MAT 142 College Mathematics

at Arizona State University

Fifth Edition

David B. Johnson
Diablo Valley College

Thomas A. Mowry
Diablo Valley College

THOMSON
BROOKS/COLE

Australia · Canada · Mexico · Singapore · Spain · United Kingdom · United States

THOMSON

BROOKS/COLE

MAT 142 College Mathematics
Johnson/Mowry

Executive Editors:
Michele Baird, Maureen Staudt &
Michael Stranz

Project Development Manager:
Linda de Stefano

Marketing Coordinators:
Lindsay Annett and Sara Mercurio

Production/Manufacturing Supervisor:
Donna M. Brown

Pre-Media Services Supervisor:
Dan Plofchan

Rights and Permissions Specialists:
Kalina Hintz and Bahman Naraghi

Cover Image
Getty Images*

Printed in the
United States of America
1 2 3 4 5 6 7 8 08 07 06

For more information, please contact Thomson Custom Solutions, 5191 Natorp Boulevard, Mason, OH 45040. Or you can visit our Internet site at www.thomsoncustom.com

The Adaptable Courseware Program consists of products and additions to existing Brooks/Cole products that are produced from camera-ready copy. Peer review, class testing, and accuracy are primarily the responsibility of the author(s).

MAT 142 College Mathematics /
Johnson/Mowry – Fifth Edition
ISBN 0-495-28420-3

International Divisions List

Asia (Including India):
Thomson Learning
(a division of Thomson Asia Pte Ltd)
5 Shenton Way #01-01
UIC Building
Singapore 068808
Tel: (65) 6410-1200
Fax: (65) 6410-1208

Australia/New Zealand:
Thomson Learning Australia
102 Dodds Street
Southbank, Victoria 3006
Australia

Latin America:
Thomson Learning
Seneca 53
Colonia Polano
11560 Mexico, D.F., Mexico
Tel (525) 281-2906
Fax (525) 281-2656

Canada:
Thomson Nelson
1120 Birchmount Road
Toronto, Ontario
Canada M1K 5G4
Tel (416) 752-9100
Fax (416) 752-8102

UK/Europe/Middle East/Africa:
Thomson Learning
High Holborn House
50-51 Bedford Row
London, WC1R 4LS
United Kingdom
Tel 44 (020) 7067-2500
Fax 44 (020) 7067-2600

Spain (Includes Portugal):
Thomson Paraninfo
Calle Magallanes 25
28015 Madrid
España
Tel 34 (0)91 446-3350
Fax 34 (0)91 445-6218

*Unless otherwise noted, all cover images used by Thomson Custom Solutions have been supplied courtesy of Getty Images with the exception of the *Earthview* cover image, which has been supplied by the National Aeronautics and Space Administration (NASA).

Preface

The goal of *Mathematics: A Practical Odyssey,* Fifth Edition, is to expose students to topics in mathematics that are usable and relevant to any educated person. It is our hope that students will encounter topics that will be useful at some time during their lives. They are also encouraged to recognize the relevance of mathematics to a well-rounded education and to appreciate the human aspect of mathematics.

New and Enhanced Features of the Fifth Edition

New Material on Voting Systems

Chapter 2 contains a new section on voting systems. The plurality, plurality-with-elimination, ranked-choice, Borda count, and pairwise comparison methods of voting are covered.

New Material on Fractal Geometry

Chapter 6 contains two new sections on fractal geometry. Several fractal shapes are studied, including the Sierpinski gasket, the Koch snowflake, and the Menger sponge. This study includes self-similarity, fractal dimension, area, and perimeter.

Concept Questions and Projects

There are a large number of new concept questions and projects.

- Concept questions focus on concepts, and usually require written responses, varying in length from a couple of sentences to a couple of paragraphs.
- Projects can be used as group or individual assignments, either in class or at home.
- Concept questions, history questions, and projects are now labeled so that they can be easily identified.
- A brief description of some of the projects is given later in this preface.

New Material on Casio Calculators

By popular demand, keystrokes for Casio graphing calculators are now included.

New Bridges Between Chapters

As much as possible, this book has been written so that its chapters are independent of each other. Sometimes, though, this independence is not desirable, because it does not allow connections to be made between seemingly unrelated topics. For this reason, the text features a number of bridges between chapters. The fifth edition contains two new bridges. (More information about bridges is given later in this preface.)

- There is a new bridge between Chapter 2 (Sets and Counting) and Chapter 6 (Geometry). That bridge discusses a connection between Pascal's Triangle and the Sierpinski Gasket.
- There is a new bridge between Chapter 6 (Geometry) and Chapter 9 (Exponential and Logarithmic Functions). That bridge discusses the use of logarithms in determining the dimension of a fractal.

Rewritten Material

- As always, examples, exercises, and articles have been updated with current real-world data and issues.
- Section 1.5 now contains a truth table for a Lewis Carroll argument containing 4 letters and 16 rows.
- Section 3.5 has new exercises about expected value and raffles.
- Section 4.1 now contains material on histograms and single-valued data classes.
- Section 4.2 now contains a formula for the location of the median.
- Section 5.4 has new exercises about lines of credit.
- Chapter 7 on Markov chains has been resequenced for better understanding and greater flexibility. Smaller problems that involve systems of two equations, and the mathematics involved in solving such systems, are covered earlier. Larger problems that involve systems of three or more equations, and the mathematics involved in solving such systems, are covered later. This gives instructors greater freedom in determining how far to go in the chapter. The elimination method, the graphing method (with a graphing calculator), and the Gauss-Jordan method (both with and without a graphing calculator) are all covered.
- Section 10.6 has new exercises on the area function.

Examples of Projects

Chapter 3—Probability

- Design a game of chance. Use probabilities and expected values to set house odds. (See page 176)
- Compare the empirically found relative frequency of tossing a coin and getting heads with the theoretical probability of getting heads. (See page 148)
- Write a research paper on the history of genetic screenings in the United States. (See page 148)
- Write a research paper on a historical figure involved in the development of probability theory. (See page 133)
- Investigate seemingly contradictory information regarding sex bias in graduate admissions and/or tuberculosis incidence in New York City and Richmond. (See page 187)

Chapter 4—Statistics

- Explore the variation in the pricing of common commodities. (See page 251)
- Conduct an opinion poll. (See page 284)
- Examine the correlation between human height and weight. (See page 295)

Chapter 5—Finance

- Use graphs to compare simple interest, interest compounded annually, and interest compounded daily. (See page 326)
- Compute the monthly payment for an adjustable rate mortgage at different points in time with varying assumptions about the interest rate. (See page 356)

- Investigate negative amortization with an adjustable rate mortgage. (See page 357)
- Buy a car. Investigate and discuss available cars, financing and affordability. (See page 357)
- Buy a home. Investigate and discuss available homes, financing affordability and real estate taxes. (See page 372)

Chapter 6—Geometry

- Verify the Pythagorean Theorem. (See page 398)
- Calculate the cost of materials to resurface a floor. (See page 398)
- Obtain conversion factors. (See page 409)
- Calculate the height of a building by using similar triangles and by using measured angles of elevation. (See page 446)
- Design your own fractal. (See page 489)
- Create your own fractal coastline. (See page 489)
- Approximate the area of a Koch snowflake. (See page 499)

Chapter 7—Markov Chains

Use Markov chains and real-world data to investigate:

- the intergenerational transition between income groups (see page 566),
- the shifting population focus on the central cities and the suburbs (see page 567), or
- the population shifts between the various regions of the United States (see page 568).

Chapter 9—Exponentials and Logarithmic Functions

- Develop the mathematical model that represents your community's population, and use that model to predict future populations. (See page 670)

Chapter 10—Calculus

- Investigate and discuss Descartes' use of language and symbols. (See page 728)
- Trace the evolution of algebra and/or of analytic geometry. (See page 728)
- Write a research paper on the contributions of Apollonius, Oresme, Descartes and Fermat to analytic geometry. (See page 728)
- Write a research paper on Fermat's last theorem. (See page 728)
- Write a research paper on a historical figure involved in the development of calculus. (See page 752)
- Write a research paper on the orbits of the planets, and the importance of this topic in motivating the invention of calculus. (See page 743)
- Use dimensional analysis to convert the distance and speed falling object formulas from the English system to the metric system. (See page 792)
- Use calculus to sketch the graph of a given polynomial function. (See page 792)
- Use calculus to find the areas of some complicated shapes. (See page 793)

Flexibility

Obviously, the book contains more material than you could ever cover in a one-semester or one-quarter course. It is written in such a way that most chapters are independent of each other. The Instructor's Manual includes a "prerequisite map" so that you can easily tell which earlier topics must be covered. It also includes sample course outlines, suggesting specific chapters to be used in forming a typical course.

The book contains a wide range of topics, varying in level of sophistication and difficulty. Chapters 1 through 6 cover topics that are not uncommon to beginning algebra-based liberal arts mathematics texts. However, because of the intermediate algebra prerequisite, the topics are covered more thoroughly, and the acquisition of problem-solving skills is emphasized. Chapters 7 through 10 are more sophisticated, covering topics not commonly found in liberal arts texts. However, the treatment is such that the material is accessible to the students who will enroll in this course.

A chapter need not be covered in its entirety; topics can easily be left out. In most cases, a chapter has a suggested core of key sections as well as a selection of optional sections, which tend to be more sophisticated. They are not labeled "optional" in the text; rather, the relationship between the sections is itemized in the prerequisite map in the Instructor's Resource Manual. The instructor determines the difficulty of the course by selecting the chapters and sections to be covered. The instructor can thus create his or her own course—one that will fit students' needs.

Bridges Between Chapters

As much as possible, this book has been written so that its chapters are independent of each other. The instructor therefore has wide latitude in selecting topics to cover and can teach a course that is responsive to the needs of his or her students and institution. Sometimes, however, this independence is not desirable, because it does not allow connections to be made between seemingly unrelated topics. For this reason, the text features a number of bridges between chapters. These bridges feature both discussion and exercises. They are clearly labeled; that is, a bridge between an earlier Chapter X and a later Chapter Y is labeled *"for those students who have completed Chapter X."*

- There is a bridge between Chapter 1 (Logic) and Chapter 2 (Sets and Counting). At the end of Section 2.1, the similarities between the respective concepts and notation of logic and sets are discussed.
- There is a bridge between Chapter 5 (Finance) and Chapter 9 (Exponential and Logarithmic Functions). That bridge discusses the relationship between the exponential growth model ($y = ae^{bt}$) and the compound interest model [$\text{FV} = \text{P}(1 + \text{i})^{\text{n}}$], and the circumstances under which one model can be used in place of the other.
- There is a bridge between Chapter 6 (Geometry) and Chapter 10 (Calculus). That bridge discusses the use of the trigonometric functions to determine the equation of a trajectory when the initial angle of elevation is other than 30°, 45°, or 60°.

Calculators

Calculators are useful and powerful tools that have become an integral part of the classroom. However, many students are unable to use their calculators effectively. Therefore, detailed instructions for scientific and graphing calculator use are included throughout the text, and several of the appendixes help students master particular aspects.

Furthermore, a number of optional subsections are included that address some of the more advanced capabilities of graphing calculators. These subsections allow instructors to incorporate technology into their class if they so desire, but they are entirely optional, and the book is in no way calculator-dependent. The graphing calculator subsections were specifically written for the following Texas Instruments calculators: TI-80, TI-81, TI-82, TI-83, TI-85, and TI-86. In addition, keystrokes for Casio graphing calculators are included.

The following graphing calculator topics are addressed:

- Chapter 3 (Probability): fractions
- Chapter 4 (Statistics): histograms, measures of central tendency and dispersion, and linear regression
- Chapter 5 (Finance): doubling time
- Chapter 7 (Matrices and Markov Chains): matrix multiplication and matrix row operations
- Chapter 8 (Linear Programming): graphing linear inequalities and the simplex method

In addition, there are four calculator-oriented appendices:

- Using a Scientific Calculator
- Using a Graphing Calculator
- Graphing with a Graphing Calculator
- Finding Points of Intersection with a Graphing Calculator

Computers

Computers are ubiquitous at the workplace, and are becoming increasingly common in the classroom. However, many students have no mathematical experience using computer software such as Microsoft Excel and Lotus 1-2-3. Therefore, a number of optional subsections are included that give instructions on the use of computerized spreadsheets. The following topics are addressed:

- Creating histograms and pie charts (Section 4.1)
- Finding the mean, median, mode and standard deviation (Section 4.3)
- Finding and drawing the line of best fit (Section 4.6)
- Creating amortization schedules (Section 5.4)

These subsections allow instructors to incorporate the computer into their class if they so desire, but they are entirely optional, and the book is in no way spreadsheet-dependent. The subsections do not assume any previous experience with computerized spreadsheets.

Course Prerequisite

Mathematics: A Practical Odyssey is written for the student who has successfully completed a course in intermediate algebra, not the student who excelled in it. It would be difficult for a student without background in intermediate algebra to succeed in a course using this book. However, some chapters are not algebra-based. These chapters require a level of critical thinking and mathematical maturity more commonly found in students who have taken intermediate algebra.

Usability

This book is user-friendly. The examples don't skip steps; key points are boxed for emphasis; step-by-step procedures are given; and there is an abundance of exposition.

Algebra Review

When appropriate, algebraic topics are reviewed, but in a very selective and focused manner. Only those topics that are used in the book are reviewed. There is no "Review of Algebra" chapter. Instead, the reviews are placed as close as possible to the topics that utilize them, usually in a Section 0 at the beginning of the chapter.

These sections are direct and to the point. They do not include any applications of the algebra; those are covered in the subsequent sections. They do not attempt to provide a thorough treatment of the algebra in question but rather focus on the algebra that will be used in the sections that follow.

Intermediate algebra courses vary significantly from school to school. Among the Section 0 topics are some that some students have never seen before, such as matrices and logarithms. In these cases, the reviews are more detailed and assume less prior knowledge.

History

The history of the subject matter is interwoven throughout most chapters. In addition, Historical Notes give in-depth biographies of the prominent people involved. It is our hope that students will see the human side of mathematics. After all, mathematics was invented by real people for real purposes and is a part of our culture. Interesting research topics are given, and writing assignments are suggested.

The exercises in this text are designed to solidify the students' understanding of the material and make them proficient in the calculations involved. It is assumed that most students who complete this course will not continue in their formal study of mathematics. Consequently, neither the exposition nor the exercises are designed to expose the students to all aspects of the topic.

Exercises

The exercises vary in difficulty. Some are exactly like the examples, and others demand more of the students. The exercises are not explicitly graded into A, B, and C categories, nor are any marked "optional"; students in this audience tend to react negatively if asked to do anything labeled in this manner. The more difficult exercises are indicated in the Instructor's Resource Manual.

The short-answer historical questions are meant to focus and reinforce the students' understanding of the historical material. They also serve to warn them that history questions may appear on exams. The essay questions can be used as an integral part of the students' grades, as background for classroom discussion, or for extra-credit work. Most are research topics and are kept as open-ended as possible.

Answers to the odd-numbered exercises are given in the back of the book, with two exceptions:

- Answers to historical questions and essay questions are not given.
- Answers are not given when the exercises instruct the students to check the answers themselves.

Error Checking

Throughout the text, there is emphasis on the importance of checking one's answers. Thus, students learn to evaluate the reasonableness of their answers, rather than accepting them at face value.

For the Instructor

Instructor's Edition—This special version of the complete student text contains a Resource Integration Guide and a complete set of answers printed in the back of the text. ISBN 0534401171

Instructor's Resource Manual—The *Instructor's Resource Manual* contains additional resources for the instructor and provides worked-out solutions to all of the problems in the text.
ISBN 0534400612

Test Bank—The *Test Bank* includes 4 tests per chapter as well as 2 final exams. The tests are made up of a combination of multiple-choice, free-response, true/false, and fill-in-the-blank questions.
ISBN 0534400620

BCA Testing for Liberal Arts Mathematics—BCA Testing is a revolutionary, Internet-ready, text-specific testing suite that allows instructors to customize exams and track student progress in an accessible, browser-based format. BCA offers full algorithmic generation of problems and free response mathematics. Results flow automatically to your gradebook and you can easily communicate to individuals, sections, or entire courses.
ISBN 0534387632

BCA Tutorial Instructor Version for Liberal Arts Mathematics—BCA Tutorial is a text-specific, interactive tutorial software program, delivered via the Web (at http://bca.brookscole.com) and is offered in both student and instructor versions. Like *BCA Testing,* it is browser-based, making it an intuitive mathematical guide even for students with little technological proficiency. So sophisticated, it's simple, *BCA Tutorial* allows students to work with real math notation in real time, providing instant analysis and feedback. The tracking program built into the instructor version of the software enables instructors to carefully monitor student progress.
ISBN 0534400639

MyCourse 2.1—Ask us about our new free online course builder! Whether you want only the easy-to-use tools to build it or the content to furnish it, Brooks/Cole offers you a simple solution for a custom course Web site that allows you to assign, track, and report on student progress; load your syllabus; and more. Contact your Brooks/Cole representative for details, or visit http://mycourse.thomsonlearning.com

CNN Today Video: Liberal Arts Mathematics

Launch your lectures with riveting footage from CNN, the world's leading 24-hour global news television network. *CNN Today Videos* allow you to integrate the news-gathering and programming power of CNN into the classroom to show students the relevance of course topics to their everyday lives. Organized by topics covered in a typical course, these videos are divided into short segments-perfect for introducing key concepts and serving as springboards for discussion. Projects built around the CNN video clips will be accessible through http://mathematics.brookscole.com. *A Thomson-Brooks/Cole exclusive. Available in North America only.*
Volume I 0534396410 Volume II 0534400663

For the Student

Student Solutions Manual—The *Student Solutions* manual provides worked out solutions to every other odd-numbered problem in the text.
ISBN 0534400604

Website—http://mathematics.brookscole.com—When you adopt a Thomson-Brooks/Cole mathematics text, you and your students will have access to a variety of teaching and learning resources. This Website features everything from book-specific resources to newsgroups. It's a great way to make teaching and learning an interactive and intriguing experience.

InfoTrac College Edition is automatically packaged FREE with every new copy of this text! *InfoTrac College Edition* is a world-class, online university library that offers the full text of articles from almost 4000 scholarly and popular publications-updated daily and going back as far as 22 years. Both adopters and their students receive unlimited access for four months.

BCA Tutorial Student Version for Liberal Arts Mathematics—This text-specific, interactive tutorial software is delivered via the Web (at http://bca.brookscole.com). It is browser-based, making it an intuitive mathematical guide, even for students with little technological proficiency. So sophisticated, it's simple. *BCA Tutorial* allows students to work with real math notation in real time, providing instant analysis and feedback. The tracking program built into the instructor version of the software enables instructors to carefully monitor student progress. *BCA Tutorial Student Version* is also available on CD Rom for those students who want to use the tutorial locally on their computers and not via the Internet.
ISBN 0534400647

Website

Located at math.brookscole.com, The Brooks/Cole Mathematics Resource Center offers book-specific student and instructor resources, as well as discipline-specific links. Student resources at this web site for this text include a Web-based version of Amortrix (the software used in Chapters 5, 7 and 8), downloadable partially completed Excel spreadsheets that are keyed to examples in this text, as well as Internet links for further research. The Website also has quizzes by chapter, whose results can be e-mailed to the instructor.

Acknowledgments

We wish to thank our wives, Gail and Paki, and our children, Lauren, Peter and Kaitlin, for their patience and support. We are also grateful to John-Paul Ramin and Bob Pirtle and all the wonderful people at Brooks/Cole who worked on this project, as well as Deann Christianson, Elaine Werner, Anne Seitz and Jean Thompson. Special thanks go to the users of the text and reviewers who evaluated the manuscript for this edition, as well as those who offered comments on previous editions.

Reviewers

Dennis Airey
Rancho Santiago College

Bruce Atkinson
Palm Beach Atlantic College

Judith Arms
University of Washington

Wayne C. Bell
Murray State University

Wayne Bishop
California State University–Los Angeles

David Boliver
Trenton State College

Barry Brunson
Western Kentucky University

Frank Burk
California State University–Chico

Laura Cameron
University of New Mexico

Jack Carter
California State University–Hayward

Timothy D. Cavanagh
University of Northern Colorado

Joseph Chavez
California State University–San Bernardino

Eric Clarkson
Murray State University

Rebecca Conti
State University of New York College at Fredonia

S. G. Crossley
University of South Alabama

Ben Divers, Jr.
Ferrum College

Al Dixon
College of the Ozarks

Joe S. Evans
Middle Tennessee State University

Hajrudin Fejzic
California State University–San Bernardino

Lloyd Gavin
California State University–Sacramento

William Greiner
McLennan Community College

Martin Haines
Olympic College

Ray Hamlett
East Central University

Virginia Hanks
Western Kentucky University

Anne Herbst
Santa Rosa Junior College

Linda Hinzman
Pasadena City College

Thomas Hull
University of Rhode Island

Robert W. Hunt
Humboldt State University

Irja Kalantari
Western Illinois University

Katalin Kolossa
Arizona State University

Lee LaRue
Paris Junior College

Thomas McCready
California State University–Chico

Vicki McMillian
Stockton State College

Narendra L. Maria
California State University–Stanislaus

John Martin
Santa Rosa Junior College

Gael Mericle
Mankato State University

Pamela G. Nelson
Panhandle State University

Carol Oelkers
Fullerton College

Michael Olinick
Middlebury College

Matthew Pickard
University of Puget Sound

John D. Putnam
University of Northern Colorado

J. Doug Richey
Northeast Texas Community College

Stewart Robinson
Cleveland State University

Eugene P. Schlereth
University of Tennessee at Chattanooga

Lawrence Somer
Catholic University of America

Michael Trapuzzano
Arizona State University

Pat Velicky
Mid-Plains Community College

Karen M. Walters
University of Northern Colorado

Dennis W. Watson
Clark College

Charles Ziegenfus
James Madison University

Contents

Logic

1

When writer Lewis Carroll took Alice on her journeys "through the looking glass" to Wonderland, she had many fantastic encounters with the hookah-smoking Caterpillar, the White Rabbit, the Red and White Queens, and the Cheshire Cat. On the surface, Carroll's writings seem to be delightful nonsense and mere children's entertainment. However, they contain hints of deeper roots and meanings. During one of her many surreal adventures, Alice came upon Tweedledum and Tweedledee. The two sarcastic twins taunted her with the following: " 'I know what you're thinking about,' said Tweedledum; 'but it isn't so, nohow.' 'Contrariwise,' continued Tweedledee, 'if it was so, it might be; and if it were so, it would be; but as it isn't, it ain't. That's logic.' "

Many people are quite surprised to learn that *Alice's Adventures in Wonderland* is as much an exercise in logic as it is a fantasy and that Lewis Carroll was actually Charles Dodgson, an Oxford mathematician. Dodgson's many writings include the whimsical *The Game of Logic* and the brilliant *Symbolic Logic,* in addition to *Alice's Adventures in Wonderland* and *Through the Looking Glass.*

Webster's dictionary defines **logic** as "the science of correct reasoning; the science which describes relationships among propositions in terms of implication, contradiction, contrariety, and conversion." In addition to being flaunted in Mr. Spock's claim that "your human emotions have drawn you to an illogical conclusion" and in Sherlock Holmes's immortal phrase "elementary, my dear Watson," logic is fundamental both to critical thinking and to problem solving. In this world of misleading commercial claims, innuendo, and political rhetoric, the ability to distinguish between valid and invalid arguments is important. Cold, emotionless Vulcans and eccentric, violin-playing detectives are not the only ones who can benefit from logic. Armed with the fundamentals of logic, we can surely join Spock and "live long and prosper!"

© The Granger Collection, NY

Logic is the science of correct reasoning. Auguste Rodin captured this ideal in his bronze sculpture *The Thinker.*

© UPI/Bettman/Corbis

In their quest for logical perfection, the Vulcans of *Star Trek* abandoned all emotion. Mr. Spock's frequent proclamation that "emotions are illogical" typified this attitude.

© The Granger Collection, NY

Using his extraordinary powers of logical deduction, Sherlock Holmes solves another mystery. "Finding the villain was elementary, my dear Watson."

1.1 Deductive versus Inductive Reasoning

Logic is the science of correct reasoning. Webster's dictionary defines **reasoning** as "the drawing of inferences or conclusions from known or assumed facts." Reasoning is an integral part of our daily lives; we take appropriate actions based on our perceptions and experiences. For instance, if the sky is heavily overcast this morning, you might assume that it will rain today and take your umbrella to work.

Problem Solving

Logic and reasoning are associated with the phrases *problem solving* and *critical thinking*. If we are faced with a problem, puzzle, or dilemma, we attempt to reason through it in hopes of arriving at a solution.

The first step in solving any problem is to define the problem in a thorough and accurate manner. Although this might sound like an obvious step, it is often overlooked. Always ask yourself, "What am I being asked to do?" Before you can solve a problem, you must understand the question. Once the problem has been defined, all known information relevant to it must be gathered, organized, and analyzed. This analysis should include a comparison of the present problem to previous ones. How is it similar? How is it different? Does a previous method of solution apply? If it seems appropriate, draw a picture of the problem; visual representations often provide insight into the interpretation of clues.

Before using any specific formula or method of solution, determine whether its use is valid for the situation at hand. A common error is to use a formula or method of solution when it does not apply. If a past formula or method of solution is appropriate, use it; if not, explore standard options and develop creative alternatives. Do not be afraid to try something different or out of the ordinary. "What if I try this . . . ?" may lead to a unique solution.

Deductive Reasoning

Once a problem has been defined and analyzed, it might fall into a known category of problems, so a common method of solution may be applied. For instance, when one is asked to solve the equation $x^2 = 2x + 1$, realizing that it is a second-degree equation (that is, a quadratic equation) leads one to put it into the standard form ($x^2 - 2x - 1 = 0$) and apply the Quadratic Formula.

EXAMPLE 1 Solve the equation $x^2 = 2x + 1$.

Solution The given equation is a second-degree equation in one variable. We know that all second-degree equations in one variable (in the form $ax^2 + bx + c = 0$) can be solved by applying the Quadratic Formula:

$$x = \frac{-b \pm \sqrt{b^2 - 4ac}}{2a}$$

Therefore, $x^2 = 2x + 1$ can be solved by applying the Quadratic Formula:

$$x^2 = 2x + 1$$
$$x^2 - 2x - 1 = 0$$
$$x = \frac{-(-2) \pm \sqrt{(-2)^2 - (4)1(-1)}}{2(1)}$$
$$x = \frac{2 \pm \sqrt{4 + 4}}{2}$$
$$x = \frac{2 \pm \sqrt{8}}{2}$$
$$x = \frac{2 \pm 2\sqrt{2}}{2}$$
$$x = \frac{2(1 \pm \sqrt{2})}{2}$$
$$x = 1 \pm \sqrt{2}$$

The solutions are $x = 1 + \sqrt{2}$ or $x = 1 - \sqrt{2}$. ■

In Example 1, we applied a general rule to a specific case; we reasoned that it was valid to apply the (general) Quadratic Formula to the (specific) equation $x^2 = 2x + 1$. This type of logic is known as **deductive reasoning**—that is, the application of a general statement to a specific instance.

Deductive reasoning and the formal structure of logic have been studied for thousands of years. One of the earliest logicians, and one of the most renowned, was Aristotle (384–322 B.C.). He was the student of the great philosopher Plato and the tutor of Alexander the Great, the conqueror of all the land from Greece to India. Aristotle's philosophy is pervasive; it influenced Roman Catholic theology through St. Thomas Aquinas and continues to influence modern philosophy. For centuries, Aristotelian logic was part of the education of lawyers and politicians and was used to distinguish valid arguments from invalid ones.

For Aristotle, logic was the necessary tool for any inquiry, and the syllogism was the sequence followed by all logical thought. A **syllogism** is an argument composed of two statements, or **premises** (the major and minor premises), followed by a **conclusion.** For any given set of premises, if the conclusion of an argument is guaranteed (that is, if it is inescapable in all instances), the argument is **valid.** If the conclusion is not guaranteed (that is, if there is at least one instance in which it does not follow), the argument is **invalid.**

Perhaps the best known of Aristotle's syllogisms is the following:

1. All men are mortal.	major premise
2. Socrates is a man.	minor premise
Therefore, Socrates is mortal.	conclusion

Applying the major premise to the minor premise, the conclusion is inescapable; the argument is valid.

Notice that the deductive reasoning used in the analysis of Example 1 has exactly the same structure as Aristotle's syllogism concerning Socrates:

1. All second-degree equations in one variable can be solved by applying the Quadratic Formula.	major premise
2. $x^2 = 2x + 1$ is a second-degree equation in one variable.	minor premise
Therefore, $x^2 = 2x + 1$ can be solved by applying the Quadratic Formula.	conclusion

Each of these syllogisms is of the following general form:

1. If *A*, then *B*.	All *A* are *B*. (major premise)
2. *x* is *A*.	We have *A*. (minor premise)
Therefore, *x* is *B*.	Therefore, we have *B*. (conclusion)

Historically, this valid pattern of deductive reasoning is known as *modus ponens.*

Deductive Reasoning and Venn Diagrams

The validity of a deductive argument can be shown by use of a Venn diagram. A **Venn diagram** is a diagram consisting of various overlapping figures contained within a rectangle (called the "universe"). To depict a statement of the form "All *A* are *B*" (or equivalently, "If *A*, then *B*"), we draw two circles, one inside the other; the inner circle represents *A*, the outer circle represents *B*. This relationship is shown in Figure 1.1.

Venn diagrams depicting "No *A* are *B*" and "Some *A* are *B*" are shown in Figures 1.2 and 1.3, respectively.

EXAMPLE 2 Construct a Venn diagram to verify the validity of the following argument:

1. All men are mortal.
2. Socrates is a man.

Therefore, Socrates is mortal.

Solution Premise 1 is of the form "All *A* are *B*" and can be represented by a diagram like that shown in Figure 1.4.

Premise 2 refers to a specific man, namely, Socrates. If we let x = Socrates, the statement "Socrates is a man" can then be represented by placing x within the circle

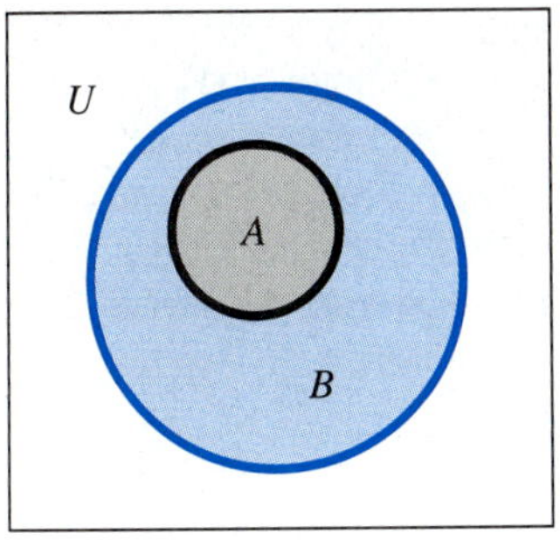

Figure 1.1
All *A* are *B*. (If *A*, then *B*.)

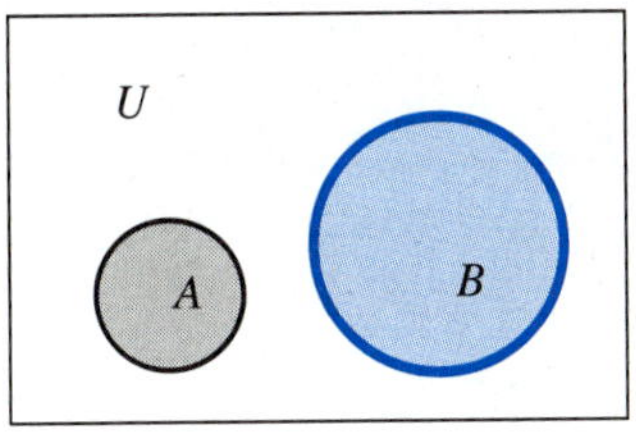

Figure 1.2
No *A* are *B*.

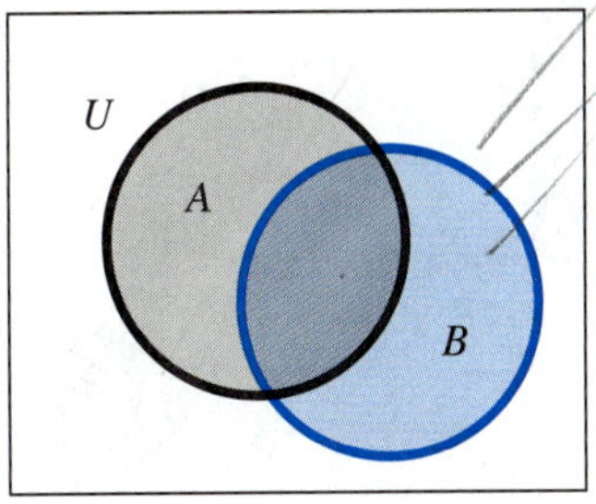

Figure 1.3
Some *A* are *B*. (At least one *A* is *B*.)

labeled "men," as shown in Figure 1.5. Because we placed *x* within the "men" circle, and all of the "men" circle is inside the "mortal" circle, the conclusion "Socrates is mortal" is inescapable; the argument is valid.

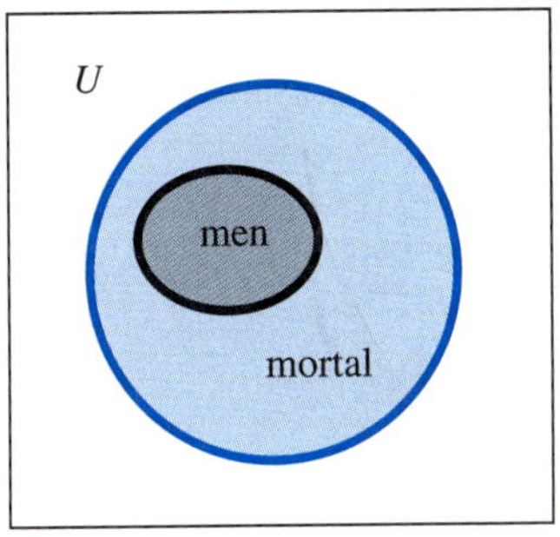

Figure 1.4

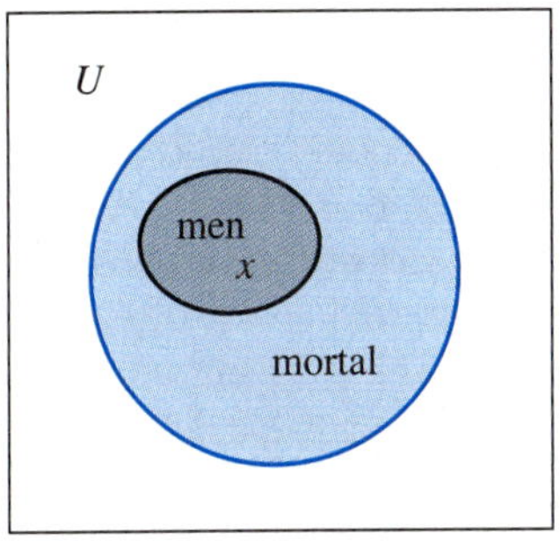

x = Socrates

Figure 1.5 ■

EXAMPLE 3 Construct a Venn diagram to determine the validity of the following argument:

1. All doctors are men.
2. My mother is a doctor.

Therefore, my mother is a man.

Solution Premise 1 is of the form "All *A* are *B*"; the argument is depicted in Figure 1.6.

No matter where *x* is placed within the "doctors" circle, the conclusion "My mother is a man" is inescapable; the argument is valid. ■

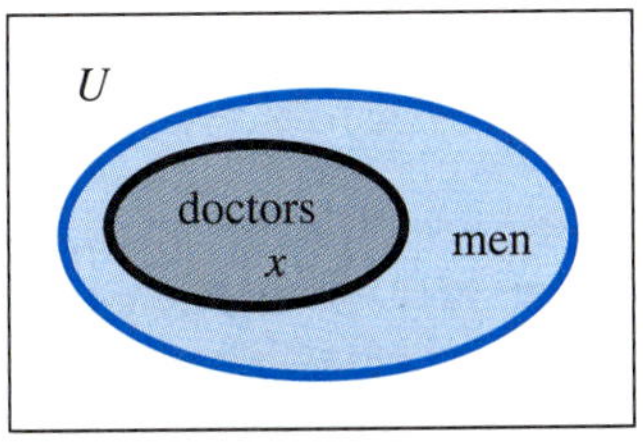

x = my mother

Figure 1.6

Saying that an argument is valid does not mean that the conclusion is true. The argument given in Example 3 *is* valid, but the conclusion is *false.* One's mother cannot be a man! Validity and truth do not mean the same thing. An argument is valid if the conclusion is inescapable, *given the premises.* Nothing is said about the truth of the premises. Thus, when examining the validity of an argument, we are not determining whether the conclusion is true or false. Saying that an argument is valid merely means that, *given the premises,* the reasoning used to obtain the conclusion is logical. However, if the premises of a valid argument are true, then the conclusion will also be true.

HISTORICAL NOTE

Aristotle

384–322 B.C.

North Wind Picture Archives

Aristotle was born in 384 B.C. in the small Macedonian town of Stagira, 200 miles north of Athens, on the shore of the Aegean Sea. Aristotle's father was the personal physician of King Amyntas II, ruler of Macedonia. When he was seventeen, Aristotle enrolled at the Academy in Athens and became a student of the famed Plato.

Aristotle was one of Plato's brightest students; he frequently questioned Plato's teachings and openly disagreed with him. Whereas Plato emphasized the study of abstract ideas and mathematical truth, Aristotle was more interested in observing the "real world" around him. Plato often referred to Aristotle as "the brain" or "the mind of the school." Plato commented, "Where others need the spur, Aristotle needs the rein."

Aristotle stayed at the Academy for twenty years, until the death of Plato. Then the king of Macedonia invited Aristotle to supervise the education of his son Alexander. Aristotle accepted the invitation and taught Alexander until he succeeded his father as ruler. At that time, Aristotle founded a school known as the Lyceum, or Peripatetic School. The school had a large library with many maps, as well as botanical gardens containing an extensive collection of plants and animals. Aristotle and his students would walk about the grounds of the Lyceum while discussing various subjects (*peripatetic* is from the Greek word meaning "to walk").

Many consider Aristotle to be a founding father of the study of biology and of science in general; he observed and classified the behavior and anatomy of hundreds of living creatures. During his many military campaigns, Alexander the Great had his troops gather specimens from distant places for Aristotle to study.

Aristotle was a prolific writer; some historians credit him with the writing of over 1,000 books. Most of his works have been lost or destroyed, but scholars have recreated some of his more influential works, including *Organon.*

Aristotle's collective works on syllogisms and deductive logic are known as *Organon,* meaning "instrument," for logic is the instrument used in the acquisition of knowledge.

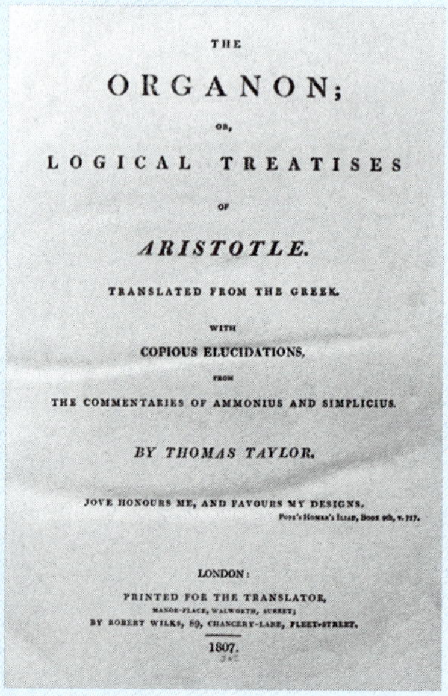
THE

ORGANON;

OR,

LOGICAL TREATISES

OF

ARISTOTLE.

TRANSLATED FROM THE GREEK.

WITH

COPIOUS ELUCIDATIONS,

FROM

THE COMMENTARIES OF AMMONIUS AND SIMPLICIUS.

BY THOMAS TAYLOR.

JOVE HONOURS ME, AND FAVOURS MY DESIGNS.

LONDON:

PRINTED FOR THE TRANSLATOR,

BY ROBERT WILKS, 89, CHANCERY-LANE, FLEET-STREET.

1807.

New York City Public Library, Photographed by David Grossman

EXAMPLE 4 Construct a Venn diagram to determine the validity of the following argument:

1. All professional wrestlers are actors.
2. The Rock is an actor.

Therefore, The Rock is a professional wrestler.

Solution Premise 1 is of the form "All A are B"; the "circle of professional wrestlers" is contained within the "circle of actors." If we let x represent The Rock, premise 2 simply requires that we place x somewhere within the actor circle; x could be placed in either of the two locations shown in Figures 1.7 and 1.8.

If x is placed as in Figure 1.7, the argument would appear to be valid; the figure supports the conclusion "The Rock is a professional wrestler." However, the placement of x in Figure 1.8 does not support the conclusion; given the premises, we can-

Even though The Rock *is* a professional wrestler, the argument used to obtain the conclusion is invalid.

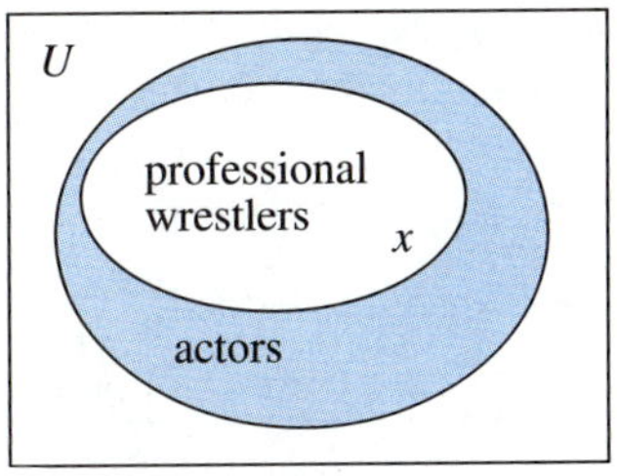

x = The Rock

Figure 1.7

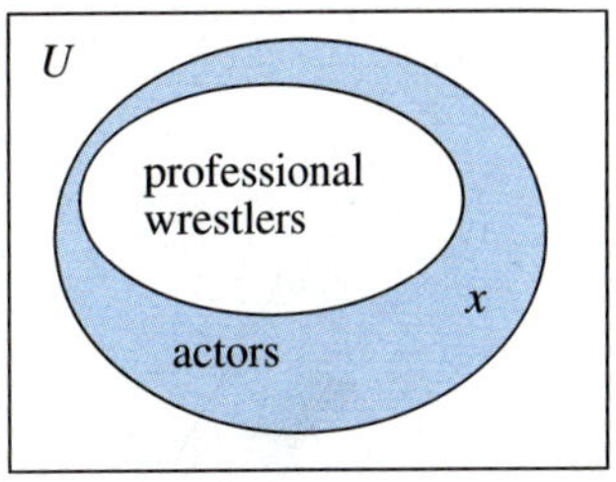

x = The Rock

Figure 1.8

not *logically* deduce that "The Rock is a professional wrestler." Since the conclusion is *not* inescapable, the argument is invalid. ■

Saying that an argument is invalid does not mean that the conclusion is false. Example 4 demonstrates that an invalid argument can have a true conclusion; even though The Rock is a professional wrestler, the argument used to obtain the conclusion is invalid. In logic, validity and truth do not have the same meaning. *Validity* refers to the process of reasoning used to obtain a conclusion; *truth* refers to conformity with fact or experience.

Venn Diagrams and Invalid Arguments

To show that an argument is invalid, you must construct a Venn diagram in which the premises are met yet the conclusion does not necessarily follow.

EXAMPLE 5 Construct a Venn diagram to determine the validity of the following argument:

1. Some plants are poisonous.
2. Broccoli is a plant.

Therefore, broccoli is poisonous.

Solution Premise 1 is of the form "Some A are B"; it can be represented by two overlapping circles (as in Figure 1.3). If we let x represent broccoli, premise 2 requires that we place x somewhere within the plant circle. If x is placed as in Figure 1.9, the argument would appear to be valid. However, if x is placed as in Figure 1.10, the conclusion does not follow. Because we can construct a Venn diagram in which the premises are met yet the conclusion does not follow (Figure 1.10), the argument is invalid.

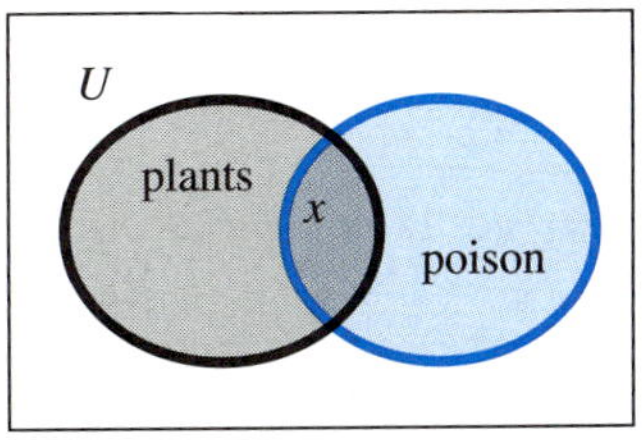

x = broccoli

Figure 1.9

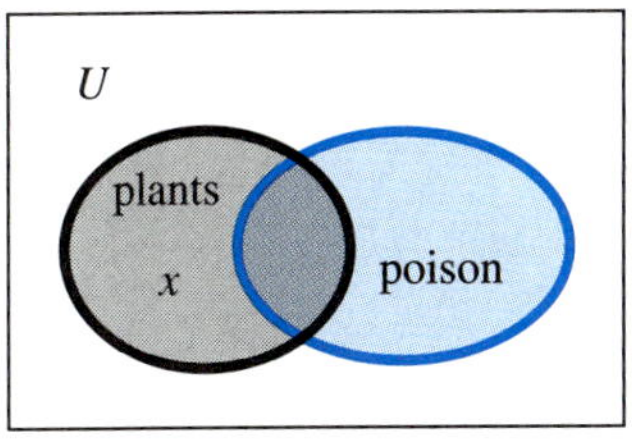

x = broccoli

Figure 1.10 ■

When analyzing an argument via a Venn diagram, you might have to draw three or more circles, as in the next example.

EXAMPLE 6 Construct a Venn diagram to determine the validity of the following argument:

1. No snake is warm-blooded.
2. All mammals are warm-blooded.

Therefore, snakes are not mammals.

Solution Premise 1 is of the form "No *A* are *B*"; it is depicted in Figure 1.11. Premise 2 is of the form "All *A* are *B*"; the "mammal circle" must be drawn within the "warm-blooded circle." Both premises are depicted in Figure 1.12.

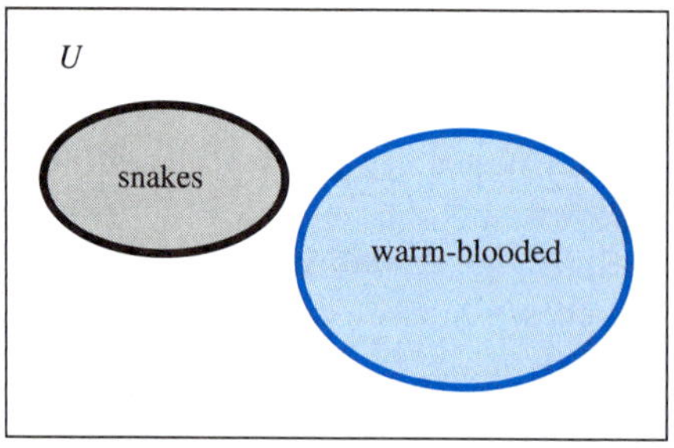

Figure 1.11

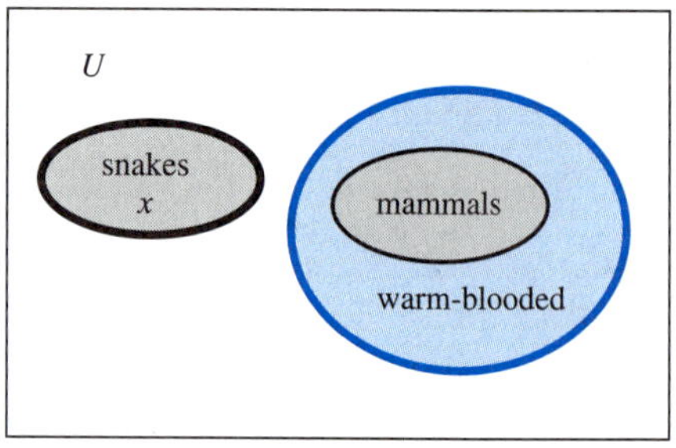

x = snake

Figure 1.12

Because we placed x (= snake) within the "snake" circle, and the "snake" circle is outside the "warm-blooded" circle, x cannot be within the "mammal" circle (which is inside the "warm-blooded" circle). Given the premises, the conclusion "Snakes are not mammals" is inescapable; the argument is valid. ■

You might have encountered Venn diagrams when you studied sets in your algebra class. The academic fields of set theory and logic are historically intertwined; set theory was developed in the late nineteenth century as an aid in the study of logical arguments. Today, set theory and Venn diagrams are applied to areas other than the study of logical arguments; we will utilize Venn diagrams in our general study of set theory in Chapter 2.

Inductive Reasoning

The conclusion of a valid deductive argument (one that goes from general to specific) is guaranteed: Given true premises, a true conclusion must follow. However, there are arguments in which the conclusion is not guaranteed even though the premises are true. Consider the following:

1. Joe sneezed after petting Frako's cat.
2. Joe sneezed after petting Paulette's cat.

Therefore, Joe is allergic to cats.

Is the conclusion guaranteed? If the premises are true, they certainly *support* the conclusion, but we cannot say with 100% certainty that Joe is allergic to cats. The conclusion is *not* guaranteed. Maybe Joe is allergic to the flea powder that the cat owners used; maybe he is allergic to the dust that is trapped in the cat's fur; or maybe he has a cold!

Reasoning of this type is called inductive reasoning. **Inductive reasoning** involves going from a series of specific cases to a general statement (see Figure 1.13). Although it may seem to follow and may in fact be true, *the conclusion in an inductive argument is never guaranteed.*

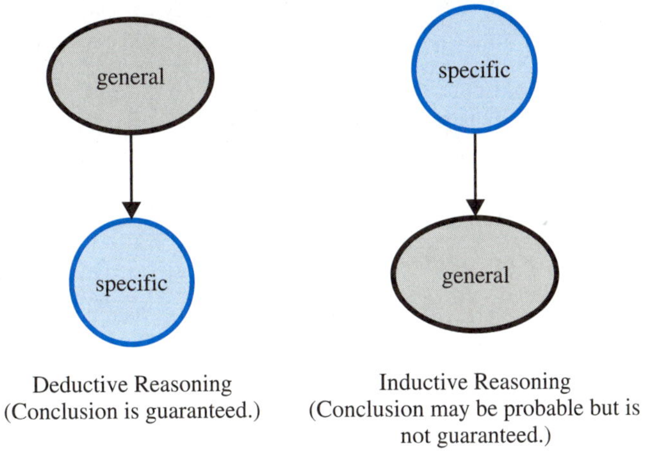

Figure 1.13

EXAMPLE 7 What is the next number in the sequence 1, 8, 15, 22, 29, . . . ?

Solution Noticing that the difference between consecutive numbers in the sequence is 7, we may be tempted to say that the next term is $29 + 7 = 36$. Is this conclusion guaranteed? No! Another sequence in which numbers differ by 7 are dates of a given day of the week. For instance, the dates of the Saturdays in the year 2005 are (January) 1, 8, 15, 22, 29, (February) 5, 12, 19, 26, ... Therefore, the next number in the sequence 1, 8, 15, 22, 29, . . . might be 5. Without further information, we cannot determine the next number in the given sequence. We can only use inductive reasoning and give one or more *possible* answers. ■

1.1 Exercises

In Exercises 1–20, construct a Venn diagram to determine the validity of the given argument.

1. a. 1. All master photographers are artists.
2. Ansel Adams is a master photographer.

Therefore, Ansel Adams is an artist.

b. 1. All master photographers are artists.
2. Ansel Adams is an artist.

Therefore, Ansel Adams is a master photographer.

2. a. 1. All Olympic gold medal winners are role models.
2. Sara Hughes is an Olympic gold medal winner.

Therefore, Sarah Hughes is a role model.

b. 1. All Olympic gold medal winners are role models.
2. Sarah Hughes is a role model.

Therefore, Sarah Hughes is an Olympic gold medal winner.

3. a. 1. All homeless people are unemployed.
2. Bill Gates is not a homeless person.

Therefore, Bill Gates is not unemployed.

b. 1. All homeless people are unemployed.
2. Bill Gates is not unemployed.

Therefore, Bill Gates is not a homeless person.

4. a. 1. All professional wrestlers are actors.
   2. Ralph Nader is not an actor.

   Therefore, Ralph Nader is not a professional wrestler.

   b. 1. All professional wrestlers are actors.
   2. Ralph Nader is not a professional wrestler.

   Therefore, Ralph Nader is not an actor.

5. 1. All pesticides are harmful to the environment.
   2. No fertilizer is a pesticide.

   Therefore, no fertilizer is harmful to the environment.

6. 1. No one who can afford health insurance is unemployed.
   2. All politicians can afford health insurance.

   Therefore, no politician is unemployed.

7. 1. No vegetarian owns a gun.
   2. All policemen own guns.

   Therefore, no policeman is a vegetarian.

8. 1. No professor is a millionaire.
   2. No millionaire is illiterate.

   Therefore, no professor is illiterate.

9. 1. All poets are loners.
   2. All loners are taxi drivers.

   Therefore, all poets are taxi drivers.

10. 1. All forest rangers are environmentalists.
    2. All forest rangers are storytellers.

    Therefore, all environmentalists are storytellers.

11. 1. Real men don't eat quiche.
    2. Clint Eastwood is a real man.

    Therefore, Clint Eastwood doesn't eat quiche.

12. 1. Real men don't eat quiche.
    2. Oscar Meyer eats quiche.

    Therefore, Oscar Meyer isn't a real man.

13. 1. All roads lead to Rome.
    2. Route 66 is a road.

    Therefore, Route 66 leads to Rome.

14. 1. All smiling cats talk.
    2. The Cheshire Cat smiles.

    Therefore, the Cheshire Cat talks.

15. 1. Some animals are dangerous.
    2. A tiger is an animal.

    Therefore, a tiger is dangerous.

16. 1. Some professors wear glasses.
    2. Mr. Einstein wears glasses.

    Therefore, Mr. Einstein is a professor.

17. 1. Some women are police officers.
    2. Some police officers ride motorcycles.

    Therefore, some women ride motorcycles.

18. 1. All poets are eloquent.
    2. Some poets are wine connoisseurs.

    Therefore, some wine connoisseurs are eloquent.

19. 1. All squares are rectangles.
    2. Some quadrilaterals are squares.

    Therefore, some quadrilaterals are rectangles.

20. 1. All squares are rectangles.
    2. Some quadrilaterals are rectangles.

    Therefore, some quadrilaterals are squares.

21. Classify each argument as deductive or inductive.

    a. 1. My television set did not work two nights ago.
    2. My television set did not work last night.

    Therefore, my television set is broken.

    b. 1. All electronic devices give their owners grief.
    2. My television set is an electronic device.

    Therefore, my television set gives me grief.

22. Classify each argument as deductive or inductive.

    a. 1. I ate a chili dog at Joe's and got indigestion.
    2. I ate a chili dog at Ruby's and got indigestion.

    Therefore, chili dogs give me indigestion.

    b. 1. All spicy foods give me indigestion.
    2. Chili dogs are spicy food.

    Therefore, chili dogs give me indigestion.

In Exercises 23–32, fill in the blank with what is most likely to be the next number. Explain (using complete sentences) the pattern generated by your answer.

23. 3, 8, 13, 18, _____
24. 10, 11, 13, 16, _____
25. 0, 2, 6, 12, _____
26. 1, 2, 5, 10, _____
27. 1, 4, 9, 16, _____
28. 1, 8, 27, 64, _____
29. 2, 3, 5, 7, 11, _____
30. 1, 1, 2, 3, 5, _____
31. 5, 8, 11, 2, _____
32. 12, 5, 10, 3, _____

In Exercises 33–36, fill in the blanks with what are most likely to be the next letters. Explain (using complete sentences) the pattern generated by your answers.

33. O, T, T, F, _____, _____
34. T, F, S, E, _____, _____

35. F, S, S, M, _____, _____

36. J, F, M, A, _____, _____

In Exercises 37–42, explain the general rule or pattern used to assign the given letter to the given word. Fill in the blank with the letter that fits the pattern.

37.

circle	square	trapezoid	octagon	rectangle
c	s	t	o	_____

38.

circle	square	trapezoid	octagon	rectangle
i	u	a	o	_____

39.

circle	square	trapezoid	octagon	rectangle
j	v	b	p	_____

40.

circle	square	trapezoid	octagon	rectangle
c	r	p	g	_____

41.

banana	strawberry	asparagus	eggplant	orange
b	z	t	u	_____

42.

banana	strawberry	asparagus	eggplant	orange
y	r	g	p	_____

43. Find two different numbers that could be used to fill in the blank.

1, 4, 7, 10, _____

Explain the pattern generated by each of your answers.

44. Find five different numbers that could be used to fill in the blank.

7, 14, 21, 28, _____

Explain the pattern generated by each of your answers.

45. Example 1 utilized the Quadratic Formula. Verify that

$$x = \frac{-b + \sqrt{b^2 - 4ac}}{2a}$$

is a solution of the equation $ax^2 + bx + c = 0$.

HINT: Substitute the fraction for x in $ax^2 + bx + c$ and simplify.

46. Example 1 utilized the Quadratic Formula. Verify that

$$x = \frac{-b - \sqrt{b^2 - 4ac}}{2a}$$

is a solution of the equation $ax^2 + bx + c = 0$.

HINT: Substitute the fraction for x in $ax^2 + bx + c$ and simplify.

47. As a review of algebra, use the Quadratic Formula to solve

$x^2 - 6x + 7 = 0$

48. As a review of algebra, use the Quadratic Formula to solve

$x^2 - 2x - 4 = 0$

Answer the following questions using complete sentences and your own words.

Concept Questions

49. Explain the difference between deductive and inductive reasoning.

50. Explain the difference between truth and validity.

51. What is a syllogism? Give an example of a syllogism that relates to your life.

History Questions

52. From the days of the ancient Greeks, the study of logic has been mandatory in what two professions? Why?

53. Who developed a formal system of deductive logic based on arguments?

54. What was the name of the school Aristotle founded? What does it mean?

55. How did Aristotle's school of thought differ from Plato's?

1.2 Symbolic Logic

The syllogism ruled the study of logic for nearly 2,000 years and was not supplanted until the development of symbolic logic in the late seventeenth century. As its name implies, symbolic logic involves the use of symbols and algebraic manipulations in logic.

Statements

All logical reasoning is based on statements. A **statement** is a sentence that is either true or false.

EXAMPLE 1 Which of the following are statements? Why or why not?

a. Apple manufactures computers.
b. Apple manufactures the world's best computers.
c. Did you buy an IBM?
d. A \$2,000 computer that is discounted 25% will cost \$1,000.
e. I am telling a lie.

Solution **a.** The sentence "Apple manufactures computers" is true; therefore, it is a statement.
b. The sentence "Apple manufactures the world's best computers" is an opinion, and as such, it is neither true nor false. It is true for some people and false for others. Therefore, it is not a statement.
c. The sentence "Did you buy an IBM?" is a question. As such, it is neither true nor false; it is not a statement.
d. The sentence "A \$2,000 computer that is discounted 25% will cost \$1,000" is false; therefore, it is a statement. (A \$2,000 computer that is discounted 25% would cost \$1,500.)
e. The sentence "I am telling a lie" is a self-contradiction, or paradox. If it were true, the speaker would be telling a lie, but in telling the truth, the speaker would be contradicting the statement that he or she was lying; if it were false, the speaker would not be telling a lie, but in not telling a lie, the speaker would be contradicting the statement that he or she was lying. The sentence is not a statement. ■

By tradition, symbolic logic uses lowercase letters as labels for statements. The most frequently used letters are p, q, r, s, and t. We can label the statement "It is snowing" as statement p in the following manner:

p: It is snowing.

If it *is* snowing, p is labeled true, whereas if it is *not* snowing, p is labeled false.

Compound Statements and Logical Connectives

It is easy to determine whether a statement such as "Charles donated blood" is true or false; either he did or he didn't. However, not all statements are so simple; some are more involved. For example, the truth of "Charles donated blood and did not wash his car, or he went to the library," depends on the truth of the individual pieces that make up the larger, compound statement. A **compound statement** is a statement that contains one or more simpler statements. A compound statement can be formed by inserting the word *not* into a simpler statement or by joining two or more statements with connective words such as *and, or, if . . . then . . . , only if,* and *if and only if.* The compound statement "Charles did *not* wash his car" is formed from the simpler statement "Charles did wash his car." The compound statement "Charles donated blood *and* did *not* wash his car, *or* he went to the library" consists of three statements, each of which may be true or false.

Figure 1.14 diagrams two equivalent compound statements.

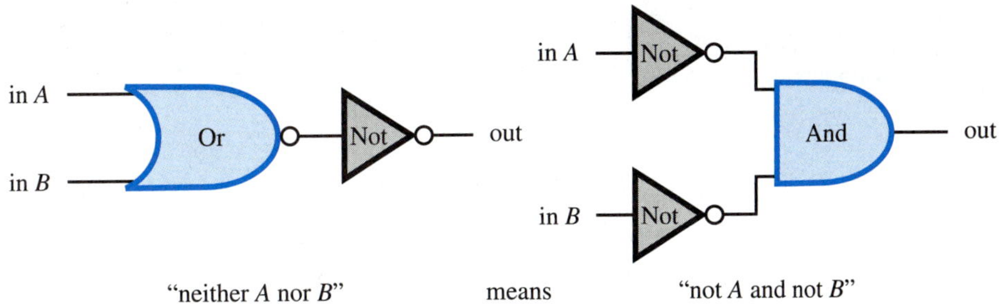

Figure 1.14 Technicians and engineers use compound statements and logical connectives to study the flow of electricity through switching circuits.

When is a compound statement true? Before we can answer this question, we must first examine the various ways in which statements can be connected. Depending on how the statements are connected, the resulting compound statement can be a *negation,* a *conjunction,* a *disjunction,* a *conditional,* or any combination thereof.

The Negation $\sim p$

The **negation** of a statement is the denial of the statement and is represented by the symbol $\sim$. The negation is frequently formed by inserting the word *not.* For example, given the statement "*p:* It is snowing," the negation would be "$\sim p$: It is not snowing." If it *is* snowing, p is true and $\sim p$ is false. Similarly, if it is *not* snowing, p is false and $\sim p$ is true. A statement and its negation always have opposite truth values; when one is true, the other is false. Because the truth of the negation depends on the truth of the original statement, a negation is classified as a compound statement.

EXAMPLE 2 Write a sentence that represents the negation of each statement:

a. The senator is a Democrat.
b. The senator is not a Democrat.
c. Some senators are Republicans.
d. All senators are Republicans.
e. No senator is a Republican.

Solution

a. The negation of "The senator is a Democrat" is "The senator is not a Democrat."

b. The negation of "The senator is not a Democrat" is "The senator is a Democrat."

c. A common error would be to say that the negation of "Some senators are Republicans" is "Some senators are not Republicans." However, "Some senators are Republicans" is not denied by "Some senators are not Republicans." The statement "Some senators *are* Republicans" implies that at least one senator is a Republican. The negation of this statement is "It is not the case that at least one senator is a Republican," or (more commonly phrased) the negation is "No senator is a Republican."

d. The negation of "All senators are Republicans" is "It is not the case that all senators are Republicans," or "There exists a senator who is not a Republican," or (more commonly phrased) "Some senators are not Republicans."

e. The negation of "No senator is a Republican" is "It is not the case that no senator is a Republican," or, in other words, "There exists at least one senator who *is* a Republican." If "some" is interpreted as meaning "at least one," the negation can be expressed as "Some senators are Republicans." ■

The words *some, all,* and *no* (or *none*) are referred to as **quantifiers.** Parts (c) through (e) of Example 2 contain quantifiers. The linked pairs of quantified statements shown in Figure 1.15 are negations of each other.

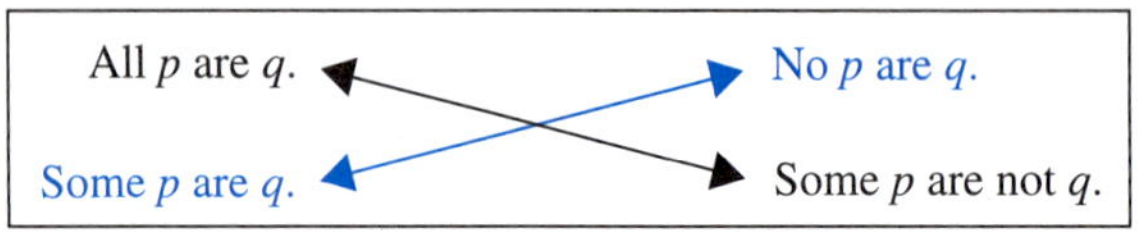

Figure 1.15
Negations of statements containing quantifiers

The Conjunction $p \wedge q$

Consider the statement "Norma Rae is a union member and she is a Democrat." This is a compound statement, because it consists of two statements—"Norma Rae is a union member" and "she (Norma Rae) is a Democrat"—and the connective word *and.* Such a compound statement is referred to as a conjunction. A **conjunction** consists of two or more statements connected by the word *and.* We use the symbol $\wedge$ to represent the word *and*; thus, the conjunction "$p \wedge q$" represents the compound statement "p and q."

EXAMPLE 3 Using the symbolic representations

p: Norma Rae is a union member.

q: Norma Rae is a Democrat.

express the following compound statements in symbolic form:

a. Norma Rae is a union member and she is a Democrat.
b. Norma Rae is a union member and she is not a Democrat.

Solution **a.** The compound statement "Norma Rae is a union member and she is a Democrat" can be represented as $p \wedge q$.
b. The compound statement "Norma Rae is a union member and she is not a Democrat" can be represented as $p \wedge \sim q$. ■

The Disjunction $p \vee q$

When statements are connected by the word *or,* a **disjunction** is formed. We use the symbol $\vee$ to represent the word *or.* Thus, the disjunction "$p \vee q$" represents the compound statement "p or q." We can interpret the word *or* in two ways. Consider the statements

p: Kaitlin is a registered Republican.

q: Paki is a registered Republican.

The statement "Kaitlin is a registered Republican or Paki is a registered Republican" can be symbolized as $p \vee q$. Notice that it is possible that *both* Kaitlin and Paki are registered Republicans. In this example, *or* includes the possibility that both things may happen. In this case, we are working with the **inclusive *or.***

Now consider the statements

p: Kaitlin is a registered Republican.

q: Kaitlin is a registered Democrat.

HISTORICAL NOTE

Gottfried Wilhelm Leibniz

1646–1716

North Wind Picture Archives

In addition to cofounding calculus (see Chapter 10), the German-born Gottfried Wilhelm Leibniz contributed much to the development of symbolic logic. A precocious child, Leibniz was self-taught in many areas. He taught himself Latin at the age of eight and began the study of Greek when he was twelve. In the process, he was exposed to the writings of Aristotle and became intrigued by formalized logic.

At the age of fifteen, Leibniz entered the University of Leipzig to study law. He received his bachelor's degree two years later, earned his master's degree the following year, and then transferred to the University of Nuremberg.

Leibniz received his doctorate in law within a year and was immediately offered a professorship but refused it, saying that he had "other things in mind." Besides law, these "other things" included politics, religion, history, literature, metaphysics, philosophy, logic, and mathematics. Thereafter, Leibniz worked under the sponsorship of the courts of various nobles, serving as lawyer, historian, and librarian to the elite. At one point, Leibniz was offered the position of librarian at the Vatican but declined the offer.

Leibniz's affinity for logic was characterized by his search for a *characteristica universalis,* or "universal character." Leibniz believed that by combining logic and mathematics, a general symbolic language could be created in which all scientific problems could be solved with a minimum of effort. In this universal language, statements and the logical relationships between them would be represented by letters and symbols. In Leibniz's words, "All truths of reason would be reduced to a kind of calculus, and the errors would only be errors of computation." In essence, Leibniz believed that once a problem had been translated into this universal language of symbolic logic, it would be solved automatically by simply applying the mathematical rules that governed the manipulation of the symbols.

Leibniz's work in the field of symbolic logic did not arouse much academic curiosity; many say that it was too far ahead of its time. The study of symbolic logic was not systematically investigated again until the nineteenth century.

In the early 1670s, Leibniz invented one of the world's first mechanical calculating machines. Leibniz's machine could multiply and divide, whereas an earlier machine invented by Blaise Pascal (see Chapter 3) could only add and subtract.

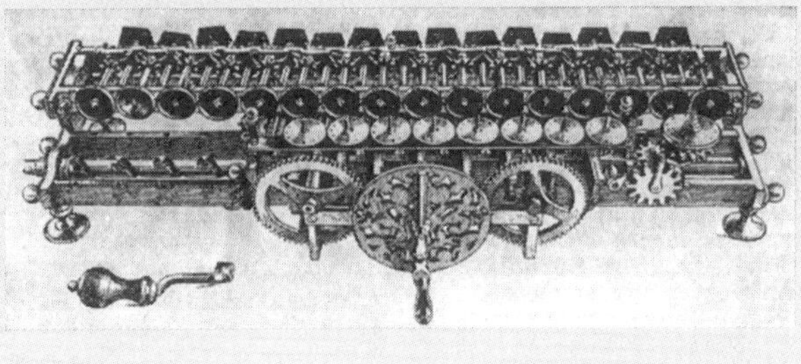

© Stock Montage

The statement "Kaitlin is a registered Republican or Kaitlin is a registered Democrat" does *not* include the possibility that both may happen; one statement *excludes* the other. When this happens, we are working with the **exclusive *or*.** In our study of symbolic logic (as in most mathematics), we will always use the *inclusive or.* Therefore, "p or q" means "p or q or both."

EXAMPLE 4 Using the symbolic representations

p: Juanita is a college graduate.

q: Juanita is employed.

express the following compound statements in words:

a. $p \vee q$ **b.** $p \wedge q$
c. $p \vee \sim q$ **d.** $\sim p \wedge q$

Solution **a.** $p \vee q$ represents the statement "Juanita is a college graduate or Juanita is employed (or both)."

b. $p \wedge q$ represents the statement "Juanita is a college graduate and Juanita is employed."

c. $p \vee \sim q$ represents the statement "Juanita is a college graduate or Juanita is not employed."

d. $\sim p \wedge q$ represents the statement "Juanita is not a college graduate and Juanita is employed." ■

The Conditional $p \rightarrow q$

Consider the statement "If it is raining, then the streets are wet." This is a compound statement because it connects two statements, namely, "it is raining" and "the streets are wet." Notice that the statements are connected with "if . . . then . . ." phrasing. Any statement of the form "if p then q" is called a **conditional** (or an **implication**); p is called the **hypothesis** (or **premise**) of the conditional, and q is called the **conclusion** of the conditional. The conditional "if p then q" is represented by the symbols "$p \rightarrow q$" (p implies q). When people use conditionals in everyday speech, they often omit the word *then,* as in "If it is raining, the streets are wet." Alternatively, the conditional "if p then q" may be phrased as "q if p" ("The streets are wet if it is raining").

EXAMPLE 5 Using the symbolic representations

p: I am healthy.

q: I eat junk food.

r: I exercise regularly.

express the following compound statements in symbolic form:

a. I am healthy if I exercise regularly.
b. If I eat junk food and do not exercise, then I am not healthy.

Solution **a.** "I am healthy if I exercise regularly" is a conditional (*if . . . then . . .*) and can be rephrased as follows:

"*If* I exercise regularly, *then* I am healthy."

Statement r is the premise. Statement p is the conclusion.

The given compound statement can be expressed as $r \rightarrow p$.

b. "If I eat junk food and do not exercise, then I am not healthy" is a conditional (*if . . . then . . .*) that contains a conjunction (*and*) and two negations (*not*):

"If I eat junk food *and* do *not* exercise, then I am *not* healthy."

The premise contains a conjunction and a negation. The conclusion contains a negation.

The premise of the conditional can be represented by $q \wedge \sim r$, while the conclusion can be represented by $\sim p$. Thus, the given compound statement has the symbolic form $(q \wedge \sim r) \rightarrow \sim p$. ■

EXAMPLE 6 Express the following statements in symbolic form:

a. All mammals are warm-blooded.

b. No snake is warm-blooded.

Solution **a.** The statement "All mammals are warm-blooded" can be rephrased as "If it is a mammal, then it is warm-blooded." Therefore, if we define p and q as

p: It is a mammal.

q: It is warm-blooded.

the statement can be expressed as $p \to q$. In general, the statement "All p are q" can be symbolized as $p \to q$.

b. The statement "No snake is warm-blooded" can be rephrased as "If it is a snake, then it is not warm-blooded." Therefore, if we define p and q as

p: It is a snake.

q: It is warm-blooded.

the statement can be expressed as $p \to \sim q$. In general, the statement "No p is q" can be symbolized as $p \to \sim q$. ■

We have seen that a statement is a sentence that is either true or false and that connecting two or more statements forms a compound statement. Figure 1.16 summarizes the logical connectives and symbols introduced in this section. The various connectives have been defined; we can now proceed in our analysis of the conditions under which a compound statement is true. This analysis is carried out in the next section.

Statement	Symbol	Read as . . .
negation	$\sim$	not
conjunction	$\wedge$	and
disjunction	$\vee$	(inclusive) or
conditional (implication)	$\to$	if . . . then . . .

Figure 1.16
Logical connectives

1.2 Exercises

1. Which of the following are statements? Why or why not?

a. His name is George Washington.

b. Abraham Lincoln was the first president of the United States.

c. Who was the first vice president of the United States?

d. Abraham Lincoln was the best president.

2. Which of the following are statements? Why or why not?

a. $3 + 5 = 6$

b. Solve the equation $2x + 5 = 3$.

c. $x^2 + 1 = 0$ has no solution.

d. $x^2 - 1 = (x + 1)(x - 1)$

e. Is $\sqrt{2}$ a rational number?

3. Determine which pairs of statements are negations of each other.

a. All of the fruits are red.

b. None of the fruits is red.

c. Some of the fruits are red.

d. Some of the fruits are not red.

4. Determine which pairs of statements are negations of each other.

a. Some of the beverages contain caffeine.

b. Some of the beverages do not contain caffeine.

c. None of the beverages contain caffeine.

d. All of the beverages contain caffeine.

5. Write a sentence that represents the negation of each statement.
   a. Her dress is not red.
   b. Some computers are priced under $100.
   c. All dogs are four-legged animals.
   d. No sleeping bag is waterproof.
6. Write a sentence that represents the negation of each statement.
   a. She is not a vegetarian.
   b. Some elephants are pink.
   c. All candy promotes tooth decay.
   d. No lunch is free.
7. Using the symbolic representations

   p: The lyrics are controversial.
   q: The performance is banned.

   express the following compound statements in symbolic form.
   a. The lyrics are controversial and the performance is banned.
   b. If the lyrics are not controversial, the performance is not banned.
   c. It is not the case that the lyrics are controversial or the performance is banned.
   d. The lyrics are controversial and the performance is not banned.
8. Using the symbolic representations

   p: The food is spicy.
   q: The food is aromatic.

   express the following compound statements in symbolic form.
   a. The food is aromatic and spicy.
   b. If the food isn't spicy, it isn't aromatic.
   c. The food is spicy and it isn't aromatic.
   d. The food isn't spicy or aromatic.
9. Using the symbolic representations

   p: A person plays the guitar.
   q: A person rides a motorcycle.
   r: A person wears a leather jacket.

   express the following compound statements in symbolic form.
   a. If a person plays the guitar or rides a motorcycle, then the person wears a leather jacket.
   b. A person plays the guitar, rides a motorcycle, and wears a leather jacket.
   c. A person wears a leather jacket and doesn't play the guitar or ride a motorcycle.
   d. All motorcycle riders wear leather jackets.
10. Using the symbolic representations

   p: The car costs $40,000.
   q: The car goes 140 mph.
   r: The car is red.

   express the following compound statements in symbolic form.
   a. All red cars go 140 mph.
   b. The car is red, goes 140 mph, and does not cost $40,000.
   c. If the car does not cost $40,000, it does not go 140 mph.
   d. The car is red and it does not go 140 mph or cost $40,000.

In Exercises 11–26, translate the sentence into symbolic form. Be sure to define each letter you use. (More than one answer is possible.)

11. All squares are rectangles.
12. No square is a triangle.
13. All whole numbers are even or odd.
14. No whole number is greater than 3 and less than 4.
15. All people born in the United States are American citizens.
16. No convicted felon is eligible to vote.
17. All muscle cars from the Sixties are polluters.
18. No electric-powered car is a polluter.
19. I do not sleep soundly if I drink coffee or eat chocolate.
20. I sleep soundly if I do not drink coffee or eat chocolate.
21. Your check is not accepted if you do not have a driver's license or a credit card.
22. Your check is accepted if you have a driver's license or a credit card.
23. If you drink and drive, you are fined or you go to jail.
24. If you are rich and famous, you have many friends and enemies.
25. You get a refund or a store credit if the product is defective.
26. The streets are slippery if it is raining or snowing.
27. Using the symbolic representations

   p: I am an environmentalist.
   q: I recycle my aluminum cans.

   express the following in words.
   a. $p \wedge q$
   b. $p \rightarrow q$
   c. $\sim q \rightarrow \sim p$
   d. $q \vee \sim p$
28. Using the symbolic representations

   p: I am innocent.
   q: I have an alibi.

   express the following in words.
   a. $p \wedge q$
   b. $p \rightarrow q$
   c. $\sim q \rightarrow \sim p$
   d. $q \vee \sim p$

29. Using the symbolic representations
p: I am an environmentalist.
q: I recycle my aluminum cans.
r: I recycle my newspapers.
express the following in words.
a. $(q \vee r) \rightarrow p$ **b.** $\sim p \rightarrow \sim(q \vee r)$
c. $(q \wedge r) \vee \sim p$ **d.** $(r \wedge \sim q) \rightarrow \sim p$

30. Using the symbolic representations
p: I am innocent.
q: I have an alibi.
r: I go to jail.
express the following in words.
a. $(p \vee q) \rightarrow \sim r$ **b.** $(p \wedge \sim q) \rightarrow r$
c. $(\sim p \wedge q) \vee r$ **d.** $(p \wedge r) \rightarrow \sim q$

Answer the following questions using complete sentences and your own words.

Concept Questions

31. What is a negation?
32. What is a conjunction?
33. What is a disjunction?
34. What is a conditional?
35. What is the difference between the inclusive *or* and the exclusive *or?*
36. Create a sentence that is a self-contradiction, or paradox, as in part (e) of Example 1.

History Questions

37. In what academic field did Gottfried Leibniz receive his degrees? Why is the study of logic important in this field?
38. Who developed a formal system of logic based on syllogistic arguments?
39. What is meant by *characteristica universalis?* Who proposed this theory?

1.3 Truth Tables

Suppose your friend Maria is a doctor, and you know that she is a Democrat. If someone told you, "Maria is a doctor and a Republican," you would say that the statement was false. On the other hand, if you were told, "Maria is a doctor or a Republican," you would say that the statement was true. Each of these statements is a compound statement—the result of joining individual statements with connective words. When is a compound statement true, and when is it false? To answer these questions, we must examine whether the individual statements are true or false and the manner in which the statements are connected.

The **truth value** of a statement is the classification of the statement as true or false and is denoted by T or F. For example, the truth value of the statement "Santa Fe is the capital of New Mexico" is T. (The statement is true.) In contrast, the truth value of "Memphis is the capital of Tennessee" is F. (The statement is false.)

A convenient way of determining whether a compound statement is true or false is to construct a truth table. A **truth table** is a listing of all possible combinations of the individual statements as true or false, along with the resulting truth value of the compound statement. As we will see, truth tables also allow us to distinguish valid arguments from invalid arguments.

The Negation $\sim p$

	p
1.	T
2.	F

Figure 1.17
Truth values for a statement p

The **negation** of a statement is the denial, or opposite, of the statement. (As was stated in the previous section, because the truth value of the negation depends on the truth value of the original statement, a negation can be classified as a compound statement.) To construct the truth table for the negation of a statement, we must first examine the original statement. A statement p may be true or false, as shown in Figure 1.17. If the statement p is true, the negation $\sim p$ is false; if p is false, $\sim p$ is true. The

	p	$\sim p$
1.	T	F
2.	F	T

Figure 1.18
Truth table for a negation $\sim p$

truth table for the compound statement $\sim p$ is given in Figure 1.18. Row 1 of the table is read "$\sim p$ is false when p is true." Row 2 is read "$\sim p$ is true when p is false."

The Conjunction $p \wedge q$

A **conjunction** is the joining of two statements with the word *and.* The compound statement "Maria is a doctor and a Republican" is a conjunction with the following symbolic representation:

p: Maria is a doctor.

q: Maria is a Republican.

$p \wedge q$: Maria is a doctor and a Republican.

The truth value of a compound statement depends on the truth values of the individual statements that make it up. How many rows will the truth table for the conjunction $p \wedge q$ contain? Because p has two possible truth values (T or F) and q has two possible truth values (T or F), we need four $(2 \cdot 2)$ rows in order to list all possible combinations of Ts and Fs, as shown in Figure 1.19.

	p	q
1.	T	T
2.	T	F
3.	F	T
4.	F	F

Figure 1.19
Truth values for two statements

For the conjunction $p \wedge q$ to be true, the components p and q must *both* be true; the conjunction is false otherwise. The completed truth table for the conjunction $p \wedge q$ is given in Figure 1.20. The symbols p and q can be replaced by any statements. The table gives the truth value of the statement "p and q," dependent upon the truth values of the individual statements "p" and "q." For instance, row 3 is read "The conjunction $p \wedge q$ is false when p is false and q is true." The other rows are read in a similar manner.

	p	q	$p \wedge q$
1.	T	T	T
2.	T	F	F
3.	F	T	F
4.	F	F	F

Figure 1.20
Truth table for a conjunction $p \wedge q$

The Disjunction $p \vee q$

A **disjunction** is the joining of two statements with the word *or.* The compound statement "Maria is a doctor or a Republican" is a disjunction (the *inclusive or*) with the following symbolic representation:

p: Maria is a doctor.

q: Maria is a Republican.

$p \vee q$: Maria is a doctor or a Republican.

Even though your friend Maria the doctor is not a Republican, the disjunction "Maria is a doctor or a Republican" is true. For a disjunction to be true, *at least one* of the components must be true. A disjunction is false only when *both* components are false. The truth table for the disjunction $p \vee q$ is given in Figure 1.21.

	p	q	$p \vee q$
1.	T	T	T
2.	T	F	T
3.	F	T	T
4.	F	F	F

Figure 1.21
Truth table for a disjunction $p \vee q$

EXAMPLE 1 Under what specific conditions is the following compound statement true? "I have a high school diploma, or I have a full-time job and no high school diploma."

Solution First, we translate the statement into symbolic form, and then we construct the truth table for the symbolic expression. Define p and q as

p: I have a high school diploma.

q: I have a full-time job.

The given statement has the symbolic representation $p \vee (q \wedge \sim p)$.

Because there are two letters, we need $2 \cdot 2 = 4$ rows. We need to insert a column for each connective in the symbolic expression $p \vee (q \wedge \sim p)$. As in algebra, we start inside any grouping symbols and work our way out. Therefore, we need a column for $\sim p$, a column for $q \wedge \sim p$, and a column for the entire expression $p \vee (q \wedge \sim p)$, as shown in Figure 1.22.

	p	q	$\sim p$	$q \wedge \sim p$	$p \vee (q \wedge \sim p)$
1.	T	T			
2.	T	F			
3.	F	T			
4.	F	F			

Figure 1.22

In the $\sim p$ column, fill in truth values that are opposite those for p. Next, the conjunction $q \wedge \sim p$ is true only when both components are true; enter a T in row 3 and Fs elsewhere. Finally, the disjunction $p \vee (q \wedge \sim p)$ is false only when both components p and $(q \wedge \sim p)$ are false; enter an F in row 4 and Ts elsewhere. The completed truth table is shown in Figure 1.23.

	p	q	$\sim p$	$q \wedge \sim p$	$p \vee (q \wedge \sim p)$
1.	T	T	F	F	T
2.	T	F	F	F	T
3.	F	T	T	T	T
4.	F	F	T	F	F

Figure 1.23
Truth table for $p \vee (q \wedge \sim p)$

As indicated in the truth table, the symbolic expression $p \vee (q \wedge \sim p)$ is true under all conditions except one: row 4; the expression is false when both p and q are false. Therefore, the statement "I have a high school diploma, or I have a full-time job and no high school diploma" is true in every case except when the speaker has no high school diploma and no full-time job. ■

If the symbolic representation of a compound statement consists of two different letters, its truth table will have $2 \cdot 2 = 4$ rows. How many rows are required if a compound statement consists of three letters—say, p, q, and r? Because each statement has two possible truth values (T and F), the truth table must contain $2 \cdot 2 \cdot 2 = 8$ rows. In general, each time a new statement is added, the number of rows doubles.

Number of Rows

If a compound statement consists of n individual statements, each represented by a different letter, the number of rows required in its truth table is 2^n.

EXAMPLE 2 Under what specific conditions is the following compound statement true? "I own a handgun, and it is not the case that I am a criminal or police officer."

Solution First, we translate the statement into symbolic form, and then we construct the truth table for the symbolic expression. Define the three simple statements as follows:

p: I own a handgun.

q: I am a criminal.

r: I am a police officer.

The given statement has the symbolic representation $p \wedge \sim(q \vee r)$. Since there are three letters, we need $2^3 = 8$ rows. We start with three columns, one for each letter. To account for all possible combinations of p, q, and r as true or false, proceed as follows:

1. Fill the first half (four rows) of column 1 with Ts and the rest with Fs, as shown in Figure 1.24(a).
2. In the next column, split each half into halves, the first half receiving Ts and the second Fs. In other words, alternate two Ts and two Fs in column 2, as shown in Figure 1.24(b).
3. Again, split each half into halves; the first half receives Ts, and the second receives Fs. Because we are dealing with the third (last) column, the Ts and Fs will alternate, as shown in Figure 1.24(c).

	p	q	r
1.	T		
2.	T		
3.	T		
4.	T		
5.	F		
6.	F		
7.	F		
8.	F		

(a)

	p	q	r
1.	T	T	
2.	T	T	
3.	T	F	
4.	T	F	
5.	F	T	
6.	F	T	
7.	F	F	
8.	F	F	

(b)

	p	q	r
1.	T	T	T
2.	T	T	F
3.	T	F	T
4.	T	F	F
5.	F	T	T
6.	F	T	F
7.	F	F	T
8.	F	F	F

(c)

Figure 1.24
Truth values for three statements

(This process of filling the first half of the first column with Ts and the second half with Fs and then splitting each half into halves with blocks of Ts and Fs applies to all truth tables.)

We need to insert a column for each connective in the symbolic expression $p \wedge \sim(q \vee r)$, as shown in Figure 1.25.

Now fill in the appropriate symbol in the column under $q \vee r$. Enter F if *both* q and r are false; enter T otherwise (that is, if at least one is true). In the $\sim(q \vee r)$ column, fill in truth values that are opposite those for $q \vee r$, as in Figure 1.26.

The conjunction $p \wedge \sim(q \vee r)$ is true only when *both* p and $\sim(q \vee r)$ are true; enter a T in row 4 and Fs elsewhere. The truth table is shown in Figure 1.27.

As indicated in the truth table, the expression $p \wedge \sim(q \vee r)$ is true only when p is true and both q and r are false. Therefore, the statement "I own a handgun, and it is not the case that I am a criminal or police officer" is true only when the speaker owns a handgun, is not a criminal, and is not a police officer—in other words, the speaker is a law-abiding citizen who owns a handgun.

	p	q	r	$q \vee r$	$\sim(q \vee r)$	$p \wedge \sim(q \vee r)$
1.	T	T	T			
2.	T	T	F			
3.	T	F	T			
4.	T	F	F			
5.	F	T	T			
6.	F	T	F			
7.	F	F	T			
8.	F	F	F			

Figure 1.25

	p	q	r	$q \vee r$	$\sim(q \vee r)$	$p \wedge \sim(q \vee r)$
1.	T	T	T	T	F	
2.	T	T	F	T	F	
3.	T	F	T	T	F	
4.	T	F	F	F	T	
5.	F	T	T	T	F	
6.	F	T	F	T	F	
7.	F	F	T	T	F	
8.	F	F	F	F	T	

Figure 1.26

	p	q	r	$q \vee r$	$\sim(q \vee r)$	$p \wedge \sim(q \vee r)$
1.	T	T	T	T	F	F
2.	T	T	F	T	F	F
3.	T	F	T	T	F	F
4.	T	F	F	F	T	T
5.	F	T	T	T	F	F
6.	F	T	F	T	F	F
7.	F	F	T	T	F	F
8.	F	F	F	F	T	F

Figure 1.27
Truth table for $p \wedge \sim(q \vee r)$

■

The Conditional $p \to q$

A **conditional** is a compound statement of the form "If p, then q" and is symbolized $p \to q$. Under what circumstances is a conditional true, and when is it false? Consider the following (compound) statement: "If you give me \$50, then I will give you a ticket to the ballet." This statement is a conditional and has the following representation:

p: You give me \$50.

q: I give you a ticket to the ballet.

$p \to q$: If you give me \$50, then I will give you a ticket to the ballet.

The conditional can be viewed as a promise: *If* you give me \$50, *then* I will give you a ticket to the ballet. Suppose you give me \$50; that is, suppose p is true. I have two options: Either I give you a ticket to the ballet (q is true), or I do not (q is false). If I do give you the ticket, the conditional $p \to q$ is true (I have kept my promise); if I do not give you the ticket, the conditional $p \to q$ is false (I have not kept my promise). These situations are shown in rows 1 and 2 of the truth table in Figure 1.28. Rows 3 and 4 require further analysis.

	p	q	$p \to q$
1.	T	T	T
2.	T	F	F
3.	F	T	?
4.	F	F	?

Figure 1.28

Suppose you do not give me \$50; that is, suppose p is false. Regardless of whether or not I give you a ticket, you cannot say that I broke my promise; that is, you cannot say that the conditional $p \to q$ is false. Consequently, since a statement is either true or false, the conditional is labeled true (by default). In other words, when the premise p of a conditional is false, it does not matter whether the conclusion q is true or false. In both cases, the conditional $p \to q$ is automatically labeled true, because it is not false.

The completed truth table for a conditional is given in Figure 1.29. Notice that the only circumstance under which a conditional is false is when the premise p is true and the conclusion q is false, as shown in row 2.

	p	q	$p \to q$
1.	T	T	T
2.	T	F	F
3.	F	T	T
4.	F	F	T

Figure 1.29

EXAMPLE 3 Under what conditions is the symbolic expression $q \to \sim p$ true?

Solution Our truth table has $2^2 = 4$ rows and contains a column for p, q, $\sim p$, and $q \to \sim p$, as shown in Figure 1.30.

In the $\sim p$ column, fill in truth values that are opposite those for p. Now, a conditional is false only when its premise (in this case, q) is true and its conclusion (in this case, $\sim p$) is false. Therefore, $q \to \sim p$ is false only in row 1; the conditional $q \to \sim p$ is true under all conditions except the condition that both p and q are true. The completed truth table is shown in Figure 1.31.

	p	q	$\sim p$	$q \to \sim p$
1.	T	T		
2.	T	F		
3.	F	T		
4.	F	F		

Figure 1.30

	p	q	$\sim p$	$q \to \sim p$
1.	T	T	F	F
2.	T	F	F	T
3.	F	T	T	T
4.	F	F	T	T

Figure 1.31
Truth table for $q \to \sim p$ ■

EXAMPLE 4 Construct a truth table for the following compound statement: "I walk up the stairs if I want to exercise or if the elevator isn't working."

Solution Rewriting the statement so the word *if* is first, we have "If I want to exercise or (if) the elevator isn't working, then I walk up the stairs."

Now we must translate the statement into symbols and construct a truth table. Define the following:

p: I want to exercise.

q: The elevator is working.

r: I walk up the stairs.

The statement now has the symbolic representation $(p \vee \sim q) \rightarrow r$. Because we have three letters, our table must have $2^3 = 8$ rows. Inserting a column for each letter and a column for each connective, we have the initial setup shown in Figure 1.32.

	p	q	r	$\sim q$	$p \vee \sim q$	$(p \vee \sim q) \rightarrow r$
1.	T	T	T			
2.	T	T	F			
3.	T	F	T			
4.	T	F	F			
5.	F	T	T			
6.	F	T	F			
7.	F	F	T			
8.	F	F	F			

Figure 1.32

In the column labeled $\sim q$, enter truth values that are the opposite of those of q. Next, enter the truth values of the disjunction $p \vee \sim q$ in column 5. Recall that a disjunction is false only when both components are false and is true otherwise. Consequently, enter Fs in rows 5 and 6 (since both p and $\sim q$ are false) and Ts in the remaining rows, as shown in Figure 1.33.

	p	q	r	$\sim q$	$p \vee \sim q$	$(p \vee \sim q) \rightarrow r$
1.	T	T	T	F	T	
2.	T	T	F	F	T	
3.	T	F	T	T	T	
4.	T	F	F	T	T	
5.	F	T	T	F	F	
6.	F	T	F	F	F	
7.	F	F	T	T	T	
8.	F	F	F	T	T	

Figure 1.33

The last column involves a conditional; it is false only when its premise is true and its conclusion is false. Therefore, enter Fs in rows 2, 4, and 8 (since $p \vee \sim q$ is true and r is false) and Ts in the remaining rows. The truth table is shown in Figure 1.34.

	p	q	r	$\sim q$	$p \vee \sim q$	$(p \vee \sim q) \rightarrow r$
1.	T	T	T	F	T	T
2.	T	T	F	F	T	F
3.	T	F	T	T	T	T
4.	T	F	F	T	T	F
5.	F	T	T	F	F	T
6.	F	T	F	F	F	T
7.	F	F	T	T	T	T
8.	F	F	F	T	T	F

Figure 1.34
Truth table for $(p \vee \sim q) \rightarrow r$

As Figure 1.34 shows, the statement "I walk up the stairs if I want to exercise or if the elevator isn't working" is true in all situations except those listed in rows 2, 4, and 8. For instance, the statement is false (row 8) when the speaker does not want to exercise, the elevator is not working, and the speaker does not walk up the stairs—in other words, the speaker stays on the ground floor of the building when the elevator is broken. ■

Equivalent Expressions

When you purchase a car, the car is either new or used. If a salesperson told you, "It is not the case that the car is not new," what condition would the car be in? This compound statement consists of one individual statement ("*p:* The car is new") and two negations:

"It is not the case that the car is not new."

$\sim$ $\sim p$

Does this mean that the car is new? To answer this question, we will construct a truth table for the symbolic expression $\sim(\sim p)$ and compare its truth values with those of the original p. Because there is only one letter, we need $2^1 = 2$ rows, as shown in Figure 1.35.

	p
1.	T
2.	F

Figure 1.35

We must insert a column for $\sim p$ and a column for $\sim(\sim p)$. Now, $\sim p$ has truth values that are opposite those of p, and $\sim(\sim p)$ has truth values that are opposite those of $\sim p$, as shown in Figure 1.36.

	p	$\sim p$	$\sim(\sim p)$
1.	T	F	T
2.	F	T	F

Figure 1.36
Truth table for $\sim(\sim p)$

Notice that the values in the column labeled $\sim(\sim p)$ are identical to those in the column labeled p. Whenever this happens, the expressions are said to be equivalent and may be used interchangeably. Therefore, the statement "It is not the case that the car is not new" is equivalent in meaning to the statement "The car is new."

Equivalent expressions are symbolic expressions that have identical truth values in each corresponding entry. The expression $p \equiv q$ is read "p is equivalent to q" or "p and q are equivalent." As we can see in Figure 1.36, an expression and its double negation are logically equivalent. This relationship can be expressed as $p \equiv \sim(\sim p)$.

EXAMPLE 5 Are the statements "If I am a homeowner, then I pay property taxes" and "I am a homeowner and I do not pay property taxes" equivalent?

Solution We begin by defining the statements:

p: I am a homeowner.

q: I pay property taxes.

$p \rightarrow q$: If I am a homeowner, then I pay property taxes.

$p \wedge \sim q$: I am a homeowner and I do not pay property taxes.

The truth table contains $2^2 = 4$ rows, and the initial setup is shown in Figure 1.37.

	p	q	$\sim q$	$p \wedge \sim q$	$p \rightarrow q$
1.	T	T			
2.	T	F			
3.	F	T			
4.	F	F			

Figure 1.37

Now enter the appropriate truth values under $\sim q$ (the opposite of q). Because the conjunction $p \wedge \sim q$ is true only when both p and $\sim q$ are true, enter a T in row 2 and Fs elsewhere. The conditional $p \rightarrow q$ is false only when p is true and q is false; therefore, enter an F in row 2 and Ts elsewhere. The completed truth table is shown in Figure 1.38.

	p	q	$\sim q$	$p \wedge \sim q$	$p \rightarrow q$
1.	T	T	F	F	T
2.	T	F	T	T	F
3.	F	T	F	F	T
4.	F	F	T	F	T

Figure 1.38

Because the entries in the columns labeled $p \wedge \sim q$ and $p \rightarrow q$ are not the same, the statements are not equivalent. "If I am a homeowner, then I pay property taxes" is *not* equivalent to "I am a homeowner and I do not pay property taxes." ■

Notice that the truth values in the columns under $p \wedge \sim q$ and $p \rightarrow q$ in Figure 1.38 are exact opposites; when one is T, the other is F. Whenever this happens, one statement is the negation of the other. Consequently, $p \wedge \sim q$ is the negation of $p \rightarrow q$ (and

HISTORICAL NOTE

George Boole

1815–1864

George Boole is called "the father of symbolic logic." Computer science owes much to this self-educated mathematician. Born the son of a poor shopkeeper in Lincoln, England, Boole had very little formal education, and his prospects for rising above his family's lower-class status were dim. Like Leibniz, he taught himself Latin; at the age of twelve, he translated an ode of Horace into English, winning the attention of the local schoolmasters. (In his day, the knowledge of Latin was a prerequisite to scholarly endeavors and to becoming a socially accepted gentleman.) After that, his academic desires were encouraged, and at the age of fifteen, he began his long teaching career. While teaching arithmetic, he studied advanced mathematics and physics.

In 1849, after nineteen years of teaching at elementary schools, Boole received his big break: He was appointed professor of mathematics at Queen's College in the city of Cork, Ireland. At last, he was able to research advanced mathematics, and he became recognized as a first-class mathematician. This was a remarkable feat, considering Boole's lack of formal training and degrees.

Boole's most influential work, *An Investigation of the Laws of Thought, on Which Are Founded the Mathematical Theories of Logic and Probabilities,* was published in 1854. In it he wrote, "There exist certain general principles founded in the very nature of language and logic that exhibit laws as identical in form as with the laws of the general symbols of algebra." With this insight, Boole had taken a big step into the world of logical reasoning and abstract mathematical analysis.

Perhaps because of his lack of formal training, Boole challenged the status quo, including the Aristotelian assumption that *all* logical arguments could be reduced to syllogistic arguments. In doing so, he employed symbols to represent concepts, as did Leibniz, but he also developed systems of algebraic manipulation to accompany these symbols. Thus, Boole's creation is a marriage of logic and mathematics. However, as is the case with almost all new theories, Boole's symbolic logic was not met with total approbation. In particular, one staunch opponent of his work was Georg Cantor, whose work on the origins of set theory and the magnitude of infinity will be investigated in Chapter 2.

In the many years since Boole's original work was unveiled, various scholars have modified, improved, generalized, and extended its central concepts. Today, Boolean algebras are the essence of computer software and circuit design. After all, a computer merely manipulates predefined symbols and conforms to a set of preassigned algebraic commands.

Through an algebraic manipulation of logical symbols, Boole revolutionized the age-old study of logic. His essay *The Mathematical Analysis of Logic* laid the foundation for his later book *An Investigation of the Laws of Thought.*

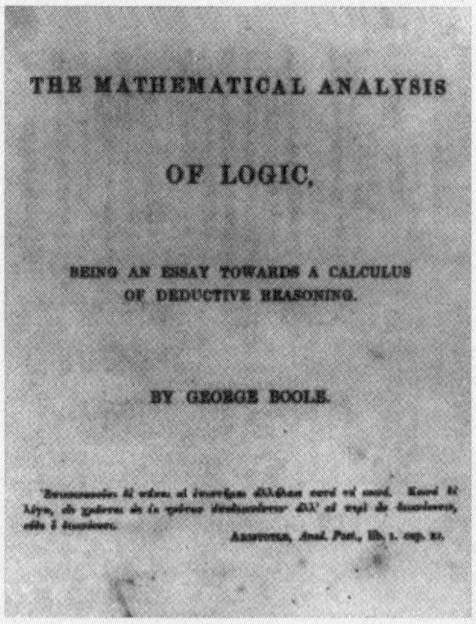

THE MATHEMATICAL ANALYSIS

OF LOGIC,

BEING AN ESSAY TOWARDS A CALCULUS OF DEDUCTIVE REASONING.

BY GEORGE BOOLE.

[illegible]

vice versa). This can be expressed as $p \wedge \sim q \equiv \sim(p \rightarrow q)$. The negation of a conditional is logically equivalent to the conjunction of the premise and the negation of the conclusion.

Statements that look or sound different may in fact have the same meaning. For example, "It is not the case that the car is not new" really means the same as "The car is new," and "It is not the case that if I am a homeowner, then I pay property taxes" actually means the same as "I am a homeowner and I do not pay property taxes." When we are working with equivalent statements, we can substitute either statement for the other without changing the truth value.

De Morgan's Laws

Earlier in this section, we saw that the negation of a negation is equivalent to the original statement; that is, $\sim(\sim p) \equiv p$. Another negation "formula" we discovered was $\sim(p \rightarrow q) \equiv p \wedge \sim q$, that is, the negation of a conditional. Can we find similar "formulas" for the negations of the other basic connectives, namely, the conjunction and the disjunction? The answer is yes, and the results are credited to the English mathematician and logician Augustus De Morgan.

De Morgan's Laws

The negation of the conjunction $p \wedge q$ is given by $\sim(p \wedge q) \equiv \sim p \vee \sim q$.

"Not p and q" is equivalent to "not p or not q."

The negation of the disjunction $p \vee q$ is given by $\sim(p \vee q) \equiv \sim p \wedge \sim q$.

"Not p or q" is equivalent to "not p and not q."

De Morgan's Laws are easily verified through the use of truth tables and will be addressed in the exercises (see Exercises 47 and 48).

EXAMPLE 6 Using De Morgan's Laws, find the negation of each of the following:

a. It is Friday and I receive a paycheck.
b. You are correct or I am crazy.

Solution **a.** The symbolic representation of "It is Friday and I receive a paycheck" is

p: It is Friday.

q: I receive a paycheck.

$p \wedge q$: It is Friday and I receive a paycheck.

Therefore, the negation is $\sim(p \wedge q) \equiv \sim p \vee \sim q$, that is, "It is not Friday or I do not receive a paycheck."

b. The symbolic representation of "You are correct or I am crazy" is

p: You are correct.

q: I am crazy.

$p \vee q$: You are correct or I am crazy.

Therefore, the negation is $\sim(p \vee q) \equiv \sim p \wedge \sim q$, that is, "You are not correct and I am not crazy." ■

As we have seen, the truth value of a compound statement depends on the truth values of the individual statements that make it up. The truth tables of the basic connectives are summarized in Figure 1.39.

Equivalent statements are statements that have the same meaning. Equivalent statements for the negations of the basic connectives are given in Figure 1.40.

	p	$\sim p$
1.	T	F
2.	F	T

Negation

	p	q	$p \wedge q$
1.	T	T	T
2.	T	F	F
3.	F	T	F
4.	F	F	F

Conjunction

	p	q	$p \vee q$
1.	T	T	T
2.	T	F	T
3.	F	T	T
4.	F	F	F

Disjunction

	p	q	$p \rightarrow q$
1.	T	T	T
2.	T	F	F
3.	F	T	T
4.	F	F	T

Conditional

Figure 1.39
Truth tables for the basic connectives

Figure 1.40
Negations of the basic connectives

1. $\sim(\sim p) \equiv p$	the negation of a negation
2. $\sim(p \wedge q) \equiv \sim p \vee \sim q$	the negation of a conjunction
3. $\sim(p \vee q) \equiv \sim p \wedge \sim q$	the negation of a disjunction
4. $\sim(p \rightarrow q) \equiv p \wedge \sim q$	the negation of a conditional

1.3 Exercises

In Exercises 1–20, construct a truth table for the symbolic expressions.

1. $p \vee \sim q$
2. $p \wedge \sim q$
3. $p \vee \sim p$
4. $p \wedge \sim p$
5. $p \rightarrow \sim q$
6. $\sim p \rightarrow q$
7. $\sim q \rightarrow \sim p$
8. $\sim p \rightarrow \sim q$
9. $(p \vee q) \rightarrow \sim p$
10. $(p \wedge q) \rightarrow \sim q$
11. $(p \vee q) \rightarrow (p \wedge q)$
12. $(p \wedge q) \rightarrow (p \vee q)$
13. $p \wedge \sim(q \vee r)$
14. $p \vee \sim(q \vee r)$
15. $p \vee (\sim q \wedge r)$
16. $\sim p \vee \sim(q \wedge r)$
17. $(\sim r \vee p) \rightarrow (q \wedge p)$
18. $(q \wedge p) \rightarrow (\sim r \vee p)$
19. $(p \vee r) \rightarrow (q \wedge \sim r)$
20. $(p \wedge r) \rightarrow (q \vee \sim r)$

In Exercises 21–36, translate the compound statement into symbolic form and then construct the truth table for the expression.

21. If it is raining, then the streets are wet.
22. If the lyrics are not controversial, the performance is not banned.
23. The water supply is rationed if it does not rain.
24. The country is in trouble if he is elected.
25. All squares are rectangles.
26. All muscle cars from the Sixties are polluters.
27. No square is a triangle.
28. No electric-powered car is a polluter.
29. Your check is accepted if you have a driver's license or a credit card.
30. You get a refund or a store credit if the product is defective.
31. If leaded gasoline is used, the catalytic converter is damaged and the air is polluted.
32. If he does not go to jail, he is innocent or has an alibi.
33. I have a college degree and I do not have a job or own a house.
34. I surf the Internet and I make purchases and do not pay sales tax.
35. If Proposition A passes and Proposition B does not, jobs are lost or new taxes are imposed.
36. If Proposition A does not pass and the legislature raises taxes, the quality of education is lowered and unemployment rises.

In Exercises 37–46, construct a truth table to determine whether the statements in each pair are equivalent.

37. The streets are wet or it is not raining.
If it is raining, then the streets are wet.
38. The streets are wet or it is not raining.
If the streets are not wet, then it is not raining.
39. He has a high school diploma or he is unemployed.
If he does not have a high school diploma, then he is unemployed.
40. She is unemployed or she does not have a high school diploma.
If she is employed, then she does not have a high school diploma.

41. If handguns are outlawed, then outlaws have handguns.
If outlaws have handguns, then handguns are outlawed.
42. If interest rates continue to fall, then I can afford to buy a house.
If interest rates do not continue to fall, then I cannot afford to buy a house.
43. If the spotted owl is on the endangered species list, then lumber jobs are lost.
If lumber jobs are not lost, then the spotted owl is not on the endangered species list.
44. If I drink decaffeinated coffee, then I do not stay awake.
If I do stay awake, then I do not drink decaffeinated coffee.
45. The plaintiff is innocent or the insurance company does not settle out of court.
The insurance company settles out of court and the plaintiff is not innocent.
46. The plaintiff is not innocent and the insurance company settles out of court.
It is not the case that the plaintiff is innocent or the insurance company does not settle out of court.
47. Using truth tables, verify De Morgan's Law

$$\sim(p \wedge q) \equiv \sim p \vee \sim q.$$

48. Using truth tables, verify De Morgan's Law

$$\sim(p \vee q) \equiv \sim p \wedge \sim q.$$

In Exercises 49–56, write the statement in symbolic form, construct the negation of the expression (in simplified symbolic form), and express the negation in words.

49. I have a college degree and I am not employed.
50. It is snowing and classes are canceled.
51. The television set is broken or there is a power outage.
52. The freeway is under construction or I do not ride the bus.
53. If the building contains asbestos, the original contractor is responsible.
54. If the legislation is approved, the public is uninformed.
55. The First Amendment has been violated if the lyrics are censored.
56. Your driver's license is taken away if you do not obey the laws.

Answer the following questions using complete sentences and your own words.

Concept Questions

57. **a.** Under what conditions is a disjunction true?
b. Under what conditions is a disjunction false?
58. **a.** Under what conditions is a conjunction true?
b. Under what conditions is a conjunction false?
59. **a.** Under what conditions is a conditional true?
b. Under what conditions is a conditional false?
60. **a.** Under what conditions is a negation true?
b. Under what conditions is a negation false?
61. What are equivalent expressions?
62. What is a truth table?
63. When constructing a truth table, how do you determine how many rows to create?

History Questions

64. Who is considered "the father of symbolic logic"?
65. Boolean algebra is a combination of logic and mathematics. What is it used for?

1.4 More on Conditionals

Conditionals differ from conjunctions and disjunctions with regard to the possibility of changing the order of the statements. In algebra, the sum $x + y$ is equal to the sum $y + x$; that is, addition is commutative. In everyday language, one realtor might say, "The house is perfect and the lot is priceless," while another says, "The lot is priceless and the house is perfect." Logically, their meanings are the same, since $(p \wedge q) \equiv (q \wedge p)$. The order of the components in a conjunction or disjunction makes no difference in regard to the truth value of the statement. This is not so with conditionals.

Variations of a Conditional

Given two statements p and q, various "if . . . then . . ." statements can be formed.

EXAMPLE 1 Using the statements

p: You are compassionate.

q: You contribute to charities.

write the sentence represented by each of the following:

a. $p \rightarrow q$ **b.** $q \rightarrow p$ **c.** $\sim p \rightarrow \sim q$ **d.** $\sim q \rightarrow \sim p$

Solution

a. $p \rightarrow q$: If you are compassionate, then you contribute to charities.
b. $q \rightarrow p$: If you contribute to charities, then you are compassionate.
c. $\sim p \rightarrow \sim q$: If you are not compassionate, then you do not contribute to charities.
d. $\sim q \rightarrow \sim p$: If you do not contribute to charities, then you are not compassionate. ■

Each part of Example 1 contains an "if . . . then . . ." statement and is called a conditional. Any given conditional has three variations: a converse, an inverse, and a contrapositive. The **converse** of the conditional "if p, then q" is the compound statement "if q, then p" That is, we form the converse of the conditional by interchanging the premise and the conclusion; $q \rightarrow p$ is the converse of $p \rightarrow q$. The statement in part (b) of Example 1 is the converse of the statement in part (a).

The **inverse** of the conditional "if p, then q" is the compound statement "if not p, then not q." We form the inverse of the conditional by negating both the premise and the conclusion; $\sim p \rightarrow \sim q$ is the inverse of $p \rightarrow q$. The statement in part (c) of Example 1 is the inverse of the statement in part (a).

The **contrapositive** of the conditional "if p, then q" is the compound statement "if not q, then not p." We form the contrapositive of the conditional by negating *and* interchanging both the premise and the conclusion; $\sim q \rightarrow \sim p$ is the contrapositive of $p \rightarrow q$. The statement in part (d) of Example 1 is the contrapositive of the statement in part (a). The variations of a given conditional are summarized in Figure 1.41. As we will see, some of these variations are equivalent, and some are not. Unfortunately, many people incorrectly treat them all as equivalent.

Name	Symbolic Form	Read As . . .
a (given) conditional	$p \rightarrow q$	If p, then q.
the converse (of $p \rightarrow q$)	$q \rightarrow p$	If q, then p.
the inverse (of $p \rightarrow q$)	$\sim p \rightarrow \sim q$	If not p, then not q.
the contrapositive (of $p \rightarrow q$)	$\sim q \rightarrow \sim p$	If not q, then not p.

Figure 1.41
Variations of a conditional

EXAMPLE 2 Given the conditional "You did not receive the proper refund if you prepared your own income tax form," write the sentence that represents each of the following:

a. the converse of the conditional
b. the inverse of the conditional
c. the contrapositive of the conditional

Solution

a. Rewriting the statement in the standard "if . . . then . . ." form, we have the conditional "If you prepared your own income tax form, then you did not receive the proper refund." The converse is formed by interchanging the premise and the conclusion. Thus, the converse is written as "If you did not receive the proper refund, then you prepared your own income tax form."

b. The inverse is formed by negating both the premise and the conclusion. Thus, the inverse is written as "If you did not prepare your own income tax form, then you received the proper refund."

c. The contrapositive is formed by negating *and* interchanging the premise and the conclusion. Thus, the contrapositive is written as "If you received the proper refund, then you did not prepare your own income tax form." ■

Equivalent Conditionals

We have seen that the conditional $p \rightarrow q$ has three variations: the converse $q \rightarrow p$, the inverse $\sim p \rightarrow \sim q$, and the contrapositive $\sim q \rightarrow \sim p$. Do any of these "if . . . then . . ." statements convey the same meaning? In other words, are any of these compound statements equivalent?

EXAMPLE 3 Determine which (if any) of the following are equivalent: a conditional $p \rightarrow q$, the converse $q \rightarrow p$, the inverse $\sim p \rightarrow \sim q$, and the contrapositive $\sim q \rightarrow \sim p$.

Solution To investigate the possible equivalencies, we must construct a truth table that contains all the statements. Because there are two letters, we need $2^2 = 4$ rows. The table must have a column for $\sim p$, one for $\sim q$, one for the conditional $p \rightarrow q$, and one for each variation of the conditional. The truth values of the negations $\sim p$ and $\sim q$ are readily entered, as shown in Figure 1.42.

	p	q	$\sim p$	$\sim q$	$p \rightarrow q$	$q \rightarrow p$	$\sim p \rightarrow \sim q$	$\sim q \rightarrow \sim p$
1.	T	T	F	F				
2.	T	F	F	T				
3.	F	T	T	F				
4.	F	F	T	T				

Figure 1.42

An "if . . . then . . ." statement is false only when the premise is true and the conclusion is false. Consequently, $p \rightarrow q$ is false only when p is T and q is F; enter an F in row 2 and Ts elsewhere in the column under $p \rightarrow q$.

Likewise, the converse $q \rightarrow p$ is false only when q is T and p is F; enter an F in row 3 and Ts elsewhere.

In a similar manner, the inverse $\sim p \rightarrow \sim q$ is false only when $\sim p$ is T and $\sim q$ is F; enter an F in row 3 and Ts elsewhere.

Finally, the contrapositive $\sim q \rightarrow \sim p$ is false only when $\sim q$ is T and $\sim p$ is F; enter an F in row 2 and Ts elsewhere.

The completed truth table is shown in Figure 1.43. Examining the entries in Figure 1.43, we can see that the columns under $p \rightarrow q$ and $\sim q \rightarrow \sim p$ are identical; each has an F in row 2 and Ts elsewhere. Consequently, a conditional and its contrapositive are equivalent: $p \rightarrow q \equiv \sim q \rightarrow \sim p$.

Likewise, we notice that $q \rightarrow p$ and $\sim p \rightarrow \sim q$ have identical truth values; each has an F in row 3 and Ts elsewhere. Thus, the converse and the inverse of a conditional are equivalent: $q \rightarrow p \equiv \sim p \rightarrow \sim q$.

	p	q	$\sim p$	$\sim q$	$p \to q$	$q \to p$	$\sim p \to \sim q$	$\sim q \to \sim p$
1.	T	T	F	F	T	T	T	T
2.	T	F	F	T	F	T	T	F
3.	F	T	T	F	T	F	F	T
4.	F	F	T	T	T	T	T	T

Figure 1.43
Truth table for a conditional and its variations

■

We have seen that different "if . . . then . . ." statements can convey the same meaning—that is, that certain variations of a conditional are equivalent (see Figure 1.44). For example, the compound statements "If you are compassionate, then you contribute to charities" and "If you do not contribute to charities, then you are not compassionate" convey the same meaning. (The second conditional is the contrapositive of the first.) Regardless of its specific contents (p, q, $\sim p$, or $\sim q$), every "if . . . then . . ." statement has an equivalent variation formed by negating *and* interchanging the premise and the conclusion of the given conditional statement.

Equivalent Statements	Symbolic Representations
a conditional and its contrapositive	$(p \to q) \equiv (\sim q \to \sim p)$
the converse and the inverse (of the conditional $p \to q$)	$(q \to p) \equiv (\sim p \to \sim q)$

Figure 1.44
Equivalent "if . . . then . . ." statements

The "Only If" Connective

Consider the statement "A prisoner is paroled only if the prisoner obeys the rules." What is the premise, and what is the conclusion? Rather than using p and q (which might bias our investigation), we define

r: A prisoner is paroled.

s: A prisoner obeys the rules.

The given statement is represented by "r only if s." Now, "r only if s" means that r can happen *only* if s happens. In other words, if s does not happen, then r does not happen, or $\sim s \to \sim r$. We have seen that $\sim s \to \sim r$ is equivalent to $r \to s$. Consequently, "r only if s" is equivalent to the conditional $r \to s$. The premise of the statement "A prisoner is paroled only if the prisoner obeys the rules" is "A prisoner is paroled," and the conclusion is "The prisoner obeys the rules."

The conditional $p \to q$ can be phrased "p only if q." Even though the word *if* precedes q, q is not the premise. *Whatever follows the connective "only if" is the conclusion of the conditional.*

EXAMPLE 4 For the compound statement "You receive a federal grant only if your artwork is not obscene," do the following:

a. Determine the premise and the conclusion.
b. Rewrite the compound statement in the standard "if . . . then . . ." form.
c. Interpret the conditions that make the statement false.

Solution **a.** Because the compound statement contains an "only if" connective, the statement that follows "only if" is the conclusion of the conditional. The premise is "You receive a federal grant." The conclusion is "Your artwork is not obscene."

b. The given compound statement can be rewritten as "If you receive a grant, then your artwork is not obscene."

c. First we define the symbols.

p: You receive a federal grant.

q: Your artwork is obscene.

Then the statement has the symbolic representation $p \to \sim q$. The truth table for $p \to \sim q$ is given in Figure 1.45.

The expression $p \to q$ is false under the conditions listed in row 1 (when p and q are both true). Therefore, the statement "You receive a federal grant only if your artwork is not obscene" is false when an artist *does* receive a federal grant *and* the artist's artwork *is* obscene. ■

	p	q	$\sim q$	$p \to \sim q$
1.	T	T	F	F
2.	T	F	T	T
3.	F	T	F	T
4.	F	F	T	T

Figure 1.45
Truth table for the conditional $p \to \sim q$

The Biconditional $p \leftrightarrow q$

What do the words *bicycle, binomial,* and *bilingual* have in common? Each word begins with the prefix *bi,* meaning "two." Just as the word *bilingual* means "two languages," the word *biconditional* means "two conditionals."

In everyday speech, conditionals often get "hooked together" in a circular fashion. For instance, someone might say, "If I am rich, then I am happy, and if I am happy, then I am rich." Notice that this compound statement is actually the conjunction (*and*) of a conditional (if rich, then happy) and its converse (if happy, then rich). Such a statement is referred to as a biconditional. A **biconditional** is a statement of the form $(p \to q) \wedge (q \to p)$ and is symbolized as $p \leftrightarrow q$. The symbol $p \leftrightarrow q$ is read "p if and only if q" and is frequently abbreviated "p iff q." A biconditional is equivalent to the conjunction of two conversely related conditionals: $p \leftrightarrow q \equiv [(p \to q) \wedge (q \to p)]$.

EXAMPLE 5 Express the biconditional "A citizen is eligible to vote if and only if the citizen is at least eighteen years old" as the conjunction of two conditionals.

Solution The given biconditional is equivalent to "If a citizen is eligible to vote, then the citizen is at least eighteen years old, *and* if a citizen is at least eighteen years old, then the citizen is eligible to vote." ■

Under what circumstances is the biconditional $p \leftrightarrow q$ true, and when is it false? To find the answer, we must construct a truth table. Utilizing the equivalence $p \leftrightarrow q \equiv [(p \to q) \wedge (q \to p)]$, we get the completed table shown in Figure 1.46. (Recall that a conditional is false only when its premise is true and its conclusion is false

	p	q	$p \to q$	$q \to p$	$(p \to q) \wedge (q \to p)$
1.	T	T	T	T	T
2.	T	F	F	T	F
3.	F	T	T	F	F
4.	F	F	T	T	T

Figure 1.46
Truth table for a biconditional $p \leftrightarrow q$

and that a conjunction is true only when both components are true.) We can see that a biconditional is true only when the two components p and q have the same truth value—that is, when p and q are both true or when p and q are both false.

Many theorems in mathematics can be expressed as biconditionals. For example, when solving a quadratic equation, we have the following: "The equation $ax^2 + bx + c = 0$ has exactly one solution if and only if the discriminant $b^2 - 4ac = 0$." Recall that the solutions of a quadratic equation are

$$x = \frac{-b \pm \sqrt{b^2 - 4ac}}{2a}$$

This biconditional is equivalent to "If the equation $ax^2 + bx + c = 0$ has exactly one solution, then the discriminant $b^2 - 4ac = 0$, and if the discriminant $b^2 - 4ac = 0$, then the equation $ax^2 + bx + c = 0$ has exactly one solution"—that is, one condition implies the other.

1.4 Exercises

In Exercises 1–2, using the given statements, write the sentence represented by each of the following.

a. $p \to q$ **b.** $q \to p$
c. $\sim p \to \sim q$ **d.** $\sim q \to \sim p$
e. Which of parts (a)–(d) are equivalent? Why?

1. p: She is a police officer.
q: She carries a gun.

2. p: I am a multimillion-dollar lottery winner.
q: I am a world traveler.

In Exercises 3–4, using the given statements, write the sentence represented by each of the following.

a. $p \to \sim q$ **b.** $\sim q \to p$
c. $\sim p \to q$ **d.** $q \to \sim p$
e. Which of parts (a)–(d) are equivalent? Why?

3. p: I watch television.
q: I do my homework.

4. p: He is an artist.
q: He is a conformist.

In Exercises 5–10, form (a) the inverse, (b) the converse, and (c) the contrapositive of the given conditional.

5. If you pass this mathematics course, then you fulfill a graduation requirement.
6. If you have the necessary tools, assembly time is less than thirty minutes.
7. The television set does not work if the electricity is turned off.
8. You do not win if you do not buy a lottery ticket.
9. You are a vegetarian if you do not eat meat.
10. If chemicals are properly disposed of, the environment is not damaged.

In Exercises 11–16, (a) determine the premise and conclusion, (b) rewrite the compound statement in the standard "if . . . then . . ." form, and (c) interpret the conditions that make the statement false.

11. I take public transportation only if it is convenient.
12. I eat raw fish only if I am in a Japanese restaurant.
13. I buy foreign products only if domestic products are not available.
14. I ride my bicycle only if it is not raining.
15. You may become a U.S. senator only if you are at least thirty years old and have been a citizen for nine years.
16. You may become the president of the United States only if you are at least thirty-five years old and were born a citizen of the United States.

In Exercises 17–22, express the given biconditional as the conjunction of two conditionals.

17. You obtain a refund if and only if you have a receipt.
18. We eat at Burger World if and only if Ju Ju's Kitsch-Inn is closed.
19. The quadratic equation $ax^2 + bx + c = 0$ has two distinct real solutions if and only if $b^2 - 4ac > 0$.
20. The quadratic equation $ax^2 + bx + c = 0$ has complex solutions if and only iff $b^2 - 4ac < 0$.
21. A polygon is a triangle iff the polygon has three sides.
22. A triangle is isosceles iff the triangle has two equal sides.

In Exercises 23–28, translate the two statements into symbolic form and use truth tables to determine whether the statements are equivalent.

23. I cannot have surgery if I do not have health insurance.
If I can have surgery, then I do have health insurance.

24. If I am illiterate, I cannot fill out an application form.
I can fill out an application form if I am not illiterate.

25. If you earn less than $12,000 per year, you are eligible for assistance.
If you are not eligible for assistance, then you earn at least $12,000 per year.

26. If you earn less than $12,000 per year, you are eligible for assistance.
If you earn at least $12,000 per year, you are not eligible for assistance.

27. I watch television only if the program is educational.
I do not watch television if the program is not educational.

28. I buy seafood only if the seafood is fresh.
If I do not buy seafood, the seafood is not fresh.

In Exercises 29–34, write an equivalent variation of the given conditional.

29. If it is not raining, I walk to work.

30. If it makes a buzzing noise, it is not working properly.

31. It is snowing only if it is cold.

32. You are a criminal only if you do not obey the law.

33. You are not a vegetarian if you eat meat.

34. You are not an artist if you are not creative.

In Exercises 35–40, determine which pairs of statements are equivalent.

35. **i.** If Proposition 111 passes, freeways are improved.
ii. If Proposition 111 is defeated, freeways are not improved.
iii. If the freeways are improved, Proposition 111 passes.
iv. If the freeways are not improved, Proposition 111 does not pass.

36. **i.** If the Giants win, then I am happy.
ii. If I am happy, then the Giants win.
iii. If the Giants lose, then I am unhappy.
iv. If I am unhappy, then the Giants lose.

37. **i.** I go to church if it is Sunday.
ii. I go to church only if it is Sunday.
iii. If I do not go to church, it is not Sunday.
iv. If it is not Sunday, I do not go to church.

38. **i.** I am a rebel if I do not have a cause.
ii. I am a rebel only if I do not have a cause.
iii. I am not a rebel if I have a cause.
iv. If I am not a rebel, I have a cause.

39. **i.** If line 34 is greater than line 29, I use Schedule X.
ii. If I use Schedule X, then line 34 is greater than line 29.
iii. If I do not use Schedule X, then line 34 is not greater than line 29.
iv. If line 34 is not greater than line 29, then I do not use Schedule X.

40. **i.** If you answer yes to all of the above, then you complete Part II.
ii. If you answer no to any of the above, then you do not complete Part II.
iii. If you completed Part II, then you answered yes to all of the above.
iv. If you did not complete Part II, then you answered no to at least one of the above.

Answer the following questions using complete sentences and your own words.

Concept Questions

41. What is a contrapositive?

42. What is a converse?

43. What is an inverse?

44. What is a biconditional?

45. How is an "if . . . then . . ." statement related to an "only if" statement?

1.5 Analyzing Arguments

Lewis Carroll's Cheshire Cat told Alice that he was mad (crazy). Alice then asked, " 'And how do you know that you're mad?' 'To begin with,' said the cat, 'a dog's not mad. You grant that?' 'I suppose so,' said Alice. 'Well, then,' the cat went on, 'you see a dog growls when it's angry, and wags its tail when it's pleased. Now *I* growl when I'm pleased, and wag my tail when I'm angry. Therefore I'm mad!' "

Does the Cheshire Cat have a valid deductive argument? Does the conclusion follow logically from the hypotheses? To answer this question, and others like it, we will utilize symbolic logic and truth tables in order to account for all possible combinations of the individual statements as true or false.

© Stock Montage

Using a logical argument, Lewis Carroll's Cheshire Cat tried to convince Alice that he was crazy. Was his argument valid?

Valid Arguments

When someone makes a sequence of statements and draws some conclusion from them, he or she is presenting an argument. An **argument** consists of two components: the initial statements, or hypotheses, and the final statement, or conclusion. When presented with an argument, a listener or reader may ask, "Does this person have a logical argument? Does his or her conclusion necessarily follow from the given statements?"

An argument is **valid** if the conclusion of the argument is guaranteed under its given set of hypotheses. (That is, the conclusion is inescapable in all instances.) For example, the argument

"All men are mortal.
Socrates is a man. } the hypotheses
Therefore, Socrates is mortal." } the conclusion

is a valid argument. Given the hypotheses, the conclusion is guaranteed. The term *valid* does not mean that all the statements are true but merely that the conclusion was reached via a proper deductive process. The argument

"All doctors are men.
My mother is a doctor. } the hypotheses
Therefore, my mother is a man." } the conclusion

is valid. Even though the conclusion is obviously false, the conclusion is guaranteed, *given the hypotheses.*

The hypotheses in a given logical argument may consist of several interrelated statements, each containing negations, conjunctions, disjunctions, and conditionals. By joining all the hypotheses in the form of a conjunction, we can form a single conditional that represents the entire argument. That is, if an argument has n hypotheses ($h_1, h_2, \ldots, h_n$) and conclusion c, the argument will have the form "if (h_1 and $h_2 \ldots$ and h_n), then c."

Church Carving May Be Original 'Cheshire Cat'

Reprinted with permission from Reuters.

LONDON — Devotees of writer Lewis Carroll believe they have found what inspired his grinning Cheshire Cat, made famous in his book "Alice's Adventures in Wonderland."

Members of the Lewis Carroll Society made the discovery over the weekend in a church at which the author's father was once rector in the Yorkshire village of Croft in northern England.

It is a rough-hewn carving of a cat's head smiling near an altar, probably dating to the 10th century. Seen from below and from the perspective of a small boy, all that can be seen is the grinning mouth.

Carroll's Alice watched the Cheshire Cat disappear "ending with the grin, which remained for some time after the rest of the head had gone."

Alice mused: "I have often seen a cat without a grin, but not a grin without a cat. It is the most curious thing I have seen in all my life."

Conditional Representation of an Argument

An argument having n hypotheses $h_1, h_2, \ldots, h_n$ and conclusion c can be represented by the conditional $[h_1 \wedge h_2 \wedge \ldots \wedge h_n] \rightarrow c$.

If the conditional representation of an argument is always true (regardless of the actual truthfulness of the individual statements), the argument is valid. If there is at least one instance in which the conditional is false, the argument is invalid.

EXAMPLE 1 Determine whether the following argument is valid:

"If he is illiterate, he cannot fill out the application.
He can fill out the application.
Therefore, he is not illiterate."

Solution First, number the hypotheses and separate them from the conclusion with a line:

1. If he is illiterate, he cannot fill out the application.
2. He can fill out the application.

Therefore, he is not illiterate.

Now use symbols to represent each different component in the statements:

p: He is illiterate.

q: He can fill out the application.

We could have defined q as "He *cannot* fill out the application" (as stated in premise 1), but it is customary to define the symbols with a positive sense. Symbolically, the argument has the form

1. $p \rightarrow \sim q$ } the hypotheses
2. q

$\therefore \sim p$ } conclusion

and is represented by the conditional $[(p \rightarrow \sim q) \wedge q] \rightarrow \sim p$. The symbol $\therefore$ is read "therefore."

To construct a truth table for this conditional, we need $2^2 = 4$ rows. A column is required for the following: each negation, each hypothesis, the conjunction of the hypotheses, the conclusion, and the conditional representation of the argument. The initial setup is shown in Figure 1.47.

			Hypothesis 1	Hypothesis 2	Column Representing All the Hypotheses	Conclusion c	Conditional Representation of the Argument
	p	q	$\sim q$	$p \to \sim q$	q	$1 \wedge 2$	$\sim p$	$(1 \wedge 2) \to c$
1.	T	T						
2.	T	F						
3.	F	T						
4.	F	F						

Figure 1.47

Fill in the truth table as follows:

~q: A negation has the opposite truth values; enter a T in rows 2 and 4 and an F in rows 1 and 3.

Hypothesis 1: A conditional is false only when its premise is true and its conclusion is false; enter an F in row 1 and Ts elsewhere.

Hypothesis 2: Recopy the q column.

$1 \wedge 2$: A conjunction is true only when both components are true; enter a T in row 3 and Fs elsewhere.

Conclusion *c*: A negation has the opposite truth values; enter an F in rows 1 and 2 and a T in rows 3 and 4.

At this point, all that remains is the final column (see Figure 1.48).

				1	2		c	
	p	q	$\sim q$	$p \to \sim q$	q	$1 \wedge 2$	$\sim p$	$(1 \wedge 2) \to c$
1.	T	T	F	F	T	F	F	
2.	T	F	T	T	F	F	F	
3.	F	T	F	T	T	T	T	
4.	F	F	T	T	F	F	T	

Figure 1.48

The last column in the truth table is the conditional that represents the entire argument. A conditional is false only when its premise is true and its conclusion is false. The only instance where the premise $(1 \wedge 2)$ is true is row 3. Corresponding to this entry, the conclusion $\sim p$ is also true. Consequently, the conditional $(1 \wedge 2) \to c$ is true in row 3. Because the premise $(1 \wedge 2)$ is false in rows 1, 2, and 4, the conditional $(1 \wedge 2) \to c$ is automatically true in those rows as well. The completed truth table is shown in Figure 1.49.

			1	2		c	
	p	q	$\sim q$	$p \to \sim q$	q	$1 \wedge 2$	$\sim p$	$(1 \wedge 2) \to c$
1.	T	T	F	F	T	F	F	T
2.	T	F	T	T	F	F	F	T
3.	F	T	F	T	T	T	T	T
4.	F	F	T	T	F	F	T	T

Figure 1.49
Truth table for the argument $[(p \to \sim q) \wedge q] \to \sim p$

The completed truth table shows that the conditional $[(p \to \sim q) \wedge q] \to \sim p$ is always true. The conditional represents the argument "If he is illiterate, he cannot fill out the application. He can fill out the application. Therefore, he is not illiterate." Thus, the argument is valid. ■

Tautologies

A **tautology** is a statement that is always true. For example, the statement

"$(a + b)^2 = a^2 + 2ab + b^2$" is a tautology.

EXAMPLE 2 Determine whether the statement $(p \wedge q) \to (p \vee q)$ is a tautology.

Solution We need to construct a truth table for the statement. Because there are two letters, the table must have $2^2 = 4$ rows. We need a column for $(p \wedge q)$, one for $(p \vee q)$, and one for $(p \wedge q) \to (p \vee q)$. The completed truth table is shown in Figure 1.50.

	p	q	$p \wedge q$	$p \vee q$	$(p \wedge q) \to (p \vee q)$
1.	T	T	T	T	T
2.	T	F	F	T	T
3.	F	T	F	T	T
4.	F	F	F	F	T

Figure 1.50
Truth table for the statement $(p \wedge q) \to (p \vee q)$

Because $(p \wedge q) \to (p \vee q)$ is always true, it is a tautology. ■

As we have seen, an argument can be represented by a single conditional. If this conditional is always true, the argument is valid (and vice versa).

Validity of an Argument

An argument having n hypotheses $h_1, h_2, \ldots, h_n$ and conclusion c is valid if and only if the conditional $[h_1 \wedge h_2 \wedge \ldots \wedge h_n] \to c$ is a tautology.

EXAMPLE 3 Determine whether the following argument is valid:
"If the defendant is innocent, the defendant does not go to jail. The defendant does not go to jail. Therefore, the defendant is innocent."

Solution Separating the hypotheses from the conclusion, we have

1. If the defendant is innocent, the defendant does not go to jail.
2. The defendant does not go to jail.

Therefore, the defendant is innocent.

Now we define symbols to represent the various components of the statements:

p: The defendant is innocent.
q: The defendant goes to jail.

Symbolically, the argument has the form

1. $p \rightarrow \sim q$
2. $\sim q$

$\therefore p$

and is represented by the conditional $[(p \rightarrow \sim q) \wedge \sim q] \rightarrow p$.

Now we construct a truth table with four rows, along with the necessary columns. The completed table is shown in Figure 1.51.

			2	1		c	
	p	q	$\sim q$	$p \rightarrow \sim q$	$1 \wedge 2$	p	$(1 \wedge 2) \rightarrow c$
1.	T	T	F	F	F	T	T
2.	T	F	T	T	T	T	T
3.	F	T	F	T	F	F	T
4.	F	F	T	T	T	F	F

Figure 1.51
Truth table for the argument $[(p \rightarrow \sim q) \wedge \sim q] \rightarrow p$

The column representing the argument has an F in row 4; therefore, the conditional representation of the argument is *not* a tautology. In particular, the conclusion does not logically follow the hypotheses when both p and q are false (row 4). The argument is not valid. Let us interpret the circumstances expressed in row 4, the row in which the argument breaks down. Both p and q are false—that is, the defendant is guilty and the defendant does *not* go to jail. Unfortunately, this situation can occur in the real world; guilty people do not *always* go to jail! As long as it is possible for a guilty person to avoid jail, the argument is invalid. ■

The following argument was presented as Example 6 in Section 1.1. In that section, we constructed a Venn diagram to show that the argument was in fact valid. We now show an alternative method; that is, we construct a truth table to determine whether the argument is valid.

EXAMPLE 4 Determine whether the following argument is valid: "No snake is warm-blooded. All mammals are warm-blooded. Therefore, snakes are not mammals."

Solution Separating the hypotheses from the conclusion, we have

1. No snake is warm-blooded.
2. All mammals are warm-blooded.

Therefore, snakes are not mammals.

These statements can be rephrased as follows:

1. If it is a snake, then it is not warm-blooded.
2. If it is a mammal, then it is warm-blooded.

Therefore, if it is a snake, then it is not a mammal.

Now we define symbols to represent the various components of the statements:

p: It is a snake.

q: It is warm-blooded.

r: It is a mammal.

Symbolically, the argument has the form

1. $p \to \sim q$
2. $r \to q$

$\therefore p \to \sim r$

and is represented by the conditional $[(p \to \sim q) \wedge (r \to q)] \to (p \to \sim r)$.

Now we construct a truth table with eight rows ($2^3 = 8$), along with the necessary columns. The completed table is shown in Figure 1.52.

						1	2		c	
	p	q	r	$\sim q$	$\sim r$	$p \to \sim q$	$r \to q$	$1 \wedge 2$	$p \to \sim r$	$(1 \wedge 2) \to c$
1.	T	T	T	F	F	F	T	F	F	T
2.	T	T	F	F	T	F	T	F	T	T
3.	T	F	T	T	F	T	F	F	F	T
4.	T	F	F	T	T	T	T	T	T	T
5.	F	T	T	F	F	T	T	T	T	T
6.	F	T	F	F	T	T	T	T	T	T
7.	F	F	T	T	F	T	F	F	T	T
8.	F	F	F	T	T	T	T	T	T	T

Figure 1.52
Truth table for the argument $[(p \to \sim q) \wedge (r \to q)] \to (p \to \sim r)$

The last column of the truth table represents the argument and contains all T's. Consequently, the conditional $[(p \to \sim q) \wedge (r \to q)] \to (p \to \sim r)$ is a tautology; the argument is valid. ■

HISTORICAL NOTE

Charles Lutwidge Dodgson

1832–1898

© Stock Montage

To those who assume that it is impossible for a person to excel both in the creative worlds of art and literature and in the disciplined worlds of mathematics and logic, the life of Charles Lutwidge Dodgson is a wondrous counterexample. Known the world over as Lewis Carroll, Dodgson penned the nonsensical classics *Alice's Adventures in Wonderland* and *Through the Looking Glass.* However, many people are surprised to learn that Dodgson (from age eighteen to his death) was a permanent resident at the University at Oxford, teaching mathematics and logic. And as if that were not enough, Dodgson is now recognized as one of the leading portrait photographers of the Victorian era.

The eldest son in a family of eleven children, Charles amused his younger siblings with elaborate games, poems, stories, and humorous drawings. This attraction to entertaining children with fantastic stories manifested itself in much of his later work as Lewis Carroll. Besides his obvious interest in telling stories, the young Dodgson was also intrigued by mathematics. At the age of eight, Charles asked his father to explain a book on logarithms. When told that he was too young to understand, Charles persisted, "But please, explain!"

The Dodgson family had a strong ecclesiastical tradition; Charles's father, great-grandfather, and great-great-grandfather were all clergymen. Following in his father's footsteps, Charles attended Christ Church, the largest and most celebrated of all the Oxford colleges. After graduating in 1854, Charles remained at Oxford, accepting the position of mathematical lecturer in 1855. However, appointment to this position was conditional upon his taking Holy Orders in the Anglican church and upon his remaining celibate. Dodgson complied and was named a deacon in 1861.

The year 1856 was filled with events that had lasting effects on Dodgson. Charles Lutwidge created his pseudonym by translating his first and middle names into Latin (Carolus Ludovic), reversing their order (Ludovic Carolus), and translating them back into English (Lewis Carroll). In this same year, Dodgson began his "hobby" of photography. He is considered by many to have been an artistic pioneer in this new field (photography was invented in 1839). Most of Dodgson's work consists of portraits that chronicle the Victorian era, and over 700 photographs taken by Dodgson have been preserved. His favorite subjects were children, especially young girls.

Dodgson's affinity for children brought about a meeting in 1856

Young Alice Liddell inspired Lewis Carroll to write *Alice's Adventures in Wonderland.* This photo is one of the many Carroll took of Alice.

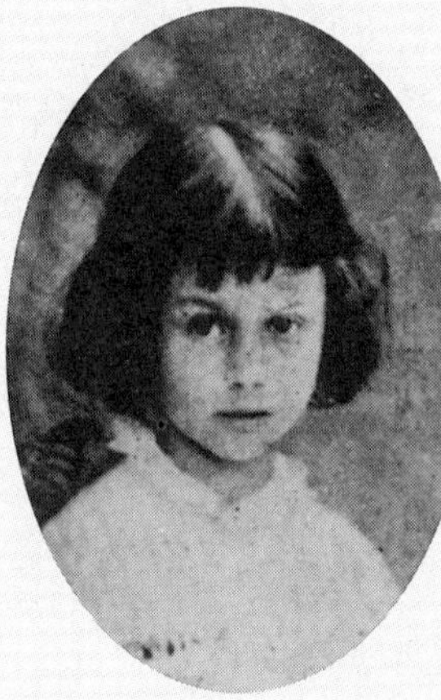

© The Granger Collection, NY

The preceding examples contained relatively simple arguments, each consisting of only two hypotheses and two simple statements (letters). In such cases, many people try to employ "common sense" to confirm the validity of the argument. For instance, the argument "If it is raining, the streets are wet. It is raining. Therefore, the streets are wet" is obviously valid. However, it might not be so simple to determine the validity of an argument that contains several hypotheses and many simple statements. Indeed, in such cases, the argument's truth table might become quite lengthy, as in the next example.

EXAMPLE 5 The following whimsical argument was written by Lewis Carroll and appeared in his 1896 book *Symbolic Logic:*

that would eventually establish his place in the history of literature. Early in the year, Dodgson met the four children of the dean of Christ Church: Harry, Lorina, Edith, and Alice Liddell. He began seeing the children on a regular basis, amusing them with stories and photographing them. Although he had a wonderful relationship with all four, Alice received his special attention.

On July 4, 1862, while rowing and picnicking with Alice and her sisters, Dodgson entertained the Liddell girls with a fantastic story of a little girl named Alice who fell into a rabbit hole. Captivated by the story, Alice Liddell insisted that Dodgson write it down for her. He complied, initially titling it *Alice's Adventure Underground.*

Dodgson's friends subsequently encouraged him to publish the manuscript, and in 1865, after editing and inserting new episodes, Lewis Carroll gave the world *Alice's Adventures in Wonderland.* Although the book appeared to be a whimsical excursion into chaotic nonsense, Dodgson's masterpiece contained many exercises in logic and metaphor. The book was a success, and in 1871 a sequel, *Through the Looking Glass,* was printed. When asked to comment on the meaning of his writings, Dodgson replied, "I'm very much afraid I didn't mean anything but nonsense! Still, you know, words mean more than we mean to express when we use them; so a whole book ought to mean a great deal more than the writer means. So, whatever good meanings are in the book, I'm glad to accept as the meaning of the book."

In addition to writing "children's stories," Dodgson wrote numerous mathematics essays and texts, including *The Fifth Book of Euclid Proved Algebraically, Formulae of Plane Trigonometry, A Guide to the Mathematical Student,* and *Euclid and His Modern Rivals.* In the field of formal logic, Dodgson's books *The Game of Logic* (1887) and *Symbolic Logic* (1896) are still used as sources of inspiration in numerous schools worldwide.

Carroll's book *The Game of Logic* presents the study of formalized logic in a gamelike fashion. After listing the "rules of the game" (complete with gameboard and markers), Carroll captures the reader's interest with nonsensical syllogisms.

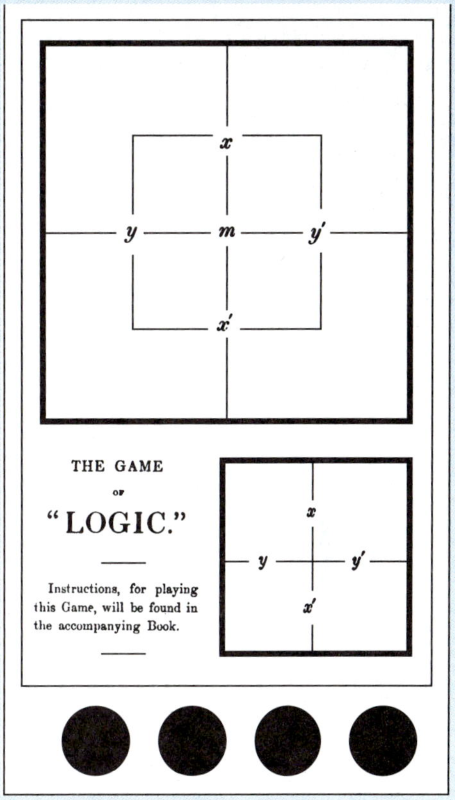

"No ducks waltz. No officers ever decline to waltz. All my poultry are ducks. Therefore, my poultry are not officers."

Construct a truth table to determine whether the argument is valid.

Solution Separating the hypotheses from the conclusion, we have

1. No ducks waltz.
2. No officers ever decline to waltz.
3. All my poultry are ducks.

Therefore, my poultry are not officers.

These statements can be rephrased as

1. If it is a duck, then it does not waltz.
2. If it is an officer, then it does not decline to waltz.
(Equivalently, "If it is an officer, then it will waltz.")
3. If it is my poultry, then it is a duck.

Therefore, if it is my poultry, then it is not an officer.

Now we define symbols to represent the various components of the statements:

p: It is a duck.

q: It will waltz.

r: It is an officer.

s: It is my poultry.

Symbolically, the argument has the form

1. $p \rightarrow \sim q$
2. $r \rightarrow q$
3. $s \rightarrow p$

$\therefore s \rightarrow \sim r$

Now we construct a truth table with sixteen rows ($2^4 = 16$), along with the necessary columns. The completed table is shown in Figure 1.53.

The last column of the truth table represents the argument and contains all T's. Consequently, the conditional $[(p \rightarrow \sim q) \wedge (r \rightarrow q) \wedge (s \rightarrow p)] \rightarrow (s \rightarrow \sim r)$ is a tautology; the argument is valid.

							1	2	3		c	
	p	q	r	s	$\sim q$	$\sim r$	$p \rightarrow \sim q$	$r \rightarrow q$	$s \rightarrow p$	$1 \wedge 2 \wedge 3$	$s \rightarrow \sim r$	$(1 \wedge 2 \wedge 3) \rightarrow c$
1.	T	T	T	T	F	F	F	T	T	F	F	T
2.	T	T	T	F	F	F	F	T	T	F	T	T
3.	T	T	F	T	F	T	F	T	T	F	T	T
4.	T	T	F	F	F	T	F	T	T	F	T	T
5.	T	F	T	T	T	F	T	F	T	F	F	T
6.	T	F	T	F	T	F	T	F	T	F	T	T
7.	T	F	F	T	T	T	T	T	T	T	T	T
8.	T	F	F	F	T	T	T	T	T	T	T	T
9.	F	T	T	T	F	F	T	T	F	F	F	T
10.	F	T	T	F	F	F	T	T	T	T	T	T
11.	F	T	F	T	F	T	T	T	F	F	T	T
12.	F	T	F	F	F	T	T	T	T	T	T	T
13.	F	F	T	T	T	F	T	F	F	F	F	T
14.	F	F	T	F	T	F	T	F	T	F	T	T
15.	F	F	F	T	T	T	T	T	F	F	T	T
16.	F	F	F	F	T	T	T	T	T	T	T	T

Figure 1.53
Truth table for the argument $[(p \rightarrow \sim q) \wedge (r \rightarrow q) \wedge (s \rightarrow p)] \rightarrow (s \rightarrow \sim r)$

1.5 Exercises

In Exercises 1–6, use the given symbols to rewrite the argument in symbolic form.

1. *p:* It is raining.
q: The streets are wet. } Use these symbols.

1. If it is raining, then the streets are wet.
2. It is raining.

Therefore, the streets are wet.

2. *p:* I have a college degree.
q: I am lazy. } Use these symbols.

1. If I have a college degree, I am not lazy.
2. I do not have a college degree.

Therefore, I am lazy.

3. *p:* It is Tuesday.
q: The tour group is in Belgium. } Use these symbols.

1. If it is Tuesday, then the tour group is in Belgium.
2. The tour group is not in Belgium.

Therefore, it is not Tuesday.

4. *p:* You are a gambler.
q: You have financial security. } Use these symbols.

1. You do not have financial security if you are a gambler.
2. You do not have financial security.

Therefore, you are a gambler.

5. *p:* You exercise regularly.
q: You are healthy. } Use these symbols.

1. You exercise regularly only if you are healthy.
2. You do not exercise regularly.

Therefore, you are not healthy.

6. *p:* The senator supports new taxes.
q: The senator is reelected. } Use these symbols.

1. The senator is not reelected if she supports new taxes.
2. The senator does not support new taxes.

Therefore, the senator is reelected.

In Exercises 7–12, use a truth table to determine the validity of the argument specified. If the argument is invalid, interpret the specific circumstances that cause it to be invalid.

7. the argument in Exercise 1
8. the argument in Exercise 2
9. the argument in Exercise 3
10. the argument in Exercise 4
11. the argument in Exercise 5
12. the argument in Exercise 6

In Exercises 13–30, define the necessary symbols, rewrite the argument in symbolic form, and use a truth table to determine whether the argument is valid. If the argument is invalid, interpret the specific circumstances that cause the argument to be invalid.

13. 1. If the Democrats have a majority, Smith is appointed and student loans are funded.
2. Smith is appointed or student loans are not funded.

Therefore, the Democrats do not have a majority.

14. 1. If you watch television, you do not read books.
2. If you read books, you are wise.

Therefore, you are not wise if you watch television.

15. 1. If you argue with a police officer, you get a ticket.
2. If you do not break the speed limit, you do not get a ticket.

Therefore, if you break the speed limit, you argue with a police officer.

16. 1. If you do not recycle newspapers, you are not an environmentalist.
2. If you recycle newspapers, you save trees.

Therefore, you are an environmentalist only if you save trees.

17. 1. All pesticides are harmful to the environment.
2. No fertilizer is a pesticide.

Therefore, no fertilizer is harmful to the environment.

18. 1. No one who can afford health insurance is unemployed.
2. All politicians can afford health insurance.

Therefore, no politician is unemployed.

19. 1. All poets are loners.
2. All loners are taxi drivers.

Therefore, all poets are taxi drivers.

20. 1. All forest rangers are environmentalists.
2. All forest rangers are storytellers.

Therefore, all environmentalists are storytellers.

21. 1. No professor is a millionaire.
2. No millionaire is illiterate.

Therefore, no professor is illiterate.

22. 1. No artist is a lawyer.
2. No lawyer is a musician.

Therefore, no artist is a musician.

23. 1. All lawyers study logic.
2. You study logic only if you are a scholar.
3. You are not a scholar.

Therefore, you are not a lawyer.

24. 1. All licensed drivers have insurance.
2. You obey the law if you have insurance.
3. You obey the law.

Therefore, you are a licensed driver.

25. If the defendant is innocent, he does not go to jail. The defendant goes to jail. Therefore, the defendant is guilty.

26. If the defendant is innocent, he does not go to jail. The defendant is guilty. Therefore, the defendant goes to jail.

27. If you are not in a hurry, you eat at Lulu's Diner. If you are in a hurry, you do not eat good food. You eat at Lulu's. Therefore, you eat good food.

28. If you give me a hamburger today, I pay you tomorrow. If you are a sensitive person, you give me a hamburger today. You are not a sensitive person. Therefore, I do not pay you tomorrow.

29. If you listen to rock and roll, you do not go to heaven. If you are a moral person, you go to heaven. Therefore, you are not a moral person if you listen to rock and roll.

30. If you follow the rules, you have no trouble. If you are not clever, you have trouble. You are clever. Therefore, you do not follow the rules.

The arguments given in Exercises 31–38 were written by Lewis Carroll and appeared in his 1896 book Symbolic Logic. *For each argument, define the necessary symbols, rewrite the argument in symbolic form, and use a truth table to determine whether the argument is valid.*

31. 1. All medicine is nasty.
2. Senna is a medicine.

Therefore, senna is nasty.

NOTE: Senna is a laxative extracted from the dried leaves of cassia plants.

32. 1. All pigs are fat.
2. Nothing that is fed on barley-water is fat.

Therefore, pigs are not fed on barley-water.

33. 1. Nothing intelligible ever puzzles me.
2. Logic puzzles me.

Therefore, logic is unintelligible.

34. 1. No misers are unselfish.
2. None but misers save eggshells.

Therefore, no unselfish people save eggshells.

35. 1. No Frenchmen like plum pudding.
2. All Englishmen like plum pudding.

Therefore, Englishmen are not Frenchmen.

36. 1. A prudent man shuns hyenas.
2. No banker is imprudent.

Therefore, no banker fails to shun hyenas.

37. 1. All wasps are unfriendly.
2. No puppies are unfriendly.

Therefore, puppies are not wasps.

38. 1. Improbable stories are not easily believed.
2. None of his stories are probable.

Therefore, none of his stories are easily believed.

Answer the following questions using complete sentences and your own words.

Concept Questions

39. What is a tautology?

40. What is the conditional representation of an argument?

41. Find a "logical" argument in a newspaper article, an advertisement, or elsewhere in the media. Analyze that argument and discuss the implications.

History Questions

42. What was Charles Dodgson's pseudonym? How did he get it? What classic "children's stories" did he write?

43. What did Charles Dodgson contribute to the study of formal logic?

44. Charles Dodgson was a pioneer in what artistic field?

45. Who was Alice Liddell?

Chapter 1 Review

TERMS

argument	deductive reasoning	invalid argument	syllogism
biconditional	disjunction	inverse	tautology
compound statement	equivalent expressions	logic	truth table
conclusion	exclusive *or*	negation	truth value
conditional	hypothesis	premise	valid argument
conjunction	implication	quantifier	Venn diagram
contrapositive	inclusive *or*	reasoning	
converse	inductive reasoning	statement	

REVIEW EXERCISES

1. Classify each argument as deductive or inductive.

a. 1. Hitchcock's "Psycho" is a suspenseful movie.
2. Hitchcock's "The Birds" is a suspenseful movie.

Therefore, all Hitchcock movies are suspenseful.

b. 1. All Hitchcock movies are suspenseful.
2. "Psycho" is a Hitchcock movie.

Therefore, "Psycho" is suspenseful.

2. Explain the general rule or pattern used to assign the given letter to the given word. Fill in the blank with the letter that fits the pattern.

day	morning	afternoon	dusk	night
y	r	f	s	____

3. Fill in the blank with what is most likely to be the next number. Explain the pattern generated by your answer.

1, 6, 11, 4, _____

In Exercises 4–9, construct a Venn diagram to determine the validity of the given argument.

4. 1. All truck drivers are union members.
2. Rocky is a truck driver.

Therefore, Rocky is a union member.

5. 1. All truck drivers are union members.
2. Rocky is not a truck driver.

Therefore, Rocky is not a union member.

6. 1. All mechanics are engineers.
2. Casey Jones is an engineer.

Therefore, Casey Jones is a mechanic.

7. 1. All mechanics are engineers.
2. Casey Jones is not an engineer.

Therefore, Casey Jones is not a mechanic.

8. 1. Some animals are dangerous.
2. A gun is not an animal.

Therefore, a gun is not dangerous.

9. 1. Some contractors are electricians.
2. All contractors are carpenters.

Therefore, some electricians are carpenters.

10. Which of the following are statements? Why or why not?

a. The Golden Gate Bridge spans Chesapeake Bay.
b. The capital of Delaware is Dover.
c. Where are you spending your vacation?
d. Hawaii is the best place to spend a vacation.

11. Determine which pairs of statements are negations of each other.

a. All of the lawyers are ethical.
b. Some of the lawyers are ethical.
c. None of the lawyers is ethical.
d. Some of the lawyers are not ethical.

12. Write a sentence that represents the negation of each statement.

a. His car is not new.
b. Some buildings are earthquakeproof.
c. All children eat candy.
d. I never cry in a movie theater.

13. Using the symbolic representations

p: The television program is educational.
q: The television program is controversial.

express the following compound statements in symbolic form.

a. The television program is educational and controversial.

b. If the television program isn't controversial, it isn't educational.
c. The television program is educational and it isn't controversial.
d. The television program isn't educational or controversial.

14. Using the symbolic representations

p: The advertisement is effective.
q: The advertisement is misleading.
r: The advertisement is outdated.

express the following compound statements in symbolic form.

a. All misleading advertisements are effective.
b. It is a current, honest, effective advertisement.
c. If an advertisement is outdated, it isn't effective.
d. The advertisement is effective and it isn't misleading or outdated.

15. Using the symbolic representations

p: It is expensive.
q: It is undesirable.

express the following in words.

a. $p \rightarrow \sim q$ **b.** $q \leftrightarrow \sim p$
c. $\sim(p \vee q)$ **d.** $(p \wedge \sim q)/(\sim p \wedge q)$

16. Using the symbolic representations

p: The movie is critically acclaimed.
q: The movie is a box office hit.
r: The movie is available on videotape.

express the following in words.

a. $(p \vee q) \rightarrow r$ **b.** $(p \wedge \sim q) \rightarrow \sim r$
c. $\sim(p \vee q) \wedge r$ **d.** $\sim r \rightarrow (\sim p \wedge \sim q)$

In Exercises 17–24, construct a truth table for the compound statement.

17. $p \vee \sim q$ **18.** $p \wedge \sim q$
19. $\sim p \rightarrow q$ **20.** $(p \wedge q) \rightarrow \sim q$
21. $q \vee \sim(p \vee r)$ **22.** $\sim p \rightarrow (q \vee r)$
23. $(q \wedge p) \rightarrow (\sim r \vee p)$ **24.** $(p \vee r) \rightarrow (q \wedge \sim r)$

In Exercises 25–28, construct a truth table to determine whether the statements in each pair are equivalent.

25. The car is unreliable or expensive.
If the car is reliable, then it is expensive.
26. If I get a raise, I will buy a new car.
If I do not get a raise, I will not buy a new car.
27. She is a Democrat or she did not vote.
She is not a Democrat and she did vote.
28. The raise is not unjustified and the management opposes it.
It is not the case that the raise is unjustified or the management does not oppose it.

In Exercises 29–34, write a sentence that represents the negation of each statement.

29. Jesse had a party and nobody came.
30. You do not go to jail if you pay the fine.
31. I am the winner or you are blind.
32. He is unemployed and he did not apply for financial assistance.
33. The selection procedure has been violated if his application is ignored.
34. The jackpot is at least $1 million.
35. Given the statements

p: You are an avid jogger.
q: You are healthy.

write the sentence represented by each of the following.

a. $p \rightarrow q$ **b.** $q \rightarrow p$
c. $\sim p \rightarrow \sim q$ **d.** $\sim q \rightarrow \sim p$
e. $p \leftrightarrow q$

36. Form (a) the inverse, (b) the converse, and (c) the contrapositive of the conditional "If he is elected, the country is in big trouble."

In Exercises 37–40, (a) determine the premise and conclusion and (b) rewrite the compound statement in the standard "if . . . then . . ." form.

37. The economy improves only if unemployment goes down.
38. The economy improves if unemployment goes down.
39. No computer is unrepairable.
40. All gemstones are valuable.

In Exercises 41 and 42, translate the two statements into symbolic form and use truth tables to determine whether the statements are equivalent.

41. If you are allergic to dairy products, you cannot eat cheese.
If you cannot eat cheese, then you are allergic to dairy products.
42. You are a fool if you listen to me.
You are not a fool only if you do not listen to me.
43. Which pairs of statements are equivalent?

i. If it is not raining, I ride my bicycle to work.
ii. If I ride my bicycle to work, it is not raining.
iii. If I do not ride my bicycle to work, it is raining.
iv. If it is raining, I do not ride my bicycle to work.

In Exercises 44–47, define the necessary symbols, rewrite the argument in symbolic form, and use a truth table to determine whether the argument is valid.

44. 1. If you do not make your loan payment, your car is repossessed.
2. Your car is repossessed.

Therefore, you did not make your loan payment.

45. 1. If you do not pay attention, you do not learn the new method.
2. You do learn the new method.

Therefore, you do pay attention.

46. 1. If you rent videocassettes, you will not go to the movie theater.
2. If you go to the movie theater, you pay attention to the movie.

Therefore, you do not pay attention to the movie if you rent videocassettes.

47. 1. If the Republicans have a majority, Farnsworth is appointed and no new taxes are imposed.
2. New taxes are imposed.

Therefore, the Republicans do not have a majority or Farnsworth is not appointed.

In Exercises 48–54, define the necessary symbols, rewrite the argument in symbolic form, and use a truth table to determine whether the argument is valid.

48. If the defendant is guilty, he goes to jail. The defendant does not go to jail. Therefore, the defendant is not guilty.

49. I will go to the concert only if you buy me a ticket. You bought me a ticket. Therefore, I will go to the concert.

50. If tuition is raised, students take out loans or drop out. If students do not take out loans, they drop out. Students do drop out. Therefore, tuition is raised.

51. If our oil supply is cut off, our economy collapses. If we go to war, our economy doesn't collapse. Therefore, if our oil supply isn't cut off, we do not go to war.

52. No professor is uneducated. No monkey is educated. Therefore, no professor is a monkey.

53. No professor is uneducated. No monkey is a professor. Therefore, no monkey is educated.

54. Vehicles stop if the traffic light is red. There is no accident if vehicles stop. There is an accident. Therefore, the traffic light is not red.

Answer the following questions using complete sentences and your own words.

Concept Questions

55. What is a statement?

56. **a.** What is a disjunction? Under what conditions is a disjunction true?
b. What is a conjunction? Under what conditions is a conjunction true?
c. What is a conditional? Under what conditions is a conditional true?
d. What is a negation? Under what conditions is a negation true?

57. What is a tautology?

58. When constructing a truth table, how do you determine how many rows to create?

History Questions

59. What role did the following people play in the development of formalized logic?
- Aristotle
- George Boole
- Augustus De Morgan
- Charles Dodgson
- Gottfried Wilhelm Leibniz

Sets and Counting

2

Recently, 1,000 college seniors were asked whether they favored increasing the state's gasoline tax to generate funds to improve highways and whether they favored increasing the state's alcohol tax to generate funds to improve the public education system. The responses were tallied, and the following results were printed in the campus newspaper: 750 favored an increase in the gasoline tax, 600 favored an increase in the alcohol tax, and 450 favored increases in both taxes. How many of these 1,000 students favored an increase in at least one of the taxes? How many favored increasing only the gasoline tax, increasing only the alcohol tax, or increasing neither tax? The mathematical tool designed to answer questions like these is the *set*. Although you might be able to answer the given questions without any formal knowledge of sets, the mental reasoning involved in obtaining your answers uses some of the basic principles of sets. (Incidentally, the answers are 900, 300, 150, and 100, respectively.)

The branch of mathematics that deals with sets is called **set theory.** Set theory can be helpful in solving both mathematical and nonmathematical problems. It is an important tool in analyzing the results of consumer surveys, marketing analyses, and political polls. Standardized admissions tests such as the Graduate Record Examination (G.R.E.) ask questions that can be answered with set theory. In this text, we use sets extensively in Chapter 3 on probability.

2.1 Sets and Set Operations

A **set** is a collection of objects or things. The objects or things in the set are called **elements** (or *members*) of the set. In our example above, we could talk about the *set* of students who favor increasing only the gasoline tax or the *set* of students who do not favor increasing either tax. In geography, we can talk about the *set* of all state capitals or the *set* of all states west of the Mississippi. It is easy to determine whether something is in these sets; for example, Des Moines is an element of the set of state capitals, whereas Dallas is not. Such sets are called **well-defined** because there is a way of determining for sure whether a particular item is an element of the set.

EXAMPLE 1 Which of the following sets are well-defined?

a. the set of all movies directed by Alfred Hitchcock
b. the set of all great rock-and-roll bands
c. the set of all possible two-person committees selected from a group of five people

Solution
a. This set is well-defined; either a movie was directed by Hitchcock, or it was not.
b. This set is *not* well-defined; membership is a matter of opinion. Some people would say that the Ramones (one of the pioneer punk bands of the late 1970s) are a member, while others might say they are not. (Note: The Ramones were inducted into the Rock and Roll Hall of Fame in 2002.)
c. This set is well-defined; either the two people are from the group of five, or they are not. ■

Notation

By tradition, a set is denoted by a capital letter, frequently one that will serve as a reminder of the contents of the set. **Roster notation** (also called *listing notation*) is a method of describing a set by listing each element of the set inside the symbols { and }, which are called *set braces.* In a listing of the elements of a set, each distinct element is listed only once, and the order of the elements doesn't matter.

The symbol $\in$ stands for the phrase *is an element of,* and $\notin$ stands for *is not an element of.* The **cardinal number** of a set A is the number of elements in the set and is denoted by $n(A)$. Thus, if R is the set of all letters in the name "Ramones," then $R = \{\text{r, a, m, o, n, e, s}\}$. Notice that m is an element of the set R, x is not an element of R, and R has 7 elements. In symbols, $\text{m} \in R$, $\text{x} \notin R$, and $n(R) = 7$.

Two sets are **equal** if they contain exactly the same elements. *The order in which the elements are listed does not matter.* If M is the set of all letters in the name "Moaners," then $M = \{\text{m, o, a, n, e, r, s}\}$. This set contains exactly the same elements as the set R of letters in the name "Ramones." Therefore, $M = R = \{\text{a, e, m, n, o, r, s}\}$.

Often it is not appropriate or not possible to describe a set in roster notation. For extremely large sets, such as the set V of all registered voters in Detroit, or for sets that contain an infinite number of elements, such as the set G of all negative real numbers, the roster method would be either too cumbersome or impossible to use. Although V could be expressed via the roster method (since each county compiles a list of all registered voters in its jurisdiction), it would take hundreds or even thousands of pages to list everyone who is registered to vote in Detroit! In the case of the set G of all negative real numbers, no list, no matter how long, is capable of listing all members of the set; there is an infinite number of negative numbers.

In such cases, it is often necessary, or at least more convenient, to use **set-builder notation**, which lists the rules that determine whether an object is

The "Ramones" or The "Moaners"? The set R of all letters in the name "Ramones" is the same as the set M of all letters in the name "Moaners." Consequently, the sets are equal; $M = R = \{a, e, m, n, o, r, s\}$. (R.I.P. Joey Ramone 1951–2001, Dee Dee Ramone 1952–2002)

an element of the set rather than the actual elements. A set-builder description of set G on page 53 is

$$G = \{x \mid x < 0 \quad \text{and} \quad x \in \Re\}$$

which is read as "the set of all x such that x is less than zero and x is a real number." A set-builder description of set V above is

$$V = \{\text{persons} \mid \text{the person is a registered voter in Detroit}\}$$

which is read as "the set of all persons such that the person is a registered voter in Detroit." In set-builder notation, the vertical line stands for the phrase "such that." Whatever is on the left side of the line is the general type of thing in the set, while the rules about set membership are listed on the right.

EXAMPLE 2 Describe each of the following in words.

a. $\{x \mid x > 0 \text{ and } x \in \Re\}$
b. {persons | the person is a living former U.S. president}
c. {women | the woman is a former U.S. president}

Solution
a. the set of all x such that x is a positive real number
b. the set of all people such that the person is a living former U.S. president
c. the set of all women such that the woman is a former U.S. president ■

The set listed in part (c) of Example 2 has no elements; there are no women who are former U.S. presidents. If we let W equal "the set of all women such that the woman is a former U.S. president," then $n(W) = 0$. A set that has no elements is called an **empty set** and is denoted by $\emptyset$ or by $\{\,\}$. Notice that since the empty set has no elements, $n(\emptyset) = 0$. In contrast, the set $\{0\}$ is not empty; it has one element, the number zero, so $n(\{0\}) = 1$.

Universal Set and Subsets

When we work with sets, we must define a universal set. For any given problem, the **universal set**, denoted by U, is the set of all possible elements of any set used in the problem. For example, when we spell words, U is the set of all letters in the alphabet. When every element of one set is also a member of another set, we say that the first set is a *subset* of the second; for instance, {p, i, n} is a subset of {p, i, n, e}. In general, we say that A is a **subset** of B, denoted by $A \subseteq B$, if for every $x \in A$ it follows that $x \in B$. Alternatively, $A \subseteq B$ if A contains no elements that are not in B. If A contains an element that is not in B, then A is not a subset of B (symbolized as $A \nsubseteq B$).

EXAMPLE 3 Let $B = \{\text{countries} \mid \text{the country has a permanent seat on the U.N. Security Council}\}$. Determine if A is a subset of B.

a. $A = \{\text{Russia, United States}\}$
b. $A = \{\text{China, Japan}\}$
c. $A = \{\text{United States, France, China, United Kingdom, Russia}\}$
d. $A = \{\ \}$

Solution We use the roster method to list the elements of set B.

$$B = \{\text{China, France, Russia, United Kingdom, United States}\}$$

a. Since every element of A is also an element of B, A is a subset of B; $A \subseteq B$.
b. Since A contains an element (Japan) that is not in B, A is not a subset of B; $A \nsubseteq B$.
c. Since every element of A is also an element of B (note that $A = B$), A is a subset of B (and B is a subset of A); $A \subseteq B$ (and $B \subseteq A$). In general, every set is a subset of itself; $A \subseteq A$ for any set A.
d. Does A contain an element that is not in B? No! Therefore, A (an empty set) is a subset of B; $A \subseteq B$. In general, the empty set is a subset of all sets; $\varnothing \subseteq A$ for any set A. ■

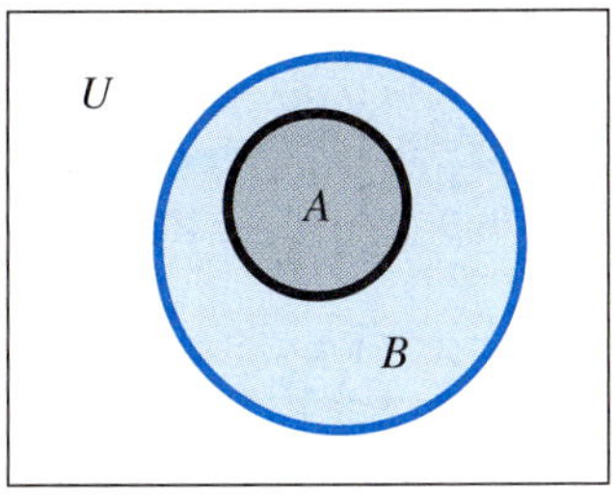

Figure 2.1
A is a subset of B. $A \subseteq B$

We can express the relationship $A \subseteq B$ visually by drawing a Venn diagram, as shown in Figure 2.1. A **Venn diagram** consists of a rectangle, representing the universal set, and various closed figures within the rectangle, each representing a set. Recall that Venn diagrams were used in Section 1.1 to determine whether an argument was valid.

If two sets are equal, they contain exactly the same elements. It then follows that each is a subset of the other. For example, if $A = B$, then every element of A is an element of B (and vice versa). In this case, A is called an **improper subset** of B. (Likewise, B is an improper subset of A.) Every set is an improper subset of itself; for example, $A \subseteq A$. On the other hand, if A is a subset of B and B contains an element not in A (that is, $A \neq B$), then A is called a **proper subset** of B. To indicate a proper subset, the symbol $\subset$ is used. While it is acceptable to write $\{1, 2\} \subseteq \{1, 2, 3\}$, the relationship of a proper subset is stressed when it is written $\{1, 2\} \subset \{1, 2, 3\}$. Notice the similarities between the subset symbols, $\subset$ and $\subseteq$, and the inequality symbols, $<$ and $\leq$, used in algebra; it is acceptable to write $1 \leq 3$, but writing $1 < 3$ is more informative.

Intersection of Sets

Sometimes an element of one set is also an element of another set; that is, the sets may overlap. This overlap is called the **intersection** of the sets. If an element is in two sets *at the same time,* it is in the intersection of the sets.

Intersection of Sets

The **intersection** of set A and set B, denoted by $A \cap B$, is

$$A \cap B = \{x \mid x \in A \quad \text{and} \quad x \in B\}$$

The intersection of two sets consists of those elements that are common to both sets.

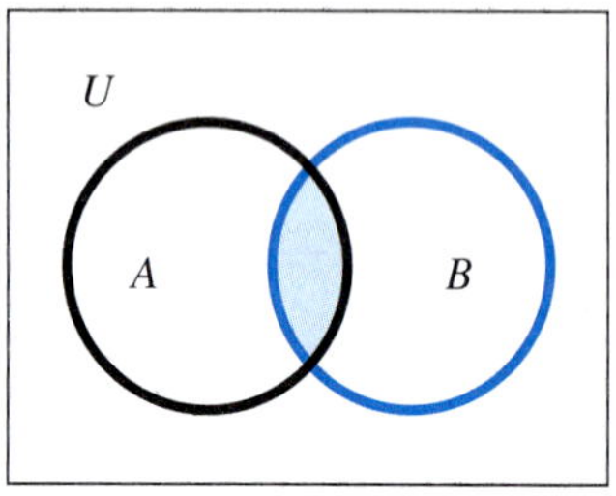

Figure 2.2
The intersection $A \cap B$

For example, given the sets $A = \{\text{Buffy, Spike, Willow, Xander}\}$ and $B = \{\text{Angel, Anya, Buffy, Giles, Spike}\}$, their intersection is $A \cap B = \{\text{Buffy, Spike}\}$.

Venn diagrams are useful in depicting the relationship between sets. The Venn diagram in Figure 2.2 illustrates the intersection of two sets; the shaded region represents $A \cap B$.

Mutually Exclusive Sets

Sometimes a pair of sets has no overlap. Consider an ordinary deck of playing cards. Let $D = \{\text{cards} \mid \text{the card is a diamond}\}$ and $S = \{\text{cards} \mid \text{the card is a spade}\}$. Certainly, *no* cards are both diamonds and spades *at the same time;* that is, $S \cap D = \emptyset$.

Two sets A and B are **mutually exclusive** (or *disjoint*) if they have no elements in common, that is, if $A \cap B = \emptyset$. The Venn diagram in Figure 2.3 illustrates mutually exclusive sets.

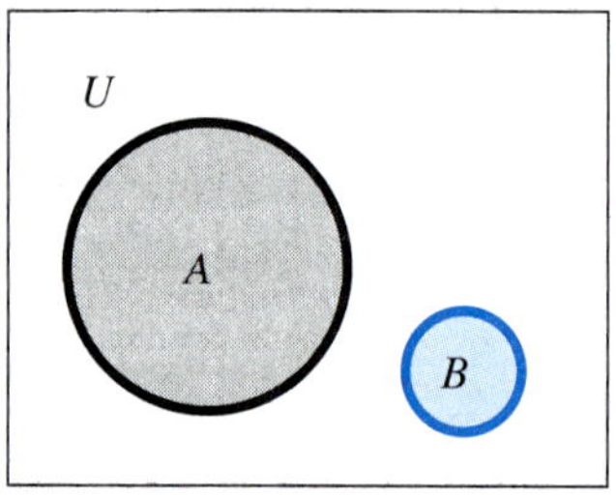

Figure 2.3
Mutually exclusive sets
($A \cap B = \emptyset$)

Union of Sets

What does it mean when we ask, "How many of the 500 college students in a transportation survey own an automobile or a motorcycle?" Does it mean "How many students own either an automobile or a motorcycle *or both*?" or does it mean "How many students own either an automobile or a motorcycle, *but not both*?" The former is called the *inclusive or,* because it includes the possibility of owning both; the latter is called the *exclusive or.* In logic and in mathematics, the word *or* refers to the *inclusive or,* unless you are told otherwise.

The meaning of the word *or* is important to the concept of union. The **union** of two sets is a new set formed by joining those two sets together, just as the union of the states is the joining together of 50 states to form one nation.

Union of Sets

The **union** of set A and set B, denoted by $A \cup B$, is

$$A \cup B = \{x \mid x \in A \quad \text{or} \quad x \in B\}$$

The union of A and B consists of all elements that are in either A or B or both, that is, all elements that are in at least one of the sets.

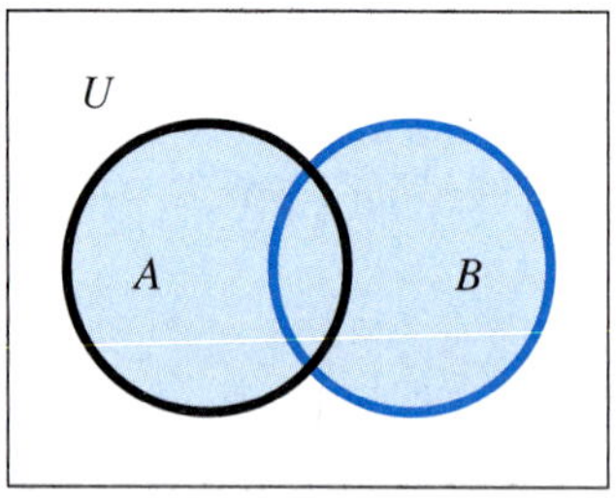

Figure 2.4
The union $A \cup B$

For example, given the sets $A = \{\text{David, Jay}\}$ and $B = \{\text{Conan, Geraldo, Oprah, Jerry}\}$, their union is $A \cup B = \{\text{Conan, David, Geraldo, Jay, Oprah, Jerry}\}$, and their intersection is $A \cap B = \emptyset$. The Venn diagram in Figure 2.4 illustrates the union of two sets; the shaded region represents $A \cup B$.

EXAMPLE 4 Given the sets $A = \{1, 2, 3\}$ and $B = \{2, 4, 6\}$, find the following.

a. $A \cap B$ (the intersection of A and B)
b. $A \cup B$ (the union of A and B)

Solution **a.** The intersection of two sets consists of those elements that are common to both sets; therefore, we have

$$\begin{aligned} A \cap B &= \{1, 2, 3\} \cap \{2, 4, 6\} \\ &= \{2\} \end{aligned}$$

b. The union of two sets consists of all elements that are in at least one of the sets; therefore, we have

$$\begin{aligned} A \cup B &= \{1, 2, 3\} \cup \{2, 4, 6\} \\ &= \{1, 2, 3, 4, 6\} \end{aligned}$$

■

Because $A \cup B$ consists of all elements that are in A or B (or both), to find $n(A \cup B)$, we add $n(A)$ plus $n(B)$. However, doing so results in an answer that might be too big; that is, if A and B have elements in common, these elements will be counted twice (once as a part of A and once as a part of B). Therefore, to find the cardinal number of $A \cup B$, we add the cardinal number of A to the cardinal number of B and then *subtract* the cardinal number of $A \cap B$ (so that the overlap is not counted twice).

Cardinal Number Formula for the Union of Sets

For any two sets A and B, the number of elements in their union is $n(A \cup B)$, where

$$n(A \cup B) = n(A) + n(B) - n(A \cap B)$$

As long as any three of the four quantities in the general formula are known, the missing quantity can be found by algebraic manipulation.

EXAMPLE 5 Given $n(U) = 169$, $n(A) = 81$, and $n(B) = 66$, find the following.

a. If $n(A \cap B) = 47$, find $n(A \cup B)$ and draw a Venn diagram depicting the composition of the universal set.
b. If $n(A \cup B) = 147$, find $n(A \cap B)$ and draw a Venn diagram depicting the composition of the universal set.

Solution **a.** We must use the Cardinal Number Formula for the Union of Sets. Substituting the three given quantities, we have

$$\begin{aligned} n(A \cup B) &= n(A) + n(B) - n(A \cap B) \\ &= 81 + 66 - 47 \\ &= 100 \end{aligned}$$

The Venn diagram in Figure 2.5 illustrates the composition of U.

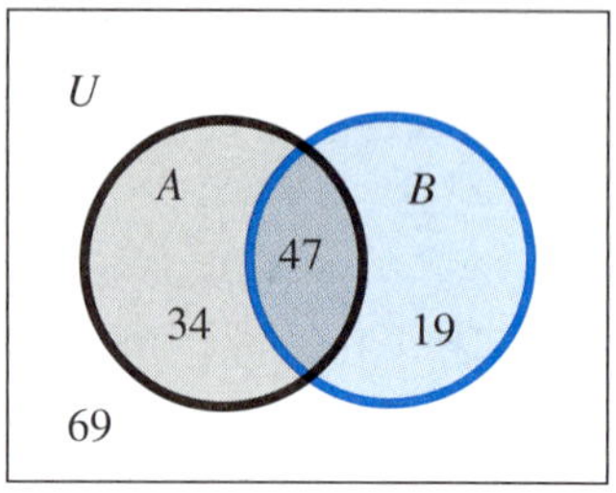

Figure 2.5

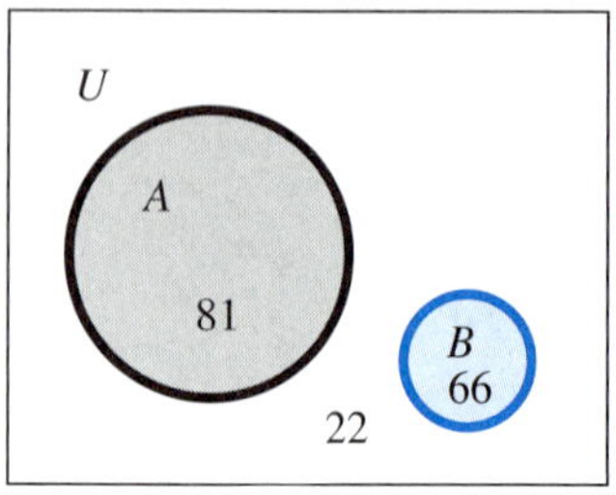

Figure 2.6

b. We must use the Cardinal Number Formula for the Union of Sets. Substituting the three given quantities, we have

$$\begin{aligned} n(A \cup B) &= n(A) + n(B) - n(A \cap B) \\ 147 &= 81 + 66 - n(A \cap B) \\ 147 &= 147 - n(A \cap B) \\ n(A \cap B) &= 147 - 147 \\ n(A \cap B) &= 0 \end{aligned}$$

Therefore, A and B have no elements in common; they are mutually exclusive. The Venn diagram in Figure 2.6 illustrates the composition of U. ■

EXAMPLE 6 A recent transportation survey of 500 college students (the universal set U) yielded the following information: 291 own an automobile (A), 179 own a motorcycle (M), and 85 own both an automobile and a motorcycle ($A \cap M$). What percent of these students own an automobile or a motorcycle?

Solution Recall that "automobile or motorcycle" means "automobile or motorcycle or both" (the inclusive *or*) and that *or* implies union. Hence, we must find $n(A \cup M)$, the cardinal number of the union of sets A and M. We are given that $n(A) = 291$, $n(M) = 179$, and $n(A \cap M) = 85$. Substituting the given values into the Cardinal Number Formula for the Union of Sets, we have

$$\begin{aligned} n(A \cup M) &= n(A) + n(M) - n(A \cap M) \\ &= 291 + 179 - 85 \\ &= 385 \end{aligned}$$

Therefore, 385 of the 500 students surveyed own an automobile or a motorcycle. Expressed as a percent, $385/500 = 0.77$; therefore, 77% of the students own an automobile or a motorcycle (or both). ■

Complement of a Set

In certain situations, it might be important to know how many things are *not* in a given set. For instance, when playing cards, you might want to know how many cards are not ranked lower than a five, or when taking a survey, you might want to know how many people did not vote for a specific proposition. The set of all elements in the universal set that are *not* in a specific set is called the *complement* of the set.

> **Complement of a Set**
>
> The **complement** of set A, denoted by A' (read "A prime" or "the complement of A"), is
>
> $$A' = \{x \mid x \in U \quad \text{and} \quad x \notin A\}$$
>
> The complement of a set consists of all elements that are in the universal set but not in the given set.

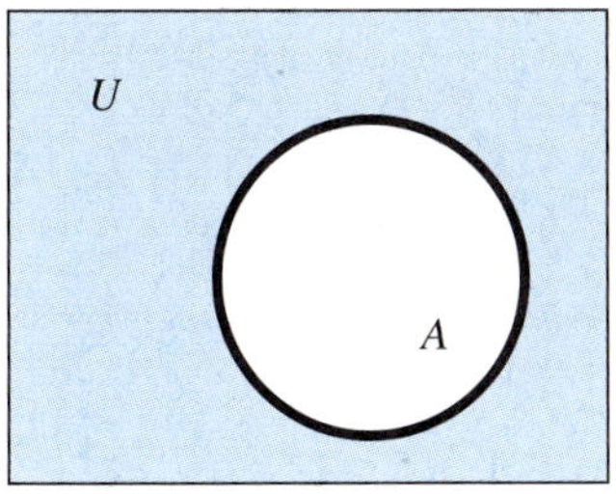

Figure 2.7
The complement A'

For example, given that $U = \{1, 2, 3, 4, 5, 6, 7, 8, 9\}$ and $A = \{1, 3, 5, 7, 9\}$, the complement of A is $A' = \{2, 4, 6, 8\}$. What is the complement of A'? Just as $-(-x) = x$ in algebra, $(A')' = A$ in set theory. The Venn diagram in Figure 2.7 illustrates the complement of set A; the shaded region represents A'.

Suppose A is a set of elements, drawn from a universal set U. If x is an element of the universal set ($x \in U$), then exactly one of the following must be true: (1) x is an element of A ($x \in A$), or (2) x is not an element of A ($x \notin A$). Since no element of the universal set can be in both A and A' at the same time, it follows that A and A' are mutually exclusive sets whose union equals the entire universal set. Therefore, the sum of the cardinal numbers of A and A' equals the cardinal number of U.

It is often quicker to count the elements that are *not* in a set than to count those that are. Consequently, to find the cardinal number of a set, we can subtract the cardinal number of its complement from the cardinal number of the universal set; that is, $n(A) = n(U) - n(A')$.

> **Cardinal Number Formula for the Complement of a Set**
>
> For any set A and its complement A',
>
> $$n(A) + n(A') = n(U)$$
>
> where U is the universal set.
> Alternatively,
>
> $$n(A) = n(U) - n(A') \quad \text{and} \quad n(A') = n(U) - n(A)$$

EXAMPLE 7 How many letters in the alphabet precede the letter w?

Solution Rather than counting all the letters that precede w, we will take a shortcut by counting all the letters that do *not* precede w. Let $L = \{\text{letters} \mid \text{the letter precedes w}\}$. Therefore, $L' = \{\text{letter} \mid \text{the letter does not precede w}\}$. Now $L' = \{\text{w, x, y, z}\}$ and $n(L') = 4$; therefore, we have

$$\begin{aligned} n(L) &= n(U) - n(L') && \text{Cardinal Number Formula for the Complement of a Set} \\ &= 26 - 4 \\ &= 22 \end{aligned}$$

There are 22 letters preceding the letter w.

Stock Montage

HISTORICAL NOTE

John Venn

1834–1923

John Venn is considered by many to be one of the originators of modern symbolic logic. Venn received his degree in mathematics from the University at Cambridge at the age of twenty-three. He was then elected a fellow of the college and held this fellowship until his death, some 66 years later. Two years after receiving his degree, Venn accepted a teaching position at Cambridge: college lecturer in moral sciences.

During the latter half of the nineteenth century, the study of logic experienced a rebirth in England. Mathematicians were attempting to symbolize and quantify the central concepts of logical thought. Consequently, Venn chose to focus on the study of logic during his tenure at Cambridge. In addition, he investigated the field of probability and published *The Logic of Chance,* his first major work, in 1866.

Venn was well read in the works of his predecessors, including the noted logicians Augustus De Morgan, George Boole, and Charles Dodgson (a.k.a. Lewis Carroll). Boole's pioneering work on the marriage of logic and algebra proved to be a strong influence on Venn; in fact, Venn used the type of diagram that now bears his name in an 1876 paper in which he examined Boole's system of symbolic logic.

Venn was not the first scholar to use the diagrams that now bear his name. Gottfried Leibniz, Leonhard Euler, and others utilized similar diagrams years before Venn did. Examining each author's diagrams, Venn was critical of their lack of uniformity. He developed a consistent, systematic explanation of the general use of geometrical figures in the analysis of logical arguments. Today, these geometrical figures are known by his name and are used extensively in elementary set theory and logic.

Venn's writings were held in high esteem. His textbooks, *Symbolic Logic* (1881) and *The Principles of Empirical Logic* (1889), were used during the late nineteenth and early twentieth centuries. In addition to his works on logic and probability, Venn also conducted much research into historical records, especially those of his college and those of his family.

Set theory and the cardinal numbers of sets are used extensively in the study of probability. Although he was a professor of logic, Venn investigated the foundations and applications of theoretical probability. Venn's first major work, *The Logic of Chance,* exhibited the diversity of his academic interests.

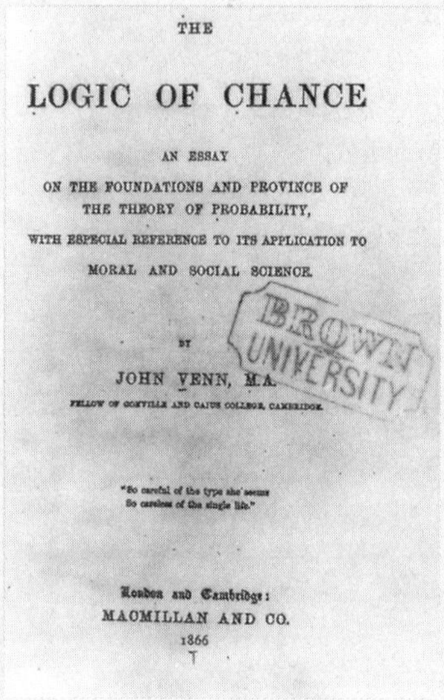
THE

LOGIC OF CHANCE

AN ESSAY

ON THE FOUNDATIONS AND PROVINCE OF
THE THEORY OF PROBABILITY,

WITH ESPECIAL REFERENCE TO ITS APPLICATION TO
MORAL AND SOCIAL SCIENCE.

BY

JOHN VENN, M.A.

FELLOW OF GONVILLE AND CAIUS COLLEGE, CAMBRIDGE.

London and Cambridge:
MACMILLAN AND CO.
1866

Brown University Library

Shading Venn Diagrams

In an effort to visualize the results of operations on sets, it may be necessary to shade specific regions of a Venn diagram. The following example shows a systematic method for shading the intersection or union of any two sets.

EXAMPLE 8 On a Venn diagram, shade in the region corresponding to the indicated set.

a. $A \cap B'$ **b.** $A \cup B'$

Solution **a.** First, draw and label two overlapping circles as shown in Figure 2.8. The two "components" of the operation $A \cap B'$ are "A" and "B'." Shade each of these components in contrasting ways; shade one of them, say A, with horizontal lines, and the

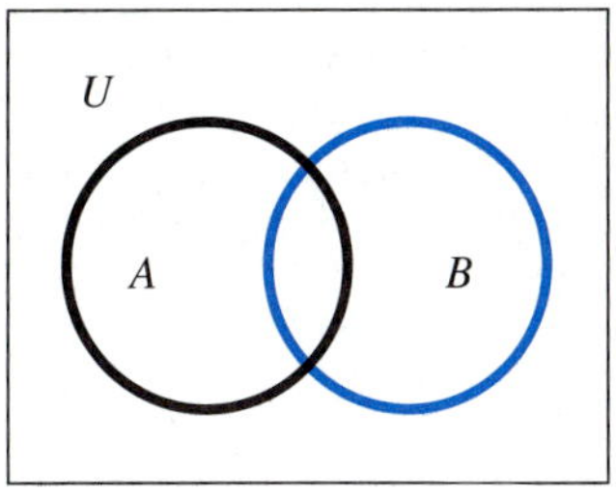

Figure 2.8

other with vertical lines as in Figure 2.9. Be sure to include a legend, or key, identifying each type of shading.

To be in the intersection of two sets, an element must be in *both* sets at the same time. Therefore, the intersection of A and B' is the region that is shaded in *both* directions (horizontal and vertical) at the same time. A final diagram depicting $A \cap B'$ is shown in Figure 2.10.

b. Refer to Figure 2.9. To be in the union of two sets, an element must be in *at least one* of the sets. Therefore, the union of A and B' consists of all regions that are shaded in *any* direction whatsoever (horizontal or vertical or both). A final diagram depicting $A \cup B'$ is shown in Figure 2.11.

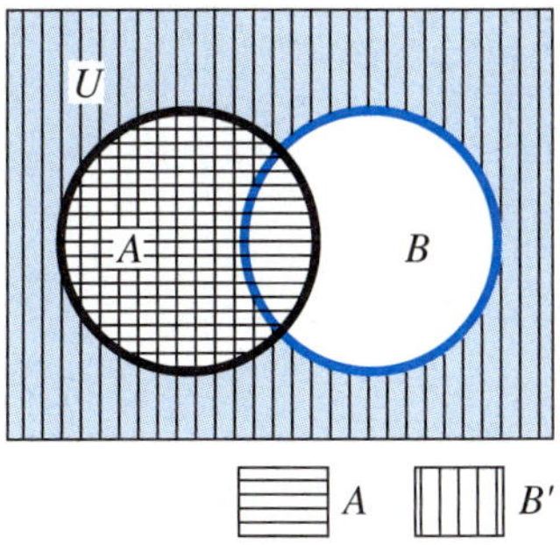

Figure 2.9

Set Theory and Logic

If you have read Chapter 1, you have probably noticed that set theory and logic have many similarities. For instance, the union symbol $\cup$ and the disjunction symbol $\vee$ have the same meaning, but they are used in different circumstances; $\cup$ goes between sets, while $\vee$ goes between logical expressions. The $\cup$ and $\vee$ symbols are similar in appearance because their usages are similar. A comparison of the terms and symbols used in set theory and logic is given in Figure 2.12.

Applying the concepts and symbols of Chapter 1, we can define the basic operations of set theory in terms of logical biconditionals. The biconditionals in Figure 2.13 are tautologies (expressions that are always true); the first biconditional is read as "x is an element of the union of sets A and B if and only if x is an element of set A or x is an element of set B."

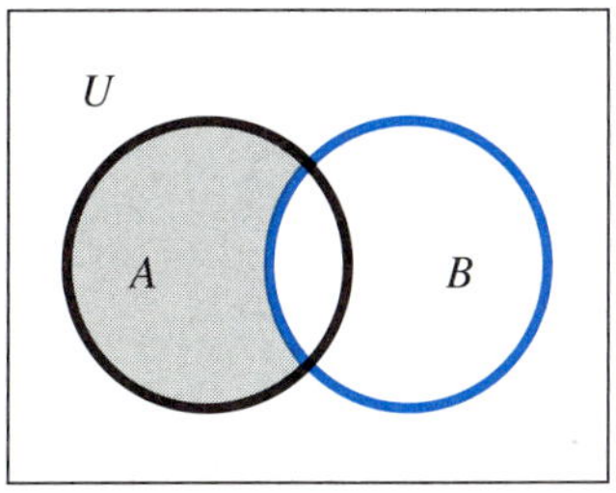

$A \cap B'$

Figure 2.10

Set Theory		Logic		Common Wording
Term	**Symbol**	**Term**	**Symbol**	
union	$\cup$	disjunction	$\vee$	or
intersection	$\cap$	conjunction	$\wedge$	and
complement	$'$	negation	$\sim$	not
subset	$\subseteq$	conditional	$\rightarrow$	if . . . then . . .

Figure 2.12
Comparison of terms and symbols used in set theory and logic

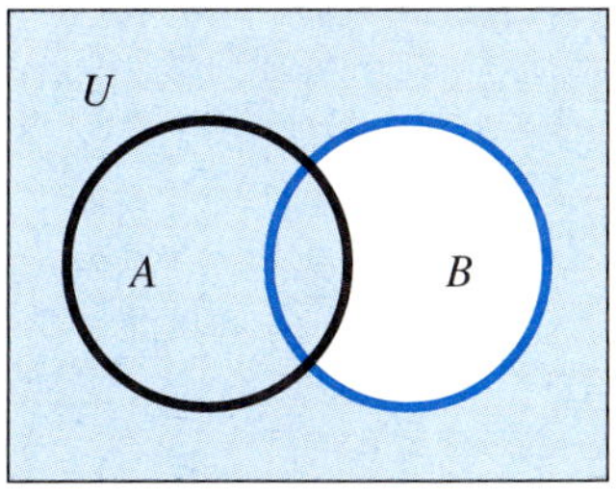

$A \cup B'$

Figure 2.11

Basic Operations in Set Theory	Logical Biconditional
union	$[x \in (A \cup B)] \leftrightarrow [x \in A \vee x \in B]$
intersection	$[x \in (A \cap B)] \leftrightarrow [x \in A \wedge x \in B]$
complement	$(x \in A') \leftrightarrow \sim (x \in A)$
subset	$(A \subseteq B) \leftrightarrow (x \in A \rightarrow x \in B)$

Figure 2.13
Set theory operations as logical biconditionals

2.1 Exercises

1. State whether the given set is well defined.
   a. the set of all black automobiles
   b. the set of all inexpensive automobiles
   c. the set of all prime numbers
   d. the set of all large numbers
2. Suppose $A = \{2, 5, 7, 9, 13, 25, 26\}$.
   a. Find $n(A)$
   b. True or false: $7 \in A$
   c. True or false: $9 \notin A$
   d. True or false: $20 \notin A$

In Exercises 3–6, list all subsets of the given set. Identify which subsets are proper and which are improper.

3. $B = \{\text{Lennon, McCartney}\}$
4. $N = \{0\}$
5. $S = \{\text{yes, no, undecided}\}$
6. $M = \{\text{classical, country, jazz, rock}\}$

In Exercises 7–10, the universal set is $U = \{0, 1, 2, 3, 4, 5, 6, 7, 8, 9\}$.

7. If $A = \{1, 2, 3, 4, 5\}$ and $B = \{4, 5, 6, 7, 8\}$, find the following.
   a. $A \cap B$ b. $A \cup B$
   c. A' d. B'
8. If $A = \{2, 3, 5, 7\}$ and $B = \{2, 4, 6, 7\}$, find the following.
   a. $A \cap B$ b. $A \cup B$
   c. A' d. B'
9. If $A = \{1, 3, 5, 7, 9\}$ and $B = \{0, 2, 4, 6, 8\}$, find the following.
   a. $A \cap B$ b. $A \cup B$
   c. A' d. B'
10. If $A = \{3, 6, 9\}$ and $B = \{4, 8\}$, find the following.
   a. $A \cap B$ b. $A \cup B$
   c. A' d. B'

In Exercises 11–16, the universal set is U = {Monday, Tuesday, Wednesday, Thursday, Friday, Saturday, Sunday}. If A = {Monday, Tuesday, Wednesday, Thursday, Friday} and B = {Friday, Saturday, Sunday}, find the indicated set.

11. $A \cap B$ 12. $A \cup B$
13. B' 14. A'
15. $A' \cup B$ 16. $A \cap B'$

In Exercises 17–26, use a Venn diagram like the one in Figure 2.14 to shade in the region corresponding to the indicated set.

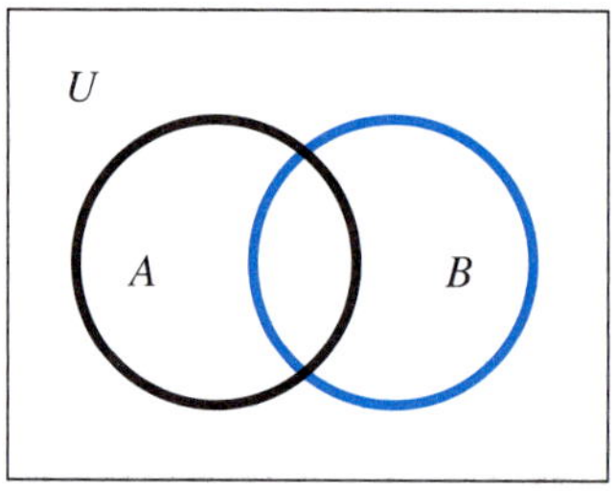

Figure 2.14

17. $A \cap B$ 18. $A \cup B$
19. A' 20. B'
21. $A \cup B'$ 22. $A' \cup B$
23. $A' \cap B$ 24. $A \cap B'$
25. $A' \cup B'$ 26. $A' \cap B'$
27. Suppose $n(U) = 150$, $n(A) = 37$, and $n(B) = 84$.
   a. If $n(A \cup B) = 100$, find $n(A \cap B)$ and draw a Venn diagram illustrating the composition of U.
   b. If $n(A \cup B) = 121$, find $n(A \cap B)$ and draw a Venn diagram illustrating the composition of U.
28. Suppose $n(U) = w$, $n(A) = x$, $n(B) = y$, and $n(A \cup B) = z$.
   a. Why must x be less than or equal to z?
   b. If $A \neq U$ and $B \neq U$, fill in the blank with the most appropriate symbol: $<$, $>$, $\leq$, or $\geq$.
   w_____z, w_____y, y_____z, x_____w
   c. Find $n(A \cap B)$ and draw a Venn diagram illustrating the composition of U.
29. In a recent transportation survey, 500 high school seniors were asked to check the appropriate box or boxes on the following form:

   ☐ I own an automobile.
   ☐ I own a motorcycle.

   The results were tabulated as follows: 91 checked the automobile box, 123 checked the motorcycle box, and 29 checked both boxes.
   a. Draw a Venn diagram illustrating the results of the survey.
   b. What percent of these students own an automobile or a motorcycle?

30. In a recent market research survey, 500 married couples were asked to check the appropriate box or boxes on the following form:

☐ We own a VCR.
☐ We own a microwave oven.

The results were tabulated as follows: 248 checked the VCR box, 314 checked the microwave oven box, and 166 checked both boxes.

a. Draw a Venn diagram illustrating the results of the survey.

b. What percent of these couples own a VCR or a microwave oven?

31. In a recent socioeconomic survey, 750 married women were asked to check the appropriate box or boxes on the following form:

☐ I have a career.
☐ I have a child.

The results were tabulated as follows: 529 checked the child box, 213 checked the career box, and 143 were blank (no boxes were checked).

a. Draw a Venn diagram illustrating the results of the survey.

b. What percent of these women had both a child and a career?

32. In a recent health survey, 750 single men in their twenties were asked to check the appropriate box or boxes on the following form:

☐ I am a member of a private gym.
☐ I am a vegetarian.

The results were tabulated as follows: 374 checked the gym box, 92 checked the vegetarian box, and 332 were blank (no boxes were checked).

a. Draw a Venn diagram illustrating the results of the survey.

b. What percent of these men were both members of a private gym and vegetarians?

For Exercises 33–36, let

$U = \{x \mid x$ is the name of one of the states in the United States$\}$

$A = \{x \mid x \in U$ and x begins with the letter A$\}$

$I = \{x \mid x \in U$ and x begins with the letter I$\}$

$M = \{x \mid x \in U$ and x begins with the letter M$\}$

$N = \{x \mid x \in U$ and x begins with the letter N$\}$

$O = \{x \mid x \in U$ and x begins with the letter O$\}$

33. Find $n(M')$.

34. Find $n(A \cup N)$.

35. Find $n(I' \cap O')$.

36. Find $n(M \cap I)$.

Exercises 37–40, let

$U = \{x \mid x$ is the name of one of the months in a year$\}$

$J = \{x \mid x \in U$ and x begins with the letter J$\}$

$Y = \{x \mid x \in U$ and x ends with the letter Y$\}$

$V = \{x \mid x \in U$ and x begins with a vowel$\}$

$R = \{x \mid x \in U$ and x ends with the letter R$\}$

37. Find $n(R')$.

38. Find $n(J \cap V)$.

39. Find $n(J \cup Y)$.

40. Find $n(V \cap R)$.

In Exercises 41–50, determine how many cards, in an ordinary deck of 52, fit the description. (If you are unfamiliar with playing cards, see the end of Section 3.1 for a description of a standard deck.)

41. spades or aces

42. clubs or twos

43. face cards or black

44. face cards or diamonds

45. face cards and black

46. face cards and diamonds

47. aces or eights

48. threes or sixes

49. aces and eights

50. threes and sixes

51. Suppose $A = \{1, 2, 3\}$ and $B = \{1, 2, 3, 4, 5, 6\}$.

a. Find $A \cap B$.

b. Find $A \cup B$.

c. In general, if $E \cap F = E$, what must be true concerning sets E and F?

d. In general, if $E \cup F = F$, what must be true concerning sets E and F?

52. Fill in the blank, and give an example to support your answer.

a. If $A \subset B$, then $A \cap B =$ _____.

b. If $A \subset B$, then $A \cup B =$ _____.

53. **a.** List all subsets of $A = \{a\}$. How many subsets does A have?

b. List all subsets of $A = \{a, b\}$. How many subsets does A have?

c. List all subsets of $A = \{a, b, c\}$. How many subsets does A have?

d. List all subsets of $A = \{a, b, c, d\}$. How many subsets does A have?

e. Is there a relationship between the cardinal number of set A and the number of subsets of set A?

f. How many subsets does $A = \{a, b, c, d, e, f\}$ have?

HINT: Use your answer to part (e).

54. Prove the Cardinal Number Formula for the Complement of a Set.

HINT: Apply the Cardinal Number Formula for the Union of Sets to A and A'.

Answer the following questions using complete sentences and your own words.

Concept Questions

55. If $A \cap B = \emptyset$, what is the relationship between sets A and B?
56. If $A \cup B = \emptyset$, what is the relationship between sets A and B?
57. Explain the difference between $\{0\}$ and $\emptyset$.
58. Explain the difference between 0 and $\{0\}$.
59. Is it possible to have $A \cap A = \emptyset$?
60. What is the difference between proper and improper subsets?
61. A set can be described by two methods: the roster method and set-builder notation. When is it advantageous to use the roster method? When is it advantageous to use set-builder notation?
62. Translate the following symbolic expressions into English sentences.
    a. $x \in (A \cap B) \leftrightarrow (x \in A \wedge x \in B)$
    b. $(x \in A') \leftrightarrow \sim (x \in A)$
    c. $(A \subseteq B) \leftrightarrow (x \in A \rightarrow x \in B)$

History Questions

63. In what academic field was John Venn a professor? Where did he teach?
64. What was one of John Venn's main contributions to the field of logic? What new benefits did it offer?

Many graduate schools require applicants to take the Graduate Record Examination (G.R.E.). This exam is intended to measure verbal, quantitative, and analytical skills that have developed throughout a person's life. Many classes and study guides are available to help people prepare for the exam. The remaining questions are typical of those found in the study guides and on the exam itself.

Exercises 65–69 refer to the following: Two collectors, John and Juneko, are each selecting a group of three posters from a group of seven movie posters: J, K, L, M, N, O, *and* P. *No poster can be in both groups. The selections made by John and Juneko are subject to the following restrictions:*

- If K is in John's group, M must be in Juneko's group.
- If N is in John's group, P must be in Juneko's group.
- J and P cannot be in the same group.
- M and O cannot be in the same group.

65. Which of the following pairs of groups selected by John and Juneko conform to the restrictions?

	John	Juneko
a.	J, K, L	M, N, O
b.	J, K, P	L, M, N
c.	K, N, P	J, M, O
d.	L, M, N	K, O, P
e.	M, O, P	J, K, N

66. If N is in John's group, which of the following could not be in Juneko's group?
    a. J **b.** K **c.** L **d.** M **e.** P
67. If K and N are in John's group, Juneko's group must consist of which of the following?
    a. J, M, and O **b.** J, O, and P
    c. L, M, and P **d.** L, O, and P
    e. M, O, and P
68. If J is in Juneko's group, which of the following is true?
    a. K cannot be in John's group.
    b. N cannot be in John's group.
    c. O cannot be in Juneko's group.
    d. P must be in John's group.
    e. P must be in Juneko's group.
69. If K is in John's group, which of the following is true?
    a. J must be in John's group.
    b. O must be in John's group.
    c. L must be in Juneko's group.
    d. N cannot be in John's group.
    e. O cannot be in Juneko's group.

2.2 Applications of Venn Diagrams

As we have seen, Venn diagrams are very useful tools for visualizing the relationships between sets. They can be used to establish general formulas involving set operations and to determine the cardinal numbers of sets. Venn diagrams are particularly useful in survey analysis.

Surveys

Surveys are often used to divide people or objects into categories. Because the categories sometimes overlap, people can fall into more than one category. Venn diagrams and the formulas for cardinal numbers can help researchers organize the data.

EXAMPLE 1 Has the advent of the DVD affected attendance at movie theaters? To study this question, Professor Redrum's film class conducted a survey of people's movie-watching habits. He had his students ask hundreds of people between the ages of sixteen and forty-five to check the appropriate box or boxes on the following form:

> ☐ I watched a movie in a theater during the past month.
> ☐ I watched a movie on a DVD during the past month.

After the professor had collected the forms and tabulated the results, he told the class that 388 people checked the theater box, 495 checked the DVD box, 281 checked both boxes, and 98 of the forms were blank. Giving the class only this information, Professor Redrum posed the following three questions.

a. What percent of the people surveyed watched a movie in a theater or on a DVD during the past month?
b. What percent of the people surveyed watched a movie in a theater only?
c. What percent of the people surveyed watched a movie on a DVD only?

Solution **a.** To calculate the desired percentages, we must determine $n(U)$, the total number of people surveyed. This can be accomplished by drawing a Venn diagram. Because the survey divides people into two categories (those who watched a movie in a theater and those who watched a movie on a DVD), we need to define two sets. Let

$$T = \{\text{people} \mid \text{the person watched a movie in a theater}\}$$
$$D = \{\text{people} \mid \text{the person watched a movie on a DVD}\}$$

Now translate the given survey information into the symbols for the sets and attach their given cardinal numbers: $n(T) = 388$, $n(D) = 495$, and $n(T \cap D) = 281$.

Our first goal is to find $n(U)$. To do so, we will fill in the cardinal numbers of all regions of a Venn diagram consisting of two overlapping circles (because we are dealing with two sets). The intersection of T and D consists of 281 people, so we draw two overlapping circles and fill in 281 as the number of elements in common (see Figure 2.15).

Because we were given $n(T) = 388$ and know that $n(T \cap D) = 281$, the difference $388 - 281 = 107$ tells us that 107 people watched a movie in a theater but did not watch a movie on a DVD. We fill in 107 as the number of people who watched a movie only in a theater (see Figure 2.16).

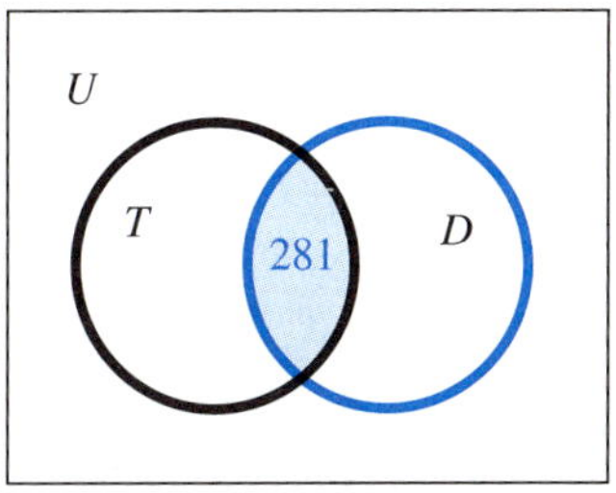

Figure 2.15

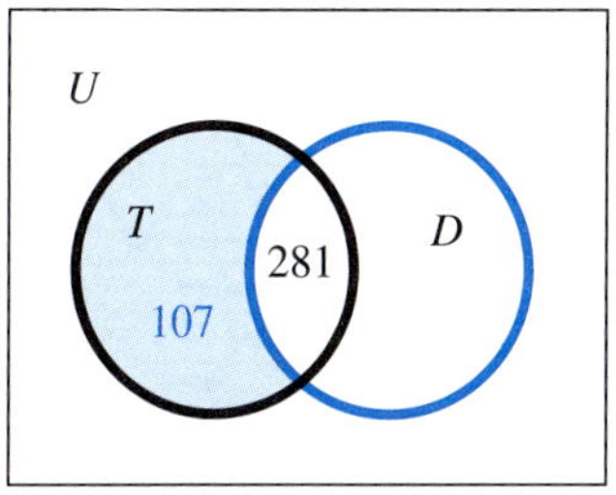

Figure 2.16

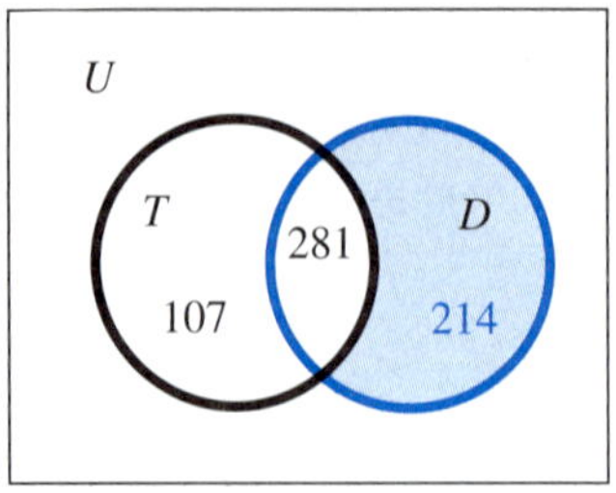

Figure 2.17

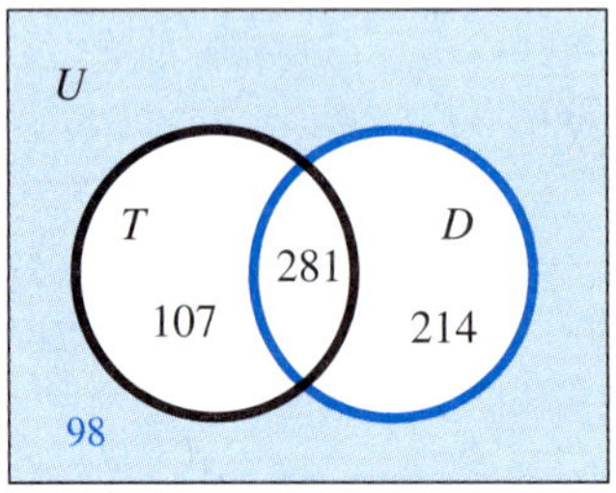

Figure 2.18

Because $n(D) = 495$, the difference $495 - 281 = 214$ tells us that 214 people watched a movie on a DVD but not in a theater. We fill in 214 as the number of people who watched a movie only on a DVD (see Figure 2.17).

The only region remaining to be filled in is the region outside both circles. This region represents people who didn't watch a movie in a theater or on a DVD and is symbolized by $(T \cup D)'$. Because 98 people didn't check either box on the form, $n[(T \cup D)'] = 98$ (see Figure 2.18).

After we have filled in the Venn diagram with all the cardinal numbers, we readily see that $n(U) = 98 + 107 + 281 + 214 = 700$. Therefore, 700 people were in the survey.

To determine what *percent* of the people surveyed watched a movie in a theater *or* on a DVD during the past month, simply divide $n(T \cup D)$ by $n(U)$:

$$\frac{n(T \cup D)}{n(U)} = \frac{107 + 281 + 214}{700} = \frac{602}{700} = 0.86$$

Therefore, exactly 86% of the people surveyed watched a movie in a theater or on a DVD during the past month.

b. To find what *percent* of the people surveyed watched a movie in a theater only, divide 107 (the number of people who watched a movie in a theater only) by $n(U)$:

$$\frac{107}{700} = 0.152857142\ldots$$

Approximately 15.3% of the people surveyed watched a movie in a theater only.

c. Because 214 people watched a movie on DVD only, $214/700 = 0.305714285\ldots$, or approximately 30.6%, of the people surveyed watched a movie on DVD only.

When you solve a cardinal number problem (a problem that asks, "How many?" or "What percent?") involving a universal set that is divided into various categories (for instance, a survey), use the following general steps.

Solving a Cardinal Number Problem

A cardinal number problem is a problem in which you are asked, "How many?" or "What percent?"

1. Define a set for each category in the universal set. If a category and its negation are both mentioned, define one set A and utilize its complement A'.
2. Draw a Venn diagram with as many overlapping circles as the number of sets you have defined.
3. Write down all the given cardinal numbers corresponding to the various given sets.
4. Starting with the innermost overlap, fill in each region of the Venn diagram with its cardinal number.

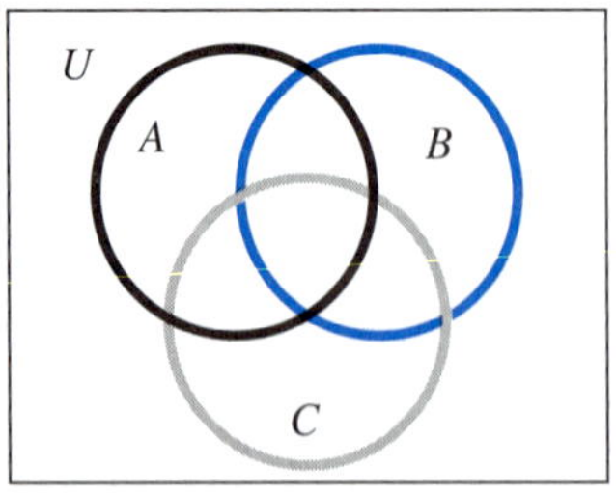

Figure 2.19

When we are working with three sets, we must account for all possible intersections of the sets. Hence, in such cases, we will use the Venn diagram shown in Figure 2.19.

EXAMPLE 2 A consumer survey was conducted to examine patterns in ownership of personal computers, cellular telephones, and DVD players. The following data were obtained: 213 people had personal computers, 294 had cellular telephones, 337 had DVD players, 109 had all three, 64 had none, 198 had cell phones and DVD players, 382 had cell phones or computers, and 61 had computers and DVD players but no cell phones.

a. What percent of the people surveyed owned a computer but no DVD player or cell phone?

b. What percent of the people surveyed owned a DVD player but no computer or cell phone?

Solution **a.** To calculate the desired percentages, we must determine $n(U)$, the total number of people surveyed. This can be accomplished by drawing a Venn diagram. Because the survey divides people into three categories (those who own a computer, those who own a cell phone, and those who own a DVD player), we need to define three sets. Let

$$C = \{\text{people} \mid \text{the person owns a personal computer}\}$$
$$T = \{\text{people} \mid \text{the person owns a cellular telephone}\}$$
$$D = \{\text{people} \mid \text{the person owns a DVD player}\}$$

Now translate the given survey information into the symbols for the sets and attach their given cardinal numbers:

213 people had computers ⟶	$n(C) = 213$
294 had cellular telephones ⟶	$n(T) = 294$
337 had DVD players ⟶	$n(D) = 337$
109 had all three (C *and* T *and* D) ⟶	$n(C \cap T \cap D) = 109$
64 had none (not C *and* not T *and* not D) ⟶	$n(C' \cap T' \cap D') = 64$
198 had cell phones and DVD players (T *and* D) ⟶	$n(T \cap D) = 198$
382 had cell phones or computers (T *or* C) ⟶	$n(T \cup C) = 382$
61 had computers and DVD players but no cell phones (C *and* D *and* not T) ⟶	$n(C \cap D \cap T') = 61$

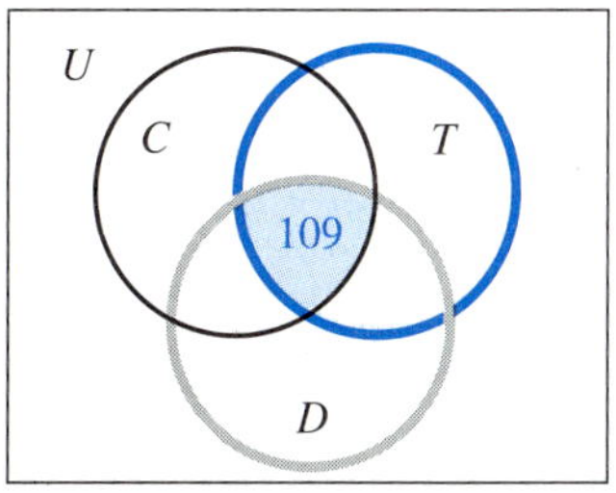

Figure 2.20

Our first goal is to find $n(U)$. To do so, we will fill in the cardinal numbers of all regions of a Venn diagram like that in Figure 2.19. We start by using information concerning membership in all three sets. Because the intersection of all three sets consists of 109 people, we fill in 109 in the region common to C and T and D (see Figure 2.20).

Next, we utilize any information concerning membership in two of the three sets. Because $n(T \cap D) = 198$, a total of 198 people are common to both T and D; some are in C, and some are not in C. Of these 198 people, 109 are in C (see Figure 2.20). Therefore, the difference $198 - 109 = 89$ gives the number not in C. Eighty-nine people are in T and D and *not* in C; that is, $n(T \cap D \cap C') = 89$. Concerning

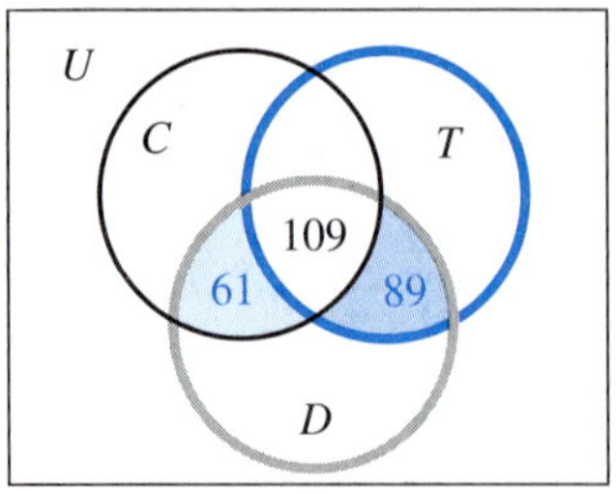

Figure 2.21

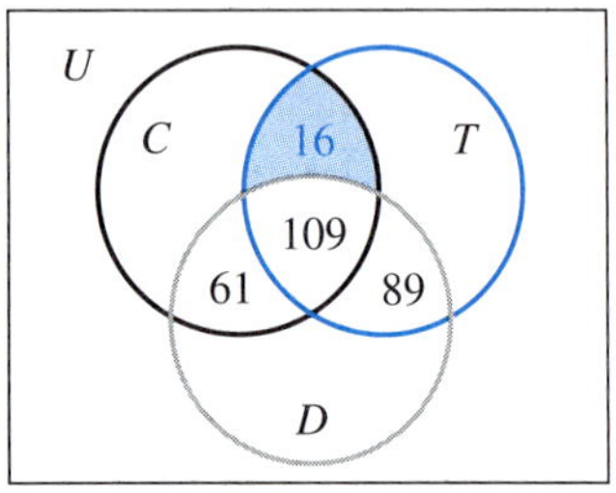

Figure 2.22

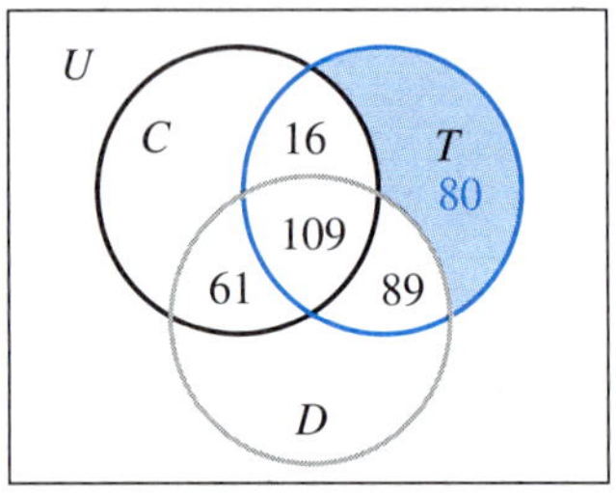

Figure 2.23

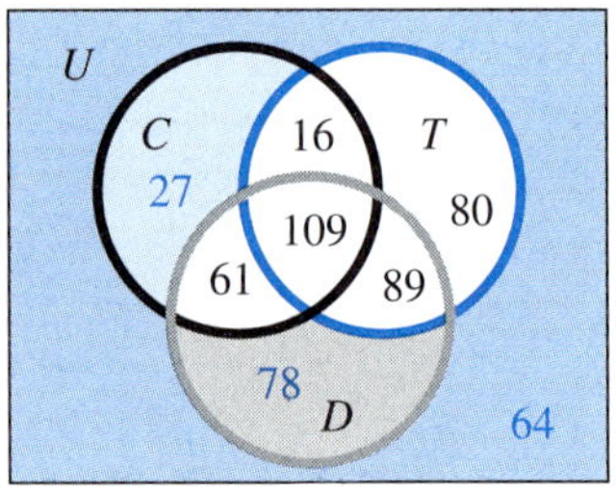

Figure 2.24

membership in the two sets C and D, we are given $n(C \cap D \cap T') = 61$. Therefore, we know that 61 people are in C and D and not in T (see Figure 2.21).

We are given $n(T \cup C) = 382$. From this number, we can calculate $n(T \cap C)$ by using the Cardinal Number Formula for the Union of Sets:

$$n(T \cup C) = n(T) + n(C) - n(T \cap C)$$
$$382 = 294 + 213 - n(T \cap C)$$
$$n(T \cap C) = 125$$

Therefore, a total of 125 people are in T and C; some are in D, and some are not in D. Of these 125 people, 109 are in D (see Figure 2.20). Therefore, the difference $125 - 109 = 16$ gives the number not in D. Sixteen people are in T and C and *not* in D; that is, $n(C \cap T \cap D') = 16$ (see Figure 2.22).

Knowing that a total of 294 people are in T (given $n(T) = 294$), we are now able to fill in the last region of T. The missing region (people in T only) has $294 - 109 - 89 - 16 = 80$ members; $n(T \cap C' \cap D') = 80$ (see Figure 2.23).

In a similar manner, we subtract the known pieces of C from $n(C) = 213$, which is given, and obtain $213 - 61 - 109 - 16 = 27$; therefore, 27 people are in C only. Likewise, to find the last region of D, we use $n(D) = 337$ (given) and obtain $337 - 89 - 109 - 61 = 78$; therefore, 78 people are in D only. Finally, the 64 people who own none of the items are placed "outside" the three circles (see Figure 2.24).

By adding up the cardinal numbers of all the regions in Figure 2.24, we find that the total number of people in the survey is 524; that is, $n(U) = 524$.

Now, to determine what *percent* of the people surveyed owned only a computer (no DVD player and no cell phone), we simply divide $n(C \cap D' \cap T')$ by $n(U)$:

$$\frac{n(C \cap D' \cap T')}{n(U)} = \frac{27}{524}$$
$$= 0.051526717\ldots$$

Approximately 5.2% of the people surveyed owned a computer and did not own a DVD player or a cellular telephone.

b. To determine what *percent* of the people surveyed owned only a DVD player (no computer and no cell phone), we divide $n(D \cap C' \cap T')$ by $n(U)$:

$$\frac{n(D \cap C' \cap T')}{n(U)} = \frac{78}{524}$$
$$= 0.148854961\ldots$$

Approximately 14.9% of the people surveyed owned a DVD player and did not own a computer or a cellular telephone. ■

De Morgan's Laws

One of the basic properties of algebra is the distributive property:

$$a(b + c) = ab + ac$$

Given $a(b + c)$, the operation outside the parentheses can be distributed over the operation inside the parentheses. It makes no difference whether you add b and c first and then multiply the sum by a or first multiply each pair, a and b, a and c, and then add their products; the same result is obtained. Is there a similar property for the complement, union, and intersection of sets?

EXAMPLE 3 Suppose $U = \{1, 2, 3, 4, 5\}$, $A = \{1, 2, 3\}$, and $B = \{2, 3, 4\}$.

a. For the given sets, does $(A \cup B)' = A' \cup B'$?
b. For the given sets, does $(A \cup B)' = A' \cap B'$?

Solution **a.** To find $(A \cup B)'$, we must first find $A \cup B$:

$$\begin{aligned} A \cup B &= \{1, 2, 3\} \cup \{2, 3, 4\} \\ &= \{1, 2, 3, 4\} \end{aligned}$$

The complement of $A \cup B$ (relative to the given universal set U) is

$$(A \cup B)' = \{5\}$$

To find $A' \cup B'$, we must first find A' and B':

$$A' = \{4, 5\} \quad \text{and} \quad B' = \{1, 5\}$$

The union of A' and B' is

$$\begin{aligned} A' \cup B' &= \{4, 5\} \cup \{1, 5\} \\ &= \{1, 4, 5\} \end{aligned}$$

Now, $\{5\} \neq \{1, 4, 5\}$; therefore, $(A \cup B)' \neq A' \cup B'$.

b. We find $(A \cup B)'$ as in part (a): $(A \cup B)' = \{5\}$. Now,

$$\begin{aligned} A' \cap B' &= \{4, 5\} \cap \{1, 5\} \\ &= \{5\} \end{aligned}$$

For the given sets, $(A \cup B)' = A' \cap B'$. ■

Part (a) of Example 3 shows that the operation of complementation *cannot* be explicitly distributed over the operation of union; that is, $(A \cup B)' \neq A' \cup B'$. However, part (b) of the example implies that there *may* be some relationship between the complement, union, and intersection of sets. The fact that $(A \cup B)' = A' \cap B'$ *for the given sets A* and *B* does not mean that it is true *for all sets A* and *B*. We will use a general Venn diagram to examine the validity of the statement $(A \cup B)' = A' \cap B'$.

When we draw two overlapping circles within a universal set, four regions are formed. Every element of the universal set U is in exactly one of the following regions, as shown in Figure 2.25:

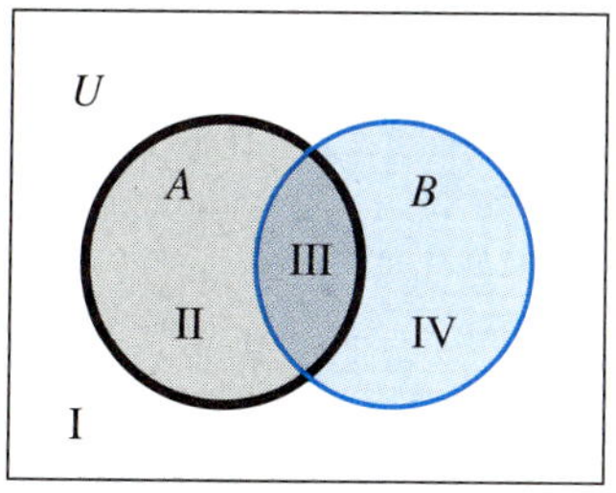

Figure 2.25

I	in neither A nor B
II	in A and not in B
III	in both A and B
IV	in B and not in A

The set $A \cup B$ consists of all elements in regions II, III, and IV. Therefore, the complement $(A \cup B)'$ consists of all elements in region I. A' consists of all elements in regions I and IV, and B' consists of the elements in regions I and II. Therefore, the elements common to both A' and B' are those in region I; that is, the set $A' \cap B'$ consists of all elements in region I. Since $(A \cup B)'$ and $A' \cap B'$ contain exactly the same elements (those in region I), the sets are equal; that is, $(A \cup B)' = A' \cap B'$ is true for all sets A and B.

The relationship $(A \cup B)' = A' \cap B'$ is known as one of **De Morgan's Laws**. Simply stated, "the complement of a union is the intersection of the complements." In a similar manner, it can be shown that $(A \cap B)' = A' \cup B'$ (see Exercise 31).

Stock Montage

HISTORICAL NOTE

Augustus De Morgan

1806–1871

Being born blind in one eye did not stop Augustus De Morgan from becoming a well-read philosopher, historian, logician, and mathematician. De Morgan was born in Madras, India, where his father was working for the East India Company. On moving to England, De Morgan was educated at Cambridge, and at the age of twenty-two, he became the first professor of mathematics at the newly opened University of London (later renamed University College).

De Morgan viewed all of mathematics as an abstract study of symbols and of systems of operations applied to these symbols. While studying the ramifications of symbolic logic, De Morgan formulated the general properties of complementation that now bear his name. Not limited to symbolic logic, De Morgan's many works include books and papers on the foundations of algebra, differential calculus, and probability. He was known to be a jovial person who was fond of puzzles, and his witty and amusing book *A Budget of Paradoxes* still entertains readers today. Besides his accomplishments in the academic arena, De Morgan was an expert flutist, spoke five languages, and thoroughly enjoyed big-city life.

Knowing of his interest in probability, an actuary (someone who studies life expectancies and determines payments of premiums for insurance companies) once asked De Morgan a question concerning the probability that a certain group of people would be alive at a certain time. In his response, De Morgan employed a formula containing the number π. In amazement, the actuary responded, "That must surely be a delusion! What can a circle have to do with the number of people alive at a certain time?" De Morgan replied that π has numerous applications and occurrences in many diverse areas of mathematics. Because it was first defined and used in geometry, people are conditioned to accept the mysterious number only in reference to a circle. However, in the history of mathematics, if probability had been systematically studied before geometry and circles, our present-day interpretation of the number π would be entirely different. In addition to his accomplishments in logic and higher-level mathematics, De Morgan introduced a convention with which we are all familiar: In a paper written in 1845, he suggested the use of a slanted line to represent a fraction, such as 1/2 or 3/4.

De Morgan was a staunch defender of academic freedom and religious tolerance. While a student at Cambridge, his application for a fellowship was refused because he would not take and sign a theological oath. Later in life, he resigned his professorship as a protest against religious bias. (University College gave preferential treatment to members of the Church of England when textbooks were selected and did not have an open policy on religious philosophy.) Augustus De Morgan was a man who was unafraid to take a stand and make personal sacrifices when it came to principles he believed in.

Gematria is a mystic pseudoscience in which numbers are substituted for the letters in a name. De Morgan's book *A Budget of Paradoxes* contains several gematria puzzles, such as, "Mr. Davis Thom found a young gentleman of the name of St. Claire busy at the Beast number: he forthwith added the letters in στκλαιρε (the Greek spelling of St. Claire) and found 666." (Verify this by using the Greek numeral system.)

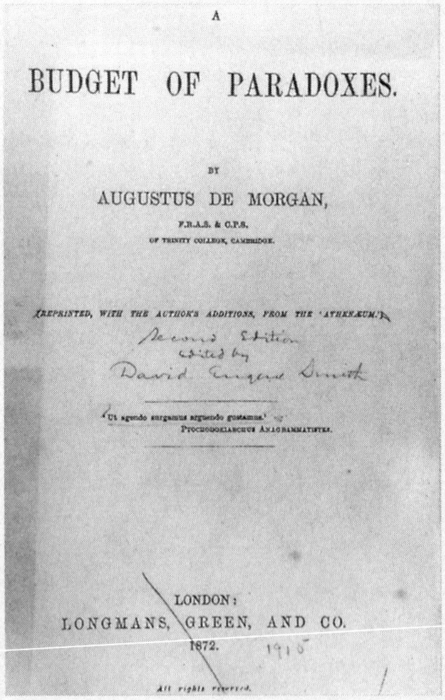
A

BUDGET OF PARADOXES.

BY

AUGUSTUS DE MORGAN,

F.R.A.S. & C.P.S.

OF TRINITY COLLEGE, CAMBRIDGE.

(REPRINTED, WITH THE AUTHOR'S ADDITIONS, FROM THE 'ATHENÆUM.')

Second Edition
edited by
David Eugene Smith

'Ut agendo surgamus arguendo gustamus.'
PROCHOOGIANCHUS ANAGRAMMATISTES.

LONDON:
LONGMANS, GREEN, AND CO.
1872. 1915

All rights reserved.

Columbia University

De Morgan's Laws

For any sets A and B,

$$(A \cup B)' = A' \cap B'$$

That is, the complement of a union is the intersection of the complements. Also,

$$(A \cap B)' = A' \cup B'$$

That is, the complement of an intersection is the union of the complements.

EXAMPLE 4 Suppose $U = \{0, 1, 2, 3, 4, 5, 6, 7, 8, 9\}$, $A = \{2, 3, 7, 8\}$, and $B = \{0, 4, 5, 7, 8, 9\}$. Use De Morgan's Law to find $(A' \cup B)'$.

Solution The complement of a union is equal to the intersection of the complements; therefore, we have

$$\begin{aligned}(A' \cup B)' &= (A')' \cap B' && \text{De Morgan's Law}\\ &= A \cap B' && (A')' = A\\ &= \{2, 3, 7, 8\} \cap \{1, 2, 3, 6\}\\ &= \{2, 3\}\end{aligned}$$

Notice that this problem could be done without using De Morgan's Law, but solving it would then involve finding first A', then $A' \cup B$, and finally $(A' \cup B)'$. This method would involve more work. (Try it!) ■

2.2 Exercises

1. A survey of 200 people yielded the following information: 94 owned a VCR, 127 owned a microwave oven, and 78 owned both. How many people owned the following?
   a. a VCR or a microwave oven
   b. a VCR but not a microwave oven
   c. a microwave oven but not a VCR
   d. neither a VCR nor a microwave oven
2. A survey of 300 workers yielded the following information: 231 belonged to a union, and 195 were Democrats. If 172 of the union members were Democrats, how many workers were in the following situations?
   a. belonged to a union or were Democrats
   b. belonged to a union but were not Democrats
   c. were Democrats but did not belong to a union
   d. neither belonged to a union nor were Democrats
3. The records of 1,492 high school graduates were examined, and the following information was obtained: 1,072 took biology, and 679 took geometry. If 271 of those who took geometry did not take biology, how many graduates took the following?
   a. both classes
   b. at least one of the classes
   c. biology but not geometry
   d. neither class
4. A department store surveyed 428 shoppers, and the following information was obtained: 214 made a purchase, and 299 were satisfied with the service they received. If 52 of those who made a purchase were not satisfied with the service, how many shoppers did the following?
   a. made a purchase and were satisfied with the service
   b. made a purchase or were satisfied with the service
   c. were satisfied with the service but did not make a purchase
   d. were not satisfied and did not make a purchase
5. In a survey, 674 adults were asked what television programs they had recently watched. The following information was obtained: 226 watched neither the

Big Game nor the New Movie, and 289 watched the New Movie. If 183 of those who watched the New Movie did not watch the Big Game, how many of the surveyed adults watched the following?

a. both programs

b. at least one program

c. the Big Game

d. the Big Game but not the New Movie

6. A survey asked 816 college freshmen whether they had been to a movie or eaten in a restaurant during the past week. The following information was obtained: 387 had been to neither a movie nor a restaurant, and 266 had been to a movie. If 92 of those who had been to a movie had not been to a restaurant, how many of the surveyed freshmen had been to the following?

a. both a movie and a restaurant

b. a movie or a restaurant

c. a restaurant

d. a restaurant but not a movie

7. A recent survey of w shoppers (that is, $n(U) = w$) yielded the following information: x shopped at Sears, y shopped at JCPenney's, and z shopped at both. How many people shopped at the following?

a. Sears or JCPenney's

b. only Sears

c. only JCPenney's

d. neither Sears nor JCPenney's

8. A recent transportation survey of w urban commuters (that is, $n(U) = w$) yielded the following information: x rode neither trains nor buses, y rode trains, and z rode only trains. How many people rode the following?

a. trains and buses **b.** only buses

c. buses **d.** trains or buses

9. A consumer survey was conducted to examine patterns in ownership of personal computers, cellular telephones, and VCRs. The following data were obtained: 313 people had personal computers, 232 had cellular telephones, 269 had VCRs, 69 had all three, 64 had none, 98 had cell phones and VCRs, 57 had cell phones but no computers or VCRs, and 104 had computers and VCRs but no cell phones.

a. What percent of the people surveyed owned a cell phone?

b. What percent of the people surveyed owned only a cell phone?

10. In a recent survey of monetary donations made by college graduates, the following information was obtained: 95 had donated to a political campaign, 76 had donated to assist medical research, 133 had donated to help preserve the environment, 25 had donated to all three, 22 had donated to none of the three, 38 had donated to a political campaign and to medical research, 46 had donated to medical research and to preserve the environment, and 54 had donated to a political campaign and to preserve the environment.

a. What percent of the college graduates donated to none of the three listed causes?

b. What percent of the college graduates donated to exactly one of the three listed causes?

11. Recently, Beck, Creed, and No Doubt had concert tours in the United States. A large group of college students was surveyed, and the following information was obtained: 825 saw Creed, 1,033 saw Beck, 1,247 saw No Doubt, 211 saw all three, 514 saw none, 240 saw only No Doubt, 677 saw No Doubt and Beck, and 201 saw Beck and Creed but not No Doubt.

a. What percent of the college students saw at least one of the bands?

b. What percent of the college students saw exactly one of the bands?

12. Dr. Hawk works in an allergy clinic, and his patients have the following allergies: 68 are allergic to dairy products, 93 are allergic to pollen, 91 are allergic to animal fur, 31 are allergic to all three, 29 are allergic only to pollen, 12 are allergic only to dairy products, 40 are allergic to dairy products and pollen.

a. What percent of Dr. Hawk's patients are allergic to animal fur?

b. What percent of Dr. Hawk's patients are allergic only to animal fur?

13. When the members of the Eye and I Photo Club discussed what type of film they had used during the past month, the following information was obtained: 77 used black and white, 24 used only black and white, 65 used color, 18 used only color, 101 used black and white or color, 27 used infrared, 9 used all three types, and 8 didn't use any film during the past month.

a. What percent of the members used only infrared film?

b. What percent of the members used at least two of the types of film?

14. After leaving the polls, many people are asked how they voted. (This is called an *exit poll.*) Concerning Propositions A, B, and C, the following information was obtained: 294 voted yes on A, 90 voted yes only on A, 346 voted yes on B, 166 voted yes only on B, 517 voted yes on A or B, 339 voted yes on C, no one voted yes on all three, and 72 voted no on all three.

a. What percent of the voters in the exit poll voted no on A?

b. What percent of the voters voted yes on more than one proposition?

15. In a recent survey, consumers were asked where they did their gift shopping. The following results were obtained: 621 shopped at Macy's, 513 shopped at Emporium, 367 shopped at Nordstrom, 723 shopped at Emporium or Nordstrom, 749 shopped at Macy's or Nordstrom, 776 shopped at Macy's or Emporium, 157 shopped at all three, 96 shopped at neither Macy's nor Emporium nor Nordstrom.

a. What percent of the consumers shopped at more than one store?

b. What percent of the consumers shopped exclusively at Nordstrom?

16. A company that specializes in language tutoring lists the following information concerning its English-speaking employees: 23 speak German, 25 speak French, 31 speak Spanish, 43 speak Spanish or French, 38 speak French or German, 46 speak German or Spanish, 8 speak Spanish, French, and German, and 7 office workers and secretaries speak English only.

a. What percent of the employees speak at least one language other than English?

b. What percent of the employees speak at least two languages other than English?

17. In a recent survey, people were asked which radio station they listened to on a regular basis. The following results were obtained: 140 listened to WOLD (oldies), 95 listened to WJZZ (jazz), 134 listened to WTLK (talk show news), 235 listened to WOLD or WJZZ, 48 listened to WOLD and WTLK, 208 listened to WTLK or WJZZ, and 25 listened to none.

a. What percent of people in the survey listened only to WTLK on a regular basis?

b. What percent of people in the survey did not listen to WTLK on a regular basis?

18. In a recent health insurance survey, employees at a large corporation were asked, "Have you been a patient in a hospital during the past year, and if so, for what reason?" The following results were obtained: 494 had an injury, 774 had an illness, 1,254 had tests, 238 had an injury and an illness and tests, 700 had an illness and tests, 501 had tests and no injury or illness, 956 had an injury or illness, and 1,543 had not been a patient.

a. What percent of the employees had been patients in a hospital?

b. What percent of the employees had tests in a hospital?

In Exercises 19 and 20, use a Venn diagram like the one in Figure 2.26.

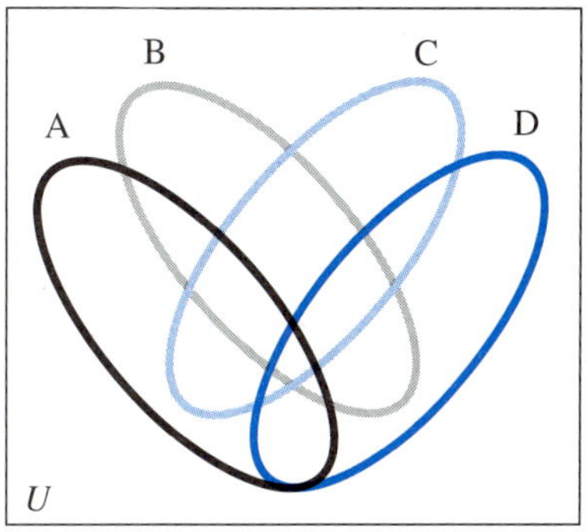

Figure 2.26

19. A survey of 136 pet owners yielded the following information: 49 own fish; 55 own a bird; 50 own a cat; 68 own a dog; 2 own all four; 11 own only fish; 14 own only a bird; 10 own fish and a bird; 21 own fish and a cat; 26 own a bird and a dog; 27 own a cat and a dog; 3 own fish, a bird, a cat, and no dog; 1 owns fish, a bird, a dog, and no cat; 9 own fish, a cat, a dog, and no bird; and 10 own a bird, a cat, a dog, and no fish. How many of the surveyed pet owners have no fish, no birds, no cats, and no dogs? (They own other types of pets.)

20. An exit poll of 300 voters yielded the following information regarding voting patterns on Propositions A, B, C, and D: 119 voted yes on A; 163 voted yes on B; 129 voted yes on C; 142 voted yes on D; 37 voted yes on all four; 15 voted yes on A only; 50 voted yes on B only; 59 voted yes on A and B; 70 voted yes on A and C; 82 voted yes on B and D; 93 voted yes on C and D; 10 voted yes on A, B, and C and no on D; 2 voted yes on A, B, and D and no on C; 16 voted yes on A, C, and D and no on B; and 30 voted yes on B, C, and D and no on A. How many of the surveyed voters voted no on all four propositions?

In Exercises 21–24, given the sets $U = \{0, 1, 2, 3, 4, 5, 6, 7, 8, 9\}$, $A = \{0, 2, 4, 5, 9\}$, and $B = \{1, 2, 7, 8, 9\}$, use De Morgan's laws to find the indicated sets.

21. $(A' \cup B)'$

22. $(A' \cap B)'$

23. $(A \cap B')'$

24. $(A \cup B')'$

In Exercises 25–30, use a Venn diagram like the one in Figure 2.27 to shade in the region corresponding to the indicated set.

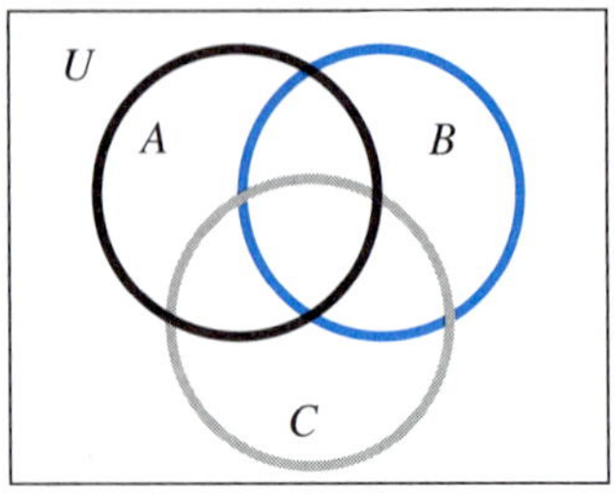

Figure 2.27

25. $A \cap B \cap C$

26. $A \cup B \cup C$

27. $(A \cup B)' \cap C$

28. $A \cap (B \cup C)'$

29. $B \cap (A \cup C')$

30. $(A' \cup B) \cap C'$

31. Using Venn diagrams, prove De Morgan's Law $(A \cap B)' = A' \cup B'$.

32. Using Venn diagrams, prove $A \cup (B \cap C) = (A \cup B) \cap (A \cup C)$.

Answer the following questions using complete sentences and your own words.

History Questions

33. What notation did De Morgan introduce in regard to fractions?

34. Why did De Morgan resign his professorship at University College?

Many graduate schools require applicants to take the Graduate Record Examination (G.R.E.). This exam is intended to measure verbal, quantitative, and analytical skills that have developed throughout a person's life. Many classes and study guides are available to help people prepare for the exam. The remaining questions are typical of those found in the study guides and on the exam itself.

Exercises 35–41 refer to the following: A nonprofit organization's board of directors, composed of four women (Angela, Betty, Carmine, and Delores) and three men (Ed, Frank, and Grant), holds frequent meetings. A meeting can be held at Betty's house, at Delores's house, or at Frank's house.

- Delores cannot attend any meetings at Betty's house.
- Carmine cannot attend any meetings on Tuesday or on Friday.
- Angela cannot attend any meetings at Delores's house.
- Ed can attend only those meetings that Grant also attends.
- Frank can attend only those meetings that both Angela and Carmine attend.

35. If all members of the board are to attend a particular meeting, under which of the following circumstances can it be held?
   a. Monday at Betty's
   b. Tuesday at Frank's
   c. Wednesday at Delores's
   d. Thursday at Frank's
   e. Friday at Betty's

36. Which of the following can be the group that attends a meeting on Wednesday at Betty's?
   a. Angela, Betty, Carmine, Ed, and Frank
   b. Angela, Betty, Ed, Frank, and Grant
   c. Angela, Betty, Carmine, Delores, and Ed
   d. Angela, Betty, Delores, Frank, and Grant
   e. Angela, Betty, Carmine, Frank, and Grant

37. If Carmine and Angela attend a meeting but Grant is unable to attend, which of the following could be true?
   a. The meeting is held on Tuesday.
   b. The meeting is held on Friday.
   c. The meeting is held at Delores's.
   d. The meeting is held at Frank's.
   e. The meeting is attended by six of the board members.

38. If the meeting is held on Tuesday at Betty's, which of the following pairs can be among the board members who attend?
   a. Angela and Frank **b.** Ed and Betty
   c. Carmine and Ed **d.** Frank and Delores
   e. Carmine and Angela

39. If Frank attends a meeting on Thursday that is not held at his house, which of the following must be true?
   a. The group can include, at most, two women.
   b. The meeting is at Betty's house.
   c. Ed is not at the meeting.
   d. Grant is not at the meeting.
   e. Delores is at the meeting.

40. If Grant is unable to attend a meeting on Tuesday at Delores's, what is the largest possible number of board members who can attend?
   a. 1 **b.** 2 **c.** 3
   d. 4 **e.** 5

41. If a meeting is held on Friday, which of the following board members *cannot* attend?
   a. Grant **b.** Delores **c.** Ed
   d. Betty **e.** Frank

2.3 Introduction to Combinatorics

If you went on a shopping spree and bought two pairs of jeans, three shirts, and two pairs of shoes, how many new outfits (consisting of a new pair of jeans, a new shirt, and a new pair of shoes) would you have? A compact disc buyers' club sends you a brochure saying that you can pick any five CDs from a group of 50 of today's hottest sounds for only $1.99. How many different combinations can you choose? Six local bands have volunteered to perform at a benefit concert, and there is some concern over the order in which the bands will perform. How many different lineups are possible? The answers to questions like these can be obtained by listing all the possibilities or by using three shortcut counting methods: the **Fundamental Principle of Counting**, **combinations**, and **permutations**. Collectively, these methods are known as **combinatorics**. (Incidentally, the answers to the questions above are 12 outfits, 2,118,760 CD combinations, and 720 lineups.) In this section, we consider the first shortcut method.

The Fundamental Principle of Counting

Daily life requires that we make many decisions. For example, we must decide what food items to order from a menu, what items of clothing to put on in the morning, and what options to order when purchasing a new car. Often, we are asked to make a series of decisions: "Do you want soup or salad? What type of dressing? What type of vegetable? What entrée? What beverage? What dessert?" These individual components of a complete meal lead to the question, "Given all the choices of soups, salads, dressings, vegetables, entrées, beverages, and desserts, what is the total number of possible dinner combinations?"

When making a series of decisions, how can you determine the total number of possible selections? One way is to list all the choices for each category and then match them up in all possible ways. To ensure that the choices are matched up in all possible ways, you can construct a **tree diagram**. A tree diagram consists of clusters of line segments, or *branches,* constructed as follows: A cluster of branches is drawn for each decision to be made such that the number of branches in each cluster equals the number of choices for the decision. For instance, if you must make two decisions and there are two choices for decision 1 and three choices for decision 2, the tree diagram would be similar to the one shown in Figure 2.28.

Although this method can be applied to all problems, it is very time consuming and impractical when you are dealing with a series of many decisions, each of which contains numerous choices. Instead of actually listing all possibilities via a tree

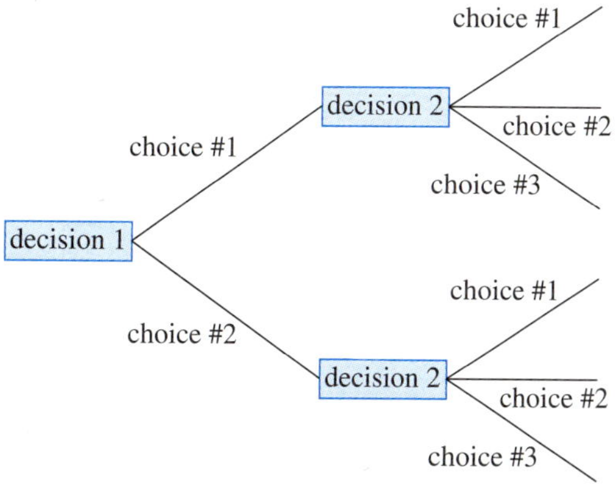

Figure 2.28

diagram, using a shortcut method might be desirable. The following example gives a clue to finding such a shortcut.

EXAMPLE 1 If you buy two pairs of jeans, three shirts, and two pairs of shoes, how many new outfits (consisting of a new pair of jeans, a new shirt, and a new pair of shoes) would you have?

Solution Because there are three categories, selecting an outfit requires a series of three decisions: You must select one pair of jeans, one shirt, and one pair of shoes. We will make our three decisions in the following order: jeans, shirt, and shoes. (The order in which the decisions are made does not affect the overall outfit.)

Our first decision (jeans) has two choices (jeans 1 or jeans 2); our tree starts with two branches, as in Figure 2.29.

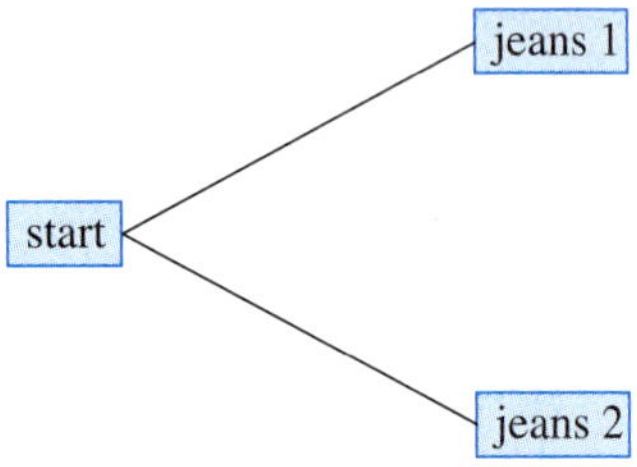

Figure 2.29

Our second decision is to select a shirt, for which there are three choices. At each pair of jeans on the tree, we draw a cluster of three branches, one for each shirt, as in Figure 2.30.

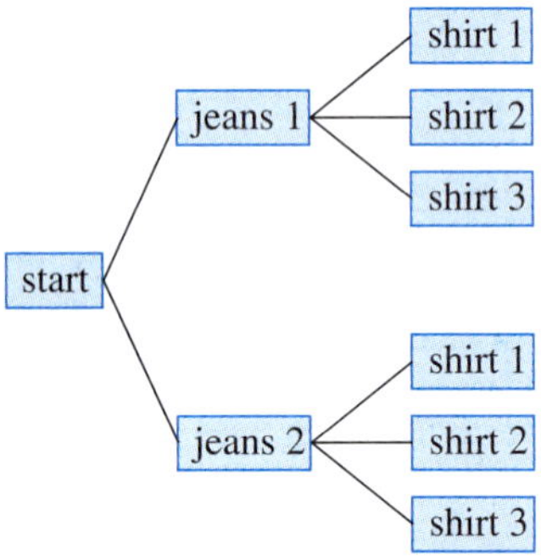

Figure 2.30

Our third decision is to select a pair of shoes, for which there are two choices. At each shirt on the tree, we draw a cluster of two branches, one for each pair of shoes, as in Figure 2.31.

We have now listed all possible ways of putting together a new outfit; twelve outfits can be formed from two pairs of jeans, three shirts, and two pairs of shoes. ■

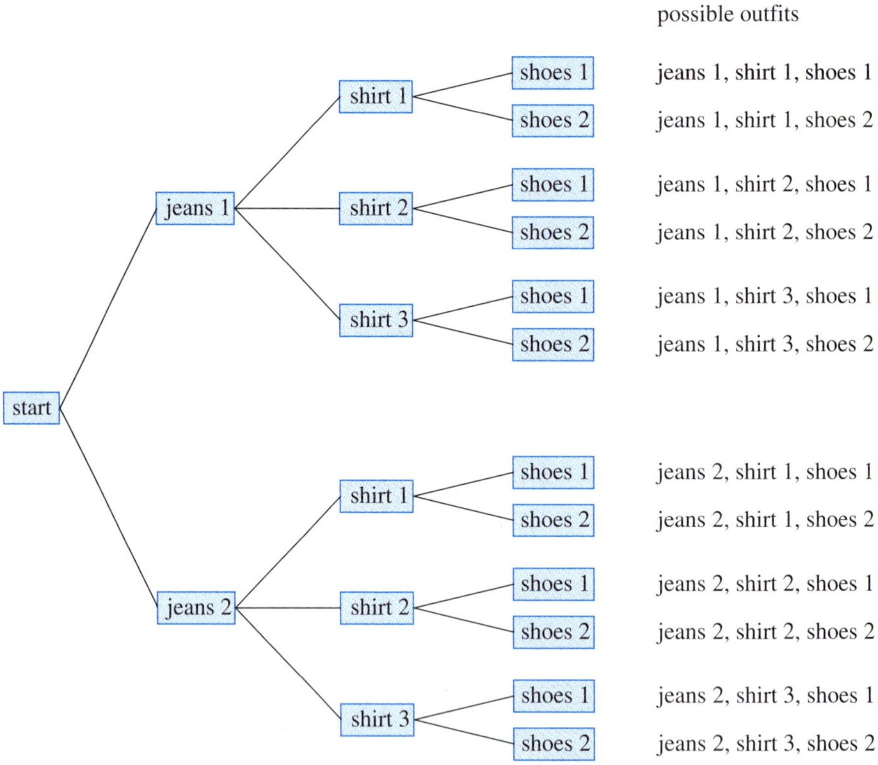

Figure 2.31

Referring to Example 1, note that each time a decision had to be made, the number of branches on the tree diagram was *multiplied* by a factor equal to the number of choices for the decision. Therefore, the total number of outfits could have been obtained by *multiplying* the number of choices for each decision:

$$2 \cdot 3 \cdot 2 = 12$$

(jeans, shirts, shoes → outfits)

The generalization of this process of multiplication is called the Fundamental Principle of Counting.

> **The Fundamental Principle of Counting**
>
> The total number of possible outcomes of a series of decisions (making selections from various categories) is found by multiplying the number of choices for each decision (or category) as follows:
>
> 1. Draw a box for each decision.
> 2. Enter the number of choices for each decision in the appropriate box and multiply.

EXAMPLE 2 A serial number consists of two consonants followed by three nonzero digits followed by a vowel (A, E, I, O, U): for example, "ST423E" and "DD666E." Determine how many serial numbers are possible given the following conditions.

a. Letters and digits cannot be repeated in the same serial number.
b. Letters and digits can be repeated in the same serial number.

Solution **a.** Because the serial number has six symbols, we must make six decisions. Consequently, we must draw six boxes:

$$\square \; \square \; \square \; \square \; \square \; \square$$

There are 21 different choices for the first consonant. Because the letters cannot be repeated, there are only 20 choices for the second consonant. Similarly, there are nine different choices for the first nonzero digit, eight choices for the second, and seven choices for the third. There are five different vowels, so the total number of possible serial numbers is

$$\boxed{21} \times \boxed{20} \times \boxed{9} \times \boxed{8} \times \boxed{7} \times \boxed{5} = 1{,}058{,}400$$

(consonants: 21, 20; nonzero digits: 9, 8, 7; vowel: 5)

There are 1,058,400 possible serial numbers when the letters and digits cannot be repeated within a serial number.

b. Because letters and digits can be repeated, the number of choices does not decrease by one each time as in part (a). Therefore, the total number of possibilities is

$$\boxed{21} \times \boxed{21} \times \boxed{9} \times \boxed{9} \times \boxed{9} \times \boxed{5} = 1{,}607{,}445$$

(consonants: 21, 21; nonzero digits: 9, 9, 9; vowel: 5)

There are 1,607,445 possible serial numbers when the letters and digits can be repeated within a serial number. ■

Factorials

EXAMPLE 3 Three students rent a three-bedroom house near campus. One of the bedrooms is very desirable (it has its own bath), one has a balcony, and one is undesirable (it is very small). In how many ways can the housemates choose the bedrooms?

Solution Three decisions must be made: who gets the room with the bath, who gets the room with the balcony, and who gets the small room. Using the Fundamental Principle of Counting, we draw three boxes and enter the number of choices for each decision. There are three choices for who gets the room with the bath. Once that decision has been made, there are two choices for who gets the room with the balcony, and finally, there is only one choice for the small room.

$$\boxed{3} \times \boxed{2} \times \boxed{1} = 6$$

There are six different ways in which the three housemates can choose the three bedrooms. ■

Combinatorics often involve products of the type $3 \cdot 2 \cdot 1 = 6$, as seen in Example 3. This type of product is called a **factorial**, and the product $3 \cdot 2 \cdot 1$ is written as 3!. In this manner, $4! = 4 \cdot 3 \cdot 2 \cdot 1$ (= 24), and $5! = 5 \cdot 4 \cdot 3 \cdot 2 \cdot 1 (= 120)$.

Factorials

If n is a positive integer, then *n factorial,* denoted by $n!$, is the product of all positive integers less than or equal to n.

$$n! = n \cdot (n-1) \cdot (n-2) \cdot \cdots \cdot 2 \cdot 1$$

As a special case, we define $0! = 1$.

Many scientific calculators have a button that will calculate a factorial. Depending on your calculator, the button will look like [x!] or [n!], and you might have to press a [shift] or [2nd] button first. For example, to calculate 6!, type the number 6, press the factorial button, and obtain 720. To calculate a factorial on most graphing calculators, do the following:

- Type the value of n. (For example, type the number 6.)
- Press the [MATH] button.
- Press the right arrow button [→] as many times as necessary to highlight [PRB].
- Press the down arrow [↓] as many times as necessary to highlight the "!" symbol, and press [ENTER].
- Press [ENTER] to execute the calculation.

To calculate a factorial on a Casio graphing calculator, do the following:

- Press the [MENU] button; this gives you access to the main menu.
- Press 1 to select the RUN mode; this mode is used to perform arithmetic operations.
- Type the value of n. (For example, type the number 6.)
- Press the [OPTN] button; this gives you access to various options displayed at the bottom of the screen.
- Press the [F6] button to see more options (i.e., [→]).
- Press the [F3] button to select probability options (i.e., [PROB]).
- Press the [F1] button to select factorial (i.e., [x!]).
- Press the [EXE] button to execute the calculation.

The factorial symbol "n!" was first introduced by Christian Kramp (1760–1826) of Strasbourg in his *Élements d'Arithmétique Universelle* (1808). Before the introduction of this "modern" symbol, factorials were commonly denoted by $\underline{n|}$. However, printing presses of the day had difficulty printing this symbol; consequently, the symbol $n!$ came into prominence because it was relatively easy for a typesetter to use.

EXAMPLE 4 Find the following values.

a. $6!$ **b.** $\dfrac{8!}{5!}$ **c.** $\dfrac{8!}{3!\cdot 5!}$

Solution **a.**
$$\begin{aligned} 6! &= 6\cdot 5\cdot 4\cdot 3\cdot 2\cdot 1 \\ &= 720 \end{aligned}$$

Therefore, $6! = 720$.

6 [x!]
6 [MATH] [PRB] [!] [ENTER]
Casio 6 [OPTN] [→] (i.e., [F6]) [PROB] (i.e., [F3]) [x!] (i.e., [F1]) [EXE]

b.
$$\begin{aligned} \frac{8!}{5!} &= \frac{8\cdot 7\cdot 6\cdot 5\cdot 4\cdot 3\cdot 2\cdot 1}{5\cdot 4\cdot 3\cdot 2\cdot 1} \\ &= \frac{8\cdot 7\cdot 6\cdot \cancel{5}\cdot \cancel{4}\cdot \cancel{3}\cdot \cancel{2}\cdot \cancel{1}}{\cancel{5}\cdot \cancel{4}\cdot \cancel{3}\cdot \cancel{2}\cdot \cancel{1}} \\ &= 8\cdot 7\cdot 6 \\ &= 336 \end{aligned}$$

Therefore, $\frac{8!}{5!} = 336$.
Using a calculator, we obtain the same result.

8 [x!] [÷] 5 [x!] [=]
8 [MATH] [PRB] [!] [÷] 5 [MATH] [PRB] [!] [ENTER]

c.
$$\begin{aligned} \frac{8!}{3!\cdot 5!} &= \frac{8\cdot 7\cdot 6\cdot 5\cdot 4\cdot 3\cdot 2\cdot 1}{(3\cdot 2\cdot 1)(5\cdot 4\cdot 3\cdot 2\cdot 1)} \\ &= \frac{8\cdot 7\cdot 6\cdot \cancel{5}\cdot \cancel{4}\cdot \cancel{3}\cdot \cancel{2}\cdot \cancel{1}}{(3\cdot 2\cdot 1)(\cancel{5}\cdot \cancel{4}\cdot \cancel{3}\cdot \cancel{2}\cdot \cancel{1})} \\ &= \frac{8\cdot 7\cdot 6}{3\cdot 2\cdot 1} \\ &= 56 \end{aligned}$$

Therefore, $\frac{8!}{3!\cdot 5!} = 56$.
Using a calculator, we obtain the same result.

8 [x!] [÷] [(] 3 [x!] [×] 5 [x!] [)] [=]
8 [MATH] [PRB] [!] [÷] [(] 3 [MATH] [PRB] [!] [×] 5 [MATH] [PRB] [!] [)] [ENTER]

■

2.3 Exercises

1. A nickel, a dime, and a quarter are tossed.
   a. Use the Fundamental Principle of Counting to determine how many different outcomes are possible.
   b. Construct a tree diagram to list all possible outcomes.
2. A die is rolled, and a coin is tossed.
   a. Use the Fundamental Principle of Counting to determine how many different outcomes are possible.
   b. Construct a tree diagram to list all possible outcomes.
3. Jamie has decided to buy either a Mega or a Better Byte personal computer. She also wants to purchase either Big Word, Word World, or Great Word word-processing software and either Big Number or Number World spreadsheet software.
   a. Use the Fundamental Principle of Counting to determine how many different packages of a computer and software Jamie has to choose from.
   b. Construct a tree diagram to list all possible packages of a computer and software.
4. Sammy's Sandwich Shop offers a soup, sandwich, and beverage combination at a special price. There are three sandwiches (turkey, tuna, and tofu), two soups (minestrone and split pea), and three beverages (coffee, milk, and mineral water) to choose from.
   a. Use the Fundamental Principle of Counting to determine how many different meal combinations are possible.
   b. Construct a tree diagram to list all possible soup, sandwich, and beverage combinations.
5. If you buy three pairs of jeans, four sweaters, and two pairs of boots, how many new outfits (consisting of a new pair of jeans, a new sweater, and a new pair of boots) will you have?
6. A certain model of automobile is available in six exterior colors, three interior colors, and three interior styles. In addition, the transmission can be either manual or automatic, and the engine can have either four or six cylinders. How many different versions of the automobile can be ordered?
7. To fulfill certain requirements for a degree, a student must take one course each from the following groups: health, civics, critical thinking, and elective. If there are four health, three civics, six critical thinking, and ten elective courses, how many different options for fulfilling the requirements does a student have?
8. To fulfill a requirement for a literature class, a student must read one short story by each of the following authors: Stephen King, Clive Barker, Edgar Allan Poe, and H. P. Lovecraft. If there are twelve King, six Barker, eight Poe, and eight Lovecraft stories to choose from, how many different combinations of reading assignments can a student choose from to fulfill the reading requirement?
9. A sporting goods store has fourteen lines of snow skis, seven types of bindings, nine types of boots, and three types of poles. Assuming that all items are compatible with each other, how many different complete ski equipment packages are available?
10. An audio equipment store has ten different amplifiers, four tuners, six turntables, eight tape decks, six compact disc players, and thirteen speakers. Assuming that all components are compatible with each other, how many different complete stereo systems are available?
11. A cafeteria offers a complete dinner that includes one serving each of appetizer, soup, entrée, and dessert for $6.99. If the menu has three appetizers, four soups, six entrées, and three desserts, how many different meals are possible?
12. A sandwich shop offers a "U-Chooz" special consisting of your choice of bread, meat, cheese, and special sauce (one each). If there are six different breads, eight meats, five cheeses, and four special sauces, how many different sandwiches are possible?
13. How many different Social Security numbers are possible? (A Social Security number consists of nine digits that can be repeated.)
14. To use an automated teller machine (ATM), a customer must enter his or her four-digit Personal Identification Number (PIN). How many different PINs are possible?
15. Every book published has an International Standard Book Number (ISBN). The number is a code used to identify the specific book and is of the form X-XXX-XXXXX-X, where X is one of digits 0, 1, 2, . . . , 9. How many different ISBNs are possible?
16. How many different Zip Codes are possible using (a) the old style (five digits) and (b) the new style (nine digits)? Why do you think the U.S. Postal Service introduced the new system?
17. Telephone area codes are three-digit numbers of the form XXX.
   a. Originally, the first and third digits were neither 0 nor 1 and the second digit was always a 0 or a 1. How many three-digit numbers of this type are possible?

b. Over time, the restrictions listed in part (a) have been altered; currently, the only requirement is that the first digit is neither 0 nor 1. How many three-digit numbers of this type are possible?

c. Why were the original restrictions listed in part (a) altered?

18. Major credit cards such as VISA and MasterCard have a sixteen-digit account number of the form XXXX-XXXX-XXXX-XXXX. How many different numbers of this type are possible?

19. The serial number on a dollar bill consists of a letter followed by eight digits and then a letter. How many different serial numbers are possible, given the following conditions?

a. Letters and digits cannot be repeated.

b. Letters and digits can be repeated.

c. The letters are nonrepeated consonants and the digits can be repeated.

20. The serial number on a new twenty-dollar bill consists of two letters followed by eight digits and then a letter. How many different serial numbers are possible, given the following conditions?

a. Letters and digits cannot be repeated.

b. Letters and digits can be repeated.

c. The first and last letters are repeatable vowels, the second letter is a consonant, and the digits can be repeated.

21. Each student at State University has a student I.D. number consisting of four digits (the first digit is nonzero, and digits may be repeated) followed by three of the letters A, B, C, D, and E (letters may not be repeated). How many different student numbers are possible?

22. Each student at State College has a student I.D. number consisting of five digits (the first digit is nonzero, and digits may be repeated) followed by two of the letters A, B, C, D, and E (letters may not be repeated). How many different student numbers are possible?

In Exercises 23–38, find the indicated value.

23. $4!$ **24.** $5!$

25. $10!$ **26.** $8!$

27. $20!$ **28.** $25!$

29. $6! \cdot 4!$ **30.** $8! \cdot 6!$

31. **a.** $\frac{6!}{4!}$ **b.** $\frac{6!}{2!}$ **32.** **a.** $\frac{8!}{6!}$ **b.** $\frac{8!}{2!}$

33. $\frac{8!}{5! \cdot 3!}$ **34.** $\frac{9!}{5! \cdot 4!}$

35. $\frac{8!}{4! \cdot 4!}$ **36.** $\frac{6!}{3! \cdot 3!}$

37. $\frac{82!}{80! \cdot 2!}$ **38.** $\frac{77!}{74! \cdot 3!}$

39. Find the value of $\frac{n!}{(n-r)!}$ when $n = 16$ and $r = 14$.

40. Find the value of $\frac{n!}{(n-r)!}$ when $n = 19$ and $r = 16$.

41. Find the value of $\frac{n!}{(n-r)!}$ when $n = 5$ and $r = 5$.

42. Find the value of $\frac{n!}{(n-r)!}$ when $n = r$.

43. Find the value of $\frac{n!}{(n-r)!r!}$ when $n = 7$ and $r = 3$.

44. Find the value of $\frac{n!}{(n-r)!r!}$ when $n = 7$ and $r = 4$.

45. Find the value of $\frac{n!}{(n-r)!r!}$ when $n = 5$ and $r = 5$.

46. Find the value of $\frac{n!}{(n-r)!r!}$ when $n = r$.

Answer the following questions using complete sentences and your own words.

Concept Questions

47. What is the Fundamental Principle of Counting? When is it used?

48. What is a factorial?

History Questions

49. Who invented the modern symbol denoting a factorial? What symbol did it replace? Why?

Many graduate schools require applicants to take the Graduate Record Examination (G.R.E.). This exam is intended to measure verbal, quantitative, and analytical skills that have developed throughout a person's life. Many classes and study guides are available to help people prepare for the exam. The remaining questions are typical of those found in the study guides and on the exam itself.

Exercises 50–54 refer to the following: In an executive parking lot, there are six parking spaces in a row, labeled 1 through 6. Exactly five cars of five different colors—black, gray, pink, white, and yellow—are to be parked in the spaces. The cars can park in any of the spaces as long as the following conditions are met:

- The pink car must be parked in space 3.
- The black car must be parked in a space next to the space in which the yellow car is parked.
- The gray car cannot be parked in a space next to the space in which the white car is parked.

50. If the yellow car is parked in space 1, how many acceptable parking arrangements are there for the five cars?

a. 1 **b.** 2 **c.** 3 **d.** 4 **e.** 5

51. Which of the following must be true of any acceptable parking arrangement?
   a. One of the cars is parked in space 2.
   b. One of the cars is parked in space 6.
   c. There is an empty space next to the space in which the gray car is parked.
   d. There is an empty space next to the space in which the yellow car is parked.
   e. Either the black car or the yellow car is parked in a space next to space 3.

52. If the gray car is parked in space 2, none of the cars can be parked in which space?
   a. 1 **b.** 3 **c.** 4 **d.** 5 **e.** 6

53. The white car could be parked in any of the spaces except which of the following?
   a. 1 **b.** 2 **c.** 4 **d.** 5 **e.** 6

54. If the yellow car is parked in space 2, which of the following must be true?
   a. None of the cars is parked in space 5.
   b. The gray car is parked in space 6.
   c. The black car is parked in a space next to the space in which the white car is parked.
   d. The white car is parked in a space next to the space in which the pink car is parked.
   e. The gray car is parked in a space next to the space in which the black car is parked.

2.4 Permutations and Combinations

The Fundamental Principle of Counting allows us to determine the total number of possible outcomes when a series of decisions (making selections from various categories) must be made. In Section 2.3, the examples and exercises involved selecting *one item each* from various categories; if you buy two pairs of jeans, three shirts, and two pairs of shoes, you will have twelve ($2 \cdot 3 \cdot 2 = 12$) new outfits (consisting of a new pair of jeans, a new shirt, and a new pair of shoes). In this section, we examine the situation when *more than one* item is selected from a category. If more than one item is selected, the selections can be made either *with* or *without* replacement.

With versus Without Replacement

Selecting items *with replacement* means that the same item *can* be selected more than once; after a specific item has been chosen, it is put back into the pool of future choices. Selecting items *without replacement* means that the same item *cannot* be selected more than once; after a specific item has been chosen, it is not replaced.

Suppose you must select a four-digit Personal Identification Number (PIN) for a bank account. In this case, the digits are selected with replacement; each time a specific digit is selected, the digit is put back into the pool of choices for the next selection. (Your PIN can be 3666; the same digit can be selected more than once.) When items are selected with replacement, we use the Fundamental Principle of Counting to determine the total number of possible outcomes; there are $10 \cdot 10 \cdot 10 \cdot 10 = 10{,}000$ possible four-digit PINs.

In many situations, items cannot be selected more than once. For instance, when selecting a committee of three people from a group of twenty, you cannot select the same person more than once. Once you have selected a specific person (say, Lauren), you do not put her back into the pool of choices. When selecting items without replacement, depending on whether the order of selection is important, *permutations* or *combinations* are used to determine the total number of possible outcomes.

Permutations

When more than one item is selected (without replacement) from a single category, and the order of selection *is* important, the various possible outcomes are called **permutations**. For example, when the rankings (first, second, and third place) in a talent contest are announced, the order of selection is important; Monte in first, Lynn in second, and Ginny in third place is different from Ginny in first, Monte in second, and Lynn in third. "Monte, Lynn, Ginny" and "Ginny, Monte, Lynn" are different permutations of the contestants. Naturally, these selections are made without replacement; we cannot select Monte for first place and reselect him for second place.

EXAMPLE 1 Six local bands have volunteered to perform at a benefit concert, but there is enough time for only four bands to play. There is also some concern over the order in which the chosen bands will perform. How many different lineups are possible?

Solution We must select four of the six bands and put them in a specific order. The bands are selected without replacement; a band cannot be selected to play and then be reselected to play again. Because we must make four decisions, we draw four boxes and put the number of choices for each decision in each appropriate box. There are six choices for the opening band. Naturally, the opening band could not be the followup act, so there are only five choices for the next group. Similarly, there are four candidates for the third group and three choices for the closing band. The total number of different lineups possible is found by multiplying the number of choices for each decision:

$$\boxed{6} \times \boxed{5} \times \boxed{4} \times \boxed{3} = 360$$

(↑ opening band under 6; ↑ closing band under 3)

With four out of six bands playing in the performance, 360 lineups are possible. *Because the order of selecting the bands is important, the various possible outcomes, or lineups, are called permutations;* there are 360 permutations of six items when the items are selected four at a time. ■

The computation in Example 1 is similar to a factorial, but the factors do not go all the way down to 1; the product $6 \cdot 5 \cdot 4 \cdot 3$ is a "truncated" (cut-off) factorial. We can change this truncated factorial into a complete factorial in the following manner:

$$6 \cdot 5 \cdot 4 \cdot 3 = \frac{6 \cdot 5 \cdot 4 \cdot 3 \cdot (2 \cdot 1)}{(2 \cdot 1)} \qquad \text{multiplying by } \frac{2}{2} \text{ and } \frac{1}{1}$$

$$= \frac{6!}{2!}$$

Notice that this last expression can be written as $\frac{6!}{2!} = \frac{6!}{(6 - 4!)}$. (Recall that we were selecting four out of six bands.) This result is generalized as follows.

Permutation Formula

The number of **permutations**, or arrangements, of r items selected without replacement from a pool of n items ($r \le n$), denoted by ${}_nP_r$, is

$${}_nP_r = \frac{n!}{(n - r)!}$$

Permutations are used whenever more than one item is selected (without replacement) from a category and the order of selection is important.

Using the notation above and referring to Example 1, note that 360 possible lineups of four bands selected from a pool of six can be denoted by ${}_6P_4 = \frac{6!}{(6-4)!} = 360$. Other notations can be used to represent the number of permutations of a group of items. In particular, the notations ${}_nP_r$, $P(n, r)$, P_r^n, and $P_{n,\,r}$ all represent the number of possible permutations (or arrangements) of r items selected (without replacement) from a pool of n items.

EXAMPLE 2 Three door prizes (first, second, and third) are to be awarded at a ten-year high school reunion. Each of the 112 attendees puts his or her name in a hat. The first name drawn wins a two-night stay at the Chat 'n' Rest Motel, the second name wins dinner for two at Juju's Kitsch-Inn, and the third wins a pair of engraved mugs. In how many different ways can the prizes be awarded?

Solution We must select 3 out of 112 people (without replacement), and the order in which they are selected *is* important. (Winning dinner is different from winning the mugs.) Hence, we must find the number of permutations of 3 items selected from a pool of 112:

$$\begin{aligned}
{}_{112}P_3 &= \frac{112!}{(112-3)!} \\
&= \frac{112!}{109!} \\
&= \frac{112 \cdot 111 \cdot 110 \cdot 109 \cdot 108 \cdot \cdots \cdot 2 \cdot 1}{109 \cdot 108 \cdot \cdots \cdot 2 \cdot 1} \\
&= 112 \cdot 111 \cdot 110 \\
&= 1{,}367{,}520
\end{aligned}$$

There are 1,367,520 different ways in which the three prizes can be awarded to the 112 people. ■

In Example 2, if you try to use a calculator to find $\frac{112!}{109!}$ directly, you will not obtain an answer. Entering 112 and pressing $\boxed{x!}$ results in a calculator error. (Try it.) Because factorials get very large very quickly, most calculators are not able to find any factorial over 69!. ($69! = 1.711224524 \times 10^{98}$.)

EXAMPLE 3 A bowling league has ten teams. How many different ways can the teams rank in the standings at the end of a tournament? (Ties are not allowed.)

Solution Because order is important, we find the number of permutations of ten items selected from a pool of ten items:

$$\begin{aligned}
{}_{10}P_{10} &= \frac{10!}{(10-10)!} \\
&= \frac{10!}{0!} \qquad \text{Recall that } 0! = 1. \\
&= \frac{10!}{1} \\
&= 3{,}628{,}800
\end{aligned}$$

In a league containing ten teams, there are 3,628,800 different standings possible at the end of a tournament. ■

Combinations

When items are selected from a group, the order of selection may or may not be important. If the order is important (as in Examples 1, 2, and 3), permutations are used to determine the total number of selections possible. What if the order of selection is *not* important? When more than one item is selected (without replacement) from a single category and the order of selection is not important, the various possible outcomes are called **combinations**.

EXAMPLE 4 Two adults are needed to chaperone a daycare center's field trip. Marcus, Vivian, Frank, and Keiko are the four managers of the center. How many different groups of chaperones are possible?

Solution In selecting the chaperones, the order of selection is *not* important; "Marcus and Vivian" is the same as "Vivian and Marcus." Hence, the permutation formula cannot be used. Because we do not yet have a shortcut for finding the total number of possibilities when the order of selection is not important, we must list all the possibilities:

Marcus and Vivian	Marcus and Frank	Marcus and Keiko
Vivian and Frank	Vivian and Keiko	Frank and Keiko

Therefore, six different groups of two chaperones are possible from the group of four managers. Because the order in which the people are selected is not important, the various possible outcomes, or groups of chaperones, are called *combinations;* there are six combinations when two items are selected from a pool of four. ■

Just as ${}_nP_r$ denotes the number of *permutations* of r elements selected from a pool of n elements, ${}_nC_r$ denotes the number of *combinations* of r elements selected from a pool of n elements. In Example 4, we found that there are six combinations of two people selected from a pool of four by listing all six of the combinations; that is, ${}_4C_2 = 6$. If we had a larger pool, listing each combination to find out how many there are would be extremely time consuming and tedious! Instead of listing, we take a different approach. We first find the number of permutations (with the permutation formula) and then alter that number to account for the distinction between permutations and combinations.

To find the number of combinations of two people selected from a pool of four, we first find the number of permutations:

$${}_4P_2 = \frac{4!}{(4-2)!} = \frac{4!}{2!} = 12$$

This figure of 12 must be altered to account for the distinction between permutations and combinations.

In Example 4, we listed combinations; one such combination was "Marcus and Vivian." If we had listed permutations, we would have had to list both "Marcus and Vivian" and "Vivian and Marcus," because the *order* of selection matters with permutations. In fact, each combination of two chaperones listed in Example 4 generates two permutations; each pair of chaperones can be given in two different orders. Thus, there are twice as many permutations of two people selected from a pool of four as there are combinations. Alternatively, there are half as many combinations of two people selected from a pool of four as there are permutations. We used the permutation formula to find that ${}_4P_2 = 12$; thus,

$${}_4C_2 = \frac{1}{2} \cdot {}_4P_2 = \frac{1}{2}(12) = 6$$

This answer certainly fits with Example 4; we listed exactly six combinations.

What if three of the four managers were needed to chaperone the daycare center's field trip? Rather than finding the number of combinations by listing each possibility, we first find the number of permutations and then alter that number to account for the distinction between permutations and combinations.

The number of permutations of three people selected from a pool of four is

$$_4P_3 = \frac{4!}{(4-3)!} = \frac{4!}{1!} = 24$$

We know that some of these permutations represent the same combination. For example, the combination "Marcus and Vivian and Keiko" generates $3! = 6$ different permutations (using initials, they are: MVK, MKV, KMV, KVM, VMK, VKM). Because each combination of three people generates six different permutations, there are one-sixth as many combinations as permutations. Thus,

$$_4C_3 = \frac{1}{6} \cdot {_4P_3} = \frac{1}{6}(24) = 4$$

This means that if three of the four managers were needed to chaperone the daycare center's field trip, there would be $_4C_3 = 4$ possible combinations.

We just saw that when two items are selected from a pool of n items, each combination of two generates $2! = 2$ permutations, so

$$_nC_2 = \frac{1}{2!} \cdot {_nP_2}$$

We also saw that when three items are selected from a pool of n items, each combination of three generates $3! = 6$ permutations, so

$$_nC_3 = \frac{1}{3!} \cdot {_nP_3}$$

More generally, when r items are selected from a pool of n items, each combination of r items generates $r!$ permutations, so

$$_nC_r = \frac{1}{r!} \cdot {_nP_r}$$

$$= \frac{1}{r!} \cdot \frac{n!}{(n-r)!} \quad \text{using the Permutation Formula}$$

$$= \frac{n!}{r! \cdot (n-r)!} \quad \text{multiplying the fractions together}$$

Combination Formula

The number of distinct **combinations** of r items selected without replacement from a pool of n items ($r \leq n$), denoted by $_nC_r$, is

$$_nC_r = \frac{n!}{(n-r)!\, r!}$$

Combinations are used whenever one or more items are selected (without replacement) from a category and the order of selection is not important.

EXAMPLE 5 A compact disc club sends you a brochure that offers any five CDs from a group of 50 of today's hottest releases. How many different selections can you make?

Solution Because the order of selection is *not* important, we find the number of combinations when five items are selected from a pool of 50:

$$
\begin{aligned}
{}_{50}C_5 &= \frac{50!}{(50-5)!\,5!} \\
&= \frac{50!}{45!\,5!} \\
&= \frac{50 \cdot 49 \cdot 48 \cdot 47 \cdot 46}{5 \cdot 4 \cdot 3 \cdot 2 \cdot 1} \\
&= 2{,}118{,}760
\end{aligned}
$$

50 [$x!$] [÷] [(] 45 [$x!$] [×] 5 [$x!$] [)] [=]

Graphing calculators have buttons that will calculate ${}_nP_r$ and ${}_nC_r$. To use them, do the following:

- Type the value of n. (For example, type the number 50.)
- Press the [MATH] button.
- Press the right arrow button [→] as many times as necessary to highlight [PRB].
- Press the down arrow button [↓] as many times as necessary to highlight the appropriate symbol— [${}_nP_r$] for permutations, [${}_nC_r$] for combinations—and press [ENTER].
- Type the value of r. (For example, type the number 5.)
- Press [ENTER] to execute the calculation.

On a Casio graphing calculator, do the following:

- Press the [MENU] button; this gives you access to the main menu.
- Press 1 to select the RUN mode; this mode is used to perform arithmetic operations.
- Type the value of n. (For example, type the number 50.)
- Press the [OPTN] button; this gives you access to various options displayed at the bottom of the screen.
- Press the [F6] button to see more options (i.e., [→]).
- Press the [F3] button to select probability options (i.e., [PROB]).
- Press the [F3] button to select combinations (i.e., [${}_nC_r$]) or the [F2] button to select permutations (i.e., [${}_nP_r$]).
- Type the value of r. (For example, type the number 5.)
- Press the [EXE] button to execute the calculation.

	50 [MATH] [PRB] [${}_nC_r$] 5 [ENTER]
Casio	50 [OPTN] [→] (i.e., [F6]) [PROB] (i.e., [F3]) [nCr] (i.e., [F3]) 5 [EXE]

In choosing five out of 50 CDs, 2,118,760 combinations are possible. ■

EXAMPLE 6 A group consisting of twelve women and nine men must select a five-person committee. How many different committees are possible if it must consist of the following?

a. three women and two men **b.** any mixture of men and women

Solution **a.** Our problem involves two categories: women and men. The Fundamental Principle of Counting tells us to draw two boxes (one for each category), enter the number of choices for each, and multiply:

the number of ways we can select three out of twelve women	×	the number of ways we can select two out of nine men	= ?

Because the order of selecting the members of a committee is not important, we will use combinations:

$$
\begin{aligned}
({}_{12}C_3) \cdot ({}_9C_2) &= \frac{12!}{(12-3)! \cdot 3!} \cdot \frac{9!}{(9-2)! \cdot 2!} \\
&= \frac{12!}{9! \cdot 3!} \cdot \frac{9!}{7! \cdot 2!} \\
&= \frac{12 \cdot 11 \cdot 10}{3 \cdot 2 \cdot 1} \cdot \frac{9 \cdot 8}{2 \cdot 1} \\
&= 220 \cdot 36 \\
&= 7{,}920
\end{aligned}
$$

[( ] 12 [x!] [÷] [( ] 9 [x!] [×] 3 [x!] [ )] [ )] [×] [( ] 9 [x!] [÷] [( ] 7 [x!] [×] 2 [x!] [ )] [ )] [=]

12 [MATH] [PRB] [nCr] 3 [×] 9 [MATH] [PRB] [nCr] 2 [ENTER]

There are 7,920 different committees consisting of three women and two men.

b. Because the gender of the committee members doesn't matter, our problem involves only one category: people. We must choose five out of the 21 people, and the order of selection is not important:

$$
\begin{aligned}
{}_{21}C_5 &= \frac{21!}{(21-5)! \cdot 5!} \\
&= \frac{21!}{16! \cdot 5!} \\
&= \frac{21 \cdot 20 \cdot 19 \cdot 18 \cdot 17}{5 \cdot 4 \cdot 3 \cdot 2 \cdot 1} \\
&= 20{,}349
\end{aligned}
$$

Calculator	21 [x!] [÷] [(] 16 [x!] [×] 5 [x!] [)] [=]
Graphing calculator	21 [MATH] [PRB] [nCr] 5 [ENTER]

There are 20,349 different committees consisting of five people. ■

EXAMPLE 7 Find the value of ${}_5C_r$ for the following values of r:

a. $r = 0$ **b.** $r = 1$ **c.** $r = 2$ **d.** $r = 3$ **e.** $r = 4$ **f.** $r = 5$

Solution

a. $${}_5C_0 = \frac{5!}{(5-0)! \cdot 0!} = \frac{5!}{5! \cdot 0!} = 1$$

b. $${}_5C_1 = \frac{5!}{(5-1)! \cdot 1!} = \frac{5!}{4! \cdot 1!} = 5$$

c. $${}_5C_2 = \frac{5!}{(5-2)! \cdot 2!} = \frac{5!}{3! \cdot 2!} = 10$$

d. $${}_5C_3 = \frac{5!}{(5-3)! \cdot 3!} = \frac{5!}{2! \cdot 3!} = 10$$

e. $${}_5C_4 = \frac{5!}{(5-4)! \cdot 4!} = \frac{5!}{1! \cdot 4!} = 5$$

f. $${}_5C_5 = \frac{5!}{(5-5)! \cdot 5!} = \frac{5!}{0! \cdot 5!} = 1$$ ■

The combinations generated in Example 7 exhibit a curious pattern. Notice that the values of ${}_5C_r$ are symmetric: ${}_5C_0 = {}_5C_5$, ${}_5C_1 = {}_5C_4$, and ${}_5C_2 = {}_5C_3$. Now examine the diagram in Figure 2.32. Each number in this "triangle" of numbers is the sum of two numbers in the row immediately above it. For example, $2 = 1 + 1$ and $10 = 4 + 6$, as shown by the inserted arrows. It is no coincidence that the values of ${}_5C_r$ found in Example 7 also appear as a row of numbers in this "magic" triangle. In fact, when the first row of the triangle is labeled the "0th" (or initial) row, the 5th row contains all the values of ${}_5C_r$ for $r = 0, 1, 2, 3, 4$, and 5. In general, the nth row of the triangle contains all the values of ${}_nC_r$ for $r = 0, 1, 2, \ldots, n$.

0th row						1					
1st row					1		1				
					↘		↙				
2nd row				1		2		1			
3rd row			1		3		3		1		
4th row		1		4		6		4		1	
					↘		↙				
5th row	1		5		10		10		5		1
and so on											

Figure 2.32

Historically, this triangular pattern of numbers is referred to as *Pascal's Triangle,* in honor of the French mathematician, scientist, and philosopher Blaise Pascal (1623–1662). Pascal is a cofounder of probability theory (see the Historical Note in Section 3.1). Although the triangle has Pascal's name attached to it, this "magic" arrangement of numbers was known to other cultures hundreds of years before Pascal's time.

The most important part of any problem involving combinatorics is deciding which counting technique (or techniques) to use. The list of general steps and the flowchart in Figure 2.33 can help you to decide which method or methods to use in a specific problem.

Which Counting Technique?

1. What is being selected?
2. If the selected items can be repeated, use the **Fundamental Principle of Counting** and multiply the number of choices for each category.
3. If there is only one category, use:
   combinations if the order of selection does not matter—that is, r items can be selected from a pool of n items in ${}_nC_r = \frac{n!}{(n-r)! \cdot r!}$ ways.
   permutations if the order of selection does matter—that is, r items can be selected from a pool of n items in ${}_nP_r = \frac{n!}{(n-r)!}$ ways.
4. If there is more than one category, use the **Fundamental Principle of Counting** with one box per category.
   a. If you are selecting one item per category, the number in the box for that category is the number of choices for that category.
   b. If you are selecting more than one item per category, the number in the box for that category is found by using step 3.

EXAMPLE 8 A standard deck of playing cards contains 52 cards.

a. How many different five-card hands containing four kings are possible?
b. How many different five-card hands containing four queens are possible?
c. How many different five-card hands containing four kings or four queens are possible?
d. How many different five-card hands containing four of a kind are possible?

Solution **a.** We use the flowchart in Figure 2.33 and answer the following questions.

Q. What is being selected?

A. Playing cards.

Q. Can the selected items be repeated?

A. No.

Q. Is there more than one category?

A. Yes: Because we must have five cards, we need four kings and one non-king. Therefore, we need two boxes:

[kings] × [non-kings]

Q. Am I selecting only one item per category?

A. *Kings*: no. Does the order of selection matter? No: Use combinations. Because there are $n = 4$ kings in the deck and we want to select $r = 4$, we

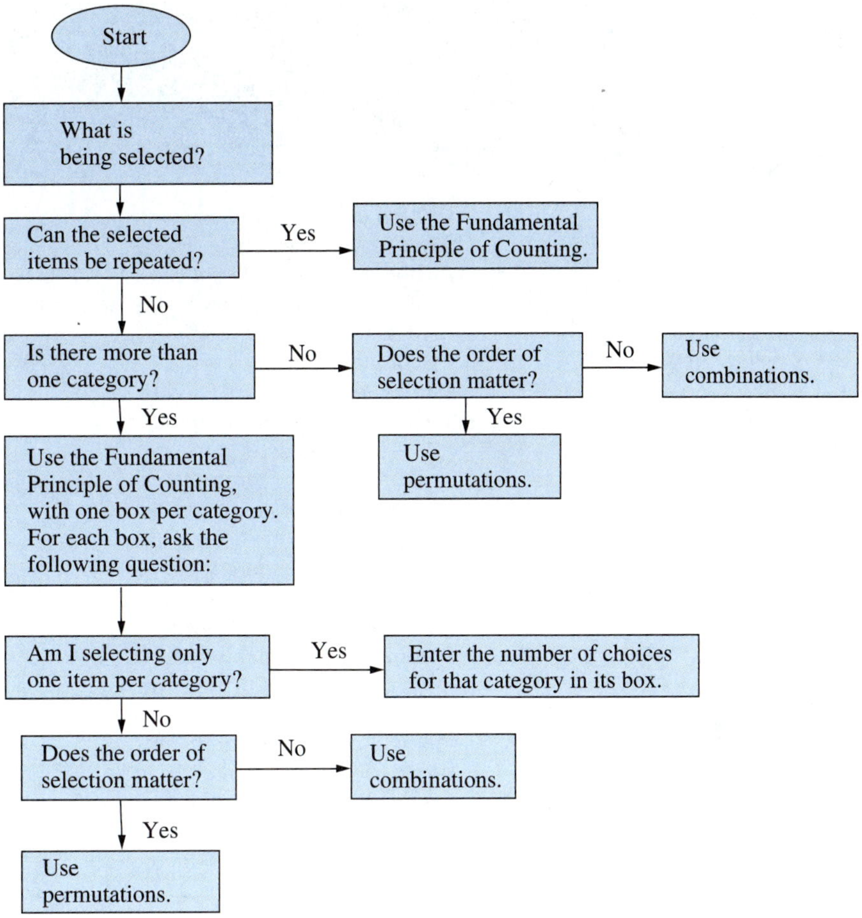

Figure 2.33
Which counting technique?

must compute ${}_4C_4$. *Non-kings*: yes. Enter the number of choices for that category: There are 48 non-kings.

$$\boxed{\text{kings}} \times \boxed{\text{non-kings}} = \boxed{{}_4C_4} \times \boxed{48}$$
$$= \frac{4!}{(4-4)! \cdot 4!} \cdot 48$$
$$= \frac{4!}{0! \cdot 4!} \cdot 48$$
$$= 1 \cdot 48$$
$$= 48$$

There are 48 different five-card hands containing four kings.

b. Using the same method as in part (a), we would find that there are 48 different five-card hands containing four queens; the number of five-card hands containing four queens is the same as the number of five-card hands containing four kings.

c. To find the number of five-card hands containing four kings or four queens, we define the following sets:

$$A = \{\text{five card hands} \mid \text{the hand contains four kings}\}$$
$$B = \{\text{five card hands} \mid \text{the hand contains four queens}\}$$

HISTORICAL NOTE

Chu Shih-chieh

CIRCA 1280–1303

Chu Shih-chieh was the last and most acclaimed mathematician of the Sung Dynasty in China. Little is known of his personal life; the actual dates of his birth and death are unknown. His work appears to have flourished during the close of the thirteenth century. It is believed that Chu Shih-chieh spent many years as a wandering scholar, earning a living by teaching mathematics to those who wanted to learn.

Two of Chu Shih-chieh's works have survived the centuries. The first, *Suan-hsüeh ch'i-meng* (*Introduction to Mathematical Studies*), was written in 1299 and contains elementary mathematics. This work was very influential in Japan and Korea, although it was lost in China until the nineteenth century.

Written in 1303, Chu's second work *Ssu-yüan yü-chien* (*Precious Mirror of the Four Elements*) contains more advanced mathematics. The topics of *Precious Mirror* include the solving of simultaneous equations and the solving of equations up to the fourteenth degree.

Of the many diagrams in *Precious Mirror,* one has special interest: the arithmetic triangle. Chu Shih-chieh's triangle contains the first eight rows of what is known in the West as Pascal's Triangle. However, Chu does not claim credit for the triangle; he refers to it as "a diagram of the *old* method for finding eighth and lower powers." "Pascal's" Triangle was known to the Chinese well over 300 years before Pascal was born!

The "Pascal" Triangle as depicted in 1303 at the front of Chu Shih-chieh's *Ssu-yüan yü-chien.* It is entitled "The Old Method Chart of the Seven Multiplying Squares" and tabulates the binomial coefficients up to the eighth power.

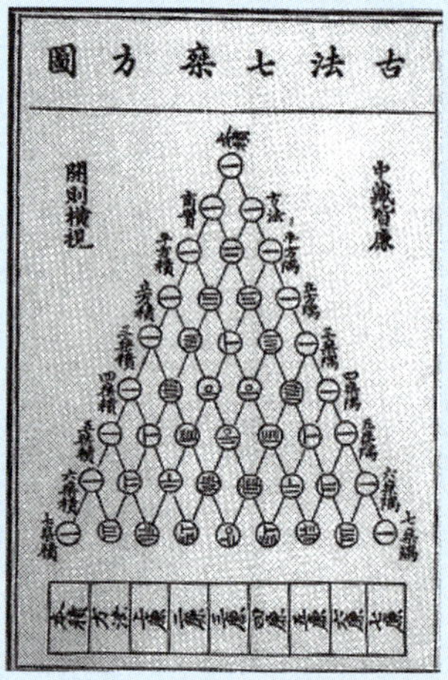

Brown University Library

Consequently,

$$A \cup B = \{\text{five card hands} \mid \text{the hand contains 4 kings } \textit{or} \text{ 4 queens}\}$$
$$A \cap B = \{\text{five card hands} \mid \text{the hand contains 4 kings } \textit{and} \text{ 4 queens}\}$$

(Recall that the union symbol, $\cup$, may be interpreted as the word "or," while the intersection symbol, $\cap$, may be interpreted as the word "and." See Figure 2.12 for a comparison of set theory and logic.)

Because there are no five-card hands that contain four kings *and* four queens, we note that $n(A \cap B) = 0$.

Using the Cardinal Number Formula for the Union of Sets, we obtain

$$\begin{aligned} n(A \cup B) &= n(A) + n(B) - n(A \cap B) \\ &= 48 + 48 - 0 \\ &= 96 \end{aligned}$$

There are 96 different five-card hands containing four kings or four queens.

d. *Four of a kind* means four cards of the same "denomination," that is, four twos, or four threes, or four fours, or . . . , or four kings, or four aces. Now, regardless of the denomination of the card, there are 48 different five-card hands that contain four of any specific denomination; there are 48 different five-card hands that contain four twos, there are 48 different five-card hands that contain four threes, there are 48 different five-card hands that contain four fours, and so on. As is shown in part (c), the word "or" implies that we *add* cardinal numbers. Consequently,

$$
\begin{aligned}
n(\text{four of a kind}) &= n(4 \text{ twos or } 4 \text{ threes or } \ldots \text{ or } 4 \text{ kings or } 4 \text{ aces}) \\
&= n(4 \text{ twos}) + n(4 \text{ threes} + \cdots + n(4 \text{ kings}) + n(4 \text{ aces}) \\
&= 48 + 48 + \cdots + 48 + 48 \quad (13 \text{ times}) \\
&= 13 \times 48 \\
&= 624
\end{aligned}
$$

There are 624 different five-card hands containing four of a kind. ■

As is shown in Example 8, there are 624 possible five-card hands that contain four of a kind. When you are dealt five cards, what is the likelihood (or probability) that you will receive one of these hands? This question, and its answer, will be explored in Section 3.4, "Combinatorics and Probability."

2.4 Exercises

In Exercises 1–12, find the indicated value:

1. a. ${}_7P_3$ **b.** ${}_7C_3$
2. a. ${}_8P_4$ **b.** ${}_8C_4$
3. a. ${}_5P_5$ **b.** ${}_5C_5$
4. a. ${}_9P_0$ **b.** ${}_9C_0$
5. a. ${}_{14}P_1$ **b.** ${}_{14}C_1$
6. a. ${}_{13}C_3$ **b.** ${}_{13}C_{10}$
7. a. ${}_{100}P_3$ **b.** ${}_{100}C_3$
8. a. ${}_{80}P_4$ **b.** ${}_{80}C_4$
9. a. ${}_xP_{x-1}$ **b.** ${}_xC_{x-1}$
10. a. ${}_xP_1$ **b.** ${}_xC_1$
11. a. ${}_xP_2$ **b.** ${}_xC_2$
12. a. ${}_xP_{x-2}$ **b.** ${}_xC_{x-2}$

13. a. Find ${}_3P_2$.
b. List all of the permutations of {a, b, c} when the elements are taken two at a time.

14. a. Find ${}_3C_2$.
b. List all of the combinations of {a, b, c} when the elements are taken two at a time.

15. a. Find ${}_4C_2$.
b. List all of the combinations of {a, b, c, d} when the elements are taken two at a time.

16. a. Find ${}_4P_2$.
b. List all of the permutations of {a, b, c, d} when the elements are taken two at a time.

17. An art class consists of 12 students. All of them must present their portfolios and explain their work to the instructor and their classmates at the end of the semester.
a. If their names are drawn from a hat to determine who goes first, second, and so on, how many presentation orders are possible?
b. If their names are put in alphabetical order to determine who goes first, second, and so on, how many presentation orders are possible?

18. An English class consists of 24 students, and three are to be chosen to give speeches in a school competition. In how many different ways can the teacher choose the team, given the following conditions?
a. The order of the speakers is important.
b. The order of the speakers is not important.

19. In how many ways can the letters in the word "STOP" be arranged? (See the photograph below.)

Exercise 19: Right Letters, Wrong Order. A woman in a passing vehicle got a chuckle out of the freshly painted 'SOTP' sign at a street corner in Anniston, Ala. A city paint crew quickly corrected the misspelling.

20. A committee of four is to be selected from a group of 15 people. How many different committees are possible, given the following conditions?
  a. There is no distinction between the responsibilities of the members.
  b. One person is the chair, and the rest are general members.
  c. One person is the chair, one person is the secretary, one person is responsible for refreshments, and one person cleans up after meetings.

21. A softball league has 14 teams. If every team must play every other team once in the first round of league play, how many games must be scheduled?

22. In a group of 19 people, each person shakes hands once with each other person in the group. How many handshakes will occur?

23. A softball league has 14 teams. How many different end-of-the-season rankings of first, second, and third place are possible (disregarding ties)?

24. Two hundred people buy raffle tickets. Three winning tickets will be drawn at random.
  a. If first prize is \$100, second prize is \$50, and third prize is \$20, in how many different ways can the prizes be awarded?
  b. If each prize is \$50, in how many different ways can the prizes be awarded?

25. A group of eight women and six men must select a four-person committee. How many committees are possible if it must consist of the following?
  a. two women and two men
  b. any mixture of men and women
  c. a majority of women

26. A group of ten seniors, eight juniors, five sophomores, and five freshmen must select a committee of four. How many committees are possible if the committee must contain the following:
  a. one person from each class
  b. any mixture of the classes
  c. exactly two seniors

Exercises 27–32 refer to a deck of 52 playing cards (jokers not allowed). If you are unfamiliar with playing cards, see the end of Section 3.1 for a description of a standard deck.

27. How many five-card poker hands are possible?

28. **a.** How many five-card poker hands consisting of all hearts are possible?
  b. How many five-card poker hands consisting of all cards of the same suit are possible?

29. **a.** How many five-card poker hands containing exactly three aces are possible?
  b. How many five-card poker hands containing three of a kind are possible?

30. **a.** How many five-card poker hands consisting of three kings and two queens are possible?
  b. How many five-card poker hands consisting of three of a kind and a pair (a *full house*) are possible?

31. How many five-card poker hands containing two pair are possible?

  HINT: You must select two of the 13 ranks, then select a pair of each, and then one of the remaining cards.

32. How many five-card poker hands containing exactly one pair are possible?

  HINT: After selecting a pair, you must select three of the remaining 12 ranks and then select one card of each.

33. A 6/53 lottery requires choosing six of the numbers 1 through 53. How many different lottery tickets can you choose? (Order is not important, and the numbers do not repeat.)

34. A 7/39 lottery requires choosing seven of the numbers 1 through 39. How many different lottery tickets can you choose? (Order is not important, and the numbers do not repeat.)

35. A 5/36 lottery requires choosing five of the numbers 1 through 36. How many different lottery tickets can you choose? (Order is not important, and the numbers do not repeat.)

36. A 6/49 lottery requires choosing six of the numbers 1 through 49. How many different lottery tickets can you choose? (Order is not important, and the numbers do not repeat.)

37. Which lottery would be easier to win, a 6/53 or a 5/36? Why?

  HINT: See Exercises 33 and 35.

38. Which lottery would be easier to win, a 6/49 or a 7/39? Why?

  HINT: See Exercises 34 and 36.

39. **a.** Find the sum of the entries in the first row of Pascal's Triangle.
b. Find the sum of the entries in the second row of Pascal's Triangle.
c. Find the sum of the entries in the third row of Pascal's Triangle.
d. Find the sum of the entries in the fourth row of Pascal's Triangle.
e. Is there a pattern to the answers to parts (a)–(d)? If so, describe the pattern you see.
f. Use the pattern described in part (e) to predict the sum of the entries in the fifth row of Pascal's Triangle.
g. Find the sum of the entries in the fifth row of Pascal's Triangle. Was your prediction in part (f) correct?
h. Find the sum of the entries in the nth row of Pascal's Triangle.

40. **a.** Add adjacent entries of the fifth row of Pascal's Triangle to obtain the sixth row.
b. Find ${}_6C_r$ for $r = 0, 1, 2, 3, 4, 5,$ and 6.

41. Use Pascal's Triangle to answer the following.
a. In which row would you find the value of ${}_4C_2$?
b. In which row would you find the value of ${}_nC_r$?
c. Is ${}_4C_2$ the second number in the fourth row?
d. Is ${}_4C_2$ the third number in the fourth row?
e. What is the location of ${}_nC_r$? Why?

42. Given the set $S = \{a, b, c, d\}$, answer the following.
a. How many one-element subsets does S have?
b. How many two-element subsets does S have?
c. How many three-element subsets does S have?
d. How many four-element subsets does S have?
e. How many zero-element subsets does S have?
f. How many subsets does S have?
g. If $n(S) = k$, how many subsets will S have?

Answer the following questions using complete sentences and your own words.

Concept Questions

43. Suppose you want to know how many ways r items can be selected from a group of n items. What determines whether you should calculate ${}_nP_r$ or ${}_nC_r$?

44. For any given values of n and r, which is larger, ${}_nP_r$ or ${}_nC_r$? Why?

Many graduate schools require applicants to take the Graduate Record Examination (G.R.E.). This exam is intended to measure verbal, quantitative, and analytical skills that have developed throughout a person's life. Many classes and study guides are available to help people prepare for the exam. The remaining questions are typical of those found in the study guides and on the exam itself.

Exercises 45–49 refer to the following: A baseball league has six teams: A, B, C, D, E, and F. All games are played at 7:30 P.M. on Fridays, and there are sufficient fields for each team to play a game every Friday night. Each team must play each other team exactly once, and the following conditions must be met:

- Team A plays team D first and team F second.
- Team B plays team E first and team C third.
- Team C plays team F first.

45. What is the total number of games that each team must play during the season?
a. 3 **b.** 4 **c.** 5 **d.** 6 **e.** 7

46. On the first Friday, which of the following pairs of teams play each other?
a. A and B; C and F; D and E
b. A and B; C and E; D and F
c. A and C; B and E; D and F
d. A and D; B and C; E and F
e. A and D; B and E; C and F

47. Which of the following teams must team B play second?
a. A **b.** C **c.** D **d.** E **e.** F

48. The last set of games could be between which teams?
a. A and B; C and F; D and E
b. A and C; B and F; D and E
c. A and D; B and C; E and F
d. A and E; B and C; D and F
e. A and F; B and E; C and D

49. If team D wins five games, which of the following must be true?
a. Team A loses five games.
b. Team A wins four games.
c. Team A wins its first game.
d. Team B wins five games.
e. Team B loses at least one game.

2.5 Voting Systems

Voting is an essential element in any democratic form of government. Whether it is the selection of leaders, the creation of laws and regulations, or deciding the outcome of issues ranging from mundane to volatile, decisions are made on the basis of the results of elections.

Maiman Rick/Corbis SyGma

Once votes have been cast, they must be counted, and a winner must be declared. Many people assume that "the candidate with the most votes wins"; however, that is not always the case. Deciding who wins an election is ultimately determined by the type of voting system that is used. "The person with the most votes wins" is a commonly used system and is called the plurality method. However, the plurality method is not the only method that is used. In this section, we will examine several common voting systems: (1) the plurality method, (2) the plurality with elimination method, (3) the ranked-choice or instant run-off method (sometimes called the Australian method), (4) the Borda count method, and (5) the pairwise comparison method.

The Plurality Method

A common way to determine the outcome of an election is to declare the candidate with the most votes the winner. This is the basis of the **plurality method of voting.**

> **Plurality Method of Voting**
>
> Each person votes for his or her favorite candidate (or choice). The candidate (or choice) who receives the most votes is declared the winner. (In case of a tie, special run-off elections may be held.)

EXAMPLE 1 City Cab is a new taxi service that will soon begin operation in a major metropolitan area. The board of directors for City Cab must purchase a fleet of new vehicles, and the vehicles must all be the same color. Four colors are available: black, white, red, and green. The five directors vote for their choice of color and the results are given in Figure 2.34. Using the plurality method of voting, which color choice wins?

Figure 2.34 Results of voting for the color of City Cab vehicles

	Director 1	Director 2	Director 3	Director 4	Director 5
Choice	green	red	red	green	red

Solution Tallying the results, we see that green received two votes and red received three. Because red received the most votes (3), red is declared the winner. ■

Referring to Example 1, it should be noted that in addition to receiving the most votes, red also received a majority of the votes, that is, more than 50% of the votes were for red. As we can see, three out of five, or 60%, of the votes were for red, while 40% were not for red.

Although the plurality method is easy to apply, this method may produce an unusual winner, as we shall see in the next example.

EXAMPLE 2 Referring to City Cab in Example 1, suppose the results of the election are those given in Figure 2.35. Using the plurality method of voting, which color choice wins?

Figure 2.35 Results of voting for the color of City Cab vehicles

	Director 1	Director 2	Director 3	Director 4	Director 5
Choice	green	red	red	black	white

Solution Once again, because red received the most votes (two), red would be declared the winner. Although red is the winner, it should be noted that red did *not* receive a *majority* of the votes. Fewer than half of the votes, that is 40%, were for red, while significantly more than half (60%) did not want the color red. This type of "winner" can lead to very difficult situations in political elections. Imagine yourself as the "winner" of a local election when more than half of the voters do *not* support you. ■

The Plurality with Elimination Method

As shown in Example 2, the plurality method of voting may produce a "winner" even though a majority of voters did not vote for the "winner." This dilemma can be avoided in several ways. One common way is to eliminate the candidate with the fewest votes and then hold another election. This is the basis of the **plurality with elimination method**. Consequently, the voters who supported the eliminated candidate now vote for someone else, that is, they must vote for their second choice. Typically, this second election will produce a winner with a majority of support. However, if a majority is still not attained, the process is repeated until a candidate obtains a majority of the votes.

> **Plurality with Elimination Method of Voting**
>
> Each person votes for his or her favorite candidate. If a candidate receives a majority of votes, that candidate is declared the winner. If no candidate receives a majority, then the candidate with the fewest votes is eliminated and a new election is held. This process continues until a candidate receives a majority of the votes.

EXAMPLE 3 The managers of We Deliver (a local shipping business) are planning a party for their 60 employees. There are three possible locations for the party: the warehouse, the park, and the beach. The employees voted on which location they preferred, and the results are given in Figure 2.36.

	Warehouse	Park	Beach
Number of votes	23	19	18

Figure 2.36
Results of voting for party location

a. Using the plurality method of voting, which location wins?
b. Using the plurality with elimination method of voting, which location wins?

Solution **a.** Because the warehouse received the most votes (23), the party will be held at the warehouse.

b. Although the warehouse received the most votes, it did not receive a majority; the warehouse received only 23/60, or approximately 38.3%, of the vote. Because the beach received the fewest number of votes (18), it is eliminated, and a new election is held. The results of the second election are given in Figure 2.37.

	Warehouse	Park	Beach
Number of votes	25	35	0

Figure 2.37
Results of the second election

Because the park received a majority of the votes ($35/60 \approx 58.3\%$), the party will be held at the park. ■

As is shown in Example 3, the outcome of an election may change depending on the type of voting system used. Therefore, it is imperative that the voters know ahead of time which system is being used!

As we have seen, the elimination of one candidate forces some people to shift their vote to their second choice in the new election. In the realm of real-world political elections, this method of elimination can be very expensive both for the local office of elections and for the candidates themselves, and it is overly time-consuming. Rather than holding a new election after eliminating the candidate with the fewest votes, a popular alternative is to have each voter rank each candidate during the first election. This is the basis of the ranked-choice or instant run-off method.

The Ranked-Choice or Instant Run-Off Method

In Example 3, the employees of We Deliver had three choices for the location of their party. If each of the 60 employees was asked to rank these choices in order of preference, in how many different ways could the locations be ranked? Because the choices are being put in an order (that is, a person must make a first choice, a second choice, and a third choice), we conclude that *permutations* can be used to determine the number of different rankings (see Section 2.4). Specifically, there are

$$ {}_3P_3 = \frac{3!}{(3-3)!} = \frac{3!}{0!} = \frac{3 \cdot 2 \cdot 1}{1} = 6 $$

different rankings (permutations) of the three locations. One such ranking is "warehouse, park, beach" or, more conveniently, WPB. Another such ranking is BPW (beach, park, warehouse). Figure 2.38 lists all possible permutations (or rankings) of the three locations.

1st choice	W	W	P	P	B	B
2nd choice	P	B	B	W	W	P
3rd choice	B	P	W	B	P	W

Figure 2.38
All permutations (or rankings) of "Warehouse, Park, Beach"

After each employee ranked the locations and the ballots were tallied, it was determined that 9 employees chose the WPB ranking, 14 chose WBP, 15 chose PBW, 4 chose PWB, 2 chose BWP, and 16 chose BPW. These results are readily portrayed in the table shown in Figure 2.39.

		Number of Ballots Cast					
		9	14	15	4	2	16
Ranked Ballot	**1st choice**	W	W	P	P	B	B
	2nd choice	P	B	B	W	W	P
	3rd choice	B	P	W	B	P	W

Figure 2.39
Voter preference table

Tables like that shown in Figure 2.39 are often called **voter preference tables**, that is, tables that list different rankings of the candidates along with the number of voters who chose each specific ranking. Tables of this nature are often used to depict the voting patterns in a ranked election.

Referring to the voter preference table in Figure 2.39, notice that the sum of the row of numbers is 60, that is, the sum of the row equals the number of people voting. In addition, to determine the number of first-choice votes for each candidate, go to the row labeled "1st choice" and observe that $9 + 14 = 23$ people had W (the warehouse) as their first choice, whereas $15 + 4 = 19$ people had P (the park) as their first choice, and $2 + 16 = 18$ had B (the beach) as their first choice. These subtotals agree with the values given in Example 3.

We now proceed as we did when using the plurality with elimination method; that is, we eliminate the candidate with the fewest first-choice votes (the beach). However, rather than holding a new election, we simply modify the original table by shifting the votes in the voter preference table upwards to fill the voids created by eliminating the beach, as shown in Figure 2.40.

		Number of Ballots Cast					
		9	14	15	4	2	16
Ranked Ballot	**1st choice**	W	W	P	P	~~B~~	~~B~~
	2nd choice	P	~~B~~	~~B~~	W	W↑	P↑
	3rd choice	~~B~~	P↑	W↑	~~B~~	P↑	W↑

Figure 2.40
Shifting votes in the voter preference table

The modified voter preference table in shown is Figure 2.41.

	Number of Ballots Cast					
	9	14	15	4	2	16
1st choice	W	W	P	P	W	P
2nd choice	P	P	W	W	P	W

Figure 2.41
Modified voter preference table

Referring to Figure 2.41, we now see that the Warehouse received $9 + 14 + 2 = 25$ first choice votes, whereas the Park received $15 + 4 + 16 = 35$ votes. Of course, these values agree with those given in Example 3.

The process described above is called the **ranked-choice or instant run-off method**. This voting system is commonly used in Australia. In March 2002, the voters of San Francisco, California, adopted the instant run-off method for the election of most political posts, including mayor, sheriff, district attorney, and many other high-visibility positions. It has been estimated that this voting method will save the City of San Francisco \$2 million per election, and voters will not have to wait several weeks for a second election. The instant run-off method of voting is summarized as follows.

Instant Run-off Method of Voting

Each voter ranks all of the candidates; that is, each voter selects his or her first choice, second choice, third choice, and so on. If a candidate receives a majority of first-choice votes, that candidate is declared the winner. If no candidate receives a majority, then the candidate with the fewest first-choice votes is eliminated and those votes are given to the next preferred candidate. If a candidate now has a majority of first-choice votes, that candidate is declared the winner. If no candidate receives a majority, this process continues until a candidate receives a majority.

EXAMPLE 4 Townsville is electing a new mayor. The candidates are Alturas (A), Bellum (B), Chan (C), and Dushay (D). The ranked ballots are tallied, and the results are summarized as shown in the voter preference table given in Figure 2.42.

		Number of Ballots Cast					
		15,527	15,287	10,023	9,105	7,978	5,660
Ranked Ballot	1st choice	C	B	C	B	A	D
	2nd choice	B	C	D	A	B	A
	3rd choice	A	D	A	D	C	C
	4th choice	D	A	B	C	D	B

Figure 2.42
Voter preference table for Mayor of Townsville

a. How many voters participated in the election?
b. Use the plurality method to determine the winner.
c. Use the instant run-off method to determine the winner.

Solution **a.** To determine the number of voters, we find the sum of the row of numbers representing the number of ballots cast for each ranking:

$$15{,}527 + 15{,}287 + 10{,}023 + 9{,}105 + 7{,}978 + 5{,}660 = 63{,}580$$

Therefore, 63,580 voters participated in the election.

b. Looking at the row of first-choice votes, we see that

A received 7,978 votes,

B received $15{,}287 + 9{,}105 = 24{,}392$ votes,

C received $15{,}527 + 10{,}023 = 25{,}550$ votes, and

D received 5,660 votes.

Because Chan (C) received the most votes (25,550), Mr. Chan becomes the new mayor of Townsville. However, it should be noted that Mr. Chan did not receive a majority of the votes; he received only $25{,}550/63{,}580 \approx 40.2\%$.

c. First, we determine the number of votes needed to obtain a majority. Because 63,580 votes were cast, a candidate needs more than half of 63,580, that is, more than $63{,}580 \div 2 = 31{,}790$ votes; in other words, a candidate needs 31,791 votes to win.

Because D received the fewest number of votes (5,660), Dushay is eliminated, and we obtain the modified voter preference table shown in Figure 2.43.

	Number of Ballots Cast					
	15,527	15,287	10,023	9,105	7,978	5,660
1st choice	C	B	C	B	A	A
2nd choice	B	C	A	A	B	C
3rd choice	A	A	B	C	C	B

Figure 2.43
Modified voter preference table for Mayor of Townsville

Examining the new row of first-choice votes, we see that

A received $7{,}978 + 5{,}660 = 13{,}638$ votes,

B received $15{,}287 + 9{,}105 = 24{,}392$ votes, and

C received $15{,}527 + 10{,}023 = 25{,}550$ votes.

Because no candidate has a majority, we eliminate the candidate with the fewest votes, that is, we eliminate Alturas and obtain the modified voter preference table shown in Figure 2.44.

	Number of Ballots Cast					
	15,527	15,287	10,023	9,105	7,978	5,660
1st choice	C	B	C	B	B	C
2nd choice	B	C	B	C	C	B

Figure 2.44
Modified voter preference table for Mayor of Townsville

Looking at the new row of first-choice votes, we see that

B received $15{,}287 + 9{,}105 + 7{,}978 = 32{,}370$ votes, and

C received $15{,}527 + 10{,}023 + 5{,}660 = 31{,}210$ votes.

Consequently, Ms. Bellum becomes the new mayor with $32{,}370/63{,}580 \approx 50.9\%$ of the vote. ■

Whereas the plurality method only utilizes a voter's first choice, the instant run-off method is designed to accommodate a voter's alternative choices. Another popular method that requires voters to rank their choices is the Borda count method.

The Borda Count Method

To win an election using the instant run-off method, a candidate must capture a majority of first-choice votes. Rather than tallying the first-choice votes only, we might want to tally the first-choice votes, the second-choice votes, the third-choice votes, and so on. This is the basis of the **Borda count method**, which is summarized in the following box.

Borda Count Method of Voting

Each voter ranks all of the candidates; that is, each voter selects his or her first choice, second choice, third choice, and so on. If there are k candidates, each candidate receives k points for each first-choice vote, $(k - 1)$ points for each second-choice vote, $(k - 2)$ points for each third-choice vote, and so on. The candidate with the most total points is declared the winner.

EXAMPLE 5 Use the Borda count method to determine the location of the party for the We Deliver employees referenced in Example 3.

Solution Recall that the employees were asked to rank the three locations (warehouse, park, and beach); the results are summarized in the voter preference table given in Figure 2.45.

		Number of Ballots Cast					
		9	14	15	4	2	16
Ranked Ballot	1st choice	W	W	P	P	B	B
	2nd choice	P	B	B	W	W	P
	3rd choice	B	P	W	B	P	W

Figure 2.45 Voter preference table

First, we tally the votes for each location as shown in Figure 2.46.

	Warehouse	Park	Beach
1st choice	9 + 14 = 23 votes	15 + 4 = 19 votes	2 + 16 = 18 votes
2nd choice	4 + 2 = 6 votes	9 + 16 = 25 votes	14 + 15 = 29 votes
3rd choice	15 + 16 = 31 votes	14 + 2 = 16 votes	9 + 4 = 13 votes

Figure 2.46 Tally of votes

Because there were $k = 3$ candidates (warehouse, park, and beach), each first-choice vote is worth 3 points, each second-choice vote is worth 2 points, and each third-choice vote is worth 1 point.

Now determine the score for each candidate by multiplying the number of votes times the appropriate number of points as shown in Figure 2.47.

	Warehouse	Park	Beach
1st choice votes (3 points each)	23 × 3 points = 69 points	19 × 3 points = 57 points	18 × 3 points = 54 points
2nd choice votes (2 points each)	6 × 2 points = 12 points	25 × 2 points = 50 points	29 × 2 points = 58 points
3rd choice votes (1 point each)	31 × 1 points = 31 points	16 × 1 points = 16 points	13 × 1 points = 13 points

Figure 2.47 Points

Summing each column, we obtain the results of the election:

$$\text{Warehouse} = 69 + 12 + 31 = 112 \text{ points}$$
$$\text{Park} = 57 + 50 + 16 = 123 \text{ points}$$
$$\text{Beach} = 54 + 58 + 13 = 125 \text{ points}$$

When the Borda count method is used, the beach is declared the winner because it has the most points (125). ■

As is illustrated by the examples referring to the We Deliver employees' party location, the outcome of the election depends on the voting system that is used. Specifically, the plurality method produced the warehouse as the winner, whereas the plurality with elimination and instant run-off methods declared the park to be the winner, and the Borda count method chose the beach. Obviously, it is very important to inform voters *before* the election as to which system will be used!

EXAMPLE 6 Referring to Example 4, use the Borda count method to determine the next Mayor of Townsville.

Solution Referring to the voter preference table given in Figure 2.42, we tally the votes as shown in Figure 2.48.

	Alturas	Bellum	Chan	Dushay
1st choice	7,978 votes	15,287 + 9,105 = 24,392 votes	15,527 + 10,023 = 25,550 votes	5,660 votes
2nd choice	9,105 + 5,660 = 14,765 votes	15,527 + 7,978 = 23,505 votes	15,287 votes	10,023 votes
3rd choice	15,527 + 10,023 = 25,550 votes	0 votes	7,978 + 5,660 = 13,638 votes	15,287 + 9,105 = 24,392 votes
4th choice	15,287 votes	10,023 + 5,660 = 15,683 votes	9,105 votes	15,527 + 7,978 = 23,505 votes

Figure 2.48
Tally of votes

Because there were $k = 4$ candidates (Alturas, Bellum, Chan, and Dushay), each first-choice vote is worth 4 points. Consequently, the second-, third-, and fourth-choice votes are worth 3, 2, and 1 points each, respectively. Now determine the score for each candidate by multiplying the number of votes times the appropriate number of points as shown in Figure 2.49.

Summing each column, we obtain the results of the election:

$$\text{Alturas} = 31{,}912 + 44{,}295 + 51{,}100 + 15{,}287 = 142{,}594 \text{ points}$$
$$\text{Bellum} = 97{,}568 + 70{,}515 + 15{,}683 = 183{,}766 \text{ points}$$
$$\text{Chan} = 102{,}200 + 45{,}861 + 27{,}276 + 9{,}105 = 184{,}442 \text{ points}$$
$$\text{Dushay} = 22{,}640 + 30{,}069 + 48{,}784 + 23{,}505 = 124{,}998 \text{ points}$$

When the Borda count method is used, Mr. Chan becomes the new mayor because he has the most points (184,442).

	Alturas	Bellum	Chan	Dushay
1st choice (4 points each)	7,978 × 4 = 31,912 points	24,392 × 4 = 97,568 points	25,550 × 4 = 102,200 points	5,660 × 4 = 22,640 points
2nd choice (3 points each)	14,765 × 3 = 44,295 points	23,505 × 3 = 70,515 points	15,287 × 3 = 45,861 points	10,023 × 3 = 30,069 points
3rd choice (2 points each)	25,550 × 2 = 51,100 votes	0 × 2 = 0 points	13,638 × 2 = 27,276 points	24,392 × 2 = 48,784 points
4th choice (1 point each)	15,287 × 1 = 15,287 votes	15,683 × 1 = 15,683 points	9,105 × 1 = 9,105 points	23,505 × 1 = 23,505 points

Figure 2.49
Points

The Pairwise Comparison Method

An election which has more than two candidates can be viewed as several mini-elections in which each possible pair of candidates compete against each other. That is, if there are three candidates, X, Y, and Z, we may consider the mini-elections or pairwise comparisons of X versus Y, X versus Z, and Y versus Z. We then determine the winner of each of these pairwise comparisons, and the candidate who wins the most of these is declared the winner of the overall election. This process is the basis of the **pairwise comparison method,** which is summarized in the following box.

Pairwise Comparison Method of Voting

Each voter ranks all of the candidates; that is, each voter selects his or her first choice, second choice, third choice, and so on. For each possible pairing of candidates, the candidate with the most votes receives 1 point; if there is a tie, each candidate receives 1/2 point. The candidate who receives the most points is declared the winner.

How many possible pairwise comparisons are there? That is, if there are k candidates and we must select two for a mini-election, how many different mini-elections are possible? The answer lies in using *combinations,* as illustrated in Section 2.4; that is, the order in which we select two candidates to compete does not matter (X versus Y is the same as Y versus X). Consequently, if there are k candidates, we must examine

$$_kC_2 = \frac{k!}{(k-2)!2!} = \frac{k(k-1)}{2}$$

possible pairwise comparisons.

EXAMPLE 7 Use the pairwise comparison method to determine the location of the party for the We Deliver employees referenced in Example 3.

Solution Recall that the employees were asked to rank the three locations (warehouse, park, and beach) and the results are summarized in the voter preference table in Figure 2.50.

		Number of Ballots Cast					
		9	14	15	4	2	16
Ranked Ballot	1st choice	W	W	P	P	B	B
	2nd choice	P	B	B	W	W	P
	3rd choice	B	P	W	B	P	W

Figure 2.50
Voter preference table

Because there are $k = 3$ "candidates," we must examine

$$_3C_2 = \frac{3!}{(3-2)!2!} = \frac{3!}{1!2!} = \frac{3 \cdot 2 \cdot 1}{1 \cdot 2 \cdot 1} = 3$$

pairwise comparisons; specifically, we investigate W versus P, W versus B, and P versus B. For each comparison, we examine each column of the voter preference table to determine which candidate is preferred.

W versus P In the first column of Figure 2.50, we see that W is preferred over P; therefore, W receives the 9 votes listed in column 1. In the second column, W is again ranked over P and receives 14 votes. In the third column, P is preferred over W, so P receives 15 votes. In the fourth column, P is again preferred over W and receives 4 votes. In the fifth column, W is preferred over P and receives 2 votes. Finally, P is preferred over W in the sixth column and receives 16 votes. Tallying these results, we obtain the following totals:

$$\text{votes for W} = 9 + 14 + 2 = 25$$
$$\text{votes for P} = 15 + 4 + 16 = 35$$

For the comparison of W versus P, voters preferred P over W by a vote of 35 to 25. Consequently, P receives 1 point.

W versus B Examining Figure 2.50, we see that the voters in the first, second, and fourth columns all preferred W over B. Consequently, W receives the 9, 14, and 4 votes listed in those columns. We also see that the voters in the third, fifth, and sixth columns all preferred B over W, and so B receives the 15, 2, and 16 votes listed in those columns. Tallying these results, we obtain the following totals:

$$\text{votes for W} = 9 + 14 + 4 = 27$$
$$\text{votes for B} = 15 + 2 + 16 = 33$$

For the comparison of W versus B, voters preferred B over W by a vote of 33 to 27. Consequently, B receives 1 point.

P versus B Figure 2.50 indicates that the voters in the first, third, and fourth columns all ranked P over B. Therefore, P receives the 9, 15, and 4 votes listed in those columns. The voters in the remaining columns (second, fifth, and sixth) preferred B over P, and B receives 14, 2, and 16 votes accordingly. Tallying these results, we obtain the following totals:

$$\text{votes for P} = 9 + 15 + 4 = 28$$
$$\text{votes for B} = 14 + 2 + 16 = 32$$

For the comparison of P versus B, voters preferred B over P by a vote of 32 to 28. Consequently, B receives 1 point.

Tallying the points, we see that B receives 2 points (B was preferred in two of the comparisons), P receives 1 point (P was preferred in one comparison), and W receives 0 points (W was not preferred in any of the comparisons). Consequently, B is declared the winner; the party will be held at the beach. ■

EXAMPLE 8 Use the pairwise comparison method to determine the new mayor of Townsville referenced in Example 4.

Solution Recall that the candidates are Alturas (A), Bellum (B), Chan (C), and Dushay (D) and that 63,580 votes were cast. The ranked ballots are tallied, and the results are summarized as shown in the voter preference table in Figure 2.51.

		Number of Ballots Cast					
		15,527	15,287	10,023	9,105	7,978	5,660
Ranked Ballot	1st choice	C	B	C	B	A	D
	2nd choice	B	C	D	A	B	A
	3rd choice	A	D	A	D	C	C
	4th choice	D	A	B	C	D	B

Figure 2.51
Voter preference table for Mayor of Townsville

Because there are $k = 4$ candidates, we must examine

$$_4C_2 = \frac{4!}{(4-2)!2!} = \frac{4!}{2! \cdot 2!} = \frac{4 \cdot 3 \cdot 2 \cdot 1}{2 \cdot 1 \cdot 2 \cdot 1} = 6$$

pairwise comparisons; specifically, we investigate A versus B, A versus C, A versus D, B versus C, B versus D, and C versus D. For each comparison, we examine each column of the voter preference table to determine which candidate is preferred.

A versus B Figure 2.51 indicates that the voters in the third, fifth, and sixth columns all chose A over B. Therefore, A receives the 10,023 + 7,978 + 5,660 = 23,661 votes listed in those columns. To determine the number of voters who prefer B over A, we could sum the votes listed in the remaining columns, or using a simpler approach, we subtract the 23,661 votes for A from the total number of votes to obtain the number of votes for B. Therefore, B receives 63,580 − 23,661 = 39,919 votes. Consequently, B is preferred and receives 1 point.

A versus C Figure 2.51 indicates that the voters in the fourth, fifth, and sixth columns all preferred A over C. Therefore, the votes are tallied as follows:

A = 9,105 + 7,978 + 5,660 = 22,743 votes
C = 63,580 − 22,743 = 40,837 votes

Consequently, C is preferred and receives 1 point.

A versus D Figure 2.51 indicates that the voters in the first, fourth, and fifth columns all preferred A over D. Therefore, the votes are tallied as follows:

$$A = 15{,}527 + 9{,}105 + 7{,}978 = 32{,}610 \text{ votes}$$
$$D = 63{,}580 - 32{,}610 = 30{,}970 \text{ votes}$$

Consequently, A is preferred and receives 1 point.

B versus C Figure 2.51 indicates that the voters in the second, fourth, and fifth columns all preferred B over C. Therefore, the votes are tallied as follows:

$$B = 15{,}287 + 9{,}105 + 7{,}978 = 32{,}370 \text{ votes}$$
$$C = 63{,}580 - 32{,}370 = 31{,}210 \text{ votes}$$

Consequently, B is preferred and receives 1 point.

B versus D Figure 2.51 indicates that the voters in the first, second, fourth, and fifth columns all preferred B over D. Therefore, the votes are tallied as follows:

$$B = 15{,}527 + 15{,}287 + 9{,}105 + 7{,}978 = 47{,}897 \text{ votes}$$
$$D = 63{,}580 - 47{,}897 = 15{,}683 \text{ votes}$$

Consequently, B is preferred and receives 1 point.

C versus D Figure 2.51 indicates that the voters in the first, second, third, and fifth columns all preferred C over D. Therefore, the votes are tallied as follows:

$$C = 15{,}527 + 15{,}287 + 10{,}023 + 7{,}978 = 48{,}815 \text{ votes}$$
$$D = 63{,}580 - 48{,}815 = 14{,}765 \text{ votes}$$

Consequently, C is preferred and receives 1 point.

Tallying the points, we see that A receives 1 point, B receives 3 points, C receives 2 points, and D receives 0 points. B has the most points (3) and so is declared the winner; Ms. Bellum is the new mayor of Townsville. ■

2.5 Exercises

1. Four candidates, Alliotti, Baker, Cruz, and Daud, are running for president of the student government. After the polls close, votes are tallied and the following results are obtained:

Candidate	Alliotti	Baker	Cruz	Daud
Number of votes	314	155	1,052	479

a. How many votes were cast?
b. Using the plurality method of voting, which candidate wins?
c. Did the winner receive a majority of the votes?

2. Five candidates, Edwards, Fischer, Gelinas, Horner, and Inclan, are running for president of the faculty senate. After the polls close, votes are tallied and the following results are obtained:

Candidate	Edwards	Fischer	Gelinas	Horner	Inclan
Number of votes	32	25	17	102	24

a. How many votes were cast?
b. Using the plurality method of voting, which candidate wins?
c. Did the winner receive a majority of the votes?

3. Three candidates, Arce, Edelstein, and Spence, are running for president of academic affairs at a local college. After the polls close, votes are tallied and the following results are obtained:

Candidate	Arce	Edelstein	Spence
Number of votes	3,021	4,198	3,132

a. How many votes were cast?
b. Using the plurality method of voting, which candidate wins?
c. Did the winner receive a majority of the votes?

4. Four candidates, Brecha, Parks, Wilcox, and Willett, are running for district supervisor. After the polls close, votes are tallied and the following results are obtained:

Candidate	Brecha	Parks	Wilcox	Willett
Number of votes	3,007	2,957	10,541	2,851

a. How many votes were cast?
b. Using the plurality method of voting, which candidate wins?
c. Did the winner receive a majority of the votes?

5. The managers of Prints Alive (a local silk-screening business) are planning a party for their 30 employees. There are three possible locations for the party: the warehouse, the park, or the beach. The employees are asked to rank these choices in order of preference, and the results are summarized in the following voter preference table.

	Number of Ballots Cast			
	6	8	11	5
1st choice	P	P	B	W
2nd choice	B	W	W	B
3rd choice	W	B	P	P

a. How many votes were cast?
b. Use the plurality method of voting to determine the winner.
c. What percent of the votes did the winner in part (b) receive?
d. Use the instant run-off method to determine the winner.
e. What percent of the votes did the winner in part (d) receive?
f. Use the Borda count method to determine the winner.
g. How many points did the winner in part (f) receive?
h. Use the pairwise comparison method to determine the winner.
i. How many points did the winner in part (h) receive?

6. The members of a local service club are volunteering to clean up and modernize the playground at one of the elementary schools in town. There are three schools: Hidden Lakes (H), Strandwood (S), and Valhalla (V). The members are asked to rank these choices in order of preference, and the results are summarized in the following voter preference table.

	Number of Ballots Cast			
	13	8	10	19
1st choice	S	V	V	H
2nd choice	H	S	H	S
3rd choice	V	H	S	V

a. How many votes were cast?
b. Use the plurality method of voting to determine the winner.
c. What percent of the votes did the winner in part (b) receive?
d. Use the instant run-off method to determine the winner.
e. What percent of the votes did the winner in part (d) receive?
f. Use the Borda count method to determine the winner.
g. How many points did the winner in part (f) receive?
h. Use the pairwise comparison method to determine the winner.
i. How many points did the winner in part (h) receive?

7. The members of a youth club are raising money so that they can attend a summer camp. There are three camps in the area: Coastline (C), Pinewood (P), and The Ranch (R). The members are asked to rank these choices in order of preference, and the results are summarized in the following voter preference table.

	Number of Ballots Cast				
	19	12	10	11	13
1st choice	C	C	P	R	R
2nd choice	P	R	C	P	C
3rd choice	R	P	R	C	P

a. How many votes were cast?
b. Use the plurality method of voting to determine the winner.
c. What percent of the votes did the winner in part (b) receive?
d. Use the instant run-off method to determine the winner.
e. What percent of the votes did the winner in part (d) receive?
f. Use the Borda count method to determine the winner.
g. How many points did the winner in part (f) receive?
h. Use the pairwise comparison method to determine the winner.
i. How many points did the winner in part (h) receive?

8. The members of a local charitable group are raising money to send a group of neighborhood children to a special event. There are three events to choose from: the circus (C), the ice show (I), and the symphony (S). The members are asked to rank these choices in order of preference, and the results are summarized in the following voter preference table.

	Number of Ballots Cast				
	6	8	7	10	15
1st choice	C	C	I	I	S
2nd choice	I	S	C	S	C
3rd choice	S	I	S	C	I

a. How many votes were cast?
b. Use the plurality method of voting to determine the winner.
c. What percent of the votes did the winner in part (b) receive?
d. Use the instant run-off method to determine the winner.
e. What percent of the votes did the winner in part (d) receive?
f. Use the Borda count method to determine the winner.
g. How many points did the winner in part (f) receive?
h. Use the pairwise comparison method to determine the winner.
i. How many points did the winner in part (h) receive?

9. Three candidates, Budd (B), Nirgiotis (N), and Shattuck (S), are running for union president. After the polls close, ranked ballots are tallied, and the results are summarized in the following voter preference table.

	Number of Ballots Cast					
	25	13	19	27	30	26
1st choice	B	B	N	N	S	S
2nd choice	N	S	S	B	B	N
3rd choice	S	N	B	S	N	B

a. How many votes were cast?
b. Use the plurality method of voting to determine the winner.
c. What percent of the votes did the winner in part (b) receive?
d. Use the instant run-off method to determine the winner.
e. What percent of the votes did the winner in part (d) receive?
f. Use the Borda count method to determine the winner.
g. How many points did the winner in part (f) receive?
h. Use the pairwise comparison method to determine the winner.
i. How many points did the winner in part (h) receive?

10. Three candidates, Maruyama (M), Peters (P), and Vilas (V), are running for district representative. After the polls close, ranked ballots are tallied, and the results are summarized in the following voter preference table.
a. How many votes were cast?
b. Use the plurality method of voting to determine the winner.

	Number of Ballots Cast					
	675	354	451	387	601	297
1st choice	M	M	P	P	V	V
2nd choice	P	V	V	M	M	P
3rd choice	V	P	M	V	P	M

c. What percent of the votes did the winner in part (b) receive?
d. Use the instant run-off method to determine the winner.
e. What percent of the votes did the winner in part (d) receive?
f. Use the Borda count method to determine the winner.
g. How many points did the winner in part (f) receive?
h. Use the pairwise comparison method to determine the winner.
i. How many points did the winner in part (h) receive?

11. Four candidates, Dolenz (D), Jones (J), Nesmith (N), and Tork (T), are running for director of public relations. After the polls close, ranked ballots are tallied, and the results are summarized in the following voter preference table.

	Number of Ballots Cast						
	225	134	382	214	81	197	109
1st choice	D	D	J	J	N	T	T
2nd choice	J	N	D	T	J	D	J
3rd choice	T	T	T	D	T	N	D
4th choice	N	J	N	N	D	J	N

a. How many votes were cast?
b. Use the plurality method of voting to determine the winner.
c. What percent of the votes did the winner in part (b) receive?
d. Use the instant run-off method to determine the winner.
e. What percent of the votes did the winner in part (d) receive?
f. Use the Borda count method to determine the winner.
g. How many points did the winner in part (f) receive?
h. Use the pairwise comparison method to determine the winner.
i. How many points did the winner in part (h) receive?

12. Four candidates, Harrison (H), Lennon (L), McCartney (M), and Starr (S), are running for regional manager. After the polls close, ranked ballots are tallied, and the results are summarized in the following voter preference table.

	Number of Ballots Cast						
	23	98	45	82	32	17	21
1st choice	H	L	L	M	M	S	S
2nd choice	L	M	S	L	L	H	M
3rd choice	S	H	M	H	S	L	H
4th choice	M	S	H	S	H	M	L

a. How many votes were cast?
b. Use the plurality method of voting to determine the winner.
c. What percent of the votes did the winner in part (b) receive?
d. Use the instant run-off method to determine the winner.
e. What percent of the votes did the winner in part (d) receive?
f. Use the Borda count method to determine the winner.
g. How many points did the winner in part (f) receive?
h. Use the pairwise comparison method to determine the winner.
i. How many points did the winner in part (h) receive?

13. Five candidates, Addley (A), Burke (B), Ciento (C), Darter (D), and Epp (E), are running for mayor. After the polls close, ranked ballots are tallied, and the results are summarized in the following voter preference table.

	Number of Ballots Cast						
	1,897	1,025	4,368	2,790	6,897	9,571	5,206
1st choice	A	A	B	C	D	D	E
2nd choice	C	D	D	E	B	E	D
3rd choice	D	C	C	D	C	B	C
4th choice	B	E	E	B	E	A	B
5th choice	E	B	A	A	A	C	A

a. How many votes were cast?
b. Use the plurality method of voting to determine the winner.
c. What percent of the votes did the winner in part (b) receive?
d. Use the instant run-off method to determine the winner.
e. What percent of the votes did the winner in part (d) receive?
f. Use the Borda count method to determine the winner.
g. How many points did the winner in part (f) receive?
h. Use the pairwise comparison method to determine the winner.
i. How many points did the winner in part (h) receive?

14. Five candidates, Fino (F), Gempler (G), Holloway (H), Isho (I), and James (J), are running for president of the Polar Bear Swim Club. After the polls close, ranked ballots are tallied, and the results are summarized in the following voter preference table.

	Number of Ballots Cast						
	12	18	29	11	21	19	18
1st choice	F	F	G	H	I	J	J
2nd choice	G	I	F	G	J	H	F
3rd choice	I	G	H	F	G	I	G
4th choice	J	H	J	I	F	G	I
5th choice	H	J	I	J	H	F	H

a. How many votes were cast?
b. Use the plurality method of voting to determine the winner.
c. What percent of the votes did the winner in part (b) receive?
d. Use the instant run-off method to determine the winner.
e. What percent of the votes did the winner in part (d) receive?
f. Use the Borda count method to determine the winner.
g. How many points did the winner in part (f) receive?
h. Use the pairwise comparison method to determine the winner.
i. How many points did the winner in part (h) receive?

15. If there are six candidates in an election and voters are asked to rank all of the candidates, how many different rankings are possible?

16. If there are eight candidates in an election and voters are asked to rank all of the candidates, how many different rankings are possible?

17. If there are six candidates in an election and voters are asked to rank all of the candidates, how many different pairwise comparisons are there?

18. If there are eight candidates in an election and voters are asked to rank all of the candidates, how many different pairwise comparisons are there?

19. In an election, there are three candidates and 25 voters.
a. What is the maximum number of points that a candidate can receive using the Borda count method?
b. What is the minimum number of points that a candidate can receive using the Borda count method?

20. In an election, there are four candidates and 75 voters.
a. What is the maximum number of points that a candidate can receive using the Borda count method?
b. What is the minimum number of points that a candidate can receive using the Borda count method?

21. In an election, there are seven candidates.
a. What is the maximum number of points that a candidate can receive using the pairwise comparison method?
b. What is the minimum number of points that a candidate can receive using the pairwise comparison method?

22. In an election, there are nine candidates.
a. What is the maximum number of points that a candidate can receive using the pairwise comparison method?
b. What is the minimum number of points that a candidate can receive using the pairwise comparison method?

Answer the following questions using complete sentences and your own words.

Concept Questions

23. What is a majority?
24. What is a ranked ballot?
25. What is a voter preference table?

2.6 Infinite Sets

> **WARNING:** Many leading nineteenth-century mathematicians and philosophers claim that the study of infinite sets may be dangerous to your mental health.

Consider the sets E and N, where $E = \{2, 4, 6, \ldots\}$ and $N = \{1, 2, 3, \ldots\}$. Both are examples of infinite sets (they "go on forever"). E is the set of all even counting numbers, and N is the set of all counting (or natural) numbers. Because every element of E is an element of N, E is a subset of N. In addition, N contains elements not in E; therefore, E is a *proper* subset of N. Which set is "bigger," E or N? Intuition may lead many people to think that N is twice as big as E because N contains all the even counting numbers *and* all the odd counting numbers. Not so! According to the work of Georg Cantor (considered by many to be the father of set theory), N and E have exactly the same number of elements! This seeming paradox, a *proper* subset that has the *same number* of elements as the set from which it came, caused a philosophic uproar in the late nineteenth century. (Hence, the warning at the beginning of this section.) To study Cantor's work (which is now accepted and considered a cornerstone in modern mathematics), we must first investigate the meaning of a one-to-one correspondence and equivalent sets.

One-to-One Correspondence

Is there any relationship between the sets $A = \{\text{one, two, three}\}$ and $B = \{\text{Pontiac, Chevrolet, Ford}\}$? Although the sets contain different types of things (numbers versus automobiles), each contains the same number of things; they are the same size. This relationship (being the same size) forms the basis of a one-to-one correspondence. A **one-to-one correspondence** between the sets A and B is a pairing up of the elements of A and B such that each element of A is paired up with exactly one element of B, and vice versa, with no element left out. For instance, the elements of A and B might be paired up as follows:

one	two	three
$\updownarrow$	$\updownarrow$	$\updownarrow$
Pontiac	Chevrolet	Ford

(Other correspondences, or matchups, are possible.) If two sets have the same cardinal number, their elements can be put into a one-to-one correspondence. Whenever a one-to-one correspondence exists between the elements of two sets A and B, the sets are **equivalent** (denoted by $A \sim B$). Hence, equivalent sets have the same number of elements.

If two sets have different cardinal numbers, it is not possible to construct a one-to-one correspondence between their elements. The sets $C = \{\text{one, two}\}$ and $B = \{\text{Pontiac, Chevrolet, Ford}\}$ do *not* have a one-to-one correspondence; no matter how their elements are paired up, one element of B will always be left over (B has more elements; it is "bigger"):

one	two	
$\updownarrow$	$\updownarrow$	
Pontiac	Chevrolet	Ford

The sets C and B are *not* equivalent.

HISTORICAL NOTE

Georg Cantor

1845–1918

Georg Ferdinand Ludwig Philip Cantor was born in St. Petersburg, Russia. His father was a stockbroker and wanted his son to become an engineer; his mother was an artist and musician. Several of Cantor's maternal relatives were accomplished musicians; in his later years, Cantor often wondered how his life would have turned out if he had become a violinist instead of pursuing a controversial career in mathematics.

Following his father's wishes, Cantor began his engineering studies at the University of Zurich in 1862. However, after one semester, he decided to study philosophy and pure mathematics. He transferred to the prestigious University of Berlin, studied under the famed mathematicians Karl Weierstrass, Ernst Kummer, and Leopold Kronecker, and received his doctorate in 1867. Two years later, Cantor accepted a teaching position at the University of Halle and remained there until he retired in 1913.

Cantor's treatises on set theory and the nature of infinite sets were first published in 1874 in *Crelle's Journal,* which was influential in mathematical circles. On their publication, Cantor's theories generated much controversy among mathematicians and philosophers. Paradoxes concerning the cardinal numbers of infinite sets, the nature of infinity, and Cantor's form of logic were unsettling to many, including Cantor's former teacher Leopold Kronecker. In fact, some felt that Cantor's work was not just revolutionary but actually dangerous.

Kronecker led the attack on Cantor's theories. He was an editor of *Crelle's Journal* and held up the publication of one of Cantor's subsequent articles for so long that Cantor refused to publish ever again in the *Journal.* In addition, Kronecker blocked Cantor's efforts to obtain a teaching position at the University of Berlin. Even though Cantor was attacked by Kronecker and his followers, others respected him. Realizing the importance of communication among scholars, Cantor founded the Association of German Mathematicians in 1890 and served as its president for many years. In addition, Cantor was instrumental in organizing the first International Congress of Mathematicians, held in Zurich in 1897.

As a result of the repeated attacks on him and his work, Cantor suffered many nervous breakdowns, the first when he was thirty-nine. He died in a mental hospital in Halle at the age of seventy-three, never having received proper recognition for the true value of his discoveries. Modern mathematicians believe that Cantor's form of logic and his concepts of infinity revolutionized all of mathematics, and his work is now considered a cornerstone in its development.

Written in 1874, Cantor's first major paper on the theory of sets, *Über eine Eigenshaft des Inbegriffes aller reellen algebraischen Zahlen* (On a Property of the System of All the Real Algebraic Numbers), sparked a major controversy concerning the nature of infinite sets. To gather international support for his theory, Cantor had his papers translated into French. This 1883 French version of Cantor's work was published in the newly formed journal *Acta Mathematica.* Cantor's works were translated into English during the early twentieth century.

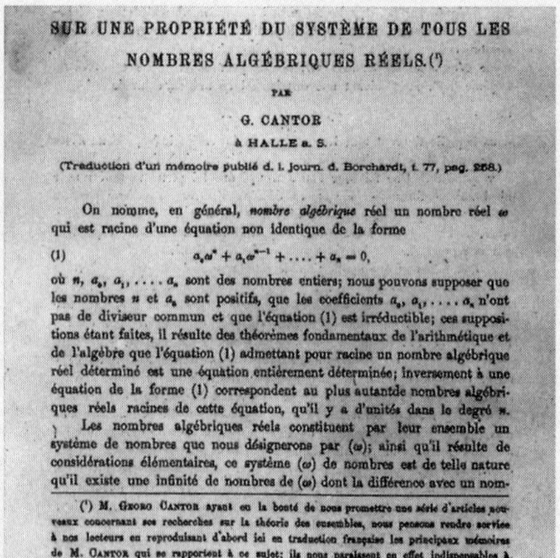

SUR UNE PROPRIÉTÉ DU SYSTÈME DE TOUS LES NOMBRES ALGÉBRIQUES RÉELS.(¹)

PAR

G. CANTOR

à HALLE a. S.

(Traduction d'un mémoire publié d. l. Journ. d. Borchardt, t. 77, pag. 258.)

On nomme, en général, *nombre algébrique* réel un nombre réel ω qui est racine d'une équation non identique de la forme

$$(1) \qquad a_0\omega^n + a_1\omega^{n-1} + \dots + a_n = 0,$$

où $n, a_0, a_1, \dots a_n$ sont des nombres entiers; nous pouvons supposer que les nombres n et a_0 sont positifs, que les coefficients $a_0, a_1, \dots a_n$ n'ont pas de diviseur commun et que l'équation (1) est irréductible; ces suppositions étant faites, il résulte des théorèmes fondamentaux de l'arithmétique et de l'algèbre que l'équation (1) admettant pour racine un nombre algébrique réel déterminé est une équation entièrement déterminée; inversement à une équation de la forme (1) correspondent au plus autantde nombres algébriques réels racines de cette équation, qu'il y a d'unités dans le degré n.

Les nombres algébriques réels constituent par leur ensemble un système de nombres que nous désignerons par (ω); ainsi qu'il résulte de considérations élémentaires, ce système (ω) de nombres est de telle nature qu'il existe une infinité de nombres de (ω) dont la différence avec un nom-

(¹) M. Georg Cantor ayant eu la bonté de nous promettre une série d'articles nouveaux concernant ses recherches sur la théorie des ensembles, nous pensons rendre service à nos lecteurs en reproduisant d'abord ici en traduction française les principaux mémoires de M. Cantor qui se rapportent à ce sujet; ils nous paraissent en effet indispensables à l'intelligence des nouveaux qui vont suivre et que l'auteur publiera de même en français. La traduction a été revue et corrigée par l'auteur. *Le rédacteur.*

Acta mathematica. 2. Imprimé 8 Juin 1883. 39

Given two sets A and B, if any one of the following statements is true, then the other statements are also true:

1. There exists a one-to-one correspondence between the elements of A and B.
2. A and B are equivalent sets.
3. A and B have the same cardinal number; that is, $n(A) = n(B)$.

EXAMPLE 1 Determine whether the sets in each of the following pairs are equivalent. If they are equivalent, list a one-to-one correspondence between their elements.

a. $A = \{\text{John, Paul, George, Ringo}\}$;
$B = \{\text{Lennon, McCartney, Harrison, Starr}\}$

b. $C = \{\alpha, \beta, \chi, \delta\}$; $D = \{\text{I, O}, \Delta\}$

c. $A = \{1, 2, 3, \ldots, 48, 49, 50\}$; $B = \{1, 3, 5, \ldots, 95, 97, 99\}$

Solution **a.** If sets have the same cardinal number, they are equivalent. Now, $n(A) = 4$ and $n(B) = 4$; therefore, $A \sim B$.

Because A and B are equivalent, their elements can be put into a one-to-one correspondence. One such correspondence follows:

$$\begin{array}{cccc} \text{John} & \text{Paul} & \text{George} & \text{Ringo} \\ \updownarrow & \updownarrow & \updownarrow & \updownarrow \\ \text{Lennon} & \text{McCartney} & \text{Harrison} & \text{Starr} \end{array}$$

b. Because $n(C) = 4$ and $n(D) = 3$, C and D are not equivalent.

c. A consists of all natural numbers from 1 to 50, inclusive. Hence, $n(A) = 50$. B consists of all odd natural numbers from 1 to 99, inclusive. Since half of the natural numbers from 1 to 100 are odd (and half are even), there are fifty $(100 \div 2 = 50)$ odd natural numbers less than 100; that is, $n(B) = 50$. Because A and B have the same cardinal number, $A \sim B$.

Many different one-to-one correspondences may be established between the elements of A and B. One such correspondence follows:

$$\begin{array}{l} A = \{1, 2, 3, \ldots, \quad n, \quad \ldots, 48, 49, 50\} \\ \qquad \updownarrow \updownarrow \updownarrow \ldots \quad \updownarrow \quad \ldots \updownarrow \updownarrow \updownarrow \\ B = \{1, 3, 5, \ldots, (2n-1), \ldots, 95, 97, 99\} \end{array}$$

That is, each natural number $n \in A$ is paired up with the odd number $(2n - 1) \in B$. The $n \leftrightarrow (2n - 1)$ part is crucial because it shows *each* individual correspondence. For example, it shows that $13 \in A$ corresponds to $25 \in B$ ($n = 13$, so $2n = 26$ and $2n - 1 = 25$). Likewise, $69 \in B$ corresponds to $35 \in A$ ($2n - 1 = 69$, so $2n = 70$ and $n = 35$). ■

As we have seen, if two sets have the same cardinal number, they are equivalent, and their elements can be put into a one-to-one correspondence. Conversely, if the elements of two sets can be put into a one-to-one correspondence, the sets have the same cardinal number and are equivalent. Intuitively, this result appears to be quite obvious. However, when Georg Cantor applied this relationship to infinite sets, he sparked one of the greatest philosophical debates of the nineteenth century.

Countable Sets

Consider the set of all counting numbers $N = \{1, 2, 3, \ldots\}$, which consists of an infinite number of elements. Each of these numbers is either odd or even. Defining O

and E as $O = \{1, 3, 5, \ldots\}$ and $E = \{2, 4, 6, \ldots\}$, we have $O \cap E = \emptyset$ and $O \cup E = N$; the sets O and E are mutually exclusive, and their union forms the entire set of all counting numbers. Obviously, N contains elements that E does not. As we mentioned earlier, the fact that E is a *proper* subset of N might lead people to think that N is "bigger" than E. In fact, N and E are the "same size"; N and E each contain the same number of elements.

Recall that two sets are equivalent and have the same cardinal number if the elements of the sets can be matched up via a one-to-one correspondence. To show the existence of a one-to-one correspondence between the elements of two sets of numbers, we must find an explicit correspondence between the general elements of the two sets. In Example 1(c), we expressed the general correspondence as $n \leftrightarrow (2n - 1)$.

EXAMPLE 2

a. Show that $E = \{2, 4, 6, 8, \ldots\}$ and $N = \{1, 2, 3, 4, \ldots\}$ are equivalent sets.
b. Find the element of N that corresponds to $1430 \in E$.
c. Find the element of N that corresponds to $x \in E$.

Solution

a. To show that $E \sim N$, we must show that there exists a one-to-one correspondence between the elements of E and N. The elements of E and N can be paired up as follows:

$$\begin{array}{rl} N = & \{1, 2, 3, 4, \ldots, n, \ldots\} \\ & \ \ \updownarrow\ \updownarrow\ \updownarrow\ \updownarrow\ \ldots\ \ \updownarrow \\ E = & \{2, 4, 6, 8, \ldots, 2n, \ldots\} \end{array}$$

Any natural number $n \in N$ corresponds with the even natural number $2n \in E$. Because there exists a one-to-one correspondence between the elements of E and N, the sets E and N are equivalent; that is, $E \sim N$.

b. $1430 = 2n \in E$, so $n = \frac{1430}{2} = 715 \in N$. Therefore, $715 \in N$ corresponds to $1430 \in E$.

c. $x = 2n \in E$, so $n = \frac{x}{2} \in N$. Therefore, $n = \frac{x}{2} \in N$ corresponds to $x = 2n \in E$. ■

We have just seen that the set of *even* natural numbers is equivalent to the set of *all* natural numbers. This equivalence implies that the two sets have the same number of elements! Although E is a proper subset of N, both sets have the same cardinal number; that is, $n(E) = n(N)$. Settling the controversy sparked by this seeming paradox, mathematicians today define a set to be an **infinite set** if it can be placed in a one-to-one correspondence with a proper subset of itself.

How many counting numbers are there? How many even counting numbers are there? We know that each set contains an infinite number of elements and that $n(N) = n(E)$, but how many is that? In the late nineteenth century, Georg Cantor defined the cardinal number of the set of counting numbers to be $\aleph_0$ (read "**aleph-null**"). Cantor utilized Hebrew letters, of which aleph, $\aleph$, is the first. Consequently, the proper response to "How many counting numbers are there?" is "There are aleph-null of them"; $n(N) = \aleph_0$. Any set that is equivalent to the set of counting numbers has cardinal number $\aleph_0$. A set is **countable** if it is finite or if it has cardinality $\aleph_0$.

Cantor was not the first to ponder the paradoxes of infinite sets. Hundreds of years before, Galileo observed that part of an infinite set contained as many elements as the whole set. In his monumental *Dialogue Concerning the Two Chief World Systems* (1632), Galileo made a prophetic observation: "There are as many (perfect) squares as there are (natural) numbers because they are just as numerous as their roots." In other words, the elements of the sets $N = \{1, 2, 3, \ldots, n, \ldots\}$ and $S = \{1^2, 2^2, 3^2, \ldots, n^2, \ldots\}$ can be put into a one-to-one correspondence ($n \leftrightarrow n^2$). Galileo pondered which of the sets (perfect squares or natural numbers) was "larger" but abandoned the subject because he could find no practical application of this puzzle.

EXAMPLE 3 Consider the following one-to-one correspondence between the set I of all integers and the set N of all natural numbers:

$$\begin{array}{l} N = \{1, 2, \ 3, \ 4, \ 5, \ldots\} \\ \quad\ \ \updownarrow\ \updownarrow\ \ \updownarrow\ \ \updownarrow\ \ \updownarrow \\ I = \{0, 1, -1, 2, -2, \ldots\} \end{array}$$

where an odd natural number n corresponds to a nonpositive integer $\frac{1-n}{2}$ and an even natural number n corresponds to a positive integer $\frac{n}{2}$.

a. Find the 613th integer; that is, find the element of I that corresponds to $613 \in N$.
b. Find the element of N that corresponds to $853 \in I$.
c. Find the element of N that corresponds to $-397 \in I$.
d. Find $n(I)$, the cardinal number of the set I of all integers.
e. Is the set of integers countable?

Solution **a.** $613 \in N$ is odd, so it corresponds to $\frac{1-613}{2} = \frac{-612}{2} = -306$. If you continued counting the integers as shown in the above correspondence, -306 would be the 613th integer in your count.

b. $853 \in I$ is positive, so

$$853 = \frac{n}{2} \qquad \text{multiplying by 2}$$

$$n = 1{,}706$$

$1{,}706 \in N$ corresponds to $853 \in I$.
This means that 853 is the 1,706th integer.

c. $-397 \in I$ is negative, so

$$-397 = \frac{1-n}{2}$$

$$-794 = 1 - n \qquad \text{multiplying by 2}$$

$$-795 = -n \qquad \text{subtracting 1}$$

$$n = 795 \qquad \text{multiplying by } -1$$

$795 \in N$ corresponds to $-397 \in I$.
This means that -397 is the 795th integer.

d. The given one-to-one correspondence shows that I and N have the same (infinite) number of elements; $n(I) = n(N)$. Because $n(N) = \aleph_0$, the cardinal number of the set of all integers is $n(I) = \aleph_0$.

e. By definition, a set is called countable if it is finite or if it has cardinality $\aleph_0$. The set of integers has cardinality $\aleph_0$, so it is countable. This means that we can "count off" all of the integers, as we did in parts (a), (b), and (c). ■

We have seen that the sets N (all counting numbers), E (all even counting numbers), and I (all integers) contain the same number of elements, $\aleph_0$. What about a set containing fractions?

EXAMPLE 4 Determine whether the set P of all positive rational numbers is countable.

Solution The elements of P can be systematically listed in a table of rows and columns as follows: All positive rational numbers whose denominator is 1 are listed in the first row, all positive rational numbers whose denominator is 2 are listed in the second row, and so on, as shown in Figure 2.52.

$\frac{1}{1}$	$\frac{2}{1}$	$\frac{3}{1}$	$\frac{4}{1}\cdots$
$\frac{1}{2}$	$\frac{2}{2}$	$\frac{3}{2}$	$\frac{4}{2}\cdots$
$\frac{1}{3}$	$\frac{2}{3}$	$\frac{3}{3}$	$\frac{4}{3}\cdots$
$\vdots$	$\vdots$	$\vdots$	$\vdots$

Figure 2.52

Each positive rational number will appear somewhere in the table. For instance, $\frac{125}{66}$ will be in row 66 and column 125. Note that not all the entries in Figure 2.52 are in lowest terms; for instance, $\frac{2}{4}$, $\frac{3}{6}$, $\frac{4}{8}$, and so on are all equal to $\frac{1}{2}$.

Consequently, in order to avoid listing the same number more than once, an entry that is not in lowest terms must be eliminated from our list. To establish a one-to-one correspondence between P and N, we can create a zigzag diagonal pattern as shown by the arrows in Figure 2.53. Starting with $\frac{1}{1}$, we follow the arrows and omit any number that is not in lowest terms (the circled numbers in Figure 2.53). In this manner, a list of all positive rational numbers with no repetitions is created. Listing the elements of P in this order, we can put them in a one-to-one correspondence with N:

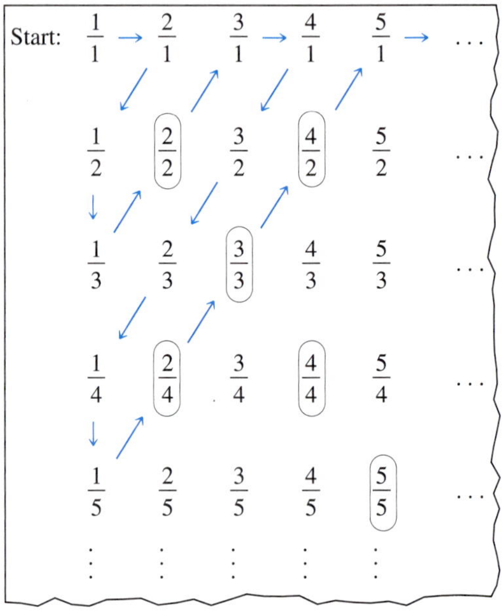

Figure 2.53
The circled rational numbers are not in lowest terms; they are omitted from the list.

$$N = \{1, 2, 3, 4, 5, 6, 7, 8, 9, 10, 11, \ldots\}$$
$$\updownarrow\ \updownarrow\ \updownarrow\ \updownarrow\ \updownarrow\ \updownarrow\ \updownarrow\ \updownarrow\ \updownarrow\ \ \updownarrow\ \ \updownarrow$$
$$P = \left\{1, 2, \frac{1}{2}, \frac{1}{3}, 3, 4, \frac{3}{2}, \frac{2}{3}, \frac{1}{4}, \frac{1}{5}, 5, \ldots\right\}$$

Any natural number n is paired up with the positive rational number found by counting through the "list" given in Figure 2.53. Conversely, any positive rational number is located somewhere in the list and is paired up with the counting number corresponding to its place in the list.

Therefore, $P \sim N$, so the set of all positive rational numbers is countable. ■

Uncountable Sets

Every infinite set we have examined so far is countable; each can be put into a one-to-one correspondence with the set of all counting numbers and consequently has cardinality $\aleph_0$. Do not be misled into thinking that all infinite sets are countable! By utilizing a "proof by contradiction," Georg Cantor showed that the infinite set $A = \{x \mid 0 \le x < 1\}$ is *not* countable. This proof involves logic that is different from what you are used to. Do not let that intimidate you.

Assume that the set $A = \{x \mid 0 \leq x < 1\}$ is countable; that is, assume $n(A) = \aleph_0$. This assumption implies that the elements of A and N can be put into a one-to-one correspondence; each $a \in A$ can be listed and counted. Because the elements of A are nonnegative real numbers less than 1, each $a_n = 0.\square\,\square\,\square\,\square\,\square\ldots$. Say, for instance, the numbers in our list are

$a_1 = 0.3750000\ldots$ the first element of A

$a_2 = 0.7071067\ldots$ the second element of A

$a_3 = 0.5000000\ldots$ the third element of A

$a_4 = 0.6666666\ldots$ and so on.

The *assumption* that A is countable implies that every element of A appears somewhere in the above list. However, we can create an element of A (call it b) that is *not* in the list. We build b according to the "diagonal digits" of the numbers in our list and the following rule: If the digit "on the diagonal" is not zero, put a 0 in the corresponding place in b; if the digit "on the diagonal" is zero, put a 1 in the corresponding place in b.

The "diagonal digits" of the numbers in our list are as follows:

$a_1 = 0.\boxed{3}750000\ldots$

$a_2 = 0.7\boxed{0}71067\ldots$

$a_3 = 0.50\boxed{0}0000\ldots$

$a_4 = 0.666\boxed{6}666\ldots$

Because the first digit on the diagonal is 3, the first digit of b is 0. Because the second digit on the diagonal is 0, the second digit of b is 1. Using all the "diagonal digits" of the numbers in our list, we obtain $b = 0.0110\ldots$. Because $0 \leq b < 1$, it follows that $b \in A$. However, the number b is not on our list of all elements of A. This is because

$b \neq a_1$ (b and a_1 differ in the first decimal place)

$b \neq a_2$ (b and a_2 differ in the second decimal place)

$b \neq a_3$ (b and a_3 differ in the third decimal place), and so on

This contradicts the assumption that the elements of A and N can be put into a one-to-one correspondence. Since the assumption leads to a contradiction, the assumption must be false; $A = \{x \mid 0 \leq x < 1\}$ is not countable. Therefore, $n(A) \neq n(N)$. That is, A is an infinite set and $n(A) \neq \aleph_0$.

An infinite set that cannot be put into a one-to-one correspondence with N is said to be **uncountable**. Consequently, an uncountable set has *more* elements than the set of all counting numbers. This implies that there are different magnitudes of infinity! To distinguish the magnitude of A from that of N, Cantor denoted the cardinality of $A = \{x \mid 0 \leq x < 1\}$ as $n(A) = c$ (c for **continuum**). Thus, Cantor showed that $\aleph_0 < c$. Cantor went on to show that A was equivalent to the entire set of all real numbers, that is $A \sim \Re$. Therefore, $n(\Re) = c$.

Although he could not prove it, Cantor hypothesized that no set could have a cardinality between $\aleph_0$ and c. This famous unsolved problem, labeled the *Continuum Hypothesis,* baffled mathematicians throughout the first half of the twentieth century. It is said that Cantor suffered a devastating nervous breakdown in 1884 when he announced that he had a proof of the Continuum Hypothesis only to declare the next day that he could show the Continuum Hypothesis to be false!

The problem was finally "solved" in 1963. Paul J. Cohen demonstrated that the Continuum Hypothesis is independent of the entire framework of set theory; that is, it can be neither proved nor disproved by using the theorems of set theory. Thus, the Continuum Hypothesis is not provable.

Although no one has produced a set with cardinality between $\aleph_0$ and c, many sets with cardinality greater than c have been constructed. In fact, modern mathematicians have shown that there are *infinitely* many magnitudes of infinity! Using subscripts, these magnitudes, or cardinalities, are represented by $\aleph_0, \aleph_1, \aleph_2, \ldots$ and have the property that $\aleph_0 < \aleph_1 < \aleph_2 < \ldots$. In this sense, the set N of all natural numbers forms the "smallest" infinite set. Using this subscripted notation, the Continuum Hypothesis implies that $c = \aleph_1$; that is, given that N forms the smallest infinite set, the set $\Re$ of all real numbers forms the next "larger" infinite set.

Points on a Line

When students are first exposed to the concept of the real number system, a number line like the one in Figure 2.54 is inevitably introduced. The real number system, denoted by $\Re$, can be put into a one-to-one correspondence with all points on a line, such that every real number corresponds to exactly one point on a line and every point on a line corresponds to exactly one real number. Consequently, any (infinite) line contains c points. What about a line segment? For example, how many points does the segment [0, 1] contain? Does the segment [0, 2] contain twice as many points as the segment [0, 1]? Once again, intuition can lead to erroneous conclusions when people are dealing with infinite sets.

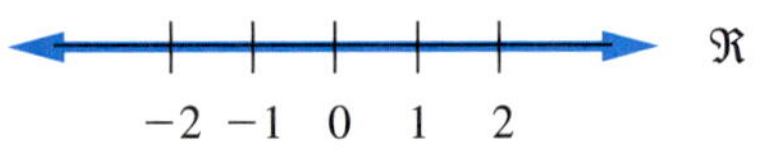

Figure 2.54

EXAMPLE 5 Show that the line segments [0, 1] and [0, 2] are equivalent sets of points.

Solution Because the segment [0, 2] is twice as long as the segment [0, 1], intuition might tell us that it contains twice as many points. Not so! Recall that two sets are equivalent (and have the same cardinal number) if their elements can be put into a one-to-one correspondence.

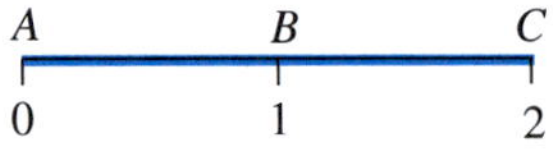

Figure 2.55

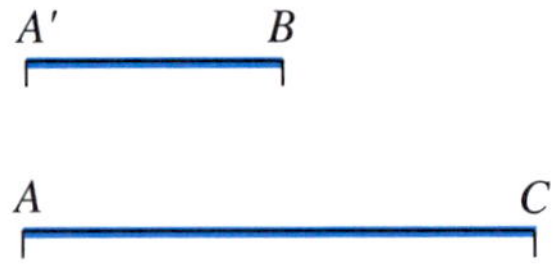

Figure 2.56

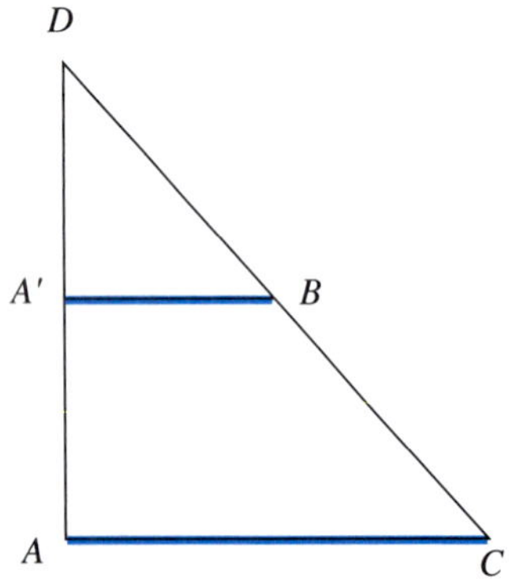

Figure 2.57

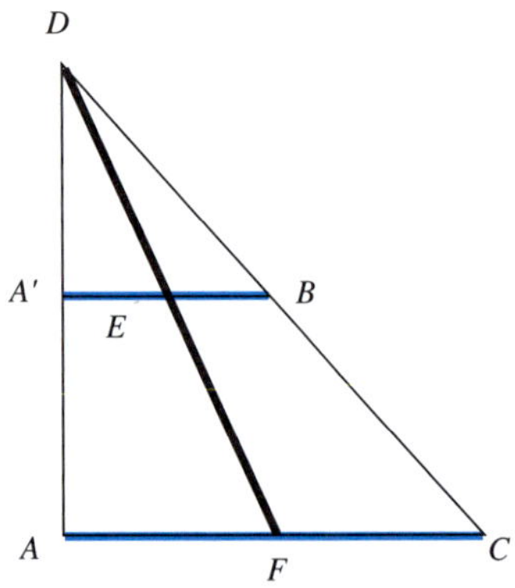

Figure 2.58

On a number line, let A represent the point 0, let B represent 1, and let C represent 2, as shown in Figure 2.55. Our goal is to develop a one-to-one correspondence between the elements of the segments AB and AC. Now draw the segments separately, with AB above AC, as shown in Figure 2.56. (To distinguish the segments from each other, point A of segment AB has been relabeled as point A'.)

Extend segments AA' and CB so that they meet at point D, as shown in Figure 2.57. Any point E on $A'B$ can be paired up with the unique point F on AC formed by the intersection of lines DE and AC, as shown in Figure 2.58. Conversely, any point F on segment AC can be paired up with the unique point E on $A'B$ formed by the intersection of lines DF and $A'B$. Therefore, a one-to-one correspondence exists between the two segments, so $[0, 1] \sim [0, 2]$. Consequently, the interval [0, 1] contains exactly the same number of points as the interval [0, 2]! ■

Even though the segment [0, 2] is twice as long as the segment [0, 1], each contains exactly the same number of points. The method used in Example 5 can be applied to any two line segments. Consequently, all line segments, regardless of their length, contain exactly the same number of points; a line segment 1 inch long has exactly the same number of points as a segment 1 mile long! Once again, it is easy to see why Cantor's work on the magnitude of infinity was so unsettling to many scholars.

Having concluded that all line segments contain the same number of points, we might ask how many points that is. What is the cardinal number? Given any line segment AB, it can be shown that $n(AB) = c$; the points of a line segment can be put into a one-to-one correspondence with the points of a line. Consequently, the interval [0, 1] contains the same number of elements as the entire real number system.

If things seem rather strange at this point, keep in mind that Cantor's pioneering work produced results that puzzled even Cantor himself. In a paper written in 1877, Cantor constructed a one-to-one correspondence between the points in a square (a two-dimensional figure) and the points on a line segment (a one-dimensional figure). Extending this concept, he concluded that a line segment and the entire two-dimensional plane contain exactly the same number of points, c. Communicating with his colleague Richard Dedekind, Cantor wrote, "I see it, but I do not believe it." Subsequent investigation has shown that the number of points contained in the interval [0, 1] is the same as the number of points contained in all of three-dimensional space! Needless to say, Cantor's work on the cardinality of infinity revolutionized the world of modern mathematics.

2.6 Exercises

In Exercises 1–10, find the cardinal numbers of the sets in each given pair to determine whether the sets are equivalent. If they are equivalent, list a one-to-one correspondence between their elements.

1. S = {Sacramento, Lansing, Richmond, Topeka}
   C = {California, Michigan, Virginia, Kansas}
2. T = {Wyoming, Ohio, Texas, Illinois, Colorado}
   P = {Cheyenne, Columbus, Austin, Springfield, Denver}
3. $R = \{a, b, c\}$; $G = \{\alpha, \beta, \chi, \delta\}$
4. W = {I, II, III}; H = {one, two}
5. $C = \{3, 6, 9, 12, \ldots, 63, 66\}$
   $D = \{4, 8, 12, 16, \ldots, 84, 88\}$
6. $A = \{2, 4, 6, 8, \ldots, 108, 110\}$
   $B = \{5, 10, 15, 20, \ldots, 270, 275\}$
7. $G = \{2, 4, 6, 8, \ldots, 498, 500\}$
   $H = \{1, 3, 5, 7, \ldots, 499, 501\}$
8. $E = \{2, 4, 6, 8, \ldots, 498, 500\}$
   $F = \{3, 6, 9, 12, \ldots, 750, 753\}$

9. $A = \{1, 3, 5, \ldots, 121, 123\}$
   $B = \{125, 127, 129, \ldots, 245, 247\}$

10. $S = \{4, 6, 8, \ldots, 664, 666\}$
    $T = \{5, 6, 7, \ldots, 335, 336\}$

11. **a.** Show that the set O of all odd counting numbers, $O = \{1, 3, 5, 7, \ldots\}$, and $N = \{1, 2, 3, 4, \ldots\}$ are equivalent sets.
    b. Find the element of N that corresponds to $1{,}835 \in O$.
    c. Find the element of N that corresponds to $x \in O$.
    d. Find the element of O that corresponds to $782 \in N$.
    e. Find the element of O that corresponds to $n \in N$.

12. **a.** Show that the set W of all whole numbers, $W = \{0, 1, 2, 3, \ldots\}$, and $N = \{1, 2, 3, 4, \ldots\}$ are equivalent sets.
    b. Find the element of N that corresponds to $932 \in W$.
    c. Find the element of N that corresponds to $x \in W$.
    d. Find the element of W that corresponds to $932 \in N$.
    e. Find the element of W that corresponds to $n \in N$.

13. **a.** Show that the set T of all multiples of 3, $T = \{3, 6, 9, 12, \ldots\}$, and $N = \{1, 2, 3, 4, \ldots\}$ are equivalent sets.
    b. Find the element of N that corresponds to $936 \in T$.
    c. Find the element of N that corresponds to $x \in T$.
    d. Find the element of T that corresponds to $936 \in N$.
    e. Find the element of T that corresponds to $n \in N$.

14. **a.** Show that the set F of all multiples of 5, $F = \{5, 10, 15, 20, \ldots\}$, and $N = \{1, 2, 3, 4, \ldots\}$ are equivalent sets.
    b. Find the element of N that corresponds to $605 \in F$.
    c. Find the element of N that corresponds to $x \in F$.
    d. Find the element of F that corresponds to $605 \in N$.
    e. Find the element of F that corresponds to $n \in N$.

15. Consider the following one-to-one correspondence between the set A of all even integers and the set N of all natural numbers:

$$\begin{array}{l} N = \{1, 2, 3, 4, 5, \ldots\} \\ \quad\;\; \updownarrow \; \updownarrow \; \updownarrow \; \updownarrow \; \updownarrow \\ A = \{0, 2, -2, 4, -4, \ldots\} \end{array}$$

where an odd natural number n corresponds to the nonpositive integer $1 - n$ and an even natural number n corresponds to the positive even integer n.
    a. Find the 345th even integer; that is, find the element of A that corresponds to $345 \in N$.
    b. Find the element of N that corresponds to $248 \in A$.
    c. Find the element of N that corresponds to $-754 \in A$.
    d. Find $n(A)$.

16. Consider the following one-to-one correspondence between the set B of all odd integers and the set N of all natural numbers:

$$\begin{array}{l} N = \{1, 2, 3, 4, 5, \ldots\} \\ \quad\;\; \updownarrow \; \updownarrow \; \updownarrow \; \updownarrow \; \updownarrow \\ B = \{1, -1, 3, -3, 5, \ldots\} \end{array}$$

where an even natural number n corresponds to the negative odd integer $1 - n$ and an odd natural number n corresponds to the odd integer n.
    a. Find the 345th odd integer; that is, find the element of B that corresponds to $345 \in N$.
    b. Find the element of N that corresponds to $241 \in B$.
    c. Find the element of N that corresponds to $-759 \in B$.
    d. Find $n(B)$.

In Exercises 17–22, show that the given sets of points are equivalent by establishing a one-to-one correspondence.

17. the line segments [0, 1] and [0, 3]
18. the line segments [1, 2] and [0, 3]
19. the circle and square shown in Figure 2.59

    HINT: Draw one figure inside the other.

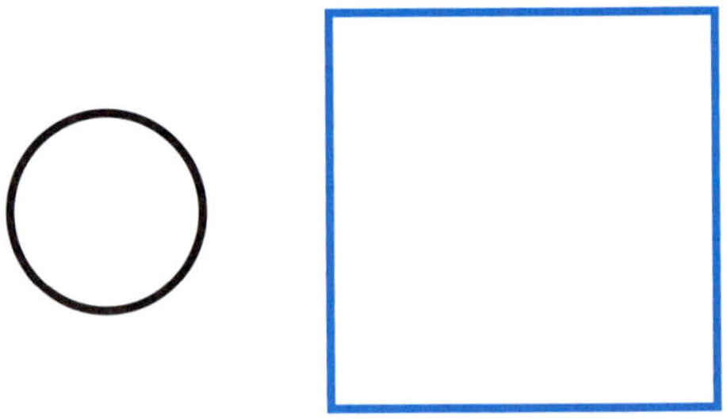

Figure 2.59

20. the rectangle and triangle shown in Figure 2.60

    HINT: Draw one figure inside the other.

Figure 2.60

21. a circle of radius 1 cm and a circle of 5 cm

HINT: Draw one figure inside the other.

22. a square of side 1 cm and a square of side 5 cm

HINT: Draw one figure inside the other.

23. Show that the set of all real numbers between 0 and 1 has the same cardinality as the set of all real numbers.

HINT: Draw a semicircle to represent the set of real numbers between 0 and 1 and a line to represent the set of all real numbers, and use the method of Example 5.

Answer the following questions using complete sentences and your own words.

Concept Questions

24. What is the cardinal number of the "smallest" infinite set? What set or sets have this cardinal number?

History Questions

25. What aspect of Georg Cantor's set theory caused controversy among mathematicians and philosophers?

26. What contributed to Cantor's breakdown in 1884?

27. Who demonstrated that the Continuum Hypothesis cannot be proven? When?

28. Write a research paper on a historical topic referred to in this section or on a related topic. Following is a list of possible topics:
- Bernhard Bolzano
- Georg Cantor
- Paul J. Cohen
- The Continuum Hypothesis
- Richard Dedekind
- Kabalah and the Hebrew alphabet
- Leopold Kronecker

Chapter 2 Review

TERMS

aleph-null
Borda count method
cardinal number
combination
combinatorics
complement
continuum
countable set
De Morgan's Laws
element
empty set
equal sets
equivalent sets
factorial
Fundamental Principle of Counting
improper subset
infinite set
instant run-off or ranked-choice method
intersection
majority
mutually exclusive
one-to-one correspondence
pairwise comparison method
permutation
plurality method
plurality with elimination method
proper subset
roster notation
set
set-builder notation
set theory
subset
tree diagram
uncountable set
union
universal set
Venn diagram
voter preference table
well-defined set

REVIEW EXERCISES

1. State whether the given set is well-defined.
a. the set of all multiples of 5.
b. the set of all difficult math problems
c. the set of all great movies
d. the set of all Oscar-winning movies

2. Given the sets
$U = \{0, 1, 2, 3, 4, 5, 6, 7, 8, 9\}$
$A = \{0, 2, 4, 6, 8\}$
$B = \{1, 3, 5, 7, 9\}$
find the following using the roster method.
a. A' **b.** B'
c. $A \cup B$ **d.** $A \cap B$

3. Given the sets $A = \{$Maria, Nobuko, Leroy, Mickey, Kelly$\}$ and $B = \{$Rachel, Leroy, Deanna, Mickey$\}$, find the following.
a. $A \cup B$ **b.** $A \cap B$

4. List all subsets of $C = \{$Dallas, Chicago, Tampa$\}$. Identify which subsets are proper and which are improper.

5. Given $n(U) = 61$, $n(A) = 32$, $n(B) = 26$, and $n(A \cup B) = 40$, do the following.
a. Find $n(A \cap B)$.
b. Draw a Venn diagram illustrating the composition of U.

6. A survey of 2,000 college seniors yielded the following information: 1,324 favored capital punishment, 937 favored stricter gun control, and 591 favored both.
   a. How many favored capital punishment or stricter gun control?
   b. How many favored capital punishment but not stricter gun control?
   c. How many favored stricter gun control but not capital punishment?
   d. How many favored neither capital punishment nor stricter gun control?

7. An exit poll yielded the following information concerning people's voting patterns on Propositions A, B, and C: 305 voted yes on A, 95 voted yes only on A, 393 voted yes on B, 192 voted yes only on B, 510 voted yes on A or B, 163 voted yes on C, 87 voted yes on all three, and 213 voted no on all three. What percent of the voters voted yes on more than one proposition?

8. Given the sets $U = \{a, b, c, d, e, f, g, h, i\}$, $A = \{b, d, f, g\}$, and $B = \{a, c, d, g, i\}$, use De Morgan's Laws to find the following.
   a. $(A' \cup B)'$ **b.** $(A \cap B')'$

9. Sid and Nancy are planning their anniversary celebration, which will include viewing an art exhibit, having dinner, and going dancing. They will go to either the Museum of Modern Art or the New Photo Gallery; dine either at Stars, at Johnny's, or at the Chelsea; and go dancing either at Le Club or at Lizards.
   a. How many different ways can Sid and Nancy celebrate their anniversary?
   b. Construct a tree diagram to list all possible ways in which Sid and Nancy can celebrate their anniversary.

10. A certain model of pickup truck is available in five exterior colors, three interior colors, and three interior styles. In addition, the transmission can be either manual or automatic, and the truck can have either two-wheel or four-wheel drive. How many different versions of the pickup truck can be ordered?

11. Each student at State University has a student I.D. number consisting of five digits (the first digit is nonzero, and digits can be repeated) followed by two of the letters *A, B, C,* and *D* (letters cannot be repeated). How many different student numbers are possible?

12. Find the value of each of the following.
   a. $(17 - 7)!$ **b.** $(17 - 17)!$
   c. $\frac{82!}{79!}$ **d.** $\frac{27!}{20!7!}$

13. In how many ways can you select three out of eleven items under the following conditions?
   a. Order of selection is not important.
   b. Order of selection is important.

14. Find the value of each of the following.
   a. ${}_{15}P_4$ **b.** ${}_{15}C_4$ **c.** ${}_{15}P_{11}$

15. A group of ten women and twelve men must select a three-person committee. How many committees are possible if it must consist of the following?
   a. one woman and two men
   b. any mixture of men and women
   c. a majority of men

16. A volleyball league has ten teams. If every team must play every other team once in the first round of league play, how many games must be scheduled?

17. A volleyball league has ten teams. How many different end-of-the-season rankings of first, second, and third place are possible (disregarding ties)?

18. Using a standard deck of 52 cards (no jokers), how many seven-card poker hands are possible?

19. Using a standard deck of 52 cards and two jokers, how many seven-card poker hands are possible?

20. A 6/42 lottery requires choosing six of the numbers 1 through 42. How many different lottery tickets can you choose?

21. What is the major difference between permutations and combinations?

22. Use Pascal's Triangle to answer the following.
   a. Which entry in which row would you find the value of ${}_7C_3$?
   b. Which entry in which row would you find the value of ${}_7C_4$?
   c. How is the value of ${}_7C_3$ related to the value of ${}_7C_4$? Why?
   d. What is the location of ${}_nC_r$? Why?

23. Given the set $S = \{a, b, c\}$, answer the following.
   a. How many one-element subsets does S have?
   b. How many two-element subsets does S have?
   c. How many three-element subsets does S have?
   d. How many zero-element subsets does S have?
   e. How many subsets does S have?
   f. How is the answer to part (e) related to $n(S)$?

24. The Metropolitan Symphonic Orchestra has volunteered to play a benefit concert with proceeds to support music programs at local high schools. The organizers need to know what music will be featured at the concert. The conductor has asked each member of the orchestra to rank the following composers: Beethoven (B), Mozart (M), Tchaikovsky (T), and Vivaldi (V). The ranked ballots have been

tallied, and the results are summarized in the following voter preference table.

	Number of Ballots Cast						
	5	7	6	12	13	17	14
1st choice	M	M	M	B	B	V	T
2nd choice	B	V	T	T	M	M	V
3rd choice	V	B	V	V	T	B	B
4th choice	T	T	B	M	V	T	M

a. How many votes were cast?
b. Use the plurality method of voting to determine the winner.
c. What percent of the votes did the winner in part (b) receive?
d. Use the instant run-off method to determine the winner.
e. What percent of the votes did the winner in part (d) receive?
f. Use the Borda count method to determine the winner.
g. How many points did the winner in part (f) receive?
h. Use the pairwise comparison method to determine the winner.
i. How many points did the winner in part (h) receive?

In Exercises 25–27, find the cardinal numbers of the sets in each given pair to determine whether the sets are equivalent. If they are equivalent, list a one-to-one correspondence between their elements.

25. $A = \{\text{I, II, III, IV, V}\}$ and $B = \{\text{one, two, three, four, five}\}$
26. $C = \{3, 5, 7, \ldots, 899, 901\}$ and $D = \{2, 4, 6, \ldots, 898, 900\}$
27. $E = \{\text{Ronald}\}$ and $F = \{\text{Reagan, McDonald}\}$
28. a. Show that the set S of perfect squares, $S = \{1, 4, 9, 16, \ldots\}$, and $N = \{1, 2, 3, 4, \ldots\}$ are equivalent sets.
b. Find the element of N that corresponds to $841 \in S$.
c. Find the element of N that corresponds to $x \in S$.
d. Find the element of S that corresponds to $144 \in N$.
e. Find the element of S that corresponds to $n \in N$.
29. Consider the following one-to-one correspondence between the set A of all integer multiples of 3 and the set N of all natural numbers:

$$\begin{array}{l} N = \{1, 2, 3, 4, 5, \ldots\} \\ \quad\;\; \updownarrow\; \updownarrow\; \updownarrow\; \updownarrow\; \updownarrow \\ A = \{0, 3, -3, 6, -6, \ldots\} \end{array}$$

where an odd natural number n corresponds to the nonpositive integer $\frac{3}{2}(1 - n)$, and an even natural number n corresponds to the positive even integer $\frac{3}{2}n$.
a. Find the element of A that corresponds to $396 \in N$.
b. Find the element of N that corresponds to $396 \in A$.
c. Find the element of N that corresponds to $-153 \in A$.
d. Find $n(A)$.
30. Show that the line segments $[0, 1]$ and $[0, \pi]$ are equivalent sets of points by establishing a one-to-one correspondence.

Answer the following questions using complete sentences and your own words.

Concept Questions

31. What is the difference between proper and improper subsets?
32. Explain the difference between $\{0\}$ and $\varnothing$.
33. What is a factorial?
34. What is the difference between permutations and combinations?
35. In voting, what is a majority?

History Questions

36. What roles did the following people play in the development of set theory and combinatorics?
- Georg Cantor
- Augustus De Morgan
- Christian Kramp
- Chu Shih-chieh
- John Venn

Probability

3

Probability theory began with a roll of the dice. The Chevalier de Méré, a seventeenth century French nobleman and generally successful gambler, had lost too much money with a certain dice bet, and he asked his friend—the French mathematician Blaise Pascal—to explain why. In answering his friend's question, Pascal laid the foundation of probability theory.

Probability theory has an interesting history filled with interesting people. Gerolamo Cardano was one of these creators of probability theory; he was also a compulsive gambler, and he may well have plagiarized much of his mathematics from one of his servants. He went on to become one of Europe's more successful physicians. Blaise Pascal was a very religious man and not a gambler. He was willing, however, to use mathematics to help his friend win. Later, Pascal gave up mathematics and the sciences and entered a convent.

Today, casinos use probability theory to set their house odds so that they are assured of making a profit. In this chapter, we will use probability theory to determine whether there are any good bets in casino games of chance.

For a long time, probability theory was viewed as a rather disreputable branch of mathematics because of its association with gambling. Gregor Mendel helped to change this view. He used probability theory to analyze the effect of randomness on heredity, thus founding the science of genetics. We will use probability theory to analyze the transmission of hair color—if you know the color of the parents' hair, you can predict the color of their child's hair. We will also analyze the transmission of inherited diseases such as sickle-cell anemia, cystic fibrosis, Huntington's disease, and Tay-Sachs disease. You should understand what probability theory predicts about such diseases because your child could acquire one of them even if you and your spouse are disease-free.

3.1 History of Probability

Probability theory is the field of mathematics that measures the likelihood of an event. It is generally considered to have originated in 1654 with the written correspondence between French mathematicians Blaise Pascal and Pierre de Fermat about several problems involving dice games. The correspondence was in response to a problem posed by Pascal's friend Antoine Gombauld, the Chevalier de Méré, a French nobleman and rather successful gambler. Gombauld had made money over the years betting that he could roll at least one six in four rolls of a single die, but he had lost money betting that he could roll at least one pair of sixes in twenty-four rolls of a pair of dice. He challenged his friend Pascal to explain why. In the ensuing communications, the two mathematicians answered Gombauld's query and laid the foundation of probability theory.

The letters between Pascal and Fermat, however, were not the first entry in the field of probability. In 1545, the Italian physician, mathematician, and gambler Gerolamo Cardano wrote the first theoretical study of probabilities and gambling, *The Book on Games of Chance.* Cardano's ideas on probabilities weren't widely recognized until after his death. Interest aroused by Pascal and Fermat's work prompted the first publication of Cardano's century-old manuscript. However, probability theory was not viewed as a serious branch of mathematics because of its involvement with gambling.

This **Bills of Mortality** (records of death) was published shortly after John Graunt's analysis.

In 1662, the Englishman John Graunt published his *Natural and Political Observations on the Bills of Mortality,* which used probabilities to analyze death records and was the first important incidence of the use of probabilities for a purpose other than gambling. English firms were selling the first life insurance policies, and they used mortality tables to make their fees appropriate to the risks involved.

Probability theory was shown to be a serious area of interest when Jacob Bernoulli's *Ars Conjectandi* (*Art of Guessing*) was published in 1713. Although its focus was on gambling, it also suggested applications of probability to government, economics, law, and genetics.

In 1812, Pierre-Simon, the Marquis de Laplace, published his *Théorie Analytique des Probabilités.* This work, written in conjunction with his astronomical studies, showed probability to be a respectable and significant field of mathematics by demonstrating its usefulness in interpreting scientific data. Laplace showed how probabilities can be used to determine the most probable true size of a quantity when repeated measurements of that quantity vary somewhat.

The work of Gregor Mendel, an Austrian monk, was especially important in the history of probability theory. In 1865, he published the results of his experiments on pea plants in the abbey garden. These results allowed for randomness in genetics and used probabilities to analyze the effect of that randomness. Ignored during his lifetime, Mendel's work is considered the foundation of genetics.

Probability theory was originally created to aid gamblers, and games of chance are the most easily and universally understood topic to which probability theory can be applied. Learning about probability and games of chance may convince you not to be a gambler. As we will see in Section 3.5, there is not a single good bet in a casino game of chance.

Roulette

Roulette is the oldest casino game still being played. Its invention has variously been credited to Pascal, the ancient Chinese, a French monk, and the Italian mathematician

© Stock Montage

Roulette quickly became a favorite game of the French upper class, as shown in this 1890 photo of Monte Carlo.

Don Pasquale. It became popular when a French policeman introduced it to Paris in 1765 in an attempt to take the advantage away from dishonest gamblers. When the Monte Carlo casino opened in 1863, roulette was the most popular game, especially among the aristocracy. The American roulette wheel has 38 numbered compartments around its circumference. Thirty-six of these compartments are numbered from 1 to 36 and are colored red or black. The remaining two are numbered 0 and 00 and are colored green. Players place their bets by putting their chips on an appropriate spot on the roulette table (see Figure 3.1). The dealer spins the wheel and then drops a ball onto the spinning wheel. The ball eventually comes to rest in one of the compartments, and that compartment's number is the winning number. (See page 168 for a photo of the roulette wheel and table.)

For example, if a player wanted to bet \$10 that the ball lands in compartment number 7, she would place \$10 worth of chips on the number 7 on the roulette table. This is a single-number bet, so house odds are 35 to 1 (see Figure 3.1 for house odds). This means that if the player wins, she wins $\$10 \cdot 35 = \$350$, and if she loses, she loses her \$10.

Similarly, if a player wanted to bet \$5 that the ball lands either on 13 or 14, he would place \$5 worth of chips on the line separating numbers 13 and 14 on the roulette table. This is a two-numbers bet, so house odds are 17 to 1. This means that if the player wins, he wins $\$5 \cdot 17 = \$85$, and if he loses, he loses his \$5.

Dice and Craps

Dice have been cast since the beginning of time, for both divination and gambling purposes. The earliest **die** (singular of *dice*) was an animal bone, usually a knucklebone or foot bone. The Romans were avid dice players. The Roman emperor

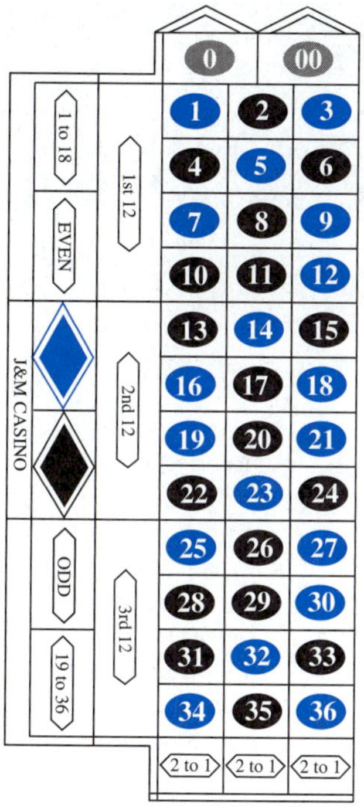

Bet	House Odds
single number	35 to 1
two numbers (“split”)	17 to 1
three numbers (“street”)	11 to 1
four numbers (“square”)	8 to 1
five numbers (“line”)	6 to 1
six numbers (“line”)	5 to 1
twelve numbers (column or section)	2 to 1
low or high (1 to 18 or 19 to 36, respectively)	1 to 1
even or odd (0 and 00 are neither even nor odd)	1 to 1
red or black	1 to 1

Figure 3.1
The roulette table and house odds for the various roulette bets. (Note: The numbers colored blue on the table shown above would be red on a real roulette table.)

Claudius I wrote a book titled *How to Win at Dice.* During the Middle Ages, dicing schools and guilds of dicers were quite popular among the knights and ladies.

Hazard, an ancestor of the dice game craps, is an English game that was supposedly invented by the Crusaders in an attempt to ward off boredom during long, drawn-out sieges. It became quite popular in England and France in the nineteenth century. The English called a throw of 2, 3, or 12 *crabs,* and it is believed that *craps* is a French mispronunciation of that term. The game came to America with the French colonization of New Orleans and spread up the Mississippi.

Brooks/Cole

A sixteenth-century two of swords.

Cards

The invention of playing cards has been credited to the Indians, the Arabs, the Egyptians, and the Chinese. During the Crusades, Arabs endured lengthy sieges by playing card games. Their European foes acquired the cards and introduced them to their homelands. Cards, like dice, were used for divination as well as gambling. In fact, the modern deck is derived from the Tarot deck, which is composed of four suits plus 22 *atouts* that are not part of any suit. Each suit represents a class of medieval society: swords represent the nobility; coins, the merchants; batons or clubs, the peasants; and cups or chalices, the church. These suits are still used in regular playing cards in southern Europe. The Tarot deck also includes a Joker and, in each suit, a king, a queen, a knight, a knave, and ten numbered cards.

Around 1500, the French dropped the knights and *atouts* from the deck and changed the suits from swords, coins, clubs, and chalices to *piques* (soldiers’

HISTORICAL NOTE

Blaise Pascal

1623–1662

© Bettman/Corbis

As a child, Blaise Pascal showed an early aptitude for science and mathematics, even though he was discouraged from studying in order to protect his poor health. Acute digestive problems and chronic insomnia made his life miserable. Few of Pascal's days were without pain.

At age sixteen, Pascal wrote a paper on geometry that won the respect of the French mathematical community and the jealousy of the prominent French mathematician René Descartes. It has been suggested that the animosity between Descartes and Pascal was in part due to their religious differences. Descartes was a Jesuit, and Pascal was a Jansenist. Although Jesuits and Jansenists were both Roman Catholics, Jesuits believed in free will and supported the sciences, while Jansenists believed in predestination and mysticism and opposed the sciences and the Jesuits.

At age nineteen, to assist his father, a tax administrator, Pascal invented the first calculating machine. Besides cofounding probability theory with Pierre de Fermat, he contributed to the advance of calculus, and his studies in physics culminated with a law on the effects of pressure on fluids that bears his name.

At age thirty-one, after a close escape from death in a carriage accident, Pascal turned his back on mathematics and the sciences and entered a Jansenist convent. Pascal came to be the greatest Jansenist, and he aroused a storm with his anti-Jesuit *Provincial Letters.* This work is still famous for its polite irony.

Pascal turned so far from the sciences that he came to believe that reason is inadequate to solve humanity's difficulties or to satisfy our hopes, and he regarded the pursuit of science as a vanity. Even so, he still occasionally succumbed to its lure. Once, to distract himself from pain, he concentrated on a geometry problem. When the pain stopped, he decided that it was a signal from God that he had not sinned by thinking of mathematics instead of his soul. This incident resulted in his only scientific work since entering the convent—and his last work. He died later that year, at age thirty-nine.

Courtesy of IBM

Pascal's calculator, the Pascaline.

pikes), *carreaux* (diamond-shaped building tiles), *trèfles* (clover leaf-shaped trefoils), and *coeurs* (hearts). In sixteenth-century Spain, *piques* were called *espados,* from which we get our term *spades.* Our diamonds are so named because of the shape of the carreaux. Clubs was an original Tarot suit, and hearts is a translation of *coeurs.*

North Wind Pictures Archives

HISTORICAL NOTE

Gerolamo Cardano

1501–1576

Gerolamo Cardano is the subject of much disagreement among historians. Some see him as a man of tremendous accomplishments, while others see him as a plagiarist and a liar. All agree that he was a compulsive gambler.

Cardano was trained as a medical doctor, but he was initially denied admission to the College of Physicians of Milan. That denial was ostensibly due to his illegitimate birth, but some suggest that the denial was in fact due to his unsavory reputation as a gambler, since illegitimacy was neither a professional nor a social obstacle in sixteenth-century Italy. His lack of professional success left him with much free time, which he spent gambling and reading. It also resulted in a stay in the poorhouse.

Cardano's luck changed when he obtained a lectureship in mathematics, astronomy, and astrology at the University of Milan. He wrote a number of books on mathematics and became famous for publishing a method of solving third-degree equations. Some claim that Cardano's mathematical success was due not to his own abilities but rather to those of Ludovico Ferrari, a servant of his who went on to become a mathematics professor.

While continuing to teach mathematics, Cardano returned to the practice of medicine. (He was finally allowed to join the College of Physicians, perhaps owing to his success as a mathematician.) Cardano wrote books on medicine and the natural sciences that were well thought of. He became one of the most highly regarded physicians in Europe and counted many prominent people among his patients. He designed a tactile system, somewhat like braille, for the blind. He also designed an undercarriage suspension device that was later adapted as a universal joint for automobiles and that is still called a *cardan* in Europe.

Cardano's investment in gambling was enormous. He not only wagered (and lost) a great deal of money but also spent considerable time and effort calculating probabilities and devising strategies. His *Book on Games of Chance* contains the first correctly calculated theoretical probabilities.

Cardano's autobiography, *The Book of My Life,* reveals a unique personality. He admitted that he loved talking about himself, his accomplishments, and his illnesses and diseases. He frequently wrote of injuries done him by others and followed these complaints with gleeful accounts of his detractors' deaths. Chapter titles include "Concerning my friends and patrons," "Calumny, defamations, and treachery of my unjust accusers," "Gambling and dicing," "Religion and piety," "The disasters of my sons," "Successes in my practice," "Things absolutely supernatural," and "Things of worth which I have achieved in various studies."

The pictures on the cards were portraits of actual people. In fourteenth-century Europe, the kings were Charlemagne (hearts), the biblical David (spades), Julius Caesar (diamonds), and Alexander the Great (clubs); the queens included Helen of Troy (hearts), Pallas Athena (spades), and the biblical Rachel (diamonds). Others honored as "queen for a day" included Joan of Arc, Elizabeth I, and Elizabeth of York, wife of Henry VII. Jacks were usually famous warriors, including Sir Lancelot (clubs) and Roland, Charlemagne's nephew (diamonds).

A modern deck of cards contains 52 cards (thirteen in each of four suits). The four suits are hearts, diamonds, clubs, and spades (♥, ♦, ♣, ♠). Hearts and diamonds are red, and clubs and spades are black. Each suit consists of cards labeled 2 through 10, followed by jack, queen, king, and ace. **Face cards** are the jack, queen, and king; and **picture cards** are the jack, queen, king, and ace.

Brooks/Cole

A sixteenth-century king of clubs.

Kathleen Olson

A modern deck of cards.

Two of the most popular card games are poker and blackjack. Poker's ancestor was a Persian game called *dsands,* which became popular in eighteenth-century Paris. It was transformed into a game called *poque,* which spread to America via the French colony in New Orleans. *Poker* is an American mispronunciation of the word *poque.*

The origins of blackjack (also known as vingt-et-un or twenty-one) are unknown. The game is called twenty-one because high odds are paid if a player's first two cards total 21 points (an ace counts 11, and the ten, jack, queen, and king each count 10). A special bonus used to be paid if those two cards were a black jack, and a black ace, hence the name *blackjack.*

3.1 Exercises

1. Roll a single die four times, and record the number of times a six comes up. Repeat this ten times. If you had made the Chevalier de Méré's favorite bet (at \$10 per game), would you have won or lost money? How much?

2. Roll a pair of dice 24 times, and record the number of times a pair of sixes comes up. Repeat this five times. If you had made the Chevalier de Méré's bet (at \$10 per game), would you have won or lost money? How much?

3. **a.** If you were to flip a pair of coins 30 times, approximately how many times do you think a pair of heads would come up? A pair of tails? One head and one tail?

b. Flip a pair of coins 30 times, and record the number of times a pair of heads comes up, the number of times a pair of tails comes up, and the number of times one head and one tail come up. How closely do the results agree with your guess?

In Exercises 4–18, use Figure 3.1 to find the outcome of the bets in roulette, given the results listed.

4. You bet \$10 on the 25.
- **a.** The ball lands on number 25.
- **b.** The ball lands on number 14.

5. You bet \$5 on 17-20 split.
- **a.** The ball lands on number 17.
- **b.** The ball lands on number 20.
- **c.** The ball lands on number 32.

6. You bet \$30 on the 22-23-24 street.
- **a.** The ball lands on number 19.
- **b.** The ball lands on number 22.
- **c.** The ball lands on number 0.

7. You bet \$20 on the 8-9-11-12 square.
- **a.** The ball lands on number 15.
- **b.** The ball lands on number 9.
- **c.** The ball lands on number 00.

8. You bet \$100 on the 0-00-1-2-3 line (the only five-number line on the table).
- **a.** The ball lands on number 29.
- **b.** The ball lands on number 2.

9. You bet \$10 on the 31-32-33-34-35-36 line.
- **a.** The ball lands on number 5.
- **b.** The ball lands on number 33.

10. You bet \$20 on the 13 through 24 section.
- **a.** The ball lands on number 00.
- **b.** The ball lands on number 15.

11. You bet \$25 on the first column.
- **a.** The ball lands on number 13.
- **b.** The ball lands on number 14.

12. You bet $30 on the low numbers.
   a. The ball lands on number 8.
   b. The ball lands on number 30.
13. You bet $50 on the odd numbers.
   a. The ball lands on number 00.
   b. The ball lands on number 5.
14. You bet $20 on the black numbers.
   a. The ball lands on number 11.
   b. The ball lands on number 12.
15. You make a $20 single-number bet on number 14, and also a $25 single-number bet on number 15.
   a. The ball lands on number 16.
   b. The ball lands on number 15.
   c. The ball lands on number 14.
16. You bet $10 on the low numbers, and also bet $20 on the 16-17-19-20 square.
   a. The ball lands on number 16.
   b. The ball lands on number 19.
   c. The ball lands on number 14.
17. You bet $30 on the 1-2 split, and also bet $15 on the even numbers.
   a. The ball lands on number 1.
   b. The ball lands on number 2.
   c. The ball lands on number 3.
   d. The ball lands on number 4.
18. You bet $40 on the 1-12 section, and also bet $10 on number 10.
   a. The ball lands on number 7.
   b. The ball lands on number 10.
   c. The ball lands on number 21.
19. How much must you bet on a single number to be able to win at least $100? (Bets must be in $1 increments.)
20. How much must you bet on a two-number split to be able to win at least $200? (Bets must be in $1 increments.)
21. How much must you bet on a twelve-number column to be able to win at least $1000? (Bets must be in $1 increments.)
22. How much must you bet on a four-number square to win at least $600? (Bets must be in $1 increments.)
23. a. How many hearts are there in a deck of cards?
   b. What fraction of a deck is hearts?
24. a. How many red cards are there in a deck of cards?
   b. What fraction of a deck is red?
25. a. How many face cards are there in a deck of cards?
   b. What fraction of a deck is face cards?
26. a. How many black cards are there in a deck of cards?
   b. What fraction of a deck is black?
27. a. How many kings are there in a deck of cards?
   b. What fraction of a deck is kings?

Answer the following questions using complete sentences and your own words.

History Questions

28. Who started probability theory? How?
29. Why was probability theory not considered a serious branch of mathematics?
30. Which authors established probability theory as a serious area of interest? What are some of the areas to which these authors applied probability theory?
31. What did Gregor Mendel do with probabilities?
32. Who was Antoine Gombauld, and what was his role in probability theory?
33. Who was Gerolamo Cardano, and what was his role in probability theory?
34. Which games of chance came to America from France via New Orleans?
35. Which implements of gambling were also used for divination?
36. What is the oldest casino game still being played?
37. How were cards introduced to Europe?
38. What is the modern deck of cards derived from?
39. What game was supposed to take the advantage away from dishonest gamblers?

Projects

40. Write a research paper on any historical topic referred to in this section or a related topic. Below is a partial list of topics.
   - the Bernoulli family (the Bernoullis were to science as the Bachs were to music)
   - Gerolamo Cardano (also called Jerome Cardan)
   - Pierre de Fermat (a lawyer whose hobby was mathematics)
   - Antoine Gombauld, the Chevalier de Méré
   - John Graunt and/or mortality tables
   - Gregor Mendel
   - Blaise Pascal
   - Don Pasquale (possibly the inventor of the roulette wheel)
   - Pierre-Simon, the Marquis de Laplace

3.2 Basic Terms of Probability

Much of the terminology and many of the computations of probability theory have their basis in set theory, because set theory contains the mathematical way of describing collections of objects and the size of those collections.

> **Basic Probability Terms**
>
> **experiment:** a process by which an observation, or **outcome,** is obtained
>
> **sample space:** the set S of all possible outcomes of an experiment
>
> **event:** any subset E of the sample space S

If a single die is rolled, the *experiment* is the rolling of the die. The possible *outcomes* are 1, 2, 3, 4, 5, and 6. The *sample space* (set of all possible outcomes) is $S = \{1, 2, 3, 4, 5, 6\}$. (The term *sample space* really means the same thing as *universal set*; the only distinction between the two ideas is that *sample space* is used only in probability theory, while *universal set* is used in any situation in which sets

Early shell games

An early certain event

are used.) There are several possible *events* (subsets of the sample space), including the following:

$E_1 = \{3\}$ "a three comes up"

$E_2 = \{2, 4, 6\}$ "an even number comes up"

$E_3 = \{1, 2, 3, 4, 5, 6\}$ "a number between 1 and 6 inclusive comes up"

Notice that an event is not the same as an outcome. An event is a subset of the sample space; an outcome is an element of the sample space. "Rolling an odd number" is an event, not an outcome. It is the set $\{1, 3, 5\}$ that is composed of three separate outcomes. Some events are distinguished from outcomes only in that set brackets are used with events and not with outcomes. For example, $\{5\}$ is an event, and 5 is an outcome; either refers to "rolling a five."

The event E_3 ("a number between 1 and 6 comes up") is called a **certain event**, since $E_3 = S$. That is, E_3 is a sure thing. "Getting 17" is an **impossible event**. No outcome in the sample space $S = \{1, 2, 3, 4, 5, 6\}$ would result in 17, so this event is the null set.

Finding Probabilities and Odds

The **probability** of an event is a measure of the likelihood that the event will occur. If a single die is rolled, the outcomes are equally likely; a three is just as likely to come up as any other number. There are six possible outcomes, so a three should come up about one out of every six rolls. That is, the probability of event E_1 ("a three comes up") is $\frac{1}{6}$. The 1 in the numerator is the number of elements in $E_1 = \{3\}$. The 6 in the denominator is the number of elements in $S = \{1, 2, 3, 4, 5, 6\}$.

If an experiment's outcomes are equally likely, then the probability of an event E is the number of outcomes in the event divided by the number of outcomes in the sample space, or $n(E)/n(S)$. (In this chapter, we discuss only experiments with equally likely outcomes.) Probability can be thought of as "success over a total."

Probability of an Event

The **probability** of an event E, denoted by $p(E)$, is

$$p(E) = \frac{n(E)}{n(S)}$$

if the experiment's outcomes are equally likely.
(*Think: success over total.*)

Many people use the words *probability* and *odds* interchangeably. However, the words have different meanings. The **odds** in favor of an event are the number of ways the event can occur compared to the number of ways the event *can fail to occur,* or "success compared to *failure*" (if the experiment's outcomes are equally likely). The odds of event E_1 ("a three comes up") are 1 to 5 (or 1:5), since a three can come up in one way and can fail to come up in five ways. Similarly, the odds of event E_3 ("a number between 1 and 6 inclusive comes up") are 6 to 0 (or 6:0), since a number between 1 and 6 inclusive can come up in six ways and can fail to come up in zero ways.

Odds of an Event

The **odds** of an event E with equally likely outcomes, denoted by $o(E)$, are given by

$$o(E) = n(E):n(E')$$

(Think: success compared with failure.)

In addition to the above meaning, the word *odds* can also refer to "house odds," which has to do with how much you will be paid if you win a bet at a casino. The odds of an event are sometimes called the **true odds** to distinguish them from the house odds.

EXAMPLE 1 A die is rolled. Find the following.

a. the probability of rolling a five
b. the odds of rolling a five
c. the probability of rolling a number below 5
d. the odds of rolling a number below 5

Solution **a.** The sample space is $S = \{1, 2, 3, 4, 5, 6\}$. $E_1 = \{5\}$ ("rolling a five"). The probability of E_1 is

$$p(E_1) = \frac{n(E_1)}{n(S)} = \frac{1}{6}$$

This means that one out of every six possible outcomes is a success (that is, a five).

b. $E_1' = \{1, 2, 3, 4, 6\}$. The odds of E_1 are

$$o(E_1) = n(E_1):n(E_1') = 1:5$$

This means that there is one possible success for every five possible failures.

c. $E_2 = \{1, 2, 3, 4\}$ ("rolling a number below 5"). The probability of E_2 is

$$p(E_2) = \frac{n(E_2)}{n(S)} = \frac{4}{6} = \frac{2}{3}$$

This means that two out of every three possible outcomes are a success.

d. $E_2' = \{5, 6\}$. The odds of E_2 are

$$o(E_2) = n(E_2):n(E_2') = 4:2 = 2:1$$

This means that there are two possible successes for every one possible failure. Notice that odds are reduced in the same manner that a fraction is reduced. ■

EXAMPLE 2 A coin is flipped. Find the following.

a. the sample space
b. the probability of event E_1, "getting heads"
c. the odds of event E_1, "getting heads"
d. the probability of event E_2, "getting heads or tails"
e. the odds of event E_2, "getting heads or tails"

Solution **a.** *Finding the sample space S*: The experiment is flipping a coin. The only possible outcomes are heads and tails. The sample space S is the set of all possible outcomes, so $S = \{h, t\}$.

b. *Finding the probability of heads*:

$$E_1 = \{h\} \text{ ("getting heads")}$$
$$p(E_1) = \frac{n(E_1)}{n(S)} = \frac{1}{2}$$

This means that one out of every two possible outcomes is a success.

c. *Finding the odds of heads*:

$$E_1' = \{t\}$$
$$o(E_1) = n(E_1) : n(E_1') = 1:1$$

This means that for every one possible success, there is one possible failure.

d. *Finding the probability of heads or tails*:

$$E_2 = \{h, t\}$$
$$p(E_2) = \frac{n(E_2)}{n(S)} = \frac{2}{2} = \frac{1}{1}$$

This means that every outcome is a success. Notice that E_2 is a certain event.

e. *Finding the odds of heads or tails*:

$$E_2' = \varnothing$$
$$o(E_2) = n(E_2) : n(E_2') = 2:0 = 1:0$$

This means that there are no possible failures. ■

Relative Frequency versus Probability

So far, we have discussed probabilities only in a theoretical way. When we found that the probability of heads was $\frac{1}{2}$, we never actually tossed a coin. It doesn't always make sense to calculate probabilities theoretically; sometimes they must be found empirically, the way a batting average is calculated. For example, in 8,389 times at bat, Babe Ruth had 2,875 hits. His batting average was $\frac{2,875}{8,389} \approx 0.343$. In other words, his probability of getting a hit was 0.343.

Sometimes a probability can be found either theoretically or empirically. We've already found that the theoretical probability of heads is $\frac{1}{2}$. We could also flip a coin a number of times and calculate (number of heads)/(total number of flips); this can be called the **relative frequency** of heads, to distinguish it from the theoretical probability of heads.

Usually, the relative frequency of an outcome is not equal to its probability, but if the number of trials is large, the two tend to be close. If you tossed a coin a couple of times, anything could happen, and the fact that the probability of heads is $\frac{1}{2}$ would have no impact on the results. However, if you tossed a coin 100 times, you would probably find that the relative frequency of heads was close to $\frac{1}{2}$. If your friend tossed a coin 1,000 times, she would probably find the relative frequency of heads to be even closer to $\frac{1}{2}$ than in your experiment. This relationship between probabilities and relative frequencies is called the **Law of Large Numbers**.

Law of Large Numbers

If an experiment is repeated a large number of times, the relative frequency of an outcome will tend to be close to the probability of that outcome.

Probabilities represent the tendency of reality, or the best guess of what will happen. At a casino, probabilities are much more useful to the casino (the "house") than to any individual gambler, because the house performs the experiment for a much larger number of trials (in other words, plays the game more often). In fact, the house plays the game so many times that the relative frequencies will be almost exactly the same as the probabilities, with the result that the house isn't gambling at all—it knows what is going to happen! Similarly, a gambler with a "system" has to play the game for a long time in order for the system to be of use.

EXAMPLE 3 If a pair of coins is flipped, find the probability of getting exactly one head.

Solution The experiment is the flipping of a pair of coins. One possible outcome is that one coin is heads and the other is tails. A second and *different* outcome is that one coin is tails and the other is heads. These two outcomes seem the same. However, if one coin was painted, it would be easy to tell them apart. Outcomes of the experiment can be described by using ordered pairs in which the first component refers to the first coin and the second component refers to the second coin. The two different ways of getting one coin heads and one coin tails are (h, t) and (t, h).

The sample space is $S = \{(h, h), (h, t), (t, h), (t, t)\}$.

The event of getting exactly one head is $E = \{(h, t), (t, h)\}$. Therefore,

$$p(E) = \frac{n(E)}{n(S)} = \frac{2}{4} = \frac{1}{2}$$

This means that if we were to toss a pair of coins a number of times, we should expect to get heads on one coin and tails on the other about half of the time. Realize that this is only a prediction, and we may in fact never get heads on one coin and tails on the other. ■

Mendel's Use of Probabilities

In his studies of flower color, Mendel pollinated peas until he produced pure-red plants (that is, plants that would produce only red-flowered offspring) and pure-white plants. He then cross-fertilized this first generation of pure-reds and pure-whites and obtained a second generation that had only red flowers. (The accepted theory of the day incorrectly predicted that he would have only pink flowers.) Finally, he cross-fertilized this second generation and obtained a third generation, approximately three-fourths of which had red flowers and one-fourth of which had white.

Mendel explained these results by postulating that there is a "determiner" responsible for flower color. These determiners are now called **genes**. Each plant has two flower color genes, one from each parent. If a plant inherits a red gene from each parent, the plant has red flowers. The plant also has red flowers if it inherits a red gene from one parent and a white gene from the other; that is, the red-flowered gene is **dominant** and the white-flowered gene is **recessive**. Which of its two genes a parent passes on to its offspring is strictly a matter of chance; Mendel successfully used probabilities to analyze this random aspect of heredity.

If we use R to stand for the red gene and w to stand for the white gene (with the capital letter indicating dominance), then the results of cross-fertilization can be described pictorially with a **Punnett square**. Figure 3.2 shows the Punnett square for the crossing of the first generation of pure-reds and pure-whites. The four possible outcomes are the same: one red gene and one white gene, resulting in a red-flowered plant.

	R	R	
w	(R, w)	(R, w)	← first parent's genes / ← offspring
w	(R, w)	(R, w)	← offspring

↑ second parent's genes

Figure 3.2

When the offspring of this experiment were cross-fertilized, Mendel found that approximately three-fourths of the offspring had red flowers and one-fourth had white. Mendel showed how to use probabilities to predict this result. Figure 3.3 shows the Punnett square for this second cross-fertilization. The sample space S consists of four outcomes:

$$S = \{(\text{R, R}), (\text{R, w}), (\text{w, R}), (\text{w, w})\}$$

	R	w	← first parent's genes
R	(R, R)	(R, w)	← offspring
w	(w, R)	(w, w)	← offspring

↑ second parent's genes

Figure 3.3

Only one of the four possible outcomes, (w, w), results in a white-flowered plant, so event E_1 that the plant has white flowers is $E_1 = \{(\text{w, w})\}$; therefore,

$$p(E_1) = \frac{n(E_1)}{n(S)} = \frac{1}{4}$$

This means that we should expect the actual relative frequency of white-flowered plants to be close to $\frac{1}{4}$.

Each of the other three outcomes, (R, R), (R, w), and (w, R), results in a red-flowered plant, because red dominates white. The event E_2 that the plant has red flowers is $E_2 = \{(\text{R, R}), (\text{R, w}), (\text{w, R})\}$; therefore,

$$p(E_2) = \frac{n(E_2)}{n(S)} = \frac{3}{4}$$

Thus, we should expect the actual relative frequency of red-flowered plants to be close to $\frac{3}{4}$.

Outcomes (R, w) and (w, R) are genetically identical; it doesn't matter which gene is inherited from which parent. For this reason, geneticists do not use the ordered-pair notation and instead refer to each of these two outcomes as "Rw." The only difficulty with this convention is that it makes the sample space appear to be S = {RR, Rw, ww}, which consists of only three elements, when in fact it consists of four elements. This distinction is important; if the sample space consisted of three equally likely elements, then the probability of a red-flowered offspring would be $\frac{2}{3}$ rather than $\frac{3}{4}$. Mendel knew that the sample space had to have four elements, because his cross-fertilization experiments resulted in a relative frequency very close to $\frac{3}{4}$, not $\frac{2}{3}$.

Ronald Fisher, a noted British statistician, used statistics to deduce that Mendel fudged his data. Mendel's relative frequencies were unusually close to the theoretical

HISTORICAL NOTE

Gregor Johann Mendel
1822–1884

Universal Press Syndicate

Johann Mendel was born to an Austrian peasant family. His interest in botany began on the family farm, where he helped his father graft fruit trees. He studied philosophy, physics, and mathematics at the University Philosophical Institute in Olmütz. He was unsuccessful in finding a job, so he quit school and returned to the farm. Depressed by the prospects of a bleak future, he became ill and stayed at home for a year.

Mendel later returned to Olmütz. After two years of study, he found the pressures of school and work to be too much, and his health again broke down. On the advice of his father and a professor, he entered the priesthood, even though he did not feel called to serve the church. His name was changed from Johann to Gregor.

Relieved of his financial difficulties, he was able to continue his studies. However, his nervous disposition interfered with his pastoral duties, and he was assigned to substitute teaching. He enjoyed this work and was popular with the staff and students, but he failed the examination for certification as a teacher. Ironically, his lowest grades were in biology. The Augustinians then sent him to the University of Vienna, where he became particularly interested in his plant physiology professor's unorthodox belief that new plant varieties can be caused by naturally arising variations. He was also fascinated by his classes in physics, where he was exposed to the physicists' experimental and mathematical approach to their subject.

After further breakdowns and failures, Mendel returned to the monastery and was assigned the low-stress job of keeping the abbey garden. There he combined the experimental and mathematical approach of a physicist with his background in biology and performed a series of experiments designed to determine whether his professor was correct in his beliefs regarding the role of naturally arising variants in plants.

Mendel studied the transmission of specific traits of the pea plant—such as flower color and stem length—from parent plant to offspring. He pollinated the plants by hand and separated

probabilities, and Fisher found that there was only about a 0.00007 chance of such close agreement. Others have suggested that perhaps Mendel did not willfully change his results but rather continued collecting data until the numbers were in close agreement with his expectations.*

Probabilities in Genetics

Cystic fibrosis is an inherited disease characterized by abnormally functioning exocrine glands that secrete a thick mucus, clogging the pancreatic ducts and lung passages. Most patients with cystic fibrosis die of chronic lung disease; until recently, most died in early childhood. This early death makes it extremely unlikely that an afflicted person would ever parent a child. Only after the advent of Mendelian genetics did it become clear how a child could inherit the disease from two healthy parents.

* R. A. Fisher, "Has Mendel's Work Been Rediscovered?" *Annals of Science* 1, 1936, pp. 115–137.

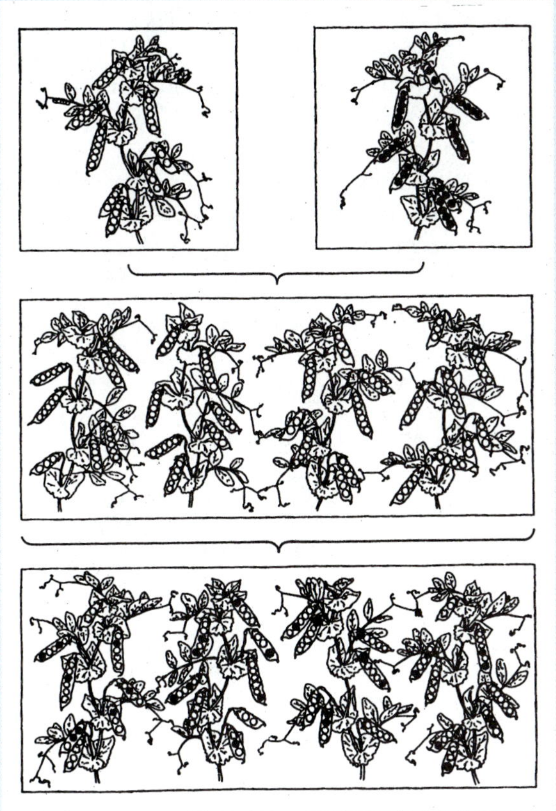

A nineteenth-century drawing illustrating Mendel's pea plants, showing the original cross, the first generation, and the second generation.

them until he had isolated each trait. For example, in his studies of flower color, he pollinated the plants until he produced pure-red plants (plants that would produce only red-flowered offspring) and pure-white plants.

At the time, the accepted theory of heredity was that of blending. In this view, the characteristics of both parents blend together to form an individual. Mendel reasoned that if the blending theory was correct, the union of a pure-red pea plant and a pure-white pea plant would result in a pink-flowered offspring. However, his experiments showed that such a union consistently resulted in red-flowered offspring.

Mendel crossbred a large number of peas that had different characteristics. In many cases, an offspring would have a characteristic of one of its parents, undiluted by that of the other parent. Mendel concluded that the question of which parent's characteristics would be passed on was a matter of chance, and he successfully used probability theory to estimate the frequency with which characteristics would be passed on. In so doing, Mendel founded modern genetics. Mendel attempted similar experiments with bees, but these experiments were unsuccessful because he was unable to control the mating behavior of the queen bee.

Mendel was ignored when he published his paper "Experimentation in Plant Hybridization." Sixteen years after his death, his work was rediscovered by three European botanists who had reached similar conclusions in plant breeding, and the importance of his work was finally recognized.

In 1989, a team of Canadian and American doctors announced the discovery of the gene that is responsible for most cases of cystic fibrosis. As a result of that discovery, a new therapy for cystic fibrosis is being developed. Researchers splice a therapeutic gene into a cold virus and administer it through an affected person's nose. When the virus infects the lungs, the gene becomes active. It is hoped that this will result in normally functioning cells, without the damaging mucus.

In April 1993, a twenty-three-year-old man with advanced cystic fibrosis became the first patient to receive this therapy. In September 1996, a British team announced that eight volunteers with cystic fibrosis had received this therapy; six were temporarily cured of the disease's debilitating symptoms. The team is now analyzing the results of a trial involving multiple doses, which could have a long-term effect. In March 1999, another British team announced a new therapy that involves administering the therapeutic gene through an aerosol spray.

Cystic fibrosis occurs in about 1 out of every 2,000 births in the Caucasian population and only in about 1 in 250,000 births in the non-Caucasian population. It is one of the most common inherited diseases in North America. One in 25 Americans carries a single gene for cystic fibrosis. Children who inherit two such genes develop the disease; that is, cystic fibrosis is recessive.

EXAMPLE 4 Each of two prospective parents carries one cystic fibrosis gene.

a. Find the probability that their child would have cystic fibrosis.
b. Find the probability that their child would be healthy (i.e., free of symptoms).
c. Find the probability that their child would be free of symptoms but could pass the cystic fibrosis gene on to his or her own child.

Solution Dominant genes are indicated by capital letters and recessive genes by lowercase letters; we will denote the recessive cystic fibrosis gene with a c and the disease-free gene with a C. Each parent is Cc and therefore does not have the disease. Figure 3.4 shows the Punnett square for the child.

	C	**c**
C	CC	Cc
c	Cc	cc

Figure 3.4

a. Cystic fibrosis is recessive, so only the cc child will have the disease; the probability of such an event is 1/4.
b. The Cc and the CC children would be healthy; thus,

$$p(\text{healthy}) = p(\text{Cc}) + p(\text{CC}) = 2/4 + 1/4 = 3/4$$

c. The Cc child would not suffer from any symptoms but could pass the cystic fibrosis gene on to his or her own child; such a person is called a **carrier**. (Both of the parents were carriers.) The probability of such an event is $\frac{2}{4} = \frac{1}{2}$. ■

Sickle-cell anemia is an inherited disease characterized by a tendency of the red blood cells to become distorted and deprived of oxygen. Although it varies in severity, the disease can be fatal in early childhood. More often, patients have a shortened life span and chronic organ damage. Newborns are now routinely screened for sickle-cell disease. The only true cure is a bone marrow transplant from a sibling without

AP/Wide World Photos

Woody Guthrie's most famous song is "This Land is Your Land." This folksinger, guitarist, and composer was a friend of Leadbelly, Pete Seeger, and Ramblin' Jack Elliott and exerted a strong influence on Bob Dylan. Guthrie died at the age of 55 of Huntington's disease.

Courtesy of Dr. Nancy Wexler

HISTORICAL NOTE

Nancy Wexler

In 1993, scientists working together at six major research centers located most genes that cause Huntington's disease. This discovery will enable people to learn whether they carry a Huntington's gene, and it will allow pregnant women to determine whether their child carries the gene. The discovery could eventually lead to a treatment.

The collaboration of research centers was organized largely by Nancy Wexler, a Columbia University professor of neuropsychology, who is herself at risk for Huntington's disease—her mother died of it 30 years ago. Dr. Wexler, President of the Hereditary Disease Foundation, has made numerous trips to study and aid the people of the Venezuelan village of Lake Maracaibo, many of whom suffer from the disease or are at risk for it. All are related to one woman who died of the disease in the early 1800s. Wexler took blood and tissue samples and gave neurological and psychoneurological tests to the inhabitants of the village. The samples and test results enabled the researchers to find the single gene that causes Huntington's disease.

In October 1993, Wexler received an Albert Lasker Medical Research Award, a prestigious honor that is often a precursor to a Nobel Prize. The award was given in recognition for her contribution to the international effort that culminated in the discovery of the Huntington's disease gene. At the awards ceremony, she explained to then first lady Hillary Rodham Clinton that her genetic heritage has made her uninsurable—she would lose her health coverage if she switched jobs. She told Mrs. Clinton that more Americans will be in the same situation as more genetic discoveries are made, unless the health care system is reformed. The then first lady incorporated this information into her speech at the awards ceremony: "It is likely that in the next years, every one of us will have a pre-existing condition and will be uninsurable. . . . What will happen as we discover those genes for breast cancer, or prostate cancer, or osteoporosis, or any of the thousands of other conditions that affect us as human beings?"

sickle-cell anemia; however, this can cause the patient's death, so it is done only under certain circumstances. There are also medications that can decrease the episodes of pain. Approximately 1 in every 500 black babies is born with sickle-cell anemia, but only 1 in 160,000 nonblack babies has the disease. This disease is **codominant**: A person with two sickle-cell genes will have the disease, while a person with one sickle-cell gene will have a mild, nonfatal anemia called **sickle-cell trait**. Approximately 8–10% of the black population has sickle-cell trait.

Huntington's disease, caused by a dominant gene, is characterized by nerve degeneration causing spasmodic movements and progressive mental deterioration. The symptoms do not usually appear until well after reproductive age has been reached; the disease usually hits people in their 40s. Death typically follows 12 to 15 years after the onset of the symptoms. No effective treatment is available, but physicians can now assess with certainty whether someone will develop the disease, and they can estimate when the disease will strike. Many of those who are at risk choose not to undergo the test, especially if they have already had children. Folk singer Arlo Guthrie is in this situation; his father, Woody Guthrie, died of Huntington's disease.

In August 1999, researchers in Britain, Germany, and the United States discovered what causes brain cells to die in people with Huntington's disease. This discovery may eventually lead to a treatment.

The row over sickle-cell

. . . Two years ago, President Nixon listed sickle-cell anemia along with cancer as diseases requiring special Federal attention. . . . Federal spending for sickle-cell anemia programs has risen from a scanty $1 million a year to $15 million for 1973. At the same time, in what can only be described as a head-long rush, at least a dozen states have passed laws requiring sickle-cell screening for blacks.

While all these efforts have been undertaken with the best intentions of both whites and blacks, in recent months the campaign has begun to stir widespread and bitter controversy. Some of the educational programs have been riddled with misinformation and have unduly frightened the black community. To quite a few Negroes, the state laws are discriminatory—and to the extent that they might inhibit childbearing, even genocidal. . . .

Parents whose children have the trait often misunderstand and assume they have the disease. In some cases, airlines have allegedly refused to hire black stewardesses who have the trait, and some carriers have been turned down by life-insurance companies—or issued policies at high-risk rates.

Because of racial overtones and the stigma that attaches to persons found to have the sickle-cell trait, many experts seriously object to mandatory screening programs. They note, for example, that there are no laws requiring testing for Cooley's anemia [or other disorders that have a hereditary basis]. . . . Moreover, there is little that a person who knows he has the disease or the trait can do about it. "I don't feel," says Dr. Robert L. Murray, a black geneticist at Washington's Howard University, "that people should be required by law to be tested for something that will provide information that is more negative than positive." . . . Fortunately, some of the mandatory laws are being repealed.

Genetic Screening

At this time, there are no conclusive tests that will tell a parent whether he or she is a cystic fibrosis carrier, nor are there conclusive tests that will tell whether a fetus has the disease. A new test resulted from the 1989 discovery of the location of most cystic fibrosis genes, but that test will detect only 85% to 95% of the cystic fibrosis genes, depending on the individual's ethnic background. The extent to which this test will be used has created quite a controversy.

Individuals who have relatives with cystic fibrosis are routinely informed about the availability of the new test. The controversial question is whether a massive genetic screening program should be instituted to identify cystic fibrosis carriers in the general population, regardless of family history. This is an important question, considering that four in five babies with cystic fibrosis are born to couples with no previous family history of the condition.

Opponents of routine screening cite a number of important concerns. The existing test is rather inaccurate; 5% to 15% of the cystic fibrosis carriers would be missed. It is not known how health insurers would use this genetic information—insurance firms could raise rates or refuse to carry people if a screening test indicated a presence of cystic fibrosis. Also, some experts question the adequacy of quality assurance for the diagnostic facilities and for the tests themselves.

Supporters of routine testing say that the individual should be allowed to decide whether to be screened. Failing to inform people denies them the opportunity to make a personal choice about their reproductive future. An individual who is found to be a carrier could choose to avoid conception, to adopt, to use artificial insemination by a donor, or to use prenatal testing to determine whether a fetus is affected—at which point the additional controversy regarding abortion could enter the picture.

The history of genetic screening programs is not an impressive one. In the 1970s, mass screening of blacks for sickle-cell anemia was instituted. This program caused unwarranted panic; those who were told they had sickle-cell trait feared that they would develop symptoms of the disease and often did not understand the probability that their children would inherit the disease (see Exercises 59 and 60). Some people with sickle-cell trait were denied health insurance and life insurance.

3.2 Exercises

In Exercises 1–14, use this information: A jar on your desk contains twelve black, eight red, ten yellow, and five green jellybeans. You pick a jellybean without looking.

1. What is the experiment?
2. What is the sample space?

In Exercises 3–14, (a) find and (b) interpret the following:

3. the probability that it is black
4. the probability that it is green
5. the probability that it is red or yellow
6. the probability that it is red or black
7. the probability that it is not yellow
8. the probability that it is not red
9. the probability that it is white
10. the probability that it is not white
11. the odds in favor of picking a black jellybean
12. the odds in favor of picking a green jellybean
13. the odds in favor of picking a red or yellow jellybean
14. the odds in favor of picking a red or black jellybean

In Exercises 15–28, one card is drawn from a well-shuffled deck of 52 cards (no jokers).

15. What is the experiment?
16. What is the sample space?

In Exercises 17–28, (a) find the probability and (b) the odds of drawing the given cards. Also, (c) interpret both the probability and the odds. (You might want to review the makeup of a deck of cards in Section 3.1.)

17. a black card
18. a heart
19. a queen
20. a two of clubs
21. a queen of spades
22. a club
23. a card below a 5 (count an ace as high)
24. a card below a 9 (count an ace as high)
25. a card above a 4 (count an ace as high)
26. a card above an 8 (count an ace as high)
27. a face card
28. a picture card

In Exercises 29–38, (a) find the probability and (b) the odds of winning the given bet in roulette. Also, (c) interpret both the probability and the odds. (You might want to review the description of the game in Section 3.1.)

29. the single-number bet
30. the two-number bet
31. the three-number bet
32. the four-number bet
33. the five-number bet
34. the six-number bet
35. the twelve-number bet
36. the low-number bet
37. the even-number bet
38. the red-number bet
39. If $p(E) = \frac{1}{5}$, find $o(E)$.
40. If $p(E) = \frac{8}{9}$, find $o(E)$.
41. If $o(E) = 3:2$, find $p(E)$.
42. If $o(E) = 4:7$, find $p(E)$.
43. If $p(E) = \frac{a}{b}$, find $o(E)$.

In Exercises 44–46, find (a) the probability and (b) the odds of winning the following bets in roulette. In finding the odds, use the formula developed in Exercise 43. Also, (c) interpret both the probability and the odds.

44. the high-number bet
45. the odd-number bet
46. the black-number bet
47. A family has two children. Using b to stand for boy and g for girl in ordered pairs, give each of the following.
    - **a.** the sample space
    - **b.** the event E that the family has exactly one daughter
    - **c.** the event F that the family has at least one daughter
    - **d.** the event G that the family has two daughters
    - **e.** $p(E)$ **f.** $p(F)$ **g.** $p(G)$
    - **h.** $o(E)$ **i.** $o(F)$ **j.** $o(G)$

    (Assume that boys and girls are equally likely.)
48. Two coins are tossed. Using ordered pairs, give the following.
    - **a.** the sample space
    - **b.** the event E that exactly one is heads
    - **c.** the event F that at least one is heads
    - **d.** the event G that two are heads
    - **e.** $p(E)$ **f.** $p(F)$ **g.** $p(G)$
    - **h.** $o(E)$ **i.** $o(F)$ **j.** $o(G)$

49. A family has three children. Using b to stand for boy and g for girl, and using ordered triples such as (b, b, g), give the following.

a. the sample space

b. the event E that the family has exactly two daughters

c. the event F that the family has at least two daughters

d. the event G that the family has three daughters

e. $p(E)$ **f.** $p(F)$ **g.** $p(G)$

h. $o(E)$ **i.** $o(F)$ **j.** $o(G)$

(Assume that boys and girls are equally likely.)

50. Three coins are tossed. Using ordered triples, give the following.

a. the sample space

b. the event E that exactly two are heads

c. the event F that at least two are heads

d. the event G that all three are heads

e. $p(E)$ **f.** $p(F)$ **g.** $p(G)$

h. $o(E)$ **i.** $o(F)$ **j.** $o(G)$

51. A couple plans on having two children.

a. Find the probability of having two girls.

b. Find the probability of having one girl and one boy.

c. Find the probability of having two boys.

d. Which is more likely: having two children of the same sex or two of different sexes? Why? - (Assume that boys and girls are equally likely.)

52. Two coins are tossed.

a. Find the probability that both are heads.

b. Find the probability that one is heads and one is tails.

c. Find the probability that both are tails.

d. Which is more likely: that the two coins match or that they don't match? Why?

53. A couple plans on having three children. Which is more likely: having three children of the same sex or of different sexes? Why? (Assume that boys and girls are equally likely.)

54. Three coins are tossed. Which is more likely: that the three coins match or that they don't match? Why?

55. A pair of dice is rolled. Using ordered pairs, give the following.

a. the sample space

HINT: S has 36 elements, one of which is (1, 1).

b. the event E that the sum is 7

c. the event F that the sum is 11

d. the event G that the roll produces doubles

e. $p(E)$ **f.** $p(F)$ **g.** $p(G)$

h. $o(E)$ **i.** $o(F)$ **j.** $o(G)$

56. Mendel found that snapdragons have no color dominance; a snapdragon with one red gene and one white gene will have pink flowers. If a pure-red snapdragon is crossed with a pure-white one, find the probability of the following.

a. a red offspring

b. a white offspring

c. a pink offspring

57. If two pink snapdragons are crossed (see Exercise 56), find the probability of the following.

a. a red offspring

b. a white offspring

c. a pink offspring

58. One parent is a cystic fibrosis carrier, and the other has no cystic fibrosis gene. Find the probability of each of the following.

a. The child would have cystic fibrosis.

b. The child would be a carrier.

c. The child would not have cystic fibrosis and not be a carrier.

d. The child would be healthy (i.e., free of symptoms).

59. If carrier-detection tests show that two prospective parents have sickle-cell trait (and are therefore carriers), find the probability of each of the following.

a. Their child would have sickle-cell anemia.

b. Their child would have sickle-cell trait.

c. Their child would be healthy (i.e., free of symptoms).

60. If carrier-detection tests show that one prospective parent is a carrier of sickle-cell anemia and the other has no sickle-cell gene, find the probability of each of the following.

a. The child would have sickle-cell anemia.

b. The child would have sickle-cell trait.

c. The child would be healthy (i.e., free of symptoms).

Tay-Sachs disease *is a recessive disease characterized by an abnormal accumulation of certain fat compounds in the spinal cord and brain, resulting in paralysis, severe mental impairment, and blindness. There is no effective treatment, and death usually occurs before the age of five. The disease occurs once in 3,600 births among Ashkenazi Jews (Jews from central and eastern Europe) but only once in 600,000 births in other populations. Carrier-detection tests and fetal-monitoring tests are available. The successful use of these tests, coupled with an aggressive counseling program, has resulted in a decrease of 90% in the incidence of this disease.*

61. If carrier-detection tests show that one prospective parent is a carrier of Tay-Sachs and the other has no Tay-Sachs gene, find the probability of each of the following.
   a. The child would have the disease.
   b. The child would be a carrier.
   c. The child would be healthy (i.e., free of symptoms).
62. If carrier-detection tests show that both prospective parents are carriers of Tay-Sachs, find the probability of each of the following.
   a. Their child would have the disease.
   b. Their child would be a carrier.
   c. Their child would be healthy (i.e., free of symptoms).
63. If a parent started to exhibit the symptoms of Huntington's disease after the birth of his or her child, find the probability of each of the following. (Assume that one parent carries a single gene for Huntington's disease and the other carries no such gene.)
   a. The child would have the disease.
   b. The child would be a carrier.
   c. The child would be healthy (i.e., free of symptoms).

Projects

64. a. In your opinion, what is the probability that the last digit of a phone number is odd? Justify your answer.
   b. Randomly choose one page from the residential section of your local phone book. Count how many phone numbers on that page have an odd last digit and how many have an even last digit. Then compute (number of phone numbers with odd last digits)/(total number of phone numbers).
   c. Is your answer to part (b) a theoretical probability or a relative frequency? Justify your answer.
   d. How do your answers to parts (a) and (b) compare? Are they exactly the same, approximately the same, or dissimilar? Discuss this comparison, taking into account the ideas of probability theory.
65. a. In your opinion, what is the probability that the first digit of a phone number is even? Justify your answer.
   b. Randomly choose one page from the residential section of your local phone book. Count how many phone numbers on that page have an even first digit and how many have an odd first digit. Then compute (number of phone numbers with even first digits)/(total number of phone numbers).
   c. Is your answer to part (b) a theoretical probability or a relative frequency? Justify your answer.
   d. How do your answers to parts (a) and (b) compare? Are they exactly the same, approximately the same, or dissimilar? Discuss this comparison, taking into account the ideas of probability theory.

Concept Questions

66. Explain how you would find the theoretical probability of rolling an even number on a single die. Explain how you would find the relative frequency of rolling an even number on a single die.
67. Give five examples of events whose probabilities must be found empirically rather than theoretically.
68. Does the theoretical probability of an event remain unchanged from experiment to experiment? Why? Does the relative frequency of an event remain unchanged from experiment to experiment? Why?
69. Consider a "weighted die"—one that has a small weight in its interior. Such a weight would cause the face closest to the weight to come up less frequently and the face farthest from the weight to come up more frequently. Would the probabilities computed in Example 1 still be correct? Why or why not? Would the definition $p(E) = \frac{n(E)}{n(S)}$ still be appropriate? Why or why not?
70. Some dice have spots that are small indentations; other dice have spots that are filled with the same material but of a different color. Which of these two types of dice is not fair? Why? What would be the most likely outcome of rolling this type of die? Why?

   HINT: 1 and 6 are on opposite faces, as are 2 and 5, and 3 and 4.
71. In the United States, 52% of the babies are boys and 48% are girls. Do these percentages contradict an assumption that boys and girls are equally likely? Why?
72. Compare and contrast theoretical probability and relative frequency. Be sure you discuss both the similarities and differences between the two.
73. Compare and contrast probability and odds. Be sure you discuss both the similarities and differences between the two.

History Questions

74. What prompted Dr. Nancy Wexler's interest in Huntington's disease?

75. What resulted from Dr. Nancy Wexler's interest in Huntington's disease?

Projects

76. **a.** Flip a coin 20 times, and compute the relative frequency of heads.

b. Repeat the experiment described in (a) 9 more times. Each time, compute the relative frequency of heads. After finishing parts (a) and (b), you will have flipped a coin 200 times, and you will have computed 10 different relative frequencies.

c. Compare your 10 relative frequencies with each other and with the theoretical probability of heads. In your discussion, use the ideas discussed under the heading "Relative Frequency versus Probability" in this section. Do not plagiarize; use your own words.

d. Combine the results of parts (a) and (b), and find the relative frequency of heads for all 200 coin tosses. Compare this relative frequency with those found in parts (a) and (b) and with the theoretical probability of heads. Be certain to incorporate the Law of Large Numbers into your discussion.

77. In the 1970s, there was a mass screening of blacks for sickle-cell anemia and a mass screening of Jews for Tay-Sachs disease. One of these was a successful program; one was not. Write a research paper on these two programs.

3.3 Basic Rules of Probability

One basic rule about probabilities is that they are always between 0 and 1 inclusive. The smallest possible probability is that of an impossible event (an event equal to the null set); that probability is 0. The largest probability is that of a certain event (an event equal to the sample space); that probability is 1. *If you ever get a negative answer or an answer greater than 1 when you calculate a probability, go back and find your error.*

Probability Rules

Rule 1	$p(\varnothing) = 0$	The probability of the null set is 0.
Rule 2	$p(S) = 1$	The probability of the sample space is 1.
Rule 3	$0 \le p(E) \le 1$	Probabilities are between 0 and 1 (inclusive).

Probability Rules 1, 2, and 3 can be formally verified as follows:

Rule 1: $$p(\varnothing) = \frac{n(\varnothing)}{n(S)} = \frac{0}{n(S)} = 0$$

Rule 2: $$p(S) = \frac{n(S)}{n(S)} = 1$$

Rule 3: E is a subset of S; therefore,

$$0 \le n(E) \le n(S)$$

$$\frac{0}{n(S)} \le \frac{n(E)}{n(S)} \le \frac{n(S)}{n(S)} \quad \text{dividing by } n(S)$$

$$0 \le p(E) \le 1$$

Mutually Exclusive Events

Two events that cannot both occur at the same time are called **mutually exclusive**. In other words, E and F are mutually exclusive if and only if $E \cap F = \varnothing$.

EXAMPLE 1 A die is rolled. Let E be the event "an even number comes up," F the event "a number greater than 3 comes up," and G the event "an odd number comes up."

a. Are E and F mutually exclusive?
b. Are E and G mutually exclusive?

Solution **a.** $E = \{2, 4, 6\}$, $F = \{4, 5, 6\}$, and $E \cap F = \{4, 6\} \neq \varnothing$ (see Figure 3.5). Therefore, E and F are *not* mutually exclusive; the number that comes up could be *both* even *and* greater than 3. In particular, it could be 4 or 6.

b. $E = \{2, 4, 6\}$, $G = \{1, 3, 5\}$, and $E \cap G = \varnothing$. Therefore, E and G *are* mutually exclusive; the number that comes up could *not* be both even and odd. ■

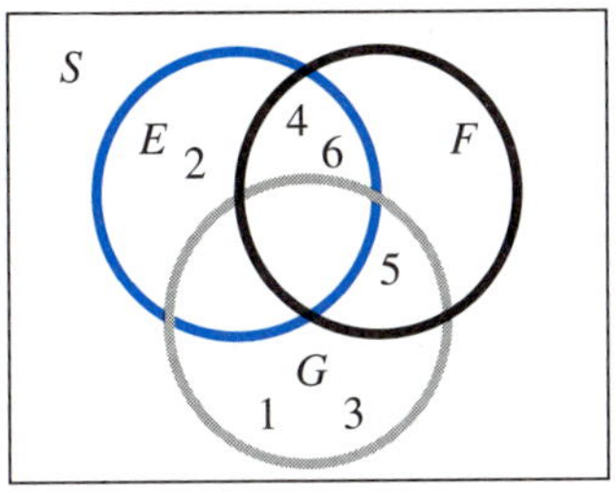

Figure 3.5

EXAMPLE 2 Let M be the event "being a mother," F the event "being a father," and D the event "being a daughter."

a. Are events M and D mutually exclusive?
b. Are events M and F mutually exclusive?

Solution **a.** M and D are mutually exclusive if $M \cap D = \varnothing$. $M \cap D$ is the set of all people who are both mothers and daughters, and that set is not empty. A person can be a mother and a daughter at the same time. M and D are not mutually exclusive because being a mother does not exclude being a daughter.

b. M and F are mutually exclusive if $M \cap F = \varnothing$. $M \cap F$ is the set of all people who are both mothers and fathers, and that set is empty. A person cannot be a mother and a father at the same time. M and F are mutually exclusive because being a mother does exclude the possibility of being a father. ■

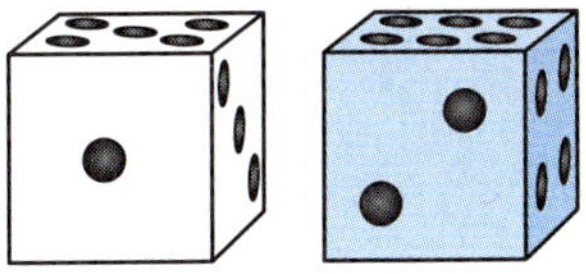

The event (1, 2)

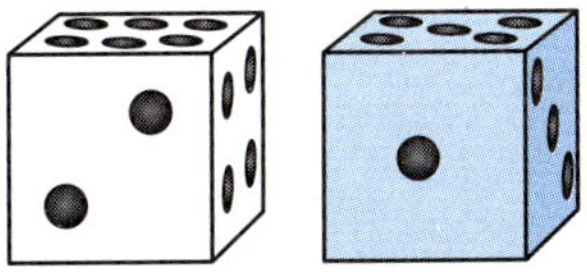

The event (2, 1)

Pair-of-Dice Probabilities

To find probabilities involving the rolling of a pair of dice, we must first determine the sample space. The *sum* can be anything from 2 to 12, but the sample space is *not* $\{2, 3, 4, 5, 6, 7, 8, 9, 10, 11, 12\}$. The sample space is the set of all outcomes, and the number 3 (for example) is not an outcome of rolling a pair of dice. Instead, the ordered pair (1, 2) is an outcome, as is (2, 1). Each of these outcomes results in a sum of 3. Of course, unless one of the dice is painted, you can't visually distinguish these two different outcomes. Figure 3.6 on page 150 lists all possible outcomes and the resulting sums. Notice that $n(S) = 6 \cdot 6 = 36$.

Figure 3.6
Outcomes of rolling two dice

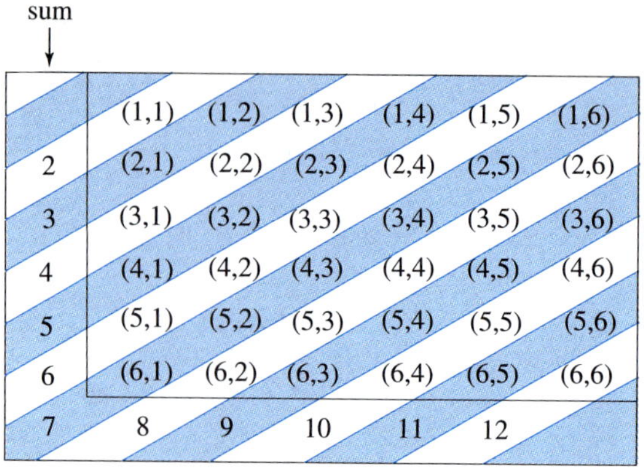

EXAMPLE 3 A pair of dice is rolled. Find the probability of each of the following events.

a. The sum is 7.
b. The sum is greater than 9.
c. The sum is even.
d. The sum is not greater than 9.
e. The sum is greater than 9 and even.
f. The sum is greater than 9 or even.

Solution **a.** To find the probability that the sum is 7, let D be the event "the sum is 7." From Figure 3.6, $D = \{(1, 6), (2, 5), (3, 4), (4, 3), (5, 2), (6, 1)\}$, so $n(D) = 6$; therefore,

$$p(D) = \frac{n(D)}{n(S)} = \frac{6}{36} = \frac{1}{6}$$

This means that if we were to roll a pair of dice a large number of times, we should expect to get a sum of 7 approximately one-sixth of the time.

> ✔ Notice that $p(D) = \frac{1}{6}$ is between 0 and 1, as are all probabilities.

b. To find the probability that the sum is greater than 9, let E be the event "the sum is greater than 9."
$E = \{(4, 6), (5, 5), (6, 4), (5, 6), (6, 5), (6, 6)\}$, so $n(E) = 6$; therefore,

$$p(E) = \frac{n(E)}{n(S)} = \frac{6}{36} = \frac{1}{6}$$

This means that if we were to roll a pair of dice a large number of times, we should expect to get a sum greater than 9 approximately one-sixth of the time.

c. Let F be the event "the sum is even."
$F = \{(1, 1), (1, 3), (2, 2), (3, 1), \ldots, (6, 6)\}$, so $n(F) = 18$ (refer to Figure 3.6); therefore,

$$p(F) = \frac{n(F)}{n(S)} = \frac{18}{36} = \frac{1}{2}$$

This means that if we were to roll a pair of dice a large number of times, we should expect to get an even sum approximately half of the time.

d. We could find the probability that the sum is not greater than 9 by counting, as in (a), (b), and (c), but the counting would be rather excessive. It is easier to use one

A Roman painting on marble of the daughters of Niobe using knucklebones as dice. This painting was found in the ruins of Herculaneum, a city that was destroyed along with Pompeii by the eruption of Vesuvius.

of the Cardinal Number Formulas from Chapter 2 on sets. The event "the sum is not greater than 9" is the complement of event E ("the sum is greater than 9") and can be expressed as E'.

$$\begin{aligned} n(E') &= n(U) - n(E) \\ &= n(S) - n(E) \\ &= 36 - 6 = 30 \end{aligned}$$

Cardinal Number Formula

"Universal set" and "sample space" represent the same idea.

$$p(E') = \frac{n(E')}{n(S)} = \frac{30}{36} = \frac{5}{6}$$

This means that if we were to roll a pair of dice a large number of times, we should expect to get a sum that's not greater than 9 approximately five-sixths of the time.

e. The event "the sum is greater than 9 and even" can be expressed as the event $E \cap F$. $E \cap F = \{(4, 6), (5, 5), (6, 4), (6, 6)\}$, so $n(E \cap F) = 4$; therefore,

$$p(E \cap F) = \frac{n(E \cap F)}{n(S)} = \frac{4}{36} = \frac{1}{9}$$

This means that if we were to roll a pair of dice a large number of times, we should expect to get a sum that's both greater than 9 and even approximately one-ninth of the time.

f. Finding the probability that the sum is greater than 9 or even by counting would require an excessive amount of counting. It is easier to use one of the Cardinal Number Formulas from Chapter 2. The event "the sum is greater than 9 or even" can be expressed as the event $E \cup F$.

$$\begin{aligned} n(E \cup F) &= n(E) + n(F) - n(E \cap F) \\ &= 6 + 18 - 4 \\ &= 20 \end{aligned}$$

Cardinal Number Formula

from parts (b), (c), and (e)

$$p(E \cup F) = \frac{n(E \cup F)}{n(S)} = \frac{20}{36} = \frac{5}{9}$$

This means that if we were to roll a pair of dice a large number of times, we should expect to get a sum that's either greater than 9 or even approximately five-ninths of the time. ■

More Probability Rules

In Example 3 above, we used some Cardinal Number Formulas from Chapter 2 to avoid excessive counting. Some people find it easier to use these rules to calculate probabilities when they are expressed in the language of probability theory.

More Probability Rules

Rule 4 $p(E \cup F) = p(E) + p(F) - p(E \cap F)$

Rule 5 If E and F are mutually exclusive, then $p(E \cup F) = p(E) + p(F)$.

Rule 6 $p(E) + p(E') = 1$ [or equivalently, $p(E) = 1 - p(E')$ or $p(E') = 1 - p(E)$]

In part (d) of Example 3, finding the probability that the sum is not greater than 9 could be done with Probability Rule 6 rather than with a Cardinal Number Formula:

$$\begin{aligned} p(E') &= 1 - p(E) \\ &= 1 - \frac{1}{6} \quad \text{from (b)} \\ &= \frac{5}{6} \end{aligned}$$

Similarly, in part (f) of Example 3, finding the probability that the sum is greater than 9 or even could be done with Probability Rule 4 rather than with a Cardinal Number Formula:

$$\begin{aligned} p(E \cup F) &= p(E) + p(F) - p(E \cap F) \\ &= \frac{1}{6} + \frac{1}{2} - \frac{1}{9} \quad \text{from (b), (c), and (e)} \\ &= \frac{3}{18} + \frac{9}{18} - \frac{2}{18} = \frac{5}{9} \end{aligned}$$

Probabilities and Venn Diagrams

Venn diagrams can be used to illustrate probabilities in the same way they are used in set theory. In this case, we label each region with its probability rather than its cardinal number.

EXAMPLE 4 Zaptronics manufactures compact discs and cases for several major record labels. A recent sampling of the products indicated that 5% have defective packaging, 3% have a defective disc, and 7% have at least one of the two defects. Find the probability that a Zaptronics product has the following.

a. both defects
b. neither defect

Solution **a.** Let P be the event "the packaging is defective" and D the event "the disc is defective." We are given $p(P) = 5\% = 0.05$, $p(D) = 3\% = 0.03$, and $p(P \cup D) = 7\% = 0.07$. We are asked to find $p(P \cap D)$. To do so, substitute into Probability Rule 4.

$$p(P \cup D) = p(P) + p(D) - p(P \cap D) \quad \text{Probability Rule 4}$$
$$0.07 = 0.05 + 0.03 - p(P \cap D) \quad \text{substituting}$$
$$0.07 = 0.08 - p(P \cap D)$$
$$p(P \cap D) = 0.01 = 1\%$$

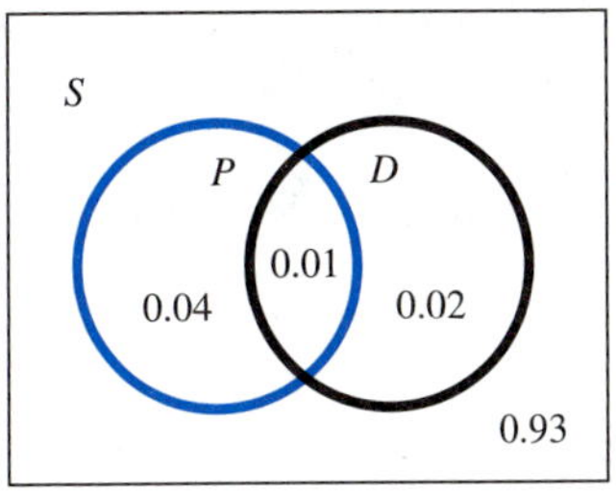

Figure 3.7

This means that 1% of Zaptronic's products have defective packaging *and* a defective disc. The Venn diagram for this problem is shown in Figure 3.7.

b. According to the Venn diagram in Figure 3.7, the probability of neither defect is $0.93 = 93\%$.

Alternatively, we are asked to find the probability that the product does not have defective packaging *and* does not have a defective disc—that is, to find $p(P' \cap D')$. This is equal to $p((P \cup D)')$, by De Morgan's law. And $(P \cup D)'$ is the complement of the event whose probability is given to us. Thus,

$$p(P' \cap D') = p((P \cup D)') \quad \text{De Morgan's law}$$
$$= 1 - p(P \cup D) \quad \text{Probability Rule 6}$$
$$= 1 - 0.07 \quad \text{substituting}$$
$$= 0.93 = 93\%$$

This means that 93% of Zaptronic's products are defect-free. ■

3.3 Exercises

In Exercises 1–10, determine whether E and F are mutually exclusive. Write a sentence justifying your answer.

1. E is the event "being a doctor," and F is the event "being a woman."

2. E is the event "it's raining," and F is the event "it's sunny."

3. E is the event "being single," and F is the event "being married."

4. E is the event "having naturally blond hair," and F is the event "having naturally black hair."

5. E is the event "having brown hair," and F is the event "having gray hair."

6. E is the event "being a plumber," and F is the event "being a stamp collector."

7. E is the event "wearing boots," and F is the event "wearing sandals."

8. E is the event "wearing shoes," and F is the event "wearing socks."

9. If a die is rolled once, E is the event "getting a four," and F is the event "getting an odd number."

10. If a die is rolled once, E is the event "getting a four," and F is the event "getting an even number."

In Exercises 11–18, a card is dealt from a complete deck of 52 playing cards (no jokers). Use probability rules (when appropriate) to find the probability that the card is as stated. (Count an ace as high.)

11. **a.** a jack and red **b.** a jack or red
c. not a red jack

12. **a.** a jack and a heart **b.** a jack or a heart
c. not a jack of hearts

13. a. a ten and a spade **b.** a ten or a spade
c. not a ten of spades

14. a. a five and black **b.** a five or black
c. not a black five

15. a. under a four
b. above a nine
c. both under a four and above a nine
d. either under a four or above a nine

16. a. above a jack
b. below a three
c. both above a jack and below a three
d. either above a jack or below a three

17. a. above a five
b. below a ten
c. both above a five and below a ten
d. either above a five or below a ten

18. a. above a seven
b. below a queen
c. both above a seven and below a queen
d. either above a seven or below a queen

In Exercises 19–26, use complements to find the probability that a card dealt from a full deck (no jokers) is as stated. (Count an ace as high.)

19. not a queen **20.** not a seven
21. not a face card **22.** not a heart
23. above a three **24.** below a queen
25. below a jack **26.** above a five

27. If $o(E) = 5:9$, find $o(E')$.

28. If $o(E) = 1:6$, find $o(E')$.

29. If $p(E) = \frac{2}{7}$, find $o(E)$ and $o(E')$.

30. If $p(E) = \frac{3}{8}$, find $o(E)$ and $o(E')$.

31. If $o(E) = a:b$, find $o(E')$.

32. If $p(E) = \frac{a}{b}$, find $o(E')$.

HINT: Use Exercise 43 from Section 3.2 and Exercise 31 above.

In Exercises 33–38, use Exercise 32 above to find the odds that a card dealt from a full deck (no jokers) is as stated.

33. not a king **34.** not an eight
35. not a face card **36.** not a club
37. above a four **38.** below a king

In Exercises 39–42, use the following information: To determine the effect their salespersons have on purchases, a department store polled 700 shoppers regarding whether or not they made a purchase and whether or not they were pleased with the service they received. Of those who made a purchase, 151 were happy with the service and 133 were not. Of those who made no purchase, 201 were happy with the service and 215 were not. Use probability rules (when appropriate) to find the probability of the event stated.

39. a. A shopper made a purchase.
b. A shopper did not make a purchase.

40. a. A shopper was happy with the service received.
b. A shopper was unhappy with the service received.

41. a. A shopper made a purchase and was happy with the service.
b. A shopper made a purchase or was happy with the service.

42. a. A shopper made no purchase and was unhappy with the service.
b. A shopper made no purchase or was unhappy with the service.

In Exercises 43–46, use the following information: A supermarket polled 1,000 customers regarding the size of their bill. The results are given in Figure 3.8.

Size of Bill	Number of Customers
below $20.00	208
$20.00–$39.99	112
$40.00–$59.99	183
$60.00–$79.99	177
$80.00–$99.99	198
$100.00 or above	122

Figure 3.8
Supermarket bills

Use probability rules (when appropriate) to find the relative frequency with which a customer's bill is as stated.

43. a. less than $40.00 **b.** $40.00 or more

44. a. less than $80.00 **b.** $80.00 or more

45. a. between $40.00 and $79.99
b. not between $40.00 and $79.99

46. a. between $20.00 and $79.99
b. not between $20.00 and $79.99

In Exercises 47–54, find the probability that the sum is as stated when a pair of dice is rolled.

47. a. 7 **b.** 9 **c.** 11

48. a. 2 **b.** 4 **c.** 6

49. a. 7 or 11 **b.** 7 or 11 or doubles

50. a. 8 or 10 **b.** 8 or 10 or doubles

51. **a.** odd and greater than 7
    b. odd or greater than 7
52. **a.** even and less than 5
    b. even or less than 5
53. **a.** even and doubles **b.** even or doubles
54. **a.** odd and doubles **b.** odd or doubles
55. Use probability rules to find the probability that a child will either have Tay-Sachs disease or be a carrier if (a) each parent is a Tay-Sachs carrier, (b) one parent is a Tay-Sachs carrier and the other parent has no Tay-Sachs gene, (c) one parent has Tay-Sachs and the other parent has no Tay-Sachs gene.
56. Use probability rules to find the probability that a child will have either sickle-cell anemia or sickle-cell trait if (a) each parent has sickle-cell trait, (b) one parent has sickle-cell trait and the other parent has no sickle-cell gene, (c) one parent has sickle-cell anemia and the other parent has no sickle-cell gene.
57. Use probability rules to find the probability that a child will neither have Tay-Sachs disease nor be a carrier if (a) each parent is a Tay-Sachs carrier, (b) one parent is a Tay-Sachs carrier and the other parent has no Tay-Sachs gene, (c) one parent has Tay-Sachs and the other parent has no Tay-Sachs gene.
58. Use probability rules to find the probability that a child will have neither sickle-cell anemia nor sickle-cell trait if (a) each parent has sickle-cell trait, (b) one parent has sickle-cell trait and the other parent has no sickle-cell gene, (c) one parent has sickle-cell anemia and the other parent has no sickle-cell gene.
59. Mary is taking two courses, photography and economics. Student records indicate that the probability of passing photography is 0.75, that of failing economics is 0.65, and that of passing at least one of the two courses is 0.85. Find the probability of the following.
    a. Mary will pass economics.
    b. Mary will pass both courses.
    c. Mary will fail both courses.
    d. Mary will pass exactly one course.
60. Alex is taking two courses, algebra and U.S. history. Student records indicate that the probability of passing algebra is 0.35, that of failing U.S. history is 0.35, and that of passing at least one of the two courses is 0.80. Find the probability of the following.
    a. Alex will pass history.
    b. Alex will pass both courses.
    c. Alex will fail both courses.
    d. Alex will pass exactly one course.
61. Of all the flashlights in a large shipment, 15% have a defective bulb, 10% have a defective battery, and 5% have both defects. If you purchase one of the flashlights in this shipment, find the probability that it has the following.
    a. a defective bulb or a defective battery
    b. a good bulb or a good battery
    c. a good bulb and a good battery
62. Of all the videotapes in a large shipment, 20% have a defective tape, 15% have a defective case, and 10% have both defects. If you purchase one of the videotapes in this shipment, find the probability that it has the following.
    a. a defective tape or a defective case
    b. a good tape or a good case
    c. a good tape and a good case
63. Verify Probability Rule 4.

    HINT: Divide the Cardinal Number Formula for the Union of Sets from Section 2.1 by $n(S)$.
64. Verify Probability Rule 5.

    HINT: Start with Probability Rule 4. Then use the fact that E and F are mutually exclusive.
65. Verify Probability Rule 6.

    HINT: Are E and E' mutually exclusive?

Answer the following questions using complete sentences and your own words.

Concept Questions

66. What is the complement of a certain event? Justify your answer.
67. Compare and contrast mutually exclusive events and impossible events. Be sure to discuss both the similarities and the differences between these two concepts.
68. Explain why it is necessary to subtract $p(E \cap F)$ in Probability Rule 4. In other words, explain why Probability Rule 5 is not true for all events.

Fractions on a Graphing Calculator

Some graphing calculators—including the TI-80, TI-82, TI-83, TI-85, TI-86, and Casio—will add, subtract, multiply, and divide fractions, and will give answers in reduced fractional form.

Reducing Fractions

The fraction 42/70 reduces to 3/5. To do this on your calculator, you must make your screen read "42/70 → Frac" ("42/70 → *b*/*c*" on a TI-80, "42⌟70" on a Casio). The way that you do this varies.

TI-80	• Type 42 ÷ 70, but do not press ENTER. This causes "42/70" to appear on the screen. • Press the FRAC button. • Use the ↓ button to highlight option 2, "→*b*/*c*." • Press ENTER. This will cause "42/70→*b*/*c*" to appear on your display. • Press ENTER and your display will read 3/5.
TI-82/TI-83	• Type 42 ÷ 70, but do not press ENTER. This causes "42/70" to appear on the screen. • Press the MATH button. • Highlight option 1, "→Frac." (Option 1 is automatically highlighted. If we were selecting a different option, we would use the ↑ and ↓ buttons to highlight it.) • Press ENTER. This causes "42/70→Frac" to appear on the screen. • Press ENTER. This causes 3/5 to appear on the screen.
TI-85/TI-86	• Type 42 ÷ 70, but do not press ENTER. This causes "42/70" to appear on the screen. • Press 2nd MATH. • Press MISC (i.e., F5). • Press MORE until "→Frac" appears. • Press →Frac (i.e., F1). This causes "42/70→Frac" to appear on the screen. • Press ENTER. This causes 3/5 to appear on the screen.
CASIO	• Type 42 $a^{b}/_{c}$ 70. • Type EXE, and your display will read "3⌟5", which should be interpreted as "3/5."

EXAMPLE 5 Use your calculator to compute

$$\frac{1}{6} + \frac{1}{2} - \frac{1}{9}$$

and give your answer in reduced fractional form.

Solution On a TI, make your screen read "1/6 + 1/2 − 1/9 → Frac" by typing

1 ÷ 6 + 1 ÷ 2 − 1 ÷ 9

and then inserting the "→Frac" command ("→b/c" on a TI-80), as described on page 156. Once you press ENTER, the screen will read "5/9."

On a Casio, make your screen read "1⌋6 + 1⌋2 − 1⌋9" by typing

1 $a^b\!/\!_c$ 6 + 1 $a^b\!/\!_c$ 2 − 1 $a^b\!/\!_c$ 9

Once you press EXE, the screen will read "5⌋9," which should be interpreted as "5/9." ■

Exercises

In Exercises 69–70, reduce the given fractions to lowest terms, both (a) by hand and (b) with a calculator. Check your work by comparing the two answers.

69. $\frac{18}{33}$ **70.** $-\frac{42}{72}$

In Exercises 71–76, perform the indicated operations, and reduce the answers to lowest terms, both (a) by hand and (b) with a calculator. Check your work by comparing the two answers.

71. $\frac{6}{15} \cdot \frac{10}{21}$

72. $\frac{6}{15} \div \frac{10}{21}$

73. $\frac{6}{15} + \frac{10}{21}$

74. $\frac{6}{15} - \frac{10}{21}$

75. $\frac{7}{6} - \frac{5}{7} + \frac{9}{14}$

76. $\frac{-8}{5} - \left(\frac{-3}{28} + \frac{5}{21}\right)$

77. How could you use your calculator to get decimal answers to the above exercises rather than fractional answers?

3.4 Combinatorics and Probability

Finding a probability involves finding the number of outcomes in an event and the number of outcomes in the sample space. So far, we have used the Probability Rules as an alternative to excessive counting. Another alternative is combinatorics—that is, permutations, combinations, and the Fundamental Counting Principle—as covered in Chapter 2. The flowchart used in Chapter 2 is summarized below.

Which Counting Technique?

1. If the problem involves more than one category, use the *Fundamental Principle of Counting* and multiply the number of choices for each category.
2. Within any one category, if the order of selection is important, use *Permutations: r* items can be selected from a group of *n* items in

$$_nP_r = \frac{n!}{(n-r)!}$$

ways.
3. Within any one category, if the order of selection is *not* important, use *Combinations: r* items can be selected from a group of *n* items in

$$_nC_r = \frac{n!}{r! \cdot (n-r)!}$$

ways.

EXAMPLE 1 A group of three people is selected at random. What is the probability that at least two of them will have the same birthday?

Solution We will assume that all birthdays are equally likely, and for the sake of simplicity, we will ignore leap year day (February 29). The experiment is to ask three people their birthdays. One possible outcome is (May 1, May 3, August 23). The sample space is the set of all possible lists of three birthdays. To find the number of elements in the sample space, we follow the flowchart from Chapter 2. The selected items are birthdays. They can be repeated (people can share the same birthday), so we must use the Fundamental Principle of Counting. We make three boxes, one for each birthday. The first birthday may be selected in any of 365 different ways, since there are 365 different days in a year. The second birthday may be selected in any of 365 ways also, as can the third birthday, because people can share the same birthday. Thus, the number of elements in the sample space is

$$n(S) = \boxed{365} \cdot \boxed{365} \cdot \boxed{365}$$

The event E is the set of all possible lists of three birthdays in which at least two of those birthdays are the same. It is rather difficult to compute $n(E)$ directly; instead, we will compute $n(E')$ and use a probability rule. E' is the set of all possible lists of three birthdays in which no two of those birthdays are the same. To find the number of elements in E', we follow the flowchart in Chapter 2. The birthdays may *not* be repeated, there is only one category (birthdays), and the order of selection does matter [(May 1, May 3, August 23) is a different list from (August 23, May 3, May 1)]. Thus, we use permutations, and the number of elements in E' is

$$\begin{aligned} n(E') &= {}_{365}P_3 = \frac{365!}{(365-3)!} \\ &= \frac{365 \cdot 364 \cdot 363 \cdot 362!}{362!} \\ &= 365 \cdot 364 \cdot 363 \quad \text{canceling} \end{aligned}$$

We are now ready to compute $p(E)$.

$$\begin{aligned} p(E) &= 1 - p(E') \quad \text{Probability Rule 6} \\ &= 1 - \frac{n(E')}{n(S)} \\ &= 1 - \frac{365 \cdot 364 \cdot 363}{365 \cdot 365 \cdot 365} \\ &= 1 - \frac{364 \cdot 363}{365 \cdot 365} \quad \text{canceling} \\ &= 0.008204 \cdots \approx 0.008 \end{aligned}$$

1 [−] 364 [×] 363 [÷] 365 [x^2] [=]

Alternatively, use the "$_nP_r$" command on a graphing calculator. From the above, $p(E) = 1 - \frac{_{365}P_3}{365^3}$. Type:

1 [−] 365 [$_nP_r$] 3 [÷] 365 [∧] 3 [ENTER]

where [$_nP_r$] refers to the "$_nP_r$" command on the "MATH" menu, discussed in Section 2.4.

This result is not at all surprising; it means that it is extremely unlikely that two or more people in a group of three share a birthday. You will, however, be surprised in the exercises when the group is increased to 30. ■

Lotteries

EXAMPLE 2 Connecticut, Missouri, Oregon, and Virginia all operate 6/44 lotteries; that is, the player selects any six of the numbers from 1 to 44. The state determines six winning numbers, usually by having a mechanical device choose six balls from a container filled with balls numbered from 1 to 44. If a player's six selections match the six winning numbers, the player wins first prize. (Smaller prizes are awarded to players who select five or four of the winning numbers.) Find the probability of winning first prize.

Many states give citizens a chance to win money in lotteries. But how great an opportunity is it?

Solution Let E represent the event "winning first prize." To find the probability, we must first find the number of outcomes in event E and the number of outcomes in the sample space S.

There is only one first-prize-winning combination, so $n(E) = 1$. However, the sample space is huge. One possible selection is

1 32 43 4 15 26

Another is

41 12 4 24 25 37

Finding $n(S)$ by counting the elements in S is impractical, so we will use combinatorics. Order does not matter, because the player can choose the six numbers in any order. Therefore, we use *combinations*:

$$n(S) = {}_{44}C_6 = \frac{44!}{6! \cdot (44 - 6)!} = 7{,}059{,}052$$

44 [x!] [÷] 6 [x!] [÷] [(] 44 [−] 6 [)] [x!] [=]

Alternatively, use the "${}_nC_r$" command on a graphing calculator. Type 44 [${}_nC_r$] 6 [ENTER]

where [${}_nC_r$] refers to the "${}_nC_r$" command on the "MATH" menu, discussed in Section 2.4.

As you can see, the sample space is amazingly huge. The probability of selecting all six winning numbers is

$$p(E) = \frac{n(E)}{n(S)} = \frac{1}{7{,}059{,}052} \approx 0.00000014$$

This means that only one out of approximately *seven million* combinations is the first-prize-winning combination. Few events in life are less likely than this. ■

Keno

The game of keno is a casino version of the lottery. In this game, the casino has a container filled with balls numbered from 1 to 80. The player buys a keno ticket, with which he selects anywhere from 1 to 15 (usually 6, 8, 9, or 10) of those 80 numbers; the player's selections are called "spots." The casino chooses 20 winning numbers, using a mechanical device

Brooks/Cole Publishing

HISTORICAL NOTE
Lotteries

Lotteries have been with us for a long time. In the Bible's Book of Numbers, Moses used a lottery to award land west of the river Jordan. In the fifteenth century, lotteries appeared in France and Belgium, when cities used lotteries to raise money to fortify their defenses and to aid the poor. The first lottery that paid cash prizes was probably started in Florence, Italy in 1530. It was so successful that many other Italian cities began to offer their own lotteries. When Italian cities united to form a nation, the first national lottery was created. *Lotto,* the Italian national lottery, continues today and is regarded as the basis for such modern gambling games as keno, state lotteries, bingo, and the illegal numbers game.

Public lotteries have a long history in the United States. The settlement of Jamestown was financed in part by an English lottery. George Washington managed a lottery that paid for a road through the Cumberland mountains. Benjamin Franklin used lotteries to finance cannons for the Revolutionary War, and John Hancock used lotteries to rebuild historic Faneuil Hall in Boston. Several universities, including Harvard, Dartmouth, Yale, and Columbia, were partly financed by lotteries.

In 1964, New Hampshire became the first state to offer a state-sponsored lottery. Since then, Americans have paid almost $428 billion for lottery tickets. In 2001 alone, Americans paid more than $39 billion for lottery tickets. Overall, 52% of this money was paid out in prizes, 35% went to the government, and 13% was spent in administering the lotteries. Of course, these numbers vary from state to state. In Delaware, only 26% was paid out in prizes, while in South Dakota, 64% was paid out in prizes.

Lottery players go berserk when the jackpots accumulate, partially because of the amazingly large winnings but also because of a lack of understanding of how unlikely it is that they will actually win. The largest cumulative jackpot was $363 million, which was split by winners in Michigan and Illinois. The largest single-winner jackpot was $197 million in Georgia in 1999.

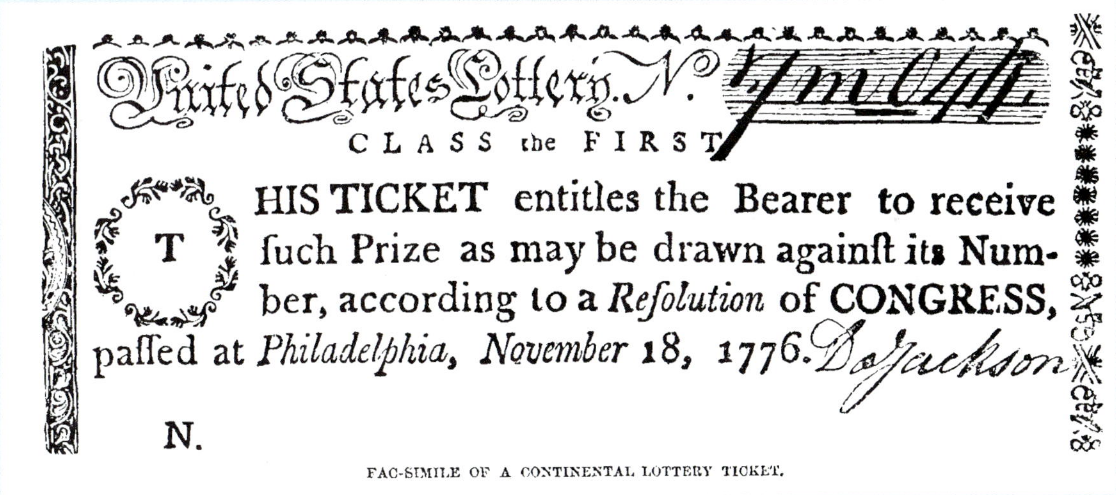
United States Lottery. No.

CLASS the FIRST

THIS TICKET entitles the Bearer to receive ſuch Prize as may be drawn againſt its Number, according to a *Reſolution* of CONGRESS, paſſed at *Philadelphia, November* 18, 1776.

N.

FAC-SIMILE OF A CONTINENTAL LOTTERY TICKET.

Pictorial History Research

The U.S. Congress operated a lottery to help fund the Revolutionary War.

to ensure a fair game. If a sufficient number of the player's spots are winning numbers, the player receives an appropriate payoff.

EXAMPLE 3 In the game of keno, if eight spots are marked, the player wins if five or more of his or her spots are selected. Find the probability of having five winning spots.

Solution The sample space S is the set of all ways that the player can select 8 numbers from the 80 numbers in the game. Order does not matter, because the player can choose the eight numbers in any order. Therefore, using combinations,

$$n(S) = {}_{80}C_8 = \frac{80!}{8! \cdot (80 - 8)!} = 28{,}987{,}537{,}150$$

Let E represent the event "having five winning spots (and 3 losing spots)." This involves two categories (winning spots and losing spots), so we will use the Fundamental Principal of Counting and multiply the number of ways of getting five winning spots and the number of ways of getting three losing spots. The casino selects 20 winning numbers, so there are

$${}_{20}C_5 = \frac{20!}{5! \cdot (20 - 5)!} = 15{,}504$$

ways of getting five winning spots. Also, the casino selects 60 losing numbers, so there are

$${}_{60}C_3 = \frac{60!}{3! \cdot (60 - 3)!} = 34{,}220$$

ways of getting three losing spots. Thus, the event consists of

$${}_{20}C_5 \cdot {}_{60}C_3 = 15{,}504 \cdot 34{,}220 = 530{,}546{,}880$$

elements, and the probability of having five winning spots in eight-spot keno is

$$p(E) = \frac{{}_{20}C_5 \cdot {}_{60}C_3}{{}_{80}C_8} = \frac{530{,}546{,}880}{28{,}987{,}537{,}150} \approx 0.018303$$

✔ In the event, there is a distinction between two categories (winning spots and losing spots); in the sample space, there is no such distinction. Thus, the numerator of

$$p(E) = \frac{{}_{20}C_5 \cdot {}_{60}C_3}{{}_{80}C_8}$$

has two parts (one for each category) and the denominator has one part. Also, the numbers in front of the Cs add correctly (20 winning spots + 60 losing spots = 80 total spots to choose from), and the numbers after the Cs add correctly (5 winning spots + 3 losing spots = 8 total spots to select).

■

Cards

One common form of poker is five-card draw, in which each player is dealt five cards. The order in which the cards are dealt is unimportant, so we compute probabilities with combinations rather than permutations.

EXAMPLE 4 Find the probability of being dealt four aces.

Solution The sample space consists of all possible five-card hands than can be dealt from a deck of 52. There are

$$_{52}C_5 = \frac{52!}{5! \cdot 47!} = 2{,}598{,}960$$

possible hands.

The event consists of all possible five-card hands that include four aces and one non-ace. This involves two categories (aces and non-aces), so we will use the Fundamental Counting Principle and multiply the number of ways of getting four aces and the number of ways of getting one non-ace. There is

$$_4C_4 = \frac{4!}{4! \cdot 0!} = 1$$

way of getting four aces, and there are

$$_{48}C_1 = \frac{48!}{1! \cdot 47!} = 48$$

ways of getting one non-ace. (These numbers could certainly be obtained with common sense rather than combinations.) Thus, the event consists of

$$_4C_4 \cdot {}_{48}C_1 = 1 \cdot 48 = 48$$

elements, and the probability of being dealt four aces is

$$p(E) = \frac{_4C_4 \cdot {}_{48}C_1}{_{52}C_5} = \frac{48}{2{,}598{,}960} \approx 0.00001847$$

✔ In the event, there is a distinction between two categories (aces and non-aces); in the sample space, there is no such distinction. Thus, the numerator of

$$p(E) = \frac{_4C_4 \cdot {}_{48}C_1}{_{52}C_5}$$

has two parts (one for each category), and the denominator has one part. Also, the numbers in front of the *C*s add correctly (4 aces + 48 non-aces = 52 cards to choose from), and the numbers after the *C*s add correctly (4 aces + 1 non-ace = 5 cards to select).

■

EXAMPLE 5 Find the probability of being dealt four of a kind.

Solution The sample space is the same as that in Example 4. The event "being dealt four of a kind" means "being dealt four twos or being dealt four threes or being dealt four fours . . . or being dealt four kings or being dealt four aces." These latter events ("four twos," "four threes," etc.) are all mutually exclusive, so we can use Probability Rule 5:

$$\begin{aligned} &p\text{ (four of a kind)} \\ &\quad = p(\text{four twos} \cup \text{four threes} \cup \cdots \cup \text{four kings} \cup \text{four aces}) \\ &\quad = p(\text{four twos}) + p(\text{four threes}) + \cdots + p(\text{four kings}) + p(\text{four aces}) \end{aligned}$$

using probability rule 5

Furthermore, these probabilities are all the same

$$p(\text{four twos}) = p(\text{four threes}) = \cdots = p(\text{four aces}) = \frac{_4C_4 \cdot {}_{48}C_1}{_{52}C_5}$$

This means that the probability of being dealt four of a kind is

$$\begin{aligned} p(\text{four of a kind}) &= p(\text{four twos}) + p(\text{four threes}) + \cdots + p(\text{four kings}) \\ &\quad + p(\text{four aces}) \\ &= \frac{{}_4C_4 \cdot {}_{48}C_1}{{}_{52}C_5} + \frac{{}_4C_4 \cdot {}_{48}C_1}{{}_{52}C_5} + \cdots + \frac{{}_4C_4 \cdot {}_{48}C_1}{{}_{52}C_5} + \frac{{}_4C_4 \cdot {}_{48}C_1}{{}_{52}C_5} \\ &= 13 \cdot \frac{{}_4C_4 \cdot {}_{48}C_1}{{}_{52}C_5} \qquad \text{there are 13 denominations (two through ace)} \\ &= 13 \cdot \frac{1 \cdot 48}{2{,}598{,}960} \qquad \text{from Example 4} \\ &= \frac{624}{2{,}598{,}960} \approx 0.0002401 \end{aligned}$$

See Example 8 in Section 2.4.

EXAMPLE 6 Find the probability of being dealt five hearts.

Solution The sample space is the same as in Example 4. The event consists of all possible five-card hands that include five hearts and no non-hearts. This involves two categories (hearts and non-hearts), so we will use the Fundamental Counting Principle and multiply the number of ways of getting five hearts and the number of ways of getting no non-hearts. There are

$${}_{13}C_5 = \frac{13!}{5! \cdot 8!} = 1{,}287$$

ways of getting five hearts, and there is

$${}_{39}C_0 = \frac{39!}{0! \cdot 39!} = 1$$

way of getting no non-hearts. Thus, the probability of being dealt five hearts is

$$p(E) = \frac{{}_{13}C_5 \cdot {}_{39}C_0}{{}_{52}C_5} = \frac{1287 \cdot 1}{2{,}598{,}960} = 0.000495198$$

✔ In the event, there is a distinction between two categories (hearts and non-hearts); in the sample space, there is no such distinction. Thus, the numerator of

$$p(E) = \frac{{}_{13}C_5 \cdot {}_{39}C_0}{{}_{52}C_5}$$

has two parts (one for each category) and the denominator has one part. Also, the numbers in front of the *C*s add correctly (13 hearts + 39 non-hearts = 52 cards to choose from), and the numbers after the *C*s add correctly (5 hearts + 0 non-hearts = 5 cards to select).

■

Notice that in Example 6, we could argue that since we're selecting only hearts, we can disregard the non-hearts. This would lead to the answer obtained in Example 6:

$$p(E) = \frac{{}_{13}C_5}{{}_{52}C_5} = \frac{1287}{2{,}598{,}960} = 0.000495198$$

However, this approach would not allow us to check our work in the manner described on page 164; the numbers in front of the *C*s don't add correctly, nor do the numbers after the *C*s.

3.4 Exercises

1. A group of 30 people is selected at random. What is the probability that at least two of them will have the same birthday?
2. A group of 60 people is selected at random. What is the probability that at least two of them will have the same birthday?
3. In 1990, California switched from a 6/49 lottery to a 6/53 lottery. Later, the state switched again to a 6/51 lottery.
   - **a.** Find the probability of winning first prize in a 6/49 lottery.
   - **b.** Find the probability of winning first prize in a 6/53 lottery.
   - **c.** Find the probability of winning first prize in a 6/51 lottery.
   - **d.** How much more probable is it that one will win the 6/49 lottery than the 6/53 lottery?
   - **e.** Why do you think California switched from a 6/49 lottery to a 6/53 lottery? And why do you think the state then switched to a 6/51 lottery? (Answer using complete sentences.)
4. Find the probability of winning second prize—that is, picking five of the six winning numbers—with a 6/53 lottery.
5. Find the probability of winning second prize (that is, picking five of the six winning numbers) with a 6/44 lottery, as played in Connecticut, Missouri, Oregon, and Virginia.
6. Find the probability of winning third prize—that is, picking four of the six winning numbers—with a 6/44 lottery.
7. Currently, the second most popular type of lottery is the 5/35 lottery. It is played in Arizona, Massachusetts, Connecticut, Iowa, South Dakota, and Nebraska.
   - **a.** Find the probability of winning first prize.
   - **b.** Find the probability of winning second prize.
8. The most popular type of lottery is the 6/49 lottery. It is currently played in Massachusetts, Maryland, New Jersey, Ohio, Pennsylvania, Washington, and Wisconsin.
   - **a.** Find the probability of winning first prize.
   - **b.** Find the probability of winning second prize.
9. The 5/39 lottery is currently played in California, Georgia, New York, Pennsylvania, and Texas.
   - **a.** Find the probability of winning first prize.
   - **b.** Find the probability of winning second prize.
10. The 6/42 lottery is currently played in Colorado, Massachusetts, Kentucky, Maine, New Hampshire, and Vermont.
    - **a.** Find the probability of winning first prize.
    - **b.** Find the probability of winning second prize.
11. There is an amazing variety of lotteries played in the United States. Currently, the following lotteries are played: 6/25, 5/26, 5/30, 6/30, 5/31, 7/31, 5/32, 5/33, 5/34, 5/35, 6/35, 5/36, 6/36, 5/37, 5/38, 4/39, 5/39, 6/39, 6/40, 6/41, 6/42, 5/44, 6/44, 6/46, 6/48, 6/49, 4/50, 6/50, 6/51, 5/52, 6/52, 6/53, and 4/100. Which is the easiest to win? Which is the hardest to win? Explain your reasoning.

    HINT: It isn't necessary to compute every single probability.
12. In the game of keno, if six spots are marked, the player wins if four or more of his or her spots are selected. Complete the following chart.

Outcome	Probability
6 winning spots	
5 winning spots	
4 winning spots	
3 winning spots	
fewer than 3 winning spots	

13. In the game of keno, if eight spots are marked, the player wins if five or more of his or her spots are selected. Complete the following chart.

Outcome	Probability
8 winning spots	
7 winning spots	
6 winning spots	
5 winning spots	
4 winning spots	
fewer than 4 winning spots	

14. In the game of keno, if nine spots are marked, the player wins if six or more of his or her spots are selected. Complete the following chart.

Outcome	Probability
9 winning spots	
8 winning spots	
7 winning spots	
6 winning spots	
5 winning spots	
fewer than 5 winning spots	

15. **a.** Find the probability of being dealt five spades when playing five-card draw poker.
    b. Find the probability of being dealt five cards of the same suit when playing five-card draw poker.
    c. When you are dealt five cards of the same suit, you have either a *flush* (if the cards are not in sequence) or a *straight flush* (if the cards are in sequence). For each suit, there are ten possible straight flushes ("ace, two, three, four, five," through "ten, jack, queen, king, ace"). Find the probability of being dealt a straight flush.
    d. Find the probability of being dealt a flush.
16. **a.** Find the probability of being dealt an "aces over kings" full house (three aces and two kings).
    b. Why are there $13 \cdot 12$ different types of full houses?
    c. Find the probability of being dealt a full house. (Round each answer off to six decimal places.)

You order twelve burritos to go from a Mexican restaurant, five with hot peppers and seven without. However, the restaurant forgot to label them. If you pick three burritos at random, find the probability of each event in Exercises 17–24.

17. All have hot peppers.
18. None has hot peppers.
19. Exactly one has hot peppers.
20. Exactly two have hot peppers.
21. At most one has hot peppers.
22. At least one has hot peppers.
23. At least two have hot peppers.
24. At most two have hot peppers.
25. Two hundred people apply for two jobs. Sixty of the applicants are women.
    a. If two persons are selected at random, what is the probability that both are women?
    b. If two persons are selected at random, what is the probability that only one is a woman?
    c. If two persons are selected at random, what is the probability that both are men?
    d. If you were an applicant and the two selected people were not of your gender, do you think that the above probabilities would indicate the presence or absence of gender discrimination in the hiring process? Why?
26. Two hundred people apply for three jobs. Sixty of the applicants are women.
    a. If three persons are selected at random, what is the probability that all are women?
    b. If three persons are selected at random, what is the probability that two are women?
    c. If three persons are selected at random, what is the probability that one is a woman?
    d. If three persons are selected at random, what is the probability that none is a woman?
    e. If you were an applicant, and the three selected people were not of your gender, should the above probabilities have an impact on your situation? Why?
27. Powerball is a complicated lottery game played in Arizona, Connecticut, the District of Columbia, Delaware, Idaho, Indiana, Iowa, Kansas, Kentucky, Louisiana, Minnesota, Missouri, Nebraska, New Hampshire, New Mexico, Oregon, Rhode Island, South Dakota, West Virginia, and Wisconsin. This popular game involves selecting any five of the numbers from 1 to 49 plus any one of the numbers from 1 to 42.
    a. Find the probability of winning first prize in Powerball.
    b. The most similar simple lottery game involves selecting any six of the numbers from 1 to 49. Find the probability of winning first prize in a 6/49 game.
    c. Which is more likely: winning first prize in Powerball or winning first prize in a 6/49 game? How many times more likely is it?
    d. Do you think that a typical lottery player would think that Powerball or a 6/49 game is easier to win? Why?
    e. Why do you think that Powerball is structured the way it is?
28. In Example 2, $n(E) = 1$ because only one of the 7,059,052 possible lottery tickets is the first prize winner. Use combinations to show that $n(E) = 1$.

Answer the following questions using complete sentences and your own words.

Concept Questions

29. Explain why, in Example 5, $p(\text{four twos}) = p(\text{four threes}) = \cdots = p(\text{four kings}) = p(\text{four aces})$.

30. Do you think a state lottery is a good thing for the state's citizens? Why? Be certain to include a discussion of both the advantages and disadvantages of a state lottery to its citizens.

31. Why are probabilities for most games of chance calculated with combinations, rather than permutations?

32. Suppose a friend or relative of yours regularly spends (and loses) a good deal of money on lotteries. How would you explain to this person why he or she loses so frequently?

33. In Example 2, $n(E) = 1$ because only one of the 7,059,052 possible lottery tickets is the first prize winner. Does this mean that it is impossible for two people to each buy a first prize winning lottery ticket? Explain.

History Questions

34. Where and how did lotteries originate?

35. Are public lotteries relative newcomers to the American scene?

3.5 Expected Value

Suppose you're playing roulette, concentrating on the \$1 single-number bet. At one point, you were \$10 ahead, but now you're \$14 behind. How much should you expect to win or lose, on average, if you were to place the bet many times?

The probability of winning a single-number bet is $\frac{1}{38}$, because there are 38 numbers on the roulette wheel and only one of them is the subject of the bet. This means that if you place the bet a large number of times, it is most likely that you will win once for every 38 times you place the bet (and lose the other 37 times). When you win, you win \$35, because the house odds are 35 to 1. When you lose, you lose \$1. Your average profit would be

$$\frac{\$35 + 37 \cdot (-\$1)}{38} = \frac{-\$2}{38} \approx -\$0.053$$

per game. This is called the *expected value* of a \$1 single-number bet, because you should expect to lose about a nickel for every dollar you bet if you play the game a long time. If you play a few times, anything could happen—you could win every single bet (though it's not likely). The house makes the bet so many times that it can be certain that its profit will be \$0.053 per dollar bet.

The standard way to find the **expected value** of an experiment is to multiply the value of each outcome of the experiment by the probability of that outcome and add the results. Here the experiment is placing a \$1 single-number bet in roulette. The outcomes are winning the bet and losing the bet. The values of the outcomes are +\$35 (if you win) and −\$1 (if you lose); the probabilities of the outcomes are $\frac{1}{38}$ and $\frac{37}{38}$, respectively (see Figure 3.9). The expected value would then be

Outcome	Value	Probability
winning	35	$\frac{1}{38}$
losing	−1	$\frac{37}{38}$

Figure 3.9

$$35 \cdot \frac{1}{38} + (-1) \cdot \frac{37}{38} \approx -\$0.053$$

It is easy to see that this calculation is algebraically equivalent to the calculation done above.

Finding an expected value of a bet is very similar to finding your average test score in a class. Suppose you're a student in a class in which you've taken

Renato Rotolo/Liaison Agency

Roulette, the oldest casino game played today, has been popular since it was introduced to Paris in 1765. Does this game have any good bets?

four tests. If your scores were 80%, 76%, 90%, and 90%, your average test score would be

$$\frac{80 + 76 + 2 \cdot 90}{4} = 84\%$$

or, equivalently,

$$80 \cdot \frac{1}{4} + 76 \cdot \frac{1}{4} + 90 \cdot \frac{2}{4}$$

The difference between finding an average test score and finding the expected value of a bet is that with the average test score you are summarizing what *has* happened, whereas with a bet you are using probabilities to project what *will* happen.

Expected Value

To find the **expected value** (or "long-term average") of an experiment, multiply the value of each outcome of the experiment by its probability and add the results.

EXAMPLE 1 By analyzing her sales records, a saleswoman has found that her weekly commissions have the following probabilities:

Commission	0	\$100	\$200	\$300	\$400
Probability	0.05	0.15	0.25	0.45	0.1

Find the saleswoman's expected commission.

Solution To find the expected commission, we multiply each possible commission by its probability and add the results. Therefore,

$$\begin{aligned}\text{expected commission} &= (0)(0.05) + (100)(0.15) + (200)(0.25) \\ &\quad + (300)(0.45) + (400)(0.1) \\ &= 240\end{aligned}$$

On the basis of her history, the saleswoman should expect to average \$240 per week in future commissions. Certainly, anything can happen in the future—she could receive a \$700 commission (it's not likely, though, because it has never happened before). ■

Why the House Wins

Four of the "best" bets that can be made in a casino game of chance are the pass, don't pass, come, and don't come bets in craps. They all have the same expected value, −\$0.014. In the long run, *there isn't a single bet in any game of chance with which you can expect to break even, let alone make a profit.* After all, the casinos are out to make money. The expected values for \$1 bets in the more common games are shown in Figure 3.10.

Game	Expected Value of \$1 Bet
baccarat	−\$0.014
blackjack	−\$0.06 to +\$0.10 (varies with strategies)
craps	−\$0.014 for pass, don't pass, come, don't come bets *only*
slot machines	−\$0.13 to ? (varies)
keno (eight-spot ticket)	−\$0.29
average state lottery	−\$0.48

Figure 3.10
Expected values of common games of chance

It is possible to achieve a positive expected value in blackjack and other card games in which a number of hands are played from the same deck without reshuffling before each hand. To do this, the player must memorize which cards have been played in previous hands. Many casinos use four decks at once to discourage memorization. Some people try to sneak small homemade computers into the casino to aid them. Naturally, the casinos forbid this.

Decision Theory

Which is the better bet—a \$1 single-number bet in roulette or a lottery ticket? Each costs \$1. The roulette bet pays \$35, but the lottery ticket might pay several million dollars. Lotteries are successful in part because the possibility of winning a large amount of money distracts people from the fact that winning is extremely unlikely. In Example 2 of Section 3.4, we found that the probability of winning first prize in many state lotteries is $\frac{1}{7{,}059{,}052} \approx 0.00000014$. At the beginning of this section, we found that the probability of winning the roulette bet is $\frac{1}{38} \approx 0.03$.

A more informed decision would take into account not only the potential winnings and losses but also their probabilities. The expected value of a bet does just that, since its calculation involves both the value and the probability of each outcome. We found that the expected value of a $1 single-number bet in roulette is about −$0.053. The expected value of the average state lottery is −$0.48 (see Figure 3.10). The roulette bet is a much better bet than is the lottery. (Of course, there is a third option, which has an even better expected value of $0.00. Not gambling!)

A decision always involves choosing between various alternatives. If you compare the expected values of the alternatives, then you are taking into account the alternatives' potential advantages and disadvantages as well as their probabilities. This form of decision making is called **decision theory**.

EXAMPLE 2 The saleswoman in Example 1 has been offered a new job that has a fixed weekly salary of $290. Financially, which is the better job?

Solution In Example 1, we found that her expected weekly commission was $240. The new job has a guaranteed weekly salary of $290. Financial considerations indicate that she should take the new job. ■

Betting Strategies

One very old betting strategy is to "cover all the numbers." In 1729, the French philosopher and writer François Voltaire organized a group that successfully implemented this strategy to win the Parisian city lottery by buying most if not all of the tickets. Their strategy was successful because, owing to a series of poor financial decisions by the city of Paris, the total value of the prizes was greater than the combined price of all of the tickets! Furthermore, there weren't a great number of tickets to buy. This strategy is still being used. (See the newspaper article on its use in 1992 in Virginia.)

A **martingale** is a gambling strategy in which the gambler doubles his or her bet after each loss. A person using this strategy in roulette, concentrating on the black numbers bet (which has 1-to-1 house odds), might lose three times before winning. This strategy would result in a net gain, as illustrated in Figure 3.11.

This seems to be a great strategy. Sooner or later the player will win a bet, and because each bet is larger than the player's total losses, he or she has to come out ahead! We will examine this strategy further in the exercises.

Bet Number	Bet	Result	Total Winnings/Losses
1	$1	lose	−$1
2	$2	lose	−$3
3	$4	lose	−$7
4	$8	win	+$1

Figure 3.11

Virginia lottery hedges on syndicate's big win

RICHMOND, VA. — Virginia lottery officials confirmed yesterday that an Australian gambling syndicate won last month's record $27 million jackpot after executing a massive block-buying operation that tried to cover all 7 million possible ticket combinations.

But lottery director Kenneth Thorson said the jackpot may not be awarded because the winning ticket may have been bought in violation of lottery rules.

The rules say tickets must be paid for at the same location where they are issued. The Australian syndicate, International Lotto Fund, paid for many of its tickets at the corporate offices of Farm Fresh Inc. grocery stores, rather than at the Farm Fresh store in Chesapeake where the winning ticket was issued, Thorson said.

"We have to validate who bought the ticket, where the purchase was made and how the purchase was made," Thorson said. "It's just as likely that we will honor the ticket as we won't honor the ticket." He said he may not decide until the end of next week.

Two Australians representing the fund, Joseph Franck and Robert Hans Roos, appeared at lottery headquarters yesterday to claim the prize.

The group succeeded in buying about 5 million of the more than 7 million possible numerical combinations before the February 15 drawing. The tactic is not illegal, although lottery officials announced new rules earlier this week aimed at making such block purchases more difficult.

The Australian fund was started last year and raised about $13 million from an estimated 2,500 shareholders who each paid a minimum of $4,000, according to Tim Phillipps of the Australian Securities Commission.

Half the money went for management expenses, much of that to Pacific Financial Resources, a firm controlled by Stefan Mandel, who won fame when he covered all the numbers in a 1986 Sydney lottery. Roos owns 10 percent of Pacific Financial Resources.

Australian Securities Commission officials said last week that the fund is under investigation for possible violations of Australian financial laws.

3.5 Exercises

In Exercises 1–10, (a) find and (b) interpret the expected value of each $1 bet in roulette.

1. the two-number bet
2. the three-number bet
3. the four-number bet
4. the five-number bet
5. the six-number bet
6. the twelve-number bet
7. the low-number bet
8. the even-number bet
9. the red-number bet
10. the black-number bet

11. Using the expected values obtained in the text and in the preceding odd-numbered exercises, determine a casino's expected net income from a 24-hour period at a single roulette table if the casino's total overhead for the table is $50 per hour and if customers place a total of $7,000 on single-number bets, $4,000 on two-number bets, $4,000 on four-number bets, $3,000 on six-number bets, $7,000 on low-number bets, and $8,000 on red-number bets.

12. Using the expected values obtained in the text and in the preceding even-numbered exercises, determine a casino's expected net income from a 24-hour period at a single roulette table if the casino's total overhead for the table is $50 per hour and if customers place a total of $8,000 on single-number bets, $3,000 on three-number bets, $4,000 on five-number bets, $4,000 on twelve-number bets, $8,000 on even-number bets, and $9,000 on black-number bets.

13. On the basis of his previous experience, the public librarian at Smallville knows that the number of books checked out by a person visiting the library has the following probabilities:

Number of Books	0	1	2	3	4	5
Probability	0.15	0.35	0.25	0.15	0.05	0.05

Find the expected number of books checked out by a person visiting this library.

14. On the basis of his sale records, a salesman knows that his weekly commissions have the following probabilities:

Commission	0	$1,000	$2,000	$3,000	$4,000
Probability	0.15	0.2	0.45	0.1	0.1

Find the salesman's expected commission.

15. Of all workers at a certain factory, the proportions earning certain hourly wages are as follows:

Hourly Wage	$8.50	$9.00	$9.50	$10.00	$12.50	$15.00
Proportion	20%	15%	25%	20%	15%	5%

Find the expected hourly wage that a worker at this factory makes.

16. Of all students at the University of Metropolis, the proportions taking certain numbers of units are as shown in Figure 3.12. Find the expected number of units that a student at U.M. takes. (Shown in Figure 3.13 at the bottom of the page.)

17. Show why the calculation at the bottom of page 167 is algebraically equivalent to the calculation in the middle of page 167.

18. In Example 1, the saleswoman's most likely weekly commission was $300. With her new job (in Example 2), she will always make $290 per week. This implies that she would be better off with the old job. Is this reasoning more or less valid than that used in Example 2? Why?

19. Maria just inherited $10,000. Her bank has a savings account that pays 4.1% interest per year. Some of her friends recommended a new mutual fund, which has been in business for three years. During its first year, the fund went up in value by 10%; during the second year, it went down by 19%; and during its third year, it went up by 14%. Maria is attracted by the mutual fund's potential for relatively high earnings but concerned by the possibility of actually losing some of her inheritance. The bank's rate is low, but it is insured by the federal government. Use decision theory to find the best investment. (Assume that the fund's past behavior predicts its future behavior.)

20. Trang has saved $8,000. It is currently in a bank savings account that pays 3.9% interest per year. He is considering putting the money into a speculative investment that would either earn 20% in one year if the investment succeeds or lose 18% in one year if it fails. At what probability of success would the speculative investment be the better choice?

21. Erica has her savings in a bank account that pays 4.5% interest per year. She is considering buying stock in a pharmaceuticals company that is developing a cure for cellulite. Her research indicates that she could earn 50% in one year if the cure is successful or lose 60% in one year if it is not. At what probability of success would the pharmaceuticals stock be the better choice?

22. Debra is buying prizes for a game at her school's fundraiser. The game has three levels of prizes, and she has already bought the second and third prizes. She wants the first prize to be nice enough to attract people to the game. The game's manufacturer has supplied her with the probabilities of winning first, second, and third prizes. Tickets cost $3 each, and she wants the school to profit an average of $1 per ticket. How much should she spend on each first prize?

Prize	Cost of Prize	Probability
1st	?	.15
2nd	$1.25	.30
3rd	$0.75	.45

23. Few students manage to complete their schooling without taking a standardized admissions test such as the Scholastic Achievement Test, or S.A.T. (used for admission to college); the Law School Admissions Test, or L.S.A.T.; and the Graduate Record Exam, or G.R.E. (used for admission to graduate school). Sometimes, these multiple-choice tests discourage guessing by subtracting points for wrong answers. In particular, a correct answer will be worth $+1$ point, and an incorrect answer on a question with 5 listed answers (a through e) will be worth $-\frac{1}{4}$ point.
   a. Find the expected value of a random guess.
   b. Find the expected value of eliminating one answer and guessing among the remaining four possible answers.
   c. Find the expected value of eliminating three answers and guessing between the remaining two possible answers.
   d. Use decision theory and your answers to parts (a), (b), and (c) to create a guessing strategy for standardized tests such as the S.A.T.

units	3	4	5	6	7	8	9	10	11	12	13	14
proportion	3%	4%	5%	6%	5%	4%	8%	12%	13%	13%	15%	12%

Figure 3.12

24. Find the expected value of a $1 bet in six-spot keno if three winning spots pays $1 (but you pay $1 to play, so you actually break even), four winning spots pays $3 (but you pay $1 to play, so you profit $2), five pays $100, and six pays $2,600. (You might want to use the probability distribution computed in Exercise 12 of Section 3.4.)

25. Find the expected value of a $1 bet in eight-spot keno if four winning spots pays $1 (but you pay $1 to play, so you actually break even), five winning spots pays $5 (but you pay $1 to play, so you profit $4), six winning spots pays $100, seven winning spots pays $1,480, and eight winning spots pays $19,000. (You might want to use the probability distribution computed in Exercise 13 of Section 3.4.)

26. Find the expected value of a $1 bet in nine-spot keno if five winning spots pays $1 (but you pay $1 to play, so you actually break even), six winning spots pays $50 (but you pay $1 to play, so you profit $49), seven pays $390, eight pays $6,000, and nine pays $25,000. (You might want to use the probability distribution computed in Exercise 14 of Section 3.4.)

27. Arizona's "Fantasy Five" is a 5/35 lottery. It differs from many other state lotteries in that its payouts are set; they do not vary with sales. To win first prize, you must select all five of the winning numbers. To win second prize, you must select any four of the five winning numbers; to win third prize, you must select any three of the five winning numbers. The first prize jackpot is $50,000 (but you pay $1 to play, so you profit $49,999). Second prize pays $500, and third prize pays $5. If you select two or fewer winning numbers, you lose your $1. Find the expected value of the Fantasy Five. (You might want to use the probabilities computed in Exercise 7 of Section 3.4.)

28. Write a paragraph in which you compare the states' fiscal policies concerning their lotteries with the casinos' fiscal policies concerning their keno games. Assume that the expected value of Arizona's Fantasy Five as described in Exercise 27 is representative of that of the other states' lotteries, and assume that the expected value of a $1 keno bet as described in Exercises 24 and/or 25 is representative of other keno bets.

29. Trustworthy Insurance Co. estimates that a certain home has a 1% chance of burning down in any one year. They calculate that it would cost $120,000 to rebuild that home. Use expected values to determine the annual insurance premium.

30. Mr. and Mrs. Trump have applied to the Trustworthy Insurance Co. for insurance on Mrs. Trump's diamond tiara. The tiara is valued at $97,500. Trustworthy estimates that the jewelry has a 2.3% chance of being stolen in any one year. Use expected values to determine the annual insurance premium.

31. The Black Gold Oil Co. is considering drilling either in Jed Clampett's back yard or his front yard. After thorough testing and analysis, they estimate that there is a 30% chance of striking oil in the back yard and a 40% chance in the front yard. They also estimate that the back yard site would either net $60 million (if oil is found) or lose $6 million (if oil is not found), and the front yard site would either net $40 million or lose $6 million. Use decision theory to determine where they should drill.

32. If in Exercise 31 Jed Clampett rejected the use of decision theory, where would he drill if he were an optimist? What would he do if he were a pessimist?

33. A community youth group is having a raffle to raise funds. Several community businesses have donated prizes. The prizes and their retail value are listed in Figure 3.13 on page 174. Each prize will be given away, regardless of the number of raffle tickets sold. Tickets are sold for $15 each. Determine the expected value of a ticket, and discuss whether it would be to your financial advantage to buy a ticket under the given circumstances.
   a. 1000 tickets are sold.
   b. 2000 tickets are sold.
   c. 3000 tickets are sold.

34. The Centerville High School PTA is having a raffle to raise funds. Several community businesses have donated prizes. The prizes and their retail value are listed in Figure 3.14 on page 174. Each prize will be given away, regardless of the number of raffle tickets sold. Tickets are sold for $30 each. Determine the expected value of a ticket, and discuss whether it would be to your financial advantage to buy a ticket under the given circumstances.
   a. 100 tickets are sold.
   b. 200 tickets are sold.
   c. 300 tickets are sold.

35. The Central State University Young Republicans Club is having a raffle to raise funds. Several community businesses have donated prizes. The prizes and their retail value are listed in Figure 3.15 on page 175. Each prize will be given away, regardless of the number of raffle tickets sold. Tickets are sold for $5 each. Determine the expected value of a ticket, and discuss whether it would be to your financial advantage to buy a ticket under the given circumstances.
   a. 1000 tickets are sold.
   b. 2000 tickets are sold.
   c. 3000 tickets are sold.

Prize	Retail Value	Number of These Prizes to Be Given Away
new car	$21,580	1
a cell phone and a one-year subscription	$940	1
a one-year subscription to an Internet service provider	$500	2
dinner for two at Spiedini's restaurant	$100	2
a one-year subscription to the local newspaper	$180	20

Figure 3.13
Use for Problem 33.

Prize	Retail Value	Number of These Prizes to Be Given Away
one week in a condo in Hawaii, and airfare for two	$2,575	1
tennis lessons for two	$500	1
a one-year subscription to an Internet service provider	$500	2
dinner for two at Haute Stuff restaurant	$120	2
a one-year membership at the Centerville Skate Park	$240	5
a copy of *Centerville Cooks*, a cookbook containing PTA members' favorite recipes	$10	20

Figure 3.14
Use for Problem 34.

Prize	Retail Value	Number of These Prizes to Be Given Away
laptop computer	$2,325	1
MP3 player	$425	2
20 CDs at Einstein Entertainment	$320	3
a giant pizza and your choice of beverage at Freddie's Pizza	$23	4
a one-year subscription to the *Young Republican Journal*	$20	15

Figure 3.15
Use for Problem 35.

36. The Southern State University Ecology Club is having a raffle to raise funds. Several community businesses have donated prizes. The prizes and their retail value are listed in Figure 3.16. Each prize will be given away, regardless of the number of raffle tickets sold. Tickets are sold for $20 each. Determine the expected value of a ticket, and discuss whether it would be to your financial advantage to buy a ticket under the given circumstances.
a. 100 tickets are sold.
b. 200 tickets are sold.
c. 300 tickets are sold.

Prize	Retail Value	Number of These Prizes to Be Given Away
one-week ecovacation to Costa Rica, including airfare for two	$2,990	1
Sierra Designs tent	$750	1
REI backpack	$355	2
Jansport daypack	$75	5
fleece jacket with Ecology Club logo	$50	10
Ecology Club T-shirt	$18	20

Figure 3.16

37. Find the expected value of the International Lotto Fund's application of the "cover all of the numbers" strategy from the newspaper article on page 171. Assume that \$5 million was spent on lottery tickets, that half of the \$13 million raised went for management expenses, and that the balance was never spent. Also assume that Virginia honors the winning ticket.

38. One application of the "cover all the numbers" strategy would be to bet \$1 on every single number in roulette.

a. Find the results of this strategy.

b. How could you use the expected value of the \$1 single-number bet ($\frac{-\$2}{38}$) to answer part (a)?

39. The application of the "cover all the numbers" strategy to a modern state lottery would involve the purchase of a large number of tickets.

a. How many tickets would have to be purchased if you were in a state that has a 6/49 lottery (the player selects 6 out of 49 numbers)?

b. How much would it cost to purchase these tickets if each costs \$1?

c. If you organized a group of 100 people to purchase these tickets and it takes 1 minute to purchase each ticket, how many days would it take to purchase the required number of tickets?

40. The application of the "cover all the numbers" strategy to Keno would involve the purchase of a large number of tickets.

a. How many tickets would have to be purchased if you were playing eight-spot keno?

b. How much would it cost to purchase these tickets if each costs \$5?

c. If it takes 5 seconds to purchase one ticket and the average keno game lasts 20 minutes, how many people would it take to purchase the required number of tickets?

41. If you had \$100 and were applying the martingale strategy to the black-number bet in roulette and you started with a \$1 bet, how many successive losses could you afford? How large would your net profit be if you lost each bet except for the last one?

42. If you had \$10,000 and were applying the martingale strategy to the black-number bet in roulette and you started with a \$1 bet, how many successive losses could you afford? How large would your net profit be if you lost each bet except for the last one?

Answer the following questions using complete sentences and your own words.

Concept Questions

43. Discuss the meaningfulness of the concept of expected value to three different people: an occasional gambler, a regular gambler, and a casino owner. To whom is the concept most meaningful? To whom is it least meaningful? Why?

44. Discuss the advantages and disadvantages of decision theory. Consider the application of decision theory to a nonrecurring situation and to a recurring situation. Also consider Jed Clampett and the Black Gold Oil Co. in Exercises 31 and 32.

45. The Historical Note in Section 3.4 states that "In 2001 alone, Americans paid more than \$39 billion for lottery tickets. Overall, 52% of this money was paid out in prizes, 35% went to the government, and 13% was spent in administering the lotteries."

a. Use this information to find the expected value of an American lottery ticket in 2001.

b. Is this method of computing an expected value more or less valid than that used in Exercises 24–26? Why?

Projects

46. Design a game of chance. Use probabilities and expected values to set the house odds so that the house will make a profit. However, be certain that your game is not so obviously pro-house that no one would be willing to play. Your project should include the following:

- a complete description of how the game is played
- a detailed mathematical analysis of the expected value of the game (or of each separate bet in the game, whichever is appropriate)
- a complete description of the bet(s) and the house odds

3.6 Conditional Probability

Public opinion polls, such as those found in newspapers and magazines and on television, frequently categorize the respondents by such groups as sex, age, race, or level of education. This is done so that the reader or listener can make comparisons and observe trends, such as "people over 40 are more likely to support the Social Security system than are people under 40." The tool that enables us to observe such trends is conditional probability.

Probabilities and Polls

In a newspaper poll concerning violence on television, 600 people were asked, "What is your opinion of the amount of violence on prime-time television—is there too much violence on television?" Their responses are indicated in Figure 3.17.

	Yes	No	Don't Know	Total
Men	162	95	23	280
Women	256	45	19	320
Total	418	140	42	600

Figure 3.17
Results of "Violence on Television" poll

Six hundred people were surveyed in this poll; that is, the sample space consists of 600 responses. Of these, 418 said they thought there was too much violence on television, so the probability of a "yes" response is $\frac{418}{600}$, or about $0.70 = 70\%$. The probability of a "no" response is $\frac{140}{600}$, or about $0.23 = 23\%$.

If we are asked to find the probability that a *woman* responded yes, we do not consider all 600 responses but instead limit the sample space to only the responses from women.

	Yes	No	Don't Know	Total
Women	256	45	19	320

The probability that a woman responded yes is $\frac{256}{320} = 0.80 = 80\%$.

Suppose we label the events in the following manner: W is the event that a response is from a woman, M is the event that a response is from a man, Y is the event

Warner Bros., Courtesy Kobal

Is there too much violence on TV?

that a response is yes, and N is the event that a response is no. Then the event that a woman responded yes would be written as

$$Y \mid W$$

The vertical bar stands for the phrase "given that"; the event $Y \mid W$ is read "a response is yes, given that the response is from a woman." The probability of this event is called a *conditional probability*:

$$p(Y \mid W) = \frac{256}{320} = \frac{4}{5} = 0.80 = 80\%$$

The numerator of this probability, 256, is the number of responses that are yes and are from women; that is, $n(Y \cap W) = 256$. The denominator, 320, is the number of responses that are from women; that is, $n(W) = 320$. A **conditional probability** is a probability whose sample space has been limited to only those outcomes that fulfill a certain condition. Because an event is a subset of the sample space, the event must also fulfill that condition. The numerator of $p(Y \mid W)$ is 256 rather than 418 even though there were 418 "yes" responses, because many of those 418 responses were made by men; we are interested only in the probability that a woman responded yes.

Conditional Probability Definition

The **conditional probability** of event A, given event B, is

$$p(A \mid B) = \frac{n(A \cap B)}{n(B)}$$

EXAMPLE 1 Using the data in Figure 3.17, find the following.

a. the probability that a response is yes, given that the response is from a man.
b. the probability that a response is from a man, given that the response is yes.
c. the probability that a response is yes and is from a man.

Solution **a.** *Finding $p(Y \mid M)$*: We are told to consider only the male responses—that is, to limit our sample space to men.

	Yes	No	Don't Know	Total
Men	162	95	23	280

$$p(Y \mid M) = \frac{n(Y \cap M)}{n(M)} = \frac{162}{280} \approx 0.58 = 58\%$$

In other words, approximately 58% of the men responded yes. (Recall that 80% of the women responded yes. This poll indicates that men and women do not have the same opinion regarding violence on television and, in particular, that a woman is more likely to oppose the violence.)

b. *Finding $p(M \mid Y)$*: We are told to consider only the "yes" responses shown in table at right.

	Yes
Men	162
Women	256
Total	418

$$p(M \mid Y) = \frac{n(M \cap Y)}{n(Y)} = \frac{162}{418} \approx 0.39 = 39\%$$

Therefore, of those who responded yes, approximately 39% were male.

c. *Finding* $p(Y \cap M)$: This is *not* a conditional probability (there is no vertical bar), so we do *not* limit our sample space.

$$p(Y \cap M) = \frac{n(Y \cap M)}{n(S)} = \frac{162}{600} = 0.27 = 27\%$$

Therefore, of all those polled, 27% were men who responded yes. ■

Notice that in Example 1, each of the three probabilities has the same numerator [$n(Y \cap M)$, the number of responses that are yes and are from men] but a different denominator. The denominator in part (a) is the number of male responses, and in part (b) it is the number of "yes" responses. In part (c), the probability is not a conditional probability, so its sample space is not limited, and the denominator is the entire original sample space of 600 responses. *If you calculate a conditional probability incorrectly, check whether you are using the correct limited sample space.*

The Product Rule

If two cards are dealt from a full deck (no jokers), how would you find the probability that both are hearts? The probability that the first card is a heart is easy to find—it's $\frac{13}{52}$, because there are 52 cards in the deck and 13 of them are hearts. The probability that the second card is a heart is more difficult to find. There are only 51 cards left in the deck (one was already dealt), but how many of these are hearts? The number of hearts left in the deck depends on the first card that was dealt. If it was a heart, then there are 12 hearts left in the deck; if it was not a heart, then there are 13 hearts left. We could certainly say that the probability that the second card is a heart, *given that the first card was a heart,* is $\frac{12}{51}$.

Therefore, the probability that the first card is a heart is $\frac{13}{52}$, and the probability that the second card is a heart, given that the first was a heart, is $\frac{12}{51}$. How do we put these two probabilities together to find the probability that *both* the first and the second cards are hearts? Should we add them? Subtract them? Multiply them? Divide them?

The answer is obtained by algebraically rewriting the Conditional Probability Definition to obtain what is called the *product rule*:

$$p(A \mid B) = \frac{n(A \cap B)}{n(B)} \qquad \text{Conditional Probability Definition}$$

$$p(A \mid B) \cdot n(B) = n(A \cap B) \qquad \text{multiplying by } n(B)$$

$$\frac{p(A \mid B) \cdot n(B)}{n(S)} = \frac{n(A \cap B)}{n(S)} \qquad \text{dividing by } n(S)$$

$$\frac{p(A \mid B)}{1} \cdot \frac{n(B)}{n(S)} = \frac{n(A \cap B)}{n(S)} \qquad \text{since } 1 \cdot n(S) = n(S)$$

$$p(A \mid B) \cdot p(B) = p(A \cap B) \qquad \text{definition of probability}$$

Product Rule

For any events A and B, the probability of A and B is

$$p(A \cap B) = p(A \mid B) \cdot p(B)$$

EXAMPLE 2 If two cards are dealt from a full deck, find the probability that both are hearts.

Solution

$$
\begin{aligned}
p(A \cap B) &= p(A \mid B) \cdot p(B) \\
p(\text{2nd heart and 1st heart}) &= p(\text{2nd heart} \mid \text{1st heart}) \cdot p(\text{1st heart}) \\
&= \frac{12}{51} \cdot \frac{13}{52} \\
&= \frac{4}{17} \cdot \frac{1}{4} \\
&= \frac{1}{17} \approx 0.06 = 6\%
\end{aligned}
$$

Therefore, there is a 6% probability that both cards are hearts. ■

Tree Diagrams

Many people find that a *tree diagram* helps them understand problems like the one in Example 2, in which an experiment is performed in stages over time. Figure 3.18 shows the tree diagram for Example 2. The first column gives a list of the possible outcomes of the first stage of the experiment; in Example 2, the first stage is dealing the first card, and its outcomes are "heart," and "not a heart." The branches leading to those outcomes represent their probabilities. The second column gives a list of the possible outcomes of the second stage of the experiment; in Example 2, the second stage is dealing the second card. A branch leading from a first-stage outcome to a second-stage outcome is the conditional probability $p(\text{2nd stage outcome} \mid \text{1st stage outcome})$.

Looking at the top pair of branches, we see that the first branch stops at "first card is a heart" and the probability is $p(\text{1st heart}) = \frac{13}{52}$. The second branch starts at "first card is a heart" and stops at "second card is a heart" and gives the conditional probability $p(\text{2nd heart} \mid \text{1st heart}) = \frac{12}{51}$. The probability that we were asked to calculate in Example 2, $p(\text{1st heart and 2nd heart})$, is that of the top limb:

$$
\begin{aligned}
p(\text{1st heart and 2nd heart}) &= p(\text{2nd heart} \mid \text{1st heart}) \cdot p\,(\text{1st heart}) \\
&= \frac{12}{51} \cdot \frac{13}{52}
\end{aligned}
$$

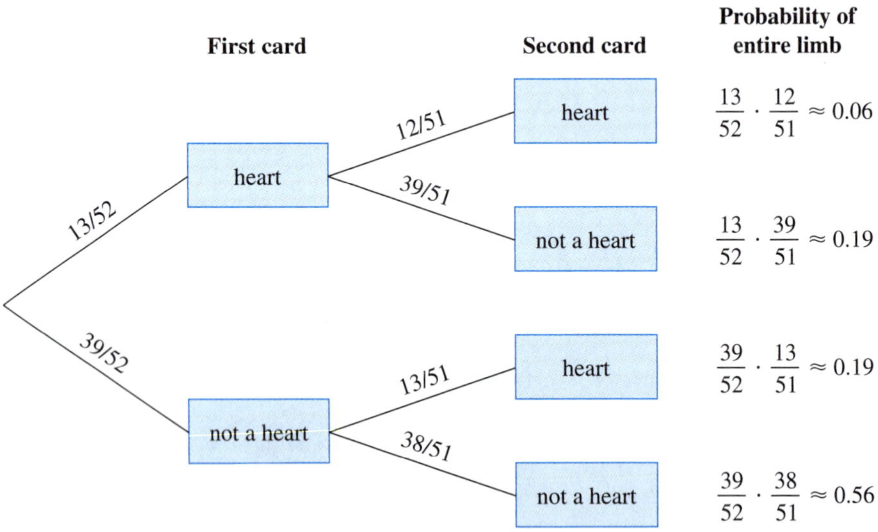

Figure 3.18

(We use the word *limb* to refer to a sequence of branches that starts at the beginning of the tree.) Notice that the sum of the probabilities of the four limbs is 1.00. Because the four limbs are the only four possible outcomes of the experiment, they must add up to 1.

Conditional probabilities always start at their condition, never at the beginning of the tree. For example, $p(\text{2nd heart} \mid \text{1st heart})$ is a conditional probability; its condition is that the first card is a heart. Thus, its branch starts at the box "first card is a heart." However, $p(\text{1st heart})$ is not a conditional probability, so it starts at the beginning of the tree. Similarly, $p(\text{1st heart and 2nd heart})$ is not a conditional probability, so it too starts at the beginning of the tree. The product rule tells us that

$$p\,(\text{1st heart and 2nd heart}) = p\,(\text{2nd heart} \mid \text{1st heart}) \cdot p(\text{1st heart})$$

That is, the product rule tells us to multiply the branches that make up the top horizontal limb. In fact, "Multiply when moving horizontally across a limb" is a restatement of the product rule.

EXAMPLE 3 Two cards are drawn from a full deck. Use the tree diagram in Figure 3.18 to find the probability that the second card is a heart.

Solution The second card can be a heart if the first card is a heart *or* if it is not. The event "the second card is a heart" is the union of the following two mutually exclusive events:

E = 1st heart and 2nd heart

F = 1st not heart and 2nd heart

We previously used the tree diagram to find that

$$p(E) = \frac{13}{52} \cdot \frac{12}{51}$$

Similarly,

$$p(F) = \frac{39}{52} \cdot \frac{13}{51}$$

Thus, we add the probabilities of limbs that result in the second card being a heart:

$$\begin{aligned} p(\text{2nd heart}) &= p(E \cup F) \\ &= p(E) + p(F) && \text{Probability Rule 5} \\ &= \frac{13}{52} \cdot \frac{12}{51} + \frac{39}{52} \cdot \frac{13}{51} = 0.25 \end{aligned}$$

■

In Example 3, the first and third limbs represent the only two ways that the second card can be a heart. These two limbs represent mutually exclusive events, so we used Probability Rule 5 [$p(E \cup F) = p(E) + p(F)$] to add their probabilities. In fact, "add when moving vertically from limb to limb" is a good restatement of Probability Rule 5.

Tree Diagram Summary

Conditional probabilities start at their condition.

Nonconditional probabilities start at the beginning of the tree.

Multiply when moving horizontally across a limb.

Add when moving vertically from limb to limb.

EXAMPLE 4 Big Fun Bicycles manufactures its product at two plants, one in Korea and one in Peoria. The Korea plant manufactures 60% of the bicycles; 4% of the Korean bikes are defective; and 5% of the Peorian bikes are defective.

a. Draw a tree diagram that shows this information.
b. Use the tree diagram to find the probability that a bike is defective and came from Korea.
c. Use the tree diagram to find the probability that a bike is defective.
d. Use the tree diagram to find the probability that a bike is defect-free.

Solution **a.** First, we need to determine which probabilities have been given and find their complements, as shown in Figure 3.19.

Probabilities Given	Complements of These Probabilities
$p(\text{Korea}) = 60\% = 0.60$	$p(\text{Peoria}) = p(\text{not Korea}) = 1 - 0.60 = 0.40$
$p(\text{defective} \mid \text{Korea}) = 4\% = 0.04$	$p(\text{not defective} \mid \text{Korea}) = 1 - 0.04 = 0.96$
$p(\text{defective} \mid \text{Peoria}) = 5\% = 0.05$	$p(\text{not defective} \mid \text{Peoria}) = 1 - 0.05 = 0.95$

Figure 3.19

The first two of these probabilities [p(Korea) and p(Peoria)] are not conditional, so they start at the beginning of the tree. The next two probabilities [$p(\text{defective} \mid \text{Korea})$ and $p(\text{not defective} \mid \text{Korea})$] are conditional, so they start at their condition (Korea). Similarly, the last two probabilities are conditional, so they start at their condition (Peoria). This placement of the probabilities yields the tree diagram in Figure 3.20.

b. The probability that a bike is defective and came from Korea is a nonconditional probability, so it starts at the beginning of the tree. Do not confuse it with the conditional probability that a bike is defective, *given that* it came from Korea, which starts at its condition (Korea). The former is the limb that goes through "Korea"

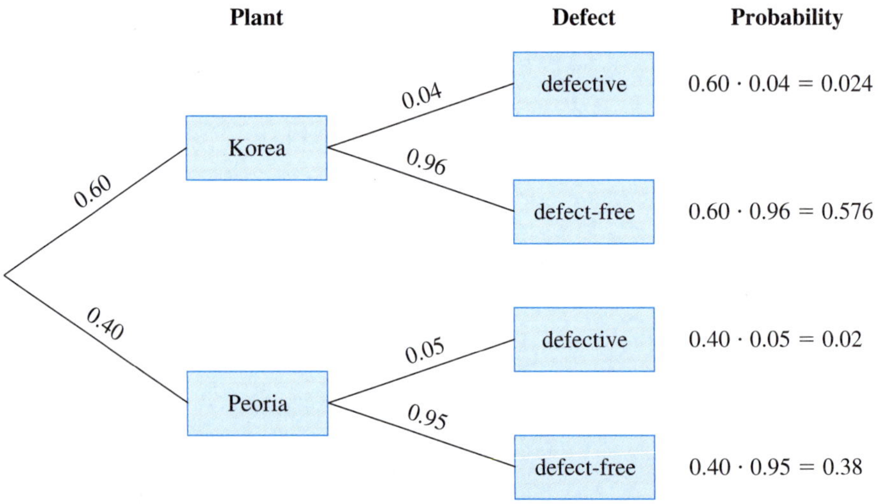

Figure 3.20
Tree diagram for Example 4

and stops at "defective"; the latter is one branch of that limb. We use the product rule to multiply when moving horizontally across a limb:

$$p(\text{defective and Korea}) = p(\text{defective} \mid \text{Korea}) \cdot p(\text{Korea})$$
$$= 0.04 \cdot 0.60 = 0.024 \qquad \text{product rule}$$

This means that 2.4% of all of Big Fun's bikes are defective bikes manufactured in Korea.

c. The event that a bike is defective is the union of two mutually exclusive events:

The bike is defective and came from Korea.

The bike is defective and came from Peoria.

These two events are represented by the first and third limbs of the tree. We use Probability Rule 5 to add when moving vertically from limb to limb:

$$p(\text{defective}) = p(\text{defective and Korea} \cup \text{defective and Peoria})$$
$$= p(\text{defective and Korea}) + p(\text{defective and Peoria})$$
$$= 0.024 + 0.02 = 0.044$$

This means that 4.4% of Big Fun's bicycles are defective.

d. The probability that a bike is defect-free is the complement of part (c).

$$p(\text{defect-free}) = p(\text{not defective}) = 1 - 0.044 = 0.956$$

Alternatively, we can find the sum of all the limbs that stop at "defect-free":

$$p(\text{defect-free}) = 0.576 + 0.38 = 0.956$$

This means that 95.6% of Big Fun's bicycles are defect-free. ■

3.6 Exercises

1. Use the data in Figure 3.17 to find the given probabilities. Also, write a sentence explaining what each means.

a. $p(N)$ **b.** $p(W)$
c. $p(N \mid W)$ **d.** $p(W \mid N)$
e. $p(N \cap W)$ **f.** $p(W \cap N)$

2. Use the data in Figure 3.17 to find the given probabilities. Also, write a sentence explaining what each means.

a. $p(Y)$ **b.** $p(M)$
c. $p(Y \mid M)$ **d.** $p(M \mid Y)$
e. $p(Y \cap M)$ **f.** $p(M \cap Y)$

In Exercises 3–6, use Figure 3.21, which gives information on the number of drivers and the number of accidents in 1987. Assume that no driver had more than one accident. (Round off to the nearest hundredth.)

3. Find the probability that a driver had an accident.

Age Group	Number of Drivers	Number of Accidents
under 20	14,100,000	5,200,000
20–24	16,900,000	5,800,000
25–34	39,800,000	9,100,000
35–44	33,200,000	5,400,000
45–54	23,400,000	2,900,000
55–64	18,500,000	2,200,000
65–74	12,500,000	1,500,000
75 and over	3,600,000	900,000
total	162,000,000	33,000,000

Figure 3.21
Source: National Safety Council's Accident Facts, 1988 edition.

4. Find the probability that a driver had an accident, given that the driver was under twenty.
5. Find the probability that a driver had an accident, given that the driver was twenty to twenty-four.
6. Find the probability that a driver had an accident, given that the driver was forty-five to fifty-four.

In Exercises 7–10, cards are dealt from a full deck of 52. Find the probabilities of the given events.

7. **a.** The first card is a club.
   b. The second card is a club, given that the first was a club.
   c. The first and second cards are both clubs.
   d. Draw a tree diagram illustrating this.
8. **a.** The first card is a king.
   b. The second card is a king, given that the first was a king.
   c. The first and second cards are both kings.
   d. Draw a tree diagram illustrating this. (Your diagram need not be a complete tree. It should have all the branches referred to in parts (a), (b), and (c), but it does not need other branches.)
9. **a.** The first card is a diamond.
   b. The second card is a spade, given that the first was a diamond.
   c. The first card is a diamond and the second is a spade.
   d. Draw a tree diagram illustrating this. (Your diagram need not be a complete tree. It should have all the branches referred to in parts (a), (b), and (c), but it does not need other branches.)
10. **a.** The first card is a jack.
   b. The second card is an ace, given that the first card was a jack.
   c. The first card is a jack and the second is an ace.
   d. Draw a tree diagram illustrating this.

*In Exercises 11 and 12, determine which probability the indicated branch in Figure 3.22 at the top of the next column refers to. For example, the branch labeled * refers to the probability p(A).*

11. **a.** the branch labeled *a*
   b. the branch labeled *b*
   c. the branch labeled *c*
12. **a.** Should probabilities (*) and (*a*) be added or multiplied? What rule tells us that? What is the result of combining them?

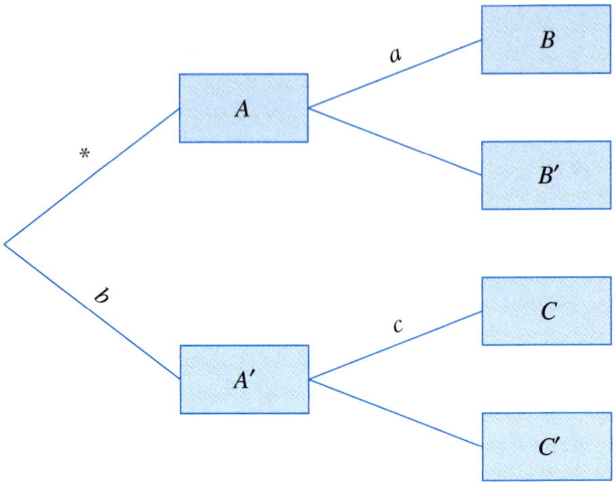

Figure 3.22

   b. Should probabilities (*b*) and (*c*) be added or multiplied? What rule tells us that? What is the result of combining them?
   c. Should the probabilities that result from parts (*a*) and (*b*) of this exercise be added or multiplied? What rule tells us that? What is the result of combining them?

In Exercises 13 and 14, a single die is rolled. Find the probabilities of the given events.

13. **a.** rolling a 6
   b. rolling a 6, given that the number rolled is even
   c. rolling a 6, given that the number rolled is odd
   d. rolling an even number, given that a 6 was rolled
14. **a.** rolling a 5
   b. rolling a 5, given that the number rolled is even
   c. rolling a 5, given that the number rolled is odd
   d. rolling an odd number, given that a 5 was rolled

In Exercises 15–18, a pair of dice is rolled. Find the probabilities of the given events.

15. **a.** The sum is 6.
   b. The sum is 6, given that the sum is even.
   c. The sum is 6, given that the sum is odd.
   d. The sum is even, given that the sum is 6.
16. **a.** The sum is 12.
   b. The sum is 12, given that the sum is even.
   c. The sum is 12, given that the sum is odd.
   d. The sum is even, given that the sum was 12.
17. **a.** The sum is 4.
   b. The sum is 4, given that the sum is less than 6.
   c. The sum is less than 6, given that the sum is 4.

18. **a.** The sum is 11.
b. The sum is 11, given that the sum is greater than 10.
c. The sum is greater than 10, given that the sum is 11.

19. A single die is rolled. Determine which of the following events is least likely and which is most likely. Do so without making any calculations. Explain your reasoning.
E_1 is the event "rolling a four."
E_2 is the event "rolling a four, given that the number rolled is even."
E_3 is the event "rolling a four, given that the number rolled is odd."

20. A pair of dice is rolled. Determine which of the following events is least likely and which is most likely. Do so without making any calculations. Explain your reasoning.
E_1 is the event "rolling a seven."
E_2 is the event "rolling a seven, given that the number rolled is even."
E_3 is the event "rolling a seven, given that the number rolled is odd."

In Exercises 21 and 22, use the following information. To determine what effect the salespeople had on purchases, a department store polled 700 shoppers as to whether or not they had made a purchase and whether or not they were pleased with the service. Of those who had made a purchase, 125 were happy with the service and 111 were not. Of those who had made no purchase, 148 were happy with the service and 316 were not.

21. Find the probability that a shopper who was happy with the service had made a purchase (round off to the nearest hundredth). What can you conclude?

22. Find the probability that a shopper who was unhappy with the service had not made a purchase. (Round off to the nearest hundredth.) What can you conclude?

In Exercises 23–26, five cards are dealt from a full deck. Find the probabilities of the given events. (Round off to four decimal places.)

23. All are spades.

24. The fifth is a spade, given that the first four were spades.

25. The last four are spades, given that the first was a spade.

26. All are the same suit.

In Exercises 27–32, round off to the nearest hundredth.

27. If three cards are dealt from a full deck, use a tree diagram to find the probability that exactly two are spades.

28. If three cards are dealt from a full deck, use a tree diagram to find the probability that exactly one is a spade.

29. If three cards are dealt from a full deck, use a tree diagram to find the probability that exactly one is an ace.

30. If three cards are dealt from a full deck, use a tree diagram to find the probability that exactly two are aces.

31. If a pair of dice is rolled three times, use a tree diagram to find the probability that exactly two throws result in sevens.

32. If a pair of dice is rolled three times, use a tree diagram to find the probability that all three throws result in sevens.

In Exercises 33–36, use the following information: A personal computer manufacturer buys 38% of its chips from Japan and the rest from America. 1.7% of the Japanese chips are defective, and 1.1% of the American chips are defective.

33. Find the probability that a chip is defective and made in Japan.

34. Find the probability that a chip is defective and made in America.

35. Find the probability that a chip is defective.

36. Find the probability that a chip is defect-free.

37. The 107th U.S. Senate convened on January 3, 2001. Data on the Senate's composition is given in Figure 3.23.

	Democrats	Republicans	Other
Male	40	46	1
Female	10	3	0

Figure 3.23

a. Find the probability that a senator is female, given that she is a Democrat.
b. Find the probability that a senator is female, given that she is a Republican.
c. What observations can you make?

38. The 107th U.S. Senate convened on January 3, 2001. Data on the Senate's composition is given in Figure 3.23.

a. Find the probability that a senator is a Democrat, given that she is female.
b. Find the probability that a senator is a Republican, given that she is female.
c. What observations can you make?

39. October 2001 data on the governors of the 50 states is given in Figure 3.24.

	Democrats	Republicans	Other
Male	17	26	2
Female	2	3	5

Figure 3.24

a. Find the probability that a governor is female, given that she is a Democrat.
b. Find the probability that a governor is female, given that she is a Republican.
c. What observations can you make?

40. October 2001 data on the governors of the 50 states is given in Figure 3.24.
a. Find the probability that a governor is a Democrat, given that she is female.
b. Find the probability that a governor is a Republican, given that she is female.
c. What observations can you make?

In Exercises 41–44, use the following information: The University of Metropolis requires its students to pass an examination in college-level mathematics before they can graduate. The students are given three chances to pass the exam; 61% pass it on their first attempt; 63% of those that take it a second time pass it then; and 42% of those that take it a third time pass it then.

41. What percent of the students pass the exam?

42. What percent of the students are not allowed to graduate because of their performance on the exam?

43. What percent of the students take the exam at least twice?

44. What percent of the students take the test three times?

45. In the game of blackjack, if the first two cards dealt to a player are an ace and either a ten, jack, queen, or king, then the player has a "blackjack," and he or she wins. Find the probability that a player is dealt a blackjack out of a full deck (no jokers).

46. In blackjack, the dealer's first card is dealt face up. If that card is an ace, then the player has the option of "taking insurance." ("Insurance" is a side bet. If the dealer has a blackjack, the player wins the insurance bet and is paid 2 to 1 odds. If the dealer does not have a blackjack, the player loses the insurance bet.) Find the probability that the dealer gets a blackjack if his or her first card is an ace.

47. In Exercise 1, find the following probabilities:
a. $p(N' \mid W)$ **b.** $p(N \mid W')$ **c.** $p(N' \mid W')$
d. Which event, $N' \mid W$, $N \mid W'$, or $N' \mid W'$, is the complement of the event $N \mid W$? Why?

48. In Exercise 2, find the following probabilities:
a. $p(Y' \mid M)$ **b.** $p(Y \mid M')$ **c.** $p(Y' \mid M')$
d. Which event, $Y' \mid M$, $Y \mid M'$, or $Y' \mid M'$, is the complement of the event $Y \mid M$? Why?

49. If A and B are arbitrary events, what is the complement of the event $A \mid B$?

50. Show that $p(A \mid B) = \frac{p(A \cap B)}{p(B)}$

HINT: Divide the numerator and denominator of the Conditional Probability Definition by $n(S)$.

51. Use Exercise 50 and appropriate answers from Exercise 1 to find $p(N \mid W)$.

52. Use Exercise 50 and appropriate answers from Exercise 2 to find $p(Y \mid M)$.

Answer the following questions using complete sentences and your own words.

Concept Questions

53. Which must be true for any events A and B?
- $p(A \mid B)$ is always greater than or equal to $p(A)$.
- $p(A \mid B)$ is always less than or equal to $p(A)$.
- Sometimes $p(A \mid B)$ is greater than or equal to $p(A)$, and sometimes $p(A \mid B)$ is less than or equal to $p(A)$, depending on the events A and B.

Answer this without making any calculations. Explain your reasoning.

54. In Exercises 3–6, why is it necessary to assume that no driver had more than one accident? Be certain to discuss what further information would be needed to answer Exercises 3–6 and how that information would affect the answers. Would they be larger or smaller than your original answers? Would they differ from your original answers by a relatively large or small amount?

HINT: In 1987, there were fewer accidents than drivers. In some future year, it may be that there are more accidents than drivers. What would this do to your answer to Exercise 3?

55. Compare and contrast the events A, $A \mid B$, $B \mid A$, and $A \cap B$. Be sure to discuss both the similarities and the differences between these events.

Projects

56. In 1973, the University of California at Berkeley admitted 1,494 of 4,321 female applicants for graduate study and 3,738 of 8,442 male applicants. (*Source:* P. J. Bickel, E. A. Hammel, and J. W. O' Connell, "Sex Bias in Graduate Admissions: Data from Berkeley," *Science,* vol. 187, 7 February 1975.)

a. Find the probability that an applicant was admitted.

b. Find the probability that an applicant was admitted, given that he was male.

c. Find the probability that an applicant was admitted, given that she was female.

d. Do these numbers indicate a bias against women?

e. Berkeley's graduate students are admitted by the department to which they apply, rather than by a campuswide admissions panel. When $p(\text{admission} \mid \text{male})$ and $p(\text{admission} \mid \text{female})$ were computed for each of the school's more than 100 departments, it was found that in four departments, $p(\text{admission} \mid \text{male})$ was greater than $p(\text{admission} \mid \text{female})$ by a significant amount, and that in six departments, $p(\text{admission} \mid \text{male})$ was less than $p(\text{admission} \mid \text{female})$ by a significant amount. Do these data indicate a bias against women?

f. What conclusions would you make and what further information would you obtain if you were an affirmative action officer for the campus?

The authors of *Sex Bias in Graduate Admissions: Data from Berkeley* attempt to explain the paradox by discussing an imaginary school with only two departments: "machismatics" and "social warfare." Machismatics admitted 200 of 400 male applicants for graduate study and 100 of 200 female applicants, while social warfare admitted 50 of 150 male applicants for graduate study and 150 of 450 female applicants. *For the school as a whole and for each of the two departments, answer the following questions.*

g. What is the probability that an applicant was admitted?

h. What is the probability that an applicant was admitted, given that he was male?

i. What is the probability that an applicant was admitted, given that she was female?

j. Do these numbers indicate a bias against women?

k. How would you explain the paradox illustrated in this problem?

57. Figure 3.25 gives information of the incidence of tuberculosis in New York City and in Richmond, Virginia in 1910.

a. Find the probability that a New York City resident died of tuberculosis.

b. Find the probability that a Caucasian New York City resident died of tuberculosis.

c. Find the probability that a non-Caucasian New York City resident died of tuberculosis.

d. Find the probability that a Richmond resident died of tuberculosis.

e. Find the probability that a Caucasian Richmond resident died of tuberculosis.

f. Find the probability that a non-Caucasian Richmond resident died of tuberculosis.

g. Which city had a more severe problem with tuberculosis?

	New York City		Richmond	
	Population	**TB Deaths**	**Population**	**TB Deaths**
Caucasian	4,675,000	8,400	81,000	130
Non-Caucasian	92,000	500	47,000	160

Figure 3.25
Source: Morris R. Cohen and Ernest Nagel, *An Introduction to Logic and Scientific Method* (New York: Harcourt Brace & Co., 1934).

3.7 Independence; Trees in Genetics

Dependent and Independent Events

Consider the dealing of two cards from a full deck. An observer who saw that the first card was a heart would be better able to predict whether the second card will be a heart than would another observer who did not see the first card. If the first card was a heart, there is one fewer heart in the deck, so it is slightly less likely that the second card will be a heart. In particular,

$$p(\text{2nd heart} \mid \text{1st heart}) = \frac{12}{51} \approx 0.24$$

whereas, as we saw in Example 3 of Section 3.6,

$$p(\text{2nd heart}) = 0.25$$

These two probabilities are different because of the effect the first card drawn has on the second. We say that the two events "first card is a heart" and "second card is a heart" are *dependent*; the result of dealing the second card depends, to some extent, on the result of dealing the first card. In general, two events E and F are **dependent** if $p(E \mid F) \neq p(E)$.

Consider two successive tosses of a single die. An observer who saw that the first toss resulted in a three would be *no better able* to predict whether the second toss will result in a three than another observer who did not observe the first toss. In particular,

$$p(\text{2nd toss is a three}) = \frac{1}{6}$$

and

$$p(\text{2nd toss is a three} \mid \text{1st toss was a three}) = \frac{1}{6}$$

These two probabilities are the same, because the first toss has no effect on the second toss. We say that the two events "first toss is a three" and "second toss is a three" are *independent*; the result of the second toss does *not* depend on the result of the first toss. In general, two events E and F are **independent** if $p(E \mid F) = p(E)$.

> **Independence/Dependence Definitions**
>
> Two events E and F are **independent** if $p(E \mid F) = p(E)$.
> *(Think: Knowing* F *does not affect* E'*s probability.)*
> Two events E and F are **dependent** if $p(E \mid F) \neq p(E)$.
> *(Think: Knowing* F *does affect* E'*s probability.)*

Many people have difficulty distinguishing between *independent* and *mutually exclusive*. (Recall that two events E and F are mutually exclusive if $E \cap F = \emptyset$, that is, if one event excludes the other.) This is probably because the relationship between mutually exclusive events and the relationship between independent events both could be described, in a very loose sort of way, by saying that "the two events have nothing to do with each other." *Never think this way;* mentally replacing "mutually exclusive" or "independent" with "having nothing to do with each other" only obscures the distinction between these two concepts. E and F are independent if knowing that F has occurred *does not* affect the probability that E will occur. E and F are dependent if knowing that F has occurred *does* affect the probability that E will occur. E and F are mutually exclusive if E and F cannot occur simultaneously.

EXAMPLE 1 Let F be the event "a person has freckles" and R the event "a person has red hair."

a. Are F and R independent?
b. Are F and R mutually exclusive?

Solution **a.** F and R are independent if $p(F \mid R) = p(F)$. With $p(F \mid R)$, we are given that a person has red hair; with $p(F)$, we are not given that information. Does knowing that a person has red hair affect the probability that the person has freckles? Yes, it does; $p(F \mid R) > p(F)$. Therefore, F and R are not independent; they are dependent.

b. F and R are mutually exclusive if $F \cap R = \emptyset$. Many people have both freckles and red hair, so $F \cap R \neq \emptyset$, and F and R are not mutually exclusive. In other words, having freckles does not exclude the possibility of having red hair; freckles and red hair can occur simultaneously. ■

EXAMPLE 2 Let T be the event "a person is tall" and R the event "a person has red hair."

a. Are T and R independent?
b. Are T and R mutually exclusive?

Solution **a.** T and R are independent if $p(T \mid R) = p(T)$. With $p(T \mid R)$, we are given that a person has red hair; with $p(T)$, we are not given that information. Does knowing that a person has red hair affect the probability that the person is tall? No, it does not; $p(T \mid R) = p(T)$, so T and R are independent.

b. T and R are mutually exclusive if $T \cap R = \emptyset$. $T \cap R$ is the event "a person is tall and has red hair." There are tall people who have red hair, so $T \cap R \neq \emptyset$, and T and R are not mutually exclusive. In other words, being tall does not exclude the possibility of having red hair; being tall and having red hair can occur simultaneously. ■

In Examples 1 and 2, we had to rely on our personal experience in concluding that knowledge that a person has red hair does affect the probability that he or she has freckles and does not affect the probability that he or she is tall. It may be the case that you have seen only one red-haired person, and she was short and without freckles. Independence is better determined by computing the appropriate probabilities than by relying on one's own personal experiences. This is especially crucial in determining the effectiveness of an experimental drug. *Double-blind* experiments, in which neither the patient nor the doctor knows whether the given medication is the experimental drug or an inert substance, are often done to ensure reliable, unbiased results.

Independence is an important tool in determining whether an experimental drug is an effective vaccine. Let D be the event that the experimental drug was administered to a patient and R the event that the patient recovered. It is hoped that $p(R \mid D) > p(R)$, that is, that the rate of recovery is greater among those who were given the drug. In this case, R and D are dependent. Independence is also an important tool in determining whether an advertisement effectively promotes a product. An ad is effective if $p(\text{consumer purchases product} \mid \text{consumer saw ad}) > p(\text{consumer purchases product})$.

EXAMPLE 3 Use probabilities to determine whether the events "thinking there is too much violence in television" and "being a man" in Example 1 of Section 3.6 are independent.

Solution Two events E and F are independent if $p(E \mid F) = p(E)$. The events "responding yes to the question on violence in television" and "being a man" are independent if $p(Y \mid M) = p(Y)$. We need to compute these two probabilities and compare them.

In Example 1 of Section 3.6, we found $p(Y \mid M) \approx 0.58$. We can use the data from the poll in Figure 3.17 to find $p(Y)$.

$$p(Y) = \frac{418}{600} \approx 0.70$$

$$p(Y \mid M) \neq p(Y)$$

The events "responding yes to the question on violence in television" and "being a man" are dependent. According to the poll, men are less likely to think that there is too much violence on television. ■

In Example 3, what should we conclude if we found that $p(Y) = 0.69$ and $p(Y \mid M) = 0.67$? Should we conclude that $p(Y \mid M) \neq p(Y)$ and that the events "thinking that there is too much violence on television" and "being a man" are dependent? Or should we conclude that $p(Y \mid M) \approx p(Y)$ and that the events are (probably) independent? In this particular case, the probabilities are relative frequencies rather than theoretical probabilities, and relative frequencies can vary. A group of 600 people was polled to determine the opinions of the entire viewing public; if the same question was asked of a different group, a somewhat different set of relative frequencies could result. While it would be reasonable to conclude that the events are (probably) independent, it would be more appropriate to include more people in the poll and make a new comparison.

Product Rule for Independent Events

The product rule says that $p(A \cap B) = p(A \mid B) \cdot p(B)$. If A and B are independent, then $p(A \mid B) = p(A)$. Combining these two equations, we get the following rule:

> **Product Rule for Independent Events**
>
> If A and B are independent events, then the probability of A and B is
>
> $$p(A \cap B) = p(A) \cdot p(B)$$

A common error that is made in computing probabilities is using the formula $p(A \cap B) = p(A) \cdot p(B)$ without verifying that A and B are independent. In fact, the Federal Aviation Administration (FAA) has stated that this is the most frequently encountered error in probabilistic analysis of airplane component failures. If it is not known that A and B are independent, you must use the Product Rule $p(A \cap B) = p(A \mid B) \cdot p(B)$.

EXAMPLE 4 If a pair of dice is tossed twice, find the probability that each toss results in a seven.

Solution In Example 3 of Section 3.3, we found that the probability of a seven is $\frac{1}{6}$. The two rolls are independent (one roll has no influence on the next), so the probability of a

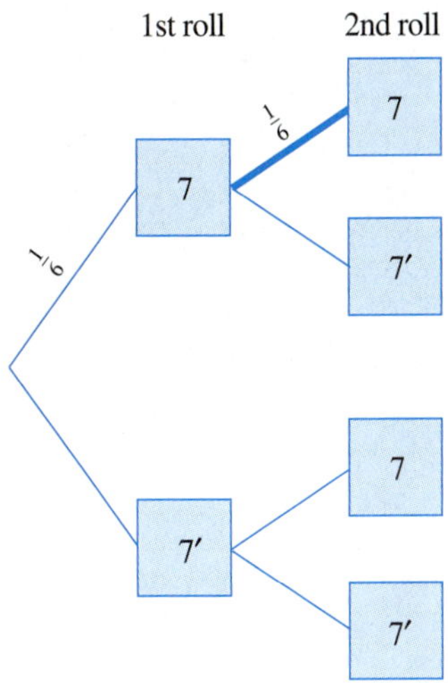

Figure 3.26

seven is $\frac{1}{6}$ regardless of what might have happened on an earlier roll; we can use the Product Rule for Independent Events:

$$
\begin{aligned}
p(A \cap B) &= p(A) \quad \cdot \quad p(B) \\
p(\text{1st is 7 and 2nd is 7}) &= p(\text{1st is 7}) \cdot p(\text{2nd is 7}) \\
&= \frac{1}{6} \quad \cdot \quad \frac{1}{6} \\
&= \frac{1}{36}
\end{aligned}
$$

See Figure 3.26. The thicker branch of the tree diagram starts at the event "1st roll is 7"and ends at the event "2nd roll is 7," so it is the conditional probability $p(\text{2nd is 7} \mid \text{1st is 7})$. However, the two rolls are independent, so $p(\text{2nd is 7} \mid \text{1st is 7}) = p(\text{2nd is 7})$. We are free to label this branch as either of these two equivalent probabilities. ■

Trees in Medicine and Genetics

Usually, medical diagnostic tests are not 100% accurate. A test might indicate the presence of a disease when the patient is in fact healthy (this is called a **false positive**), or it might indicate the absence of a disease when the patient does in fact have the disease (a **false negative**). Probability trees can be used to determine the probability that a person whose test results were positive actually has the disease.

EXAMPLE 5 Medical researchers have recently devised a diagnostic test for "white lung" (an imaginary disease caused by the inhalation of chalk dust). Teachers are particularly susceptible to this disease; studies have shown that half of all teachers are afflicted with it. The test correctly diagnoses the presence of white lung in 99% of the persons who have it and correctly diagnoses its absence in 98% of the persons who do not have it. Find the probability that a teacher whose test results are positive actually has white lung and the probability that a teacher whose test results are negative does not have white lung.

Solution First, we determine which probabilities have been given and find their complements, as shown in Figure 3.27. We use + to denote the event that a person receives a positive diagnosis and − to denote the event that a person receives a negative diagnosis.

Probabilities Given	Complements of Those Probabilities
$p(\text{ill}) = 0.50$	$p(\text{healthy}) = p(\text{not ill}) = 1 - 0.50 = 0.50$
$p(- \mid \text{healthy}) = 98\% = 0.98$	$p(+ \mid \text{healthy}) = 1 - 0.98 = 0.02$
$p(+ \mid \text{ill}) = 99\% = 0.99$	$p(- \mid \text{ill}) = 1 - 0.99 = 0.01$

Figure 3.27

The first two of these probabilities [$p(\text{ill})$ and $p(\text{healthy})$] are not conditional, so they start at the beginning of the tree. The next two probabilities [$p(- \mid \text{healthy})$ and $p(+ \mid \text{healthy})$] are conditional, so they start at their condition (healthy). Similarly, the last two probabilities are conditional, so they start at their condition (ill). This placement of the probabilities yields the tree diagram in Figure 3.28 on page 192.

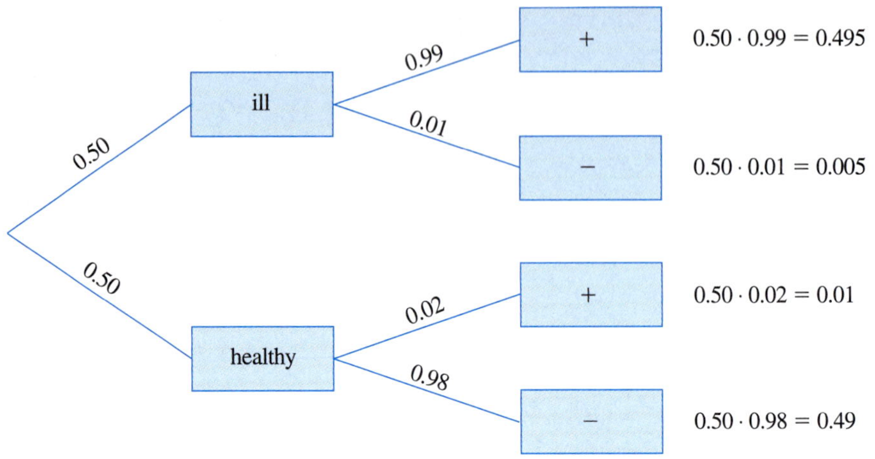

Figure 3.28

The four probabilities to the right of the tree are

$p(\text{ill and } +) = 0.495$
$p(\text{ill and } -) = 0.005$
$p(\text{healthy and } +) = 0.01$
$p(\text{healthy and } -) = 0.49$

We are to find the probability that a teacher whose test results are positive actually has white lung; that is, we are to find $p(\text{ill} \mid +)$. Thus, we are given that the test results are positive, and we need only consider branches that involve positive test results: $p(\text{ill and } +) = 0.495$ and $p(\text{healthy and } +) = 0.01$. The probability that a teacher whose test results are positive actually has white lung is

$$p(\text{ill} \mid +) = \frac{0.495}{0.495 + 0.01} = 0.9801\ldots \approx 98\%$$

The probability that a teacher whose test results are negative does not have white lung is

$$p(\text{healthy} \mid -) = \frac{0.49}{0.49 + 0.005} = 0.98989\ldots \approx 99\%$$

The probabilities show that this diagnostic test works well in determining whether a teacher actually has white lung. In the exercises, we will see how well it would work with schoolchildren. ■

EXAMPLE 6 Mr. and Mrs. Smith each had a sibling who died of cystic fibrosis. The Smiths are considering having a child. They have not been tested to determine whether they are carriers. What is the probability that their child would have cystic fibrosis?

Solution In our previous examples involving trees, we were given probabilities; the conditional or nonconditional status of those probabilities helped us to determine the physical layout of the tree. In this example, we are not given any probabilities. To determine the physical layout of the tree, we have to separate what we know to be true from what is only possible or probable. The tree focuses on what is possible or probable. We know that both Mr. and Mrs. Smith had a sibling who died of cystic fibrosis, and it is possible that their child would inherit that disease. The tree's branches will represent the series of possible events that could result in the Smith child inheriting cystic fibrosis.

What events must take place if the Smith child is to inherit the disease? First, the grandparents would have to have had cystic fibrosis genes. Next, the Smiths themselves would have to have inherited those genes from their parents. Finally, the Smith child would have to inherit those genes from his or her parents.

Cystic fibrosis is recessive, which means that a person can inherit it only if he or she receives two cystic fibrosis genes, one from each parent. Mr. and Mrs. Smith each had a sibling who had cystic fibrosis, so each of the four grandparents must have been a carrier. They could not have actually had the disease because the Smiths would have known, and we would have been told.

We now know the physical layout of our tree. The four grandparents were definitely carriers. Mr. and Mrs. Smith were possibly carriers. The Smith child will possibly inherit the disease. The first set of branches will deal with Mr. and Mrs. Smith, and the second set will deal with the child.

The Smith child will not have cystic fibrosis unless Mr. and Mrs. Smith are both carriers. Figure 3.29 shows the Punnett square for Mr. and Mrs. Smith's possible genetic configuration.

	C	**c**	
	C	**c**	← one grandparent's genes
C	CC	Cc	← possible offspring
c	Cc	cc	← possible offspring

↑ other grandparent's genes

Figure 3.29

Neither Mr. Smith nor Mrs. Smith has the disease, so we can eliminate the cc possibility. Thus, the probability that Mr. Smith is a carrier is $\frac{2}{3}$, as is the probability that Mrs. Smith is a carrier. Furthermore, these two events are independent, since the Smiths are (presumably) unrelated.

Using the Product Rule for Independent Events, we have

$$p(\text{Mr. S is a carrier and Mrs. S is a carrier})$$
$$= p(\text{Mr. S is a carrier}) \cdot p(\text{Mrs. S is a carrier}) = \frac{2}{3} \cdot \frac{2}{3} = \frac{4}{9}$$

The same Punnett square tells us that the probability that their child will have cystic fibrosis, given that the Smiths are both carriers, is $\frac{1}{4}$. Letting B be the event that both parents are carriers and F the event that the child has cystic fibrosis, we obtain the tree diagram in Figure 3.30.

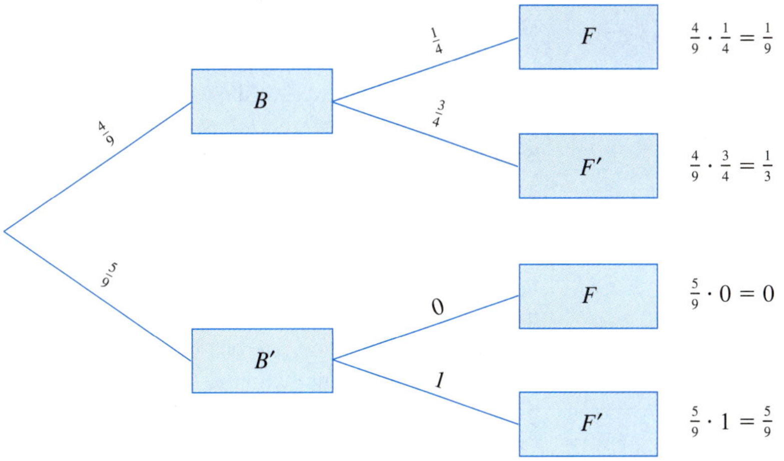

Figure 3.30

The probability that the Smiths' child would have cystic fibrosis is $\frac{1}{9}$. ■

Notice that the tree in Figure 3.30 could have been drawn differently, as shown in Figure 3.31. It is not necessary to draw a branch going from the B' box to the F box, since the child cannot have cystic fibrosis if both parents are not carriers.

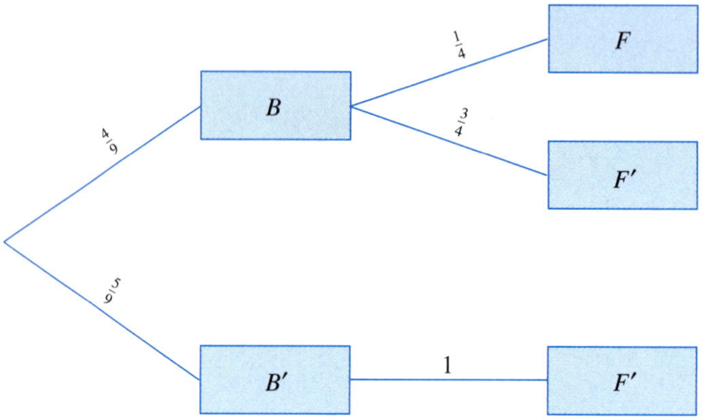

Figure 3.31
An alternative to the tree in Figure 3.30

Hair Color

Like cystic fibrosis, hair color is inherited, but the method by which it is transmitted is more complicated than the method by which an inherited disease is transmitted. This more complicated method of transmission allows for the possibility that a child might have hair that is colored differently from that of other family members.

Hair color is determined by two pairs of genes, one pair that determines placement on a blond/brown/black spectrum and one pair that determines the presence or absence of red pigment. These two pairs of genes are independent.

Melanin is a brown pigment that affects the color of hair (as well as the colors of eyes and skin). The pair of genes that controls the brown hair colors does so by determining the amount of melanin in the hair. This gene has three forms, each traditionally labeled with an M (for melanin): M^{Bd}, or blond (a light melanin deposit); M^{Bw}, or brown (a medium melanin deposit); and M^{Bk}, or black (a heavy melanin deposit). Everyone has two of these genes, and their combination determines the brown aspect of hair color, as illustrated in Figure 3.32.

Genes	Hair Color
$M^{Bd}M^{Bd}$	blond
$M^{Bd}M^{Bw}$	light brown
$M^{Bd}M^{Bk}$ $M^{Bw}M^{Bw}$	medium brown
$M^{Bw}M^{Bk}$	dark brown
$M^{Bk}M^{Bk}$	black

Figure 3.32

Genes	Blond ($M^{Bd}M^{Bd}$)	Light Brown ($M^{Bd}M^{Bw}$)	Medium Brown ($M^{Bd}M^{Bk}$,$M^{Bw}M^{Bw}$)	Dark Brown ($M^{Bw}M^{Bk}$)	Black ($M^{Bk}M^{Bk}$)
R^-R^-	blond	light brown	medium brown	dark brown	black
R^+R^-	strawberry blond	reddish brown	chestnut	shiny dark brown	shiny black
R^+R^+	bright red	dark red	auburn	glossy dark brown	glossy black

Figure 3.33

The hair colors in Figure 3.32 are altered by the presence of red pigment, which is determined by another pair of genes. This gene has two forms: R^-, or no red pigment, and R^+, or red pigment. Because everyone has two of these genes, there are three possibilities for the amount of red pigment: R^-R^-, R^+R^-, and R^+R^+. The amount of red pigment in a person's hair is independent of the brownness of his or her hair.

The actual color of a person's hair is determined by the interaction of these two pairs of genes, as shown in Figure 3.33.

EXAMPLE 7 The Rosses are going to have a child. Mr. Ross has blond hair, and Mrs. Ross has reddish brown hair. Find their child's possible hair colors and the probabilities of each possibility.

Solution The parent with blond hair has genes $M^{Bd}M^{Bd}$ and R^-R^-. The parent with reddish brown hair has genes $M^{Bd}M^{Bw}$ and R^+R^-. We need to use two Punnett squares, one for the brownness of the hair (Figure 3.34) and one for the presence of red pigment (Figure 3.35).

	M^{Bd}	M^{Bd}
M^{Bd}	$M^{Bd}M^{Bd}$	$M^{Bd}M^{Bd}$
M^{Bw}	$M^{Bd}M^{Bw}$	$M^{Bd}M^{Bw}$

Figure 3.34
Brownness

	R^-	R^-
R^+	R^+R^-	R^+R^-
R^-	R^-R^-	R^-R^-

Figure 3.35
Red pigment

$$p(M^{Bd}M^{Bd}) = \tfrac{2}{4} = \tfrac{1}{2}$$
$$p(M^{Bd}M^{Bw}) = \tfrac{2}{4} = \tfrac{1}{2}$$

$$p(R^+R^-) = \tfrac{2}{4} = \tfrac{1}{2}$$
$$p(R^-R^-) = \tfrac{2}{4} = \tfrac{1}{2}$$

We will use a tree diagram to determine the possible hair colors and their probabilities (see Figure 3.36 on page 196).

The Rosses' child could have strawberry blond, blond, reddish brown, or light brown hair. The probability of each color is $\frac{1}{4}$. Notice that there is a 50% probability that the child will have hair that is colored differently from that of either parent.

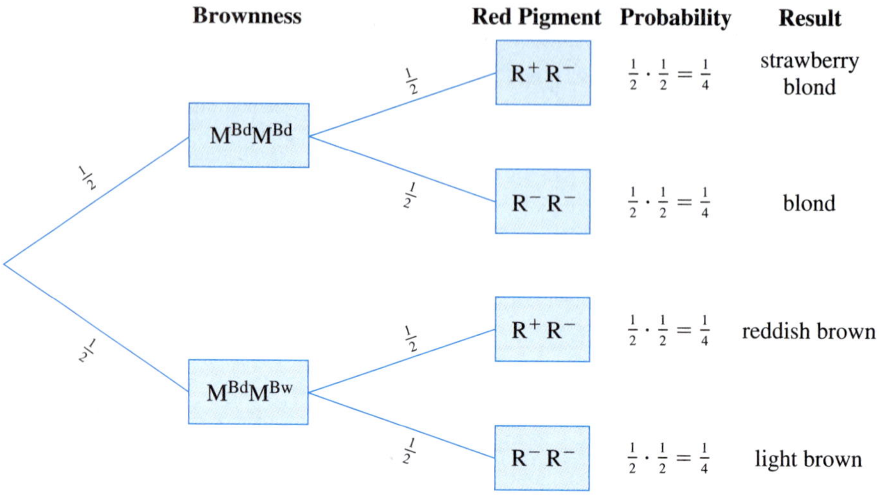

Figure 3.36

3.7 Exercises

In Exercises 1–8, use your own personal experience with the events described to determine whether (a) E *and* F *are independent and (b)* E *and* F *are mutually exclusive. (Where appropriate, these events are meant to be simultaneous; for example, in Exercise 2, "*E *and* F*" would mean "it's simultaneously raining and sunny.") Write a sentence justifying each of your answers.*

1. *E* is the event "being a doctor," and *F* is the event "being a woman."
2. *E* is the event "it's raining," and *F* is the event "it's sunny."
3. *E* is the event "being single," and *F* is the event "being married."
4. *E* is the event "having naturally blond hair," and *F* is the event "having naturally black hair."
5. *E* is the event "having brown hair," and *F* is the event "having gray hair."
6. *E* is the event "being a plumber," and *F* is the event "being a stamp collector."
7. *E* is the event "wearing shoes," and *F* is the event "wearing sandals."
8. *E* is the event "wearing shoes," and *F* is the event "wearing socks."

In Exercises 9–10, use probabilities, rather than your own personal experience, to determine whether (a) E and F are independent and (b) E and F are mutually exclusive. (c) Interpret your answers to parts (a) and (b).

9. If a die is rolled once, *E* is the event "getting a 4," and *F* is the event "getting an odd number."
10. If a die is rolled once, and *E* is the event "getting a 4," and *F* is the event "getting an even number."
11. Determine whether the events "responding yes to the question on violence in television" and "being a woman" in Example 1 of Section 3.6 are independent.
12. Determine whether the events "having a defect" and "being manufactured in Peoria" in Example 4 of Section 3.6 are independent.

In Exercises 13 and 14, use the following information: To determine what effect salespeople had on purchases, a department store polled 700 shoppers as to whether or not they made a purchase and whether or not they were pleased with the service. Of those who made a purchase, 125 were happy with the service and 111 were not. Of those who made no purchase, 148 were happy with the service and 316 were not.

13. Are the events "being happy with the service" and "making a purchase" independent? What conclusion can you make?
14. Are the events "being unhappy with the service" and "not making a purchase" independent? What conclusion can you make?
15. A personal computer manufacturer buys 38% of its chips from Japan and the rest from America. Of the Japanese chips, 1.7% are defective, whereas 1.1% of

the American chips are defective. Are the events "defective" and "Japanese-made" independent? What conclusion can you draw? (See Exercises 33–36 in Section 3.6.)

16. A skateboard manufacturer buys 23% of its ball bearings from a supplier in Akron, 38% from one in Atlanta, and the rest from one in Los Angeles. Of the ball bearings from Akron, 4% are defective; 6.5% of those from Atlanta are defective; and 8.1% of those from Los Angeles are defective.
a. Find the probability that a ball bearing is defective.
b. Are the events "defective" and "from the Los Angeles supplier" independent?
c. Are the events "defective" and "from the Atlanta supplier" independent?
d. What conclusion can you draw?

In Exercises 17–20, you may wish to use Exercise 50 in Section 3.6.

17. Sixty percent of all computers sold last year were HAL computers, 4% of all computer users quit using computers, and 3% of all computer users used HAL computers and then quit using computers. Are the events "using a HAL computer" and "quit using computers" independent? What conclusion can you make?

18. Ten percent of all computers sold last year were Peach computers, 4% of all computer users quit using computers, and 0.3% of all computer users used Peach computers and then quit using computers. Are the events "using a Peach computer" and "quit using computers" independent? What conclusion can you make?

19. Forty percent of the country used SmellSoGood deodorant, 10% of the country quit using deodorant, and 4% of the country used SmellSoGood deodorant and then quit using deodorant. Are the events "used SmellSoGood" and "quit using deodorant" independent? What conclusion can you make?

20. Nationwide, 20% of all TV viewers use VideoLink cable, 8% of all TV viewers switched from cable to a satellite service, and 7% of all TV viewers used VideoLink cable and then switched from cable to a satellite service. Are the events "used VideoLink cable" and "switched from cable to satellite" independent? What conclusion can you make?

21. Suppose that the space shuttle has three separate computer control systems—the main system and two backup duplicates of it. The first backup would monitor the main system and kick in if the main system failed. Similarly, the second backup would monitor the first. We can assume that a failure of one system is independent of a failure of another system, since the systems are separate. The probability of failure for any one system on any one mission is known to be 0.01.
a. Find the probability that the shuttle is left with no computer control system on a mission.
b. How many backup systems does the space shuttle need if the probability that the shuttle is left with no computer control system on a mission must be $\frac{1}{\text{billion}}$?

22. Use the information in Example 5 to find $p(\text{healthy} \mid +)$, the probability that a teacher receives a false positive.

23. Use the information in Example 5 to find $p(\text{ill} \mid -)$, the probability that a teacher receives a false negative.

24. Overwhelmed with their success in diagnosing white lung in teachers, public health officials decided to administer the test to all schoolchildren, even though only one child in 1,000 has contracted the disease. Recall from Example 5 that the test correctly diagnoses the presence of white lung in 99% of the persons who have it and correctly diagnoses its absence in 98% of the persons who do not have it.
a. Find the probability that a schoolchild whose test results are positive actually has white lung.
b. Find the probability that a schoolchild whose test results are negative does not have white lung.
c. Find the probability that a schoolchild whose test results are positive does not have white lung.
d. Find the probability that a schoolchild whose test results are negative actually has white lung.
e. Which of these events is a false positive? Which is a false negative?
f. Which of these probabilities would you be interested in if you or one of your family members tested positive?
g. Discuss the usefulness of this diagnostic test, both for teachers (as in Example 5) and for schoolchildren.

25. In 1996, the Centers for Disease Control estimated that 1,000,000 of the 261,000,000 residents of the United States are HIV-positive. (HIV is the virus that is believed to cause AIDS.) The SUDS diagnostic test correctly diagnoses the presence of AIDS/HIV 99.9% of the time and correctly diagnoses its absence 99.6% of the time.
a. Find the probability that a person whose test results are positive actually has HIV.
b. Find the probability that a person whose test results are negative does not have HIV.

c. Find the probability that a person whose test results are positive does not have HIV.

d. Find the probability that a person whose test results are negative actually has HIV.

e. Which of the probabilities would you be interested in if you or someone close to you tested positive?

f. Which of these events is a false positive? Which is a false negative?

g. Discuss the usefulness of this diagnostic test.

h. It has been proposed that all immigrants to the United States should be tested for HIV before being allowed into the country. Discuss this proposal.

26. Assuming that the SUDS test cannot be made more accurate, what changes in the circumstances described in Exercise 25 would increase the usefulness of the SUDS diagnostic test? Give a specific example of this change in circumstances, and demonstrate how it would increase the test's usefulness by computing appropriate probabilities.

27. Compare and contrast the circumstances and the probabilities in Example 5 and in Exercises 24, 25, and 26. Discuss the difficulties in using a diagnostic test when that test is not 100% accurate.

28. Use the information in Example 6 to find the following:

a. the probability that the Smiths' child is a cystic fibrosis carrier

b. the probability that the Smiths' child is healthy (i.e., has no symptoms)

c. the probability that the Smiths' child is healthy and not a carrier

HINT: You must consider three possibilities: both of the Smiths are carriers, only one of the Smiths is a carrier, and neither of the Smiths is a carrier.

29. a. Two first cousins marry. Their mutual grandfather's sister died of cystic fibrosis. They have not been tested to see whether they are carriers. Find the probability that their child will have cystic fibrosis. (Assume that no other grandparents are carriers.)

b. Two unrelated people marry. Each had a grandparent whose sister died of cystic fibrosis. They have not been tested to see whether they are carriers. Find the probability that their child will have cystic fibrosis. (Assume that no other grandparents are carriers.)

30. It is estimated that one in twenty-five Americans is a cystic fibrosis carrier. Find the probability that a randomly selected American couple's child will have cystic fibrosis (assuming that they are unrelated).

31. In 1989, researchers announced a new carrier-detection test for cystic fibrosis. However, it was discovered in 1990 that the test will detect the presence of the cystic fibrosis gene in only 85% of cystic fibrosis carriers. If two unrelated people are in fact carriers of cystic fibrosis, find the probability that they both test positive. Find the probability that they don't both test positive.

32. Ramon del Rosario's mother's father died of Huntington's disease. His mother died in childbirth, before symptoms of the disease would have appeared in her. Find the probability that Ramon will have the disease. (Huntington's disease is discussed on page 143 in Section 3.2.)

33. Albinism is a recessive disorder that blocks the normal production of pigmentation. The typical albino has white hair, white skin, and pink eyes and is visually impaired. Mr. Jones is an albino, and although Ms. Jones is normally pigmented, her brother is an albino. Neither of Ms. Jones' parents are albinos. Find the probability that their child will be an albino.

34. Find the probability that the Joneses child will not be an albino but will be a carrier. (See Exercise 33.)

35. If the Joneses' first child is an albino, find the probability that their second child will be too. (See Exercise 33.)

36. The Donohues are going to have a child. She has shiny black hair, and he has bright red hair. Find their child's possible hair colors and the probabilities of each possibility.

37. The Yorks are going to have a child. She has black hair, and he has dark red hair. Find their child's possible hair colors and the probabilities of each possibility.

38. The Eastwoods are going to have a child. She has chestnut hair ($M^{Bd}M^{Bk}$), and he has dark brown hair. Find their child's possible hair colors and the probabilities of each possibility.

39. The Wilsons are going to have a child. She has strawberry blond hair, and he has shiny dark brown hair. Find their child's possible hair colors and the probabilities of each possibility.

40. The Breuners are going to have a child. She has blond hair, and he has glossy dark brown hair. Find their child's possible hair colors and the probabilities of each possibility.

41. The Landres are going to have a child. She has chestnut hair ($M^{Bd}M^{Bk}$), and he has shiny dark brown hair. Find their child's possible hair colors and the probabilities of each possibility.

42. The Hills are going to have a child. She has reddish brown hair, and he has strawberry blond hair. Find their child's possible hair colors and the probabilities of each possibility.

43. Recall from Section 3.1 that Antoine Gombauld, the Chevalier de Méré, had made money over the years by betting with even odds that he could roll at least one six in four rolls of a single die. This problem finds the probability of winning that bet, and the expected value of the bet.
   a. Find the probability of rolling a six in one roll of one die.
   b. Find the probability of not rolling a six in one roll of one die.
   c. Find the probability of never rolling a six in four rolls of a die.

   HINT: This would mean that the first roll is not a six *and* the second roll is not a six *and* the third is not a six *and* the fourth is not a six.

   d. Find the probability of rolling at least one six in four rolls of a die.

   HINT: Use complements.

   e. Find the expected value of this bet if \$1 is wagered.

44. Recall from Section 3.1 that Antoine Gombauld, the Chevalier de Méré, had lost money over the years by betting with even odds that he could roll at least one pair of sixes in 24 rolls of a pair of dice and that he could not understand why. This problem finds the probability of winning that bet and the expected value of the bet.
   a. Find the probability of rolling a double six in one roll of a pair of dice.
   b. Find the probability of not rolling a double six in one roll of a pair of dice.
   c. Find the probability of never rolling a double six in 24 rolls of a pair of dice.
   d. Find the probability of rolling at least one double six in 24 rolls of a pair of dice.
   e. Find the expected value of this bet if \$1 is wagered.

45. Use Exercise 50 of Section 3.6 to explain the calculations of $p(\text{ill} \mid +)$ and $p(\text{healthy} \mid -)$ in Example 5 of this section.

46. Dr. Wellby's patient exhibits symptoms associated with acute neural toxemia (an imaginary disease), but the symptoms can have other, innocuous causes. Studies show that only 25% of those who exhibit the symptoms actually have acute neural toxemia (ANT). A diagnostic test correctly diagnoses the presence of ANT in 88% of the persons who have it and correctly diagnoses its absence in 92% of the persons who do not have it. ANT can be successfully treated, but the treatment causes side effects in 2% of the patients. If left untreated, 90% of those with ANT die; the rest recover fully. Dr. Wellby is considering ordering a diagnostic test for her patient and treating the patient if test results are positive, but she is concerned about the treatment's side effects.
   a. Dr. Wellby could choose to test her patient and administer the treatment if the results are positive. Find the probability that her patient's good health will return under this plan.
   b. Dr. Wellby could choose to avoid the treatment's side effects by not administering the treatment. (This also implies not testing the patient.) Find the probability that her patient's good health will return under this plan.
   c. Find the probability that the patient's good health will return if he undergoes treatment regardless of the test's outcome.
   d. Based on the probabilities, should Dr. Wellby order the test and treat the patient if test results are positive?

47. Probability theory began when Antoine Gombauld, the Chevalier de Méré, asked his friend Blaise Pascal why he had made money over the years betting that he could roll at least one six in four rolls of a single die but he had lost money betting that he could roll at least one pair of sixes in 24 rolls of a pair of dice. Use decision theory and the results of Exercises 43 and 44 to answer Gombauld.

Answer the following questions using complete sentences and your own words.

Concept Questions

48. In Example 5, we are given that $p(- \mid \text{healthy}) = 98\%$ and $p(+ \mid \text{ill}) = 99\%$, and we computed that $p(\text{ill} \mid +) \approx 98\%$ and that $p(\text{healthy} \mid -) \approx 99\%$. Which of these probabilities would be most important to a teacher who was diagnosed as having white lung? Why?

49. Are the melanin hair color genes (M^{Bd}, M^{Bw}, and M^{Bk}) dominant, recessive, or codominant? Are the redness hair color genes (R^+ and R^-) dominant, or recessive, or codominant? Why?

50. Compare and contrast the concepts of independence and mutual exclusivity. Be sure to discuss both the similarities and the differences between these two concepts.

Chapter 3 Review

TERMS

carrier
certain event
codominant gene
conditional probability
decision theory
dependent event
dominant gene
event
expected value
experiment
false negative
false positive
impossible event
independent events
Law of Large Numbers
mutually exclusive events
odds
outcome
probability
Punnett square
recessive gene
relative frequency
sample space

REVIEW EXERCISES

1. What role did the following people play in the development of probability theory?
   - Jacob Bernoulli
   - Gerolamo Cardano
   - Pierre de Fermat
   - Antoine Gombauld, the Chevalier de Méré
   - John Graunt
   - Pierre-Simon, the Marquis de Laplace
   - Gregor Mendel
   - Blaise Pascal
2. A card is dealt from a well-shuffled deck of 52 cards. Find and interpret the probability and the odds of being dealt each of the following.
   a. a red card **b.** a queen
   c. a club **d.** the queen of clubs
   e. a queen or a club **f.** not a queen
3. Find and interpret the probability and the odds of winning the following bets in roulette. (You might want to review the game of roulette in Section 3.1.)
   a. the three-number bet
   b. the twelve-number bet
   c. the even-number bet
4. Three coins are tossed. Find each of the following.
   a. the experiment
   b. the sample space
   c. the event E that exactly two are tails
   d. the event F that two or more are tails
   e. the probability of E and the odds of E
   f. the probability of F and the odds of F
5. A pair of dice is tossed. Find the probability of rolling each of the following.
   a. a seven **b.** an eleven
   c. a seven, an eleven, or doubles
   d. a number that's both odd and greater than 8
   e. a number that's either odd or greater than 8
   f. a number that's neither odd nor greater than 8
6. Three cards are dealt from a deck of 52. Find the probability of each of the following.
   a. All three are hearts.
   b. Exactly two are hearts.
   c. At least two are hearts.
   d. The first is an ace of hearts, the second a two of hearts, and the third a three of hearts.
7. A pair of dice is rolled three times. Find the probability of each of the following.
   a. All three are sevens.
   b. Exactly two are sevens.
   c. At least two are sevens.
8. Gregor's Garden Corner buys 40% of its plants from the Green Growery and the balance from Herb's Herbs. Of the plants from the Green Growery, 20% must be returned, and 10% of those from Herb's Herbs must be returned.
   a. Find the probability that a plant must be returned.
   b. Are the events "a plant must be returned" and "a plant was from Herb's Herbs" independent?
9. A long-stemmed pea is dominant over a short-stemmed pea. A pea with one long-stemmed gene and one short-stemmed gene is crossed with a pea with two short-stemmed genes. Find the following.
   a. the probability that the offspring will be long-stemmed
   b. the probability that the offspring will be short-stemmed
10. Cystic fibrosis is caused by a recessive gene. If two cystic fibrosis carriers produce a child, find the probability of each of the following.
   a. The child will have cystic fibrosis.
   b. The child will be a carrier.
   c. The child will neither have the disease nor be a carrier.

11. Sickle-cell anemia is caused by a codominant gene. If two sickle-cell carriers produce a child, find the probability of each of the following.
   a. The child will have the disease.
   b. The child will be a carrier (i.e., have sickle-cell trait).
   c. The child will neither have the disease nor be a carrier.
12. Huntington's disease is caused by a dominant gene. If one parent has Huntington's disease, find the probability of each of the following. (Assume that the affected parent inherited a single gene for Huntington's disease.)
   a. The child will have the disease.
   b. The child will be healthy.
13. Tay-Sachs disease is caused by a recessive gene. If one parent is a Tay-Sachs carrier and the other has no Tay-Sachs gene, find the probability of each of the following.
   a. The child will have the disease.
   b. The child will be a carrier.
   c. The child will neither have the disease nor be a carrier.
14. Mr. Moody's brother died of cystic fibrosis, as did his wife's brother. Find the probability that the Moody's child will have cystic fibrosis.
15. Jock O'Neill, a sportscaster, and Trudy Bell, a member of the state assembly, are both running for governor of the state of Erehwon. A recent telephone poll asked 800 randomly selected voters for whom they planned to vote. The results of this poll are shown in Figure 3.37.
   a. Find the probability that an urban resident supports O'Neill and the probability that an urban resident supports Bell.
   b. Find the probability that a rural resident supports O'Neill and the probability that a rural resident supports Bell.
   c. Find the probability that an O'Neill supporter lives in an urban area and the probability that an O'Neill supporter lives in a rural area.
   d. Find the probability that a Bell supporter lives in an urban area and the probability that a Bell supporter lives in a rural area.
   e. Where are O'Neill supporters more likely to live? Where are Bell supporters more likely to live?
   f. Which candidate do the urban residents tend to prefer? the rural residents?
   g. Are the events "supporting O'Neill" and "living in an urban area" independent?
   h. On the basis of the poll, who is ahead in the gubernatorial race?
16. Are the following events independent or dependent? Are they mutually exclusive?
   a. "It's springtime" and "it's sunny."
   b. "It's springtime" and "it's Monday."
   c. "It's springtime" and "it's autumn."
   d. "The first card dealt is an ace" and "the second card dealt is an ace."
   e. "The first roll of the dice results in a seven" and "the second roll results in a seven."
17. a. Find the probability of choosing all nine winning spots in nine-spot keno.
   b. Find the probability of choosing eight winning spots in nine-spot keno.
18. Find the expected value of the low-number bet in roulette.
19. Find the expected value of a $1 bet in nine-spot keno if five winning spots break even, six winning spots pay $50, seven pay $390, eight pay $6,000, and nine pay $25,000.
20. Use decision theory to compare a $1 low-number bet in roulette with a $1 bet in nine-spot keno. (See Exercises 18 and 19.)
21. Find the probability of being dealt a pair of tens and three jacks when playing five-card draw poker.

Concept Questions

22. What is a conditional probability?
23. What is meant by independence?
24. Why are probabilities always between 0 and 1 (inclusive)?
25. Give an example of a permutation and a similar example of a combination.
26. Give an example of two events that are mutually exclusive and an example of two events that are not mutually exclusive.

	Jock O'Neill	Trudy Bell	Undecided
Urban Residents	266	184	22
Rural Residents	131	181	16

Figure 3.37

Statistics

4

Statistics are everywhere. The news, whether reported in a newspaper, on television, or over the radio, includes statistics of every kind. When shopping for a new car, you will certainly examine the statistics (average miles per gallon, acceleration times, braking distances, and so on) of the various makes and models you're considering. Statistics abound in government studies, and their interpretation affects us all. Industry is driven by statistics; they are essential to the direction of quality control, marketing research, productivity, and many other factors. Sporting events are laden with statistics concerning the past performance of the teams and players.

A person who understands the nature of statistics is equipped to see beyond the short-term and individual perspective. He or she is also better prepared to deal with those who use statistics in misleading ways. To many people, the word *statistics* conjures up an image of an endless list of facts and figures. Where do they come from? What do they mean? In this chapter, you will learn to handle basic statistical problems and expand your knowledge of the meanings, uses, and misuses of statistics.

4.1 Population, Sample, and Data

The field of **statistics** can be defined as the science of collecting, organizing, and summarizing data in such a way that valid conclusions and meaningful predictions can be drawn from them. The first part of this definition, "collecting, organizing, and summarizing data," applies to **descriptive statistics.** The second part, "drawing valid conclusions and making meaningful predictions," describes **inferential statistics.**

Population versus Sample

Who will become the next president of the United States? During election years, political analysts spend a lot of time and money trying to determine what percent of the vote each candidate will receive. However, because there are over 175 million registered voters in the United States, it would be virtually impossible to contact each and every one of them and ask, "Who do you plan on voting for?" Consequently, analysts select a smaller group of people, determine their intended voting patterns, and project their results onto the entire body of all voters.

Because of time and money constraints, it is very common for researchers to study the characteristics of a small group in order to estimate the characteristics of a larger group. In this context, the set of all objects under study is called the **population,** and any subset of the population is called a **sample** (see Figure 4.1).

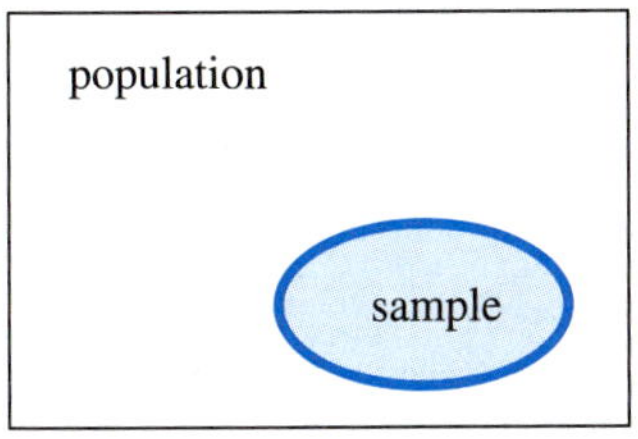

Figure 4.1

When we are studying a large population, we might not be able to collect data from every member of the population, so we collect data from a smaller, more manageable sample. Once we have collected these data, we can summarize by calculating various descriptive statistics, such as the average value. Inferential statistics, then, deals with drawing conclusions (hopefully, valid ones!) about the population, based on the descriptive statistics of the sample data.

Sample data are collected and summarized to help us draw conclusions about the population. A good sample is representative of the population from which it was taken. Obviously, if the sample is not representative, the conclusions concerning the population might not be valid. The most difficult aspect of inferential statistics is obtaining a representative sample. Remember that conclusions are only as reliable as the sampling process and that information will usually change from sample to sample.

Frequency Distributions

The first phase of any statistical study is the collection of data. Each element in a set of data is referred to as a **data point.** When data are first collected, the data points might show no apparent patterns or trends. To summarize the data and detect any trends, we must organize the data. This is the second phase of descriptive statistics. The most common way to organize raw data is to create a **frequency distribution,** a table that lists each data point along with the number of times it occurs (its **frequency**).

The composition of a frequency distribution is often easier to see if the frequencies are converted to percents, especially if large amounts of data are being summarized. The **relative frequency** of a data point is the frequency of the data point expressed as a percent of the total number of data points (that is, made *relative* to the total). The relative frequency of a data point is found by dividing its frequency by the total number of data points in the data set. Besides listing the frequency of each data point, a frequency distribution should also contain a column that gives the relative frequencies.

EXAMPLE 1 While bargaining for their new contract, the employees of 2 Dye 4 Clothing asked their employers to provide daycare service as an employee benefit. Examining the

AP/Wide World Photos/Gianni Foggi

What person who lived in the twentieth century do you admire most? The top ten responses from a Gallup poll taken on Dec. 20–21, 1999, are as follows: (1) Mother Teresa, (2) Martin Luther King, Jr., (3) John F. Kennedy, (4) Albert Einstein, (5) Helen Keller, (6) Franklin D. Roosevelt, (7) Billy Graham, (8) Pope John Paul II, (9) Eleanor Roosevelt, and (10) Winston Churchill.

personnel files of the company's 50 employees, the management recorded the number of children under six years of age that each employee was caring for. The following results were obtained:

0	2	1	0	3	2	0	1	1	0
0	1	1	2	4	1	0	1	1	0
2	1	0	0	3	0	0	1	2	1
0	0	2	4	1	1	0	1	2	0
1	1	0	3	5	1	2	1	3	2

Organize the data by creating a frequency distribution.

Solution First, we list each different number in a column, putting them in order from smallest to largest (or vice versa). Then we use tally marks to count the number of times each data point occurs. The frequency of each data point is shown in the third column of Figure 4.2.

Number of Children under Six	Tally	Frequency	Relative Frequency
0	卌 卌 卌 I	16	$\frac{16}{50} = 0.32 = 32\%$
1	卌 卌 卌 III	18	$\frac{18}{50} = 0.36 = 36\%$
2	卌 IIII	9	$\frac{9}{50} = 0.18 = 18\%$
3	IIII	4	$\frac{4}{50} = 0.08 = 8\%$
4	II	2	$\frac{2}{50} = 0.04 = 4\%$
5	I	1	$\frac{1}{50} = 0.02 = 2\%$
		$n = 50$	total = 100%

Figure 4.2

To get the relative frequencies, we divide each frequency by 50 (the total number of data points) and change the resulting decimal to a percent, as shown in the fourth column of Figure 4.2.

✔ Adding the frequencies, we see that there is a total of $n = 50$ data points in the distribution. This is a good way to monitor the tally process.

The raw data has now been organized and summarized. At this point, we can see that about one-third of the employees have no need for child care (32%), while the remaining two-thirds (68%) have at least one child under six years of age who would benefit from company-sponsored day care. The most common trend (that is, the data point with the highest relative frequency for the 50 employees) is having one child (36%). ■

Grouped Data

When raw data consists of only a few distinct values (for instance, the data in Example 1, which consisted of only the numbers 0, 1, 2, 3, 4, and 5), we can easily organize the data and determine any trends by listing each data point along with its frequency and relative frequency. However, when the raw data consists of many nonrepeated data points, listing each one separately does not help us see any trends the data set might contain. In such cases, it is useful to group the data into intervals or classes and then determine the frequency and relative frequency of each group rather than of each data point.

EXAMPLE 2 Keith Reed is an instructor for an acting class offered through a local arts academy. The class is open to anyone who is at least sixteen years old. Forty-two people are enrolled; their ages are as follows:

26	16	21	34	45	18	41	38	22
48	27	22	30	39	62	25	25	38
29	31	28	20	56	60	24	61	28
32	33	18	23	27	46	30	34	62
49	59	19	20	23	24			

Organize the data by creating a frequency distribution.

Solution This example is quite different from Example 1. Example 1 had only six different data values, whereas this example has many. Listing each distinct data point and its frequency might not summarize the data well enough for us to draw conclusions. Instead, we will work with grouped data.

First, we find the largest and smallest values (62 and 16). Subtracting, we find the range of ages to be $62 - 16 = 46$ years. In working with grouped data, it is customary to create between four and eight groups of data points. We arbitrarily choose six groups, with the first group beginning at the smallest data point, 16. To find the beginning of the second group (and hence the end of the first group), divide the range by the number of groups, round off this answer to be consistent with the data, and then add the result to the smallest data point:

$$46 \div 6 = 7.6666666\ldots \approx 8 \quad \text{(this is the width of each group)}$$

The beginning of the second group is $16 + 8 = 24$, so the first group consists of people from sixteen up to (but not including) twenty-four years of age.

In a similar manner, the second group consists of people from twenty-four up to (but not including) thirty-two ($24 + 8 = 32$) years of age. The remaining groups are formed and the ages tallied in the same way. The frequency distribution is shown in Figure 4.3.

x = Age	Tally	Frequency	Relative Frequency
$16 \le x < 24$	卌 卌 I	11	$\frac{11}{42} \approx 26\%$
$24 \le x < 32$	卌 卌 III	13	$\frac{13}{42} \approx 31\%$
$32 \le x < 40$	卌 II	7	$\frac{7}{42} \approx 17\%$
$40 \le x < 48$	III	3	$\frac{3}{42} \approx 7\%$
$48 \le x < 56$	II	2	$\frac{2}{42} \approx 5\%$
$56 \le x < 64$	卌 I	6	$\frac{6}{42} \approx 14\%$
		$n = 42$	total = 100%

Figure 4.3

Now that the data has been organized, we can observe various trends: Ages from twenty-four to thirty-two are most common (31% is the highest relative frequency), and ages from forty-eight to fifty-six are least common (5% is the lowest). Also, over half the people enrolled (57%) are from sixteen to thirty-two years old. ■

When we are working with grouped data, we can choose the groups in any desired fashion. The method used in Example 2 might not be appropriate in all situations. For example, we used the smallest data point as the beginning of the first group, but we could have begun the first group at an even smaller number. The box at the top of the next page gives a general method for constructing a frequency distribution.

Histograms

When data are grouped in intervals, they can be depicted by a **histogram,** a bar chart that shows how the data are distributed in each interval. To construct a histogram, mark off the class limits on a horizontal axis. If each interval has equal width, we draw two

Constructing a Frequency Distribution

1. If the raw data consists of many different values, create intervals and work with grouped data. If not, list each distinct data point. [When working with grouped data, choose from four to eight intervals. Divide the range (high minus low) by the desired number of intervals, round off this answer to be consistent with the data, and then add the result to the lowest data point to find the beginning of the second group.]
2. Tally the number of data points in each interval or the number of times each individual data point occurs.
3. List the frequency of each interval or each individual data point.
4. Find the relative frequency by dividing the frequency of each interval or each individual data point by the total number of data points in the distribution. The resulting decimal can be expressed as a percent.

vertical axes; the axis on the left exhibits the frequency of an interval, and the axis on the right gives the corresponding relative frequency. We then draw a rectangle above each interval; the height of the rectangle corresponds to the number of data points contained in the interval. The vertical scale on the right gives the percentage of data contained in each interval. The histogram depicting the distribution of the ages of the people in Keith Reed's acting class (Example 2) is shown in Figure 4.4.

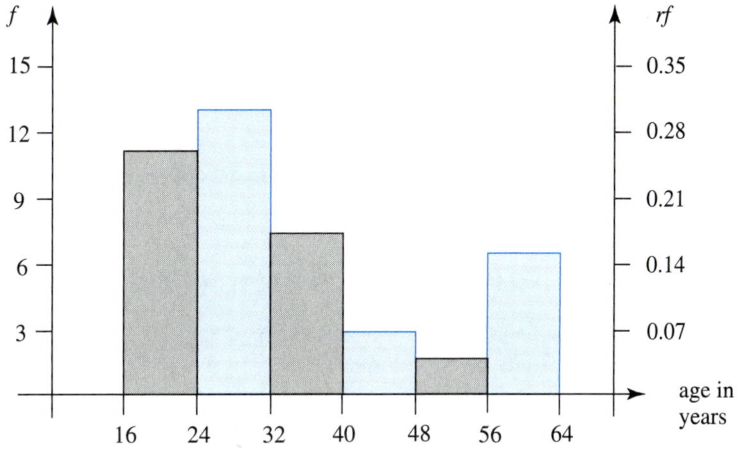

Figure 4.4
Ages of the people in Keith Reed's acting class

What happens if the intervals do not have equal width? For instance, suppose the ages of the people in Keith Reed's acting class are those given in the frequency distribution shown in Figure 4.5 on the next page.

With frequency and relative frequency as the vertical scales, the histogram depicting this new distribution is given in Figure 4.6 on the next page.

Does the histogram in Figure 4.6 give a truthful representation of the distribution? No; the rectangle over the interval from 45 to 65 appears to be larger than the rectangle over the interval from 30 to 45, yet the interval from 45 to 65 contains less data than the interval from 30 to 45. This is misleading; rather than comparing the heights of the rectangles, our eyes naturally compare the areas of the rectangles. Therefore, to make an accurate comparison, *the areas of the rectangles must correspond to the*

x = Age	Frequency	Relative Frequency	Class Width
$15 \le x < 20$	4	$\frac{4}{42} \approx 10\%$	5
$20 \le x < 25$	11	$\frac{11}{42} \approx 26\%$	5
$25 \le x < 30$	6	$\frac{6}{42} \approx 14\%$	5
$30 \le x < 45$	11	$\frac{11}{42} \approx 26\%$	15
$45 \le x < 65$	10	$\frac{10}{42} \approx 24\%$	20
	$n = 42$	total = 100%	

Figure 4.5

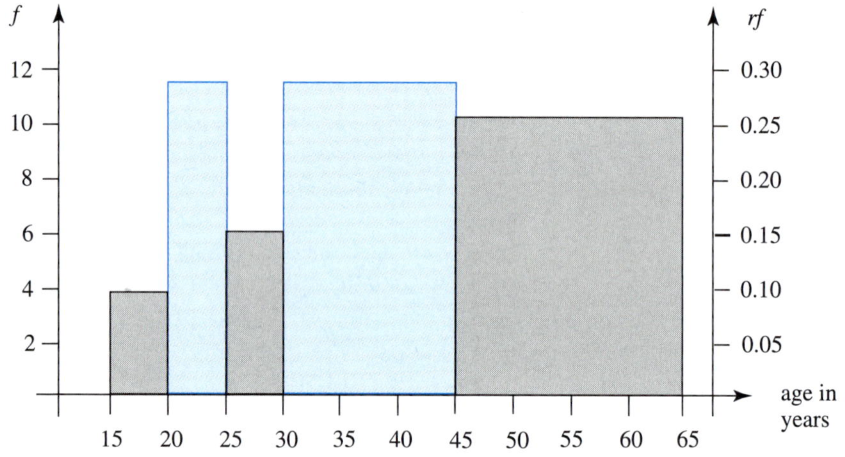

Figure 4.6
Why is this histogram misleading?

relative frequencies of the intervals. This is accomplished by utilizing the **density** of each interval.

Histograms and Relative Frequency Density

Density is a ratio. In science, density is used to determine the concentration of weight in a given volume: Density = weight/volume. For example, the density of water is 62.4 pounds per cubic foot. In statistics, density is used to determine the concentration of data in a given interval: Density = (percent of total data)/(size of an interval). Because relative frequency is a measure of the percentage of data within an interval, we shall calculate the relative frequency density of an interval to determine the concentration of data within the interval.

Definition of Relative Frequency Density

Given a set of n data points, if an interval contains f data points, then the **relative frequency density** (*rfd*) of the interval is

$$rfd = \frac{f/n}{\Delta x}$$

where Δx is the width of the interval.

For example, if the interval $20 \leq x < 25$ contains 11 out of 42 data points, then the relative frequency density of the interval is

$$rfd = \frac{f/n}{\Delta x}$$

$$= \frac{11/42}{5} = 0.052380952\ldots$$

If a histogram is constructed using relative frequency density as the vertical scale, the area of a rectangle will correspond to the relative frequency of the interval, as shown in Figure 4.7.

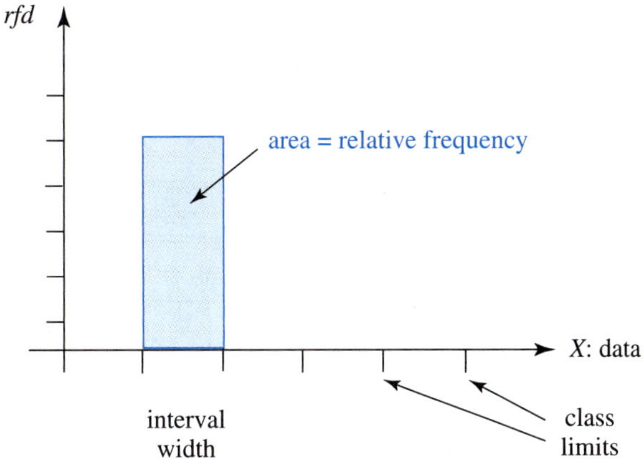

$$\begin{aligned} \text{area} &= \text{base} \cdot \text{height} \\ &= \Delta x \cdot rfd \\ &= \Delta x \cdot \frac{f/n}{\Delta x} \\ &= f/n \\ &= \text{relative frequency} \end{aligned}$$

Figure 4.7

Adding a new column to the frequency distribution given in Figure 4.5, we obtain the relative frequency densities shown in Figure 4.8.

We now construct a histogram using relative frequency density as the vertical scale. The histogram depicting the distribution of the ages of the people in Keith

x = Age	Frequency	Relative Frequency	Class Width	Relative Frequency Density
$15 \leq x < 20$	4	$\frac{4}{42} \approx 10\%$	5	$\frac{4}{42} \div 5 \approx 0.020$
$20 \leq x < 25$	11	$\frac{11}{42} \approx 26\%$	5	$\frac{11}{42} \div 5 \approx 0.052$
$25 \leq x < 30$	6	$\frac{6}{42} \approx 14\%$	5	$\frac{6}{42} \div 5 \approx 0.028$
$30 \leq x < 45$	11	$\frac{11}{42} \approx 26\%$	15	$\frac{11}{42} \div 15 \approx 0.017$
$45 \leq x < 65$	10	$\frac{10}{42} \approx 24\%$	20	$\frac{10}{42} \div 20 \approx 0.012$
	$n = 42$	total = 100%		

Figure 4.8

Reed's acting class (using the frequency distribution in Figure 4.8) is shown in Figure 4.9.

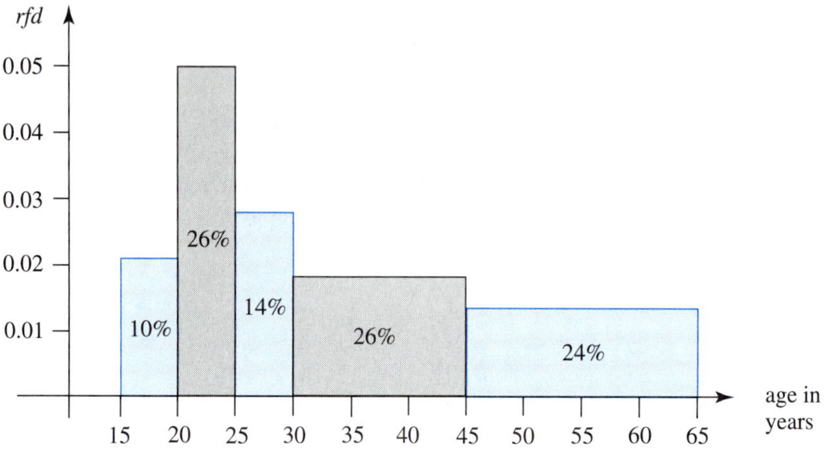

Figure 4.9
Ages of the people in Keith Reed's acting class

Comparing the histograms in Figures 4.6 and 4.9, we see that using relative frequency density as the vertical scale (rather than frequency) gives a more truthful representation of a distribution when the interval widths are unequal.

EXAMPLE 3 To study the output of a machine that fills bags with corn chips, a quality control engineer randomly selected and weighed a sample of 200 bags of chips. The frequency distribution in Figure 4.10 summarizes the data. Construct a histogram for the weights of the bags of corn chips.

x = Weight (ounces)	f = Number of Bags
$15.3 \le x < 15.5$	10
$15.5 \le x < 15.7$	24
$15.7 \le x < 15.9$	36
$15.9 \le x < 16.1$	58
$16.1 \le x < 16.3$	40
$16.3 \le x < 16.5$	20
$16.5 \le x < 16.7$	12

Figure 4.10

Solution Because each interval has the same width ($\Delta x = 0.2$), we construct a combined frequency and relative frequency histogram. The relative frequencies are given in Figure 4.11.

We now draw coordinate axes with appropriate scales and rectangles (Figure 4.12).

x	f	$rf = f/n$
$15.3 \le x < 15.5$	10	0.05
$15.5 \le x < 15.7$	24	0.12
$15.7 \le x < 15.9$	36	0.18
$15.9 \le x < 16.1$	58	0.29
$16.1 \le x < 16.3$	40	0.20
$16.3 \le x < 16.5$	20	0.10
$16.5 \le x < 16.7$	12	0.06
	$n = 200$	sum = 1.00

Figure 4.11

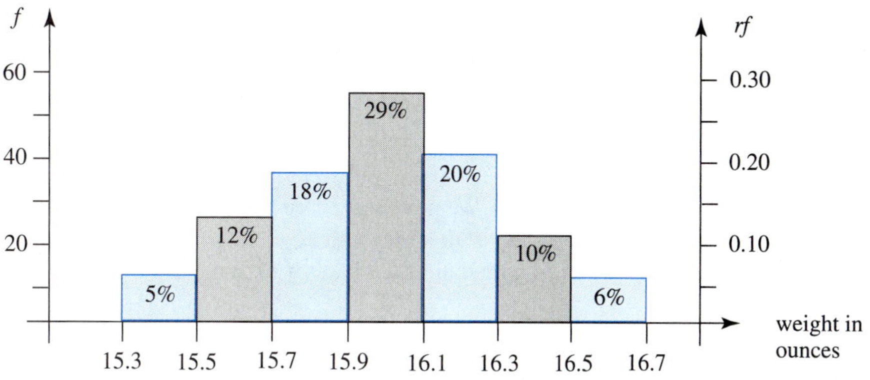

Figure 4.12
Weights of bags of corn chips

Notice the (near) symmetry of the histogram. We will study this type of distribution in more detail in Section 4.4. ■

Histograms and Single-Valued Classes

The histograms that we have constructed so far have all utilized intervals of grouped data. For instance, the first group of ages in Keith Reed's acting class was from 16 to 24 years old ($16 \le x < 24$). However, if a set of data consists of only a few distinct values, it may be advantageous to consider each distinct value to be a "class" of data; that is, we utilize single-valued classes of data.

EXAMPLE 4 A sample of high school seniors was asked, "How many television sets are in your house?" The frequency distribution in Figure 4.13 summarizes the data. Construct a histogram using single-valued classes of data.

Number of Television sets	Frequency
0	2
1	13
2	18
3	11
4	5
5	1

Figure 4.13

Solution Rather than using intervals of grouped data, we use a single value to represent each class. Because each class has the same width ($\Delta x = 1$), we construct a combined frequency and relative frequency histogram. The relative frequencies are given in Figure 4.14.

Number of Television Sets	Frequency	Relative Frequency
0	2	2/50 = 4%
1	13	13/50 = 26%
2	18	18/50 = 36%
3	11	11/50 = 22%
4	5	5/50 = 10%
5	1	1/50 = 2%
	$n = 50$	total = 100%

Figure 4.14

We now draw coordinate axes with appropriate scales and rectangles. In working with single valued classes of data, it is common to write the single value at the midpoint of the base of each rectangle as shown in Figure 4.15.

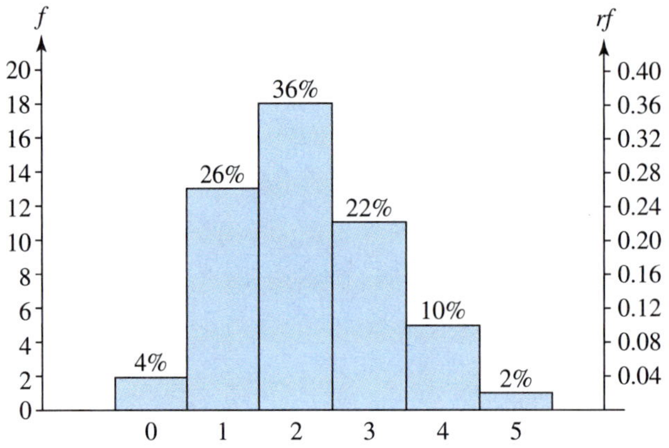

Figure 4.15
Number of television sets

■

Pie Charts

Many statistical studies involve **categorical data**—that which is grouped according to some common feature or quality. One of the easiest ways to summarize categorical data is through the use of a **pie chart.** A pie chart shows how various categories of a set of data account for certain proportions of the whole.

Financial incomes and expenditures are invariably shown as pie charts, as in Figure 4.16.

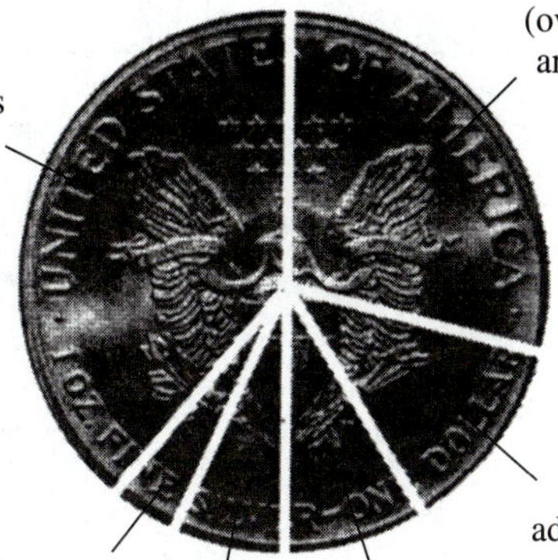

Figure 4.16
How a typical medical dollar is spent

To draw the "slice" of the pie representing the relative frequency (percentage) of the category, the appropriate central angle must be calculated. Since a complete circle comprises 360 degrees, we obtain the required angle by multiplying 360° times the relative frequency of the category.

EXAMPLE 5 What type of academic degree do you hope to earn? The different types of degrees and the number of each type of degree conferred in the United States during the 1999–2000 academic year is given in Figure 4.17. Construct a pie chart to summarize the data.

Type of Degree	Frequency (thousands)
associate's	565
bachelor's	1,238
master's	457
doctorate	45

Figure 4.17
Academic degrees conferred (1999–2000) *Source:* National Center for Education Statistics

Solution Find the relative frequency of each category and multiply it by 360° to determine the appropriate central angle. The necessary calculations are shown in Figure 4.18.

Now use a protractor to lay out the angles and draw the "slices." The name of each category can be written directly on the slice, or, if the names are too long, a legend consisting of various shadings may be used. Each slice of the pie should contain its relative frequency, expressed as a percent. Remember, the whole reason for constructing a pie chart is to convey information visually; pie charts should enable the reader to instantly compare the relative proportions of categorical data. See Figure 4.19.

Type of Degree	Frequency (thousands)	Relative Frequency	Central Angle
associate's	565	$\frac{565}{2,305} \approx 0.245$	$0.245 \times 360° = 88.2°$
bachelor's	1,238	$\frac{1,238}{2,305} \approx 0.537$	$0.537 \times 360° = 193.3°$
master's	457	$\frac{457}{2,305} \approx 0.198$	$0.198 \times 360° = 71.3°$
doctorate	45	$\frac{45}{2,305} \approx 0.020$	$0.020 \times 360° = 7.2°$
	$n = 2,305$	sum = 1.000	total = 360°

Figure 4.18

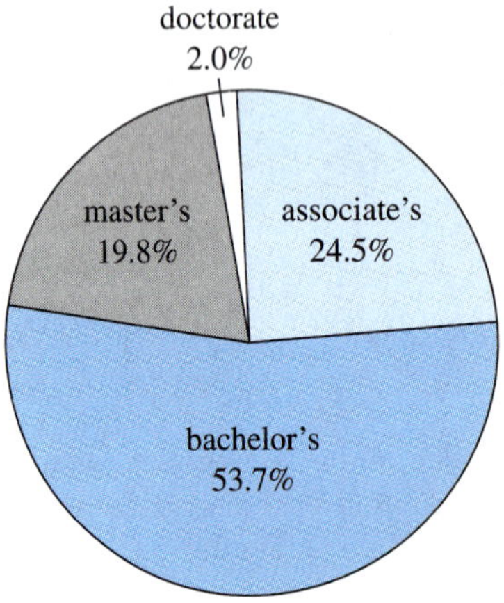

Figure 4.19
Academic degrees conferred (1999–2000)

4.1 Exercises

1. To study the library habits of students at a local college, 30 randomly selected students were surveyed to determine the number of times they had been to the library during the last week. The following results were obtained:

1	5	2	1	1	4	2	1	5	4
5	2	5	1	2	3	4	1	1	2
3	5	4	1	2	2	4	5	1	2

   a. Organize the given data by creating a frequency distribution.
   b. Construct a pie chart to represent the data.
   c. Construct a histogram using single-valued classes of data.

2. To study the eating habits of students at a local college, 30 randomly selected students were surveyed to determine the number of times they had purchased food at the school cafeteria during the last week. The following results were obtained:

2	3	1	2	2	3	2	1	2	3
1	1	5	5	4	3	5	2	2	2
3	2	3	4	1	4	1	2	3	3

   a. Organize the given data by creating a frequency distribution.
   b. Construct a pie chart to represent the data.
   c. Construct a histogram using single-valued classes of data.

3. To study the composition of families in Manistee, Michigan, 40 randomly selected married couples were surveyed to determine the number of children in each family. The following results were obtained:

2	1	3	0	1	0	2	5	1	2	0	2	2	1
4	3	1	1	3	4	1	1	0	2	0	0	2	2
1	0	3	1	1	3	4	2	1	3	0	1		

   a. Organize the given data by creating a frequency distribution.
   b. Construct a pie chart to represent the data.
   c. Construct a histogram using single-valued classes of data.

4. To study the spending habits of shoppers in Orlando, Florida, 50 randomly selected shoppers at a mall were surveyed to determine the number of

credit cards they carried. The following results were obtained:

2 5 0 4 2 1 0 6 3 5 4 3 4 0
5 2 5 2 0 2 5 0 2 5 2 5 4 3
5 6 1 0 6 3 5 3 4 0 5 2 2 5
2 0 2 0 4 2 1 0

a. Organize the given data by creating a frequency distribution.
b. Construct a pie chart to represent the data.
c. Construct a histogram using single-valued classes of data.

5. The speeds, in miles per hour, of 40 randomly monitored cars on Interstate 40 near Winona, Arizona, were as follows:

66 71 76 61 73 78 74 67 80 63 69
78 66 70 77 60 72 58 65 70 64 75
80 75 62 67 72 59 74 65 54 69 73
79 64 68 57 51 68 79

a. Organize the given data by creating a frequency distribution. (Group the data into six intervals.)
b. Construct a histogram to represent the data.

6. The weights, in pounds, of 35 packages of ground beef at the Cut Above Market were as follows:

1.0 1.9 2.5 1.2 2.0 0.7 1.3 2.4 1.1
3.3 2.4 0.8 2.3 1.7 1.0 2.8 1.4 3.0
0.9 1.1 1.4 2.2 1.5 3.2 2.1 2.7 1.8
1.6 2.3 2.6 1.3 2.9 1.9 1.2 0.5

a. Organize the given data by creating a frequency distribution. (Group the data into six intervals.)
b. Construct a histogram to represent the data.

7. To examine the effects of a new registration system, a campus newspaper asked freshmen how long they had to wait in a registration line. The frequency distribution in Figure 4.20 summarizes the responses. Construct a histogram to represent the data.

x = Time in Minutes	Number of Freshmen
$0 \le x < 10$	101
$10 \le x < 20$	237
$20 \le x < 30$	169
$30 \le x < 40$	79
$40 \le x < 50$	51
$50 \le x < 60$	63

Figure 4.20

8. The frequency distribution shown in Figure 4.21 lists the annual salaries of the managers at Universal Manufacturing of Melonville. Construct a histogram to represent the data.

x = Salary (in thousands)	Number of Managers
$\$30 \le x < 40$	6
$40 \le x < 50$	12
$50 \le x < 60$	10
$60 \le x < 70$	5
$70 \le x < 80$	7
$80 \le x < 90$	3

Figure 4.21

9. The frequency distribution shown in Figure 4.22 lists the hourly wages of the workers at Universal Manufacturing of Melonville. Construct a histogram to represent the data.

x = Hourly Wage	Number of Employees
$\$4.00 \le x < 5.50$	21
$5.50 \le x < 7.00$	35
$7.00 \le x < 8.50$	42
$8.50 \le x < 10.00$	27
$10.00 \le x < 11.50$	18
$11.50 \le x < 13.00$	9

Figure 4.22

10. To study the output of a machine that fills boxes with cereal, a quality control engineer weighed 150 boxes of Brand X cereal. The frequency distribution in Figure 4.23 summarizes her findings. Construct a histogram to represent the data.

x = Weight (in ounces)	Number of Boxes
$15.3 \le x < 15.6$	13
$15.6 \le x < 15.9$	24
$15.9 \le x < 16.2$	84
$16.2 \le x < 16.5$	19
$16.5 \le x < 16.8$	10

Figure 4.23

11. The ages of the nearly 4 million women who gave birth in the United States in 1997 are given in Figure 4.24. Construct a histogram to represent the data.

Age	Number of Women
$15 \le x < 20$	486,000
$20 \le x < 25$	948,000
$25 \le x < 30$	1,075,000
$30 \le x < 35$	891,000
$35 \le x < 40$	410,000
$40 \le x < 45$	77,000
$45 \le x < 50$	4,000

Figure 4.24
Ages of women giving birth in 1997
Source: U.S. Bureau of the Census

12. The age composition of the population of the United States in the year 2000 is given in Figure 4.25. Replace the interval "85 and over" with the interval $85 \le x \le 100$ and construct a histogram to represent the data.

Age	Number of People (in thousands)
$0 < x < 5$	19,176
$5 \le x < 10$	20,550
$10 \le x < 15$	20,528
$15 \le x < 25$	39,184
$25 \le x < 35$	39,892
$35 \le x < 45$	44,149
$45 \le x < 55$	37,678
$55 \le x < 65$	24,275
$65 \le x < 85$	30,752
85 and over	4,240

Figure 4.25
Age composition of the population of the United States in the year 2000
Source: U.S. Bureau of the Census

In Exercises 13 and 14, use the age composition of the 14,980,000 students enrolled in institutions of higher education in the United States during 2000, as given in Figure 4.26.

Age of Males	Number of Students
$14 \le x < 18$	94,000
$18 \le x < 20$	1,551,000
$20 \le x < 22$	1,420,000
$22 \le x < 25$	1,091,000
$25 \le x < 30$	865,000
$30 \le x < 35$	521,000
35 and over	997,000
total	6,539,000

Figure 4.26
Age composition of students in higher education
Source: U.S. National Center for Education Statistics

Age of Females	Number of Students
$14 \le x < 18$	78,000
$18 \le x < 20$	1,907,000
$20 \le x < 22$	1,597,000
$22 \le x < 25$	1,305,000
$25 \le x < 30$	1,002,000
$30 \le x < 35$	664,000
35 and over	1,888,000
total	8,441,000

Figure 4.26 ***Continued***

13. Using the data in Figure 4.26, replace the interval "35 and over" with the interval $35 \le x \le 60$ and construct a histogram to represent the male data.
14. Using the data in Figure 4.26, replace the interval "35 and over" with the interval $35 \le x \le 60$ and construct a histogram to represent the female data.

15. The frequency distribution shown in Figure 4.27 lists the ages of 200 randomly selected students who received a bachelor's degree at State University last year. Where possible, determine what percent of the graduates had the following ages:
   a. less than twenty-three
   b. at least thirty-one
   c. at most twenty
   d. not less than nineteen
   e. at least nineteen but less than twenty-seven
   f. not between twenty-three and thirty-five

Age	Number of Students
$10 \le x < 15$	1
$15 \le x < 19$	4
$19 \le x < 23$	52
$23 \le x < 27$	48
$27 \le x < 31$	31
$31 \le x < 35$	16
$35 \le x < 39$	29
39 and over	19

Figure 4.27

16. The frequency distribution shown in Figure 4.28 lists the number of hours per day a randomly selected sample of teenagers spent watching television. Where possible, determine what percent of the teenagers spent the following number of hours watching television:
   a. less than 4 hours
   b. at least 5 hours
   c. at least 1 hour
   d. less than 2 hours
   e. at least 2 hours but less than 4 hours
   f. more than 3.5 hours

Hours per Day	Number of Teenagers
$0 \le x < 1$	18
$1 \le x < 2$	31
$2 \le x < 3$	24
$3 \le x < 4$	38
$4 \le x < 5$	27
$5 \le x < 6$	12
$6 \le x < 7$	15

Figure 4.28

17. Figure 4.29 lists the top five reasons given by patients for emergency room visits in 1999. Construct a pie chart to represent the data.

Reason	Number of Patients (thousands)
stomach pain	6,610
chest pain	5,608
fever	4,678
headache	2,809
cough	2,596

Figure 4.29
Source: U.S. Department of Health and Human Services

18. Figure 4.30 lists the world's top six countries as tourist destinations in 2000. Construct a pie chart to represent the data.

Country	Number of Arrivals (millions)
France	75.5
Spain	50.9
United States	48.2
Italy	41.2
China	31.2
United Kingdom	25.2

Figure 4.30
Source: World Tourism Organization

19. Figure 4.31 lists the race of new AIDS cases in the United States reported in 1999.

a. Construct a pie chart to represent the male data.

b. Construct a pie chart to represent the female data.

c. Construct a pie chart to represent the total data.

Race	Male	Female
White	12,775	1,908
Black	14,885	6,743
Hispanic	6,085	1,608
Asian	297	62
Native American	132	41

Figure 4.31
New AIDS cases in the United States in 1999
Source: U.S. Department of Health and Human Services

20. Figure 4.32 lists the types of accidental deaths in the United States in 2000. Construct a pie chart to represent the data.

Type of Accident	Number of Deaths
motor vehicle	43,000
falls	16,200
poison	12,100
drowning	3,900
fire	3,600
choking	3,400
firearms	600

Figure 4.32
Types of accidental deaths in 2000
Source: National Safety Council

21. Figure 4.33 lists some common specialties of physicians in the United States in 1999.

a. Construct a pie chart to represent the male data.

b. Construct a pie chart to represent the female data.

c. Construct a pie chart to represent the total data.

Specialty	Male	Female
internal medicine	94,244	34,465
family practice	50,176	18,887
general surgery	35,572	3,739
pediatrics	31,050	28,499
psychiatry	27,790	11,266
obstetric/gynecology	26,158	13,205

Figure 4.33
Physicians by Gender and Specialty in 1999 *Source:* American Medical Association

22. Figure 4.34 lists the major metropolitan areas on intended residence for immigrants admitted to the United States in 1998. Construct a pie chart to represent the data.

Metropolitan Area	Number of Immigrants
New York, NY	81,417
Los Angeles, CA	60,220
Chicago, IL	31,033
Miami, FL	29,242
Washington DC	25,639

Figure 4.34
Immigrants and areas of residence in 1998
Source: Immigration and Naturalization Service

Answer the following questions using complete sentences and your own words.

Concept Questions

23. Explain the meanings of the terms *population* and *sample.*

24. The cholesterol levels of the 800 residents of Land-o-Lakes, Wisconsin, were recently collected and organized in a frequency distribution. Do these data represent a sample or a population? Explain your answer.

25. Explain the difference between frequency, relative frequency, and relative frequency density. What does each measure?

26. When is relative frequency density used as the vertical scale in constructing a histogram? Why?

27. In some frequency distributions, data are grouped in intervals; in others, they are not.

a. When should data be grouped in intervals?

b. What are the advantages and disadvantages of using grouped data?

Technology and Statistical Graphs

In Example 2 of this section, we created a frequency distribution and a histogram for the ages of the students in an acting class. Much of this work can be done on a computer or a graphing calculator.

Histograms and Pie Charts on a Computerized Spreadsheet

A **spreadsheet** is a large piece of paper marked off in rows and columns. Accountants use spreadsheets to organize numerical data and perform computations. A **computerized spreadsheet,** such as Microsoft Excel or Lotus 1-2-3, is a computer program that mimics the appearance of a paper spreadsheet. It frees the user from performing any computations; instead, it allows the user merely to give instructions on how to perform those computations. The instructions in this subsection were specifically written for Microsoft Excel; however, all computerized spreadsheets work somewhat similarly.

When you start a computerized spreadsheet, you see something that looks like a table waiting to be filled in. The rows are labeled with numbers and the columns with letters, as shown in Figure 4.35.

	A	B	C	D	E
1					
2					
3					
4					
5					

FIGURE 4.35

The individual boxes are called **cells.** The cell in column A row 1 is called cell A1; the cell below it is called cell A2, because it is in column A row 2.

A computerized spreadsheet is an ideal tool to use in creating a histogram or a pie chart. We will illustrate this process by preparing both a histogram and a pie chart for the ages of the students in Keith Reed's acting class, as discussed in Example 2.

Entering the Data

Step 1 *Label the columns.* Use the mouse and/or the arrow buttons to move to cell A1, type in "age of student" and press "return" or "enter." (If there were other data, we could enter it in other columns. For example, if the students' names were included, we could type "name of student" in cell A1, and "age of student" in cell B1.) The columns' widths can be adjusted—using your mouse, click on the right edge of the "A" label at the top of the first column, and move it.

Step 2 *Enter the students' ages in column A.* Move to cell A2, type in "26" and press "return" or "enter." Move to cell A3, type in "16" and press "return" or "enter." In a similar manner, enter all of the ages. You can enter the ages in any order; Excel will sort the data for you. After you complete this step, your spreadsheet should look like that in Figure 4.36 (except it should go down a lot further). You can move up and down in the spreadsheet by clicking on the arrows on the upper-right and lower-right corners of the spreadsheet.

	A	B	C	D
1	age of student			
2	26			
3	16			
4	21			
5	34			
6	45			
7	18			
8	41			

FIGURE 4.36

Step 3 *Save the spreadsheet.* Use your mouse to select "File" at the very top of the screen. Then pull your mouse down until "Save As" is highlighted, and let go. Your instructor may give you further instructions on where and how to save your spreadsheet.

Preparing the Data for a Chart

Step 1 *Enter the group boundaries.* Excel uses "bin numbers" rather than group boundaries. A group's bin number is the highest number that should be included in that group. In Example 2, the first group was $16 \leq x < 24$. The highest age that should be included in this group is 23, so the group's bin number is 23. Be careful; this group's bin number is not 24, since people who are 24 years old should be included in the second group.

- If necessary, use the up arrow in the upper-right corner of the spreadsheet to scroll to the top of the spreadsheet.
- Type "bin numbers" in cell B1.
- Determine each group's bin number, and enter them in column B.

After you complete this step, your spreadsheet should look like that in Figure 4.37.

	A	B	C	D
1	age of student	bin numbers		
2	26	23		
3	16	31		
4	21	39		
5	34	47		
6	45	55		
7	18	63		
8	41			

FIGURE 4.37

Step 2 *Have Excel determine the frequencies.*

- Use your mouse to select "Tools" at the very top of the screen. Then pull your mouse down until "Data Analysis" is highlighted, and let go.
- If "Data Analysis" is not listed under "Tools," then
  - Select "Tools" at the top of the screen, pull down until "Add-Ins" is highlighted, and let go.
  - Select "Analysis ToolPak-VBA."
  - Use your mouse to press the "OK" button.
  - Select "Tools" at the top of the screen, pull down until "Data Analysis" is highlighted, and let go.
- In the "Data Analysis" box that appears, use your mouse to highlight "Histogram" and press the "OK" button.
- In the "Histogram" box that appears, use your mouse to click on the white rectangle that follows "Input Range," and then use your mouse to draw a box around all of the ages. (To draw the box, move your mouse to cell A2, press the mouse button, and move the mouse down until all of the entered ages are enclosed in a box.) This should cause "$A$2:$A$43" to appear in the Input Range rectangle.
- Use your mouse to click on the white rectangle that follows "Bin Range", and then use your mouse to draw a box around all of the bin numbers. This should cause "$B$2:$B$7" to appear in the Bin Range rectangle.
- Be certain that there is a dot in the button to the left of "Output Range." Then click on the white rectangle that follows "Output Range," and use your mouse to click on cell C1. This should cause "$C$1" to appear in the Output Range rectangle.
- Use your mouse to press the "OK" button.

After you complete this step, your spreadsheet should look like that in Figure 4.38 on page 222.

	A	B	C	D
1	age of student	bin numbers	Bin	frequency
2	26	23	23	11
3	16	31	31	13
4	21	39	39	7
5	34	47	47	3
6	45	55	55	2
7	18	63	63	6
8	41		More	0

FIGURE 4.38

Step 3 *Prepare labels for the chart.*

- In column E, list the group boundaries.
- In column F, list the frequencies.

After you complete this step, your spreadsheet should look like that in Figure 4.39 (the first few columns are not shown).

C	D	E	F
Bin	frequency	group boundaries	frequency
23	11	16-24	11
31	13	24-32	13
39	7	32-40	7
47	3	40-48	3
55	2	48-56	2
63	6	56-64	6
More	0		

FIGURE 4.39

Drawing a Histogram

Step 1 *Use the Chart Wizard to draw the histogram.*

- Use your mouse to press the "Chart Wizard" button at the top of the spreadsheet. (It looks like a histogram and it might have a magic wand.)
- Select "Column" and then an appropriate style.
- Press the "Next" button.

- Use your mouse to click on the white rectangle that follows "Data range," and then use your mouse to draw a box around all of the group boundaries and frequencies from step 3 in "Preparing the data for a chart". This should cause "=Sheet1!$E$2:$F$7" to appear in the Data range rectangle.
- Press the "Next" button.
- Under "Chart Title" type an appropriate title.
- After "Category (X)" type an appropriate title for the x-axis, such as "students' ages."
- After "Value (Y)" type an appropriate title for the y-axis, such as "frequencies."
- Press the "Finish" button and the histogram will appear.

Step 2 *Save the spreadsheet.* Use your mouse to select "File" at the very top of the screen. Then pull your mouse down until "Save" is highlighted, and let go.

Step 3 *Print the histogram.*

- Click on the chart, and it will be surrounded by a thicker border than before.
- Use your mouse to select "File" at the very top of the screen. Then pull your mouse down until "Print" is highlighted, and let go.
- Respond appropriately to the "Print" box that appears. Usually, it is sufficient to press a "Print" button or an "OK" button, but the actual way that you respond to the "Print" box depends on your printer.

Drawing a Pie Chart

Step 1 *Delete old charts.* If you still have a histogram on your screen, you must delete it. Click on it once and press the delete button on your keyboard.

Step 2 *Use the Chart Wizard to draw the pie chart.*

- Use your mouse to press the "Chart Wizard" button at the top of the spreadsheet. (It looks like a histogram and it might have a magic wand.)
- Select "Pie" and then an appropriate style.
- Press the "Next" button.
- Use your mouse to click on the white rectangle that follows "Data range," and then use your mouse to draw a box around all of the group boundaries and frequencies from step 3 in "Preparing the data for a chart." This should cause "=Sheet1!$E$2:$F$7" to appear in the Data range rectangle.
- Press the "Next" button.
- Under "Chart Title" type an appropriate title.
- Press the "Finish" button and the pie chart will appear.

Step 3 *Save the spreadsheet.* Use your mouse to select "File" at the very top of the screen. Then pull your mouse down until "Save" is highlighted, and let go.

Step 4 *Print the pie chart.*

- Click on the chart, and it will be surrounded by a thicker border than before.
- Use your mouse to select "File" at the very top of the screen. Then pull your mouse down until "Print" is highlighted, and let go.
- Respond appropriately to the "Print" box that appears. Usually, it is sufficient to press a "Print" button or an "OK" button, but the actual way that you respond to the "Print" box depends on your printer.

Histograms on a Graphing Calculator

Entering the Data

To enter the data from Example 2, do the following.

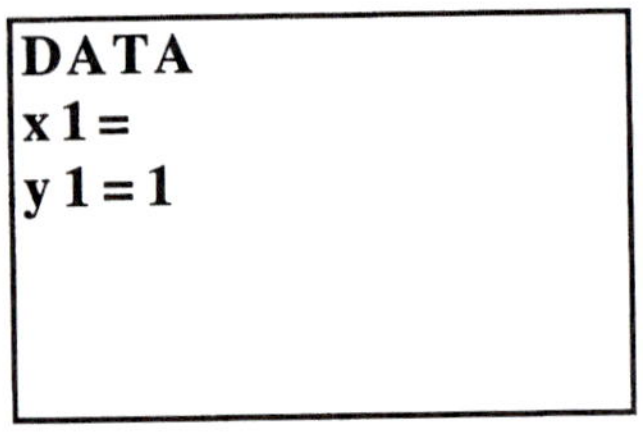

Figure 4.40
A TI-81's data screen

Entering the Data on a TI-81:

- *Put the calculator into statistics mode* by pressing 2nd STAT.
- *Set the calculator up for entering the data* by scrolling to the right, selecting the "DATA" menu, and selecting "Edit" from the "DATA" menu. This should result in the screen shown in Figure 4.40. (If your screen is messier, clear it by returning to the "DATA" menu and selecting "ClrStat.")
- *Enter the students' ages* (in any order) as x_1, x_2, etc. The corresponding y's are the frequencies; leave them as 1s. When completed, your screen should have a blank following "x $_{43}$="; this indicates that 42 entries have been made and the calculator is ready to receive the 43rd. This allows you to check whether you have left any entries out.

Note: If some data points frequently recur, you can enter the data points as x's and their frequencies as y's, rather than reentering a data point each time it recurs.

- Press 2nd QUIT.

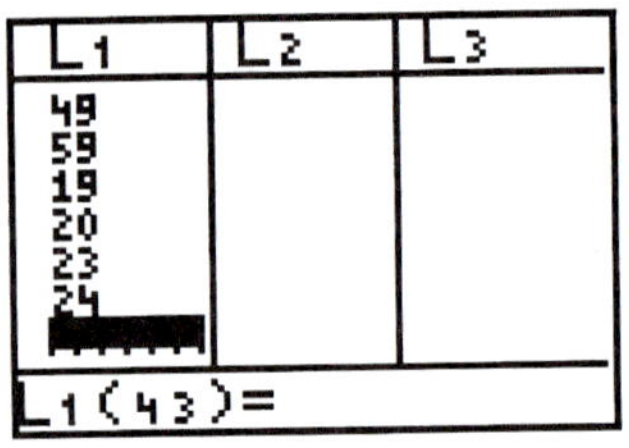

Figure 4.41
A TI-82's list screen

Entering the Data on a TI-80/82/83/86:

- *Put the calculator into statistics mode* by pressing STAT (TI-86: 2nd STAT).
- *Set the calculator up for entering the data* by selecting "Edit" from the "EDIT" menu (TI-86: select "Edit" by pressing F2), and the "list screen" appears, as shown in Figure 4.41. (A TI-86's list screen says "xStat," "yStat," and "fStat" instead of "L_1," "L_2," "L_3," respectively.) If data already appear in a list (as they do in list L_1 in Figure 4.41), use the arrow buttons to highlight the name of the list (i.e., "L_1" or "xStat") and press CLEAR ENTER.
- *Enter the students' ages* in list L_1 (TI-86: in list xStat), in any order, using the arrow buttons and the ENTER button. When completed, your screen should look similar to that in Figure 4.41. Notice the "L_1(43)=" at the bottom of the screen; this indicates that 42 entries have been made, and the calculator is ready to receive the 43rd. This allows you to check whether you have left any entries out.

Note: If some data points frequently recur, you can enter the data points in list L_1 and their frequencies in list L_2 (TI-86: in list fStat), rather than reentering a data point each time it recurs.

- Press 2nd QUIT.

Entering the Data on a TI-85:

- *Put the calculator into statistics mode* by pressing STAT.
- *Set the calculator up for entering the data* by pressing EDIT (i.e., F2). Under most circumstances, the calculator will respond by automatically naming the list of x-coordinates "xStat" and the list of y-coordinates "yStat," as shown in Figure 4.42. (You can rename these lists by typing over the calculator-selected names.) Press ENTER twice to indicate that these two lists are named to your satisfaction.

Figure 4.42

Figure 4.43

- Your calculator is now ready for entering the data. The screen should look like Figure 4.43. [If your screen is messier, you can clear it by pressing CLRxy (i.e., F5)].
- *Enter the students' ages* (in any order) as x_1, x_2, etc. The corresponding *y*s are the frequencies; leave them as 1s. When completed, your screen should have a blank following "$x_{43}=$"; this indicates that 42 entries have been made, and the calculator is ready to receive the 43rd. This allows you to check if you have left any entries out.

Note: If some data points frequently recur, you can enter the data points as *x*s and their frequencies as *y*s, rather than reentering a data point each time it recurs.

- Press 2nd QUIT.

Entering the Data on a Casio:

- *Put the calculator into statistics mode* by pressing MENU, highlighting STAT, and pressing EXE.
- *Enter the data* in List 1 in any order, using the arrow buttons and the EXE button. See Figure 4.44. If data already appear in a list and you want to erase it, use the arrow keys to move to that list, press F6 and then F4 (i.e., DEL-A, which stands for "delete all"), and then press F1 (i.e., YES). Notice that the entries are numbered; this allows you to check whether you have left any entries out.

Figure 4.44
A Casio's list screen

Drawing a Histogram

Once the data are entered, you can draw a histogram.

Drawing a Histogram on a TI-81:

- Press Y= and clear any functions that may appear. Erase the "Y= " screen by pressing 2nd QUIT.
- *Enter the group boundaries* by pressing RANGE, entering the left boundary of the first group as xmin (16 for this problem), the right boundary of the last group plus 1 as xmax ($64 + 1 = 65$ for this problem), and the group width as xscl (8 for this problem). (The calculator will create histograms only with equal group widths.) Enter 0 for ymin and the largest frequency for ymax. (You may guess; it's easy to change it later if you guess wrong.)

- *Draw a histogram* by pressing [2nd], [STAT], scrolling to the right and selecting the "DRAW" menu, and selecting "Hist" from the "DRAW" menu. When "Hist" appears on the screen, press [ENTER]. If some of the bars are too long or too short for the screen, alter ymin accordingly.
- Press an arrow button to read off ordered pairs. By placing the mark on top of a bar, you can approximate its frequency.

Drawing a Histogram on a TI-80/82/83:

- Press [Y=] and clear any functions that may appear.
- *Enter the group boundaries* by pressing [WINDOW], entering the left boundary of the first group as xmin (16 for this problem), the right boundary of the last group plus 1 as xmax (64 + 1 = 65 for this problem), and the group width as xscl (8 for this problem). (The calculator will create histograms only with equal group widths.) Enter 0 for ymin, and the largest frequency for ymax. (You may guess; it's easy to change it later if you guess wrong.)
- *Set the calculator up to draw a histogram* by pressing [2nd] [STAT PLOT] and selecting "Plot 1." Turn the plot on and select the histogram icon.
- Tell the calculator to put the data entered in list L_1 on the *x*-axis by selecting "L_1" for "Xlist," and to consider each entered data point as having a frequency of 1 by selecting "1" for "Freq."

Note: If some data points frequently recur and you entered their frequencies in list L_2, then select "L_2" rather than "1" for "Freq" by typing [2nd] [L_2].

- *Draw a histogram* by pressing [GRAPH]. If some of the bars are too long or too short for the screen, alter ymin accordingly.
- Press [TRACE] to find out the left and right boundaries and the frequency of the bars, as shown in Figure 4.45. Use the arrow buttons to move from bar to bar.
- Press [2nd] [STAT PLOT], select "Plot 1" and turn the plot off, or else the histogram will appear on future graphs.

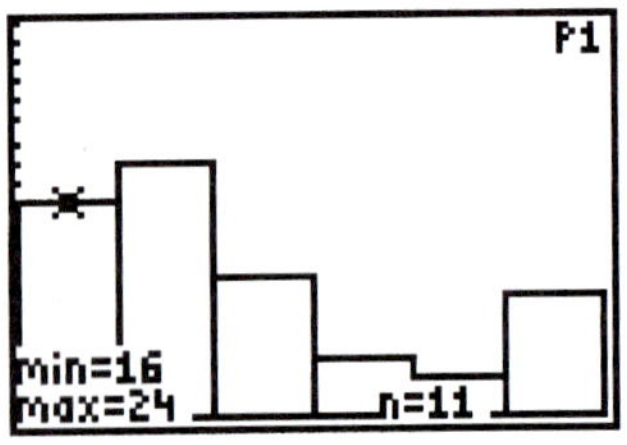

Figure 4.45
The first bar's boundaries are 16 and 24; its frequency is 11.

Drawing a Histogram on a TI-85:

- Put the calculator into graphing mode, press [y(x) =], and clear any functions that may appear.
- *Enter the group boundaries* by pressing [2nd] [RANGE], entering the left boundary of the first group as xmin (16 for this problem), the right boundary of the last group plus 1 as xmax (64 + 1 = 65 for this problem), and the group width as xscl (8 for this problem). (The calculator will create histograms only with equal group widths.) Enter 0 for ymin, and the largest frequency for ymax. (You may guess; it's easy to change it later if you guess wrong.) Quit the graphing mode by pressing [2nd] [QUIT].
- *Put the calculator back into statistics mode* by pressing [STAT], and then press [DRAW]. If this immediately produces a drawing, clear it by pressing [CLDRW].
- *Draw the histogram* by pressing [HIST]. If some of the bars are too long or too short for the screen, alter ymax accordingly. If the button labels obscure the histogram, press [EXIT] once.
- Press [CLDRW], or else the histogram will appear on future graphs.

Drawing a Histogram on a TI-86:

- Put the calculator into graphing mode, press [y(x) =], and clear any functions that may appear.
- *Enter the group boundaries* by pressing [2nd] [WIND], entering the left boundary of the first group as xmin (16 for this problem), the right boundary of the last group plus

1 as xmax (64 + 1 = 65 for this problem), and the group width as xscl (8 for this problem). (The calculator will create histograms only with equal group widths.) Enter 0 for ymin, and the largest frequency for ymax. (You may guess; it's easy to change it later if you guess wrong.) Quit the graphing mode by pressing [2nd] [QUIT].

- *Put the calculator back into statistics mode* by pressing [2nd] [STAT].
- *Set the calculator up to draw a histogram* by pressing [PLOT] (i.e., [F3]) and then [Plot1] (i.e., [F1]). Turn the plot on, press the down arrow until the symbol after "Type=" is flashing, and press [HIST] (i.e., [F4]).
- Tell the calculator to put the data entered in xStat on the x-axis by selecting "xStat" for "Xlist," and to consider each entered data point as having a frequency of 1 by typing "1" for "Freq."

Note: If some data points frequently recur and you entered their frequencies in list fStat, then press [fStat] (i.e., [F3]) rather than "1" for "Freq."

- Press [2nd] [QUIT], and put the calculator back into statistics mode by pressing [2nd] [STAT].
- *Draw the histogram* by pressing [DRAW]. If some of the bars are too long or too short for the screen, alter ymax accordingly. If the button labels obscure the histogram, press [EXIT] once.
- Press [PLOT] and then [Plot1] and turn the plot off, or else the histogram will appear on future graphs.

Drawing a Histogram on a Casio:

- Press [GRPH] (i.e., [F1]).
- Press [SET] (i.e., [F6]).
- Make the resulting screen, which is labeled "StatGraph1," read as follows:

Graph Type	:Hist
Xlist	:List1
Frequency	: 1

To make the screen read as described above, do the following:

- Use the down arrow button to scroll down to "Graph Type."
- Press [F6].
- Press [HIST] (i.e., [F1]).
- In a similar manner, change "Xlist" and "Frequency" if necessary.

- Press [EXE] and return to List1.
- Press [GPH1] (i.e., [F1]) and make the resulting screen, which is labeled "Set Interval," read as follows:

Start: 16 "Start" refers to the beginning of the first group

Pitch: 8 "Pitch" refers to the width of each group

- Press [DRAW] (i.e., [F6]), and the calculator will display the histogram.
- Press [SHIFT] and then [TRCE] (i.e., [F1]) to find out the left boundaries and the frequency of the bars. Use the arrow buttons to move from bar to bar. See Figure 4.46.

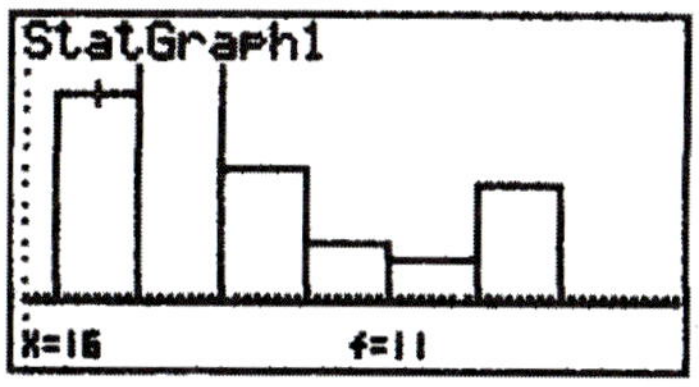

Figure 4.46
The first bar's left boundary is 16; its frequency is 11.

4.1 Exercises

28. In July 2002, *Money* magazine reported the best savings yields in the country for money-market accounts, bond funds, and certificates of deposit.

a. Construct a histogram to represent the data. (Group the data into six intervals.)

b. What observations does the histogram enable you to make?

3.00%	3.00%	3.00%	2.80%	2.79%	1.93%	1.90%	1.88%	1.83%	1.79%
1.62%	1.46%	1.43%	1.43%	1.41%	5.42%	5.28%	5.25%	5.20%	5.15%
3.24%	3.23%	2.98%	2.86%	2.75%	2.90%	2.76%	2.75%	2.70%	3.30%
3.25%	3.21%	3.20%	5.62%	5.50%	5.50%	5.39%			

29. In a 1996 article, *Money* magazine claimed that "state regulation (of the insurance industry) is a joke."* The article "ranks the states by the percentage of legislators with links to insurers who sit on committees focusing on insurance issues."

a. Construct a histogram to represent the data for all 50 states (group the data into five intervals).

b. What observations does the histogram enable you to make?

Mississippi	40%	Louisiana	38%	Arkansas	37%
Virginia	35%	North Carolina	33%	Missouri	32%
Florida	31%	Alabama	26%	Georgia	25%
Texas	25%	West Virginia	24%	Indiana	21%
Ohio	18%	Minnesota	18%	Iowa	17%
North Dakota	17%	Pennsylvania	15%	Arizona	14%
Wyoming	14%	Illinois	14%	Kentucky	13%
Idaho	13%	Delaware	13%	Utah	12%
Kansas	12%	Wisconsin	10%	Hawaii	10%
New Mexico	10%	Rhode Island	9%	Maryland	9%
New Jersey	8%	New York	8%	South Carolina	8%
Washington	8%	Tennessee	8%	Maine	8%
Oklahoma	8%	New Hampshire	7%	Connecticut	5%

(Unlisted states were ranked 0%.)

**Source: Money,* August 1996.

4.2 Measures of Central Tendency

Who is the best running back in professional football? How much does a typical house cost? What is the most popular television program? The answers to questions like these have one thing in common: They are based on averages. To compare the capabilities of athletes, we compute their average performances. This computation usually involves the ratio of two totals, such as (total yards gained)/ (total number of carries) = average gain per carry. In real estate, the average-price house is found by listing the prices of all houses for sale (from lowest to highest) and selecting the price in the middle. Television programs are rated by the average number of households tuned in to each particular program.

Rather than listing every data point in a large distribution of numbers, people tend to summarize the data by selecting a representative number, calling it the average. Three figures—the *mean,* the *median,* and the *mode*—describe the "average" or "center" of a distribution of numbers. These averages are known collectively as the **measures of central tendency.**

The Mean

The **mean** is the average people are most familiar with; it can also be the most misleading. Given a collection of n data points, $x_1, x_2, \ldots, x_n$, the mean is found by adding up the data and dividing by the number of data points:

$$\text{the mean of } n \text{ data points} = \frac{x_1 + x_2 + \cdots + x_n}{n}$$

If the data are collected from a sample, then the mean is denoted by $\bar{x}$ (read "x bar"); if the data are collected from an entire population, then the mean is denoted by μ (lowercase Greek letter "mu"). Unless stated otherwise, we will assume that the data represent a sample, so the mean will be symbolized by $\bar{x}$.

Mathematicians have developed a type of shorthand, called **summation notation,** to represent the sum of a collection of numbers. The Greek letter Σ ("sigma") corresponds to the letter S and represents the word *sum.* Given a group of data points $x_1, x_2 \ldots, x_n$, we use the symbol Σx to represent their sum; that is, $\Sigma x = x_1 + x_2 + \cdots + x_n$.

In 2001, Barry Bonds of the San Francisco Giants hit 73 home runs and set an all-time record in professional baseball. On average, how many home runs per season did Bonds hit? See Exercise 11.

Definition of the Mean

Given a sample of n data points, $x_1, x_2, \ldots, x_n$, the **mean,** denoted by $\bar{x}$, is

$$\bar{x} = \frac{\Sigma x}{n} \quad \text{or} \quad \bar{x} = \frac{\text{the sum of the data points}}{\text{the number of data points}}$$

Many scientific calculators have statistical functions built into them. These functions allow you to enter the data points and press the "x-bar" button to obtain the mean. Consult your manual for specific instructions.

EXAMPLE 1 Vince Fratangelo wanted to investigate the price of gasoline at various service stations in his area. He surveyed seven stations and recorded the price per gallon of self-serve premium unleaded (92 octane) fuel. The raw data are as follows:

$1.399   1.349   1.299   1.429   1.399   1.379   1.259

Find the mean price per gallon of self-serve premium unleaded gasoline.

Solution

$$\begin{aligned}\bar{x} &= \frac{\Sigma x}{n} \\ &= \frac{\$1.399 + 1.349 + 1.299 + 1.429 + 1.399 + 1.379 + 1.259}{7} \\ &= \frac{\$9.513}{7} \\ &= \$1.359 \text{ per gallon}\end{aligned}$$

■

"Bob and Ruth! Come on in. ... Have you met Russell and Bill, our 1.5 children?"

If a sample of 460 families have 690 children altogether, then the mean number of children per family is $\overline{x} = 1.5$.

EXAMPLE 2 In 2001, the U.S. Bureau of Labor Statistics tabulated a survey of workers' ages and wages. The frequency distribution in Figure 4.47 summarizes the age distribution of workers who received minimum wage ($5.15 per hour). Find the mean age of a worker receiving minimum wage.

y = Age	Number of Workers
$16 \le y < 20$	640,000
$20 \le y < 25$	660,000
$25 \le y < 35$	372,000
$35 \le y < 45$	276,000
$45 \le y < 55$	171,000
$55 \le y < 65$	111,000
	$n = 2{,}230{,}000$

Figure 4.47
Source: Bureau of Labor Statistics, U.S. Department of Labor

Solution To find the mean age of the workers, we must add up the ages of all the workers and divide by 2,230,000. However, because we are given grouped data, the ages of the individual workers are unknown to us. In this situation, we use the midpoint of each interval as the representative of the interval; consequently, our answer is an approximation.

To find the midpoint of an interval, add the endpoints and divide by 2. For instance, the midpoint of the interval $16 \le y < 20$ is

$$\frac{16 + 20}{2} = 18$$

We can then say that each of the 640,000 people in the first interval is approximately eighteen years old. Adding up these workers' ages, we obtain

$$\begin{aligned} &18 + 18 + 18 + \cdots + 18 \quad \text{(six hundred forty thousand times)} \\ &\qquad = (640{,}000)(18) \\ &\qquad = 11{,}520{,}000 \text{ years} \end{aligned}$$

If we let f = frequency and x = the midpoint of an interval, the product $f \cdot x$ gives us the total age of the workers in an interval. The results of this procedure for all the workers are shown in Figure 4.48.

y = Age	f = Frequency	x = Midpoint	$f \cdot x$
$16 \le y < 20$	640,000	18	11,520,000
$20 \le y < 25$	660,000	22.5	14,850,000
$25 \le y < 35$	372,000	30	11,160,000
$35 \le y < 45$	276,000	40	11,040,000
$45 \le y < 55$	171,000	50	8,550,000
$55 \le y < 65$	111,000	60	6,660,000
	$n = 2{,}230{,}000$		$\Sigma(f \cdot x) = 63{,}780{,}000$

Figure 4.48

Because $f \cdot x$ gives the sum of the ages of the workers in an interval, the symbol $\Sigma(f \cdot x)$ represents the sum of all the $(f \cdot x)$; that is, $\Sigma(f \cdot x)$ represents the sum of the ages of *all* workers.

The mean age is found by dividing the sum of the ages of all the workers by the number of workers:

$$\begin{aligned} \bar{x} &= \frac{\Sigma(f \cdot x)}{n} \\ &= \frac{63{,}780{,}000}{2{,}230{,}000} \\ &= 28.60089 \text{ years} \end{aligned}$$

The mean age of the workers earning minimum wage is approximately 28.6 years.

✔ One common mistake that is made in working with grouped data is to forget to multiply the midpoint of an interval by the frequency of the interval. Another common mistake is to divide by the number of intervals instead of by the total number of data points.

The procedure for calculating the mean when working with grouped data (as illustrated in Example 2) is summarized in the following box.

Calculating the Mean: Grouped Data

Given a frequency distribution containing several groups of data, the mean $\bar{x}$ can be found by using the following formula:

$$\bar{x} = \frac{\Sigma(f \cdot x)}{n}$$ where x = the midpoint of a group, f = the frequency of the group, and $n = \Sigma f$

EXAMPLE 3 Ten college students were comparing their wages earned at part-time jobs. Nine earned \$10.00 per hour working at jobs ranging from waiting on tables to working in a bookstore. The tenth student earned \$200.00 per hour modeling for a major fashion magazine. Find the mean wage of the ten students.

Solution The data point \$10.00 is repeated nine times, so we multiply it by its frequency. Thus, the mean is as follows:

$$\begin{aligned}\bar{x} &= \frac{\Sigma(f \cdot x)}{n}\\ &= \frac{(9 \cdot 10) + (1 \cdot 200)}{10}\\ &= \frac{290}{10}\\ &= 29\end{aligned}$$

The mean wage of the students is \$29.00 per hour. ■

Example 3 seems to indicate that the average wage of the ten students is \$29.00 per hour. Is that a reasonable figure? If nine out of ten students earn \$10.00 per hour, can we justify saying that their average wage is \$29.00? Of course not! Even though the mean wage *is* \$29.00, it is not a convincing "average" for this specific group of data. The mean is inflated because one student made \$200.00 per hour. This wage is called an **outlier** (or **extreme value**) because it is significantly different from the rest of the data. Whenever a collection of data has extreme values, the mean can be greatly affected and might not be an accurate measure of the average.

The Median

The **median** is the "middle value" of a distribution of numbers. To find it, we first put the data in numerical order. (If a number appears more than once, we include it as many times as it occurs.) If there is an odd number of data points, the median is the middle data point; if there is an even number of data points, the median is defined to be the mean of the two middle values. In either case, the median separates the distribution into two equal parts. Thus, the median can be viewed as an "average." (The word *median* is also used to describe the strip that runs down the middle of a freeway; half the freeway is on one side, and half is on the other. This common usage is in keeping with the statistical meaning.)

EXAMPLE 4 Find the median of the following sets of raw data.

a. 2 8 3 12 6 2 11
b. 2 8 3 12 6 2 11 8

Solution **a.** First, we put the data in order from smallest to largest. Because there is an odd number of data points, ($n = 7$), we pick the middle one:

2 2 3 6 8 11 12
↑ middle value (under 6)

The median is 6. (Notice that 6 is the fourth number in the list.)

b. We arrange the data first. Because there is an even number of data points, ($n = 8$), we pick the two middle values and find their mean:

2 2 3 6 8 8 11 12
↑ ↑ (under 6 and 8)

$$\frac{(6 + 8)}{2} = 7$$

Therefore, the median is 7. (Notice that 7 is halfway between the fourth and fifth numbers in the list.) ■

In Example 4, we saw that when $n = 7$, the median was the fourth number in the list, and that when $n = 8$, the median was halfway between the fourth and fifth numbers in the list. Consequently, the *location* of the median depends on n, the number of numbers in the set of data. The formula

$$L = \frac{n + 1}{2}$$

can be used to find the location, L, of the median; when $n = 7$,

$$L = \frac{7 + 1}{2} = 4$$

(the median is the fourth number in the list), and when $n = 8$,

$$L = \frac{8 + 1}{2} = 4.5$$

(the median is halfway between the fourth and fifth numbers in the list).

Location of the Median

Given a sample of n data points, the location of the median can be found by using the following formula:

$$L = \frac{n + 1}{2}$$

Once the data has been arranged from smallest to largest, the median is the Lth number in the list.

EXAMPLE 5 Find the median wage for the ten students in Example 3.

Solution First, we put the ten wages in order:

10 10 10 10 10 10 10 10 10 200

Because there are $n = 10$ data points, the location of the median is

$$L = \frac{10 + 1}{2} = 5.5$$

That is, the median is halfway between the fifth and sixth numbers. To find the median, we add the fifth and sixth numbers and divide by 2. Now, the fifth number is 10 and the sixth number is 10, so the median is

$$\frac{10 + 10}{2} = 10$$

Therefore, the median wage is \$10.00. This is a much more meaningful "average" than the mean of \$29.00. ■

Doctors' average income in U.S. is now $177,000

Associated Press. Reprinted with permission.

WASHINGTON— The average income of the nation's physicians rose to \$177,400 in 1992, up 4 percent from the year before, the American Medical Association said yesterday. . . .

The average income figures, compiled annually by the AMA and based on a telephone survey of more than 4,100 physicians, ranged from a low of \$111,800 for general practitioners and family practice doctors to a high of \$253,300 for radiologists.

Physicians' median income was \$148,000 in 1992, or 6.5 percent more than a year earlier. Half the physicians earned more than that and half earned less.

The average is pulled higher than the median by the earnings of the highest paid surgeons, anesthesiologists and other specialists at the top end of the scale . . . .

Here are the AMA's average and median net income figures by specialty for 1992:

General/family practice: \$111,800, \$100,000.
Internal medicine: \$159,300, \$130,000.
Surgery: \$244,600, \$207,000.
Pediatrics: \$121,700, \$112,000.
Obstetrics/gynecology: \$215,100, \$190,000.
Radiology: \$253,300, \$240,000.
Psychiatry: \$130,700, \$120,000.
Anesthesiology: \$228,500, \$220,000.
Pathology: \$189,800, \$170,000.
Other: \$165,400, \$150,000.

Figure 4.49

If a collection of data contains extreme values, the median, rather than the mean, is a better indicator of the "average" value. For instance, in discussions of real estate, the median is usually used to express the "average" price of a house. (Why?) In a similar manner, when the incomes of professionals are compared, the median is a more meaningful representation. The discrepancy between mean (average) income and median income is illustrated in the news article shown in Figure 4.49. Although the article is dated (it's from 1992), it is very informative as to the differences between the mean and the median.

EXAMPLE 6 The students in Ms. Kahlo's art class were asked how many siblings they had. The frequency distribution in Figure 4.50 summarizes the responses. Find the median number of siblings.

Number of Siblings	Number of Responses
0	2
1	8
2	5
3	6

Figure 4.50

Solution The frequency distribution indicates that two students had no (0) siblings, eight students had one (1) sibling, five students had two (2) siblings, and six students had three (3) siblings. Therefore, there were $n = 2 + 8 + 5 + 6 = 21$ students in the class; consequently, there are 21 data points. Listing the data in order, we have

$$0, 0, 1, \ldots 1, 2, \ldots 2, 3, \ldots 3$$

Because there are $n = 21$ data points, the location of the median is

$$L = \frac{21 + 1}{2} = 11$$

That is, the median is the eleventh number. Because the first ten numbers are 0s and 1s ($2 + 8 = 10$), the eleventh number is a 2. Consequently, the median number of siblings is 2. ■

The Mode

The third measure of central tendency is the **mode.** The mode is the most frequent number in a collection of data; that is, it is the data point with the highest frequency. Because it represents the most common number, the mode can be viewed as an average. A distribution of data can have more than one mode or none at all.

EXAMPLE 7 Find the mode(s) of the following sets of raw data:

a. 4 10 1 8 5 10 5 10
b. 4 9 1 10 1 10 4 9
c. 9 6 1 8 3 10 3 9

Solution
a. The mode is 10, because it has the highest frequency (3).
b. There is no mode, because each number has the same frequency (2).
c. The distribution has two modes—namely, 3 and 9—each with a frequency of 2. A distribution that has two modes is called *bimodal.* ■

In summarizing a distribution of numbers, it is most informative to list all three measures of central tendency. It helps to avoid any confusion or misunderstanding in situations in which the word *average* is used. In the hands of someone with questionable intentions, numbers can be manipulated to mislead people. In his book *How to Lie with Statistics,* Darrell Huff states, "The secret language of statistics, so appealing in a fact-minded culture, is employed to sensationalize, inflate, confuse, and oversimplify. Statistical methods and statistical terms are necessary in reporting the mass data of social and economic trends, business conditions, 'opinion' polls, the census. But without writers who use the words with honesty and understanding, and readers who know what they mean, the result can only be semantic nonsense." An educated public should always be on the alert for oversimplification of data via statistics. The next time someone mentions an "average," ask "Which one?" Although people might not intentionally try to mislead you, their findings can be misinterpreted if you do not know the meaning of their statistics and the method by which the statistics were calculated.

4.2 Exercises

In Exercises 1–4, find the mean, median, and mode of the given set of raw data.

1. 9 12 8 10 9 11 12
15 20 9 14 15 21 10

2. 20 25 18 30 21 25 32 27
32 35 19 26 38 31 20 23

3. 1.2 1.8 0.7 1.5 1.0 0.7 1.9 1.7 1.2
0.8 1.7 1.3 2.3 0.9 2.0 1.7 1.5 2.2

4. 0.07 0.02 0.09 0.04 0.10 0.08 0.07 0.13
0.05 0.04 0.10 0.07 0.04 0.01 0.11 0.08

5. Find the mean, median, and mode of each set of data.

a. 9 9 10 11 12 15

b. 9 9 10 11 12 102

c. How do your answers for parts (a) and (b) differ (or agree)? Why?

6. Find the mean, median, and mode for each set of data.

a. 80 90 100 110 110 140

b. 10 90 100 110 110 210

c. How do your answers for parts (a) and (b) differ (or agree)? Why?

7. Find the mean, median, and mode of each set of data.

a. 2 4 6 8 10 12

b. 102 104 106 108 110 112

c. How are the data in part (b) related to the data in part (a)?

d. How do your answers for parts (a) and (b) compare?

8. Find the mean, median, and mode of each set of data.

a. 12 16 20 24 28 32

b. 600 800 1,000 1,200 1,400 1,600

c. How are the data in part (b) related to the data in part (a)?

d. How do your answers for parts (a) and (b) compare?

9. Jerry Rice holds the all-time record in professional football for scoring touchdowns. The number of touchdown receptions (TDs) for each of his seasons is given in Figure 4.51 (*Source:* cbs.sportsline.com). Find the mean, median, and mode of the number of touchdown receptions per year by Rice.

Year	TDs
1985	3
1986	15
1987	22
1988	9
1989	17
1990	13
1991	14
1992	10

Year	TDs
1993	15
1994	13
1995	15
1996	8
1997	1
1998	9
1999	5
2000	7
2001	9

Figure 4.51

10. Wayne Gretzky, known as "The Great One," holds the all-time record in professional hockey for scoring goals. The number of goals for each of his seasons is given in Figure 4.52 (*Source: The World Almanac*). Find the mean, median, and mode of the number of goals per season by Gretzky.

Season	Goals
1979–80	51
1980–81	55
1981–82	92
1982–83	71
1983–84	87
1984–85	73
1985–86	52
1986–87	62
1987–88	40
1988–89	54

Season	Goals
1989–90	40
1990–91	41
1991–92	31
1992–93	16
1993–94	38
1994–95	11
1995–96	23
1996–97	25
1997–98	23
1998–99	9

Figure 4.52

11. Barry Bonds of the San Francisco Giants set an all-time record in professional baseball by hitting 73 home runs in one season (2001). The number of home runs (HR) for each of his seasons in professional baseball is given in Figure 4.53. Find the mean, median, and mode of the number of home runs hit per year by Bonds.

Year	HR
1986	16
1987	25
1988	24
1989	19
1990	33
1991	25
1992	34
1993	46
1994	37

Year	HR
1995	33
1996	42
1997	40
1998	37
1999	34
2000	49
2001	73
2002	46

Figure 4.53

12. Michael Jordan has been recognized as an extraordinary player in professional basketball, especially in terms of the number of points per game he has scored. Jordan's average number of points per game (PPG) for each of his seasons is given in Figure 4.54. Find the mean, median, and mode of the average points per game made per season by Jordan.

Season	PPG
1984–85	28.2
1985–86	22.7
1986–87	37.1
1987–88	35.0
1988–89	32.5
1989–90	33.6
1990–91	31.5

Season	PPG
1991–92	30.1
1992–93	32.6
1994–95	26.9
1995–96	30.4
1996–97	29.6
1997–98	28.7
2001–02	22.9

Figure 4.54

13. The frequency distribution in Figure 4.55 lists the results of a quiz given in Professor Gilbert's statistics class. Find the mean, median, and mode of the scores.

Score	Number of Students
10	3
9	10
8	9
7	8
6	10
5	2

Figure 4.55
Quiz scores in Professor Gilbert's statistics class

14. Todd Booth, an avid jogger, kept detailed records of the number of miles he ran per week during the past year. The frequency distribution in Figure 4.56 (see page 238) summarizes his records. Find the mean, median, and mode of the number of miles per week that Todd ran.

Miles Run per Week	Number of Weeks
0	5
1	4
2	10
3	9
4	10
5	7
6	3
7	4

Figure 4.56
Miles run by Todd Booth

15. To study the output of a machine that fills boxes with cereal, a quality control engineer weighed 150 boxes of Brand X cereal. The frequency distribution in Figure 4.57 summarizes his findings. Find the mean weight of the boxes of cereal.

x = Weight (in ounces)	Number of Boxes
$15.3 \le x < 15.6$	13
$15.6 \le x < 15.9$	24
$15.9 \le x < 16.2$	84
$16.2 \le x < 16.5$	19
$16.5 \le x \le 16.8$	10

Figure 4.57
Amount of Brand X cereal per box

16. To study the efficiency of its new price-scanning equipment, a local supermarket monitored the amount of time its customers had to wait in line. The frequency distribution in Figure 4.58 summarizes the findings. Find the mean amount of time spent in line.

17. Katrina must take five exams in a math class. If her scores on the first four exams are 71, 69, 85, and 83, what score does she need on the fifth exam for her overall mean to be
a. at least 70?
b. at least 80?
c. at least 90?

18. Eugene must take four exams in a geography class. If his scores on the first three exams are 91, 67, and 83, what score does he need on the fourth exam for his overall mean to be
a. at least 70?
b. at least 80?
c. at least 90?

x = Time (in minutes)	Number of Customers
$0 \le x < 1$	79
$1 \le x < 2$	58
$2 \le x < 3$	64
$3 \le x < 4$	40
$4 \le x \le 5$	35

Figure 4.58
Time spent waiting in a supermarket checkout line

19. The mean salary of 12 men is \$58,000, and the mean salary of 8 women is \$42,000. Find the mean salary of all 20 people.

20. The mean salary of 12 men is \$52,000, and the mean salary of 4 women is \$84,000. Find the mean salary of all 16 people.

21. Maria drove from Chicago, Illinois, to Milwaukee, Wisconsin, a distance of 90 miles, at a mean speed of 60 miles per hour. On her return trip, the traffic was much heavier, and her mean speed was 45 miles per hour. Find Maria's mean speed for the round trip.

HINT: Divide the total distance by the total time.

22. Sully drove from Atlanta, Georgia, to Birmingham, Alabama, a distance of 150 miles, at a mean speed of 50 miles per hour. On his return trip, the traffic was much lighter, and his mean speed was 60 miles per hour. Find Sully's mean speed for the round trip.

HINT: Divide the total distance by the total time.

23. The mean age of a class of 25 students is 23.4 years. How old would a 26th student have to be for the mean age of the class to be 24.0 years?

24. The mean age of a class of 15 students is 18.2 years. How old would a 16th student have to be for the mean age of the class to be 21.0 years?

25. The mean salary of eight employees is \$40,000, and the median is \$42,000. The highest-paid employee gets a \$6,000 raise.
a. What is the new mean salary of the eight employees?
b. What is the new median salary of the eight employees?

26. The mean salary of ten employees is \$32,000, and the median is \$30,000. The highest-paid employee gets a \$5,000 raise.

a. What is the new mean salary of the ten employees?

b. What is the new median salary of the ten employees?

27. The number of civilians holding various federal government jobs and their mean monthly earnings for May 1999 are given in Figure 4.59.

Department	Number of Civilian Workers	Mean Monthly Earnings
State Department	25,067	\$4,701.12
Justice Department	123,779	\$4,252.77
Congress	16,930	\$3,800.87
Department of the Navy	191,001	\$4,691.25
Department of the Air Force	162,389	\$3,267.01
Department of the Army	233,841	\$2,897.44

Figure 4.59
Monthly earnings for civilian jobs (May 1999)
Source: U.S. Office of Personnel Management

a. Find the mean monthly earnings of all civilians employed by the Navy, Air Force, and Army.

b. Find the mean monthly earnings of all civilians employed by the State Department, Justice Department, and Congress.

28. The ages of the nearly 4 million women who gave birth in the United States in 1997 are given in Figure 4.60. Find the mean age of these women.

Age	Number of Women
$15 \le x < 20$	486,000
$20 \le x < 25$	948,000
$25 \le x < 30$	1,075,000
$30 \le x < 35$	891,000
$35 \le x < 40$	410,000
$40 \le x < 45$	77,000
$45 \le x < 50$	4,000

Figure 4.60
Ages of women giving birth in 1997
Source: U.S. Bureau of the Census

29. The age composition of the population of the United States in the year 2000 is given in Figure 4.61.

a. Find the mean age of all people in the United States under the age of eighty-five.

b. Replace the interval "85 and over" with the interval $85 \le X \le 100$ and find the mean age of all people in the United States.

Age	Number of People (in thousands)
$0 < x < 5$	19,176
$5 \le x < 10$	20,550
$10 \le x < 15$	20,528
$15 \le x < 25$	39,184
$25 \le x < 35$	39,892
$35 \le x < 45$	44,149
$45 \le x < 55$	37,678
$55 \le x < 65$	24,275
$65 \le x < 85$	30,752
85 and over	4,240

Figure 4.61
Age composition of the population of the United States in the year 2000
Source: U.S. Bureau of the Census

In Exercises 30 and 31, use the age composition of the 14,980,000 students enrolled in institutions of higher education in the United States during 2000, as given in Figure 4.62.

Age of Males	Number of Students
$14 \le x < 18$	94,000
$18 \le x < 20$	1,551,000
$20 \le x < 22$	1,420,000
$22 \le x < 25$	1,091,000
$25 \le x < 30$	865,000
$30 \le x < 35$	521,000
35 and over	997,000
total	6,539,000

Figure 4.62
Age composition of students in higher education
Source: U.S. National Center for Education Statistics

Age of Females	Number of Students
$14 \le x < 18$	78,000
$18 \le x < 20$	1,907,000
$20 \le x < 22$	1,597,000
$22 \le x < 25$	1,305,000
$25 \le x < 30$	1,002,000
$30 \le x < 35$	664,000
35 and over	1,888,000
total	8,441,000

Figure 4.62 *Continued*

30. a. Find the mean age of all male students in higher education under thirty-five.

b. Replace the interval "35 and over" with the interval $35 \le x \le 60$ and find the mean age of all male students in higher education.

31. a. Find the mean age of all female students in higher education under thirty-five.

b. Replace the interval "35 and over" with the interval $35 \le x \le 60$ and find the mean age of all female students in higher education.

Answer the following questions using complete sentences and your own words.

Concept Questions

32. What are the three measures of central tendency? Briefly explain the meaning of each.

33. Suppose the mean of Group I is A and the mean of Group II is B. We combine Groups I and II to form Group III. Is the mean of Group III equal to $\frac{A+B}{2}$? Explain.

34. Why do we use the midpoint of an interval when calculating the mean of grouped data?

4.3 Measures of Dispersion

To settle an argument over who was the better bowler, George and Danny agreed to bowl six games, and whoever had the highest "average" would be considered best. Their scores were as shown in Figure 4.63.

George	185	135	200	185	250	155
Danny	182	185	188	185	180	190

Figure 4.63

Each bowler then arranged his scores from lowest to highest and computed the mean, median, and mode:

George

135 155 185 185 200 250

$$\text{mean} = \frac{\text{sum of scores}}{6} = \frac{1{,}110}{6} = 185$$

$$\text{median} = \text{middle score} = \frac{185 + 185}{2} = 185$$

$$\text{mode} = \text{most common score} = 185$$

Danny

180 182 185 185 188 190

$$\text{mean} = \frac{1{,}110}{6} = 185$$

$$\text{median} = \frac{185 + 185}{2} = 185$$

$$\text{mode} = 185$$

Much to their surprise, George's mean, median, and mode were exactly the same as Danny's! Using the measures of central tendency alone to summarize their performances, the bowlers appear identical. Even though their averages were identical, however, their performances were not; George was very erratic, while Danny was very consistent. Who is the better bowler? On the basis of high score, George is better. On the basis of consistency, Danny is better.

George and Danny's situation points out a fundamental weakness in using only the measures of central tendency to summarize data. In addition to finding the averages of a set of data, the consistency, or spread, of the data should also be taken into account. This is accomplished by using **measures of dispersion,** which determine how the data points differ from the average.

Deviations

It is clear from George and Danny's bowling scores that it is sometimes desirable to measure the relative consistency of a set of data. Are the numbers consistently bunched up? Are they erratically spread out? To measure the dispersion of a set of data, we need to identify an average or typical distance between the data points and the mean. The difference between a single data point x and the mean $\bar{x}$ is called the **deviation from the mean** (or simply the **deviation**) and is given by $(x - \bar{x})$. A data point that is close to the mean will have a small deviation, whereas data points far from the mean will have large deviations, as shown in Figure 4.64.

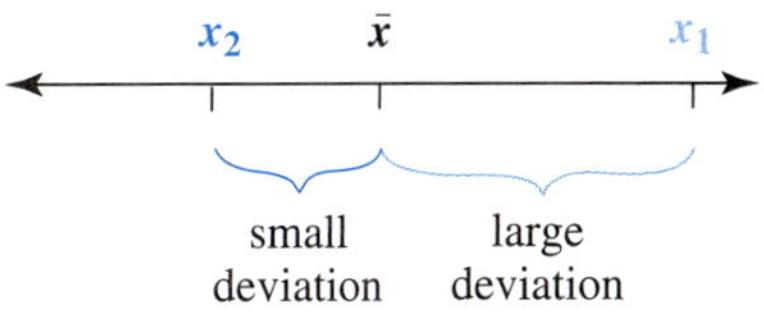

Figure 4.64

To find the typical deviation of the data points, you might be tempted to add up all the deviations and divide by the total number of data points, thus finding the "average" deviation. Unfortunately, this process leads nowhere. To see why, we will find the mean of the deviations of George's bowling scores.

EXAMPLE 1 George bowled six games, and his scores were 185, 135, 200, 185, 250, and 155. Find the mean of the scores, the deviation of each score, and the mean of the deviations.

Solution

$$\bar{x} = \frac{\text{sum of scores}}{6} = \frac{1{,}110}{6} = 185$$

The mean score is 185.

To find the deviations, subtract the mean from each score, as shown in Figure 4.65.

Score (x)	Deviation $(x - 185)$
135	−50
155	−30
185	0
185	0
200	15
250	65
	sum = 0

Figure 4.65

$$\text{mean of the deviations} = \frac{\text{sum of deviations}}{6} = \frac{0}{6} = 0$$

The mean of the deviations is zero. ■

In Example 1, the sum of the deviations of the data is zero. This *always* happens; that is, $\Sigma\,(x - \bar{x}) = 0$ for any set of data. The negative deviations are the "culprits"—they will always cancel out the positive deviations. Therefore, to use deviations to study the spread of the data, we must modify our approach and convert the negatives into positives. We do this by squaring each deviation.

Variance and Standard Deviation

Before proceeding, we must be reminded of the difference between a population and a sample. A population is the universal set of all possible items under study; a sample is any group or subset of items selected from the population. (Samples are used to study populations.) In this context, George's six bowling scores represent a sample, not a population; because we do not know the scores of *all* the games George has ever bowled, we are limited to a sample. Unless otherwise specified, we will consider any given set of data to represent a sample, not an entire population.

To measure the typical deviation contained within a set of data points, we must first find the **variance** of the data. Given a sample of n data points, the variance of the data is found by squaring each deviation, adding the squares, and then dividing the sum by the number $(n - 1)$.*

Sample Variance Deviation

Given a sample of n data points, $x_1, x_2, \ldots, x_n$, the **variance** of the data, denoted by s^2, is

$$s^2 = \frac{\Sigma(x - \bar{x})^2}{n - 1}$$

The variance of a sample is found by dividing the sum of the squares of the deviations by $n - 1$. The symbol s^2 is a reminder that the deviations have been squared.

Because we are working with n data points, you might wonder why we divide by $n - 1$ rather than by n. The answer lies in the study of inferential statistics. Recall that inferential statistics deal with the drawing of conclusions concerning the nature of a population based on observations made within a sample. Hence, the variance of a sample can be viewed as an estimate of the variance of the population. However, because the population will vary more than the sample (a population has more data points), dividing the sum of the squares of the sample deviations by n would underestimate the true variance of the entire population. To compensate for this underestimation, statisticians have determined that dividing the sum of the squares of the deviations by $n - 1$ rather than by n produces the best estimate of the true population variance.

Variance is the tool with which we can obtain a measure of the typical deviation contained within a set of data. However, because the deviations have been squared, we must perform one more operation to obtain the desired result: We must take the square root. The square root of variance is called the **standard deviation** of the data.

---

* If the n data points represent the entire population, the population variance, denoted by σ^2, is found by squaring each deviation, adding the squares, and then dividing the sum by n.

Standard Deviation Definition

Given a sample of n data points, $x_1, x_2, \ldots, x_n$, the **standard deviation** of the data, denoted by s, is

$$s = \sqrt{\text{variance}}$$

To find the standard deviation of a set of data, first find the variance and then take the square root of the variance.

EXAMPLE 2 George bowled six games, and his scores were 185, 135, 200, 185, 250, and 155. Find the standard deviation of his scores.

Solution To find the standard deviation, we must first find the variance. The mean of the six data points is 185. The necessary calculations for finding variance are shown in Figure 4.66.

Data (x)	Deviation ($x - 185$)	Deviation Squared $(x - 185)^2$
135	-50	$(-50)^2 = 2{,}500$
155	-30	$(-30)^2 = 900$
185	0	$(0)^2 = 0$
185	0	$(0)^2 = 0$
200	15	$(15)^2 = 225$
250	65	$(65)^2 = 4{,}225$
		sum $= 7{,}850$

Figure 4.66

$$\text{variance} = \frac{\text{sum of the squares of the deviations}}{n - 1}$$

$$s^2 = \frac{7{,}850}{6 - 1}$$

$$= \frac{7{,}850}{5}$$

$$= 1{,}570$$

The variance is $s^2 = 1{,}570$. Taking the square root, we have

$$s = \sqrt{1{,}570}$$

$$= 39.62322551\ldots$$

It is customary to round off s to one place more than the original data. Hence, the standard deviation of George's bowling scores is $s = 39.6$ points. ■

Because they give us information concerning the spread of data, variance and standard deviation are called **measures of dispersion.** Standard deviation (and variance) is a relative measure of the dispersion of a set of data; the larger the standard deviation, the more spread out the data. Consider George's standard deviation of 39.6. This appears to be high, but what exactly constitutes a "high" standard deviation? Unfortunately, because it is a relative measure, there is no hard-and-fast distinction between a "high" and a "low" standard deviation.

By itself, the standard deviation of a set of data might not be very informative, but standard deviations are very useful in comparing the relative consistencies of two sets of data. Given two groups of numbers of the same type (for example, two sets of bowling scores, two sets of heights, or two sets of prices), the set with the lower standard deviation contains data that are more consistent, whereas the data with the higher standard deviation are more spread out. Calculating the standard deviation of Danny's six bowling scores, we find $s = 3.7$. Since Danny's standard deviation is less than George's, we infer that Danny is more consistent. If George's standard deviation is less than 39.6 the next time he bowls six games, we would infer that his game has become more consistent (the scores would not be spread out as far).

Alternate Methods for Finding Variance

The procedure for calculating variance is very direct: First find the mean of the data, then find the deviation of each data point, and finally divide the sum of the squares of the deviations by $(n - 1)$. However, using the Sample Variance Definition to find the variance can be rather tedious. Fortunately, many scientific calculators are programmed to find the variance (and standard deviation) if you just push a few buttons. Consult your manual to utilize the statistical capabilities of your calculator.

If your calculator doesn't have built-in statistical functions, you might still be able to take a shortcut in calculating variance. Instead of using the Definition of Variance (as in Example 2), we can use an alternate formula that contains the two sums Σx and Σx^2, where Σx represents the sum of the data and Σx^2 represents the sum of the squares of the data, as shown in the following box.

Alternate Formula for Sample Variance

Given a sample of n data points, $x_1, x_2, \ldots, x_n$, the **variance** of the data, denoted by s^2, can be found by

$$s^2 = \frac{1}{(n-1)}\left[\Sigma x^2 - \frac{(\Sigma x)^2}{n}\right]$$

Note: Σx^2 means "square each data point, then add"; $(\Sigma x)^2$ means "add the data points, then square."

Although we will not prove it, this Alternate Formula for Sample Variance is algebraically equivalent to the Sample Variance Definition; given any set of data, either method will produce the same answer. At first glance, the Alternate Formula might appear to be more difficult to use than the Definition. Don't be fooled by its appearance! As we will see, the Alternate Formula is relatively quick and easy to apply.

EXAMPLE 3 Using the Alternate Formula for Sample Variance, find the standard deviation of George's bowling scores as given in Example 2.

Solution Recall that George's scores were 185, 135, 200, 185, 250, and 155. To find the standard deviation, we must first find the variance.

The Alternate Formula for Sample Variance requires that we find the sum of the data and the sum of the squares of the data. These calculations are shown in

Figure 4.67. Applying the Alternate Formula for Sample Variance, we have

$$s^2 = \frac{1}{(n-1)}\left[\Sigma x^2 - \frac{(\Sigma x)^2}{n}\right]$$
$$= \frac{1}{6-1}\left[213{,}200 - \frac{(1{,}110)^2}{6}\right]$$
$$= \frac{1}{5}[213{,}200 - 205{,}350]$$
$$= \frac{7{,}850}{5} = 1{,}570$$

x	x^2
135	18,225
155	24,025
185	34,225
185	34,225
200	40,000
250	62,500
$\Sigma x = 1{,}110$	$\Sigma x^2 = 213{,}200$

Figure 4.67

The variance is $s^2 = 1{,}570$. (Note that this is the same as the variance calculated in Example 2 using the Definition.)

Taking the square root, we have

$$s = \sqrt{1{,}570}$$
$$= 39.62322551\ldots$$

Rounded off, the standard deviation of George's bowling scores is $s = 39.6$ points. ■

When we are working with grouped data, the individual data points are unknown. In such cases, the midpoint of each interval should be used as the representative value of the interval.

EXAMPLE 4 In 2001, the U.S. Bureau of Labor Statistics tabulated a survey of workers' ages and wages. The frequency distribution in Figure 4.68 summarizes the age distribution of workers who received minimum wage ($5.15 per hour). Find the standard deviation of the ages of these workers.

y = Age	Number of Workers
$16 \le y < 20$	640,000
$20 \le y < 25$	660,000
$25 \le y < 35$	372,000
$35 \le y < 45$	276,000
$45 \le y < 55$	171,000
$55 \le y < 65$	111,000
	$n = 2{,}230{,}000$

Figure 4.68
Source: Bureau of Labor Statistics, U.S. Department of Labor

Solution Because we are given grouped data, the first step is to determine the midpoint of each interval. We do this by adding the endpoints and dividing by 2.

To utilize the Alternate Formula for Sample Variance, we must find the sum of the data and the sum of the squares of the data. The sum of the data is found by multiplying each midpoint by the frequency of the interval and adding the results; that is, $\Sigma(f \cdot x)$. The sum of the squares of the data is found by squaring each midpoint, multiplying by the corresponding frequency, and adding; that is, $\Sigma(f \cdot x^2)$. The calculations are shown in Figure 4.69 (see page 246).

y = Age	f = Frequency	x = Midpoint	$f \cdot x$	$f \cdot x^2$
$16 \le y < 20$	640,000	18	11,520,000	207,360,000
$20 \le y < 25$	660,000	22.5	14,850,000	334,125,000
$25 \le y < 35$	372,000	30	11,160,000	334,800,000
$35 \le y < 45$	276,000	40	11,040,000	441,600,000
$45 \le y < 55$	171,000	50	8,550,000	427,500,000
$55 \le y < 65$	111,000	60	6,660,000	399,600,000
	$n = 2{,}230{,}000$		$\Sigma(f \cdot x) = 63{,}780{,}000$	$\Sigma(f \cdot x^2) = 2{,}144{,}985{,}000$

Figure 4.69

Applying the Alternate Formula for Sample Variance, we have

$$\begin{aligned} s^2 &= \frac{1}{n-1}\left[\Sigma(f \cdot x^2) - \frac{(\Sigma f \cdot x)^2}{n}\right] \\ &= \frac{1}{2{,}230{,}000 - 1}\left[2{,}144{,}985{,}000 - \frac{(63{,}780{,}000)^2}{2{,}230{,}000}\right] \\ &= \frac{1}{2{,}229{,}999}[2{,}144{,}985{,}000 - 1{,}824{,}165{,}202] \\ &= \frac{320{,}819{,}798}{2{,}229{,}999} \end{aligned}$$

The variance is $s^2 = 143.8654448$. Taking the square root, we have

$$\begin{aligned} s &= \sqrt{143.8654448} \\ &= 11.99439222\ldots \end{aligned}$$

Rounded off, the standard deviation of the ages of the workers receiving minimum wage is $s = 12.0$ years. ■

The procedure for calculating variance when working with grouped data (as illustrated in Example 4) is summarized in the following box.

Alternate Formula for Sample Variance: Grouped Data

Given a frequency distribution containing several groups of data, the variance s^2 can be found by

$$s^2 = \frac{1}{(n-1)}\left[\Sigma(f \cdot x^2) - \frac{(\Sigma f \cdot x)^2}{n}\right]$$

where x = the midpoint of a group, f = the frequency of the group, and $n = \Sigma f$.

To obtain the best analysis of a collection of data, we should use the measures of central tendency and the measures of dispersion in conjunction with each other. The most common way to combine these measures is to determine what percent of the data lies within a specified number of standard deviations of the mean. The phrase "one standard deviation of the mean" refers to all numbers within the interval

$[\bar{x} - s, \bar{x} + s]$, that is, all numbers that differ from $\bar{x}$ by at most s. Likewise, "two standard deviations of the mean" refers to all numbers within the interval $[\bar{x} - 2s, \bar{x} + 2s]$. One, two, and three standard deviations of the mean are shown in Figure 4.70.

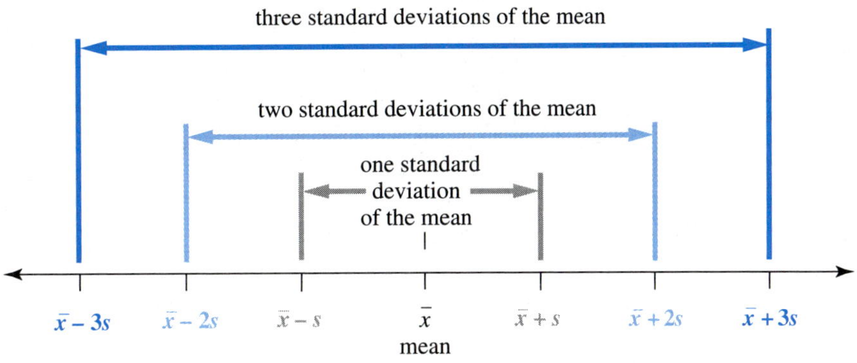

Figure 4.70

EXAMPLE 5 Jim Courtright surveyed the price of unleaded gasoline (self-serve) at the gas stations in his neighborhood. He found the following prices, in dollars per gallon:

1.269 1.249 1.199 1.299 1.269 1.239 1.349 1.289 1.259

What percent of the data lies within one standard deviation of the mean?

Solution We must first find the mean $\bar{x}$ and the standard deviation s.

Summing the nine data points, we have $\Sigma x = 11.421$. Summing the squares of the data points, we have $\Sigma x^2 = 14.507249$. The mean is

$$\bar{x} = \frac{(11.421)}{9} = 1.269.$$

Using the Alternate Formula for Sample Variance, we have

$$\begin{aligned} s^2 &= \frac{1}{(n-1)}\left[\Sigma x^2 - \frac{(\Sigma x)^2}{n}\right] \\ &= \frac{1}{9-1}\left[14.507249 - \frac{(11.421)^2}{9}\right] \\ &= \frac{1}{8}[14.507249 - 14.493249] \\ &= \frac{0.014}{8} \end{aligned}$$

The variance is $s^2 = 0.00175$. Taking the square root, we have

$$\begin{aligned} s &= \sqrt{0.00175} \\ &= 0.041833001\ldots \end{aligned}$$

The standard deviation of the prices is 0.0418 (dollars).

To find one standard deviation of the mean, we add and subtract the standard deviation to and from the mean:

$$\begin{aligned} [\bar{x} - s, \bar{x} + s] &= [1.269 - 0.0418, 1.269 + 0.0418] \\ &= [1.2272, 1.3108] \end{aligned}$$

Arranging the data from smallest to largest, we see that seven of the nine data points are between 1.2272 and 1.3108:

1.199 1.239 1.249 1.259 1.269 1.269 1.289 1.299 1.349

$\bar{x} - s$ (1.2272) ↑ ———— ↑ $\bar{x}$ ———— ↑ $\bar{x} + s$ (1.3108)

Therefore, $\frac{7}{9}$=0.77777777 or 78% of the data lie within one standard deviation of the mean. ■

4.3 Exercises

1. Perform each task, given the following sample data:

   3 8 5 3 10 13

   a. Use the Sample Variance Definition to find the variance and standard deviation of the data.
   b. Use the Alternate Formula for Sample Variance to find the variance and standard deviation of the data.
2. Perform each task, given the following sample data:

   6 10 12 12 11 17 9

   a. Use the Sample Variance Definition to find the variance and standard deviation of the data.
   b. Use the Alternate Formula for Sample Variance to find the variance and standard deviation of the data.
3. Perform each task, given the following sample data:

   10 10 10 10 10 10

   a. Find the variance of the data.
   b. Find the standard deviation of the data.
4. Find the mean and standard deviation of each set of data.
   a. 2 4 6 8 10 12
   b. 102 104 106 108 110 112
   c. How are the data in (b) related to the data in (a)?
   d. How do your answers for (a) and (b) compare?
5. Find the mean and standard deviation of each set of data.
   a. 12 16 20 24 28 32
   b. 600 800 1,000 1,200 1,400 1,600
   c. How are the data in (b) related to the data in (a)?
   d. How do your answers for (a) and (b) compare?
6. Find the mean and standard deviation of each set of data.
   a. 50 50 50 50 50
   b. 46 50 50 50 54
   c. 5 50 50 50 95
   d. How do your answers for (a), (b), and (c) compare?
7. Joey and Dee Dee bowled five games at the Rock 'n' Bowl Lanes. Their scores are given in Figure 4.71.
   a. Find the mean score of each bowler. Who has the highest mean?
   b. Find the standard deviation of each bowler's scores.
   c. Who is the more consistent bowler? Why?

Joey	144	171	220	158	147
Dee Dee	182	165	187	142	159

Figure 4.71
Bowling scores

8. Paki surveyed the price of unleaded gasoline (self-serve) at gas stations in Novato and Lafayette. The raw data, in dollars per gallon, are given in Figure 4.72.
   a. Find the mean price in each city. Which city has the lowest mean?
   b. Find the standard deviation of prices in each city.
   c. Which city has more consistently priced gasoline? Why?

Novato	1.309	1.289	1.339	1.309	1.259	1.239
Lafayette	1.329	1.269	1.189	1.349	1.289	1.229

Figure 4.72
Price (in dollars) of one gallon of unleaded gasoline

9. Barry Bonds of the San Francisco Giants set an all-time record in professional baseball by hitting 73 home runs in one season (2001). The number of home runs (HR) for each of his seasons in professional baseball is given in Figure 4.73. Find the standard deviation of the number of home runs hit per year by Bonds.

Year	HR
1986	16
1987	25
1988	24
1989	19
1990	33
1991	25
1992	34
1993	46
1994	37

Year	HR
1995	33
1996	42
1997	40
1998	37
1999	34
2000	49
2001	73
2002	46

Figure 4.73

10. Michael Jordan has been recognized as an extraordinary player in professional basketball, especially in terms of the number of points per game he has scored. Jordan's average number of points per game (PPG) for each of his seasons is given in Figure 4.74. Find the standard deviation of the average points per game made per season by Jordan.

Season	PPG
1984–85	28.2
1985–86	22.7
1986–87	37.1
1987–88	35.0
1988–89	32.5
1989–90	33.6
1990–91	31.5

Season	PPG
1991–92	30.1
1992–93	32.6
1994–95	26.9
1995–96	30.4
1996–97	29.6
1997–98	28.7
2001–02	22.9

Figure 4.74

11. The Truly Amazing Dudes are a group of comic acrobats. The heights (in inches) of the ten acrobats are as follows:

68 50 70 67 72 78 69 68 66 67

a. Find the mean and standard deviation of the heights.

b. What percent of the data lies within one standard deviation of the mean?

c. What percent of the data lies within two standard deviations of the mean?

Is your height or weight "average"? These characteristics can vary considerably within any specific group of people. The mean is used to represent the average, and the standard deviation is used to measure the "spread" of a collection of data.

12. The weights (in pounds) of the ten Truly Amazing Dudes are as follows:

152 196 144 139 166 83 186 157 140 138

a. Find the mean and standard deviation of the weights.

b. What percent of the data lies within one standard deviation of the mean?

c. What percent of the data lies within two standard deviations of the mean?

13. The normal monthly rainfall in Seattle, Washington, is given in Figure 4.75 on page 250 (*Source:* U.S. Department of Commerce).

Month	Jan.	Feb.	Mar.	Apr.	May	June
Inches	5.4	4.0	3.8	2.5	1.8	1.6

Month	July	Aug.	Sept.	Oct.	Nov.	Dec.
Inches	0.9	1.2	1.9	3.3	5.7	6.0

Figure 4.75

a. Find the mean and standard deviation of the monthly rainfall in Seattle.
b. What percent of the year will the monthly rainfall be within one standard deviation of the mean?
c. What percent of the year will the monthly rainfall be within two standard deviations of the mean?

14. The normal monthly rainfall in Phoenix, Arizona, is given in Figure 4.76 (*Source:* U.S. Department of Commerce).

Month	Jan.	Feb.	Mar.	Apr.	May	June
Inches	0.7	0.7	0.9	0.2	0.1	0.1

Month	July	Aug.	Sept.	Oct.	Nov.	Dec.
Inches	0.8	1.0	0.9	0.7	0.7	1.0

Figure 4.76

a. Find the mean and standard deviation of the monthly rainfall in Phoenix.
b. What percent of the year will the monthly rainfall be within one standard deviation of the mean?
c. What percent of the year will the monthly rainfall be within two standard deviations of the mean?

15. The frequency distribution in Figure 4.77 lists the results of a quiz given in Professor Gilbert's statistics class.
a. Find the mean and standard deviation of the scores.

Score	Number of Students	Score	Number of Students
10	5	7	8
9	10	6	3
8	6	5	2

Figure 4.77
Quiz scores in Professor Gilbert's statistics class

b. What percent of the data lies within one standard deviation of the mean?
c. What percent of the data lies within two standard deviations of the mean?
d. What percent of the data lies within three standard deviations of the mean?

16. Amy surveyed the prices for a quart of a certain brand of motor oil. The sample data, in dollars per quart, is summarized in Figure 4.78.

Price per Quart	Number of Stores
0.99	2
1.09	5
1.19	10
1.29	13
1.39	9
1.49	3

Figure 4.78
Price (in dollars) for a quart of motor oil

a. Find the mean and the standard deviation of the prices.
b. What percent of the data lies within one standard deviation of the mean?
c. What percent of the data lies within two standard deviations of the mean?
d. What percent of the data lies within three standard deviations of the mean?

17. To study the output of a machine that fills boxes with cereal, a quality control engineer weighed 150 boxes of Brand X cereal. The frequency distribution in Figure 4.79 summarizes his findings. Find the standard deviation of the weight of the boxes of cereal.

x = Weight (in ounces)	Number of Boxes
$15.3 \le x < 15.6$	13
$15.6 \le x < 15.9$	24
$15.9 \le x < 16.2$	84
$16.2 \le x < 16.5$	19
$16.5 \le x < 16.8$	10

Figure 4.79
Amount of Brand X cereal per box

18. To study the efficiency of its new price-scanning equipment, a local supermarket monitored the amount of time its customers had to wait in line. The frequency distribution in Figure 4.80 summarizes the findings. Find the standard deviation of the amount of time spent in line.

x = Time (in minutes)	Number of Customers
$0 \le x < 1$	79
$1 \le x < 2$	58
$2 \le x < 3$	64
$3 \le x < 4$	40
$4 \le x \le 5$	35

Figure 4.80
Time spent waiting in a supermarket checkout line

19. The ages of the nearly 4 million women who gave birth in the United States in 1997 are given in Figure 4.81. Find the standard deviation of the ages of these women.

Age	Number of Women
$15 \le x < 20$	486,000
$20 \le x < 25$	948,000
$25 \le x < 30$	1,075,000
$30 \le x < 35$	891,000
$35 \le x < 40$	410,000
$40 \le x < 45$	77,000
$45 \le x < 50$	4,000

Figure 4.81
Ages of women giving birth in 1997
Source: U.S. Bureau of the Census

20. The age composition of the population of the United States in the year 2000 is given in Figure 4.82. Replace the interval "85 and over" with the interval $85 \le x \le 100$ and find the standard deviation of the ages of all people in the United States.

Age	Number of People (in thousands)
$0 < x < 5$	19,176
$5 \le x < 10$	20,550
$10 \le x < 15$	20,528
$15 \le x < 25$	39,184
$25 \le x < 35$	39,892
$35 \le x < 45$	44,149
$45 \le x < 55$	37,678
$55 \le x < 65$	24,275
$65 \le x < 85$	30,752
85 and over	4,240

Figure 4.82
Age composition of the population of the United States in the year 2000
Source: U.S. Bureau of the Census

Answer the following questions using complete sentences and your own words.

Concept Questions

21. **a.** When studying the dispersion of a set of data, why are the deviations from the mean squared?
b. What effect does squaring have on a deviation that is less than 1?
c. What effect does squaring have on a deviation that is greater than 1?
d. What effect does squaring have on the data's units?
e. Why is it necessary to take a square root when calculating standard deviation?

22. Why do we use the midpoint of an interval when calculating the standard deviation of grouped data?

Projects

23. The purpose of this project is to explore the variation in the pricing of a common commodity. Go to several different stores that sell food (the more the better) and record the price of one gallon of whole milk.
a. Compute the mean, median, and mode of the data.
b. Compute the standard deviation of the data.
c. What percent of the data lie within one standard deviation of the mean?

d. What percent of the data lie within two standard deviations of the mean?

e. What percent of the data lie within three standard deviations of the mean?

24. The purpose of this project is to explore the variation in the pricing of a common commodity. Go to several different gas stations (the more the better) and record the price of one gallon of premium gasoline (91 octane).

a. Compute the mean, median, and mode of the data.

b. Compute the standard deviation of the data.

c. What percent of the data lie within one standard deviation of the mean?

d. What percent of the data lie within two standard deviations of the mean?

e. What percent of the data lie within three standard deviations of the mean?

Technology and Measures of Central Tendency and Dispersion

In Examples 1, 2, and 3 of this section, we found the mean, median, mode, variance, and standard deviation of George's bowling scores. This work can be done quickly and easily on either a computerized spreadsheet or a graphing calculator.

Measures of Central Tendency and Dispersion on a Computerized Spreadsheet

Step 1 *Enter the data* from Example 1 of this section as discussed in Section 4.1.

Step 2 *Have Excel compute the mean, standard deviation, etc.*

- Use your mouse to select "Tools" at the very top of the screen. Then pull your mouse down until "Data Analysis" is highlighted, and let go. (If "Data Analysis" is not listed under "Tools," then follow the instructions given in Section 4.1.)
- In the "Data Analysis" box that appears, use your mouse to highlight "Descriptive Statistics" and press the "OK" button.
- In the "Descriptive Statistics" box that appears, use your mouse to click on the white rectangle that follows "Input Range," and then use your mouse to draw a box around all of the bowling scores. (To draw the box, move your mouse to cell A2, press the mouse button, and move the mouse down until all of the entered scores are enclosed in a box.) This should cause "$A$2:$A$7" to appear in the Input Range rectangle.
- Be certain that there is a dot in the button to the left of "Output Range." Then click on the white rectangle that follows "Output Range," and use your mouse to click on cell B1. This should cause "$B$1" to appear in the Output Range rectangle.
- Select "Summary Statistics."
- Use your mouse to press the "OK" button.

After you complete this step, your spreadsheet should look like that in Figure 4.83.

Measures of Central Tendency and Dispersion on a Graphing Calculator

Calculating the Mean, the Variance, and the Standard Deviation

On a TI-81:

- Enter the data from Example 1 of this section as discussed in Section 4.1.
- Press [2nd] [STAT].
- Select "1-Var" from the "CALC" menu.
- When "1-Var" appears on the screen, press [ENTER].

	A	B	C	D
1	george scores	*Column 1*		
2	185	Mean	185	
3	135	Standard Error	16.1761141	
4	200	Median	185	
5	185	Mode	185	
6	250	Standard Deviation	39.6232255	
7	155	Sample Variance	1570	
8		Kurtosis	0.838675	
		Skewness	0.60763436	
		Range	115	
		Minimum	135	
		Maximum	250	
		Sum	1110	
		Count	6	
		Confidence Level (95%)	41.5819571	

Figure 4.83

On a TI-80/82/83:

- Enter the data from Example 1 of this section in list L_1 as discussed in Section 4.1.
- Press STAT.
- Select "1-Var Stats" from the "CALC" menu.
- When "1-Var" appears on the screen, press 2nd L_1 ENTER.

On a TI-85:

- Enter the data from Example 1 of this section as discussed in Section 4.1.
- Press CALC (i.e., 2nd M1).
- The calculator will respond with the list names you selected previously. Press ENTER twice to indicate that these two names are correct, or change them if they are incorrect.
- Press 1-VAR (i.e. F1).

On a TI-86:

- Enter the data from Example 1 of this section as discussed in Section 4.1.
- Press 2nd QUIT, and put the calculator back into statistics mode by pressing 2nd STAT.

On a Casio:

- Enter the data from Example 1 of this section as described in Section 4.1.
- Press CALC (i.e., F2).
- Press 1VAR (i.e., F1).

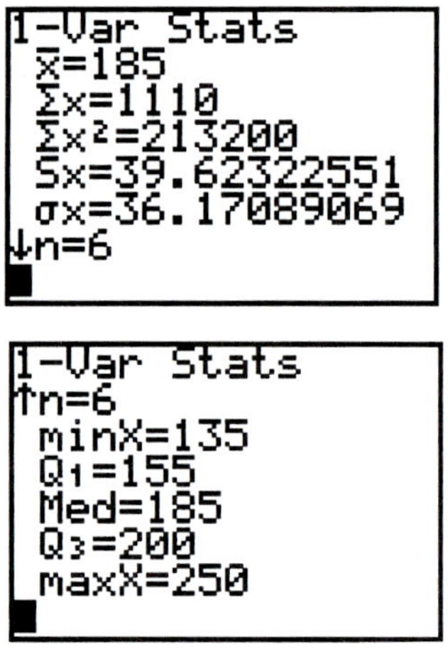

Figure 4.84

The above steps will result in the first screen in Figure 4.84. This screen gives the mean, the sample standard deviation (S_x on a TI, $\chi\sigma$n-1 on a Casio), the population standard deviation (σ_x on a TI, $\chi\sigma$n on a Casio), and the number of data points (n). The second screen, available only on the TI-82/83, and the Casio, can be obtained by pressing the down arrow. It gives the minimum and maximum data points (minX and maxX, respectively) as well as the median (Med).

Calculating the Sample Variance

The above work does not yield the sample variance. To find it, follow these steps:

- Quit the statistics mode.
- Get S_x on the screen:

TI-81:	Press [VARS] and select "S_x" from the "XY" menu.
TI-80/82/83:	Press [VARS], select "Statistics," and then select "S_x" from the "X/Y" menu.
TI-85:	Press [2nd] [VARS], press [MORE] until the "STAT" option appears and select that option, and then select "S_x."
TI-86:	Press [2nd] [STAT], press [VARS] (i.e., [F5]) and then [S_x] (i.e., [F3]).

- Once S_x is on the screen, square it by pressing [x^2] [ENTER]. The variance is 1,570.

Calculating the Mean, the Variance, and the Standard Deviation with Grouped Data

To calculate the mean and standard deviation from the frequency distribution that utilizes grouped data, follow the same steps except:

On a TI-81 or TI-85:

- Enter the midpoints of the classes as x's.
- Enter the frequencies of those classes as the corresponding y's.

On a TI-80/82/83:

- Enter the midpoints of the classes in list L_1.
- Enter the frequencies of those classes in list L_2. Each frequency must be less than 100.
- After "1-Var Stats" appears on the screen, press [2nd] [L_1] [,] [2nd] [L_2] [ENTER].

On a TI-86:

- Enter the midpoints of the classes in list xStat.
- Enter the frequencies of those classes in list fStat.
- After "1-Var Stats" appears on the screen, press [2nd] [LIST], and then [NAMES] (i.e., [F3]).
- Press [xStat] [,] [fStat], and the screen will read "OneVar xStat,fStat."
- Press [ENTER].

Exercises

25. In a 1996 article,* *Money* magazine claimed that "state regulation (of the insurance industry) is a joke." The article "ranks the states by the percentage of legislators with links to insurers who sit on committees focusing on insurance issues."

a. Find the mean and population standard deviation (σ) of these data for all 50 states.

b. Interpret the results of (a).

Mississippi	40%	Louisiana	38%	Arkansas	37%
Virginia	35%	North Carolina	33%	Missouri	32%
Florida	31%	Alabama	26%	Georgia	25%
Texas	25%	West Virginia	24%	Indiana	21%
Ohio	18%	Minnesota	18%	Iowa	17%
North Dakota	17%	Pennsylvania	15%	Arizona	14%
Wyoming	14%	Illinois	14%	Kentucky	13%
Idaho	13%	Delaware	13%	Utah	12%
Kansas	12%	Wisconsin	10%	Hawaii	10%
New Mexico	10%	Rhode Island	9%	Maryland	9%
New Jersey	8%	New York	8%	South Carolina	8%
Washington	8%	Tennessee	8%	Maine	8%
Oklahoma	8%	New Hampshire	7%	Connecticut	5%

(Unlisted states were ranked 0%.)

* *Source: Money,* August 1996.

4.4 The Normal Distribution

Sets of data may exhibit various trends or patterns. Figure 4.85 shows a histogram of the weights of bags of corn chips. Notice that most of the data is near the "center" and that the data tapers off at either end. Furthermore, the histogram is nearly symmetric; it is almost the same on both sides. This type of distribution (nearly symmetric, with most of the data in the middle) occurs quite often in many different situations. In order to study the composition of such distributions, statisticians have created an ideal **bell-shaped curve** describing a **normal distribution,** as shown in Figure 4.86 (see page 256).

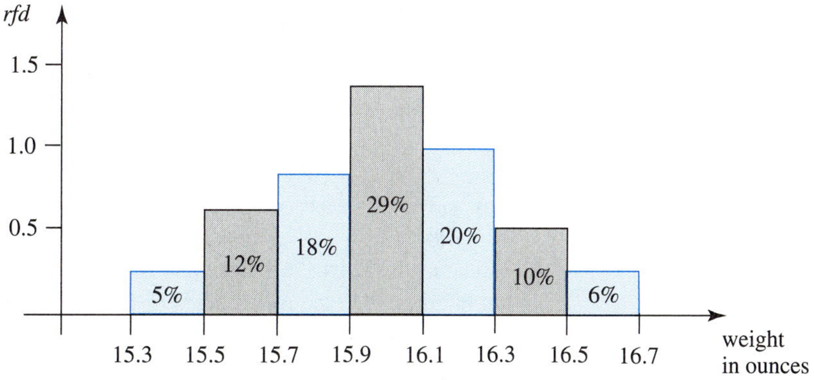

Figure 4.85
Weights of bags of corn chips

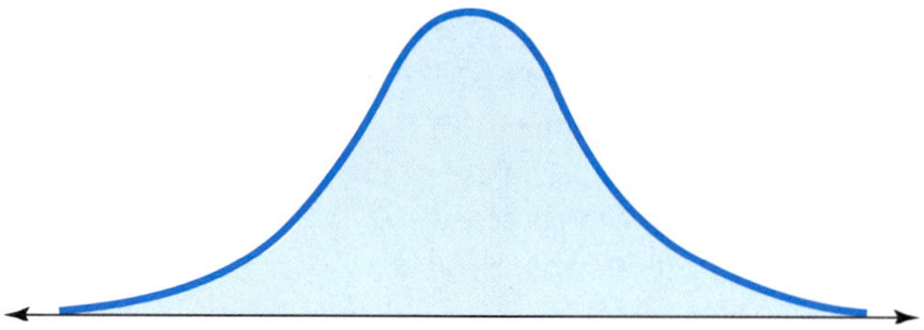

Figure 4.86
Normal distribution

Before we can study the characteristics and applications of a normal distribution, we must make a distinction between different types of variables.

Discrete versus Continuous Variables

The number of children in a family is variable, because it varies from family to family. In listing the number of children, only whole numbers (0, 1, 2, and so on) can be used. In this respect, we are limited to a collection of discrete, or separate, values. A variable is **discrete** if there are "gaps" between each possible variable value. Consequently, any variable that involves counting is discrete.

On the other hand, a person's height or weight doesn't have such a restriction. When someone grows, he or she does not instantly go from 67 inches to 68 inches; a person grows continuously from 67 inches to 68 inches, attaining all possible values in between. For this reason, height is called a continuous variable. A variable is **continuous** if it can assume *any value* in an interval of real numbers. Consequently, any variable that involves measurement is continuous; someone might claim to be 67 inches tall and to weigh 152 pounds, but the true values might be 67.13157 inches and 151.87352 pounds. Heights and weights are expressed (discretely) as whole numbers solely for convenience; most people do not have rulers or bathroom scales that allow them to obtain measurements that are accurate to ten or more decimal places!

Normal Distributions

The collection of all possible values that a discrete variable can assume forms a countable set. For instance, we can list all the possible numbers of children in a family. In contrast, a continuous variable will have an uncountable number of possibilities because it can assume any value in an interval. For instance, the weights (a continuous variable) of bags of corn chips could be *any* value x such that $15.3 \le x \le 16.7$.

When we sample a continuous variable, some values may occur more often than others. As we can see in Figure 4.85, the weights are "clustered" near the center of the histogram, with relatively few located at either end. If a continuous variable has a symmetric distribution such that the highest concentration of values is at the center and the lowest is at both extremes, the variable is said to have a **normal distribution** and is represented by a smooth, continuous, bell-shaped curve like that in Figure 4.86.*

The normal distribution, which is found in a wide variety of situations, has two main qualities: (1) the frequencies of the data points nearer the center or "average"

* This is an informal definition only. The formal definition of a normal distribution involves the number pi; the natural exponential e^x, and the mean and variance of the distribution.

are increasingly higher than the frequencies of data points far from the center, and (2) the distribution is symmetric (one side is a mirror image of the other). *Because of these two qualities, the mean, median, and mode of a normal distribution all coincide at the center of the distribution.*

Just like any other collection of numbers, the spread of normal distribution is measured by its standard deviation. It can be shown that for any normal distribution, slightly more than two-thirds of the data (68.26%) will lie within one standard deviation of the mean, 95.44% will lie within two standard deviations, and virtually all the data (99.74%) will lie within three standard deviations of the mean. Recall that μ (the Greek letter "mu") represents the mean of a population and σ (the Greek letter "sigma") represents the standard deviation of the population. The spread of a normal distribution, with μ and σ used to represent the mean and standard deviation, is shown in Figure 4.87.

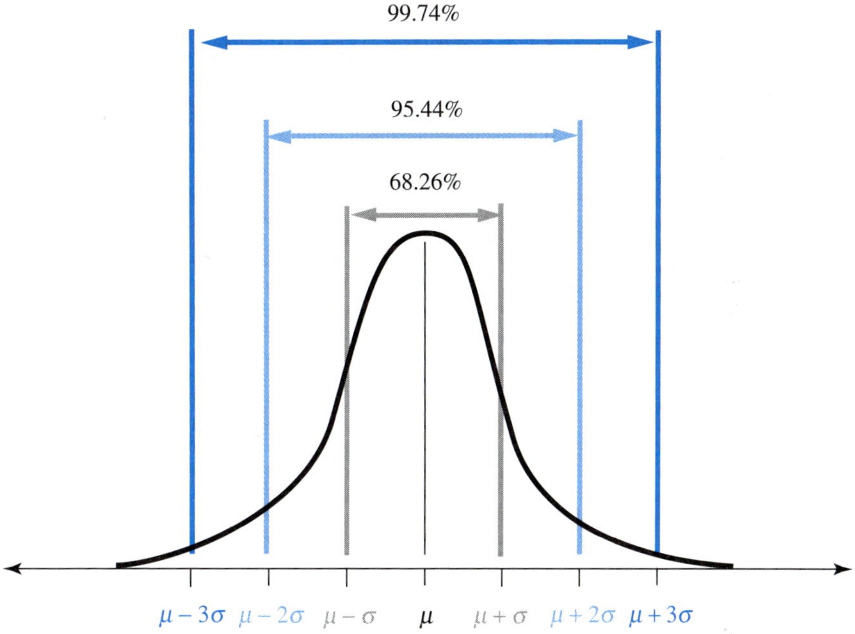

Figure 4.87
The spread of a normal distribution

EXAMPLE 1 The heights of a large group of people are assumed to be normally distributed. Their mean height is 66.5 inches, and the standard deviation is 2.4 inches. Find and interpret the intervals representing one, two, and three standard deviations of the mean.

Solution The mean is $\mu = 66.5$, and the standard deviation is $\sigma = 2.4$.

Step 1 *One standard deviation of the mean:*

$$\begin{aligned}\mu \pm 1\sigma &= 66.5 \pm 1(2.4)\\ &= 66.5 \pm 2.4\\ &= [64.1, 68.9]\end{aligned}$$

Therefore, approximately 68% of the people are between 64.1 and 68.9 inches tall.

Step 2 *Two standard deviations of the mean:*

$$\begin{aligned}\mu \pm 2\sigma &= 66.5 \pm 2(2.4)\\ &= 66.5 \pm 4.8\\ &= [61.7, 71.3]\end{aligned}$$

Therefore, approximately 95% of the people are between 61.7 and 71.3 inches tall.

Step 3 *Three standard deviations of the mean:*

$$\begin{aligned}\mu \pm 3\sigma &= 66.5 \pm 3(2.4)\\ &= 66.5 \pm 7.2\\ &= [59.3, 73.7]\end{aligned}$$

Nearly all of the people (99.74%) are between 59.3 and 73.7 inches tall. ■

In Example 1, we found that virtually all the people under study were between 59.3 and 73.7 inches tall. A clothing manufacturer might want to know what percent of these people are shorter than 66 inches or what percent are taller than 73 inches. Questions like these can be answered by using probability and a normal distribution. (We will do this in Example 6.)

Probability, Area, and Normal Distributions

In Chapter 3, we mentioned that relative frequency is really a type of probability. If 3 out of every 100 people have red hair, you could say that the relative frequency of red hair is $\frac{3}{100}$ (or 3%), or you could say that the probability of red hair $p(x = \text{red hair})$ is 0.03. Therefore, to find out what percent of the people in a population are taller than 73 inches, we need to find $p(x > 73)$, the probability that x is greater than 73, where x represents the height of a randomly selected person.

Recall that a sample space is the set S of all possible outcomes of a random experiment. Consequently, the probability of a sample space must always equal 1; that is, $p(S) = 1$ (or 100%). If the sample space S has a normal distribution, its outcomes and their respective probabilities can be represented by a bell curve.

Recall that when constructing a histogram, relative frequency density (*rfd*) was used to measure the heights of the rectangles. Consequently, the *area* of a rectangle gave the relative frequency (percent) of data contained in an interval. In a similar manner, we can imagine a bell curve being a histogram composed of infinitely many "skinny" rectangles, as in Figure 4.88.

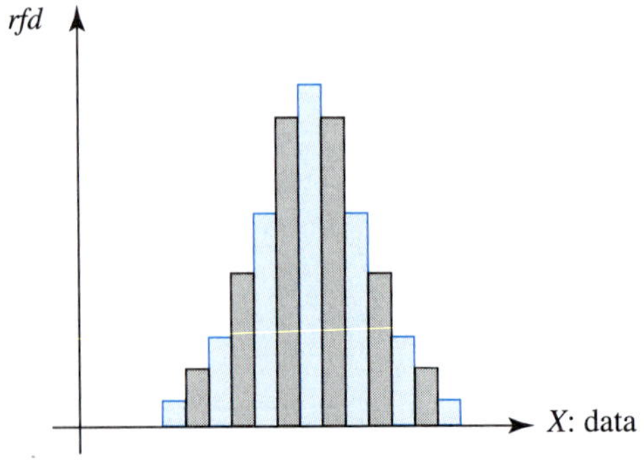

Figure 4.88

For a normal distribution, the outcomes nearer the center of the distribution occur more frequently than those at either end; the distribution is denser in the middle and sparser at the extremes. This difference in density is taken into account by consideration of the area under the bell curve; the center of the distribution is denser, contains more area, and has a higher probability of occurrence than the extremes. Consequently, we use the area under the bell curve to represent the probability of an outcome. Because $p(S) = 1$, we define the entire area under the bell curve to equal 1.

Because a normal distribution is symmetric, 50% of the data will be greater than the mean, and 50% will be less. (The mean and the median coincide in a symmetric distribution.) Therefore, the probability of randomly selecting a number x greater than the mean is $p(x > \mu) = 0.5$, and that of selecting a number x less than the mean is $p(x < \mu) = 0.5$, as shown in Figure 4.89.

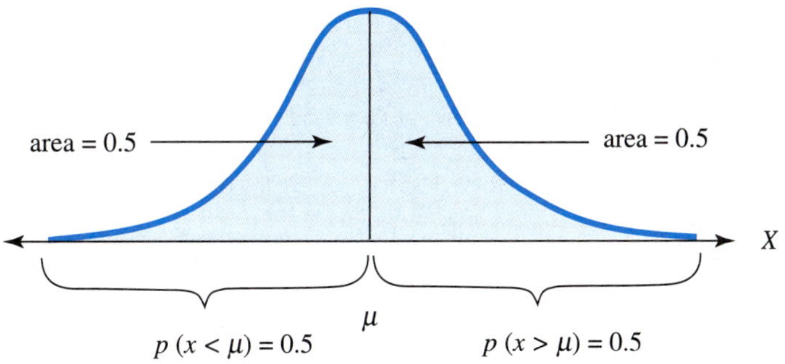

Figure 4.89

To find the probability that a randomly selected number x is between two values (say a and b), we must determine the area under the curve from a to b; that is, $p(a < x < b) =$ area under the bell curve from $x = a$ to $x = b$, as shown in Figure 4.90(a). Likewise, the probability that x is greater than or less than any specific number is given by the area of the tail, as shown in Figure 4.90(b). To find probabilities involving data that are normally distributed, we must find the area of the appropriate region under the bell curve.

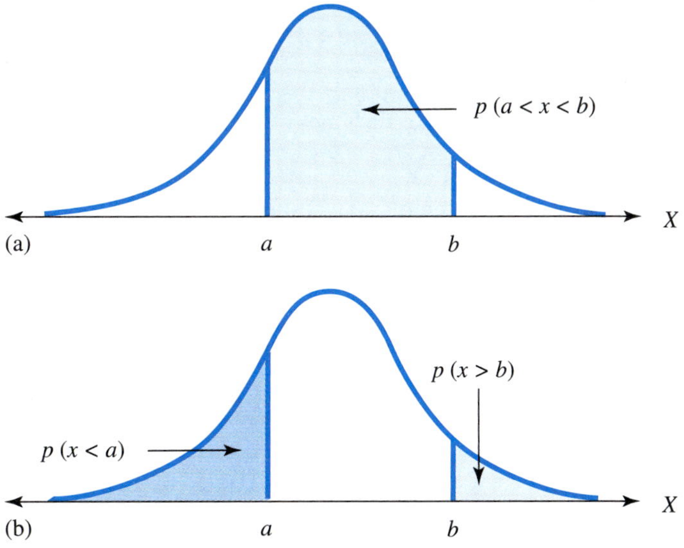

Figure 4.90

© Stock Montage

HISTORICAL NOTE

Carl Friedrich Gauss

1777–1855

Dubbed "the Prince of Mathematics," Carl Gauss is considered by many to be one of the greatest mathematicians of all time. At the age of three, Gauss is said to have discovered an arithmetic error in his father's bookkeeping. The child prodigy was encouraged by his teachers and excelled throughout his early schooling. When he was fourteen, Gauss was introduced to Ferdinand, the Duke of Brunswick. Impressed with the youth, the duke gave Gauss a yearly stipend and sponsored his education for many years.

In 1795, Gauss enrolled at Göttingen University, where he remained for three years. While at Göttingen, Gauss had complete academic freedom; he was not required to attend lectures, he had no required conferences with professors or tutors, and he did not take exams. Much of his time was spent studying independently in the library. For reasons unknown to us, Gauss left the university in 1798 without a diploma. Instead, he sent his dissertation to the University of Helmstedt and in 1799 was awarded his degree without the usual oral examination.

In 1796, Gauss began his famous mathematical diary. Discovered 40 years after his death, the 146 sometimes cryptic entries exhibit the diverse range of topics that Gauss pondered and pioneered. The first entry was Gauss's discovery (at the age of nineteen) of a method for constructing a 17-sided polygon with a compass and a straightedge. Other entries include important results in number theory, algebra, calculus, analysis, astronomy, electricity, magnetism, the foundations of geometry (see Section 6.7), and probability.

At the dawn of the nineteenth century, Gauss began his lifelong study of astronomy. On January 1, 1801, the Italian astronomer Giuseppe Piazzi discovered Ceres, the first of the known planetoids (minor planets or asteroids). Piazzi and others observed Ceres for 41 days, until it was lost behind the sun. Because of his interest in the mathematics of astronomy, Gauss turned his attention to Ceres. Working with a minimum amount of data, he successfully calculated the orbit of Ceres. At the end of the year, the planetoid was rediscovered in exactly the spot that Gauss had predicted!

To obtain the orbit of Ceres, Gauss utilized his method of least squares, a technique for dealing with experimental error. Letting x represent the error between an experimentally obtained value and the true value it represents, Gauss's theory involved minimizing x^2—that is, obtaining the least square of the error. Theorizing that the probability of a small error was higher than that of a large error, Gauss subsequently developed the normal distribution, or bell-shaped curve, to explain the probabilities of the random errors. Because of his pioneering efforts, some mathematicians refer to the normal distribution as the Gaussian distribution.

In 1807, Gauss became director of the newly constructed observatory at Göttingen; he held the position until his death some 50 years later.

Published in 1809, Gauss's *Theoria Motus Corporum Coelestium (Theory of the Motion of Heavenly Bodies)* contained rigorous methods of determining the orbits of planets and comets from observational data via the method of least squares. It is a landmark in the development of modern mathematical astronomy and statistics.

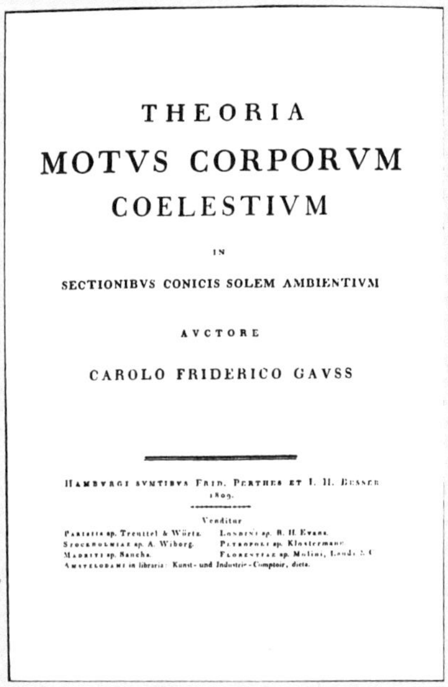

THEORIA

MOTVS CORPORVM

COELESTIVM

IN

SECTIONIBVS CONICIS SOLEM AMBIENTIVM

AVCTORE

CAROLO FRIDERICO GAVSS

HAMBVRGI SVMTIBVS FRID. PERTHES ET I. H. BESSER

1809.

Venditur

Parisiis ap. Treuttel & Würtz. Londini ap. R. H. Evans.

Stockholmiae ap. A. Wiborg. Petropoli ap. Klostermann.

Madriti ap. Sancha. Florentiae ap. Molini, Landi & C.

Amstelodami in libraria: Kunst- und Industrie-Comptoir, dicta.

© Stock Montage

The Standard Normal Distribution

All normal distributions share the following features: they are symmetric, bell-shaped curves, and virtually all the data (99.74%) lie within three standard deviations of the mean. Depending on whether the standard deviation is large or small, the bell curve will be either flat and spread out or peaked and narrow, as shown in Figure 4.91.

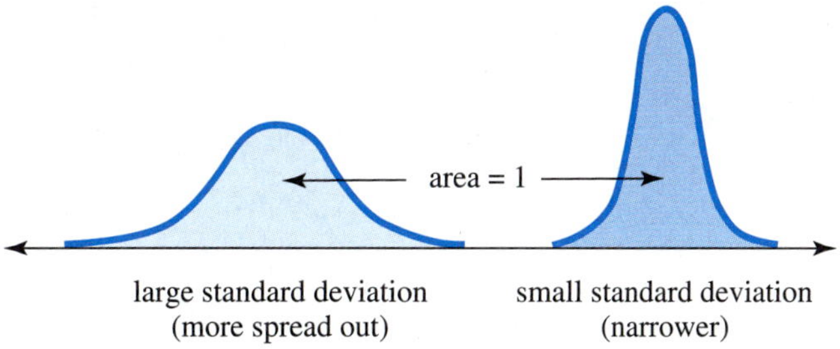

Figure 4.91

To find the area under any portion of any bell curve, mathematicians have devised a means of comparing the proportions of any curve with the proportions of a special curve defined as "standard." To find probabilities involving normally distributed data, we utilize the bell curve associated with the standard normal distribution.

The **standard normal distribution** is the normal distribution whose mean is 0 and standard deviation is 1, as shown in Figure 4.92. The standard normal distribution is also called the ***z*-distribution;** we will always use the letter z to refer to the standard normal. By convention, we will use the letter x to refer to any other normal distribution.

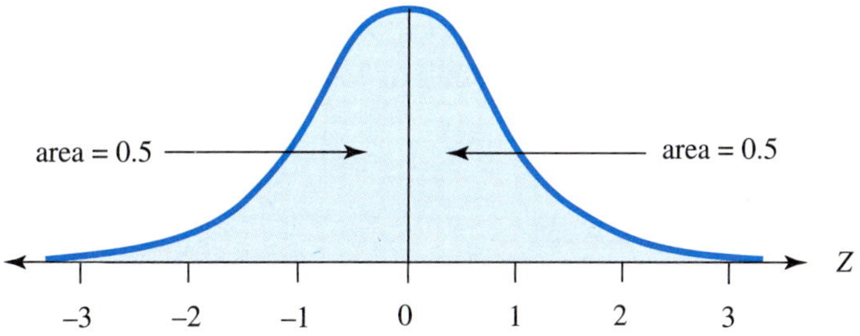

Figure 4.92
The standard normal distribution (mean = 0; standard deviation = 1)

Tables have been developed for finding areas under the standard normal curve using the techniques of calculus. A few statistical calculators will also give these areas. We will use the table in Appendix F to find $p(0 < z < z^*)$, the probability that z is between 0 and a positive number z^*, as shown in Figure 4.93(a). The table in Appendix F is known as the **body table** because it gives the probability of an interval located in the middle, or body, of the bell curve.

The tapered end of a bell curve is known as a **tail.** To find the probability of a tail—that is, to find $p(z > z^*)$ or $p(z < z^*)$ where z^* is a positive real number— subtract the probability of the corresponding body from 0.5, as shown in Figure 4.93(b).

EXAMPLE 2 Find the following probabilities (that is, the areas), where z represents the standard normal distribution.

a. $p(0 < z < 1.25)$ **b.** $p(z > 1.87)$

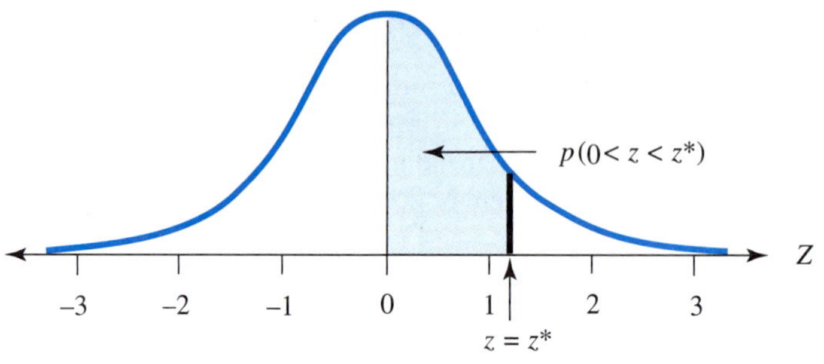

(a) Area found by using the body table (Appendix F)

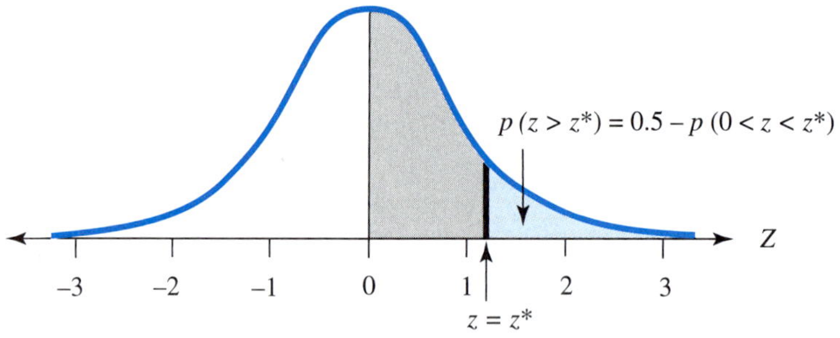

(b) Area of a tail, found by subtracting the corresponding body area from 0.5

Figure 4.93

Solution **a.** As a first step, it is always advisable to draw a picture of the z-curve and shade in the desired area. We will use the body table directly, because we are working with a central area (see Figure 4.94).

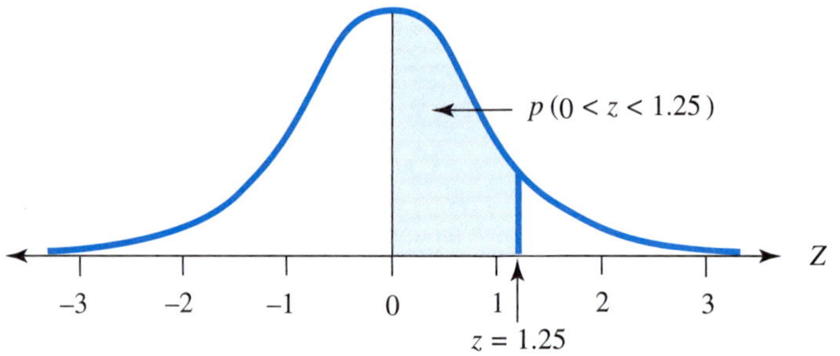

Figure 4.94

The z-numbers are located along the left edge and the top of the table. Locate the whole number and the first-decimal-place part of the number (1.2) along the left edge; then locate the second-decimal-place part of the number (0.05) along the top. The desired probability (area) is found at the intersection of the row and column of the two parts of the z-number. Thus, $p(0 < z < 1.25) = 0.3944$, as shown in Figure 4.95.

z^*	0.00	0.01	0.02	0.03	0.04	0.05	0.06	0.07	0.08	0.09
∘										
∘										
∘										
1.1	0.3643	0.3665	0.3686	0.3708	0.3729	0.3749	0.3770	0.3790	0.3810	0.3830
1.2	0.3849	0.3869	0.3888	0.3907	0.3925	0.3944	0.3962	0.3980	0.3997	0.4015
1.3	0.4032	0.4049	0.4066	0.4082	0.4099	0.4115	0.4131	0.4147	0.4162	0.4177
∘										
∘										
∘										

Figure 4.95

Hence, we could say that about 39% of the z-distribution lies between $z = 0$ and $z = 1.25$.

b. To find the area of a tail, we subtract the corresponding body area from 0.5, as shown in Figure 4.96. Therefore,

$$\begin{aligned} p(z > 1.87) &= 0.5 - p(0 < z < 1.87) \\ &= 0.5 - 0.4692 \\ &= 0.0308 \end{aligned}$$

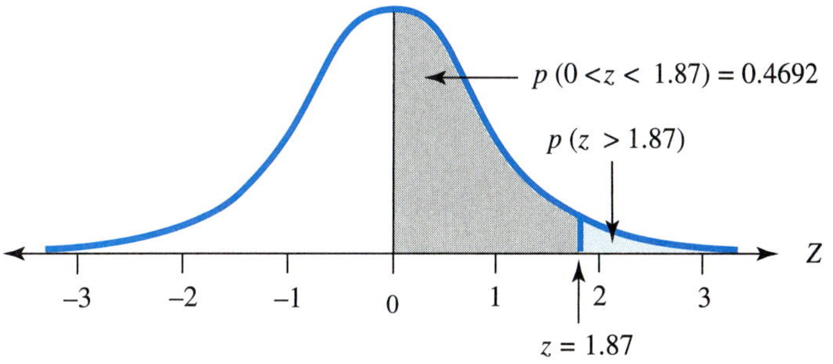

Figure 4.96 ■

The body table can also be used to find areas other than those given explicitly as $p(0 < z < z^*)$ and $p(z > z^*)$ where z^* is a positive number. By adding or subtracting two areas, we can find probabilities of the type $p(a < z < b)$, where a and b are positive or negative numbers, and probabilities of the type $p(z < c)$, where c is a positive or negative number.

EXAMPLE 3 Find the following probabilities (the areas), where z represents the standard normal distribution.

a. $p(0.75 < z < 1.25)$ **b.** $p(-0.75 < z < 1.25)$

Solution **a.** Because the required region, shown in Figure 4.97, doesn't begin exactly at $z = 0$, we cannot look up the desired area directly in the body table. Whenever z is between two nonzero numbers, we will take an indirect approach to finding the required area.

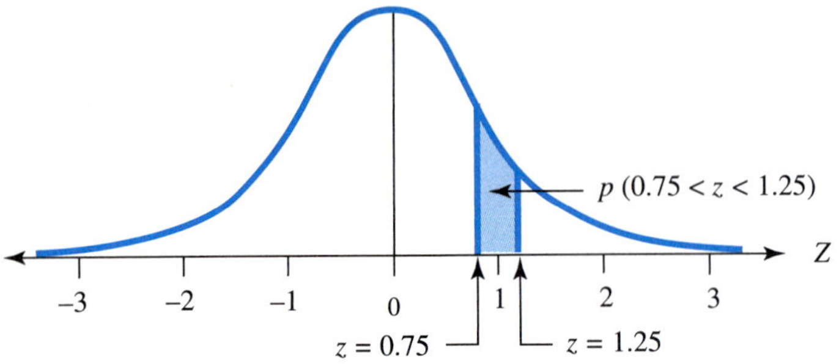

Figure 4.97

The total area under the curve from $z = 0$ to $z = 1.25$ can be divided into two portions: the area under the curve from 0 to 0.75 and the area under the curve from 0.75 to 1.25.

To find the area of the "strip" between $z = 0.75$ and $z = 1.25$, we *subtract* the area of the smaller body (from $z = 0$ to $z = 0.75$) from that of the larger body (from $z = 0$ to $z = 1.25$), as shown in Figure 4.98.

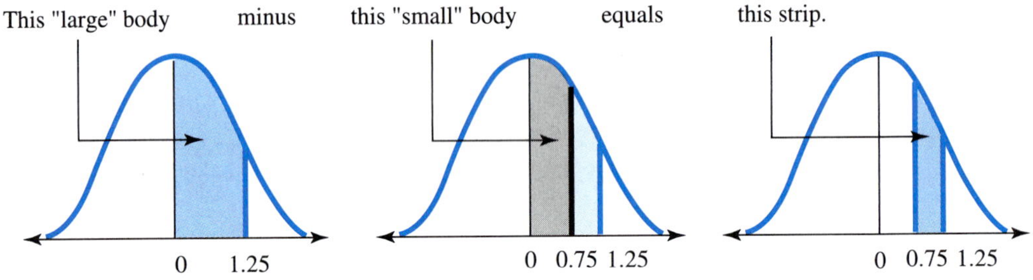

Figure 4.98
Area of a strip

$$\text{area of strip} = \text{area of large body} - \text{area of small body}$$

$$\begin{aligned} p(0.75 < z < 1.25) &= p(0 < z < 1.25) - p(0 < z < 0.75) \\ &= 0.3944 - 0.2734 \\ &= 0.1210 \end{aligned}$$

Therefore, $p(0.75 < z < 1.25) = 0.1210$. Hence, we could say that about 12.1% of the z-distribution lies between $z = 0.75$ and $z = 1.25$.

b. The required region, shown in Figure 4.99, can be divided into two regions: the area from $z = -0.75$ to $z = 0$ and the area from $z = 0$ to $z = 1.25$.

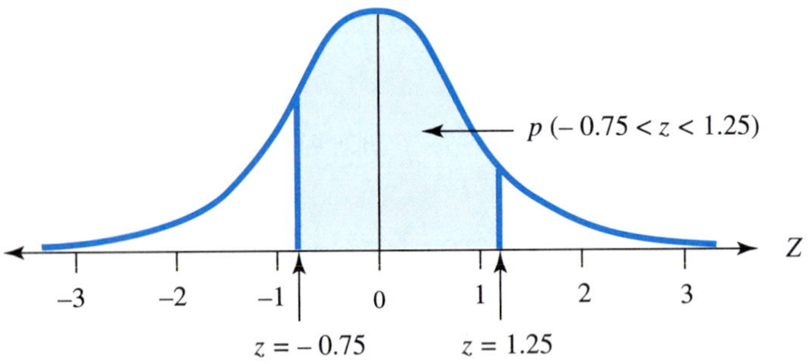

Figure 4.99

To find the total area of the region between $z = -0.75$ and $z = 1.25$, we *add* the area of the "left" body (from $z = -0.75$ to $z = 0$) to the area of the "right" body (from $z = 0$ to $z = 1.25$), as shown in Figure 4.100.

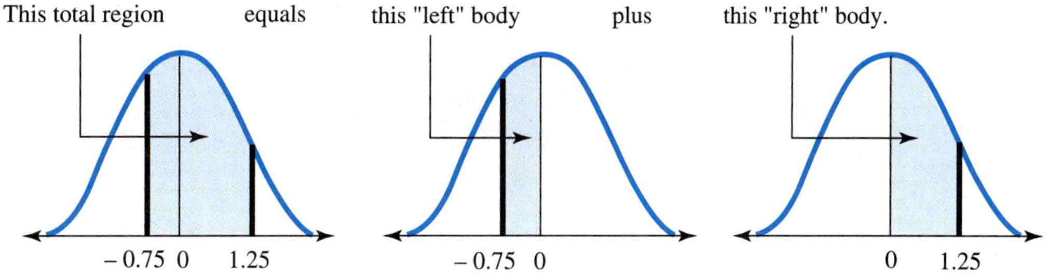

Figure 4.100

This example is different from our previous examples in that it contains a negative z-number. A glance at the tables reveals that negative numbers are not included! However, recall that normal distributions are symmetric. Therefore, the area of the body from $z = -0.75$ to $z = 0$ is the same as that from $z = 0$ to $z = 0.75$; that is, $p(-0.75 < z < 0) = p(0 < z < 0.75)$. Therefore,

$$\begin{aligned}
\text{total area of region} &= \text{area of left body} + \text{area of right body} \\
p(-0.75 < z < 1.25) &= p(-0.75 < z < 0) + p(0 < z < 1.25) \\
&= p(0 < z < 0.75) + p(0 < z < 1.25) \\
&= 0.2734 + 0.3944 \\
&= 0.6678
\end{aligned}$$

Therefore, $p(-0.75 < z < 1.25) = 0.6678$. Hence, we could say that about 66.8% of the z-distribution lies between $z = -0.75$ and $z = 1.25$. ■

EXAMPLE 4 Find the following probabilities (the areas), where z represents the standard normal distribution.

a. $p(z < 1.25)$ **b.** $p(z < -1.25)$

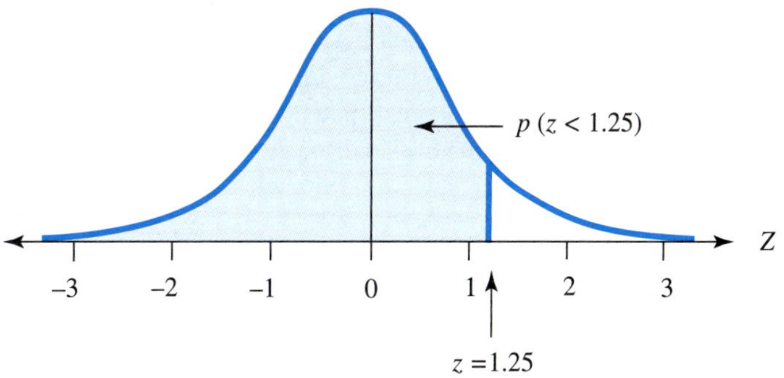

Figure 4.101

Solution **a.** The required region is shown in Figure 4.101. Because 50% of the distribution lies to the left of 0, we can add 0.5 to the area of the body from $z = 0$ to $z = 1.25$:

$$\begin{aligned} p(z < 1.25) &= p(z < 0) + p(0 < z < 1.25) \\ &= 0.5 + 0.3944 \\ &= 0.8944 \end{aligned}$$

Therefore, $p(z < 1.25) = 0.8944$. Hence, we could say that about 89.4% of the z-distribution lies to the left of $z = 1.25$.

b. The required region is shown in Figure 4.102. Because a normal distribution is symmetric, the area of the left tail ($z < -1.25$) is the same as the area of the corresponding right tail ($z > 1.25$).

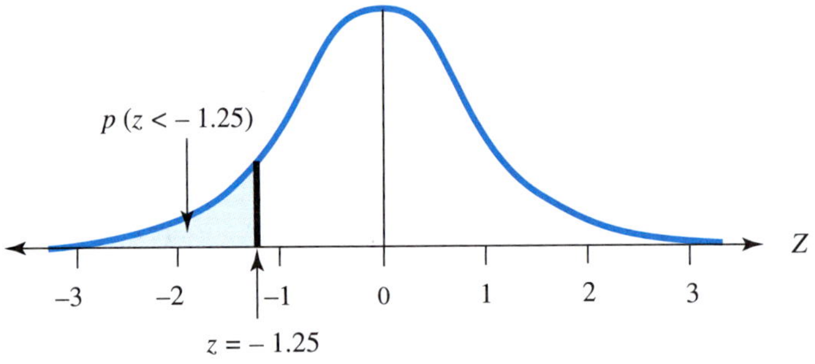

Figure 4.102

Therefore,

$$\begin{aligned} p(z < -1.25) &= p(z > 1.25) \\ &= 0.5 - p(0 < z < 1.25) \\ &= 0.5 - 0.3944 \\ &= 0.1056 \end{aligned}$$

Hence, we could say that about 10.6% of the z-distribution lies to the left of $z = -1.25$. ■

Converting to the Standard Normal

Weather forecasters in the United States usually report temperatures in degrees Fahrenheit. Consequently, if a temperature is given in degrees Celsius, most people would convert it to Fahrenheit in order to judge how hot or cold it was. A similar situation arises when we are working with a normal distribution. Suppose we know that a large set of data is normally distributed with a mean value of 68 and a standard deviation of 4. What percent of the data will lie between 65 and 73? We are asked to find $p(65 < x < 73)$.

Converting a Normal Distribution into the Standard Normal z

Every number x in a given normal distribution has a corresponding number z in the standard normal distribution. The **z-number** that corresponds to the number x is

$$z = \frac{x - \mu}{\sigma}$$

where μ is the mean and σ the standard deviation of the given normal distribution.

To find this probability, we must first convert the given normal distribution to the standard normal distribution and then look up the approximate z-numbers.

The body table (Appendix F) applies to the standard normal z-distribution. When we are working with any other normal distribution (denoted by X), we must first convert the x-distribution into the standard normal z-distribution. This conversion is done with the help of the following rule.

Given a number x, its corresponding z-number counts the number of standard deviations the number lies from the mean. For example, suppose the mean and standard deviation of a normal distribution are $\mu = 68$ and $\sigma = 4$. The z-number corresponding to $x = 78$ is

$$z = \frac{x - \mu}{\sigma} = \frac{78 - 68}{4} = 2.5$$

This implies that $x = 78$ lies two and one-half standard deviations above the mean, 68. Similarly, for $x = 65$,

$$z = \frac{65 - 68}{4} = -0.75$$

Therefore, $x = 65$ lies three-quarters of a standard deviation below the mean, 68.

EXAMPLE 5 Suppose a population is normally distributed with a mean of 24.6 and a standard deviation of 1.3. What percent of the data will lie between 25.3 and 26.8?

Solution We are asked to find $p(25.3 < x < 26.8)$, the area of the region shown in Figure 4.103. Because we need to find the area of the strip between 25.3 and 26.8, we must find the body of each and subtract, as in part (a) of Example 3.

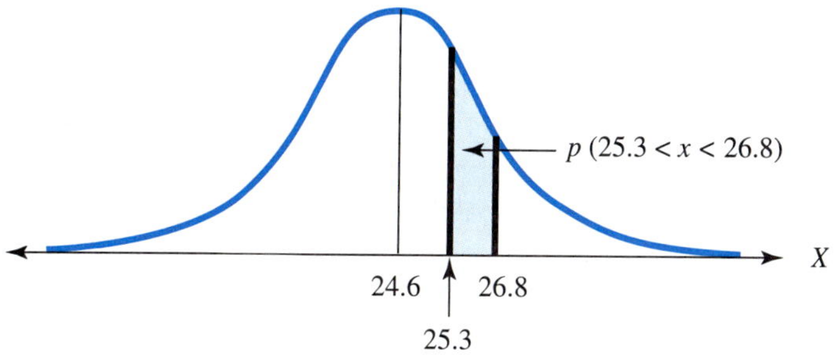

Figure 4.103

Using the Conversion Formula $z = (x - \mu)/\sigma$ with $\mu = 24.6$ and $\sigma = 1.3$, we first convert $x = 25.3$ and $x = 26.8$ into their corresponding z-numbers.

Converting $x = 25.3$	Converting $x = 26.8$
$z = \frac{x - \mu}{\sigma}$	$z = \frac{x - \mu}{\sigma}$
$= \frac{25.3 - 24.6}{1.3}$	$= \frac{26.8 - 24.6}{1.3}$
$= 0.5384615$	$= 1.6923077$
$= 0.54$	$= 1.69$

Therefore,

$$\begin{aligned} p(25.3 < x < 26.8) &= p(0.54 < z < 1.69) \\ &= p(0 < z < 1.69) - p(0 < z < 0.54) \\ &= 0.4545 - 0.2054 \quad \text{using the body table} \\ &= 0.2491 \end{aligned}$$

Assuming a normal distribution, approximately 24.9% of the data will lie between 25.3 and 26.8. ■

EXAMPLE 6 The heights of a large group of people are assumed to be normally distributed. Their mean height is 68 inches, and the standard deviation is 4 inches. What percentage of these people are the following heights?

a. taller than 73 inches **b.** between 60 and 75 inches

Solution **a.** Let x represent the height of a randomly selected person. We need to find $p(x > 73)$, the area of a tail, as shown in Figure 4.104.

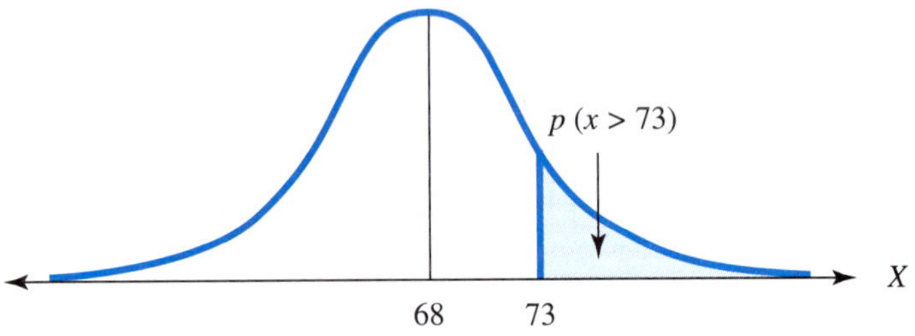

Figure 4.104

First, we must convert $x = 73$ to its corresponding z-number. Using the Conversion Formula with $x = 73$, $\mu = 68$, and $\sigma = 4$, we have

$$\begin{aligned} z &= \frac{x - \mu}{\sigma} \\ &= \frac{73 - 68}{4} \\ &= 1.25 \end{aligned}$$

Therefore,

$$\begin{aligned} p(x > 73) &= p(z > 1.25) \\ &= 0.5 - p(0 < z < 1.25) \\ &= 0.5 - 0.3944 \\ &= 0.1056 \end{aligned}$$

Approximately 10.6% of the people will be taller than 73 inches.

b. We need to find $p(60 < x < 75)$, the area of the central region shown in Figure 4.105. Notice that we will be adding the areas of the two bodies.

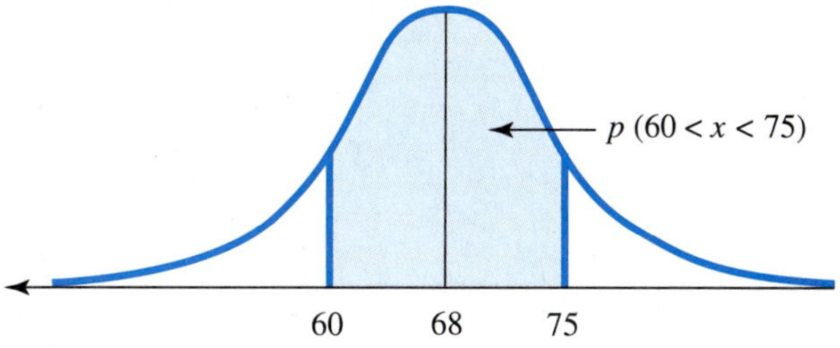

Figure 4.105

First, we convert $x = 60$ and $x = 75$ to their corresponding z-numbers:

$$
\begin{aligned}
p(60 < x < 75) &= p\left(\frac{60-68}{4} < z < \frac{75-68}{4}\right) && \text{using the Conversion Formula} \\
&= p(-2.00 < z < 1.75) \\
&= p(-2.00 < z < 0) + p(0 < z < 1.75) && \text{expressing the area as two bodies} \\
&= p(0 < z < 2.00) + p(0 < z < 1.75) && \text{using symmetry} \\
&= 0.4772 + 0.4599 && \text{using the body table} \\
&= 0.9371
\end{aligned}
$$

Approximately 93.7% of the people will be between 60 and 75 inches tall. ■

All the preceding examples involved finding probabilities that contained only the strict $<$ or $>$ inequalities, never $\leq$ or $\geq$ inequalities; the endpoints were never included. What if the endpoints are included? How does $p(a < x < b)$ compare with $p(a \leq x \leq b)$? Because probabilities for continuous data are found by determining *area* under a curve, including the endpoints does not affect the probability! The probability of a single point $p(x = a)$ is 0, because there is no "area" over a single point. (We obtain an area only when we are working with an interval of numbers.) Consequently, if x represents continuous data, then $p(a \leq x \leq b) = p(a < x < b)$; it makes no difference whether the endpoints are included.

EXAMPLE 7 Tall Dudes is a clothing store that specializes in fashions for tall men. Its informal motto is "Our customers are taller than 80% of the rest." Assuming the heights of men to be normally distributed with a mean of 67 inches and a standard deviation of 5.5 inches, find the heights of Tall Dudes' clientele.

Solution Let c = the height of the shortest customer at Tall Dudes, and let x represent the height of a randomly selected man. We are given that the heights of all men are normally distributed with $\mu = 67$ and $\sigma = 5.5$.

Assuming Tall Dudes' clientele to be taller than 80% of all men implies that $x < c$ 80% of the time and $x > c$ 20% of the time. Hence, we can say that the probability of selecting someone shorter than the shortest tall dude is $p(x < c) = 0.80$ and that the probability of selecting a tall dude is $p(x > c) = 0.20$, as shown in Figure 4.106 (see page 270).

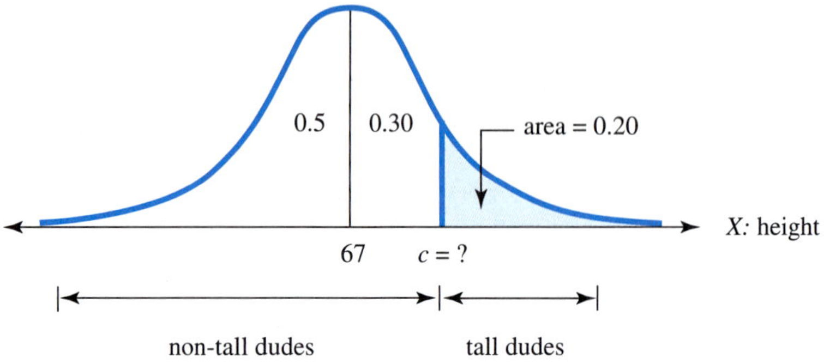

Figure 4.106

We are given that the area of the right tail is 0.20 (and the right body is 0.30), and we need to find the appropriate cutoff number c. This is exactly the reverse of all the previous examples, in which we were given the cutoff numbers and asked to find the area. Thus, our goal is to find the z-number that corresponds to a body of area 0.30 and convert it into its corresponding x-number.

When we scan through the *interior* of the body table, the number closest to the desired area of 0.30 is 0.2995, which is the area of the body when $z = 0.84$. This means that $p(0 < z < 0.84) = 0.2995$, as shown in Figure 4.107.

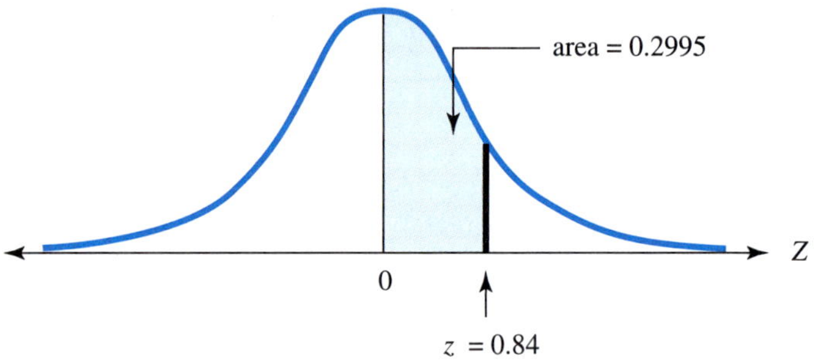

Figure 4.107

Therefore, the number c that we are seeking lies 0.84 standard deviations above the mean. All that remains is to convert $z = 0.84$ into its corresponding x-number by substituting $x = c$, $z = 0.84$, $\mu = 67$, and $\sigma = 5.5$ into the Conversion Formula:

$$z = \frac{x - \mu}{\sigma}$$

$$0.84 = \frac{c - 67}{5.5}$$

$$(5.5)(0.84) = c - 67$$

$$4.62 = c - 67$$

$$c = 71.62 \quad (\approx 72 \text{ inches, or 6 feet})$$

Therefore, Tall Dudes caters to men who are at least 71.62 inches, or about 6 feet tall. ■

4.4 Exercises

1. What percent of the standard normal z-distribution lies between the following values?
   a. $z = 0$ and $z = 1$
   b. $z = -1$ and $z = 0$
   c. $z = -1$ and $z = 1$ (*Note:* This interval represents one standard deviation of the mean.)
2. What percent of the standard normal z-distribution lies between the following values?
   a. $z = 0$ and $z = 2$
   b. $z = -2$ and $z = 0$
   c. $z = -2$ and $z = 2$ (*Note:* This interval represents two standard deviations of the mean.)
3. What percent of the standard normal z-distribution lies between the following values?
   a. $z = 0$ and $z = 3$
   b. $z = -3$ and $z = 0$
   c. $z = -3$ and $z = 3$ (*Note:* This interval represents three standard deviations of the mean.)
4. What percent of the standard normal z-distribution lies between the following values?
   a. $z = 0$ and $z = 1.5$
   b. $z = -1.5$ and $z = 0$
   c. $z = -1.5$ and $z = 1.5$ (*Note:* This interval represents one and one-half standard deviations of the mean.)
5. A population is normally distributed with mean 24.7 and standard deviation 2.3.
   a. Find the intervals representing one, two, and three standard deviations of the mean.
   b. What percentage of the data lies in each of the intervals in part (a)?
   c. Draw a sketch of the bell curve.
6. A population is normally distributed with mean 18.9 and standard deviation 1.8.
   a. Find the intervals representing one, two, and three standard deviations of the mean.
   b. What percentage of the data lies in each of the intervals in part (a)?
   c. Draw a sketch of the bell curve.
7. Find the following probabilities.
   a. $p(0 < z < 1.62)$ b. $p(1.30 < z < 1.84)$
   c. $p(-0.37 < z < 1.59)$ d. $p(z < -1.91)$
   e. $p(-1.32 < z < -0.88)$ f. $p(z < 1.25)$
8. Find the following probabilities.
   a. $p(0 < z < 1.42)$ b. $p(1.03 < z < 1.66)$
   c. $p(-0.87 < z < 1.71)$ d. $p(z < -2.06)$
   e. $p(-2.31 < z < -1.18)$ f. $p(z < 1.52)$
9. Find c such that each of the following is true.
   a. $p(0 < z < c) = 0.1331$
   b. $p(c < z < 0) = 0.4812$
   c. $p(-c < z < c) = 0.4648$
   d. $p(z > c) = 0.6064$
   e. $p(z > c) = 0.0505$
   f. $p(z < c) = 0.1003$
10. Find c such that each of the following is true.
    a. $p(0 < z < c) = 0.3686$
    b. $p(c < z < 0) = 0.4706$
    c. $p(-c < z < c) = 0.2510$
    d. $p(z > c) = 0.7054$
    e. $p(z > c) = 0.0351$
    f. $p(z < c) = 0.2776$
11. A population X is normally distributed with mean 250 and standard deviation 24. For each of the following values of x, find the corresponding z-number. Round off your answers to two decimal places.
    a. $x = 260$ b. $x = 240$
    c. $x = 300$ d. $x = 215$
    e. $x = 321$ f. $x = 197$
12. A population X is normally distributed with mean 72.1 and standard deviation 9.3. For each of the following values of x, find the corresponding z-number. Round off your answers to two decimal places.
    a. $x = 90$ b. $x = 80$
    c. $x = 75$ d. $x = 70$
    e. $x = 60$ f. $x = 50$
13. A population is normally distributed with mean 36.8 and standard deviation 2.5. Find the following probabilities.
    a. $p(36.8 < x < 39.3)$ b. $p(34.2 < x < 38.7)$
    c. $p(x < 40.0)$ d. $p(32.3 < x < 41.3)$
    e. $p(x = 37.9)$ f. $p(x > 37.9)$
14. A population is normally distributed with mean 42.7 and standard deviation 4.7. Find the following probabilities.
    a. $p(42.7 < x < 47.4)$ b. $p(40.9 < x < 44.1)$
    c. $p(x < 50.0)$ d. $p(33.3 < x < 52.1)$
    e. $p(x = 45.3)$ f. $p(x > 45.3)$
15. The mean weight of a box of cereal filled by a machine is 16.0 ounces, with a standard deviation of 0.3 ounce. If the weights of all the boxes filled by the machine are normally distributed, what percent of the boxes will weigh the following amounts?
    a. less than 15.5 ounces
    b. between 15.8 and 16.2 ounces
16. The amount of time required to assemble a component on a factory assembly line is normally distributed with a mean of 3.1 minutes and a standard deviation of 0.6 minute. Find the probability that a randomly selected employee will take the given amount of time to assemble the component.

a. more than 4.0 minutes
b. between 2.0 and 2.5 minutes

17. The time it takes an acrylic paint to dry is normally distributed. If the mean is 2 hours 36 minutes with a standard deviation of 24 minutes, find the probability that the drying time will be as follows.
a. less than 2 hours 15 minutes
b. between 2 and 3 hours

HINT: Convert everything to minutes (or to hours).

18. The shrinkage in length of a certain brand of blue jeans is normally distributed with a mean of 1.1 inches and a standard deviation of 0.2 inch. What percent of this brand of jeans will shrink the following amounts?
a. more than 1.5 inches
b. between 1.0 and 1.25 inches

19. The mean volume of a carton of milk filled by a machine is 1.0 quart, with a standard deviation of 0.06 quart. If the volumes of all the cartons are normally distributed, what percent of the cartons will contain the following amounts?
a. at least 0.9 quart
b. at most 1.05 quarts

20. The amount of time between taking a pain reliever and getting relief is normally distributed with a mean of 23 minutes and a standard deviation of 4 minutes. Find the probability that the time between taking the medication and getting relief is as follows.
a. at least 30 minutes
b. at most 20 minutes

21. The results of a statewide exam for assessing the mathematics skills of realtors were normally distributed with a mean score of 72 and a standard deviation of 12. The realtors who scored in the top 10% are to receive a special certificate, while those in the bottom 20% will be required to attend a remedial workshop.
a. What score does a realtor need in order to receive a certificate?
b. What score will dictate that the realtor attend the workshop?

22. Professor Harde assumes that exam scores are normally distributed and wants to grade "on the curve." The mean score was 58, with a standard deviation of 16.
a. If she wants 14% of the students to receive an A, find the minimum score to receive an A.
b. If she wants 19% of the students to receive a B, find the minimum score to receive a B.

23. The time it takes an employee to package the components of a certain product is normally distributed with $\mu = 8.5$ minutes and $\sigma = 1.5$ minutes. To boost productivity, management has decided to give special training to the 34% of employees who took the greatest amount of time to package the components. Find the amount of time taken to package the components that will indicate that an employee should get special training.

24. The time it takes an employee to package the components of a certain product is normally distributed with $\mu = 8.5$ and $\sigma = 1.5$ minutes. As an incentive, management has decided to give a bonus to the 20% of employees who took the shortest amount of time to package the components. Find the amount of time taken to package the components that will indicate that an employee should get a bonus.

Answer the following questions using complete sentences and your own words.

Concept Questions

25. What are the characteristics of a normal distribution?
26. Are all distributions of data normally distributed? Support your answer with an example.
27. Why is the total area under a bell curve equal to 1?
28. Why are there no negative z-numbers in the body table?
29. When converting an x-number to a z-number, what does a negative z-number tell you about the location of the x-number?
30. Is it logical to assume that the heights of all high school students in the United States are normally distributed? Explain.
31. Is it reasonable to assume that the ages of all high school students in the United States are normally distributed? Explain.

History Questions

32. Who is known as "the Prince of Mathematics"? Why?
33. What mathematician was instrumental in the creation of the normal distribution? What application prompted this person to create the normal distribution?

4.5 Polls and Margin of Error

One of the most common applications of statistics is the evaluation of the results of surveys and public opinion polls. Most editions of the daily newspaper contain the results of at least one poll. Headlines announce the attitude of the nation toward a myriad of topics ranging from the actions of politicians to controversial current issues. (Figure 4.108). How are these conclusions reached? What do they mean? How valid are they? In this section, we investigate these questions and obtain results concerning the "margin of error" associated with the reporting of "public opinion."

Americans Divided Over Abortion Debate

Similar percentages call themselves pro-choice and pro-life

by Lydia Saad

Three decades of extensive polling on the abortion issue have shown that Americans hold a complex set of opinions about the morality and legality of terminating a woman's pregnancy. However, when asked in a new Gallup poll to sum up their abortion views according to the labels favored by activists on each side, the public is almost evenly split on the issue, with 48% currently calling themselves "pro-choice" and 42% identifying themselves as "pro-life."

Where Gallup does find significant differences in views on abortion is between people belonging to different political parties, and between those who hold different levels of religious commitment. Democrats and independents are much more likely than are Republicans to consider themselves pro-choice, with slightly more than half of Democrats and independents calling themselves pro-choice, compared to only 38% of Republicans. The differences are even stronger according to the religious commitment of respondents. Among those who say religion is very important in their lives, more than half identify themselves as pro-life. However, among those for whom religion is only fairly important or not important at all, the pro-choice position is the dominant view.

For results based on the sample of national adults (N=1, 014) surveyed April 30–May 2, 1999, the margin of sampling error is ±3 percentage points.

Figure 4.108

Sampling and Inferential Statistics

The purpose of conducting a survey or poll is to obtain information about a population—for example, adult Americans. Because there are approximately 200 million Americans over the age of eighteen, it would be very difficult, time-consuming, and expensive to contact every one of them. The only realistic alternative is to poll a sample and use the science of inferential statistics to draw conclusions about the population as a whole. Different samples have different characteristics depending on, among other things, the age, sex, education, and locale of the people in the sample. Therefore, it is of the utmost importance that a sample be representative of the population. Obtaining a representative sample is the most difficult aspect of inferential statistics.

Another problem facing pollsters is determining *how many* people should be selected for the sample. Obviously, the larger the sample, the more likely it will reflect the population. However, larger samples cost more money, so a limited budget will limit the sample size. Conducting surveys can be very costly, even for a small to moderate sample. For example, a survey conducted in 1989 by the Gallup Organization that contacted 1,005 adults and 500 teenagers would have cost \$100,000 (the pollsters donated their services for this survey). The results of this poll indicated that Americans thought the "drug crisis" was the nation's top problem (stated by 27% of the adults and 32% of the teenagers).

After a sample has been selected and its data analyzed, information about the sample is generalized to the entire population. Because 27% of the 1,005 adults in a poll stated that the drug crisis was the nation's top problem, we would like to conclude that 27% of *all* adults have the same belief. Is this a valid generalization? That is, how confident is the pollster that the feelings of the people in the sample reflect those of the population?

Sample Proportion versus Population Proportion

If x members (for example, people, automobiles, households) in a sample of size n have a certain characteristic, then the proportion of the sample, or **sample proportion,** having this characteristic is given by $\frac{x}{n}$. For instance, in a sample of $n = 70$ automobiles, if $x = 14$ cars have a defective fan switch, then the proportion of the sample having a defective switch is $\frac{14}{70} = 0.2$, or 20%. The true proportion of the entire population, or **population proportion,** having the characteristic is represented by the letter P. A sample proportion $\frac{x}{n}$ is an estimate of the population proportion P.

Sample proportions $\frac{x}{n}$ vary from sample to sample; some will be larger than P, and some will be smaller. Of the 1,005 adults in the Gallup Poll sample mentioned above, 27% viewed the drug crisis as the nation's top problem. If a different sample of 1,005 had been chosen, 29% might have had this view. If still another 1,005 had been selected, this view might have been shared by only 25%. We will assume that the sample proportions $\frac{x}{n}$ are normally distributed around the population proportion P. The set of all sample proportions, along with their probabilities of occurring, can be represented by a bell curve like the one in Figure 4.109.

In general, a sample estimate is not 100% accurate; although a sample proportion might be close to the true population proportion, it will have an error term associated with it. The difference between a sample estimate and the true (population) value is called the **error of the estimate.** We can use a bell curve (like the one in Figure 4.109) to predict the probable error of a sample estimate.

Before developing this method of predicting the error term, we need to introduce some special notation. The symbol z_α (read "z alpha") will be used to represent the

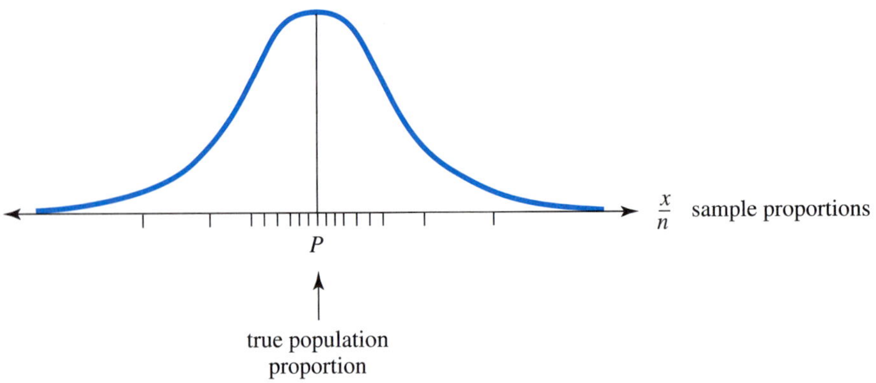

FIGURE 4.109
Sample proportions normally distributed around the true (population) proportion

positive z-number that has a right body of area α. That is, z_α is the number such that $p(0 < z < z_\alpha) = \alpha$, as shown in Figure 4.110.

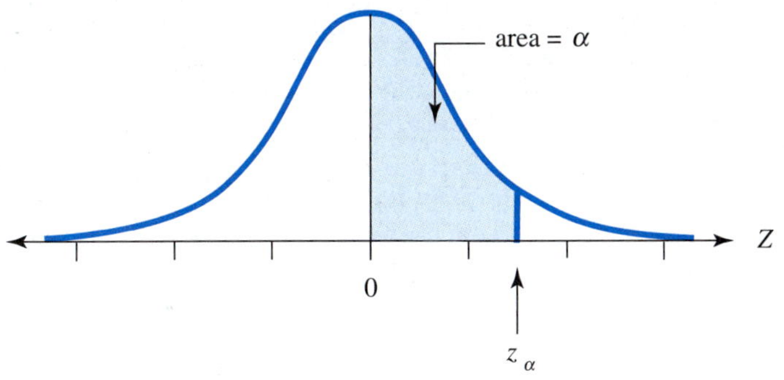

Figure 4.110
$p(0 < z < z_\alpha) = \alpha$

EXAMPLE 1 Use the body table to find the following values:

a. $z_{0.3925}$ **b.** $z_{0.475}$ **c.** $z_{0.45}$ **d.** $z_{0.49}$

Solution

a. $z_{0.3925}$ represents the z-number that has a body of area 0.3925. Looking at the interior of the body table, we find 0.3925 and see that it corresponds to the z-number 1.24—that is, $p(0 < z < 1.24) = 0.3925$. Therefore, $z_{0.3925} = 1.24$.

b. In a similar manner, we find that a body of area 0.4750 corresponds to $z = 1.96$. Therefore, $z_{0.475} = 1.96$.

c. Looking through the interior of the table, we cannot find a body of area 0.45. However, we do find a body of 0.4495 (corresponding to $z = 1.64$) and a body of 0.4505 (corresponding to $z = 1.65$). Because the desired body is *exactly halfway* between the two listed bodies, we use a z-number that is exactly halfway between the two listed z-numbers, 1.64 and 1.65. Therefore, $z_{0.45} = 1.645$.

d. We cannot find a body of the desired area, 0.49, in the interior of the table. The closest areas are 0.4898 (corresponding to $z = 2.32$) and 0.4901 (corresponding to $z = 2.33$). Because the desired area (0.49) is *closer* to 0.4901, we use $z = 2.33$. Therefore, $z_{0.49} = 2.33$. ■

Margin of Error

Sample proportions $\frac{x}{n}$ vary from sample to sample; some will have a small error, and some will have a large error. Knowing that sample estimates have inherent errors, statisticians make predictions concerning the largest possible error associated with a sample estimate. This error is called the **margin of error** of the estimate and is denoted by **MOE.** Because the margin of error is a prediction, we cannot guarantee that it is absolutely correct; that is, the probability that a prediction is correct might be 0.95, or it might be 0.75.

In the field of inferential statistics, the probability that a prediction is correct is referred to as the **level of confidence** of the prediction. For example, we might say that we are 95% confident that the maximum error of an opinion poll is plus or minus 3 percentage points; that is, if 100 samples were analyzed, 95 of them would have proportions that differ from the true population proportion by an amount less than or equal to 0.03, and 5 of the samples would have an error greater than 0.03.

Assuming that sample proportions are normally distributed around the population proportion (as in Figure 4.109), we can use the z-distribution to determine the margin of error associated with a sample proportion. In general, the margin of error depends on the sample size and the level of confidence of the estimate.

HISTORICAL NOTE

George H. Gallup 1901–1984

© Corbis/Bettman

To many people, the name *Gallup* is synonymous with opinion polls. George Horace Gallup, the founder of the American Institute of Public Opinion, began his news career while attending the University of Iowa. During his junior year as a student of journalism, Gallup became the editor of his college newspaper, the *Daily Iowan.* After receiving his bachelor's degree in 1923, Gallup remained at the university nine years as an instructor of journalism.

In addition to teaching, Gallup continued his own studies of human nature and public opinion. Interest in how the public reacts to advertisements and perceives various issues of the day led Gallup to combine his study of journalism with the study of psychology. In 1925, he received his master's degree in psychology. Gallup's studies culminated in 1928 with his doctoral thesis, *A New Technique for Objective Methods for Measuring Reader Interest in Newspapers.* Gallup's new technique of polling the public was to utilize a stratified sample, that is, a sample that closely mirrors the composition of the entire population. Gallup contended that a stratified sample of 1,500 people was sufficient to obtain reliable estimates. For his pioneering work in this new field, Gallup was awarded his Ph.D. in journalism in 1928.

Gallup founded the American Institute of Public Opinion in 1935 with the stated purpose "to impartially measure and report public opinion on political and social issues of the day without regard to the rightness or wisdom of the views expressed." His first triumph was his prediction of the winner of the 1936 presidential election between Franklin D. Roosevelt and Alfred Landon. While many, including the prestigious *Literary Digest,* predicted Landon to win, Gallup correctly predicted Roosevelt as the winner.

Gallup Polls have correctly predicted all presidential elections since, with the exception of the 1948 race between Thomas Dewey and Harry S Truman. Much to his embarrassment, Gallup predicted Dewey as the winner. Truman won the election with 49.9% of the vote, while Gallup had predicted that he would receive only 44.5%. Gallup's explanation was that he had ended his poll too far in advance of election day and had disregarded the votes of those who

Margin of Error Formula

Given a sample size n, the **margin of error,** denoted by **MOE,** for a poll involving sample proportions is

$$\text{MOE} = \frac{z_{\alpha/2}}{2\sqrt{n}}$$

where α represents the level of confidence of the poll. That is, the probability is α that the sample proportion has an error of at most MOE.

EXAMPLE 2 Assuming a 90% level of confidence, find the margin of error associated with each of the following:

a. sample size $n = 275$ **b.** sample size $n = 750$

Solution **a.** The margin of error depends on two things: the sample size and the level of confidence. For a 90% level of confidence, $\alpha = 0.90$. Hence, $\frac{\alpha}{2} = 0.45$, and $z_{\alpha/2} = z_{0.45} = 1.645$.

The Gallup Organization predicted that Thomas Dewey would win the 1948 presidential election. Much to Gallup's embarrassment, Harry S. Truman won the election and triumphantly displayed a newspaper containing Gallup's false prediction. With the exception of this election, Gallup Polls have correctly predicted every presidential election since 1936.

were undecided. Of his error, Gallup said, "We are continually experimenting and continually learning."

Although some people criticize the use of polls, citing their potential influence and misuse, Gallup considered the public opinion poll to be "one of the most useful instruments of democracy ever devised." Answering the charge that he and his polls influenced elections, Gallup retorted, "One might as well insist that a thermometer makes the weather!" In addition, Gallup confessed that he had not voted in a presidential election since 1928. Above all, Gallup wanted to ensure the impartiality of his polls.

Besides polling people regarding their choices in presidential campaigns, Gallup was the first pollster to ask the public to rate a president's performance and popularity. Today, these "presidential report cards" are so common we may take them for granted. In addition to presidential politics, Gallup also dealt with sociological issues, asking questions such as "What is the most important problem facing the country?"

Polling has become a multimillion-dollar business. In 1983, the Gallup Organization had revenues totaling $6.7 million. Today, Gallup Polls are syndicated in newspapers across the country and around the world.

Substituting this value and $n = 275$ into the MOE formula, we have the following:

$$\begin{aligned} \text{MOE} &= \frac{z_{\alpha/2}}{2\sqrt{n}} \\ &= \frac{1.645}{2\sqrt{275}} \\ &= 0.049598616\ldots \\ &\approx 0.050 \qquad \text{rounding off to three decimal places} \\ &= 5.0\% \end{aligned}$$

When we are polling a sample of 275 people, we can say that we are 90% confident that the maximum possible error in the sample proportion will be plus or minus 5.0 percentage points.

b. For a 90% level of confidence, $\alpha = 0.90$, $\frac{\alpha}{2} = 0.45$, and $z_{\alpha/2} = z_{0.45} = 1.645$. Substituting this value and $n = 750$ into the MOE formula, we have the following:

$$\begin{aligned} \text{MOE} &= \frac{z_{\alpha/2}}{2\sqrt{n}} \\ &= \frac{1.645}{2\sqrt{750}} \\ &= 0.030033453\ldots \\ &\approx 0.030 && \text{rounding off to three decimal places} \\ &= 3.0\% \end{aligned}$$

When we are polling a sample of 750 people, we can say that we are 90% confident that the maximum possible error in the sample proportion will be plus or minus 3.0 percentage points. ■

If we compare the margins of error in parts (a) and (b) of Example 2, we notice that by increasing the sample size (from 275 to 750), the margin of error was reduced (from 5.0 to 3.0 percentage points). Intuitively, this should make sense; a larger sample gives a better estimate (has a smaller margin of error).

EXAMPLE 3 To obtain an estimate of the proportion of all Americans who think the president is doing a good job, a random sample of 500 Americans is surveyed, and 345 respond, "The president is doing a good job."

a. Determine the sample proportion of Americans who think the president is doing a good job.

b. Assuming a 95% level of confidence, find the margin of error associated with the sample proportion.

Solution **a.** $n = 500$ and $x = 345$. The sample proportion is

$$\frac{x}{n} = \frac{345}{500} = 0.69$$

Sixty-nine percent of the sample think the president is doing a good job.

b. We must find MOE when $n = 500$ and $\alpha = 0.95$. Because $\alpha = 0.95$, $\frac{\alpha}{2} = 0.475$ and $z_{\alpha/2} = z_{0.475} = 1.96$.
Therefore,

$$\begin{aligned} \text{MOE} &= \frac{z_{\alpha/2}}{2\sqrt{n}} \\ &= \frac{1.96}{2\sqrt{500}} \\ &= 0.043826932\ldots \\ &\approx 0.044 && \text{rounding off to three decimal places} \\ &= 4.4\% \end{aligned}$$

The margin of error associated with the sample proportion is plus or minus 4.4 percentage points. We are 95% confident that 69% ($\pm$4.4%) of all Americans think the president is doing a good job. In other words, on the basis of our sample proportion $\left(\frac{x}{n}\right)$ of 69%, we predict (with 95% certainty) that the true population proportion (P) is somewhere between 64.6% and 73.4%. ■

EXAMPLE 4 The article shown in Figure 4.111 was released by the Gallup Organization in June 1999.

a. The poll states that 79% of the Americans questioned said they favored the registration of all firearms. Assuming a 95% level of confidence (the most commonly used level of confidence), find the margin of error associated with the survey.

b. Assuming a 98% level of confidence, find the margin of error associated with the survey.

Solution **a.** We must find MOE when $n = 1{,}022$ and $\alpha = 0.95$. Because $\alpha = 0.95$, $\alpha/2 = 0.475$ and $z_{\alpha/2} = z_{0.475} = 1.96$.
Therefore,

$$\begin{aligned} \text{MOE} &= \frac{z_{\alpha/2}}{2\sqrt{n}} \\ &= \frac{1.96}{2\sqrt{1{,}022}} \\ &= 0.0306549511\ldots \\ &\approx 0.031 \qquad \text{rounding off to three decimal places} \end{aligned}$$

The margin of error associated with the survey is plus or minus 3.1%. We are 95% confident that 79% $\pm 3.1\%$ of all Americans favor the registration of all firearms.

POLL RELEASES
June 16, 1999

Americans Support Wide Variety of Gun Control Measures

Eight out of ten favor registration of all firearms

by Frank Newport

PRINCETON, NJ — Motivated in part by the tragic school shootings in Colorado and Georgia this spring, the idea of instituting new and more stringent gun control laws has become one of the most hotly debated issues on the current national agenda. A new Gallup poll shows that Americans strongly support most of the specific types of gun control measures now being debated in Congress, and that support for the idea of a general requirement that all firearms be registered is now higher than it was as recently as last fall.

The "registration of all firearms" proposition now receives somewhat stronger support than in the past. The current higher level of support—79%—may reflect the influence of this spring's tragic school shootings in Colorado and in Georgia.

There are some differences by party in support for the measure, but even among Republicans, support is strong, at 71%, compared to 88% Democrats. Women are also somewhat more likely than men to support registration of all firearms, by an 88% to 69% margin.

An interesting finding: more than sixty years ago, in 1938, Gallup asked Americans about their support for registration of all "pistols and revolvers" and found 84% support, suggesting that the current sentiment in favor of many of these gun measures does not necessarily represent a wholesale change in the attiudes of the public over the ensuing decades.

The results above are based on telephone interviews with a randomly selected national sample of 1,022 adults, 18 years and older, conducted June 11-13, 1999.

Figure 4.111

b. We must find MOE when $n = 1{,}022$ and $\alpha = 0.98$. Because $\alpha = 0.98$, $\alpha/2 = 0.49$ and $z_{\alpha/2} = z_{0.49} = 2.33$.
Therefore,

$$\begin{aligned} \text{MOE} &= \frac{z_{\alpha/2}}{2\sqrt{n}} \\ &= \frac{1.96}{2\sqrt{1{,}022}} \\ &= 0.0364418551\ldots \\ &\approx 0.036 \qquad \text{rounding off to three decimal places} \end{aligned}$$

The margin of error associated with the survey is plus or minus 3.6%. We are 98% confident that 79% ±3.6% of all Americans favor the registration of all firearms. ■

If we compare the margins of error found in parts (a) and (b) of Example 4, we notice that as the level of confidence went up (from 95% to 98%) the margin of error increased (from 3.1% to 3.6%). Intuitively, if we want to be more confident in our predictions, we should give our prediction more leeway (a larger margin of error).

When the results of the polls are printed in a newspaper, the sample size, level of confidence, margin of error, date of survey, and location of survey may be given as a footnote, as shown in Figure 4.112.

POLL RELEASES
April 9, 1999

Americans Oppose General Legalization of Marijuana
But Support Use for Medicinal Purposes
by David Moore

By a large majority, Americans continue to oppose the general legalization of marijuana, but by an even larger majority they would support the drug's use for medicinal purposes. These findings, from a Gallup poll conducted March 19-21, follow the announcement several days earlier from the Institute of Medicine, an affiliate of the National Academy of Sciences, that marijuana's active ingredients can ease the pain, nausea and vomiting caused by cancer and AIDS.

According to the poll, just 29% of respondents support general legalization of marijuana, while 69% are opposed. These figures represent a slight gain for the proposal, with support higher than last year by five percentage points. In four previous polls dating back to 1979, support has varied from 23% to 28%. The earliest poll asking about this issue was conducted in 1969, when just 12% supported legalization.

Despite this opposition to the general legalization of the drug, by a three-to-one margin Americans would support making marijuana available to doctors, so it could be prescribed to reduce pain and suffering. In six states, voters have already approved marijuana for medicinal use, although the drug remains banned by federal law. With the responses to both questions taken into account, the poll shows that 28% of Americans support legalization of marijuana for whatever reason, 25% oppose it even for medicinal purposes, while 43% support it for medicinal purposes but not for general use.

Survey Methods

The results are based on telephone interviews with a randomly selected national sample of 1,018 adults, 18 years and older, conducted March 19-21, 1999. For results based on this sample, one can say with 95 percent confidence that the maximum error attributable to sampling and other random effects is plus or minus 3 percentage points. In addition to sampling error, question wording and practical difficulties in conducting surveys can introduce error or bias into the findings of public opinion polls.

Figure 4.112

EXAMPLE 5 Verify the margin of error stated in the article shown in Figure 4.112.

Solution The footnote to the article states that for a sample size of 1,018 and a 95% level of confidence, the margin of error is $\pm 3\%$, that is, MOE = 0.03 for $n = 1{,}018$ and $\alpha = 0.95$. Because $\alpha = 0.95$, $\frac{\alpha}{2} = 0.475$ and $z_{\alpha/2} = z_{0.475} = 1.96$. Therefore,

$$\begin{aligned} \text{MOE} &= \frac{z_{\alpha/2}}{2\sqrt{n}} \\ &= \frac{1.96}{2\sqrt{1{,}018}} \\ &= 0.0307151179\ldots \\ &\approx 0.03 \qquad \text{rounding off to two decimal places} \end{aligned}$$

Therefore, the stated margin of error of $\pm 3\%$ is accurate. ■

News articles do not always mention the level of confidence of a survey. However, if the sample sizes and margin of error are given, the level of confidence can be determined, as shown in Example 6.

EXAMPLE 6 Do you prefer paper or plastic shopping bags? The news article shown below in Figure 4.113 presents the results of a Gallup poll pertaining to this question. Find the level of confidence of the poll.

Shoppers' replies to the big question

COX NEWS SERVICE

WASHINGTON —

Paper or plastic?

A new Gallup Poll has revealed that 48 percent of American consumers answer "paper" and 37 percent say "plastic" when a grocery store checkout clerk asks what kind of bag they want.

Pollsters telephoned 1,021 adults to find the answer to this seemingly universal question, within a margin of error of plus or minus 3 percent.

Of the respondents, 83 percent said their stores offered a choice of sack types. About 10 percent said their grocers had only paper bags, and about 7 percent said their grocers provided only plastic.

About one in eight grocery shoppers—13 percent—uses both paper and plastic bags, the poll said. Fewer than 2 percent answered anything other than paper or plastic or both, including those who bring reusable cloth bags to the grocery store.

The study found that 53 percent of men use paper bags, as compared with 43 percent of women. Shoppers under age 55 and those with incomes of $25,000 or more are also more likely to prefer paper.

Figure 4.113

Solution We are given $n = 1{,}021$ and MOE = 0.03. To find α, the level of confidence of the poll, we must first find $z_{\alpha/2}$ and the area of the bodies, as shown in Figure 4.114.

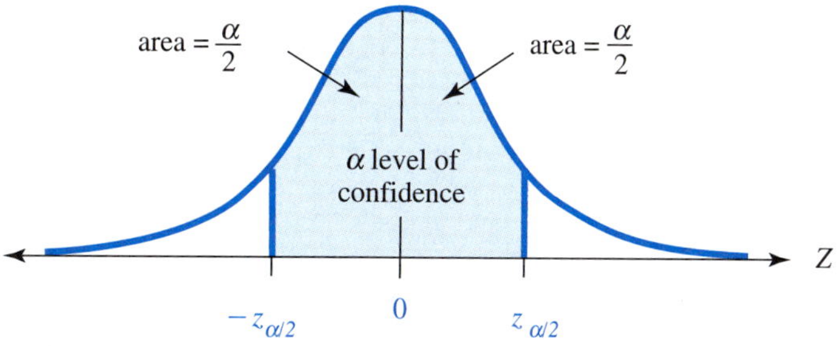

Figure 4.114

Substituting the given values into the MOE formula, we have

$$\text{MOE} = \frac{z_{\alpha/2}}{2\sqrt{n}}$$

$$0.03 = \frac{z_{\alpha/2}}{2\sqrt{1{,}021}}$$

$$z_{\alpha/2} = 0.03(2\sqrt{1{,}021}) \quad \text{multiplying each side by } 2\sqrt{1{,}021}$$

$$z_{\alpha/2} = 1.917185437\ldots$$

$$z_{\alpha/2} \approx 1.92 \quad \text{rounding off to two decimal places}$$

Using the body table, we can find the area under the bell curve between $z = 0$ and $z = 1.92$; that is, $p(0 < z < 1.92) = 0.4726$.

Therefore, $\frac{\alpha}{2} = 0.4726$, and multiplying by 2, we have $\alpha = 0.9452$. Thus, the level of confidence is $\alpha = 0.9452$ (or 95%), as shown in Figure 4.115. ■

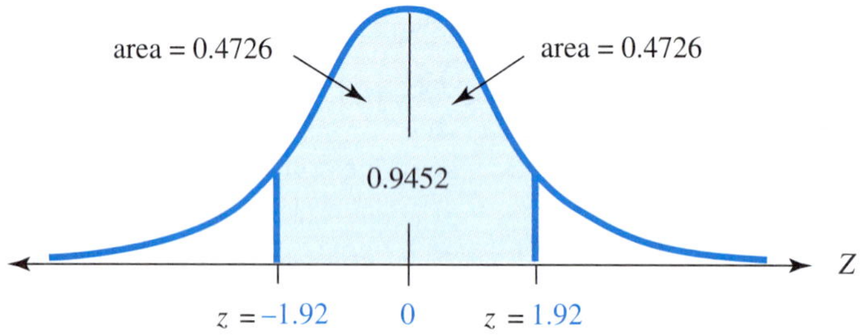

Figure 4.115

4.5 Exercises

In Exercises 1–4, use the body table to find the specified z-number.

1. **a.** $z_{0.2517}$ **b.** $z_{0.1217}$ **c.** $z_{0.4177}$ **d.** $z_{0.4960}$
2. **a.** $z_{0.0199}$ **b.** $z_{0.2422}$ **c.** $z_{0.4474}$ **d.** $z_{0.4936}$
3. **a.** $z_{0.4250}$ **b.** $z_{0.4000}$ **c.** $z_{0.3750}$ **d.** $z_{0.4950}$
4. **a.** $z_{0.4350}$ **b.** $z_{0.4100}$ **c.** $z_{0.2750}$ **d.** $z_{0.4958}$
5. Find the z-number associated with a 92% level of confidence.
6. Find the z-number associated with a 97% level of confidence.
7. Find the z-number associated with a 75% level of confidence.
8. Find the z-number associated with an 85% level of confidence.

In Exercises 9–22, round off your answers (sample proportions and margins of error) to three decimal places (a tenth of a percent).

9. The Gallup Poll in Example 5 states that 29% of respondents support general legalization of marijuana. For each of the following levels of confidence, find the margin of error associated with the sample.
   a. a 90% level of confidence
   b. a 98% level of confidence
10. The Gallup Poll in Example 6 states that 48% of the Americans questioned say "paper" when a grocery store checkout clerk asks what kind of bag they want. For each of the following levels of confidence, find the margin of error associated with the sample.
   a. an 80% level of confidence
   b. a 99% level of confidence
11. A survey asked, "How important is it to you to buy products that are made in America?" Of the 600 Americans surveyed, 450 responded, "It is important." For each of the following levels of confidence, find the sample proportion and the margin of error associated with the poll.
   a. a 90% level of confidence
   b. a 95% level of confidence

12. In the survey in Exercise 11, 150 of the 600 Americans surveyed responded, "It is not important." For each of the following levels of confidence, find the sample proportion and the margin of error associated with the poll.
   a. an 85% level of confidence
   b. a 98% level of confidence
13. A survey asked, "Have you ever bought a lottery ticket?" Of the 2,710 Americans surveyed, 2,141 said yes, and 569 said no.*
   a. Determine the sample proportion of Americans who have purchased a lottery ticket.
   b. Determine the sample proportion of Americans who have not purchased a lottery ticket.
   c. With a 90% level of confidence, find the margin of error associated with the sample proportions.
14. A survey asked, "Which leg do you put into your trousers first?" Of the 2,710 Americans surveyed, 1,138 said left, and 1,572 said right.*
   a. Determine the sample proportion of Americans who put their left leg into their trousers first.
   b. Determine the sample proportion of Americans who put their right leg into their trousers first.
   c. With a 90% level of confidence, find the margin of error associated with the sample proportions.
15. A survey asked, "Do you prefer showering or bathing?" Of the 1,220 American men surveyed, 1,049 preferred showering, and 171 preferred bathing. In contrast, 1,043 of the 1,490 American women surveyed preferred showering, and 447 preferred bathing.*
   a. Determine the sample proportion of American men who prefer showering.
   b. Determine the sample proportion of American women who prefer showering.
   c. With a 95% level of confidence, find the margin of error associated with the sample proportions.
16. A survey asked, "Do you like the way you look in the nude?" Of the 1,220 American men surveyed, 830 said yes, and 390 said no. In contrast, 328 of the 1,490 American women surveyed said yes, and 1,162 said no.*
   a. Determine the sample proportion of American men who like the way they look in the nude.
   b. Determine the sample proportion of American women who like the way they look in the nude.
   c. With a 95% level of confidence, find the margin of error associated with the sample proportions.
17. A survey asked, "Do you know kids who carry weapons in school?" Of the 2,035 teenage Americans surveyed, 1,160 said yes, and 875 said no.†
   a. Determine the sample proportion of teenage Americans who know kids who carry weapons in school.
   b. Determine the sample proportion of teenage Americans who do not know kids who carry weapons in school.
   c. With a 95% level of confidence, find the margin of error associated with the sample proportions.
18. A survey asked, "Should schools make condoms available to students?" Of the 2,035 teenage Americans surveyed, 1,221 said yes, and 814 said no.†
   a. Determine the sample proportion of teenage Americans who think that schools should make condoms available to students.
   b. Determine the sample proportion of teenage Americans who think that schools should not make condoms available to students.
   c. With a 90% level of confidence, find the margin of error associated with the sample proportions.
19. A survey asked, "Can you imagine a situation in which you might become homeless?" Of the 2,503 Americans surveyed, 902 said yes.‡
   a. Determine the sample proportion of Americans who can imagine a situation in which they might become homeless.
   b. With a 90% level of confidence, find the margin of error associated with the sample proportion.
   c. With a 98% level of confidence, find the margin of error associated with the sample proportion.
   d. How does your answer to part (c) compare to your answer to part (b)? Why?
20. A survey asked, "Do you think that homeless people are responsible for the situation they are in?" Of the 2,503 Americans surveyed, 1,402 said no.‡

* Data from Poretz and Sinrod, *The First Really Important Survey of American Habits* (Los Angeles: Price Stern Sloan Publishing, 1989).

† Data from Lynn Minton, "What Kids Say," *Parade Magazine,* August 1, 1993: 4–6.

‡ Data from Mark Clements, "What Americans Say about the Homeless," *Parade Magazine,* Jan. 9, 1994: 4–6.

a. Determine the sample proportion of Americans who think that homeless people are not responsible for the situation they are in.
b. With an 80% level of confidence, find the margin of error associated with the sample proportion.
c. With a 95% level of confidence, find the margin of error associated with the sample proportion.
d. How does your answer to part (c) compare to your answer to part (b)? Why?

21. A sample consisting of 430 men and 765 women was asked various questions pertaining to international affairs. With a 95% level of confidence, find the margin of error associated with the following samples.
a. the male sample
b. the female sample
c. the combined sample

22. A sample consisting of 942 men and 503 women was asked various questions pertaining to the nation's economy. For a 95% level of confidence, find the margin of error associated with the following samples.
a. the male sample
b. the female sample
c. the combined sample

23. A poll pertaining to environmental concerns had the following footnote: "Based on a sample of 1,763 adults, the margin of error is plus or minus 2.5 percentage points." Find the level of confidence of the poll.

HINT: See Example 6.

24. A poll pertaining to educational goals had the following footnote: "Based on a sample of 2,014 teenagers, the margin of error is plus or minus 2 percentage points." Find the level of confidence of the poll.

HINT: See Example 6.

25. A recent poll pertaining to educational reforms involved 640 men and 820 women. The margin of error for the combined sample is 2.6%. Find the level of confidence for the entire poll.

HINT: See Example 6.

26. A recent poll pertaining to educational reforms involved 640 men and 820 women. The margin of error is 3.9% for the male sample and 3.4% for the female sample. Find the level of confidence for the male portion of the poll and for the female portion of the poll.

HINT: See Example 6.

Answer the following questions using complete sentences and your own words.

Concept Questions

27. What is a sample proportion? How is it calculated?
28. What is a margin of error? How is it calculated?
29. If the sample size is increased in a survey, would you expect the margin of error to increase or decrease? Why?

History Questions

30. Who founded the American Institute of Public Opinion? When? In what two academic fields did this person receive degrees?

Projects

31. The purpose of this project is to conduct an opinion poll. Select a topic that is relevant to you and/or your community. Create a multiple-choice question to gather people's opinions concerning this issue. For example:

> If you could vote today, how would you vote on Proposition X?
> i. support ii. oppose iii. undecided

a. Ask 50 people your question, and record their responses. Calculate the sample proportion for each category of response. Use a 95% level of confidence, and calculate the margin of error for your survey.
b. Ask 100 people your question, and record their responses. Calculate the sample proportion for each category of response. Use a 95% level of confidence, and calculate the margin of error for your survey.
c. How does the margin of error in part (b) compare to the margin of error in part (a)?

4.6 Linear Regression

When x and y are variables and m and b are constants, the equation $y = mx + b$ has infinitely many solutions of the form (x, y). A specific ordered pair (x_1, y_1) is a solution of the equation if $y_1 = mx_1 + b$. Because every solution of the given equation lies on a straight line, we say that x and y are *linearly related.*

If we are given two ordered pairs (x_1, y_1) and (x_2, y_2), we should be able to "work backwards" and find the equation of the line passing through them; assuming that x and y are linearly related, we can easily find the equation of the line passing through the points (x_1, y_1) and (x_2, y_2). The process of finding the equation of a line passing through given points is known as **linear regression;** the equation thus found is called the **mathematical model** of the linear relationship. Once the model has been constructed, it can be used to make predictions concerning the values of x and y.

EXAMPLE 1 Charlie is planning a family reunion and wants to place an order for custom T-shirts from Prints Alive (the local silk-screen printer) to commemorate the occasion. He has ordered shirts from Prints Alive on two previous occasions; on one occasion, he paid \$164 for 24 shirts; on another, he paid \$449 for 84. Assuming a linear relationship between the cost of T-shirts and the number ordered, predict the cost of ordering 100 shirts.

Solution Letting $x =$ the number of shirts ordered and $y =$ the total cost of the shirts, the given data can be expressed as two ordered pairs: $(x_1, y_1) = (24, 164)$ and $(x_2, y_2) = (84, 449)$. We must find $y = mx + b$, the equation of the line passing through the two points.

First, we find m, the slope:

$$\begin{aligned} m &= \frac{y_2 - y_1}{x_2 - x_1} \\ &= \frac{449 - 164}{84 - 24} \\ &= \frac{285}{60} \\ &= 4.75 \end{aligned}$$

Now we use one of the ordered pairs to find b, the y-intercept. Either point will work; we will use $(x_1, y_1) = (24, 164)$.

The slope-intercept form of a line is $y = mx + b$. Solving for b, we obtain

$$\begin{aligned} b &= y - mx \\ &= 164 - 4.75(24) \\ &= 164 - 114 \\ &= 50 \end{aligned}$$

Therefore, the equation of the line is $y = 4.75x + 50$. We use this linear model to predict the cost of ordering $x = 100$ T-shirts.

$$\begin{aligned} y &= 4.75x + 50 \\ &= 4.75(100) + 50 \\ &= 475 + 50 \\ &= 525 \end{aligned}$$

We predict that it will cost \$525 to order 100 T-shirts. ■

Linear Trends and Line of Best Fit

Example 1 illustrates the fact that two points determine a unique line. To find the equation of the line, we must find the slope and y-intercept. If we are given more than two points, the points might not be collinear. When collecting real-world data, this is usually the case. However, after plotting the scatter of points on an x-y coordinate system, it may appear that they "almost" fit on a line. If a sample of ordered pairs tend to "go in the same general direction," we say that they exhibit a **linear trend.** See Figure 4.116.

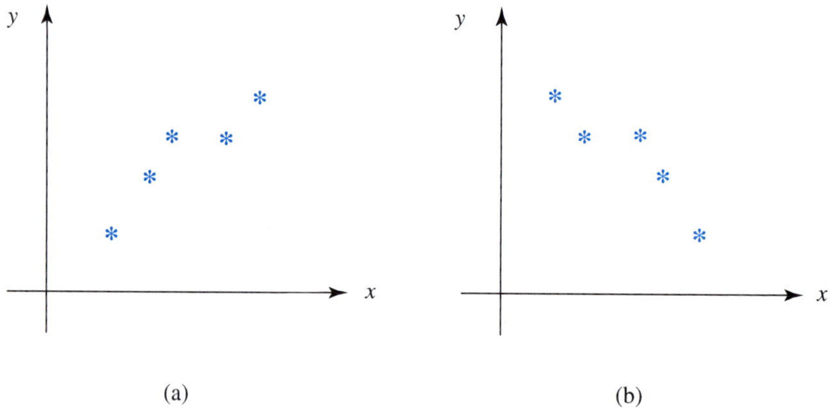

Figure 4.116
Ordered pairs that exhibit a linear trend

When a scatter of points exhibits a linear trend, we construct the line that best approximates the trend. This line is called the **line of best fit** and is denoted by $\hat{y} = mx + b$. The "hat" over the y indicates that the calculated value of y is a prediction based on linear regression. See Figure 4.117.

To calculate the slope and y-intercept of the line of best fit, mathematicians have developed formulas based on the method of least squares. (See the Historical Note on Carl Gauss in Section 4.4.)

Recall that the symbol Σ means "sum." Therefore, Σx represents the sum of the x-coordinates of the points and Σy the sum of the y-coordinates. To find Σxy, multiply the x- and y-coordinates of each point and sum the results.

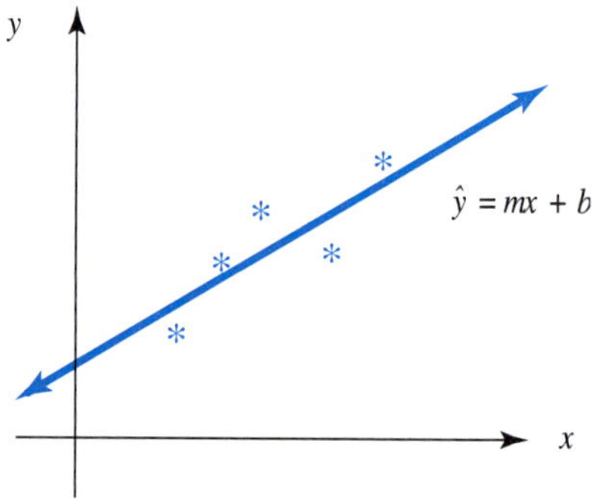

Figure 4.117
The line of best fit $y = mx + b$

Line of Best Fit

Given a sample of n ordered pairs $(x_1, y_1), (x_2, y_2), \ldots, (x_n, y_n)$.
The **line of best fit** (the line that best represents the data) is denoted by $\hat{y} = mx + b$, where the slope m and y-intercept b are given by

$$m = \frac{n(\Sigma xy) - (\Sigma x)(\Sigma y)}{n(\Sigma x^2) - (\Sigma x)^2} \quad \text{and} \quad b = \bar{y} - m\bar{x}$$

$\bar{x}$ and $\bar{y}$ denote the means of the x- and y-coordinates, respectively.

EXAMPLE 2 Given: the ordered pairs (5, 14), (9, 17), (12, 16), (14, 18), and (17, 23).

a. Find the equation of the line of best fit.

b. Plot the given data and sketch the graph of the line of best fit on the same coordinate system.

Solution **a.** Organize the data in a table and compute the appropriate sums, as shown in Figure 4.118.

(x, y)	x	x^2	y	xy
(5, 14)	5	25	14	$5 \cdot 14 = 70$
(9, 17)	9	81	17	$9 \cdot 17 = 153$
(12, 16)	12	144	16	$12 \cdot 16 = 192$
(14, 18)	14	196	18	$14 \cdot 18 = 252$
(17, 23)	17	289	23	$17 \cdot 23 = 391$
$n = 5$ ordered pairs	$\Sigma x = 57$	$\Sigma x^2 = 735$	$\Sigma y = 88$	$\Sigma xy = 1{,}058$

Figure 4.118

First, we find the slope:

$$m = \frac{n(\Sigma xy) - (\Sigma x)(\Sigma y)}{n(\Sigma x^2) - (\Sigma x)^2}$$

$$= \frac{5(1{,}058) - (57)(88)}{5(735) - (57)^2}$$

$$= 0.643192488\ldots$$

[( ] 5 [×] 1058 [−] 57 [×] 88 [ )] [÷] [( ] 5 [×] 735 [−] 57 [x^2] [ )] [=]

For graphing calculators, see the instructions on page 296.

Once m has been calculated, we store it in the memory of our calculator. We will need it to calculate b, the y-intercept.

$$b = \bar{y} - m\bar{x}$$

$$= \left(\frac{88}{5}\right) - 0.643192488\left(\frac{57}{5}\right)$$

$$= 10.26760564\ldots$$

88 [÷] 5 [−] [(] [RCL] [×] 57 [÷] 5 [)] [=]

Therefore, the line of best fit, $\hat{y} = mx + b$, is

$$\hat{y} = 0.643192488x + 10.26760564$$

Rounding off to one decimal place, we have

$$\hat{y} = 0.6x + 10.3$$

b. To graph the line, we need to plot two points. One point is the y-intercept $(0, b) = (0, 10.3)$. To find another point, we pick an appropriate value for x—say, $x = 18$—and calculate $\hat{y}$:

$$\begin{aligned}\hat{y} &= 0.6x + 10.3\\ &= 0.6(18) + 10.3\\ &= 21.1\end{aligned}$$

Therefore, the point $(x, \hat{y}) = (18, 21.1)$ is on the line of best fit.

Plotting (0, 10.3) and (18, 21.1), we construct the line of best fit; it is customary to use asterisks (*) to plot the given ordered pairs as shown in Figure 4.119.

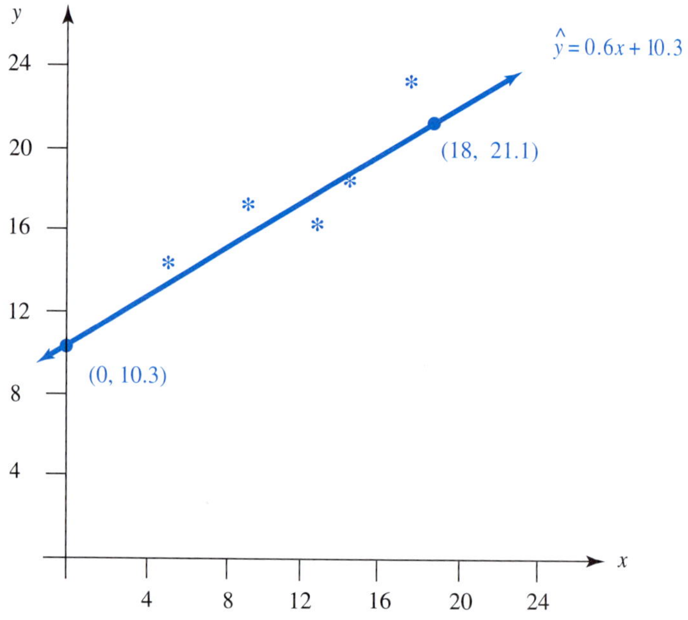

Figure 4.119

Coefficient of Linear Correlation

Given a sample of n ordered pairs, we can always find the line of best fit. Does the line accurately portray the data? Will the line give accurate predictions? To answer these questions, we must consider the relative strength of the linear trend exhibited by the given data. If the given points are close to the line of best fit, there is a strong linear relation between x and y; the line will generate good predictions. If the given points are widely scattered about the line of best fit, there is a weak linear relation, and predictions based on it are probably not reliable.

One way to measure the strength of a linear trend is to calculate the **coefficient of linear correlation,** denoted by r. The formula for calculating r is shown in the box.

Coefficient of Linear Correlation

Given a sample of n ordered pairs, $(x_1, y_1), (x_2, y_2), \ldots, (x_n, y_n)$.

The **coefficient of linear correlation,** denoted by r, is given by

$$r = \frac{n(\Sigma xy) - (\Sigma x)(\Sigma y)}{\sqrt{n(\Sigma x^2) - (\Sigma x)^2}\sqrt{n(\Sigma y^2) - (\Sigma y)^2}}$$

The calculated value of r is always between -1 and 1, inclusive; that is, $-1 \leq r \leq 1$. If the given ordered pairs lie perfectly on a line whose slope is *positive,* then the calculated value of r will equal 1 (think 100% perfect with positive slope). In this case, both variables have the same behavior: As one increases (or decreases), so will the other. On the other hand, if the data points fall perfectly on a line whose slope is *negative,* the calculated value of r will equal -1 (think 100% perfect with negative slope). In this case, the variables have opposite behavior: As one increases, the other decreases, and vice versa. See Figure 4.120.

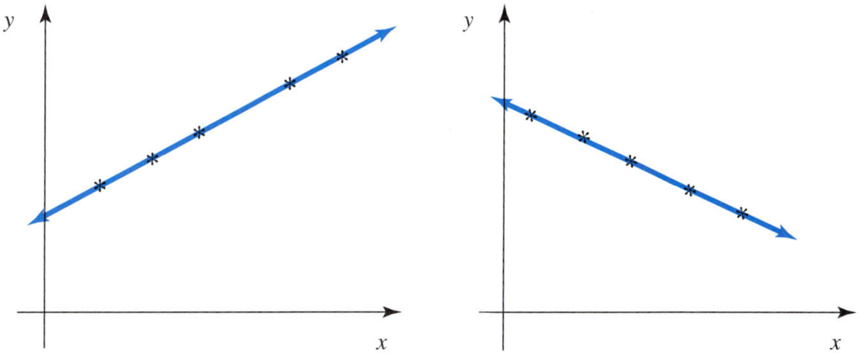

$r = 1$
Perfect, positive linear relation
(a)

$r = -1$
Perfect, negative linear relation
(b)

Figure 4.120

If the value of r is close to 0, there is little or no *linear* relation between the variables. This does not mean that the variables are not related. It merely means that no *linear* relation exists; the variables might be related in some nonlinear fashion, as shown in Figure 4.121.

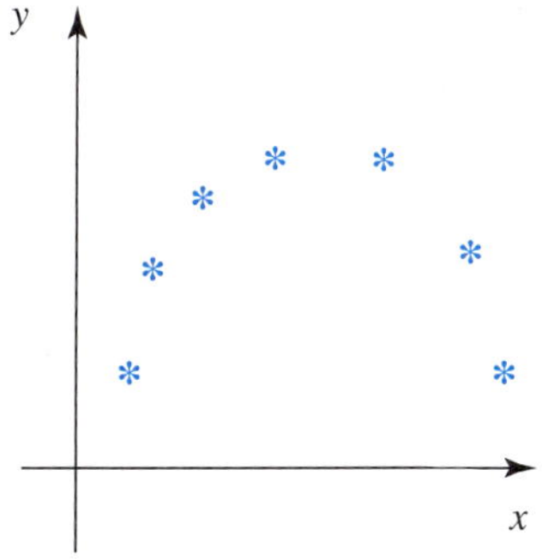

r close to 0

Figure 4.121

In summary, the closer r is to 1 or -1, the stronger the linear relation between x and y; the line of best fit will generate reliable predictions. The closer r is to 0, the weaker the linear relation; the line of best fit will generate unreliable predictions. If r is positive, the variables have a direct relationship (as one increases, so does the other); if r is negative, the variables have an inverse relationship (as one increases, the other decreases).

EXAMPLE 3 Calculate the coefficient of linear correlation for the ordered pairs given in Example 2.

Solution The ordered pairs are (5, 14), (9, 17), (12, 16), (14, 18), and (17, 23). We add a y^2 column to the table in Example 2, as shown in Figure 4.122.

(x, y)	x	x^2	y	y^2	xy
(5, 14)	5	25	14	196	$5 \cdot 14 = 70$
(9, 17)	9	81	17	289	$9 \cdot 17 = 153$
(12, 16)	12	144	16	256	$12 \cdot 16 = 192$
(14, 18)	14	196	18	324	$14 \cdot 18 = 252$
(17, 23)	17	289	23	529	$17 \cdot 23 = 391$
$n = 5$ ordered pairs	$\Sigma x = 57$	$\Sigma x^2 = 735$	$\Sigma y = 88$	$\Sigma y^2 = 1{,}594$	$\Sigma xy = 1{,}058$

Figure 4.122

Now use the formula to calculate r:

$$r = \frac{n(\Sigma xy) - (\Sigma x)(\Sigma y)}{\sqrt{n(\Sigma x^2) - (\Sigma x)^2}\sqrt{n(\Sigma y^2) - (\Sigma y)^2}}$$

$$= \frac{5(1{,}058) - (57)(88)}{\sqrt{5(735) - (57)^2}\sqrt{5(1{,}594) - (88)^2}}$$

$$= 0.883062705\ldots$$

(5 × 1058 − 57 × 88) ÷ ((5 × 735 − 57 x^2) $\sqrt{x}$ × (5 × 1594 − 88 x^2) $\sqrt{x}$) =

The coefficient of linear correlation is reasonably close to 1, so the line of best fit will generate reasonably reliable predictions. (Notice that the data points in Figure 4.119 are fairly close to the line of best fit.) ■

EXAMPLE 4 Unemployment and personal income are undoubtedly related; we would assume that as the national unemployment rate increases, total personal income would decrease. Figure 4.123 gives the unemployment rate and the total personal income for the United States for various years.

a. Use linear regression to predict the total personal income of the United States if the unemployment rate is 5.0%.
b. Use linear regression to predict the unemployment rate if the total personal income of the United States is \$10 billion.
c. Are the predictions in parts (a) and (b) reliable? Why or why not?

Year	Unemployment Rate (percent)	Total Personal Income (billions)
1975	8.5	\$1.3
1980	7.1	2.3
1985	7.2	3.4
1990	5.6	4.8
1995	5.6	6.1
2000	4.0	8.3

Figure 4.123

Sources: Bureau of Labor Statistics, U.S. Dept. of Labor and Bureau of Economic Analysis, U.S. Dept. of Commerce

Solution **a.** Letting x = unemployment rate, and y = total personal income, we have $n = 6$ ordered pairs, as shown in Figure 4.124.

x (percent)	y (billions)
8.5	1.3
7.1	2.3
7.2	3.4
5.6	4.8
5.6	6.1
4.0	8.3
$\Sigma x = 38$	$\Sigma y = 26.2$

Figure 4.124

Using a calculator, we find the following sums:

$$\Sigma x^2 = 253.22 \qquad \Sigma y^2 = 147.68 \qquad \Sigma xy = 146.1$$

First we find the slope:

$$\begin{aligned} m &= \frac{n(\Sigma xy) - (\Sigma x)(\Sigma y)}{n(\Sigma x^2) - (\Sigma x)^2} \\ &= \frac{6(146.1) - (38)(26.2)}{6(253.22) - (38)^2} \\ &= -1.579925651\ldots \end{aligned}$$

Now we calculate b, the y-intercept:

$$\begin{aligned} b &= \bar{y} - m\bar{x} \\ &= \left(\frac{26.2}{6}\right) - (-1.579925651)\left(\frac{38}{6}\right) \\ &= 14.37286246\ldots \end{aligned}$$

Therefore, the line of best fit, $\hat{y} = mx + b$, is

$$\hat{y} = -1.579925651x + 14.37286246$$

Rounding off to two decimal places (one more than the data), we have

$$\hat{y} = -1.58x + 14.37$$

Now, substituting $x = 5.0$ (5% unemployment) into the equation of the line of best fit, we have

$$\begin{aligned}\hat{y} &= -1.58x + 14.37\\ &= -1.58(5.0) + 14.37\\ &= -7.9 + 14.37\\ &= 6.47\end{aligned}$$

If the unemployment rate is 5.0%, we predict that the total personal income of the United States will be approximately \$6.47 billion.

b. To predict the unemployment rate when the total personal income is \$10 billion, we let $y = 10$, substitute into $\hat{y}$, and solve for x:

$$\begin{aligned}10 &= -1.58x + 14.37\\ 1.58x + 10 &= 14.37 && \text{adding 1.58x to both sides}\\ 1.58x &= 14.37 - 10 && \text{subtracting 10 from both sides}\\ 1.58x &= 4.37\\ x &= \frac{4.37}{1.58} && \text{dividing by 1.58}\\ &= 2.765822785\ldots\end{aligned}$$

We predict that the unemployment rate will be approximately 2.8% when the total personal income is \$10 billion.

c. To investigate the reliability of our predictions (the strength of the linear trend), we must calculate the coefficient of linear correlation:

$$r = \frac{n(\Sigma xy) - (\Sigma x)(\Sigma y)}{\sqrt{n(\Sigma x^2) - (\Sigma x)^2}\sqrt{n(\Sigma y^2) - (\Sigma y)^2}}$$

$$r = \frac{6(146.1) - (38)(26.2)}{\sqrt{6(253.22) - (38)^2}\sqrt{6(147.68) - (26.2)^2}}$$

$$= -0.9704385162\ldots$$

Because r is extremely close to -1, we conclude that our predictions are very reliable; the linear relationship between x and y is extremely high. Furthermore, since r is negative, we know that y (total personal income) decreases as x (unemployment rate) increases. ■

4.6 Exercises

1. A set of $n = 6$ ordered pairs has the following sums:

$$\Sigma x = 64 \quad \Sigma x^2 = 814 \quad \Sigma y = 85$$
$$\Sigma y^2 = 1{,}351 \quad \Sigma xy = 1{,}039$$

a. Find the line of best fit.

b. Predict the value of y when $x = 11$.

c. Predict the value of x when $y = 19$.

d. Find the coefficient of linear correlation.

e. Are the predictions in parts (b) and (c) reliable? Why or why not?

2. A set of $n = 8$ ordered pairs has the following sums:

$$\Sigma x = 111 \quad \Sigma x^2 = 1{,}869 \quad \Sigma y = 618$$
$$\Sigma y^2 = 49{,}374 \quad \Sigma xy = 7{,}860$$

a. Find the line of best fit.
b. Predict the value of y when $x = 8$.
c. Predict the value of x when $y = 70$.
d. Find the coefficient of linear correlation.
e. Are the predictions in parts (b) and (c) reliable? Why or why not?

3. A set of $n = 5$ ordered pairs has the following sums:

$$\Sigma x = 37 \quad \Sigma x^2 = 299 \quad \Sigma y = 38$$
$$\Sigma y^2 = 310 \quad \Sigma xy = 279$$

a. Find the line of best fit.
b. Predict the value of y when $x = 5$.
c. Predict the value of x when $y = 7$.
d. Find the coefficient of linear correlation.
e. Are the predictions in parts (b) and (c) reliable? Why or why not?

4. Given: the ordered pairs (4, 40), (6, 37), (8, 34), and (10, 31).
a. Find and interpret the coefficient of linear correlation.
b. Find the line of best fit.
c. Plot the given ordered pairs and sketch the graph of the line of best fit on the same coordinate system.

5. Given: the ordered pairs (5, 5), (7, 10), (8, 11), (10, 15), and (13, 16).
a. Plot the ordered pairs. Do the ordered pairs exhibit a linear trend?
b. Find the line of best fit.
c. Predict the value of y when $x = 9$.
d. Plot the given ordered pairs and sketch the graph of the line of best fit on the same coordinate system.
e. Find the coefficient of linear correlation.
f. Is the prediction in part (c) reliable? Why or why not?

6. Given: the ordered pairs (5, 20), (6, 15), (10, 14), (12, 15), and (13, 10).
a. Plot the ordered pairs. Do the ordered pairs exhibit a linear trend?
b. Find the line of best fit.
c. Predict the value of y when $x = 8$.
d. Plot the given ordered pairs and sketch the graph of the line of best fit on the same coordinate system.
e. Find the coefficient of linear correlation.
f. Is the prediction in part (c) reliable? Why or why not?

7. Given: the ordered pairs (2, 6), (3, 12), (6, 15), (7, 4), (10, 6), and (11, 12).
a. Plot the ordered pairs. Do the ordered pairs exhibit a linear trend?
b. Find the line of best fit.
c. Predict the value of y when $x = 8$.
d. Plot the given ordered pairs and sketch the graph of the line of best fit on the same coordinate system.
e. Find the coefficient of linear correlation.
f. Is the prediction in part (c) reliable? Why or why not?

8. The average hourly wage and the unemployment rate in the United States are given in Figure 4.125.

Year	Average Hourly Wage	Unemployment Rate
1992	$10.57	7.5%
1993	10.83	6.9
1994	11.12	6.1
1995	11.43	5.6
1996	11.82	5.4
1997	12.28	4.9
1998	12.77	4.5

FIGURE 4.125
Average hourly wage and unemployment rates
Source: Bureau of Labor Statistics

a. Letting $x =$ the average hourly wage and $y =$ the unemployment rate, plot the data. Do the data exhibit a linear trend?
b. Find the line of best fit.
c. Predict the unemployment rate when the average hourly wage is $12.00.
d. Predict the average hourly wage when the unemployment rate is 7.0%.
e. Find the coefficient of linear correlation.
f. Are the predictions in parts (c) and (d) reliable? Why or why not?

9. The average hourly wage and the average tuition at public four-year institutions of higher education in the United States are given in Figure 4.126.

Year	Average Hourly Wage	Average Tuition at 4-year Institutions
1992	$10.57	$2,349
1993	10.83	2,537
1994	11.12	2,681
1995	11.43	2,811
1996	11.82	2,975
1997	12.28	3,111
1998	12.77	3,247

FIGURE 4.126
Average hourly wage and average tuition
Sources: Bureau of Labor Statistics and National Center for Education Statistics

a. Letting x = the average hourly wage and y = the average tuition, plot the data. Do the data exhibit a linear trend?
b. Find the line of best fit.
c. Predict the average tuition when the average hourly wage is $12.50.
d. Predict the average hourly wage when the average tuition is $3,000.
e. Find the coefficient of linear correlation.
f. Are the predictions in parts (c) and (d) reliable? Why or why not?

10. The average number of barrels (in hundred thousands) of crude oil exported and imported per day by the United States is given in Figure 4.127.
a. Letting x = the number of barrels exported per day and y = the number of barrels imported per day, plot the data. Do the data exhibit a linear trend?
b. Find the line of best fit.
c. Predict the number of barrels imported per day when there are 900,000 barrels exported per day.
d. Predict the number of barrels exported per day when there are 9,000,000 barrels imported per day.
e. Find the coefficient of linear correlation.
f. Are the predictions in parts (c) and (d) reliable? Why or why not?

Year	Barrels Exported (per day)	Barrels Imported (per day)
1982	8.2	51.1
1984	7.2	54.4
1986	7.9	62.2
1988	8.2	74.0
1990	8.6	80.2
1992	9.5	78.9
1994	9.4	90.0
1996	9.8	94.0
1998	9.5	107.1

FIGURE 4.127
Barrels of crude oil exported and imported per day (hundred thousands)
Source: U.S. Department of Energy

11. The numbers of marriages and divorces (in millions) in the United States are given in Figure 4.128.
a. Letting x = the number of marriages and y = the number of divorces in a year, plot the data. Do the data exhibit a linear trend?
b. Find the line of best fit.
c. Predict the number of divorces in a year when there are 2,750,000 marriages.
d. Predict the number of marriages in a year when there are 1,500,000 divorces.
e. Find the coefficient of linear correlation.
f. Are the predictions in parts (c) and (d) reliable? Why or why not?

year	1965	1970	1975	1980	1985	1990	2000
marriages	1.800	2.158	2.152	2.413	2.425	2.448	2.329
divorces	0.479	0.708	1.036	1.182	1.187	1.175	1.135

Figure 4.128
Number of marriages and divorces (millions)
Source: National Center for Health Statistics

Year	Median Price	Mortgage Rate
1995	$110,500	7.85%
1996	115,800	7.71
1997	121,800	7.68
1998	128,400	7.10
1999	136,900	7.26

FIGURE 4.129
Median-priced home and average mortgage rates
Source: National Association of Realtors

12. The median-priced home and average mortgage rate in the United States are given in Figure 4.129.
   a. Letting $x =$ the median price of a home and $y =$ the average mortgage rate, plot the data. Do the data exhibit a linear trend?
   b. Find the line of best fit.
   c. Predict the average mortgage rate if the median price of homes is $135,000.
   d. Predict the median-priced home if the average mortgage rate is 7.25%.
   e. Find the coefficient of linear correlation.
   f. Are the predictions in parts (c) and (d) reliable? Why or why not?

Answer the following questions using complete sentences and your own words.

Concept Questions

13. What is a line of best fit? How do you find it?
14. How do you measure the strength of a linear trend?
15. What is a positive linear relation? Give an example.
16. What is a negative linear relation? Give an example.

Projects

17. Measure the heights and weights of 10 people. Let $x =$ height and $y =$ weight.
   a. Plot the ordered pairs. Do the ordered pairs exhibit a linear trend?
   b. Use the data to find the line of best fit.
   c. Find the coefficient of linear correlation.
   d. Will your line of best fit produce reliable predictions? Why or why not?

Technology and Linear Regression

In Example 2 of this section, we computed the slope and y-intercept of the line of best fit for the five ordered pairs (5, 14), (9, 17), (12, 16), (14, 18), and (17, 23). These calculations can be tedious when done by hand, even with only five data points. In the real world, there are always a large number of data points, and the calculations are always done with the aid of technology.

Linear Regression on a Computerized Spreadsheet

Computerized spreadsheets can compute the slope and y-intercept of the line of best fit, graph the line, and create a scatter diagram.

Step 1 *Enter the data* from Example 2 of this section as discussed in Section 4.1. Put the x-coordinates in column A and the y-coordinates in column B.

Step 2 *Use the Chart Wizard to draw the scatter diagram.*
- Use your mouse to press the "Chart Wizard" button at the top of the spreadsheet. (It looks like a histogram and it might have a magic wand.)
- Select "XY (Scatter)" and choose an appropriate style.
- Press the "Next" button.
- Use your mouse to click on the white rectangle that follows "Data range," and then use your mouse to draw a box around all of the x-coordinates and y-coordinates from step 1.

- Press the "Next" button.
- Under "Chart Title" type an appropriate title.
- After "Category (X)" type an appropriate title for the x-axis.
- After "Category (Y)" type an appropriate title for the y-axis.
- Press the "Finish" button and the scatter diagram will appear.

Step 3 *Save the spreadsheet.* Use your mouse to select "File" at the very top of the screen. Then pull your mouse down until "Save" is highlighted, and let go.

Step 4 *Compute and draw the line of best fit.*

- Click on the chart, and it will be surrounded by a thicker border than before.
- Use your mouse to select "Chart" at the very top of the screen. Then pull your mouse down until "Trendline" is highlighted, and let go.
- Press the "Linear" button.
- Click on "Options."
- Select "Display Equation on Chart."
- Select "Display R-Squared value on Chart.
- Press the "OK" button and the scatter diagram, the line of best fit, the equation of the line of best fit, and r^2 (the square of the correlation coefficient) appear. If the equation is in the way, you can click on it and move it to a better location.
- Use the square root button on your calculator to find r, the correlation coefficient.

Step 5 *Save the spreadsheet.*

Step 6 *Print the scatter diagram and line of best fit.*

- Click on the diagram, and it will be surrounded by a thicker border than before.
- Use your mouse to select "File" at the very top of the screen. Then pull your mouse down until "Print" is highlighted, and let go.
- Respond appropriately to the "Print" box that appears. Usually, it is sufficient to press a "Print" button or an "OK" button, but the actual way that you respond to the "Print" box depends on your printer.

Linear Regression on a Graphing Calculator

Graphing calculators can draw a scatter diagram, compute the slope and y-intercept of the line of best fit, and graph the line.

Entering the Data

On a TI-81:

```
DATA
x1=
y1=1
```

Figure 4.130

- *Put the calculator into statistics mode* by pressing 2nd STAT.
- *Set the calculator up for entering the data* from Example 2 by scrolling to the right and selecting the "DATA" menu, and selecting "Edit" from the "DATA" menu. This should result in the screen shown in Figure 4.130. (If your screen is messier, clear it by returning to the "DATA" menu and selecting "ClrStat."
- Enter 5 for x_1, 14 for y_1, 9 for x_2, etc.

On a TI-80/82/83/86:

- *Put the calculator into statistics mode* by pressing STAT (TI-86: 2nd STAT).
- *Set the calculator up for entering the data* from Example 2 by selecting "Edit" from the "EDIT" menu (TI-86: Select "Edit" by pressing F2), and the "List Screen" appears, as shown in Figure 4.131. (A TI-86's list screen says "xStat," "yStat," and "fStat" instead of "L_1," "L_2," and "L_3", respectively.) If data already appear in a list (as they do in lists L_1 and L_2 in Figure 4.131) and you want to clear it, use the arrow buttons to highlight the name of the list and press CLEAR ENTER.
- Use the arrow buttons and the ENTER button to enter the x-coordinates in list L_1 and the corresponding y-coordinates in list L_2. (TI-86: Enter the x-coordinates in list xStat and the corresponding y-coordinates in list yStat. Also enter a frequency of 1 in list fStat, for each of the entries in list xStat.) When done, your screen should look like Figure 4.131. (TI-86 screens should also have a list of 1s in the last column.)

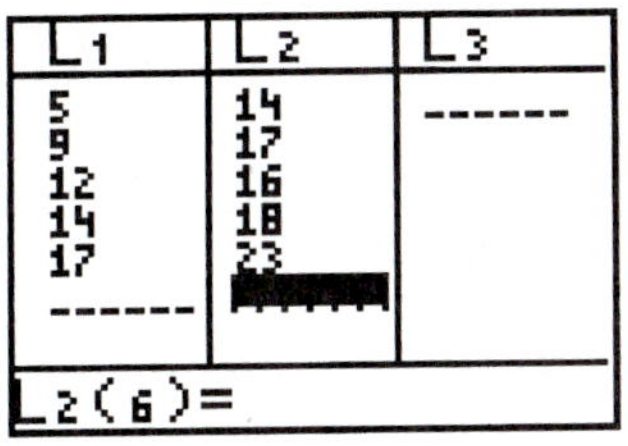

Figure 4.131

On a TI-85:

- *Put the calculator into statistics mode* by pressing STAT.
- *Set the calculator up for entering the data* from Example 2 by pressing EDIT (i.e., F2).
- The calculator will respond by automatically naming the list of x-coordinates "xStat" and the list of y-coordinates "yStat," as shown in Figure 4.132. (You can rename these lists by typing over the calculator-selected names.) Press ENTER twice to indicate that these two lists are named to your satisfaction.
- Enter 5 for x_1, 14 for y_1, 9 for x_2, etc.

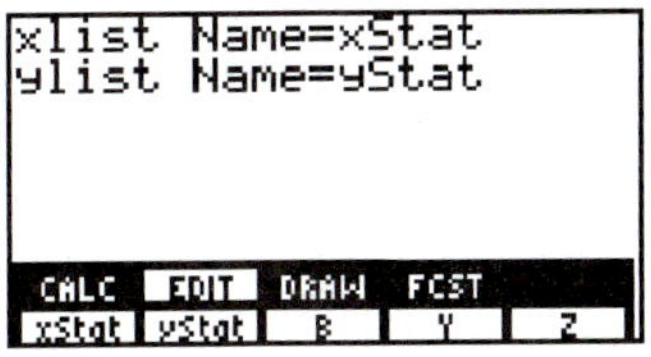

Figure 4.132

On a Casio:

- *Put the calculator into statistics mode* by pressing MENU, highlighting STAT, and pressing EXE.
- Use the arrow buttons and the EXE button to enter the x-coordinates from Example 2 in List 1 and the corresponding y-coordinates in List 2. When you are done, your screen should look like Figure 4.133. If data already appears in a list and you want to erase it, use the arrow keys to move to that list, press F6 and then F4 (i.e., DEL-A, which stands for "delete all") and then F1 (i.e., YES).

Figure 4.133
A Casio's list screen

Finding the Equation of the Line of Best Fit

Once the data have been entered, it's easy to find the equation. Recall from Example 2 that the equation is $y = 0.64319248x + 10.2676056$.

On a TI-80:

- Press STAT, scroll to the right and select the "CALC" menu, and select "LinReg(ax + b)" from the "CALC" menu.
- When "LinReg(ax + b)" appears on the screen, press 2nd L_1 , 2nd L_2 ENTER.
- The slope (labeled "a" rather than "m"), the *y*-intercept, and the correlation coefficient will appear on the screen.

On a TI-81:

- Press 2nd STAT and select "LinReg" from the "CALC" menu.
- When "LinReg" appears on the screen, press ENTER.
- The slope (confusingly labeled "b" rather than "m"), the *y*-intercept (confusingly labeled "a" rather than "b"), and the correlation coefficient (labeled "r") will appear on the screen.

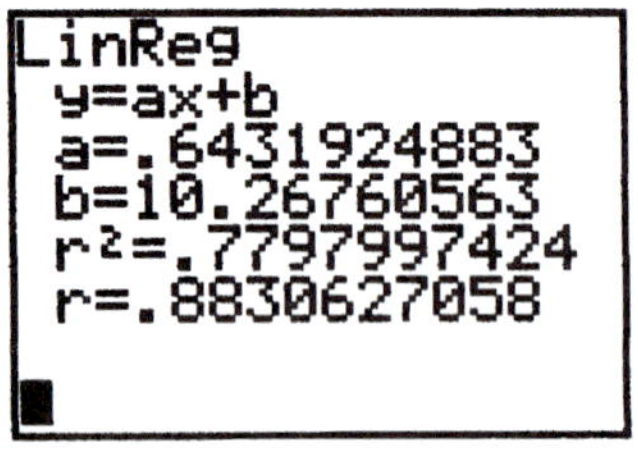

Figure 4.134
A TI-83's equation of the line of best fit

On a TI-82/83:

- Press STAT, scroll to the right and select the "CALC" menu, and select "LinReg(ax + b)" from the "CALC" menu.
- When "LinReg(ax + b)" appears on the screen, press ENTER.
- The slope (labeled "a" rather than "m"), the *y*-intercept, and the correlation coefficient will appear on the screen. (See Figure 4.134.)

Note: The TI-83 does not display the correlation coefficient unless you tell it to. To do so, press 2nd CATALOG, scroll down and select "DiagnosticOn," and press ENTER.

On a TI-85:

- Press CALC (i.e., 2nd M1).
- The calculator will respond with the list names you selected above. Press ENTER twice to indicate that these two names are correct, or change them if they are incorrect.
- Select linear regression by pressing LINR (i.e., F2).
- The slope (confusingly labeled "b" rather than "m"), the *y*-intercept (confusingly labeled "a" rather than "b"), the correlation coefficient (labeled "corr"), and the number of data points (labeled "n") will appear on the screen.

On a TI-86:

- Press EXIT.
- Press 2nd STAT and select "CALC" by pressing F1.
- Select linear regression by pressing LinR (i.e., F3).
- When "LinR" appears on the screen, press ENTER.
- The slope (confusingly labeled "b" rather than "m"), the *y*-intercept (confusingly labeled "a" rather than "b"), the correlation coefficient (labeled "corr"), and the number of data points (labeled "n") will appear on the screen. (Use the ↓ button to see it all.)

On a Casio:

- Press CALC (i.e., F2) to calculate the equation.
- Press REG (i.e., F3); "REG" stands for "regression."
- Press X (i.e., F1) for linear regression. (See Figure 4.135.)

Figure 4.135
A Casio's equation of the line of best fit

Drawing a Scatter Diagram and the Line of Best Fit

Once the equation has been found, the line can be graphed.

On a TI-80:

- *Quit the statistics mode* by pressing 2nd QUIT.
- Press WINDOW and enter appropriate values. For this problem, let x and y range from 0 to 25. Leave the "WINDOW" screen by pressing 2nd QUIT.
- *Set the calculator up to draw a scatter diagram* by pressing 2nd STAT PLOT and selecting "Plot 1." Turn the plot on and select the scatter icon.
- Tell the calculator to put the data entered in list L_1 on the x-axis by selecting "L_1" for "XL," and to put the data entered in list L_2 on the y-axis by selecting "L_2" for "YL."
- *Obtain a scatter diagram* by pressing GRAPH.
- If you don't want the data points displayed on the line of best fit, press 2nd STAT PLOT and turn off plot 1.
- *Quit the statistics mode* by pressing 2nd QUIT.
- *Enter the equation of the line of best fit* by pressing Y= VARS, selecting "Statistics," scrolling to the right to select the "EQ" menu, and selecting "REGEQ" (for regression equation).
- *Graph the line of best fit* by pressing GRAPH.
- Press TRACE to read off the data points as well as points on the line of best fit. Use the up and down arrows to switch between them.

On a TI-81:

- *Quit the statistics mode* by pressing 2nd QUIT.
- Press RANGE and enter appropriate values. For this problem, let x and y range from 0 to 25. Leave the "RANGE" screen by pressing 2nd QUIT.
- Press Y= and clear any functions that may appear.
- *Enter the equation of the line of best fit* for Y_1 by pressing VARS, selecting "LR" (for linear regression), and selecting "RegEQ" (for regression equation).
- Press GRAPH to draw the line of best fit.
- *Draw a scatter diagram* by pressing 2nd STAT, scrolling to the right and selecting the "DRAW" menu, and selecting "Scatter" from the "DRAW" menu. When "Scatter" appears on the screen, press ENTER. (Unfortunately, the data points tend to be hard to see.)
- Press 2nd DRAW and select "ClrDraw," or else the scatter diagram will appear on future graphs.

On a TI-82/83:

- Press Y=, and clear any functions that may appear.
- Set the calculator up to draw a scatter diagram by pressing 2nd STAT PLOT and selecting "Plot 1." Turn the plot on and select the scatter icon.
- Tell the calculator to put the data entered in list L_1 on the x-axis by selecting "L_1" for "Xlist," and to put the data entered in list L_2 on the y-axis by selecting "L_2" for "Ylist."
- Automatically set the range of the window and obtain a scatter diagram by pressing ZOOM and selecting option 9: "ZoomStat."
- If you don't want the data points displayed on the line of best fit, press 2nd STAT PLOT and turn off plot 1.
- Quit the statistics mode by pressing 2nd QUIT.
- Enter the equation of the line of best fit by pressing Y= VARS, selecting "Statistics," scrolling to the right to select the "EQ" menu, and selecting "RegEQ" (for regression equation).
- Automatically set the range of the window and obtain a scatter diagram by pressing ZOOM and selecting option 9: "ZoomStat." (See Figure 4.136.)
- Press TRACE to read off the data points as well as points on the line of best fit. Use the up and down arrows to switch between data points and points on the line of best fit. Use the left and right arrows to move left and right on the graph.

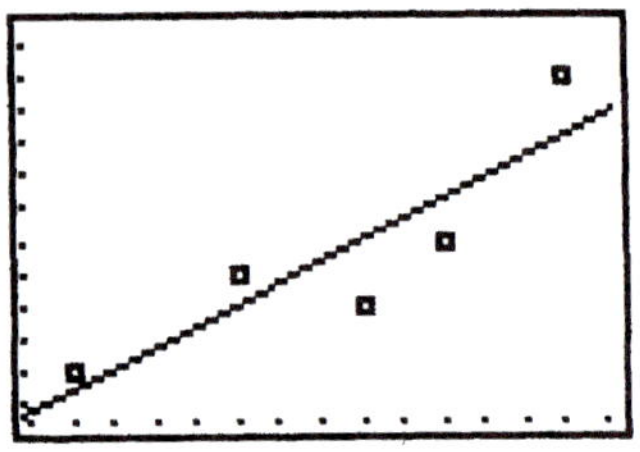

Figure 4.136
A TI-83's graph of the line of best fit and the scatter diagram

On a TI-85/86:

- Quit the statistics mode by pressing 2nd QUIT.
- Put the calculator into graphing mode, press y(x)=, and clear any functions that may appear. Press RANGE or WIND and enter appropriate values. For this problem, let x and y range from 0 to 25. Quit the graphing mode by pressing 2nd QUIT.
- Put the calculator into statistics mode by pressing STAT (TI-86: 2nd STAT).
- Press DRAW. If this immediately produces a drawing, clear it by pressing CLDRW. (You may need to press MORE to find CLDRW.)
- Press either SCAT (for a scatter diagram) or DRREG (for the regression line) or both. (You may need to press MORE to find DRREG.) If the button labels obscure the line, press EXIT once.
- Press CLDRW, or else the scatter diagram will appear on future graphs.

On a Casio:

- Press SHIFT QUIT to return to List 1 and List 2.
- Press GRPH (i.e., F1).
- Press SET (i.e., F6).
- Make the resulting screen, which is labeled "StatGraph1," read as follows:

Graph Type	:Scatter
Xlist	:List1
Ylist	:List2
Frequency	:1

To make the screen read as described above,

- Use the down arrow button to scroll down to "Graph Type."
- Press Scat (i.e., F1).
- In a similar manner, change "Xlist," "Ylist," and "Frequency" if necessary.

- Press SHIFT QUIT to return to List1 and List2.
- Press GRPH (i.e., F1).
- Press GPH1 (i.e., F1) to obtain the scatter diagram.
- Press X (i.e., F1).
- Press DRAW (i.e., F6), and the calculator will display the scatter diagram as well as the line of best fit. (See Figure 4.137.)

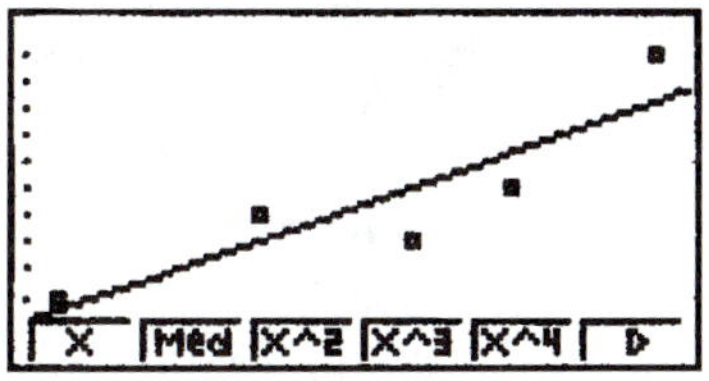

Figure 4.137
A Casio's graph of the line of best fit and the scatter diagram

Chapter 4 Review

TERMS

bell-shaped curve
categorical data
coefficient of linear correlation
continuous variable
data point
density
descriptive statistics
deviation
discrete variable
error of the estimate
extreme value
frequency
frequency distribution
histogram
inferential statistics
level of confidence
linear regression
linear trend
line of best fit
margin of error (MOE)
mathematical model
mean
measures of central tendency
measures of dispersion
median
mode
normal distribution
outlier
pie chart
population
population proportion
relative frequency
relative frequency density
sample
sample proportion
standard deviation
standard normal distribution
statistics
summation notation
variance
z-number

REVIEW EXERCISES

1. Find (a) the mean, (b) the median, (c) the mode, and (d) the standard deviation of the following set of raw data:

   5 8 10 4 8 10 6 8 7 5

2. To study the composition of families in Winslow, Arizona, 40 randomly selected married couples were surveyed to determine the number of children in each family. The following results were obtained:

   3 1 0 4 1 3 2 2 0 2 0 2 2 1
   4 3 1 1 3 4 2 1 3 0 1 0 2 5
   1 2 3 0 0 1 2 3 1 2 0 2

   a. Organize the given data by creating a frequency distribution.
   b. Find the mean number of children per family.
   c. Find the median number of children per family.
   d. Find the mode number of children per family.
   e. Find the standard deviation of the number of children per family.
   f. Construct a histogram using single-valued classes of data.

3. The frequency distribution in Figure 4.138 lists the number of hours per day that a randomly selected sample of teenagers spent watching television. Where possible, determine what percent of the

teenagers spent the following number of hours watching television.

a. less than 4 hours
b. not less than 6 hours
c. at least 2 hours
d. less than 2 hours
e. at least 4 hours but less than 8 hours
f. more than 3.5 hours

x = Hours per Day	Frequency
$0 \le x < 2$	23
$2 \le x < 4$	45
$4 \le x < 6$	53
$6 \le x < 8$	31
$8 \le x \le 10$	17

Figure 4.138

4. To study the efficiency of its new oil-changing system, a local service station monitored the amount of time it took to change the oil in customers' cars. The frequency distribution in Figure 4.139 summarizes the findings.

a. Find the mean number of minutes to change the oil in a car.
b. Find the standard deviation of the amount of time to change the oil in a car.
c. Construct a histogram to represent the data.

x = Time (in minutes)	Number of Customers
$3 \le x < 6$	18
$6 \le x < 9$	42
$9 \le x < 12$	64
$12 \le x < 15$	35
$15 \le x \le 18$	12

Figure 4.139

5. If your scores on the first four exams (in this class) are 74, 65, 85, and 76, what score do you need on the next exam for your overall mean to be at least 80?

6. The mean salary of 12 men is \$37,000, and the mean salary of 8 women is \$28,000. Find the mean salary of all 20 people.

7. Timo and Henke golfed five times during their vacation. Their scores are given in Figure 4.140.

a. Find the mean score of each golfer. Who has the lowest mean?
b. Find the standard deviation of each golfer's scores.
c. Who is the more consistent golfer? Why?

Timo	103	99	107	93	92
Henke	101	92	83	96	111

Figure 4.140

8. Suzanne surveyed the prices for a quart of a certain brand of motor oil. The sample data, in dollars per quart, are summarized in Figure 4.141.

a. Find the mean and standard deviation of the prices.
b. What percentage of the data lies within one standard deviation of the mean?
c. What percentage of the data lies within two standard deviations of the mean?
d. What percentage of the data lies within three standard deviations of the mean?

Price per Quart	Number of Stores
0.99	2
1.09	3
1.19	7
1.29	10
1.39	14
1.49	4

Figure 4.141

9. Classify the following types of data as discrete, continuous, or neither.

a. weights of motorcycles
b. colors of motorcycles
c. number of motorcycles
d. ethnic background of students
e. number of students
f. amounts of time spent studying

10. What percentage of the standard normal z-distribution lies in the following intervals?

a. between $z = 0$ and $z = 1.75$
b. between $z = -1.75$ and $z = 0$
c. between $z = -1.75$ and $z = 1.75$

11. A large group of data is normally distributed with mean 78 and standard deviation 7.
   a. Find the intervals that represent one, two, and three standard deviations of the mean.
   b. What percentage of the data lies in each interval in part (a)?
   c. Draw a sketch of the bell curve.
12. The time it takes a latex paint to dry is normally distributed. If the mean is $3\frac{1}{2}$ hours with a standard deviation of 45 minutes, find the probability that the drying time will be as follows.
   a. less than 2 hours 15 minutes
   b. between 3 and 4 hours

   HINT: Convert everything to hours (or to minutes).
13. All incoming freshmen at a major university are given a diagnostic mathematics exam. The scores are normally distributed with a mean of 420 and a standard deviation of 45. If the student scores less than a certain score, he or she will have to take a review course. Find the cutoff score at which 34% of the students would have to take the review course.
14. Find the specified z-number.
   a. $z_{0.4441}$ b. $z_{0.4500}$ c. $z_{0.1950}$ d. $z_{0.4975}$
15. A survey asked, "Do you think that the president is doing a good job?" Of the 1,200 Americans surveyed, 800 responded yes. For each of the following levels of confidence, find the sample proportion and the margin of error associated with the poll.
   a. a 90% level of confidence
   b. a 95% level of confidence
16. A survey asked, "Do you support capital punishment?" Of the 1,000 Americans surveyed, 750 responded no. For each of the following levels of confidence, find the sample proportion and the margin of error associated with the poll.
   a. an 80% level of confidence
   b. a 98% level of confidence
17. A sample consisting of 580 men and 970 women was asked various questions pertaining to international affairs. For a 95% level of confidence, find the margin of error associated with the following samples.
   a. the male sample
   b. the female sample
   c. the combined sample
18. A poll pertaining to environmental concerns had the following footnote: "Based on a sample of 1,098 adults, the margin of error is plus or minus 2 percentage points." Find the level of confidence of the poll.
19. A set of $n = 5$ ordered pairs has the following sums:

$$\Sigma x = 66 \quad \Sigma x^2 = 1{,}094 \quad \Sigma y = 273$$
$$\Sigma y^2 = 16{,}911 \quad \Sigma xy = 4{,}272$$

   a. Find the line of best fit.
   b. Predict the value of y when $x = 16$.
   c. Predict the value of x when $y = 57$.
   d. Find the coefficient of linear correlation.
   e. Are the predictions in parts (b) and (c) reliable? Why or why not?
20. Given the ordered pairs (5, 38), (10, 30), (20, 33), (21, 25), (24, 18), and (30, 20),
   a. Plot the ordered pairs. Do the ordered pairs exhibit a linear trend?
   b. Find the line of best fit.
   c. Predict the value of y when $x = 15$.
   d. Plot the given ordered pairs and sketch the graph of the line of best fit on the same coordinate system.
   e. Find the coefficient of linear correlation.
   f. Is the prediction in part (c) reliable? Why or why not?
21. The value (in billions of dollars) of agricultural exports and imports in the United States is given in Figure 4.142.

Year	Agricultural Exports (billion dollars)	Agricultural Imports (billion dollars)
1990	40.4	22.7
1991	37.8	22.7
1992	42.6	24.5
1993	42.9	24.6
1994	44.0	26.6
1995	54.7	29.9
1996	59.9	32.6
1997	57.4	35.8
1998	53.7	37.0

FIGURE 4.142
Agricultural exports and imports (billion dollars)
Source: U.S. Department of Agriculture

   a. Letting x = the value of agricultural exports and y = the value of agricultural imports, plot the data. Do the data exhibit a linear trend?
   b. Find the line of best fit.

c. Predict the value of agricultural imports when the value of agricultural exports is 50.0 billion dollars.
d. Predict the value of agricultural exports when the value of agricultural imports is 35.0 billion dollars.
e. Find the coefficient of linear correlation.
f. Are the predictions in parts (c) and (d) reliable? Why or why not?

Answer the following questions using complete sentences and your own words.

Concept Questions

22. What are the three measures of central tendency? Briefly explain the meaning of each.
23. What does standard deviation measure?
24. What is a normal distribution? What is the standard normal distribution?
25. What is a margin of error? How is it related to a sample proportion?
26. How do you measure the strength of a linear trend?

History Questions

27. What role did the following people play in the development of statistics?
    - George Gallup
    - Carl Friedrich Gauss

Finance

5

Today, almost everyone borrows money, whether in the form of a student loan, a car loan, a credit card loan, or a home loan. Almost everyone saves money, through either a savings account at a bank, a money market account, or some form of annuity. To make educated decisions about such matters, you should understand the mathematics of finance.

Webster's Dictionary defines **finance** as "the system that includes the circulation of money, the granting of credit, the making of investments, and the provision of banking facilities." In this chapter, we investigate the mathematics of two aspects of finance: loans and nonspeculative investments—such as savings accounts, money market accounts, certificates of deposit, and annuities. (Speculative investments are investments that involve some risk, such as stocks and real estate.)

Loans and investments are very similar financial transactions; both involve the flow of money from one party to another, the return of the money to its source, and the payment of a fee to the source for the use of the money. When you make a deposit in your savings account, you view the transaction as an investment, but the bank views it as a loan; you are lending the bank your money, which they will lend to another customer, perhaps to buy a house. When you borrow money to buy a car, you view the transaction as a loan, but the bank views it as an investment; the bank is investing its money in you in order to make a profit.

In this chapter, we explore some of the different forms of loans and investments, along with the various ways in which they are calculated, so that you can make the best choice when you have to make a financial decision. We investigate the different types of automobile loans available and discuss how home loans work. We also show how you can accumulate an incredibly large amount of money by saving only $50 a month.

5.1 Simple Interest

When a lender lends money to a borrower, the lender usually expects to be repaid more than was lent. The original amount of money lent is called the **principal,** or **present value;** if some of the principal has been paid back, then the portion that remains unpaid is called the **outstanding principal,** or **balance.** The total amount of money the lender is paid back is called the **future value;** this includes the original amount lent and the lender's profit, or **interest.** How much interest will be paid depends on the **interest rate** (usually expressed as a percent per year); the **term,** or the length of time before the debt is repaid; and how the interest is calculated.

Many short-term loans and some other forms of investment are calculated at simple interest. **Simple interest** means that the amount of interest is computed as a percent per year of the principal.

Simple Interest Formula

The simple interest I on a principal P at an annual rate of interest r for t years is

$$I = Prt$$

The future value FV is the total of the principal and the interest; therefore,

$$\begin{aligned} FV &= P + I \\ &= P + Prt \\ &= P(1 + rt) \end{aligned}$$

Simple Interest Future Value Formula

The future value FV of a principal P at an annual rate of interest r for t years is

$$FV = P(1 + rt)$$

One of the most common uses of simple interest is a short-term (i.e., a year or less) simple interest loan that requires a single lump sum payment at the end of the term. Businesses routinely obtain these loans to purchase equipment or inventory, to pay operating expenses, or to pay taxes.

EXAMPLE 1 South Face, a mountaineering and camping store, borrowed \$340,000 at 5.1% for 120 days to purchase inventory for the Christmas season. Find the interest on the loan.

Solution We are given $P = 340{,}000$ and $r = 5.1\% = 0.051$. Using dimensional analysis (as discussed in Appendix E) to convert days to years, we get

$$t = 120 \text{ days} = 120 \cancel{\text{days}} \cdot \frac{1 \text{ year}}{365 \cancel{\text{days}}} = \frac{120}{365} \text{ years}$$

Using the Simple Interest Formula, we have

$$\begin{aligned} I &= Prt \\ &= 340{,}000 \cdot 0.051 \cdot \frac{120}{365} \end{aligned}$$

$$= 5700.8219\ldots$$
$$\approx \$5{,}700.82$$

This means that South Face has agreed to pay its lender \$340,000 plus \$5,700.82 interest at the end of 120 days. ■

There is an important distinction between the variables *FV, I, P, r,* and *t*: *FV, I,* and *P* measure amounts of money, while *r* and *t* do not. For example, consider the interest rate *r* and the interest *I*. People frequently confuse these two. However, the interest rate *r* is a percent and the interest *I* is an amount of money; in Example 1, the interest rate is 5.1% and the interest is \$5,700.82. To emphasize this distinction, we will always use capital letters for variables that measure amounts of money and lowercase letters for other variables. Using this notation may help you avoid substituting 5.1% = 0.051 for *I* when it should be substituted for *r*.

In Example 1, we naturally used 365 days per year, but some institutions traditionally count a year as 360 days and a month as 30 days (especially if that tradition works in their favor). This is a holdover from the days before calculators and computers; the numbers were simply easier to work with. Also, we used normal round-off rules to round 5700.8219. . . to 5,700.82; some institutions round off some interest calculations in their favor. In this book, we will count a year as 365 days and use normal round-off rules (unless stated otherwise).

The issue of how a financial institution rounds off its interest calculations might seem unimportant; after all, we're talking about a difference of a fraction of a penny. However, consider one classic form of computer crime: the round-down fraud, performed on a computer system that processes a large number of accounts. Frequently, such systems use the normal round-off rules in their calculations and keep track of the difference between the theoretical account balance if no rounding off is done and the actual account balance with rounding off. Whenever that difference reaches or exceeds 1¢, the extra penny is deposited in (or withdrawn from) the account. A fraudulent computer programmer can write the program so that the extra penny is deposited in *his* or *her* account. This fraud is difficult to detect because the accounts appear to be balanced. While each individual gain is small, the total gain can be quite large if a large number of accounts is processed on a regular basis.

EXAMPLE 2 To pay its taxes, Espree Clothing borrowed \$185,000 at 7.3% for four months. Find the future value of the loan.

Solution We are given $P = 185{,}000$ and $r = 7.3\% = 0.073$. Using dimensional analysis to convert months to years, we get

$$t = 4 \text{ months} = 4 \text{ months} \cdot \frac{1 \text{ year}}{12 \text{ months}} = \frac{4}{12} \text{ years.}$$

Using the Simple Interest Future Value Formula, we have

$$FV = P(1 + rt)$$
$$= 185{,}000\left(1 + .073 \cdot \frac{4}{12}\right)$$
$$= 189{,}501.6666\ldots$$
$$\approx \$189{,}501.67$$

This means that Espree has agreed to pay their lender \$189,501.67 at the end of four months.

185000 [×] [(] 1 [+] .073 [×] 4 [÷] 12 [)] [=]

With a graphing calculator, press [ENTER] instead of [=].

✔ Notice that the answer is somewhat higher than the original $185,000, as it should be.

Technically, Example 2 should specify which months are involved, because months vary from 28 to 31 days in length. The four-month time span from January through April does not have the same number of days as the four-month time span from March through June and thus would generate a slightly different amount of interest. The calculation in Example 2 gives a very good approximation of the future value after *any* four-month time span, but it is *only* an approximation.

A written contract signed by the lender and the borrower is called a **loan agreement** or a **note.** The **maturity value** of the note (or just the *value* of the note) refers to the note's future value. Thus, the value of the note in Example 2 was $189,501.67.

EXAMPLE 3 Find the amount of money that must be invested now at a $5\frac{3}{4}\%$ interest rate so that it will be worth $1,000 in two years.

Solution We are asked to find the present value, or principal *P*, that will generate a future value of $1,000.

We know that $FV = 1{,}000$, $r = 5\frac{3}{4}\% = 0.0575$, and $t = 2$. Using the Future Value Formula, we have

$$FV = P(1 + rt)$$
$$1{,}000 = P(1 + .0575 \cdot 2)$$
$$P = \frac{1{,}000}{1 + 0.0575 \cdot 2} \qquad \text{dividing}$$
$$= 896.86099\ldots$$
$$\approx \$896.86$$

1000 [÷] [(] 1 [+] .0575 [×] 2 [)] [=]

With a graphing calculator, press [ENTER] instead of [=].

✔ Notice that the answer is somewhat smaller than $1000, as it should be.

National Debt

For every year from 1970 through 1998 (and for most years since 1931), the U.S. federal budget called for deficit spending, that is, spending more money than is received. Deficit spending grew tremendously during the Reagan and Bush (senior) presidencies, but it ceased during the Clinton years. The proposed 2003 budget (see Figures 5.1(a) and (b)) calls for deficit spending of \$109 billion. The total national debt held by the public in mid-2002 was \$6,151 billion; this amounts to a debt of more than \$21,000 per person in the United States. President Bush's 2003 proposed budget calls for spending \$180 billion for interest on the national debt.

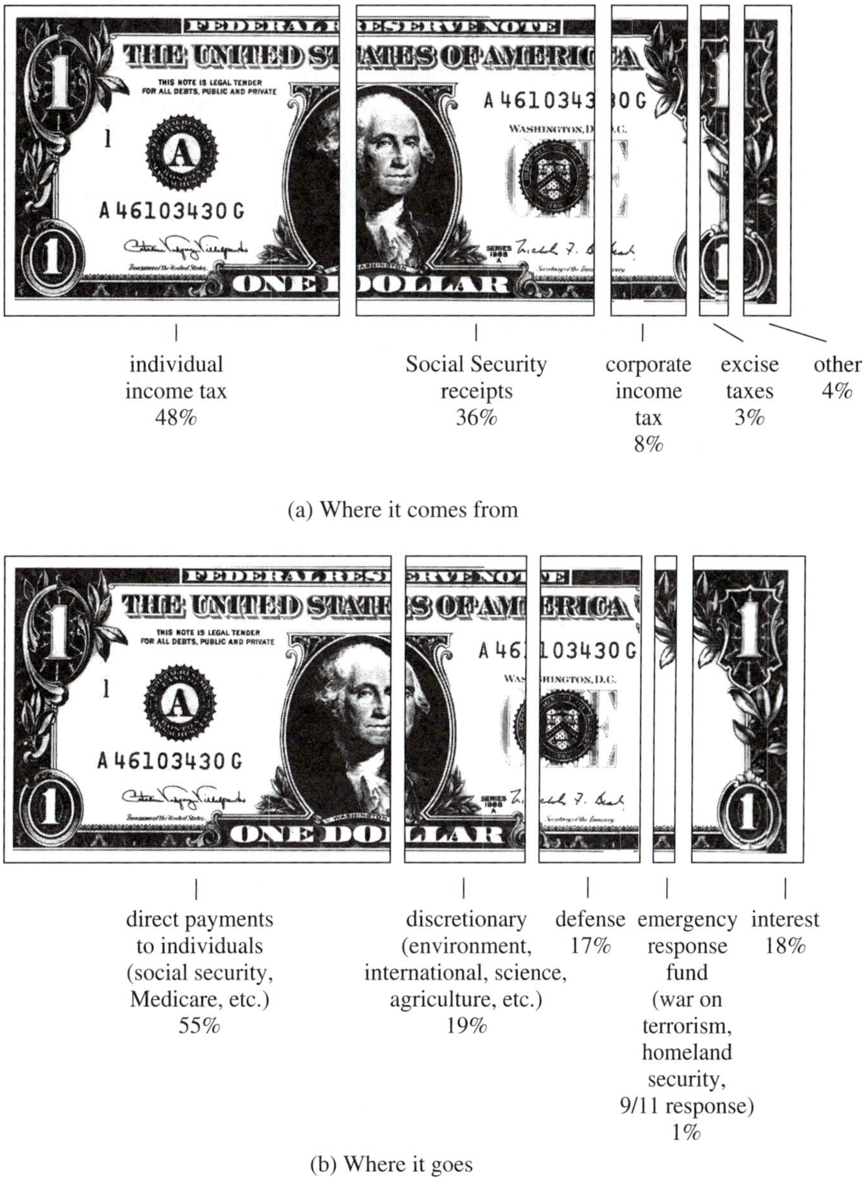

Figure 5.1
The federal government dollar (2003 estimate)

Office of Management and Budget, *Budget of the United States Government, FY2003*

EXAMPLE 4 Find the simple interest rate that was paid on the national debt in 2003.

Solution From the information given on page 309, we know that $I = \$180$ billion, $P = \$6{,}151$ billion, and $t = 1$ year. Using the Simple Interest Formula, we have

$$I = Prt$$
$$\$180 \text{ billion} = \$6{,}151 \text{ billion} \cdot r \cdot 1$$
$$\text{r} = \$\ 180 \text{ billion}/\$\ 6{,}151 \text{ billion} = 0.02926\ldots \approx 2.9\%$$

■

Add-on Interest

Purchases made at car dealerships, appliance stores, and furniture stores can be financed through the store itself. Frequently, this type of loan involves **add-on interest,** which consists of a simple interest charge on the loan amount, distributed equally over each payment.

EXAMPLE 5 Flo and Eddie Turkel purchase $1,300 worth of kitchen appliances at Link's Appliance Mart. They put $200 down and agree to pay the balance at a 10% add-on rate for two years. Find their monthly payment.

Solution We are given $P = \text{loan amount} = 1{,}300 - 200 = 1{,}100$, $r = 10\% = 0.10$, and $t = 2$ years. The total amount due is the future value.

$$FV = P(1 + rt)$$
$$= 1{,}100\,(1 + 0.10 \cdot 2)$$
$$= 1{,}320$$

The total amount due is spread out over 24 monthly payments; therefore,

$$\text{monthly payment} = \frac{1{,}320}{24} = \$55$$

■

Credit Card Finance Charge

Credit cards have become part of the American way of life. Purchases made with a credit card are subject to a finance charge, but there is frequently a grace period, and no finance charge is assessed if full payment is received by the payment due date. One of the most common methods of calculating credit card interest is the **average daily balance** method. To find the average daily balance, the balance owed on the account is found for each day in the billing period, and the results are averaged. Simple interest is then charged on the result.

EXAMPLE 6 The activity on Eric Johnson's Visa account for one billing is shown below. Find the average daily balance and the finance charge if the billing period is October 15 through November 14, the previous balance was $346.57, and the annual interest rate is 21%.

October 21	payment	$50.00
October 23	restaurant	$42.14
November 7	clothing	$18.55

HISTORICAL NOTE

Credit Card History

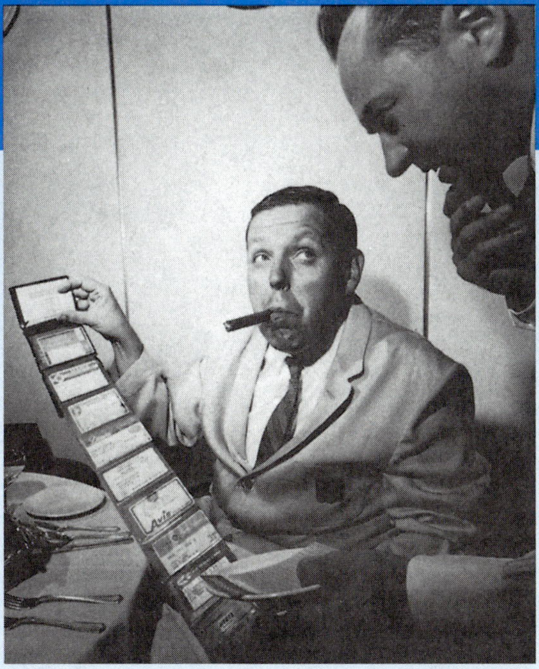
AP/Wide World Photos

Credit cards were first used in the United States in 1915, when Western Union issued a metal card to some of its regular customers. Holders of these cards were allowed to defer their payments and were assured of prompt and courteous service. Shortly thereafter, several gasoline companies, hotels, department stores, and railroads issued credit cards to their preferred customers.

The use of credit cards virtually ceased during World War II owing to government restrictions on credit. In 1950, a New York lawyer established the Diners' Club after being embarrassed when he lacked sufficient cash to pay a dinner bill. A year later, the club had billed more than $1 million; Carte Blanche and American Express soon followed. These "travel and entertainment" cards were and are attractive to the public because they provide a convenient means of paying restaurant, hotel, and airline bills; they eliminate both the possibility of a nonlocal check being refused and the need to carry a large amount of cash.

The first bank card was issued in 1951 by Franklin National Bank in New York. Within a few years, 100 other banks had followed suit. Because these bank cards were issued by individual banks, they were accepted for use only in relatively small geographical areas. In 1965, the California-based Bank of America began licensing other banks (both in the United States and abroad) to issue BankAmericards. Other banks formed similar groups, which allowed customers to use the card out of state.

In 1970, the Bank of America transferred administration of its bank card program to a new company owned by the banks that issued the card. In 1977, the card was renamed Visa. MasterCard evolved from the merger of several regional bank card associations. In 1998 there were over 800 million Visa cardholders, and nearly $1.4 trillion in products and services were purchased using Visa cards.

The first Diners' Club card

Diner's Club Corp.

Solution To find the average daily balance, we need the balance for each day in the billing period and the number of days at that balance, as shown in Figure 5.2. The average daily balance is then the weighted average of each daily balance, with each balance weighted to reflect the number of days at that balance.

Time Interval	Days	Daily Balance
October 15–20	6	$346.57
October 21–22	2	$346.57 − $50.00 = $296.57
October 23–November 6	15	$296.57 + $42.14 = $338.71
November 7–14	8	$338.71 + $18.55 = $357.26

Figure 5.2

$$\text{average daily balance} = \frac{6 \cdot 346.57 + 2 \cdot 296.57 + 15 \cdot 338.71 + 8 \cdot 357.26}{6 + 2 + 15 + 8}$$
$$= 342.29967\ldots$$
$$\approx \$342.30$$

(6 × 346.57 + 2 × 296.57 + 15 × 338.71 + 8 × 357.26) ÷ (6 + 2 + 15 + 8) =

With a graphing calculator, type ENTER instead of =.

The finance charge consists of simple interest on the average daily balance. We know that $P = 342.29967\ldots$, $r = 21\% = 0.21$, and $t = 31$ days $= \frac{31}{365}$ year. Using the Simple Interest Formula, we have the following result:

$$I = Prt$$
$$= 342.29967 \cdot 0.21 \cdot \frac{31}{365}$$
$$= 6.1051258$$
$$\approx \$6.11$$

■

5.1 Exercises

In Exercises 1–6, (a) find and (b) interpret the simple interest of the given loan amounts.

1. loan amount of \$2,000 at 8% for 3 years
2. loan amount of \$35,037 at 6% for 2 years
3. loan amount of \$420 at $6\frac{3}{4}$% for 9 months
4. loan amount of \$8,950 at $9\frac{1}{2}$% for 10 months
5. loan amount of \$1,410 at $12\frac{1}{4}$% for 325 days
6. loan amount of \$5,682 at $11\frac{3}{4}$% for 278 days

In Exercises 7–10, (a) find and (b) interpret the future value of the given present values.

7. present value of \$3,670 deposited at $2\frac{3}{4}$% for 7 years
8. present value of \$4,719 deposited at 4.1% for 11 years
9. present value of \$12,430 deposited at $5\frac{7}{8}$% for 2 years 3 months
10. present value of \$172.39 deposited at 6% for 3 years 7 months

In Exercises 11–16, (a) find and (b) interpret the maturity value of the given loan amounts.

11. \$1,400 borrowed at $7\frac{1}{8}$% for 9 months
12. \$3,250 borrowed at $8\frac{1}{2}$% for 1 year 1 month
13. \$5,900 borrowed at $14\frac{1}{2}$% for 112 days
14. \$2,720 borrowed at $12\frac{3}{4}$% for 275 days
15. \$16,500 borrowed at $11\frac{7}{8}$% from April 1 to July 10 of the same year
16. \$2,234 borrowed at $12\frac{1}{8}$% from March 10 to December 20 of the same year

In Exercises 17–22, (a) find and (b) interpret the present value of the given future value.

17. future value of \$8,600 at $9\frac{1}{2}$% simple interest for 3 years
18. future value of \$420 at $5\frac{1}{2}$% simple interest for 2 years
19. future value of \$1,112 at $3\frac{5}{8}$% simple interest for 1 year 11 months
20. future value of \$5,750 at $4\frac{7}{8}$% simple interest for 2 years 2 months
21. future value of \$1,311 at $6\frac{1}{2}$% simple interest for 317 days
22. future value of \$4,200 at $6\frac{3}{4}$% simple interest for 509 days
23. How much must be deposited now at 6% interest so that in 1 year and 8 months an account will contain \$3,000?

24. How much must be deposited now at $5\frac{7}{8}\%$ interest so that in 2 years and 7 months an account will contain \$1,900?

25. Fred Murtz just received his income tax refund of \$1,312.82, and he needs \$1,615 to buy a new stereo system. If his money can earn $6\frac{7}{8}\%$ interest, how long must he invest his tax refund?

26. Sam Spade inherited \$7,000. He wants to buy a used car, but the type of car he wants typically sells for around \$8,000. If his money can earn $6\frac{1}{2}\%$ interest, how long must he invest his money?

27. Alice Cohen buys a two-year-old Honda from a car dealer for \$9,000. She puts \$500 down and finances the rest through the dealer at 13% add-on interest. If she agrees to make 36 monthly payments, find the size of each payment.

28. Sven Lundgren buys a three-year-old Chevrolet Celebrity from a car dealer for \$4,600. He puts \$300 down and finances the rest through the dealer at 12.5% add-on interest. If he agrees to make 24 monthly payments, find the size of each payment.

29. Ray and Teresa Martinez buy a bedroom set at Fowler's Furniture for \$3,700. They put \$500 down and finance the rest through the store at 9.8% add-on interest. If they agree to make 36 monthly payments, find the size of each payment.

30. Helen and Dick Davis buy a refrigerator at Appliance Barn for \$1,200. They put \$100 down and finance the rest through the store at 11.6% add-on interest. If they agree to make 36 monthly payments, find the size of each payment.

31. The activity on Stuart Ratner's Visa account for one billing period is shown below. Find the average daily balance and the finance charge if the billing period is April 11 through May 10, the previous balance was \$126.38, and the annual interest rate is 18%.

April 15	payment	\$15.00
April 22	record store	\$25.52
May 1	clothing	\$32.18

32. The activity on Marny Zell's MasterCard account for one billing period is shown below. Find the average daily balance and the finance charge if the billing period is June 26 through July 25, the previous balance was \$396.68, and the annual interest rate is 19.5%.

June 30	payment	\$100.00
July 2	gasoline	\$ 36.19
July 10	restaurant	\$ 53.00

33. The activity on Denise Helling's Sears account for one billing period is shown here. Find the average daily balance and the finance charge if the billing period is March 1 through March 31, the previous balance was \$157.14, and the annual interest rate is 21%.

March 5	payment	\$25.00
March 17	tools	\$36.12

34. The activity on Charlie Wilson's Visa account for one billing period is shown below. Find the average daily balance and the finance charge if the billing period is November 11 through December 10, the previous balance was \$642.38, and the annual interest rate is 20%.

November 15	payment	\$150.00
November 28	office supplies	\$ 23.82
December 1	toy store	\$312.58

35. Donovan and Pam Hamilton bought a house from Edward Gurney for \$162,500. Typically, a purchaser will make a down payment to the seller of 10% to 20% of the purchase price and borrow the rest from a bank. However, the Hamiltons did not have sufficient savings for a 10% down payment, and Mr. Gurney was a motivated seller. In lieu of a 10% down payment, Mr. Gurney accepted a 5% down payment at the time of the sale and a promissory note from the Hamiltons for an additional 5%, due in four years. The note required the Hamiltons to make monthly interest payments to Mr. Gurney at 10% interest until the note expired. The Hamiltons obtained a loan from their bank for the remaining 90% of the purchase price. The bank in turn paid the sellers the remaining 90% of the purchase price, less a sales commission (6% of the purchase price) paid to the sellers' and the buyers' real estate agents.
   a. Find the Hamiltons' down payment.
   b. Find the amount that the Hamiltons borrowed from their bank.
   c. Find the amount that the Hamiltons borrowed from Mr. Gurney.
   d. Find the Hamiltons' monthly interest payment to Mr. Gurney.
   e. Find Mr. Gurney's total income from all aspects of the down payment (including the down payment, the amount borrowed under the promissory note, and the monthly payments required by the promissory note).
   f. Find Mr. Gurney's total income from the Hamiltons' bank.
   g. Find Mr. Gurney's total income from all aspects of the sale.

36. George and Peggy Fulwider bought a house from Sally Sinclair for \$233,500. Typically, a purchaser

will make a down payment to the seller of 10% to 20% of the purchase price and borrow the rest from a bank. However, the Fulwiders did not have sufficient savings for a 10% down payment, and Ms. Sinclair was a motivated seller. In lieu of a 10% down payment, Ms. Sinclair accepted a 5% down payment at the time of the sale and a promissory note from the Fulwiders for an additional 5%, due in four years. The note required the Fulwiders to make monthly interest payments to Ms. Sinclair at 10% interest until the note expired. The Fulwiders obtained a loan from their bank for the remaining 90% of the purchase price. The bank in turn paid the sellers the remaining 90% of the purchase price, less a sales commission (6% of the purchase price) paid to the sellers' and the buyers' real estate agents.

a. Find the Fulwiders' down payment.

b. Find the amount that the Fulwiders borrowed from their bank.

c. Find the amount that the Fulwiders borrowed from Ms. Sinclair.

d. Find the Fulwiders' monthly interest payment to Ms. Sinclair.

e. Find Ms. Sinclair's total income from all aspects of the down payment (including the down payment, the amount borrowed under the promissory note, and the monthly payments required by the promissory note).

f. Find Ms. Sinclair's income from the Fulwiders' bank.

g. Find Ms. Sinclair's total income from all aspects of the sale.

37. The Bushes bought a house from the Clintons for $389,400. In lieu of a 20% down payment, the Clintons accepted a 10% down payment at the time of the sale and a promissory note from the Bushes for an additional 10%, due in four years. The Bushes also agreed to make monthly interest payments to the Clintons at 11% interest until the note expires. The Bushes obtained a loan from their bank for the remaining 80% of the purchase price. The bank in turn paid the sellers the remaining 80% of the purchase price, less a sales commission (6% of the purchase price) paid to the sellers' and the buyers' real estate agents.

a. Find the Bushes' down payment.

b. Find the amount that the Bushes borrowed from their bank.

c. Find the amount that the Bushes borrowed from the Clintons.

d. Find the Bushes' monthly interest payment to the Clintons.

e. Find the Clintons' total income from all aspects of the down payment (including the down payment, the amount borrowed under the promissory note, and the monthly payments required by the promissory note).

f. Find the Clintons' income from the Bushes' bank.

g. Find the Clintons' total income from all aspects of the sale.

38. Sam Needham bought a house from Sheri Silva for $238,300. In lieu of a 20% down payment, Ms. Silva accepted a 10% down payment at the time of the sale and a promissory note from Mr. Needham for an additional 10%, due in four years. Mr. Needham also agreed to make monthly interest payments to Ms. Silva at 9% interest until the note expires. Mr. Needham obtained a loan from his bank for the remaining 80% of the purchase price. The bank in turn paid Ms. Silva the remaining 80% of the purchase price, less a sales commission (6% of the purchase price) paid to the sellers' and the buyers' real estate agents.

a. Find Mr. Needham's down payment.

b. Find the amount that Mr. Needham borrowed from his bank.

c. Find the amount that Mr. Needham borrowed from Ms. Silva.

d. Find Mr. Needham's monthly interest payment to Ms. Silva.

e. Find Ms. Silva's total income from all aspects of the down payment (including the down payment, the amount borrowed under the promissory note, and the monthly payments required by the promissory note).

f. Find Ms. Silva's income from Mr. Needham's bank.

g. Find Ms. Silva's total income from all aspects of the sale.

Answer the following questions using complete sentences and your own words.

Concept Questions

39. Could Exercises 1–22 all be done with the Simple Interest Formula? How? Could Exercises 1–22 all be done with the Simple Interest Future Value Formula? How? Why do we have both formulas?

40. Which is always higher, future value or present value? Why?

History Questions

41. What was the first credit card?
42. What was the first post–World War II credit card?
43. Who created the first post–World War II credit card?
44. What event prompted the creation of the first post–World War II credit card?
45. What was the first interstate bank card?

5.2 Compound Interest

Many forms of investment (including savings accounts) earn **compound interest,** in which case interest is periodically paid on the existing account balance, which includes both the original principal and previous interest payments. This form of interest results in significantly higher earnings over a long period of time. The **compounding period** (usually annually, semiannually, quarterly, monthly, or daily) refers to the frequency at which interest is computed and deposited. The effects of the compounding period, the interest rate, and the time interval can be much more dramatic with compound interest than with simple interest. It is important that you understand these effects in order to make wise financial decisions.

EXAMPLE 1 One thousand dollars is deposited in an account where it earns 8% interest compounded quarterly. Find the account balance (future value) after six months, using the Simple Interest Future Value Formula to compute the balance at the end of each compounding period.

Solution Because interest is compounded quarterly in this case, interest is computed and deposited every quarter of a year.

At the end of the first quarter, $P = 1{,}000$, $r = 8\% = 0.08$, and $t =$ one quarter or $\frac{1}{4}$ year.

$$\begin{aligned} FV &= P(1 + rt) \\ &= 1{,}000\left(1 + 0.08 \cdot \frac{1}{4}\right) \\ &= 1{,}000(1 + 0.02) \\ &= \$1{,}020 \end{aligned}$$

At the end of the second quarter, $P = 1{,}020$ (the new principal), $r = 0.08$, and $t = \frac{1}{4}$ (the length of time, in years, that the new principal has been in the account).

$$\begin{aligned} FV &= P(1 + rt) \\ &= 1{,}020\left(1 + 0.08 \cdot \frac{1}{4}\right) \\ &= 1{,}020(1 + 0.02) \\ &= \$1{,}040.40 \end{aligned}$$

At the end of six months, the account balance is $1,040.40. ■

This process would become tedious if we were asked to compute the balance after 20 years. Therefore, compound interest problems are usually solved with their own formula.

Notice that for each quarter's calculation in Example 1, we multiplied the annual rate of 8% (0.08) by the time one quarter ($\frac{1}{4}$ year) and got 2% (0.02). This 2% is the

quarterly rate (or, more generally, the **periodic rate**). A periodic rate is any rate that is prorated in this manner from an annual rate.

If i is the periodic interest rate, then the future value (the account balance in the case of an investment or the maturity value in the case of a loan) at the end of the first period is

$$FV = P(1 + i)$$

Because this is the account balance at the beginning of the second period, it becomes the new principal (or present value). Substituting the new $P(1 + i)$ for P, we have an account balance at the end of the second period of

$$\begin{aligned} FV &= [P(1 + i)] \cdot (1 + i) \\ &= P(1 + i)^2 \end{aligned}$$

$P(1 + i)^2$ becomes the account balance at the beginning of the third period, and the future value at the end of the third period is

$$\begin{aligned} FV &= [P(1 + i)^2] \cdot (1 + i) \quad \text{substituting } P(1 + i)^2 \text{ for } P \\ &= P(1 + i)^3 \end{aligned}$$

We can generalize this procedure by replacing the exponent with n to represent any number of periods. Thus, we get the following Compound Interest Formula.

Compound Interest Formula

At the end of n periods, the future value FV of an initial principal P subject to compound interest at a periodic interest rate i for n periods is

$$FV = P(1 + i)^n$$

Notice that we have maintained our variables tradition—i and n do not measure amounts of money, so they are not capital letters. We now have three interest-related variables:

- r, the annual interest rate (not an amount of money)
- i, the periodic interest rate (not an amount of money)
- I, the interest itself (an amount of money)

EXAMPLE 2 Recompute Example 1 using the Compound Interest Formula.

Solution Compounding quarterly, we have $P = 1{,}000$ and $i = 1/4\text{th of } 8\% = \frac{0.08}{4} = 0.02$. Using dimensional analysis (as discussed in Appendix E) to convert years to quarters, we get

$$n = \frac{1}{2}\text{ year} = \frac{1}{2}\,\cancel{\text{year}} \cdot \left(\frac{4\text{ quarters}}{1\,\cancel{\text{year}}}\right) = 2\text{ quarters}$$

(i and n are *periodic* figures)

$$\begin{aligned} FV &= P(1 + i)^n \\ &= 1{,}000(1 + 0.02)^2 \\ &= \$1{,}040.40 \end{aligned}$$

1000 [×] [(] 1 [+] .02 [)] [x^2] [=]

With a graphing calculator, press [ENTER] instead of [=].

This is a slightly easier calculation than the one in Example 1. ■

Comparing the future value derived from compound interest in Example 2 with the future value derived from simple interest, we find little difference. Computing future value at simple interest, we have $P = 1{,}000$, $r = 8\% = 0.08$, and $t = 6$ months $= \frac{1}{2}$ year (r and t are *annual* figures).

$$
\begin{aligned}
FV &= P(1 + rt) \\
&= 1{,}000\left(1 + 0.08 \cdot \frac{1}{2}\right) \\
&= \$1{,}040.00
\end{aligned}
$$

In this instance, compound interest gives us an extra 40¢. As we will see, however, the difference between simple interest and compound interest isn't always so insignificant.

EXAMPLE 3 In 1777, Jacob DeHaven, a wealthy Pennsylvania merchant, responded to a desperate plea from George Washington when it looked as if the Revolutionary War was about to be lost. He lent the Continental Congress \$450,000 to rescue Washington's troops at Valley Forge. When the war was over, Mr. DeHaven unsuccessfully tried to collect what was owed him. He died penniless in 1812. In 1990, his descendants sued the U.S. government for repayment of the original amount plus interest at the then prevailing rate of 6%.* How much did the government owe his descendants on the 1990 anniversary of the loan, if the interest was

a. compounded monthly? **b.** simple interest?

Solution **a.** To find the future value at monthly compounded interest, we have $P = 450{,}000$ and $i = 1/12$th of $6\% = 0.06/12$. Using dimensional analysis to convert years to months, we get

$$n = 213 \text{ years} = 213 \cancel{\text{years}} \cdot \frac{12 \text{ months}}{1 \cancel{\text{year}}} = 2556 \text{ months}$$

(i and n are *periodic* figures)

$$
\begin{aligned}
FV &= P(1+i)^n \\
&= 450{,}000(1+0.06/12)^{2556} \\
&= 1.547627234\ldots \times 10^{11} \\
&\approx \$154{,}762{,}723{,}400
\end{aligned}
$$

* *Source: New York Times,* May 27, 1990.

450000 [×] [(] 1 [+] 0.06 [÷] 12 [)] [y^x] 2556 [=]

With a graphing calculator, press [∧] instead of [y^x] and [ENTER] instead of [=].

Notice how much easier this method is than the method of Example 1, where we would have had to make 2,556 calculations!

b. To find the future value at simple interest, we have $P = 450{,}000$, $r = 6\% = 0.06$, and $t = 213$ years (r and t are *annual* figures).

$$\begin{aligned} FV &= P(1 + rt) \\ &= 450{,}000(1 + 0.06 \cdot 213) \\ &= \$6{,}201{,}000 \end{aligned}$$

■

In Example 3, the future value with compound interest is almost *25,000 times* the future value with simple interest. Over longer periods of time, compound interest is immensely more profitable to the investor than simple interest, because compound interest pays interest on interest. Similarly, compounding more frequently (daily rather than quarterly, for example) is more profitable to the investor.

The effects of the size of the time interval and the compounding period can be seen in Figures 5.3–5.5 on page 319.

EXAMPLE 4 **a.** Find the future value of \$2,500 deposited in an account where it earns 10.3% interest compounded daily for 15 years.

b. Find the interest earned.

Solution **a.** Compounding daily, we have $P = 2{,}500$, $i = \frac{1}{365}$ of $10.3\% = \frac{0.103}{365}$, and

$$n = 15 \text{ years} = 15 \text{ years} \cdot \frac{365 \text{ days}}{1 \text{ year}} = 5{,}475 \text{ days.}$$

$$\begin{aligned} FV &= P(1 + i)^n \\ &= 2{,}500\left(1 + \frac{0.103}{365}\right)^{5475} \\ &= 11{,}717.3749\ldots \\ &\approx \$11{,}717.37 \end{aligned}$$

2500 [×] [(] 1 [+] .103 [÷] 365 [)] [y^x] 5475 [=]

With a graphing calculator, press [∧] instead of [y^x] and [ENTER] instead of [=].

Future Value of $1,000 at 10% Interest		
	in 1 year	in 10 years
Simple interest	$1,100.00	$2,000.00
Compounded annually	$1,100.00	$2,593.74
Compounded quarterly	$1,103.81	$2,685.06
Compounded monthly	$1,104.71	$2,707.04
Compounded daily	$1,105.16	$2,717.91

Figure 5.3

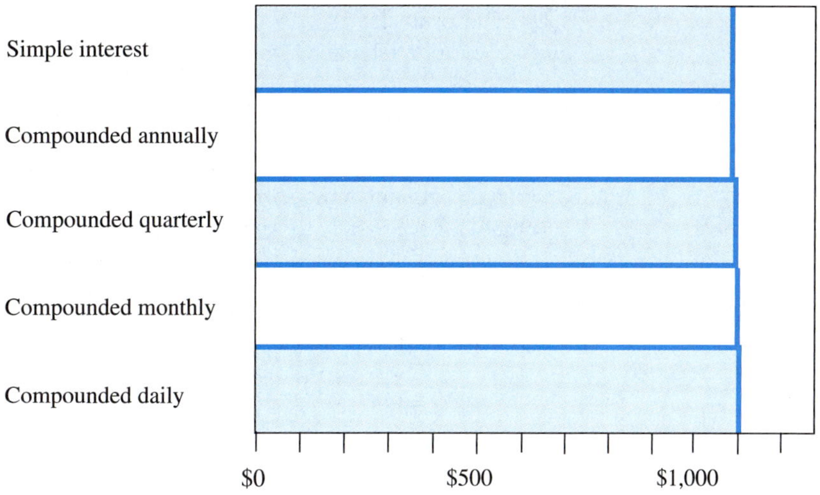

Figure 5.4
Future value of $1,000 invested at 10% interest in 1 year

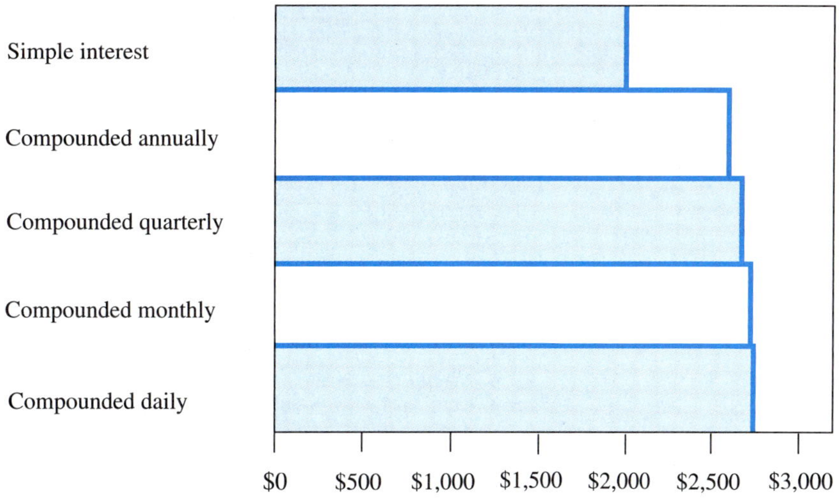

Figure 5.5
Future value of $1,000 invested at 10% interest in 10 years

WARNING: If you compute $\frac{0.103}{365}$ separately, you will get a long decimal. Do not round off that decimal, because the resulting answer could be inaccurate. Doing the calculation all at once (as shown above) avoids this difficulty.

When a calculation involves a large exponent, different calculators will give slightly different answers. This difference will rarely be more than a few pennies.

b. The principal is \$2,500, and the total of principal and interest is \$11,717.37; therefore, the interest is

$$\$11{,}717.37 - \$2{,}500 = \$9{,}217.37$$

■

EXAMPLE 5 Find the amount of money that must be invested now at $7\frac{3}{4}\%$ interest compounded annually for the money to be worth \$2,000 in 3 years.

Solution The question actually asks us to find the present value, or principal P, that will generate a future value of \$2,000. We have $FV = 2{,}000$, $i = 7\frac{3}{4}\% = 0.0775$, and $n = 3$.

$$\begin{aligned} FV &= P(1 + i)^n \\ 2{,}000 &= P(1 + 0.0775)^3 \\ P &= \frac{2{,}000}{1.0775^3} \\ &= 1{,}598.7412\ldots \\ &\approx \$1{,}598.74 \end{aligned}$$

■

Annual Yield

Which investment is more profitable: one that pays 5.8% compounded daily or one that pays 5.9% compounded quarterly? It's difficult to tell; certainly, 5.9% is a better rate than 5.8%, but compounding daily is better than compounding quarterly. The two rates cannot be directly compared because of their different compounding frequencies. The way to tell which is the better investment is to find the annual yield of each.

The **annual yield** (or just **yield**) of a compound interest deposit is the *simple interest rate* that has the same future value the compound rate would have in one year. The annual yields of two different investments can be compared, because they are both simple interest rates. Annual yield provides the consumer with a uniform basis for comparison. The annual yield should be slightly higher than the compound rate, because compound interest is slightly more profitable than simple interest over a short period of time. The compound rate is sometimes called the **nominal rate** to distinguish it from the yield (here, *nominal* means *named* or *stated*).

Because the annual yield is the simple interest rate that has the same future value as the compound rate in one year, the formula for the future value at the simple interest rate will equal the formula for the future value at the compound rate (with r being the annual yield).

$$FV\,(\text{compound interest}) = FV\,(\text{simple interest})$$
$$P\,(1 + i)^n = P\,(1 + rt)$$

EXAMPLE 6 Find the annual yield of \$2,500 deposited in an account where it earns 10.3% interest compounded daily for 15 years.

Solution

Compound Interest

$P = 2{,}500$

$i = \frac{1}{365}$ of $10.3\% = \frac{0.103}{365}$

$n = 1$ year $= 365$ days

(i and n are *periodic* figures)

Simple Interest

$P = 2{,}500$

$r =$ unknown annual yield

$t = 1$ year

(r and t are *annual* figures)

$$P\,(1 + i)^n = P(1 + rt)$$

$$2{,}500\left(1 + \frac{0.103}{365}\right)^{365} = 2{,}500\,(1 + \mathrm{r} \cdot 1)$$

$$\left(1 + \frac{0.103}{365}\right)^{365} = 1 + \mathrm{r} \quad \text{dividing by 2,500}$$

$$\mathrm{r} = \left(1 + \frac{0.103}{365}\right)^{365} - 1 \quad \text{simplifying}$$

$$= 0.10847527\ldots = 10.847527\ldots\%$$

$$\approx 10.85\%$$

✔ Notice that the annual yield is slightly higher than the compound rate, as it should be.

By tradition, we round the answer to the nearest hundredth of a percent, so the annual yield is 10.85%. This means that in one year's time, 10.3% interest compounded daily has the same effect as 10.85% simple interest. For any period of time longer than a year, 10.3% compounded daily will yield *more* interest than would 10.85% simple interest because compound interest pays interest on interest. ■

Notice that in Example 6, the principal of \$2,500 canceled out. If the principal were \$2 or \$2 million, it would still cancel out, and the annual yield of 10.3% compounded daily would still be 10.85%. The principal doesn't matter in computing the annual yield. Also notice that the 15 years did not enter into the calculation; annual yield is *always* based on a *one-year* period, so that we can compare various investment terms to a standard.

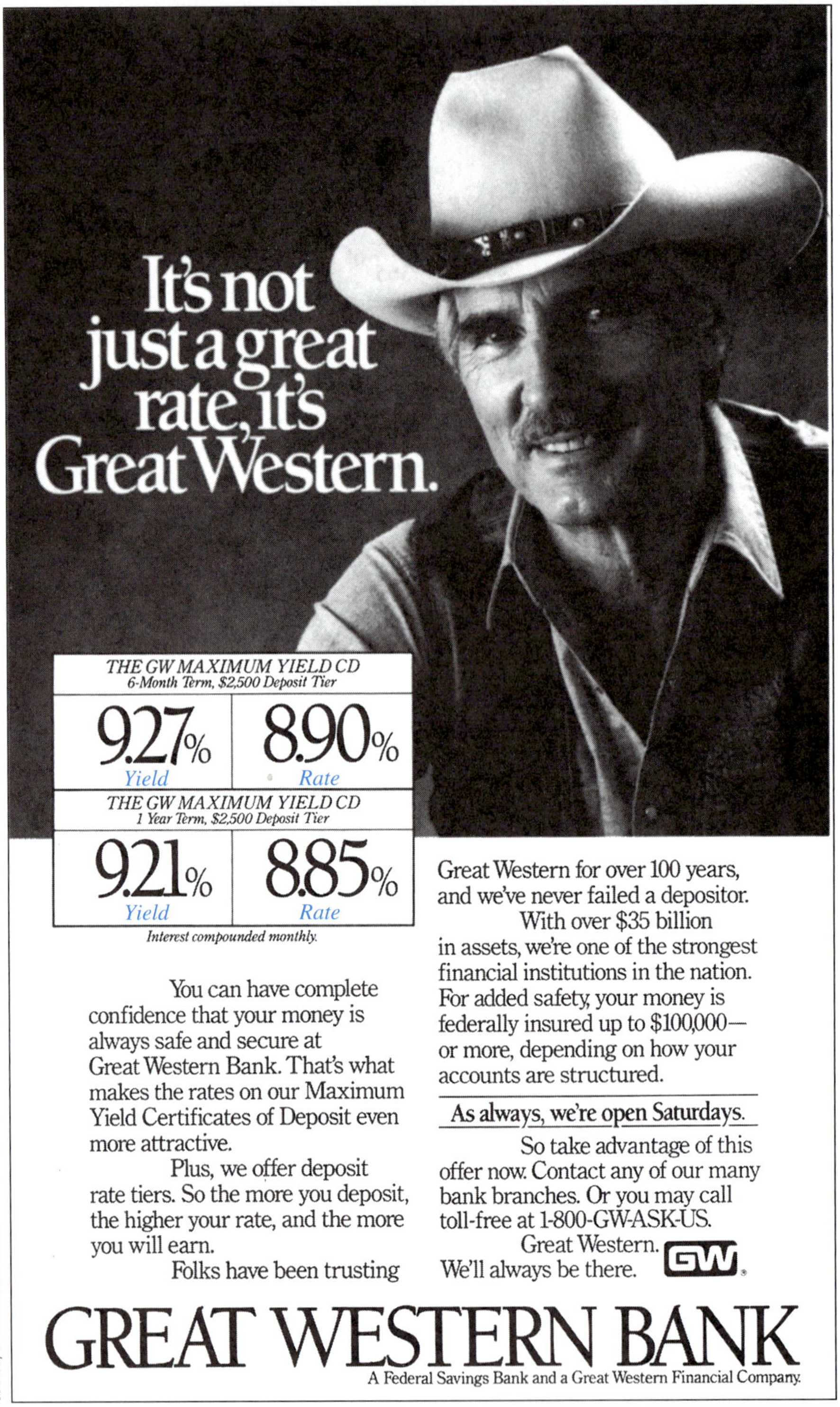

Courtesy of Great Western Bank

Most financial institutions advertise their annual yields as well as their interest rates.

EXAMPLE 7 Find the annual yield corresponding to a nominal rate of 8.4% compounded monthly.

Solution We are told neither the principal nor the time, but (as discussed above), these variables don't affect the annual yield.

Compounding monthly, we have $i = \frac{1}{12}$ of 8.4% $= \frac{0.084}{12}$, $n = 1$ year $= 12$ months, and $t = 1$ year.

$$FV\text{(compounded monthly)} = FV\text{(simple interest)}$$
$$P(1+i)^n = P(1+rt)$$
$$(1+i)^n = 1+rt \quad \text{dividing by } P$$
$$\left(1+\frac{0.084}{12}\right)^{12} = 1 + \text{r} \cdot 1$$
$$\text{r} = \left(1+\frac{0.084}{12}\right)^{12} - 1$$
$$\text{r} = 0.0873107\ldots = 8.73107\ldots\%$$
$$\approx 8.73\%$$

✔ Notice that the annual yield is slightly higher than the compound rate, as it should be.

■

In Example 7, we found that 8.4% compounded monthly generates an annual yield of 8.73%. This means that 8.4% compounded monthly has the same effect as 8.73% simple interest in one year's time. Furthermore, as Figure 5.6 indicates, 8.4% compounded monthly has the same effect as 8.73% *compounded annually for any time period.*

For a Principal of $1,000	After 1 Year	After 10 Years
FV at 8.4% compounded monthly	$1,087.31	$2,309.60
FV at 8.73% simple interest	$1,087.30	$1,873.00
FV at 8.73% compounded annually	$1,087.30	$2,309.37

Figure 5.6

Notice that all three rates have the same future value after one year (the 1¢ difference is due to rounding off the annual yield to 8.73%). However, after 10 years, the simple interest has fallen way behind, while the 8.4% compounded monthly and the 8.73% compounded annually remain the same (except for the round-off error). This always happens. The annual yield is the simple interest rate that has the same future value the compound rate would have *in one year;* it is also the *annually compounded rate* that has the same future value the nominal rate would have *after any amount of time.*

An annual yield formula does exist, but the annual yield can be calculated efficiently without it, as shown above. The formula will be developed in the exercises.

5.2 Exercises

In Exercises 1–6, find the periodic rate that corresponds to the given compound rate if the rate is compounded (a) quarterly, (b) monthly, (c) daily, (d) biweekly (every two weeks), and (e) semimonthly (twice a month). (Do not round off the periodic rate.)

1. 12% **2.** 6% **3.** 3.1%
4. 6.8% **5.** 9.7% **6.** 10.1%

In Exercises 7–10, find the number of periods that corresponds to the given time span if a period is (a) a quarter of a year, (b) a month, and (c) a day. (Ignore leap years.)

7. $8\frac{1}{2}$ years **8.** $9\frac{3}{4}$ years
9. 30 years **10.** 45 years

In Exercises 11–16, (a) find and (b) interpret the future value of the given amount.

11. $3,000 at 6% compounded annually for 15 years
12. $7,300 at 7% compounded annually for 13 years
13. $5,200 at $6\frac{3}{4}$% compounded quarterly for $8\frac{1}{2}$ years
14. $36,820 at $7\frac{7}{8}$% compounded quarterly for 4 years
15. $1,960 at $4\frac{1}{8}$% compounded daily for 17 years (ignore leap years)
16. $12,350 at 6% compounded daily for 10 years (ignore leap years)

In Exercises 17–22, find and interpret the annual yield corresponding to the given nominal rate.

17. 8% compounded monthly
18. $5\frac{1}{2}$% compounded quarterly
19. $4\frac{1}{4}$% compounded daily
20. $12\frac{5}{8}$% compounded daily
21. 10% compounded (a) quarterly, (b) monthly, (c) daily
22. $12\frac{1}{2}$% compounded (a) quarterly, (b) monthly, (c) daily

In Exercises 23–26, (a) find and (b) interpret the present value that will generate the given future value.

23. $1,000 at 8% compounded annually for 7 years
24. $9,280 at $9\frac{3}{4}$% compounded monthly for 2 years, 3 months
25. $3,758 at $11\frac{7}{8}$% compounded monthly for 17 years 7 months
26. $4,459 at $10\frac{3}{4}$% compounded quarterly for 4 years

For Exercises 27–31, note the following information. A certificate of deposit (or CD) is an agreement between a bank and a saver in which the bank guarantees an interest rate and the saver commits to leaving his or her deposit in the account for an agreed-upon period of time.

27. First National Bank offers two-year CDs at 9.12% compounded daily, and Citywide Savings offers two-year CDs at 9.13% compounded quarterly. Compute the annual yield for each institution and determine which is more advantageous for the consumer.
28. National Trust Savings offers five-year CDs at 8.25% compounded daily, and Bank of the Future offers five-year CDs at 8.28% compounded annually. Compute the annual yield for each institution and determine which is more advantageous for the consumer.

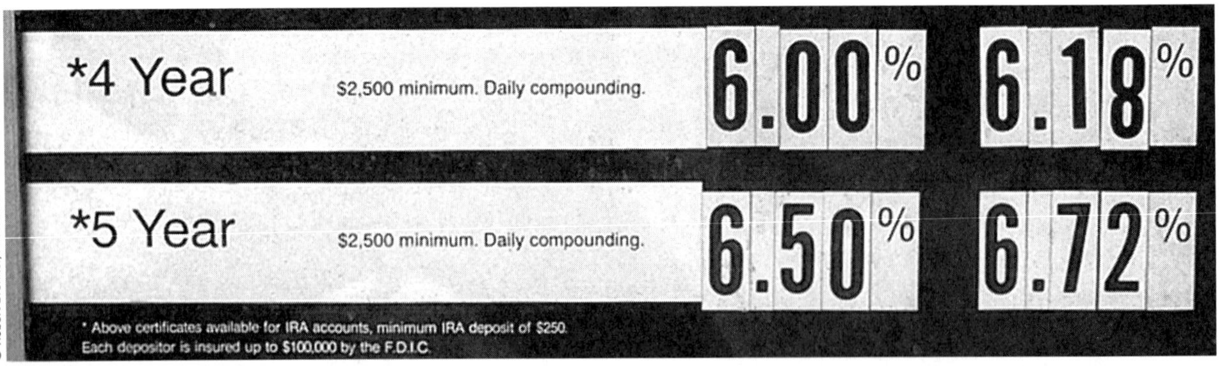

29. Verify the annual yield for the five-year CD quoted in the bank sign at the bottom of page 324.

30. Verify the yield for the one-year CDs quoted in the Jean Lafitte Savings Bank advertisement below.

Worrying what to do with your money?
Bury it in Jean Lafitte Savings Bank's 1-year "High Seas" CDs!
You'll rest in peace.

Rate:	8.7% compounded monthly
Yield:	10.16%

31. Verify the yield for the two-year CDs quoted in the Cole Younger Savings Bank advertisement below.

It's not just a great deal –
it's highway robbery!
2-year CDs from
Cole Younger Savings Bank:

9.3% interest, compounded daily
10.74% yield

32. Verify the "yield" for money market accounts quoted in the Great Western Bank advertisement on page 322.

33. When Jason Levy was born, his grandparents deposited \$3,000 into a special account for Jason's college education, earning $6\frac{1}{2}\%$ interest compounded daily.

a. How much will be in the account when Jason is eighteen?

b. If, on becoming eighteen, Jason arranged for the monthly interest to be sent to him, how much would he receive each 30-day month?

34. When Alan Cooper was born, his grandparents deposited \$5,000 into a special account for Alan's college education. The account earned $7\frac{1}{4}\%$ interest compounded daily.

a. How much will be in the account when Alan is eighteen?

b. If, on becoming eighteen, Alan arranged for the monthly interest to be sent to him, how much would he receive each 30-day month?

For Exercises 35–38, note the following information: An Individual Retirement Account (or IRA) is an account in which the saver does not pay income tax on the amount deposited but is not allowed to withdraw the money until retirement. (The saver pays income tax at that point, but his or her tax bracket is much lower then.)

35. At age twenty-seven, Lauren Johnson deposited \$1,000 into an IRA, where it earns $7\frac{7}{8}\%$ compounded monthly. What will it be worth when she retires at sixty-five?

36. At age thirty-six, Dick Shoemaker deposited \$2,000 into an IRA, where it earns $8\frac{1}{8}\%$ compounded semiannually. What will it be worth when he retires at sixty-five?

37. Jane Brecha wants to have an IRA that will be worth \$100,000 when she retires at age sixty-five.

a. How much must she deposit at age thirty-five at $8\frac{3}{8}\%$ compounded daily?

b. If, at age sixty-five, she arranges for the monthly interest to be sent to her, how much will she receive each 30-day month?

38. James Magee wants to have an IRA that will be worth \$150,000 when he retires at age sixty-five.

a. How much must he deposit at age twenty-six at $6\frac{1}{8}\%$ compounded daily?

b. If, at age sixty-five, he arranges for the monthly interest to be sent to him, how much will he receive each 30-day month?

39. In August 2002, Bank of the West offered six-month CDs at 1.64% interest compounded daily.

a. Find the CD's annual yield.

b. How much would a \$1,000 CD be worth at maturity?

c. How much interest would you earn?

d. What percent of the original \$1,000 is this interest?

e. The answer to part (d) is not the same as that of part (a). Why?

f. The answer to part (d) is close to, but not exactly half, that of part (a). Why?

40. In August 2002, Bank of the West offered 6-month CDs at 1.64% interest compounded daily, and 1-year CDs at 1.93% interest compounded daily. Maria Ruiz bought a six-month \$2,000 CD, even though she knew she wouldn't need the money for at least a year, because it was predicted that interest rates would rise.

a. Find the future value of Maria's CD.

b. Six months later, Maria's CD came to term, and in the intervening time, interest rates had risen. She reinvested the principal and interest from her first CD in a second six-month CD that paid 2.05% interest compounded daily. Find the future value of Maria's second CD.

c. Would Maria have been better off if she had bought a one-year CD in August of 2002?

d. If Maria's second CD paid 2.25% interest compounded daily, rather than 2.05%, would she be better off with the two six-month CDs or the one-year CD?

41. Develop a formula for the annual yield of a compound interest rate.

HINT: Follow the procedure given in Example 7, but use the letters i and n in place of numbers.

In Exercises 42–46, use the formula found in Exercise 41 to compute the annual yield corresponding to the given nominal rate.

42. $9\frac{1}{2}\%$ compounded monthly

43. $7\frac{1}{4}\%$ compounded quarterly

44. $12\frac{3}{8}\%$ compounded daily

45. $5\frac{5}{8}\%$ compounded (a) semiannually, (b) quarterly, (c) monthly, (d) daily, (e) biweekly, (f) semimonthly

46. $10\frac{1}{2}\%$ compounded (a) semiannually, (b) quarterly, (c) monthly, (d) daily, (e) biweekly, (f) semimonthly

Answer the following questions using complete sentences and your own words.

Concept Questions

47. Why is there no work involved in finding the annual yield of a given simple interest rate?

48. Why is there no work involved in finding the annual yield of a given compound interest rate, when that rate is compounded annually?

49. Which should be higher, the annual yield of a given rate compounded quarterly or compounded monthly? Explain why, *without* performing any calculations or referring to any formulas.

50. Why should the annual yield of a given compound interest rate be higher than the compound rate? Why should it be only slightly higher? Explain why, *without* performing any calculations or referring to any formulas.

51. Explain the difference between simple interest and compound interest.

52. *Money Magazine* and other financial publications regularly list the top-paying money-market funds, the top-paying bond funds, and the top-paying CDs, and their yields. Why do they list yields rather than interest rates and compounding periods?

53. Equal amounts are invested in two different accounts. One account pays simple interest, and the other pays compound interest at the same rate. When will the future values of the two accounts be the same?

54. Suppose you invest some money in a new account that pays 5% interest compounded annually and you do not make any further deposits into that account. Which of the following must be true?

- The account grows by the same dollar amount in the second year as it did in the first year.
- The account grows by a larger dollar amount in the second year than it did in the first year.
- The account grows by a smaller dollar amount in the second year than it did in the first year.

Why?

55. *For those who have completed Section 1.1:* Is the logic used in deriving the Compound Interest Formula inductive or deductive? Why?

Project

56. Suppose you have $1,000 invested at 5% annual interest and you do not make any further deposits into that account. Let x measure years after you invested the money, and let y measure the future value of the account. Draw a graph that shows the relationship between x and y, for $0 \leq x \leq 5$, if the interest is

a. simple interest

b. compounded annually

c. compounded daily

d. Discuss the differences between the three graphs. In your discussion, address the difference in their shapes, where they coincide, which graph is above the others, and which is below the others.

Doubling Time on a Graphing Calculator

Simple interest is a very straightforward concept. If an account earns 5% simple interest, then 5% of the principal is paid for each year that principal is in the account. In one year, the account earns 5% interest; in two years, it earns 10% interest; in three years, it earns 15% interest, and so on.

It is not nearly so easy to get an intuitive grasp of compound interest. If an account earns 5% interest compounded daily, then it does not earn only 5% interest in one year, and it does not earn only 10% interest in two years.

Annual yield is one way of gaining an intuitive grasp of compound interest. If an account earns 5% interest compounded daily, then it will earn 5.13% interest in one year (because the annual yield is 5.13%), but it does not earn merely $2 \cdot 5.13\% = 10.26\%$ interest in two years.

Doubling time is another way of gaining an intuitive grasp of compound interest. **Doubling time** is the amount of time it takes for an account to double in value; that is, it's the amount of time it takes for the future value to become twice the principal. To find the doubling time for an account that earns 5% interest compounded daily, substitute $2P$ for the future value and solve the resulting equation.

$$FV = P(1 + i)^n \quad \text{Compound Interest Future Value Formula}$$

$$2P = P\left(1 + \frac{0.05}{365}\right)^n \quad \text{substituting}$$

$$2 = \left(1 + \frac{0.05}{365}\right)^n \quad \text{dividing by } P$$

How do you solve this for n? One method involves logarithms; you may have learned that method in Intermediate Algebra. Another method involves the graphing calculator.

EXAMPLE 8 Use a graphing calculator to find the doubling time for an account that earns 5% interest compounded daily.

Solution

Step 1 *Use the calculator to graph two different equations:*

$$y = 2 \quad \text{and} \quad y = \left(1 + \frac{0.05}{365}\right)^x$$

Follow the procedure discussed in Appendix C. Fill in the "Y=" screen, as shown in Figure 5.7(a). If you use the "ZoomStandard" command, you will get the graph shown in Figure 5.7(b).

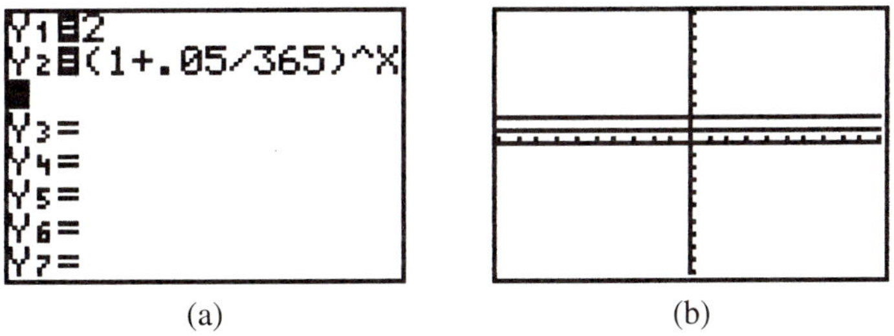

Figure 5.7

Step 2 *Adjust the calculator's screen so that it shows the point at which these two equations intersect.* To do this, we must determine appropriate values of $x$min, $x$max, $y$min, and $y$max.

In the equation $y = (1 + \frac{0.05}{365})^x$, x measures time and y measures money. Neither can be negative, so set $x$min and $y$min to 0.

We can get a rough idea of how big x should be by finding the doubling time for simple interest (a much easier calculation):

$$FV = P(1 + rt) \quad \text{Simple Interest Future Value Formula}$$
$$2P = P(1 + 0.05t) \quad \text{substituting } 2P \text{ for } FV$$
$$2 = 1 + 0.05t \quad \text{dividing by } P$$
$$1 = 0.05t$$
$$t = \frac{1}{0.05} = 20$$

The doubling time for 5% *simple* interest is 20 years. Simple interest isn't as productive as compound interest, so the doubling time for 5% *compound* interest will be less than 20 years. In the equation $y = (1 + \frac{0.05}{365})^x$, x measures time in days, and 20 years = 20 · 365 = 7,300 days, so set $x$max to 7,300.

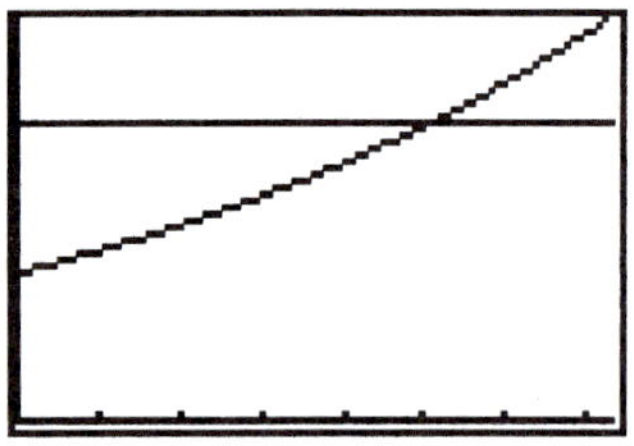

Figure 5.8

The value of $y$max should correspond to $x$max = 7,300:

$$y = \left(1 + \frac{0.05}{365}\right)^{7300}$$
$$= 2.71809\ldots$$
$$\approx 2.7$$

Set $y$max to 2.7.

Step 3 *Press the* GRAPH *button to obtain the screen shown in Figure 5.8.* Notice that we can now see the point of intersection.

Step 4 *Use the calculator to find the point of intersection.* Follow the procedure discussed in Appendix D and summarized below.

TI-80/81:	Use the "zoom" and "trace" commands
TI-82/83:	Select the "intersect" option from the "CALC" menu
TI-85/86:	Select "MATH" from the "GRAPH" menu
CASIO:	Press G-Solv and then ISCT

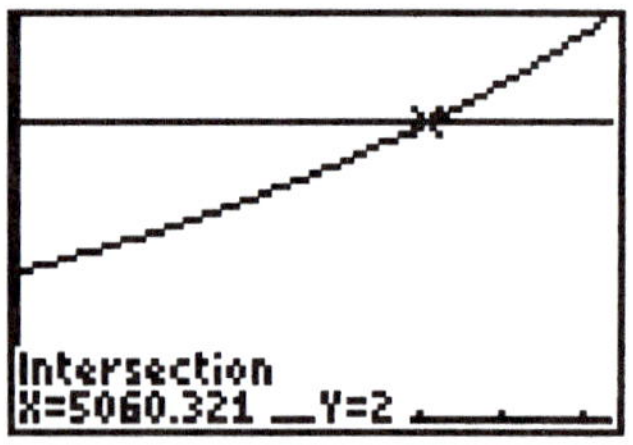

Figure 5.9

Figure 5.9 shows the result of this work. The intersection is at (5,060.321, 2). This means that it takes 5,060.321 days (or about 5060.321/365 = 13.86389315 . . . ≈ 13.9 years) for money invested at 5% interest compounded daily to double.

✔ The solution can be checked by substituting 5060.321 for x in the equation

$$2 = \left(1 + \frac{0.05}{365}\right)^x$$

The right side is

$$\left(1 + \frac{0.05}{365}\right)^x = \left(1 + \frac{0.05}{365}\right)^{5060.321}$$
$$= 2.000000005$$

This solution is mathematically (almost) correct, as shown in the preceding box. Practically speaking, however, the ".321" part of the solution doesn't make sense. If interest is compounded daily, then at the end of each day, your account is credited with that day's interest. After 5,060 days, your account would contain slightly less than twice the original principal. After 5,060.321 days, your account balance would not have changed, since interest won't be credited until the end of the day. After 5,061 days, your account would contain slightly more than twice the original principal. The doubling time is 5,061 days $\approx$ 13.9 years. ■

Exercises

57. The solution in Example 8 didn't quite check; we found that the y-value at the point of intersection is 2.000000005, not 2. Why is there a discrepancy?

58. If $1,000 is deposited into an account that earns 5% interest compounded daily, the doubling time is approximately 5,061 days.
  a. Find the amount in the account after 5,061 days.
  b. Find the amount in the account after $2 \cdot 5{,}061$ days.
  c. Find the amount in the account after $3 \cdot 5{,}061$ days.
  d. Find the amount in the account after $4 \cdot 5{,}061$ days.
  e. What conclusion can you make?

59. Do the following. (Give the number of periods and the number of years, rounded to the nearest hundredth.)
  a. Find the doubling time corresponding to 5% interest compounded annually.
  b. Find the doubling time corresponding to 5% interest compounded quarterly.
  c. Find the doubling time corresponding to 5% interest compounded monthly.
  d. Find the doubling time corresponding to 5% interest compounded daily.
  e. Discuss the effect of the compounding period on doubling time.

60. Do the following. (Give the number of periods and the number of years, rounded to the nearest hundredth.)
  a. Find the doubling time corresponding to 6% interest compounded annually.
  b. Find the doubling time corresponding to 7% interest compounded annually.
  c. Find the doubling time corresponding to 10% interest compounded annually.
  d. Discuss the effect of the interest rate on doubling time.

61. If you invest $10,000 at 8.125% interest compounded daily, how long will it take for you to accumulate $15,000? How long will it take for you to accumulate $100,000? (Give the number of periods and the number of years, rounded to the nearest hundredth.)

62. If you invest $15,000 at $9\frac{3}{8}\%$ interest compounded daily, how long will it take for you to accumulate $25,000? How long will it take for you to accumulate $100,000? (Give the number of periods and the number of years, rounded to the nearest hundredth.)

63. If you invest $20,000 at $6\frac{1}{4}\%$ interest compounded daily, how long will it take for you to accumulate $30,000? How long will it take for you to accumulate $100,000? (Give the number of periods and the number of years, rounded to the nearest hundredth.)

5.3 Annuities

Many people have long-term financial goals and limited means with which to accomplish them. Your goal might be to save $3,000 over the next four years for your college education, to save $10,000 over the next 10 years for the down payment on a home, to save $30,000 over the next 18 years to finance your new baby's college education, or to save $300,000 over the next 40 years for your retirement. It seems incredible, but each of these goals can be achieved by saving only $50 a month (if interest rates are favorable). All you need to do is start an annuity.

An **annuity** is a sequence of equal, regular payments into an account where each payment receives compound interest. Because most annuities involve relatively small periodic payments, they're affordable for the average person. Over longer periods of time, the payments themselves start to amount to a significant sum, but it's really the power of compound interest that makes annuities so amazing. If you pay \$50 a month into an annuity for the next 40 years, then your total payment is

$$\frac{\$50}{\cancel{\text{month}}} \cdot \frac{12 \cancel{\text{ months}}}{\cancel{\text{year}}} \cdot 40 \cancel{\text{ years}} = \$24{,}000$$

However, if the annuity pays 10% interest compounded monthly, after 40 years the account will contain over \$316,000!

A **Christmas club** is an annuity that is set up to save for Christmas shopping. A Christmas club participant makes regular equal deposits, and the deposits and the resulting interest are released to the participant in December when the money is needed. Christmas clubs are different from other annuities in that they span a short amount of time—a year at most—and thus earn only a small amount of interest. (People set them up to be sure they're putting money aside rather than to generate interest.) Our first few examples deal with Christmas clubs, because their short time span enables us to see how an annuity actually works.

Calculating Short-Term Annuities

EXAMPLE 1 On August 12, Patty Leitner joined a Christmas club through her bank. For the next 3 months, she would deposit \$200 at the beginning of each month. The money would earn $8\frac{3}{4}\%$ interest compounded monthly, and on December 1, she could withdraw her money for shopping. Use the Compound Interest Formula to find the future value of her account.

Solution We are given $P = 200$ and $i = \frac{1}{12}$ of $8\frac{3}{4}\% = \frac{0.0875}{12}$. First, calculate the future value of the first payment (made on September 1), using $n = 3$ (it will receive interest during September, October, and November).

$$\begin{aligned} FV &= P(1 + i)^n \\ &= 200\left(1 + \frac{0.0875}{12}\right)^3 \\ &= 204.40698\ldots \\ &\approx \$204.41 \end{aligned}$$

Next, calculate the future value of the second payment (made on October 1), using $n = 2$ (it will receive interest during October and November).

$$\begin{aligned} FV &= P(1 + i)^n \\ &= 200\left(1 + \frac{0.0875}{12}\right)^2 \\ &= 202.9273\ldots \\ &\approx \$202.93 \end{aligned}$$

To calculate the future value of the third payment (made on November 1), use $n = 1$ (it will receive interest during November).

$$\begin{aligned} FV &= P(1 + i)^n \\ &= 200\left(1 + \frac{0.0875}{12}\right)^1 \\ &= 201.45833\ldots \\ &\approx \$201.46 \end{aligned}$$

The payment schedule and interest earned are illustrated in Figure 5.10. The future value of Patty's annuity is the sum of the future values of each payment:

$$\begin{aligned} FV &\approx \$204.41 + \$202.93 + \$201.46 \\ &= \$608.80 \end{aligned}$$

Patty's deposits will total \$600.00; therefore, she will earn \$8.80 interest on her deposits. (See Figure 5.10.)

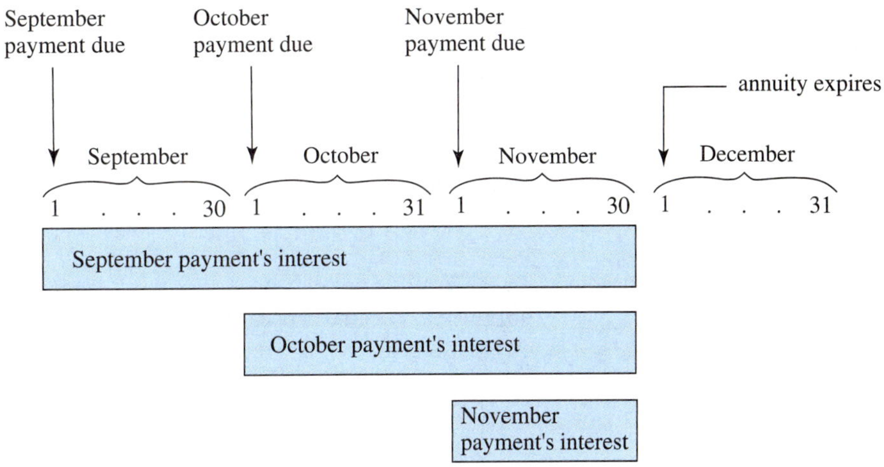

Figure 5.10

The **payment period** of an annuity is the time between payments; in Example 1, the payment period was one month. The **term** is the time from the beginning of the first payment period to the end of the last period; the term of Patty's Christmas club was three months. When an annuity has **expired** (that is, when its term is over), the entire account or any portion of it can be withdrawn. Naturally, any portion not withdrawn will continue to receive interest.

Most annuities are **simple;** that is, their compounding period is the same as their payment period (for example, if payments are made monthly, then interest is compounded monthly). In this book, we will work only with simple annuities. An **annuity due** is an annuity for which each payment is due at the beginning of its time period; Patty's annuity in Example 1 was an annuity due, because the payments were due at the *beginning* of each month. An **ordinary annuity** is an annuity for which each payment is due at the end of its time period; as the name implies, this form of annuity is more typical. As we will see in the next example, the difference is one of accounting.

EXAMPLE 2 Dan Bach also joined a Christmas club. His was like Patty's except that his payments were due at the end of each month and his first payment was due September 30. Use the Compound Interest Formula to find the future value of his account.

Solution This is an *ordinary* annuity because payments are due at the *end* of each month; interest is compounded monthly. From Example 1, we know that $P = 200$ and $i = \frac{1}{12}$ of $8\frac{3}{4}\% = \frac{0.0875}{12}$.

To calculate the future value of the first payment (made on September 30), use $n = 2$ (this payment will receive interest during October and November).

$$\begin{aligned} FV &= P(1 + i)^n \\ &= 200\left(1 + \frac{0.0875}{12}\right)^2 \\ &= 202.9273\ldots \\ &\approx \$202.93 \end{aligned}$$

To calculate the future value of the second payment (made on October 31), use $n = 1$ (this payment will receive interest during November).

$$\begin{aligned} FV &= P(1 + i)^n \\ &= 200\left(1 + \frac{0.0875}{12}\right)^1 \\ &= 201.45833\ldots \\ &\approx \$201.46 \end{aligned}$$

To calculate the future value of the third payment (made on November 30), note that no interest is earned, because the payment is due November 30 and the annuity expires December 1. Therefore,

$$FV = \$200$$

Dan's payment schedule and interest payments are illustrated in Figure 5.11.

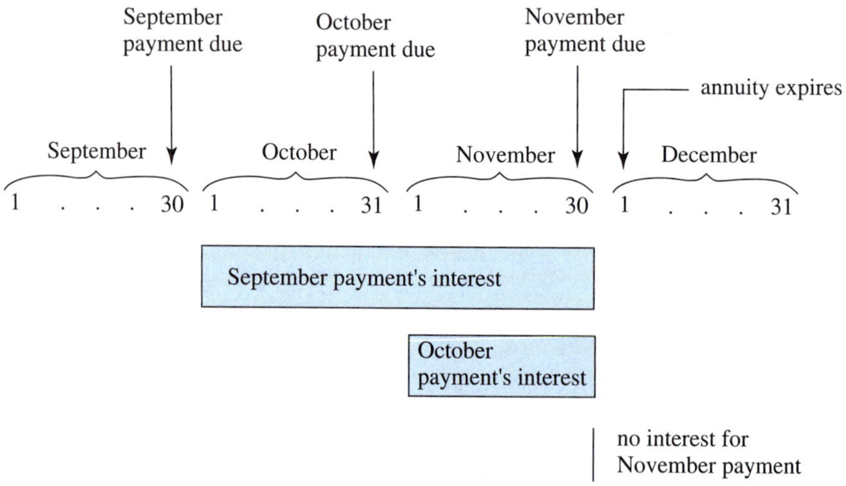

Figure 5.11

The future value of Dan's annuity is the sum of the future values of each payment:

$$\begin{aligned} FV &\approx \$200 + \$201.46 + \$202.93 \\ &= \$604.39 \end{aligned}$$

Dan earned \$4.39 interest on his deposits. ■

In Examples 1 and 2, why did Patty earn more interest than Dan? The reason is that each of her payments was made a month earlier and therefore received an extra month's interest. In fact, we could find the future value of Patty's account by giving Dan's future value one more month's interest:

$$604.39 \cdot \left(1 + \frac{0.0875}{12}\right)^1 = 608.80$$

$$\text{Dan's } FV \cdot \text{extra interest} = \text{Patty's } FV$$

The difference between an ordinary annuity and an annuity due is strictly an accounting difference, because any ordinary annuity in effect will become an annuity due if you leave all the funds in the account for one extra period.

Calculating Long-Term Annuities

The procedure followed in Examples 1 and 2 reflects what actually happens with annuities, and it works fine for a small number of payments. However, most annuities are long-term. If we were to calculate the future value of an annuity with a term of 30 years of monthly payments, the above procedure would be very laborious! Long-term annuities should be calculated using a new formula.

For an *ordinary* annuity with payment "pymt," periodic rate i, and a term of n payments, the first payment receives interest for $n - 1$ periods (payment is made at the end of the first period, so it receives no interest for that period). Its future value is then

$$FV\,(\text{first pymt}) = \text{pymt}(1 + i)^{n-1}$$

The last payment receives no interest (under the annuity), because it is due at the end of the last period and the annuity expires the next day. Its future value is simply

$$FV\,(\text{last pymt}) = \text{pymt}$$

The next-to-last payment receives one period's interest, so its future value is

$$FV\,(\text{next-to-last pymt}) = \text{pymt}(1 + i)^1$$

The future value of the annuity is the sum of all these future values of individual payments:

$$FV = \text{pymt} + \text{pymt}(1 + i)^1 + \text{pymt}(1 + i)^2 + \cdots + \text{pymt}(1 + i)^{n-1}$$

To get a shortcut formula from all this, we can first multiply each side of the equation above by $(1 + i)$ and then subtract the original equation from the result. This allows a lot of canceling.

$$\begin{aligned} & FV(1 + i) = \quad \text{pymt}(1 + i) + \text{pymt}(1 + i)^2 + \ldots + \text{pymt}(1 + i)^{n-1} + \text{pymt}(1 + i)^n \\ \text{minus: } & FV = \text{pymt} + \text{pymt}(1 + i)^1 + \text{pymt}(1 + i)^2 + \ldots + \text{pymt}(1 + i)^{n-1} \\ \hline \text{equals: } & FV(1 + i) - FV = \text{pymt}(1 + i)^n - \text{pymt} && \text{subtracting} \\ & FV(1 + i - 1) = \text{pymt}[(1 + i)^n - 1] && \text{factoring} \\ & FV(i) = \text{pymt}[(1 + i)^n - 1] \\ & FV = \text{pymt}\,\frac{(1 + i)^n - 1}{i} && \text{dividing by } i \end{aligned}$$

This is the future value of the ordinary annuity.

Ordinary Annuity Formula

The future value FV of an ordinary annuity with payment size "pymt," periodic rate i, and a term of n payments is

$$FV(\text{ordinary}) = \text{pymt}\,\frac{(1+i)^n - 1}{i}$$

As we saw in Examples 1 and 2, the difference between an ordinary annuity and an annuity due is that with an annuity due, each payment receives one more period's interest (because each payment is made at the beginning of the period). Thus, the future value of an annuity due is the future value of an ordinary annuity plus one more period's interest.

$$\begin{aligned} FV\,(\text{of annuity due}) &= FV\,(\text{of ordinary annuity}) \cdot (1+i) \\ &= \text{pymt}\,\frac{(1+i)^n - 1}{i}(1+i) \end{aligned}$$

Annuity Due Formula

The future value FV of an annuity due with payment size "pymt," periodic rate i, and a term of n payments is

$$\begin{aligned} FV(\text{due}) &= FV(\text{ordinary}) \cdot (1+i) \\ &= \text{pymt}\,\frac{(1+i)^n - 1}{i}(1+i) \end{aligned}$$

Tax-Deferred Annuities

A **tax-deferred annuity (TDA)** is an annuity that is set up to save for retirement. Money is automatically deducted from the participant's paychecks until retirement, and the federal (and perhaps state) tax deduction is computed *after* the annuity payment has been deducted, resulting in significant tax savings. In some cases, the employer also makes a regular contribution to the annuity.

The following example involves a long-term annuity. Usually, the interest rate of a long-term annuity varies somewhat from year to year. In this case, calculations must be viewed as predictions, not guarantees.

EXAMPLE 3 Jim Moran just got a new job, and he immediately set up a TDA to save for retirement. He arranged to have \$200 taken out of each of his monthly checks, which will earn $8\frac{3}{4}\%$ interest. Owing to the tax-deferring effect of the TDA, his take-home pay went down by only \$115. Jim just had his thirtieth birthday, and his ordinary annuity will come to term when he is sixty-five. Find the future value of the annuity.

Solution This is an ordinary annuity, with pymt = 200, $i = \frac{1}{12}$ of $8\frac{3}{4}\% = \frac{0.0875}{12}$, and $n = 35 \text{ years} = 35 \text{ years} \cdot \frac{12 \text{ months}}{1 \text{ year}} = 420$ monthly payments.

$$
\begin{aligned}
FV &= \text{pymt}\,\frac{(1+i)^n - 1}{i} \\
&= 200\,\frac{\left(1 + \frac{0.0875}{12}\right)^{420} - 1}{\frac{0.0875}{12}} \\
&= \$552{,}539.96
\end{aligned}
$$

Because $\frac{0.0875}{12}$ occurs twice in the calculation, compute it first and put it into your calculator's memory. (See Appendixes A and B for memory instructions.) After storing the periodic rate, type

(1 + RCL) y^x 420 − 1 = ÷ RCL × 200 =

(1 + ALPHA A) ^ 420 − 1 ENTER ÷ ALPHA A × 200 ENTER

■

How much of the future value of the annuity in Example 3 is interest? Jim made 420 payments of \$200 each, so he paid a total of 420 · \$200 = \$84,000. The interest is then \$552,539.96 − \$84,000 = \$468,539.96. The interest is almost six times as large as the total of Jim's payments! The magnitude of the earnings illustrates the amazing power of annuities and the effect of compound interest over a long period of time.

EXAMPLE 4 Find the future value of the annuity in Example 3 if Jim's payments are made at the beginning of each period.

Solution

$$
\begin{aligned}
FV(\text{due}) &= FV(\text{ordinary}) \cdot (1 + i) \\
&= 552{,}539.96 \cdot \left(1 + \frac{0.0875}{12}\right) \quad \text{from Example 3} \\
&\approx \$556{,}568.90
\end{aligned}
$$

■

Sinking Funds

A **sinking fund** is an annuity in which the future value is a specific amount of money that will be used for a certain purpose, such as a child's education or the down payment on a home.

EXAMPLE 5 Tom and Sue Pegnim want to set up a sinking fund to save for their new baby's college education. How much would they have to have deducted from their biweekly paycheck to have \$30,000 in 18 years, at $9\frac{1}{4}\%$ interest? Assume that the account is an ordinary annuity.

Solution This is an ordinary annuity, with $i = \frac{1}{26}$ of $9\frac{1}{4}\% = \frac{0.0925}{26}$, $n = 18$ years $= 18 \text{ years} \cdot \frac{26 \text{ periods}}{1 \text{ year}} = 468$ periods, and $FV = 30{,}000$.

$$FV = \text{pymt}\,\frac{(1+i)^n - 1}{i}$$

$$30{,}000 = \text{pymt}\,\frac{\left(1 + \dfrac{0.0925}{26}\right)^{468} - 1}{\dfrac{0.0925}{26}}$$

We want to find the value of "pymt." To do so, we must divide 30,000 by the fraction on the right side of the equation. Because the fraction is so complicated, it's best to first calculate the fraction and then multiply its reciprocal by 30,000.

First calculate $\frac{0.0925}{26}$ and put it into your calculator's memory. Then type

[( ] 1 [+] [RCL] [ )] [y^x] 468 [−] 1 [=] [÷] [RCL] [=] [1/x] [×] 30000 [=]

calculating the fraction (from [( ] through [=]); reciprocal ([1/x])

[( ] 1 [+] [ALPHA] [A] [ )] [^] 468 [−] 1 [ENTER] [÷] [ALPHA] [A] [ENTER]

calculating the fraction

[x^{-1}] [×] 30000 [ENTER]

reciprocal ([x^{-1}])

This calculation gives pymt = 24.995038; Tom and Sue would need to have only $25.00 taken out of each biweekly paycheck to save $30,000 in 18 years. Notice that to have exactly $30,000 saved, they would need to deduct exactly $24.995038 from their paychecks. They will actually have a few dollars more than $30,000 because the paycheck deduction was rounded up. ■

EXAMPLE 6 If, in Example 5, Tom and Sue Pegnim have exactly $25.00 deducted from each paycheck, find the following.

a. the actual future value of their annuity
b. the portion of the future value that is their contribution
c. the total interest

Solution **a.** This is an ordinary annuity, with pymt = 25.00, $i = \frac{1}{26}$ of $9\frac{1}{4}\% = \frac{0.0925}{26}$, and $n = 18$ years $= 18 \text{ years} \cdot \frac{26 \text{ periods}}{1 \text{ year}} = 468$ periods.

$$FV = \text{pymt}\,\frac{(1+i)^n - 1}{i}$$

$$= 25\,\frac{\left(1 + \frac{0.0925}{26}\right)^{468} - 1}{\frac{0.0925}{26}}$$

$$= 30{,}005.956\ldots$$

$$\approx \$30{,}005.96$$

b. Their contribution is 468 payments of \$25 each = 468 · \$25 = \$11,700.
c. The interest is then \$30,005.96 − \$11,700.00 = \$18,305.96.

The interest is more than one and a half times as large as the total of the Pegnims' payments! While this isn't quite as incredible as Jim Moran's interest, which was almost six times the total of his payments (see Examples 3 and 4), it is still a very attractive investment. ■

Present Value of an Annuity

The **present value of an annuity** refers to the lump sum that can be deposited at the beginning of the annuity's term, at the same interest rate and with the same compounding period, that would yield the same amount as the annuity. This value can help the saver understand his or her options; it refers to an alternative way of saving the same amount of money in the same amount of time. It is called the present value because it refers to the single action the saver can take *in the present* (that is, at the beginning of the annuity's term) that would have the same effect as the annuity.

EXAMPLE 7 What lump sum deposit would Tom and Sue Pegnim from Examples 5 and 6 have to make (at the same interest rate) to have the same amount of money after 18 years as their annuity would give them?

Solution We are asked to find the present value of their annuity. Because we have already found the future value, what remains is a compound interest problem.

Interest is compounded biweekly; from Example 6, we know that $i = \frac{1}{26}$ of $9\frac{1}{4}\% = \frac{0.0925}{26}$, $n = 468$, and $FV = 30{,}005.956$.

$$FV = P(1+i)^n$$

$$30{,}005.956 = P\left(1 + \frac{0.0925}{26}\right)^{468}$$

$$P = \frac{30{,}005.956}{\left(1 + \frac{0.0925}{26}\right)^{468}}$$

$$= 5{,}693.6452\ldots$$

$$\approx \$5{,}693.65$$

30005.956 [÷] [(] 1 [+] .0925 [÷] 26 [)] [y^x] 468 [=]

With a graphing calculator, replace [y^x] with [^] and [=] with [ENTER].

This means that the Pegnims would have to deposit \$5,693.65 at their baby's birth to save as much money as the annuity would yield. The Pegnims chose an annuity over a lump sum deposit because they couldn't afford to tie up almost \$5,700 for 18 years but could afford to deduct \$25 out of each paycheck. ■

EXAMPLE 8 Find the present value of an ordinary annuity that has \$200 monthly payments for 25 years if the account receives $10\frac{1}{2}\%$ interest.

Solution We could find the future value of the annuity and then find the lump sum deposit whose future value matches it, as we did in the previous example. However, it's simpler to do the calculation all at once. The key is to realize that the future value of the lump sum must equal the future value of the annuity:

$$\text{future value of lump sum} = \text{future value of annuity}$$

$$P(1+i)^n = \text{pymt}\,\frac{(1+i)^n - 1}{i}$$

Lump Sum Compounded Monthly

$P = ?$

$$i = \frac{1}{12} \text{ of } 10\frac{1}{2}\% = \frac{0.105}{12}$$

$$n = 25 \text{ years} = 25 \text{ years} \cdot \frac{12 \text{ months}}{1 \text{ year}} = 300 \text{ months}$$

Annuity (Ordinary Annuity)

$\text{pymt} = 200$

$$i = \frac{0.105}{12}$$

$n = 300$ months

Note that i and n are the same with the annuity and the lump sum.

$$P(1+i)^n = \text{pymt}\,\frac{(1+i)^n - 1}{i}$$

$$P\left(1 + \frac{0.105}{12}\right)^{300} = 200\,\frac{\left(1+\frac{0.105}{12}\right)^{300} - 1}{\frac{0.105}{12}}$$

First, calculate the right side, as with any annuity calculation. Then divide by $(1 + \frac{0.105}{12})^{300}$ to find P.

$$P = 21{,}182.363\ldots \approx \$21{,}182.36$$

This means that an investor would have to make a lump sum deposit of more than \$21,000 to have as much money after 25 years as with monthly \$200 annuity payments. ■

Present Value of Annuity Formula

The present value of an annuity is the lump sum *P*, where

$$FV(\text{lump sum}) = FV(\text{annuity})$$

$$P(1 + i)^n = \text{pymt}\,\frac{(1 + i)^n - 1}{i}$$

There is a special formula for finding the present value of an ordinary annuity, but computations can be done quite efficiently without it. The formula will be developed in the exercises.

5.3 Exercises

In Exercises 1–4, (a) find and (b) interpret the future value of the given annuity.

1. ordinary annuity, \$120 monthly payment, $5\frac{3}{4}\%$ interest, 1 year

2. ordinary annuity, \$175 monthly payment, $6\frac{1}{8}\%$ interest, 11 years

3. annuity due, \$100 monthly payment, $5\frac{7}{8}\%$ interest, 4 years

4. annuity due, \$150 monthly payment, $6\frac{1}{4}\%$ interest, 13 years

5. On February 8, Bert Sarkis joined a Christmas club. His bank will automatically deduct \$75 from his checking account at the end of each month and deposit it into his Christmas club account, where it will earn 7% interest. The account comes to term on December 1. Find the following.

a. the future value of the account
b. Bert's total contribution to the account
c. the total interest

6. On March 19, Rachael Westlake joined a Christmas club. Her bank will automatically deduct \$110 from her checking account at the end of each month and deposit it into her Christmas club account, where it will earn $6\frac{7}{8}\%$ interest. The account comes to term on December 1. Find the following.

a. the future value of the account
b. Rachael's total contribution to the account
c. the total interest

7. On February 23, Ginny Deus joined a Christmas club. Her bank will automatically deduct \$150 from her checking account at the beginning of each month and deposit it into her Christmas club account, where it will earn $7\frac{1}{4}\%$ interest. The account comes to term on December 1. Find the following.

a. the future value of the account
b. Ginny's total contribution to the account
c. the total interest

8. On January 19, Lynn Knight joined a Christmas club. Her bank will automatically deduct \$100 from her checking account at the beginning of each month and deposit it into her Christmas club account, where it will earn 6% interest. The account comes to term on December 1. Find the following.

a. the future value of the account
b. Lynn's total contribution to the account
c. the total interest

9. Pat Gilbert recently set up a TDA to save for her retirement. She arranged to have \$175 taken out of each of her monthly checks; it will earn $10\frac{1}{2}\%$ interest. She just had her thirty-ninth birthday, and her ordinary annuity comes to term when she is sixty-five. Find the following.

a. the future value of the account
b. Pat's total contribution to the account
c. the total interest

10. Dick Eckel recently set up a TDA to save for his retirement. He arranged to have \$110 taken out of each of his biweekly checks; it will earn $9\frac{7}{8}\%$ interest. He just had his twenty-ninth birthday, and his ordinary annuity comes to term when he is sixty-five. Find the following.
    a. the future value of the account
    b. Dick's total contribution to the account
    c. the total interest
11. Sam Whitney recently set up a TDA to save for his retirement. He arranged to have \$290 taken out of each of his monthly checks; it will earn 11% interest. He just had his forty-fifth birthday, and his ordinary annuity comes to term when he is sixty-five. Find the following.
    a. the future value of the account
    b. Sam's total contribution to the account
    c. the total interest
12. Art Dull recently set up a TDA to save for his retirement. He arranged to have \$50 taken out of each of his biweekly checks; it will earn $9\frac{1}{8}\%$ interest. He just had his thirtieth birthday, and his ordinary annuity comes to term when he is sixty-five. Find the following.
    a. the future value of the account
    b. Art's total contribution to the account
    c. the total interest

In Exercises 13–18, (a) find and (b) interpret the present value of the given annuity.

13. the annuity in Exercise 1
14. the annuity in Exercise 2
15. the annuity in Exercise 5
16. the annuity in Exercise 6
17. the annuity in Exercise 9
18. the annuity in Exercise 10

In Exercises 19–24, find the monthly payment that will yield the given future value.

19. \$100,000 at $9\frac{1}{4}\%$ interest for 30 years; ordinary annuity
20. \$45,000 at $8\frac{7}{8}\%$ interest for 20 years; ordinary annuity
21. \$250,000 at $10\frac{1}{2}\%$ interest for 40 years; ordinary annuity
22. \$183,000 at $8\frac{1}{4}\%$ interest for 25 years; ordinary annuity
23. \$250,000 at $10\frac{1}{2}\%$ interest for 40 years; annuity due
24. \$183,000 at $8\frac{1}{4}\%$ interest for 25 years; annuity due
25. Mr. and Mrs. Gonsales set up a TDA to save for their retirement: \$100 will be deducted from each of Mr. Gonsales's biweekly paychecks, which will earn $8\frac{1}{8}\%$ interest.
    a. Find the future value of their ordinary annuity if it comes to term after they retire in $35\frac{1}{2}$ years.
    b. After retiring, Mr. and Mrs. Gonsales convert their annuity to a savings account, which earns 6.1% interest compounded monthly. At the end of each month, they withdraw \$650 for living expenses. Complete the chart in Figure 5.12 for their postretirement account:

Month Number	Account Balance (beginning of month)	Interest for the Month	With-drawal	Account Balance (end of month)
1				
2				
3				
4				
5				

Figure 5.12

26. Mr. and Mrs. Jackson set up a TDA in order to save for their retirement. They agreed to have \$125 deducted from each of Mrs. Jackson's biweekly paychecks, which will earn $7\frac{5}{8}\%$ interest.
    a. Find the future value of their ordinary annuity if it comes to term after they retire in $32\frac{1}{2}$ years.
    b. After retiring, the Jacksons convert their annuity to a savings account, which earns 6.3% interest compounded monthly. At the end of each month, they withdraw \$700 for living expenses. Complete the chart in Figure 5.13 for their post-retirement account.

Month Number	Account Balance (beginning of month)	Interest for the Month	With-drawal	Account Balance (end of month)
1				
2				
3				
4				
5				

Figure 5.13

27. Jeanne and Harold Kimura want to set up a TDA that will generate sufficient interest at maturity to meet their living expenses, which they project to be \$950 per month.

   a. Find the amount needed at maturity to generate \$950 per month interest, if they can get $6\frac{1}{2}\%$ interest compounded monthly.
   b. Find the monthly payment they would have to put into an ordinary annuity to obtain the future value found in (a) if their money earns $8\frac{1}{4}\%$ and the term is 30 years.

28. Susan and Bill Stamp want to set up a TDA that will generate sufficient interest at maturity to meet their living expenses, which they project to be \$1,200 per month.

   a. Find the amount needed at maturity to generate \$1,200 per month interest if they can get $7\frac{1}{4}\%$ interest compounded monthly.
   b. Find the monthly payment they would have to put into an ordinary annuity to obtain the future value found in (a) if their money earns $9\frac{3}{4}\%$ and the term is 25 years.

29. When Shannon Pegnim was 14, she got an after-school job at a local pet shop. Her parents told her that if she put some of her earnings into an IRA, they would contribute an equal amount to her IRA. (An IRA, or Individual Retirement Account, is an annuity that is set up to save for retirement. IRAs differ from TDAs in that an IRA allows the participant to contribute money whenever he or she wants, whereas a TDA requires the participant to have a specific amount deducted from each of his or her paychecks.) That year, and every year thereafter, she deposited \$1,000 into her IRA. When she became 25 years old, her parents stopped contributing, but Shannon increased her annual deposit to \$2,000 and continued depositing that amount annually until she retired at age 65. Her IRA paid 8.5% interest. Find the following.

   a. the future value of the account
   b. Shannon's and her parents' total contributions to the account
   c. the total interest
   d. the future value of the account if Shannon waited until she was 19 before she started her IRA
   e. the future value of the account if Shannon waited until she was 24 before she started her IRA

30. When Bo McSwine was 16, he got an after-school job at his parents' barbecue restaurant. His parents told him that if he put some of his earnings into an IRA, they would contribute an equal amount to his IRA. That year, and every year thereafter, he deposited \$900 into his IRA. When he became 21 years old, his parents stopped contributing, but Bo increased his annual deposit to \$1,800, and continued depositing that amount annually until he retired at age 65. His IRA paid 7.75% interest. Find the following.

   a. the future value of the account
   b. Bo's and his parents' total contributions to the account
   c. the total interest
   d. the future value of the account if Bo waited until he was 18 before he started his IRA
   e. the future value of the account if Bo waited until he was 25 before he started his IRA

31. If Shannon Pegnim from Exercise 29 started her IRA at age 35 rather than 14, how big of an annual contribution would she have had to have made in order to have the same amount saved at age 65?

32. If Bo McSwine from Exercise 30 started his IRA at age 35 rather than 16, how big of an annual contribution would he have had to have made in order to have the same amount saved at age 65?

33. Toni Torres wants to save \$1,200 over the next two years to use as a down payment on a new car. If her bank offers her 9% interest, what monthly payment would she need to put into an ordinary annuity in order to reach her goal?

34. Fred and Melissa Furth's daughter Sally will be a freshman in college in six years. To help cover their extra expenses, the Furths decide to set up a sinking fund of \$12,000. If the account pays 7.2% interest and they want to make quarterly payments, find the size of each payment.

35. Anne Geyer buys some land in Utah. She agrees to pay the seller a lump sum of \$65,000 in five years. Until then, she will make monthly simple interest payments to the seller at 11% interest.

   a. Find the amount of each interest payment.
   b. Anne sets up a sinking fund to save the \$65,000. Find the size of her semiannual payments if her payments are due at the end of every six-month period and her money earns $8\frac{3}{8}\%$ interest.
   c. Prepare a table showing the amount in the sinking fund after each deposit.

36. Chrissy Fields buys some land in Oregon. She agrees to pay the seller a lump sum of \$120,000 in six years. Until then, she will make monthly simple interest payments to the seller at 12% interest.

a. Find the amount of each interest payment.

b. Chrissy sets up a sinking fund to save the \$120,000. Find the size of her semiannual payments if her money earns $10\frac{3}{4}\%$ interest.

c. Prepare a table showing the amount in the sinking fund after each deposit.

37. Develop a formula for the present value of an ordinary annuity by solving the Present Value of Annuity Formula for P and simplifying.

38. Use the formula developed in Exercise 37 to find the present value of the annuity in Exercise 2.

39. Use the formula developed in Exercise 37 to find the present value of the annuity in Exercise 1.

40. Use the formula developed in Exercise 37 to find the present value of the annuity in Exercise 6.

41. Use the formula developed in Exercise 37 to find the present value of the annuity in Exercise 5.

Answer the following questions using complete sentences and your own words.

Concept Questions

42. *For those who have completed Section 1.1:* Is the logic used in deriving the Ordinary Annuity Formula inductive or deductive? Why? Is the logic used in deriving the relationship

$$FV(\text{due}) = FV(\text{ordinary}) \cdot (1 + i)$$

inductive or deductive? Why?

43. Write a paragraph or two in which you explain the difference between an ordinary annuity and an annuity due.

44. Compare and contrast an annuity with a lump-sum investment that receives compound interest. Be sure to discuss both the similarity and the difference between these two concepts, as well as the advantages and disadvantages of each.

45. Which is always higher—the present value of an annuity or the future value? Why?

46. If you want to retire on \$2,000 a month for 25 years, do you need to save \$2,000 a month for 25 years before retiring? Why or why not?

Annuities on a Graphing Calculator

EXERCISES

47. a. Use the method discussed in Section 5.2 on doubling time to find how long it takes for an annuity to have a balance of \$500,000, if the ordinary annuity requires monthly \$200 payments that earn 5% interest. Notice that payments are made on a monthly basis, so the answer must be a whole number of months.

b. State what values you used for *x*min, *x*max, *x*scl, *y*min, *y*max, and *y*scl. Explain how you arrived at those values.

c. Draw a freehand sketch of the graph obtained on your calculator.

48. Analyze the effect of the interest rate on annuities by recalculating Exercise 47 with interest rates of 6%, 8%, and 10%. Discuss the impact of the increased rate.

49. Analyze the effect of the payment period on annuities by recalculating Exercise 47 with twice-monthly payments of \$100 each that earn 5% interest. Discuss the impact of the altered period.

5.4 Amortized Loans

An **amortized loan** is a loan for which the loan amount, plus interest, is paid off in a series of regular equal payments. In section 5.1, we looked at one type of amortized loan: the add-on interest loan. A second type of amortized loan is the simple interest amortized loan. There is an important difference between these two types of loans that any potential borrower should be aware of: *The payments are smaller with a simple interest amortized loan than with an add-on interest loan* (assuming, naturally, that the loan amounts, interest rates, and number of payments are the same).

A **simple interest amortized loan** is really a type of annuity; specifically, it is an ordinary annuity whose future value is the loan amount plus compound interest. This raises a puzzling question. How can a simple interest amortized loan be defined as an annuity whose future value is the loan amount plus *compound* interest and still be called a *simple* interest loan? We will see.

The following formula for a simple interest amortized loan is based on the definition above, which requires that the future value of the ordinary annuity equal the future value of the loan amount.

Simple Interest Amortized Loan Formula

$$\text{future value of annuity} = \text{future value of loan amount}$$

$$\text{pymt}\,\frac{(1+i)^n - 1}{i} = P(1+i)^n$$

where "pymt" is the loan payment, i is the periodic interest rate, n is the number of periods, and P is the present value or loan amount.

Algebraically, the above formula could be used to determine any one of the four unknowns (pymt, i, n, and P) if the other three are known. In Section 5.3, we used it to find the present value P of an annuity. In this section, we use it to find the size of the payment of a simple interest amortized loan.

EXAMPLE 1 Heidi Ochikubo buys a car for \$13,518.77. She makes a \$1,000 down payment and finances the balance through a four-year simple interest amortized loan from her bank. She is charged 12% interest. Find the following.

a. the size of her monthly payment
b. the total interest for the loan

Solution **a.** We are given $P = 13{,}518.77 - 1{,}000.00 = 12{,}518.77$, $i = \frac{1}{12}$ of $12\% = 0.01$, and $n = 4 \text{ years} = 4 \text{ years} \cdot \frac{12 \text{ months}}{\text{years}} = 48 \text{ months}$.

$$\text{future value of annuity} = \text{future value of loan amount}$$

$$\text{pymt}\,\frac{(1+i)^n - 1}{i} = P(1+i)^n$$

$$\text{pymt}\,\frac{(1+0.01)^{48} - 1}{0.01} = 12{,}518.77\,(1+0.01)^{48}$$

To find "pymt," we need to divide the right side by the fraction on the left side or, equivalently, multiply by its reciprocal. First, find the fraction on the left side, as with any annuity calculation; then multiply the right side by the fraction's reciprocal.

[(] 1 [+] .01 [)] [y^x] 48 [−] 1 [=] [÷] .01 [=]

fraction on left side

[$1/x$] [×] 12518.77 [×] [(] 1 [+] .01 [)] [y^x] 48 [=]

reciprocal — right side

With a graphing calculator, replace [y^x] with [^], [$1/x$] with [x^{-1}] and [=] with [ENTER]

We get pymt = \$329.667 . . . ≈ \$329.67

✔ Heidi borrowed \$12,518.77, and each payment includes principal and interest; therefore, the payment must be larger than 12,518.77/48 = \$260.81.

b. Heidi has agreed to make 48 payments of \$329.67 each, for a total of 48 · \$329.67 = \$15,824.16. Of this, \$12,518.77 is principal, so the balance of \$15,824.16 − \$12,518.77 = \$3,305.39 is interest. ■

Amortization Schedules

Each payment of an amortized loan includes principal and interest. A *simple interest* amortized loan is so named because the interest portion of each payment is simple interest on the outstanding principal. An **amortization schedule** is a list of several periods of payments showing the principal and interest portions of those payments and the outstanding principal (or balance) after each payment is made.

Most lending agencies provide the borrower with an amortization schedule on an annual basis. The data on the amortization schedule are important to the borrower for two reasons. First, the borrower needs to know the total interest paid, for tax purposes. (Interest paid on a home loan is usually deductible from the borrower's income tax, and interest paid on a loan by a business is usually deductible.) Second, the borrower needs the data if he or she is considering paying off the loan early. Such prepayment could save money, because an advance payment would be all principal and would not include any interest; however, the lending institution might charge a **prepayment penalty** that would absorb some or all of the interest savings.

EXAMPLE 2 Prepare an amortization schedule for the first two months of Heidi Ochikubo's simple interest amortized loan in Example 1.

Solution From Example 1, we have loan amount = \$12,518.77, paid in 48 monthly payments of \$329.67.

For any simple interest loan, the interest portion of each payment is simple interest on the outstanding principal or balance, so we use the Simple Interest Formula $I = Prt$ to compute the interest. Recall that r and t are annual figures.

For each payment, $r = 12\% = 0.12$ and $t = 1$ month $= \frac{1}{12}$ year.

For payment number 1, $P = \$12{,}518.77$ (the amount borrowed).

$$\begin{aligned} I &= Prt \\ &= 12{,}518.77 \cdot 0.12 \cdot \frac{1}{12} \\ &= 125.1877 \\ &\approx \$125.19 \end{aligned}$$

The principal portion of payment number 1 is

$\$329.67 - \$125.19 = \$204.48$ payment minus interest portion

The outstanding principal or balance after payment is

$\$12{,}518.77 - \$204.48 = \$12{,}314.29$ previous principal minus principal portion

For payment number 2, $P = \$12{,}314.29$ (the outstanding principal).

$$\begin{aligned} I &= Prt \\ &= 12{,}314.29 \cdot 0.12 \cdot \frac{1}{12} \\ &= 123.1429 \\ &\approx \$123.14 \end{aligned}$$

The principal portion of payment number 2 is

$\$329.67 - \$123.14 = \$206.53$

The balance after payment is

$\$12{,}314.29 - \$206.53 = \$12{,}107.76$

The chart in Figure 5.14 is the amortization schedule for the first two months of Heidi's loan. Notice that the principal portion increased and the interest portion decreased after the first payment. This pattern continues throughout the life of the loan, and the final payment is mostly principal. After each payment, the balance is somewhat smaller, so the interest on the balance is somewhat smaller as well.

Payment Number	Principal Portion	Interest Portion	Total Payment	Balance
0	—	—	—	\$12,518.77
1	\$204.48	\$125.19	\$329.67	\$12,314.29
2	\$206.53	\$123.14	\$329.67	\$12,107.76

Figure 5.14

■

EXAMPLE 3 Comp-U-Rent needs to borrow \$60,000 to increase its inventory of rental computers. The company is confident that the expanded inventory will generate sufficient extra income to allow it to pay off the loan in a short amount of time, so it wishes to borrow the money for only three months. First National Bank offered a simple interest amortized loan at $8\frac{3}{4}\%$ interest, and World Bank offered an add-on interest amortized loan at the same interest rate.

a. Find what the monthly payment would be with First National Bank.
b. Find what the monthly payment would be with World Bank.
c. Prepare an amortization schedule for the better offer.

Solution **a.** First National Bank offered a simple interest amortized loan; we have $P =$ \$60,000, $i = \frac{1}{12}$ of $8\frac{3}{4}\% = \frac{0.0875}{12}$, and $n = 3$ months.

$$\text{future value of annuity} = \text{future value of loan amount}$$

$$\text{pymt}\,\frac{(1+i)^n - 1}{i} = P(1+i)^n$$

$$\text{pymt}\,\frac{\left(1+\dfrac{0.0875}{12}\right)^3 - 1}{\dfrac{0.0875}{12}} = 60{,}000\left(1+\frac{0.0875}{12}\right)^3$$

First, compute the fraction on the left side. Then find "pymt" by taking the reciprocal of that fraction and multiplying it by the right side.

We get

$$\begin{aligned}\text{pymt} &= 20{,}292.376\\ &\approx \$\,20{,}292.38\end{aligned}$$

✔ Comp-U-Rent borrowed \$60,000, and each payment includes principal and interest; therefore, the payment must be larger than \$60,000/3 = \$20,000.

b. World Bank offered an add-on interest amortized loan; we have $P =$ \$60,000, $r = 8\frac{3}{4}\% = 0.0875$, and $t = 3$ months $= \frac{1}{4}$ year. The total amount due is

$$\begin{aligned}FV &= P(1+rt)\\ &= \$60{,}000\left(1 + 0.0875\cdot\frac{1}{4}\right)\\ &\approx \$61{,}312.50\end{aligned}$$

There are three monthly payments; therefore, monthly payment $= \frac{\$61{,}312.50}{3} =$ \$20,437.50.

c. The better loan offer is the simple interest amortized loan through First National Bank, because the monthly payments are \$20,292.38, compared to \$20,437.50 for the add-on interest loan at the same interest rate with World Bank.

For each payment, $r = 8\frac{3}{4}\% = 0.0875$ and $t = 1$ month $= \frac{1}{12}$ year.

For payment number 1, $P = \$60{,}000$ (the amount borrowed).

$$\begin{aligned} I &= Prt \\ &= 60{,}000 \cdot 0.0875 \cdot \frac{1}{12} \\ &\approx \$437.50 \end{aligned}$$

The principal portion of payment number 1 is

$\$20{,}292.38 - \$437.50 = \$19{,}854.88$ payment minus interest portion

The balance after payment is

$\$60{,}000.00 - \$19{,}854.88 = \$40{,}145.12$ previous principal minus principal portion

For payment number 2, $P = \$40{,}145.12$.

$$\begin{aligned} I &= Prt \\ &= 40{,}145.12 \cdot 0.0875 \cdot \frac{1}{12} \\ &= 292.72483\ldots \\ &\approx \$292.72 \end{aligned}$$

The principal portion of payment number 2 is

$$\$20{,}292.38 - \$292.72 = \$19{,}999.66$$

The balance after payment is

$$\$40{,}145.12 - \$19{,}999.66 = \$20{,}145.46$$

For payment number 3, $P = \$20{,}145.46$.

$$\begin{aligned} I &= Prt \\ &= 20{,}145.46 \cdot 0.0875 \cdot \frac{1}{12} \\ &= 146.89398\ldots \\ &\approx \$146.89 \end{aligned}$$

The principal portion of payment number 3 is

$$\$20{,}292.38 - \$146.89 = \$20{,}145.49$$

The balance after payment is

$$\$20{,}145.46 - \$20{,}145.49 = -\$0.03$$

Negative three cents can't be right. After the final payment is made, the balance *must be* \$0.00. The discrepancy arises from the fact that we rounded off the payment size from \$20,292.376 to \$20,292.38. If there were some way the borrower could make a monthly payment that is not rounded off, the calculation above would have yielded an amount due of \$0.00. To repay the exact amount owed, we must compute the last payment differently.

The principal portion of payment number 3 *must* be \$20,145.46, because that is the balance, and this payment is the only chance to pay it. The payment must also include \$146.89 interest, as calculated above. Thus, the last payment is the sum of the principal due and the interest on that principal:

$$\$20{,}145.46 + \$146.89 = \$20{,}292.35$$

The balance after payment is then \$20,145.46 − \$20,145.46 = \$0.00, as it should be.

The amortization schedule is given in Figure 5.15.

Payment Number	Principal Portion	Interest Portion	Total Payment	Balance
0	—	—	—	\$60,000.00
1	\$19,854.88	\$437.50	\$20,292.38	\$40,145.12
2	\$19,999.66	\$292.72	\$20,292.38	\$20,145.46
3	\$20,145.46	\$146.89	\$20,292.35	\$0.00

Figure 5.15

Amortization Schedule Steps

For each payment, list the payment number, principal portion, interest portion, total payment and balance.

For each payment:

Step 1. Find the interest on the balance—use the simple interest formula.

For each payment except the last:

Step 2. The principal portion is the payment minus the interest portion.

Step 3. The new balance is the previous balance minus the principal portion.

For the last payment:

Step 4. The principal portion is the previous balance.

Step 5. The total payment is the sum of the principal portion and interest portion.

Payment Number	Principal Portion	Interest Portion	Total Payment	Balance
0	—	—	—	loan amount
first through next-to-last	total payment minus interest portion	simple interest on previous balance; use $I = Prt$	use Simple Interest Amortized Loan Formula	previous balance minus this payment's principal portion
last	previous balance	simple interest on previous balance; use $I = Prt$	principal portion plus interest portion	\$0.00

In preparing an amortization schedule, the new balance is the previous balance minus the principal portion. This means that only the principal portion of a payment goes toward paying off the loan; the interest portion is the lender's fee for the use of its money. In Example 3, the interest portion of Comp-U-Rent's first loan payment was $437.50, and the principal portion was $19,854.88, so almost all of that payment went toward paying off the loan. However, if Comp-U-Rent's loan were for 30 years rather than three months, the interest portion of the first payment would remain at $437.50, but the principal portion would be $34.52, and very little of that payment would go toward paying off the loan.

It is extremely depressing for first-time home purchasers to discover how little of their initial payment actually goes toward paying off the loan. The homeowner has a few alternatives: paying off the loan early, finding a new loan with a better interest rate, or "creatively altering" the traditional structure of a simple interest amortized loan. Popular examples of the third alternative include altering the payment period from monthly to biweekly and altering the loan's duration from 30 years to 20 or 15. These alternatives are discussed in the exercises.

Another example of altering the structure of the loan is making one extra payment each year. This extra payment is all principal, and it creates some very complex accounting that is beyond the scope of this book; it is a viable alternative only if the loan does not have a prepayment penalty. An informed source in the banking world has told us that if the borrower makes one extra payment each year for the life of a 30-year loan, then the loan will be paid off approximately 10 years early!

Finding an Unpaid Balance

Prepaying a loan has some real advantages. Any additional amount paid beyond the required monthly principal and interest must be classified by the lender as payment toward the principal. The size of the monthly payment is unaffected by a prepayment, but the interest portion of future monthly payments will be decreased because less principal is owed. Furthermore, the loan will be paid off early for the same reason.

Prepaying also has some disadvantages. Obviously, the borrower must have sufficient savings to afford prepayments. Using part of one's savings for prepayment means that that money is no longer earning interest. Also, one's income tax would go up as a result of prepayment if the interest portion of each payment is deductible.

If a borrower is thinking about paying off his or her loan early, several things should be considered. Some lenders charge a prepayment penalty, an extra fee assessed if the loan is prepaid. The lender can do this only if the loan agreement allows it; when you sign a loan agreement, you should determine whether a prepayment penalty is included. If a borrower is considering prepaying a loan, he or she should weigh the advantage of not making the monthly payments against the interest he or she could earn by investing the money that will be used to prepay. This involves knowing the **unpaid balance.** If an amortization schedule has not already been prepared, the borrower can easily compute the unpaid balance by subtracting the current value of the annuity from the current value of the loan.

A common error in using this formula is to let n equal the number of periods in the entire life of the loan, rather than the number of periods from the beginning of the loan until the time of prepayment.

Unpaid Balance Formula

unpaid balance = current value of loan amount − current value of annuity

$$= P(1 + i)^n - \text{pymt}\,\frac{(1 + i)^n - 1}{i}$$

where "pymt" is the loan payment, i is the periodic interest rate, n is the number of periods *from the beginning of the loan to the present,* and P is the loan amount.

EXAMPLE 4 Ten years ago, Rob and Shelly Golumb bought a home for $140,000. They paid the sellers a 20% down payment and obtained a simple interest amortized loan for the balance from their bank at $10\frac{3}{4}\%$ for 30 years.

The Granger Collection

Rob and Shelly Golumb won $100,000 on a TV game show!

Recently, they won $100,000 on a TV game show. They are considering paying off their home loan with that money, and they want to know how much of their winnings would be left.

a. Find the size of the Golumb's monthly payment.
b. Find the unpaid balance of their loan.
c. Find what portion of their winnings would be left after prepaying their loan.
d. Find the amount of interest they would save by prepaying.

Solution **a.** *Finding their monthly payment* We are given that down payment = 20% of \$140,000 = \$28,000, P = loan amount = \$140,000 − \$28,000 = \$112,000, i = 1/12th of $10\frac{3}{4}\%$ = 0.1075/12, and n = 30 years = 30 years · (12 months)/(1 year) = 360 months.

$$\text{future value of annuity} = \text{future value of loan amount}$$

$$\text{pymt}\,\frac{(1+i)^n - 1}{i} = P(1+i)^n$$

$$\text{pymt}\,\frac{(1+0.1075/12)^{360} - 1}{0.1075/12} = 112{,}000\,(1+0.1075/12)^{360}$$

Computing the fraction on the left side and multiplying its reciprocal by the right side, we get

$$\text{pymt} = 1{,}045.4991\ldots \approx \$1{,}045.50$$

b. *Finding the unpaid balance* The Golumbs have made payments on their loan for 10 years = 120 months, so n = 120 (not 360!).

$$\begin{aligned}\text{unpaid balance} &= \text{current value of loan amount} - \text{current value of annuity}\\ &= P(1+i)^n - \text{pymt}\,\frac{(1+i)^n - 1}{i}\\ &= 112{,}000\,(1+0.1075/12)^{120}\\ &\quad - 1{,}045.50\,\frac{(1+0.1075/12)^{120} - 1}{0.1075/12}\\ &= 102{,}981.4237\ldots \approx \$102{,}981.42\end{aligned}$$

First, compute .1075/12 and put it in memory. Then type

112000 [×] [(] 1 [+] [RCL] [)] [y^x] 120 [−] 1045.50 [×] [(] [(] 1 [+] [RCL] [)] [y^x] 120 [−] 1 [)] [÷] [RCL] [=]

112000 [×] [(] 1 [+] [ALPHA] [A] [)] [∧] 120 [−] 1045.50 [×] [(] [(] 1 [+] [ALPHA] [A] [)] [∧] 120 [−] 1 [)] [÷] [ALPHA] [A] [ENTER]

This means that after 10 years of payments on a loan of \$112,000, the Golumbs still owe \$102,981.42!

c. *Finding the portion of their winnings that would be left* The Golumbs won \$100,000, so they could not afford to prepay the loan unless they used more than their game show winnings.

d. *Finding the amount of interest they would save by prepaying* If they prepaid, they would not have to make their remaining 240 monthly payments of \$1,045.50 each, a total of

$$240 \cdot \$1{,}045.50 = \$250{,}920.00$$

Of this amount, \$102,981.42 is principal, and the interest they would save by prepaying is

$$\$250{,}920.00 - \$102{,}981.42 = \$147{,}938.58$$

They would have to weigh this savings and the extra money they would have from not making monthly payments against the interest they would receive from investing all or part of their winnings. ■

5.4 Exercises

In the following exercises, all loans are simple interest amortized loans with monthly payments, unless labeled otherwise.

In Exercises 1–6, find (a) the monthly payment and (b) the total interest for the given simple interest amortized loan.

1. \$5,000 at $9\frac{1}{2}\%$ for 4 years
2. \$8,200 at $10\frac{1}{4}\%$ for 6 years
3. \$10,000 at $6\frac{1}{8}\%$ for 5 years
4. \$20,000 at $7\frac{3}{8}\%$ for $5\frac{1}{2}$ years
5. \$155,000 at $9\frac{1}{2}\%$ for 30 years
6. \$289,000 at $10\frac{3}{4}\%$ for 35 years
7. Wade Ellis buys a new car for \$16,113.82. He puts 10% down and obtains a simple interest amortized loan for the balance at $11\frac{1}{2}\%$ interest for four years.
   a. Find his monthly payment.
   b. Find the total interest.
   c. Prepare an amortization schedule for the first two months of the loan.
8. Guy de Primo buys a new car for \$9,837.91. He puts 10% down and obtains a simple interest amortized loan for the balance at $8\frac{7}{8}\%$ for four years.
   a. Find his monthly payment.
   b. Find the total interest.
   c. Prepare an amortization schedule for the first two months of the loan.
9. Chris Burditt bought a house for \$212,500. He put 20% down and obtained a simple interest amortized loan for the balance at $10\frac{7}{8}\%$ for 30 years.
   a. Find his monthly payment.
   b. Find the total interest.
   c. Prepare an amortization schedule for the first two months of the loan.
   d. Most lenders will approve a home loan only if the total of all the borrower's monthly payments, including the home loan payment, is no more than 38% of the borrower's monthly income. How much must Mr. Burditt make to qualify for the loan?
10. Shirley Trembley bought a house for \$187,600. She put 20% down and obtained a simple interest amortized loan for the balance at $6\frac{3}{8}\%$ for 30 years.
   a. Find her monthly payment.
   b. Find the total interest.
   c. Prepare an amortization schedule for the first two months of the loan.
   d. Most lenders will approve a home loan only if the total of all the borrower's monthly payments, including the home loan payment, is no more than 38% of the borrower's monthly income. How much must Ms. Trembley make to qualify for the loan?
11. Ray and Helen Lee bought a house for \$189,500. They put 10% down, borrowed 80% from their bank for 30 years at 11.5%, and convinced the seller to take a second mortgage for the remaining 10%. That 10% is due in full in five years (this is called a *balloon payment*), and the Lees agreed to make monthly interest-only payments to the seller at 12% simple interest in the interim.
   a. Find the Lees' down payment.
   b. Find the amount that the Lees borrowed from their bank.
   c. Find the amount that the Lees borrowed from the seller.
   d. Find the Lees' monthly payment to the bank.
   e. Find the Lees' monthly interest payment to the seller.
   f. If the Lees save for their balloon payment with a sinking fund, find the size of the necessary monthly payments into that fund if their money earns 6% interest.
   g. Find the Lees' total monthly payment for the first five years.
   h. Find the Lees' total monthly payment for the last 25 years.
12. Jack and Laurie Worthington bought a house for \$163,700. They made a 10% down payment, borrowed 80% from their bank for 30 years at 12%, and

convinced the seller to take a second mortgage for the remaining 10%. That 10% is due in full in five years, and the Worthingtons agreed to make monthly interest-only payments to the seller at 12% simple interest in the interim.

a. Find the Worthingtons' down payment.
b. Find the amount that the Worthingtons borrowed from their bank.
c. Find the amount that the Worthingtons borrowed from the seller.
d. Find the Worthingtons' monthly payment to the bank.
e. Find the Worthingtons' monthly interest payment to the seller.
f. If the Worthingtons save for their balloon payment with a sinking fund, find the size of the necessary monthly payments into that fund if their money earns 7% interest.
g. Find the Worthingtons' total monthly payment for the first five years.
h. Find the Worthingtons' total monthly payment for the last 25 years.

13. Dennis Lamenti wants to buy a new car that costs \$15,829.32. He has two possible loans in mind. One loan is through the car dealer; it is a four-year add-on interest loan at $7\frac{3}{4}\%$ and requires a down payment of \$1,000. The second loan is through his bank; it is a four-year simple interest amortized loan at $8\frac{7}{8}\%$ and requires a 10% down payment.

a. Find the monthly payment for each loan.
b. Find the total interest paid for each loan.
c. Which loan should Dennis choose? Why?

14. Barry Wood wants to buy a used car that costs \$4,000. He has two possible loans in mind. One loan is through the car dealer; it is a three-year add-on interest loan at 6% and requires a down payment of \$300. The second loan is through his credit union; it is a three-year simple interest amortized loan at 9.5% and requires a 10% down payment.

a. Find the monthly payment for each loan.
b. Find the total interest paid for each loan.
c. Which loan should Barry choose? Why?

15. Investigate the effect of the term on simple interest amortized automobile loans by finding the monthly payment and the total interest for a loan of \$11,000 at $9\frac{7}{8}\%$ interest for the following terms.

a. 3 years b. 4 years c. 5 years

16. Investigate the effect of the interest rate on simple interest amortized automobile loans by finding the monthly payment and the total interest for a four-year loan of \$12,000 at the following interest rates.

a. $8\frac{1}{2}\%$ b. $8\frac{3}{4}\%$ c. 9% d. 10%

17. Investigate the effect of the interest rate on home loans by finding the monthly payment and the total interest for a 30-year loan of \$100,000 at the following interest rates.

a. 6% b. 7% c. 8%
d. 9% e. 10% f. 11%

18. Some lenders are now offering 15-year home loans. Investigate the effect of the term on home loans by finding the monthly payment and total interest for a loan of \$100,000 at 10% for the following terms.

a. 30 years b. 15 years

19. Verify (a) the monthly payments and (b) the interest savings in the Continental Savings advertisement shown below.

Continental Savings of America

20. The home loan in Exercise 19 presented two options. The 30-year option required a smaller monthly payment. A consumer who chooses the

30-year option could take the savings in the monthly payment generated by that option and invest that savings in an annuity. At the end of 15 years, the annuity might be large enough to pay off the 30-year loan. Determine whether this is a wise plan, if the annuity's interest rate is 7.875%. (Disregard the tax ramifications of this approach.)

21. Some lenders are now offering loans with biweekly payments rather than monthly payments. Investigate the effect of this option on home loans by finding the payment and total interest on a 30-year loan of $100,000 at 10% interest if payments are made (a) monthly and (b) biweekly.

22. Pool-N-Patio World needs to borrow $75,000 to increase its inventory for the upcoming summer season. The owner is confident that he will sell most, if not all, of the new inventory during the summer, so he wishes to borrow the money for only four months. His bank has offered him a simple interest amortized loan at $7\frac{3}{4}\%$ interest.

a. Find the size of the monthly bank payment.

b. Prepare an amortization schedule for all four months of the loan.

23. Slopes R Us needs to borrow $120,000 to increase its inventory of ski equipment for the upcoming season. The owner is confident that she will sell most, if not all, of the new inventory during the winter, so she wishes to borrow the money for only five months. Her bank has offered her a simple interest amortized loan at $8\frac{7}{8}\%$ interest.

a. Find the size of the monthly bank payment.

b. Prepare an amortization schedule for all five months of the loan.

24. The owner of Video Extravaganza is opening a second store and needs to borrow $93,000. Her success with the first store has made her confident that she will be able to pay off her loan quickly, so she wishes to borrow the money for only four months. Her bank has offered her a simple interest amortized loan at $9\frac{1}{8}\%$ interest.

a. Find the size of the monthly bank payment.

b. Prepare an amortization schedule for all four months of the loan.

25. The Green Growery Nursery needs to borrow $48,000 to increase its inventory for the upcoming summer season. The owner is confident that he will sell most, if not all, of the new plants during the summer, so he wishes to borrow the money for only four months. His bank has offered him a simple interest amortized loan at $9\frac{1}{4}\%$ interest.

a. Find the size of the monthly bank payment.

b. Prepare an amortization schedule for all four months of the loan.

*For Exercises 26–28, note the following information. A **line of credit** is an agreement between a bank and a borrower where the borrower can borrow any amount of money (up to a mutually agreed-upon maximum) at any time, simply by writing a check. Typically, monthly simple interest payments are required, and the borrower is free to make principal payments as frequently or infrequently as he or she wants. Usually, a line of credit is secured by the title to the borrower's house, and the interest paid to the bank by the borrower is deductible from the borrower's income taxes.*

26. Kevin and Roxanne Gahagan did not have sufficient cash to pay their income taxes. However, they had previously set up a line of credit with their bank. On April 15, they wrote a check to the I.R.S. on their line of credit for $6,243. The line's interest rate is 5.75%.

a. Find the size of the required monthly interest payment.

b. The Gahagans decided that it would be in their best interests to get this loan paid off in eight months. Find the size of the monthly principal-plus-interest payment that would accomplish this.

HINT: In effect, the Gahagans are converting the loan to an amortized loan.

c. Prepare an amortization schedule for all eight months of the loan.

d. Find the amount of line of credit interest that the Gahagans could deduct from their taxes next year.

27. James and Danna Wright did not have sufficient cash to pay their income taxes. However, they had previously set up a line of credit with their bank. On April 15, they wrote a check to the I.R.S. on their line of credit for $10,288. The line's interest rate is 4.125%.

a. Find the size of the required monthly interest payment.

b. The Wrights decided that it would be in their best interests to get this loan paid off in six months. Find the size of the monthly principal-plus-interest payment that would accomplish this.

HINT: In effect, the Wrights are converting the loan to an amortized loan.

c. Prepare an amortization schedule for all six months of the loan.

d. Find the amount of line of credit interest that the Wrights could deduct from their taxes next year.

28. Homer Simpson did not have sufficient cash to pay his income taxes. However, he had previously set up a line of credit with his bank. On April 15, he wrote a check to the I.R.S. on his line of credit for $12,982. The line's interest rate is 8.25%.

a. Find the size of the required monthly interest payment.

b. Homer decided that it would be in his best interests to get this loan paid off in seven months. Find the size of the monthly principal-plus-interest payment that would accomplish this.

HINT: In effect, Simpson is converting the loan to an amortized loan.

c. Prepare an amortization schedule for all seven months of the loan.

d. Find the amount of line of credit interest that Simpson could deduct from his taxes next year.

In Exercises 29 and 30, you are asked to prepare an amortization schedule for an add-on interest loan. This schedule should have the same data as does an amortization schedule for a simple interest amortized loan, but the computational procedure is different. Use the information in Section 5.1 on add-on interest loans to help you determine this procedure.

29. Prepare an amortization schedule for the first three months of each of Dennis Lamenti's two possible loans in Exercise 13. By comparing the schedules, what advantages can you see in one loan over the other?

30. Prepare an amortization schedule for the first two months of each of Barry Wood's two possible loans in Exercise 14. By comparing the schedules, what advantages can you see in one loan over the other?

31. Wade Ellis buys a new car for $16,113.82. He puts 10% down and obtains a simple interest amortized loan for the balance at $11\frac{1}{2}$% interest for four years. Three years and two months later, he sells his car. Find the unpaid balance on his loan.

32. Guy de Primo buys a new car for $9,837.91. He puts 10% down and obtains a simple interest amortized loan for the balance at $10\frac{7}{8}$% interest for four years. Two years and six months later, he sells his car. Find the unpaid balance on his loan.

33. Gary Kersting buys a house for $212,500. He puts 20% down and obtains a simple interest amortized loan for the balance at $10\frac{7}{8}$% interest for 30 years. Eight years and two months later, he sells his house. Find the unpaid balance on his loan.

34. Shirley Trembley buys a house for $187,600. She puts 20% down and obtains a simple interest amortized loan for the balance at $11\frac{3}{8}$% interest for 30 years. Ten years and six months later, she sells her house. Find the unpaid balance on her loan.

35. Harry and Natalie Wolf have a three-year-old loan with which they purchased their house; their interest rate is $13\frac{3}{8}$%. Since they obtained this loan, interest rates have dropped, and they can now get a loan for $8\frac{7}{8}$% through their credit union. Because of this, the Wolfs are considering refinancing their home. Each loan is a 30-year simple interest amortized loan, and neither has a prepayment penalty. The existing loan is for $152,850, and the new loan would be for the current amount due on the old loan.

a. Find their monthly payment with the existing loan.

b. Find the loan amount for their new loan.

c. Find the monthly payment with their new loan.

d. Find the total interest they will pay if they do *not* get a new loan.

e. Find the total interest they will pay if they *do* get the new loan.

f. Should the Wolfs refinance their home? Why or why not?

36. Russ and Roz Rosow have a 10-year-old loan with which they purchased their house; their interest rate is $10\frac{5}{8}$%. Since they obtained this loan, interest rates have dropped, and they can now get a loan for $9\frac{1}{4}$% through their credit union. Because of this, the Rosows are considering refinancing their home. Each loan is a 30-year simple interest amortized loan, and neither has a prepayment penalty. The existing loan is for $112,000, and the new loan would be for the current amount due on the old loan.

a. Find the monthly payment with their existing loan.

b. Find the loan amount for their new loan.

c. Find the monthly payment with their new loan.

d. Find the total interest they will pay if they do *not* get a new loan.

e. Find the total interest they will pay if they *do* get the new loan.

f. Should the Rosows refinance their home? Why or why not?

37. Michael and Lynn Sullivan have a 10-year-old loan for \$187,900 with which they purchased their home. They just sold their highly profitable import-export business and are considering paying off their home loan. Their loan is a 30-year simple interest amortized loan at $10\frac{1}{2}\%$ and has no prepayment penalty.

a. Find the size of their monthly payment.

b. Find the unpaid balance of the loan.

c. Find the amount of interest they would save by prepaying.

d. The Sullivans decided that if they paid off their loan, they would deposit the equivalent of half their monthly payment into an annuity. If the ordinary annuity pays 9% interest, find its future value after 20 years.

e. The Sullivans decided that if they did not pay off their loan, they would deposit an amount equivalent to their unpaid balance into an account that pays $9\frac{3}{4}\%$ interest compounded monthly. Find the future value of this account after 20 years.

f. Should the Sullivans prepay their loan? Why or why not?

38. Charlie and Ellen Wilson have a 25-year-old loan for \$47,000 with which they bought their home. The Wilsons have retired and are living on a fixed income, so they are contemplating paying off their home loan. Their loan is a 30-year simple interest amortized loan at $4\frac{1}{2}\%$ and has no prepayment penalty. They also have savings of \$73,000, which they have invested in a certificate of deposit currently paying $8\frac{1}{4}\%$ simple interest. Should they pay off their home loan? Why or why not?

Answer the following questions using complete sentences and your own words.

Concept Questions

39. In Example 3, an add-on interest amortized loan required larger payments than did a simple interest amortized loan at the same interest rate. What is there about the structure of an add-on interest loan that makes its payments larger than those of a simple interest amortized loan?

HINT: The interest portion is a percent of what quantity?

40. Why are the computations for the last period of an amortization schedule different from those for all preceding periods?

41. Give two different situations in which a borrower would need the information contained in an amortization schedule.

42. A borrower would need to know the unpaid balance of a loan if the borrower were to prepay the loan. Give three different situations in which it might be in a borrower's best interests to prepay a loan.

HINT: See some of the preceding exercises.

43. If you double the period of an amortized loan, does your monthly payment halve? Why or why not?

44. If you double the loan amount of an amortized loan, does your monthly payment double? Why or why not?

Projects

For Exercises 45–47, note the following information. An ***adjustable rate mortgage*** *(or ARM) is, as the name implies, a mortgage where the interest rate is allowed to change; as a result, the payment changes too. At first, an ARM costs less than a fixed rate mortgage—its initial interest rate is usually two or three percentage points lower. As time goes by, the rate is adjusted; as a result, it may or may not continue to hold this advantage.*

45. Trustworthy Savings offers a 30-year adjustable rate mortgage with an initial rate of 5.375%. The rate and the required payment are adjusted annually. Future rates are set at 2.875 percentage points above the 11th District Federal Home Loan Bank's cost of funds. Currently, that cost of funds is 4.839%. The loan's rate is not allowed to rise more than 2 percentage points in any one adjustment, nor is it allowed to rise above 11.875%. Trustworthy Savings also offers a 30-year fixed rate mortgage with an interest rate of 7.5%.

a. Find the monthly payment for the fixed rate mortgage on a loan amount of \$100,000.

b. Find the monthly payment for the ARM's first year on a loan amount of \$100,000.

c. How much would the borrower save in the mortgage's first year by choosing the adjustable rather than the fixed rate mortgage?

d. Find the unpaid balance at the end of the ARM's first year.

e. Find the interest rate and the value of n for the ARM's second year, if the 11th District Federal Home Loan Bank's cost of funds does not change during the loan's first year.

f. Find the monthly payment for ARM's second year, if the 11th District Federal Home Loan Bank's cost of funds does not change during the loan's first year.

g. How much would the borrower save in the mortgage's first two years by choosing the adjustable rather than the fixed rate mortgage, if the cost of funds does not change?
h. Discuss the advantages and disadvantages of an adjustable rate mortgage.

46. American Dream Savings Bank offers a 30-year adjustable rate mortgage with an initial rate of 4.25%. The rate and the required payment are adjusted annually. Future rates are set at 3 percentage points above the one-year Treasury bill rate, which is currently 5.42%. The loan's rate is not allowed to rise more than 2 percentage points in any one adjustment, nor is it allowed to rise above 10.25%. American Dream Savings Bank also offers a 30-year fixed rate mortgage with an interest rate of 7.5%.
a. Find the monthly payment for the fixed rate mortgage on a loan amount of $100,000.
b. Find the monthly payment for the ARM's first year on a loan amount of $100,000.
c. How much would the borrower save in the mortgage's first year by choosing the adjustable rather than the fixed rate mortgage?
d. Find the unpaid balance at the end of the ARM's first year.
e. Find the interest rate and the value of n for the ARM's second year, if the one-year Treasury bill rate does not change during the loan's first year.
f. Find the monthly payment for the ARM's second year, if the Treasury bill rate does not change during the loan's first year.
g. How much would the borrower save in the mortgage's first two years by choosing the adjustable rather than the fixed rate mortgage, if the Treasury bill rate does not change?
h. Discuss the advantages and disadvantages of an adjustable rate mortgage.

47. Bank Two offers a 30-year adjustable rate mortgage with an initial rate of 3.95%. This initial rate is in effect for the first six months of the loan, after which it is adjusted on a monthly basis. The monthly payment is adjusted annually. Future rates are set at 2.45 percentage points above the 11th District Federal Home Loan Bank's cost of funds. Currently, that cost of funds is 4.839%.
a. Find the monthly payment for the ARM's first year on a loan amount of $100,000.
b. Find the unpaid balance at the end of the ARM's first six months.
c. Find the interest portion of the seventh payment, if the cost of funds does not change.
d. Usually, the interest portion is smaller than the monthly payment, and the difference is subtracted from the unpaid balance. However, the interest portion found in part (c) is *larger* than the monthly payment found in part (a), and the difference is *added* to the unpaid balance found in part (b). Why would this difference be added to the unpaid balance? What effect will this have on the loan?
e. The situation described in part (d) is called **negative amortization.** Why?
f. What is there about the structure of Bank Two's loan that allows negative amortization?

48. This is an exercise in buying a car. It involves choosing a car and selecting the car's financing. Write a paper describing all of the following points.
a. You might not be in a position to buy a car now. If that is the case, fantasize about your future. What job do you have? How long have you had that job? What is your salary? If you're married, does your spouse work? Do you have a family? What needs will your car fulfill? Make your fantasy realistic. Briefly describe what has happened between the present and your future fantasy. (If you are in a position to buy a car now, discuss these points on a more realistic level.)
b. Go shopping for a car. Look at new cars, used cars, or both. Read newspaper and magazine articles about your choices (see, for example, *Consumer Reports, Motor Trend,* and *Road and Track*). Discuss in detail the car you selected and why you did so. How will your selection fulfill your (projected) needs? What do the newspaper and magazines say about your selection? Why did you select a new or a used car?
c. Go shopping for financing. Many banks, savings and loans, and credit unions have information on car loans available on request. Get all of the information you need from at least two lenders, but do not bother the staff unnecessarily. You might be able to find the necessary information in the newspaper. Perform all appropriate computations yourself—do not have the lenders tell you the payment size. Summarize the appropriate data (including the down payment, payment size, interest rate, duration of loan, type of loan, and loan fees) in your paper, and discuss which loan you would choose. Explain how you would be able to afford your purchase.

Amortization Schedules on a Computer

Interest paid on a home loan is deductible from the borrower's income taxes, and interest paid on a loan by a business is usually deductible. A borrower with either of these types of loans needs to know the total interest paid on the loan during the final year. The way to determine the total interest paid during a given year is to prepare an amortization schedule for that year. Typically, the lender provides the borrower with an amortization schedule, but it's not uncommon for this schedule to arrive after taxes are due. In this case, the borrower must either do the calculation personally or pay taxes without the benefit of the mortgage deduction and then file an amended set of tax forms after the amortization schedule has arrived.

Computing a year's amortization schedule is rather tedious, and neither a scientific calculator nor a graphing calculator offers relief. The best tool for the job is a computer, combined either with the Amortrix computer program (available with this book) or with a computerized spreadsheet. Each is discussed below.

Amortization Schedules and Amortrix The Amortrix computer program, which is available with this book, will enable you to compute an amortization schedule for any time period. Amortrix is available for both Macintosh and Windows-based computers. Also, a somewhat different version of Amortrix is available over the web (web address: math.brookscole.com). The web version utilizes a different interface than that discussed below, so the following discussion must be used as a general guide rather than specific instructions if you use the web version of Amortrix.

When you start Amortrix, a main menu appears. Use the mouse to click on the "Amortization Schedule" option. Once you're into the "Amortization Schedule" part of the program, type the loan amount (P). Follow the instructions that appear on the screen to enter the annual interest rate, the payment period, and the number of payment periods (n). The area labeled "Expressionist" is a calculator; you may use it to compute the number of payment periods—just use the mouse to press Expressionist's buttons.

If you need to change any of the above information, either use the mouse to click on the appropriate box or type the letter in brackets in that box's label (for example, type "A" for the box labeled "Loan [A]mount").

Once all of the necessary information is entered, use the mouse to click on the box labeled "[C]reate Table" (or just type "C"). This causes the schedule for the first 12 payments to appear; to see the schedule for other payments, use the arrow buttons in the lower right corner of the screen. An Amortrix amortization schedule is shown in Figure 5.16. You can print a copy of a portion of the table by clicking on the box labeled "[P]rint Table."

The program will *not* correctly compute the last loan payment; it will compute the last payment in the same way it computes all other payments rather than in the way shown earlier in this section. You will have to correct this last payment if you use the computer to prepare an amortization schedule for a time period that includes the last payment.

Amortization Schedules and Computerized Spreadsheets A **spreadsheet** is a large piece of paper marked off in rows and columns. Accountants use spreadsheets to organize numerical data and perform computations. A **computerized spreadsheet,** such as Microsoft Excel or Lotus 1-2-3, is a computer program that mimics the appearance of a paper spreadsheet. It frees the user from performing any computations; you can just give instructions on how to perform those computations.

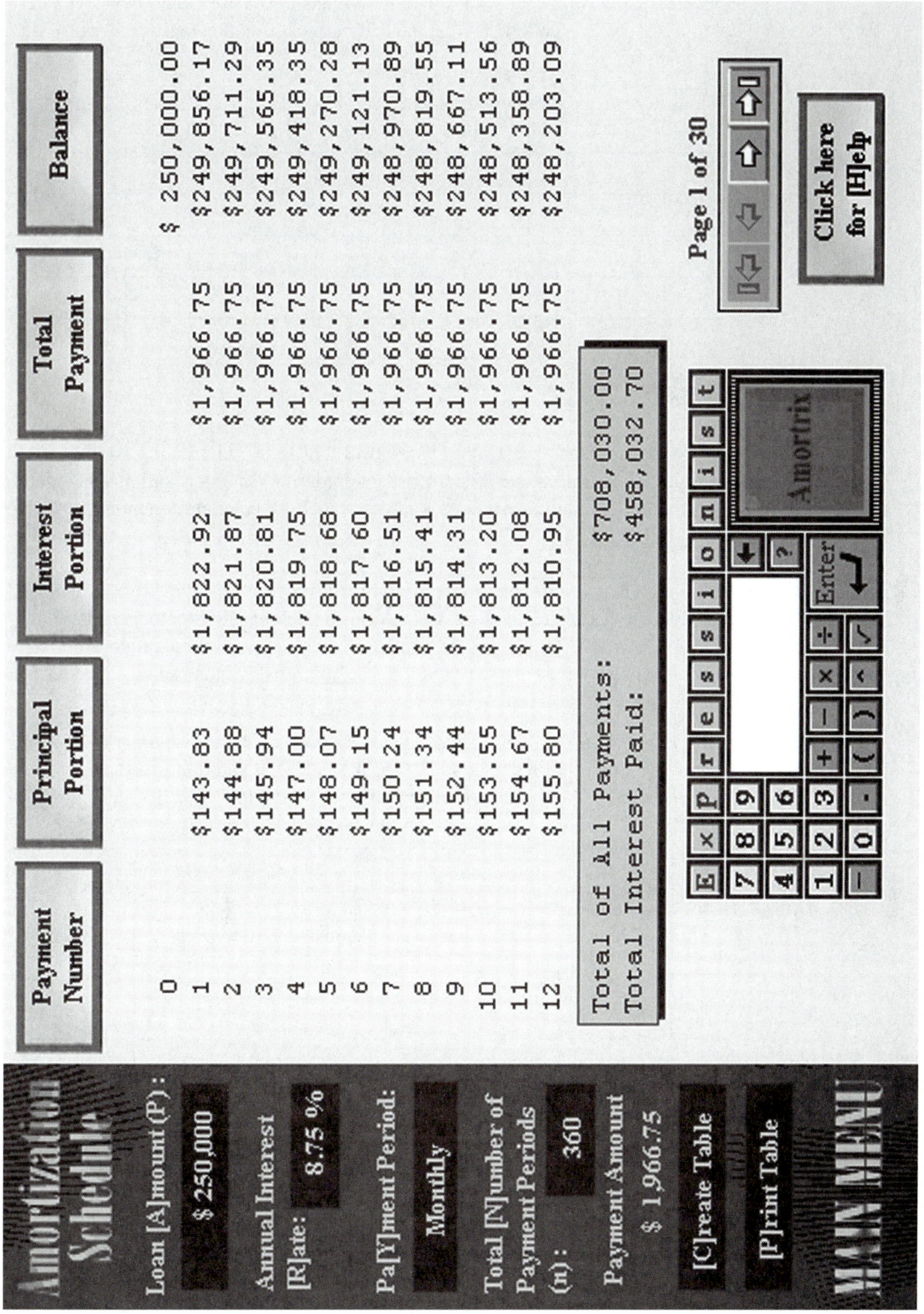

Figure 5.16
Amortrix amortization table

When you start a computerized spreadsheet, you see something that looks like a table waiting to be filled in. The rows are labeled with numbers and the columns with letters, as shown in Figure 5.17. The individual boxes are called **cells.** The cell in column A, row 1, is called cell A1; the cell below it is called cell A2, because it is in column A, row 2.

	A	B	C	D	E
1					
2					
3					
4					
5					

Figure 5.17

A computerized spreadsheet is an ideal tool to use in creating an amortization schedule. We will illustrate this process by preparing the amortization schedule for a four-month, \$175,000 loan at 7.75% interest; such a loan requires a monthly payment of \$44,458.65.

Step 1 *Label the columns.* Use the mouse and/or the arrow buttons to move to cell A1, type in "Payment Number," and press "return" or "enter." Move to cell B1, type in "Principal Portion," and press "return" or "enter." Type the remaining amortization schedule labels in cells C1, D1, and E1. The columns' widths can be adjusted—your instructor will tell you how.

Step 2 *Fill in row 2 with payment 0 information.* Move to cell A2, type in "0," and press "return" or "enter." Move to cell E2, type in "175000," and press "return" or "enter." After you complete this step, your spreadsheet should look like that in Figure 5.18.

	A	B	C	D	E
1	Payment Number	Principal Portion	Interest Portion	Total Payment	Balance
2	0				175000
3					
4					
5					
6					

Figure 5.18

Step 3 *Fill in row 3 with instructions on how to compute payment 1 information.*

- Move to cell D3 and type in "44458.65," the total payment.
- Column A should eventually contain the numbers 0, 1, 2, 3, and 4. Since each of these numbers is one more than the previous number, the instructions in cell A3

should say to add 1 to the contents in cell A2. To do this, type "= A2 + 1" or "+ A2 + 1" and press "return" or "enter." Notice that these computational instructions begin with either the "=" symbol or the "+" symbol. Computational instructions are always preceded by one of these symbols. (Your software may accept only one of these symbols.)

- Cell C3 should contain instructions on computing interest on the previous balance, using the Simple Interest Formula $I = Prt$.

$$\begin{aligned} I &= P \cdot r \cdot t \\ &= (\text{the previous balance}) \cdot (0.0775) \cdot (1/12) \\ &= \text{E2} * .0775/12 \quad \text{computers use * for multiplication} \end{aligned}$$

  Move to cell C3, type in either "= E2 * .0775 / 12" or "+ E2 * .0775 / 12," and press "return" or "enter."

- Cell B3 should contain instructions on computing the payment's principal portion.

$$\begin{aligned} \text{principal portion} &= \text{payment} - \text{interest portion} \\ &= \text{D3} - \text{C3} \end{aligned}$$

  Move to cell B3 and type in either "= D3 − C3" or "+ D3 − C3."

- Cell E3 should contain instructions on computing the new balance.

$$\begin{aligned} \text{new balance} &= \text{previous balance} - \text{principal portion} \\ &= \text{E2} - \text{B3} \end{aligned}$$

  Move to cell E3 and type in either "= E2 − B3" or "+ E2 − B3."

After you complete this step, your spreadsheet should look like that in Figure 5.19.

Figure 5.19

	A	B	C	D	E
1	Payment Number	Principal Portion	Interest Portion	Total Payment	Balance
2	0				175000
3	1	43328.44167	1130.20833	44458.65	131671.5583
4					
5					

Notice that the cells in row 3 show more decimal places than is appropriate. We will take care of this later.

Step 4 *Fill in rows 4 and 5 with instructions on how to compute payment 2 and payment 3 information.* Payment 2 and payment 3 computations are just like payment 1 computations, so all we have to do is copy the payment 1 instructions from row 3 and paste them into rows 4 and 5.

To copy the payment 1 instructions from row 3, do the following.

- Use your mouse to move to cell A3.
- Press the mouse's button, move the mouse to the right until cells B3, C3, D3, and E3 are highlighted, and let go of the button. Be careful that you do not highlight any other cells.

- Use the mouse to move to the word "Edit" at the very top of the screen.
- Press the mouse's button, move the mouse down until the word "Copy" is highlighted, and let go of the button.

To paste the payment 1 instructions into row 4, do the following.

- Use your mouse to move to cell A4.
- Press the mouse's button and move the mouse to the right and down until cells A4, B4, C4, D4, E4, A5, B5, C5, D5, and E5 are highlighted.
- Use the mouse to move to the word "Edit" at the very top of the screen.
- Press the mouse's button, move the mouse down until the word "Paste" is highlighted, and let go of the button.

After you complete this step, your spreadsheet should look like that in Figure 5.20.

	A	B	C	D	E
1	Payment Number	Principal Portion	Interest Portion	Total Payment	Balance
2	0				175000
3	1	43328.44167	1130.20833	44458.65	131671.5583
4	2	43608.27119	850.378814	44458.65	88063.28715
5	3	43889.90794	568.742063	44458.65	44173.37921
6					

Figure 5.20

Note: Most amortization schedules involve more than four payments. In this case, the step 4 instructions apply to all payments other than the first and last.

Step 5 *Fill in row 6 with instructions on how to compute information on the last payment.* Except for the principal portion and the total payment, the last payment computations are just like previous payment computations, so we can do more copying and pasting.

- Copy the instructions from cell A5 and paste them into cell A6.
- Copy the instructions from cell C5 and paste them into cell C6.
- Copy the instructions from cell E5 and paste them into cell E6.
- Cell B6 should contain the principal portion of the last payment, which is equal to the previous balance. Move to cell B6, type in either "= E5" or "+ E5," and press "return" or "enter."
- Cell D6 should contain instructions on computing the last total payment.

$$\text{total payment} = \text{principal portion} + \text{interest portion}$$
$$= B6 + C6$$

Move to cell D6 and type in either "= B6 + C6" or "+ B6 + C6."

After you complete this step, your spreadsheet should look like that in Figure 5.21.

	A	B	C	D	E
1	Payment Number	Principal Portion	Interest Portion	Total Payment	Balance
2	0				175000
3	1	43328.44167	1130.20833	44458.65	131671.5583
4	2	43608.27119	850.378814	44458.65	88063.28715
5	3	43889.90794	568.742063	44458.65	44173.37921
6	4	44173.37921	285.286407	44458.665617	0

Figure 5.21

Step 6 *Format the contents of columns B, C, D, and E in currency style.* Your instructor will tell you how. After this formatting, your spreadsheet should look like that in Figure 5.22.

	A	B	C	D	E
1	Payment Number	Principal Portion	Interest Portion	Total Payment	Balance
2	0				$175,000.00
3	1	$43,328.44	$1,130.21	$44,458.65	$131,671.56
4	2	$43,608.27	$850.38	$44,458.65	$88,063.29
5	3	$43,889.91	$568.74	$44,458.65	$44,173.38
6	4	$44,173.38	$285.29	$44,458.67	$0

Figure 5.22

Finding the Total Interest Paid Once an amortization schedule is prepared, it is a simple matter to find the total interest paid—just find the sum of the interest portions of the appropriate payments. If you are using Amortrix, compute this sum with your own calculator or Expressionist. If you are using a computerized spreadsheet, you can use the spreadsheet to compute the sum. In the example illustrated in Figure 5.19, the total interest paid is the sum of cells C3 through C6. To find this quantity, go to cell C7 (or any other empty cell), type " = sum(C3:C6)" or "@sum(C3:C6)" and press "return" or "enter." With some software, it's possible to find this sum by going to cell C7 and using the mouse to press the "Σ" button.

Exercises

In Exercises 49–56, do the following.

a. *Use a computer to prepare an amortization schedule for the given loan's first year.*

b. *Find the amount that could be deducted from the borrower's taxable income (i.e., find the total interest paid) in the loan's first year if the first payment was made in January and if the interest is in fact deductible.*

c. *Find the unpaid balance at the beginning of the loan's last year.*

d. *Use a computer to prepare an amortization schedule for the given loan's last year. If you are using a* *computerized spreadsheet, this will entail using the solution to part (c) in place of the loan amount, and starting with payment numbers more appropriate than 0 and 1.*

e. *Find the amount that could be deducted from the borrower's taxes (i.e., the total interest paid) in the loan's last year.*

Answers to parts b, c, and e are given in the back of the book. Your answers may vary from those given by a few cents.

49. the loan in Exercise 7 of Exercises 5.4

50. the loan in Exercise 8 of Exercises 5.4

51. the loan in Exercise 9 of Exercises 5.4

52. the loan in Exercise 10 of Exercises 5.4

53. a 30-year simple interest amortized home loan for \$350,000 at $14\frac{1}{2}\%$ interest

54. a five-year simple interest amortized car loan for \$29,500 at $13\frac{1}{4}\%$ interest

55. a five-year simple interest amortized car loan for \$10,120 at $7\frac{7}{8}\%$ interest

56. a 30-year simple interest amortized home loan for \$109,000 at $9\frac{1}{2}\%$ interest

57. Use a computer to prepare an amortization schedule for a simple interest amortized loan of \$32,440 at 5.25% interest for one year.

5.5 Annual Percentage Rate on a Graphing Calculator

A **simple interest loan** is any loan for which the interest portion of each payment is simple interest on the outstanding principal. A *simple interest amortized loan* fulfills this requirement; in fact, we compute an amortization schedule for a simple interest amortized loan by finding the simple interest on the outstanding principal.

EXAMPLE 1 Chris Dant paid \$18,327 for a new car. Her dealer offered her a five-year add-on interest loan at 8.5% interest.

a. Find the size of her monthly payment.

b. Find the principal portion and the interest portion of each payment.

c. Determine if this loan is a simple interest loan.

Solution **a.** *Finding the monthly payment.* We are given $P = 18{,}327$, $r = 0.085$, and $t = 5$ years. The total interest charge is simple interest on the loan amount:

$$\begin{aligned} I &= Prt \\ &= 18{,}327 \cdot 0.085 \cdot 5 \\ &= 7{,}788.975 \\ &\approx \$7{,}788.98 \end{aligned}$$

The total of principal and interest is then

$$\mathrm{P} + \mathrm{I} = 18{,}327 + 7{,}788.98 = \$26{,}115.98$$

This amount is equally distributed over 60 monthly payments, so the monthly payment is

$$\frac{26{,}115.98}{60} = 435.26633\ldots \approx \$435.27$$

b. *Finding the principal and interest portions.* The total amount due consists of \$18,327 in principal and \$7,788.98 in interest. The total amount due is equally distributed over

60 monthly payments, so the principal and interest are also equally distributed over 60 monthly payments. The principal portion of each payment is then

$$\frac{18{,}327}{60} = \$305.45$$

and the interest portion is

$$\frac{7{,}788.98}{60} = 129.81633\ldots \approx \$129.82$$

c. *Determining whether the loan is a simple interest loan.* The loan is a simple interest loan if the interest portion of each payment is simple interest on the outstanding principal. For the first payment, the outstanding principal is \$18,327; simple interest on this amount is

$$\begin{aligned} I &= Prt \\ &= 18{,}327 \cdot 0.085 \cdot \frac{1}{12} \\ &= 129.81625 \approx \$129.82 \end{aligned}$$

In part (b), we found that the interest portion of each payment is also \$129.82; thus, for the first payment, the interest portion is simple interest on the outstanding principal.

For the second payment, the outstanding principal is

$$18{,}327 - 305.45 = 18{,}021.55$$

Simple interest on this amount is

$$\begin{aligned} I &= Prt \\ &= 18{,}021.55 \cdot 0.085 \cdot \frac{1}{12} \\ &= 127.65264\ \ldots \approx \$127.65 \end{aligned}$$

However, the interest portion of each payment is \$129.82; thus, for the second payment, the interest portion is *more than* simple interest on the outstanding principal. The loan is *not* a simple interest loan; it requires higher interest payments than would a simple interest loan at the same rate. ■

Whenever a loan is not a simple interest loan, the Truth in Lending Act requires the lender to disclose the annual percentage rate to the borrower. The **annual percentage rate (A.P.R.) of an add-on interest loan** is the simple interest rate that makes the dollar amounts the same if the loan is recomputed as a simple interest amortized loan.

EXAMPLE 2 Find the A.P.R. of the add-on interest loan in Example 1.

Solution Substitute the dollar amounts into the Simple Interest Amortized Loan Formula and solve for the interest rate *i*.

$$\text{pymt}\,\frac{(1+i)^n - 1}{i} = P(1+i)^n$$

$$435.27\,\frac{(1+i)^{60} - 1}{i} = 18{,}327(1+i)^{60}$$

Solving this equation without the aid of a graphing calculator would be difficult if not impossible. To solve it with a graphing calculator:

- Enter $435.27 \dfrac{(1 + x)^{60} - 1}{x}$ for Y_1
- Enter $18,327\ (1 + x)^{60}$ for Y_2
- Graph the two functions
- Find the point at which the two functions intersect, as discussed in Appendix D and summarized here.

TI-80/81:	use ZOOM and TRACE buttons
TI-82/83:	select the "intersect" option from the "CALC" menu
TI-85/86:	select "MATH" from the "GRAPH" menu, and then "ISECT" from the "GRAPH MATH" menu
Casio:	press G-Solv and then ISCT

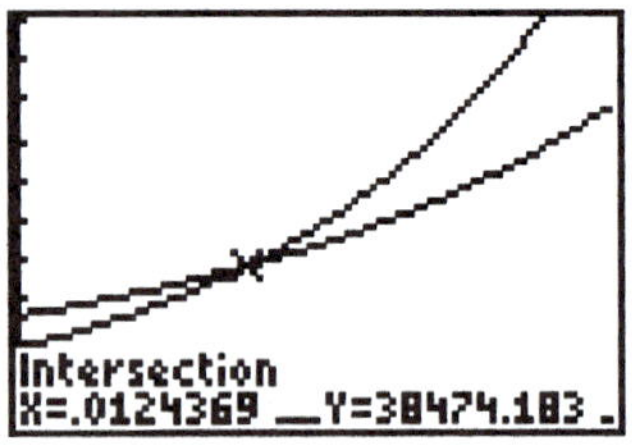

Figure 5.23

The graph is shown in Figure 5.23. The intersection is at $i = 0.0124369$. This is a monthly rate; the corresponding annual rate is $12 \cdot 0.0124369 = 0.1492428 \approx 14.92\%$. The 8.5% add-on interest loan in Example 1 has an A.P.R. of 14.92%. This means that the add-on interest loan requires the same monthly payment that a 14.92% simple interest amortized loan would have.

The Truth in Lending Act, which requires a bank to divulge its loan's A.P.R., allows a tolerance of one-eighth of 1% (= 0.125%) in the claimed A.P.R. Thus, the dealership would be legally correct if it stated that the A.P.R. was between $14.92428\% - 0.125\% = 14.79928\%$ and $14.92428\% + 0.125\% = 15.04928\%$. ■

Finance Charges

Sometimes **finance charges** other than the interest portion of the monthly payment are associated with a loan; these charges must be paid when the loan agreement is

Courtesy of LBJ Library

HISTORICAL NOTE

Truth in Lending Act

The **Truth in Lending Act** was signed by President Lyndon Johnson in 1968. Its original intent was to promote credit shopping by requiring lenders to use uniform language and calculations and to make full disclosure of credit charges so that consumers could shop for the most favorable credit terms. The Truth in Lending Act is interpreted by the Federal Reserve Board's *Regulation Z.*

During the 1970s, Congress amended the act many times, and corresponding changes were made in Regulation Z. These changes resulted in a huge increase in the length and complexity of the law. Its scope now goes beyond disclosure to grant significant legal rights to the borrower.

signed. For example, a **point** is a finance charge that is equal to 1% of the loan amount; a **credit report fee** pays for a report on the borrower's credit history, including any late or missing payments and the size of all outstanding debts; an **appraisal fee** pays for the determination of the current market value of the property (the auto, boat, or home) to be purchased with the loan. The Truth in Lending Act requires the lender to inform the borrower of the total finance charge, which includes the interest, the points, and some of the fees (see Figure 5.24). Most of the finance charges must be paid before the loan is awarded, so in essence the borrower must pay money now in order to get more money later. The law says that this means the lender is not really borrowing as much as he or she thinks. According to the law, the actual amount loaned is the loan amount minus all points and those fees included in the finance charge. The **A.P.R. of a simple interest amortized loan** is the rate that reconciles the payment and this actual loan amount.

Costs Included in the Prepaid Finance Charges	
2 points	\$2,415.32
prorated interest	\$1,090.10
prepaid mortgage insurance	\$ 434.50
loan fee	\$1,242.00
document preparation fee	\$ 80.00
tax service fee	\$ 22.50
processing fee	\$ 42.75
subtotal	\$5,327.17

Other Costs Not Included in the Finance Charge	
appraisal fee	\$ 70.00
credit report	\$ 60.00
closing fee	\$ 670.00
title insurance	\$ 202.50
recording fee	\$ 20.00
notary fee	\$ 20.00
tax and insurance escrow	\$ 631.30
subtotal	\$1,673.80

Figure 5.24
Sample portion of a federal truth-in-lending disclosure statement (loan amount = \$120,765.90)

EXAMPLE 3 Glen and Tanya Hansen bought a home for \$140,000. They paid the sellers a 20% down payment and obtained a simple interest amortized loan for the balance from their bank at $10\frac{3}{4}\%$ for 30 years. The bank in turn paid the sellers the remaining 80% of the purchase price, less a 6% sales commission paid to the sellers' and the buyers' real estate agents. (The transaction is illustrated in Figure 5.25.) The bank charged the Hansens two points plus fees totaling \$3,247.60; of these fees, \$1,012.00 were included in the finance charge.

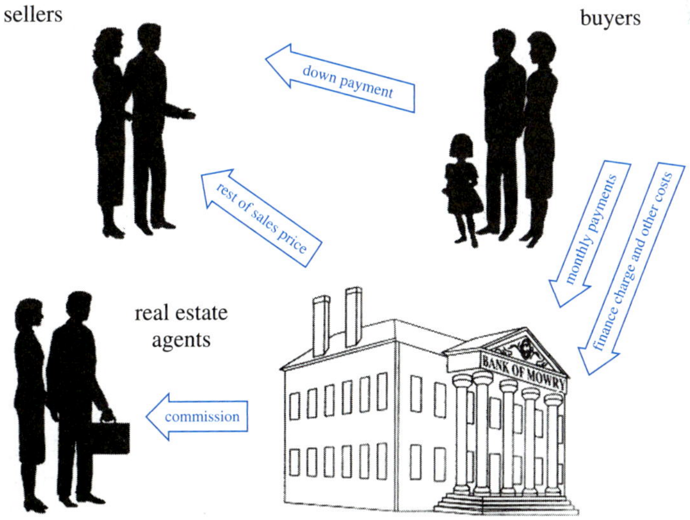

Figure 5.25

a. Find the size of the Hansens' monthly payment.
b. Find the total interest paid.
c. Compute the total finance charge.
d. Find the A.P.R.

Solution **a.** *Finding their monthly payment* We are given down payment = 20% of \$140,000 = \$28,000, P = loan amount = \$140,000 − \$28,000 = \$112,000, $i = \frac{1}{12}$ of $10\frac{3}{4}\%$ $= \frac{0.1075}{12}$, and n = 30 years = 30 years · (12 months)/(1 year) = 360 months.

$$\text{future value of annuity} = \text{future value of loan amount}$$

$$\text{pymt}\,\frac{(1+i)^n - 1}{i} = P(1+i)^n$$

$$\text{pymt}\,\frac{\left(1+\dfrac{0.1075}{12}\right)^{360} - 1}{\dfrac{0.1075}{12}} = 112{,}000\left(1+\frac{0.1075}{12}\right)^{360}$$

Computing the fraction on the left side and multiplying its reciprocal by the right side, we get

$$\text{pymt} = 1{,}045.4991 \approx \$1{,}045.50$$

b. *Finding the total interest paid* The total interest paid is the total amount paid minus the amount borrowed. The Hansens agreed to make 360 monthly payments of \$1045.50 each, for a total of 360 · \$1,045.50 = \$376,380.00. Of this, \$112,000 is principal; therefore, the total interest is

$$376{,}380 - 112{,}000 = \$264{,}380$$

c. *Computing their total finance charge*

2 points = 2% of \$112,000 =	\$ 2,240	
included fees =	\$ 1,012	
total interest paid =	\$264,380	from part (b)
total finance charge =	\$267,632	

d. *Finding the A.P.R.* The A.P.R. is the simple interest rate that makes the dollar amounts the same if the loan is recomputed using the legal loan amount (loan amount less points and fees) in place of the actual loan amount. The legal loan amount is

$$P = \$112{,}000 - \$3252 = \$108{,}748$$

Substitute the dollar amounts into the simple interest amortized loan formula and solve for the interest rate i.

$$\text{pymt}\,\frac{(1+i)^n - 1}{i} = P(1+i)^n \qquad \text{Simple Interest Amortized Loan Formula}$$

$$1{,}045.50\,\frac{(1+i)^{360} - 1}{i} = 108{,}748\,(1+i)^{360} \qquad \text{substituting}$$

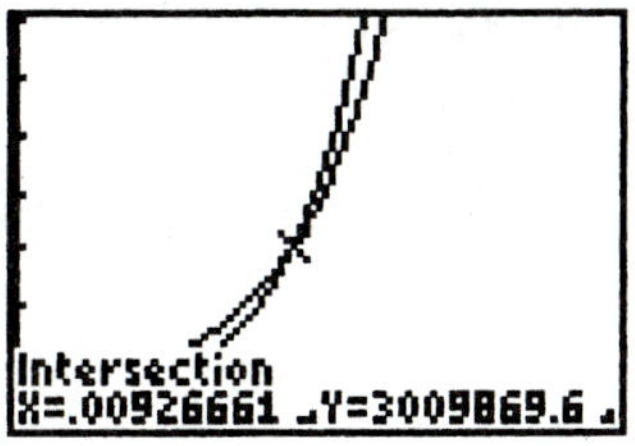

Figure 5.26

To solve this with a graphing calculator:

- Enter $1{,}045.50 \dfrac{(1 + x)^{360} - 1}{x}$ for Y_1.
- Enter $108{,}748(1 + x)^{360}$ for Y_2.
- Graph the two functions.
- Find the point at which the two functions intersect.

The graph is shown in Figure 5.26. The intersection is at $i = 0.00926661$. This is a monthly rate; the corresponding annual rate is $12 \cdot 0.00926661 = 0.11119932 \approx 11.12\%$.

The A.P.R. is 11.12%. This means that the $10\frac{3}{4}\%$ loan requires the same monthly payment as a 11.12% loan with no points or fees would have. ■

If you need to obtain a loan, it is not necessarily true that the lender with the lowest interest rate will give you the least expensive loan. One lender may charge more points or higher fees than does another lender. One lender may offer an add-on interest loan, while another lender offers a simple interest amortized loan. These differences can have a significant impact on the cost of a loan. If two lenders offer loans at the same interest rate but differ in any of these ways, that difference will be reflected in the A.P.R. "Lowest interest rate" does *not* mean "least expensive loan," but "lowest A.P.R." *does* mean "least expensive loan."

A.P.R. Steps

1. *Compute the payment.*
   - **a.** Use the methods of Section 5.1 with an add-on interest loan.
   - **b.** Use the methods of Section 5.4 with a simple interest amortized loan.
2. *Use the Simple Interest Amortized Loan Formula to compute the A.P.R.*
   - **a.** Substitute the payment from step 1 for pymt.
   - **b.** If there are points and fees, substitute the legal loan amount (legal loan amount = loan amount less points and those fees included in the finance charge) for *P*.
   - **c.** Use your graphing calculator to find the periodic interest rate.
   - **d.** Convert the periodic interest rate to an annual rate. This annual rate is the A.P.R.
3. *Determine whether the claimed A.P.R. is legally correct.* If the lender claims a certain A.P.R., the Truth in Lending Act allows a tolerance of one-eighth of 1% (= 0.125%).

Estimating Prepaid Finance Charges

As we have seen, a borrower usually must pay an assortment of fees when obtaining a home loan. These fees can be quite substantial, and they can vary significantly from lender to lender. For example, the total fees in Figure 5.24 were \$7,000.97; the Hansens' total fees in Example 3 were \$5,487.60. When shopping for a home loan, a borrower should take these fees into consideration.

All fees must be disclosed to the borrower when he or she signs the loan papers. Furthermore, the borrower must be given an estimate of the fees when he or she applies for the loan. Before an application is made, the borrower can use a loan's A.P.R. to obtain a reasonable approximation of those fees that are included in the finance charge (for a fixed rate loan). This would allow the borrower to make a more educated decision in selecting a lender.

EXAMPLE 4 Felipe and Delores Lopez are thinking of buying a home for \$152,000. A potential lender advertises an 80%, 30-year simple interest amortized loan at $7\frac{1}{2}\%$ interest with an A.P.R. of 7.95%.

a. Find the size of the Lopezes' monthly payment.
b. Use the A.P.R. to approximate the fees included in the finance charge.

Solution **a.** *Finding the monthly payment* The loan amount is 80% of \$152,000 = \$121,600, $i = \frac{1}{12}$ of $7.5\% = \frac{0.075}{12}$, and $n = 30$ years $= 360$ months.

$$\text{future value of annuity} = \text{future value of loan amount}$$

$$\text{pymt}\,\frac{(1+i)^n - 1}{i} = P(1+i)^n$$

$$\text{pymt}\,\frac{\left(1+\frac{0.075}{12}\right)^{360} - 1}{\frac{0.075}{12}} = 121{,}600\left(1+\frac{0.075}{12}\right)^{360}$$

Computing the fraction on the left side and multiplying its reciprocal by the right side, we get

$$\text{pymt} = 850.2448 \approx \$850.25$$

b. *Approximating the fees included in the finance charge* We approximate the fees by computing the legal loan amount (loan amount less points and fees), using the A.P.R. as the interest rate and the payment computed in part (a) as "pymt." Therefore, $i = \frac{1}{12}$ of $7.95\% = \frac{0.0795}{12}$, pymt $= 850.25$.

$$\text{future value of annuity} = \text{future value of loan amount}$$

$$\text{pymt}\,\frac{(1+i)^n - 1}{i} = P(1+i)^n$$

$$850.25\,\frac{\left(1+\frac{0.0795}{12}\right)^{360} - 1}{\frac{0.0795}{12}} = \text{P}\left(1+\frac{0.0795}{12}\right)^{360}$$

Computing the left side, and dividing by $(1 + \frac{0.0795}{12})^{360}$, we get

$$P = 116{,}427.6304 = \$116{,}427.63$$

This is the legal loan amount, that is, the loan amount less points and fees. The Lopezes were borrowing \$121,600, so that leaves \$121,600 − \$116,427.63 = \$5,172.37 in points and fees. This is an estimate of the points and fees included in the finance charge, such as a loan fee, a document preparation fee, and a processing fee; it does not include fees such as an appraisal fee, a credit report fee, and a title insurance fee. ■

5.5 Exercises

1. Wade Ellis buys a new car for \$16,113.82. He puts 10% down and obtains a simple interest amortized loan for the balance at $11\frac{1}{2}\%$ interest for four years. If loan fees included in the finance charge total \$814.14, find the A.P.R.
2. Guy de Primo buys a new car for \$9,837.91. He puts 10% down and obtains a simple interest amortized loan for the balance at $10\frac{7}{8}\%$ interest for four years. If loan fees included in the finance charge total \$633.87, find the A.P.R.
3. Chris Burditt bought a house for \$212,500. He put 20% down and obtained a simple interest amortized loan for the balance at $10\frac{7}{8}\%$ interest for 30 years. If Chris paid 2 points and \$4,728.60 in fees, \$1,318.10 of which are included in the finance charge, find the A.P.R.
4. Shirley Trembley bought a house for \$187,600. She put 20% down and obtained a simple interest amortized loan for the balance at $11\frac{3}{8}\%$ for 30 years. If Shirley paid 2 points and \$3,427.00 in fees, \$1,102.70 of which are included in the finance charge, find the A.P.R.
5. Jennifer Tonda wants to buy a used car that costs \$4,600. The used car dealer has offered her a four-year add-on interest loan that requires no down payment at 8% annual interest, with an A.P.R. of $14\frac{1}{4}\%$.
   a. Find the monthly payment.
   b. Verify the A.P.R.
6. Melody Shepherd wants to buy a used car that costs \$5,300. The used car dealer has offered her a four-year add-on interest loan that requires a \$200 down payment at 7% annual interest, with an A.P.R. of 10%.
   a. Find the monthly payment.
   b. Verify the A.P.R.
7. Anne Scanlan is buying a used car that costs \$10,340. The used car dealer has offered her a five-year add-on interest loan at 9.5% interest, with an A.P.R. of 9.9%. The loan requires a 10% down payment.
   a. Find the monthly payment.
   b. Verify the A.P.R.
8. Stan Loll bought a used car for \$9,800. The used car dealer offered him a four-year add-on interest loan at 7.8% interest, with an A.P.R. of 8.0%. The loan requires a 10% down payment.
   a. Find the monthly payment.
   b. Verify the A.P.R.
9. Susan Chin is shopping for a car loan. Her savings and loan offers her a simple interest amortized loan for four years at 9% interest. Her bank offers her a simple interest amortized loan for four years at 9.1% interest. Which is the less expensive loan?
10. Stephen Tamchin is shopping for a car loan. His credit union offers him a simple interest amortized loan for 4 years at 7.1% interest. His bank offers him a simple interest amortized loan for four years at 7.3% interest. Which is the less expensive loan?
11. Ruben Lopez is shopping for a home loan. Really Friendly Savings and Loan offers him a 30-year simple interest amortized loan at 9.2% interest, with an A.P.R. of 9.87%. The Solid and Dependable Bank offers him a 30-year simple interest amortized loan at 9.3% interest, with an A.P.R. of 9.80%. Which loan would have the lower payments? Which loan would be the least expensive, taking into consideration monthly payments, points, and fees? Justify your answers.
12. Keith Moon is shopping for a home loan. Sincerity Savings offers him a 30-year simple interest amortized loan at 8.7% interest, with an A.P.R. of 9.12%. Pinstripe National Bank offers him a 30-year simple interest amortized loan at 8.9% interest, with an A.P.R. of 8.9%. Which loan would have the lower payments? Which loan would be the least expensive, taking into consideration monthly payments, points, and fees? Justify your answers.
13. The Nguyens are thinking of buying a home for \$119,000. A potential lender advertises an 80%, 30-year simple interest amortized loan at $8\frac{1}{4}\%$ interest, with an A.P.R. of 9.23%. Use the A.P.R. to approximate the fees included in the finance charge.
14. Ellen Taylor is thinking of buying a home for \$126,000. A potential lender advertises an 80%, 30-year simple interest amortized loan at $10\frac{3}{4}\%$ interest, with an A.P.R. of 11.57%. Use the A.P.R. to approximate the fees included in the finance charge.
15. James Magee is thinking of buying a home for \$124,500. Bank of the Future advertises an 80%, 30-year simple interest amortized loan at $9\frac{1}{4}\%$ interest, with an A.P.R. of 10.23%. R.T.C. Savings and Loan advertises an 80%, 30-year simple interest amortized loan at 9% interest with an A.P.R. of 10.16%.
    a. Find James's monthly payment if he borrows through Bank of the Future.
    b. Find James's monthly payment if he borrows through R.T.C. Savings and Loan.
    c. Use the A.P.R. to approximate the fees included in the finance charge by Bank of the Future.

d. Use the A.P.R. to approximate the fees included in the finance charge by R.T.C. Savings and Loan.

e. Discuss the advantages of each of the two loans. Who would be better off with the Bank of the Future loan? Who would be better off with the R.T.C. loan?

16. Holly Kresch is thinking of buying a home for \$263,800. State Bank advertises an 80%, 30-year simple interest amortized loan at $6\frac{1}{4}\%$ interest, with an A.P.R. of 7.13%. Boonville Savings and Loan advertises an 80%, 30-year simple interest amortized loan at $6\frac{1}{2}\%$ interest with an A.P.R. of 7.27%.

a. Find Holly's monthly payment if she borrows through State Bank.

b. Find Holly's monthly payment if she borrows through Boonville Savings and Loan.

c. Use the A.P.R. to approximate the fees included in the finance charge by State Bank.

d. Use the A.P.R. to approximate the fees included in the finance charge by Boonville Savings and Loan.

e. Discuss the advantages of the two loans. Who would be better off with the State Bank loan? Who would be better off with the Boonville loan?

Answer the following questions using complete sentences and your own words.

Concept Questions

17. If the A.P.R. of a simple interest amortized home loan is equal to the loan's interest rate, what conclusions could you make about the loan's required fees and points?

18. Compare and contrast the annual percentage rate of a loan with the annual yield of a compound interest rate. Be sure to discuss both the similarity and the difference between these two concepts.

Projects

19. This is an exercise in buying a home. It involves choosing a home and selecting the home's financing. Write a paper describing all of the following points.

a. You might not be in a position to buy a home now. If this is the case, fantasize about your future. What job do you have? How long have you had that job? What is your salary? If you're married, does your spouse work? Do you have a family? What needs will your home fulfill? Make your future fantasy realistic. Briefly describe what has happened between the present and your future fantasy. (If you are in a position to buy a home now, discuss these points on a more realistic level.)

b. Go shopping for a home. Look at houses, or condominiums, or both. Look at new homes, or used homes, or both. (Used homes can easily be visited by going to an "open house," where the owners are gone and the real estate agent allows interested parties to inspect the home. Open houses are probably listed in your local newspaper.) Read appropriate newspaper and magazine articles (for example, in the "Real Estate" section of your local newspaper). Discuss in detail the home you selected and why you did so. How will your selection fulfill your (projected) needs? Why did you select a house or a condominium? Why did you select a new or a used home? Explain your choice of location, house size, and features of the home.

c. Go shopping for financing. Many banks, savings and loans, and credit unions have information on home loans available on request. Get all of the information you need from at least two lenders, but do not bother the staff unnecessarily. You might be able to find the necessary information in the newspaper. Perform all appropriate computations yourself—do not have commercial lenders tell you the payment size. Summarize the appropriate data in your paper and discuss which loan you would choose. Include in your discussion the down payment, the presence or absence of a prepayment penalty, the duration of the loan, the interest rate, the A.P.R., the payment size, and other terms of the loan. Use the A.P.R. to approximate the fees included in the finance charge; use Figure 5.24 to approximate the fees that are not included in the finance charge.

d. Also discuss the real estate taxes (your instructor will provide you with information on the local tax rate) and the effect of your purchase on your income taxes (interest paid on a home loan is deductible from the borrower's income taxes).

e. Most lenders will approve a home loan only if the total of all the borrower's monthly payments, including the home loan payment, real estate taxes, credit card payments, and car loan payments, is no more than 38% of the borrower's monthly income. Discuss your ability to qualify for the loan.

5.6 Payout Annuities

The annuities that we have discussed are all savings instruments. In Section 5.3, we defined an annuity as a sequence of equal, regular payments into an account where each payment receives compound interest. A saver who utilizes such an annuity can accumulate a sizable sum.

Annuities can be payout instruments rather than savings instruments. After you retire, you may wish to have part of your savings sent to you each month for living expenses. You may wish to receive equal, regular payments from an account where each payment has earned compound interest. Such an annuity is called a **payout annuity.** Payout annuities are also used to pay for a child's college education.

Calculating Short-Term Payout Annuities

EXAMPLE 1 On November 1, Debra Landre will make a deposit at her bank that will be used for a payout annuity. For the next three months, commencing on December 1, she will receive a payout of $1,000 per month. Use the Compound Interest Formula to find how much money she must deposit on November 1 if her money earns 10% compounded monthly.

Solution First, calculate the principal necessary to receive $1,000 on December 1. Use $FV = 1{,}000$ and $n = 1$ (interest is earned for one month).

$$FV = P(1 + i)^n$$

$$1{,}000 = \mathrm{P}\left(1 + \frac{0.10}{12}\right)^1$$

$$\mathrm{P} = 1{,}000 \div \left(1 + \frac{0.10}{12}\right)^1 = 991.7355\ldots \approx \$991.74$$

Next, calculate the principal necessary to receive $1,000 on January 1. Use $n = 2$ (interest is earned for two months).

$$FV = P(1 + i)^n$$

$$1{,}000 = \mathrm{P}\left(1 + \frac{0.10}{12}\right)^2$$

$$\mathrm{P} = 1{,}000 \div \left(1 + \frac{0.10}{12}\right)^2 = 983.5393\ldots \approx \$983.54$$

Now calculate the principal necessary to receive $1,000 on February 1. Use $n = 3$ (interest is earned for three months).

$$FV = P(1 + i)^n$$

$$1{,}000 = \mathrm{P}\left(1 + \frac{0.10}{12}\right)^3$$

$$\mathrm{P} = 1{,}000 \div \left(1 + \frac{0.10}{12}\right)^3 = 975.4109\ldots \approx \$975.41$$

Debra must deposit the sum of the above three amounts if she is to receive three monthly payouts of $1,000 each. Her total principal must be

$$\$991.74 + \$983.54 + \$975.41 = \$2{,}950.69$$

✔ If Debra's principal received no interest, then she would need $3 \cdot \$1{,}000 = \$3{,}000$. Since her principal does receive interest, she needs slightly less than $3,000.

■

Comparing Payout Annuities and Savings Annuities

Payout annuities and savings annuities are similar but not identical. It is important to understand their differences before we proceed. The following example is the "savings annuity" version of Example 1; that is, it is the savings annuity most similar to the payout annuity discussed in Example 1.

EXAMPLE 2 On November 1, Debra Landre set up an ordinary annuity with her bank. For the next three months she will make a payment of $1,000 per month. Each of those payments will receive compound interest. At the end of the three months (i.e., on February 1), she can withdraw her three $1,000 payments plus the interest that they will have earned. Use the Compound Interest Formula to find how much money she can withdraw.

Solution Her first payment of $1,000 would be due on November 30. It would earn interest for two months (December and January).

$$FV = P(1 + i)^n$$

$$\text{FV} = 1{,}000\left(1 + \frac{0.10}{12}\right)^2 \approx \$1{,}016.74$$

Her second payment of $1,000 would be due on December 31. It would earn interest for one month (January).

$$FV = P(1 + i)^n$$

$$\text{FV} = 1{,}000\left(1 + \frac{0.10}{12}\right)^1 \approx \$1{,}008.33$$

Her third payment would be due on January 31. It would earn no interest (since the annuity expires February 1), so its future value is $1,000. On February 1, Debra could withdraw

$$\$1{,}016.74 + \$1{,}008.33 + \$1{,}000 = \$3{,}025.07$$

■

Naturally, we could have found the future value of Debra's ordinary annuity more easily with the Ordinary Annuity Formula.

$$FV(\text{ordinary}) = \text{pymt}\,\frac{(1 + i)^n - 1}{i}$$

$$= 1{,}000\,\frac{\left(1 + \frac{0.10}{12}\right)^3 - 1}{\frac{0.10}{12}} \approx \$3{,}025.07$$

The point of the method shown in Example 2 is to illustrate the differences between a savings annuity and a payout annuity.

Calculating Long-Term Payout Annuities

The procedure used in Example 1 reflects what actually happens with payout annuities, and it works fine for a small number of payments. However, most annuities are long-term. In the case of a savings annuity, we do not need to calculate the future value of each individual payment, as we did in Example 2; instead, we can use the Ordinary Annuity Formula. We need such a formula for payout annuities. We can find a formula if we look more closely at how the payout annuity from Example 1 compares with the savings annuity from Example 2.

In Example 2, we found that the future value of an ordinary annuity with three \$1,000 payments is

$$FV = 1{,}000\left(1 + \frac{0.10}{12}\right)^2 + 1{,}000\left(1 + \frac{0.10}{12}\right)^1 + 1{,}000$$

In Example 1, we found that the total principal necessary to generate three \$1,000 payouts is

$$P = 1{,}000 \div \left(1 + \frac{0.10}{12}\right)^1 + 1{,}000 \div \left(1 + \frac{0.10}{12}\right)^2 + 1{,}000 \div \left(1 + \frac{0.10}{12}\right)^3$$

This can be rewritten, using exponent laws, as

$$P = 1{,}000\left(1 + \frac{0.10}{12}\right)^{-1} + 1{,}000\left(1 + \frac{0.10}{12}\right)^{-2} + 1{,}000\left(1 + \frac{0.10}{12}\right)^{-3}$$

This is quite similar to the future value of the ordinary annuity—the only difference is the exponents. If we multiply each side by $(1 + \frac{0.10}{12})^3$, even the exponents will match.

$$\begin{aligned} P\left(1 + \frac{0.10}{12}\right)^3 = {} & 1{,}000\left(1 + \frac{0.10}{12}\right)^{-1}\left(1 + \frac{0.10}{12}\right)^3 \\ & + 1{,}000\left(1 + \frac{0.10}{12}\right)^{-2}\left(1 + \frac{0.10}{12}\right)^3 \\ & + 1{,}000\left(1 + \frac{0.10}{12}\right)^{-3}\left(1 + \frac{0.10}{12}\right)^3 \end{aligned}$$

$$P\left(1 + \frac{0.10}{12}\right)^3 = 1{,}000\left(1 + \frac{0.10}{12}\right)^2 + 1{,}000\left(1 + \frac{0.10}{12}\right)^1 + 1{,}000\left(1 + \frac{0.10}{12}\right)^0$$

The right side of the above equation is the future value of an ordinary annuity, so we can use the Ordinary Annuity Formula to rewrite it.

$$P\left(1 + \frac{0.10}{12}\right)^3 = 1{,}000\,\frac{\left(1 + \frac{0.10}{12}\right)^3 - 1}{\frac{0.10}{12}}$$

If we generalize by replacing $\frac{0.10}{12}$ with i, 3 with n, and 1,000 with pymt, we have our Payout Annuity Formula.

Payout Annuity Formula

$$P(1 + i)^n = \text{pymt}\,\frac{(1 + i)^n - 1}{i}$$

where P is the total principal necessary to generate n payouts, "pymt" is the size of the payout, and i is the periodic interest rate.

We've seen this formula before. In Section 5.3, we used it to find the present value P of a savings annuity. In Section 5.4, we used it to find the payment of a simple interest amortized loan. In this section, we use it to find the required principal for a payout annuity. This is a versatile formula.

The following example involves a long-term annuity. Usually, the interest rate of a long-term annuity varies somewhat from year to year. In this case, calculations must be viewed as predictions, not guarantees.

EXAMPLE 3 Fabiola Macias is about to retire, so she is setting up a payout annuity with her bank. She wishes to receive a payout of \$1,000 per month for the next 25 years. Use the Payout Annuity Formula to find how much money she must deposit if her money earns 10% compounded monthly.

Solution We are given that pymt = 1,000, $i = \frac{1}{12}$ of 10% $= \frac{0.10}{12}$, and $n = 25 \cdot 12 = 300$.

$$P(1 + i)^n = \text{pymt}\,\frac{(1 + i)^n - 1}{i}$$

$$P\left(1 + \frac{0.10}{12}\right)^{300} = 1{,}000\,\frac{\left(1 + \frac{0.10}{12}\right)^{300} - 1}{\frac{0.10}{12}}$$

To find P, we need to calculate the right side and then divide by the $(1 + \frac{0.10}{12})^{300}$ from the left side.

We get 110,047.23005 ≈ \$110,047.23. This means that if Fabiola deposits \$110,047.23, she will receive monthly payouts of \$1,000 each for 25 years, or a total of $25 \cdot 12 \cdot 1{,}000 = \$300{,}000$.

✔ If Fabiola's principal received no interest, then she would need $300 \cdot \$1{,}000 = \$300{,}000$. Since her principal does receive compound interest for a long time, she needs significantly less than \$300,000.

■

If Fabiola Macias in Example 3 were like most people, she wouldn't have \$110,047.23 in savings when she retires, so she wouldn't be able to set up a payout annuity for herself. However, if she had set up a savings annuity 30 years before retirement, she could have saved that amount by making monthly payments of only \$48.68. This is something that almost anyone can afford, and it's a wonderful deal. Thirty years of monthly payments of \$48.68, while you're working, can generate 25 years of monthly payments of \$1,000 when you're retired. In the exercises, we will explore this combination of a savings annuity and a payout annuity.

Payout Annuities with Inflation

The only trouble with Fabiola's retirement payout annuity in Example 3 is that she is ignoring inflation. In 25 years, she will still be receiving $1,000 a month, but her money won't buy as much as it does today. Fabiola would be better off if she allowed herself an annual **cost-of-living adjustment (C.O.L.A.).**

EXAMPLE 4 After retiring, Fabiola Macias set up a payout annuity with her bank. For the next 25 years, she will receive payouts that start at $1,000 per month and then receive an annual C.O.L.A. of 3%. Find the size of her monthly payout for

a. the first year **b.** the second year
c. the third year **d.** the 25th year

Solution **a.** During the first year, no adjustment is made, so she will receive $1,000 per month.

b. During the second year, her monthly payout of $1,000 will increase 3%, so her new monthly payout will be

$$1{,}000 \cdot (1 + .03) = \$1{,}030$$

c. During the third year, her monthly payout of $1,030 will increase 3%, so her new monthly payout will be

$$1{,}030 \cdot (1 + .03) = \$1{,}060.90$$

Since the 1,030 in the above calculation came from computing $1{,}000 \cdot (1 + .03)$, we could rewrite this calculation as

$$1000 \cdot (1 + .03)^2 = \$1{,}060.90$$

d. By the 25th year, she will have received twenty-four 3% increases, so her monthly payout will be

$$1{,}000 \cdot (1 + .03)^{24} = 2{,}032.7941\ldots \approx \$2{,}032.79$$

■

If a payout annuity is to have automatic annual cost-of-living adjustments, the following formula should be used to find the principal that must be deposited. The C.O.L.A. is an annual one, so all other figures must also be annual figures; in particular, r is the *annual* interest rate, t is the duration of the annuity in *years,* and we use an *annual* payout.

Annual Payout Annuity with C.O.L.A. Formula

A payout annuity of t years, where the payouts receive an annual C.O.L.A., requires a principal of

$$P = (\text{pymt})\,\frac{1 - \left(\dfrac{1+c}{1+r}\right)^t}{r - c}$$

where "pymt" is the annual payout for the first year, c is the annual C.O.L.A. rate, and r is the annual rate at which interest is earned on the principal.

EXAMPLE 5 After retiring, Sam Needham set up a payout annuity with his bank. For the next 25 years, he will receive annual payouts then start at $12,000 and then receive an annual C.O.L.A. of 3%. Use the Annual Payout Annuity with C.O.L.A. Formula to find how much money he must deposit if his money earns 10% interest per year.

Solution We are given that the annual payout is pymt = 12,000, $r = 10\% = 0.10$, $c = 3\% = 0.03$, and $t = 25$.

$$P = (\text{pymt})\frac{1 - \left(\frac{1+c}{1+r}\right)^t}{r - c}$$

$$= (12{,}000)\frac{1 - \left(\frac{1+0.03}{1+0.10}\right)^{25}}{0.10 - 0.03}$$

$$= (12{,}000)\frac{1 - \left(\frac{1.03}{1.10}\right)^{25}}{0.10 - 0.03}$$

$$= 138{,}300.4587 \approx \$138{,}300.46$$

1 [−] [(] 1.03 [÷] 1.10 [)] [y^x] 25 [=] [÷] [(] .10 [−] .03 [)] [×] 12000 [=]

With a graphing calculator, press [^] rather than [y^x] and [ENTER] rather than [=].

This means that if Sam deposits $138,300.46 now, he will receive

$12,000 in one year
$\$12{,}000 \cdot (1 + .03)^1 = \$12{,}360$ in two years

$\$12{,}000 \cdot (1 + .03)^2 = \$12{,}730.80$ in three years
$\$12{,}000 \cdot (1 + .03)^{24} = \$24{,}393.53$ in 25 years

✔ If Sam's principal received no interest, he would need 25 · $12,000 = $300,000. Since his principal does receive compound interest for a long time, he needs significantly less than $300,000.

■

Compare Sam's payout annuity in Example 5 with Fabiola's in Example 3. Sam receives annual payouts that start at $12,000 and slowly increase to $24,393. Fabiola receives exactly $1,000 per month (or $12,000 per year) for the same amount of time. Sam was required to deposit $138,300, and Fabiola was required to deposit $110,047.23.

5.6 Exercises

1. Suzanne Miller is planning for her retirement, so she is setting up a payout annuity with her bank. She wishes to receive a payout of $1,200 per month for 20 years.
   a. How much money must she deposit if her money earns 8% interest compounded monthly?
   b. Find the total amount that Suzanne will receive from her payout annuity.
2. James Magee is planning for his retirement, so he is setting up a payout annuity with his bank. He wishes to receive a payout of $1,100 per month for 25 years.
   a. How much money must he deposit if his money earns 9% interest compounded monthly?
   b. Find the total amount that James will receive from his payout annuity.
3. Dean Gooch is planning for his retirement, so he is setting up a payout annuity with his bank. He wishes to receive a payout of $1,300 per month for 25 years.
   a. How much money must he deposit if his money earns 7.3% interest compounded monthly?
   b. Find the total amount that Dean will receive from his payout annuity.
4. Holly Krech is planning for her retirement, so she is setting up a payout annuity with her bank. She wishes to receive a payout of $1,800 per month for 20 years.
   a. How much money must she deposit if her money earns 7.8% interest compounded monthly?
   b. Find the total amount that Holly will receive from her payout annuity.
5. **a.** How large a monthly payment must Suzanne Miller (from Exercise 1) make if she saves for her payout annuity with an ordinary annuity, which she sets up 30 years before her retirement? (The two annuities pay the same interest rate.)
   b. Find the total amount that Suzanne will pay into her ordinary annuity, and compare it with the total amount that she will receive from her payout annuity.
6. **a.** How large a monthly payment must James Magee (from Exercise 2) make if he saves for his payout annuity with an ordinary annuity, which he sets up 25 years before his retirement? (The two annuities pay the same interest rate.)
   b. Find the total amount that James will pay into his ordinary annuity, and compare it with the total amount that he will receive from his payout annuity.
7. **a.** How large a monthly payment must Dean Gooch (from Exercise 3) make if he saves for his payout annuity with an ordinary annuity, which he sets up 30 years before his retirement? (The two annuities pay the same interest rate.)
   b. How large a monthly payment must he make if he sets the ordinary annuity up 20 years before his retirement?
8. **a.** How large a monthly payment must Holly Krech (from Exercise 4) make if she saves for her payout annuity with an ordinary annuity, which she sets up 30 years before her retirement? (The two annuities pay the same interest rate.)
   b. How large a monthly payment must she make if she sets the ordinary annuity up 20 years before her retirement?
9. Lily Chang is planning for her retirement, so she is setting up a payout annuity with her bank. For 20 years, she wishes to receive annual payouts that start at $14,000 and that receive an annual C.O.L.A. of 4%.
   a. How much money must she deposit if her money earns 8% interest per year?
   b. How large will Lily's first annual payout be?
   c. How large will Lily's second annual payout be?
   d. How large will Lily's last annual payout be?
10. Wally Brown is planning for his retirement, so he is setting up a payout annuity with his bank. For 25 years, he wishes to receive annual payouts that start at $16,000 and that receive an annual C.O.L.A. of 3.5%.
    a. How much money must he deposit if his money earns 8.3% interest per year?
    b. How large will Wally's first annual payout be?
    c. How large will Wally's second annual payout be?
    d. How large will Wally's last annual payout be?
11. Oshri Karmon is planning for his retirement, so he is setting up a payout annuity with his bank. He is now thirty years old, and he will retire when he is sixty. He wants to receive annual payouts for 25 years, and he wants those payouts to receive an annual C.O.L.A. of 3.5%.
    a. He wants his first payout to have the same purchasing power as does $13,000 today. How big should that payout be if he assumes inflation of 3.5% per year?
    b. How much money must he deposit when he is sixty if his money earns 7.2% interest per year?

c. How large a monthly payment must he make if he saves for his payout annuity with an ordinary annuity? (The two annuities pay the same interest rate.)

d. How large a monthly payment would he make if he waits until he is forty before starting his ordinary annuity?

12. Shelly Franks is planning for her retirement, so she is setting up a payout annuity with her bank. She is now thirty-five years old, and she will retire when she is sixty-five. She wants to receive annual payouts for 20 years, and she wants those payouts to receive an annual C.O.L.A. of 4%.

a. She wants her first payout to have the same purchasing power as does $15,000 today. How big should that payout be if she assumes inflation of 4% per year?

b. How much money must she deposit when she is sixty-five if her money earns 8.3% interest per year?

c. How large a monthly payment must she make if she saves for her payout annuity with an ordinary annuity? (The two annuities pay the same interest rate.)

d. How large a monthly payment would she make if she waits until she is forty before starting her ordinary annuity?

In Exercises 13–16, use the Annual Payout Annuity with C.O.L.A. Formula to find the deposit necessary to receive monthly payouts with an annual cost-of-living adjustment. In order to use the formula, all figures must be annual figures, including the payout and the annual rate. You can adapt the formula for monthly payouts by using

- *the future value of a one-year ordinary annuity in place of the annual payout, where "pymt" is the monthly payout, and*
- *the annual yield of the given compound interest rate in place of the annual rate r.*

13. Fabiola Macias is about to retire, so she is setting up a payout annuity with her bank. She wishes to receive a monthly payout for the next 25 years, where the payout starts at $1,000 per month and receives an annual C.O.L.A. of 3%. Her money will earn 10% compounded monthly.

a. The annual payout is the future value of a one-year ordinary annuity. Find this future value.

b. The annual rate r is the annual yield of 10% interest compounded monthly. Find this annual yield (*do not* round it off).

c. Use the Annual Payout Annuity with C.O.L.A. Formula to find how much money she must deposit.

d. Fabiola could have saved for her payout annuity with an ordinary annuity. If she had started doing so 30 years ago, what would the required monthly payments have been? (The two annuities pay the same interest rate.)

14. Gary Kersting is about to retire, so he is setting up a payout annuity with his bank. He wishes to receive a monthly payout for the next 20 years, where the payout starts at $1,300 per month and receives an annual C.O.L.A. of 4%. His money will earn 8.7% compounded monthly.

a. The annual payout is the future value of a one-year ordinary annuity. Find this future value.

b. The annual rate r is the annual yield of 8.7% interest compounded monthly. Find this annual yield (*do not* round it off).

c. Use the Annual Payout Annuity with C.O.L.A. Formula to find how much money he must deposit.

d. Gary could have saved for his payout annuity with an ordinary annuity. If he had started doing so 25 years ago, what would the required monthly payments have been? (The two annuities pay the same interest rate.)

15. Conrad von Schtup is about to retire, so he is setting up a payout annuity with his bank. He wishes to receive a monthly payout for the next 23 years, where the payout starts at $1,400 per month and receives an annual C.O.L.A. of 5%. His money will earn 8.9% compounded monthly.

a. The annual payout is the future value of a one-year ordinary annuity. Find this future value.

b. The annual rate r is the annual yield of 8.9% interest compounded monthly. Find this annual yield (*do not* round it off).

c. Use the Annual Payout Annuity with C.O.L.A. Formula to find how much money he must deposit.

d. Conrad could have saved for his payout annuity with an ordinary annuity. If he had started doing so 20 years ago, what would the required monthly payments have been? (The two annuities pay the same interest rate.)

16. Mitch Martinez is about to retire, so he is setting up a payout annuity with his bank. He wishes to receive a monthly payout for the next 30 years, where the payout starts at $1,250 per month and receives an annual C.O.L.A. of 4%. His money will earn 7.8% compounded monthly.

a. The annual payout is the future value of a one-year ordinary annuity. Find this future value.
b. The annual rate r is the annual yield of 7.8% interest compounded monthly. Find this annual yield.
c. Use the Annual Payout Annuity with C.O.L.A. Formula to find how much money he must deposit.
d. Mitch could have saved for his payout annuity with an ordinary annuity. If he had started doing so 20 years ago, what would the required monthly payments have been? (The two annuities pay the same interest rate.)

17. Bob Pirtle won $1 million in a state lottery. He was surprised to learn that he will not receive a check for $1 million. Rather, for 20 years, he will receive an annual check from the state for $50,000. The state finances this series of checks by buying Bob a payout annuity. Find what the state pays for Bob's payout annuity if the interest rate is 8%.
18. John-Paul Ramin won $2.3 million in a state lottery. He was surprised to learn that he will not receive a check for $2.3 million. Rather, for 20 years, he will receive an annual check from the state for $\frac{1}{20}$ of his winnings. The state finances this series of checks by buying John-Paul a payout annuity. Find what the state pays for John-Paul's payout annuity if the interest rate is 7.2%.

Answer the following questions using complete sentences and your own words.

Concept Questions

19. Compare and contrast a savings annuity with a payout annuity. How do they differ in purpose? How do they differ in structure? How do their definitions differ?

HINT: Compare Examples 1 and 2.

20. Under what circumstances would a savings annuity and a payout annuity be combined?

Projects

21. This is an exercise in saving for your retirement. Write a paper describing all of the following points.
    a. Go to the library and find out what the annual rate of inflation has been for each of the last ten years. Use the average of these figures as a prediction of the future annual rate of inflation.
    b. Estimate the total monthly expenses you would have if you were retired now. Include housing, food, and utilities.
    c. Use parts (a) and (b) to predict your total monthly expenses when you retire, assuming that you retire at age sixty-five.
    d. Plan on financing your monthly expenses with a payout annuity. How much money must you deposit when you are sixty-five, if your money earns 7.5% interest per year?
    e. How large a monthly payment must you make if you save for your payout annuity with an ordinary annuity, starting now? (The two annuities pay the same interest rate.)
    f. How large a monthly payment must you make if you save for your payout annuity with an ordinary annuity, starting ten years from now?
    g. How large a monthly payment must you make if you save for your payout annuity with an ordinary annuity, starting 20 years from now?

Chapter 5 Review

TERMS

add-on interest loan
adjustable rate mortgage
amortization schedule
amortized loan
annual percentage rate (A.P.R.) of an add-on interest loan
annual percentage rate of a simple interest amortized loan
annual yield
annuity
annuity due versus ordinary annuity
average daily balance
balance
compounding period
compound interest
computerized spreadsheet
cost-of-living adjustment (C.O.L.A.)
doubling time
finance
finance charges
future value
interest
interest rate
loan agreement

maturity value
negative amortization
nominal rate
note
outstanding principal
payment period
payout annuity
periodic interest rate
point
prepayment penalty
present value
present value of an annuity
principal
simple interest
simple interest amortized loan
simple interest loan
sinking fund
spreadsheet
tax-deferred annuity (TDA)
term
Truth in Lending Act
unpaid balance
yield

FORMULAS

Simple Interest: $I = Prt$

Simple Interest Future Value: $FV = P(1 + rt)$

Compound Interest: $FV = P(1 + i)^n$

Annual Yield:
FV(compound interest) = FV(simple interest)
(Find the simple interest rate.)

Ordinary Annuity:
$$FV(\text{ordinary}) = \text{pymt}\,\frac{(1+i)^n - 1}{i}$$

Annuity Due: $FV(\text{due}) = FV(\text{ordinary}) \cdot (1 + i)$

Present Value of Annuity:
FV(lump sum) = FV(annuity)
(Find the lump sum.)

Simple Interest Amortized Loan:
FV(ordinary annuity) = FV(compound interest)
(Find "pymt.")

Unpaid Balance:
$$\begin{aligned}\text{unpaid balance} &= \text{current value of loan amount} \\ &\quad - \text{current value of annuity} \\ &= P(1+i)^n - \text{pymt}\,\frac{(1+i)^n - 1}{i}\end{aligned}$$
(n is the number of periods *from the beginning of the loan to the present.*)

Payout Annuity: $P(1+i)^n = \text{pymt}\,\dfrac{(1+i)^n - 1}{i}$

Annual Payout Annuity with C.O.L.A.:
$$P = (\text{annual pymt})\,\frac{1 - \left(\dfrac{1+c}{1+r}\right)^t}{r - c}$$

REVIEW EXERCISES

1. Find the interest earned by a deposit of \$8,140 at $9\frac{3}{4}\%$ simple interest for 11 years.
2. Find the interest earned by a deposit of \$8,140 at $9\frac{3}{4}\%$ interest compounded monthly for 11 years.
3. Find the maturity value of a loan of \$3,550 borrowed at $12\frac{1}{2}\%$ simple interest for 1 year and 2 months.
4. Find the future value of \$3,550 deposited at $12\frac{1}{2}\%$ interest compounded monthly for 1 year and 2 months.
5. Lynn Knight inherited \$7,000. She wants to buy a used car, but the type of car she wants typically sells for around \$8,000. If her money can earn $7\frac{1}{2}\%$ simple interest, how long must she invest her money?

6. The activity on Sue Washburn's MasterCard account for one billing period is shown below. Find the average daily balance and the finance charge if the billing period is August 26 through September 25, the previous balance was \$3,472.38, and the annual interest rate is $19\frac{1}{2}\%$.

August 30	payment	\$100.00
September 2	gasoline	\$ 34.12
September 10	restaurant	\$ 62.00

7. George and Martha Simpson bought a house from Sue Sanchez for \$205,500. In lieu of a 20% down payment, Ms. Sanchez accepted 5% down at the time of the sale and a promissory note from the Simpsons for the remaining 15%, due in eight years. The Simpsons also agreed to make monthly interest payments to Ms. Sanchez at 12% simple interest until the note expires. The Simpsons borrowed the remaining 80% of the purchase price from their bank. The bank paid that amount, less a commission of 6% of the purchase price, to Ms. Sanchez.
   a. Find the Simpsons' monthly interest-only payment to Ms. Sanchez.
   b. Find Ms. Sanchez's total income from all aspects of the down payment.
   c. Find Ms. Sanchez's total income from all aspects of the sale of the house, including the down payment.
8. Extremely Trustworthy Savings offers five-year CDs at 7.63% compounded annually, and Bank of the South offers five-year CDs at 7.59% compounded daily. Compute the annual yield for each institution and determine which offering is more advantageous for the consumer.
9. Tien Ren Chiang wants to have an IRA that will be worth \$250,000 when he retires at age sixty-five.
   a. How much must he deposit at age twenty-five at $10\frac{1}{8}\%$ compounded quarterly?
   b. If at age sixty-five he arranges for the monthly interest to be sent to him, how much would he receive each month? (Assume that he will continue to receive $10\frac{1}{8}\%$ interest, compounded monthly.)
10. Find the future value of an ordinary annuity with monthly payments of \$200 that earns $6\frac{1}{8}\%$ interest, after 11 years.
11. Find the present value of the annuity in Exercise 10.
12. Find the future value of an annuity due with monthly payments of \$200 that earns $6\frac{1}{8}\%$ interest, after 11 years.
13. Matt and Leslie Silver want to set up a TDA that will generate sufficient interest on maturity to meet their living expenses, which they project to be \$1,300 per month.
   a. Find the amount needed at maturity to generate \$1,300 per month interest if they can get $8\frac{1}{4}\%$ interest compounded monthly.
   b. Find the monthly payment they would have to make into an ordinary annuity to obtain the future value found in part (a) if their money earns $9\frac{3}{4}\%$ and the term is 30 years.
14. Ben Suico buys a new car for \$13,487.31. He puts 10% down and obtains a simple interest amortized loan for the rest at $10\frac{7}{8}\%$ interest for five years.
   a. Find his monthly payment.
   b. Find the total interest.
   c. Prepare an amortization schedule for the first two months of the loan.
   d. If loan fees included in the finance charge total \$633.87, verify the lender's statement that the A.P.R. is 11.4%.
   e. Mr. Suico decides to sell his car two years and six months after he bought it. Find the unpaid balance on his loan.
15. Scott Frei wants to buy a used car that costs \$6,200. The used car dealer has offered him a four-year add-on interest loan that requires a \$200 down payment at 9.9% annual interest with an A.P.R. of 10%.
   a. Find the monthly payment.
   b. Verify the A.P.R.
16. Susan and Steven Tamchin are thinking of buying a home for \$198,000. A potential lender advertises an 80%, 30-year simple interest amortized loan at $8\frac{1}{2}\%$ interest, with an A.P.R. of 9.02%.
   a. Find the size of the Tamchins' monthly payment.
   b. Use the A.P.R. to approximate the fees included in the finance charge.
17. Fred Rodgers is planning for his retirement, so he is setting up a payout annuity with his bank. He wishes to receive a payout of \$1,700 per month for 25 years.
   a. How much money must he deposit if his money earns 6.1% interest compounded monthly?
   b. How large a monthly payment would Fred have made if he saved for his payout annuity with an ordinary annuity, set up 30 years before his retirement? (The two annuities pay the same interest rate.)
   c. Find the total amount that Fred will pay into his ordinary annuity and the total amount that he will receive from his payout annuity.
18. Sue West is planning for her retirement, so she is setting up a payout annuity with her bank. She is now thirty years old, and she will retire when she

is sixty. She wants to receive annual payouts for 25 years, and she wants those payouts to have an annual C.O.L.A. of 4.2%.

a. She wants her first payout to have the same purchasing power as does $17,000 today. How big should that payout be if she assumes inflation of 4.2% per year?

b. How much money must she deposit when she is sixty if her money earns 8.3% interest per year?

c. How large a monthly payment must she make if she saves for her payout annuity with an ordinary annuity? (The two annuities pay the same interest rate.)

Geometry

6

Let no one unversed in geometry enter this door.
Inscription to Plato's Academy, 387 B.C.

Geometry is one of the oldest branches of mathematics; its roots include two very practical human endeavors: surveying and astronomy. From the beginning of recorded time, people have attempted to measure the land and the heavens. The word *geometry* literally means "earth measurement"; it is derived from the Greek words *geos*, meaning "earth," and *metros*, meaning "measure."

Ancient geometry consisted of a fragmented assortment of examples, guesses, tricks, and rule-of-thumb procedures. It was an empirical subject in which approximate answers were usually good enough for practical purposes. A typical problem might have been the determination of the area of a triangular field of grain. The Babylonians and Egyptians were masters at using this collection of "how to's."

As a result of commerce, travel, and warfare, the empirical geometry of the Egyptians reached the ancient Greeks; however, the Greeks' pursuit of absolute truth through abstract thought set them apart from their predecessors. Beginning with Thales of Miletus, the Greeks insisted that geometric statements be established by *deductive reasoning* rather than by trial and error. The orderly development of theorems by *proof* was characteristic of Greek mathematics. This was an entirely new concept in the realm of mathematics. Consequently, the Greeks no longer concerned themselves with triangular fields of grain but with "triangles" and the characteristics associated with "triangularity." To this end, Greek scholars set out to systematically develop the relationships between all known mathematics of the day. The culmination of this effort was the writing of the monumental *Elements* by Euclid in the fourth century B.C.

Conceptually, geometry underwent little change over the next 2,000 years. The seventeenth century witnessed the marriage of geometry and algebra, the so-called *analytic geometry* of Descartes and Fermat (see Chapter 10). Euclid's iron grip on the bases of geometry was finally challenged in the nineteenth century, when non-Euclidean geometries were established. In the twentieth century, there was a proliferation of many types of new geometries. Just as the union of geometry and

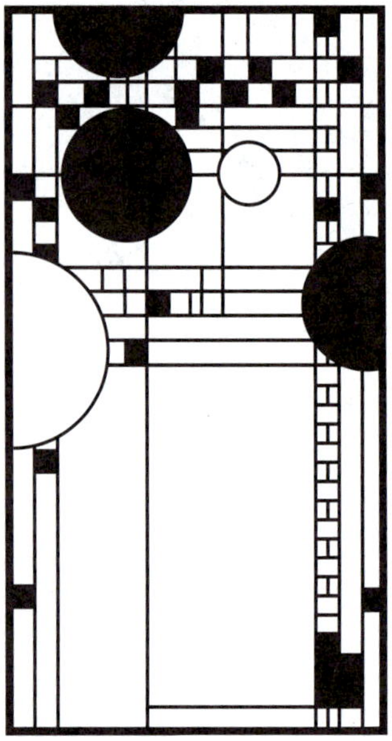

Frank Lloyd Wright Foundation

Architect Frank Lloyd Wright used simple geometric shapes to design this stained glass window in 1911.

algebra heralded the new analytic geometry, the modern-day union of geometry and computers led to the new fractal geometry of Benoit Mandelbrot.

In this chapter, we review the standard concepts and calculations of length, area, and volume and present a brief history of the major accomplishments of ancient, analytic, and modern geometries.

6.1 Perimeter and Area

Geometric shapes have intrigued people throughout history. Art, religion, science, engineering, architecture, psychology, and advertising are just a few of the many areas that make use of triangles, rectangles, squares, circles, cubes, pyramids, cones, spheres, and a host of other forms. Having seen many of these shapes in nature, geometers have constructed ideal representations of them and have developed various formulas for measuring their lengths (one-dimensional), areas (two-dimensional), and volumes (three-dimensional). In this section, we examine some features of the most commonly encountered two-dimensional figures.

Polygons

Two-dimensional figures can be classified by the number of sides they have. A **polygon** is a many-sided figure. A pentagon is a five-sided figure, a hexagon is a six-sided figure, and an octagon is an eight-sided figure. However, the names of polygons do not necessarily end with *-gon*. Although a three-sided figure could be called a *trigon*, we prefer *triangle*. Likewise, a four-sided figure is referred to as a quadrilateral rather than a quadragon.

Our study of polygons focuses on finding the distance around a figure and the amount of space enclosed within the figure. Some of the polygons we will examine are shown in Figure 6.1. The symbol ⊾ represents an angle of 90° (90 degrees = a square corner).

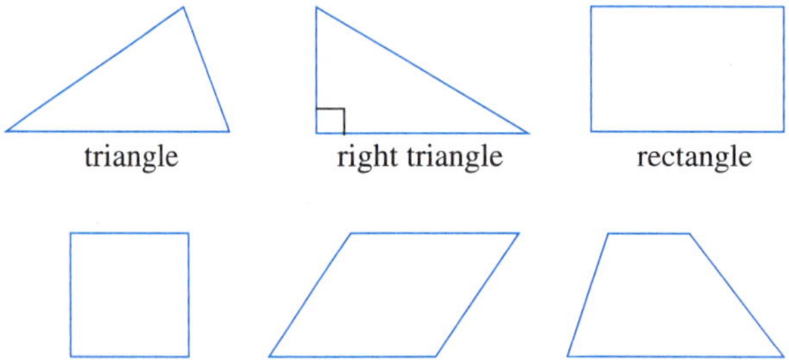

Figure 6.1
Common polygons

The **perimeter** of (or distance around) a two-dimensional figure is the sum of the lengths of its sides. As shown in Figure 6.2, the perimeter of a rectangular scarf 18 inches wide and 2 feet long is 7 feet. (We must first convert 18 inches into 1.5 feet.)

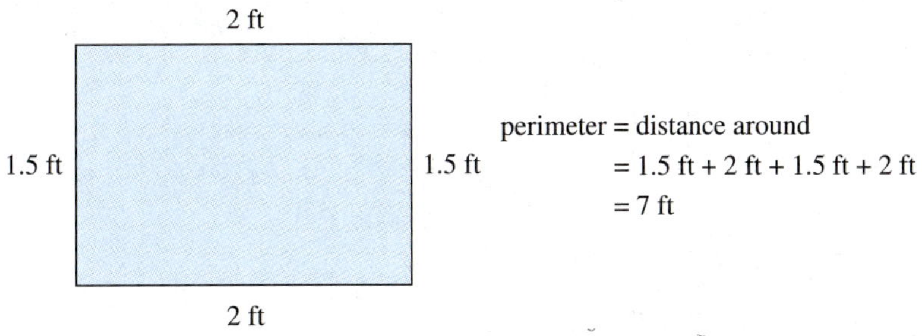

Figure 6.2

The **area** of a two-dimensional figure is the number of square units (for example, square inches, square miles) it takes to fill the interior of that figure. As shown in Figure 6.3, a rectangular rug 6 feet long and 3 feet 6 inches wide has an area of 21 square feet (or 21 ft^2).

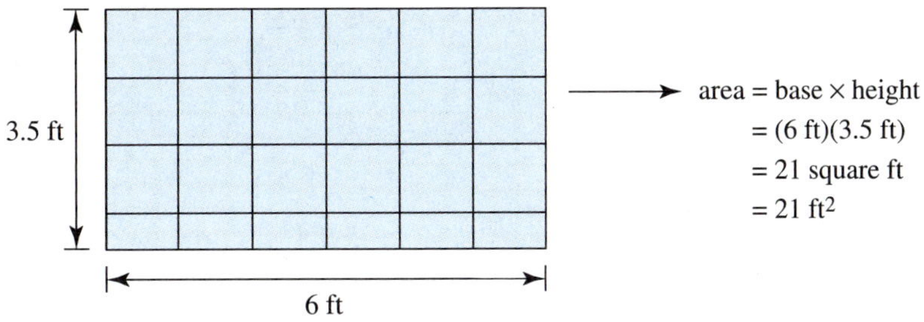

Figure 6.3

It has been suggested that the concepts of square units and area have their origins in the weaving of fabric. Single strands of yarn have a linear, or one-dimensional, measurement, such as inches or feet. However, when an equal number of "horizontal" and "vertical" strands are woven together on a loom, a square figure is formed. Therefore, a natural way to measure the amount of cloth created from the strands is to employ units consisting of squares (square feet). (See Figure 6.4.)

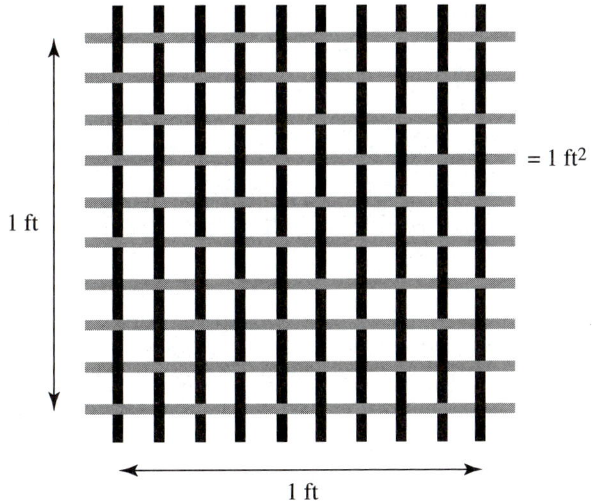

Figure 6.4

It is very easy to find the area of a rectangle or square. On the basis of these quadrilaterals, we can also find the areas of triangles, trapezoids, and parallelograms. A **trapezoid** is a quadrilateral with one pair of parallel sides; a **parallelogram** is a quadrilateral with two pairs of parallel sides. The area of a **triangle, rectangle,** parallelogram, or trapezoid can be found by use of the appropriate formula given in Figure 6.5, with A = area, b = base, h = height, and b_1 and b_2 = bases.

Why does the area of a triangle equal one-half the product of the base times the height? The answer lies in the formula for the area of a rectangle. A triangle can be

Area Formulas for Common Polygons

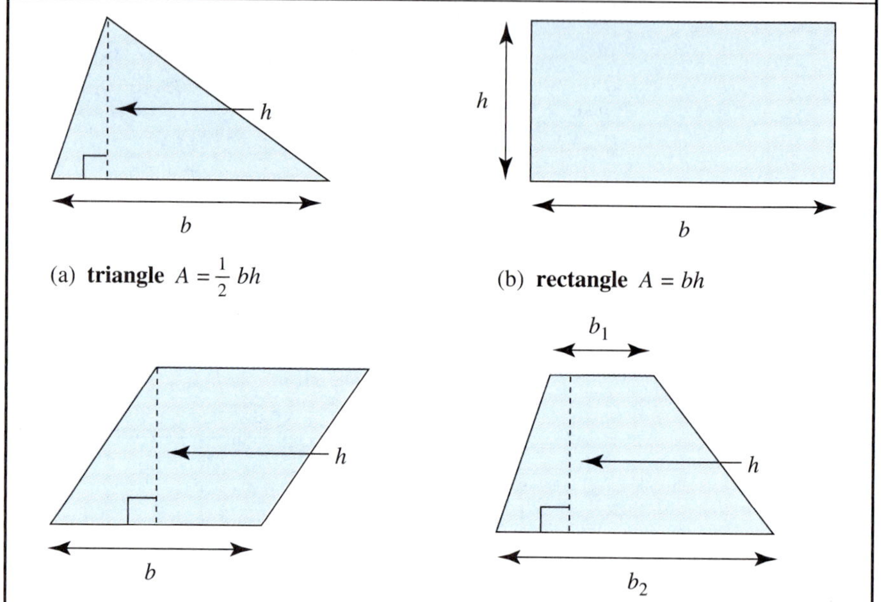

(a) **triangle** $A = \frac{1}{2}bh$ (b) **rectangle** $A = bh$

(c) **parallelogram** $A = bh$ (d) **trapezoid** $A = \frac{1}{2}(b_1 + b_2)h$

Figure 6.5

divided into two smaller triangles. If copies of these smaller triangles are then "added on" to the original triangle, a rectangle of area $b \cdot h$ can be formed, as shown in Figure 6.6. Because the area of the original triangle is half that of the rectangle, we have the desired result.

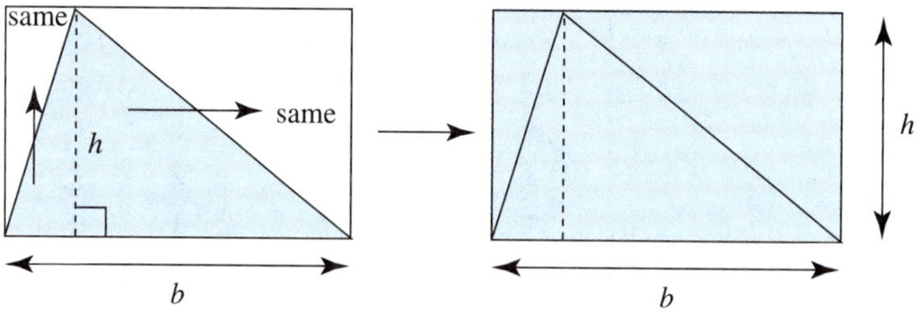

Figure 6.6
$A_{\text{triangle}} = \frac{1}{2}A_{\text{rectangle}} = \frac{1}{2}bh$

Why does the area of a parallelogram equal the product of the base times the height? Once again, the answer lies in the formula for the area of a rectangle. A parallelogram can be rearranged to form a rectangle of area $b \cdot h$, as shown in Figure 6.7. That is, the area of the parallelogram is the same as the area of the rectangle, and we have the desired result.

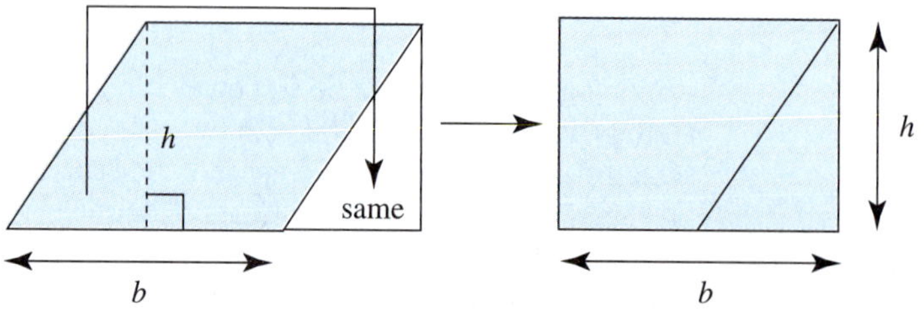

Figure 6.7
$A_{\text{parallelogram}} = A_{\text{rectangle}} = bh$

Why does the area of a trapezoid equal one-half the product of the sum of the bases times the height? Yet again, the answer lies in the formula for the area of a rectangle! The two triangular "tips" of the trapezoid can be cut off and rearranged to form a rectangle, as shown in Figure 6.8. The base of the rectangle equals the average of the two bases of the trapezoid; that is, $b_{rectangle} = (b_1 + b_2)/2$. Thus, we have the desired result.

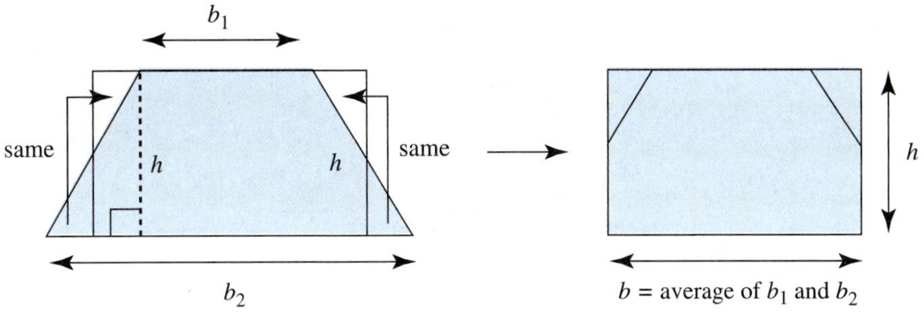

Figure 6.8

$$A_{trapezoid} = A_{rectangle} = \left(\frac{b_1 + b_2}{2}\right)(h)$$

Heron's Formula for the Area of a Triangle

If the height of triangle is not known, we cannot use the common formula $A = \frac{1}{2}bh$; an alternative method must be used. When the lengths of all three sides are known, the "semiperimeter" can be used to find the area of the triangle. As the name implies, the semiperimeter is half the perimeter. The formula is referred to as Heron's Formula, in honor of the ancient Greek mathematician Heron of Alexandria (circa A.D. 75).

Heron's Formula for the Area of a Triangle

$$s = \frac{1}{2}(a + b + c)$$

$$A = \sqrt{s(s - a)(s - b)(s - c)}$$

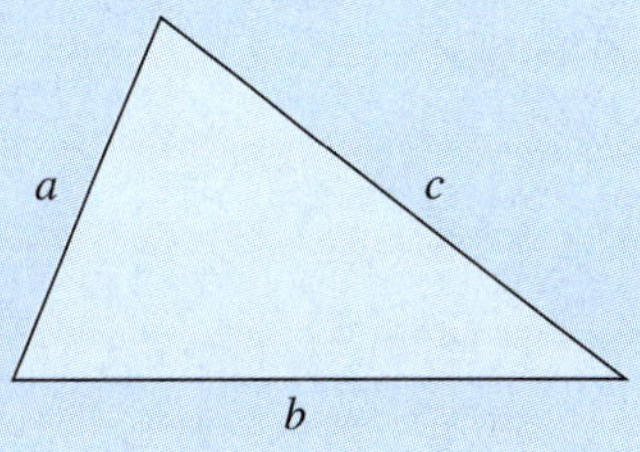

EXAMPLE 1 Assuming the same growing conditions, which of the fields shown in Figures 6.9 and 6.10 would produce more grain?

a. **Figure 6.9**

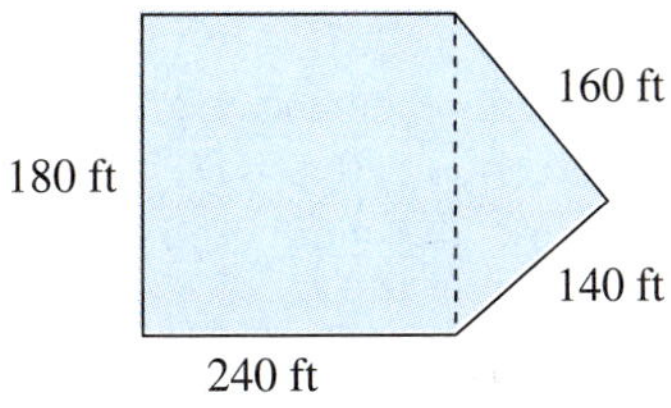

b. **Figure 6.10**

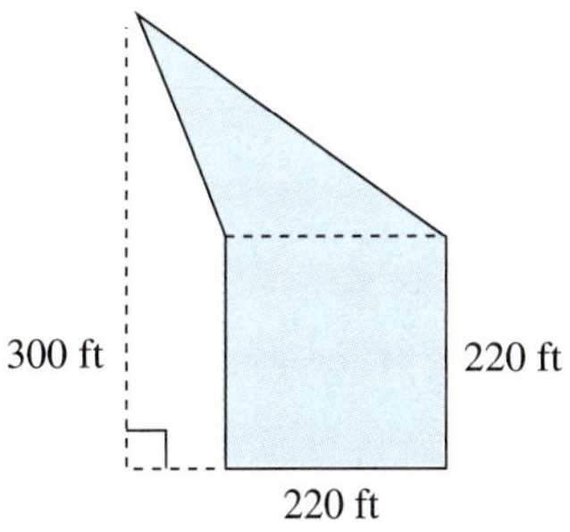

Solution The field with the larger area would produce more grain.

Finding the Area of the Field in Part (a) The field consists of a rectangle and a triangle, as shown in Figure 6.11. To find the total area, we must find the area of each separate shape and add the results together.

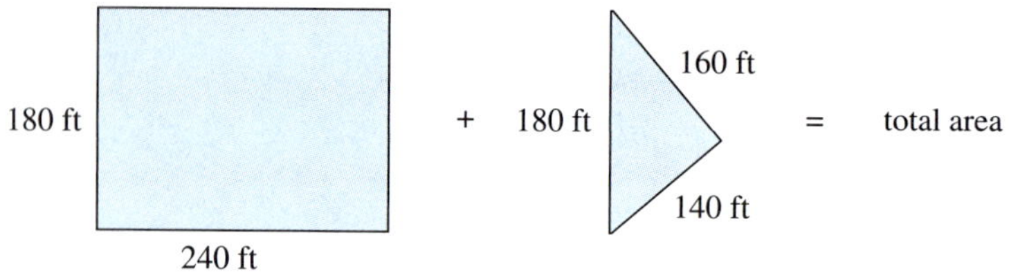

Figure 6.11

$$A_{\text{rectangle}} + A_{\text{triangle}} = A_{\text{total}}$$

The area of the rectangle is

$$A_{\text{rectangle}} = bh = (240 \text{ ft})(180 \text{ ft}) = 43{,}200 \text{ ft}^2$$

To find the area of the triangle, we use the semiperimeter, because the lengths of all three sides are known:

$$s = \frac{1}{2}(a + b + c) = \frac{1}{2}(180 \text{ ft} + 160 \text{ ft} + 140 \text{ ft}) = 240 \text{ ft}$$

Using Heron's Formula, we have

$$\begin{aligned} A_{\text{triangle}} &= \sqrt{240(240 - 180)(240 - 160)(240 - 140)} \\ &= \sqrt{(240 \text{ ft})(60 \text{ ft})(80 \text{ ft})(100 \text{ ft})} \\ &= \sqrt{115{,}200{,}000 \text{ ft}^4} \\ &= 10{,}733.12629\ldots \text{ ft}^2 \\ &\approx 10{,}733 \text{ ft}^2 \\ A_{\text{total}} &= A_{\text{rectangle}} + A_{\text{triangle}} = 43{,}200 \text{ ft}^2 + 10{,}733 \text{ ft}^2 \\ &= 53{,}933 \text{ ft}^2 \end{aligned}$$

The total area of the field in part (a) is 53,933 square feet.

Finding the Area of the Field in Part (b) The field in part (b) is composed of a square and a triangle. The lengths of two of the sides of the triangle are not given, so the semiperimeter cannot be used. However, we do know that the base of the triangle is 220 feet, and we can find its height. The total height of the figure is 300 feet, and the height of the square is 220 feet; we deduce that the height of the triangle is 300 ft − 220 ft = 80 ft. (See Figure 6.12.) Now the area of the square is

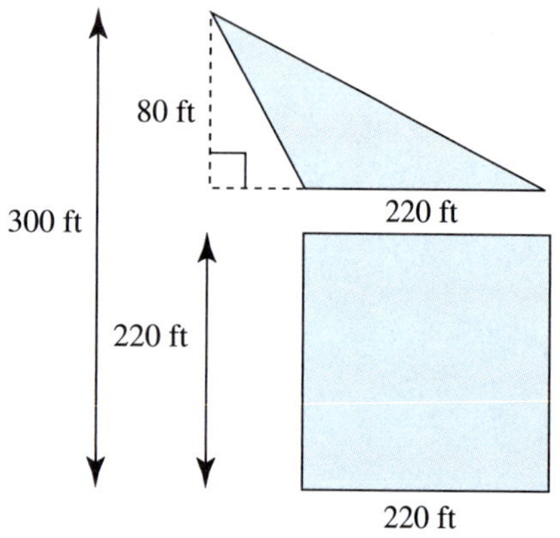

$$A_{\text{square}} = b^2 = (220 \text{ ft})^2 = 48{,}400 \text{ ft}^2$$

Figure 6.12

The area of the triangle is

$$A_{\text{triangle}} = \frac{1}{2}bh = \frac{1}{2}(220 \text{ ft})(80 \text{ ft}) = 8{,}800 \text{ ft}^2$$

$$A_{\text{total}} = A_{\text{square}} + A_{\text{triangle}} = 48{,}400 \text{ ft}^2 + 8{,}800 \text{ ft}^2$$
$$= 57{,}200 \text{ ft}^2$$

The total area of the field in part (b) is 57,200 square feet.

Assuming the same growing conditions, the field in part (b) would produce more grain, because it has a larger area than the field in part (a). ■

EXAMPLE 2 A local art supply store has donated two custom-size canvases to be used for murals in a youth center. Before they can be used, they must be treated with a special primer. Of the two canvases shown in Figures 6.13 and 6.14, which requires more primer?

a. Figure 6.13

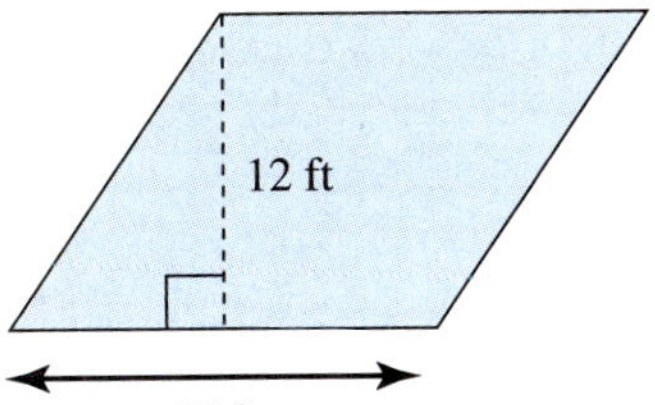

b. Figure 6.14

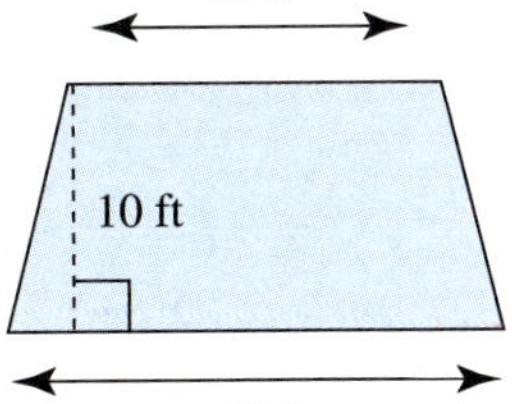

Solution The canvas with the larger area requires more primer.

Finding the Area of the Parallelogram in Part (a)

$$A_{\text{parallelogram}} = bh = (15 \text{ ft})(12 \text{ ft}) = 180 \text{ ft}^2$$

The canvas in part (a) contains 180 square feet of material to be primed.

Finding the Area of the Trapezoid in Part (b)

$$A_{\text{trapezoid}} = \frac{1}{2}(b_1 + b_2)h$$
$$= \frac{1}{2}(20 \text{ ft} + 16 \text{ ft})(10 \text{ ft})$$
$$= \frac{1}{2}(36 \text{ ft})(10 \text{ ft})$$
$$= 180 \text{ ft}^2$$

The canvas in part (b) also contains 180 square feet of material to be primed. Because the figures have the same area, each would require the same amount of primer. ■

Right Triangles

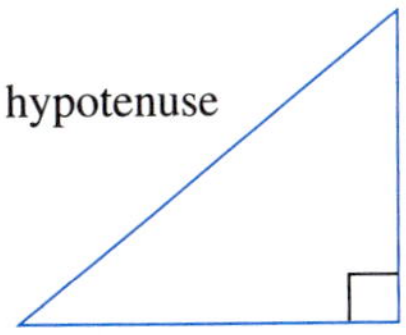

Figure 6.15
Right triangle

If one of the angles of a triangle is a right angle (a square corner, or 90°), the triangle is called a **right triangle.** The side opposite the right angle is called the **hypotenuse** and is labeled c, while the remaining two sides are called the **legs** and are labeled a and b, as shown in Figure 6.15.

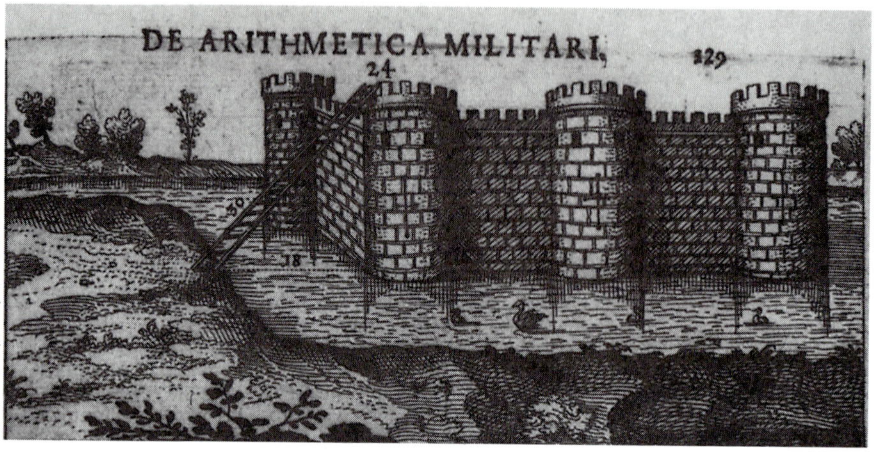

An early military handbook utilized the Pythagorean Theorem. Because $18^2 + 24^2 = 30^2$, a 30-foot ladder is required to reach the top of a 24-foot wall from the opposite side of an 18-foot moat.

A special relationship exists between the hypotenuse and the legs of a right triangle. Over 2,000 years ago, early geometers observed that if the longest side of a right triangle (the hypotenuse) was squared, the number obtained was always the same as the sum of the squares of the two other sides (the legs). This observation was proved to be true for *all* right triangles. It is referred to as the **Pythagorean Theorem,** in honor of the ancient Greek mathematician Pythagoras of Samos (circa 580 B.C.), although the result was known to earlier peoples.

Pythagorean Theorem

For any right triangle, the square of the hypotenuse equals the sum of the squares of the legs.

$$c^2 = a^2 + b^2$$

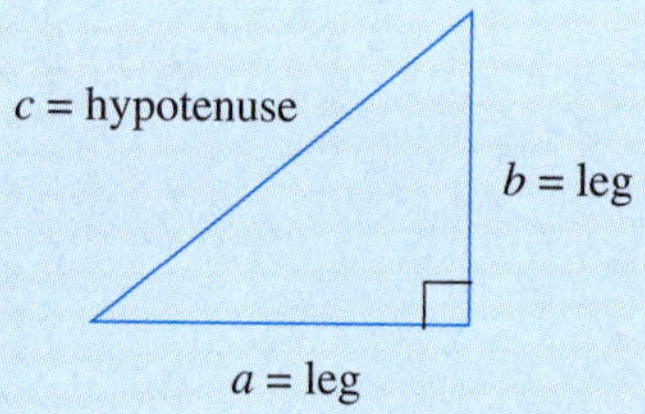

The converse of this theorem is also true; if the sides of a triangle satisfy the relationship $c^2 = a^2 + b^2$, then the triangle is a right triangle.

EXAMPLE 3 The length of the hypotenuse of a right triangle is 25 feet, and the length of one of the legs is 7 feet. (See Figure 6.16.) Find the area of the triangle.

Solution To find the area of a triangle, we must know either a base and its corresponding (perpendicular) height, or all three sides. In this case, we must find the missing side of the given triangle. Because we have a right triangle, we can apply the Pythagorean Theorem to find the missing leg, b:

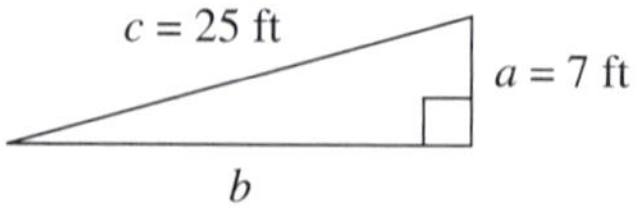

Figure 6.16

$$a^2 + b^2 = c^2$$
$$b^2 = c^2 - a^2$$
$$b^2 = (25)^2 - (7)^2$$
$$b^2 = 625 - 49 = 576$$
$$b = \sqrt{576} = 24 \text{ ft}$$

Now we find the area of the triangle:

$$\begin{aligned} A &= \frac{1}{2}bh \\ &= \frac{1}{2}(24\text{ ft})(7\text{ ft}) \\ &= 84\text{ ft}^2 \end{aligned}$$

The area of the triangle is 84 square feet. (We obtain the same answer if we use Heron's Formula.) ■

Circles

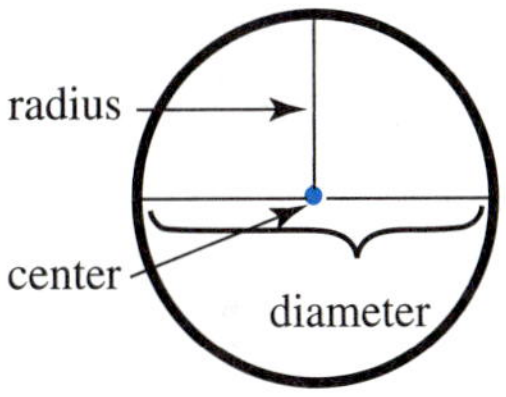

Figure 6.17

Many people consider the circle a perfect geometric figure; it has no beginning or end, it has total symmetry, and it motivated the invention of the famous irrational (nonfraction) number π **(pi).** The early Greeks defined a **circle** as the set of all points in a plane equidistant from a fixed point. The fixed point is called the **center** of the circle; a line segment from the center to any point on the circle is called a **radius;** and any line segment connecting two points on the circle and passing through the center is called a **diameter.** (The length of a diameter is twice the length of a radius.) See Figure 6.17.

Recall that the distance around a figure is called its *perimeter*; however, as a special case, the distance around a circle is called its **circumference.** Thousands of years ago, geometers observed a curious relationship: If they measured the circumference C of *any* circle (probably by using a string or a rope), they found that it was always a little bit longer than 3 times the diameter d of the circle, as shown in Figure 6.18. In other words, the ratio of circumference to diameter is constant: circumference/diameter = constant. This constant number, which is a little larger than 3, is represented by the Greek letter π (read "pi"; rhymes with *sly*). Hence,

$$\frac{\text{circumference}}{\text{diameter}} = \pi \qquad \text{or} \qquad \text{circumference} = \pi \cdot \text{diameter}$$

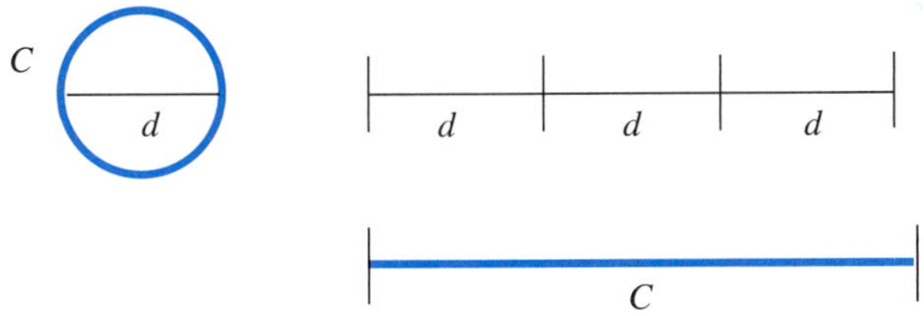

Figure 6.18

Mathematicians have shown that π is irrational; that is, it cannot be written exactly as a fraction, and its decimal expansion never repeats or terminates. Computers have calculated π to hundreds of thousands of decimal places, and we have

$$\pi \approx 3.14159265358979323846264338327 9\ldots$$

Various approximations of π have been used throughout history (some of them are investigated in Sections 6.3 and 6.4). Today, the most commonly used classroom estimates are 3.1416 and $\frac{22}{7}$. When π is needed in calculations, simply press the appropriate button on your calculator. If your calculator doesn't have a $\boxed{\pi}$ button, use $\pi = 3.1416$.

Pi is used to calculate the circumference and the area of a circle. We will investigate the origins of the area formula in Section 6.3.

Circumference and Area of a Circle

The **circumference** C of a circle of radius r is

$$C = 2\pi r$$

The **area** A of a circle of radius r is

$$A = \pi r^2$$

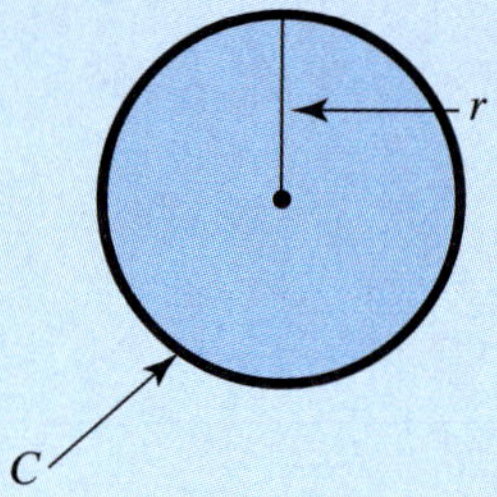

EXAMPLE 4 A Norman window consists of a rectangle with a semicircle mounted on top. For the window shown in Figure 6.19, find the following.

a. the area **b.** the perimeter

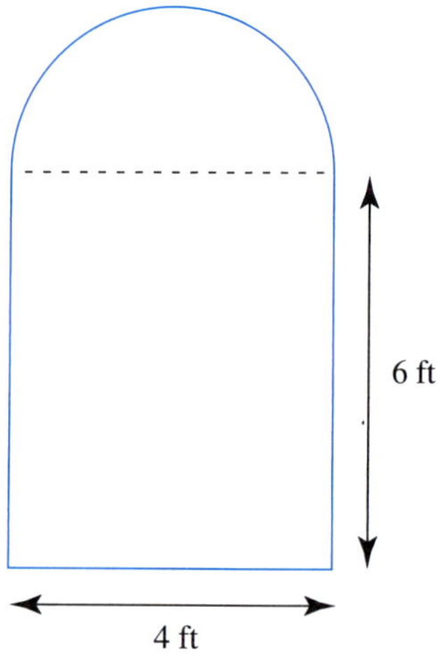

Figure 6.19

Solution **a.** The area of the entire window is the sum of the areas of the rectangle and the semicircle. A semicircle is half a circle, so the semicircular region has area equal to one-half the area of a complete circle. The base of the rectangle is the same as the diameter of the semicircle. Hence, the diameter is 4 feet, and the radius is 2 feet (see Figure 6.20).

$$\begin{aligned} A_{\text{total}} &= A_{\text{rectangle}} + A_{\text{semicircle}} \\ &= bh + \frac{1}{2}(\pi r^2) \\ &= (4\text{ ft})(6\text{ ft}) + \frac{1}{2}\pi(2\text{ ft})^2 \\ &= 24\text{ ft}^2 + 2\pi\text{ ft}^2 \\ &= (24 + 2\pi)\text{ ft}^2 \\ &= 30.28318531\ldots\text{ ft}^2 \end{aligned}$$

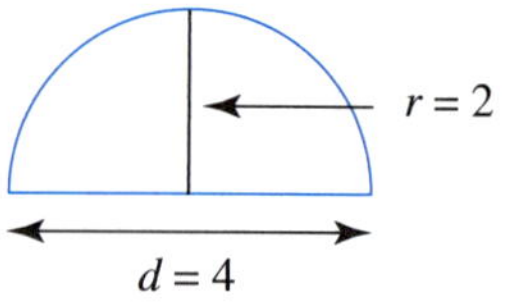

Figure 6.20

24 [+] 2 [×] [π] [=]

With a graphing calculator, press [ENTER] instead of [=].

Thus, the area of the Norman window is approximately 30.3 square feet.

b. The perimeter of the window consists of a semicircle (one-half the circumference of a circle), one horizontal line segment, and two vertical line segments.

$$P = \frac{1}{2}C + b + 2h$$
$$= \frac{1}{2}(\pi \cdot 4\text{ ft}) + 4\text{ ft} + 2(6\text{ ft})$$
$$= 2\pi\text{ ft} + 16\text{ ft}$$
$$= 22.28318531\ldots\text{ ft}$$

Thus, the perimeter of the Norman window is approximately 22.3 feet. ■

6.1 Exercises

When necessary, round off answers to one decimal place. In Exercises 1–8, find the area of each figure.

1.

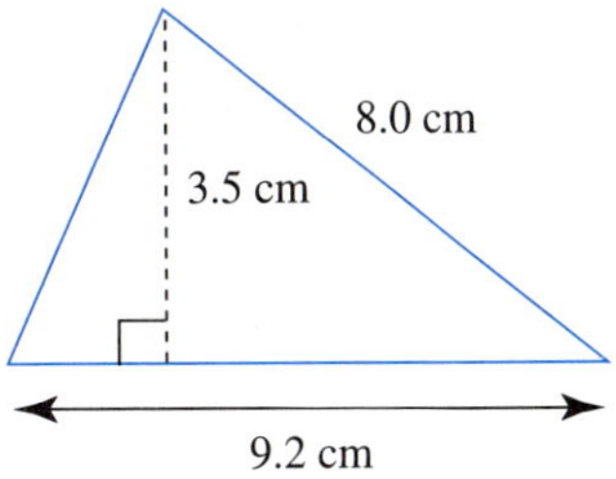

2.

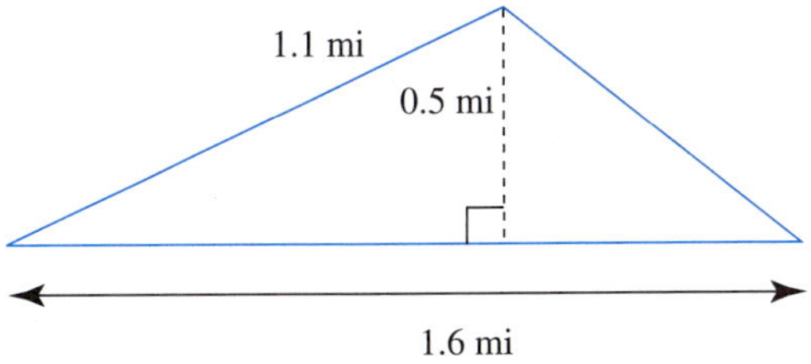

3.

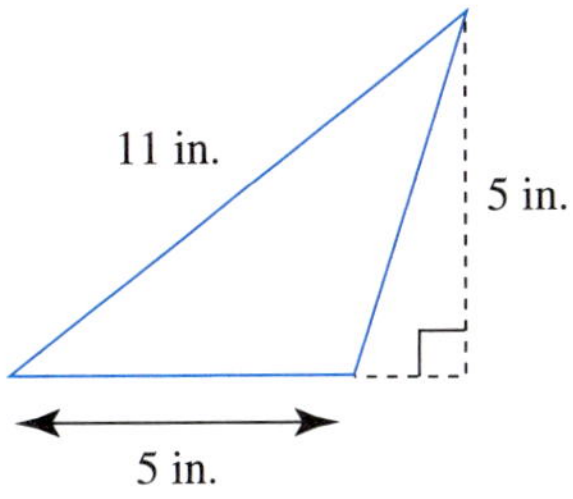

4.

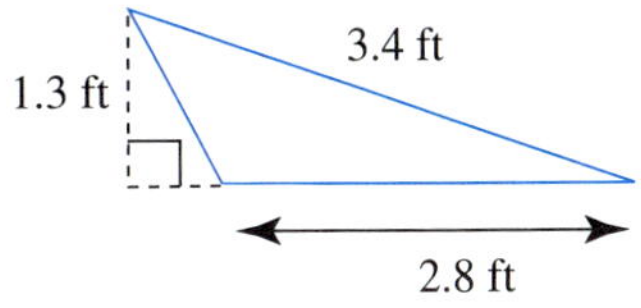

5.

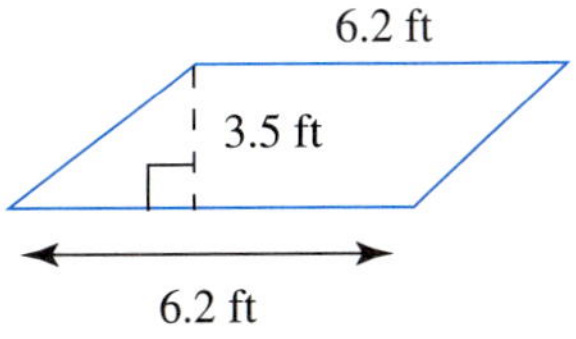

6.

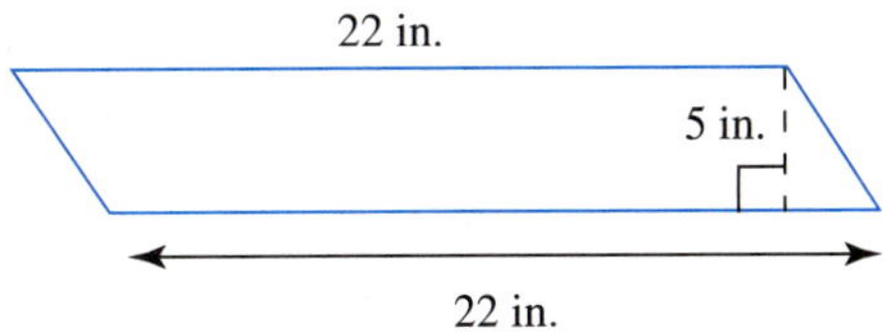

7.

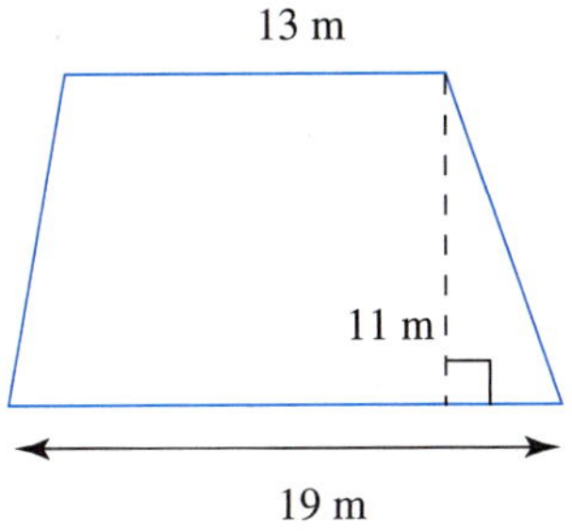

8.

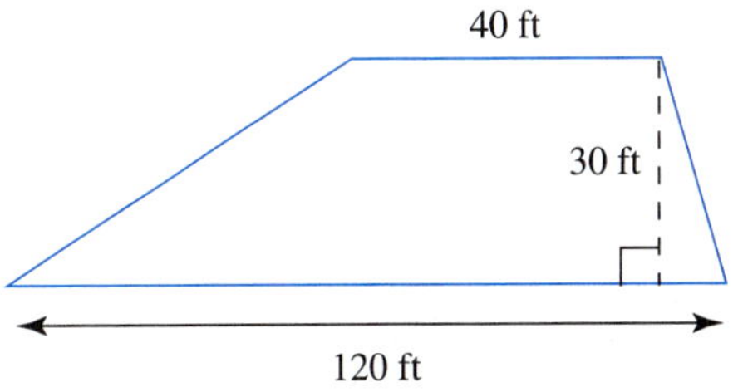

In Exercises 9 and 10, find (a) the area and (b) the circumference of the circle.

9.

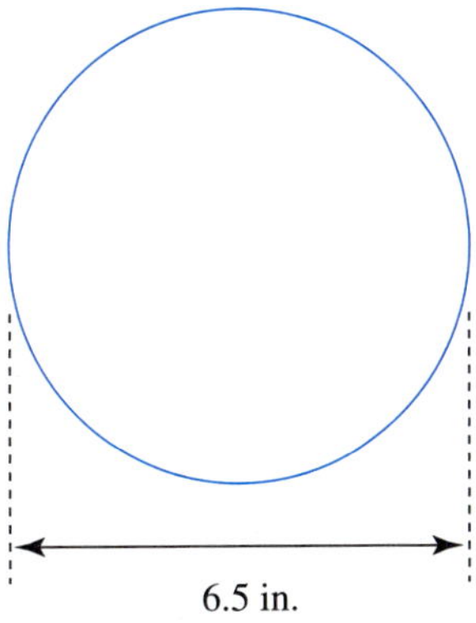

10.

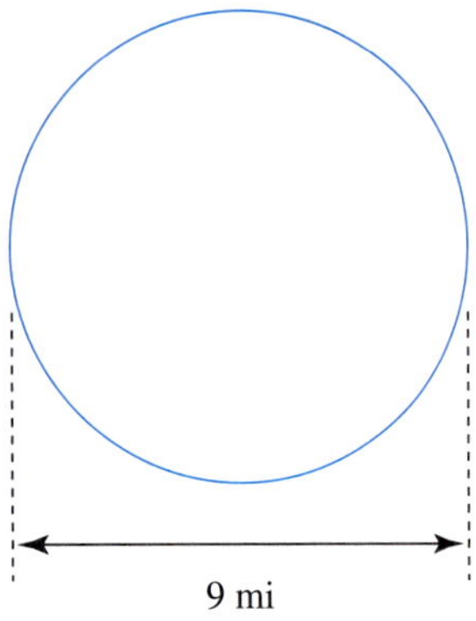

In Exercises 11–20, find (a) the area and (b) the perimeter of each figure.

11.

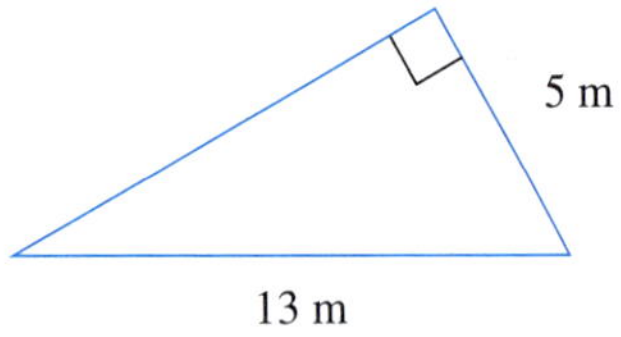

12.

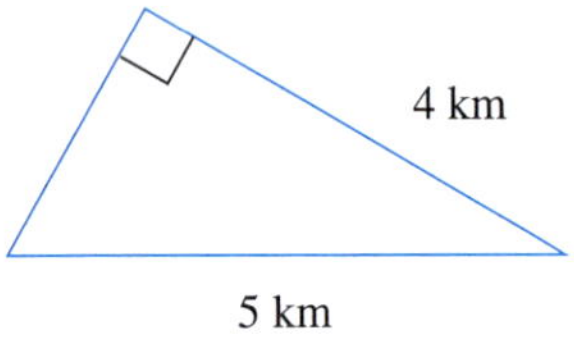

13.

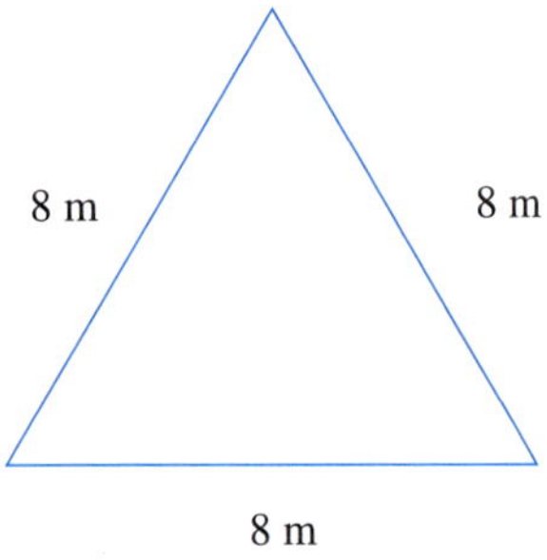

14.

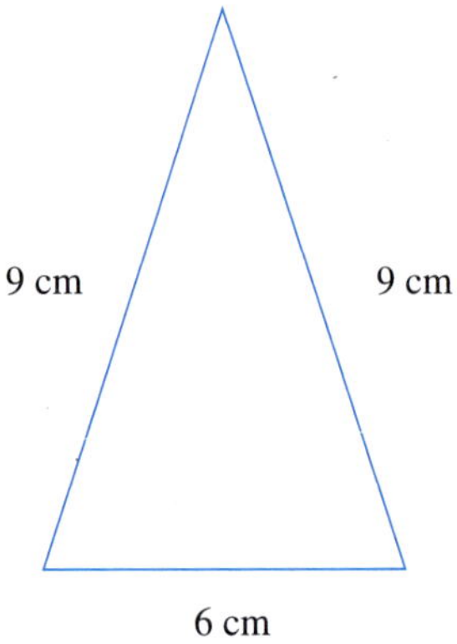

15.

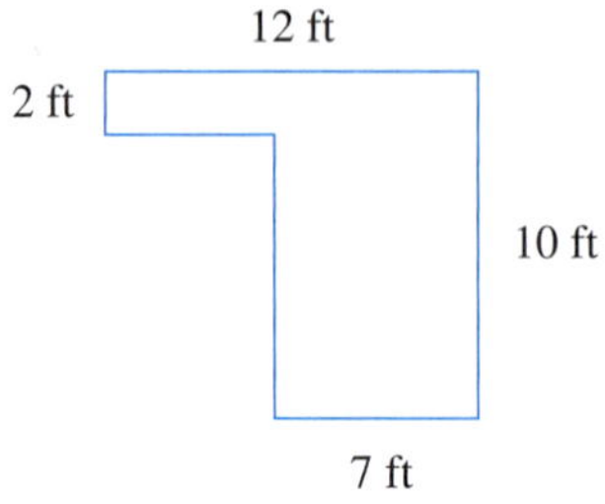

16.

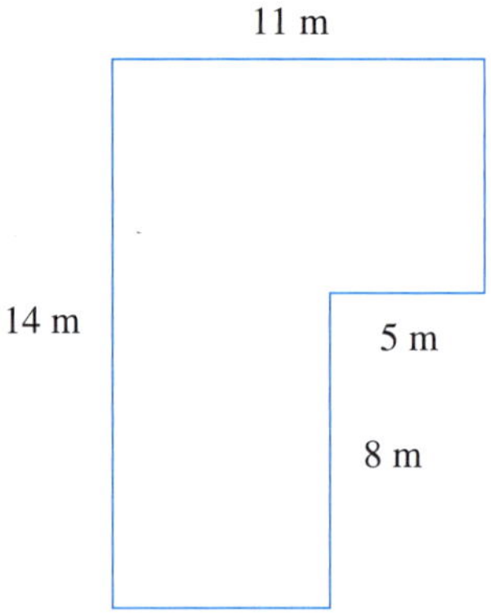

17.

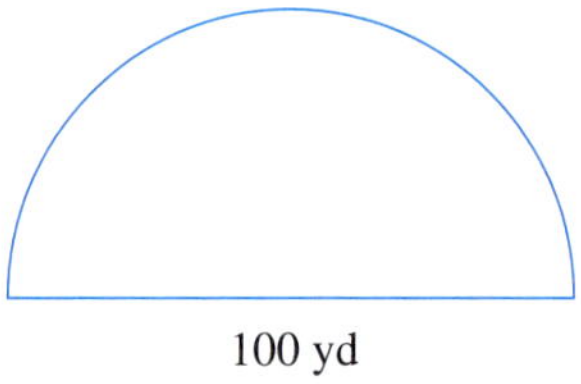

18.

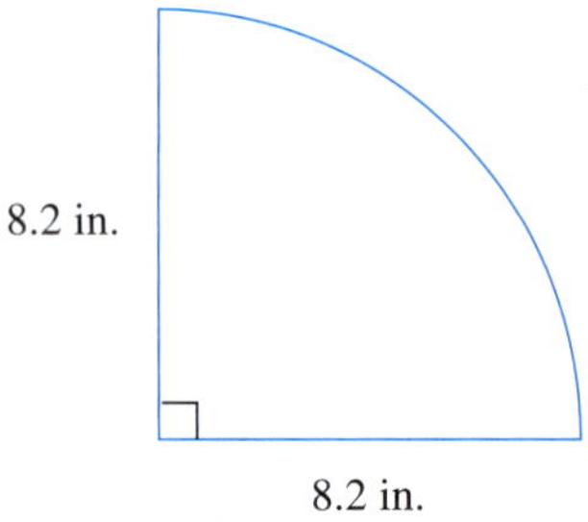

19.

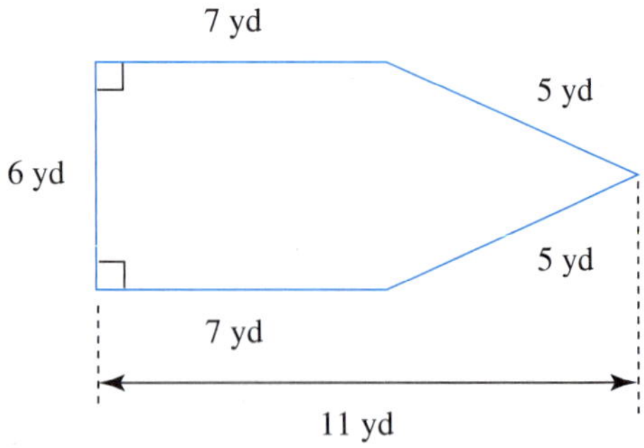

20.

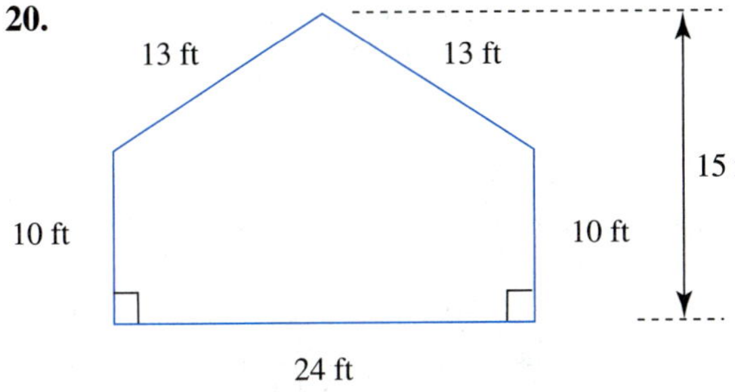

In Exercises 21 and 22, find (a) the area and (b) the perimeter of each Norman window.

HINT: See Example 4.

21.

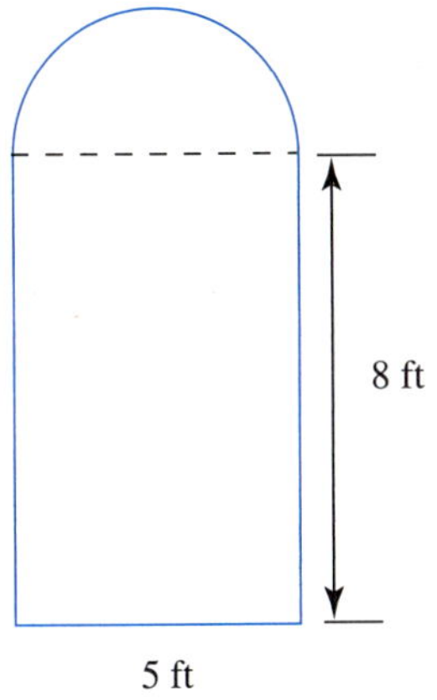

22.

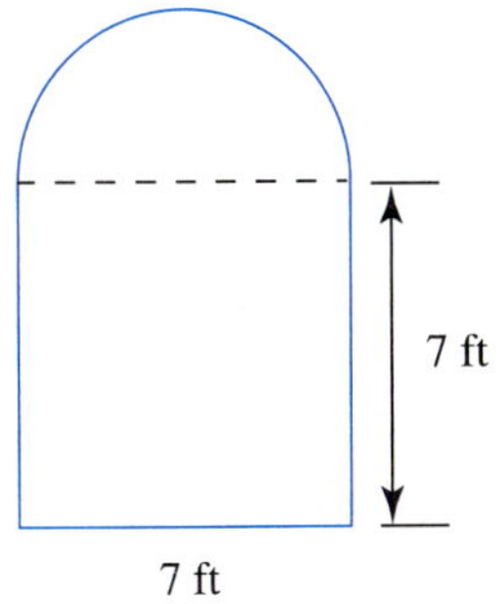

23. A circular swimming pool has diameter 50 feet and is centered in a fenced-in square region measuring 80 feet by 80 feet. A concrete sidewalk 5 feet wide encircles the pool, and the rest of the region is grass, as shown.

a. Find the surface area of the water.
b. Find the area of the concrete sidewalk.
c. Find the area of the grass.

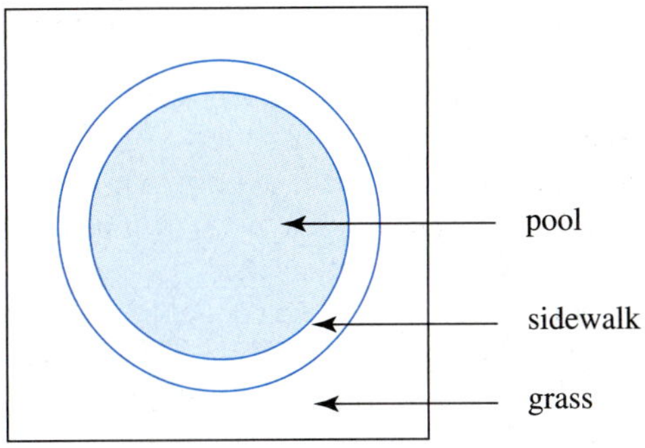

24. A rectangular swimming pool 30 feet by 15 feet is surrounded by a uniform concrete sidewalk 5 feet wide. The pool is positioned inside a fenced-in circular grassy region of diameter 60 feet, as shown.

a. Find the surface area of the water.
b. Find the area of the concrete sidewalk.
c. Find the area of the grass.

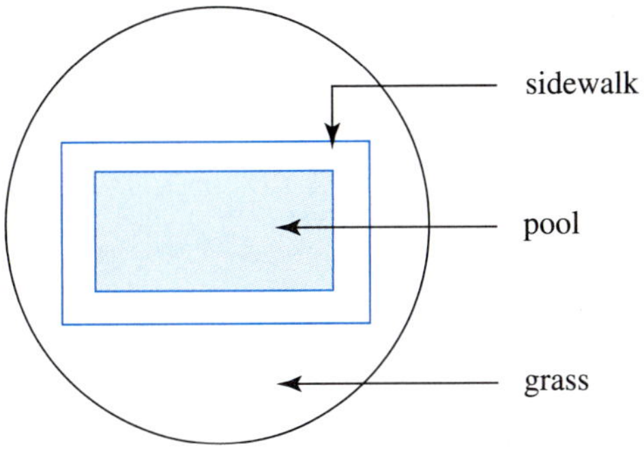

25. The perimeter of a square window is 18 feet 8 inches. Find the area of the window (a) in square inches and (b) in square feet.

26. A square window has an area of 729 square inches. Find the perimeter of the window.

27. The circumference of a circular window is 10 feet. Find the area of the window.

28. A circular window has an area of 10 square feet. Find the circumference of the window.

29. You jog $\frac{3}{4}$ mile due north, then jog $1\frac{1}{2}$ miles due east, and then return to your starting point via a straight-line path. How many miles have you jogged?

30. You walk 100 yards due south, then 120 yards due west, and then 30 yards due north. How far are you from your starting point?

31. A 10-foot ladder leans against a wall. If the base of the ladder is 6 feet from the wall, how far up the wall does the ladder reach?

32. A ladder is leaning against a building. If the bottom of the ladder is $7\frac{1}{2}$ feet from the wall and the top of the ladder is 10 feet above the ground, how long is the ladder?

33. An oval athletic field is the union of a rectangle and semicircles at opposite ends, as shown below.
    a. Find the total area of the field.
    b. Find the distance around the field.

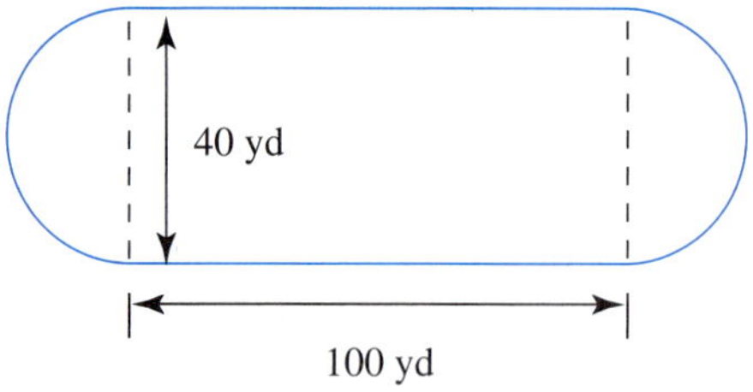

34. An oval athletic field is the union of a square and semicircles at opposite ends, as shown below. If the total area of the field is 1,300 square yards, find the dimensions of the square.

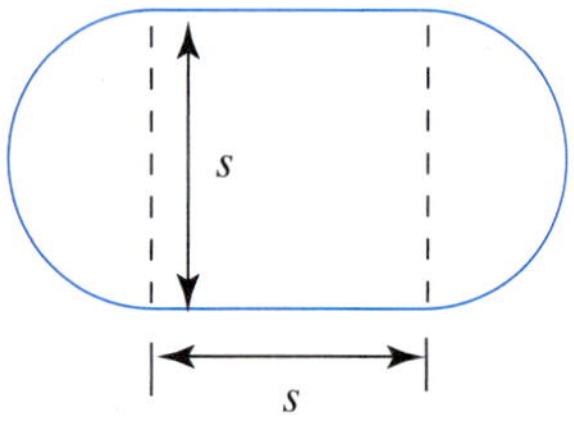

35. The lengths of the edges of a triangular canvas are 100 feet, 140 feet, and 180 feet. You want to waterproof the tarp. If one can of waterproofing will treat 1,000 square feet, how many cans do you need?

36. Steve Loi wants to fertilize his backyard. The yard is a triangle whose sides are 110 feet, 120 feet, and 150 feet. If one bag of fertilizer will cover 1,200 square feet, how many bags does he need?

37. Reddie's Pizza Bash is famous for its Loads-o-Meat Deluxe Pizza. It comes in three sizes: small (13 inches) for \$11.75, large (16 inches) for \$14.75, and super (19 inches) for \$22.75. Which size is the best deal?

    HINT: Find the price per square inch.

38. The best-selling pizza at Magic Mushroom Pizza Deli is the Vegetarian Surprise. It comes in three sizes: small (12 inches) for \$11.25, large (14 inches) for \$14.75, and super (18 inches) for \$20.75. Which size is the best deal?

    HINT: Find the price per square inch.

Answer the following questions using complete sentences and your own words.

Projects

39. The purpose of this project is to verify the Pythagorean Theorem.
    a. On a sheet of paper that has a square corner, measure 3 inches from the corner along the edge of the paper and put a dot. Now measure 4 inches from the same corner along the other edge of the paper and put a dot. Use the Pythagorean Theorem to calculate the distance between the dots. Now, measure the distance between the dots. How does your measurement compare to your calculation?
    b. Repeat part (a) using 6 and 8 inches.
    c. Repeat part (a) using $7\frac{1}{2}$ and 10 inches.
    d. Repeat part (a) using 3 and 4 centimeters.
    e. Repeat part (a) using 5 and 12 centimeters.
    f. Repeat part (a) using 7 and 24 centimeters.

40. The purpose of this project is to calculate the cost of materials to resurface a floor.
    a. Pick a rectangular room in your apartment or house. Measure the dimensions of the floor, and calculate its area in square feet.
    b. Convert your answer from part (a) to square yards. (*HINT:* How many square feet are in 1 square yard?)
    c. Suppose you want to install a new floor covering in your selected room. Of the three materials, carpet, vinyl, and wood, which do you think would be most expensive? Least expensive?
    d. Go to a local carpet and floor covering store, and select a style of carpet for your room. What is the unit price of the product? Use the unit price to calculate the cost of the carpet for your room.
    e. Instead of carpet, select a vinyl product for the floor of your room. What is the unit price of the product? Use the unit price to calculate the cost of the vinyl for the floor of your room.
    f. Instead of carpet or vinyl, select a wood floor covering for your room. What is the unit price of the product? Use the unit price to calculate the cost of the wood floor for your room.
    g. How do your answers to parts (d)–(f) compare with your answer to part (c)?

6.2 Volume and Surface Area

Having examined the perimeter (one-dimensional measurement) and area (two-dimensional measurement) of common geometric figures in Section 6.1, we now explore the geometry of three-dimensional figures. In particular, we investigate the calculation of volume and surface area and apply the results to a variety of situations.

Problem Solving

Outdoors Unlimited, a camping supply store, is having a sale on its demonstration models, discontinued lines, and irregular items. Two water containers are on sale—one cylindrical and the other rectangular, as shown in Figure 6.21. You want to purchase the container that holds the most water, but unfortunately the original information on the capacity of each container has been lost. Which container would you choose? To compare the capacities of the two vessels, you must find their volumes.

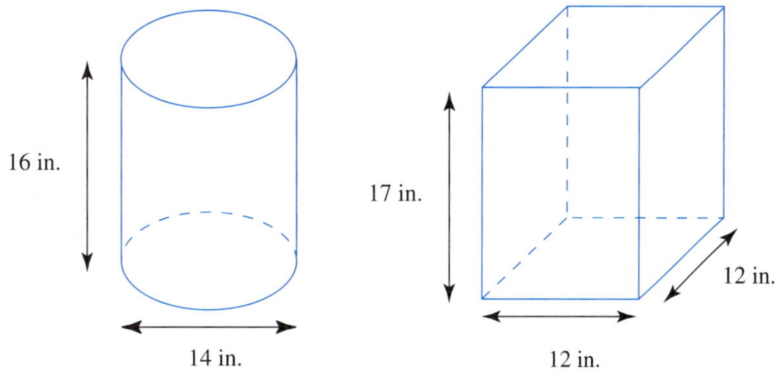

Figure 6.21

Volume is a measure of the amount of space occupied by a three-dimensional object. Volume was originally investigated in the course of bartering and commerce. When grain, wine, and oil were traded, various vessels or containers were used as standards. However, different cultures used different standard vessels, so elaborate conversion systems were necessary. Pints, quarts, gallons, hogsheads, cords, pecks, and bushels are the modern-day legacy of early vessel measurement. Today, the fundamental figure used in the calculation of volume is the cube; volume is expressed in terms of "cubic" units such as cubic feet (ft^3) and cubic centimeters (cc). (See Figure 6.22.)

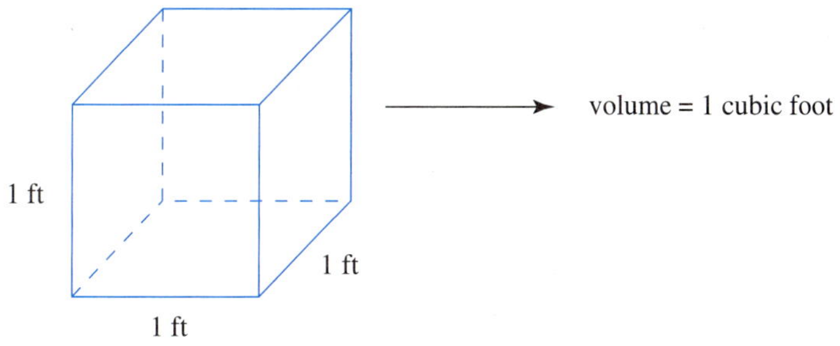

Figure 6.22

The volume of a rectangular box is found by multiplying its length times its width times its height; that is, $V = L \cdot w \cdot h$. Notice that $(L \cdot w)$ equals the area of the (rectangular) base of the box. Consequently, the volume of the box can be found by multiplying the area of the base times the height of the box, as shown in Figure 6.23.

$$\text{volume} = L \cdot w \cdot h$$
$$V = (L \cdot w) \cdot h$$
$$V = A_{\text{base}} \cdot h$$

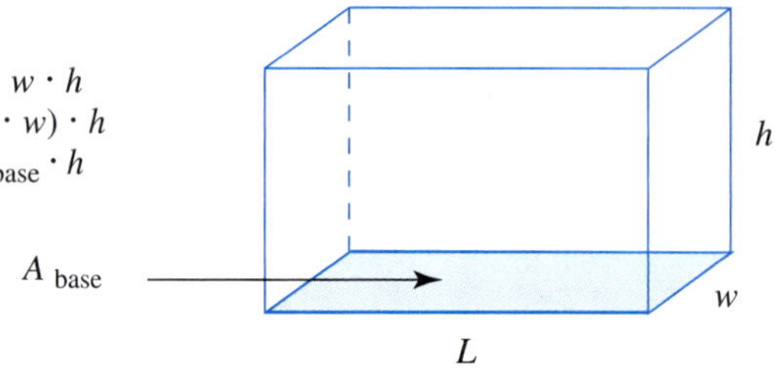

Figure 6.23

A rectangular box has the property that every cross section, or slice, taken parallel to the base produces an identical rectangle. It is precisely this property that allows us to calculate the volume of the box simply by multiplying the area of the base times the height. In general, the volume of any figure that has *identical* cross sections (same shape and size) from top to bottom is found by multiplying the area of the base times the height of the figure.

Volume of a Figure Having Identical Cross Sections

If a three-dimensional figure has **height** h and identical cross sections from top to bottom, each of **area** A, the **volume** V of the figure is

$$V = A \cdot h$$

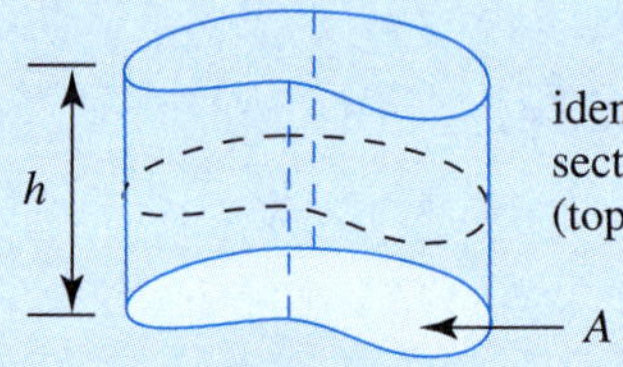

EXAMPLE 1 Referring to the cylindrical and rectangular water containers shown in Figure 6.21, determine which holds the most water.

Solution The holding capacity of each container is determined by its volume, so we must find the volume of each figure.

Finding the Volume of the Rectangular Container The rectangular box has identical cross sections from top to bottom; each cross section is a 12-inch-by-12-inch square.

$$\begin{aligned} V_{\text{box}} &= A_{\text{base}} \cdot h \\ &= A_{\text{square}} \cdot h \\ &= (b^2) \cdot h \\ &= (12 \text{ in.})^2 \cdot (17 \text{ in.}) \\ &= 2{,}448 \text{ in.}^3 \end{aligned}$$

The rectangular container holds 2,448 cubic inches of water.

Finding the Volume of the Cylindrical Container The **cylinder** has identical cross sections from top to bottom; each is a circle of radius 7 inches (half the diameter).

$$
\begin{aligned}
V_{\text{cylinder}} &= A_{\text{base}} \cdot h \\
&= A_{\text{circle}} \cdot h \\
&= (\pi r^2) \cdot h \\
&= \pi \cdot (7 \text{ in.})^2 \cdot (16 \text{ in.}) \\
&= 2{,}463.00864 \ldots \text{ in.}^3
\end{aligned}
$$

π × 7 x^2 × 16 =

With a graphing calculator, press ENTER instead of =.

The cylindrical container holds approximately 2,463 cubic inches of water. Therefore, the cylindrical container holds about 15 cubic inches more than the rectangular container. ■

Surface Area

The **surface area** of a three-dimensional figure is the sum total of the areas of all the surfaces that compose the figure. Although a formula for the surface area of a rectangular box could be given, it is not necessary, because we already know how to find the area of each rectangular face. Nonrectangular objects, however, deserve more attention. Cylinders and spheres have specific formulas requiring more analysis.

EXAMPLE 2 Referring to the cylindrical and rectangular water containers shown in Figure 6.21, determine which has the greater surface area.

Solution **Finding the Surface Area of the Rectangular Container** The surface area of a box is composed of six pieces: the top, the bottom, and the four sides. In this case, the top and bottom are identical squares; each has area b^2. The sides are identical rectangles; each has area $b \cdot h$.

$$
\begin{aligned}
A_{\text{box}} &= 2 \cdot A_{\text{square}} + 4 \cdot A_{\text{rectangle}} \\
&= 2(b^2) + 4(b \cdot h) \\
&= 2(12 \text{ in.})^2 + 4(12 \text{ in.})(17 \text{ in.}) \\
&= 288 \text{ in.}^2 + 816 \text{ in.}^2 \\
&= 1{,}104 \text{ in.}^2
\end{aligned}
$$

The surface area of the rectangular container is 1,104 square inches.

Finding the Surface Area of the Cylindrical Container The surface area of a cylinder is composed of three pieces: the top, the bottom, and the curved side. The top and bottom are circles, each with an area of πr^2. We find the area of the curved side by cutting it from top to bottom and rolling it out flat. The result is a rectangle with length equal to the circumference of the cylinder, as shown in Figure 6.24.

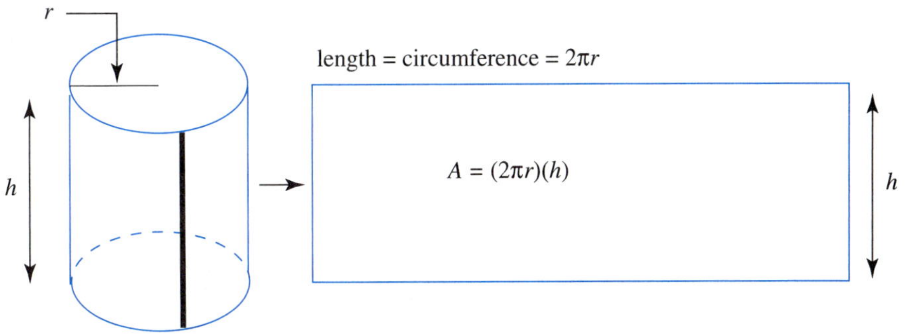

Figure 6.24

$$\begin{aligned} A_{\text{cylinder}} &= 2 \cdot A_{\text{circle}} + A_{\text{side}} \\ &= 2(\pi r^2) + (2\pi r) \cdot h \\ &= 2\pi \cdot (7 \text{ in.})^2 + 2\pi \cdot (7 \text{ in.})(16 \text{ in.}) \\ &= 98\pi \text{ in.}^2 + 224\pi \text{ in.}^2 \\ &= 322\pi \text{ in.}^2 \\ &= 1{,}011.592834 \ldots \text{ in.}^2 \end{aligned}$$

The surface area of the cylindrical container is approximately 1,012 square inches. Therefore, the surface area of the rectangular container is about 92 square inches more than that of the cylindrical container. ■

Comparing Examples 1 and 2, we see that even though the volume of the cylinder is greater than that of the box, the surface area of the cylinder is less than that of the box. In essence, the cylinder is more "efficient" due to its curved surface.

Spheres

How much larger is a basketball than a soccer ball? Depending on which measurement you use for the comparison, different answers are possible. The "size" of a ball can be given in terms of its diameter (one-dimensional measurement), its surface area (two-dimensional measurement), or its volume (three-dimensional measurement).

A **sphere** is the three-dimensional counterpart of a circle; it is the set of all points *in space* equidistant from a fixed point. The fixed point is called the **center** of the sphere, a line segment connecting the center and any point on the sphere is a **radius,** and a line segment connecting any two points of the sphere *and* passing through the center is a **diameter.** The length of a diameter is twice that of a radius. (See Figure 6.25.)

The ancient Greeks knew the following formulas for determining the volume and surface area of a sphere.

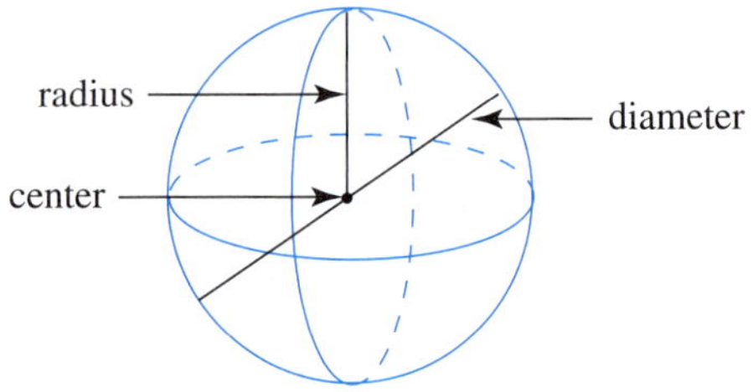

Figure 6.25

Volume and Surface Area of a Sphere

The **volume** V of a sphere of radius r is

$$V = \frac{4}{3}\pi r^3$$

The **surface area** A of a sphere of radius r is

$$A = 4\pi r^2$$

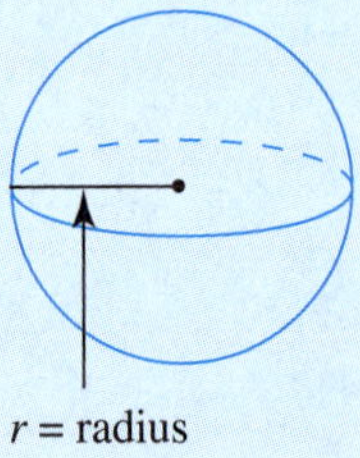

EXAMPLE 3 A regulation basketball has a circumference of 30 inches; the circumference of a standard soccer ball is 27 inches. (Each ball is a sphere.)

a. Find and compare the volumes of the two types of balls.
b. Find and compare the surface areas of the two types of balls.

Solution **a.** To find the volume of a sphere, we must know its radius. We are given the circumference, so we can find the radius as follows:

$$C = 2\pi r$$

$$r = \frac{C}{2\pi} \quad \text{dividing by } 2\pi$$

Finding the Volume of the Basketball The circumference of a basketball is $C = 30$ inches, so we have

$$r_{\text{basketball}} = \frac{C}{2\pi} = \frac{30 \text{ in.}}{2\pi} = \frac{15}{\pi} \text{ in.}$$

Rather than using a calculator at this point, we substitute this exact value directly into the volume formula and then use a calculator:

$$\begin{aligned} V_{\text{basketball}} &= \frac{4}{3}\pi r^3 \\ &= \frac{4}{3}\pi\left(\frac{15}{\pi} \text{ in.}\right)^3 \\ &= 455.9453264\ldots \text{ in.}^3 \end{aligned}$$

4 [÷] 3 [×] [π] [×] [(] 15 [÷] [π] [)] [y^x] 3 [=]

4 [÷] 3 [×] [π] [×] [(] 15 [÷] [π] [)] [∧] 3 [ENTER]

The volume of the basketball is approximately 456 cubic inches.

Finding the Volume of the Soccer Ball The circumference of a soccer ball is $C = 27$ inches; thus,

$$r_{\text{soccer ball}} = \frac{C}{2\pi} = \frac{27 \text{ in.}}{2\pi} = \frac{13.5 \text{ in.}}{\pi}$$

Therefore,

$$\begin{aligned} V_{\text{soccer ball}} &= \frac{4}{3}\pi r^3 \\ &= \frac{4}{3}\pi\left(\frac{13.5 \text{ in.}}{\pi}\right)^3 \\ &= 332.3841429\ldots \text{ in.}^3 \end{aligned}$$

The volume of the soccer ball is approximately 332 cubic inches.

To compare the volumes, consider the ratio of the larger ball to the smaller ball:

$$\begin{aligned} \frac{V_{\text{basketball}}}{V_{\text{soccer ball}}} &= \frac{456 \text{ in.}^3}{332 \text{ in.}^3} \approx 1.37 \\ V_{\text{basketball}} &\approx 1.37 V_{\text{soccer ball}} \\ V_{\text{basketball}} &\approx 137\% \text{ of } V_{\text{soccer ball}} \end{aligned}$$

Therefore, the volume of the basketball is 37% more than the volume of the soccer ball.

b. To find the surface area of each ball, substitute the radii found in part (a) into the surface area formula.

Finding the Surface Area of the Basketball Recall that $r_{\text{basketball}} = \frac{15}{\pi}$ in.

$$\begin{aligned} A &= 4\pi r^2 \\ &= 4\pi\left(\frac{15}{\pi}\text{ in.}\right)^2 \\ &= 286.4788976\ldots\text{ in.}^2 \end{aligned}$$

4 [×] [π] [×] [(] 15 [÷] [π] [)] [x^2] [=]

With a graphing calculator, press [ENTER] instead of [=].

The surface area of the basketball is approximately 286 square inches.

Finding the Surface Area of the Soccer Ball Recall that $r_{\text{soccer ball}} = \frac{13.5}{\pi}$ in.

$$\begin{aligned} A &= 4\pi r^2 \\ &= 4\pi\left(\frac{13.5}{\pi}\text{ in.}\right)^2 \\ &= 232.047907\ldots\text{ in.}^2 \end{aligned}$$

The surface area of the soccer ball is approximately 232 square inches.

As with the volume comparison, use the ratio of the larger ball to the smaller ball to compare their surface areas:

$$\begin{aligned} \frac{A_{\text{basketball}}}{A_{\text{soccer ball}}} &= \frac{286\text{ in.}^2}{232\text{ in.}^2} \approx 1.23 \\ A_{\text{basketball}} &\approx 1.23\,A_{\text{soccer ball}} \\ A_{\text{basketball}} &\approx 123\%\text{ of }A_{\text{soccer ball}} \end{aligned}$$

Therefore, the surface area of the basketball is 23% more than the surface area of the soccer ball. ■

Even though the circumference of a basketball is 11% more than that of a soccer ball ($C_{\text{basketball}}/C_{\text{soccer ball}} = 30/27 = 1.11$), the surface area is 23% more, and the volume is 37% more. These comparisons differ because of the dimensions of the measurements; C is one-dimensional, A is two-dimensional, and V is three-dimensional. Each measurement involves the multiplication of an additional value of r.

Cones and Pyramids

Two other common three-dimensional objects are the cone and the pyramid. A **cone** has a circular base, whereas the base of a **pyramid** is a polygon. In either case, cross sections parallel to the base all have the same shape as the base, but they differ in size;

starting at the base, the cross sections get progressively smaller until they reach a single point at the top of the figure. For both a cone and a pyramid, the volume is one-third the product of the area of the base times the height.

> **Volume of a Cone or a Pyramid**
>
> If a cone or pyramid has height h and if the area of the base is A, the **volume** V of the figure is
>
> $$v = \frac{1}{3} A \cdot h$$
>
> 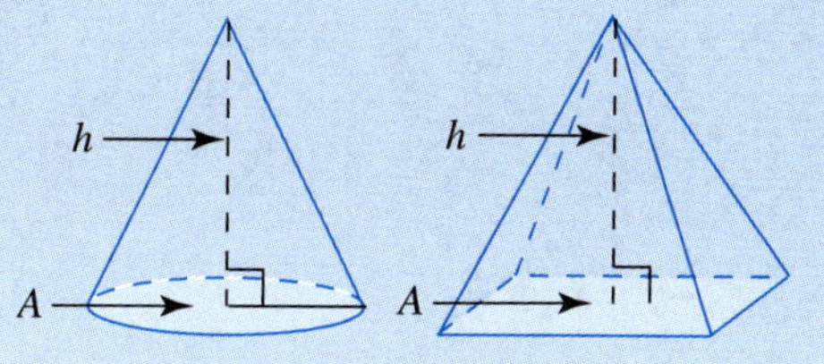
>

EXAMPLE 4 Find and compare the volumes of the cone and pyramid shown in Figure 6.26.

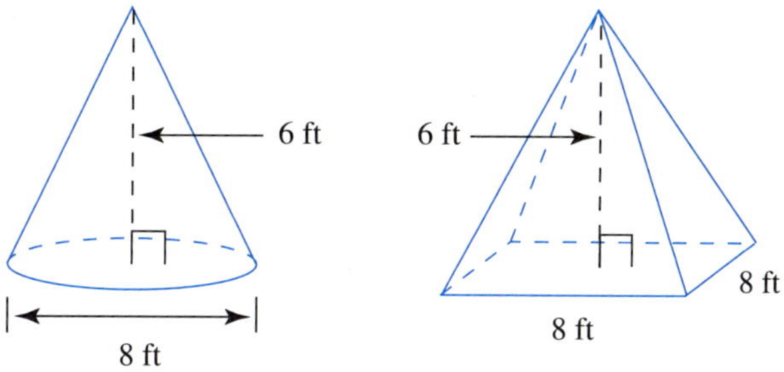

Figure 6.26

Solution Finding the Volume of the Cone We have $r = 8\text{ ft}/2 = 4\text{ ft}$ and $h = 6\text{ ft}$.

$$\begin{aligned} V &= \frac{1}{3} A \cdot h \\ &= \frac{1}{3}(\pi r^2)h \\ &= \frac{1}{3}\pi(4\text{ ft})^2(6\text{ ft}) \\ &= 32\pi\text{ ft}^3 \\ &= 100.5309649\ldots\text{ ft}^3 \end{aligned}$$

The volume of the cone is approximately 101 cubic feet.

Finding the Volume of the Pyramid

$$\begin{aligned} V &= \frac{1}{3} A \cdot h \\ &= \frac{1}{3}(b^2)h \\ &= \frac{1}{3}(8\text{ ft})^2(6\text{ ft}) \\ &= 128\text{ ft}^3 \end{aligned}$$

The volume of the pyramid is 128 cubic feet.

Comparing the Larger Volume to the Smaller Volume We have

$$\frac{V_{\text{pyramid}}}{V_{\text{cone}}} = \frac{128 \text{ ft}^3}{101 \text{ ft}^3} \approx 1.27$$

$$V_{\text{pyramid}} \approx 1.27\ V_{\text{cone}}$$

$$V_{\text{pyramid}} \approx 127\,\% \text{ of } V_{\text{cone}}$$

Therefore, the volume of the pyramid is 27% larger than the volume of the cone. ■

6.2 Exercises

When necessary, round off answers to two decimal places.

In Exercises 1–6, find (a) the volume and (b) the surface area of each figure.

1.

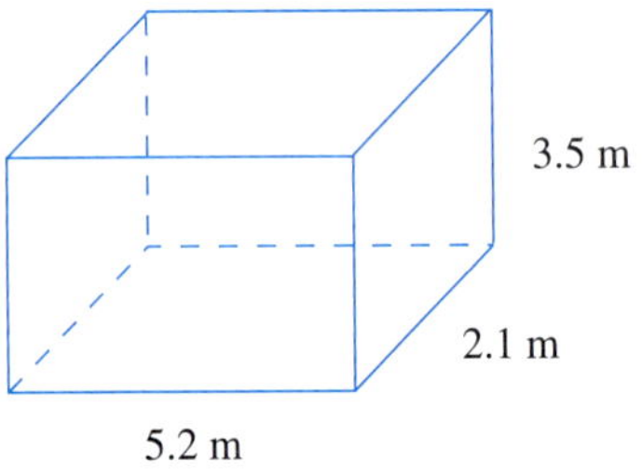

2.

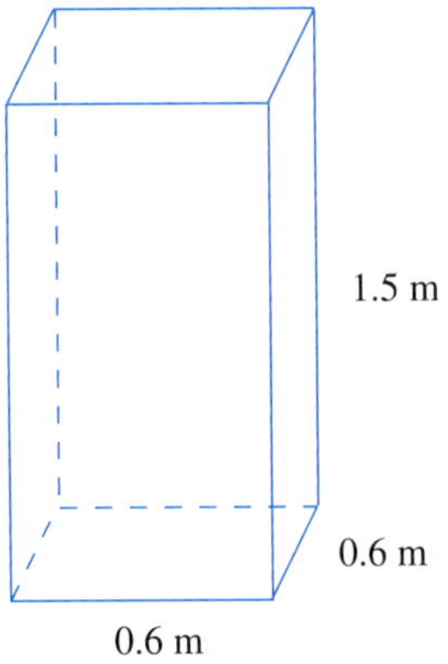

3.

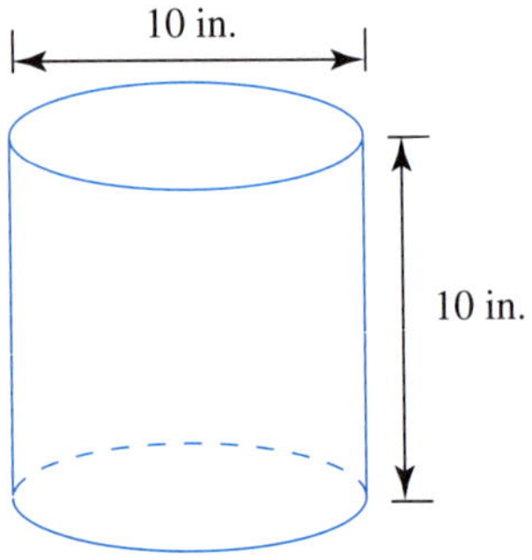

4.

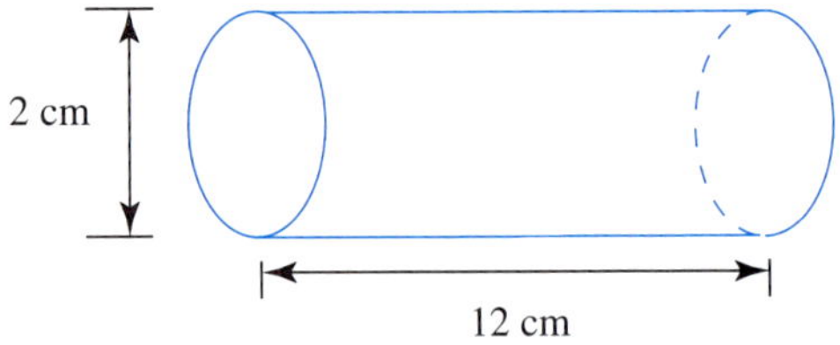

5.

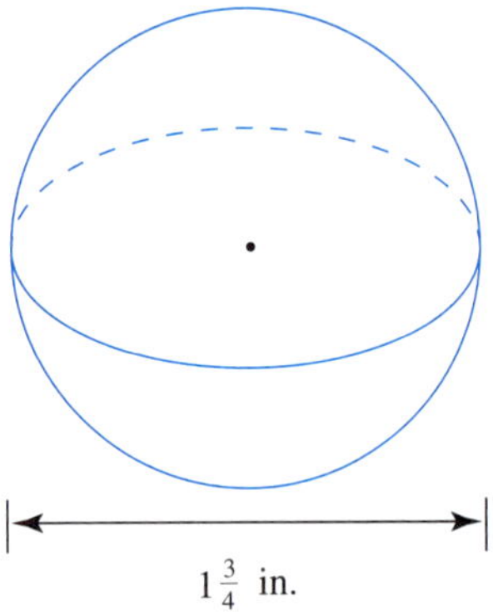

6.

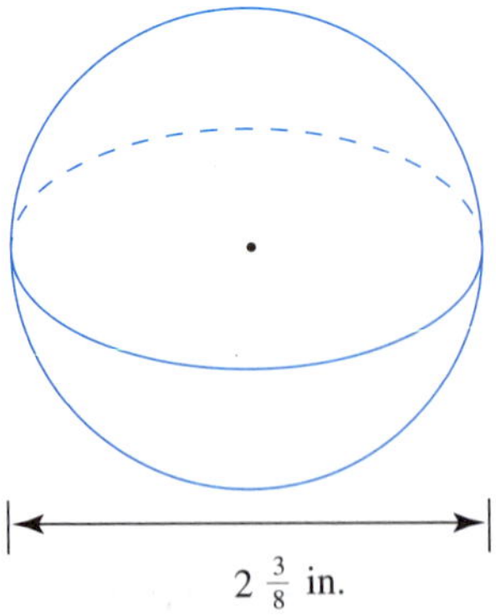

In Exercises 7–14, find the volume of each figure. All dimensions are given in feet.

7.

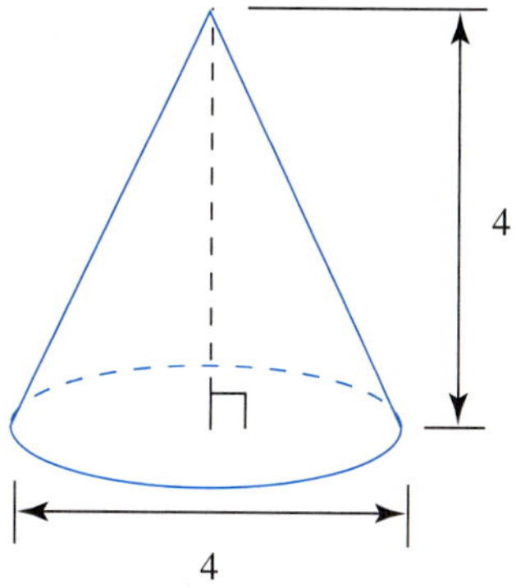

8.

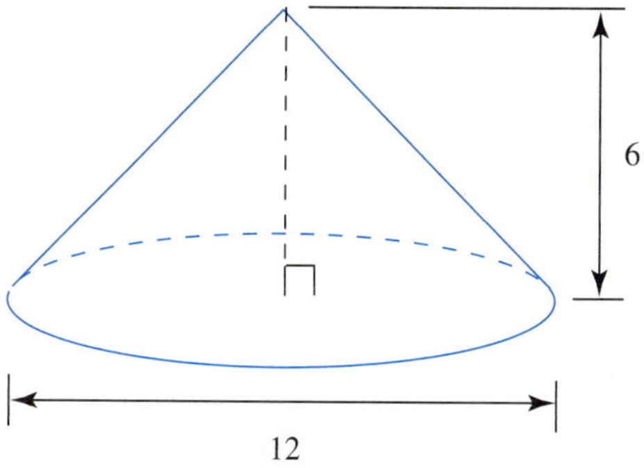

9.

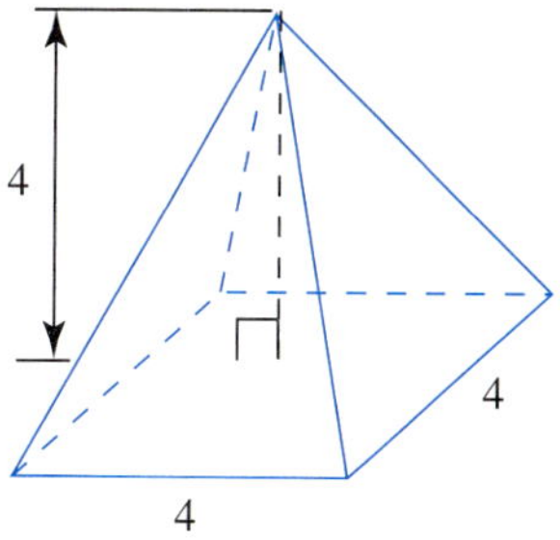

10.

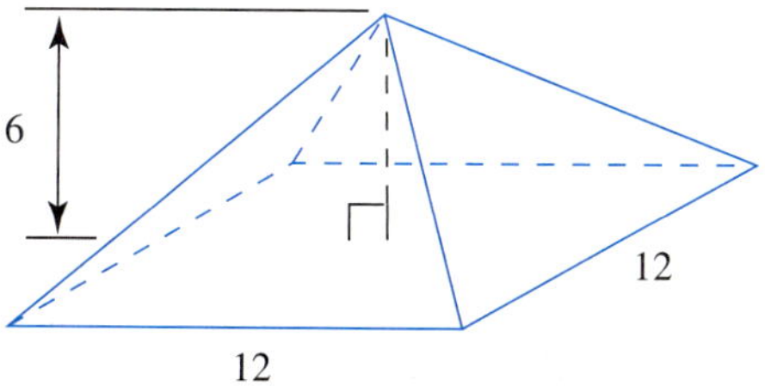

11.

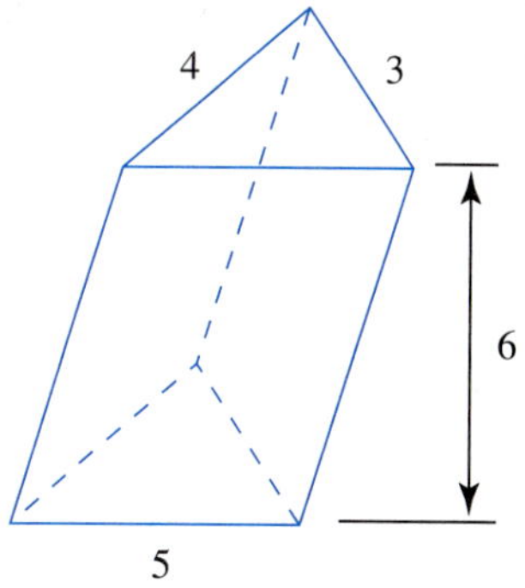

12.

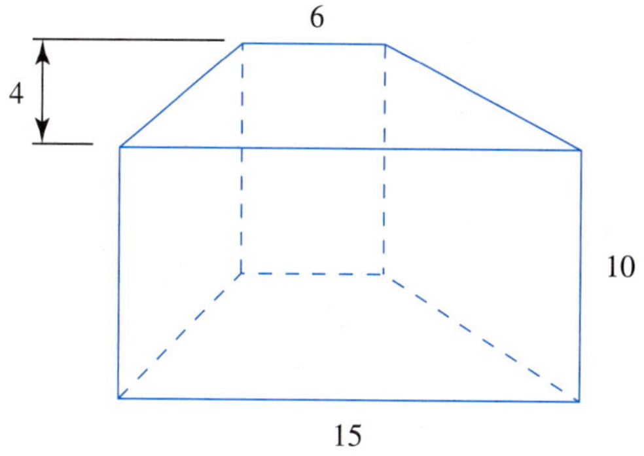

13.

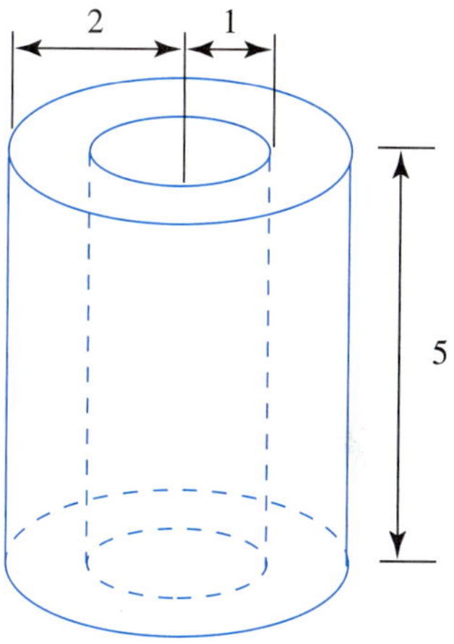

14.

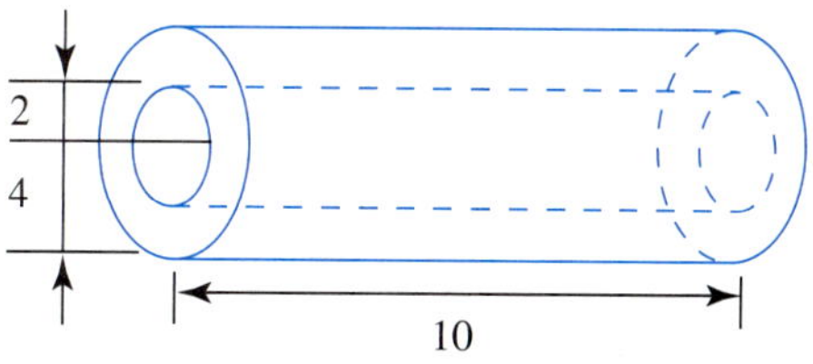

15. From a 10-inch-by-16-inch piece of cardboard, 1.5-inch-square corners are cut out, as shown on page 408, and the resulting flaps are folded up to form an open box. Find (a) the volume and (b) the external surface area of the box.

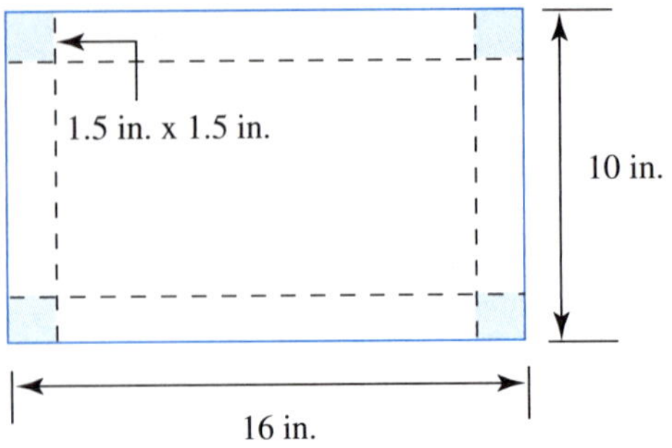

16. From a 24-inch-square piece of cardboard, square corners are to be cut out as shown, and the resulting flaps folded up to form an open box. Find the volume of the resulting box if (a) 4-inch-square corners are cut out and (b) 8-inch-square corners are cut out.

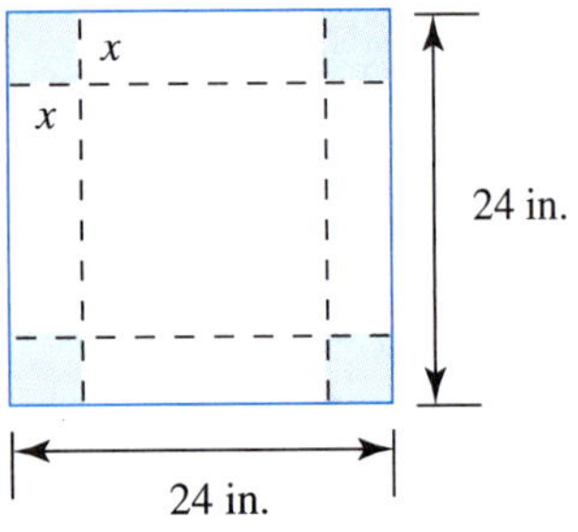

17. A grain silo consists of a cylinder with a hemisphere on top. Find the volume of a silo in which the cylindrical part is 50 feet tall and has a diameter of 20 feet, as shown.

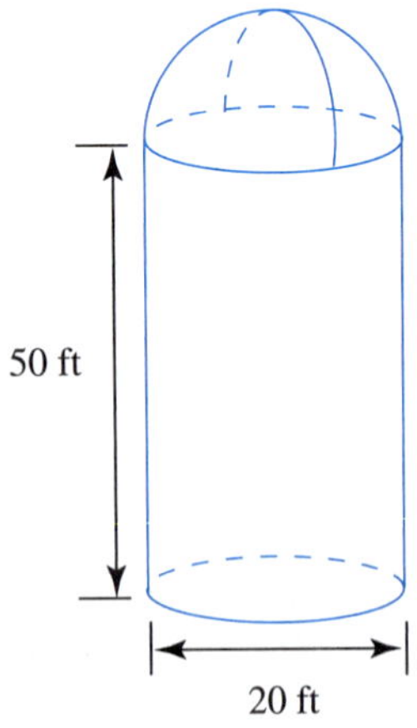

18. A propane gas tank consists of a cylinder with a hemisphere at each end. Find the volume of the tank if the overall length is 10 feet and the diameter of the cylinder is 4 feet, as shown.

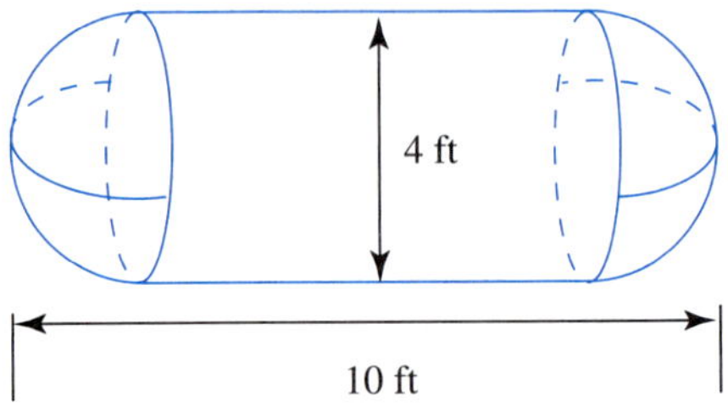

19. A regulation baseball (hardball) has a circumference of 9 inches; a regulation softball has a circumference of 12 inches.
  a. Find and compare the volumes of the two types of balls.
  b. Find and compare the surface areas of the two types of balls.

20. A regulation tennis ball has a diameter of $2\frac{1}{2}$ inches; a Ping-Pong ball has a diameter of $1\frac{1}{2}$ inches.
  a. Find and compare the volumes of the two types of balls.
  b. Find and compare the surface areas of the two types of balls.

21. The diameter of the earth is approximately 7,920 miles; the diameter of the moon is approximately 2,160 miles. How many moons could fit inside the earth?

HINT: Compare their volumes.

22. The diameter of the earth is approximately 7,920 miles; the diameter of the planet Jupiter (the largest planet in our solar system) is approximately 88,640 miles. How many earths could fit inside Jupiter?

HINT: Compare their volumes.

23. The diameter of the earth is approximately 7,920 miles; the diameter of the planet Pluto (the smallest) is approximately 1,500 miles. How many Plutos could fit inside the earth?

HINT: Compare their volumes.

24. The diameter of the planet Jupiter (the largest planet in our solar system) is approximately 88,640 miles; the diameter of the planet Pluto (the smallest) is approximately 1,500 miles. How many Plutos could fit inside Jupiter?

HINT: Compare their volumes.

In Exercises 25 and 26, use the following information: Firewood is measured and sold by the cord. *A* cord *of wood is a rectangular pile of wood 4 feet wide, 4 feet high, and 8 feet long, as shown in Figure 6.27. Therefore, one cord = 128 cubic feet.*

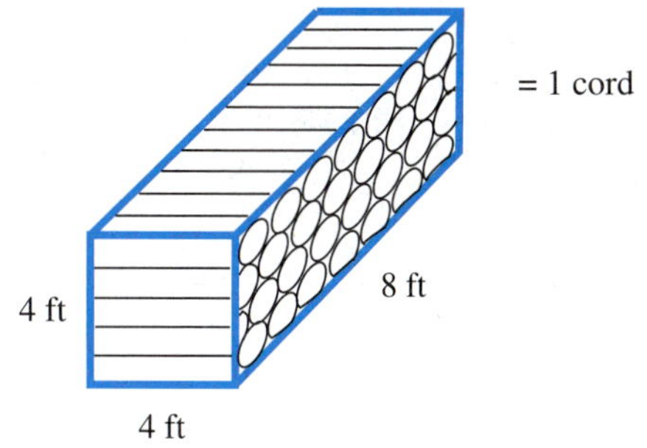

Figure 6.27

25. A cord of seasoned almond wood costs \$190. You paid \$190 for a pile that was 4 feet wide, 2 feet high, and 10 feet long. Did you get an honest deal? If not, what should the cost have been?

26. A cord of seasoned oak wood costs \$160. You paid \$640 for a pile that was 4 feet wide, 6 feet high, and 20 feet long. Did you get an honest deal? If not, what should the cost have been?

27. Tennis balls are packaged three to a cylindrical can. If the diameter of a tennis ball is $2\frac{1}{2}$ inches, find the volume of the can.

28. Golf balls are packaged three to a rectangular box. If the diameter of a golf ball is $1\frac{3}{4}$ inches, find the volume of the box.

29. Ron Thiele bought an older house and wants to put in a new concrete driveway. The driveway will be 36 feet long, 9 feet wide, and 6 inches thick. Concrete (a mixture of sand, gravel, and cement) is measured by the cubic yard. One sack of dry cement mix costs \$7.30, and it takes four sacks to mix up 1 cubic yard of concrete. How much will it cost Ron to buy the cement?

30. Marcus Robinson bought an older house and wants to put in a new concrete patio. The patio will be 18 feet long, 12 feet wide, and 3 inches thick. Concrete is measured by the cubic yard. One sack of dry cement mix costs \$7.30, and it takes four sacks to mix up 1 cubic yard of concrete. How much will it cost Marcus to buy the cement?

31. The Great Pyramid of Cheops (erected about 2600 B.C.) originally had a square base measuring 756 feet by 756 feet and was 480 feet high. (It has since been eroded and damaged.) Find the volume of the original Great Pyramid.

32. The student union at the University of Utopia (U^2) is a pyramid with a 150-foot square base and a height of 40 feet. Find the volume of the student union at U^2.

33. A water storage tank is an upside-down cone, as shown. If the diameter of the circular top is 12 feet and the length of the sloping side is 10 feet, how much water will the tank hold?

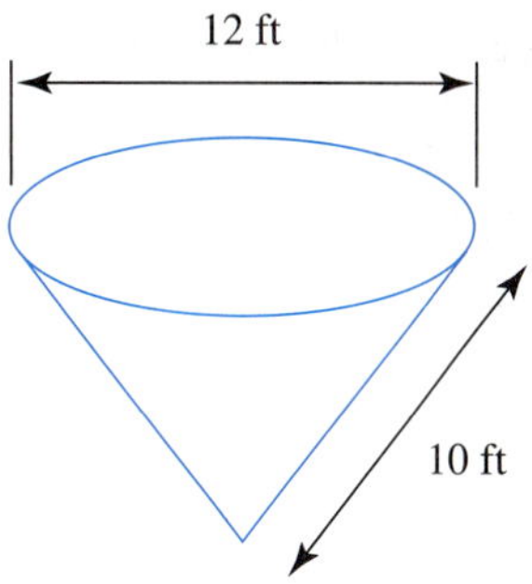

34. The diameter of a conical paper cup is 3.5 inches, and the length of the sloping side is 4.55 inches, as shown. How much water will the cup hold?

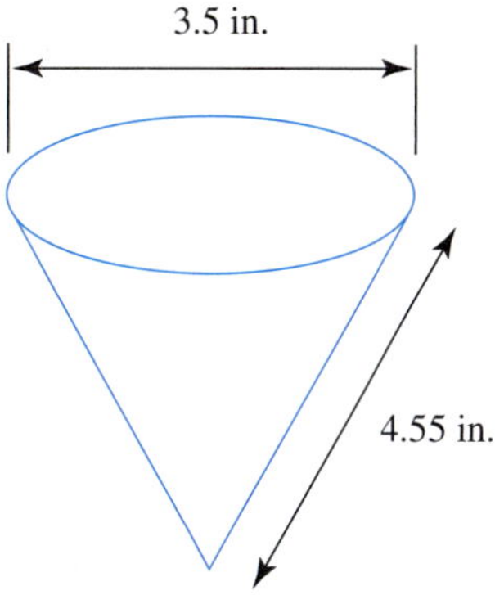

Answer the following questions using complete sentences and your own words.

Projects

35. How do you convert fluid ounces to cubic inches or vice versa? The purpose of this project is to obtain appropriate conversion factors.

- **a.** Select a canned food product or beverage such as a can of soup, soda, juice, or vegetable. According to the label, how many fluid ounces of product does the can contain?
- **b.** Measure the height and diameter of the can in inches. Use these measurements to calculate the volume of the can in cubic inches.
- **c.** Divide the volume calculated in part (b) by the amount given in part (a). This ratio will give the number of cubic inches per fluid ounce.

d. Your answer to part (c) should be close to 1.8; that is, a commonly used conversion factor is 1 fluid ounce = 1.8047 cubic inches. How accurate is your calculation?

e. If a can contains 32 fluid ounces of juice, how large is the can in terms of cubic inches?

f. Now measure the height and diameter of the can in centimeters. Use these measurements to calculate the volume of the can in cubic centimeters.

g. Divide the volume calculated in part (f) by the amount given in part (a). This ratio will give the number of cubic centimeters per fluid ounce.

h. Your answer to part (g) should be close to 29.6; that is, a commonly used conversion factor is 1 fluid ounce = 29.5735 cubic centimeters. How accurate is your calculation?

i. If a can contains 32 fluid ounces of juice, how large is the can in terms of cubic centimeters?

6.3 Egyptian Geometry

The mathematics of the early Egyptians was very practical in its content and use. Mathematics was required in the surveying of land, the construction of buildings, the creation of calendars, and the recordkeeping necessary for commerce. All these activities contributed to the advancement of Egyptian civilization.

The Metropolitan Museum of Art

Found in the Tomb of Menna, this Egyptian wall painting depicts a harvest scene. Notice the use of knotted ropes to measure a field of grain.

Repeated surveying was a necessity for the Egyptians because of the periodic flooding of the Nile. The floodbanks were very fertile, so this land was most valuable and was taxed accordingly. However, when the river flooded its banks, all landmarks and boundaries would be washed away, making it imperative to be able to lay out geometric figures of the correct size and shape.

The Egyptian surveyors used ropes with equally spaced knots to measure distance. In fact, surveyors were referred to as "rope stretchers." These knotted ropes were also used to construct right angles. A right triangle is formed when sides of lengths 3, 4, and 5 are used. A rope with twelve equally spaced knots can easily be stretched to form such a triangle, as shown in Figure 6.28.

EXAMPLE 1 Which of the configurations of knotted ropes shown in Figures 6.29 and 6.30 would form a right triangle?

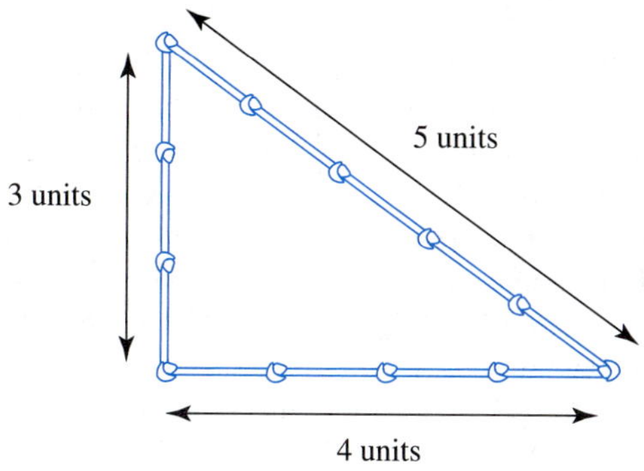

Figure 6.28

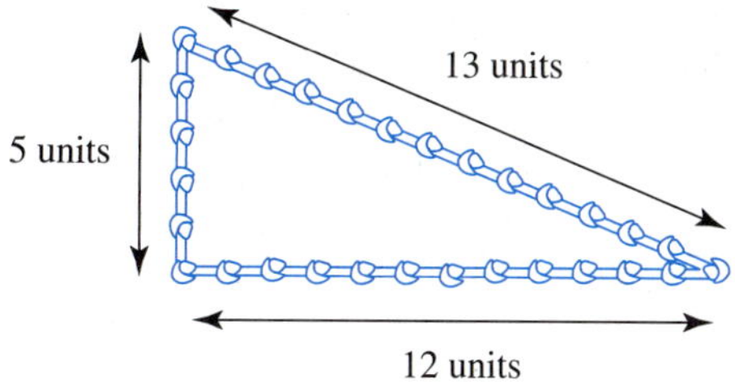

Figure 6.29

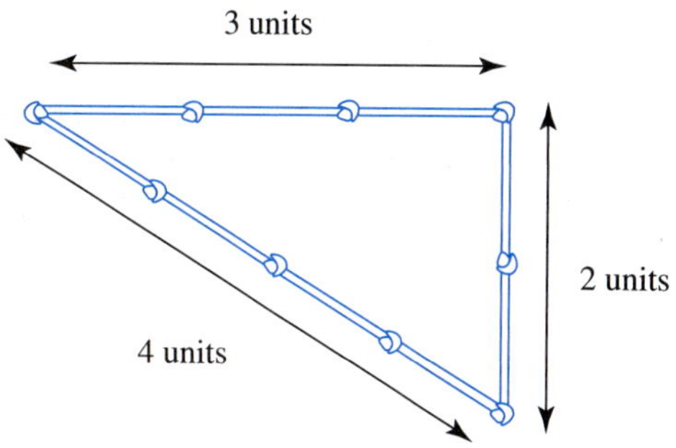

Figure 6.30

Solution **a.** Counting the number of segments on each side, we find the sides to be of lengths 5, 12, and 13 units. If the triangle is a right triangle, the sides will satisfy the Pythagorean theorem, $a^2 + b^2 = c^2$. Because 13 is the largest value, we let $c = 13$ and check the theorem:

$$\begin{aligned} a^2 + b^2 &= 5^2 + 12^2 \\ &= 25 + 144 \\ &= 169 \\ &= 13^2 = c^2 \end{aligned}$$

The sides satisfy the Pythagorean Theorem; therefore, a right angle is formed at the vertex opposite the side of length 13.

b. The sides of the triangle have lengths 2, 3, and 4. Let $a = 2$ and $b = 3$ (the two shortest sides) and check the Pythagorean Theorem:

$$\begin{aligned} a^2 + b^2 &= 2^2 + 3^2 \\ &= 4 + 9 \\ &= 13 \\ &\neq 4^2 = c^2 \end{aligned}$$

The lengths do not satisfy the Pythagorean Theorem; no right angle is formed. ■

Units of Measurement

The basic unit used by the ancient Egyptians for measuring length was the cubit. A *cubit* (*cubitum* is Latin for "elbow") is the distance from a person's elbow to the end of the middle finger. Just as a yard can be subdivided into smaller units of feet and inches, a cubit can be subdivided into smaller units of *palms* and *fingers*. One "royal" cubit (the unit used in official land measurement) equals seven palms, whereas one "common" cubit equals six palms; a royal cubit is longer than a common cubit. (In this book, we will take each cubit to be a royal cubit.) In either case, one palm equals four fingers. Since a cubit is a relatively small length, it is an inconvenient unit to use when measuring large distances. Consequently, the Egyptians defined a *khet* to equal 100 cubits; khets were used when land was surveyed.

The basic unit of area used by the Egyptians was the setat. A *setat* is equal to one square khet; a square whose sides each measure one khet (100 cubits) has an area of exactly one setat (or 10,000 square cubits). (See Figure 6.31.) One setat is approximately two-thirds of an acre.

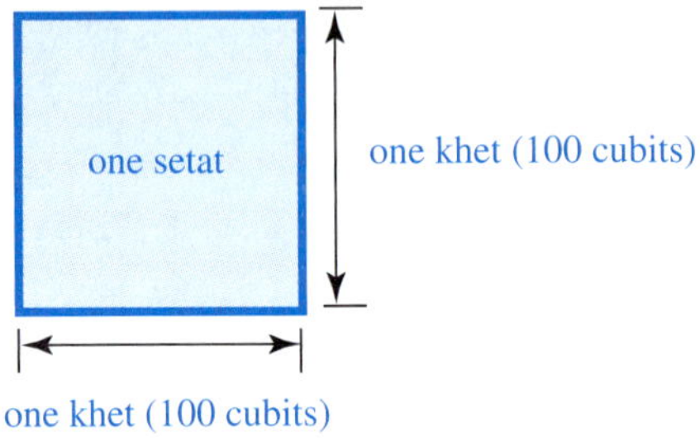

Figure 6.31
1 setat = 1 square khet
= 10,000 square cubits

If lengths are measured in terms of feet, volume is calculated in terms of cubic feet. Since the Egyptians measured lengths in terms of cubits, we would expect volume to be expressed in terms of cubic cubits. However, the basic unit of volume used by the Egyptians was the *khar*; one khar equals two-thirds of a cubic cubit (or 1 cubic cubit $= \frac{3}{2}$ khar). The Egyptian units of measurement are summarized in the following box.

Egyptian Units of Measurement

1 cubit = 7 palms
1 palm = 4 fingers
1 khet = 100 cubits
1 setat = 1 square khet = 10,000 square cubits
1 khar $= \frac{2}{3}$ cubic cubit

Empirical Geometry (If It Works, Use It)

The body of mathematical knowledge possessed by the Egyptians was organized and presented quite differently from ours today. Whereas contemporary mathematics deals with general formulas, theorems, and proofs, most of the surviving written records of the Egyptians consist of single problems without mention of a general formula. The Egyptians lacked a formal system of variables and algebra; they were unable to write a formula like $A = Lw$. Consequently, their work resembles a verbal narration on how to solve a specific problem or puzzle.

A specific example of this single-problem approach comes from what is now called the Moscow Papyrus, an Egyptian papyrus dating from about 1850 B.C. and containing 25 mathematical problems. It now resides in the Museum of Fine Arts in Moscow. Problem 14 of the Moscow Papyrus deals with finding the volume of a truncated pyramid (that is, a pyramid with its top cut off). The problem translates as follows: "Example of calculating a truncated pyramid. If you are told: a truncated pyramid of 6 for the vertical height by 4 on the base by 2 on the top: You are to square this 4; result 16. You are to double 4; result 8. You are to square this 2; result 4. You are to add the 16 and the 8 and the 4; result 28. You are to take $\frac{1}{3}$ of 6; result 2. You are to take 28 twice; the result 56. See, it is of 56. You will find it right."

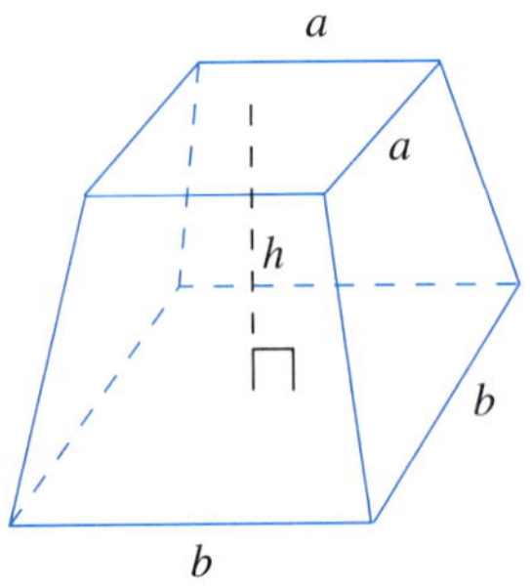

Figure 6.32

The corresponding modern-day formula for finding the volume of a truncated pyramid like the one shown in Figure 6.32 is $V = \frac{h}{3}(a^2 + ab + b^2)$, where h is the height and a and b are the lengths of the sides of the square top and square base. The following line-by-line comparison shows the relationship between the Egyptians' work and the modern-day formula.

Problem from Moscow Papyrus	Modern Counterparts
Example of calculating a truncated pyramid.	$V = \frac{h}{3}(a^2 + ab + b^2)$
If you are told:	Given
a truncated pyramid of 6 for the vertical height	$h = 6$,
by 4 on the base	$b = 4$, and
by 2 on the top:	$a = 2$.
You are to square this 4; result 16.	$b^2 = 16$
You are to double 4; result 8.	$ab = 8$
You are to square this 2; result 4.	$a^2 = 4$
You are to add the 16 and the 8 and the 4; result 28.	$a^2 + ab + b^2 = 28$
You are to take $\frac{1}{3}$ of 6; result 2.	$\frac{h}{3} = 2$
You are to take 28 twice;	$\frac{h}{3}(a^2 + ab + b^2)$
the result 56.	$V = 56$

EXAMPLE 2 Which of the following structures has the greater storage capacity?

a. a truncated pyramid 15 cubits high with a 6-cubit-by-6-cubit top and an 18-cubit-by-18-cubit bottom (Figure 6.33)

b. a regular pyramid 20 cubits tall with an 18-cubit-by-18-cubit base (Figure 6.34)

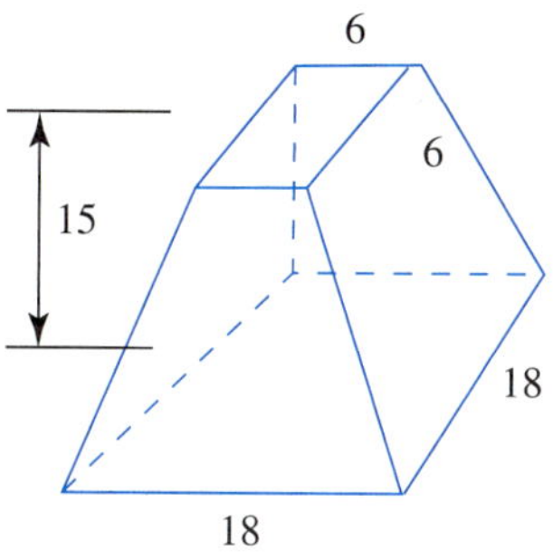

Figure 6.33

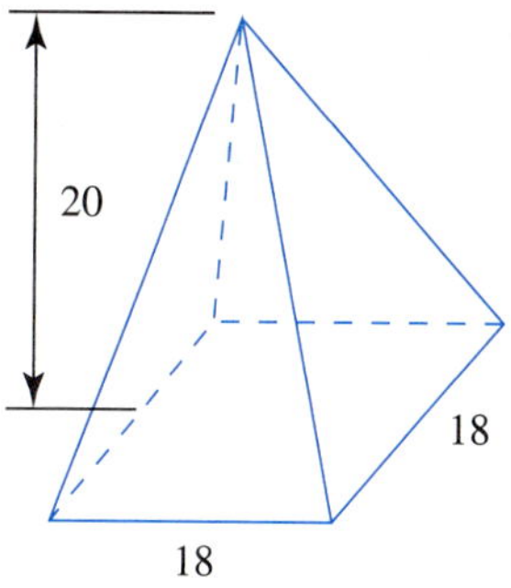

Figure 6.34

Solution Because volume is a measure of the amount of space within a three-dimensional figure, it is used to compare the storage capacities of structures.

a. To find the volume of the truncated pyramid, we use $h = 15$ cubits, $a = 6$ cubits, and $b = 18$ cubits:

$$
\begin{aligned}
V &= \frac{h}{3}(a^2 + ab + b^2) \\
&= \frac{15 \text{ cubits}}{3}[(6 \text{ cubits})^2 + (6 \text{ cubits})(18 \text{ cubits}) + (18 \text{ cubits})^2] \\
&= (5 \text{ cubits})(36 \text{ cubits}^2 + 108 \text{ cubits}^2 + 324 \text{ cubits}^2) \\
&= (5 \text{ cubits})(468 \text{ cubits}^2) \\
&= 2{,}340 \text{ cubits}^3 \\
&= (2{,}340 \text{ cubits}^3)\left(\frac{\frac{3}{2}\text{ khar}}{1 \text{ cubit}^3}\right) \quad \text{converting cubic cubits to khar} \\
&= 3{,}510 \text{ khar}
\end{aligned}
$$

The volume of the truncated pyramid is 3,510 khar.

b. To find the volume of the pyramid, we use $h = 20$ cubits and $b = 18$ cubits. The pyramid is not truncated, so $a = 0$:

$$\begin{aligned} V &= \frac{h}{3}(b^2) \\ &= \frac{20 \text{ cubits}}{3}(18 \text{ cubits})^2 \\ &= 2{,}160 \text{ cubits}^3 \\ &= (2{,}160 \text{ cubits}^3)\left(\frac{\frac{3}{2} \text{ khar}}{1 \text{ cubit}^3}\right) \quad \text{converting cubic cubits to khar} \\ &= 3{,}240 \text{ khar} \end{aligned}$$

The volume of the regular pyramid is 3,240 khar.

Even though the two figures have the same base and the regular pyramid is taller, the truncated pyramid has a larger volume and thus a greater storage capacity. This result is due to the fact that the faces of the truncated pyramid have a steeper slope than those of the regular pyramid. ■

Because of its single-problem approach, Egyptian geometry consisted of a fragmented assortment of guesses, tricks, and rule-of-thumb procedures. It was an empirical subject in which approximate answers were usually good enough for practical purposes. This does not mean that the Egyptians weren't concerned with accuracy. The Great Pyramid of Cheops is a testament to the Egyptians' ability to calculate, measure, and construct with a high degree of precision.

The Great Pyramid of Cheops

Built around 2600 B.C. with an original height of 481.2 feet, the Great Pyramid is the most awesome of all ancient structures. The blocks of limestone composing the pyramid have a combined weight of over 5 million tons. The base is almost a perfect square; the average length of the four sides is 755.78 feet, with a maximum discrepancy of only 4.5 inches (a relative error of only 0.05%)! In addition, the sides of the pyramid are oriented to correspond to the four major compass headings of north, south, east, and west. This positioning has an error of less than 1 degree.

Because of its near perfection, some people believe that the pyramid contains mystic puzzles and the answers to the riddles of the universe. For instance, if you divide the semiperimeter of the base (the distance halfway around, or two times the length of a side) by the height of the pyramid, an interesting result is obtained:

$$\frac{\text{semiperimeter}}{\text{height}} = \frac{2(755.78)}{481.2} = 3.141230258\ldots$$

The comparison of this number to the modern-day approximation of π is remarkable (π is approximately 3.141592654 . . .). Was this planned, or is it merely coincidental? The debate continues.

As surveyors, astronomers, and architects, the early Egyptians were undoubtedly concerned with the mathematics of a circle. By taking careful measurements, they knew that the circumference of a circle was proportional to its diameter—that is, that circumference/diameter = constant. Today, we call this constant π. On the basis of a contemporary interpretation of an ancient document known as the Rhind Papyrus, it appears that the Egyptians would have concluded that $\pi = \frac{256}{81}$ (≈ 3.16).

© Michael Justice/Liaison Agency

The precision of the Great Pyramid of Cheops is a testament to the ancient Egyptians' geometric ability. Some people believe that it also contains mystic puzzles.

The Rhind Papyrus

Most of our knowledge of early Egyptian mathematics was obtained from two famous papyri: the Moscow (or Golenischev) Papyrus and the Rhind Papyrus. In 1858, the Scottish lawyer and amateur archeologist Henry Rhind was vacationing in Luxor, Egypt. While investigating the buildings and tombs of Thebes, he came across an old rolled-up papyrus. The papyrus was written by the scribe Ahmes around 1650 B.C. and contained 84 mathematical problems and their solutions. We know that the papyrus is a copy of an older original, for it begins: "The entrance into the knowledge of all existing things and all obscure secrets. This book was copied in the year 33, the fourth month of the inundation season, under the King of Upper and Lower Egypt 'A-user-Re,' endowed with life, in likeness to writings made of old in the time of the King of Upper and Lower Egypt Ne-mat'et-Re. It is the scribe Ahmes who copies this writing." The reference to the ruler Ne-mat'et-Re places the original script around 2000 B.C. Of particular interest is *Problem 48*, in which the area of a circle (and hence π) is investigated.

Problem 48 of the Rhind Papyrus concludes that "*the area of a circle of diameter 9 is the same as the area of a square of side 8*." Today, we know that this is not exactly true; a circle of diameter 9 has area $A = \pi(4.5)^2 \approx 63.6$, whereas a square of side 8 has area $A = 8^2 = 64$. As we can see, the geometry of the early Egyptians was not perfect. But in many practical applications, the answers were close enough for their intended use.

The reasoning behind the conclusion of Problem 48 went like this: Construct a square whose sides are 9 units each, and inscribe a circle in the square. Thus, the diameter of the circle is 9 units. Now divide the sides of the square into three equal segments of 3 units each. Remove the triangular corners, as shown in Figure 6.35, and an irregular octagon is formed. The area of this octagon is then used as an approximation of the area of the circle: $A_{\text{circle}} \approx A_{\text{octagon}}$.

Notice that the area of the octagon is equal to the area of the large square minus the area of the four triangular corners. Rearranging these triangles, we can see that the

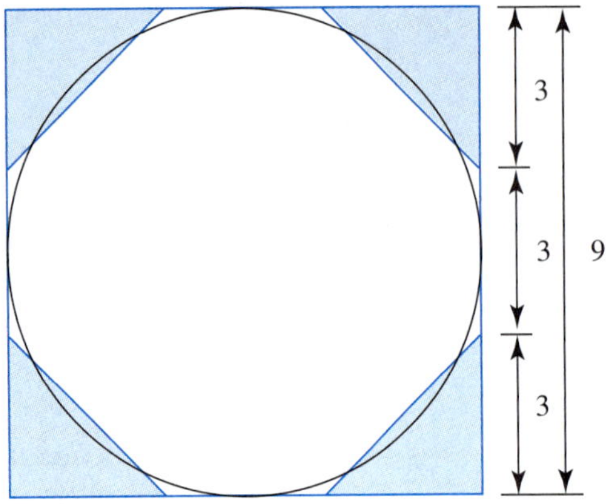

Figure 6.35

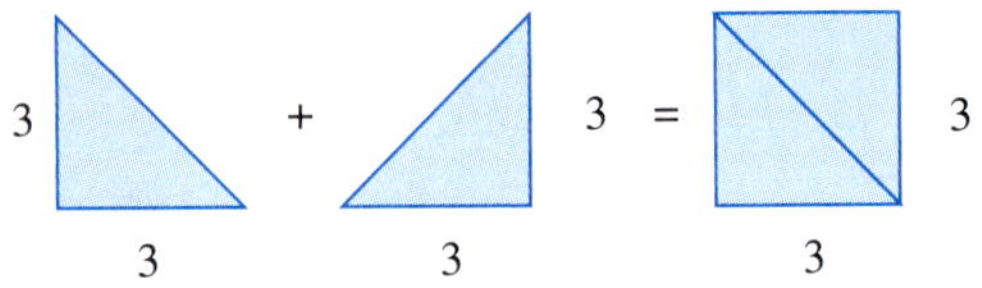

Figure 6.36

area of two triangular corners is the same as the area of one square of side 3 units, as shown in Figure 6.36.

Therefore,

$$
\begin{aligned}
A_{\text{circle}} \approx A_{\text{octagon}} &= A_{\text{square of side 9}} - 4A_{\text{triangle}} \\
&= A_{\text{square of side 9}} - 2A_{\text{square of side 3}} \\
&= (9)^2 - 2(3)^2 \\
&= 81 - 18 \\
&= 63
\end{aligned}
$$

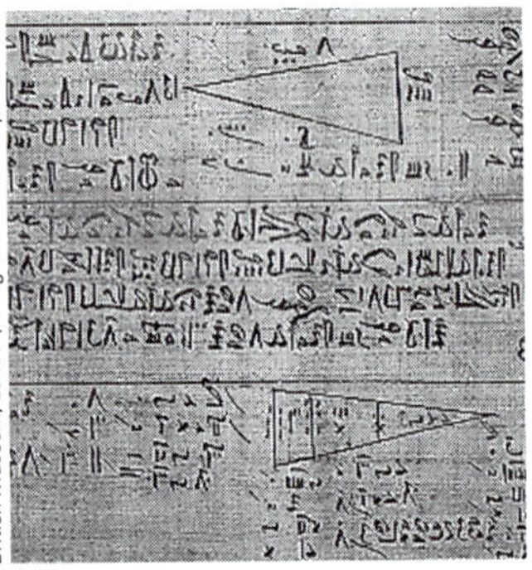

The Rhind Papyrus is one of the world's oldest known mathematics textbooks. Notice the triangles in this portion of the 18-foot-long papyrus.

Therefore, $A_{\text{circle}} \approx 63$ square units. The Egyptians realized that the area of the circle was not 63, but close to it. They mistakenly thought that the area of the circle must be exactly 64, because 64 is close to 63 and 64 ($= 8^2$) is the area of a square with whole-number sides. (This 4,000-year-old "logic" doesn't seem reasonable today.) Thus, the ancient mathematicians reached their conclusion that

$$A_{\text{circle of diameter 9}} = A_{\text{square of side 8}}$$

Two approximations were used in Problem 48:

$$A_{\text{circle of diameter 9}} \approx A_{\text{octagon}}$$
$$A_{\text{octagon}} = 63 \approx 64 = A_{\text{square of side 8}}$$

Of these two estimates, the first (Figure 6.35) was an underestimate and the second an overestimate; thus, the net "canceling" of errors contributed to the accuracy of their overall calculation. The Egyptians' geometry was not perfect, but it did give reasonable answers.

Pi and the Area of a Circle

The Egyptians obtained the "formula" $C = \pi d$ by observing that the ratio circumference/diameter is a constant. How did they obtain the "formula" for the area of a circle? (The Egyptians did not work with actual formulas; instead, they used verbal examples, as we saw with truncated pyramids.) The Egyptians might have used a method of rearrangement to find the area of any given circle. The method of rearrangement says that if a region is cut up into smaller pieces and rearranged to form a new shape, the new shape will have the same area as the original figure. We will apply this method to a circle, as the Egyptians might have done, and see what happens.

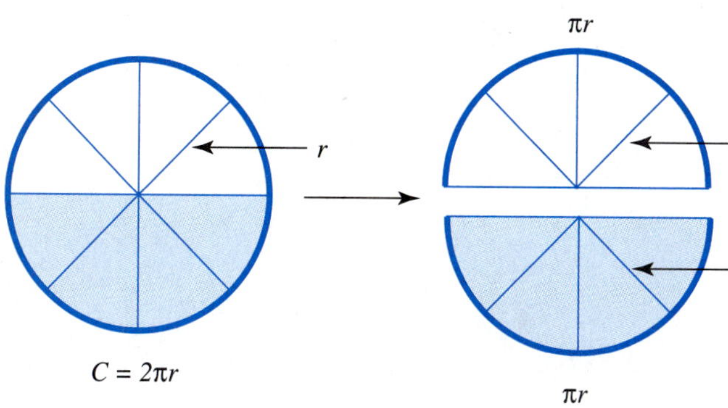

Figure 6.37

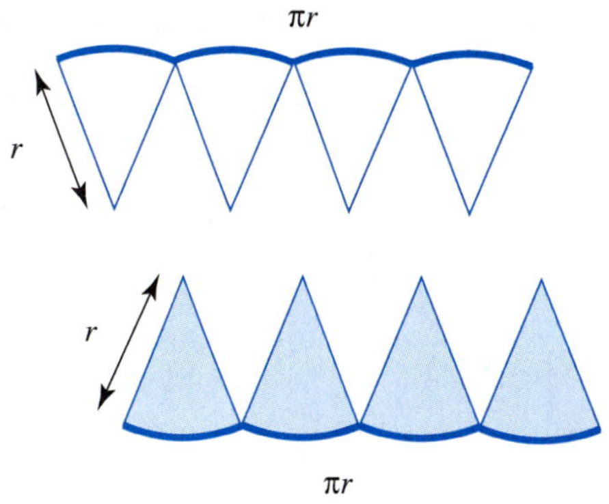

Figure 6.38

Suppose a circle of radius r is cut in half, and each semicircular region is then cut up into an equal number of uniform slices (like a pie). The circumference of the entire circle is known to be $2\pi r$, so the length of each semicircle is πr, as shown in Figure 6.37.

Now rearrange the slices as shown in Figure 6.38. The total area of these slices is still the same as the area of the circle. These pieces are then joined together as in Figure 6.39.

If we continue this process, cutting up the circle into more and more slices that get thinner and thinner, the rearranged figure becomes quite rectangular, as shown in Figure 6.40.

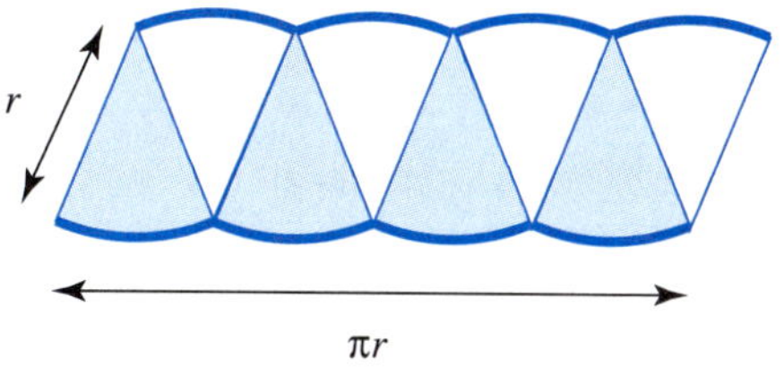

Figure 6.39

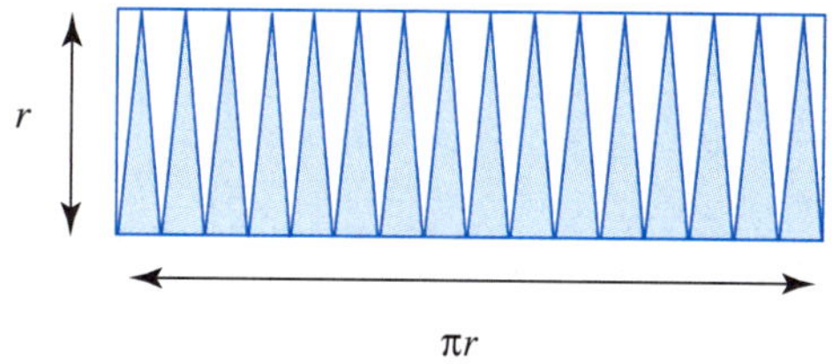

Figure 6.40

Rather than having a bumpy top and bottom, as in Figure 6.39, the top and bottom of this new figure are fairly straight owing to the arcs of the circle becoming so short that they appear to be straight line segments.

Using the formula for the area of a rectangle (the rearranged figure), we have $A = bh = (\pi r)(r) = \pi r^2$. Because the area of the rearranged figure is πr^2, the area of the original circle is also πr^2.

EXAMPLE 3 Combining the result of Problem 48 of the Rhind Papyrus with the method of rearrangement of area, find the Egyptian approximation of π.

Solution Problem 48 of the Rhind Papyrus states that the area of a circle of diameter 9 (or radius $\frac{9}{2}$) is equal to the area of a square of side 8; that is,

$$A_{\text{circle of diameter 9}} = A_{\text{square of side 8}} = 64$$

Using rearrangement of area, we also have

$$A_{\text{circle of diameter 9}} = \pi\left(\frac{9}{2}\right)^2$$

Equating these two expressions, we have

$$\pi\left(\frac{9}{2}\right)^2 = 64$$

$$\frac{81}{4}\pi = 64$$

$$\pi = 64\left(\frac{4}{81}\right)$$

$$\pi = \frac{256}{81} \ (= 3.160493827 \ldots)$$

This is thought to be the procedure that the Egyptians used to obtain their value of π. This approximation is remarkably close to our modern-day value! ■

EXAMPLE 4 For a circle of radius 5 inches, do the following:

a. Use $\pi = \frac{256}{81}$ (the Egyptian approximation of pi) to find the area of the circle.
b. Use the value of π contained in a scientific calculator to find the area of the circle.
c. Find the error of the Egyptian calculation relative to the calculator value.

Solution **a.**

$$\begin{aligned} A &= \pi r^2 \\ &= \left(\frac{256}{81}\right)(5 \text{ in.})^2 \\ &= 79.01234568 \ldots \text{ in.}^2 \end{aligned}$$

Using the Egyptian value, we find that the area of the circle is approximately 79.0 square inches.

b.

$$\begin{aligned} A &= \pi r^2 \\ &= \pi(5 \text{ in.})^2 \\ &= 25\pi \text{ in.}^2 \\ &= 78.53981634 \ldots \text{ in.}^2 \end{aligned}$$

Using the calculator value, we find that the area of the circle is approximately 78.5 square inches. Note that if both decimals were rounded off to the nearest whole unit, they would yield the same answer.

c. To find the error of the Egyptian calculation relative to the calculator value, we must find the difference of the calculations and divide this difference by the calculator value:

$$\begin{aligned} A_{\text{Egyptian}} - A_{\text{calculator}} &= 79.01234568 \text{ in.}^2 - 78.53981634 \text{ in.}^2 \\ &= 0.47252934 \text{ in.}^2 \end{aligned}$$

$$\begin{aligned} \text{error relative to calculator} &= \frac{0.47252934 \text{ in.}^2}{78.53981634 \text{ in.}^2} \\ &= 0.00601643 \ldots \\ &= 0.6\% \end{aligned}$$

The error of the Egyptian calculation relative to the calculator value is 0.6%. ■

6.3 Exercises

In Exercises 1–4, determine whether each configuration of knotted ropes would form a right triangle.

1. 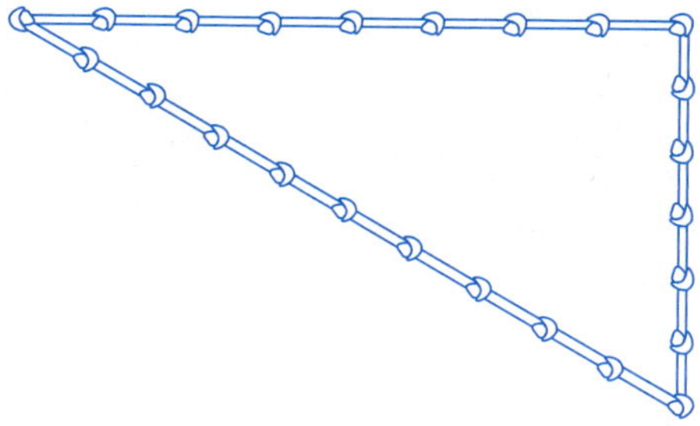

2. 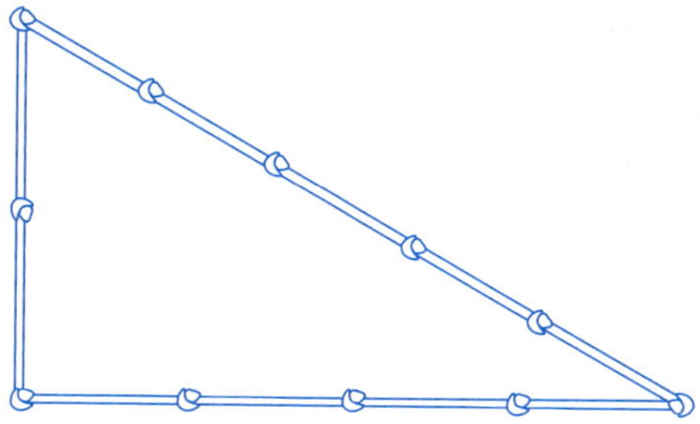

3. 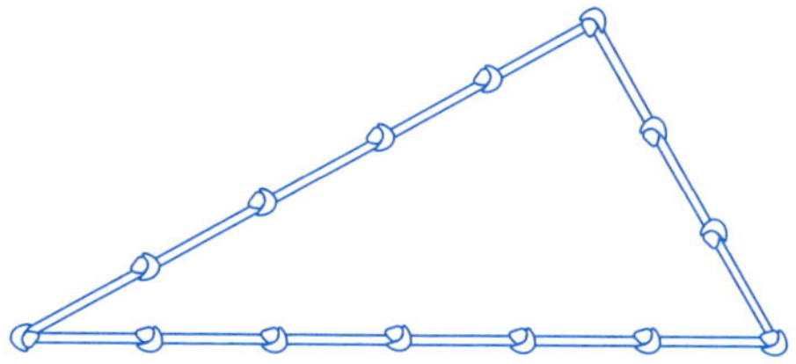

4. 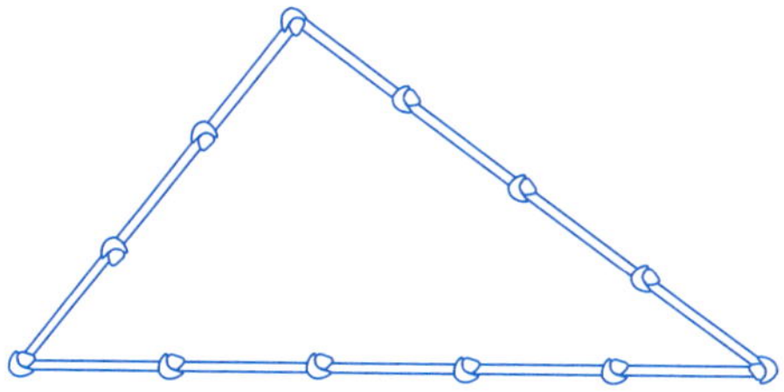

5. Which of the following structures has the larger storage capacity?
   a. a truncated pyramid that is 12 cubits high with a 2-cubit-by-2-cubit top and a 15-cubit-by-15-cubit bottom
   b. a regular pyramid that is 18 cubits tall with a 15-cubit-by-15-cubit base

6. Which of the following structures has the larger storage capacity?
   a. a truncated pyramid that is 33 cubits high with a 6-cubit-by-6-cubit top and a 30-cubit-by-30-cubit bottom
   b. a regular pyramid that is 50 cubits tall with a 30-cubit-by-30-cubit base

In Exercises 7 and 8, fill in the blanks as in the problem from the Moscow Papyrus (the Egyptian method for finding the volume of a truncated pyramid, immediately preceding Example 2).

7. "If you are told: a truncated pyramid of 9 for the vertical height by 6 on the base by 3 on the top: You are to square this 6; result _____. You are to triple 6; result _____. You are to square this 3; result _____. You are to add the _____ and the _____ and the _____; result _____. You are to take $\frac{1}{3}$ of _____; result _____. You are to take _____ three times; the result _____. See, it is of _____. You will find it right."

8. "If you are told: a truncated pyramid of 12 for the vertical height by 10 on the base by 4 on the top: You are to square this 10; result _____. You are to quadruple 10; result _____. You are to square this 4; result _____. You are to add the _____ and the _____ and the _____; result _____. You are to take $\frac{1}{3}$ of _____; result _____. You are to take _____ four times; the result _____. See, it is of _____. You will find it right."

9. Following the method of Problem 48 of the Rhind Papyrus, find a square, with a whole number as the length of its side, that has the same (approximate) area as a circle of diameter 24 units.

   HINT: Divide the sides of a 24-by-24 square into three equal segments.

10. Following the method of Problem 48 of the Rhind Papyrus, find a square, with a whole number as the length of its side, that has the same (approximate) area as a circle of diameter 42 units.

   HINT: Divide the sides of a 42-by-42 square into three equal segments.

11. Using the result of Exercise 9 and the method of Example 3, obtain an approximation of π.

12. Using the result of Exercise 10 and the method of Example 3, obtain an approximation of π.

13. For a circle of radius 3 palms, do the following.
   a. Use $\pi = \frac{256}{81}$ (the Egyptian approximation of pi) to find the area of the circle.
   b. Use the value of π contained in a scientific calculator to find the area of the circle.
   c. Find the error of the Egyptian calculation relative to the calculator value.

14. For a circle of radius 6 palms, do the following.
   a. Use $\pi = \frac{256}{81}$ (the Egyptian approximation of pi) to find the area of the circle.

b. Use the value of π contained in a scientific calculator to find the area of the circle.
c. Find the error of the Egyptian calculation relative to the calculator value.

15. For a sphere of radius 3 palms, do the following.
a. Use $\pi = \frac{256}{81}$ (the Egyptian approximation of pi) to find the volume of the sphere.
b. Use the value of π contained in a scientific calculator to find the volume of the sphere.
c. Find the error of the Egyptian calculation relative to the calculator value.

16. For a sphere of radius 6 palms, do the following.
a. Use $\pi = \frac{256}{81}$ (the Egyptian approximation of pi) to find the volume of the sphere.
b. Use the value of π contained in a scientific calculator to find the volume of the sphere.
c. Find the error of the Egyptian calculation relative to the calculator value.

17. Find the perimeter of a triangle having sides with the following measurements: 2 cubits, 5 palms, 3 fingers; 3 cubits, 3 palms, 2 fingers; 4 cubits, 4 palms, 3 fingers.

18. It has been suggested that one palm equals 7.5 centimeters. Using a standard equivalence, one centimeter equals 0.3937 inch.
a. How many inches are in one royal cubit?
b. How many inches are in one common cubit?
c. Measure your own personal cubit. How does it compare to the royal and common cubits?

19. Use part (a) of Exercise 18 to determine which is larger: one khar or one cubic foot.

20. A rectangular field measures 50 cubits by 200 cubits. Find the area of this field in setats.

21. A triangular field has sides that measure 150 cubits, 200 cubits, and 250 cubits. Find the area of this field in setats.

22. A rectangular room is 20 cubits long, 15 cubits wide, and 8 cubits high. Find the volume of this room in khar.

23. The area of a square is 4 setats. Find the length of a side of this square. Express your answer in the following units.
a. khet
b. cubits

24. The volume of a cube is 12 khar. Find the length, in cubits, of an edge of this cube.

25. Problem 41 of the Rhind Papyrus pertains to finding the volume of a cylindrical granary. The diameter of the granary is 9 cubits, and its height is 10 cubits.
a. Calculate the volume of this granary, in khar, using the Egyptian method (Problem 48 of the Rhind Papyrus) to calculate the area of a circle.
b. Calculate the volume of this granary, in khar, using the conventional formula $A = \pi r^2$ to calculate the area of a circle.
c. Find the error of the Egyptian calculation of volume relative to the conventional calculation.

Answer the following questions using complete sentences and your own words.

History Questions

26. List four activities in which the ancient Egyptians used mathematics and, in so doing, contributed to the advancement of their civilization.

27. How did the Egyptians construct right angles?

28. Who wrote the Rhind Papyrus?

29. What does Problem 48 of the Rhind Papyrus state?

30. Write a research paper on a historical topic referred to in this section or on a related topic. Following is a list of possible topics:
- ancient agriculture and the Nile River (How did agriculture lead to mathematical and scientific advancement?)
- early astronomy (What motivated the early astronomers? What is the connection between early astronomy and mathematics?)
- early calendars (What purpose did calendars serve? Who created them?)
- the Great Pyramid of Cheops (How was it constructed? What was its purpose?)
- the history of π (Trace its history and applications.)
- the Moscow Papyrus (What does it contain? When was it translated?)
- the Rhind Papyrus (Is it intact? What does it contain? When was it translated?)
- the Rosetta Stone (How was it used in mathematics? In politics? In science? In archeology?)

6.4 The Greeks

The primary question was not "What do we know," but "How do we know it."

—ARISTOTLE TO THALES

Many people accept common knowledge without questioning it, especially if they have firsthand experience of it. However, as Aristotle's comment to Thales points out, the ancient Greek scholars were not satisfied with mere facts; they were forever asking "why" in their search for absolute truth in the world around them. Whereas the Egyptians constructed a right angle by stretching a knotted rope to form a triangle with sides of 3, 4, and 5 units, the Greeks wanted to know *why* this method worked. The Greeks were not content to accept a claim simply because experience indicated that it worked. They wanted proof, and they obtained proof through the systematic application of logic and deductive reasoning.

Thales of Miletus

The empirical geometry of the Egyptians reached the ancient Greeks through commerce, travel, and warfare. One of the first Greek scholars to study this geometry was Thales of Miletus (625–547 B.C.). Thales is known as the first of the Seven Sages of Greece. (The Seven Sages of Greece were famous for their practical knowledge. In addition to Thales, the sages were Solon of Athens, Bias of Priene, Chilo of Sparta, Cleobulus of Rhodes, Periander of Corinth, and Pittacus of Mitylene.)

A successful businessman, statesman, and philosopher, Thales had occasion to travel to Egypt and consequently learned of Egyptian geometry. While in Egypt, he won the respect of the pharaoh by calculating the height of the Great Pyramid of Cheops. He did this by measuring the shadows of the pyramid and of his walking staff. Knowing the height of his staff, he used a proportion to calculate the height of the pyramid. (See Section 10.0.)

In the history of mathematics, Thales is credited with being the first to prove that corresponding sides of similar triangles are proportional. If the three angles of one triangle are equal to the three angles of another triangle, the triangles are **similar.** Consequently, similar triangles have the same shape but may differ in size.

The Sides of Similar Triangles Are Proportional

If the three angles of one triangle equal the three angles of another triangle, then

$$\frac{a}{d} = \frac{b}{e} = \frac{c}{f}$$

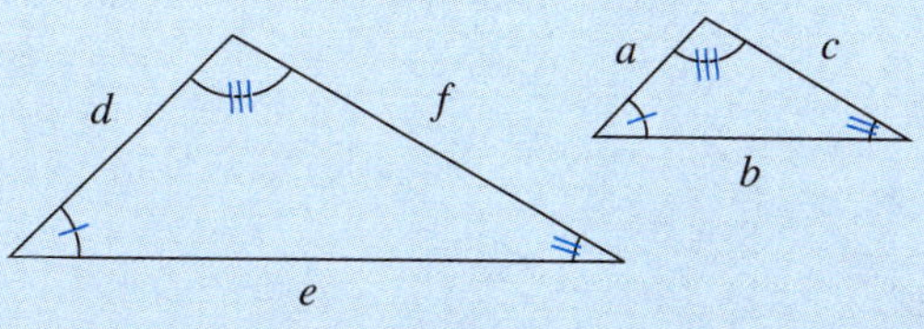

EXAMPLE 1 For the two triangles in Figure 6.41, find the lengths of the unknown sides.

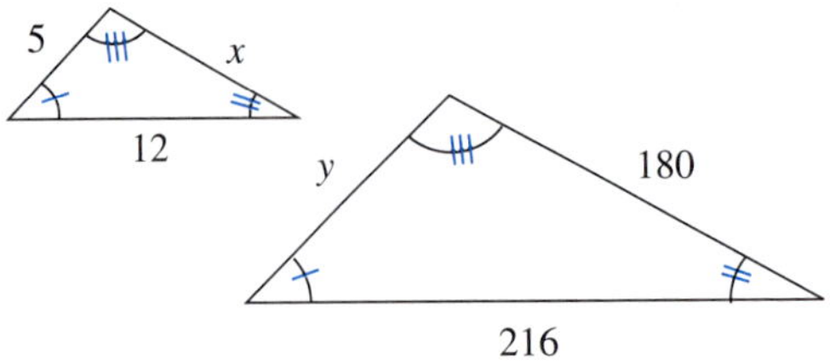

Figure 6.41

Solution Because the triangles are similar (they have equal angles), their sides are proportional.

Comparing the small triangle to the large triangle, we have

$$\frac{x}{180} = \frac{12}{216}$$

$$x = \frac{12 \cdot 180}{216} \quad \text{multiplying both sides by 180}$$

$$x = 10$$

Comparing the large triangle to the small triangle, we have

$$\frac{y}{5} = \frac{216}{12}$$

$$y = \frac{216 \cdot 5}{12} \quad \text{multiplying both sides by 5}$$

$$y = 90$$

■

Pythagoras of Samos

The name Pythagoras is synonymous with the famous formula that relates the lengths of the sides of a right triangle, $a^2 + b^2 = c^2$. However, the life and teachings of this famous Greek mathematician, philosopher, and mystic are shrouded in mystery, legend, and conjecture.

Pythagoras of Samos

Born on the Aegean island of Samos about 500 B.C., Pythagoras traveled and studied extensively as a young man. His travels took him to Egypt, Phoenicia, and Babylonia. Returning to Samos when he was fifty, Pythagoras found his homeland under the rule of the tyrant Polycrates. For an unknown reason, Pythagoras was banned from Samos and migrated to Croton, in the south of present-day Italy, where he founded a school. His students and disciples, known as Pythagoreans, eventually came to hold considerable social power.

Pythagoras had his students concentrate on four subjects: the theory of numbers, music, geometry, and astronomy. To him, these subjects constituted the core of knowledge necessary for an educated person. This core of four subjects, later known as the *quadrivium*, persisted until the Middle Ages. At that time, three other subjects, known as the *trivium*, were added to the list: logic, grammar, and rhetoric. These seven areas of study came to be known as the seven liberal arts and formed the proper course of study for all educated people.

The Pythagoreans were a secretive sect. Their motto, "All is number," indicates their belief that numbers have a mystical quality and are the essence of the universe. Numbers had a mystic aura about them, and all things had a numerical representation. For instance, the number 1 stood for reason (the number of absolute truth), 2 stood for woman (the number of opinion), 3 stood for man (the number of harmony), 4 stood for justice (the product of equal terms, $4 = 2 \times 2 = 2 + 2$), 5 stood for marriage (the sum of man and woman, $5 = 2 + 3$), and 6 stood for creation (from which man and woman came, $6 = 2 \times 3$). In general, odd numbers were masculine, and even numbers were feminine. Indeed, the Pythagorean doctrine was a strange mixture of cosmic philosophy and number mysticism. The Pythagoreans' beliefs were augmented by many rites and taboos, which included refusing to eat beans (they were sacred), refusing to pick up a fallen object, and refusing to stir a fire with an iron.

In the realm of astronomy, Pythagoras had a curious theory. Being geocentric, he believed the earth to be the center of the universe. The sun, moon, and five known planets circled the earth, each traveling on its own crystal sphere. Because of the friction of these gigantic bodies whirling about, each body would produce a unique tone based on its distance from the earth. The combined effect of the seven bodies was the harmonious celestial music of the gods, which Pythagoras was the only mortal blessed to hear!

Near the end of his life, Pythagoras and his followers became more political. Their power and influence were felt in Croton and the other Greek cities of southern Italy. Legend has it that this power and the Pythagoreans' sense of autocratic supremacy initiated a conflict with the local peoples and governments that ultimately led to his death.

Historians do not agree on the exact circumstances of Pythagoras's death, but it is known that during a violent revolt against the Pythagoreans, the locals set fire to the school. Some say that Pythagoras died in the flames. Others say that he escaped the inferno but was chased to the edge of a bean field. Rather than trample the sacred beans, Pythagoras allowed the crowd to overtake and kill him. Some people believe that Pythagoras was murdered by a disgruntled disciple. In death, as in life, Pythagoras is shrouded in mystery.

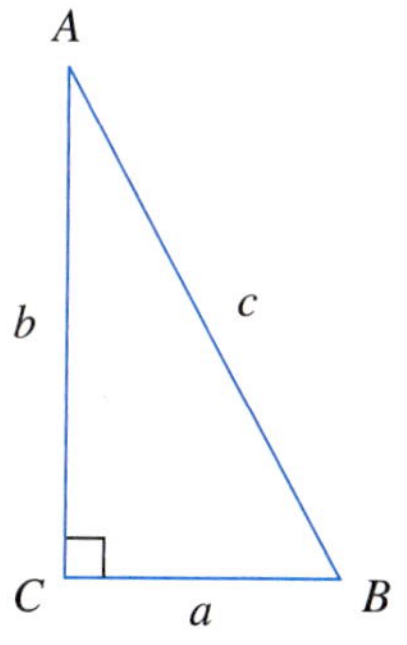

Figure 6.42

Although Pythagoras is given credit for proving the theorem that bears his name, there is no hard evidence that he did so. All his teachings were verbal; there are no written records of his actual work. Furthermore, his disciples had to swear not to reveal any of his doctrines to outsiders.

Since the days of Pythagoras, many proofs of "his" theorem have been put forth. Most involve the method of rearrangement of area. A few years before being elected the twentieth president of the United States, James Garfield produced an original proof. His method, shown below, involved the creation of a trapezoid that consisted of three right triangles.

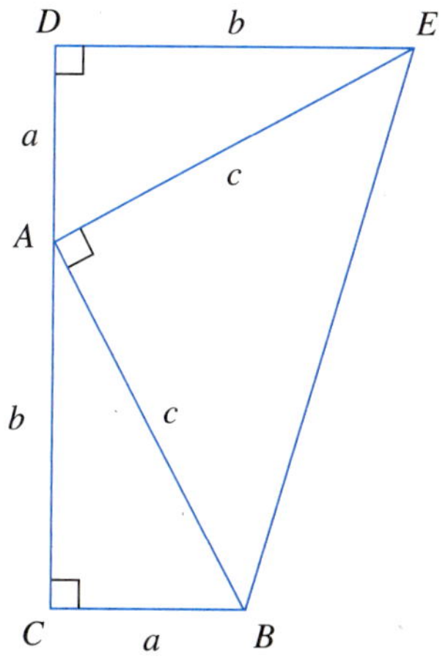

Figure 6.43

James Garfield's Proof of the Pythagorean Theorem

Given: the right triangle ABC, as shown in Figure 6.42.

Extend segment CA a distance of a, and call this point D.

Construct a perpendicular segment of length b at D.

The endpoint of this segment is E. (See Figure 6.43.)

Triangles ABC and EAD are congruent, so $\angle EAB$ is a right angle. (Why?)

Figure $BCDE$ is a trapezoid with parallel bases CB and DE.

The area of $BCDE$ can be found by using the formula for the area of a trapezoid, or it can be found by adding the areas of the three right triangles. Apply both methods and set their results equal as follows:

$$\text{area of trapezoid} = (\text{average base})(\text{height}) = \left(\frac{a+b}{2}\right)(a+b)$$

$$\text{area of } \Delta\, ABC = \frac{1}{2}(\text{base})(\text{height}) = \frac{1}{2}ab$$

$$\text{area of } \Delta\, EAD = \frac{1}{2}(\text{base})(\text{height}) = \frac{1}{2}ab$$

$$\text{area of } \Delta\, EAB = \frac{1}{2}(\text{base})(\text{height}) = \frac{1}{2}c^2$$

$$\text{area of trapezoid} = \text{sum of the areas of the three triangles}$$

$$\left(\frac{a+b}{2}\right)(a+b) = \left(\frac{1}{2}ab\right) + \left(\frac{1}{2}ab\right) + \left(\frac{1}{2}c^2\right)$$

$$(a+b)(a+b) = ab + ab + c^2 \quad \text{multiplying each side by 2}$$

$$a^2 + ab + ab + b^2 = ab + ab + c^2 \quad \text{multiplying } (a+b)(a+b)$$

$$a^2 + b^2 = c^2 \quad \text{subtracting } 2ab \text{ from each side}$$

Thus, given any right triangle with legs a and b and hypotenuse c, we have shown that $a^2 + b^2 = c^2$.

Euclid of Alexandria

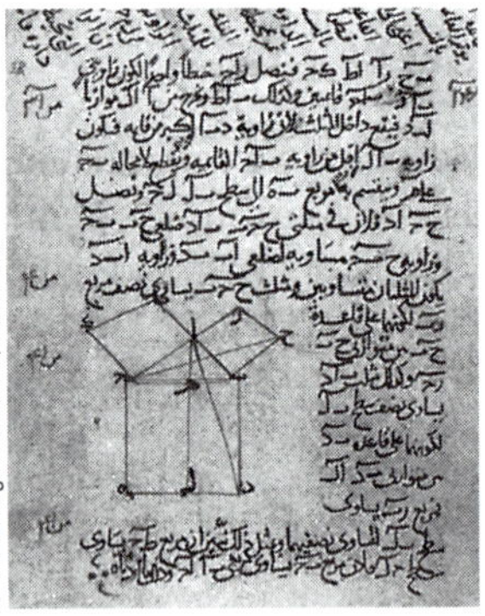

Euclid's proof of the Pythagorean Theorem as it appeared in an Arabic translation of *Elements*.

The most widely published and read book in the history of the printed word is, of course, the Bible. And what do you suppose the second most widely published book is? Many people are quite surprised to discover that the number 2 all-time best-seller is a mathematics textbook! Written by the Greek mathematician Euclid of Alexandria around 300 B.C., *Elements* was hand-copied by scribes for nearly 2,000 years. Since the coming of the printing press, over 1,000 editions have been printed, the first in 1482. *Elements* consists of 13 "books," or units. Although most people associate Euclid with geometry, *Elements* also contains number theory and elementary algebra, all the mathematics known to the scholars of the day.

Euclid's monumental contribution to the world of mathematics was his method of organization, not his discovery of new theorems. He took all the mathematical knowledge that had been compiled since the days of Thales (300 years earlier) and organized it in such a way that every result followed logically from its predecessors. To begin this chain of proof, Euclid had to start with a handful of assumptions, or things that cannot be proved. These assumptions are called *axioms*, or *postulates*, and are accepted without proof. By carefully choosing five geometric postulates, Euclid proceeded to prove 465 results, many of which were quite complicated and not at all intuitively obvious. The beauty of his work is that so much was proved from so few assumptions.

Euclid's Postulates of Geometry

1. A straight line segment can be drawn from any point to any other point.
2. A (finite) straight line segment can be extended continuously into an (infinite) straight line.
3. A circle may be described with any center and any radius.

4. All right angles are equal to one another.
5. Given a line and a point not on that line, there is one and only one line through the point parallel to the original line.

Euclid of Alexandria

The only "controversial" item on this list is postulate 5, the so-called Parallel Postulate. This proved to be the unraveling of Euclidean geometry, and 2,000 years later non-Euclidean geometries were born. Non- Euclidean geometries will be discussed in Section 6.6.

The rigor of Euclid's work was unmatched in the ancient world. *Elements* was written for mature, sophisticated thinkers. Even today, the real significance of Euclid's work lies in the superb training it gives in logical thinking. Abraham Lincoln knew of the rigor of Euclid's *Elements*. At the age of forty, while still a struggling lawyer, Lincoln mastered the first six books in the *Elements* on his own, solely as training for his mind.

Teachers of mathematics invariably encounter resistance from students when abstract mathematics with little apparent application is presented. Even in the days of Euclid, students were dismayed at the complexity of the proofs and asked the question so often echoed by today's students: "Why do we have to learn this stuff?" According to legend, one of Euclid's beginning students asked, "What shall I gain by learning these things?" Euclid quickly summoned his servant, responding, "Give him a coin, since he must make gain out of what he learns!" A second legend handed down through the years claims that King Ptolemy once asked Euclid if there was a shorter way of learning geometry than the study of his *Elements*, to which Euclid replied, "There is no royal road to geometry." Sorry, no shortcuts today!

Deductive Proof

To investigate the Greeks' method of geometric deduction, we will examine congruent triangles. **Congruent triangles** are triangles that have exactly the same shape and size; their corresponding angles and sides are equal, as shown in Figure 6.44. "Triangle ABC is congruent to triangle DEF" (denoted by $\Delta ABC \cong \Delta DEF$) means that all six of the following statements are true:

$$\angle A = \angle D \qquad \angle B = \angle E \qquad \angle C = \angle F$$
$$AB = DE \qquad BC = EF \qquad CA = FD$$

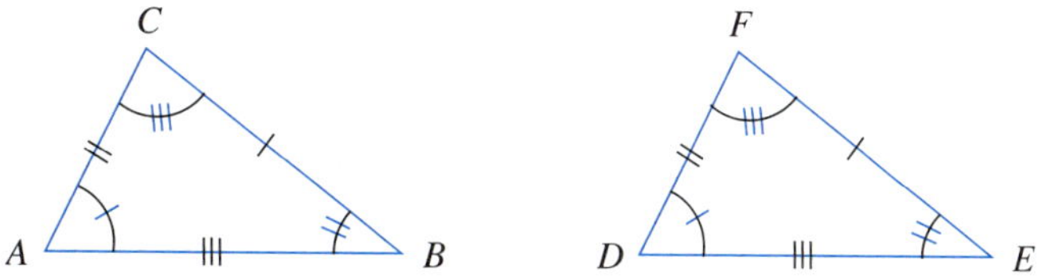

Figure 6.44

HISTORICAL NOTE

Archimedes of Syracuse

(287–212 B.C.)

Archimedes demonstrated that a very heavy object could be raised with relatively little effort by using a series of pulleys (a compound pulley) or by using levers. He was so confident of his principles of levers and pulleys that he boasted that he could move anything. One of Archimedes' many famous quotes is "Give me a place to stand and I will move the earth!"

Aware of his genius, King Hieron turned to Archimedes on many occasions. One amusing legend concerns the problem of the king's gold crown. After giving a goldsmith a quantity of gold for the creation of an elaborate crown, King Hieron suspected the smith of keeping some of the gold and replacing it with an equal weight of silver. Since he had no proof, Hieron turned to Archimedes. While pondering the king's problem, Archimedes submerged himself in a full tub of water to bathe. As he climbed in, he noticed the water overflowing the tub. In a flash of brilliance, he discovered the solution to the king's problem. Overjoyed with his idea, Archimedes is rumored to have run down the street, totally naked, shouting, "Eureka! Eureka!" ("I have found it! I have found it!")

What Archimedes had found was that because silver is less dense than gold, an equal weight of silver would have a greater volume than an equal weight of gold and consequently would displace more water. When the crown and a piece of pure gold of the same weight were immersed in water, the smith's fate was sealed. Even though the crown and the pure gold had the same weight, the crown displaced more water, thus proving that it had been adulterated.

In 215 B.C., the Romans laid siege to Syracuse on the island of Sicily. The Roman commander Marcellus had no idea of the fierce resistance his troops

A seventeenth-century engraving depicting Archimedes' invention of a system of mirrors designed to focus the sun's rays on attacking ships to set them afire. The focusing properties of these shapes will be discussed in Section 6.6.

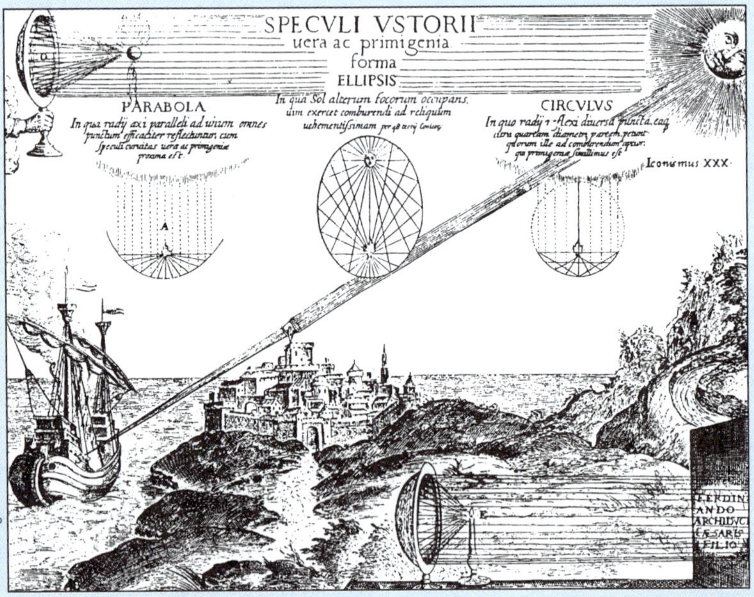

Euclid proved that two triangles are congruent—that is, $\Delta ABC \cong \Delta DEF$—in any of the following three circumstances:

1. SSS: If the three sides of one triangle equal the three sides of the other triangle, then the triangles are congruent. (The corresponding angles will automatically be equal.) See Figure 6.45.
2. SAS: If two sides and the included angle of one triangle equal two sides and the included angle of the other triangle, then the triangles are congruent. (The other side and angles will automatically be equal.) See Figure 6.46.
3. ASA: If two angles and the included side of one triangle equal two angles and the included side of the other triangle, then the triangles are congruent. (The other angle and sides will automatically be equal.) See Figure 6.47.

North Wind Picture Archives

were to face. The attack was from the sea. A native of Syracuse, Archimedes (then seventy-two years old) designed and helped to build a number of ingenious weapons to defend his home. He designed huge catapults that hurled immense boulders at the Roman ships. These catapults were set to throw the projectiles at different ranges so that no matter where the ships were, they were always under fire. When a ship managed to come in close, the defenders lowered large hooks over the city walls, grabbed the ship, lifted it into the air with pulleys, and tossed it back into the sea.

Needless to say, the Greek defenders put up a good fight. They fought so well that the siege lasted nearly three years. However, one day, while the people of Syracuse were feasting and celebrating a religious festival, Roman sympathizers within the city informed the attackers of weaknesses in the city's defense, and a bloody sack began.

Marcellus had given strict orders to take Archimedes alive; no harm was to come to him or his house. At the time, Archimedes was studying the drawings of circles he had made in the sand. Preoccupied, he did not notice the Roman soldier standing next to him until the soldier cast a shadow on his drawings in the sand. The agitated mathematician called out, "Don't disturb my circles!" At that, the insulted soldier used his sword and killed Archimedes of Syracuse.

The Roman commander Marcellus so grieved at the loss of Archimedes that when he learned of the mathematician's wish for the design of his tombstone, he fulfilled it.

Archimedes' tombstone

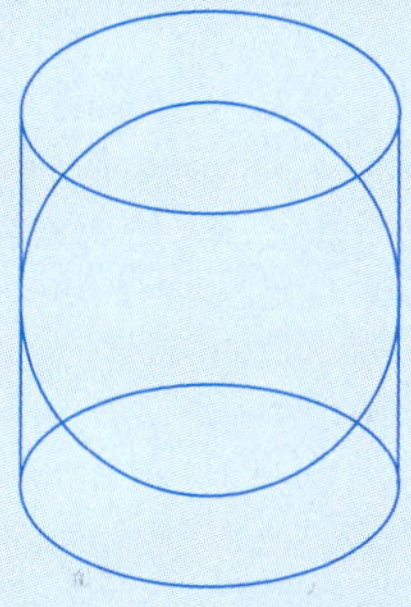

2 : 3

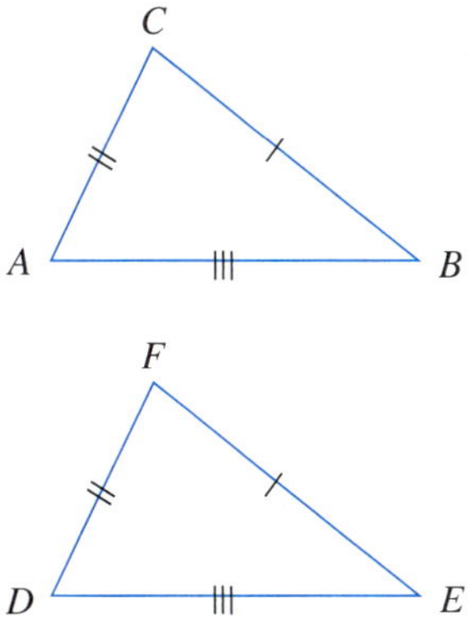

Figure 6.45
SSS implies $\Delta ABC \cong \Delta DEF$

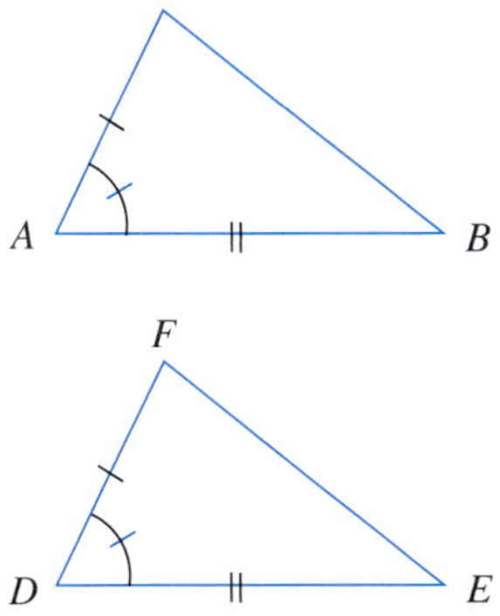

Figure 6.46
SAS implies $\Delta ABC \cong \Delta DEF$

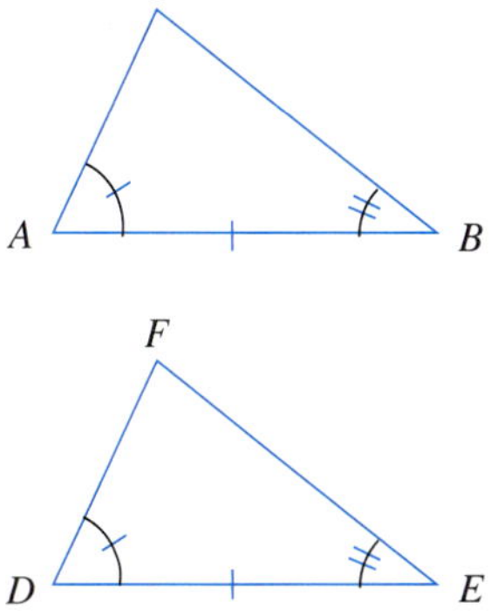

Figure 6.47
ASA implies $\Delta ABC \cong \Delta DEF$

When proving a result in geometry, we use a two-column method. The first column is a list of each valid statement used in the proof, starting with the given statements and progressing to the conclusion. The second column gives the reason for or justification of each statement in the first column.

EXAMPLE 2 Given: $CA = DB$

$\angle CAB = \angle DBA$

Prove: $CB = DA$

(See Figure 6.48.)

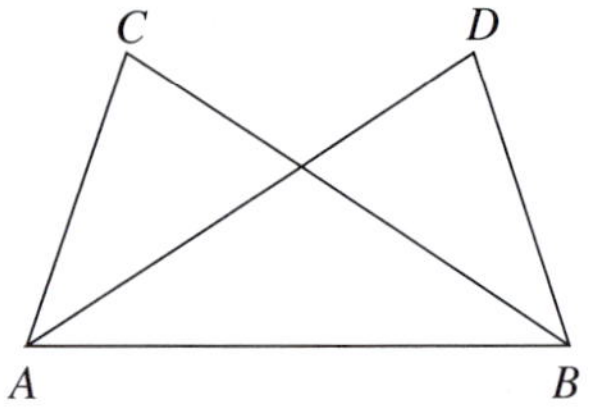

Figure 6.48

Solution

Statements	Reasons
1. $CA = DB$	1. Given
2. $\angle CAB = \angle DBA$	2. Given
3. $AB = AB$	3. Anything equals itself.
4. $\Delta CAB \cong \Delta DBA$	4. SAS
5. $CB = DA$	5. Corresponding parts of congruent triangles are equal.

■

Archimedes of Syracuse

Many scholars consider Archimedes the father of physics; he was the first scientific engineer, for he combined mathematics and the deductive logic of Euclid with the scientific method of experimentation. Born around 287 B.C. in the Greek city of Syracuse on the island of Sicily, Archimedes gave the world many labor-saving devices, such as pulleys, levers, and a water pump that utilized a screw-shaped cylinder. In addition, he created formidable weapons that were used in the defense of Syracuse, developed the fundamental concept of buoyancy, and wrote many treatises on mathematics and geometry, including ones on the calculation of π and on the surface area and volume of spheres, cylinders, and cones.

To calculate the value of π to any desired level of accuracy, Archimedes approximated a circle by using a regular polygon—a polygon whose sides all have the same length. His method was to inscribe a regular polygon of n sides inside a circle and determine the perimeter of the polygon. Because it was inside the circle, the perimeter of the polygon was smaller than the circumference of the circle. Then he would circumscribe an n-sided polygon around the circle and find its perimeter, which of course was larger than the circumference, as shown in Figure 6.49. By increasing n (the number of sides of the polygon) from 6 to 12 to 24 and so on, he got polygons whose perimeters were progressively closer and closer to the circumference of the circle. Using a polygon of 96 sides, he obtained

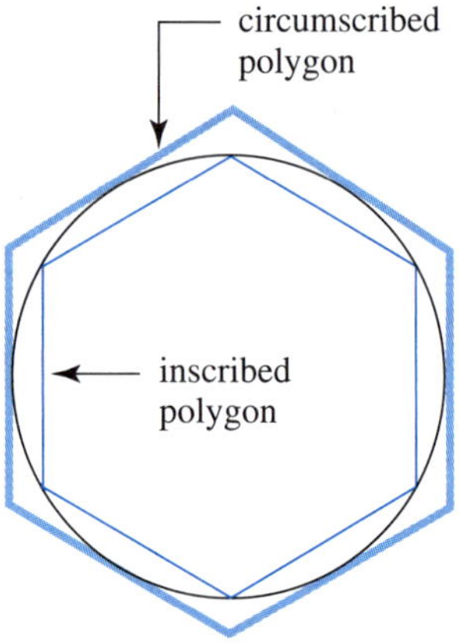

Figure 6.49

$$3\frac{10}{71} < \pi < 3\frac{1}{7}$$

This upper limit of $\frac{22}{7}$ is still used today as a common approximation for π.

EXAMPLE 3 A regular hexagon is inscribed in a circle of radius r, while another regular hexagon is circumscribed around the same circle.

a. Using the perimeter of the inscribed hexagon as an approximation of the circumference of the circle, obtain an estimate of π.

b. If the perimeter of the circumscribed hexagon has sides of length $s = (2\sqrt{3}/3)r$, obtain an estimate of π.

Solution

a. Let s = length of a side of the inscribed hexagon; hence, $P_{\text{hexagon}} = 6s$. A regular hexagon is composed of six equilateral triangles; that is, triangles in which all sides are equal. Therefore, $s = r$, as shown in Figure 6.50.

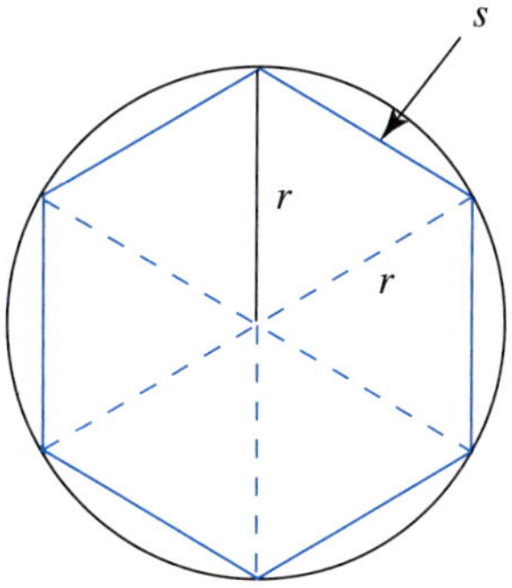

Figure 6.50
Inscribed hexagon

$$
\begin{aligned}
C_{\text{circle}} &\approx P_{\text{inscribed hexagon}} \\
2\pi r &\approx 6s \\
2\pi r &\approx 6r && \text{substituting } s \text{ for } r \\
\pi &\approx \frac{6r}{2r} && \text{dividing by } 2r \\
\pi &\approx 3
\end{aligned}
$$

Therefore, π is approximately 3. Note that an **inscribed polygon** will underestimate the circumference, so our approximation of π is too small.

b. Let s = length of a side of the circumscribed hexagon; hence, $P = 6s$, as shown in Figure 6.51. We are given that $s = (2\sqrt{3}/3)r$.

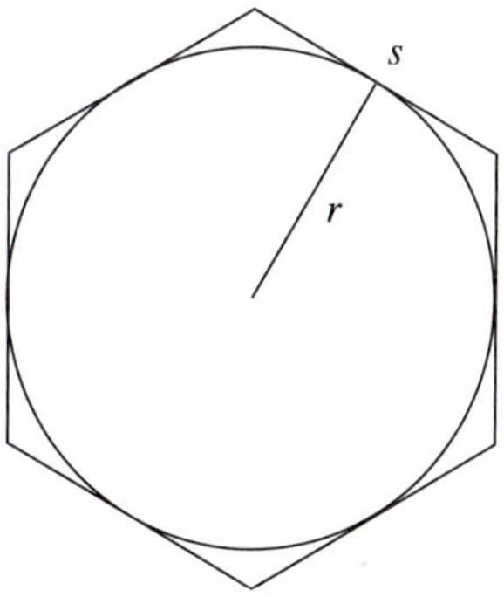

Figure 6.51
Circumscribed hexagon

$$
\begin{aligned}
C_{\text{circle}} &\approx P_{\text{circumscribed hexagon}} \\
2\pi r &\approx 6s \\
2\pi r &\approx 6\left(\frac{2\sqrt{3}}{3}r\right) && \text{substituting } s = (2\sqrt{3}/3)r \\
2\pi r &\approx 4\sqrt{3}r \\
\pi &\approx \frac{4\sqrt{3}r}{2r} && \text{dividing by } 2r \\
\pi &\approx 2\sqrt{3} \\
\pi &\approx 3.4641016
\end{aligned}
$$

Therefore, π is approximately 3.4641016. Note that a **circumscribed polygon** will overestimate the circumference, so our approximation of π is too big.

Combining (b) with (a), we have

$$3 < \pi < 3.4641016$$

■

Of all the work Archimedes did in the field of solid geometry, he was proudest of one particular achievement. Archimedes found that if a sphere is inscribed in a cylinder, the volume of the sphere will be two-thirds the volume of the cylinder. In other words, the ratio of their volumes is 2:3. The result pleased him so much that he wanted a diagram of a sphere inscribed in a cylinder and the ratio 2:3 to be the only markings on his tombstone.

6.4 Exercises

In Exercises 1–4, the given triangles are similar. Find the lengths of the missing sides.

1.

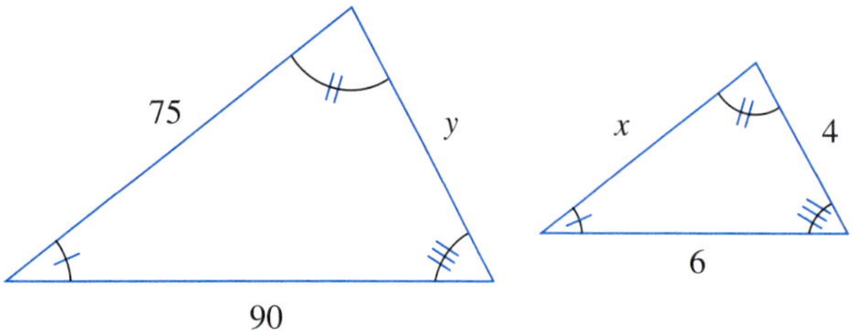

2.

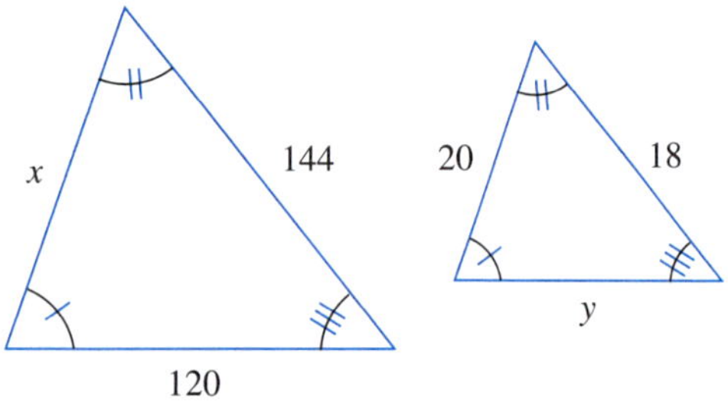

3.

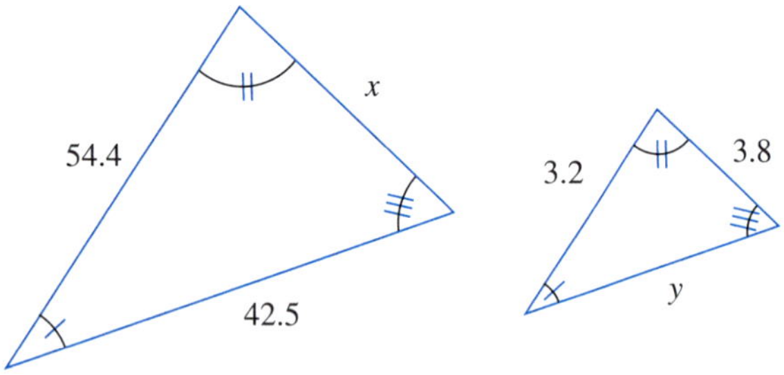

4.

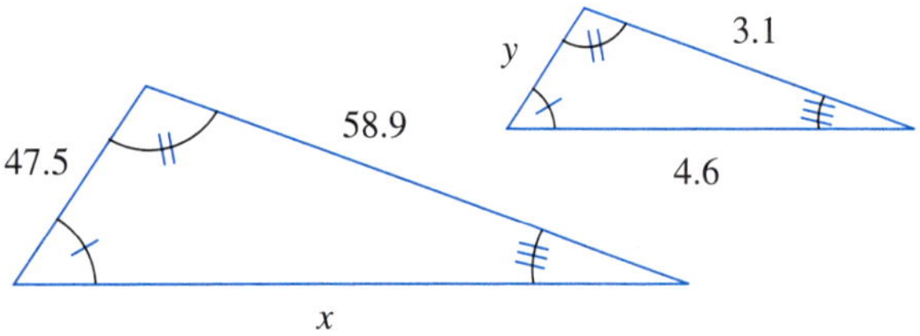

5. A 6-foot-tall man casts a shadow of 3.5 feet at the same instant that a tree casts a shadow of 21 feet. How tall is the tree?

6. A 5.4-foot-tall woman casts a shadow of 2 feet at the same instant that a telephone pole casts a shadow of 9 feet. How tall is the pole?

7. Use the following diagrams to show that $a^2 + b^2 = c^2$.

HINT: The area of square #1 = the area of square #2. (Why?)

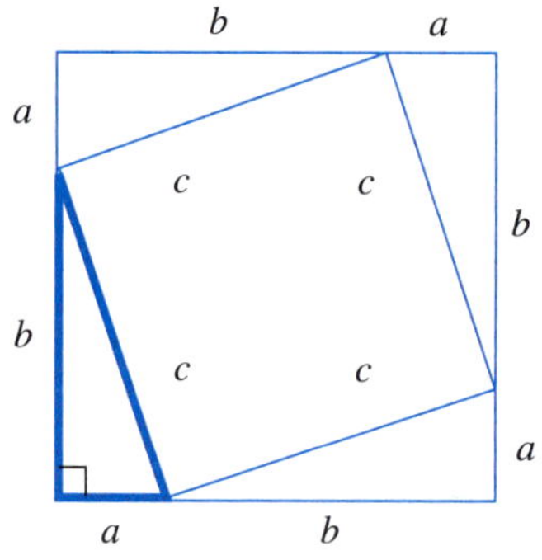

Square #1

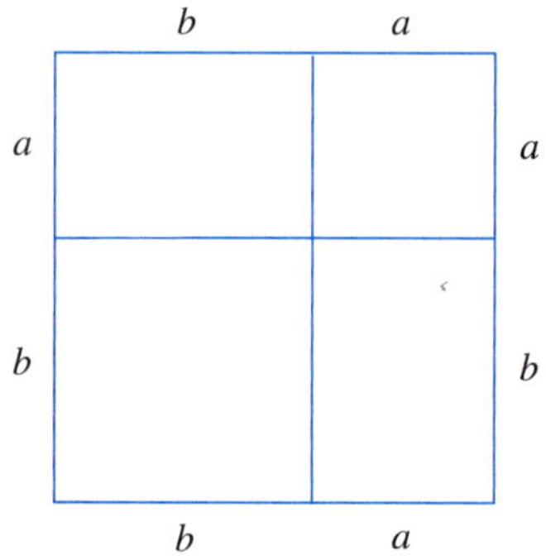

Square #2

8. Use the following diagram and the method of rearrangement to show that $a^2 + b^2 = c^2$.

HINT: Find the area of the figure in two ways, as was done in Garfield's proof.

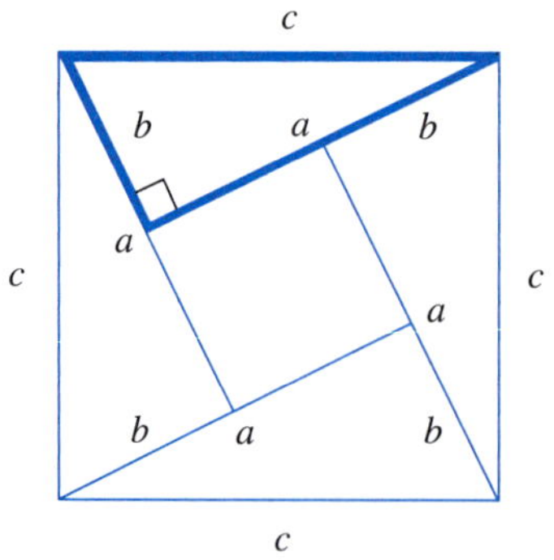

9. Find the length of the longest object that will fit inside a rectangular box 4 feet long, 3 feet wide, and 2 feet high.

HINT: Draw a picture and use the Pythagorean Theorem.

10. Find the length of the longest object that will fit inside a cube 2 feet long, 2 feet wide, and 2 feet high.

HINT: Draw a picture and use the Pythagorean Theorem.

11. Find the length of the longest object that will fit inside a cylinder that has a radius of $2\frac{1}{4}$ inches and is 6 inches high.

HINT: Draw a picture and use the Pythagorean Theorem.

12. Find the length of the longest object that will fit inside a cylinder that has a radius of 1 inch and is 2 inches high.

HINT: Draw a picture and use the Pythagorean Theorem.

In Exercises 13–18, use the two-column method to prove the indicated result.

13. Given: $AD = CD$
$AB = CB$
Prove: $\angle DBA = \angle DBC$

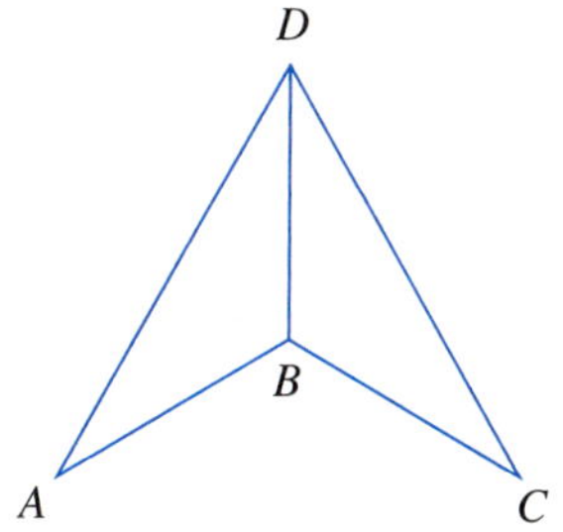

14. Given: $CA = DB$
$CB = DA$
Prove: $\angle ACB = \angle ADB$

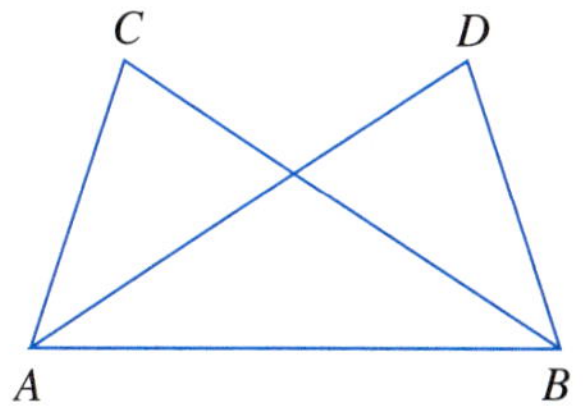

15. Given: $AD = BD$
$\angle ADC = \angle BDC$
Prove: $AC = BC$

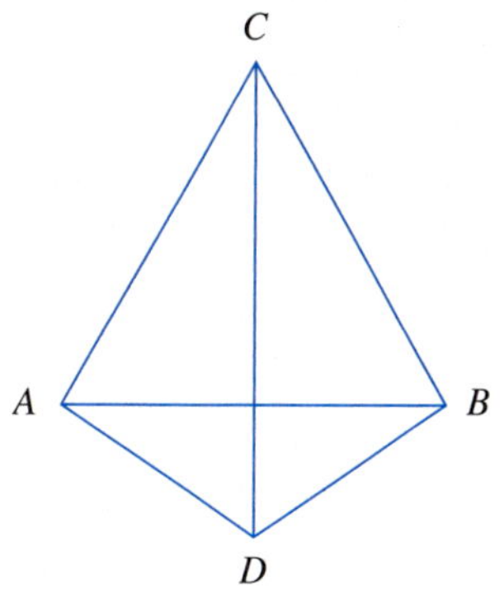

16. Given: $\angle ACD = \angle BCD$
$\angle ADC = \angle BDC$
Prove: $AC = BC$

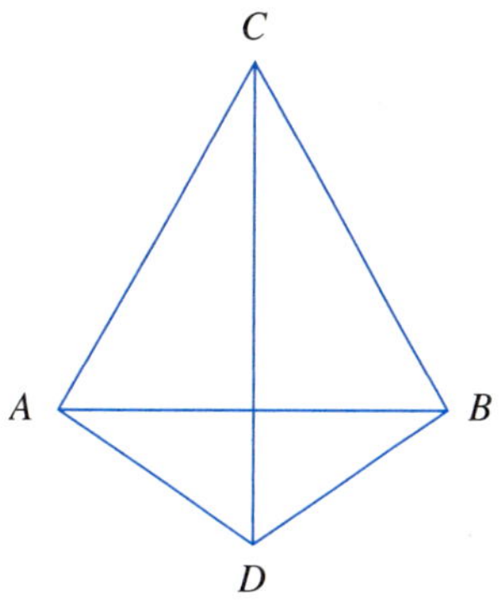

17. Given: $AE = CE$
$AB = CB$
Prove: $\angle ADB = \angle CDB$

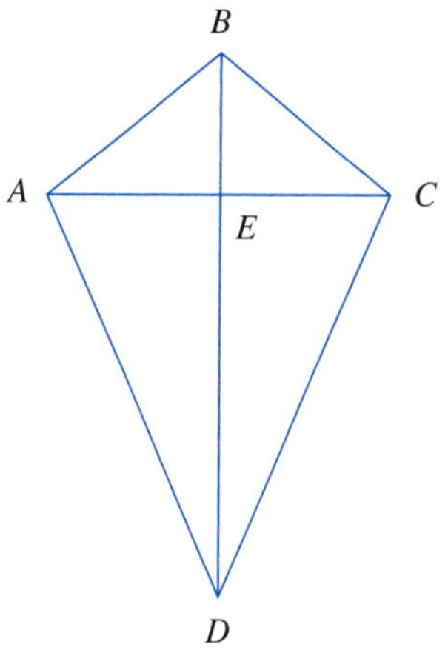

18. Given: $AE = CE$
$\angle AEB = \angle CEB$
Prove: $AD = CD$
(See figure at top of next page.)

Figure for Exercise 18 below.

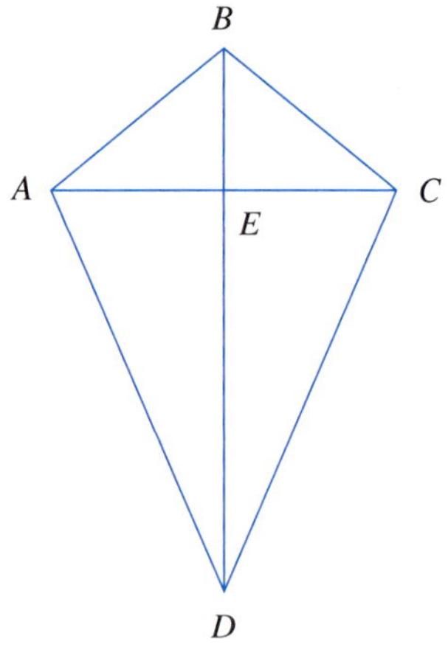

In Exercises 19–22, follow the method of Example 3.

19. A regular octagon (eight-sided polygon) is inscribed in a circle of radius r, while another regular octagon is circumscribed around the same circle.

a. The length of a side of the inscribed octagon is $s = \sqrt{2 - \sqrt{2}}r$. Using the perimeter of this polygon as an approximation of the circumference of the circle, obtain an estimate of π.

b. The length of a side of the circumscribed octagon is $s = 2(\sqrt{2} - 1)r$. Using the perimeter of this polygon as an approximation of the circumference of the circle, obtain an estimate of π.

20. A regular dodecagon (12-sided polygon) is inscribed in a circle of radius r, while another regular dodecagon is circumscribed around the same circle.

a. The length of a side of the inscribed dodecagon is $s = [(\sqrt{6} - \sqrt{2})/2]r$. Using the perimeter of this polygon as an approximation of the circumference of the circle, obtain an estimate of π.

b. The length of a side of the circumscribed dodecagon is $s = [(\sqrt{6} - \sqrt{2})/(\sqrt{2} + \sqrt{3})]r$. Using the perimeter of this polygon as an approximation of the circumference of the circle, obtain an estimate of π.

21. A regular 16-sided polygon is inscribed in a circle of radius r, while another regular 16-sided polygon is circumscribed around the same circle.

a. The length of a side of the inscribed polygon is $s = \sqrt{2 - \sqrt{2 + \sqrt{2}}}r$. Using the perimeter of this polygon as an approximation of the circumference of the circle, obtain an estimate of π.

b. The length of a side of the circumscribed polygon is

$$s = \left(\frac{2\sqrt{2 - \sqrt{2}}}{2 + \sqrt{2 + \sqrt{2}}}\right)r$$

Using the perimeter of this polygon as an approximation of the circumference of the circle, obtain an estimate of π.

22. A regular 24-sided polygon is inscribed in a circle of radius r, while another 24-sided polygon is circumscribed around the same circle.

a. The length of a side of the inscribed polygon is $s = [\sqrt{8 - \sqrt{24} - \sqrt{8}})/2]r$. Using the perimeter of this polygon as an approximation of the circumference of the circle, obtain an estimate of π.

b. The length of a side of the circumscribed polygon is

$$s = \left(\frac{2\sqrt{8 - \sqrt{24} - \sqrt{8}}}{\sqrt{8} + \sqrt{24} + \sqrt{8}}\right)r$$

Using the perimeter of this polygon as an approximation of the circumference of the circle, obtain an estimate of π.

23. If a sphere is inscribed in a cylinder, the diameter of the sphere equals the height of the cylinder. Using this relationship, show Archimedes' favorite result, that is, that the volumes of the sphere and the cylinder are in the ratio 2:3.

Answer the following questions using complete sentences and your own words.

History Questions

24. What is the connection between Thales and Egypt?

25. What was the motto of the Pythagoreans?

26. Which president of the United States developed an original proof of the Pythagorean Theorem?

27. What is the Parallel Postulate?

28. How did the geometry of the Greeks differ from that of the Egyptians?

29. Write a research paper on a historical topic referred to in this section or on a related topic. Following is a list of possible topics:

- Alexandria as a center of knowledge (Why and when did Alexandria flourish as a center of knowledge in the ancient world?)
- ancient war machines (How were mathematics and physics used in the design and construction of ancient weapons of war?)
- Archimedes
- classic construction problems (What do the problems "trisecting an angle," "squaring a circle," and "doubling a cube" have in common? Why are these problems special in the history of geometry?)

- Eratosthenes of Cyrene (How did he calculate the earth's circumference? What did he contribute to the making of maps?)
- Euclid
- Eudoxus of Cnidos (What is the "Golden Section" or "Golden Ratio"?)
- Heron of Alexandria (What is Heron's Formula?)
- Hippocrates of Chios (What are the lunes of Hippocrates?)
- Pappus of Alexandria (What did Pappus's Mathematical Collection contain?)
- pulleys and levers (How do pulleys and levers reduce the force required to lift an object?)
- Pythagoras and the Pythagoreans
- Thales of Miletus (How did he calculate the height of the Great Pyramid of Cheops?)
- Zeno of Elea (What are Zeno's paradoxes?)

Projects

30. The purpose of this project is to calculate the height of a tall building by using similar triangles. Choose a tall building and measure the length of its shadow. Now have an assistant measure the length of your shadow. The top of the building, the base of the building, and the tip of the shadow of the building form a triangle, say, ΔABC. Likewise, the top of your head, your heels, and the tip of your shadow form a triangle, say, ΔDEF. (See Figure 6.52.) Assuming that rays of sunlight are parallel, we conclude that ΔABC and ΔDEF are similar. Use the ratio of your height to the length of your shadow to calculate the height of the building.

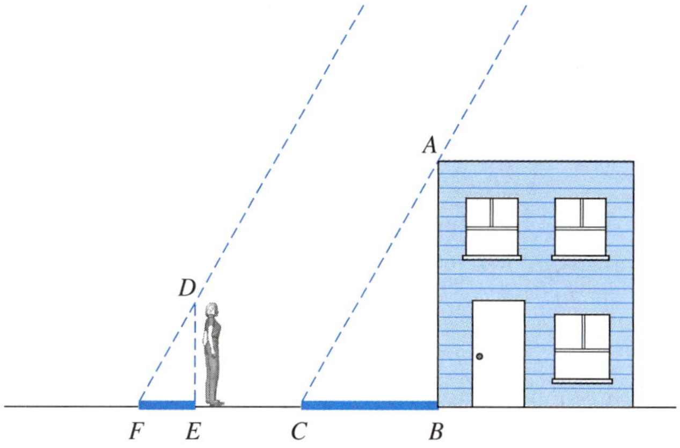

Figure 6.52

6.5 Right Triangle Trigonometry

Plane geometry includes the study of many simple figures, such as circles, triangles, rectangles, squares, parallelograms, and trapezoids. Triangles are of special interest; they are the simplest figures that can be drawn with straight sides. **Trigonometry** is the branch of mathematics that details the study of triangles. The word *trigonometry* literally means "triangle measurement"; it is derived from the Greek words *trigon*, meaning "triangle," and *metros*, meaning "measure." Although modern trigonometry has numerous applications that do not involve triangles, trigonometry's roots lie in the study of right triangles; for example, we will use trigonometry to calculate the length of a cable that runs from the ground to the top of a warehouse, as shown in Figure 6.53.

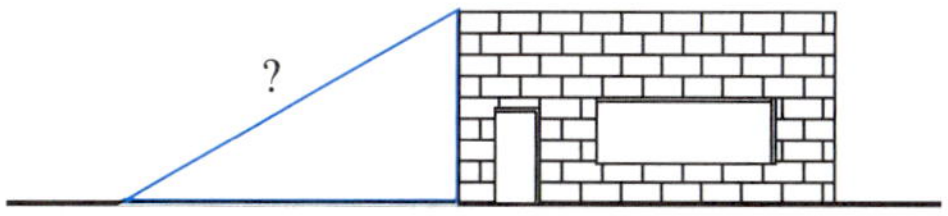

Figure 6.53

Angle Measurement

A triangle has three sides and three angles. The sides can be measured in terms of any linear unit, such as inches, feet, centimeters, or meters. How are angles measured? The two most common units are *degrees* and *radians*. We will consider only degrees in this textbook.

A *right angle* is an angle that makes a square corner. By definition, a right angle is said to have a measure of *90 degrees*, denoted 90°. See Figure 6.54(a). Consequently, a straight line can be viewed as an angle of measure 180°, as shown in Figure 6.54(b).

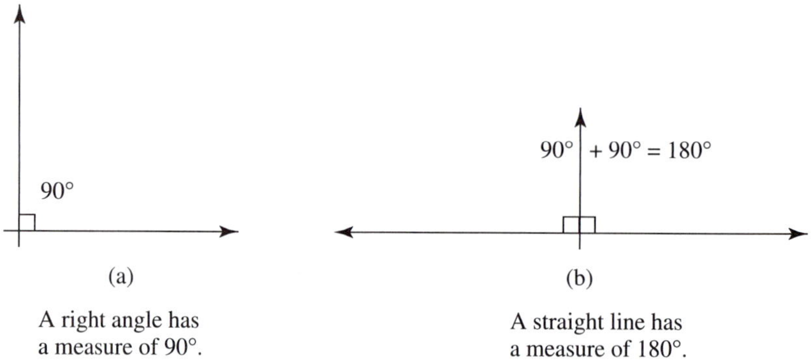

Figure 6.54

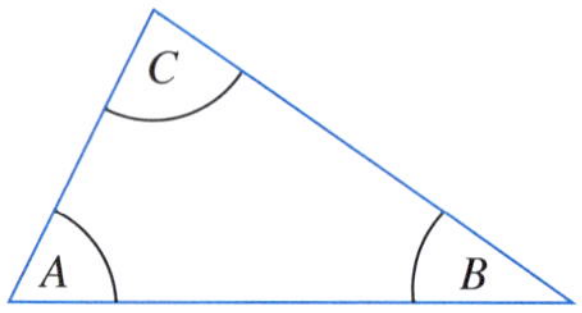

Figure 6.55

If a right angle is "cut up" into 90 equal pieces, each "slice" would represent an angle of measure 1°. From a previous math course, you may recall that for any triangle, the sum of the measures of the three angles is 180°. Why is this true? Suppose the angles of a triangle are *A*, *B*, and *C*, as shown in Figure 6.55.

Now construct two copies of the triangle and place them next to the original, as shown in Figure 6.56.

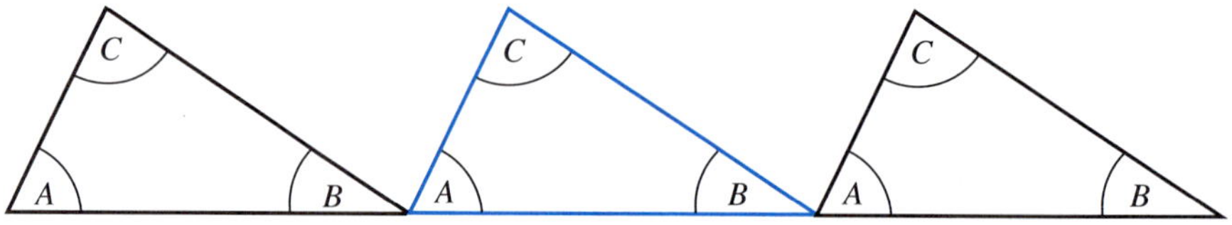

Figure 6.56

If the triangle on the right is flipped backwards and upside down, it will fit perfectly into the "notch" between the other two triangles, as shown in Figure 6.57.

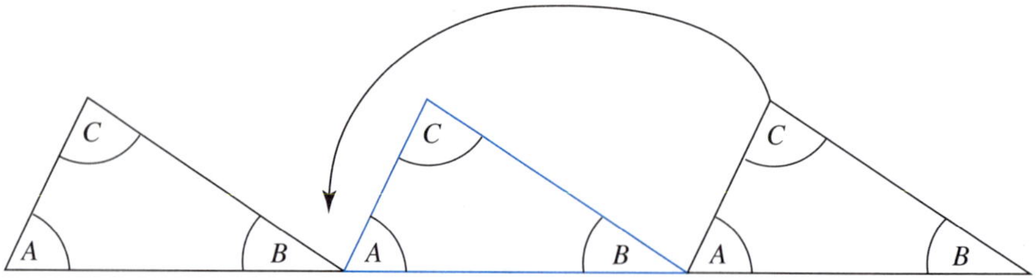

Figure 6.57

Notice that the three angles A, B, and C now form a straight line, as shown on the bottom edge of Figure 6.58; consequently, the sum of the angles equals 180°.

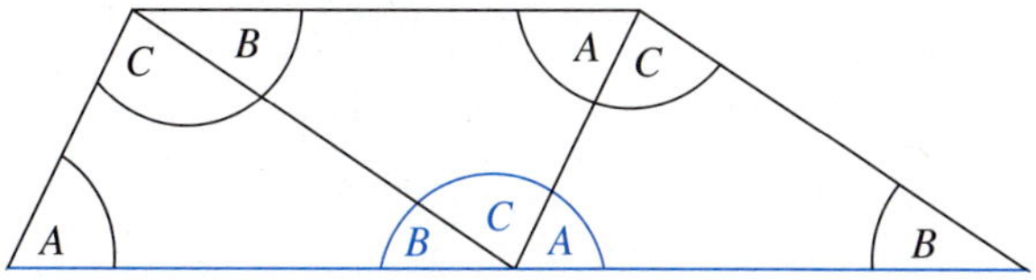

$$A + B + C = 180°$$

Figure 6.58

Special Triangles

There are two special triangles that frequently arise in trigonometry: the 30°-60°-90° triangle and the 45°-45°-90° triangle. The sides of each of these triangles have special relationships.

Each angle of a square has a measure of 90°. If a square with sides of length x is cut in two along one of its diagonals, each resulting right triangle will have two equal sides (of length x) and two 45° angles. See Figure 6.59.

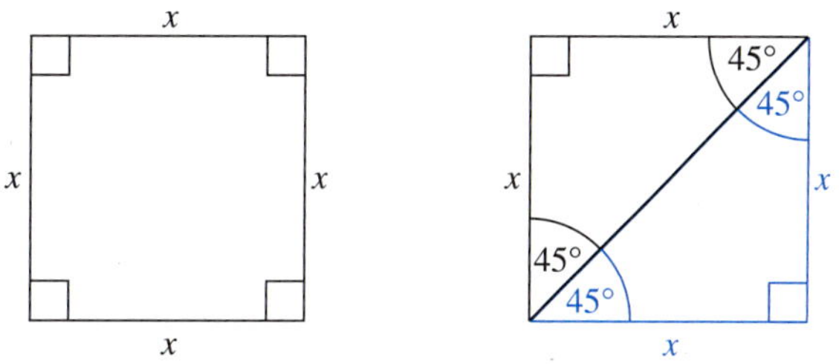

Figure 6.59

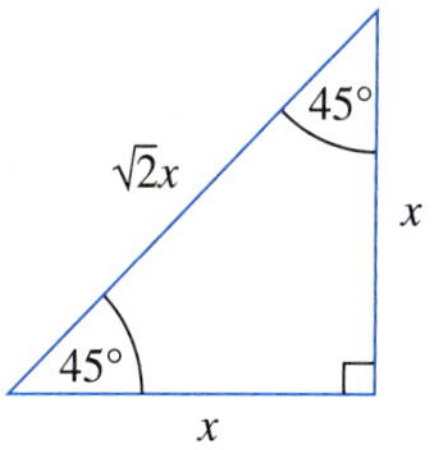

Figure 6.60
Isosceles right triangle

A triangle of this type is called an *isosceles right triangle*. (An isosceles triangle is a triangle that has two equal sides.) Applying the Pythagorean theorem, we find the length of the hypotenuse as follows:

$$c^2 = a^2 + b^2 \qquad c = \text{hypotenuse, } a \text{ and } b = \text{legs}$$
$$c^2 = x^2 + x^2 \qquad \text{substituting } a = x \text{ and } b = x$$
$$c^2 = 2x^2 \qquad \text{combining like terms}$$

Now take the (positive) square root of each side and simplify:

$$\sqrt{c^2} = \sqrt{2x^2}$$
$$c = \sqrt{2}x$$

Therefore, the hypotenuse has length $\sqrt{2}x$. See Figure 6.60.

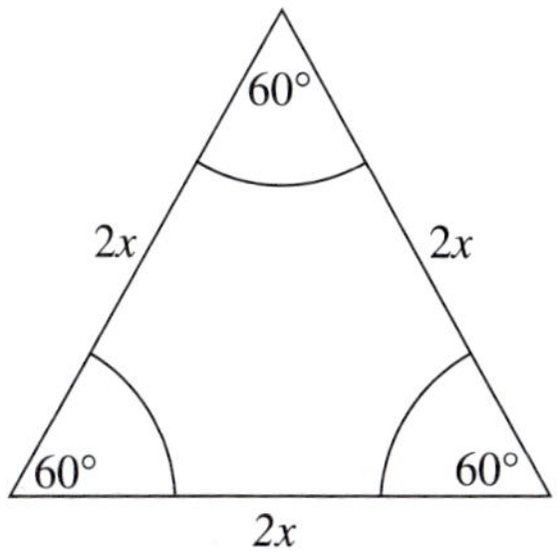

Figure 6.61
Equilateral triangle

An *equilateral triangle* is a triangle in which all three sides have the same length; consequently, all three angles have the same measure. Since the sum of the three angles must equal 180°, it follows that each angle in an equilateral triangle has a measure of 60°. For future convenience, suppose each side has length $2x$, as shown in Figure 6.61.

Suppose one of the angles of an equilateral triangle is bisected (cut into two equal pieces); the bisected 60° angle will generate two 30° angles. Consequently, two right

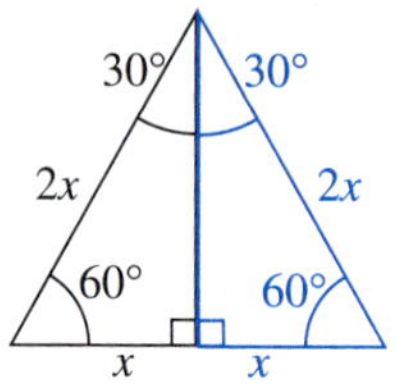

Figure 6.62
Bisected equilateral triangle

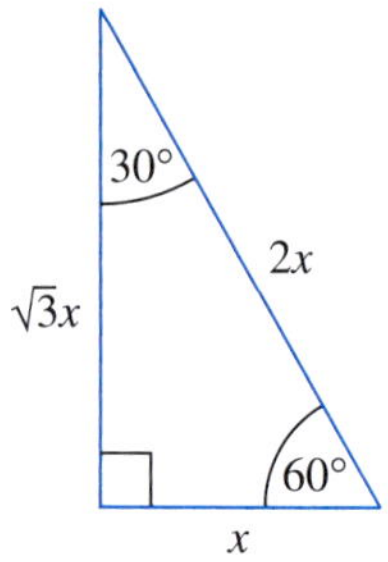

Figure 6.63
30°-60°-90° triangle

triangles are created. In addition, the side opposite the bisected angle will be cut into two equal segments of length x. See Figure 6.62.

Applying the Pythagorean theorem to one of these right triangles (called a *30°-60°-90° triangle*), we find the length of the missing (vertical) leg as follows:

$$\begin{aligned} a^2 + b^2 &= c^2 && a \text{ and } b = \text{legs, } c = \text{hypotenuse} \\ x^2 + b^2 &= (2x)^2 && \text{substituting } a = x \text{ and } c = 2x \\ x^2 + b^2 &= 4x^2 && \text{multiplying} \\ b^2 &= 3x^2 && \text{subtracting } x^2 \text{ from both sides} \end{aligned}$$

Now take the (positive) square root of each side and simplify:

$$\begin{aligned} \sqrt{b^2} &= \sqrt{3x^2} \\ b &= \sqrt{3}x \end{aligned}$$

Therefore, the leg has length $\sqrt{3}x$. See Figure 6.63.

Trigonometric Ratios

An *acute angle* is an angle whose measure is between 0° and 90°. Let θ (the Greek letter theta) equal the measure of one of the acute angles of a right triangle. Using the abbreviations "opp," "adj," and "hyp" to represent, respectively, the side opposite angle θ, the side adjacent to angle θ, and the hypotenuse, a right triangle can be labeled, as in Figure 6.64.

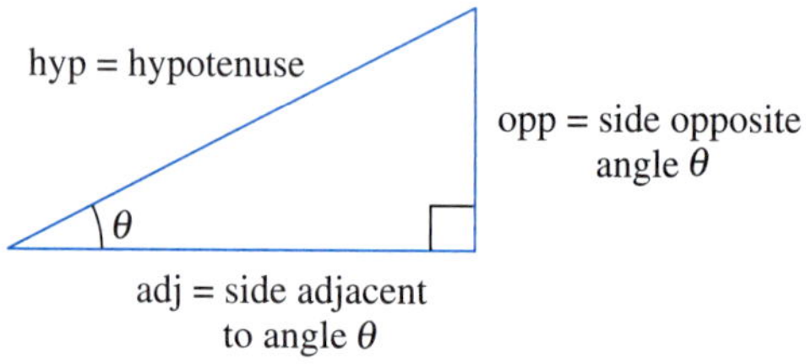

Figure 6.64

In Section 6.4, we saw that for similar triangles (triangles that have the same shape but different size), corresponding sides are proportional; that is, ratios of corresponding sides are constant. Consequently, no matter how large or how small a right triangle is, the ratio of its sides is determined solely by the size of the acute angles. Three specific ratios of sides form the basis of trigonometry; these ratios are called the *sine, cosine*, and *tangent* and are defined as follows.

Trigonometric Ratios

Let θ equal the measure of one of the acute angles of a right triangle.

1. The **sine of θ**, denoted **$\sin\theta$**, is the ratio of the side opposite θ compared to the hypotenuse: $\sin\theta = \dfrac{\text{opp}}{\text{hyp}}$
2. The **cosine of θ**, denoted **$\cos\theta$**, is the ratio of the side adjacent θ compared to the hypotenuse: $\cos\theta = \dfrac{\text{adj}}{\text{hyp}}$
3. The **tangent of θ**, denoted **$\tan\theta$**, is the ratio of the side opposite θ compared to the side adjacent θ: $\tan\theta = \dfrac{\text{opp}}{\text{adj}}$

The following example examines the special isosceles right triangle.

EXAMPLE 1 Find the sine, cosine, and tangent of 45°.

Solution Recall the general isosceles right triangle shown in Figure 6.60 and reproduced here. The length of the side opposite 45° is x, as is the length of the side adjacent 45°; the length of the hypotenuse is $\sqrt{2}x$. Applying the trigonometric ratios to the angle of measure 45°, we have the following:

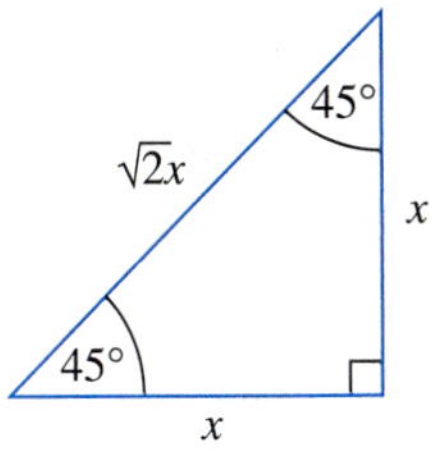

$$\sin 45° = \frac{\text{opp}}{\text{hyp}} = \frac{x}{\sqrt{2}x} = \frac{1}{\sqrt{2}}$$

To rationalize the denominator, multiply by $\frac{\sqrt{2}}{\sqrt{2}}$.

Therefore, $\sin 45° = \frac{\sqrt{2}}{2}$.

$$\cos 45° = \frac{\text{adj}}{\text{hyp}} = \frac{x}{\sqrt{2}x} = \frac{1}{\sqrt{2}}$$

To rationalize the denominator, multiply by $\frac{\sqrt{2}}{\sqrt{2}}$.

Therefore, $\cos 45° = \frac{\sqrt{2}}{2}$.

$$\tan 45° = \frac{\text{opp}}{\text{adj}} = \frac{x}{x} = 1$$

Therefore, $\tan 45° = 1$. ■

Trigonometric Ratios for 45°

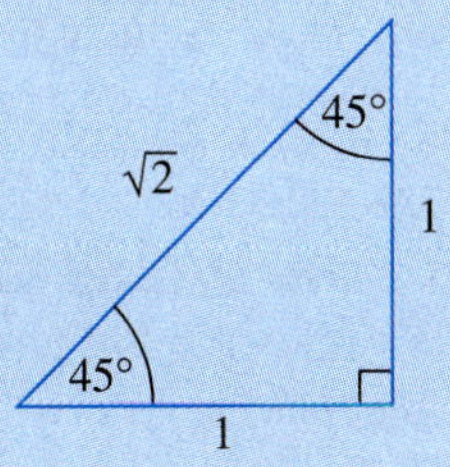

$$\sin 45° = \frac{1}{\sqrt{2}} = \frac{\sqrt{2}}{2}$$

$$\cos 45° = \frac{1}{\sqrt{2}} = \frac{\sqrt{2}}{2}$$

$$\tan 45° = \frac{1}{1} = 1$$

Notice that in Example 1, the value of x has no influence on the trigonometric ratios; x always "cancels" in both the numerator and denominator. When working with a 45° angle, an "easy" triangle to remember is the triangle in which $x = 1$. The results are summarized below.

The following example examines the special 30°-60°-90° triangle.

EXAMPLE 2
a. Find the sine, cosine, and tangent of 30°.
b. Find the sine, cosine, and tangent of 60°.

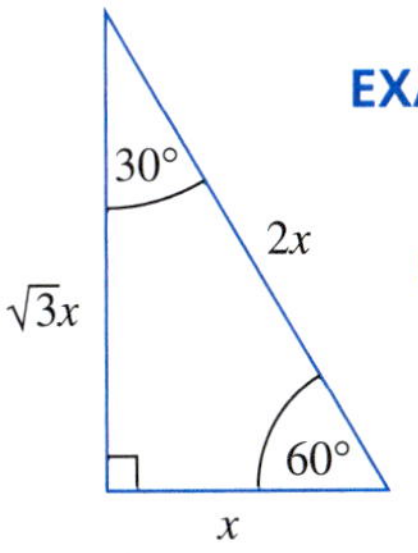

Solution **a.** Recall the general 30°-60°-90° triangle shown in Figure 6.63 and reproduced here. The length of the side opposite 30° is x, the length of the side adjacent 30° is $\sqrt{3}x$, and the length of the hypotenuse is $2x$. Applying the trigonometric ratios to the angle of measure 30°, we have the following:

$$\sin 30° = \frac{\text{opp}}{\text{hyp}} = \frac{x}{2x} = \frac{1}{2}$$

Therefore, $\sin 30° = \frac{1}{2}$.

$$\cos 30° = \frac{\text{adj}}{\text{hyp}} = \frac{\sqrt{3}x}{2x} = \frac{\sqrt{3}}{2}$$

Therefore, $\cos 30° = \frac{\sqrt{3}}{2}$.

$$\tan 30° = \frac{\text{opp}}{\text{adj}} = \frac{x}{\sqrt{3}x} = \frac{1}{\sqrt{3}}$$

To rationalize the denominator, multiply by $\frac{\sqrt{3}}{\sqrt{3}}$.

Therefore, $\tan 30° = \frac{\sqrt{3}}{3}$.

b. Applying the trigonometric ratios to the angle of measure 60°, we have the following:

$$\sin 60° = \frac{\text{opp}}{\text{hyp}} = \frac{\sqrt{3}x}{2x} = \frac{\sqrt{3}}{2}$$

Therefore, $\sin 60° = \frac{\sqrt{3}}{2}$.

$$\cos 60° = \frac{\text{adj}}{\text{hyp}} = \frac{x}{2x} = \frac{1}{2}$$

Therefore, $\cos 60° = \frac{1}{2}$.

$$\tan 60° = \frac{\text{opp}}{\text{adj}} = \frac{\sqrt{3}x}{x} = \frac{\sqrt{3}}{1}$$

Therefore, $\tan 60° = \sqrt{3}$. ■

Notice that in Example 2, the value of x has no influence on the trigonometric ratios; x always "cancels" in both the numerator and denominator. When working with a 30° or 60° angle, an "easy" triangle to remember is the triangle in which $x = 1$. The results are summarized below.

Trigonometric Ratios for 30° and 60°

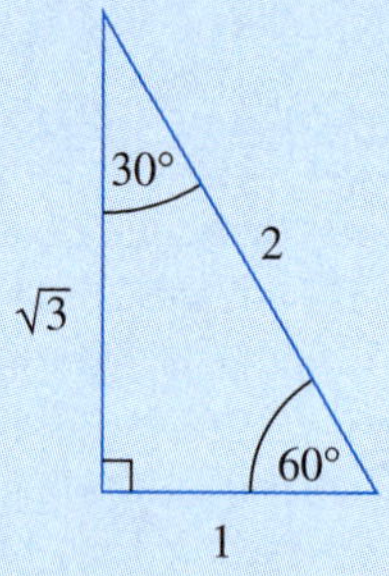

$\sin 30° = \frac{1}{2}$ | $\sin 60° = \frac{\sqrt{3}}{2}$

$\cos 30° = \frac{\sqrt{3}}{2}$ | $\cos 60° = \frac{1}{2}$

$\tan 30° = \frac{1}{\sqrt{3}} = \frac{\sqrt{3}}{3}$ | $\tan 60° = \sqrt{3}$

Before we investigate acute angles other than the special ones (30°, 45°, and 60°), we first consider an application of trigonometric ratios.

EXAMPLE 3 A cable runs from the top of a warehouse to a point on the ground 54.0 feet from the base of the building. If the cable makes an angle of 30° with the ground (as shown in Figure 6.65), calculate

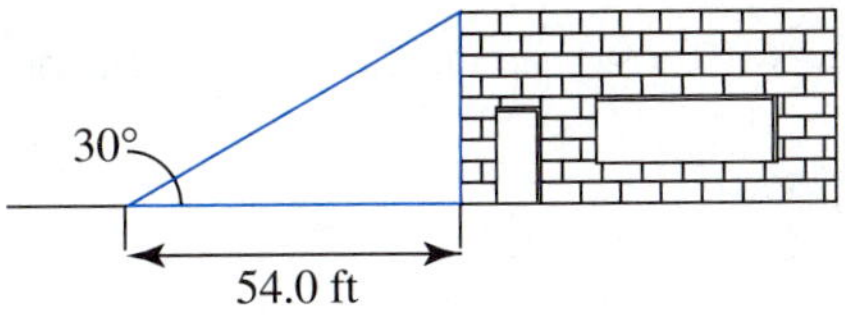

Figure 6.65

a. the height of the warehouse
b. the length of the cable

Solution Let h = height of the warehouse and c = length of the cable. Assuming that the building makes a right angle with the ground, we have the right triangle shown in Figure 6.66.

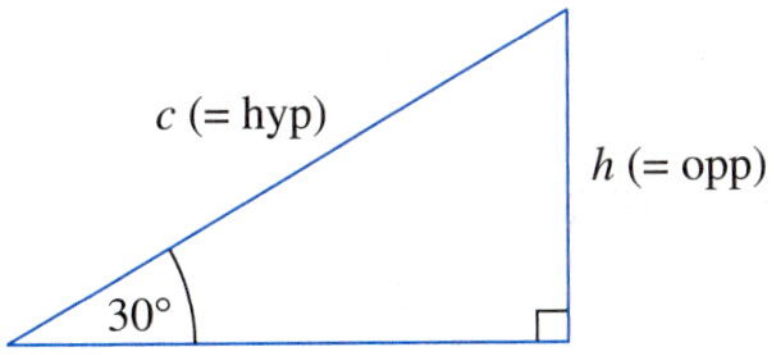

Figure 6.66

a. The unknown height of the building is the side opposite the given angle of 30°; it is labeled *opp*. Similarly, the known side of 54.0 feet is adjacent to the angle; it is labeled *adj*. By comparing *opp* to *adj*, we use the tangent ratio to find the height of the building:

$$\tan\theta = \frac{\text{opp}}{\text{adj}} \qquad \text{definition of the tangent ratio}$$

$$\tan 30^\circ = \frac{h}{54.0} \qquad \text{substituting } \theta = 30^\circ, \text{opp} = h, \text{adj} = 54.0$$

$$h = 54.0 \tan 30^\circ \qquad \text{multiplying both sides by 54.0}$$

However, we know that $\tan 30^\circ = \frac{\sqrt{3}}{3}$. Substituting, we obtain

$$h = 54.0\left(\frac{\sqrt{3}}{3}\right) = 18\sqrt{3} = 31.1769\ldots$$

The height of the warehouse is approximately 31.2 feet.

b. The unknown length of the cable is the hypotenuse; it is labeled *hyp*. By comparing *adj* (which is known) to *hyp*, we use the cosine ratio to find the length of the cable:

$$\cos\theta = \frac{\text{adj}}{\text{hyp}} \qquad \text{definition of the cosine ratio}$$

$$\cos 30^\circ = \frac{54.0}{c} \qquad \text{substituting } \theta = 30^\circ, \text{adj} = 54.0, \text{hyp} = c$$

$$c(\cos 30^\circ) = 54.0 \qquad \text{multiplying both sides by } c$$

$$c = \frac{54.0}{\cos 30^\circ} \qquad \text{dividing both sides by } \cos 30^\circ$$

However, we know that $\cos 30° = \frac{\sqrt{3}}{2}$. Substituting, we obtain

$$c = \frac{54.0}{\frac{\sqrt{3}}{2}}$$

$$c = \left(\frac{54.0}{1}\right)\left(\frac{2}{\sqrt{3}}\right) \qquad \text{dividing by a fraction: invert and multiply}$$

$$c = \frac{108}{\sqrt{3}} \cdot \frac{\sqrt{3}}{\sqrt{3}} \qquad \text{rationalizing the denominator}$$

$$c = \frac{108\sqrt{3}}{3} = 36\sqrt{3} = 62.3538\ldots$$

The length of the cable is approximately 62.4 feet. ■

Using a Calculator

So far, we have found trigonometric ratios of only three angles: 30°, 45°, and 60°. Historically, these were the "easiest" owing to their association with special triangles. How do we find the trigonometric ratios for *any* acute angle? In the past, mathematicians developed extensive tables that gave the desired values; today, we use calculators.

When using a calculator, the proper unit of angle measure must be selected. In this textbook, we measure angles in terms of degrees; the calculator must be in the "degree mode." If your calculator has a DRG button, press it as many times as necessary so that the display reads "D" or "DEG." (*DRG* signifies *D*egrees-*R*adians-*G*radiens.) Some scientific calculators have a reference chart printed near the display window; press MODE and the appropriate symbol to select "degrees." If you have a graphing calculator, press the MODE button and use the arrow buttons to select "degrees."

EXAMPLE 4 Use a calculator to find the following values.

a. sin 17° **b.** cos 25.3° **c.** tan 83.45°

Solution **a.** Locate the SIN button. With most scientific calculators, you enter the measure of the angle first, then press SIN; with most graphing calculators, you press SIN first, type the measure of the angle, and then press ENTER. In any case, we find that sin 17° = 0.292371704723. . . .

17 SIN
SIN 17 ENTER

b. Using the COS button, we find that cos 25.3° = 0.9040825497. . . .

25.3 [COS]

[COS] 25.3 [ENTER]

c. Using the [TAN] button, we find that tan 83.45° = 8.70930765678. . . .

83.45 [TAN]

[TAN] 83.45 [ENTER]

Finding an Acute Angle

If you know the measure of an acute angle, you can use a calculator to find the angle's trigonometric ratios. A calculator can also be used to work a problem in reverse; if you know a trigonometric ratio, a calculator can find the measure of the desired acute angle.

Specifically, let $a = opp$, $b = adj$, and $c = hyp$ be the lengths of the sides of a right triangle, where θ equals the measure of the acute angle opposite side a, as shown in Figure 6.67.

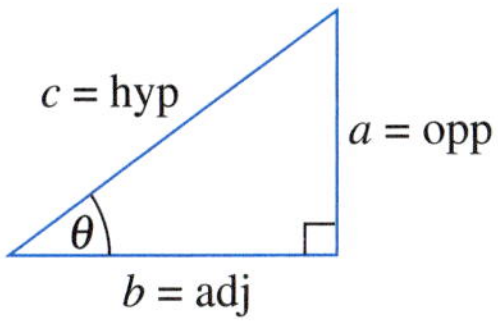

Figure 6.67

To find θ, note that $\sin\theta = opp/hyp = a/c$; alternatively, we can say that θ is *the angle whose sine is* a/c. This description of θ can be symbolized as $\theta = \sin^{-1}(a/c)$; that is, θ is the "inverse sine" of the ratio (a/c). Similar statements can be made concerning the cosine and tangent. These results are summarized below.

Inverse Trigonometric Ratios

If a, b, and c are the lengths of the sides of a right triangle, and θ is the acute angle shown in the figure to the right, then

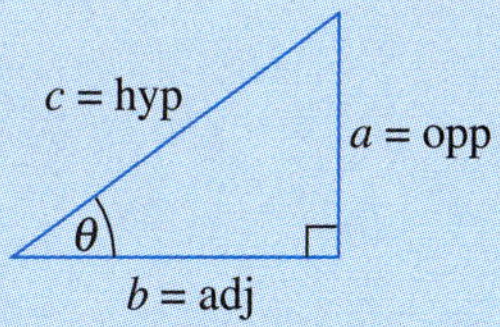

$\theta = \sin^{-1}\left(\frac{a}{c}\right)$ means θ is *the angle whose sine is* $\frac{a}{c}$; that is, $\sin\theta = \frac{a}{c}$.

$\theta = \cos^{-1}\left(\frac{b}{c}\right)$ means θ is *the angle whose cosine is* $\frac{b}{c}$; that is, $\cos\theta = \frac{b}{c}$.

$\theta = \tan^{-1}\left(\frac{a}{b}\right)$ means θ is *the angle whose tangent is* $\frac{a}{b}$, that is, $\tan\theta = \frac{a}{b}$.

EXAMPLE 5 Use a calculator to find the measure of angle θ in each of the following right triangles:

a.

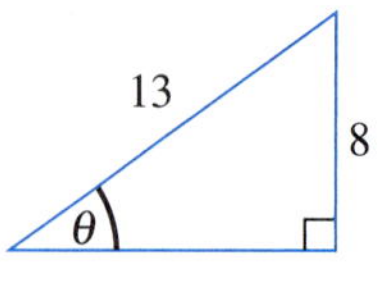

b.

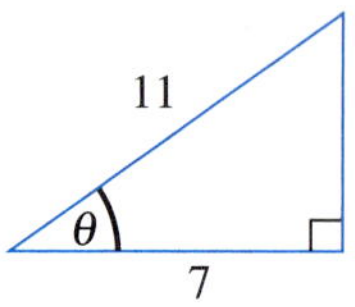

c.

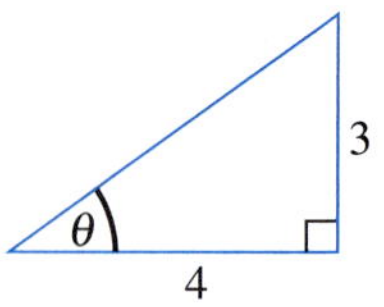

Solution **a.** Applying the sine ratio, we have $\sin\theta = \dfrac{8}{13}$, or equivalently, $\theta = \sin^{-1}\left(\dfrac{8}{13}\right)$. On a calculator, pressing [INV] [SIN] or [2nd] [SIN] or [shift] [SIN] will summon the "inverse sine." Using a calculator, we obtain $\theta = 37.97987244\ldots°$, or $\theta \approx 37.98°$.

8 [÷] 13 [=] [2nd] [SIN⁻¹]
[2nd] [SIN⁻¹] [(] 8 [÷] 13 [)] [ENTER]

b. Applying the cosine ratio, we have $\cos\theta = \frac{7}{11}$, or equivalently, $\theta = \cos^{-1}\left(\dfrac{7}{11}\right)$. Using a calculator, we obtain $\theta = 50.47880364\ldots°$, or $\theta \approx 50.48°$.

7 [÷] 11 [=] [2nd] [COS⁻¹]
[2nd] [COS⁻¹] [(] 7 [÷] 11 [)] [ENTER]

c. Applying the tangent ratio, we have $\tan\theta = \frac{3}{4}$, or equivalently, $\theta = \tan^{-1}(\frac{3}{4})$. Using a calculator, we obtain $\theta = 36.86989765\ldots°$, or $\theta \approx 36.87°$.

3 [÷] 4 [=] [2nd] [TAN⁻¹]
[2nd] [TAN⁻¹] [(] 3 [÷] 4 [)] [ENTER]

■

Angles of Elevation and Depression

If an object is above your eye level, you must raise your eyes to focus on the object. The angle measured from a horizontal line to the object is called the **angle of elevation.** If an object is below your eye level, you must lower your eyes to focus on the object. The angle measured from a horizontal line to the object is called the **angle of depression.** (See Figure 6.68.)

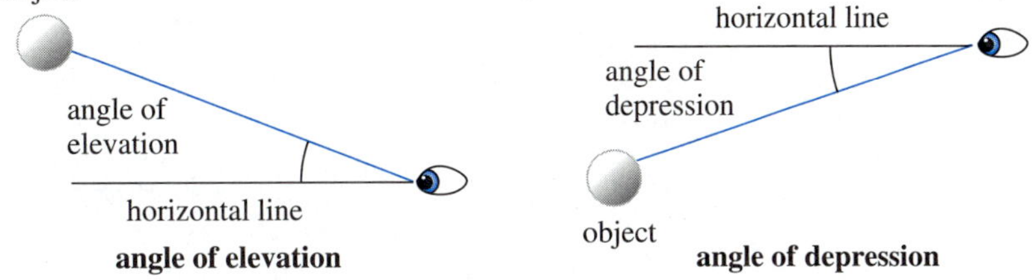

Figure 6.68

EXAMPLE 6 A communications antenna is on top of a building. You are 100.0 feet from the building. Using a transit (a surveyor's instrument), you measure the angle of elevation of the bottom of the antenna as 20.6° and the angle of elevation of the top of the antenna as 27.3°. How tall is the antenna?

Solution Let h = height of the antenna, and x = height of the building. The given information is depicted by the two (superimposed) right triangles, shown in Figure 6.69.

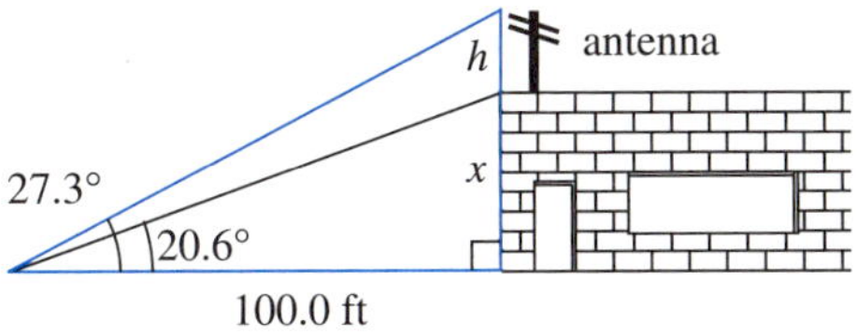

Figure 6.69

Referring to the large right triangle, the side opposite the 27.3° angle is $(x + h)$, and the side adjacent is 100.0 feet. Using the tangent ratio (*opp*/*adj*), we have

$$\tan 27.3° = \frac{x + h}{100.0}$$

Solving for h, we obtain

$$h = 100.0\,(\tan 27.3°) - x \qquad \textbf{(1)}$$

Using the tangent ratio for the small right triangle, we have

$$\tan 20.6° = \frac{x}{100.0}$$

Hence,

$$x = 100.0\,(\tan 20.6°) \qquad \textbf{(2)}$$

Substituting equation (2) into equation (1), we obtain

$$h = 100.0\,(\tan 27.3°) - 100.0\,(\tan 20.6°)$$

or

$$h = 100.0\,(\tan 27.3° - \tan 20.6°)$$ factoring 100.0 from each term

Using a calculator, we have

$$h = 14.0263198\ldots$$

The antenna is approximately 14.0 feet tall. ■

6.5 Exercises

In Exercises 1–8, use trigonometric ratios to find the unknown sides and angles in each right triangle. Do not use a calculator.

1.

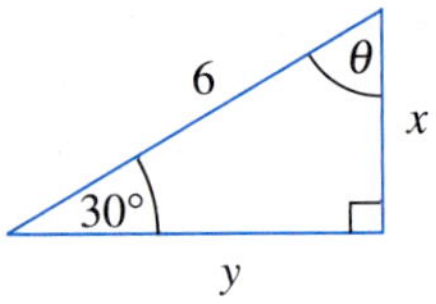

2.

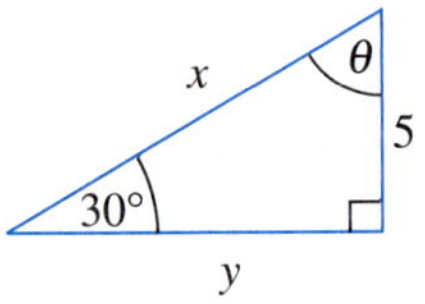

3.

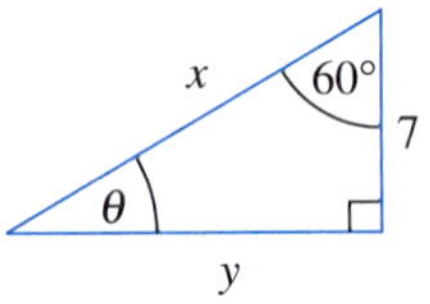

4.

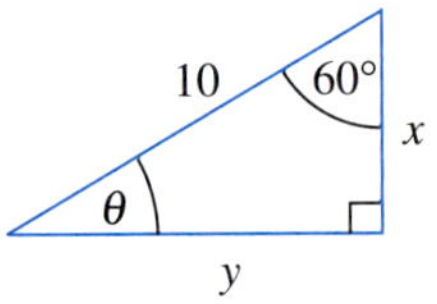

5.

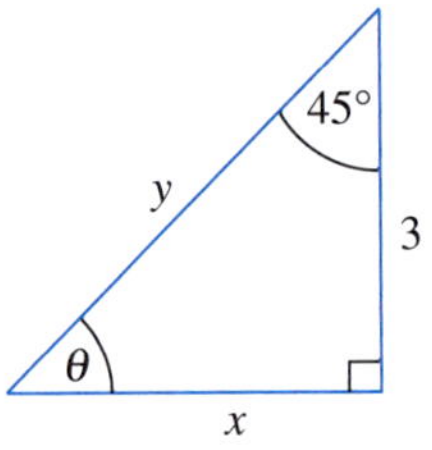

6.

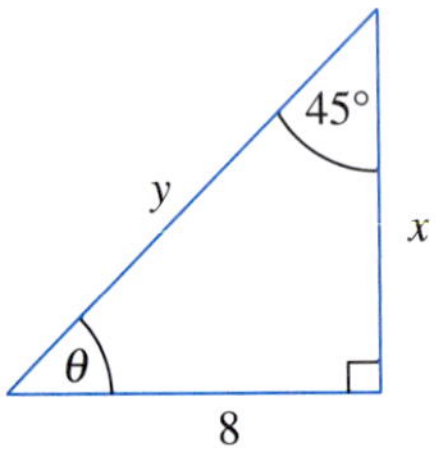

7.

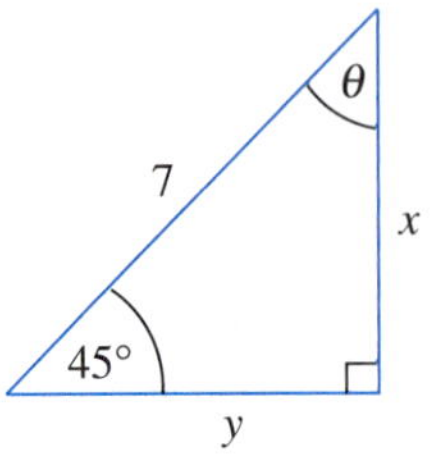

8.

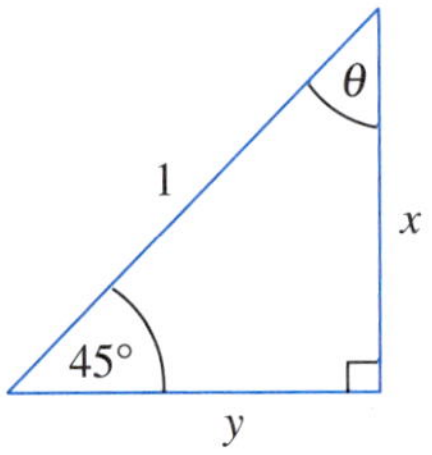

In Exercises 9–26, use the given information to draw a right triangle labeled like the one shown in Figure 6.70. Use trigonometric ratios and a calculator to find the unknown sides and angles. Round off sides and angles to the same number of decimal places as the given sides and angles; in Exercises 21–26, round off angles to one decimal place.

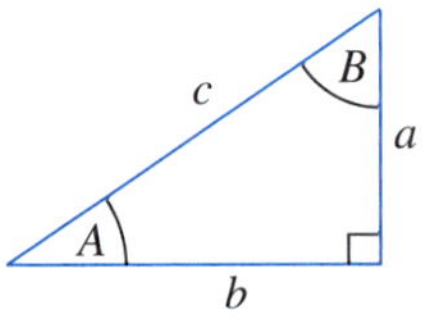

Figure 6.70

9. $a = 12.0$ and $A = 37°$
10. $a = 23.0$ and $A = 74°$
11. $b = 5.6$ and $A = 54.3°$
12. $b = 19.5$ and $A = 20.1°$
13. $c = 0.92$ and $B = 49.9°$
14. $c = 0.086$ and $B = 18.7°$
15. $a = 1{,}546$ and $B = 9.15°$
16. $a = 2{,}666$ and $B = 81.07°$
17. $c = 54.40$ and $A = 53.125°$
18. $c = 81.60$ and $A = 32.375°$
19. $b = 1.002$ and $B = 5.00°$
20. $b = 1.321$ and $B = 3.00°$
21. $a = 15.0$ and $c = 23.0$
22. $a = 6.0$ and $c = 17.0$
23. $a = 6.0$ and $b = 7.0$
24. $a = 52.1$ and $b = 29.6$
25. $b = 0.123$ and $c = 0.456$
26. $b = 0.024$ and $c = 0.067$

27. A cable runs from the top of a building to a point on the ground 66.5 feet from the base of the building. If the cable makes an angle of 43.9° with the ground (as shown in Figure 6.71), find

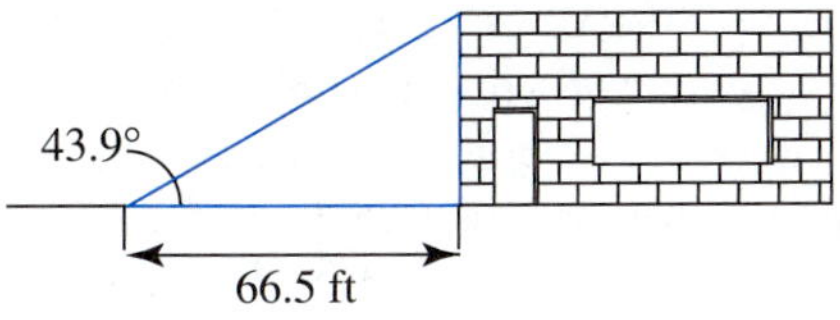

Figure 6.71

a. the height of the building
b. the length of the cable
Round off answers to the nearest tenth of a foot.

28. A support cable runs from the top of a telephone pole to a point on the ground 42.7 feet from its base. If the cable makes an angle of 29.6° with the ground (as shown in Figure 6.72), find

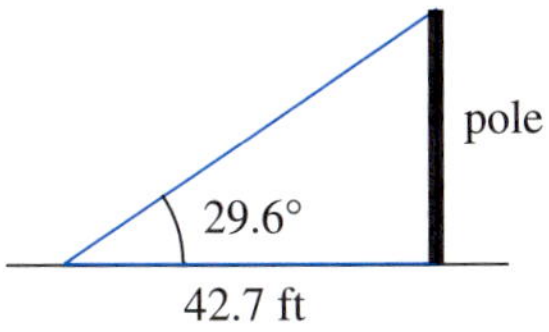

Figure 6.72

a. the height of the pole
b. the length of the cable
Round off answers to the nearest tenth of a foot.

29. A 20-foot-high diving tower is built on the edge of a large swimming pool. From the tower, you measure the angle of depression of the far edge of the pool. If the angle is 7.6°, how wide (to the nearest foot) is the pool?

30. You are at the top of a lighthouse and measure the angle of depression of an approaching ship. If the lighthouse is 90 feet tall and the angle of depression of the ship is 4.12°, how far from the lighthouse is the ship? Round your answer to the nearest multiple of ten.

31. The "pitch" of a roof refers to the vertical rise measured against a standard horizontal distance of 12 inches. If the pitch of a roof is 4 in 12 (the roof rises 4 inches for every 12 horizontal inches), find the acute angle the roof makes with a horizontal line. Round your answer to one decimal place.

32. If the pitch of a roof is 3 in 12, find the acute angle the roof makes with a horizontal line. Round your answer to one decimal place. See Exercise 31.

33. A bell tower is known to be 48.5 feet tall. You measure the angle of elevation of the top of the tower. How far from the tower (to the nearest tenth of a foot) are you if the angle is
a. 15.4°?
b. 61.2°?

34. You are standing on your seat in the top row of an outdoor sports stadium, 200 feet above ground level. You see your car in the parking lot below and measure its angle of depression. How far (to the nearest foot) from the stadium is your car if the angle is
a. 73.5°?
b. 22.9°?

35. A billboard is on top of a building. You are 125.0 feet from the building. You measure the angle of elevation of the bottom of the billboard as 33.4° and the angle of elevation of the top of the billboard as 41.0°. How tall is the billboard? Round your answer to one decimal place.

36. The distance between two buildings is 180 feet. You are standing on the roof of the taller building. You measure the angle of depression of the top and bottom of the shorter building. If the angles are 36.7° and 71.1°, respectively, find
a. the height of the taller building
b. the height of the shorter building
Round your answers to the nearest foot.

37. You are hiking along a river and see a tall tree on the opposite bank. You measure the angle of elevation of the top of the tree and find it to be 61.0°. You then walk 50 feet directly away from the tree and measure the angle of elevation. If the second measurement is 49.5° (see Figure 6.73), how tall is the tree? Round your answer to the nearest foot.

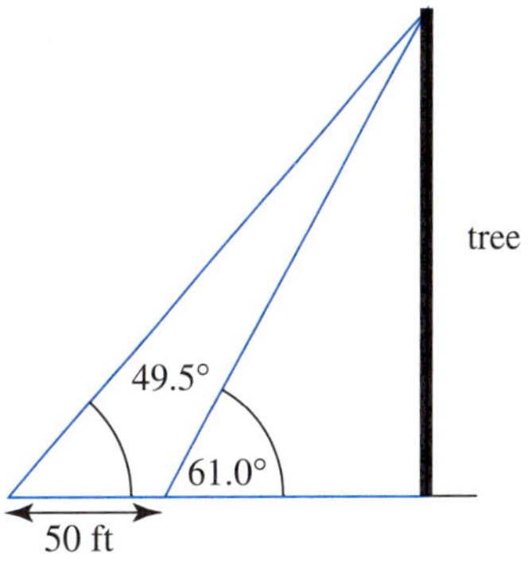

Figure 6.73

38. While sightseeing in Washington D.C., you visit the Washington Monument. From an unknown distance, you measure the angle of elevation of the top of the monument. You then move 100 feet backward (directly away from the monument) and measure the angle of elevation of the top of the monument. If the angles are 61.6° and 54.2°, respectively, how tall (to the nearest foot) is the Washington Monument?

39. While sightseeing in St. Louis, you visit the Gateway Arch. Standing near the base of the structure, you measure the angle of elevation of the top of the arch. You then move 120 feet backward and measure the angle of elevation of the top of the arch. If the angles are 73.72° and 64.24°, respectively, how tall (to the nearest foot) is the Gateway Arch?

Jefferson National Expansion Memorial/National Park Service

40. While sightseeing in Seattle, you visit the Space Needle. From a point 175 feet from the structure, you measure the angle of elevation of the top of the structure. If the angle is 73.87°, how tall (to the nearest foot) is the Space Needle?

The Granger Collection, NY

41. The Sears Tower (in Chicago) is the tallest building in the United States. From a point 900 feet from the building, you measure the angle of elevation of the top of the building. If the angle is 58.24°, how tall (to the nearest foot) is the Sears Tower?

42. A tree is growing on a hillside. From a point 100 feet downhill from the base of the tree, the angle of elevation to the base of the tree is 28.3° with an additional 12.5° to the top of the tree. How tall is the tree?

43. A tree is growing on a hillside. From a point 100 feet uphill from the base of the tree, the angle of depression to the top of the tree is 28.3° with an additional 12.5° to the base of the tree. How tall is the tree?

44. Two observers are 500 feet apart. Each measures the angle of elevation of a hot air balloon that lies in a vertical plane passing through their locations; the angles are 53° and 42°. Find the height of the balloon if the observers are on

a. opposite sides of the balloon. See Figure 6.74.
b. the same side of the balloon. See Figure 6.75.

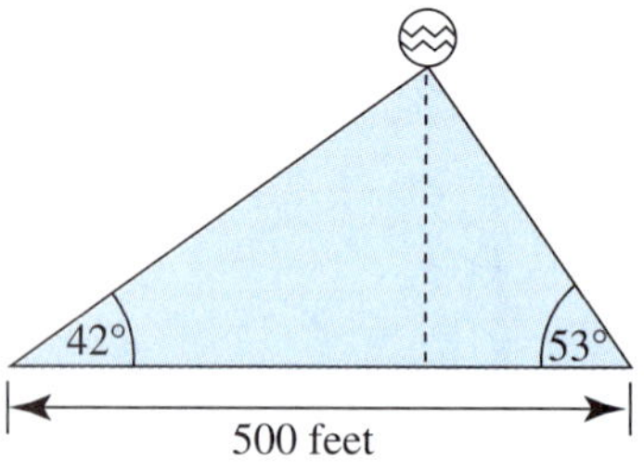

Figure 6.74

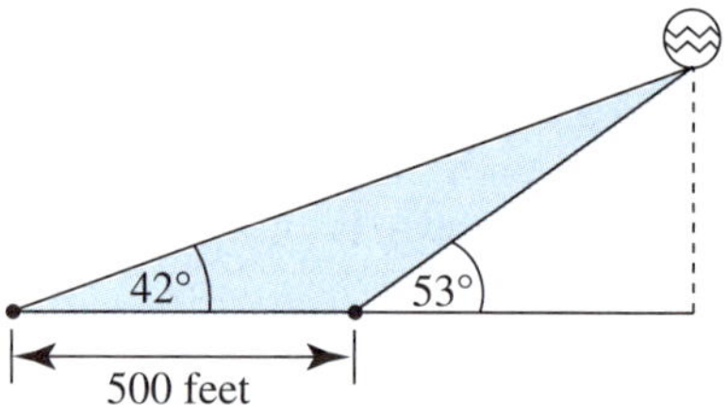

Figure 6.75

Projects

45. The purpose of this project is to calculate the height of a tall building by measuring an angle of elevation.

a. First, we must construct a device to measure an angle of elevation. You will need a protractor, a plastic drinking straw, a bead (or button), and a piece of thread. Attach the straw to the straight edge of the protractor so that the straw lines up with the 0° marks on the protractor. Next, tie the thread to the bead and attach the thread to the center mark of the protractor as shown in Figure 6.76. Note that when the straw is horizontal, the thread is positioned over the 90° mark on the scale. To measure an angle of elevation, look through the straw and focus on the top of the object being measured. Allow the thread to hang vertically, and have an assistant read the larger of the two angles indicated by the thread. (One scale will give an angle greater than 90°, the other less than 90°.) Subtract 90° from the larger angle, and you will obtain the angle of elevation. The angle of elevation indicated by the device shown in Figure 6.76 is $115° - 90° = 25°$.

b. Choose a tall building. Pick an accessible point away from the building, and measure the angle of elevation of the top of the building. Now use a tape measure to determine the distance from your location to the base of the building. Set up the appropriate trigonometric ratio, and calculate the height of the building.

c. Pick a second location, and repeat part (b).

d. Your answers to parts (b) and (c) should be roughly the same. Are they? If not, give possible reasons for the discrepancy.

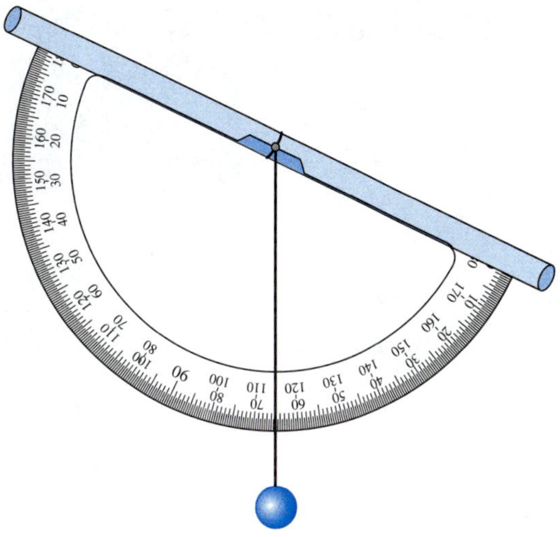

Figure 6.76

6.6 Conic Sections and Analytic Geometry

In order to seek truth it is necessary once in the course of our life to doubt as far as possible all things.

—RENÉ DESCARTES

It is easy to take our system of mathematical notation and algebraic manipulation for granted or even to regard the study of mathematics as an inconvenience and doubt its relevance to our individual lives and goals. René Descartes, one of the great philosophers and mathematicians of the seventeenth century, felt the same way. In doubting the "system" of mathematics of his predecessors, Descartes contributed to the creation of one of the major foundations of modern mathematics: analytic geometry. Simply put, **analytic geometry** is the marriage of algebra and geometry.

Most mathematics from ancient times through the Middle Ages consisted of a verbal description of a geometric method for solving a single problem. (The empirical geometry of the Egyptians is a case in point.) In the Europe of the Middle Ages, "advanced" mathematics was available only to those who knew Latin, as all major works were written in this "language of scholars." Rather than relying on language (which differs from culture to culture), Descartes pioneered the modern notion of using single letters to represent known and unknown quantities. He used the last letters of the alphabet—x, y, and z—to represent variables and the first letters—a, b, and c—to represent constants.

The Greeks verbally defined a parabola to be "the set of all points in a plane equidistant from a given line and a point not on the line"; modern mathematicians, by contrast, favor the equation $y = ax^2 + bx + c$. When we say that the graph of the equation $y = ax^2 + bx + c$ is a parabola, we are operating in the realm of analytic geometry. No doubt your initial exposure to this field was in your beginning or intermediate algebra course, when you were asked to graph an equation such as $y = 2x + 1$ or $y = x^2$.

HISTORICAL NOTE

Hypatia

A.D. 370–415

Hypatia, one of the first women to be recognized for her mathematical accomplishments, lived in Alexandria, one of the largest and most academically prominent cities on the Mediterranean Sea. This Hellenistic city was famed for its university and its library, said to be the largest of its day.

Hypatia's early environment was filled with intellectual challenge and stimulation. Her father, Theon, was a professor of mathematics and director of the museum and library at the University of Alexandria. He gave his daughter a classic education in the arts, literature, mathematics, science, and philosophy. In addition to her studies at the University of Alexandria, Hypatia traveled the Mediterranean. While in Athens, she attended a school conducted by the famed writer Plutarch.

On her return to Alexandria, Hypatia continued in her father's footsteps: She lectured on mathematics and philosophy at the university and directed the museum and library. Her lectures were well received by enthusiastic students and scholars alike; many considered Hypatia to be an oracle. Although none of her writings remains intact, historians attribute several mathematical treatises to Hypatia, including commentaries on the astronomical works of Diophantus and Ptolemy, on the conics of Apollonius, and on the geometry of Euclid. In addition to her insightful lectures and mathematical works, letters written by Hypatia's contemporaries credit her with the invention of devices used in the study of astronomy.

At this time, Alexandria was part of the Roman Empire and was undergoing a power struggle between Christians and pagans (worshippers of Greek and Roman gods). Bishop Cyril was using his position in the Christian church to usurp the power of the Alexandrian government, which was under the rule of the Roman prefect Orestes. Orestes was known to have attended many of Hypatia's lectures and was believed to have been Hypatia's lover. Since Hypatia was a symbol of classic Greek culture, Bishop Cyril associated her with paganism and viewed her as a threat to his quest for Christian power.

In a frenzied attempt to eradicate the pagan influences of Alexandria, a Christian mob incited by Bishop Cyril attacked Hypatia while she was riding in her chariot. She was stripped naked and dragged through the streets, then tortured and murdered.

Hypatia was the last symbol of the ancient culture of Alexandria. She preserved and carried forward the knowledge and wisdom of the Golden Age of Greek civilization. It is tragic that her intelligence and devotion led to her violent death.

Analytic geometry was not created overnight. Many individuals from diverse cultures made significant contributions. Some historians take the easy way out and say that Descartes invented it, but the story of the creation of analytic geometry is much more complicated than that. Because of its intimate association with the creation of calculus, the development of analytic geometry is discussed in Chapter 10.

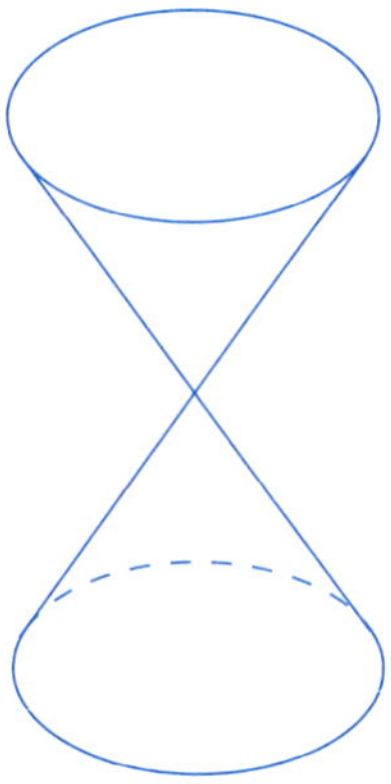

Figure 6.77
A double cone

Conic Sections

A **conic section** is the figure formed when a plane intersects a cone. What do the cross sections of a cone look like? The ancient Greeks studied this problem extensively. Recall that their approach was verbal, involving statements such as "a parabola is the set of all points in a plane equidistant from a given line and a point not on the line" or "a circle is the set of all points in a plane equidistant from a given point." When we study the various conic sections, we use a "double cone" like the one shown in Figure 6.77. Depending on the inclination of the plane of intersection, a circle, a parabola, ellipse, or hyperbola is formed. (See Figure 6.78.)

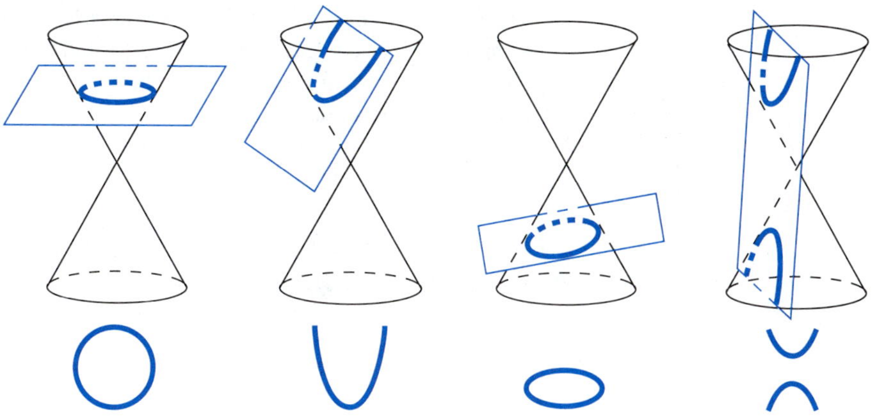

Figure 6.78

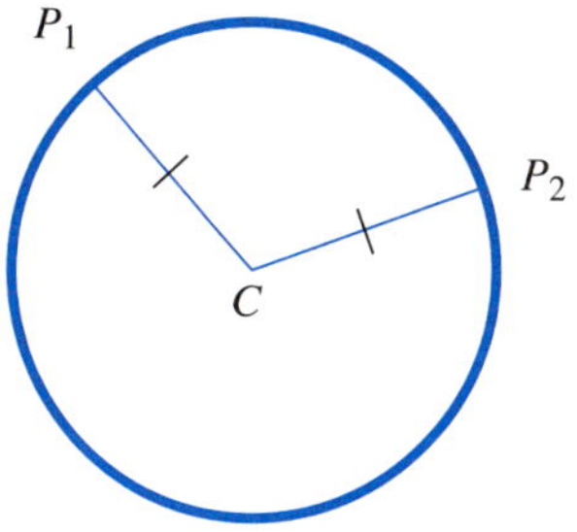

Figure 6.79

The Circle

When a cross section is taken parallel to the base of the cone, as in Figure 6.78(a), a **circle** is obtained. The Greeks defined a circle as "the set of all points in a plane equidistant from a given fixed point." That is, no matter which point P on the circle you select, its distance from the fixed point (the center) is always the same, as shown in Figure 6.79.

Using modern algebra, a circle can also be described as the graph of an algebraic equation, as shown in Figure 6.80.

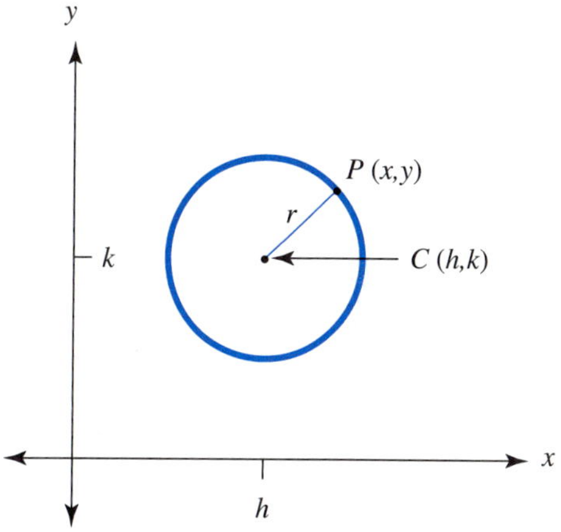

Figure 6.80
$(x - h)^2 + (y - k)^2 = r^2$

Equation of a Circle

A **circle** of radius r, centered at the point (h, k), is the set of all points satisfying the equation

$$(x - h)^2 + (y - k)^2 = r^2$$

EXAMPLE 1 Find the center and radius of the circle $x^2 + y^2 + 6x - 8y + 21 = 0$.

Solution To find the coordinates of the center (that is, to find h and k), we must put the given equation in the form $(x - h)^2 + (y - k)^2 = r^2$. The expressions $(x - h)^2$ and $(y - k)^2$ are called *perfect squares*; each is the square of a binomial.

First, group the terms containing x and those containing y and put the constant on the right side of the equation:

$$x^2 + y^2 + 6x - 8y + 21 = 0$$
$$(x^2 + 6x) + (y^2 - 8y) = -21$$

The given expression $(x^2 + 6x)$ is *not* a perfect square; we must add the appropriate term in order to complete the square. To find this missing term, take half the coefficient of x and square it; we must add $(\frac{6}{2})^2 = 9$. Likewise, we must add $(\frac{-8}{2})^2 = 16$ to complete the square of y.

$$(x^2 + 6x + \boxed{9}) + (y^2 - 8y + \boxed{16}) = -21 + \boxed{9} + \boxed{16}$$
$$(x + 3)^2 + (y - 4)^2 = 4$$
$$[x - (-3)]^2 + (y - 4)^2 = 2^2$$

Therefore, the center is the point $C(-3, 4)$, and the radius is $r = 2$, as shown in Figure 6.81.

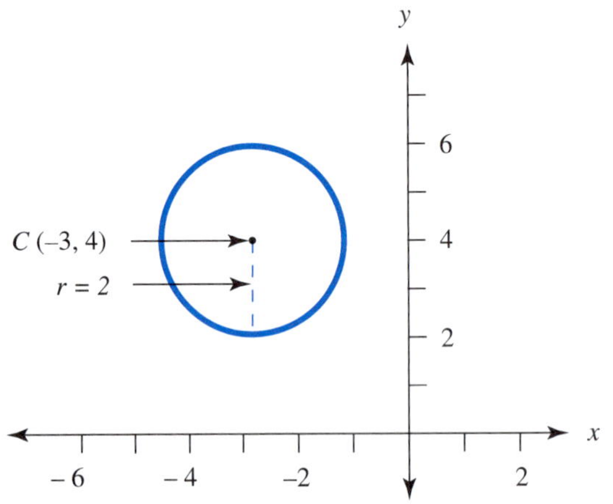

Figure 6.81
$x^2 + y^2 + 6x - 8y + 21 = 0$

■

The Parabola

When a cross section passes through the base of the cone and intersects only one portion, as in Figure 6.78(b), the resulting figure is a **parabola.** The Greeks defined a parabola as "the set of all points in a plane equidistant from a given line and a given point not on the line." No matter which point P on the parabola you select, its distance from the given line (called the *directrix*) is the same as its distance from the given point (called the *focus*), as shown in Figure 6.82. Every parabola can be divided into two equal pieces. The line that cuts the parabola into symmetric halves is called the **line of symmetry.** The point at which the line of symmetry intersects the parabola is called the **vertex** of the parabola.

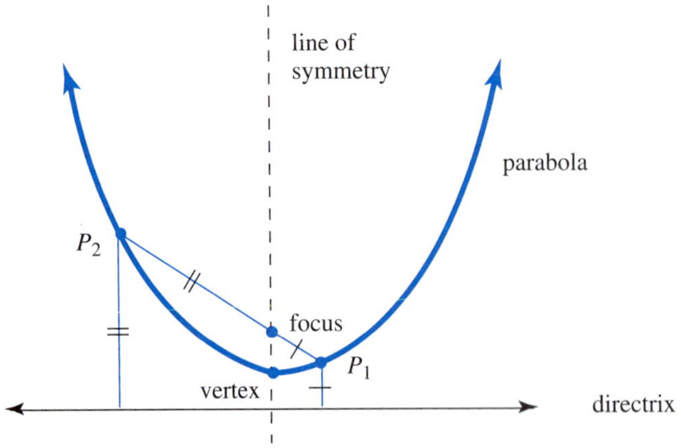

Figure 6.82

Parabolas have many applications. In particular, satellite dish antennas and the dishes used with sports microphones are designed so that their cross sections are parabolas; that is, they are parabolic reflectors. When incoming radio waves, microwaves, or sound waves hit the dish, they are reflected to the focus, where they are collected and converted to electronic pulses. The focusing ability of a parabola is also applied to the construction of mirrors used in telescopes and in solar heat collection. (See Figure 6.83.)

Our study of parabolas concentrates on locating the focus. In this respect, we will consider only a parabola whose vertex is at the origin (0, 0) of a rectangular coordinate system.

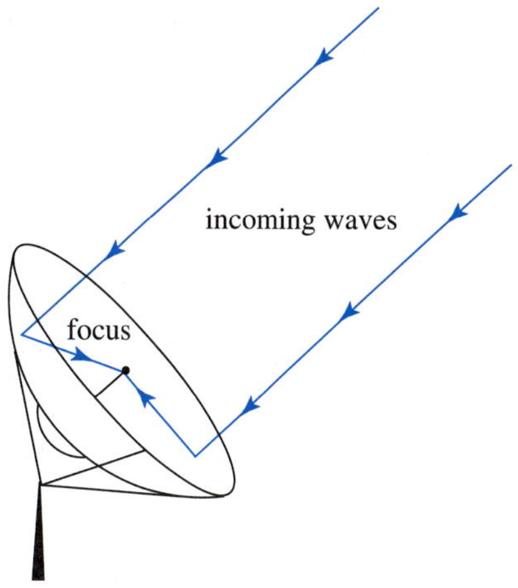

Figure 6.83

A Parabola and Its Focus

A **parabola,** with vertex at the point (0, 0) and **focus** at the point (0, p), is the set of all points (x, y) satisfying the equation

$$4py = x^2$$

A parabola and its focus, located on a rectangular coordinate system, are shown in Figure 6.84.

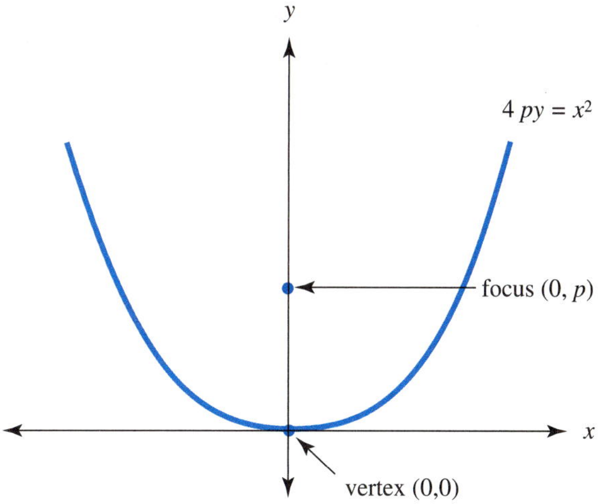

Figure 6.84
A parabola and its focus

EXAMPLE 2 A Peace Corps worker has found a way for villagers to obtain Mylar-coated parabolic dishes that will concentrate the sun's rays for the solar heating of water. If a dish is 7 feet wide and 1.5 feet deep, where should the water container be placed?

Solution The water will be heated most rapidly if the container is placed at the focus. (The incoming solar radiation will be reflected by the parabolic dish and concentrated at the focus of the parabola.)

Draw a parabola with its vertex at the origin. As Figure 6.85 demonstrates, the given dimensions tell us that the point $Q(3.5, 1.5)$ is on the parabola.

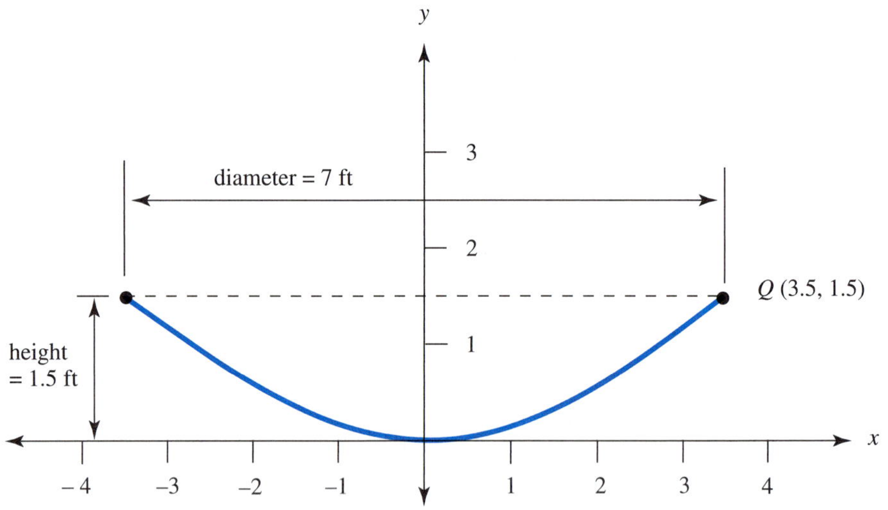

Figure 6.85

Substituting $x = 3.5$ and $y = 1.5$ into the general equation $4py = x^2$, we can find p, the location of the focus:

$$\begin{aligned} 4py &= x^2 \\ 4p(1.5) &= (3.5)^2 \\ 6p &= 12.25 \\ p &= 2.04166666\ldots \\ p &\approx 2 \end{aligned}$$

The focus is located at the point (0, 2). Therefore, the water container should be placed 2 feet above the bottom of the parabolic dish. ■

Besides collecting incoming waves, parabolic reflectors also concentrate outgoing waves. The reflectors in flashlights and automotive headlights are parabolic. If the bulb is positioned at the focus, then the outgoing light will be more intensely concentrated in a forward beam.

The Ellipse

When a cross section of a cone is not parallel to the base and does not pass through the base, as in Figure 6.78(c), the resulting figure is an **ellipse.** An ellipse is a symmetric oval. The Greeks defined an ellipse as "the set of all points in a plane the *sum* of whose distances from two given points is constant." That is, if C_1 and C_2 represent the given points (called the **foci**), then no matter which point P on the ellipse you select, the distance from P to C_1 plus the distance from P to C_2 is always the same, as shown in Figure 6.86.

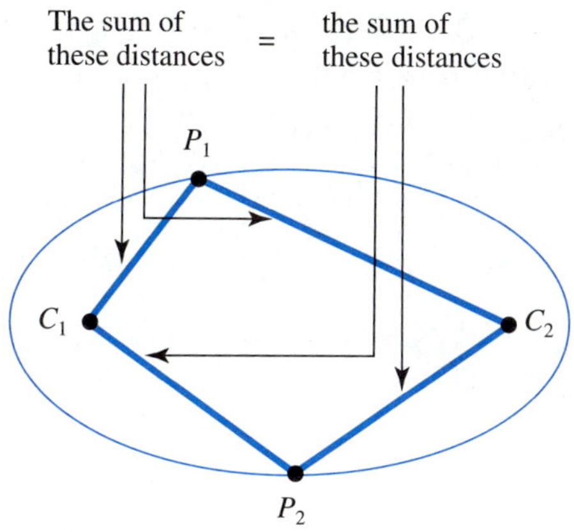

Figure 6.86
Ellipse

Ellipses are commonly used in astronomy. The planets revolve about the sun, each following its own elliptical path with the sun at one of the foci. The elliptical motion of the planets was first hypothesized by Johann Kepler in 1609.

An ellipse has an interesting reflective property. Anything that is emitted from one focus and strikes the ellipse will be directed to the other focus. For example, suppose you had an elliptical billiard table with a hole located at one of the foci. A cue ball located at the other focus would always go in the hole, regardless of the direction in which it was shot. This principle is used in the design of whispering galleries. If the walls and ceiling of a room are elliptical, a person standing at one focus will be able to hear the whispering of someone standing at the other focus; the sound waves are reflected and directed from one focus to the other. The world's most famous whispering gallery is located in St. Paul's Cathedral in London, England, which was completed in 1710. Other galleries are located in the U.S. Capitol in Washington, D.C., in the Mormon Tabernacle in Salt Lake City, Utah, and in Gloucester Cathedral in Gloucester, England (built in the eleventh century).

Although they can be located anywhere in a plane, we will consider only ellipses centered at the origin of a rectangular coordinate system. (See Figure 6.87.) (If the center is not at the origin, the method of completing the square can be used to locate the center, as in Example 1.)

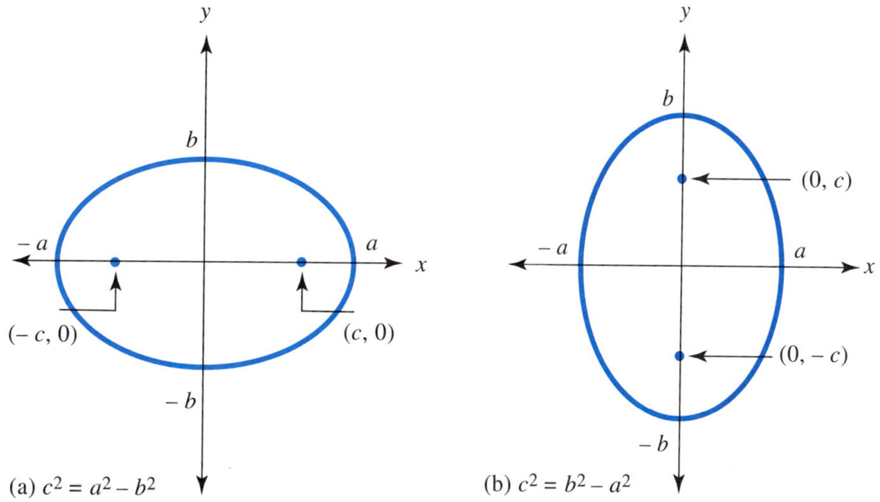

Figure 6.87

> **Equation of an Ellipse**
>
> An **ellipse,** centered at the origin, is the set of all points (x, y) satisfying the equation
>
> $$\frac{x^2}{a^2} + \frac{y^2}{b^2} = 1 \quad \text{(}a\text{ and }b\text{ are positive constants)}$$
>
> The **foci** are located on the longer axis of the ellipse and are found by solving the equation c^2 = difference between a^2 and b^2.

EXAMPLE 3 Sketch the graph and find the foci of the ellipse $9x^2 + 25y^2 = 225$.

Solution First, we put the equation in the general formula $x^2/a^2 + y^2/b^2 = 1$. To obtain this form, we divide each side of the equation by 225 so that the right side will equal 1:

$$9x^2 + 25y^2 = 225$$

$$\frac{9x^2}{225} + \frac{25y^2}{225} = \frac{225}{225} \quad \text{dividing by 225}$$

$$\frac{x^2}{25} + \frac{y^2}{9} = 1 \quad \text{simplifying the fractions}$$

(Note that $a^2 = 25$ and $b^2 = 9$.) We can now sketch the ellipse by finding the x- and y-intercepts.

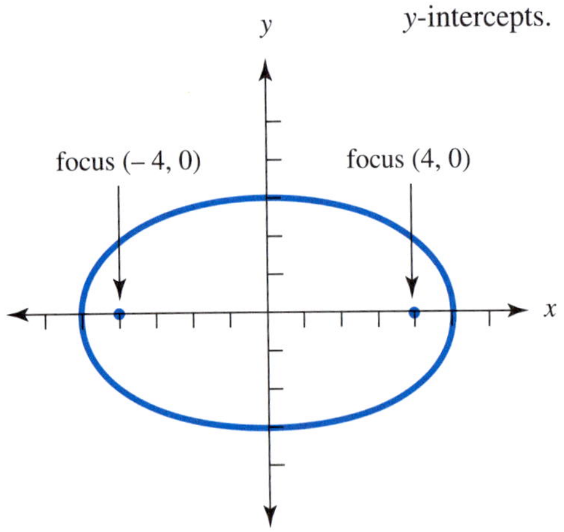

Figure 6.88
$9x^2 + 25y^2 = 225$

Finding the x-Intercepts:

(Substitute $y = 0$ and solve for x.)

$$\frac{x^2}{25} + \frac{0^2}{9} = 1$$

$$\frac{x^2}{25} = 1$$

$$x^2 = 25$$

$$x = \pm 5$$

Finding the y-Intercepts:

(Substitute $x = 0$ and solve for y.)

$$\frac{0^2}{25} + \frac{y^2}{9} = 1$$

$$\frac{y^2}{9} = 1$$

$$y^2 = 9$$

$$y = \pm 3$$

Finding the Foci:

$$c^2 = a^2 - b^2$$

$$= 25 - 9$$

$$= 16$$

Therefore, $c = \pm 4$.

Plotting the intercepts and connecting them with a smooth oval, we sketch the ellipse and its foci as shown in Figure 6.88. ■

The Hyperbola

When a cross section intersects both portions of the double cone, as in Figure 6.78(d), the resulting figure is a **hyperbola.** A hyperbola consists of two separate, symmetric branches. The Greeks defined a hyperbola as "the set of all points in a plane the *difference* of whose distances from two given points is constant." That is, if C_1 and C_2 represent the given points (called the **foci**), then no matter which point P on the hyperbola you select, subtracting the smaller distance (between the point and each focus) from the larger distance always results in the same value, as shown in Figure 6.89.

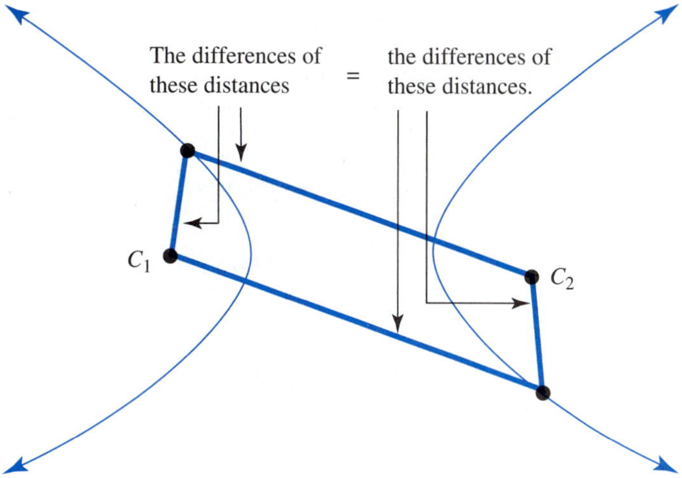

Figure 6.89
Hyperbola

As with the ellipse, we will consider only hyperbolas centered at the origin of a rectangular coordinate system, as shown in Figure 6.90.

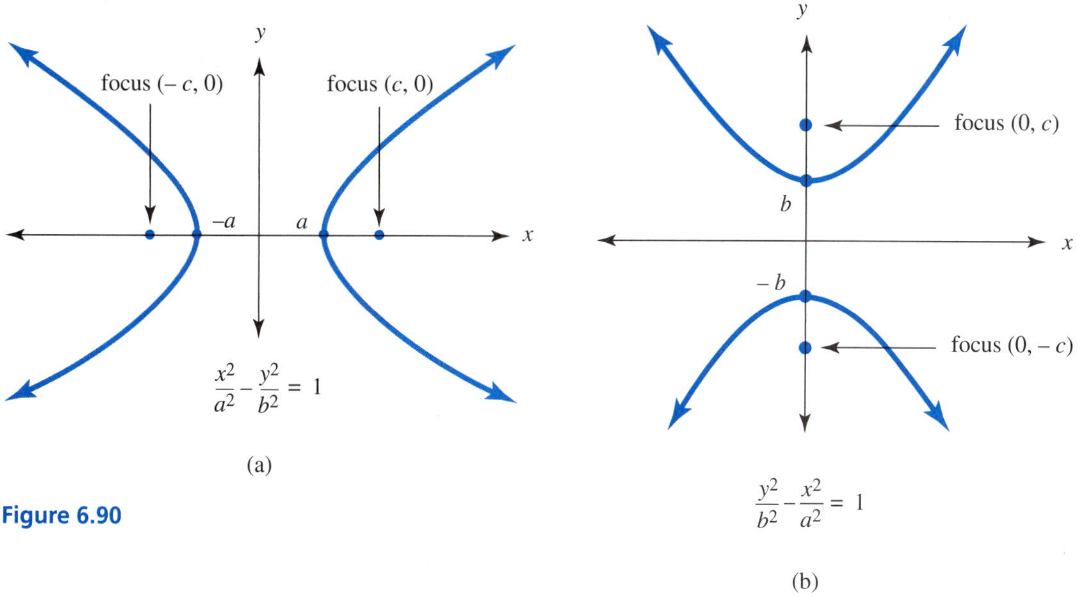

Figure 6.90

Equation of a Hyperbola

A **hyperbola,** centered at the origin, is the set of all points (x, y) satisfying either the equation

$$\frac{x^2}{a^2} - \frac{y^2}{b^2} = 1 \quad \text{or} \quad \frac{y^2}{b^2} - \frac{x^2}{a^2} = 1 \qquad \text{(}a\text{ and }b\text{ are positive constants)}$$

The **foci** are found by solving the equation $c^2 = a^2 + b^2$.

EXAMPLE 4 Sketch the graph and find the foci of the hyperbola $y^2 - 4x^2 = 4$.

Solution First, put the equation in the general form, with the right side of the equation equal to 1:

$$y^2 - 4x^2 = 4$$

$$\frac{y^2}{4} - \frac{4x^2}{4} = \frac{4}{4} \quad \text{dividing by 4}$$

$$\frac{y^2}{4} - \frac{x^2}{1} = 1 \quad \text{simplifying the fractions}$$

(Note that $a^2 = 1$ and $b^2 = 4$.)

Finding the x-Intercepts:

(Substitute $y = 0$ and solve for x.)

$$\frac{0^2}{4} - \frac{x^2}{1} = 1$$

$$-\frac{x^2}{1} = 1$$

$$x^2 = -1$$

no solution; no x-intercepts

Finding the y-Intercepts:

(Substitute $x = 0$ and solve for y.)

$$\frac{y^2}{4} - \frac{0^2}{1} = 1$$

$$\frac{y^2}{4} = 1$$

$$y^2 = 4$$

$$y = \pm 2$$

Because there are no x-intercepts, the branches of the hyperbola open up and down. (Whenever the y^2 term comes first in the general equation, the branches of the hyperbola will open up and down.)

The values of a and b determine the shape of the hyperbola. Locate the points $a = \pm 1$ on the x-axis. Locate the y-intercepts $b = \pm 2$.

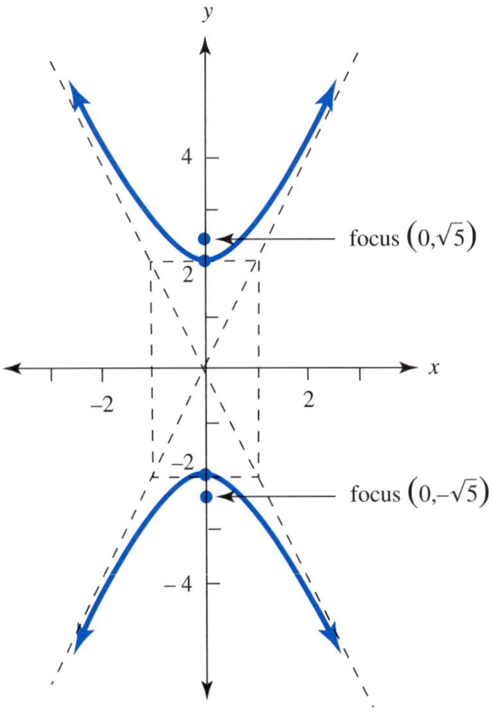

Figure 6.91
$y^2 - 4x^2 = 4$

Using a dotted line, draw the rectangle (with sides parallel to the axes) formed by these points. Lightly draw in the diagonals of this rectangle. Now draw in the two branches of the hyperbola, as shown in Figure 6.91.

Finding the Foci:

$$c^2 = a^2 + b^2$$
$$= 1 + 4 = 5$$

Therefore, $c = \pm\sqrt{5}$ (≈ 2.24). The foci are the points $(0, \sqrt{5})$ and $(0, -\sqrt{5})$. ■

The most common application of the hyperbola is in radio-assisted-navigation, or LORAN (LOng-RAnge Navigation). Various national and international transmitters continually emit radio signals at regular time intervals. By measuring the *difference* in reception times of a pair of these signals, the navigator of a ship at sea knows that the vessel is somewhere on the hyperbola that has the transmitters as foci. Using a second pair of transmitters, the navigator knows that the ship is somewhere on a second hyperbola. By plotting both hyperbolas on a map, the navigator knows that the ship is located at the intersection of the two hyperbolas, as shown in Figure 6.92.

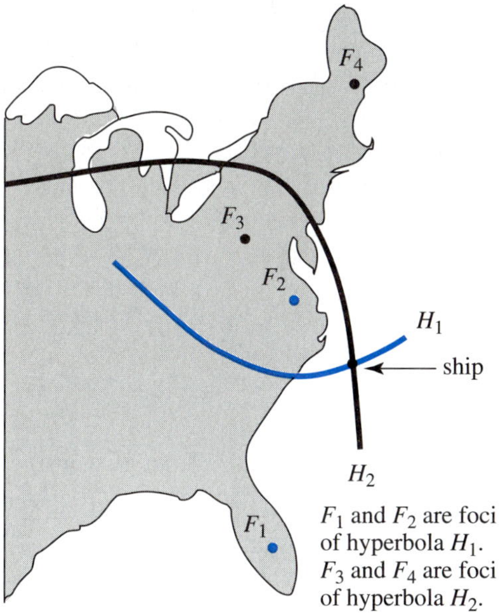

Figure 6.92

6.6 Exercises

In Exercises 1–8, find the center and radius of the given circle and sketch its graph.

1. $x^2 + y^2 = 1$
2. $x^2 + y^2 = 4$
3. $x^2 + y^2 - 4x - 5 = 0$
4. $x^2 + y^2 + 8y + 12 = 0$
5. $x^2 + y^2 - 10x + 4y + 13 = 0$
6. $x^2 + y^2 + 2x - 6y - 15 = 0$
7. $x^2 + y^2 - 10x - 10y + 25 = 0$
8. $x^2 + y^2 + 8x + 8y + 16 = 0$

In Exercises 9–12, sketch the graph of the given parabola and find its focus.

9. $y = x^2$
10. $y = 2x^2$
11. $y = \frac{1}{2}x^2$
12. $y = \frac{1}{4}x^2$

13. A Mylar-coated parabolic reflector dish (used for the solar heating of water) is 9 feet wide and $1\frac{3}{4}$ feet deep. Where should the water container be placed to heat the water most rapidly?

14. A satellite dish antenna has a parabolic reflector dish that is 18 feet wide and 4 feet deep. Where is the focus located?

15. A flashlight has a diameter of 3 inches. If the reflector is 1 inch deep, where should the light bulb be located to concentrate the light in a forward beam?

16. An automotive headlight has a diameter of 7 inches. If the reflector is 3 inches deep, where should the light bulb be located to concentrate the light in a forward beam?

In Exercises 17–24, sketch the graph and find the foci of the given ellipse.

17. $4x^2 + 9y^2 = 36$
18. $x^2 + 4y^2 = 16$
19. $25x^2 + 4y^2 = 100$
20. $x^2 + 9y^2 = 9$
21. $16x^2 + 9y^2 = 144$
22. $4x^2 + y^2 = 36$
23. $4x^2 + y^2 - 8x - 4y + 4 = 0$

HINT: Complete the square for x and for y.

24. $25x^2 + 9y^2 + 150x - 90y + 225 = 0$

HINT: Complete the square for x and for y.

25. An elliptical room is designed to function as a whispering gallery. If the room is 30 feet long and 24 feet wide, where should two people stand to optimize the whispering effect?

26. An elliptical billiard table is 8 feet long and 5 feet wide. Where are the foci located?

27. During the earth's elliptical orbit around the sun (the sun is located at one of the foci), the earth's greatest distance from the sun is 94.51 million miles, and the shortest distance is 91.40 million miles. Find the equation (in the form $x^2/a^2 + y^2/b^2 = 1$) of the earth's elliptical orbit.

HINT: Use $a + c =$ greatest distance and $a - c =$ shortest distance, as shown in Figure 6.93, to solve for a and then c; then find b.

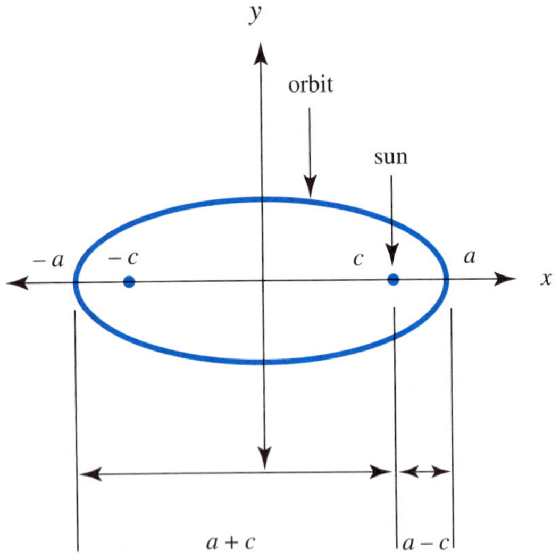

Figure 6.93

28. During Mercury's elliptical orbit around the sun (the sun is located at one of the foci), the planet's greatest distance from the sun is 43.38 million miles, and the shortest distance is 28.58 million miles. Find the equation (in the form $x^2/a^2 + y^2/b^2 = 1$) of Mercury's elliptical orbit.

HINT: Use $a + c =$ greatest distance and $a - c =$ shortest distance, as shown in Figure 6.93, to solve for a and c; then find b.

In Exercises 29–36, sketch the graph and find the foci of the given hyperbola.

29. **a.** $x^2 - y^2 = 1$
b. $y^2 - x^2 = 1$
30. **a.** $x^2 - y^2 = 4$
b. $y^2 - x^2 = 4$
31. $4x^2 - 9y^2 = 36$
32. $x^2 - 4y^2 = 16$
33. $25y^2 - 4x^2 = 100$
34. $y^2 - 9x^2 = 9$
35. $x^2 - 4y^2 - 6x + 5 = 0$

HINT: Complete the square for x and for y.

36. $9y^2 - x^2 - 36y - 2x + 26 = 0$

HINT: Complete the square for x and for y.

Answer the following questions using complete sentences and your own words.

History Questions

37. Who is one of the first women to be mentioned in the history of mathematics? Why was she murdered?

38. Write a research paper on a historical topic referred to in this section or on a related topic. Following is a list of possible topics:

- Apollonius of Perga (What contributions did he make to the study of conic sections?)
- Hypatia (Discuss her accomplishments.)
- René Descartes (What contributions did he make to the study of conic sections?)
- Edmond Halley (What is Halley's comet? How did Halley predict its arrival?)
- Johann Kepler (How did he develop his theory of elliptical orbits?)
- LORAN (How is LORAN used in navigation? When and where did it originate?)
- parabolic reflectors and antennas (How have parabolic reflectors changed the dissemination of information around the world?)

6.7 Non-Euclidean Geometry

Mathematics is that subject in which we do not know what we are talking about, or whether what we are saying is true.

—Bertrand Russell

From the time of Thales of Miletus, mathematics has been based on the axiomatic system of deduction and proof. Euclid's monumental work, *Elements*, was the epitome of this effort. By the careful choice of five geometric postulates (things that are assumed to be true), Euclid proceeded to build the entire structure of all geometric knowledge of the day. The ancient Greeks' attitude was "the truth of these five axioms is obvious; therefore, everything that follows from them is true also." Over 2,000 years and many headaches, heartbreaks, and humiliations later, the attitude of a current mathematician is more "*if* we *assume* these axioms to be valid, then everything that follows from them is valid also." While the Greeks considered geometry (and all of mathematics) to be a system of absolute truth, Bertrand Russell's twentieth-century comment implies that results that are based on *assumptions* cannot be considered true or false; the most one can hope for is *consistency*.

The Parallel Postulate

Euclid's empire was based on his perception of reality. This reality in turn consisted of five axiomatic concepts so simple and so obvious that no one could deny that they were in fact true. After all, reality is what is confirmed by experience, and experience had never shown the following statements to be false:

1. A straight line can be drawn from any point to any other point.
2. A (finite) straight line segment can be extended continuously in an (infinite) straight line.
3. A circle may be described with any center and any radius.
4. All right angles are equal to one another.
5. Given a line and a point not on that line, there is one and only one line through the point parallel to the original line.

Taking the axiomatic method to heart, scholars scrutinized even the five "obvious" postulates for any possible overlap or dependence. Could one of the five be deduced logically from the others? The search was on for the smallest possible set of postulates that could serve as the basis of all geometry.

The most likely candidate for a dependent postulate was number 5, the so-called **Parallel Postulate** (see Figure 6.94). From the beginning, it aroused the curiosity and inquiry of mathematicians the world over. Many "proofs" of this postulate were put forth, but each was dismissed upon further scrutiny by later investigations. Each "proof" had its flaw, its logical demise. Even Euclid himself had tried to prove it but couldn't, and he consequently called it a postulate (it was obviously true).

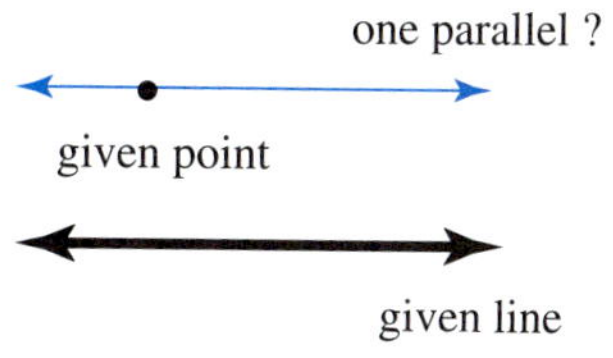

Figure 6.94

The more mathematicians studied the postulate, the more elusive it became. Indeed, the competition to prove the Parallel Postulate became so intense that the French mathematician d'Alembert called it "the scandal of elementary geometry." On one occasion, the highly admired Joseph-Louis Lagrange presented a paper on the postulate to the French Academy. However, halfway through his presentation, he stopped, with the comment "I must meditate further on this." He put the paper away and never spoke of it again.

Girolamo Saccheri

In all the furor to prove the Parallel Postulate, one particular attempt is worth mentioning. In 1733, a Jesuit priest named Girolamo Saccheri published *Euclides ab Omni Naevo Vindicatus* (*Euclid Vindicated of Every Blemish*). This work contained Saccheri's "proof" of the Parallel Postulate. Saccheri was a professor of mathematics, logic, and philosophy at the University of Pavia, Italy. He is said to have had an incredible memory; commonly mentioned was his ability to play three games of chess simultaneously, without looking at the chessboards!

Saccheri was the first to examine the consequences of assuming the Parallel Postulate to be false; that is, he attempted a proof by contradiction. *Reductio ad absurdum*, or proof by contradiction, is an indirect means of proving a statement. Simply put, to prove something, you assume it to be false. If you can show that this assumption leads to a contradiction, then the assumed falsity of the statement is itself false, implying that the statement is true. This is an acceptable line of reasoning that has been used since the days of classic Greek logic. Euclid himself used reductio ad absurdum in *Elements*, though not in reference to the Parallel Postulate.

In Saccheri's quest to prove the Parallel Postulate, he assumed it to be false, thereby hoping to obtain some sort of logical contradiction. The denial of the Parallel Postulate consists of two alternatives:

1. Given a line and a point not on the line, there are no lines through the point parallel to the original line (parallels don't exist).
2. Given a line and a point not on the line, there are at least two lines through the point parallel to the original line.

Surely, each of these assumptions would lead to a contradiction. To his surprise, Saccheri couldn't find one! How could this be? Were the foundations of Euclid's world built on rock or on sand? Rather than accepting the implications of his work, Saccheri manufactured a spurious contradiction in order to be able to proclaim support of the Parallel Postulate.

Saccheri's work had little impact at the time. The implication that the Parallel Postulate was independent of the other four postulates (that is, that it could not be proved) was too far out of line for the thinking of the day. Mathematicians were so convinced of the dependence of the Parallel Postulate that little interest was given to any other alternative. However, less than 100 years later, three men (Carl Gauss, Janos Bolyai, and Nikolai Lobachevsky) independently resurrected the denial of the Parallel Postulate and took roads similar to the one that had led Saccheri to question the "truth" of Euclidean geometry.

Carl Friedrich Gauss

Carl Friedrich Gauss

Carl Friedrich Gauss, the "Prince of Mathematicians," first became interested in the study of parallels in 1792, at the age of fifteen (see Historical Note on page 260.). Although he never published any theory that went counter to Euclid's Parallel Postulate, his diaries and letters to close friends over the years affirmed his realization that geometries based on axioms different from Euclid's could exist.

Having failed to prove the Parallel Postulate, Gauss began to ponder its sanctity. According to his diaries, he put forth an axiom that contradicted Euclid's; specifically, he assumed that more than one parallel could be drawn through a point not on a line. Rather than trying to obtain a contradiction as Saccheri had attempted (and thus prove the Parallel Postulate via reductio ad absurdum), Gauss began to see that a geometry that was at odds with Euclid's but internally consistent could be developed. However, Gauss never went public; he didn't want to face the inevitable public controversy.

Janos Bolyai

Janos Bolyai was a Hungarian army officer and the son of a respected mathematician. From an early age, Janos was exposed to the world of mathematics. Janos's father, Wolfgang, had studied with Gauss at Göttingen, and their sporadic correspondence lasted a lifetime. In fact, in their correspondence, Gauss had pointed out the fallacy in the elder Bolyai's "proof" of the Parallel Postulate.

In 1817, Janos entered the Imperial Engineering Academy in Vienna at the age of fifteen. After completing his studies in 1823, he embarked upon a military career. An expert fencer with numerous successful duels to his credit, Bolyai was also an accomplished violinist.

The early teachings of his father prompted Bolyai to continue his study of mathematics. However, when Janos informed his father of his interest in the study of parallels, the elder Bolyai wrote, "Do not waste an hour's time on that problem. It does not lead to any result; instead it will come to poison all your life." Ignoring his father's decree, the younger Bolyai pursued the elusive Parallel Postulate.

After several failed attempts to prove the Parallel Postulate, Bolyai, too, proceeded down the path of denying Euclid's assumption of the existence of a unique parallel. In a manner not unlike that of Gauss, even though he was unaware of Gauss's work, Bolyai assumed that more than one parallel existed through a point not on a line. He developed several theorems in this new, consistent geometry. He was so excited over his discovery that he wrote, "Out of nothing I have created a strange new world."

Hastening to publish his work, the younger Bolyai published his theory as a 24-page appendix to the elder Bolyai's two-volume *Tentamen Juventutem Studiosam in Elementa Matheseos Purae* (*An Attempt to Introduce Studious Youth to the Elements of Pure Mathematics*). Although the book is dated 1829, it was actually printed in 1832.

When Wolfgang Bolyai sent a copy to his old friend Carl Gauss, the reply was less than heartening to Janos. Gauss responded, "If I begin by saying that I dare not praise this work, you will of course be surprised for a moment; but I cannot do otherwise. To praise it would amount to praising myself. For the entire content of the work, the approach which your son has taken, and the results to which he is led, coincide almost exactly with my own meditations which have occupied my mind for the past thirty or thirty-five years. It was my plan to put it all down on paper eventually, so that at least it would not perish with me. So I am greatly surprised to be spared this effort, and am overjoyed that it happens to be the son of my old friend who outstrips me in such a remarkable way."

Not seeing the true compliment that the great Gauss had bestowed on him, Bolyai feared that his work was being stolen. His feelings were compounded by the total lack of interest on the part of other mathematicians, and deep periods of depression set in. Janos Bolyai never published again.

Nikolai Lobachevsky

Nikolai Lobachevsky

Labeling it "imaginary geometry," Nikolai Ivanovitch Lobachevsky published the first complete text on non-Euclidean geometry in 1829. Many have since heralded him as the Copernicus of geometry. Just as Copernicus had challenged the long-established theory that the earth was the center of the universe, Lobachevsky's alternative view of geometry was in direct contradiction to the long-revered work of Euclid. Referring to Lobachevsky, Einstein said, "He dared to challenge an axiom."

КАЗАНСКІЙ
ВѢСТНИКЪ,
ИЗДАВАЕМЫЙ
ПРИ
ИМПЕРАТОРСКОМЪ
КАЗАНСКОМЪ УНИВЕРСИТЕТѢ.
ЧАСТЬ XXV.
КНИЖКА II. и III.
ФЕВРАЛЬ и МАРТЪ,
1829 года.
Печатано въ Университетской Типографіи.

Lobachevsky's *On the Foundations of Geometry* was published in this 1829 issue of the academic journal of the University of Kazan.

Lobachevsky spent most of his life at the University of Kazan, near Siberia. Founded by Czar Alexander I in 1804, Kazan was Europe's easternmost center of higher education. Among its first students, Lobachevsky received his master's degree in mathematics and physics in 1811 at the age of eighteen. He remained at Kazan teaching courses for civil servants until 1816, when he was promoted to full professorship.

Geometry received Nikolai's special attention. Being inquisitive, Lobachevsky made several attempts to prove the Parallel Postulate; he failed at each. He then proceeded to examine the consequences of substituting an alternative to Euclid's postulate of a unique parallel; Lobachevsky assumed that more than one parallel could be drawn through a point. Living in distant isolation from the learning capitals of Europe, Lobachevsky was unaware of the similar approaches taken by Gauss and Bolyai.

Lobachevsky's idea of a geometry based on an axiom in opposition to Euclid began to take form in 1823, when he drew up an outline for a geometry course he was teaching. In 1826, he gave a lecture and presented a paper incorporating his belief in the feasibility of a geometry based on axioms different from Euclid's. Although this paper has been lost (as have so many in the history of mathematics), it was the first recorded attempt to breach Euclid's bastion. In 1829, the monthly academic journal of the University of Kazan printed a series of his works titled *On the Foundations of Geometry*; this publication is considered by many to be the official birth of the radically new, non-Euclidean geometry.

As with anything new and at odds with the status quo, Lobachevsky's work was not embraced with open arms. The St. Petersburg Academy rejected it for publication in its scholarly journal and printed an uncomplimentary review. In contrast to Gauss, who did not have the courage to print, and Bolyai, who did not have the fortitude to face his opponents, Lobachevsky remained undaunted. He proceeded in his work, expanding *Foundations* into *New Elements of Geometry, with a Complete Theory of Parallels*, which was also published in Kazan's academic journal of 1835. Shortly thereafter, his work began to be recognized outside of Kazan; he was published in Moscow, Paris, and Berlin.

In 1846, when Gauss received a copy of Lobachevsky's latest book, *Geometrical Investigations on the Theory of Parallels* (which contained only 61 pages), he wrote to a colleague, "I have had occasion to look through again that little volume by Lobachevsky. You know that for fifty-four years now I have held the same conviction. I have found in Lobachevsky's work nothing that is new to me, but the development is made in a way different from that which I have followed, and certainly by Lobachevsky in a skillful way and a truly geometrical spirit." However, Gauss did not give public approval to Lobachevsky's work.

As is the case with those who pioneer ideas and art forms that are incomprehensible to the world at large, the radicals of geometry, Lobachevsky and Bolyai, never received full recognition of the value of their work during their lifetimes. However, they opened the door for a host of new ideas in geometry and in the axiomatic system in general.

Bernhard Riemann

Bernhard Riemann

The groundbreaking work of Lobachevsky and Bolyai was ignored for many years. The mathematician who finally convinced the academic world of the merits of non-Euclidean geometry (that is, geometries based on the denial of Euclid's Parallel Postulate) was born in 1826, at the time when Lobachevsky and Bolyai were initially presenting their ideas. Although his life was cut short (he died of tuberculosis at the age of thirty-nine), Georg Friedrich Bernhard Riemann's contributions to the world of modern mathematics were monumental.

At the age of nineteen, Bernhard Riemann enrolled at the University of Göttingen with the intention of pursuing the study of theology and philosophy. However, he became so interested in the study of mathematics that he devoted his life entirely to that field. As a graduate student at Göttingen, Riemann studied under Gauss, who was unusually impressed with his protégé's abilities. In his report to the faculty, Gauss said that Riemann's dissertation came from "a creative, active, truly mathematical mind, and of a gloriously fertile originality." Coming from the Prince of Mathematicians, that was a compliment indeed!

After receiving his degree, Riemann wanted to stay at Göttingen as a member of the faculty. To prove himself, he first had to deliver a "probationary" lecture to his former teachers. When he submitted three possible topics for his oration, Gauss chose the third, "On the Hypotheses That Underlie the Foundation of Geometry." Delivered in 1854 and published in 1868 (two years after his death), Riemann's lecture is considered by many to be one of the highlights of modern mathematical history.

Unlike Gauss, Lobachevsky, and Bolyai (each of whom assumed that through a point not on a line more than one parallel existed), Riemann denied the existence of all parallel lines! He philosophized that we could just as well assume that all lines eventually intersect as assume that some (parallel lines) do not intersect.

Geometric Models

One reason for the initial resistance to non-Euclidean geometry was practical experience. In our perception of the "flatness" of the immediate world in which we live, we naturally accept the Parallel Postulate as true. It is easy to envision a rectangular grid of city blocks with straight streets that do not intersect. However, our world is not flat; the earth is spherical. An age-old problem associated with map making has been the projection of the curved earth (a globe) into a plane (a flat map) and vice versa. (What happens if we try to peel the "skin" off of a globe or stretch a piece of rectangular graph paper around a sphere?)

Granted, non-Euclidean geometries run counter to common sense. To understand them, we need to draw pictures; we must create models. Each model will have its own unique definition of a line and a plane. In all cases, **parallel lines** are lines in a plane that do not intersect one another. Geometries can be categorized by three types, depending on what is assumed concerning parallel lines:

1. *Euclidean geometry*: Through a point not on a given line, there exists exactly one line parallel to the given line (Euclid's Parallel Postulate).
2. *Lobachevskian geometry*: Through a point not on a given line, there exist more than one line parallel to the given line.
3. *Riemannian geometry*: Through a point not on a given line, there exists no line parallel to the given line.

Because they are based on postulates that contradict Euclid's Parallel Postulate, Lobachevskian and Riemannian geometries are called non-Euclidean geometries.

The model that is used to express Euclidean geometry is the plane. Lying in a flat plane, lines extend indefinitely into space and have infinite length; they do not "wrap" around the earth. See Figure 6.95.

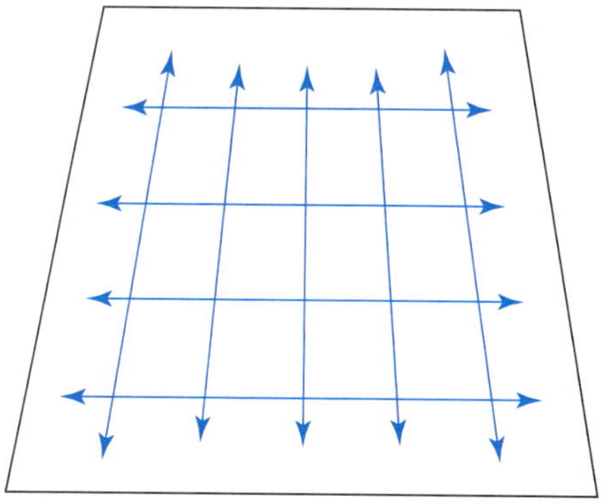

Figure 6.95
The Euclidean planar model

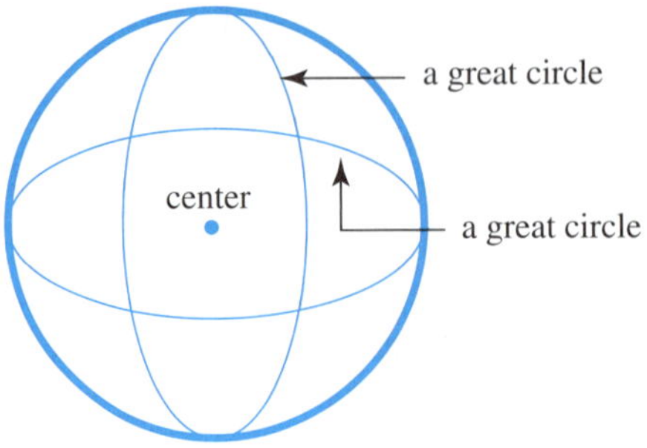

Figure 6.96
The Riemannian spherical model

In contrast, the model used to express Riemannian geometry is the sphere. Using this model, we define lines to be the great circles that encompass the sphere. A **great circle** is a circle whose center lies at the center of the sphere, as shown in Figure 6.96. No matter how they are drawn, each pair of great circles will always intersect in two points. Consequently, parallel lines do not exist! Riemannian geometry is important in navigation, because the shortest distance between two points on a sphere is the path along a great circle.

Figure 6.97
A Poincaré plane (disk)

A model that is used to express Lobachevskian geometry is the interior of a circle. This model was described by the Frenchman Henri Poincaré (1854–1912). The Poincaré model defines a plane to be all points "inside" a circle. (The points *on* the circle are excluded.) This region is also called a **disk;** it is shown in Figure 6.97.

Poincaré's model of a Lobachevskian geometry defines a line to be either of the following:

1. a diameter of the disk
2. a circular arc connecting two points on the boundary of the disk*

The two types of Poincaré lines are shown in Figure 6.98. Because the boundary of the disk is excluded, a (Poincaré) line has no endpoints.

Using Poincaré's disk model, we can construct more than one parallel line through a given point, as shown in Figure 6.99. Notice that lines *CD* and *EF* pass through the point *Q* and that each line is parallel to line *AB*. (*CD* and *AB* are parallel because they do not intersect. Likewise, *EF* and *AB* are parallel because they do not intersect.)

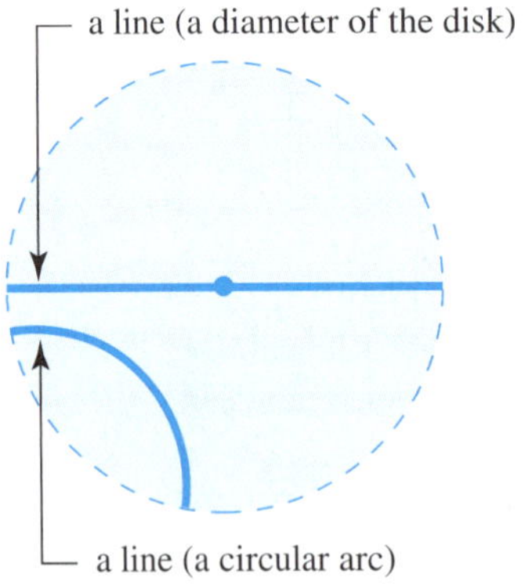

Figure 6.98
The Poincaré disk model for Lobachevskian geometry

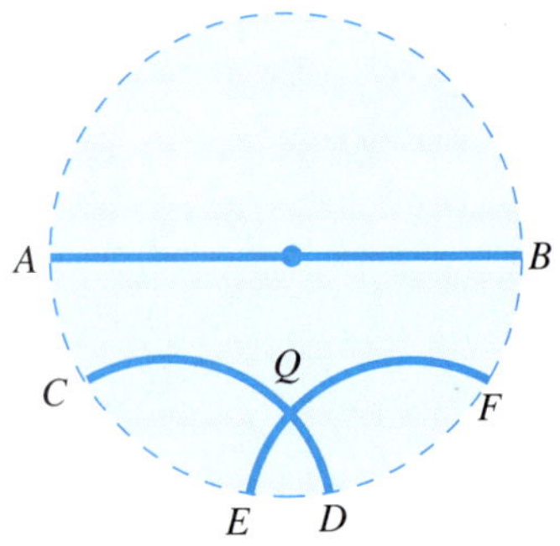

Figure 6.99

* Technically, the arc is the intersection of the disk with an orthogonal circle. Orthogonal circles are circles whose radii are perpendicular at the points where the circles intersect.

A Comparison of Triangles

Owing to the difference in the number of parallel lines, non-Euclidean geometries have properties that differ from those of Euclidean geometry. As you know, the sum of the angles of a triangle always *equals* 180° in Euclidean geometry, as shown in Figure 6.100(a). In contrast, when a triangle is drawn on a sphere, the sum of the angles is *greater* than 180°, as shown in Figure 6.100(b). Finally, the sum of the angles of a triangle drawn on a Poincaré disk is *less* than 180°, as shown in Figure 6.100(c).

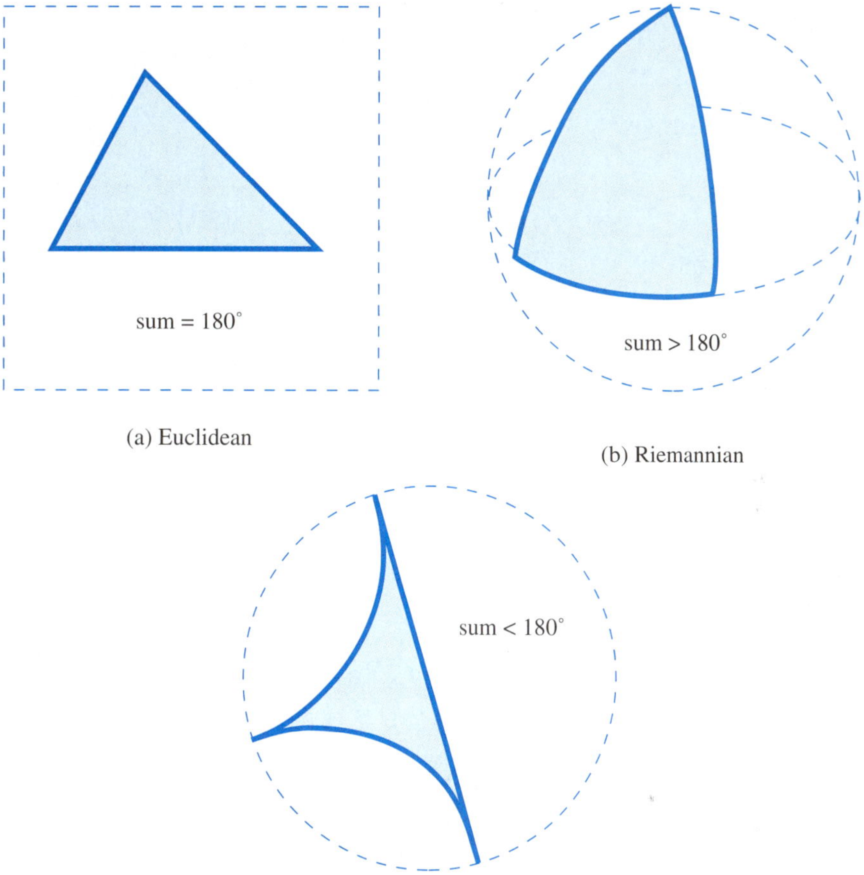

(a) Euclidean

(b) Riemannian

(c) Lobachevskian

Figure 6.100
The angles of a triangle

Geometry in Art

The study of geometry has always been an integral part of an education in art. Whether an artist leans toward realism or surrealism, a fundamental knowledge of geometric relationships is essential. No artist has been more successful in combining art and geometry than the Dutch artist Maurits Cornelis Escher (1898–1972). Escher is known for his repetitious plane-filling patterns, which often depict the metamorphosis of one figure into another.

The precise geometric appearance of Escher's work is not accidental; he possessed the knowledge, the talent, and the creativity to manipulate perspective and dimension in both Euclidean and non-Euclidean geometry. In fact, several of Escher's images were based on the non-Euclidean geometry of a Poincaré disk. A sketch from one of his many notebooks (Figure 6.101) shows Escher's study of a disk in his preparation to work in non-Euclidean space. Notice the triangles formed by the circular arcs.

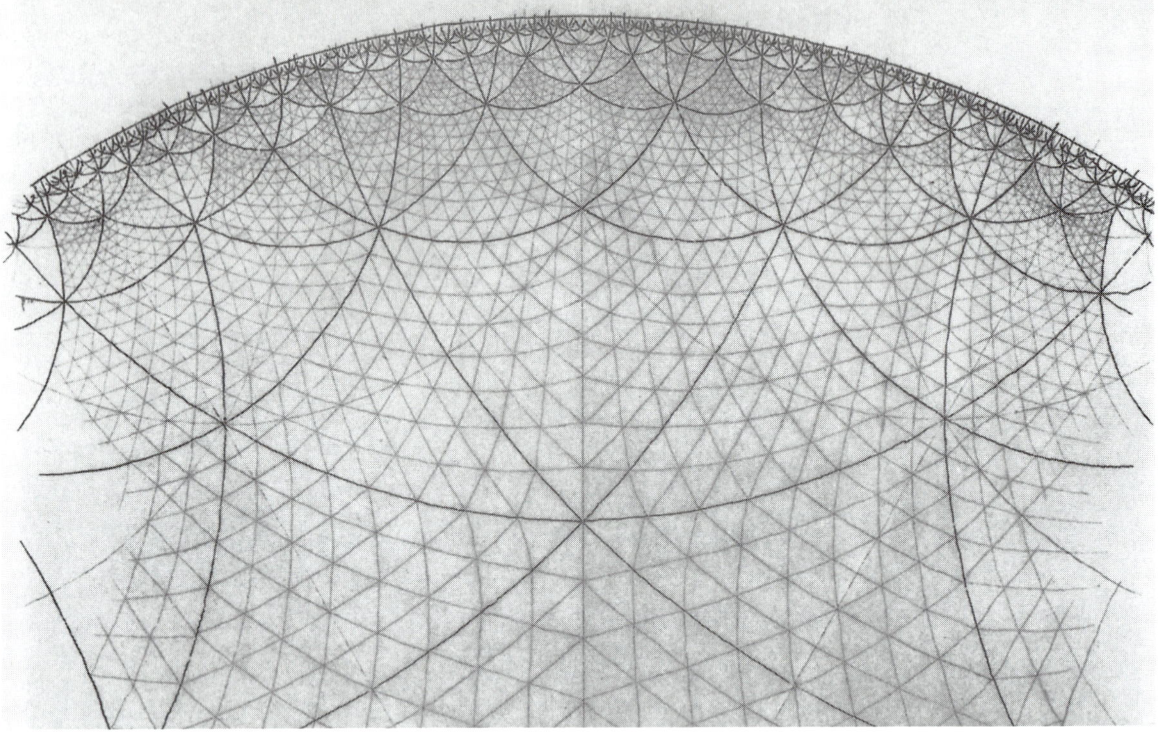

Figure 6.101
Escher drew this sketch of a Poincaré disk before creating the work of art shown in Figure 6.103.

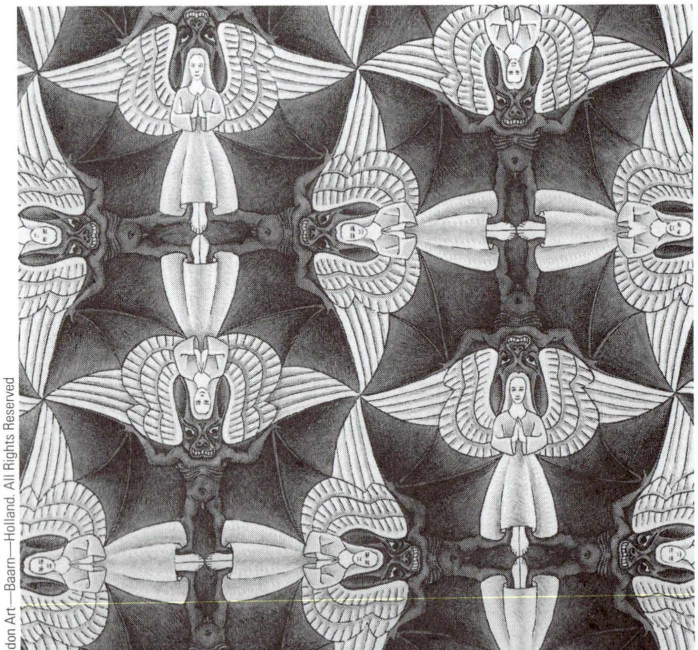

Figure 6.102
A repetitive pattern that completely fills a plane is known as a tiling; these complementary angels and devils form a tiling of the Euclidian plane.

Escher often created the same image on different geometric models. For instance, after he developed a pattern of complementary angels and devils on a "flat" Euclidean plane (Figure 6.102), Escher then employed Lobachevskian geometry to map the pattern onto a disk (Figure 6.103) and Riemannian geometry to transpose the pattern onto a sphere (Figure 6.104).

Figure 6.103
Utilizing Lobachevskian geometry, Escher tiled a Poincaré disk with the angels and devils of Figure 6.102.

Figure 6.104
Utilizing Riemannian geometry, Escher transposed the angels and devils of Figure 6.102 onto a sphere; the design is carved in solid maple.

6.7 Exercises

In Exercises 1–6, answer the questions for the planar model of Euclidean geometry.

1. Through a point not on a given line, how many parallels to the given line are there?
2. What can be said about the sum of the angles of a triangle?
3. How many right angles can a triangle have?
4. If two different lines are each parallel to a third line, are they necessarily parallel to each other?
5. In how many points can a pair of distinct lines intersect?
6. Do lines have finite or infinite length?

In Exercises 7–12, answer the questions for the spherical model of Riemannian geometry.

7. Through a point not on a given line, how many parallels to the given line are there?
8. What can be said about the sum of the angles of a triangle?
9. How many right angles can a triangle have?
10. Do lines have finite or infinite length?
11. In how many points can a pair of distinct lines intersect?
12. What is the minimum number of sides required to form a polygon?

In Exercises 13–18, answer the questions for Poincaré's model of Lobachevskian geometry.

13. Through a point not on a given line, how many parallels to the given line are there?
14. What can be said about the sum of the angles of a triangle?
15. How many right angles can a triangle have?
16. If two different lines are each parallel to a third line, are they necessarily parallel to each other?
17. In how many points can a pair of distinct lines intersect?
18. Do lines have finite or infinite length?

Answer the following questions using complete sentences and your own words.

History Questions

19. What is Euclid's Parallel Postulate?
20. What was "the scandal of elementary geometry"?
21. Who was the first mathematician to examine the consequences of assuming the Parallel Postulate to be false? What were his findings? What was his reaction?
22. Who was the first mathematician to affirm the existence of a non-Euclidean geometry? Did he publish his work? Why or why not?
23. Who stopped publishing his mathematical ideas after he had published his work on non-Euclidean geometry? Why did he stop?
24. Who is called the Copernicus of geometry? Why?
25. Who introduced a geometry in which parallel lines do not exist?
26. Write a research paper on a historical topic referred to in this section or on a related topic. Following is a list of possible topics:
    - Janos Bolyai (What influence did he have on the development of non-Euclidean geometry?)
    - M. C. Escher (How did he combine art and geometry?)
    - Carl Friedrich Gauss (What influence did he have on the development of non-Euclidean geometry?)
    - Immanuel Kant (How did his philosophy affect the development of non-Euclidean geometry?)
    - Felix Klein (What model did he create to represent Lobachevskian geometry?)
    - Adrien-Marie Legendre (How did he attempt to prove the Parallel Postulate?)
    - Nikolai Lobachevsky (Why is he given most of the credit for the creation of non-Euclidean geometry?)
    - Henri Poincaré (What did he contribute to the study of geometry?)
    - Georg Friedrich Bernhard Riemann (What did he contribute to the study of geometry?)
    - Girolamo Saccheri (What is the connection between Saccheri quadrilaterals and non-Euclidean geometry?)

6.8 Fractal Geometry

The classic geometry of the Greeks is not the only geometry. In Section 6.7, we learned of the non-Euclidean geometries created by Gauss, Bolyai, Lobachevsky, and Riemann. Intuitively, Euclid's geometry seems most natural to us, probably because we seem to live on a plane. However, it could be argued that Riemannian geometry is more natural because we actually live on a sphere, not a plane.

The shapes of the classic geometry of the Greeks—the triangle, square, rectangle, circle, ellipse, parabola and hyperbola—do not seem natural to us. Although these idealized shapes occasionally exist in nature, we more commonly see complex, less perfect shapes such as those of mountains and plants, rather than the simple, Euclidean shapes of circles and rectangles. As mathematician Benoit Mandelbrot said, "Clouds are not spheres, mountains are not cones, coastlines are not circles, and bark is not smooth, nor does lightning travel in a straight line" (Benoit Mandelbrot, *The Fractal Geometry of Nature*).

Amazingly, the fern and the mountain in Figure 6.105 are neither photographs nor drawings; they are shapes formed with fractal geometry, a new alternative geometry. In many respects, fractal geometry has more in common with nature than does the classic geometry of the Greeks.

"Zabriski; Point," a fractal forgery of a mirage.

Figure 6.105
A fractal forgery of a fern. Notice its self-similarity.

The Sierpinski Gasket

The Sierpinski gasket is one of the older examples of fractal geometry. It was first studied by Polish mathematician Waclaw Sierpinski in 1915. To understand this shape, we will create one. Take the following steps, which are illustrated in Figure 6.106:

Step 1 Draw a triangle and fill in its interior.

Step 2 Put a hole in the triangle:

- Find the midpoints of the sides of the triangle.
- Connect the midpoints and form a new triangle.
- Form a hole by removing the new triangle.

This leaves three filled in triangles, each a smaller version of the original triangle from step 1.

Step 3 Put a hole in each of the three triangles from step 2 by applying the procedure from step 2 to each triangle.

Step 4 Put a hole in each of the triangles from step 3 by applying the procedure from step 2 to each triangle.

Step 5 (and beyond) Put a hole in each of the triangles from the previous step by applying the procedure from step 2 to each triangle.

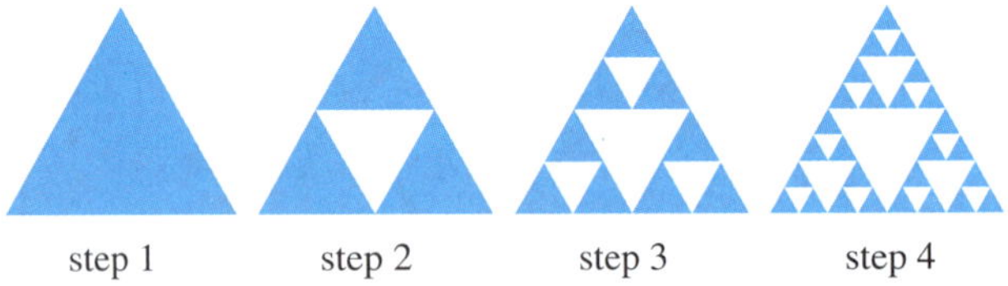

Figure 6.106
Creating a Sierpinski gasket

Figure 6.107
Step 5 of a Sierpinski gasket

The Sierpinski gasket (Figure 6.107) is the result of continuing the process described above through steps 6 and 7 and 8 and beyond without stopping.

After several steps, it becomes impossible to visually tell the difference between the result of one step and the result of the next step. That is, you probably couldn't tell the difference between the result of stopping at step 10 and the result of stopping at step 11 unless you used a magnifying glass. However, neither of these shapes would be a Sierpinski gasket; the only way to get a Sierpinski gasket is to continue the process without stopping, forming more and more tinier triangles, each with holes and even tinier triangles inside it.

Self-Similarity

If you did use a magnifying glass to enlarge a portion of a Sierpinski gasket, what would you see? The enlarged portion would look exactly the same as the nonenlarged portion, as shown in Figure 6.108. This is called *self-similarity*. A shape has **self-similarity** if parts of that shape appear within itself at different scales. Fractal geometry utilizes shapes that have self-similarity. Some shapes have **exact self-similarity** (in which the enlarged portion is exactly the same as the nonenlarged portion), and some shapes have **approximate self-similarity.**

Notice that the fern in Figure 6.105 has self-similarity. One branch of the fern looks just like the fern itself. And that one branch consists of many smaller branches, each of which looks just like the fern itself. Approximate self-similarity is a common

Figure 6.108
The self-similarity of a Sierpinski gasket

Figure 6.109
The approximate self-similarity of a cauliflower. One of these pictures is a blow-up of the other. Which is which?

Figure 6.110

feature of shapes that exist in nature. For example, cauliflower has a self-similar structure (see Figure 6.109 and 6.110), as do some trees. The fact that fractal geometry utilizes self-similarity is one reason why fractal geometry generates shapes that could exist in the real world.

Self-similarity can be found in art, too. For example, consider M. C. Escher's *Smaller and Smaller* (Figure 6.111), and Jonathan Swift's *On Poetry. A Rhapsody*. These two works certainly exhibit self-similarity.

So Nat'ralists observe, a Flea
Hath smaller fleas that on him prey;
And these have smaller Fleas to bite 'em
And so proceed ad infinitum.

—JONATHAN SWIFT

ON POETRY. A RHAPSODY

Recursive Processes

Step 2 of building a Sierpinski gasket gives a procedure for putting a hole in a triangle. Steps 3 and beyond instruct us to apply that procedure to the result, over and over again. Such a process, in which a procedure is applied and the same procedure is applied to the result, over and over again, is called a **recursive process.**

M. C. Escher's *Smaller and Smaller.*
Figure 6.111

Two of the key elements of fractal geometry are self-similarity and recursive processes. We will not attempt a more formal definition of fractal geometry, since it goes beyond the scope of this text. Benoit Mandelbrot, who coined the term *fractal*, described a fractal as a mathematical set or object whose form is extremely irregular and/or fragmented at all scales. Mandelbrot derived the term *fractal* from the Latin *fractus* (to break, to create irregular fragments).

The Skewed Sierpinski Gasket

The skewed Sierpinski gasket is a simple variation of the Sierpinski gasket. The steps are the same as before, except for a twist.

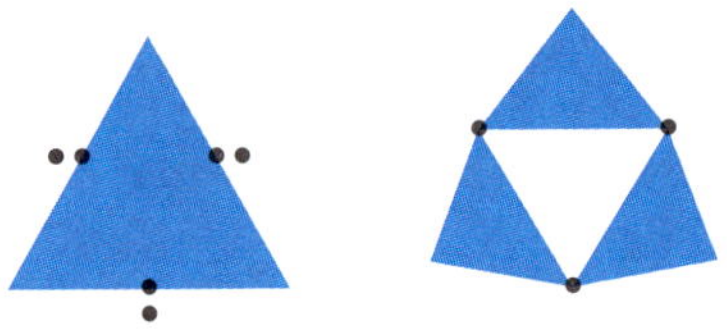

Figure 6.112

Step 1 Draw a triangle and fill in its interior (as before).

Step 2 Put a twisted hole in the triangle: (See Figure 6.112.)

- Find the midpoints of the sides of the triangle (as before).
- For each midpoint, find a nearby point (this is the new twist).
- Connect the nearby points and the vertices of the original triangle.
- Form a twisted hole by removing the center triangle.

Step 3 Put a twisted hole in each of the three triangles from step 2 by applying the procedure from step 2 to each triangle.

Step 4 Put a twisted hole in each of the triangles from step 3 by applying the procedure from step 2 to each triangle.

Step 5 (and beyond) Put a twisted hole in each of the triangles from the previous step by applying the procedure from step 2 to each triangle. Continue this process indefinitely.

Figure 6.113 shows a skewed Sierpinski gasket. Surprisingly, this fractal looks somewhat like a mountain! It has both the shape and the texture of a real mountain. Fractal geometry generates shapes that could exist in the real world.

The Sierpinski Gasket and the Chaos Game

There is an interesting connection between probability (as studied in Chapter 3) and the Sierpinski gasket. This connection involves the chaos game, which is similar to a game of chance.

Figure 6.113
A skewed Sierpinski gasket and a real mountain

Chaos Game Equipment:

- a sketch of a triangle with the vertices labeled *A*, *B*, and *C*. (See Figure 6.114.)
- a die
  - If you roll a 1 or a 2, vertex *A* "wins."
  - If you roll a 3 or a 4, vertex *B* "wins."
  - If you roll a 5 or a 6, vertex *C* "wins."

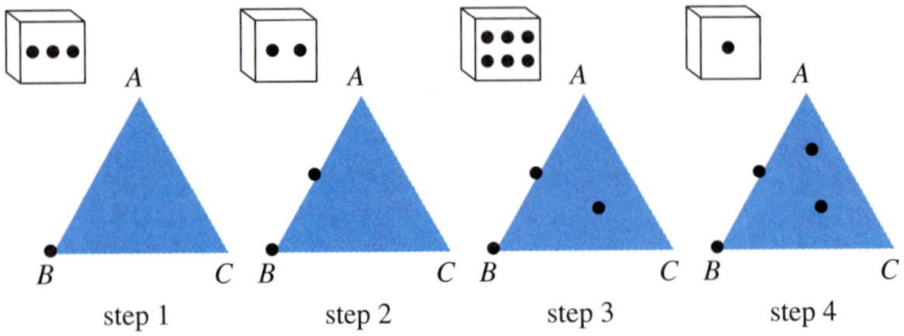

Figure 6.114
The beginnings of the chaos game

Chaos Game Procedure:

Step 1 Roll the die. Put a dot at the winning vertex. For example, if you roll a 3, put a dot at vertex B. (See Figure 6.114.)

Step 2 Roll the die. Put a dot halfway between the last dot and the winning vertex. For example, if you roll a 2, put a dot halfway between vertex *B* (the last dot) and vertex *A* (the winning vertex). (See Figure 6.114.)

Step 3 Roll the die. Put a dot halfway between the last dot and the winning vertex, as determined by the die. For example, if you roll a 6, put a dot halfway between the last dot and vertex *C* (the winning vertex). (See Figure 6.114.)

Step 4 (and beyond) Continue rolling the die. Each time put a dot halfway between the last dot and the winning vertex, as determined by the die. For example, if you roll a 1, put a dot halfway between the last dot and vertex *A* (the winning vertex). (See Figure 6.114.) Then roll again.

Figure 6.115(a) shows the chaos game after 100 rolls. Not surprisingly, it's just a scattering of dots. Figure 6.115(d) shows the chaos game after 10,000 rolls. Quite surprisingly, it looks like the Sierpinski gasket! After all, the chaos game is a sequence of random events, so you would think you would get a random result, and if you play the game twice, you would think you'd get two different results. But this is not the case; no matter how many times you play the chaos game, if you roll the die enough, you get the same result: the Sierpinski gasket.

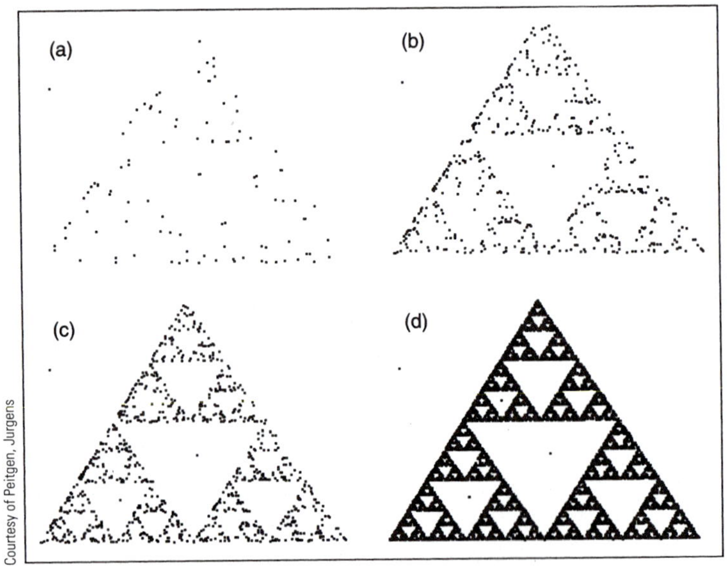

Courtesy of Peitgen, Jurgens

Figure 6.115
The chaos game after (a) 100 rolls, (b) 500 rolls, (c) 1,000 rolls, and (d) 10,000 rolls

Pascal's Triangle and the Sierpinski Gasket *(for those who have read Section 2.4: Permutations and Combinations.)*

There is an interesting connection between Pascal's triangle (as studied in Section 2.4) and the Sierpinski gasket. If you color the odd numbers of the first eight rows of Pascal's triangle black and the even numbers white, the result looks like the result of step 3 in creating the Sierpinski gasket. This is shown in Figure 6.116.

Figure 6.117 shows the result of coloring the first 64 rows of Pascal's triangle. Is this a Sierpinski gasket? (See Exercise 21.)

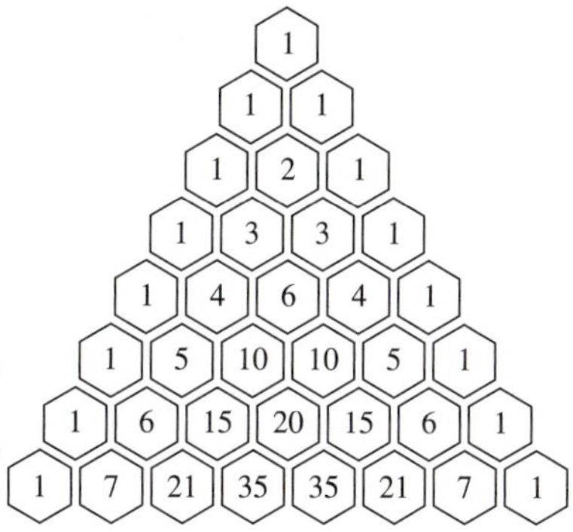

the first eight rows of Pascal's Triangle

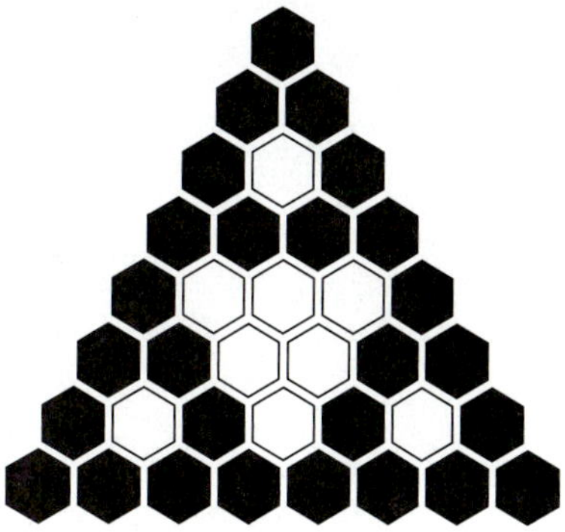

the first eight rows of Pascal's Triangle with the odd numbers colored black and the even numbers colored white

Figure 6.116

Figure 6.117 The first 64 rows of Pascal's Triangle with the odd numbers colored black and the even numbers colored white

The Koch Snowflake

The Koch snowflake is another example of fractal geometry. It was first studied by Swedish mathematician Helge von Koch in 1904. To create this shape, follow the following steps, which are illustrated in Figure 6.118.

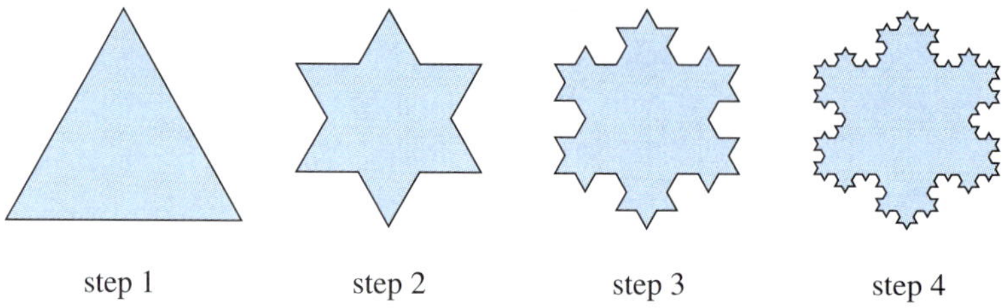

Figure 6.118
Creating a Koch snowflake

Step 1 Draw an equilateral triangle (that is, a triangle whose sides are equal and whose angles are equal).

Step 2 In the middle of each side of the equilateral triangle, attach a smaller equilateral triangle (pointing outwards). The smaller equilateral triangle's sides should be one third as long as the original larger equilateral triangle's sides. This forms a Star of David, with 12 sides.

Step 3 In the middle of each side of the Star of David, attach an equilateral triangle (pointing outwards). This equilateral triangle's sides should be one third as long as the Star of David's sides.

Step 4 (and beyond) Alter the result of the previous step by attaching an equilateral triangle to the middle of each side. Continue this process indefinitely. The resulting curve, shown in Figure 6.119, is called a Koch snowflake. Notice its similarity to a real snowflake, such as that in Figure 6.120.

Figure 6.119
A Koch snowflake

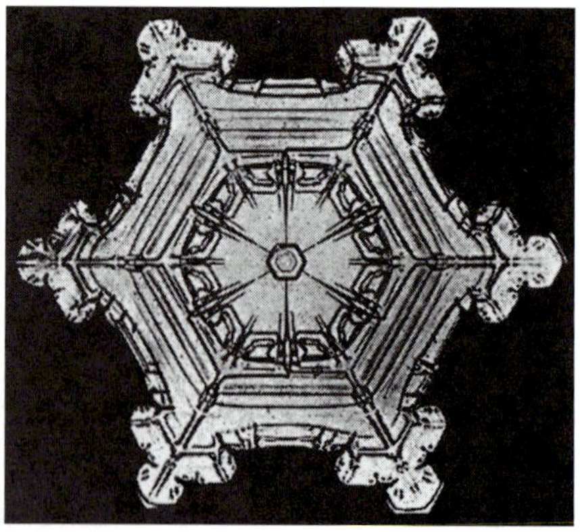

Dover Publications

Figure 6.120
Notice the similarity between the Koch snowflake and the real snowflake shown here.

The Menger Sponge

The Menger sponge is closely related to the Sierpinski gasket. To create this shape, follow the following steps, which are illustrated in Figure 6.121.

Step 1 Draw a cube.

Step 2 Remove some of the cube:

- Subdivide each of the cube's faces into nine equal squares. This subdivides the cube itself into 27 small cubes (nine in the front, nine in the middle, and nine in the back).
- Remove the small cube at the center of each face. There are six such small cubes.
- Remove the small cube at the very center of the original cube.

Step 3 A number of small cubes remain at the end of step 2. Remove some of each of these cubes by applying the procedure from step 2 to them.

Step 4 (and beyond) A number of cubes remain at the end of the previous step. Remove some of each of these cubes by applying the procedure from step 2 to them. Continue this process indefinitely. The resulting shape is called a Menger sponge. (See Figure 6.122.)

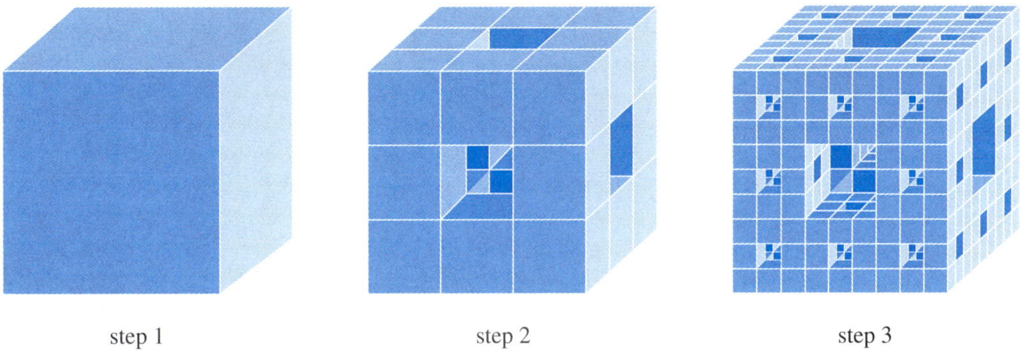

Figure 6.121
Creating a Menger sponge

Figure 6.122
A Menger sponge

Dimension

You're probably familiar with the term *dimension*. A box has three dimensions: up/down, back/forth, and left/right. A rectangle has two dimensions: up/down and left/right. A line has one dimension: left/right. A circle (one that's not filled in) also has one dimension: clockwise/counterclockwise. See Figure 6.123.

What is the dimension of a piece of paper? Many would say it has two dimensions: up/down and left/right. Actually, a piece of paper has a third dimension, which is thickness. Tissue paper is very thin, typing paper is somewhat thicker, and card stock is thicker still. It's just that we tend to forget about this third dimension, the thickness, because it's so small relative to the paper's length and width. The only way a piece of paper would have two dimensions would be if it were infinitely thin. And paper isn't infinitely thin.

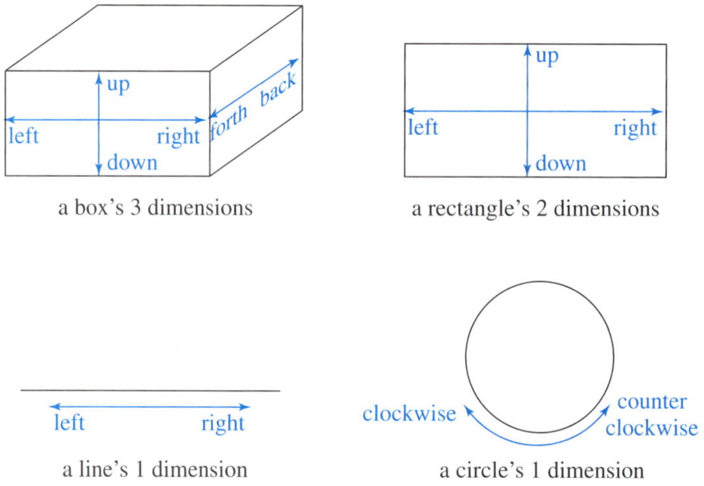

Figure 6.123

The Dimension of a Fractal

What is the dimension of a Koch snowflake? Many would say that, like a circle, it has one dimension: clockwise/counterclockwise. But a Koch snowflake wiggles around a lot; it's filled with little tiny crinkles. And those crinkles seem to make it somewhat thicker than a circle. The fact that a piece of paper has some thickness makes the paper's dimension 3 rather than 2. Does the fact that the Koch snowflake has some thickness affect its dimension?

The dimension of a fractal is certainly less obvious than the dimension of a box, a rectangle, or a line. Because of this, it must be determined with a new approach. We'll start by applying this new approach to a square.

If we take a square and double both its length and its width, we get four new squares, each equal to the original square. (See Figure 6.124.) We say that the scale factor is 2, because we multiplied the length and width by 2.

What happens if we apply a scale factor of 3 to the original square? If we multiply both the length and the width by 3, we get nine new squares, each equal to the original square. (See Figure 6.125.)

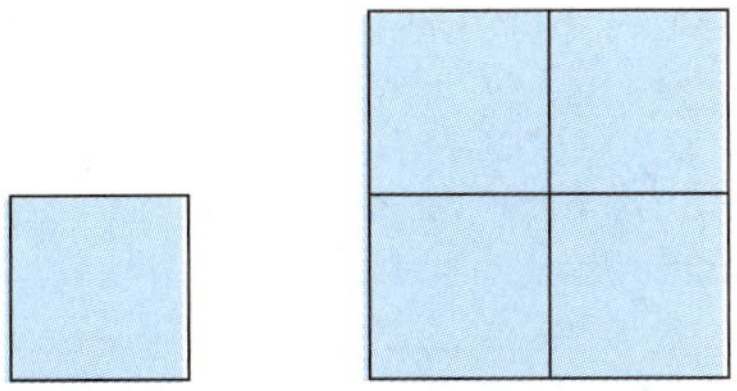

the original square

double its length and width and get 4 new squares, each equal to the original square

Figure 6.124

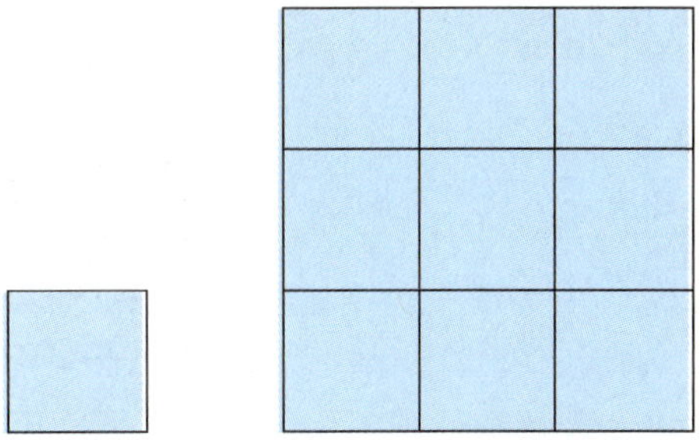

the original square

triple its length and width and get 9 new squares, each equal to the original square

Figure 6.125

In each case, the scale factor, raised to the second power, gives the number of new squares:

- When the scale factor is 2, we get $2^2 = 4$ new squares.
- When the scale factor is 3, we get $3^2 = 9$ new squares.

What happens with cubes? If we apply a scale factor of 2 to a cube, then we get eight new cubes, each equal to the original cube. And if we apply a scale factor of 3, then we get 27 new cubes, each equal in size to the original cube. See Figure 6.126.

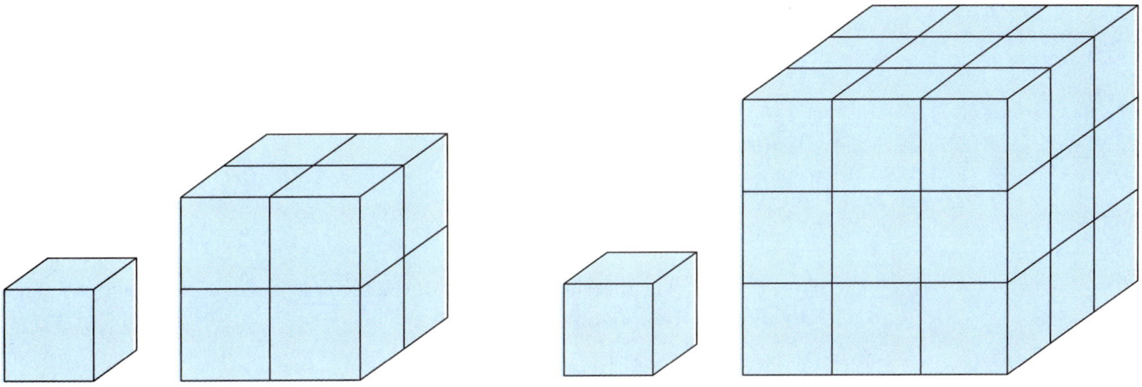

the original cube

apply a scale factor of 2 and get 8 new cubes, each equal to the original cube

the original cube

apply a scale factor of 3 and get 27 new cubes, each equal to the original cube

Figure 6.126

With the square, the scale factor raised to the *second* power gives the number of new squares. But with the cube, the scale factor raised to the *third* power gives the number of new cubes:

- When the scale factor is 2, we get $2^3 = 8$ new cubes.
- When the scale factor is 3, we get $3^3 = 27$ new cubes.

Clearly, there is a relationship between the scale factor, the number of new objects, and the dimension. The scale factor, raised to the power of the dimension, equals the number of new objects. It turns out that this relationship can be applied to fractals just as well as it can be applied to squares and cubes.

Dimension Formula

The dimension of an object is the number d that satisfies the equation

$$s^d = n$$

where s is the scale factor and n is the number of new objects that result, where each new object is equal in size to the original object.

EXAMPLE 1 Find the dimension of a Koch snowflake.

Solution We will start by looking at one piece of a Koch snowflake. If we apply a scale factor of $s = 3$ to this piece, we get a larger piece of a Koch snowflake. This larger piece has $n = 4$ parts, each of which is equal in size to the original piece. See Figure 6.127.

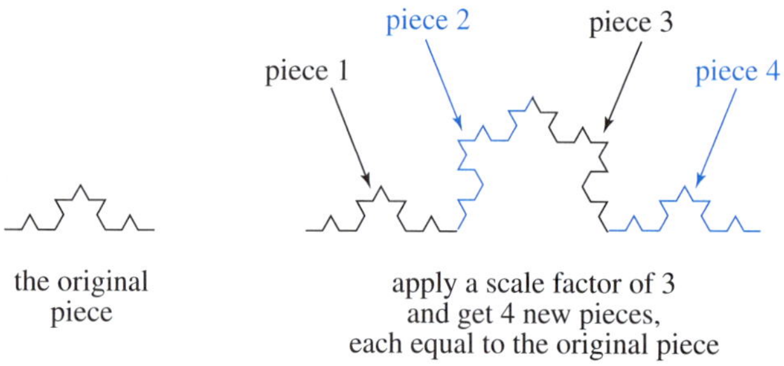

Figure 6.127

This means that the dimension of the piece of a Koch snowflake is the number d that solves the equation $3^d = 4$:

$s^d = n$ the dimension formula

$3^d = 4$ substituting

If we apply a scale factor of $s = 3$ to an entire Koch snowflake, rather than to just one piece of the snowflake, each piece turns into $n = 4$ pieces, each of which is equal to the original piece. So the dimension of the entire snowflake is the same as that of the piece of a snowflake. It is the number d that solves the equation $3^d = 4$.

What number would d be? Solving this equation involves mathematics that will be covered in Chapter 9. Instead, we'll approximate d to the nearest tenth, using a trial and error process.

Clearly, $d \neq 1$, because $3^1 = 3 \neq 4$. Also, $d \neq 2$, because $3^2 = 9 \neq 4$. Substituting 1 for d gives a result that's too small, and substituting 2 for d gives a result that's too big. Instead, d is a number between 1 and 2. We'll approximate d to the nearest tenth, using a trial and error process. Using a calculator, we get the following results:

$$3^{1.1} = 3.348\ldots$$
$$3^{1.2} = 3.737\ldots$$
$$3^{1.3} = 4.171\ldots$$

3 [y^x] 1.3 [=]
3 [∧] 1.3 [ENTER]

Substituting 1.2 for d gives a result that's a little too small, and substituting 1.3 gives a result that's a little too big. This means that d is between 1.2 and 1.3. And since 4.171 . . . is closer to 4 than is 3.737 . . . , we can tell that d is closer to 1.3 than to 1.2. Thus, $d \approx 1.3$.

So the dimension of a Koch snowflake is a little bigger than the dimension of a circle. (The Koch snowflake has dimension $d \approx 1.3$, while the circle has dimension $d = 1$.) The little tiny crinkles do make the snowflake somewhat thicker than a circle. And this extra thickness makes the snowflake's dimension somewhat bigger than that of a circle. But it's quite strange that a fractal can have a dimension that is not a whole number. ■

In discussing squares and cubes, it didn't matter what scale factor we used. We got the same result for the dimension regardless of whether we used $s = 2$ or $s = 3$. In fact, we would get the same result for any scale factor. But in Example 1, we deliberately used a scale factor of 3 because that scale factor gave us a larger piece of a Koch snowflake with four parts, *each of which is equal to the original piece*. If we used a scale factor of 2, we would still get a larger piece of a Koch snowflake with four parts, *but those parts would not be equal to the original piece.* They would be too small to equal the original piece. See Figure 6.128. With fractals, it's important to carefully choose a scale factor so that each of the new pieces is equal in size to the original piece.

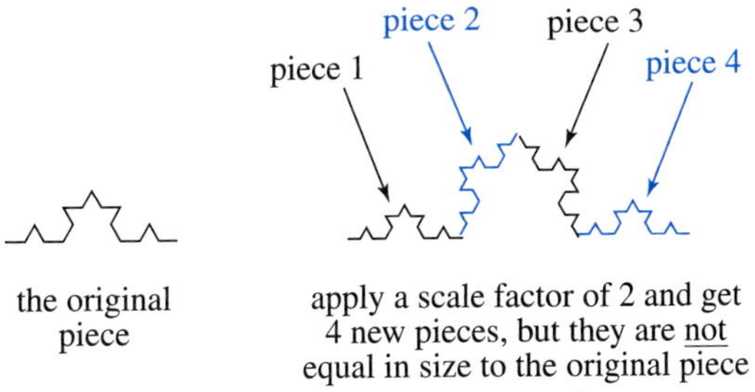

Figure 6.128

HISTORICAL NOTE

Benoit Mandelbrot

1924–

W. H. Freeman and Company

Benoit Mandelbrot was born in Warsaw, Poland. His father made his living buying and selling clothes. His mother was a doctor. When Mandelbrot was 11, he emigrated with his family to France. There, his uncle, who was a professor of mathematics at the Collège de France, took responsibility for his education. After World War II started, poverty and the need to survive kept Mandelbrot away from school. As a result, he educated himself. He now attributes much of his success to this self-education. It allowed him to think in different ways and to avoid the standard modes of thought encouraged by a conventional education.

After obtaining a master's degree in aeronautics from the California Institute of Technology and a Ph.D. at the Université de Paris, Mandelbrot went to the Institute for Advanced Study at Princeton. Later, he became an IBM Fellow at their Watson Research Center in New York. IBM allowed him great latitude in choosing the direction of his research. At IBM, he started his famous work on fractals, which led to his being called "the father of fractal geometry." Also at IBM, he developed one particularly important fractal called the Mandelbrot set, and he developed some of the first computer programs that print graphics.

In part because of his unique education, he decided early on to make contributions to many different branches of science. His academic posts indicate this; they include professor of mathematics at Yale, Harvard, and the École Polytechnique and visiting professor at Harvard (first in economics and later in applied mathematics), Yale (engineering), and the Einstein College of Medicine (physiology). He has received many honors and prizes, including the Barnard Medal for Meritorious Service to Science, the Franklin Medal, the Alexander von Humbolt Prize, the Steinmetz Medal, the Science for Art Prize, the Harvey Prize for Science and Technology, the Nevada Medal, the Wolf Prize in physics, and the Honda Prize.

One of the most beautiful and complex fractals is the Mandelbrot set.

W. H. Freeman and Company

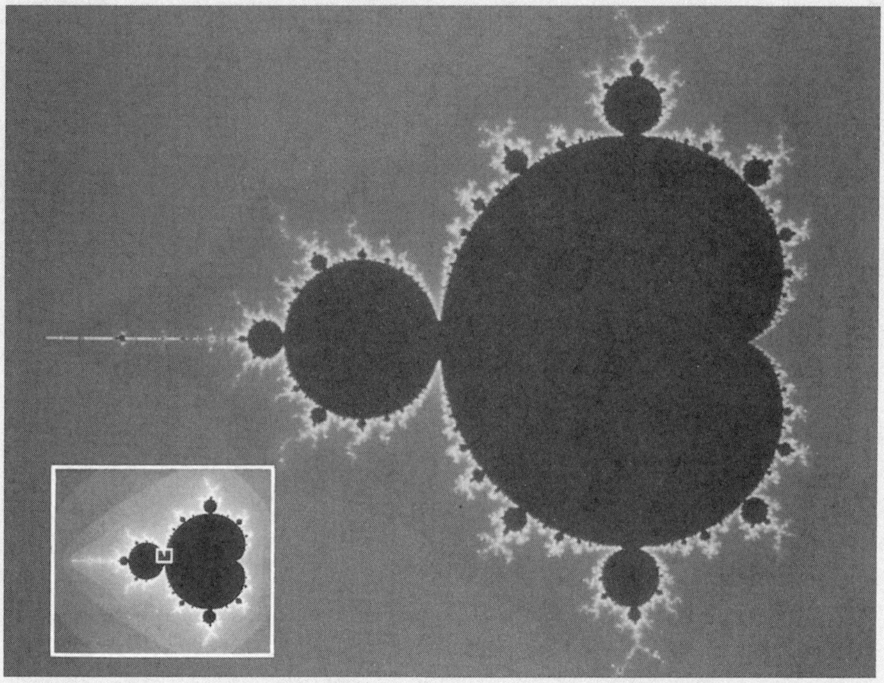

Springer-Verlag, Berlin

The Mandelbrot set (above) and the first of a series of successive close-ups (below).

Springer-Verlag, Berlin

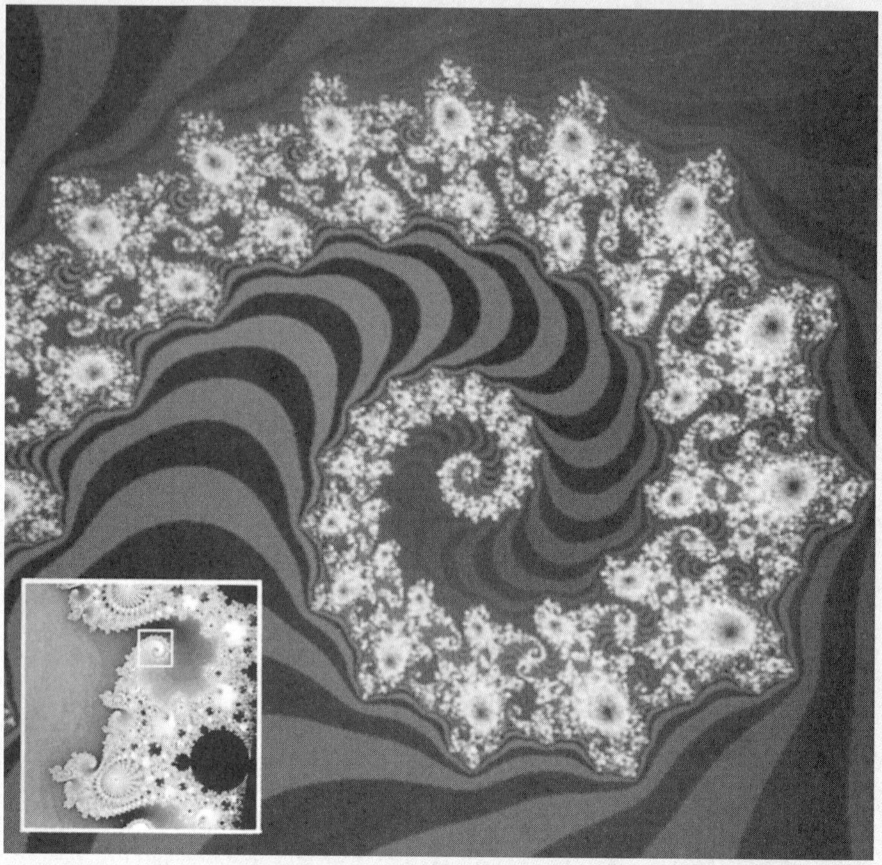

Springer-Verlag, Berlin

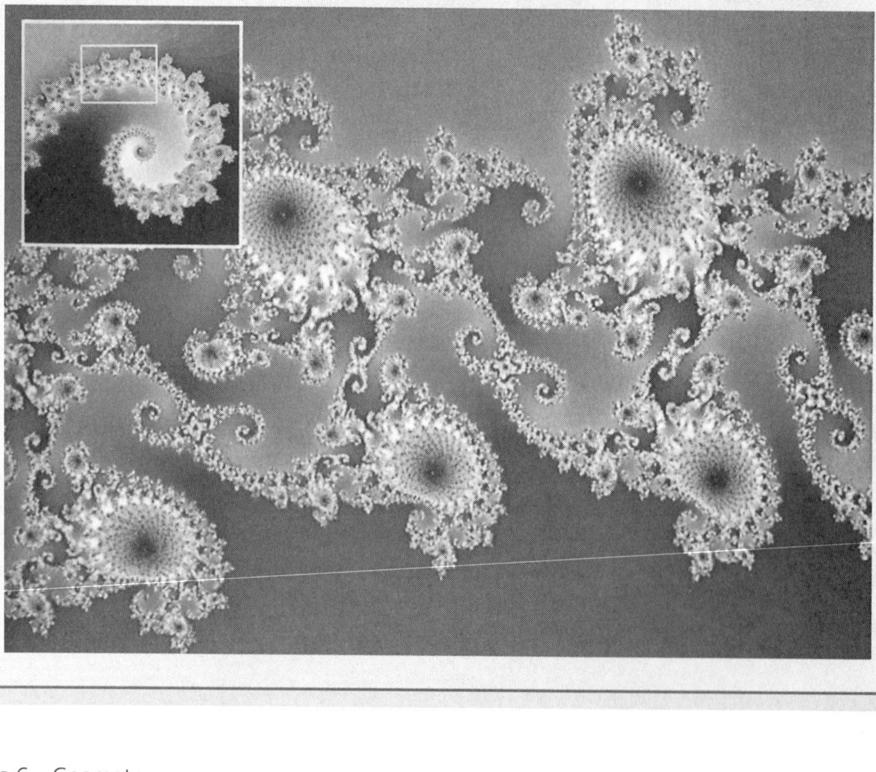

Springer-Verlag, Berlin

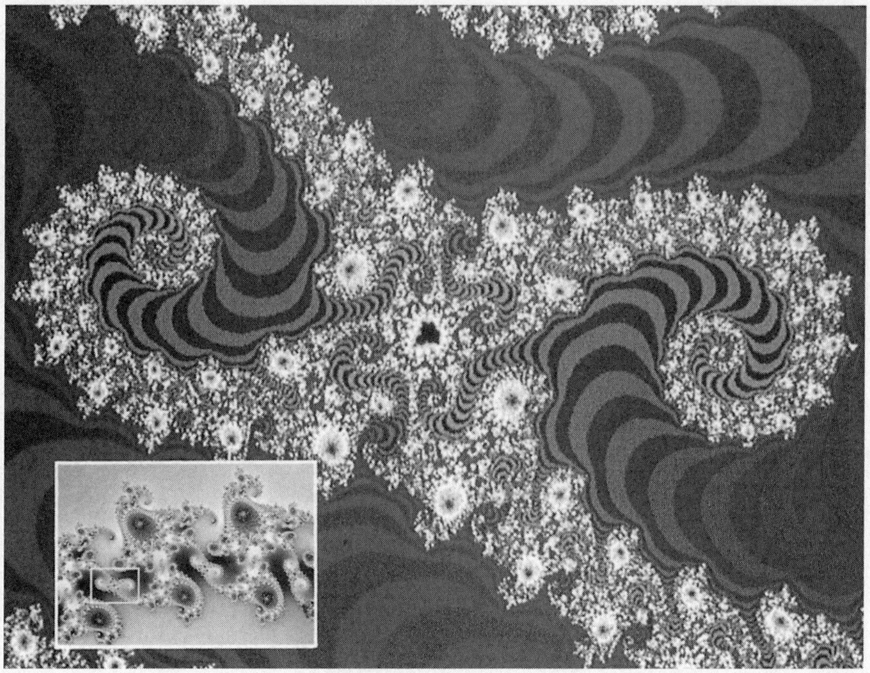

Springer-Verlag, Berlin

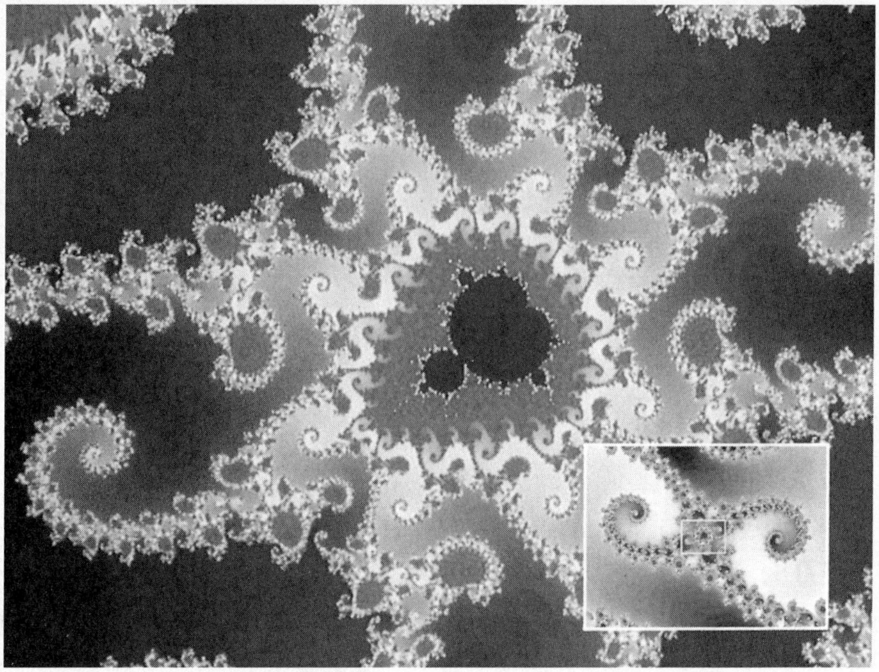

Springer-Verlag, Berlin

The next-to-last in our series of blow-ups. Notice the Mandelbrot set at the center of this image. What an extreme example of self-similarity!

Springer-Verlag, Berlin

The last in our series of blow-ups. This image also contains a Mandelbrot set near its center. Even more self-similarity!

Some Applications of Fractals

A country's coastline has self-similarity. If you took two photographs of an uninhabited coastline, one from an elevation of 1,000 feet and the other from an elevation of 5,000 feet, you would have difficulty determining which photo is which. This self-similarity means that fractal geometry can be used to analyze the complexity of coastlines. In fact, geographers classify coastlines according to their fractal dimension. South Africa's coastline is quite smooth, with a fractal dimension close to 1. On the other hand, Norway's coastline is heavily indented by glacier-carved fjords; its fractal dimension is about 1.5.

It's interesting to note that a Koch snowflake is sometimes called a Koch island, because the snowflake's outline looks somewhat like the coastline of an island. (See Exercise 34.)

Geologists have found that the fractal dimension of geologic features such as mountains, coastlines, faults, valleys, and river courses are related to the conditions under which they were formed and the processes to which they have been subjected. The fractal dimension of a mountain's contour line will be higher if the mountain is made of easily eroded rock or if it is in an area of high rainfall.

Scientists have also found that the fractal dimension of the surfaces of small particles such as soot and pharmaceuticals is related to their physical characteristics. For example, the tendency of soot to adhere to lung surfaces and the rate of absorption of ingested pharmaceuticals are both related to their fractal dimensions.

6.8 Exercises

1. *The Sierpinski gasket*

a. Using a straightedge, carefully and accurately draw five equilateral triangles, each on a separate piece of paper. Make each side 1 foot in length (or, optionally, make your triangle as large as the paper will allow and label each side as being 1 foot in length).

b. With the first equilateral triangle, complete the first step of a Sierpinski gasket. With the second equilateral triangle, complete the first two steps of a Sierpinski gasket. Continue in this fashion until you have completed the first five steps with the fifth equilateral triangle. With each triangle, shade in the appropriate regions.

2. *The Sierpinski carpet.* The Sierpinski carpet is closely related to the Sierpinski gasket and the Menger sponge. It is the square version of the Sierpinski gasket and the flat version of the Menger sponge. To create this shape, use the following steps. (See Figure 6.129.)

Step 1 Draw a square.

Step 2 Subdivide the square into nine equal squares. Remove the central square.

Step 3 The original square now has a square hole in its center, surrounded by eight other squares. Subdivide each of these eight squares into nine smaller squares. Remove each of the central squares.

Step 4 (and beyond) Continue this process indefinitely.

a. Using a straightedge, carefully and accurately draw four squares, each on a separate piece of paper. Make each side 1 foot in length (or, optionally, make your triangle as large as the paper will allow and label each side as being 1 foot in length).

b. With the first square, complete the first step of a Sierpinski carpet. With the second square, complete the first two steps of a Sierpinski carpet. Continue in this fashion until you have completed the first four steps with the fourth square. With each square, shade in the appropriate regions.

3. *The Mitsubishi Gasket.* Carefully and accurately complete the first four steps of a Mitsubishi gasket.

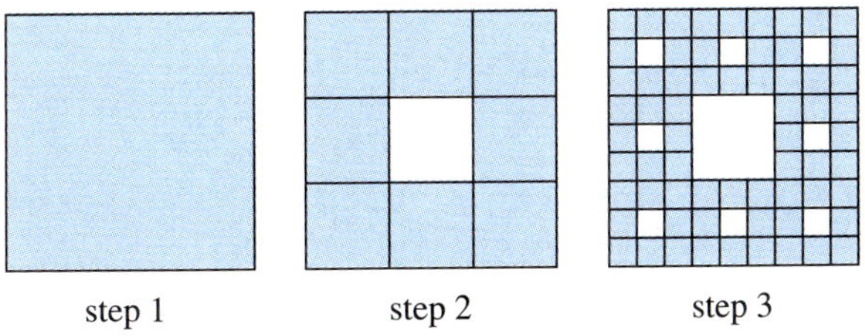

Figure 6.129
Creating a Sierpinski carpet

To create this shape, use the following steps. (See Figure 6.130.)

Figure 6.130
Creating a Mitsubishi gasket

Step 1 Draw a triangle and fill in its interior.

Step 2 Put three holes in the triangle:

- Find the points that divide the sides of the triangle into thirds.
- Form nine new triangles by connecting those points with lines parallel to the sides.
- Form three holes by removing three of these new triangles, as shown in Figure 6.130.

This leaves six filled in triangles, each a smaller version of the original triangle from step 1.

Step 3 Put three holes in each of the six triangles from step 2 by applying the procedure from step 2 to each triangle.

Step 4 Put three holes in each of the triangles from step 3 by applying the procedure from step 2 to each triangle.

Step 5 (and beyond) Put three holes in each of the triangles from the previous step by applying the procedure from step 2 to each triangle.

4. *The square snowflake.* Carefully and accurately complete the first four steps of a square snowflake. To create this shape, use the following steps. (See Figure 6.131.)

Step 1 Draw a square.

Step 2 Divide each of the square's straight lines into fourths. In a clockwise order:

- Leave the first fourth alone.
- Replace the second fourth with a square that's on the outer side of the line.
- Replace the third fourth with a square that's on the inner side of the line.
- Leave the last fourth alone.

Apply the above procedure to each of the square's straight lines. *Be sure to do the above in a clockwise order!*

Step 3 Apply the procedure from step 2 to each of the straight lines in step 2.

Step 4 (and beyond) Apply the procedure from step 2 to each of the previous step's straight lines.

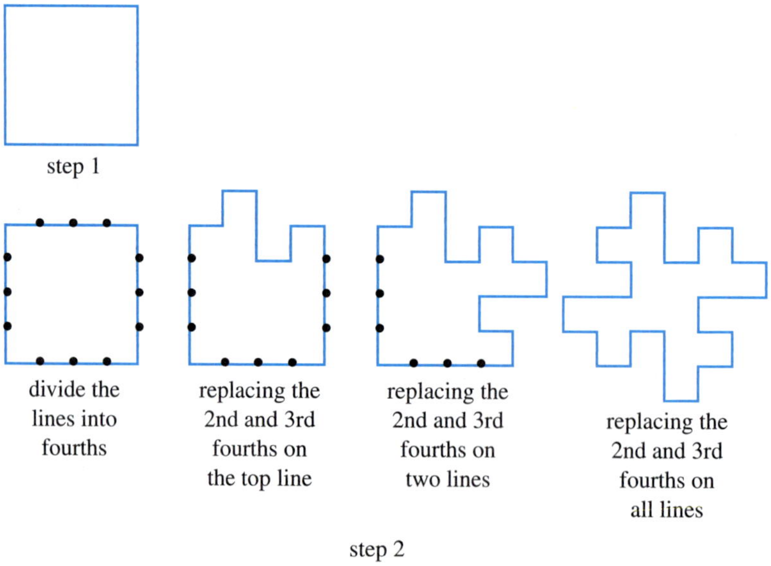

Figure 6.131
Creating a square snowflake

5. Find the exact dimension of a fractal that has $n = 25$ and $s = 5$. Do not approximate.
6. Find the exact dimension of a fractal that has $n = 5$ and $s = 25$. Do not approximate.
7. Find the exact dimension of a fractal that has $n = 3$ and $s = 27$. Do not approximate.
8. Find the exact dimension of a fractal that has $n = 2$ and $s = 32$. Do not approximate.
9. Find the approximate dimension of the Sierpinski gasket. Round off to the nearest tenth. What does this number say about the fractal?
10. Find the approximate dimension of the Sierpinski carpet from Exercise 2. Round off to the nearest tenth. What does this number say about the fractal?
11. Find the approximate dimension of the Mitsubishi gasket from Exercise 3. Round off to the nearest tenth. What does this number say about the fractal?
12. Find the approximate dimension of the Menger sponge. Round off to the nearest tenth. What does this number say about the fractal?
13. Find the approximate dimension of the square snowflake from Exercise 4. Round off to the nearest tenth. What does this number say about the fractal?
14. Of the assigned problems from 9 through 13, which fractal is the least complex? Which is the most complex? Why?
15. Use a scale factor of 4 to find the dimension of a cube. What do you observe?
16. Use a scale factor of 5 to find the dimension of a square. What do you observe?
17. Use scale factors to find the dimension of a rectangle.
18. Play the chaos game through 40 rolls of a die. Label the points "P_1," "P_2," "P_3," etc. Discuss your results.

Answer the following using complete sentences and your own words.

Concepts Questions

19. When playing the chaos game, why is it impossible to ever land in the hole formed in step 2 of the Sierpinski gasket process? Why is this an important consideration in seeing that the chaos game yields the Sierpinski gasket?
20. Figure 6.115(d) shows the result of the chaos game after 10,000 rolls. Is this a Sierpinski gasket, or is it the result of one of the steps in forming a Sierpinski gasket or is it merely similar to the Sierpinski gasket or one of the steps in forming the gasket? Explain your answer.
21. *For those who have read section 2.4: Permutations and Combinations*. Figure 6.117 shows the result of coloring the first 64 rows of Pascal's triangle. Is this a Sierpinski gasket, or is it the result of one of the steps in forming a Sierpinski gasket? Explain your answer. If it is the result of one of the steps in forming a Sierpinski gasket, determine which step.
22. Describe the self-similarity of the given fractal. Also determine whether the self-similarity is exact or approximate.
    a. the Sierpinski gasket
    b. the skewed Sierpinski gasket
    c. the Koch snowflake
    d. the Menger sponge
23. Describe the self-similarity for the fractal described in each of the assigned exercises (see Exercises 2–4). In each case, describe the self-similarity. Also determine whether the self-similarity is exact or approximate.
24. Discuss the recursive nature of:
    a. the Sierpinski gasket
    b. the skewed Sierpinski gasket
    c. the Koch snowflake
    d. the Menger sponge
25. Discuss the recursive nature of the fractal described in each of the assigned exercises (see Exercises 2–4).
26. Name three naturally occurring objects that have self-similarity. Do these objects have exact or approximate self-similarity?
27. Which type of self-similarity is most typically found in nature: exact or approximate? Which type is most typically found in fractal geometry?
28. Name a naturally occurring object that has a recursive nature. Describe its recursive nature.
29. In fractal geometry, self-similarity occurs at ever-smaller scales. That is, parts of a shape appear within itself at infinitely many different scales. Is this the case with the self-similarity found in nature? Why?
30. Describe two applications of fractal dimension.

History Questions

31. Who is known as the father of fractal geometry?
32. What was different about the education of the father of fractal geometry? What impact did this difference have on him?

Projects

33. Design your own fractal. Discuss its self-similarity and its recursive nature. Find its dimension
34. *Create your own coastline.* Draw a line segment. Roll a die. If either a 1, 2, or 3 comes up, apply step 2 of

the Koch snowflake procedure to that line segment, with the triangle(s) pointing upwards. If a 4, 5, or 6 comes up, apply step 2 of the Koch snowflake procedure to that line segment, with the triangle(s) pointing downwards. Continue this process until your fractal resembles a coastline. Include in your work the final sketch and a description of exactly how you created that shape. In your description, list each roll of the die and describe how that roll changed your image.

6.9 The Perimeter and Area of a Fractal

There is something both puzzling and useful about the perimeter and area of a fractal. We're going to find the perimeter of a Koch snowflake. To do so, we'll first find the perimeter of the first few steps of the process of creating a Koch snowflake. Then we'll use that work to find the perimeter of a Koch snowflake.

The Perimeter of the Koch Snowflake

EXAMPLE 1 Find the perimeter P_1 of the equilateral triangle formed in step 1 of the process of creating a Koch snowflake, if each side is of length 1 foot. (See Figure 6.132.)

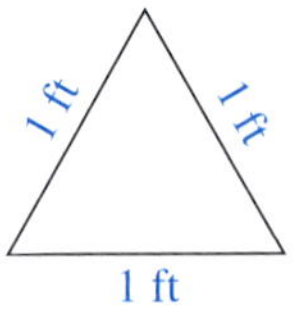

Figure 6.132

Solution The perimeter is just the sum of the lengths of the sides, so it is

$$P_1 = 1 \text{ ft} + 1 \text{ ft} + 1 \text{ ft} = 3 \text{ ft}$$

Alternatively, there are three sides, each of length 1 foot, so the perimeter is

$$P_1 = (3 \cancel{\text{sides}}) \cdot (1 \text{ foot}/\cancel{\text{side}}) = 3 \text{ ft}$$

(Notice that we are using dimensional analysis, as discussed in Appendix E.) We call this P_1 as a memory device; P stands for perimeter, and the subscript 1 indicates that it is the perimeter of the triangle formed in step one. ■

EXAMPLE 2 Find the perimeter P_2 of the Star of David that is formed in step 2 of the process of creating a Koch snowflake. (See Figure 6.133.)

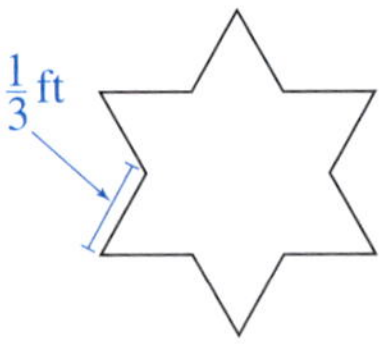

Figure 6.133

Solution The sides of the new smaller triangles are one third as long as the sides of the larger equilateral triangle from Example 1, so each side is

$$1/3 \cdot 1 \text{ ft} = 1/3 \text{ ft}$$

There are 12 sides, so the perimeter is

$$P_2 = (12 \cancel{\text{sides}}) \cdot (1/3 \text{ ft}/\cancel{\text{side}}) = 4 \text{ ft} \quad \text{using dimensional analysis}$$ ■

EXAMPLE 3 Find the perimeter P_3 of the shape that is formed in step 3 of the process of creating a Koch snowflake. (See Figure 6.134.)

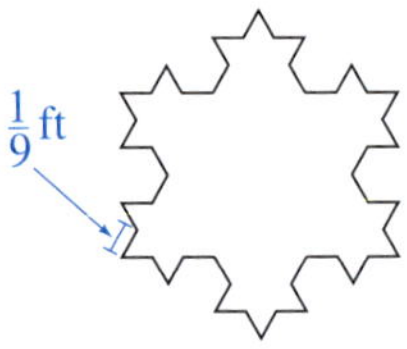

Figure 6.134

Solution The sides of the new smaller triangles are one-third as long as the sides of the triangle from Example 2, so each side is

$$1/3 \cdot 1/3 \text{ ft} = 1/9 \text{ ft}$$

There are 48 sides (count them!), so the perimeter is

$$P_3 = (48 \cancel{\text{sides}}) \cdot (1/9 \text{ ft}/\cancel{\text{side}})$$
$$= 48/9 \text{ ft} = 16/3 \text{ ft} \quad \text{using dimensional analysis}$$

The results of Examples 1, 2, and 3 are summarized in Figure 6.135.

Step	Picture	Number of Sides	Length of Each Side	Perimeter
1		3 sides	1 ft/side	$P_1 = 3$ sides $\cdot$ 1 ft/side $= 3$ ft
2		12 sides	1/3 ft/side	$P_2 = 12$ sides $\cdot$ 1/3 ft/side $= 4$ ft
3		48 sides	1/9 ft/side	$P_3 = 48$ sides $\cdot$ 1/9 ft/side $= 16/3$ ft

Figure 6.135 ■

EXAMPLE 4 Find the perimeter P of the Koch snowflake.

Solution To find the perimeter P, we have to generalize from the solutions of Examples 1, 2, and 3. There are several patterns to observe.

First notice that the number of sides is increasing by a factor of 4. In Figure 6.135, under "Number of Sides," each entry is 4 times the previous entry. In particular, $4 \cdot 3$ sides $= 12$ sides and $4 \cdot 12$ sides $= 48$ sides. This should make sense, because in each step in the Koch snowflake procedure, you replace every with a four-sided 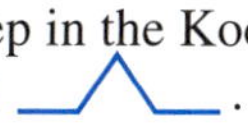.

This means that every single side is replaced with four sides. So naturally, the number of sides increases by a factor of 4. (The word *factor* means "multiplier," so saying that the number of sides increases by a factor of 4 means that the number of sides is multiplied by 4.)

The second pattern to observe is that the length of each side is decreasing by a factor of 1/3. In Figure 6.135, under "Length of Each Side," each entry is 1/3 times the previous entry. In particular, $1/3 \cdot 1$ ft $= 1/3$ ft and $1/3 \cdot 1/3$ ft $= 1/9$ ft. This too should make sense, because each step in the Koch snowflake procedure makes each new side 1/3 as long as each side from the previous step.

The last pattern to observe is that the perimeter is increasing by a factor of 4/3. In Figure 6.135, under "Perimeter," each entry is 4/3 times the previous entry. In particular, $4/3 \cdot 3$ ft $= 4$ ft and $4/3 \cdot 4$ ft $= 16/3$ ft. This should also make sense, because there are four times as many sides, and each side is 1/3 as long, so the perimeter increases by a factor of $4 \cdot 1/3 = 4/3$.

These observations are summarized and extended in Figure 6.136 on page 492.

Notice that for each step number, the perimeter is 3 times 4/3 to a power, where the power is 1 less than the step number. In other words, if we let n be the step number, then the perimeter at step n is

$$P_n = 3 \cdot (4/3)^{n-1}$$

Let's check this formula by using it to find P_5:

$$P_5 = 3 \cdot (4/3)^{5-1} = 3 \cdot (4/3)^4$$

This checks with the last entry in Figure 6.136.

Step	Picture	Number of Sides	Length of Each Side	Perimeter
1	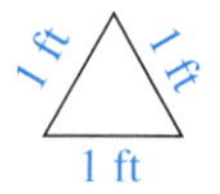	3 sides	1 ft/side	$P_1 = 3$ ft
2	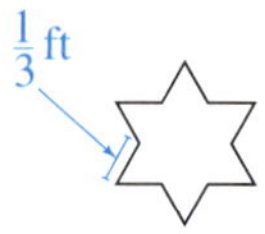	$4 \cdot 3$ sides $= 12$ sides	$1/3 \cdot 1$ ft/side $= 1/3$ ft/side	$P_2 = 4/3 \cdot P_1$ $= 4/3 \cdot 3$ ft
3	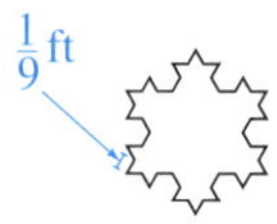	$4 \cdot 12$ sides $= 48$ sides	$1/3 \cdot 1/3$ ft/side $= 1/9$ ft/side	$P_3 = 4/3 \cdot P_2$ $= 4/3 \cdot 4/3 \cdot 3$ ft $= (4/3)^2 \cdot 3$ ft
4		$4 \cdot 48$ sides $= 192$ sides	$1/3 \cdot 1/9$ ft/side $= 1/27$ ft/side	$P_4 = 4/3 \cdot P_3$ $= 4/3 \cdot (4/3)^2 \cdot 3$ ft $= (4/3)^3 \cdot 3$ ft
5		$4 \cdot 192$ sides $= 768$ sides	$1/3 \cdot 1/27$ ft/side $= 1/81$ ft/side	$P_5 = 4/3 \cdot P_4$ $= 4/3 \cdot (4/3)^3 \cdot 3$ ft $= (4/3)^4 \cdot 3$ ft

Figure 6.136

The Koch snowflake is the result of continuing the process illustrated in Figure 6.136 indefinitely. That means that the perimeter of the Koch snowflake is the result of multiplying previous perimeters by 4/3, indefinitely. At step 100, the perimeter is huge:

$$P_{100} = 3 \cdot (4/3)^{100-1} = 3 \cdot (4/3)^{99} \text{ ft}$$

With a calculator, we find that this is approximately 7×10^{12} ft = 7,000,000,000,000 ft. At step 1,000, the perimeter is incredibly huge:

$$P_{1000} = 3 \cdot (4/3)^{1000-1} = 3 \cdot (4/3)^{999} \text{ ft}$$

On the calculator, this is approximately 2×10^{125} ft. (Remember, 2×10^{125} is a 2 followed by 125 zeros!)

3 [×] [(] 4 ÷ 3 [)] [y^x] 999 [=]

3 [×] [(] 4 ÷ 3 [)] [^] 999 [ENTER]

We must conclude that the perimeter of the Koch snowflake is infinite! This is hard to make sense of, because the perimeter just doesn't look that long. But remember, after several steps, it becomes impossible to visually tell the difference between the result of one step and the result of the next step. So you can't visually tell the difference between the result of step 5 (which has perimeter $P_5 \approx 9.48$ ft) and a Koch snowflake (which has infinite perimeter). Most of the perimeter is in all the little tiny crinkles. ■

The Area of the Koch Snowflake

EXAMPLE 5 Find the area A_1 of the equilateral triangle formed in step 1 of the process of creating a Koch snowflake, if each side is of length 1 foot. (See Figure 6.137.)

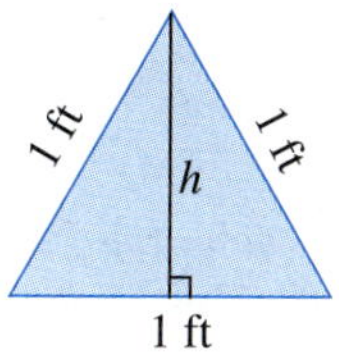

Figure 6.137

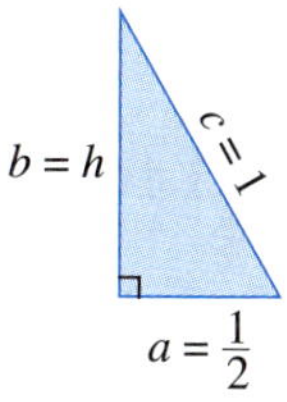

Figure 6.138

Solution To find the area, we need the base and the height. It's easy to see that the base is 1 foot, but we have to calculate the height h. The height h splits the equilateral triangle into two identical triangles, one on the each side of the height. Because the two triangles are identical, their bases are equal, so each must be 1/2 ft. We can apply the Pythagorean Theorem to either of these triangles, because they are right triangles. (See Figure 6.138.)

$$a^2 + b^2 = c^2 \quad \text{Pythagorean Theorem}$$
$$(1/2)^2 + h^2 = 1^2 \quad \text{substituting}$$
$$1/4 + h^2 = 1$$
$$h^2 = 1 - 1/4 = 3/4$$
$$h = \sqrt{\frac{3}{4}} = \frac{\sqrt{3}}{\sqrt{4}} = \frac{\sqrt{3}}{2} \text{ ft}$$

This is the height of the right triangle, but it is also the height of the triangle formed in step 1 of the Koch snowflake process. This means that the area of that triangle is

$$A_1 = \frac{1}{2}bh = \frac{1}{2} \cdot 1 \text{ ft} \cdot \frac{\sqrt{3}}{2} \text{ ft} = \frac{\sqrt{3}}{4} \text{ sq ft}$$

■

EXAMPLE 6 Find the area A_2 of the Star of David formed in step 2 of the process of creating a Koch snowflake. (See Figure 6.139.)

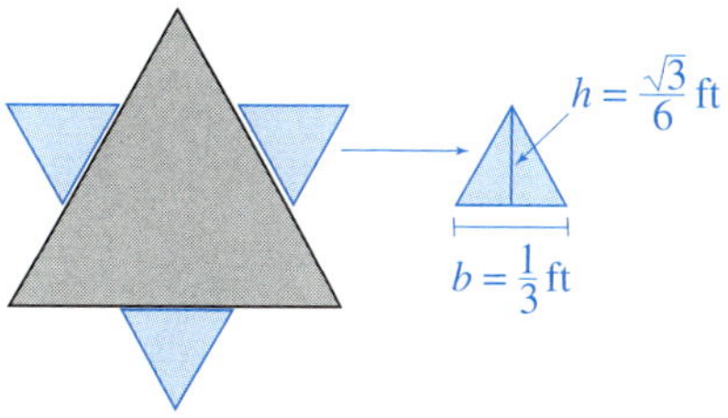

Figure 6.139

Solution The area is the sum of the area of the original equilateral triangle plus the areas of the three new smaller triangles. The original equilateral triangle's area was found in Example 5. It is

$$A_1 = \frac{\sqrt{3}}{4} \text{ sq ft}$$

The new smaller triangles are identical, so they have the same area. In step 2, all sides are 1/3 of what they were in step 1, so the base is $b = 1/3 \cdot 1$ ft $= 1/3$ ft, and the height is $h = 1/3 \cdot \sqrt{3}/2 = \sqrt{3}/6$ ft. This means that the area of each smaller triangle is

$$A = \frac{1}{2}bh = \frac{1}{2} \cdot \frac{1}{3} \text{ ft} \cdot \frac{\sqrt{3}}{6} \text{ ft} = \frac{\sqrt{3}}{36} \text{ sq ft}$$

There are three of these new smaller triangles. Thus, the area A_2 of the Star of David is

$$A_2 = (\text{area of original equilateral triangle}) + 3 \cdot (\text{area of new smaller triangle})$$
$$= \frac{\sqrt{3}}{4} + 3 \cdot \frac{\sqrt{3}}{36} \text{ sq ft}$$
$$= \frac{\sqrt{3}}{4} + \frac{\sqrt{3}}{12} \text{ sq ft} \quad \text{canceling}$$

■

EXAMPLE 7 Find the area A_3 of the shape formed in step 3 of the process of creating a Koch snowflake. (See Figure 6.140.)

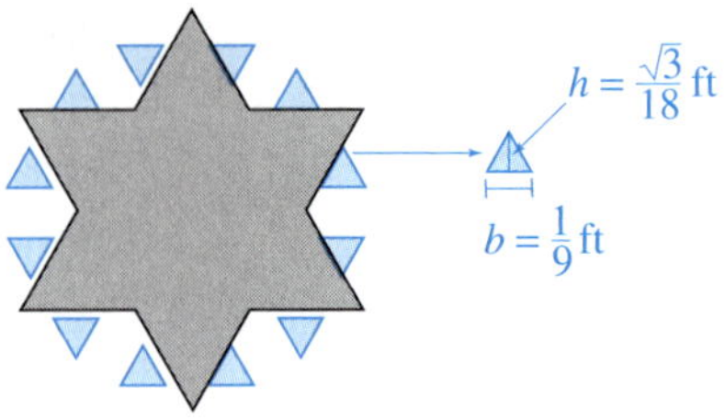

Figure 6.140

Solution The area is the sum of the area of the Star of David from Example 6 plus the areas of the new smaller triangles. The Star of David's area is

$$A_2 = \frac{\sqrt{3}}{4} + \frac{\sqrt{3}}{12} \text{ sq ft}$$

The new smaller triangles are identical, so they have the same area. The base is $b = 1/3 \cdot 1/3$ ft $= 1/9$ ft, and the height is $h = 1/3 \cdot \sqrt{3}/6 = \sqrt{3}/18$ ft. This means that the area of each smaller triangle is

$$A = \frac{1}{2}bh = \frac{1}{2} \cdot \frac{1}{9} \text{ ft} \cdot \frac{\sqrt{3}}{18} \text{ ft} = \frac{\sqrt{3}}{324} \text{ sq ft}$$

Also, there are 12 new smaller triangles (count them!). Thus, the area A_3 is

$$\begin{aligned} A_3 &= (\text{area of Star of David}) + 12 \cdot (\text{area of new smaller triangle}) \\ &= \frac{\sqrt{3}}{4} + \frac{\sqrt{3}}{12} + 12 \cdot \frac{\sqrt{3}}{324} \text{ sq ft} \\ &= \frac{\sqrt{3}}{4} + \frac{\sqrt{3}}{12} + \frac{\sqrt{3}}{27} \text{ sq ft} \qquad \text{canceling} \end{aligned}$$

■

While there are a number of patterns that we could observe, these patterns lead to a more complicated result than do the perimeter patterns discussed in Example 4. Utilizing these patterns to compute the exact area of the Koch snowflake is beyond the scope of this text. In Exercise 11, we will use Figure 6.140 to approximate the area.

Notice that the result of each step of the process of creating a Koch snowflake can be completely enclosed in a square that is 2 feet on each side, with plenty of room left over, as shown in Figure 6.141. This means that the area of the Koch snowflake is significantly less than the area of the square, which is 4 square feet.

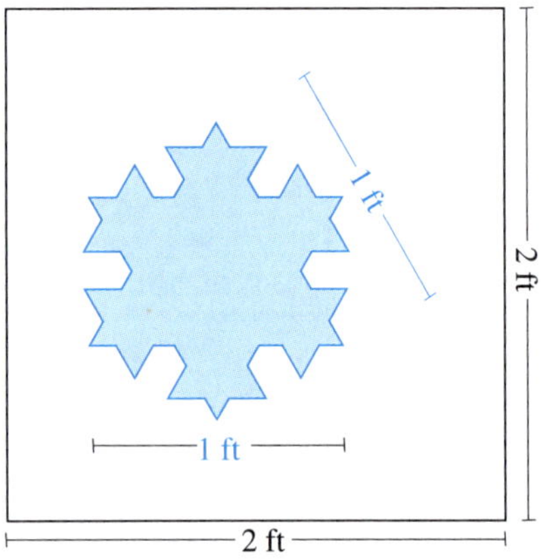

Figure 6.141

The Perimeter and Area of a Fractal

At the beginning of this section, we said, "There is something both puzzling and useful about the perimeter and area of a fractal." Now that we've found the perimeter and area of the Koch snowflake, we can discuss the puzzling and useful aspect of this topic.

The Koch snowflake has an infinite perimeter but an area that is significantly less than 4 square feet. This means that the Koch snowflake fits an infinitely long curve (the perimeter) inside a small space. The Sierpinski gasket also has an infinite perimeter but an area of zero! (The area is zero because of all of the holes—They just take over the Sierpinski gasket.) So the Sierpinski gasket also fits an infinitely long curve inside a small space. The Menger sponge has an infinite surface area but a volume of zero! This means that the Menger sponge fits an infinitely large surface inside a small space.

An Application of Fractals

It is certainly puzzling that a Koch snowflake has an infinite perimeter but a small area, and it's certainly puzzling that a Menger sponge has an infinite surface area but a volume of zero. But why are these facts useful?

Consider the human body's system of veins, arteries, and capillaries. This system takes up a small part of the body's space, but its length is incredibly large; laid end to end, the system of veins and arteries would be more than 40,000 miles long. Like a Koch snowflake, the length is huge, but it fits inside a relatively small space.

Also, consider the human kidney. Its surface area is practically infinite (and for good reason, since the kidney's surface area determines how much blood it can cleanse), but its volume is small—the kidney is only 4 inches long.

Scientists use fractal geometry to analyze these various parts of the body. It may well be that fractal geometry will help us better understand the functioning of the human body.

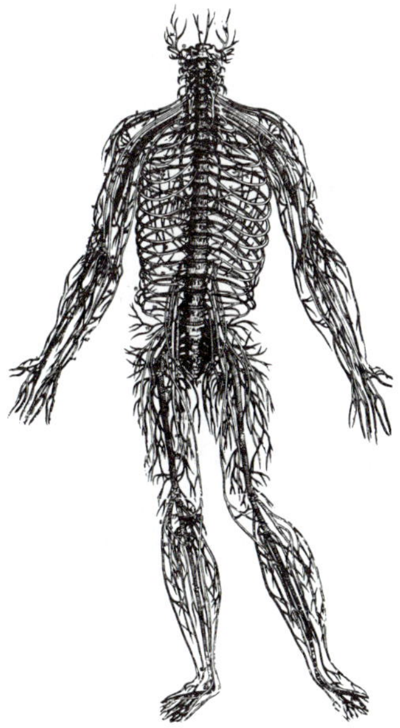

A. Vesalvas

The human body's system of veins, arteries and capillaries.

6.9 Exercises

1. *Finding the perimeter of a Sierpinski gasket.* (By "perimeter," we mean the total distance around all of the filled in regions, not the distance around the outside only.)

   a. Use your drawings from Exercise 1 of Section 6.8 to explain each of the calculations in the chart in Figure 6.142 on page 496, where "triangle" refers to a filled-in triangle, not a triangular hole.

Step	Number of Triangles	Length of Each Side	Perimeter of One Triangle	Total Perimeter of All Triangles
1	1	1 ft	$3 \cdot 1$ ft = 3 ft	$1 \cdot 3$ ft = 3 ft
2	3*	$1/2 \cdot 1$ ft = 1/2 ft	$3 \cdot 1/2$ ft = 3/2 ft	$3 \cdot 3/2$ ft = 9/2 ft
3				
4				
5				
6				

*3 not 4, because only three triangles are filled in.

Figure 6.142

b. Use your drawings from Exercise 1 of Section 6.8 to complete the chart in Figure 6.142. Include units.

c. By what factor does the number of triangles increase? Explain why your answer makes sense.

d. By what factor does the length of each side decrease? Explain why your answer makes sense.

e. By what factor does the perimeter of one triangle change? Explain why your answer makes sense. Is the perimeter of one triangle increasing or decreasing? Why?

f. By what factor does the total perimeter of all triangles change? Explain why your answer makes sense. Is the total perimeter of one triangle increasing or decreasing? Why?

g. Use your answers to parts (a)–(f) to find a formula for the total perimeter of all triangles at step n.

h. Use your answer to part (g) and your calculator to find the perimeter of the Sierpinski gasket. Explain your work. (By "perimeter," we mean the total distance around all of the filled in regions, not the distance around the outside only.)

2. *Finding the area of a Sierpinski gasket.* (By "area" we mean the total area of all the filled in regions.)

a. Use your drawings in Exercise 1 of Section 6.8 to complete the chart in Figure 6.143, where "triangle" refers to a filled in triangle, not to a triangular hole. Include units.

b. By what factor does the number of triangles increase? Explain why your answer makes sense.

c. By what factor does the length of the base decrease? Explain why your answer makes sense.

d. By what factor does the length of the height decrease? Explain why your answer makes sense.

e. By what factor does the area of each triangle decrease? Explain why your answer makes sense.

f. By what factor does the total area of all triangles change? Explain why your answer makes sense. Is the total area increasing or decreasing? Why?

g. Use your answers to parts (a)–(f) to find a formula for the total area of all triangles at step n.

h. Use your answer to part (g) and your calculator to find the area of the Sierpinski gasket. (By "area," we mean the total area of all of the filled in regions.) Explain your work.

Step	Number of Triangles	Length of the Base	Length of the Height	Area of one Triangle	Total area of all Triangles
1	1				
2	3*				
3					
4					
5					
6					

*3 not 4, because only three triangles are filled in.

Figure 6.143

Step	Number of New Squares*	Length of Each Side	Perimeter of One New Square*	Total Perimeter of All New Squares*	Total Perimeter of All Squares*
1	1	1 ft	4·1 ft = 4 ft	1·4 ft = 4 ft	4 ft
2	1	1/3·1 ft = 1/3 ft	4·1/3 ft = 4/3 ft	1·4/3 ft = 4/3 ft	4 ft + 4/3 ft
3	8				
4					
5					
6					

* "New square" refers to the squares that are both new to this step and that contribute to the total perimeter. " All squares" refers to the squares that are new to this step as well as those from previous steps.

Figure 6.144

3. *Finding the perimeter of a Sierpinski carpet.* (See Exercise 2 in Section 6.8 for a description of this fractal.) (By "perimeter," we mean the total distance around all of the filled in regions.)
   - **a.** Use your drawings in Exercise 2 of Section 6.8 to explain each of the calculations in the chart in Figure 6.144. These calculations are illustrated in Figure 6.145.
   - **b.** Use your drawings from Exercise 2 of Section 6.8 to complete the chart in Figure 6.144. Include units.
   - **c.** By what factor does the number of squares increase (starting at step 2)? Explain why your answer makes sense.
   - **d.** By what factor does the length of each side decrease? Explain why your answer makes sense.
   - **e.** By what factor does the perimeter of one new square change? Explain why your answer makes sense. Is the perimeter of one new square increasing or decreasing? Why?
   - **f.** By what factor does the total perimeter of all new squares change? Explain why your answer makes sense. Is the total perimeter of all new squares increasing or decreasing? Why?
   - **g.** Use your answers to parts (a)–(f) to find a formula for the total perimeter of all new squares at step n.

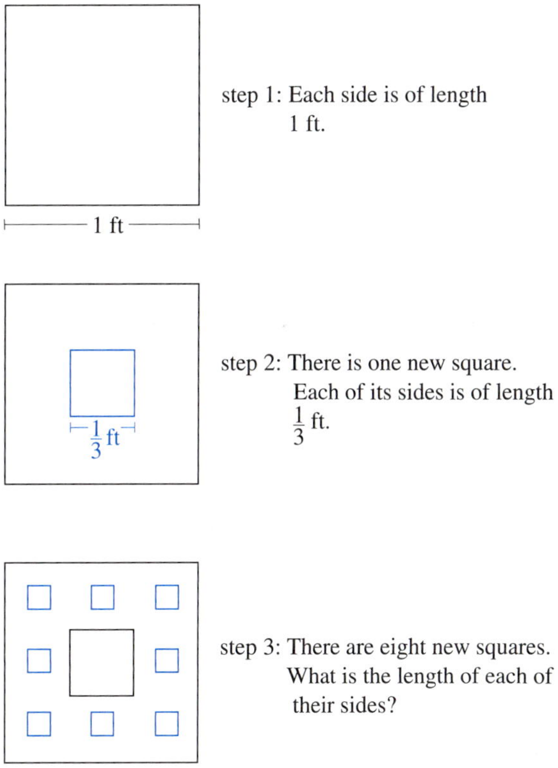

Finding the perimeter and area of a Sierpinski carpet.

Figure 6.145

h. Use your answer to part (g) and your calculator to determine what happens to the total perimeter of all new squares as we move from step to step.
i. Use your answer to part (h) to find the perimeter of the Sierpinski carpet. Explain your work. (By "perimeter," we mean the total distance around all of the filled in regions, not the distance around the outside only.)

(*HINT*: How must the total perimeter of all squares compare with the total perimeter of all new squares?)

4. *Finding the area of a Sierpinski carpet.* (See Exercise 2 in Section 6.8 for a description of this fractal.) (By "area," we mean the total area of all the filled in regions.)
   a. Use your drawings in Exercise 2 of Section 6.8 to explain each of the calculations in the chart in Figure 6.146. These calculations are illustrated in Figure 6.145 on page 497.
   b. Use your drawings from Exercise 2 of Section 6.8 to complete the chart in Figure 6.146.
   c. By what factor does the number of squares increase (starting at step 2)? Explain why your answer makes sense.
   d. By what factor does the length of each side decrease? Explain why your answer makes sense.
   e. By what factor does the area of one square change? Explain why your answer makes sense. Is the area of one square increasing or decreasing? Why?
   f. By what factor does the total area of all new squares change? Explain why your answer makes sense. Is the total area of all new squares increasing or decreasing? Why?
   g. By what factor is the total area after removing all new squares changing? Is the total area after removing all new squares increasing or decreasing? Why?
   h. Use your answers to parts (a)–(g) to find a formula for the total area after removing all new squares at step n.
   i. Use your answer to part (h) and your calculator to find the area of the Sierpinski carpet. Explain your answer.
5. *The Mitsubishi Gasket.* (See Exercise 3 in Section 6.8 for a description of this fractal.)
   a. Find a formula for the total perimeter of step n of the process described in Exercise 3, Section 6.8.
   b. Use the formula from part (a) to find the perimeter of the Mitsubishi gasket.
   c. Find a formula for the total area of step n of the above process.
   d. Use the formula from part (c) to find the area of the Mitsubishi gasket.

Step	Number of New Squares	Length of Each Side	Area of One New Square	Total Area of All New Squares*	Total area after Removing All New Squares*
1	1	1 ft	$(1\text{ ft})^2$ $= 1$ sq ft	$1 \cdot 1$ sq ft $= 1$ sq ft	1 sq ft – 0 = 1 sq ft (Nothing is removed in step 1.)
2	1	$1/3 \cdot 1$ ft = 1/3 ft	$(1/3\text{ ft})^2$ $= 1/9$ sq ft	$1 \cdot 1/9$ sq ft $= 1/9$ sq ft	1 sq ft – 1/9 sq ft = 8/9 sq ft
3	8				
4					
5					
6					

* "New Squares" refers to the squares that are both new to this step and that contribute to the area.

Figure 6.146

6. *A square snowflake.* (See Exercise 4 in Section 6.8 for a description of this fractal.)
   a. Find a formula for the total perimeter of step n of the process described in Exercise 4, Section 6.8.
   b. Use the formula from part (a) to find the perimeter of a square snowflake.
   c. Find a formula for the total area of step n of the above process. (*HINT:* It's an incredibly easy formula. Don't make it hard.)
   d. Use the formula from part (c) to find the area of a square snowflake.
7. *For those who have read Section 1.1: Deductive versus Inductive Reasoning.* In Example 4, we discussed the perimeter of the results of the first few steps of the process of creating a Koch snowflake, and we concluded that $P_4 = 64/9$ and $P_5 = 256/27$. In arriving at these conclusions, some of the reasoning was inductive and some was deductive. Which reasoning was inductive? Which was deductive? Why?
8. *For those who have read Section 6.5: Right Triangle Trigonometry.* In part (b) of Example 1, we found the base and height of a triangle. Discuss how to use the information developed in Section 6.5, under *Special Triangles*, to find this base and height.

Answer the questions 9 and 10 using complete sentences and your own words.

Concept Questions

9. Describe an application of fractal area, perimeter, and volume.

Projects

10. Design your own fractal. Discuss its self-similarity and its recursive nature. Find its perimeter and area (or its surface area and volume, if appropriate).
11. *Approximating the area of a Koch snowflake.*
    a. Use Example 7 to complete the chart in Figure 6.147. Include units.
    b. Expand the chart to include steps 4–8.
    c. Use your calculator and the "Total Area" entries in the chart you completed in part (b) to compute decimal approximations of A_1 through A_8. Round off to four decimal places.
    d. So far, we haven't found the area of the Koch snowflake. Instead, we've found the areas of the first few steps of creating a Koch snowflake. Would the area of the Koch snowflake be larger or smaller than A_8? Why? Use the result of part (c) to estimate the area of the Koch snowflake.

Step	Number of New Triangles	Base of New Triangle	Height of New Triangle	Area of each New Triangle	Area of all New Triangles	Total Area
1	1	1 ft	$\frac{\sqrt{3}}{2}$ ft	$A_1 = \frac{1}{2} \cdot 1 \cdot \frac{\sqrt{3}}{2}$ $= \frac{\sqrt{3}}{4}$ sq ft	$1 \cdot \frac{\sqrt{3}}{4}$ sq ft $= \frac{\sqrt{3}}{4}$ sq ft	A_1 $= \frac{\sqrt{3}}{4}$ sq ft
2	3	$\frac{1}{3} \cdot 1$ ft $= \frac{1}{3}$ ft	$\frac{1}{3} \cdot \frac{\sqrt{3}}{2}$ ft $= \frac{\sqrt{3}}{6}$ ft	$\frac{1}{2} \cdot \frac{1}{3} \cdot \frac{\sqrt{3}}{6}$ $= \frac{\sqrt{3}}{36}$ sq ft	$3 \cdot \frac{\sqrt{3}}{36}$ sq ft $= \frac{\sqrt{3}}{12}$ sq ft	$A_2 =$ $\frac{\sqrt{3}}{4} + \frac{\sqrt{3}}{12}$ sq ft
3						

Figure 6.147

Chapter 6 Review

TERMS

analytic geometry
angle of depression
angle of elevation
area
center
chaos game
circle
circumference
circumscribed polygon
cone
congruent triangles
conic section
cosine
cylinder
diameter
dimension
disk
ellipse
foci
focus
fractal
great circle
height
hyperbola

hypotenuse
inscribed polygon
leg
line of symmetry
Menger sponge
parabola
parallel lines
Parallel Postulate
parallelogram
perimeter
π (pi)
polygon
pyramid
Pythagorean Theorem
radius
rectangle
recursive process
right triangle
self-similarity (exact self-similarity and approximate self-similarity)
Sierpinski gasket
similar triangles
sine
skewed Sierpinski gasket
sphere
surface area
tangent
trapezoid
triangle
trigonometry
vertex
volume

REVIEW EXERCISES

1. Find the area and perimeter of the accompanying figure.

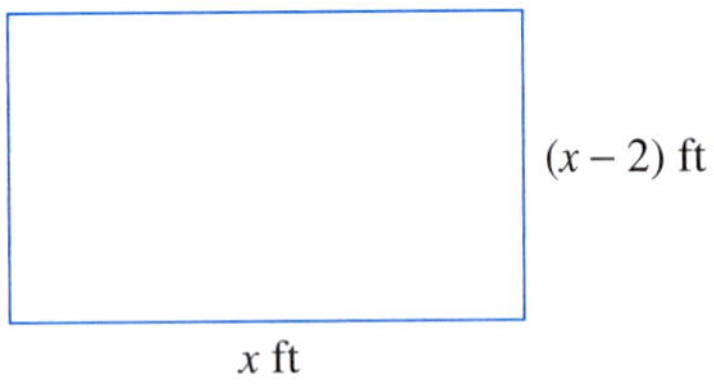

2. Find the area and perimeter of the accompanying figure.

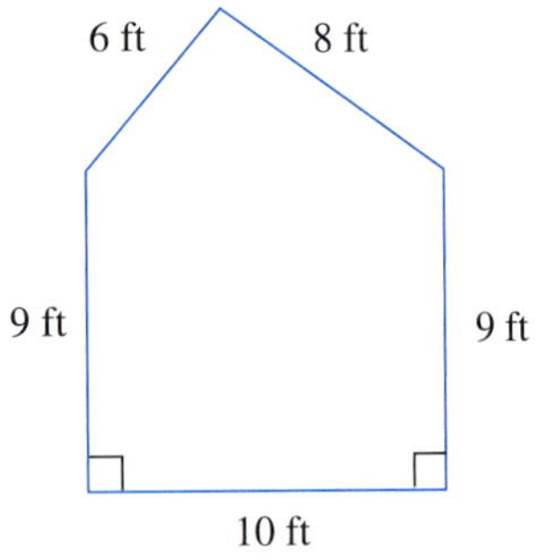

3. Find the area and perimeter of the accompanying figure.

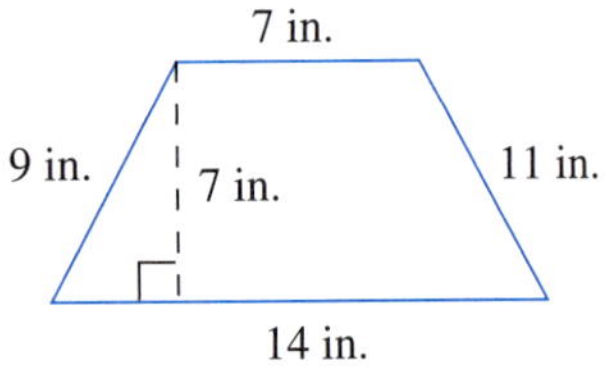

4. Find the area and perimeter of the accompanying figure.

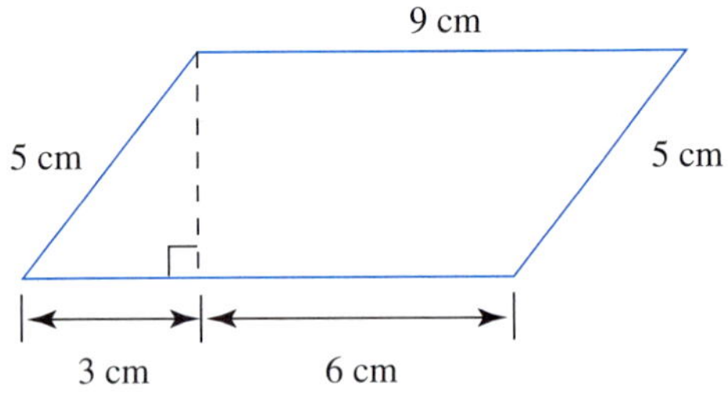

5. Find the area and perimeter of the accompanying figure.

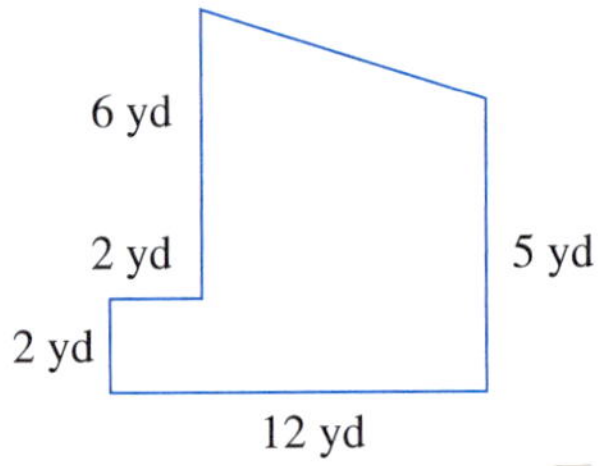

6. Find the volume of the accompanying figure.

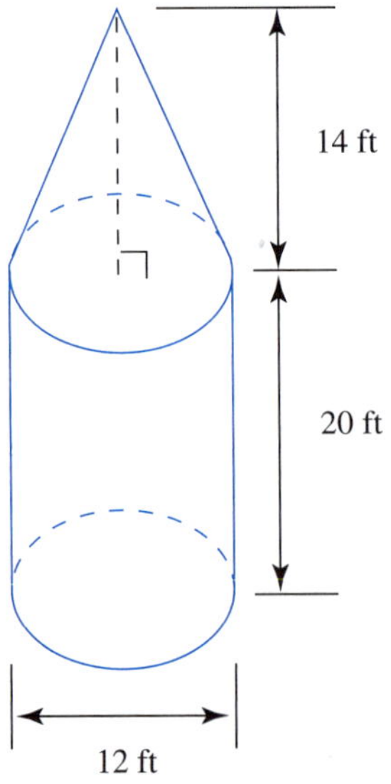

7. Find the volume and surface area of the accompanying figure.

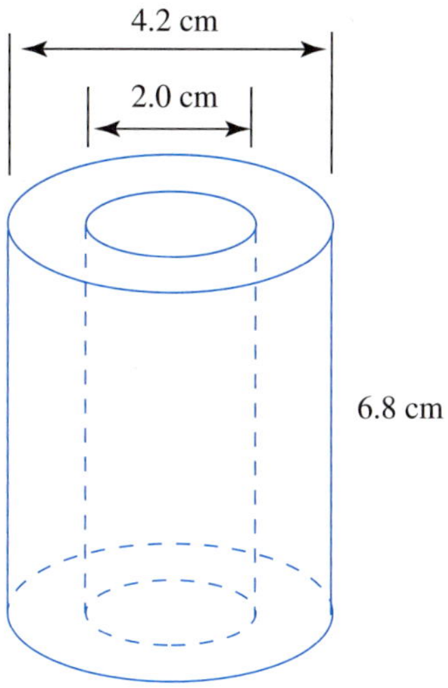

8. Find the volume and surface area of the accompanying figure.

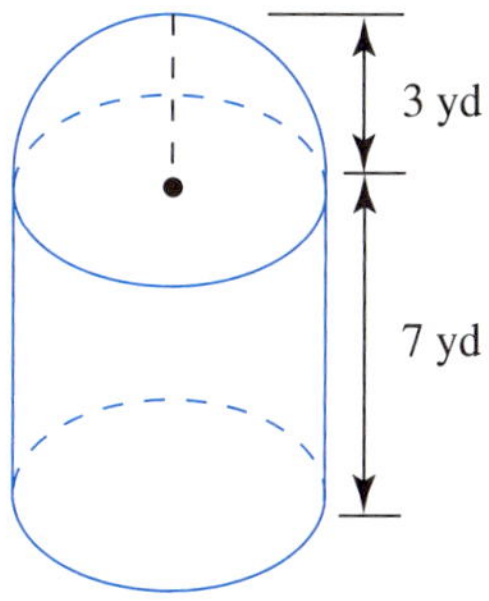

9. Find the volume and surface area of the accompanying figure.

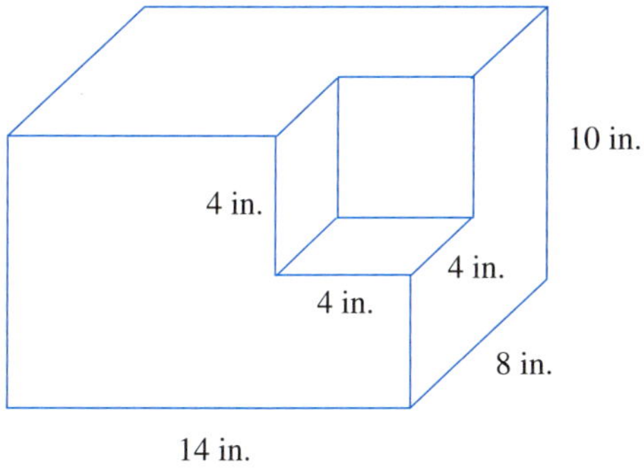

10. You jog $1\frac{3}{4}$ miles due south, then jog $\frac{1}{2}$ mile due west, and then return to your starting point via a straight-line path. How many miles have you run?

11. Tony Endres wants to fertilize his backyard. The yard is a triangle whose sides are 100 feet, 130 feet, and 160 feet. If one bag of fertilizer will cover 800 square feet, how many bags does he need?

12. An oval athletic field is the combination of a square and semicircles at opposite ends, as shown below. If the total area of the field is 1,200 square yards, find the area of the square.

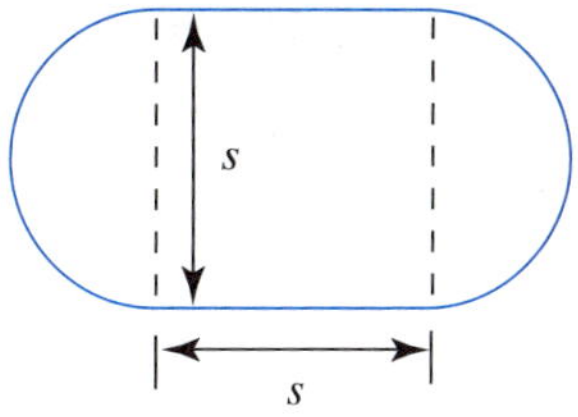

13. From a 12-inch-by-18-inch piece of cardboard, 2-inch-square corners are cut out as shown, and the resulting flaps are folded up to form an open box. Find (a) the volume and (b) the external surface area of the box.

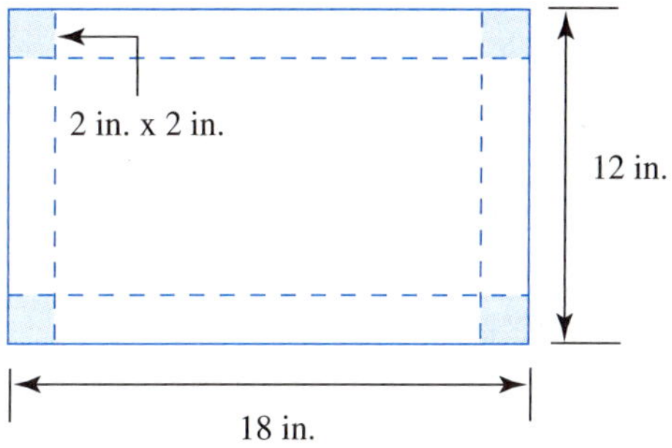

14. A regulation baseball (hardball) has a circumference of 9 inches. A regulation tennis ball has a diameter of $2\frac{1}{2}$ inches.
   a. Find and compare the volumes of the two types of balls.
   b. Find and compare the surface areas of the two types of balls.

15. Ted Nirgiotis wants to construct a plexiglass greenhouse in the shape of a pyramid. The base is 13 feet square, and the pyramid is 8 feet high.
   a. Find the volume of the greenhouse.
   b. Find the surface area of the greenhouse. (The floor is not included.)

16. For a circle of radius 1 cubit, do the following:
   a. Use $\pi = \frac{256}{81}$ (the Egyptian approximation of pi) to find the area of the circle.
   b. Use the value of π contained in a scientific calculator to find the area of the circle.
   c. Find the error of the Egyptian calculation relative to the calculator value.

17. For a sphere of radius 1 cubit, do the following.
  a. Use $\pi = \frac{256}{81}$ (the Egyptian approximation of pi) to find the volume of the sphere.
  b. Use the value of π contained in a scientific calculator to find the volume of the sphere.
  c. Find the error of the Egyptian calculation relative to the calculator value.

18. Which of the following structures has the larger storage capacity?
  a. a truncated pyramid 30 cubits high with a 6-cubit-by-6-cubit top and a 40-cubit-by-40-cubit bottom
  b. a regular pyramid 50 cubits tall with a 40-cubit-by-40-cubit base

19. Following the method of Problem 48 of the Rhind Papyrus, find a square, with a whole number as the length of its side, that has the same (approximate) area as a circle of diameter 18 units.

  HINT: Divide the sides of an 18-by-18 square into three equal segments.

20. Use the result of Exercise 19 to obtain an approximation of π.

21. Find the length of the longest object that will fit inside a rectangular trunk 5 feet long, 3.5 feet wide, and 2.5 feet high.

22. Find the length of the longest object that will fit inside a garbage can (a cylinder) that has a diameter of 2.5 feet and a height of 4 feet.

23. Use the given figures to prove the Pythagorean Theorem.

  HINT: Use the method of rearrangement and label the sides of each square in the figure on the right.

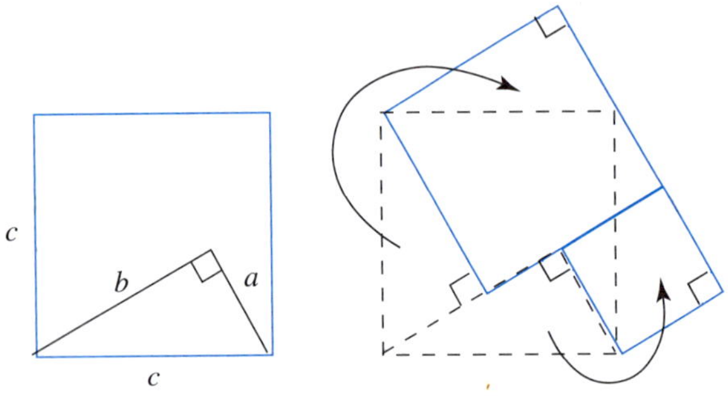

In Exercises 24 and 25, use the two-column method to prove the indicated result.

24. Given: $AD = CD$
  $AB = CB$
  Prove: $AE = CE$

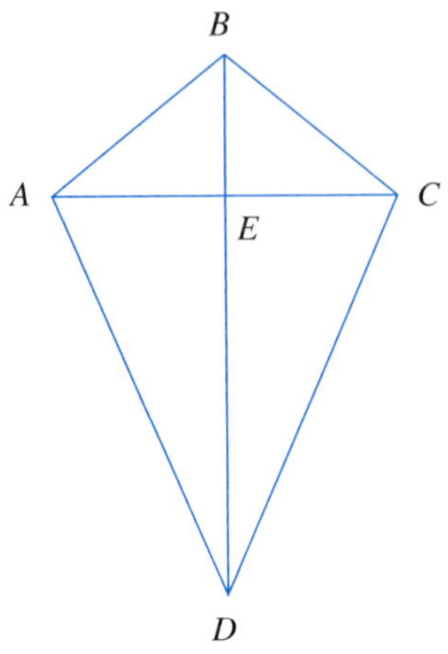

25. Given: $\angle CBA = \angle DAB$
  $BC = AD$
  Prove: $\angle CAD = \angle DBC$

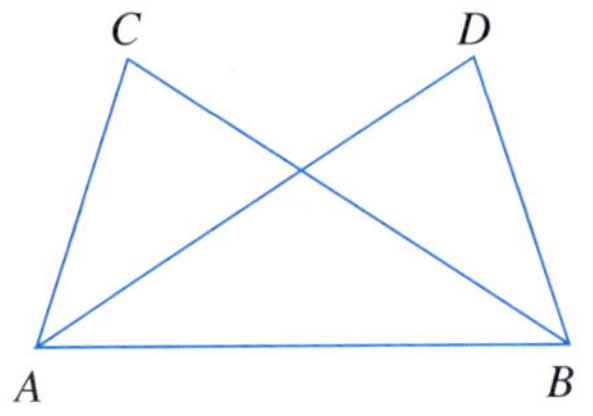

26. A regular octagon (eight-sided polygon) is circumscribed around a circle of radius r. The length of a side of the octagon is $s = 2(\sqrt{2} - 1)r$. Using the perimeter of the octagon as an approximation of the circumference of the circle, obtain an estimate of π.

27. A regular dodecagon (12-sided polygon) is inscribed in a circle of radius r. The length of a side of the dodecagon is $s = [(\sqrt{6} - \sqrt{2})/2]r$. Using the perimeter of the dodecagon as an approximation of the circumference of the circle, obtain an estimate of π.

28. Find the center and radius and sketch the graph of the circle given by

$$x^2 + y^2 - 2x - 2y + 1 = 0$$

29. Find the foci and sketch the graph of the hyperbola given by each of the following equations.
  a. $9x^2 - 4y^2 = 36$
  b. $4y^2 - 9x^2 = 36$

30. A Mylar-coated parabolic reflector is used to solar-heat water. The disk is 8 feet wide and 2 feet deep. Where should the water container be placed to heat the water most rapidly?

31. Find the foci and sketch the graph of the ellipse given by

$$9x^2 + 4y^2 = 36$$

32. An elliptical room is designed to function as a whispering gallery. If the room is 40 feet long and 20 feet wide, where should two people stand to optimize the whispering effect?

In Exercises 33 and 34, draw and label a right triangle like the one shown, and use trigonometric ratios and a calculator to find the unknown sides and angles. Round off your answers to one decimal place.

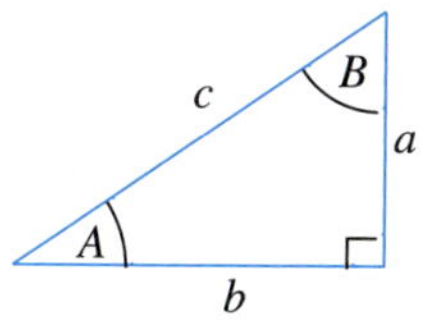

33. $A = 25.4°$ and $c = 56.1$ ft.

34. $a = 45.2$ cm and $b = 67.8$ cm

35. An electrician is at the top of a 50.0-foot utility pole. She measures the angle of depression of the base of another utility pole on the opposite side of a river. If the angle is 13.2°, how far apart are the poles?

36. While sightseeing in Paris, you visit the Eiffel Tower. From an unknown distance, you measure the angle of elevation of the top of the tower. You then move 120 feet backwards (directly away from the tower) and measure the angle of elevation of the top of the tower. If the angles are 79.7° and 73.1°, respectively, how tall (to the nearest foot) is the Eiffel tower?

37. **a.** What is the Parallel Postulate?

b. What are the two alternatives if we assume the Parallel Postulate to be false?

c. What geometric models can be used to describe a geometry based on each of the alternatives listed in part (b)?

d. Draw a triangle using each of the geometric models listed in part (c).

In Exercises 38 and 39, answer each question for (a) Euclidean, (b) Lobachevskian, and (c) Riemannian geometry.

38. Through a point not on a given line, how many parallels to the given line are there?

39. What can be said about the sum of the angles of a triangle?

40. *The Box Fractal.* To create this shape, use the following steps:

Step 1 Draw a square and fill in its interior.

Step 2 Remove some of the square:
- Subdivide the square into nine equal squares.
- Remove the four small squares at the middle of each edge.

This leaves five filled in squares, one at each corner and one at the very center. See Figure 6.148.

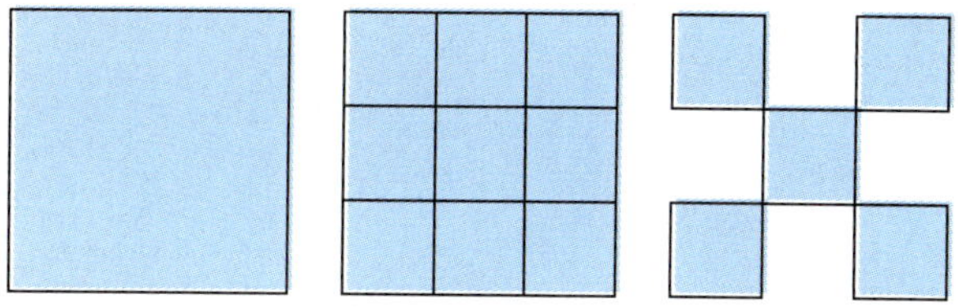

Figure 6.148
Creating a box fractal

Step 3 Remove some of the five squares from step 2 by applying the procedure from step 2 to each filled in square.

Step 4 Remove some of the squares from step 3 by applying the procedure from step 2 to each filled in square.

Step 5 (and beyond) Remove some of the five squares from the previous step by applying the procedure from step 2 to each filled in square.

a. Carefully and accurately complete the first five steps of the box fractal.

b. Find the total perimeter of steps 1–5 of the above process.

c. Find a formula for the total perimeter of step n of the above process.

d. Use the formula from part (c) to find the perimeter of the box fractal.

e. Find a formula for the total area of step n of the above process.

f. Use the formula from part (e) to find the area of the box fractal.

g. Find the dimension of the box fractal.

h. Discuss the self-similarity of the box fractal.

i. Discuss the recursive nature of the box fractal.

Answer the following questions using complete sentences and your own words.

History Questions

41. What roles did the following people play in the development and application of geometry?

- Archimedes of Syracuse
- Euclid of Alexandria
- Hypatia
- Johann Kepler
- Pythagoras of Samos
- Girolamo Saccheri
- Janos Bolyai
- Carl Friedrich Gauss
- Nikolai Lobachevsky
- Bernhard Riemann
- Thales of Miletus
- Benoit Mandelbrot

Matrices and Markov Chains

7

Markov chains are used to analyze trends and predict the future. They have many applications in business, sociology, the physical sciences, and biology. For example, in business, Markov chains can be used to predict the future success of a product. In sociology, Markov chains can be used to predict the eventual outcome of trends, such as the shrinking of the American middle class and the growth of the suburbs at the expense of central cities and rural areas. In this chapter, we investigate these two sociological problems and also see how businesses use Markov chains.

The method of Markov chains consists of using probabilistic information in the analysis of current trends in order to predict their outcomes. Probability trees are used to analyze trends, and matrices are used to summarize the probability trees and simplify the calculations. As an alternative to probability trees, matrices make the work easier. (A **matrix** is a rectangular arrangement of numbers enclosed by brackets, such as $\begin{bmatrix} 3 & -2 \\ 0 & 1 \end{bmatrix}$).

The method of Markov chains also involves solving systems of equations. Some of these systems are easily solved with the elimination method learned in intermediate algebra, but you may find that you prefer to use the Gauss-Jordan method, especially when the system involves more than two equations. The Gauss-Jordan method is based on the elimination method, but it uses matrices rather than equations. This use of matrices not only abbreviates the amount of writing and work involved in solving systems of equations, but also opens the door for the use of technology in the form of either a graphing calculator or a computer.

Matrices will be used throughout this chapter. In Section 7.0, we will review matrix terminology and notation and matrix multiplication. In Section 7.1, we will introduce Markov chains and discuss the application of matrix multiplication to Markov chains. In section 7.2 we will review the elimination method of solving systems of equations. In Section 7.3 we will further explore Markov chains and discuss the use of systems of equations in Markov Chains. In Section 7.4 we will discuss the use of the elimination method and the Gauss-Jordan method in solving larger systems of equations. In Section 7.5 we will continue to explore Markov chains and discuss the use of systems of equations in Markov chains.

7.0 Review of Matrices

Terminology and Notation

A **matrix** (plural *matrices*) is a rectangular arrangement of numbers enclosed by brackets. For example,

$$\begin{bmatrix} 3 & -2 & 0 \\ 1 & -27 & 5 \end{bmatrix}$$

is a matrix. Each number in the matrix is called an **element** or an **entry** of the matrix; 3 is an element of the above matrix, as is -27. The horizontal listings of elements are called **rows;** the vertical listings are called **columns.** The first row of the matrix

$$\begin{bmatrix} 3 & -2 & 0 \\ 1 & -27 & 5 \end{bmatrix} \leftarrow \text{first row}$$

$$\uparrow \text{ last column}$$

is "3 -2 0," and the last column is "0 5." The **dimensions** of a matrix with m rows and n columns are $m \times n$ (read "m by n"). The matrix above has two rows and three columns, so its dimensions are 2×3; that is, it is a 2×3 matrix.

The dimensions of a matrix are like the dimensions of a room. If a room is 7 feet on one side and 10 feet on the other, then the dimensions of that room are 7×10. We do not actually multiply 7 times 10 to determine the dimensions (although we would to find the area). Likewise, the dimensions of the above matrix are 2×3; we do not actually multiply 2 times 3 and get 6. If we did, we would have the *number of elements* in the matrix rather than the *dimensions* of the matrix.

A matrix with only one row is called a **row matrix.** A matrix with only one column is called a **column matrix.** A matrix is **square** if it has as many rows as columns. For example,

$\begin{bmatrix} 3 & -2 & 0 \end{bmatrix}$ is a row matrix.

$\begin{bmatrix} 26 \\ 89 \end{bmatrix}$ is a column matrix.

$\begin{bmatrix} 0 & -1 \\ 5 & 3 \end{bmatrix}$ is a square matrix.

$\begin{bmatrix} 3 & -4 & 5 \\ 0 & 18 & 2 \end{bmatrix}$ is neither a row matrix, a column matrix, nor a square matrix.

A matrix is usually labeled with a capital letter, and the entries of that matrix are labeled with the same letter in lowercase with a double subscript. The first subscript refers to the entry's row; the second subscript refers to its column. For example, b_{23} refers to the element of matrix B that is in the second row and the third column. Similarly, if

$$A = \begin{bmatrix} 5 & 7 \\ 9 & 11 \end{bmatrix}$$

then $a_{11} = 5$, $a_{12} = 7$, $a_{21} = 9$, and $a_{22} = 11$.

Matrix Multiplication

In your intermediate algebra class, you might have learned to add, subtract, and multiply matrices and to take the determinant of a matrix. In this chapter, we will be studying applications that involve the multiplication of matrices but not their addition or subtraction or the taking of determinants. Thus, we limit our discussion to matrix multiplication.

Matrix multiplication is a process that at first seems rather strange and obscure. Its usefulness will become apparent when we apply it to probability trees in the method of Markov chains later in this chapter.

EXAMPLE 1 If $A = [3 \;\; -2]$ and $B = \begin{bmatrix} 1 \\ 0 \end{bmatrix}$, find the product AB.

Solution

$$\text{AB} = [3 \;\; -2]\begin{bmatrix} 1 \\ 0 \end{bmatrix} = ?$$

To find AB,

- multiply the first element of A by the first element of B,
- then multiply the second element of A by the second element of B,
- and then add the results:

$$AB = \begin{bmatrix} \boxed{3} & \boxed{-2} \end{bmatrix}\begin{bmatrix} \boxed{1} \\ \boxed{0} \end{bmatrix} = \begin{bmatrix} \boxed{3 \cdot 1} + \boxed{-2 \cdot 0} \end{bmatrix} = [3]$$

EXAMPLE 2 If $A = \begin{bmatrix} 3 & -2 \\ 4 & 1 \end{bmatrix}$ and $B = \begin{bmatrix} 1 & 5 \\ 0 & -7 \end{bmatrix}$, find the product AB.

Solution

$$AB = \begin{bmatrix} 3 & -2 \\ 4 & 1 \end{bmatrix}\begin{bmatrix} 1 & 5 \\ 0 & -7 \end{bmatrix} = ?$$

We start with row 1 of A and column 1 of B. We multiply corresponding elements together, add the products, and put the sum in row 1, column 1 of the product matrix, as we did in Example 1.

$$AB = \begin{bmatrix} \boxed{3} & \boxed{-2} \\ 4 & 1 \end{bmatrix}\begin{bmatrix} \boxed{1} & 5 \\ \boxed{0} & -7 \end{bmatrix} = \begin{bmatrix} \boxed{3 \cdot 1} + \boxed{-2 \cdot 0} & ? \\ ? & ? \end{bmatrix} = \begin{bmatrix} \boxed{3} & ? \\ ? & ? \end{bmatrix}$$

We continue in this manner until each row of A has been multiplied by each column of B. Thus, we now multiply row 1 of A and column 2 of B and put the result in row 1, column 2 of the product matrix:

$$AB = \begin{bmatrix} \boxed{3} & \boxed{-2} \\ 4 & 1 \end{bmatrix}\begin{bmatrix} 1 & \boxed{5} \\ 0 & \boxed{-7} \end{bmatrix} = \begin{bmatrix} 3 & \boxed{3 \cdot 5} + \boxed{-2 \cdot -7} \\ ? & ? \end{bmatrix} = \begin{bmatrix} 3 & \boxed{29} \\ ? & ? \end{bmatrix}$$

We're finished with row 1; what remains is to multiply row 2 of A first by column 1 of B and then by column 2. Multiplying row 2 of A by column 1 of B gives

$$AB = \begin{bmatrix} 3 & -2 \\ \boxed{4} & \boxed{1} \end{bmatrix}\begin{bmatrix} \boxed{1} & 5 \\ \boxed{0} & -7 \end{bmatrix} = \begin{bmatrix} 3 & 29 \\ \boxed{4 \cdot 1} + \boxed{1 \cdot 0} & ? \end{bmatrix} = \begin{bmatrix} 3 & 29 \\ \boxed{4} & ? \end{bmatrix}$$

Finally, we multiply row 2 of A by column 2 of B and put the result in row 2, column 2 of the product matrix:

$$AB = \begin{bmatrix} 3 & -2 \\ \boxed{4} & \boxed{1} \end{bmatrix}\begin{bmatrix} 1 & \boxed{5} \\ 0 & \boxed{-7} \end{bmatrix} = \begin{bmatrix} 3 & 29 \\ 4 & \boxed{4 \cdot 5} + \boxed{1 \cdot -7} \end{bmatrix} = \begin{bmatrix} 3 & 29 \\ 4 & \boxed{13} \end{bmatrix}$$

The product of A and B is

$$\text{AB} = \begin{bmatrix} 3 & 29 \\ 4 & 13 \end{bmatrix}$$

■

A matrix is really just a table. Consider the table in Figure 7.1, which gives prices of CDs and DVDs.

The table could be rewritten as the matrix

$$\begin{bmatrix} 14 & 19 \\ 20 & 25 \end{bmatrix}$$

	Sale Price	**Regular Price**
CDs	\$14	\$19
DVDs	\$20	\$25

Figure 7.1

Matrix multiplication is a natural way to manipulate data in tables. If we were to purchase three CDs and two DVDs, we would calculate the sale price as

$$3 \cdot \$14 + 2 \cdot \$20 = \$82$$

and the regular price as

$$3 \cdot \$19 + 2 \cdot \$25 = \$107$$

This calculation is matrix multiplication; it is the same as

$$\begin{bmatrix} 3 & 2 \end{bmatrix}\begin{bmatrix} 14 & 19 \\ 20 & 25 \end{bmatrix} = \begin{bmatrix} 3 \cdot 14 + 2 \cdot 20 & 3 \cdot 19 + 2 \cdot 25 \end{bmatrix}$$
$$= \begin{bmatrix} 82 & 107 \end{bmatrix}$$

It is not always possible to multiply two matrices together. In the previous matrix multiplication problem if the first matrix had one extra element, we wouldn't be able to multiply.

$$\begin{bmatrix} 3 & 2 & 77 \end{bmatrix}\begin{bmatrix} 14 & 19 \\ 20 & 25 \end{bmatrix} = [3 \cdot 14 + 2 \cdot 20 + 77 \cdot ? \quad 3 \cdot 19 + 2 \cdot 25 + 77 \cdot ?]$$

It's easy to tell when two matrices cannot be multiplied: If you run out of elements to pair together when you're multiplying the first row by the first column, then the two matrices cannot be multiplied. In the above problem, we paired 3 and 14, and 2 and 20, but we had nothing to pair with 77. This tells us that the two matrices cannot be multiplied together.

Some people prefer to determine whether it's possible to multiply two matrices before they actually start multiplying. In the above problem, we could not multiply a 1×3 matrix by a 2×2 matrix.

$$\begin{bmatrix} 3 & 2 & 77 \end{bmatrix}\begin{bmatrix} 14 & 19 \\ 20 & 25 \end{bmatrix}$$

$1 \times 3 \qquad 2 \times 2$

↑ ↑

different, so we can't multiply

Previously, when 77 wasn't there, we *were* able to multiply.

$$\begin{bmatrix} 3 & 2 \end{bmatrix} \begin{bmatrix} 14 & 19 \\ 20 & 25 \end{bmatrix}$$

$$\underset{\uparrow}{1 \times 2} \quad \underset{\uparrow}{2 \times 2}$$

the same, so we can multiply

EXAMPLE 3 $A = \begin{bmatrix} 3 & -2 \\ 4 & 1 \end{bmatrix}$, $B = \begin{bmatrix} 1 & 0 \end{bmatrix}$, and $C = \begin{bmatrix} 1 \\ 0 \end{bmatrix}$.

a. Find AB, if it exists.
b. Find AC, if it exists.

Solution **a.** $AB = \begin{bmatrix} 3 & -2 \\ 4 & 1 \end{bmatrix} \begin{bmatrix} 1 & 0 \end{bmatrix} = \begin{bmatrix} 3 \cdot 1 + -2 \cdot ? & 3 \cdot 0 + -2 \cdot ? \\ 4 \cdot 1 + \quad 1 \cdot ? & 4 \cdot 0 + \quad 1 \cdot ? \end{bmatrix}$

The product AB does not exist, because we ran out of elements to pair together. Alternatively, A is a 2×2 matrix, and B is a 1×2 matrix.

$$\begin{bmatrix} 3 & -2 \\ 4 & 1 \end{bmatrix} \begin{bmatrix} 1 & 0 \end{bmatrix}$$

$$\underset{\uparrow}{2 \times 2} \quad \underset{\uparrow}{1 \times 2}$$

different, so we can't multiply

b. $AC = \begin{bmatrix} 3 & -2 \\ 4 & 1 \end{bmatrix} \begin{bmatrix} 1 \\ 0 \end{bmatrix} = \begin{bmatrix} 3 \cdot 1 + -2 \cdot 0 \\ 4 \cdot 1 + \quad 1 \cdot 0 \end{bmatrix} = \begin{bmatrix} 3 \\ 4 \end{bmatrix}$

The product of A and C is

$$\begin{bmatrix} 3 \\ 4 \end{bmatrix}$$

If we had checked the dimensions in advance, we would have found that A and C can be multiplied:

$$\begin{bmatrix} 3 & -2 \\ 4 & 1 \end{bmatrix} \begin{bmatrix} 1 \\ 0 \end{bmatrix}$$

$$\underset{\uparrow}{2 \times 2} \quad \underset{\uparrow}{2 \times 1}$$

the same, so we can multiply

This dimension check also gives the dimensions of the product matrix. The matching twos "cancel" and leave a 2×1 matrix. ■

Matrix Multiplication

If matrix A is an $m \times n$ matrix and matrix B is an $n \times p$ matrix, then the product AB exists and is an $m \times p$ matrix.

(Think: $m \times n \cdot n \times p$ yields $m \times p$.)

To find the entry in the ith row and the jth column of the product matrix AB, multiply each element of A's ith row by the corresponding element of B's jth column and add the results.

Properties of Matrix Multiplication

When you multiply numbers together, you use certain properties so automatically that you don't even realize you are using them. In particular, multiplication of real numbers is ***commutative;*** that is, $ab = ba$ (the order in which you multiply doesn't matter). Multiplication of real numbers is also ***associative;*** that is, $a(bc) = (ab)c$ (if you multiply three numbers together, it doesn't matter which two you multiply first).

Matrix multiplication is certainly different from real number multiplication. It's important to know whether matrix multiplication is commutative and associative, because people tend to use those properties without being aware of doing so.

EXAMPLE 4 Determine whether matrix multiplication is commutative by finding the product BA for the matrices B and A given in Example 2 and comparing BA with AB.

Solution

$$BA = \begin{bmatrix} 1 & 5 \\ 0 & -7 \end{bmatrix}\begin{bmatrix} 3 & -2 \\ 4 & 1 \end{bmatrix} = \begin{bmatrix} 1\cdot 3 + 5\cdot 4 & 1\cdot -2 + 5\cdot 1 \\ 0\cdot 3 + -7\cdot 4 & 0\cdot -2 + -7\cdot 1 \end{bmatrix}$$

$$= \begin{bmatrix} 23 & 3 \\ -28 & -7 \end{bmatrix}$$

The product of B and A is $BA = \left[\begin{smallmatrix} 23 & 3 \\ -28 & -7 \end{smallmatrix}\right]$.

In Example 2, we found that $AB = \left[\begin{smallmatrix} 3 & 29 \\ 4 & 13 \end{smallmatrix}\right]$.

Clearly, AB is not the same as BA; that is, $AB \neq BA$. ■

Matrix multiplication is not commutative; in general, $AB \neq BA$. This means that you have to be careful to multiply in the right order—it's easy to be careless and find BA when you're asked to find AB.

EXAMPLE 5 Check the associative property for matrix multiplication by finding $A(BC)$ and $(AB)C$ for matrices A, B, and C given below.

$$A = \begin{bmatrix} 2 & -3 \end{bmatrix} \qquad B = \begin{bmatrix} 4 & 0 & -3 \\ 2 & -1 & 5 \end{bmatrix} \qquad C = \begin{bmatrix} -3 & 5 \\ 2 & 0 \\ 1 & -1 \end{bmatrix}$$

Solution To find $A(BC)$, we first find BC and then multiply it by A. Notice that B's dimensions are 2×3 and C's dimensions are 3×2. This tells us that B and C can be multiplied and that the product will be a 2×2 matrix.

$$BC = \begin{bmatrix} 4 & 0 & -3 \\ 2 & -1 & 5 \end{bmatrix}\begin{bmatrix} -3 & 5 \\ 2 & 0 \\ 1 & -1 \end{bmatrix}$$

$$= \begin{bmatrix} 4\cdot -3 + 0\cdot 2 + -3\cdot 1 & 4\cdot 5 + 0\cdot 0 + -3\cdot -1 \\ 2\cdot -3 + -1\cdot 2 + 5\cdot 1 & 2\cdot 5 + -1\cdot 0 + 5\cdot -1 \end{bmatrix}$$

$$= \begin{bmatrix} -15 & 23 \\ -3 & 5 \end{bmatrix}$$

To find $A(BC)$, we place A on the *left* of BC and multiply. Notice that A's dimensions are 1×2 and BC's dimensions are 2×2, so A and BC can be multiplied, and the product will be a 1×2 matrix.

$$\begin{aligned} \text{A(BC)} &= [2 \ \ -3]\begin{bmatrix} -15 & 23 \\ -3 & 5 \end{bmatrix} \\ &= [2 \cdot -15 + -3 \cdot -3 \quad 2 \cdot 23 + -3 \cdot 5] \\ &= [-21 \quad 31] \end{aligned}$$

Now that we've found $A(BC)$, we need to find $(AB)C$ and then see whether the products are the same. To find $(AB)C$, we first find AB and then multiply it by C. Notice that A's dimensions are 1×2 and B's dimensions are 2×3, so A and B can be multiplied.

$$\begin{aligned} \text{AB} &= [2 \ \ -3]\begin{bmatrix} 4 & 0 & -3 \\ 2 & -1 & 5 \end{bmatrix} \\ &= [2 \cdot 4 + -3 \cdot 2 \quad 2 \cdot 0 + -3 \cdot -1 \quad 2 \cdot -3 + -3 \cdot 5] \\ &= [2 \quad 3 \ \ -21] \end{aligned}$$

To find $(AB)C$, we place AB on the *left* of C and multiply.

$$\begin{aligned} \text{(AB)C} &= [2 \ \ 3 \ \ -21]\begin{bmatrix} -3 & 5 \\ 2 & 0 \\ 1 & -1 \end{bmatrix} \\ &= [2 \cdot -3 + 3 \cdot 2 + -21 \cdot 1 \quad 2 \cdot 5 + 3 \cdot 0 + -21 \cdot -1] \\ &= [-21 \quad 31] \end{aligned}$$

Therefore, $A(BC) = (AB)C$. ■

In Example 5, $A(BC) = (AB)C$, even though the work involved in finding $A(BC)$ was different from the work involved in finding $(AB)C$. This always happens; $A(BC) = (AB)C$, provided that the dimensions of A, B, and C are such that they can be multiplied together. *Matrix multiplication is associative.*

Identity Matrices

EXAMPLE 6

If $I = \begin{bmatrix} 1 & 0 \\ 0 & 1 \end{bmatrix}$ and $A = \begin{bmatrix} 2 & 3 \\ 4 & 5 \end{bmatrix}$, find IA.

$$\begin{aligned} \text{IA} &= \begin{bmatrix} 1 & 0 \\ 0 & 1 \end{bmatrix}\begin{bmatrix} 2 & 3 \\ 4 & 5 \end{bmatrix} \\ &= \begin{bmatrix} 1 \cdot 2 + 0 \cdot 4 & 1 \cdot 3 + 0 \cdot 5 \\ 0 \cdot 2 + 1 \cdot 4 & 0 \cdot 3 + 1 \cdot 5 \end{bmatrix} \\ &= \begin{bmatrix} 2 & 3 \\ 4 & 5 \end{bmatrix} \end{aligned}$$

■

In Example 6, notice that $IA = A$. If you were to multiply in the opposite order, you would find that $AI = A$. The matrix I is similar to the number 1, because $I \cdot A = A \cdot I = A$, just as $1 \cdot a = a \cdot 1 = a$.

The matrix $\begin{bmatrix} 1 & 0 \\ 0 & 1 \end{bmatrix}$ is called an identity matrix, because multiplying this matrix by any other 2×2 matrix A yields a product *identical* to A. The matrix

HISTORICAL NOTE

Arthur Cayley & James Joseph Sylvester

1821–1895 **1814–1897**

The theory of matrices was a product of the unique partnership of Arthur Cayley and James Sylvester. They met in their twenties and were friends, colleagues, and co-authors for the rest of their lives.

Cayley's mathematical ability was recognized at an early age, and he was encouraged to study the subject. He graduated from Cambridge University at the top of his class. After graduation, Cayley was awarded a three-year fellowship that allowed him to do as he pleased. During this time, he made several trips to Europe, where he spent his time taking walking tours, mountaineering, painting, reading novels, and studying architecture, as well as reading and writing mathematics. He wrote twenty-five papers in mathematics, papers that were well received by the mathematical community.

When Cayley's fellowship expired, he found that no position as a mathematician was open to him unless he entered the clergy, so he left mathematics and prepared for a legal career. When he was admitted to the bar, he met James Joseph Sylvester.

Sylvester's mathematical ability had also been recognized at an early age. He studied mathematics at the University of London at the age of fourteen, under Augustus De Morgan (see Chapter 2 for more information on this famous mathematician). He entered Cambridge University at the age of seventeen and won several prizes. However, Cambridge would not award him his degrees because he was Jewish. He completed his bachelor's and master's degrees at Trinity College in Dublin. Many years later, Cambridge changed its discriminatory policy and gave Sylvester his degrees.

Sylvester taught science at University College in London for two years, found that he didn't like teaching science, and quit. He went to the United States, where he got a job teaching mathematics at the University of Virginia. After three months, he quit when the administration refused to discipline a student who had insulted him. After several unsuccessful attempts to obtain a teaching position, he returned to England and worked for an insurance firm as an actuary, retaining his interest in mathematics only through tutoring. Florence Nightingale was one of his private pupils. When Sylvester became thoroughly bored with insurance work, he studied for a legal career and met Cayley.

Cayley and Sylvester revived and intensified each other's interest in mathematics, and each started to write mathematics again. During his fourteen years spent practicing law,

$\begin{bmatrix} 1 & 0 & 0 \\ 0 & 1 & 0 \\ 0 & 0 & 1 \end{bmatrix}$ is also called an identity matrix, because multiplying this 3×3 matrix by any other 3×3 matrix A yields a product identical to A. An **identity matrix** is a square matrix I that has ones for each entry in the **diagonal** (the diagonal that starts in the upper left corner) and zeros for all other entries. The product of any matrix A and an identity matrix is always A, as long as the dimensions of the two matrices are such that they can be multiplied.

Properties of Matrix Multiplication

1. *There is no commutative property:* In general, $AB \neq BA$. You must be careful about the order in which you multiply.
2. *Associative property:* $A(BC) = (AB)C$, provided the dimensions of A, B, and C are such that they can be multiplied together.
3. *Identity property:* An **identity matrix** is a square matrix I that has ones for each entry in the diagonal (the diagonal that starts in the upper left corner) and zeros for all other entries. If I and A have the same dimensions, then $IA = AI = A$.

Cayley wrote almost 300 papers. Each frequently expressed gratitude to the other for assistance and inspiration. In one of his papers, Sylvester wrote that "the theorem above enunciated was in part suggested in the course of a conversation with Mr. Cayley (to whom I am indebted for my restoration to the enjoyment of mathematical life)." In another, he said, "Mr. Cayley habitually discourses pearls and rubies."

Cayley joyfully departed from the legal profession when Cambridge offered him a professorship in mathematics, even though his income suffered as a result. He was finally able to spend his life studying, teaching, and writing mathematics. He became quite famous as a mathematician, writing almost 1,000 papers in algebra and geometry, often in collaboration with Sylvester. Many of these papers are pioneering works of scholarship. Cayley also played an important role in changing Cambridge's policy that had prohibited the admission of women as students.

Sylvester was repeatedly honored for his pioneering work in algebra. He left the law but was unable to obtain a professorship in mathematics at a prominent institution until late in his life. At the age of sixty-two, he accepted a position at the newly founded Johns Hopkins University in Baltimore as its first professor of mathematics. While there, he founded the *American Journal of Mathematics,* introduced graduate work in mathematics into American universities, and generally stimulated the development of mathematics in America. He also arranged for Cayley to spend a semester at Johns Hopkins as guest lecturer. At the age of seventy, Sylvester returned to England to become Savilian Professor of Geometry at Oxford University.

Cayley and Sylvester were responsible for the theory of matrices, including the operation of matrix multiplication. Sixty-seven years after the invention of matrix theory, Heisenberg recognized it as the perfect tool for his revolutionary work in quantum mechanics. The work of Cayley and Sylvester in algebra became quite important for modern physics, particularly in the theory of relativity. Cayley also wrote on non-Euclidean geometry (see Chapter 6 for information on non-Euclidean geometry).

Arthur Cayley

James Sylvester

7.0 Exercises

In Exercises 1–10, (a) find the dimensions of the given matrix and (b) determine whether the matrix is a row matrix, a column matrix, a square matrix, or none of these.

1. $A = \begin{bmatrix} 5 & 0 \\ 22 & -3 \\ 18 & 9 \end{bmatrix}$

2. $B = \begin{bmatrix} 1 & 13 & 207 \\ -4 & 8 & 100 \\ 0 & 1 & 5 \end{bmatrix}$

3. $C = \begin{bmatrix} 23 \\ 41 \end{bmatrix}$

4. $D = \begin{bmatrix} 2 & 0 & 19 & -3 \\ 62 & 13 & 44 & 1 \\ 5 & 5 & 30 & 12 \\ 0 & 0 & 0 & 0 \end{bmatrix}$

5. $E = [3 \quad 0]$

6. $F = \begin{bmatrix} -2 & 10 \\ 4 & -3 \end{bmatrix}$

7. $G = \begin{bmatrix} 12 & -11 & 5 \\ -9 & 4 & 0 \\ 1 & 9 & 5 \end{bmatrix}$

8. $H = \begin{bmatrix} 1 & 0 \\ 0 & -1 \end{bmatrix}$

9. $J = \begin{bmatrix} 5 \\ -3 \\ 11 \end{bmatrix}$

10. $K = \begin{bmatrix} 2 & -5 & 13 & 0 \\ -1 & 4 & 3 & 6 \\ 8 & -10 & 4 & 0 \end{bmatrix}$

In Exercises 11–20, find the indicated elements of the matrices given in Exercises 1–10.

11. a_{21}

    HINT: You're asked to find the entry in row 2, column 1 of the matrix A, which is given in Exercise 1.

12. b_{23}
13. c_{21}
14. d_{34}
15. e_{11}
16. f_{22}
17. g_{12}
18. h_{21}
19. j_{21}
20. k_{23}

In Exercises 21–30, find the indicated products (if they exist) of the matrices given in Exercises 1–10.

21. **a.** AC
    b. CA
22. **a.** AE
    b. EA
23. **a.** AD
    b. DA
24. **a.** GH
    b. HG
25. **a.** CG
    b. GC
26. **a.** DK
    b. KD
27. **a.** JB
    b. BJ
28. **a.** HK
    b. KH
29. **a.** AF
    b. FA
30. **a.** AB
    b. BA

31. Use matrix multiplication and Figure 7.2 to find the sale price and the regular price of purchasing five CDs and three tapes.

	Sale Price	Regular Price
CDs	\$14	\$19
DVDs	\$20	\$25

Figure 7.2

32. Use matrix multiplication and Figure 7.2 to find the sale price and the regular price of purchasing four CDs and one tape.
33. Use matrix multiplication and Figure 7.3 to find the price of purchasing two slices of pizza and one cola at Blondie's and at SliceMan's.

	Blondie's Price	SliceMan's Price
Pizza	\$1.25	\$1.30
Cola	\$ .95	\$1.10

Figure 7.3

34. Use matrix multiplication and Figure 7.3 to find the price of purchasing six slices of pizza and two colas at Blondie's and at SliceMan's.
35. Jim, Eloise, and Sylvie are enrolled in the same four courses: English, Math, History, and Business. English and Math are 4-unit courses, while History and Business are 3-unit courses. Their grades are given in Figure 7.4.

	English	Math	History	Business
Jim	A	B	B	C
Eloise	C	A	A	C
Sylvie	B	A	B	A

Figure 7.4

Counting A's as 4 grade points, B's as 3 grade points, and C's as 2 grade points, use matrix multiplication to compute Jim's, Eloise's, and Sylvie's grade point averages (GPAs).

HINT: Start by computing Jim's total grade points without using matrices. Then set up two matrices, one that is a matrix version of Figure 7.4 with grade points rather than letter grades and another that gives the four courses' units. Set up these two matrices so that their product mirrors your by-hand calculation of Jim's total grade points. Finally, multiply by 1 over the number of units taken so that the result is GPA, not total grade points.

36. Dave, Jay, Johnny, and Larry are each enrolled in the same four courses: Communications, Broadcasting, Marketing, and Contracts. Communications and Contracts are 4-unit courses, while Broadcasting and Marketing are 3-unit courses. Their grades are given in Figure 7.5.

	Communi-cations	**Broad-casting**	**Marketing**	**Contracts**
Dave	A	B	C	A
Jay	A	B	C	B
Johnny	A	A	A	A
Larry	B	A	B	C

Figure 7.5

Counting A's as 4 grade points, B's as 3 grade points, and C's as 2 grade points, use matrix multiplication to compute Dave's, Jay's, Johnny's, and Larry's grade point averages. (See Exercise 35.)

37. Verify the associative property by finding $A(BC)$ and $(AB)C$.

$$\text{A} = \begin{bmatrix} -4 & 5 \\ 2 & 3 \end{bmatrix} \qquad \text{B} = \begin{bmatrix} -3 & 0 & -1 \\ 1 & 4 & -2 \end{bmatrix}$$

$$C = \begin{bmatrix} 0 \\ 5 \\ -1 \end{bmatrix}$$

38. Verify the associative property by finding $A(BC)$ and $(AB)C$.

$$\text{A} = \begin{bmatrix} 2 & 0 \\ -1 & 1 \\ 3 & 2 \end{bmatrix} \qquad \text{B} = \begin{bmatrix} 5 & 3 & -2 \\ 2 & 0 & 1 \end{bmatrix}$$

$$C = \begin{bmatrix} 1 \\ 2 \\ 0 \end{bmatrix}$$

In Exercises 39–44, find the product (if it exists).

39. $\begin{bmatrix} 1 & 0 \\ 0 & 1 \end{bmatrix} \cdot \begin{bmatrix} 3 & -2 \\ 4 & 0 \end{bmatrix}$

40. $\begin{bmatrix} 1 & 0 & 0 \\ 0 & 1 & 0 \\ 0 & 0 & 1 \end{bmatrix} \cdot \begin{bmatrix} -4 & 5 & 2 \\ 8 & -1 & 9 \\ -2 & 27 & 4 \end{bmatrix}$

41. $\begin{bmatrix} 1 & 0 & 0 \\ 0 & 1 & 0 \\ 0 & 0 & 1 \end{bmatrix} \cdot \begin{bmatrix} 4 & 1 & -1 \\ 5 & 12 & 3 \end{bmatrix}$

42. $\begin{bmatrix} 27 & 19 \\ 42 & 25 \end{bmatrix} \cdot \begin{bmatrix} 1 & 0 \\ 0 & 1 \end{bmatrix}$

43. $\begin{bmatrix} 19 & 7 & 34 \\ 74 & 0 & -11 \\ 13 & -2 & 44 \end{bmatrix} \cdot \begin{bmatrix} 1 & 0 & 0 \\ 0 & 1 & 0 \\ 0 & 0 & 1 \end{bmatrix}$

44. $\begin{bmatrix} 1 & 0 \\ 0 & 1 \end{bmatrix} \cdot \begin{bmatrix} 6 & 18 & -3 \\ 0 & 1 & 5 \\ -14 & 5 & -2 \end{bmatrix}$

Answer the following questions using complete sentences and your own words.

Concept Questions

45. Describe the two ways of determining whether two matrices can be multiplied.

46. Why is an identity matrix so named?

47. Give an example of two matrices A and B such that $AB = BA$.

48. Give an example of two matrices A and B such that the product AB exists but the product BA does not exist.

49. What must be true about matrices A and B if the products AB and BA both exist?

50. Make up a realistic word problem that involves matrix multiplication, as Exercises 31–36 do.

History Questions

51. Compare and contrast Arthur Cayley and James Joseph Sylvester. Be sure to discuss both the similarities and the differences between these two people. In particular, compare and contrast why each turned his back on mathematics, why each returned to mathematics, their relative success in schooling, and their relative success in employment.

52. Why did Cambridge University not award Sylvester his degrees?

53. Why was no position as a mathematician open to Cayley early in his career?

Matrix Multiplication on a Graphing Calculator

The TI-81, TI-82, TI-83, TI-85, TI-86, and Casio graphing calculators can multiply matrices (the TI-80 cannot). Consider the matrices

$$A = \begin{bmatrix} 1 & 2 \\ 3 & 4 \end{bmatrix} \quad \text{and} \quad B = \begin{bmatrix} 1 & 5 \\ 0 & -7 \end{bmatrix}$$

The TI-81, TI-82, and TI-83 require that you name these matrices "[A]" and "[B]" with brackets; the TI-85 and TI-86 require that you name them "A" and "B" without brackets. The Casio requires that you name them "Mat A" and "Mat B."

TI-85/86 users When your calculator is in "ALPHA" mode, some buttons will generate a letter of the alphabet. If your calculator has a flashing "A" on the screen, then it is in "ALPHA" mode. One way to put your calculator in "ALPHA" mode is to press the [ALPHA] button. However, some other commands will automatically put your calculator in "ALPHA" mode. Don't confuse the [2nd] and [ALPHA] buttons. The [2nd] button refers to the mathematical labels above and to the *left* of the buttons.

Entering a Matrix To enter matrix A on a TI, select "EDIT" from the "MATRIX" menu and then matrix A from the "MATRX EDIT" menu:

TI-81/82/83	■ Press [MATRX] ■ Use the [→] button to highlight "EDIT" ■ Highlight option 1 "[A]"* ■ Press [ENTER]
TI-85/86	■ Press [2nd] [MATRX] ■ Press [EDIT]. (i.e. [F2]) ■ Press [A] [MATRX]†
Casio	■ Press [MENU] ■ Use the arrow buttons to highlight "MAT" ■ Press [EXE] ■ Use the arrow buttons to highlight "Mat A"

Then enter the dimensions of our 2×2 matrix A by typing

2 [ENTER] 2 [ENTER]

To enter the elements, type

1 [ENTER] 2 [ENTER] 3 [ENTER] 4 [ENTER]

---

*Option 1 is automatically highlighted. If we were selecting some other option, we would use the [↑] and [↓] buttons to highlight it.

†The [LOG] button becomes the [A] button when the calculator is in "ALPHA" mode. Usually this requires preceding the [A] button with the [ALPHA] button. However, the TI-85/86 was automatically placed in "ALPHA" mode when you pressed [EDIT], as shown by the flashing "A" on the screen.

In a similar manner, enter matrix B. When you're done entering, type 2nd QUIT.

Note that the calculator uses double subscript notation. When we entered the "4" on page 516, the calculator showed that this was entered as a_{22}.

Viewing a Matrix To view matrix A, type the following:

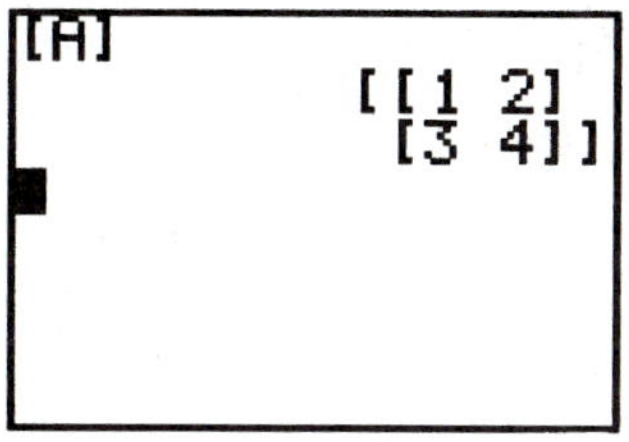

FIGURE 7.6
Matrix A

TI-81	Type 2nd [A] ENTER, and matrix [A] will appear on the screen, as shown in Figure 7.6
TI-82/83	▪ Type MATRX ▪ Highlight option 1: "[A]" ▪ Press ENTER, and "[A]" will appear on the screen ▪ Press ENTER, and the matrix itself will appear on the screen, as shown in Figure 7.6
TI-85/86	Type ALPHA A ENTER and matrix A will appear on the screen, as shown in Figure 7.6
Casio	▪ Press MENU ▪ Use the arrow buttons to highlight "MAT" and press EXE ▪ Use the arrow buttons to highlight "Mat A" ▪ Press EXE, and the matrix will appear on the screen, similar to Figure 7.6

Multiplying Two Matrices To calculate AB, make the screen read "A * B" or "[A] * [B]" or "MatA × MatB" by typing this:

TI-81	2nd [A] × 2nd [B] ENTER
TI-82/83	MATRX 1 × MATRX 2 ENTER
TI-85/86	ALPHA A × ALPHA B ENTER
Casio	▪ Press MENU ▪ Use the arrow buttons to highlight "RUN" and press EXE ▪ Press OPTN ▪ Press MAT (i.e., F2) ▪ Make the screen read "Mat A × Mat B" by typing MAT ALPHA A × MAT ALPHA B ▪ Press EXE

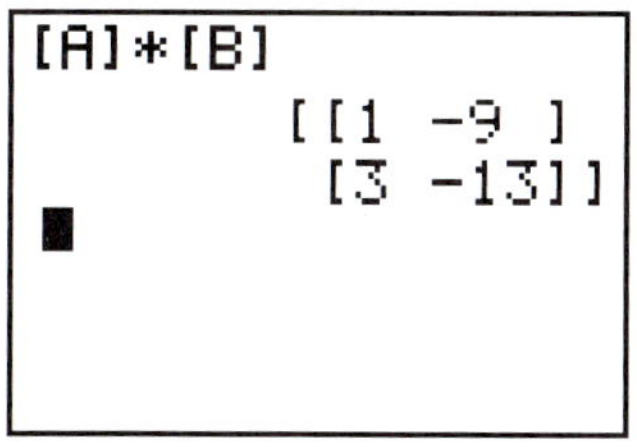

FIGURE 7.7

The result on a TI-83 is shown in Figure 7.7; other calculators' results are similar.

Exercises

In Exercises 54–62, use the following matrices.

$$A = \begin{bmatrix} 5 & -7 \\ 3 & 9 \end{bmatrix} \quad B = \begin{bmatrix} 7 & 9 \\ -8 & 0 \end{bmatrix}$$

$$C = \begin{bmatrix} 8 & -12 & 13 \\ 52 & 17 & -31 \\ 72 & 28 & -15 \end{bmatrix}$$

$$D = \begin{bmatrix} 14 & -22 & 53 \\ 94 & -15 & -35 \\ 83 & 0 & 7 \end{bmatrix} \quad E = \begin{bmatrix} 5 & -31 \\ 83 & -33 \\ 60 & 0 \end{bmatrix}$$

$$F = \begin{bmatrix} 93 & -11 & 39 \\ 53 & 66 & 83 \end{bmatrix}$$

In Exercises 54–55, find the indicated product (a) by hand and (b) with a graphing calculator. Check your work by comparing the solutions to parts (a) and (b); answers are not given in the back of the book.

54. AB

55. BA

In Exercises 56–62, find the indicated product with a graphing calculator. (When a matrix is too big for the screen, use the arrow keys to view it.)

56. a. CD
b. DC

57. a. EF
b. FE

58. a. AB
b. BA

59. a. $(BF)C$
b. $B(FC)$

60. a. $(DE)A$
b. $D(EA)$

61. a. C^2
b. C^5

62. a. D^2
b. D^6

7.1 Markov Chains

Suppose we repeatedly observe the status of some characteristic at successive points in time, where the observations' outcomes are elements of some finite set. For example, a college student might observe her class standing at the beginning of each school year. These observations' outcomes are limited to elements of the finite set {freshman, sophomore, junior, senior, quit, graduated}. If the student is a freshman in 2000, then she could be a freshman or a sophomore or quit in 2001, each with a certain probability. If the probability of making a certain observation at a certain time depends only on the outcome of the immediately preceding observation, then these observations and their probabilities form a **Markov chain.** Our college student's observations would form a Markov chain; the probability that she graduates in 2004 depends only on her status in 2003 and not on her status in any previous year.

Markov chains were developed in the early 1900s by Andrei Markov, a Russian mathematician. They have many applications in the physical sciences, business, sociology, and biology. In business, Markov chains are used to analyze data on customer satisfaction with a product, to assess the effect of the product's advertising, and to predict what portion of the market the product will eventually command. In sociology, Markov chains are used to analyze sociological trends, such as the shrinking of the American middle class and the growth of the suburbs at the expense of central cities and rural areas, and to predict the eventual outcome of such trends. Markov chains are also used to predict the weather, to analyze genetic inheritance, and to predict the daily fluctuation in a stock's price.

EXAMPLE 1 Bif is a laundry detergent. The company that makes Bif has just launched a new advertising campaign. A market analysis indicates that, as a result of this campaign, 40% of the consumers who currently *do not* use Bif will buy it the next time they buy a detergent. Another market analysis has studied customer loyalty toward Bif. It indicates that 80% of the consumers who currently *do* use Bif will buy it again the next time they buy a detergent. Rewrite these data in probability form and find the complements of these events.

Solution We are given

$$p(\text{next purchase is Bif} \mid \text{current purchase is Bif}) = 0.8$$
$$p(\text{next purchase is Bif} \mid \text{current purchase is not Bif}) = 0.4$$

We can compute the complements:

$$p(\text{next purchase is not Bif} \mid \text{current purchase is Bif}) = 1 - 0.8 = 0.2$$
$$p(\text{next purchase is not Bif} \mid \text{current purchase is not Bif}) = 1 - 0.4 = 0.6$$

These relationships are summarized in Figure 7.8.

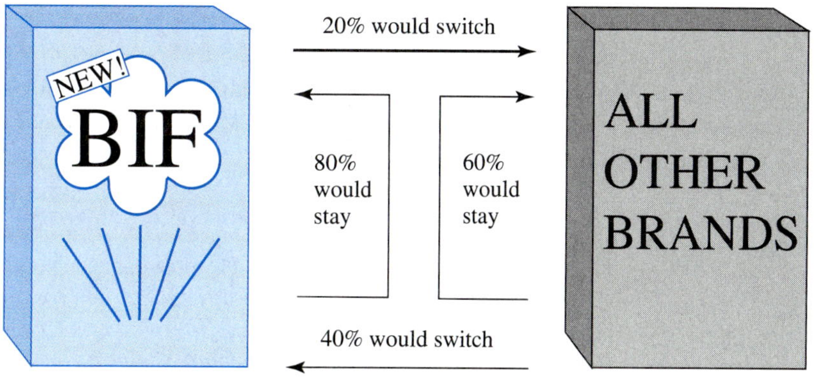

FIGURE 7.8

■

Notice that Example 1 describes a Markov chain:

- The observations are made at successive points in time (shopping trips at which a detergent is purchased).
- The observations' outcomes are elements of the finite set {Bif is purchased, something other than Bif is purchased}, or {B, B′} for short.
- The probability of making a certain observation at a certain time depends only on the outcome of the immediately preceding observation (the probability that the next purchase is Bif depends only on the current purchase, not on any previous purchases).

The observations' outcomes are called **states.** The states in Example 1 are "a consumer purchases Bif" ("Bif" or B for short) and "a consumer purchases something other than Bif" ("not Bif" or B′ for short). All states manifest themselves as both **current states** and **following states.** The current states in Example 1 are "a consumer *currently* purchases Bif" and "a consumer *currently* purchases something other than

Bif," and the next following states are "a consumer will purchase Bif *the next time he or she purchases a detergent*" and "a consumer will purchase something other than Bif *the next time he or she purchases a detergent.*"

Transition Matrices

A **transition matrix** is a matrix whose entries are the probabilities of passing from current states to following states. A transition matrix has a row and a column for each state. The rows refer to current manifestations of those states, while the columns refer to later (or following) manifestations of those states. A transition matrix is read like a chart; an entry in a certain row and column represents the probability of making a transition from the current state represented by that row to the following state represented by that column.

EXAMPLE 2 Write the transition matrix T for the data in Example 1.

Solution Our states are "Bif" and "not Bif" (or just B and B′); therefore, our transition matrix T will be the following 2×2 matrix:

first following state

$$T = \begin{array}{c} \begin{matrix} \text{B} & \text{B}' \end{matrix} \\ \begin{bmatrix} 0.8 & 0.2 \\ 0.4 & 0.6 \end{bmatrix} \end{array} \begin{matrix} \text{B} \\ \text{B}' \end{matrix}$$

current state

The entry 0.4 is the probability that a consumer makes a transition from buying something other than Bif (current state is B′) to buying Bif (first following state is B). ■

We can make several observations about transition matrices.

1. *A transition matrix must be square*, because there is one row and one column for each state. If you create a transition matrix that is not square, go back and see which state you did not list as both a current state and a following state.
2. *Each entry in a transition matrix must be between* 0 *and* 1 *(inclusive)*, because the entries are probabilities. If you create a transition matrix that has an entry that is less than 0 or greater than 1, go back and find your error.
3. *The sum of the entries of any row must be* 1, because the entries of a row are the probabilities of changing from the state represented by that row to *any* of the possible following states. If you create a transition matrix that has a row that does not add to 1, go back and find your error.

Transition Matrix Observations

1. A transition matrix must be square.
2. Each entry in a transition matrix must be between 0 and 1 (inclusive).
3. The sum of the entries of any row must be 1.

The transition matrix in Example 2 shows the transitional trends in the consumers' selection of their next detergent purchases. What effect will these trends have on Bif's success as a product? Will Bif's market share increase or decrease?

Probability Matrices

EXAMPLE 3 A marketing analysis for Bif shows that Bif currently commands 25% of the market. Write this, and its complement, in probability form.

Solution We are given

$$p(\text{current purchase is Bif}) = .25$$

Its complement is

$$p(\text{current purchase is not Bif}) = 1 - .25 = .75$$ ■

A **probability matrix** is a row matrix in which each entry is the probability of a possible state. The columns of a probability matrix must be labeled in the same way as the rows and columns of the transition matrices.

EXAMPLE 4 Write the probability matrix P for the data in Example 3.

Solution A probability matrix has one row, with a column for each state; our states are "Bif" and "not Bif," so our probability matrix will be a 1×2 matrix.

$$\begin{matrix} & \text{B} & \text{B}' \\ \text{p} = & [.25 & .75] \end{matrix}$$ ■

We can make several observations about probability matrices, which parallel the observations made earlier about transition matrices.

Probability Matrix Observations

1. A probability matrix is a row matrix.
2. Each entry in a probability matrix must be between 0 and 1 (inclusive).
3. The sum of the entries of the row must be 1.

Using Markov Chains to Predict the Future

EXAMPLE 5 Use a probability tree to predict Bif's market share after the first following purchase. In other words, find $p(\text{1st following purchase is Bif})$ and $p(\text{1st following purchase is not Bif})$.

Solution This can be computed with the tree in Figure 7.9 on page 591. We want $p(\text{1st following purchase is Bif})$, which is the sum of the probabilities of the limbs that stop at Bif. Similarly, $p(\text{1st following purchase is not Bif})$ is the sum of the probabilities of the limbs that stop at not Bif.

$$\left.\begin{aligned} p(\text{1st following purchase is Bif}) &= (0.25 \cdot 0.8) + (0.75 \cdot 0.4) \\ &= 0.2 + 0.3 = 0.5 \\ p(\text{1st following purchase is not Bif}) &= (0.25 \cdot 0.2) + (0.75 \cdot 0.6) \\ &= 0.05 + 0.45 = 0.5 \end{aligned}\right\} \text{from Figure 7.9}$$

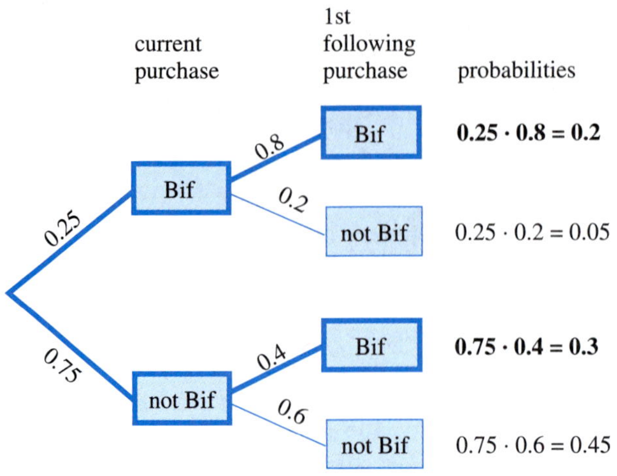

FIGURE 7.9

In other words, after the first following purchase, Bif will command 50% of the market (up from a previous 25%). Remember, though, that this is only a prediction, based on the assumption that current trends continue unchanged. ■

Notice that the calculations done in Example 5 are the same ones that are done in computing the product of the probability matrix P and the transition matrix T:

$$\begin{aligned} PT &= [0.25 \quad 0.75] \cdot \begin{bmatrix} 0.8 & 0.2 \\ 0.4 & 0.6 \end{bmatrix} \\ &= [0.25 \cdot 0.8 + 0.75 \cdot 0.4 \quad 0.25 \cdot 0.2 + 0.75 \cdot 0.6] \\ &= [0.2 + 0.3 \quad 0.05 + 0.45] \\ &= [0.5 \quad 0.5] \end{aligned}$$

Recall that matrix multiplication is not commutative; that is, $PT \neq TP$. You must be careful to multiply in the correct order.

This parallelism between trees and matrices is illustrated in Figure 7.10.

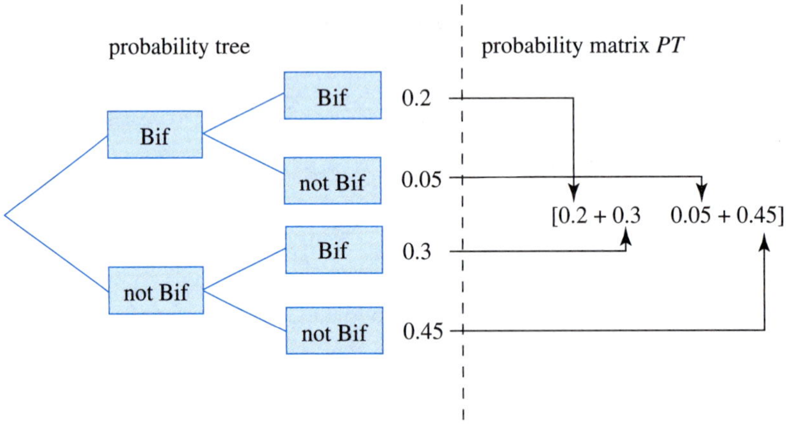

FIGURE 7.10

EXAMPLE 6 Use a probability tree to predict Bif's market share after the second following purchase.

Solution This can be computed with the tree shown in Figure 7.11. This tree is the tree from Example 5, altered to include one more set of branches. We want p(2nd following purchase is Bif), which is the sum of the probabilities of the limbs that stop at Bif under "2nd following purchase."

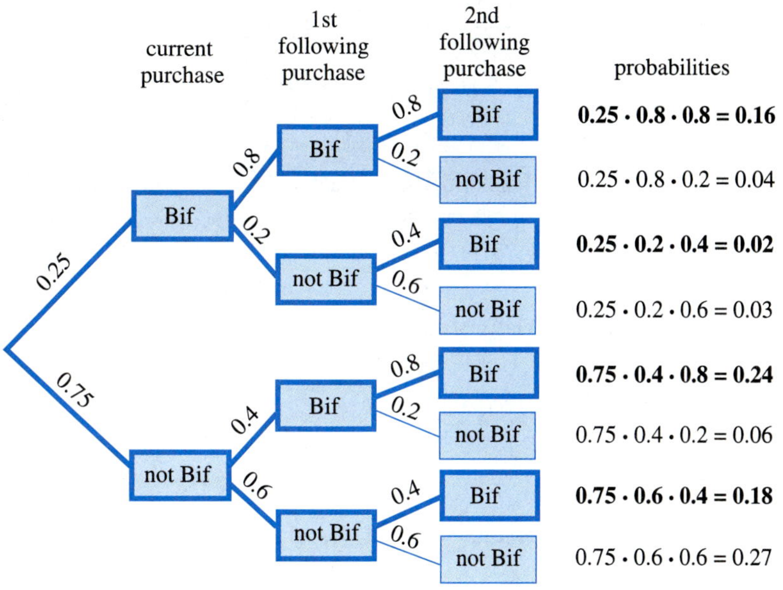

FIGURE 7.11

Therefore,

$$\begin{aligned} p(\text{2nd following purchase is Bif}) &= (0.25 \cdot 0.8 \cdot 0.8) + (0.25 \cdot 0.2 \cdot 0.4) \\ &\quad + (0.75 \cdot 0.4 \cdot 0.8) + (0.75 \cdot 0.6 \cdot 0.4) \\ &= 0.6 \end{aligned}$$

In other words, after the second following purchase, Bif will command 60% of the market. Remember, though, that this is only a prediction, based on the assumption that current trends continue unchanged. ■

Notice that the calculation done in Example 6 could be rewritten in a factored form:

$$\begin{aligned} &(\boxed{0.25 \cdot 0.8} \cdot 0.8) + (\boxed{0.25 \cdot 0.2} \cdot 0.4) + (\boxed{0.75 \cdot 0.4} \cdot 0.8) + (\boxed{0.75 \cdot 0.6} \cdot 0.4) \\ &= (\boxed{0.25 \cdot 0.8} + \boxed{0.75 \cdot 0.4})(0.8) + (\boxed{0.25 \cdot 0.2} + \boxed{0.75 \cdot 0.6})(0.4) \quad \text{factoring} \\ &= (\quad \boxed{0.5} \quad)(0.8) + (\quad \boxed{0.5} \quad)(0.4) \end{aligned}$$

from PT   from PT

Part of this ($\boxed{0.25 \cdot 0.8} + \boxed{0.75 \cdot 0.4} = \boxed{0.5}$ and $\boxed{0.25 \cdot 0.2} + \boxed{0.75 \cdot 0.6} = \boxed{0.5}$) is the same as the calculation we did in finding PT. And the whole thing is the same calculation that is done in computing row 1, column 1 of the product of the probability matrix PT and the transition matrix T:

$$\begin{aligned} PT^2 &= \quad PT \quad \cdot \quad T \\ &= [0.5 \quad 0.5] \cdot \begin{bmatrix} 0.8 & 0.2 \\ 0.4 & 0.6 \end{bmatrix} \\ &= [0.5 \cdot 0.8 + 0.5 \cdot 0.4 \qquad 0.5 \cdot 0.2 + 0.5 \cdot 0.6] \\ &= [0.6 \quad 0.4] \end{aligned}$$

If this parallelism between probability trees and matrices continues, then p(2nd following purchase is not Bif), when computed from the above tree, should match the second entry in PT^2. From the tree:

p(2nd following purchase is not Bif) =

$$(\boxed{0.25 \cdot 0.8} \cdot 0.2) + (\boxed{0.25 \cdot 0.2} \cdot 0.6) + (\boxed{0.75 \cdot 0.4} \cdot 0.2) + (\boxed{0.75 \cdot 0.6} \cdot 0.6)$$

$$= (\boxed{0.25 \cdot 0.8} + \boxed{0.75 \cdot 0.4})(0.2) + (\boxed{0.25 \cdot 0.2} + \boxed{0.75 \cdot 0.6})(0.6) \quad \text{factoring}$$

$$= (\boxed{0.5})(0.2) + (\boxed{0.5})(0.6)$$

This does indeed match the second entry of PT^2. Predictions of Bif's future market shares can be made either with trees or with matrices. The matrix work is certainly simpler, both to set up and to calculate.

EXAMPLE 7 Use matrices to predict Bif's market share after the third following purchase.

Solution Bif's market share will be an entry in PT^3. One way to do this calculation is to observe that

$$PT^3 = P(T^2T) = (PT^2)T \quad \text{associative property}$$

and to use the previously calculated PT^2:

$$PT^3 = (PT^2)T = [0.6 \quad 0.4] \cdot \begin{bmatrix} 0.8 & 0.2 \\ 0.4 & 0.6 \end{bmatrix} = [0.64 \quad 0.36]$$

Thus, p(3rd following purchase is Bif) = 0.64. [Also, p(3rd following purchase is not Bif) = 0.36.] This means that three purchases after the time when the market analysis was done, 64% of the detergents purchased will be Bif (*if current trends continue*). In other words, Bif's market share will be 64%. Bif will be doing quite well—recall that it had a market share of only 25% at the beginning of the new advertising campaign. ■

Predictions with Markov Chains

1. *Create the probability matrix P.* P is a row matrix whose entries are the initial probabilities of the states.
2. *Create the transition matrix T.* T is the square matrix whose entries are the probabilities of passing from current states to first following states. The rows refer to the current states, and the columns refer to the next following states.
3. *Calculate PT^n.* The matrix PT^n is a row matrix whose entries are the probabilities of the nth following states. Be careful that you multiply in the correct order: $PT^n \neq T^nP$.

EXAMPLE 8 In sociology, a family is considered to belong to the lower class, the middle class, or the upper class, depending on the family's total annual income. Sociologists have found that the strongest determinant of an individual's class is the class of his or her parents.

a. Convert the data given on the next page (U.S. Bureau of the Census; *Statistical Abstract of the United States, 1996*) on family incomes in 1994 into a probability matrix.

Class	Family Income	Percent of Population
lower class	under $15,000	16%
middle class	$15,000–$74,999	67%
upper class	$75,000 or more	17%

b. Census data suggest the following (illustrated in Figure 7.12):

- Of those individuals whose parents belong to the lower class, 21% will become members of the middle class, and 1% will become members of the upper class.
- Of those individuals whose parents belong to the middle class, 6% will become members of the lower class, and 4% will become members of the upper class.
- Of those individuals whose parents belong to the upper class, 1% will become members of the lower class, and 10% will become members of the middle class.

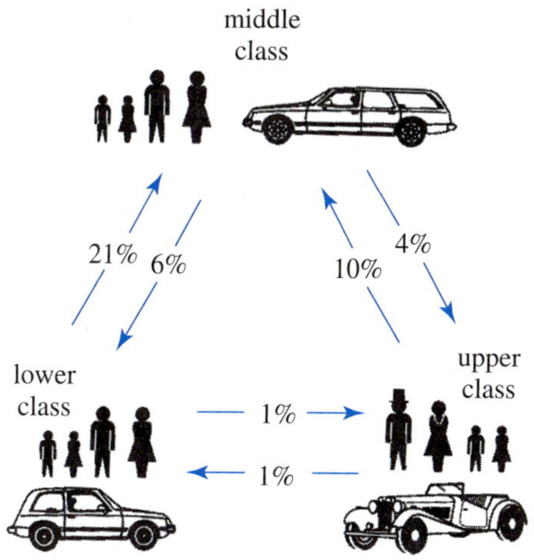

FIGURE 7.12

Convert this information into a transition matrix.

c. Predict the percent of U.S. families in the lower, middle, and upper classes after one generation.

d. Predict the percent of U.S. families in the lower, middle, and upper classes after two generations.

Solution **a.** *Creating the probability matrix P* Recall that a probability matrix has one row, with a column for each state. Our states are "lower," "middle," and "upper"; our probability matrix will be a 1×3 matrix.

$$P = \begin{matrix} \text{low} & \text{mid} & \text{up} \\ [0.16 & 0.67 & 0.17] \end{matrix}$$

b. *Creating the transition matrix T* Because we have three classes, the transition matrix T will be a 3×3 matrix, with rows referring to the current state, or class of the parents, and columns referring to the following state, or class of the child. We are given the following portion of T:

class of child

$$T = \begin{bmatrix} ? & 0.21 & 0.01 \\ 0.06 & ? & 0.04 \\ 0.01 & 0.10 & ? \end{bmatrix} \begin{matrix} \text{low} \\ \text{mid} \\ \text{up} \end{matrix}$$

(columns: low, mid, up; rows: class of parents)

Filling in the blanks is easy. If 21% of the children of lower-class parents become members of the middle class and 1% become members of the upper class, that leaves 78% to remain members of the lower class. [Alternatively, the entries in any row must add to 1, so the missing entry is $1 - (0.21 + 0.01) = 0.78$.] Similar calculations fill in the other two blanks. Thus, we have

class of child

$$T = \begin{bmatrix} 0.78 & 0.21 & 0.01 \\ 0.06 & 0.90 & 0.04 \\ 0.01 & 0.10 & 0.89 \end{bmatrix} \begin{matrix} \text{low} \\ \text{mid} \\ \text{up} \end{matrix}$$

(columns: low, mid, up) class of parents

c. *Predicting the distribution after one generation* We are asked to calculate PT^1, the probabilities of the first following generation.

$$PT^1 = PT = [0.16 \quad 0.67 \quad 0.17] \cdot \begin{bmatrix} 0.78 & 0.21 & 0.01 \\ 0.06 & 0.90 & 0.04 \\ 0.01 & 0.10 & 0.89 \end{bmatrix}$$

$$= [0.1667 \quad 0.6536 \quad 0.1797]$$

$$\approx \begin{matrix} \text{low} & \text{mid} & \text{up} \\ [0.17 & 0.65 & 0.18] \end{matrix}$$

This means that, if current trends continue, in one generation's time the lower class will grow from 16% of the population to approximately 17%, the middle class will shrink from 67% to 65%, and the upper class will grow from 17% to 18%.

✓ Notice that the row adds to 1, as it should.

d. *Predicting the distribution after two generations* We are asked to calculate PT^2, the probabilities of the second following generation. The easiest way to do this calculation is to observe that $PT^2 = (PT)T$ and to use the previously calculated PT. Our answers will be more accurate if we do not round off PT before multiplying it by T.

$$PT^2 = (PT)T = [0.1667 \quad 0.6536 \quad 0.1797] \cdot \begin{bmatrix} 0.78 & 0.21 & 0.01 \\ 0.06 & 0.90 & 0.04 \\ 0.01 & 0.10 & 0.89 \end{bmatrix}$$

$$= [0.171039 \quad 0.641217 \quad 0.187744]$$

$$\approx \begin{matrix} \text{low} & \text{mid} & \text{up} \\ [0.17 & 0.64 & 0.19] \end{matrix}$$

This means that, if current trends continue, in two generations' time the lower class will grow from 16% of the population to approximately 17%, the middle class will shrink from 67% to 64%, and the upper class will grow from 17% to 19%.

✓ Notice that the row adds to 1, as it should.

■

HISTORICAL NOTE

Andrei Andreevich Markov

1856–1922

Novosti Information Agency

Andrei Markov

Andrei Markov lived most of his life in St. Petersburg, Russia. During his lifetime, St. Petersburg was the capital of czarist Russia, a major seaport and commercial center, as well as an international center of literature, theater, music, and ballet. Markov's family belonged to the upper class; his father worked for the forestry department and managed a private estate.

In high school, Markov showed a talent for mathematics but generally was not a good student. He studied mathematics at St. Petersburg University, where he received his bachelor's, master's, and doctor's degrees. He then went on to teach at St. Petersburg University. The head of the mathematics department, P. L. Chebyshev, was a famous mathematician and statistician. Markov became a consistent follower of Chebyshev's ideas. Nominated by Chebyshev, he was elected to the prestigious St. Petersburg Academy of Sciences.

Markov's research was primarily in the areas of statistics, probability theory, calculus, and number theory. His most famous work, on Markov chains, was motivated solely by theoretical concerns. In fact, he never wrote about their applications other than in a linguistic analysis of Pushkin's *Eugene Onegin*.

During the early twentieth century, Markov participated in the liberal movement that climaxed in the Russian Revolution. When the czar overruled the election of author and revolutionary Maxim Gorky to the St. Petersburg Academy of Sciences, Markov wrote letters of protest to academic and state officials. When the czar dissolved the Duma (an elected assembly), Markov denounced the czarist government. When the government celebrated the three-hundredth anniversary of the House of Romanov (the czars' family) Markov organized a celebration of the two-hundredth anniversary of the publishing of Jacob Bernoulli's book on probabilities, *Ars Conjectandi* (see Section 3.1). After the czar finally abdicated, Markov asked the Academy to send him to teach mathematics at a secondary school in a small country town in the center of Russia. He returned to St. Petersburg after a winter of famine. Soon after his return, his health declined rapidly, and he died.

The first page of Markov's *Calculus of Probabilities*

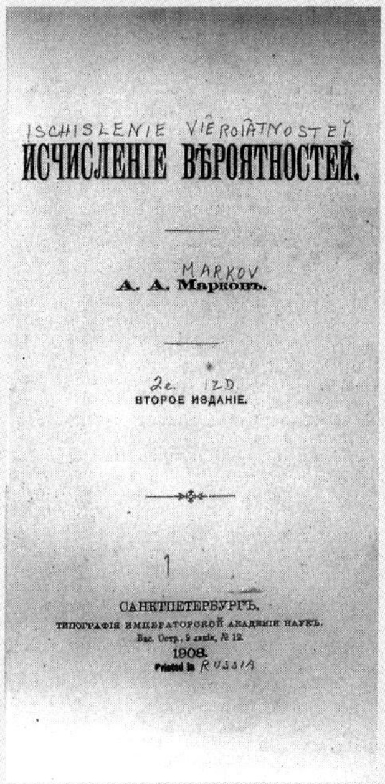

Brown University

7.1 Exercises

In Exercises 1–6, (a) rewrite the given data (and, if appropriate, their complements) in probability form and (b) convert these probabilities into a probability matrix.

1. A marketing analysis shows that KickKola currently commands 14% of the cola market.
2. A marketing analysis shows that SoftNWash currently commands 26% of the fabric softener market.
3. A census report shows that currently, 32% of the residents of Metropolis own their own home and that 68% rent.
4. A survey shows that 23% of the shoppers in seven midwestern states regularly buy their groceries at Safe Shop, 29% regularly shop at PayNEat, and the balance shop at any one of several smaller markets.

5. Silver's Gym currently commands 48% of the health club market in Metropolis, Fitness Lab commands 37%, and ThinNFit commands the balance of the market.
6. Smallville has three Chinese restaurants: Asia Gardens, Chef Chao's, and Chung King Village. Currently, Asia Gardens gets 41% of the business, Chef Chao's gets 33%, and Chung King Village gets the balance.

In Exercises 7–12, (a) rewrite the given data in probability form and (b) convert the data into a transition matrix.

7. A marketing analysis for KickKola indicates that 12% of the consumers who do not currently drink KickKola will purchase KickKola the next time they buy a cola (in response to a new advertising campaign) and that 63% of the consumers who currently drink KickKola will purchase it the next time they buy a cola.
8. A marketing analysis for SoftNWash indicates that 9% of the consumers who do not currently use SoftNWash will purchase SoftNWash the next time they buy a fabric softener (in response to a free sample sent to selected consumers) and that 29% of the consumers who currently use SoftNWash will purchase it the next time they buy a fabric softener.
9. The Metropolis census report shows that 12% of the renters plan to buy a home in the next 12 months and that 3% of the homeowners plan to sell their home and rent instead.
10. The survey for Safe Shop indicates that 8% of the consumers who currently shop at Safe Shop will purchase their groceries at PayNEat the next time they shop and that 5% will switch to some other store. Also, 12% of the consumers who currently shop at PayNEat will purchase their groceries at Safe Shop the next time they shop, and 2% will switch to some other store. In addition, 13% of the consumers who currently shop at neither store will purchase their groceries at Safe Shop the next time they shop, and 10% will shop at PayNEat.
11. An extensive survey of Metropolis's gym users indicates that 71% of the current members of Silver's will continue their annual membership when it expires, 12% will quit and join Fitness Lab, and the rest will quit and join ThinNFit. Fitness Lab has been unable to keep its equipment in good shape, and as a result 32% of its members will defect to Silver's and 34% will leave for ThinNFit. ThinNFit's members are quite happy, and as a result, 96% plan to renew their annual membership, with half of the balance planning to move to Silver's and half to Fitness Lab.
12. Chung King Village recently mailed a coupon to all residents of Smallville offering two dinners for the price of one. As a result, 67% of those who normally eat at Asia Gardens plan on trying Chung King Village within the next month when they eat Chinese food, and 59% of those who normally eat at Chef Chao's plan on trying Chung King Village. Also, all of Chung King Village's normal customers will return to take advantage of the special. And 30% of those who normally eat at Asia Gardens will eat there within the next month because of its convenient location. Chef Chao's has a new chef who isn't doing very well, and as a result, only 15% of those who normally eat there are planning on returning the next time they eat Chinese food.
13. Use the information in Exercises 1 and 7 to predict KickKola's market share at the following times.
    - **a.** the first following purchase, using probability trees
    - **b.** the second following purchase, using probability trees
    - **c.** the first following purchase, using matrices
    - **d.** the second following purchase, using matrices
    - **e.** Discuss the advantages and disadvantages of using trees, as well as those of using matrices.

    NOTE: The answers to (c) and (d) are not given in the back of the book, but they are the same as those of (a) and (b).
14. Use the information in Exercises 2 and 8 to predict SoftNWash's market share at the following times.
    - **a.** the first following purchase, using probability trees
    - **b.** the second following purchase, using probability trees
    - **c.** the first following purchase, using matrices
    - **d.** the second following purchase, using matrices
    - **e.** Discuss the advantages and disadvantages of using trees, as well as those of using matrices.
15. Use the information in Exercises 5 and 11 to predict the market shares of Silver's Gym, Fitness Lab, and ThinNFit at the following times.
    - **a.** in one year, using probability trees
    - **b.** in one year, using matrices
    - **c.** in two years, using matrices
    - **d.** in three years

    NOTE: The answer to (b) is not given in the back of the book, but it is the same as that of (a).

16. Use the information in Exercises 6 and 12 to predict the market shares of Asia Gardens, Chef Chao's, and Chung King Village at the following times.
  a. in one month, using probability trees
  b. in one month, using matrices
  c. in two months, using matrices
  d. in three months

17. Sierra Cruiser currently commands 41% of the mountain bike market. A nationwide survey performed by *Get Out of My Way* magazine indicates that 31% of the bike owners who do not currently own a Sierra Cruiser will purchase a Sierra Cruiser the next time they buy a mountain bike and that 12% of the bikers who currently own a Sierra Cruiser will purchase one the next time they buy a mountain bike. If the average customer buys a new mountain bike every two years, predict Sierra Cruiser's market share in four years.

18. A marketing analysis shows that Clicker Pens currently commands 46% of the pen market. The analysis also indicates that 46% of the consumers who do not currently own a Clicker pen will purchase a Clicker the next time they buy a pen and that 37% of the consumers who currently own a Clicker will not purchase one the next time they buy a pen. If the average consumer buys a new pen every three weeks, predict Clicker's market share after six weeks.

Projects

19. a. The Census Bureau classifies all residents of the United States as residents of a central city, or a suburb, or a nonmetropolitan area. In the 1990 census, the bureau reported that the central cities held 77.8 million people, the suburbs 114.9 million people, and the nonmetropolitan areas 56.0 million people. Compute the proportion of U.S. residents for each category and present those proportions as a probability matrix. (Round off to three decimal places.)

b. The Census Bureau (in Current Population Reports, P20-463, *Geographic Mobility, March 1990 to March 1991*) also reported the information in Figure 7.13 and Figure 7.14 regarding migration between these three areas from 1990 to 1991. (Numbers are in thousands.)

Use the data in the table and the data in part (a) to compute the probabilities that a central city resident will move to a suburb or to a nonmetropolitan area, the probabilities that a suburban resident will move to a central city or to a nonmetropolitan area, and the probabilities that a nonmetropolitan resident will move to a central city or to a suburb. Present these probabilities in the form of a transition matrix. (Round off to three decimal places.)

	Moved to		
Moved from	**Central City**	**Suburb**	**Nonmetropolitan Area**
Central City	(x)	4,946	736
Suburb	2,482	(x)	964
Nonmetropolitan Area	741	1,075	(x)

Figure 7.13

NOTE: Persons moving from one central city to another were not considered. This is represented in the above chart by "(x)."

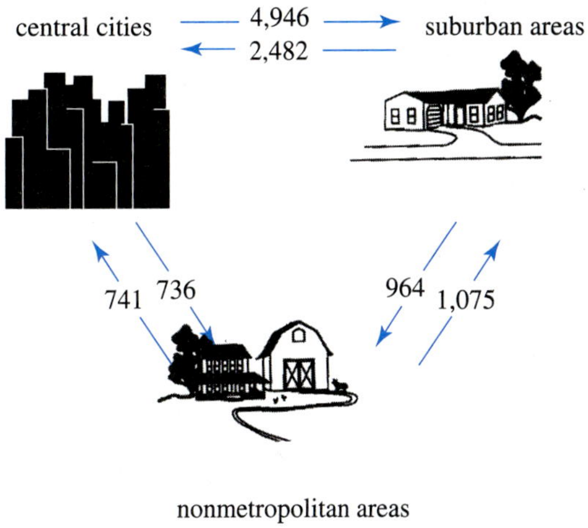

Figure 7.14
Movers between cities, suburbs, and nonmetropolitan areas, and net change due to migration, 1990–91 (numbers in thousands)

c. Predict the percent of U.S. residents living in a central city, a suburb, and a nonmetropolitan area in 1991.

d. Predict the percent of U.S. residents residing in a central city, a suburb, and a nonmetropolitan area in 1992.

e. Which prediction is the stronger one—that in part (c) or that in part (d)? Why?

20. a. The Census Bureau (in *Statistical Abstract of the United States: 2000*) reported the 2000 national and regional populations in Figure 7.15.

Region	Population (in thousands)
United States	281,422
Northeast	53,594
Midwest	64,393
South	100,237
West	63,198

Figure 7.15

Compute the proportion of U.S. residents of each region and present those proportions as a probability matrix. (Round off to three decimal places).

b. The Census Bureau (in Current Population Reports P20-*538 Geographical Mobility: March 1999 to March 2000*) also reported the information in Figures 7.16 and 7.17 regarding movement between regions from 1999 to 2000 (numbers are in thousands).

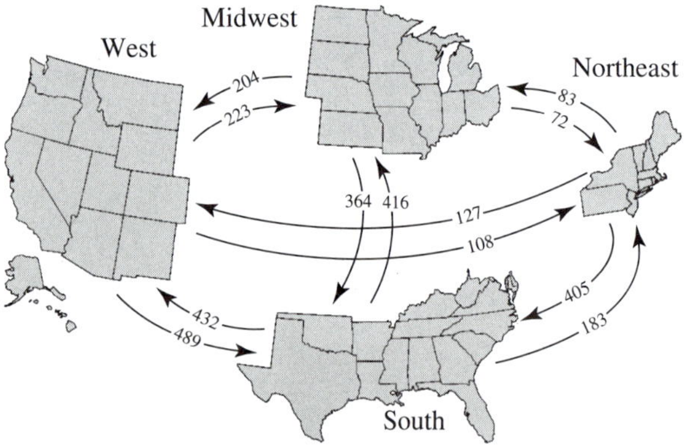

Figure 7.16
Movers between regions, 1999–2000 (numbers in thousands)

Region Moved from	Region Moved to			
	Northeast	Midwest	South	West
Northeast	(*x*)	83	405	127
Midwest	72	(*x*)	364	204
South	183	416	(*x*)	432
West	108	223	489	(*x*)

Figure 7.17

NOTE: The Census data involved interregional movement; thus, persons moving from one part of a region to another part of the same region were not considered. This is represented in the above chart by "(*x*)."

Compute the probabilities that a resident of any one region move to any of the other regions. Present these probabilities in the form of a transition matrix (round off to three decimal places). Assume that the 2000–2001 probabilities are equal to the corresponding 1999–2000 probabilities.

c. Predict the percent of U.S. residents who resided in the Northeast, the Midwest, the South, and the West in 2001.

d. Predict the percent of U.S. residents who resided in the Northeast, the Midwest, the South, and the West in 2002.

e. Which prediction is the stronger one—that in part (c) or that in part (d)? Why?

21. a. In 1981, Neil Sampson, executive vice president of the National Association of Conservation Districts, compiled data on the shifting land use pattern in the United States. [*Source*: R. Neil Sampson, *Farmland or Wasteland: A Time to Choose* (Rodale Press, 1981).] He stated that in 1977, there were 413 million acres of cropland, 127 million acres of land that had a high or medium potential for conversion to cropland (for example, grasslands or forests), and 856 million acres of land that had little or no potential for conversion to cropland (such as urban land or land with heavily deteriorated soil). See Figure 7.18. Compute the proportion of land in each category and present those proportions as a probability matrix (round off to three decimal places).

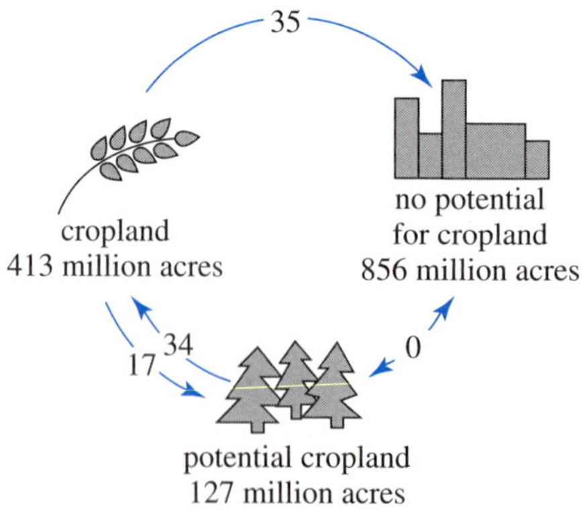

Figure 7.18

b. He also estimated that between 1967 and 1977, there were shifts in land use (in millions of acres) as shown in the accompanying chart.

Previous Use of Land	New Use of Land		
	Cropland	**Potential Cropland**	**Noncropland**
Cropland	(x)	17	35
Potential Cropland	34	(x)	0^*
Noncropland	0	0^*	(x)
*Unknown but assumed to cancel each other out.			

Use the data contained in the chart and the data in part (a) to compute the probabilities that an acre of land of any of the three categories will shift to another use (in 10 years' time). Present these probabilities in the form of a transition matrix (round off to three decimal places).

c. Predict the amount of land that was cropland, potential cropland, and noncropland in 1987, assuming that the trend described above continues.

d. Predict the amount of land that was cropland, potential cropland, and noncropland in 1997, assuming that the trend described above continues.

Answer the following questions using complete sentences and your own words.

Concept Questions

22. Are PT, PT^2, and PT^3 transition matrices or probability matrices? Why?

23. Are T^2 and T^3 transition matrices or probability matrices? Why?

24. What are the differences between a transition matrix and a probability matrix?

25. What information would a marketing analyst need to predict the future market share of a product? How would he or she obtain such information?

26. When a prediction is made using Markov chains, that prediction is based on a number of assumptions, one involving the trend, one involving the data summarized in the probability matrix P, and one involving the data summarized in the transition matrix T. What are these assumptions? What would make the assumptions invalid?

27. Compare and contrast the tree method of Markov chains with the matrix method of Markov chains. What are the advantages of the two different methods?

28. Choose one of Exercises 17–21 and discuss why the given data describes a Markov chain.

29. What is the difference between a current state and a following state?

History Questions

30. In creating Markov chains, was Markov motivated by theoretical concerns or by specific applications?

31. Was Markov a supporter of the czar or of the revolution?

32. How did Markov celebrate the 300th anniversary of the House of Romanov?

33. What did Markov do when the czar abdicated?

7.2 Systems of Linear Equations

Linear Equations

To graph the equation $2x + 3y = 8$, first solve the equation for y.

$$2x + 3y = 8$$
$$3y = -2x + 8$$
$$y = -\frac{2}{3}x + \frac{8}{3}$$

Comparing this equation with the slope-intercept formula $y = mx + b$, we find that this is the equation of a line with slope $m = -\frac{2}{3}$ and y-intercept $b = \frac{8}{3}$, as shown in Figure 7.19.

The graph of the equation $5x + 7y = 12$ is also a line, because we could follow the above procedure and rewrite the equation in the form $y = mx + b$. In fact, the graph of any equation of the form $ax + by = c$ is a line. (If b were 0, the above procedure wouldn't work, but the graph would still be a line.) For this reason, an equation of the form $ax + by = c$ is called a *linear equation*.

To graph an equation of the form $ax + by + cz = d$, we would need a z-axis in addition to the usual x- and y-axes; the graph of such an equation is always a plane. Despite its geometrical appearance, however, such an equation is called a linear equation owing to its algebraic similarity to $ax + by = c$. A **linear equation** is an equation that can be written in the form $ax + by = c$, the form $ax + by + cz = d$, or a similar form with more unknowns. In section 7.3, we will use linear equations to solve Markov chain problems.

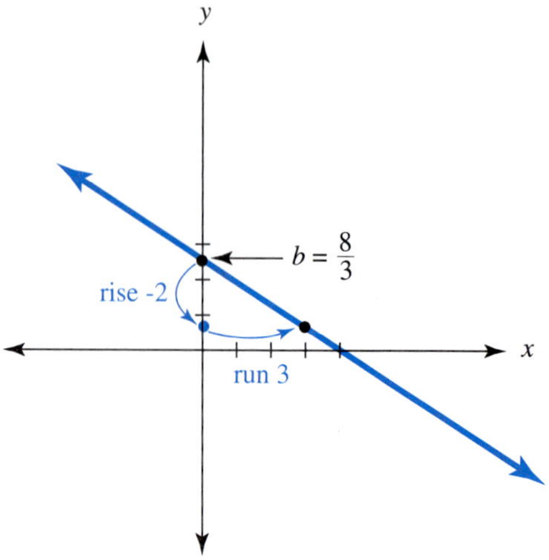

Figure 7.19
The graph of $y = -\frac{2}{3}x + \frac{8}{3}$

Systems of Equations

A **system of equations** is a set of more than one equation. **Solving a system of equations** means finding all ordered pairs (x, y) [or ordered triples (x, y, z) and so on] that will satisfy each equation in the system. Geometrically, this means finding all points that are on the graph of each equation in the system.

The system

$$\begin{aligned} x + y &= 3 \\ 2x + 3y &= 8 \end{aligned}$$

is a system of linear equations because it is a set of two equations and each equation is of the form $ax + by = c$. The ordered pair $(1, 2)$ is a solution to the system, because $(1, 2)$ satisfies each equation:

$$\begin{aligned} x + y &= 1 + 2 = 3 \\ 2x + 3y &= 2 \cdot 1 + 3 \cdot 2 = 8 \end{aligned}$$

Does this system have any other solutions? If we solve each equation for y, we get

$$\begin{aligned} y &= -1x + 3 \\ y &= \frac{-2}{3}x + \frac{8}{3} \end{aligned}$$

The first equation is the equation of a line with slope -1 and y-intercept 3; the second is the equation of a line with slope $\frac{-2}{3}$ and y-intercept $\frac{8}{3}$. Because the slopes are different, the lines are not parallel and intersect in only one point. Thus, (1, 2) is the only solution to the system; it is the point at which the two lines intersect. This system is illustrated in Figure 7.20.

Not all systems have only one solution. A system with two linear equations in two unknowns describes a pair of lines. The two lines can be parallel and not intersect (in which case the system has no solution), or the two lines can be the same (in which case the system has an infinite number of solutions), as shown in Figure 7.21.

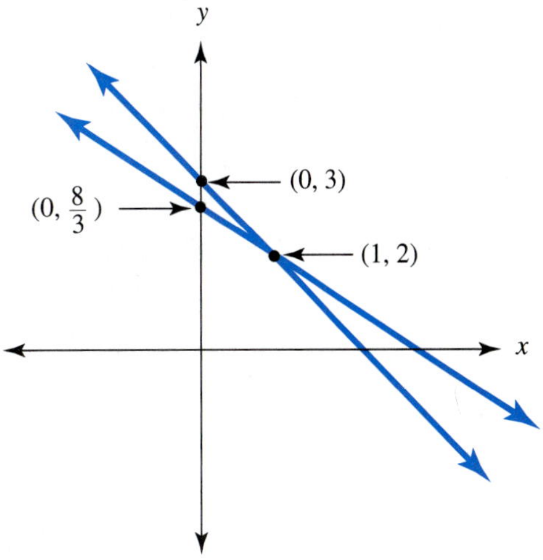

Figure 7.20

	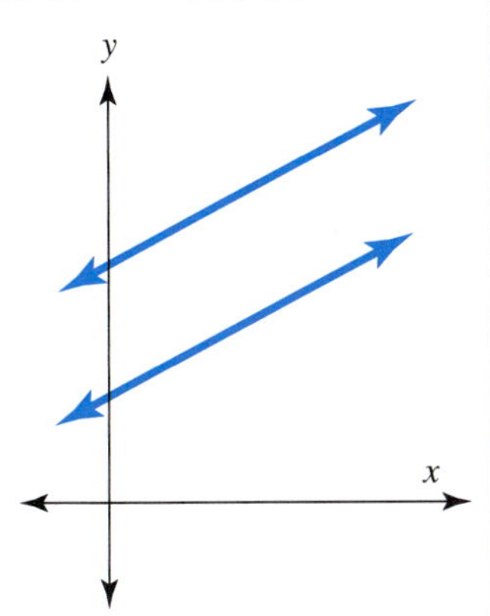	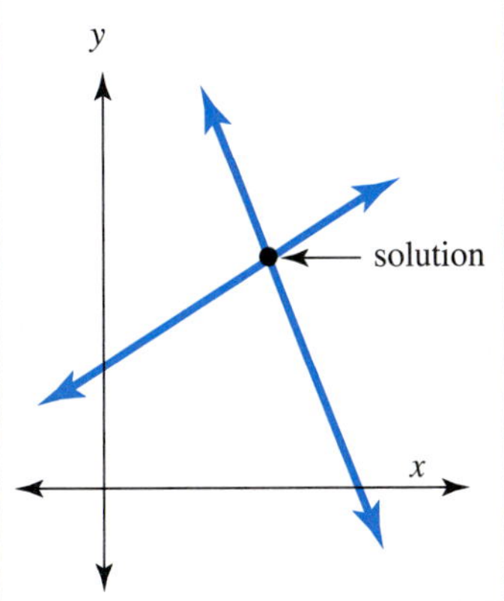	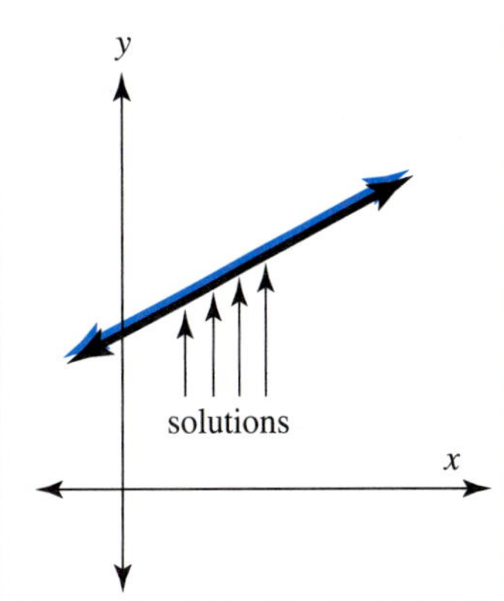
the lines:	parallel and nonintersecting	intersecting	the same
the number of solutions:	0	1	infinite

Figure 7.21

A system of two linear equations in *three* unknowns (x, y, and z) describes a set of two planes. Such a system can have no solution or an infinite number of solutions, but it cannot have only one solution, as illustrated in Figure 7.22 on page 534. In fact, *any system of linear equations that has fewer equations than unknowns cannot have a unique solution.*

A system of *three* linear equations in three unknowns describes a set of three planes. Such a system can have no solution, only one solution, or an infinite number of solutions, as illustrated in Figure 7.23 on page 534. In fact, *any system of linear equations that has as many equations as unknowns can have either no solution, one unique solution, or an infinite number of solutions.*

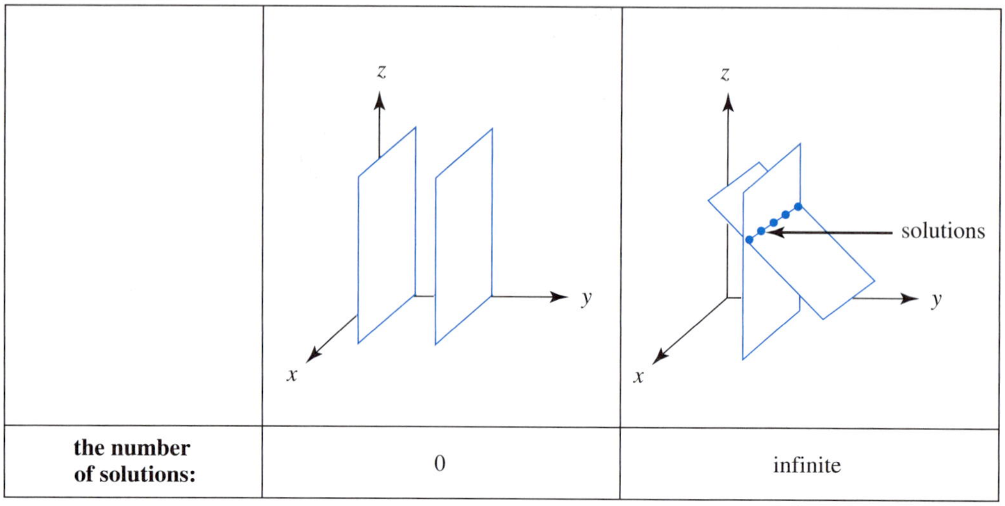

Figure 7.22
A system of 2 linear equations in 3 unknowns can have no solutions or an infinite number of solutions.

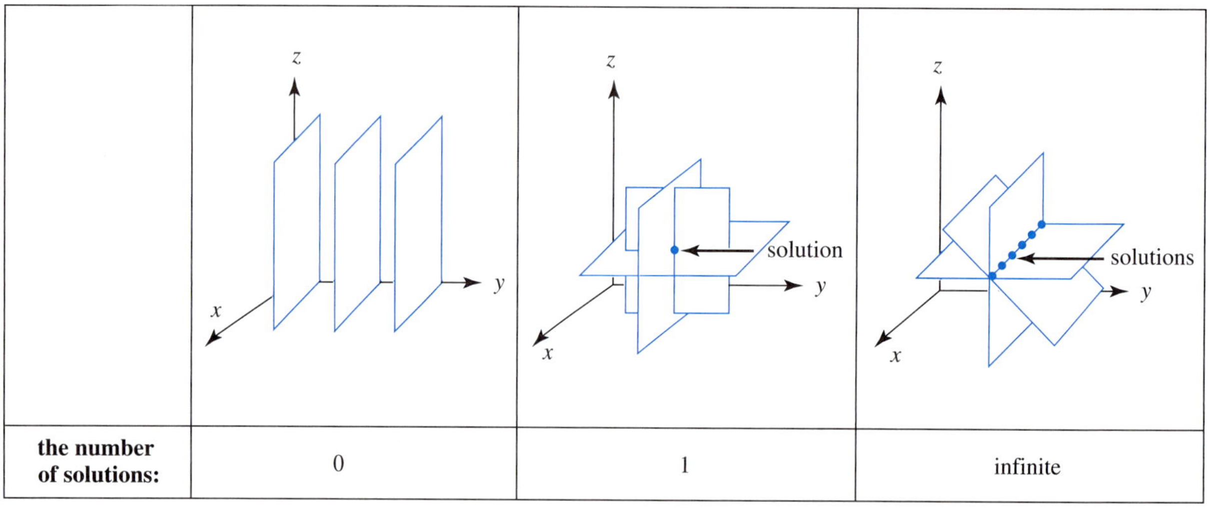

Figure 7.23
A system of 3 linear equations in 3 unknowns can have no solutions, 1 solution, or an infinite number of solutions.

EXAMPLE 1 How many solutions could the following system have?

$$\begin{aligned} 3x - y \quad + z \quad &= 5 \\ 2x + y \quad - 3z \quad &= 1 \\ 20x + 10y - 30z &= 10 \end{aligned}$$

Solution This system seems to have as many equations as unknowns; however, this is not the case. The third equation is 10 times the second equation, so its presence is unnecessary. The above system is equivalent to the system

$$\begin{aligned} 3x - y + z &= 5 \\ 2x + y - 3z &= 1 \end{aligned}$$

This system has three unknowns but only two *unique* equations, so it can have either no solution or an infinite number of solutions. It cannot have one solution. ■

Number of Solutions of a System of Linear Equations

A system that has:	will have:
fewer equations than unknowns	either no solution or an infinite number of solutions
as many unique equations as unknowns	either no solution, one solution, or an infinite number of solutions

Solving Systems: The Elimination Method

Our work with Markov chains will involve solving systems of equations. Many methods are used to solve systems of equations. Perhaps the easiest method for solving a small system is the elimination method (also called the addition method). With this method, you add together equations or multiples of equations in such a way that a variable is eliminated. You probably saw this method in your intermediate algebra class.

EXAMPLE 2 Use the elimination method to solve the following system:

$$3x - 2y = 5$$
$$2x + 3y = 4$$

Solution We'll eliminate x if we multiply the first equation by 2 and the second equation by -3 and add the results:

$$\begin{aligned} 6x - 4y &= 10 && \text{2 times equation 1} \\ + \ -6x - 9y &= -12 && \text{-3 times equation 2} \\ \hline -13y &= -2 \\ y &= \frac{2}{13} \end{aligned}$$

We can find x by substituting $y = \frac{2}{13}$ into either of the original equations:

$$\begin{aligned} 3x - 2y &= 5 \\ 3x - 2 \cdot \frac{2}{13} &= 5 \\ 3x - \frac{4}{13} &= \frac{65}{13} && \text{rewriting 5 with a common denominator} \\ 3x &= \frac{69}{13} \\ x &= \frac{23}{13} \end{aligned}$$

Thus, the solution to the system of equations is the ordered pair $\left(\frac{23}{13}, \frac{2}{13}\right)$. ■

✔ We can easily check our work by substituting the solution back into the original equations.

$$3x - 2y = 3 \cdot \frac{23}{13} - 2 \cdot \frac{2}{13} = \frac{69}{13} - \frac{4}{13} = \frac{65}{13} = 5 \checkmark$$

$$2x + 3y = 2 \cdot \frac{23}{13} + 3 \cdot \frac{2}{13} = \frac{46}{13} + \frac{6}{13} = \frac{52}{13} = 4 \checkmark$$

Naturally, these calculations can be performed on a calculator.

7.2 Exercises

In Exercises 1–6, determine whether the given ordered pair or ordered triple solves the given system of equations.

1. (4, 1)
$3x - 5y = 7$
$2x + 2y = 10$

2. (7, −2)
$2x - 4y = 30$
$x + y = 5$

3. (−5, 3)
$3x + y = 4$
$10x - 4y = -62$

4. (−1, −2)
$2x + 2y = -5$
$3x - 4y = 2$

5. (4, −1, 2)
$2x + 3y - z = 3$
$x + y - z = 5$
$10x - 2y = 3$

6. (0, 5, −1)
$3x - 2y + z = -11$
$2x + 4y + z = 19$
$x - z = 1$

In Exercises 7–12, do the following: **a.** *Find each line's slope and y intercept.* **b.** *Use the slopes and y intercepts to determine whether the given system has no solution, one solution, or an infinite number of solutions. Do not actually solve the system.*

7. $5x + 2y = 4$
$6x - 19y = 72$

8. $3x + 132y = 19$
$45x + 17y = 4$

9. $4x + 3y = 12$
$8x + 6y = 24$

10. $19x - 22y = 1$
$190x - 220y = 10$

11. $x + y = 7$
$3x + 2y = 8$
$2x + 2y = 14$

12. $3x - y = 12$
$2x + 3y = 5$
$5x + 2y = 17$

In Exercises 13–18, determine whether the given system could have a single solution. Do not actually solve the system.

13. $3x - 2y + 5z = 1$
$2x + y = 2$
$5x + 7y - z = 0$

14. $8x - 4y + 2z = 10$
$3x + y + z = 1$
$4x - 2y + z = 5$

15. $x + y + z = 1$
$2x + 2y + 2z = 2$
$3x - y + 10z = 45$

16. $x + y + z = 34$
$5x - y + 2z = 9$
$3x + 3y + 3z = 102$

17. $x + 2y + 3z = 4$
$5x - y = 2$

18. $9x - 21y = 476$
$x + 3y + z = 12$

Solve the systems in Exercises 19–24 with the elimination method. Check your answers by substituting them back in. (Answers are not given at the back of the book.)

19. $2x + 3y = 5$
$4x - 2y = 2$

20. $5x + 3y = 11$
$2x + 7y = 16$

21. $3x - 7y = 27$
$4x - 5y = 23$

22. $5x - 2y = -23$
$x + 2y = 5$

23. $5x - 9y = -12$
$3x + 7y = -1$

24. $5x - 12y = 9$
$3x + 3y = 2$

Answer the following questions using complete sentences and your own words.

Concept Questions

25. By sketching lines, determine whether a system of three equations in two unknowns could have no solution, one solution, or an infinite number of solutions. What if the three equations were unique?
26. By sketching planes, determine whether a system of four equations in three unknowns could have no solution, one solution, or an infinite number of solutions. What if the four equations were unique?
27. Solving a system of two equations in two unknowns is the same as finding all points that are on the graph of both of the two equations. Why do we not solve systems by graphing them (by hand) and locating the points that are on both graphs?
28. What does the graph of a linear equation in two unknowns look like? What does the graph of a linear equation in three unknowns look like? Why is a linear equation in three unknowns called a linear equation?
29. Suppose you have a linear equation in two unknowns, and you multiply both sides of that equation by a constant (not zero). How would the graph of the new equation compare to that of the original equation? Would they be parallel lines, perpendicular lines, the same lines, or what? Why?

Solving Systems of Two Linear Equations in Two Unknowns on a Graphing Calculator

A system of two linear equations in two unknowns can be solved nicely on a graphing calculator.

Solving Systems of Two Linear Equations in Two Unknowns on a Graphing Calculator

- Solve each equation for y.
- Enter the results into the graphing calculator as Y_1 and Y_2.
- Graph the two equations on the graphing calculator.
- Use the calculator to find the point of intersection. (The details are given in Appendices C and D.)
- Check your solution by substituting it into each of the original equations.

EXAMPLE 3 Use a graphing calculator to solve the following system:

$$3x - 2y = 5$$
$$2x + 3y = 4$$

Solution

Step 1 *Solve each equation for y.*

$$3x - 2y = 5 \Rightarrow -2y = -3x + 5 \Rightarrow \frac{-2y}{-2} = \frac{-3x + 5}{-2} \Rightarrow y = \frac{3}{2}x - \frac{5}{2}$$

$$2x + 3y = 4 \Rightarrow 3y = -2x + 4 \Rightarrow \frac{3y}{3} = \frac{-2x + 4}{3} \Rightarrow y = -\frac{2}{3}x + \frac{4}{3}$$

Step 2 *Enter the results into the graphing calculator as* Y_1 and Y_2.

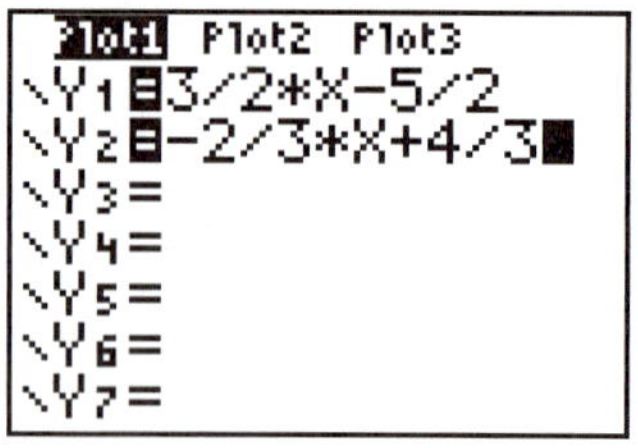

Figure 7.24
A TI-83's "Y=" screen
(Other calculators are similar.)

Step 3 *Graph the two equations on the graphing calculator.*

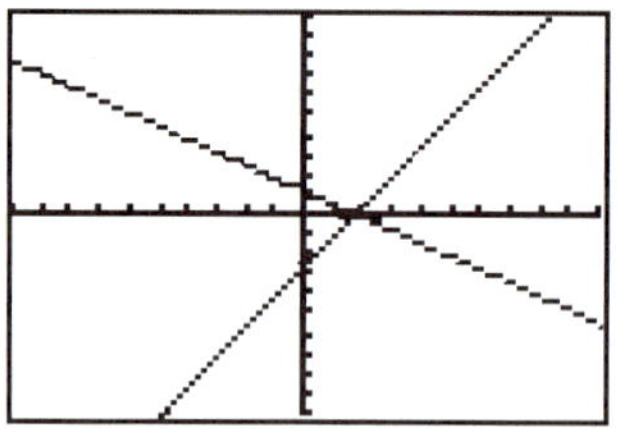

Figure 7.25
The graph of the system
$3x - 2y = 5$
$2x + 3y = 4$

Step 4 *Use the calculator to find the point of intersection.*

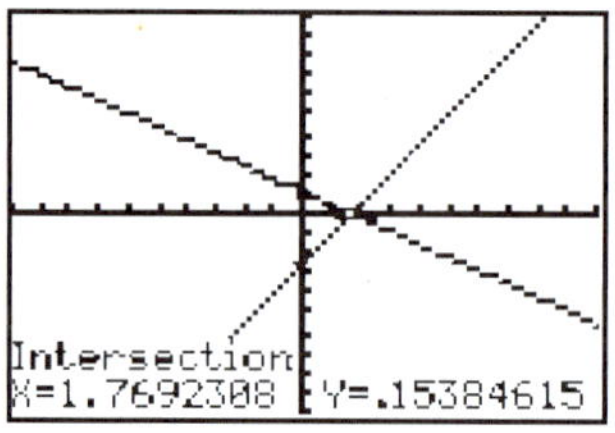

Figure 7.26
The solution of the system
$3x - 2y = 5$
$2x + 3y = 4$

The system's solution is approximately (1.7692308, 0.15384615).

Step 5 *Check your solution by substituting it into each of the original equations.*

$$3x - 2y = 3 \cdot 1.769230 - 2 \cdot 0.15384615 = 4.9999167 \approx 5$$
$$2x + 3y = 2 \cdot 1.769230 + 3 \cdot 0.15384615 = 3.99994445 \approx 4$$

The solution checks, but not perfectly. This is because the calculator rounded off the point of intersection in step 4.

Exercises

Solve the systems in Exercises 30–35 with a graphing calculator. Check your answers by substituting them back in. (Answers are not given at the back of the book.)

30. Exercise 20
31. Exercise 21
32. Exercise 22
33. Exercise 23
34. Exercise 24
35. Exercise 19

Concept Questions

36. Describe how you could use a graphing calculator to solve a system of three equations in two unknowns.

37. Why could you not use a graphing calculator to solve a system of two equations in three unknowns?

7.3 Long-Range Predictions with Markov Chains

In Section 7.1, we were given Bif's current market share (25%), and we predicted Bif's market share after one, two, and three purchases (that is, we found the probabilities of the first, second, and third following states) by computing PT, PT^2, and PT^3. In a similar way, we could predict Bif's market share after four purchases by computing PT^4. These predictions are shown in Figure 7.27. It appears that the trend in increasing market share created by Bif's new advertising campaign and customer satisfaction with the product will stabilize, and that Bif will ultimately command close to 70% of the market. See Figure 7.28 on page 540.

Bif's Projected Market Share	Bif	Not Bif	Probability Matrix
current purchase	25%	75%	P
1st following purchase	50%	50%	$PT = PT^1$
2nd following purchase	60%	40%	PT^2
3rd following purchase	64%	36%	PT^3
4th following purchase	65.6%	34.4%	PT^4

Figure 7.27

If Bif had started out with a much smaller initial market share—say, only 10% rather than 25%—a similar pattern would emerge, as shown in Figure 7.29.

Bif's Projected Market Share	Bif	Not Bif	Probability Matrix
current purchase	10%	90%	P
1st following purchase	44%	56%	$PT = PT^1$
2nd following purchase	57.6%	42.4%	PT^2
3rd following purchase	63.04%	36.96%	PT^3
4th following purchase	65.22%	34.78%	PT^4

Figure 7.29

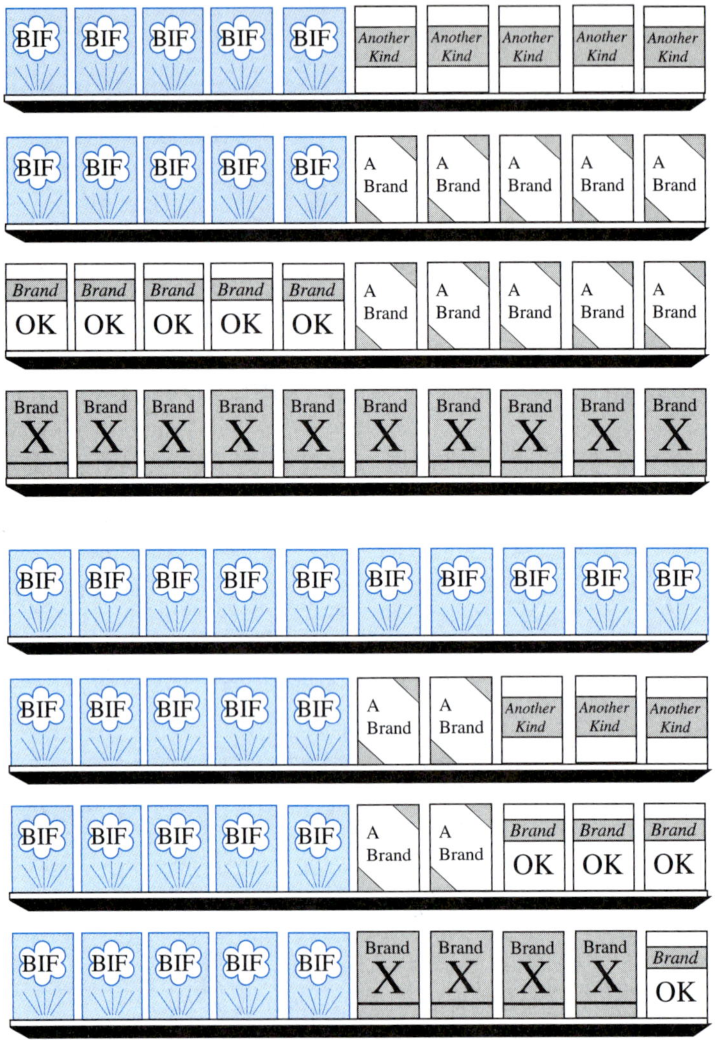

Figure 7.28 As a result of a new advertising campaign and customers' satisfaction with the product, Bif is predicted to increase its market share from 25% . . .

. . . to 65.6% after the fourth following purchase.

$$P = [0.1 \quad 0.9]$$

$$T = \begin{bmatrix} 0.8 & 0.2 \\ 0.4 & 0.6 \end{bmatrix} \quad \text{as before}$$

Surprisingly, it appears that Bif's new advertising campaign and customer satisfaction with the product would have the same ultimate effect regardless of Bif's initial market share. Under certain circumstances, PT^n will stabilize in a way that is unaffected by the value of P. Our goal in this section is to find the level at which the trend will stabilize. The **equilibrium matrix L,** the matrix at which the trend stabilizes, is a probability matrix $L = PT^n$ such that all following probability matrices are equal. In other words, multiplying the equilibrium matrix L by T would have no effect, and LT would equal L. In fact, we find the equilibrium matrix by solving the matrix equation $LT = L$.

EXAMPLE 1 Make a long-range forecast for Bif's market share, given the transition data from Section 7.1.

Solution The long-range forecast is the probability matrix at which the trend in Bif's changing market share stabilizes; that is, it is the equilibrium matrix L. We find L by solving the matrix equation $LT = L$.

$L = [x, \; y]$ a row matrix because L is a probability matrix

$T = \begin{bmatrix} 0.8 & 0.2 \\ 0.4 & 0.6 \end{bmatrix}$ from Section 7.1

$$\underset{L}{[x \quad y]} \cdot \underset{T}{\begin{bmatrix} 0.8 & 0.2 \\ 0.4 & 0.6 \end{bmatrix}} = \underset{L}{[x \quad y]}$$

$[0.8x + 0.4y \quad 0.2x + 0.6y] = [x \quad y]$ multiplying

This yields the following system:

$$0.8x + 0.4y = x$$
$$0.2x + 0.6y = y$$

Combining like terms gives

$$-0.2x + 0.4y = 0$$
$$0.2x - 0.4y = 0$$

These two equations are equivalent (we can multiply one equation by -1 to get the other), so we can discard one. This system has one unique equation and two unknowns, and we found in Section 7.2 that a system with fewer equations than unknowns will have either no solution or an infinite number of solutions. This leaves us in a difficult situation, because we are looking for a single solution. However, recall that $L = [x \quad y]$ is a probability matrix, so its entries must add to 1. Thus, $x + y$ is a second equation, the system now has as many unique equations as unknowns, and we can try to find a single solution to the system

$$0.2x - 0.4y = 0$$
$$x + y = 1$$

This small system can easily be solved with either the elimination method or a graphing calculator.

Proceeding with the elimination method, we multiply the second equation by 0.4. The resulting system is

$$0.2x - 0.4y = 0$$
$$0.4x + 0.4y = 0.4$$

Adding gives

$$0.6x = 0.4$$
$$\frac{0.6x}{0.6} = \frac{0.4}{0.6}$$
$$x = \frac{2}{3}$$

Substituting into $x + y = 1$ gives

$$\frac{2}{3} + y = 1$$
$$y = \frac{1}{3}$$

Thus, $L = [x \quad y] = [\frac{2}{3} \;\; \frac{1}{3}]$, and Bif's market share should stabilize at $\frac{2}{3}$, or approximately 67%.

As a check, multiply L by T and see if you get L.

✔ $$LT = \left[\frac{2}{3} \;\; \frac{1}{3}\right] \cdot \begin{bmatrix} 0.8 & 0.2 \\ 0.4 & 0.6 \end{bmatrix}$$
$$\approx [0.6666667 \quad 0.3333333]$$
$$\approx \left[\frac{2}{3} \;\; \frac{1}{3}\right]$$

■

Long-Range Predictions with Markov Chains

1. *Create the transition matrix T.* T is the square matrix, discussed in Section 7.1, whose entries are the probabilities of passing from current states to next following states. The rows refer to the current states, and the columns refer to the next following states.
2. *Create the equilibrium matrix L.* The equilibrium matrix L is the long-range prediction. L is the matrix that solves the matrix equation $LT = L$. If there are two states, then $L = [x \quad y]$.
3. *Find and simplify the system of equations described by $LT = L$.*
4. *Discard any redundant equations and include the equation $x + y = 1$.*
5. *Solve the resulting system.* Use elimination or a graphing calculator.
6. ✔ *Check your work by verifying that $LT = L$.*

7.3 Exercises

In Exercises 1–4, find the equilibrium matrix L by solving LT = L for L.

1. $T = \begin{bmatrix} 0.1 & 0.9 \\ 0.2 & 0.8 \end{bmatrix}$ **2.** $T = \begin{bmatrix} 0.5 & 0.5 \\ 0.6 & 0.4 \end{bmatrix}$

3. $T = \begin{bmatrix} 0.3 & 0.7 \\ 0.8 & 0.2 \end{bmatrix}$ **4.** $T = \begin{bmatrix} 0.6 & 0.4 \\ 0.25 & 0.75 \end{bmatrix}$

5. A marketing analysis shows that 12% of the consumers who do not currently drink KickKola will purchase KickKola the next time they buy a cola and that 63% of the consumers who currently drink KickKola will purchase it the next time they buy a cola. Make a long-range prediction of KickKola's ultimate market share, assuming that current trends continue. (See Exercise 13 in Section 7.1.)

6. A marketing analysis shows that 9% of the consumers who do not currently use SoftNWash will purchase SoftNWash the next time they buy a fabric softener and that 29% of the consumers who currently use SoftNWash will purchase it the next time they buy a fabric softener. Make a long-range prediction of SoftNWash's ultimate market share, assuming that current trends continue. (See Exercise 14 in Section 7.1.)

7. A census report shows that 32% of the residents of Metropolis own their own home and that 68% rent. The report shows that 12% of the renters plan to buy a home in the next 12 months and that 3% of the home owners plan to sell their home and rent instead. Make a long-range prediction of the percent of Metropolis residents who will own their own home and the percent who will rent. Give two assumptions on which this prediction is based. (See Exercises 3 and 9 in Section 7.1.)

8. A marketing analysis indicates that 46% of the consumers who do not currently own a Clicker pen will purchase a Clicker the next time they buy a pen and that 37% of the consumers who currently own a Clicker will not purchase one the next time they buy a pen. If the average consumer buys a new pen every three weeks, make a long-range prediction of Clicker's ultimate market share, assuming that current trends continue. (See Exercise 18 in Section 7.1.)

9. A nationwide survey performed by *Get Out of My Way* magazine indicates that 31% of the bike owners who do not currently own a Sierra Cruiser will purchase a Sierra Cruiser the next time they buy a mountain bike and that 12% of the bikers who currently own a Sierra Cruiser will purchase one the next time they buy a mountain bike. If the average customer buys a new mountain bike every two years, make a long-range prediction of Sierra Cruiser's ultimate market share, assuming that current trends continue. (See Exercise 17 in Section 7.1.)

7.4 Solving Larger Systems of Equations

We will discuss two different methods for solving larger systems of equations (i.e., systems of more than two equations with more than two unknowns): the elimination method and the Gauss-Jordan method.

Solving Larger Systems with the Elimination Method

Solving a system of three equations in three unknowns with the elimination method is quite similar to solving a system of two equations in two unknowns. There are a few more steps because there are more variables to eliminate.

Solving a System of Three Equations in Three Unknowns with the Elimination Method

1. Combine two equations so that one variable is eliminated.
2. Combine two different equations so that the same variable is eliminated.
3. Steps 1 and 2 result in a system of two equations in two unknowns. Solve that system with the elimination method, as discussed in Section 7.2.
4. Substitute back into various equations to find the values of the remaining variables.
5. Check your answer by substituting it into the original equations.

EXAMPLE 1 Use the elimination method to solve the following system:

$$5x - 7y + 2z = 25$$
$$3x + 2y - 5z = -16$$
$$7x + 3y - 9z = -25$$

Solution

Step 1 *Combine two equations so that one variable is eliminated.*
We'll combine the first two equations, and eliminate z. To do this, multiply the first equation by 5 and the second by 2, and add the results.

$$\begin{array}{rl} 25x - 35y + 10z = 125 & \text{5 times equation 1} \\ +\ 6x + 4y - 10z = -32 & \text{2 times equation 2} \\ \hline 31x - 31y \qquad = 93 & \end{array}$$

Step 2 *Combine two different equations so that the same variable is eliminated.*
We'll combine the last two equations and eliminate z. To do this, multiply the second equation by 9 and the third by -5, and add the results.

$$\begin{array}{rl} 27x + 18y - 45z = -144 & \text{9 times equation 2} \\ +\ -35x - 15y + 45z = 125 & -5 \text{ times equation 3} \\ \hline -8x + 3y \qquad = -19 & \end{array}$$

Step 3 *Steps 1 and 2 result in a system of two equations in two unknowns. Solve that system with the elimination method, as discussed in Section 7.2.* We are left with a smaller system to solve:

$$\begin{array}{rl} 31x - 31y = 93 & \text{the result of step 1} \\ -8x + 3y = -19 & \text{the result of step 2} \end{array}$$

First, simplify the first equation of this new system by dividing through by 31:

$$\begin{array}{rl} x - y = 3 & \text{dividing the first equation by 31} \\ -8x + 3y = -19 & \text{the second equation} \end{array}$$

We'll eliminate y if we multiply the first equation by 3 and add the result to the second equation.

$$\begin{array}{rl} 3x - 3y = 9 & \text{3 times equation 1} \\ +\ -8x + 3y = -19 & \text{equation 2} \\ \hline -5x \qquad = -10 & \\ -5x/-5 = -10/-5 & \\ x = 2 & \end{array}$$

Step 4 *Substitute back into various equations to find the values of the remaining variables.*
We can find y by substituting $x = 2$ into $x - y = 3$

$$\begin{array}{rl} 2 - y = 3 & \text{substituting} \\ -y = 1 & \text{solving for } y \\ y = -1 & \end{array}$$

We can find z by substituting $x = 2$ and $y = -1$ into $5x - 7y + 2z = 25$

$$\begin{array}{rl} 5x - 7y + 2z = 25 & \\ 5 \cdot 2 - 7 \cdot (-1) + 2z = 25 & \text{substituting} \\ 17 + 2z = 25 & \text{solving for } z \\ 2z = 8 & \\ z = 8/2 = 4 & \end{array}$$

The system's solution is $(x, y, z) = (2, -1, 4)$.

Step 5 *Check your answer by substituting it into the original equations.* It's not necessary to substitute this solution into the original system's first equation, because in step 4 we just finished forcing that equation to work.

$$3x + 2y - 5z = -16 \quad \Rightarrow \quad 3 \cdot 2 + 2 \cdot (-1) - 5 \cdot 4 = -16 \ \checkmark$$
$$7x + 3y - 9z = -25 \quad \Rightarrow \quad 7 \cdot 2 + 3 \cdot (-1) - 9 \cdot 4 = -25 \ \checkmark$$

■

Solving Systems: The Gauss-Jordan Method

You might find that you prefer to use the Gauss-Jordan method rather than the elimination method, especially when the system involves more than two equations. The Gauss-Jordan method is based on the elimination method, but it uses matrices to abbreviate the amount of writing and work involved and to allow for the use of technology, in the form of either a graphing calculator or a computer. We will illustrate the Gauss-Jordan method with a small system, but keep in mind that the Gauss-Jordan method is really more appropriate for use with larger systems. A small system simply provides an easy introduction to the method.

EXAMPLE 2 Use the Gauss-Jordan method to solve the following system:

$$\begin{aligned} x + y &= 3 \\ 2x + 3y &= 8 \end{aligned}$$

Solution

Step 1 *Write the system in such a way that all the variables are on the left side of the equations.* Write in all the coefficients, including coefficients of 1.

$$\begin{aligned} 1x + 1y &= 3 \\ 2x + 3y &= 8 \end{aligned}$$

Step 2 *Rewrite the system in matrix form.* This is done by eliminating the letters, addition symbols, and equal symbols and putting the remaining numbers into a matrix.

$$\begin{matrix} x & y & \end{matrix}$$
$$\begin{bmatrix} 1 & 1 & 3 \\ 2 & 3 & 8 \end{bmatrix}$$

Notice that we've labeled the first two columns x and y. This is an optional way of keeping track of the origins of the numbers.

The goal of the remaining steps is to rewrite the matrix in the following form:

$$\begin{matrix} x & y & \end{matrix}$$
$$\begin{bmatrix} 1 & 0 & ? \\ 0 & 1 & ? \end{bmatrix}$$

A matrix written in this form is called an **augmented identity matrix,** because it is an identity matrix augmented by one extra column. When this goal is achieved, we will have our solution, because the form above is the matrix equivalent of the system

$$\begin{aligned} 1x + 0y &= ? \\ 0x + 1y &= ? \end{aligned}$$

which simplifies to the system

$$x = ?$$
$$y = ?$$

To achieve this goal, we use the following **row operations,** each of which is the matrix equivalent of a tool used in the elimination method.

Row Operations

1. *Multiply or divide a row by any number (except zero).* This is equivalent to multiplying or dividing an equation by a number.
2. *Add one row to another row or add a multiple of one row to another row.* This is equivalent to adding one equation to another equation or adding a multiple of an equation to another equation.
3. *Interchange any two rows.* This is equivalent to interchanging two equations.

Note: Each of the following steps in the Gauss-Jordan method is accompanied by the equivalent step in the elimination method. Following the Gauss-Jordan steps is easier if you see them as the equivalent of the elimination method steps with many of the symbols left out.

Gauss-Jordan Method	Elimination Method
$\begin{matrix} x & y & \\ \end{matrix}$ $\begin{bmatrix} 1 & 1 & 3 \\ 2 & 3 & 8 \end{bmatrix}$	$\begin{aligned} 1x + 1y &= 3 \\ 2x + 3y &= 8 \end{aligned}$

Step 3 *Use the row operations to rewrite the matrix so its first column is "1 0."* In other words, we want the matrix to become

$$\begin{bmatrix} 1 & ? & ? \\ 0 & ? & ? \end{bmatrix}$$

The first column already has 1 as the first entry; what remains is to use the row operations to change 2 to 0. We will use row operation 2 to accomplish this, adding −2 times the first row to the second row (or "−2R1 + R2").

$$\begin{array}{ccc} -2 \cdot 1 & -2 \cdot 1 & -2 \cdot 3 \\ +\quad 2 & +\quad 3 & +\quad 8 \\ \hline 0 & 1 & 2 \end{array} \qquad \begin{array}{rcl} -2 \cdot 1x + -2 \cdot 1y &=& -2 \cdot 3 \\ +\quad 2x + \quad 3y &=& +\quad 8 \\ \hline 0x + \quad 1y &=& 2 \end{array}$$

If we replace the original second row with this result, then the resulting matrix will have the desired column.

$\begin{matrix} x & y & \\ \end{matrix}$ $\begin{bmatrix} 1 & 1 & 3 \\ 0 & 1 & 2 \end{bmatrix}$	$\begin{aligned} 1x + 1y &= 3 \\ 0x + 1y &= 2 \end{aligned}$

Step 4 *Use the row operations to rewrite the matrix so its second column is "0 1."* In other words, we want the matrix to become

$$\begin{bmatrix} 1 & 0 & ? \\ 0 & 1 & ? \end{bmatrix}$$

The second column already has 1 as the second entry; what remains is to use the row operations to change the first entry (also 1) to 0. We will accomplish this by adding the first row to −1 times the second row (or "R1 + −1R2").

$$\begin{array}{ccc} 1 & 1 & 3 \\ + -1\cdot 0 & + -1\cdot 1 & + -1\cdot 2 \\ \hline 1 & 0 & 1 \end{array} \qquad \begin{array}{rcr} 1x + \quad 1y & = & 3 \\ + -1\cdot 0x + -1\cdot 1y & = & + -1\cdot 2 \\ \hline 1x + \quad 0y & = & 1 \end{array}$$

If we replace the first row with this result, then the resulting matrix will have the desired column.

$$\begin{array}{c} \begin{matrix} x & y & \end{matrix} \\ \begin{bmatrix} 1 & 0 & 1 \\ 0 & 1 & 2 \end{bmatrix} \end{array} \qquad \begin{aligned} 1x + 0y &= 1 \\ 0x + 1y &= 2 \end{aligned}$$

Step 5 *Read off the solution.* The matrix

$$\begin{bmatrix} 1 & 0 & 1 \\ 0 & 1 & 2 \end{bmatrix}$$

is just shorthand for the system

$$\begin{aligned} 1x + 0y &= 1 \\ 0x + 1y &= 2 \end{aligned}$$

This system simplifies to

$$\begin{aligned} x &= 1 \\ y &= 2 \end{aligned}$$

The solution is the ordered pair (1, 2).

Alternatively, the solution can be read from the matrix without reverting to the system of equations. We can find the value of the variable heading each column by reading down the column, turning at 1, and stopping at the end of the row.

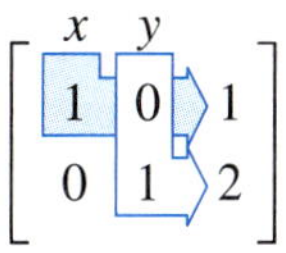

Step 6 ✔ *Check the solution by substituting it into the original system.*

$$\begin{aligned} x + y &= 1 + 2 = 3 \\ 2x + 3y &= 2\cdot 1 + 3\cdot 2 = 8 \end{aligned}$$

■

Notice that in Example 3, steps 3 and 4 are really the same steps applied to different columns. In each of these two steps, we rewrote a column as a column in the identity matrix $\begin{bmatrix} 1 & 0 \\ 0 & 1 \end{bmatrix}$. This process is called **pivoting.** Also notice that, in the execution of steps 3 and 4, the columns already had 1 in the desired location; if that had not been the case, more work would have been necessary, as shown in the next example. The following procedure (a more complete version of the steps used above) will be used on all problems.

Gauss-Jordan Steps

1. *Write the system with all the variables on the left side of the equations and with all coefficients showing.*
2. *Rewrite the system in matrix form.* Eliminate the letters, addition symbols, and equal symbols. Keep all signs for negative numbers.
3. *Use the row operations to rewrite the matrix as an augmented identity matrix.* (This is called *pivoting*.) Decimals can be used rather than fractions as long as you round *each* calculation off to four decimal places, to keep some accuracy.
4. *Read off the solution.* Find the value of the variable heading a column by reading down the column, turning at 1, and stopping at the end of the row.
5. *Check the solution by substituting it into the original system.*

There are many different ways to execute step 3. Some people prefer to do each problem with a routine procedure that always works, while others prefer to do each problem differently, finding clever shortcuts that minimize the calculations. If you are among the former, the following procedure is recommended.

Routine Pivoting Procedure

1. *Change the appropriate entry of the first column into 1 by multiplying or dividing the row containing that entry by a number.*
2. *Change each of the other entries in the column into 0 by adding a multiple of the row with 1 in it to the row in which you wish to have a zero.*
3. *Repeat the above steps with the next column.*

EXAMPLE 3 Use the Gauss-Jordan method to solve the following system:

$$2x + 3y = 14$$
$$3x = 2y + 8$$

Note: While it would probably be easier to solve this problem with the elimination method, the point of this example is to illustrate the Gauss-Jordan method.

Solution

Step 1 *Write the system with all the variables on the left side of the equations and with all coefficients showing.*

$$2x + 3y = 14$$
$$3x - 2y = 8$$

Step 2 *Rewrite the system in matrix form.* Eliminate the letters, addition symbols, and equal symbols.

$$\begin{array}{c} \begin{matrix} x & \;\; y & \;\;\;\;\; \end{matrix} \\ \begin{bmatrix} 2 & 3 & 14 \\ 3 & -2 & 8 \end{bmatrix} \end{array}$$

Step 3 *Use the row operations to rewrite the matrix as an augmented identity matrix.* (We will use the Routine Pivoting Procedure given on the previous page.)

Starting with the first column: *Change the appropriate entry into 1 by multiplying or dividing the row containing that entry by a number.* In other words, we pivot on 2.

$$\begin{array}{cc} x & y & \\ \end{array}$$
$$\begin{bmatrix} 2 & 3 & 14 \\ 3 & -2 & 8 \end{bmatrix} \text{ pivoting on 2}$$

We need to change 2 into 1, so we divide the first row by 2 (or "R1 ÷ 2").

$$\begin{array}{cc} x & y \end{array}$$
$$\begin{bmatrix} 1 & 1.5 & 7 \\ 3 & -2 & 8 \end{bmatrix}$$

Next, change each of the other entries in the column into zeros by adding a multiple of the row with 1 in it to the row in which we wish to have 0. We need to change 3 into 0, so we add −3 times the first row (the row with 1) to the second row (the row in which we wish to have 0) and place the result in the second row (or "−3R1 + R2:R2").

$$\begin{bmatrix} 1 & 1.5 & 7 \\ 0 & -6.5 & -13 \end{bmatrix}$$

Now that we're finished with the first column, we move on to the second column and execute the same steps.

Change the appropriate entry into 1 by multiplying or dividing the row containing that entry by a number.

$$\begin{bmatrix} 1 & 1.5 & 7 \\ 0 & -6.5 & -13 \end{bmatrix} \qquad \text{pivoting on } -6.5$$

We need to pivot on −6.5, so we divide the second row by −6.5 (or "R2 ÷ −6.5").

$$\begin{bmatrix} 1 & 1.5 & 7 \\ 0 & 1 & 2 \end{bmatrix}$$

Next, change each of the other entries in the column into zeros by adding a multiple of the row with 1 in it to the row in which we wish to have 0. We need to change 1.5 into 0, so we add the first row to −1.5 times the second row and place the result in the first row (or "R1 + −1.5R2 : R1").

$$\begin{array}{cc} x & y \end{array}$$
$$\begin{bmatrix} 1 & 0 & 4 \\ 0 & 1 & 2 \end{bmatrix}$$

We have finished step 3; we've used the row operations to rewrite the matrix as the identity matrix with an extra column at the end.

Step 4 *Read off the solution.* The matrix represents the system

$$1x + 0y = 4$$
$$0x + 1y = 2$$

which simplifies to the system

$$x = 4$$
$$y = 2$$

As a shortcut, we can find the value of the variable heading a column by reading down the column, turning at 1, and stopping at the end of the row. The value of x is 4, and the value of y is 2. Either way, our solution is (4, 2).

$$\begin{array}{c} \begin{matrix} x & y & \end{matrix} \\ \left[\begin{matrix} 1 & 0 & 4 \\ 0 & 1 & 2 \end{matrix}\right] \end{array}$$

Step 5 ✔ *Check the solution by substituting it into the original system.*

$$2x + 3y = 2 \cdot 4 + 3 \cdot 2 = 14$$
$$3x - 2y = 3 \cdot 4 - 2 \cdot 2 = 8$$

■

In the preceding two examples, the Gauss-Jordan method seems to involve much more effort than the elimination method, because we gave a very detailed explanation. In the next example, we will leave out the details and allow you to see the advantages of the Gauss-Jordan method.

EXAMPLE 4 Use the Gauss-Jordan method to solve the following system:

$$2x + z = y + 5$$
$$x + 2y + z = 3$$
$$4x - 3z = 0$$

Solution

Step 1 *Write the system with all the variables on the left side of the equations and with all coefficients showing.*

$$2x - 1y + 1z = 5$$
$$1x + 2y + 1z = 3$$
$$4x + 0y - 3z = 0$$

Step 2 *Rewrite the system in matrix form.* Eliminate the letters, addition symbols, and equal symbols.

$$\begin{array}{c} \begin{matrix} x & y & z & \end{matrix} \\ \left[\begin{matrix} 2 & -1 & 1 & 5 \\ 1 & 2 & 1 & 3 \\ 4 & 0 & -3 & 0 \end{matrix}\right] \end{array}$$

Step 3 *Use the row operations to rewrite the matrix as an augmented identity matrix.* The goal of this step is to rewrite the matrix in the following form:

$$\begin{array}{c} \begin{matrix} x & y & z & \end{matrix} \\ \left[\begin{matrix} 1 & 0 & 0 & ? \\ 0 & 1 & 0 & ? \\ 0 & 0 & 1 & ? \end{matrix}\right] \end{array}$$

We rewrite the matrix so that the first column is "1 0 0." The Routine Pivoting Procedure specifies that we start this by dividing the first row by 2, but it's easier to use row operation 3 and interchange the first two rows. (Remember that this is equivalent to switching the order of the first two equations.) The Routine Pivoting Procedure isn't necessarily the most efficient procedure.

pivot on 1

$$\begin{array}{c} \\ \end{array}\begin{bmatrix} x & y & z & \\ 1 & 2 & 1 & 3 \\ 2 & -1 & 1 & 5 \\ 4 & 0 & -3 & 0 \end{bmatrix}$$

$-2R1 + R2{:}R2$

$$\begin{bmatrix} 1 & 2 & 1 & 3 \\ 0 & -5 & -1 & -1 \\ 4 & 0 & -3 & 0 \end{bmatrix}$$

$-4R1 + R3{:}R3$

$$\begin{bmatrix} 1 & 2 & 1 & 3 \\ 0 & -5 & -1 & -1 \\ 0 & -8 & -7 & -12 \end{bmatrix}$$

Now that we're finished with the first column, we move on to the second column and execute the same steps.

pivot on -5

$$\begin{bmatrix} 1 & 2 & 1 & 3 \\ 0 & -5 & -1 & -1 \\ 0 & -8 & -7 & -12 \end{bmatrix}$$

$R2 \div -5$

$$\begin{bmatrix} 1 & 2 & 1 & 3 \\ 0 & 1 & 0.2 & 0.2 \\ 0 & -8 & -7 & -12 \end{bmatrix}$$

$R1 - 2R2{:}R1$
$8R2 + R3{:}R3$

$$\begin{bmatrix} 1 & 0 & 0.6 & 2.6 \\ 0 & 1 & 0.2 & 0.2 \\ 0 & 0 & -5.4 & -10.4 \end{bmatrix}$$

Now that we're finished with the second column, we move on to the third column and execute the same steps.

pivot on -5.4.

$$\begin{bmatrix} 1 & 0 & 0.6 & 2.6 \\ 0 & 1 & 0.2 & 0.2 \\ 0 & 0 & -5.4 & -10.4 \end{bmatrix}$$

$R3 \div -5.4$

$$\begin{bmatrix} 1 & 0 & 0.6 & 2.6 \\ 0 & 1 & 0.2 & 0.2 \\ 0 & 0 & 1 & 1.9259 \end{bmatrix}$$

rounding to four decimal places

$R1 - 0.6R3{:}R1$
$R2 - 0.2R3{:}R2$

$$\begin{bmatrix} 1 & 0 & 0 & 1.4445 \\ 0 & 1 & 0 & -0.1852 \\ 0 & 0 & 1 & 1.9259 \end{bmatrix}$$

rounding to four decimal places

We have finished step 3; we've used the row operations to rewrite the matrix as an augmented identity matrix.

Step 4 *Read off the solution.* The above matrix is just shorthand for the system

$$\begin{aligned} 1x + 0y + 0z &= 1.4445 \\ 0x + 1y + 0z &= -0.1852 \\ 0x + 0y + 1z &= 1.9259 \end{aligned}$$

This system simplifies to

$$\begin{aligned} x &= 1.4445 \\ y &= -0.1852 \\ z &= 1.9259 \end{aligned}$$

The (approximate) solution is the ordered triple (1.4445, −0.1852, 1.9259). Alternatively, the solution can be read from the matrix:

$$\begin{array}{ccc} x & y & z & \\ \left[\begin{array}{ccc} 1 & 0 & 0 \\ 0 & 1 & 0 \\ 0 & 0 & 1 \end{array}\right. & & & \left.\begin{array}{c} 1.4445 \\ -0.1852 \\ 1.9259 \end{array}\right] \end{array}$$

Step 5 *Check the solution by substituting it into the original system.*

$$\begin{aligned} 2x - y + z &= 2(1.4445) - (-0.1852) + (1.9259) = 5.0001 \\ x + 2y + z &= (1.4445) + 2(-0.1852) + (1.9259) = 3 \\ 4x - 3z &= 4(1.4445) - 3(1.9259) = 0.0003 \end{aligned}$$

Note that our solution doesn't check perfectly; we should get 5, 3 and 0 when we substitute the solution into the three equations. This discrepancy is a result of rounding off. ■

7.4 Exercises

In Exercises 1–6, solve the given system with the elimination method.

1. $5x + y - z = 17$
$2x + 5y + 2z = 0$
$3x + y + z = 11$

2. $9x - 2y + 4z = 29$
$2x + 3y - 4z = 3$
$x + y + z = 1$

3. $x + y + z = 14$
$3x - 2y + z = 3$
$5x + y + 2z = 29$

4. $2x + y - z = 4$
$3x + 2y - 7z = 0$
$5x - 3y + 2z = 20$

5. $x - y + 4z = -13$
$2x - z = 12$
$3x + y = 25$

6. $2x + 2y + z = 5$
$3x - y = 5$
$2y + 7z = -9$

Rewrite the systems of equations in Exercises 7–14 in matrix form.

7. $2x + 7y = 11$
$3x - 2y = 15$

8. $4x + 5y = 6$
$3x + 2y = 9$

9. $2x = y + 5$
$3x - y = 41$

10. $2x - 19y = 55$
$3y = 2x + 8$

11. $2x + 3y - 7z = 53$
$5x - 2y + 12z = 19$
$x + y + z = 55$

12. $3x - y + z = 22$
$5x + y + 10z = 38$
$18x + 10y - 121z = 15$

13. $5x + z = 2$
$3x - y = 15$
$2x + 2y + 2z = 53$

14. $8x + 2y = 43$
$2x - z = 15$
$5x + y + 3z = 0$

In Exercises 15–20, use the row operations to change the indicated column into an identity matrix column.

15. $\begin{bmatrix} 2 & 6 & 10 \\ -2 & 1 & 4 \end{bmatrix}$ Change the first column.

16. $\begin{bmatrix} 4 & 12 & 40 \\ -2 & 1 & 4 \end{bmatrix}$ Change the first column.

17. $\begin{bmatrix} 1 & 3 & 12 \\ 0 & 5 & 7 \end{bmatrix}$ Change the second column.

18. $\begin{bmatrix} 1 & 4 & 3 \\ 0 & 2 & -4 \end{bmatrix}$ Change the second column.

19. $\begin{bmatrix} 2 & 2 & 4 & 12 \\ 2 & -1 & 4 & 3 \\ 1 & 2 & -9 & 2 \end{bmatrix}$ Change the first column.

20. $\begin{bmatrix} 4 & 8 & 12 & 4 \\ 2 & 2 & 3 & 4 \\ -1 & 4 & -2 & 2 \end{bmatrix}$ Change the first column.

In Exercises 21–24, find the solution that corresponds to the given matrix.

21. $\begin{bmatrix} 1 & 0 & 9 \\ 0 & 1 & 2 \end{bmatrix}$

22. $\begin{bmatrix} 1 & 0 & 0.7654 \\ 0 & 1 & -3.7990 \end{bmatrix}$

23. $\begin{bmatrix} 1 & 0 & 0 & 0.5281 \\ 0 & 1 & 0 & -0.6205 \\ 0 & 0 & 1 & 0 \end{bmatrix}$

24. $\begin{bmatrix} 1 & 0 & 0 & 3.4076 \\ 0 & 1 & 0 & 1.2066 \\ 0 & 0 & 1 & -0.9934 \end{bmatrix}$

Use the Gauss-Jordan method to solve the systems in Exercises 25–36. If you use decimals, round off each calculation to four decimal places. Check your answers by substituting them back in. (Answers are not *given at the back of the book.)*

25. $x + y = 3$
$2x - 3y = -4$

26. $x - y = 2$
$5x + 2y = 17$

27. $2x + 4y = 12$
$3x - y = 4$

28. $3x - 9y = 15$
$2x + 11y = 10$

29. $5x + 2y = 19$
$3x - 4y = -25$

30. $4x + 9y = 35$
$3x + 2y = 12$

31. $5x + y - z = 17$
$2x + 5y + 2z = 0$
$3x + y + z = 11$

32. $9x - 2y + 4z = 29$
$2x + 3y - 4z = 3$
$x + y + z = 1$

33. $x + y + z = 14$
$3x - 2y + z = 3$
$5x + y + 2z = 29$

34. $2x + y - z = 4$
$3x + 2y - 7z = 0$
$5x - 3y + 2z = 20$

35. $x - y + 4z = -13$
$2x - z = 12$
$3x + y = 25$

36. $2x + 2y + z = 5$
$3x - y = 5$
$2y + 7z = -9$

The systems in Exercises 37–40 have more equations than unknowns. Use the Gauss-Jordan method to solve the given system.

HINT: Temporarily ignore any one of the given equations and solve the resulting system. Then substitute the resulting solution into the ignored equation.

37. $3x - 5y + z = 12$
$2x + y + z = 3$
$5x - 4y + z = 0$
$x + y + z = 4$

38. $x + 2y - 3z = 4$
$5x + 2y + z = 12$
$3x - y = 7$
$4x + z = 12$

39. $x + y = 4$
$3x + z = 5$
$2x - y = 8$
$x + 2y + 3z = -17$

40. $5x - 4y + 2z = 8$
$3x + 2y + z = 1$
$4x - y - z = 12$
$8x + 10y + 2z = 18.$

If a system has fewer equations than unknowns, it may have an infinite number of solutions. In Exercises 41–44, use the Gauss-Jordan method to find three solutions for each system. Solutions are not given in the back of the book because there are an infinite number of them. Check your solutions by substituting them back into the equations.

HINT: After completing the Gauss-Jordan work, rewrite the resulting matrix as a system of equations. Then substitute numbers for z and find the resulting values of x and y.

41. $x + y + z = 3$
$2x - y + z = 2$

42. $3x + y + 5z = 12$
$4x - 12y = 9$

43. $6x + 10y = 22$
$3x - 2z = 5$

44. $x + 2y + 3z = 5$
$4x - 3y - z = 2$

Answer the following question using complete sentences and your own words.

Concept Question

45. Compare and contrast the elimination method and the Gauss-Jordan method. How are they similar? How are they different? What advantages does the elimination method have? What advantages does the Gauss-Jordan method have?

Technology and the Row Operations

One advantage of the Gauss-Jordan method over the elimination method is that it will solve larger systems of equations in a more systematic and efficient manner. Another advantage is that it is amenable to the use of technology and thus affords relief from tedious calculations. Some graphing calculators can perform the row operations on a matrix, as can a computer combined with the Amortrix computer program (which is available with this book). A computer provides an easier interface than does a graphing calculator, but a graphing calculator may be more readily available to you. Each is discussed below.

The Row Operations and Amortrix

The Amortrix computer program, which is available with this book, will enable you to perform the row operations on a matrix easily and quickly. Amortrix is available for both Macintosh and Windows-based computers. Also, a somewhat different version of Amortrix is available over the Web (see the preface for the web address). The web version utilizes a different interface than that discussed below, so the following discussion must be used as a general guide rather than specific instructions if you use the web version of Amortrix.

We will discuss how to use Amortrix to solve the following system:

$$3.1x + 2.7y - 4.9z = 5.3$$
$$6.9x + 4.2y - 3.7z = 9.2$$
$$4.1x - 1.7y + 9.3z = 7.7$$

This system can be rewritten as the following matrix:

$$\begin{bmatrix} 3.1 & 2.7 & -4.9 & 5.3 \\ 6.9 & 4.2 & -3.7 & 9.2 \\ 4.1 & -1.7 & 9.3 & 7.7 \end{bmatrix}$$

When you start Amortrix, a main menu appears. Use the mouse to click on the "Matrix Row Operations" option. Once you're into the "Matrix Row Operations" part of the program, press the "Create a New Matrix" button. The program will ask for the dimensions of the matrix you want to work on and will tell you how to make a response. Once you've responded, the program will display a matrix of the appropriate dimensions. Use the mouse to click on the "Edit Values" button and follow the instructions that appear on the screen to enter the above numbers.

To pivot on 3.1, first divide row 1 by 3.1. To do this, use the mouse to click on the "ROW 1" label to the left of the matrix. This causes two boxes to appear—one that allows you to multiply the row by a number and one that allows you to divide the row by a number. Use the mouse to press the division box, click on the white box, type in "3.1", and press "return" or "enter". When you press the "OK" box, the computer will execute the row operation. See Figure 7.30 on page 556.

To multiply row 1 by -6.9 and add the result to row 2, use the mouse to click first on the "ROW 1" label and then the "ROW 2" label to the left of the matrix. *When adding a multiple of one row to another row, always click first on the row to be multiplied.* This causes two boxes to appear, one that allows you to add two rows and one that allows you to swap two rows. Use the mouse to press the "add" box, click on the first white box, type in "-6.9" and press "return" or "enter". When you press the "OK" box, the computer will execute the row operation. In a similar manner, multiply row 1 by -4.1 and add the result to row 3. This completes the first pivot. See Figure 7.31 on page 557.

At this point, you're ready to move on to the second column. The next step is to divide row two by the entry in row 2 column 2. The computer shows that entry as being -1.8097. However, this number is rounded off, and if you type -1.8097 in the white box, the row operation will result in 1.0054 appearing in row 2 column 2, rather than the desired 1. To avoid this, use the mouse to click on the -1.8097 in row 2, column 2; the computer will divide by the *nonrounded* version of this number, and the row operation will result in a 1 in row 2, column 2.

In a similar manner, continue until you obtain the solution.

Row Operations on a TI Graphing Calculator

The Texas Instruments graphing calculators can perform the row operations. (The TI-80 cannot.)

First, enter the matrix

$$A = \begin{bmatrix} 1 & 2 \\ 3 & 4 \end{bmatrix}$$

as matrix A in the manner discussed in Section 7.0. When you're done entering, type 2nd QUIT. Then duplicate the following instructions on your calculator, and make sure that you get the correct result.

Interchanging Two Rows The "RowSwap" command interchanges two rows. This and other row operations are available under the "MATRX" menu. To swap rows 1 and 2 of matrix A, make the screen read, "RowSwap([A],1,2)" ["rSwap(A,1,2)" on a TI-85/86]. To achieve this, do the following. (Continue reading on page 558.)

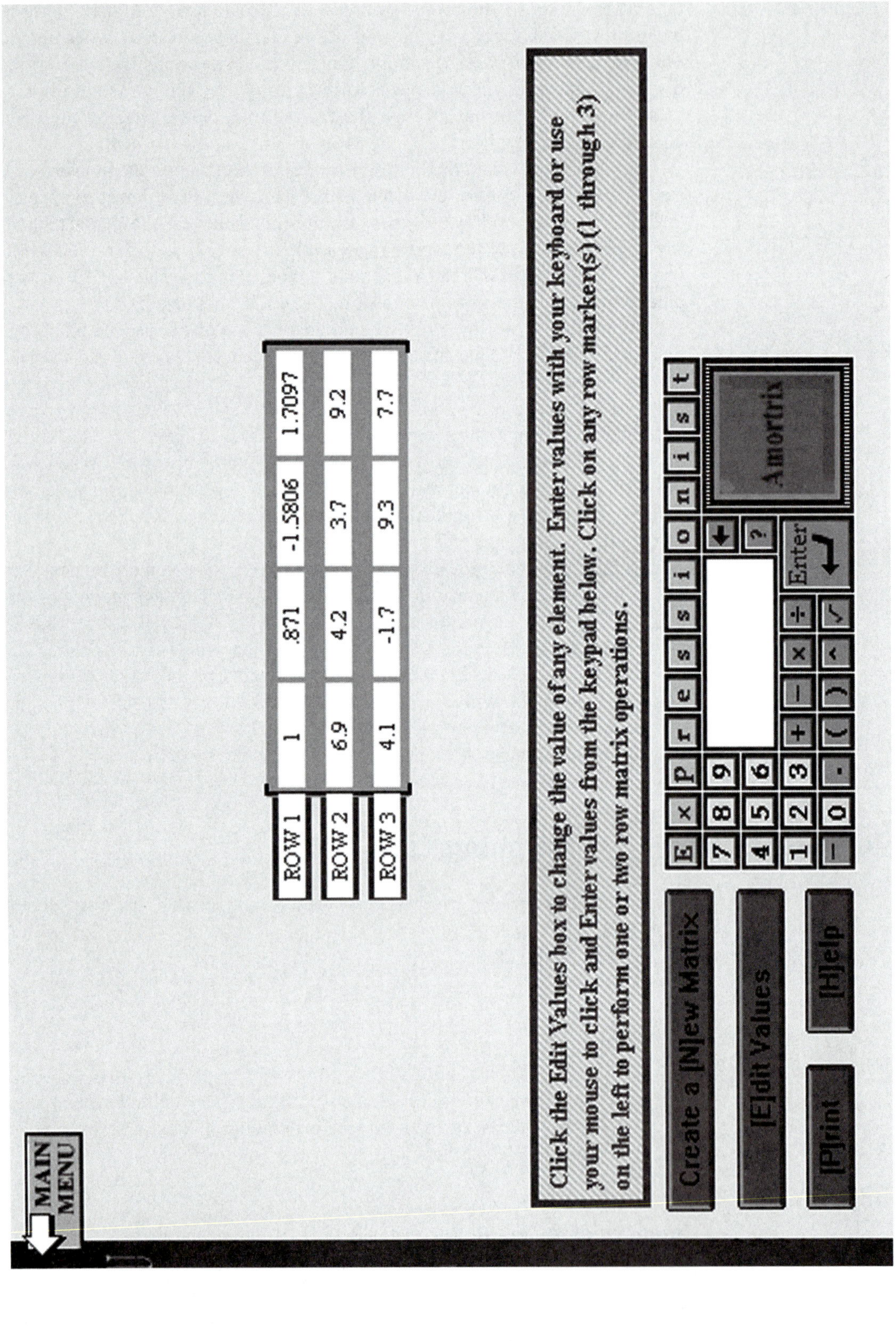

Figure 7.30 Using Amortrix to solve a system of equations.

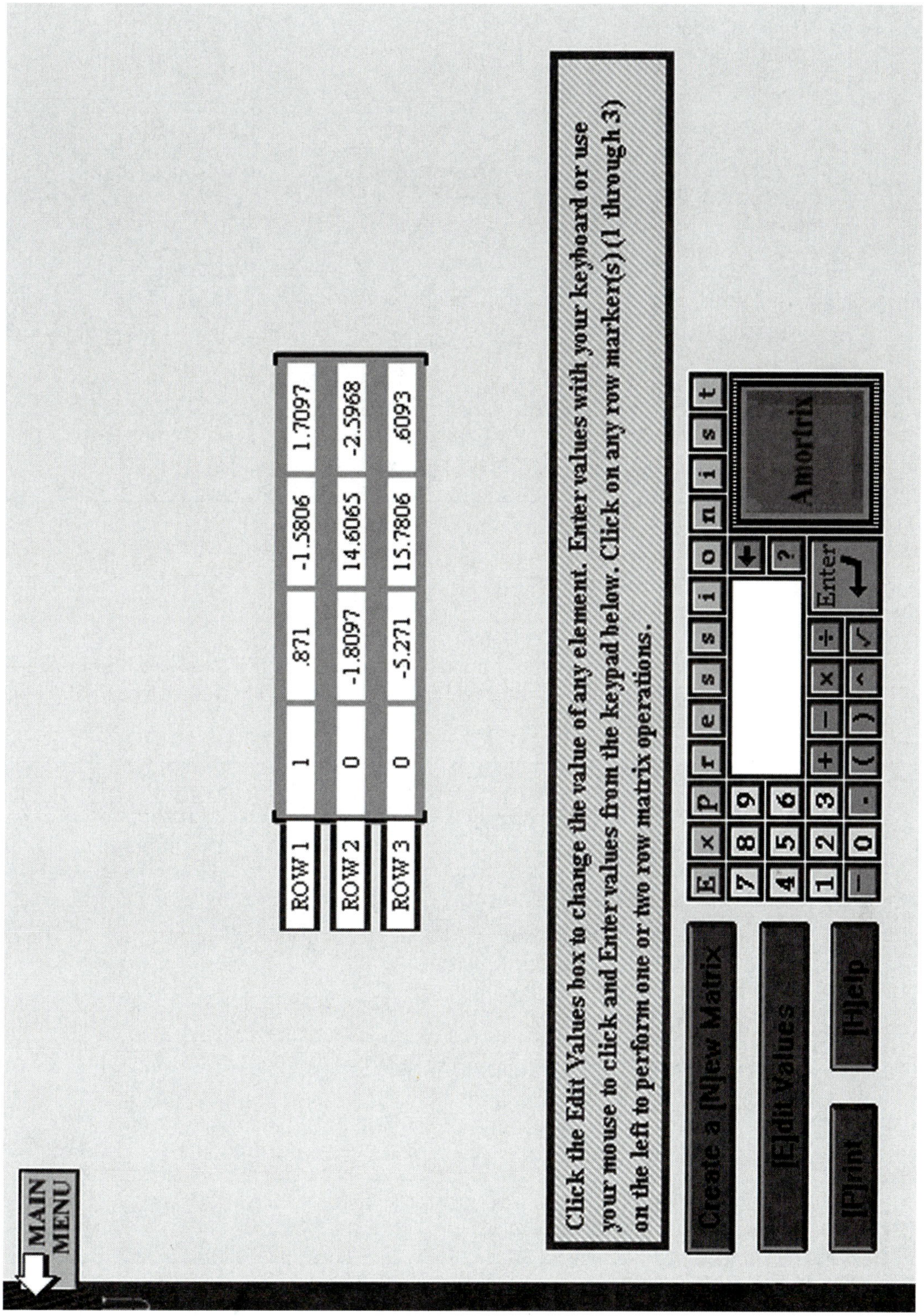

Figure 7.31 Using Amortrix to solve a system of equations.

TI-81 ■ Press MATRX ■ Highlight option 1: "RowSwap(" ■ Press ENTER	Press 2nd [A]	Press ALPHA , 1 ALPHA , 2) ENTER
TI-82/83 ■ Press MATRX ■ Scroll to "MATH" ■ Highlight the "RowSwap(" option (it's down the page) ■ Press ENTER	■ Press MATRX ■ Highlight option 1: "[A]" ■ Press ENTER	Press , 1 , 2) ENTER
TI-85/86 ■ Press 2nd MATRX ■ Press OPS (i.e. F4) ■ Select "rSwap" by pressing MORE F2	Press ALPHA A	Press , 1 , 2) ENTER
This generates "RowSwap(" or "rSwap("	*This generates "[A]" or "A"*	*This generates ",1,2)"*

The result should be $\begin{bmatrix} 3 & 4 \\ 1 & 2 \end{bmatrix}$.

Adding Two Rows To add rows 1 and 2 of matrix A and place the result in row 2, make the screen read, "Row+([A],1,2)" ["rAdd(A,1,2)" on a TI-85/86]. To achieve this, do the following.

TI-81 ■ Press MATRX ■ Highlight "Row+(" ■ Press ENTER	Press 2nd [A]	Press ALPHA , 1 ALPHA , 2) ENTER
TI-82/83 ■ Press MATRX ■ Scroll to "MATH" ■ Highlight "Row+(" ■ Press ENTER	■ Press MATRX ■ Highlight option 1: "[A]" ■ Press ENTER	Press , 1 , 2) ENTER
TI-85/86 ■ Press 2nd MATRX ■ Press OPS (i.e. F4) ■ Select "rAdd" (i.e. MORE F3)	Press ALPHA A	Press , 1 , 2) ENTER
This generates "Row+(" or "r Add("	*This generates "[A]" or "A"*	*This generates ",1,2)"*

The result should be $\begin{bmatrix} 1 & 2 \\ 4 & 6 \end{bmatrix}$.

Multiplying or Dividing a Row by a Number To multiply row 1 of matrix A by 5, make the screen read, "∗Row(5,[A],1" ["multR(5,A,1) on a TI-85/86].

The result should be $\begin{bmatrix} 5 & 10 \\ 3 & 4 \end{bmatrix}$.

To divide a row by a number, multiply by the number's reciprocal. For example, divide a row by 5 by multiplying it by $\frac{1}{5}$.

Adding a Multiple of One Row to Another Row To multiply row 1 by 5, add the result to row 2, and place the result in row 2, make the screen read, "∗Row+(5,[A],1,2)" ["mRAdd(5,A,1,2)" on a TI-85-86].

The result should be $\begin{bmatrix} 1 & 2 \\ 8 & 14 \end{bmatrix}$.

Solving a System on a Graphing Calculator With this experience, you should be able to use your graphing calculator to solve the following system:

$$3.1x + 2.7y - 4.9z = 5.3$$
$$6.9x + 4.2y + 3.7z = 9.2$$
$$4.1x - 1.7y + 9.3z = 7.7$$

This system can be rewritten as the following matrix:

$$\begin{bmatrix} 3.1 & 2.7 & -4.9 & 5.3 \\ 6.9 & 4.2 & 3.7 & 9.2 \\ 4.1 & -1.7 & 9.3 & 7.7 \end{bmatrix}$$

Enter this matrix into your calculator. Check the matrix to make sure that you made no entry errors. (Use the [→] button to view the right end of the matrix.)

To pivot on 3.1, select the row operation, "∗Row("[TI-85/86: "multR("]. Use that command to divide row 1 by 3.1 (i.e., multiply it by 1/3.1).

After performing this or any other row operation, *you must store the resulting matrix or it will be lost*! To store the result as matrix B:

TI-81	Type	[STO] [2nd] [[B]] [ENTER]
TI-82/83	Type	[STO] [MATRX] 2 [ENTER]
TI-85/86	Type	[STO] [B] [ENTER]

Store the result of the next row operation as matrix C, and then alternate between B and C in storing later results. This keeps the previous matrix available, in case you made any errors. It also keeps the original matrix available, as matrix A.

Figure 7.32

Maintaining Accuracy with Decimals It is important to maintain accuracy by not rounding off until the end of the problem. For example, part way through solving the above system, one obtains the matrix shown in Figure 7.32.

It would be tempting to divide row 2 by −1.8096; unfortunately, doing so results in an inaccurate final solution. Instead, divide row 2 by −1.80967741935; this maintains as much accuracy as possible.

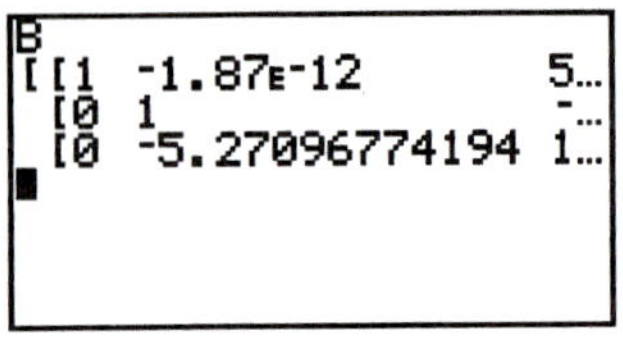

Figure 7.33

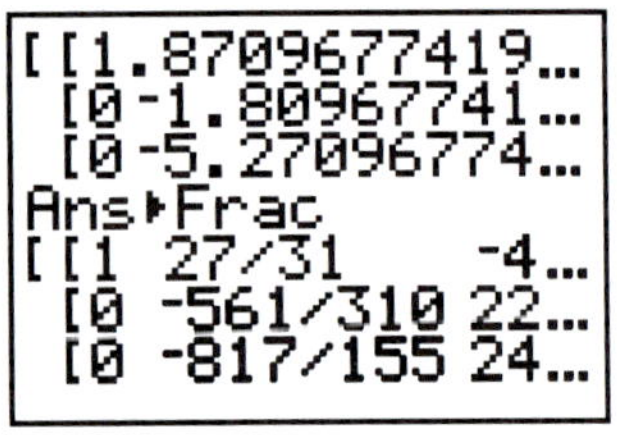

Figure 7.34

A similar issue arises a little further along, when one obtains the matrix shown in Figure 7.33.

The "$-1.87E - 12$" in row 1, column 2 is the calculator's version of -1.87×10^{-12}. This number is not the desired 0; however, it is awfully close to 0. This occurs because the calculator has of necessity rounded off the results of various previous calculations.

Maintaining Accuracy with Fractions It is important to maintain accuracy with this work. With decimals, that involves not rounding off until the end of the problem and treating numbers like "-1.87E-12" as if they were zero. Another, possibly easier way to maintain accuracy is to use fractions via the "→ Frac" command (described in Section 3.3) after each row operation. With this approach, we obtain the matrix in Figure 7.34 rather than that shown in Figure 7.32. And rather than dividing row 2 by -1.80967741935 (and possibly miscopying this number, which would mess everything up), we would multiply row 2 by $-310/561$, the reciprocal of $-561/310$.

Augmented Identity Matrices The TI-83, TI-85, and TI-86 have a "rref" command that automatically rewrites a matrix as an augmented identity matrix. To do this, make the screen read "rref [A]" on a TI-83, or "rrefA" on a TI-85/86.

TI-83	Press MATRX, scroll to "MATH" and select "rref("	Press MATRX 1 ENTER
TI-85/86	Press 2nd MATRX, select "OPS" (i.e. F4) and then "rref" (i.e. F5)	Press ALPHA A ENTER
	This generates "rref(" or "rref"	*This generates "[A]" or "A"*

Row Operations on a Casio Graphing Calculator

The Casio graphing calculator can perform the row operations.

First, enter the matrix

$$A = \begin{bmatrix} 1 & 2 \\ 3 & 4 \end{bmatrix}$$

as matrix A in the manner discussed in Section 7.0. Then duplicate the following instructions on your calculator, and make sure that you get the correct result.

Interchanging Two Rows The "Swap" button interchanges two rows. To swap rows 1 and 2 of matrix A, do the following:

- Press R-OP (i.e., F1).
- Press Swap (i.e., F1).
- Press 1 EXE. This selects one of the rows to be swapped.
- Press 2 EXE. This selects the other row to be swapped.

The result should be $\begin{bmatrix} 3 & 4 \\ 1 & 2 \end{bmatrix}$. The Casio automatically stores this as matrix A.

Adding Two Rows To add rows 1 and 2 of the swapped version of matrix A and place the result in row 2, do the following:

- Press [Rw+] (i.e., [F4]).
- Press 1 [EXE]. This selects a row that will be added.
- Press 2 [EXE]. This selects the row that will be added and replaced.

The result should be $\begin{bmatrix} 3 & 4 \\ 4 & 6 \end{bmatrix}$. The Casio automatically stores this as matrix A.

Multiplying or Dividing a Row by a Number To multiply row 1 of the latest version of matrix A, do the following:

- Press [×Rw] (i.e., [F2]).
- Press 5 [EXE]. This is the multiplier.
- Press 1 [EXE]. This selects the row that will be multiplied and replaced.

The result should be $\begin{bmatrix} 15 & 20 \\ 4 & 6 \end{bmatrix}$. The Casio automatically stores this as matrix A.

To divide a row by a number, multiply by the number's reciprocal. For example, divide a row by 5 by multiplying it by 1/5.

Adding a Multiple of One Row to Another Row To multiply row 1 of the latest version of matrix A by 5, add the result to row 2, and place the result in row 2, do the following:

- Press [×Rw+] (i.e., [F3]).
- Press 5 [EXE]. This is the multiplier.
- Press 1 [EXE]. This selects the row to be multiplied.
- Press 2 [EXE]. This selects the row that will be added and replaced.

The result should be $\begin{bmatrix} 15 & 20 \\ 79 & 106 \end{bmatrix}$. The Casio automatically stores this as matrix A.

Solving a System on a Graphing Calculator With this experience, you should be able to use your graphing calculator to solve the following system:

$$3.1x + 2.7y - 4.9z = 5.3$$
$$6.9x + 4.2y + 3.7z = 9.2$$
$$4.1x - 1.7y + 9.3z = 7.7$$

This system can be rewritten as the following matrix:

$$\begin{bmatrix} 3.1 & 2.7 & -4.9 & 5.3 \\ 6.9 & 4.2 & 3.7 & 9.2 \\ 4.1 & -1.7 & 9.3 & 7.7 \end{bmatrix}$$

Enter this matrix into your calculator. Check the matrix to make sure that you made no entry errors.

To pivot on 3.1, select the row operation "xRw." Use that command to divide row 1 by 3.1 (i.e., multiply it by 1/3.1).

After performing this or any other row operation, the Casio automatically replaces the previous matrix with the new matrix. This means that, if you make an error in performing a row operation, you have to start all over again. The only recourse to this is to start by storing the original matrix as both matrix *A* and matrix *B*. Perform all of the row operations on one of these matrices, and if you make an error, you still have the original matrix stored. You might also consider storing other intermediate results as matrix C, matrix D, and so on. This is time consuming, but not nearly as time consuming as starting all over because of an error.

To store matrix A as matrix B (and thus have two copies of the same matrix), do the following:

- Press MENU.
- Use the arrow buttons to highlight "RUN" and press EXE.
- Press OPTN.
- Press MAT (i.e., F2).
- Type Mat ALPHA A → Mat ALPHA B

Exercises

46. Use the course software or a graphing calculator to finish solving the system discussed in this section. Check your answer with your calculator; it is *not* given in the back of the book. (Your answer should check, but not perfectly, since the program rounded off its calculations.)

Use the course software or a graphing calculator to solve the systems in Exercises 47–58. Check your solutions with your calculator; they are not *given in the back of the book. Your answer should consist of a list of the row operations used in the order in which you used them, in addition to the final solution of the problem.*

47. $x + y + z = 2.35$
$-2x + 3y - 3z = 6.2$
$1.2x - 0.4y + 2.1z = -2.64$

48. $4.6x - 7.2y + 9.8z = 16.36$
$3.2x + 6.2y + 5.3z = 105.7$
$9.5x - 3.6y - 2.5z = 5.58$

49. $5.1x - 3.2y + 9.8z = 15$
$7.3x - 4.6y - 1.2z = 5$
$8.0x + y + 2z = 14$

50. $18x - 22y + 53z = 9$
$-3x + 28y - 10z = 0$
$5x + y - 10z = 33$

51. $x = 2y - z$
$3y + 2z - 15x = 0$
$9x + 2y + 7z = 29$

52. $y - z = 20x$
$8x + 2y - 5z = 19$
$37x + 22y - 55z = 39$

53. $1.3x + 5.3y - 8.9z + 5.2w = 2.8$
$4.7x - 5.5y - 3.8z - 7.3w = 5.0$
$5.3x - 1.0y - 3.3z + 8.9w = 8.3$
$7.4x - 3.2y + 9.9z + 5.7w = 82.9$

54. $38x + 39y + 18z + 32w = 449$
$-57x + 48y + 29z - 12w = 39$
$-38x + 30y - 94z + 42w = 93$
$42x + 15y - 34z + 12w = 38$

55. $3.9x_1 + 4.9x_2 - 3.9x_3 - 4.2x_4 = 4.2$
$4.7x_1 + 3.7x_2 + 3.8x_3 + 9.7x_4 = 1.1$
$7.0x_1 - 7.2x_2 + 3.9x_3 + 3.6x_4 = 5$
$8.5x_1 + 6.3x_2 - 9.0x_3 - 3.7x_4 = 0$

56. $63x_1 + 95x_2 - 36x_3 - 99x_4 = 148$
$74x_1 - 77x_2 + 38x_3 + 96x_4 = 95$
$12x_1 - 6x_2 + x_3 = 5$
$84x_1 + 96x_2 - 46x_3 - 145x_4 = 80$

Projects

57. $6.3x_1 + 5.9x_2 - 3.9x_3 - 4.7x_4 + 9.1x_5 = 71.8$
$6.4x_1 + 9.2x_2 + 5.1x_3 + 2.2x_4 - 7.6x_5 = 81.1$
$3.4x_1 - 7.0x_2 + 2.9x_3 + 3.5x_4 + 4.2x_5 = 15$
$5.7x_1 + 3.6x_2 - 9.0x_3 - 2.4x_4 + 3.3x_5 = 100$
$3.6x_1 + 4.3x_2 - 5.7x_3 - 6.4x_4 - 2.8x_5 = 53$

58. $63x_1 + 85x_2 - 26x_3 - 89x_4 + 72x_5 = 148$
$32x_1 - 54x_2 + 95x_3 + 146x_4 - 32x_5 = 95$
$12x_1 - 6x_2 + x_3 + 12x_4 = 5$
$84x_1 + 96x_2 - 46x_3 - 145x_4 = 80$
$472x_1 - 59x_2 + 98x_3 + 16x_4 - 72x_5 = 83$

7.5 More on Markov Chains

In this section, we will continue to make long-range predictions with Markov chains, as we did in Section 7.3. However, the problems in this section will involve more than two unknowns. The resulting systems can be solved by using either the elimination method or the Gauss-Jordan method.

EXAMPLE 1 In sociology, a family is considered to belong to the lower class, the middle class, or the upper class, depending on the family's total annual income. Sociologists have found that the strongest determinant of an individual's class is the class of his or her parents. Census data suggest the following (see Figure 7.35).

- Of those individuals whose parents belong to the lower class, 21% will become members of the middle class, and 1% will become members of the upper class.
- Of those individuals whose parents belong to the middle class, 6% will become members of the lower class, and 4% will become members of the upper class.
- Of those individuals whose parents belong to the upper class, 1% will become members of the lower class, and 10% will become members of the middle class.

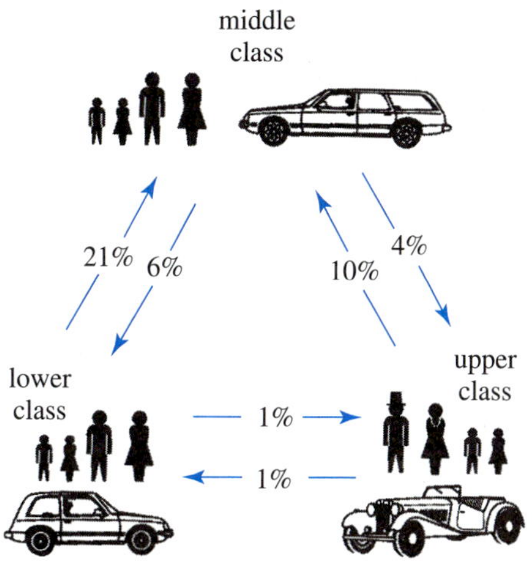

Figure 7.35

Make a long-range prediction of the levels at which the lower, middle, and upper classes will stabilize.

Solution We are asked to find the equilibrium matrix L.

Step 1 *Create the transition matrix T.* We did this in Section 7.1.

$$T = \begin{bmatrix} 0.78 & 0.21 & 0.01 \\ 0.06 & 0.90 & 0.04 \\ 0.01 & 0.10 & 0.89 \end{bmatrix}$$

Step 2 *Create the equilibrium matrix L.* L is a row matrix, and in this problem L has three entries—one for lower class, one for middle class, and one for upper class.

$$L = [x \quad y \quad z]$$

Step 3 *Find and simplify the system of equations described by LT = L.*

$$\begin{array}{ccc} L & T & = \quad L \end{array}$$

$$[x \quad y \quad z] \cdot \begin{bmatrix} 0.78 & 0.21 & 0.01 \\ 0.06 & 0.90 & 0.04 \\ 0.01 & 0.10 & 0.89 \end{bmatrix} = [x \quad y \quad z]$$

$$[0.78x + 0.06y + 0.01z \quad 0.21x + 0.90y + 0.10z \quad 0.01x + 0.04y + 0.89z] = [x \quad y \quad z]$$

This matrix equation describes the following system:

$$\begin{aligned} 0.78x + 0.06y + 0.01z &= x \\ 0.21x + 0.90y + 0.10z &= y \\ 0.01x + 0.04y + 0.89z &= z \end{aligned}$$

We can simplify this system by combining like terms.

$$\begin{aligned} -0.22x + 0.06y + 0.01z &= 0 \\ 0.21x - 0.10y + 0.10z &= 0 \\ 0.01x + 0.04y - 0.11z &= 0 \end{aligned}$$

We can further simplify the system by multiplying by 100.

$$\begin{aligned} -22x + 6y + 1z &= 0 \\ 21x - 10y + 10z &= 0 \\ 1x + 4y - 11z &= 0 \end{aligned}$$

Step 4 *Discard any redundant equations and include the equation* $x + y + z = 1$. The third equation is the negative of the sum of the first two equations, so its presence is redundant. Thus, any one of the three equations can be dropped. $L = [x \quad y \quad z]$ is a probability matrix, so the sum of its entries must be 1; that is, $x + y + z = 1$.

$$-22x + 6y + 1z = 0 \qquad \textbf{(1)}$$
$$21x - 10y + 10z = 0 \qquad \textbf{(2)}$$
$$x + y + z = 1 \qquad \textbf{(3)}$$

Step 5 *Solve the resulting system.* Because this is a larger system, we'll use the Gauss-Jordan method and round off all calculations to four decimal places. (Alternatively, we could use the elimination method.)

$$\begin{bmatrix} 1 & 1 & 1 & 1 \\ -22 & 6 & 1 & 0 \\ 21 & -10 & 10 & 0 \end{bmatrix}$$ putting equation (3) first, to take advantage of the 1

22R1 + R2:R2

−21R1 + R3:R3

$$\begin{bmatrix} 1 & 1 & 1 & 1 \\ 0 & 28 & 23 & 22 \\ 0 & -31 & -11 & -21 \end{bmatrix}$$

R2 ÷ 28:R2

$$\begin{bmatrix} 1 & 1 & 1 & 1 \\ 0 & 1 & 0.8214 & 0.7857 \\ 0 & -31 & -11 & -21 \end{bmatrix}$$

R1 − R2:R1

31R2 + R3:R3

$$\begin{bmatrix} 1 & 0 & 0.1786 & 0.2143 \\ 0 & 1 & 0.8214 & 0.7857 \\ 0 & 0 & 14.4634 & 3.3567 \end{bmatrix}$$

R3 ÷ 14.4634:R3

$$\begin{bmatrix} 1 & 0 & 0.1786 & 0.2143 \\ 0 & 1 & 0.8214 & 0.7857 \\ 0 & 0 & 1 & 0.2321 \end{bmatrix}$$

R1 − 0.1786R3:R1

R2 − 0.8214R3:R2

$$\begin{bmatrix} 1 & 0 & 0 & 0.1728 \\ 0 & 1 & 0 & 0.5951 \\ 0 & 0 & 1 & 0.2321 \end{bmatrix}$$

The resulting solution is

$$L = [x \quad y \quad z] = [0.1728 \quad 0.5951 \quad 0.2321] \approx [0.17 \quad 0.60 \quad 0.23]$$

Step 6 ✔ *Check your work by verifying that $LT = L$.*

$$\underset{L}{[0.1728 \quad 0.5951 \quad 0.2321]} \cdot \underset{T}{\begin{bmatrix} 0.78 & 0.21 & 0.01 \\ 0.06 & 0.90 & 0.04 \\ 0.01 & 0.10 & 0.89 \end{bmatrix}} = L$$

$$= [0.172811 \quad 0.595088 \quad 0.232101]$$

$$\approx [0.1728 \quad 0.5951 \quad 0.2321]$$

This means that, if current trends continue, the lower class will eventually stabilize at 17% of the population, the middle class will stabilize at 60%, and the upper class will stabilize at 23%. (Recall that in 1994, the lower class was 16%, the middle class was 67%, and the upper class was 17%, according to the Census Bureau.) It is important to remember that this prediction is based on the assumption that current trends continue.

Note that the solution didn't check perfectly; this is because we rounded to four decimal places in our calculations. If we had used fractions rather than decimals (and possibly a TI graphing calculator's "→Frac" command, the solution would check perfectly. ■

7.5 Exercises

In Exercises 1–2, find the equilibrium matrix L by solving $LT = L$ for L.

1. $T = \begin{bmatrix} 0.3 & 0.2 & 0.5 \\ 0.1 & 0.8 & 0.1 \\ 0.4 & 0.3 & 0.3 \end{bmatrix}$

2. $T = \begin{bmatrix} 0.4 & 0.3 & 0.3 \\ 0.2 & 0.7 & 0.1 \\ 0.3 & 0.3 & 0.4 \end{bmatrix}$

3. An extensive survey of Metropolis's gym users indicates that 71% of the current members of Silver's Gym will continue their annual membership when it expires, 12% will quit and join Fitness Lab, and the rest will quit and join ThinNFit. Fitness Lab has been unable to keep its equipment in good shape, and as a result, 32% of its members will defect to Silver's and 34% will leave for ThinNFit. ThinNFit's members are quite happy, and as a result, 96% plan on renewing their annual memberships, with half of the balance planning on moving to Silver's and half to Fitness Lab. Make a long-range prediction of the ultimate market shares of the three health clubs, assuming that current trends continue. (See Exercise 15 in Section 7.1.)

4. Smallville has three Chinese restaurants: Asia Gardens, Chef Chao's, and Chung King Village. Chung King Village recently mailed a coupon to all residents of Smallville offering two dinners for the price of one. As a result, 67% of those who normally eat at Asia Gardens plan to try Chung King Village within the next month, and 59% of those who normally eat at Chef Chao's plan to try Chung King Village. Also, all of Chung King Village's normal customers will return to take advantage of the special. And 30% of those who normally eat at Asia Gardens will eat there within the next month because of its convenient location. Chef Chao's has a new chef who isn't doing very well, and as a result, only 15% of those who normally eat there are planning to return next time. Make a long-range prediction of the ultimate market shares of the three restaurants, assuming that current trends continue. (See Exercise 16 in Section 7.1.)

5. A survey shows that 23% of the shoppers in seven midwestern states regularly buy their groceries at Safe Shop, 29% regularly shop at PayNEat, and the balance shop at any one of several smaller markets. The survey indicates that 8% of the consumers who currently shop at Safe Shop will purchase their groceries at PayNEat the next time they shop and that 5% will switch to some other store. Also, 12% of the consumers who currently shop at PayNEat will purchase their groceries at Safe Shop the next time they shop, and 2% will switch to some other store. In addition, 13% of the consumers who currently shop at neither store will purchase their groceries at Safe Shop the next time they shop, and 10% will shop at PayNEat. Predict the percent of midwestern shoppers who will be regular Safe Shop customers and the percent who will be regular PayNEat customers as a result of this trend. (See Exercises 4 and 10 in Section 7.1.)

6. In October 1990, a New York Times/CBS News poll contained the information shown in Figure 7.36. For example, 78% of those registered as a Democrat reported that they would vote for a Democrat in the next election.
   a. Complete the chart.
   b. What assumption must be made in completing the chart?
   c. If the poll were interpreted as predicting a change in registered party affiliation, make a long-range prediction of the percentage of voters who will be registered as Democrats, as Republicans, and as Independents.
   d. Do you think that the prediction in part (b) would be a good prediction? Why?

Projects

7. In May 1992, the New York Times reported on a University of Michigan study on intergenerational transition between income groups. That study contained the information shown in Figure 7.37. For example, 30% of those whose fathers were in the lowest income group stayed in the lowest income group.
   a. Make a long-range prediction of the percent of males in the various income groups.
   b. The predictions in part (a) are not 10%, 40%, 40%, and 10%. Should they be? Explain.

      HINT: Does "the bottom 10%" mean the same income bracket for the son as it does for the father?
   c. What assumptions are made in the prediction in part (a)? Consider the fact that some men have no sons, some have one son, and some have

	Next Vote Will Be for a Democrat	**Next Vote Will Be for a Republican**	**Next Vote Will Be for an Independent**
Registered as a Democrat	78%	6%	?
Registered as a Republican	6%	75%	?
Registered as an Independent	30%	29%	?

Figure 7.36
Source: *New York Times*, October 12, 1990, page A21

	Son's Income in Bottom 10%	Son's Income From 10% to 50%	Son's Income From 50% to 90%	Son's Income in Top 10%
Father's Income in Bottom 10%	30%	52%	17%	1%
Father's Income From 10% to 50%	10%	48%	38%	4%
Father's Income From 50% to 90%	4%	38%	48%	10%
Father's Income in Top 10%	1%	17%	52%	30%

Figure 7.37

Source: *New York Times*, May 18, 1992, page D5. Data has been adjusted so that pecentages add to 100%.

many sons. Also consider the fact that only males are involved in the study.

8. a. The Census Bureau classifies all residents of the United States as residents of a central city, or a suburb, or a nonmetropolitan area. In the 1990 census, the bureau reported that the central cities held 77.8 million people, the suburbs 114.9 million people, and the nonmetropolitan areas 56.0 million people. The Census Bureau (in Current Population Reports, P20–463, *Geographic Mobility*: *March 1990 to March 1991*) also reported the information in the accompanying table and in Figure 7.38 regarding migration between these three areas from 1990 to 1991 (numbers are in thousands).

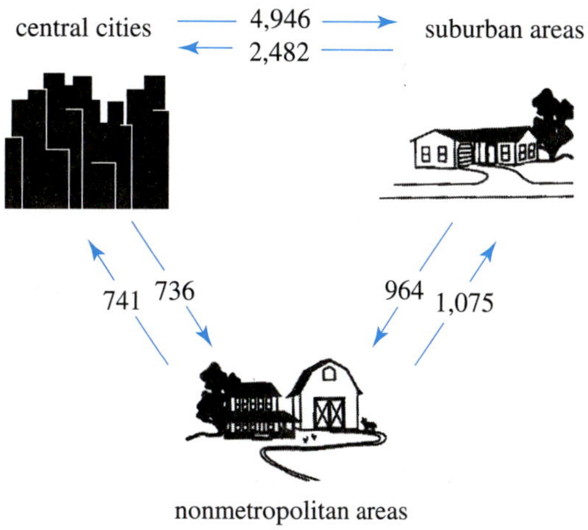

Figure 7.38
Movers between cities, suburbs, and nonmetropolitan areas, and net change due to migration, 1990–1991 (numbers in thousands)

	Moved to		
Moved From	**Central City**	**Suburb**	**Nonmetropolitan Area**
Central City	(x)	4,946	736
Suburb	2,482	(x)	964
Nonmetropolitan Area	741	1,075	(x)

NOTE: Persons moving from one central city to another were not considered. This is represented in the above chart by "(x)".

Compute the probabilities that a central city resident will move to a suburb or to a nonmetropolitan area, the probabilities that a suburban resident will move to a central city or to a nonmetropolitan area, and the probabilities that a nonmetropolitan resident will move to a central city or to a suburb. Present these probabilities in the form of a transition matrix. (Round off to three decimal places.) (See Exercise 19(b) in Section 7.1.)

b. Make a long-range prediction of the ultimate percent of U.S. residents who will reside in a central city, a suburb, and a nonmetropolitan area, assuming that the trend indicated by the above data continues.

9. The Census Bureau (in *Statistical Abstract of the United States*: *2000*) reported the 2000 national and regional populations in Figure 7.39.

Region	Population (in thousands)
United States	281,422
Northeast	53,594
Midwest	64,393
South	100,237
West	63,198

Figure 7.39

The Census Bureau (in *Current Population Reports P20–538, Geographical Mobility*: *March 1999 to March 2000*) also reported the information in Figures 7.40 and 7.41 regarding the movement between the regions from 1999 to 2000 (numbers are in thousands).

a. Compute the probabilities that a resident of any one region will move to any of the other regions. Present these probabilities in the form of a transition matrix. (Round off to three decimal places.) (See Exercise 20(b) in Section 7.1.)

b. Make a long-range prediction of the ultimate percent of U.S. residents who will reside in the Northeast, the Midwest, the South, and the West, assuming that the trend described by the above data continues.

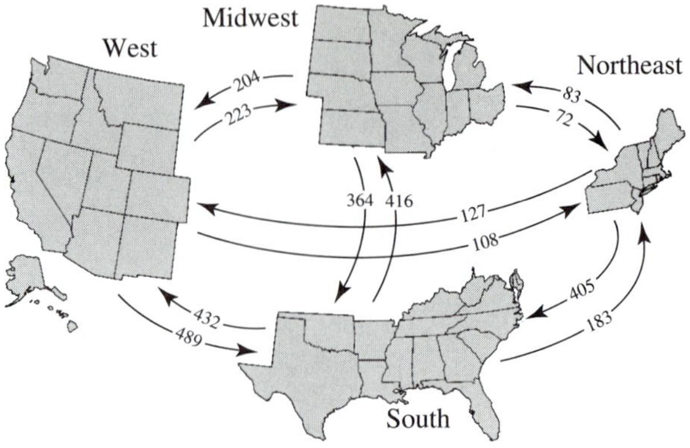

Figure 7.41

10. Seven different makes of automobile are available for purchase in the country of Outer Moldavia: the Asian-manufactured Toyonda and Nissota, the American Reo, Henry J, and DeSoto, and the European Hugo. The Moldavian Department of Motor Vehicles kept records of the sale of all new and used autos. These records show that 60% of those people who sold their old Toyonda replaced it with a new Toyonda, 10%

	Car Purchased					
Car Sold	**Toyonda**	**Nissota**	**Reo**	**Henry J**	**DeSoto**	**Hugo**
Toyonda	60%	10%	12%	8%	7%	3%
Nissota	9%	64%	7%	8%	9%	3%
Reo	19%	10%	54%	9%	7%	1%
Henry J	5%	6%	5%	69%	12%	3%
DeSoto	3%	5%	13%	16%	61%	2%
Hugo	18%	15%	18%	16%	17%	16%

Figure 7.42

Region Moved from	Region Moved to			
	Northeast	Midwest	South	West
Northeast	(x)	83	405	127
Midwest	72	(x)	364	204
South	183	416	(x)	432
West	108	223	489	(x)

Figure 7.40

replaced it with a Nissota, 12% switched to a Reo, 8% switched to a Henry J, 7% bought a DeSoto, and 3% actually bought a Hugo. Similar statistics on the other brands are shown in Figure 7.42. If this trend continues, what percent of the Moldavian market will each brand of automobile command?

11. **a.** Use the information on United States class structure in Example 8 in Section 7.1 to calculate PT, PT^2, PT^3, etc., and list the first entry of each of those calculations, rounded to two decimal places. Continue in this fashion until your result agrees with the result of Example 2 in this section.

NOTE: Use some form of technology.

b. For which value of n did PT^n approximate the result of Example 2 with two-decimal-place accuracy?

c. Is it reasonable to approximate the equilibrium matrix L with two-decimal-place accuracy by calculating PT, PT^2, PT^3, and so on until you get two successive answers that round to the same result? Why?

Answer the following questions using complete sentences.

Concept Questions

12. What information would a marketing analyst need to make a long-range prediction of the future market share of a product? How would he or she obtain such information?

13. How could a marketing analyst predict when the equilibrium matrix will be achieved?

Chapter 7 Review

TERMS

associative property	dimensions	Markov chain	solving a system of equations
augmented identity matrix	element	matrix	square matrix
column	entry	pivoting	state
column matrix	equilibrium matrix	probability matrix	system of equations
commutative property	following state	row	transition matrix
current state	identity matrix	row matrix	
diagonal	linear equation	row operations	

REVIEW EXERCISES

In Exercises 1–4, determine whether the given matrix is a row matrix, a column matrix, a square matrix, or none of these; also give the dimensions of the matrix.

1. $\begin{bmatrix} 5 & 7 \\ 3 & 9 \\ 1 & 0 \end{bmatrix}$

2. $\begin{bmatrix} 6 & 3 & 8 \\ 1 & 6 & 3 \\ 4 & 9 & 6 \end{bmatrix}$

3. $\begin{bmatrix} 8 \\ 2 \\ 4 \\ 7 \end{bmatrix}$

4. $\begin{bmatrix} 4 & 67 & 43 & 27 \end{bmatrix}$

In Exercises 5–9, find the indicated product, if it exists.

5. $\begin{bmatrix} 1 & 6 \\ 8 & 4 \end{bmatrix} \cdot \begin{bmatrix} -3 & 6 \\ 0 & -5 \end{bmatrix}$

6. $\begin{bmatrix} 6 & 3 & 5 \\ 0 & 2 & 5 \end{bmatrix} \cdot \begin{bmatrix} -3 & 6 \\ 0 & -5 \end{bmatrix}$

7. $\begin{bmatrix} -3 & 6 \\ 0 & -5 \end{bmatrix} \cdot \begin{bmatrix} 6 & 3 & 5 \\ 0 & 2 & 5 \end{bmatrix}$

8. $\begin{bmatrix} 5 & 3 & -9 \\ 1 & 0 & 5 \end{bmatrix} \cdot \begin{bmatrix} 7 & 4 & 6 & 9 \\ 3 & 5 & -7 & 2 \\ 1 & 6 & 9 & 3 \end{bmatrix}$

9. $\begin{bmatrix} 7 & 4 & 6 & 9 \\ 3 & 5 & -7 & 2 \\ 1 & 6 & 9 & 3 \end{bmatrix} \cdot \begin{bmatrix} 5 & 3 & -9 \\ 1 & 0 & 5 \end{bmatrix}$

10. Johnco Industries is planning a new promotional campaign for its "Veg-O-Slicer." Use matrix multiplication and Figure 7.43 to find the cost of three newspaper ads and four television ads in New York City and the cost of a similar ad campaign in Washington, D.C.

	Newspaper Ad	Television Ad
New York City	\$5,000	\$7,000
Washington, D.C.	\$5,500	\$6,200

Figure 7.43

11. Is matrix multiplication commutative? Is it associative?

Solve the systems in Exercises 12 and 13 with the elimination method or a graphing calculator. Check your answers by substituting them back in. (Answers are not *given at the back of the book.)*

12. $3x + 5y = -14$
    $4x - 7y = 36$
13. $5x - 7y = -29$
    $2x + 4y = 2$

Solve the systems in Exercises 14–16 with the Gauss-Jordan method or the elimination method. Check your answers by substituting them back in. (Answers are not *given at the back of the book.)*

14. $3x + 5y = -14$
    $4x - 7y = 36$
15. $5x - 7y = -29$
    $2x + 4y = 2$
16. $7x - 8y + 3z = -43$
    $5x + 3y - z = 24$
    $11x - 5y + 12z = 0$
17. Department of Motor Vehicles records indicate that 12% of the automobile owners in the state of Jefferson own a Toyonda. A recent survey of Jefferson automobile owners, commissioned by Toyonda Motors, shows that 62% of the Toyonda owners would buy a Toyonda for their next car and that 16% of those automobile owners who do not own a Toyonda would buy a Toyonda for their next car. The same survey indicates that automobile owners buy a new car an average of once every three years.
    a. Predict Toyonda Motors' market share after three years.
    b. Predict Toyonda Motors' market share after six years.
    c. Make a long-range prediction of Toyonda Motors' market share.
    d. What assumptions are these predictions based on?
18. Currently, 23% of the residences in Foxtail County are apartments, 18% are condominiums or townhouses, and the balance are single-family houses. The Foxtail County Contractors' Association (FCCA) has commissioned a survey that shows that 3% of the apartment residents plan to move to a condominium or townhouse within the next two years and that 5% plan to move to a single-family house. The same survey indicates that 1% of those who currently reside in a condominium or townhouse plan to move to an apartment and that 11% plan to move to a single-family house; also, 2% of the single-family house dwellers plan to move to an apartment, and 4% plan to move to a condominium or townhouse.
    a. What recommendation should the FCCA make regarding the construction of apartments, condominiums, and single-family houses in the next four years?
    b. What long-term recommendations should the FCCA make?
    c. What important factors does the survey ignore?

History Questions

19. Who invented the theory of matrices?
20. Who invented Markov chains?

Linear Programming

8

Linear programming is a method of solving problems that involve a quantity to be maximized or minimized when that quantity is subject to certain restrictions. Linear programming was invented in the 1940s by George Dantzig, as a result of an Air Force research project concerned with computing the most efficient and economical way to distribute personnel, weapons, and supplies to the various fronts during World War II. The word *programming* in the term "linear programming" means creating a plan or procedure that solves a problem; it is not a reference to computer programming.

Since its conception, linear programming has been extremely successful. The petroleum industry was the first civilian industry to use it. Now refineries use linear programming methods to blend gasoline, decide what crude oil to buy, and determine what products to produce. The steel industry uses it to evaluate ores and to determine what products to produce and when to build new furnaces. Airlines use it to minimize costs related to the scheduling of flights, subject to constraints such as the amount of time a pilot or a crew may fly. The federal energy authorities use linear programming to explore policy alternatives. Various businesses and government agencies use it to determine the best way to control water and air pollution, assign personnel to jobs, and achieve racial balance in schools. Supermarket chains use it to determine which warehouses should ship which products to the stores. Investment companies use it to create portfolios with the best mix of stocks and bonds.

In this chapter, we focus on how small businesses can use linear programming to determine the best way to allocate their limited resources. For example, a craftsman who manufactures coffee tables and end tables by hand would have to decide how many tables of each type he should make. He would need to consider his limited resources of time and money as well as his desire to increase profit. Linear programming analyzes the effect of such limitations on the craftsman's business and helps him make an informed decision.

In large corporations, managers must decide how to allocate their limited resources (such as raw materials, labor, and machinery) to maximize their profit and meet other objectives. You might assume that such a corporation's resources are not really limited—that a manager could just determine a production level for each of her products and obtain the resources needed to meet those production

AP/World Wide Photos

The method of linear programming resulted from 1940s research to determine the least expensive way to distribute personnel and equipment to the various fronts of World War II.

levels. However, this is frequently not the case. Sufficient resources might not be available, or they might be limited by their location or their cost. Resources are usually limited, and their allocation is a difficult decision. Linear programming analyzes the effect of such limitations and helps corporate managers make informed decisions.

Corporate managers have so many different products and resources to consider that the necessary calculations must be done by computer. It is estimated that linear programming accounts for at least as much computer time as does payroll or inventory control. And some experts claim that linear programming is the most widely used form of modern mathematics.

8.0 Review of Linear Inequalities

A **linear equation** in two variables x and y is an equation that can be written in the form $ax + by = c$, where a, b, and c are constants. Such an equation is referred to as a *linear* equation because its graph is a line. We can graph most linear equations by solving for y and using the **slope-intercept formula** $y = mx + b$ to find the slope m and the y-intercept b of the line.

A **linear inequality** in two variables x and y is an inequality that can be written in the form $ax + by < c$ (or with $>$, $\leq$, or $\geq$ instead of $<$). In other words, a linear inequality is the result of replacing a linear equation's equals symbol with an inequality symbol.

Our goal in this chapter is to **optimize** a quantity—that is, to maximize or minimize a quantity. A craftsman who manufactures coffee tables and end tables by hand would want to know how to invest his resources of time and money to maximize his profit. His resources are restricted; he wants to work *no more than* 40 hours each week, and he has *at most* $1,000 to spend on materials each week. These restrictions are described mathematically with inequalities, because they involve the phrases "no more than" and "at most." To analyze the craftsman's restricted resources and the effect of these restrictions on his profit, we must be able to graph the inequalities that describe the restrictions.

Each restriction in a linear programming problem must be expressible as a linear inequality; this is the meaning of the word *linear* in the name "linear programming." In this section, we discuss the graphing of linear inequalities. We can graph a linear inequality by solving the inequality for y, graphing the line described by the associated equation, and then shading the region to one side of that line.

EXAMPLE 1 Graph the linear inequality $2x + y \le 6$.

Solution

Step 1 *Solve the linear inequality for y.*

$$2x + y \le 6 \rightarrow y \le -2x + 6$$

Step 2 *Graph the line.* The equation associated with the inequality is $y = -2x + 6$. Comparing this equation with the slope-intercept formula $y = mx + b$, we find that this is the equation of a line with slope $m = -2$ and y-intercept $b = 6$. Because the slope is -2 and because slope means "rise over run," we have

$$\frac{\text{rise}}{\text{run}} = -2 = \frac{-2}{1}$$

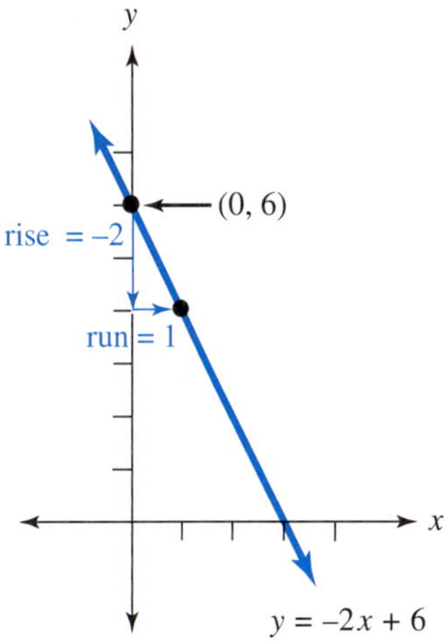

Figure 8.1

To graph the line, we place a point at 6 on the y-axis (since the y-intercept is 6) and then from that point rise -2 (that is, move two units down) and run 1 (that is, move one unit to the right). This takes us to a new point. We connect the points with a line. Because "=" is a part of "≤," any point on the line $y = -2x + 6$ must also be a point on the graph of the inequality $y \le -2x + 6$. We show this by using a solid line, as in Figure 8.1. (If our inequality were $y < -2x + 6$, then a point on the line would *not* be a point on the graph of the inequality. We would show this by using a dashed line.)

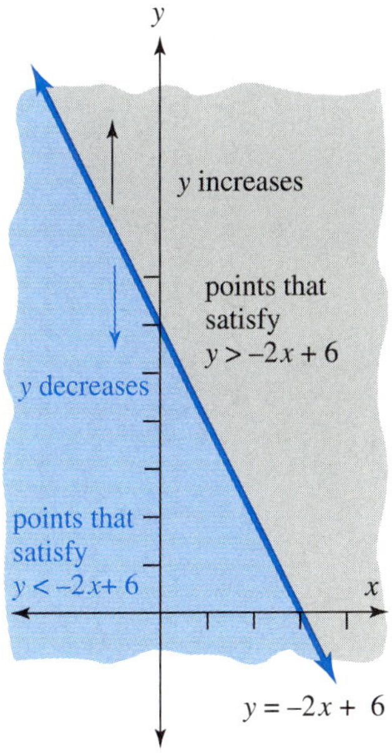

Figure 8.2

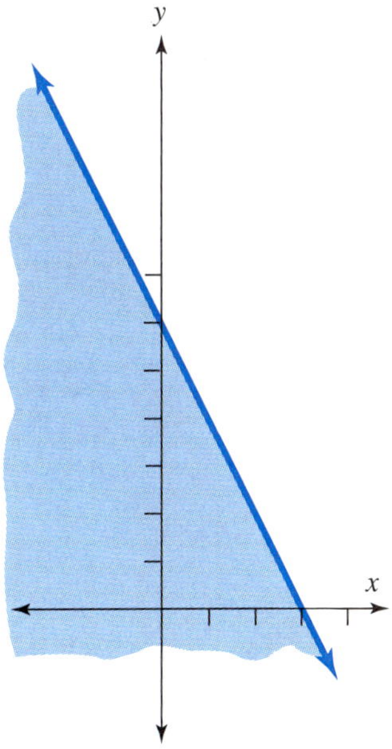

region of solutions of $y \le -2x + 6$ (or equivalently of $2x + y \le 6$)

Figure 8.3

Step 3 *Shade in one side of the line.* Two types of points satisfy the inequality $y \le -2x + 6$: points that satisfy $y = -2x + 6$ and points that satisfy $y < -2x + 6$.

Points that satisfy $y = -2x + 6$ were graphed in step 2, when we graphed the line. Points that satisfy *y is less than* $-2x + 6$ are the points *below* the line, because values of y decrease if we move down and increase if we move up, as shown in Figure 8.2.

Thus, to graph the inequality $y \le -2x + 6$, we make the line solid and shade in the region *below* the line, as shown in Figure 8.3. The solution of $y \le -2x + 6$ is the set of all points on or below the line; any point on or below the line will successfully substitute into the inequality, and any point above the line will not. This region is called the **region of solutions** of the inequality. ■

EXAMPLE 2 Graph $3x - 2y < 12$.

Solution

Step 1 *Solve the linear inequality for y.*

$$\begin{aligned} 3x - 2y &< 12 \\ -2y &< -3x + 12 \\ \frac{-2y}{-2} &> \frac{-3x + 12}{-2} && \text{multiplying or dividing by a negative reverses the direction of an inequality} \\ y &> \frac{-3x}{-2} + \frac{12}{-2} && \text{distributing } -2 \\ y &> \frac{3}{2}x - 6 \end{aligned}$$

Step 2 *Graph the line*. The associated equation is $y = \frac{3}{2}x - 6$, which is a line with slope $m = \frac{3}{2}$ and y-intercept $b = -6$.

$$m = \frac{3}{2} \to \frac{\text{rise}}{\text{run}} = \frac{3}{2}$$

To graph the line, we place a point at -6 on the y-axis and from that point rise 3 and run 2. Because "=" is *not* part of " > ," a point on the line $y = \frac{3}{2}x - 6$ is *not* a point on the graph of the inequality $y > \frac{3}{2}x - 6$. We show this by using a dashed line.

Step 3 *Shade in one side of the line*. Because values of y *increase* if we move upward and because we want to graph where y *is greater than* $\frac{3}{2}x - 6$, we shade in the region above the dashed line. The region of solutions of the inequality is the set of all points above (but not on) the line, as shown in Figure 8.4.

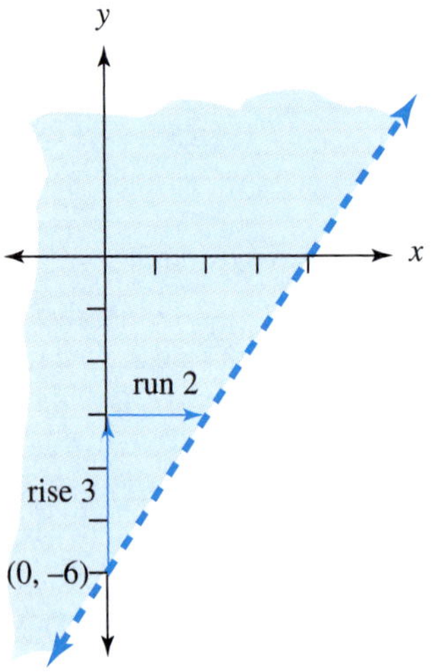

region of solutions of $y > \frac{3}{2}x - 6$
(or equivalently of $3x - 2y < 12$)

Figure 8.4

■

EXAMPLE 3 Graph $x \geq 3$.

Solution

Step 1 *Solve the linear inequality for y.* This can't be done, since there is no y in the inequality $x \geq 3$.

Step 2 *Graph the line.* The associated equation is $x = 3$. Because any point with an x-coordinate of 3 satisfies this equation, the graph of $x = 3$ is a vertical line through (3, 0) and (3, 1) and (3, 2). Also, "=" is part of "≥," so any point on the line $x = 3$ is a point on the graph of the inequality $x \geq 3$. We show this by using a solid line.

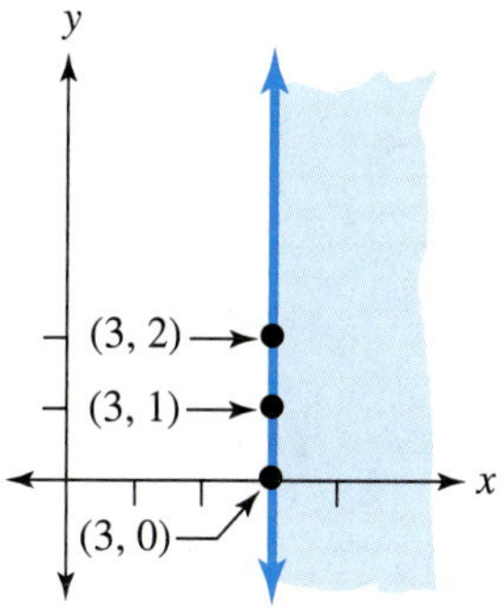

region of solutions of $x \geq 3$

Figure 8.5

Step 3 *Shade in one side of the line.* Values of x increase if we move to the right and decrease if we move to the left. The points on the line are the points where x equals 3, and the points to the right of the line are the points where x is greater than 3. Thus, the region of solutions of the inequality is the set of all points on or to the right of the line, as shown in Figure 8.5. ■

Graphing the Region of Solutions of a Linear Inequality

1. *Solve the linear inequality for y.* This puts the inequality in slope-intercept form. If the inequality has no y, then solve the inequality for x.
2. *Graph the line.* The line is described by the equation associated with the inequality.

If the equation is:	then the line:
in slope-intercept form ($y = mx + b$) . . .	.has slope m and y-intercept b.
in the form $x = a$	.is a vertical line through $(a, 0)$.

If the inequality is:	then the line:
≤ or ≥	.is part of the region of solutions —use a solid line.
< or >	.is not part of the region of solutions—use a dashed line.

3. *Shade in one side of the line.*
   y increases as you move up, and decreases as you move down.
   x increases as you move to the right, and decreases as you move to the left.

Systems of Linear Inequalities

The inequalities we deal with in this chapter come from restrictions. Usually, a linear programming problem involves more than one restriction and therefore more than one inequality. A **system of linear inequalities** is a set of more than one linear inequality. The **region of solutions** of a system of linear inequalities is the set of all points that simultaneously satisfy each inequality in the system.

To graph the region of solutions of a system of linear inequalities, we graph each inequality on the same axes and shade in the intersection of their solutions.

EXAMPLE 4 Graph the following system of inequalities:

$$x + y \geq 3$$
$$-x + 2y \leq 0$$

Solution With the aid of the following chart, we can graph the two inequalities on the same axes, as shown in Figure 8.6.

Original Inequality	Slope-Intercept Form	Associated Equation	Graph of the Inequality
$x + y \geq 3$	$y \geq -x + 3$	$y = -1x + 3$	all points on or above the line with slope -1 and y-intercept 3
$-x + 2y \leq 0$	$2y \leq x \rightarrow y \leq \frac{1}{2}x$	$y = \frac{1}{2}x + 0$	all points on or below the line with slope $\frac{1}{2}$ and y-intercept 0

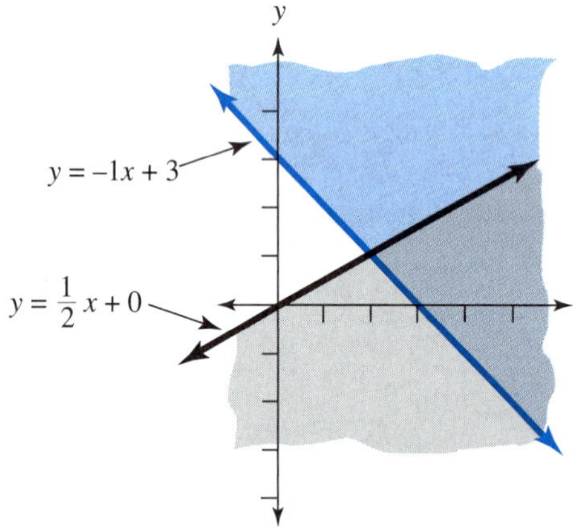

$x + y \geq 3$ and $-x + 2y \leq 0$ graphed on the same axes

Figure 8.6

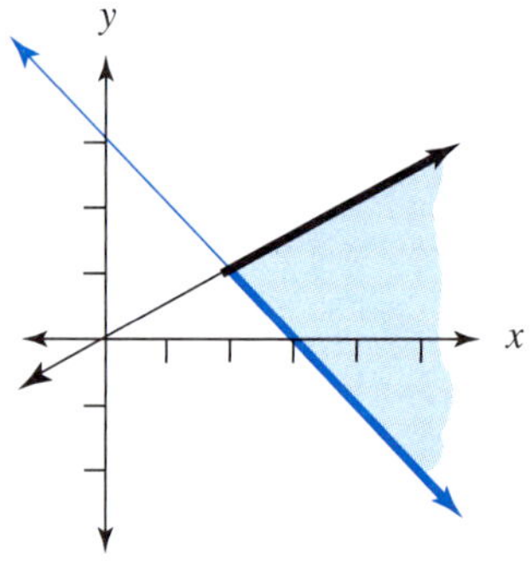

region of solutions of the system
$x + y \geq 3$
$-x + 2y \leq 0$

Figure 8.7

The region of solutions of the system is the set of all points that satisfy both the first *and* the second inequality. As we saw in Chapter 2, *intersection* is the set-theoretical word that corresponds to *and*, so the region of solutions is the intersection of the graphs of the two inequalities. This is the region shaded in Figure 8.7; it includes the solid lines bounding that region. ■

EXAMPLE 5 Graph the following system of inequalities:

$$2x + y \leq 8$$
$$x + 2y \leq 10$$
$$x \geq 0$$
$$y \geq 0$$

Solution With the aid of the following chart, we can graph the two inequalities on the same axes.

Original Inequality	Slope-Intercept Form	Associated Equation	Graph of the Inequality
$2x + y \le 8$	$y \le -2x + 8$	$y = -2x + 8$	all points on or below the line with slope -2 and y-intercept 8
$x + 2y \le 10$	$2y \le -x + 10 \rightarrow$ $y \le -\frac{1}{2}x + 5$	$y = -\frac{1}{2}x + 5$	all points on or below the line with slope $-\frac{1}{2}$ and y-intercept 5
$x \ge 0$	(not applicable)	$x = 0$	all points on or to the right of the y-axis
$y \ge 0$	$y \ge 0$	$y = 0$	all points on or above the x-axis

The inequalities $x \ge 0$ and $y \ge 0$ appear frequently in linear programming problems. They tell us that our graph is in the first quadrant.

The region of solutions of the system is the intersection of the graph of each individual inequality. It is shown in Figure 8.8.

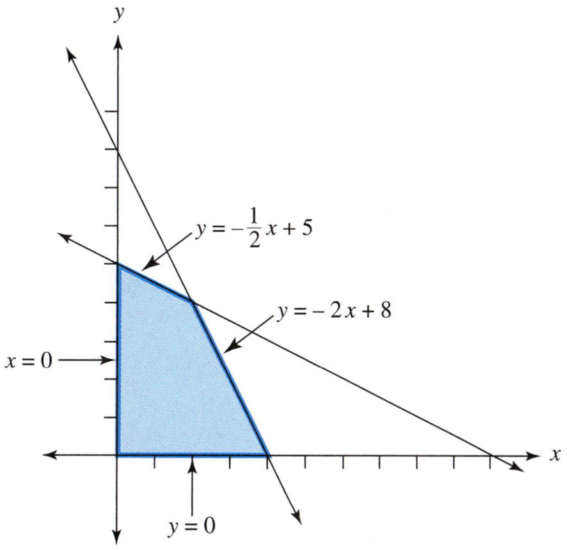

region of solutions of the system
$2x + y \le 8$
$x + 2y \le 10$
$x \ge 0$
$y \ge 0$

Figure 8.8

The region of solutions in Example 4 is different from that of Example 5 in that the former is not totally enclosed. For that reason, it is called an **unbounded region.** Regions that are totally enclosed, like that in Example 5, are called **bounded regions.** See Figure 8.9 on page 578. When we graph a system of inequalities as part of a linear programming problem, we must analyze unbounded regions differently than we do bounded regions.

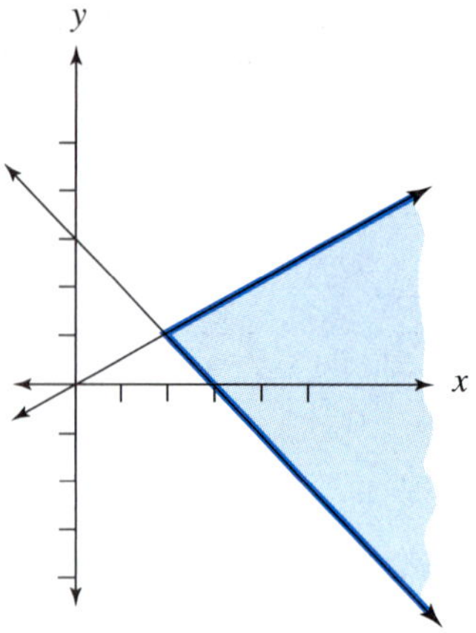

region of solutions
from Example 4;
an unbounded region

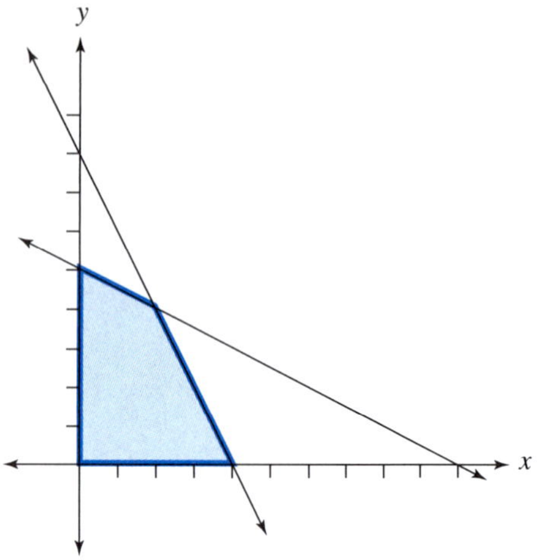

region of solutions
from Example 5;
a bounded region

Figure 8.9

Finding Corner Points

Our work in linear programming will involve graphing the region of solutions of a system of linear inequalities, as we did in Example 5. It will also involve finding the region's corner points. A **corner point** is a point that is at a corner of the region of solutions.

EXAMPLE 6 Find the corner points of the region of solutions in Example 5.

Solution The region has four corner points, as shown in Figure 8.10.

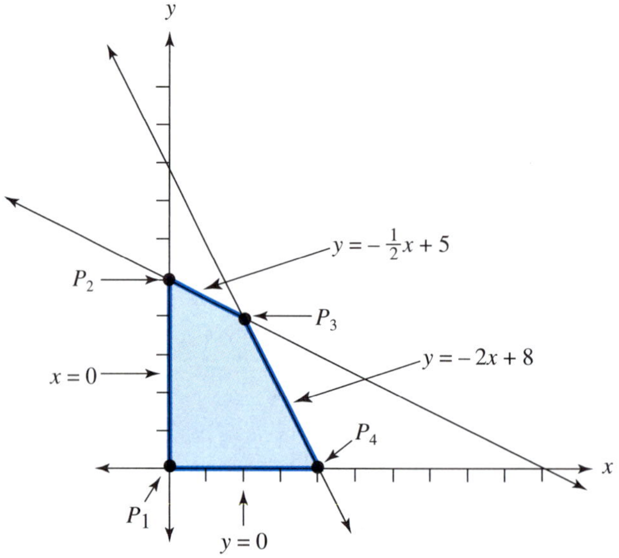

Figure 8.10

- P_1 at the origin
- P_2 where $y = -\frac{1}{2}x + 5$ intersects $x = 0$ (the y-axis)
- P_3 where $y = -\frac{1}{2}x + 5$ intersects $y = -2x + 8$
- P_4 where $y = -2x + 8$ intersects $y = 0$ (the x-axis)

There is no work to do to find points P_1 and P_2. Point P_1 is clearly (0, 0). Point P_2 is the y-intercept of $y = -\frac{1}{2}x + 5$; that y-intercept is $b = 5$, so P_2 is (0, 5).

We can find P_3 by solving the system of equations

$$y = -\tfrac{1}{2}x + 5$$
$$y = -2x + 8$$

We can solve this system with the elimination method, with the Gauss-Jordan method from Chapter 7, or with the substitution method you might have encountered in intermediate algebra. To proceed with the elimination method, multiply the second equation by -1 and add the results.

$$\begin{array}{rrl} y = & -\tfrac{1}{2}x + 5 & \\ + \ -y = & 2x - 8 & \text{-1 times the second equation} \\ \hline 0 = & \tfrac{3}{2}x - 3 & \end{array}$$

$$\frac{3}{2}x = 3$$

$$\frac{2}{3} \cdot \frac{3}{2}x = \frac{2}{3} \cdot 3$$

$$x = 2$$

We can find y by substituting $x = 2$ into either of the original equations.

$$y = -2x + 8$$
$$y = -2 \cdot 2 + 8 = 4$$

Thus, P_3 is (2, 4).

We can find P_4 by solving the system of equations

$$y = -2x + 8$$
$$y = 0 \qquad \text{the equation of the x-axis}$$

To proceed with the elimination method, multiply the second equation by -1 and add the results.

$$\begin{array}{rrl} y = & -2x + 8 & \\ + \ -y = & 0 & \text{-1 times the second equation} \\ \hline 0 = & -2x + 8 & \end{array}$$

$$2x = 8$$
$$x = 4$$

Since P_4 is on the x-axis, its y-coordinate must be 0. Thus, P_4 is (4, 0). The region of solutions' four corner points are

- P_1 at (0, 0)
- P_2 at (0, 5)
- P_3 at (2, 4)
- P_4 at (4, 0) ■

8.0 Exercises

In Exercises 1–8, graph the region of solutions of the given linear inequality.

1. $3x + y < 4$
2. $8x + y > 2$
3. $4x - 3y \leq 9$
4. $5x - 2y \geq 6$
5. $x \geq 4$
6. $x \leq -3$
7. $y \leq -4$
8. $y \geq -2$

In Exercises 9–21, (a) graph the region of solutions of the given system of linear inequalities, (b) determine whether the region of solutions is bounded or unbounded, and (c) find all of the region's corner points.

9. $y > 2x + 1$
   $y \leq -x + 4$
10. $y < -2x + 6$
    $y \geq -x + 7$
11. $2x + 3y < 17$
    $3x - y \geq -2$
12. $5x - y \geq 7$
    $2x - 3y < -5$
13. $x + 2y \leq 4$
    $3x - 2y \leq -12$
    $x - y < -7$
14. $x - y + 1 \geq 0$
    $3x + 2y + 8 \geq 0$
    $3x - y < 6$
15. $2x + 5y \leq 70$
    $5x + y \leq 60$
    $x \geq 0$
    $y \geq 0$
16. $x + 20y \leq 460$
    $21x + y \leq 861$
    $x \geq 0$
    $y \geq 0$
17. $15x + 22y \leq 510$
    $35x + 12y \leq 600$
    $x + y > 10$
    $x \geq 0$
    $y \geq 0$
18. $3x + 20y \leq 2{,}200$
    $19x + 9y \leq 2{,}755$
    $2x + y \leq 120$
    $x \geq 0$
    $y \geq 0$
19. $0.50x + 1.30y \leq 2.21$
    $6x + y \leq 9$
    $0.7x + 0.6y < 3.00$
    $x \geq 0$
    $y \geq 0$
20. $3.70x + 0.30y \leq 1.17$
    $0.10x + 2.20y \leq 0.47$
    $x \geq 0$
    $y \geq 0$
21. $x - 2y + 16 \geq 0$
    $3x + y \leq 30$
    $x + y \leq 14$
    $x \geq 0$
    $y \geq 0$

Concept Questions

22. Make up a system of inequalities with a region of solutions that is bounded. Do not use any of the systems discussed in the text.
23. Make up a system of inequalities with a region of solutions that is unbounded. Do not use any of the systems discussed in the text.
24. Make up a system of inequalities with a region of solutions that is the triangle with corner points (0, 0), (10, 0), and (0, 20).
25. Make up a system of inequalities with a region of solutions that has corner points (0, 0), (20, 0), (10, 40), and (0, 50).

Answer the following questions with complete sentences and your own words.

26. Why do we describe the solution of a system of linear inequalities with a graph, rather than a list of points?
27. List three industries that routinely use linear programming, and give examples of how they use it.
28. Why do large corporations have to be concerned about "limited" resources?

Graphing Linear Inequalities on a Graphing Calculator

A Texas Instruments graphing calculator can perform all of the specific tasks that are part of graphing the region of solutions of a system of linear inequalities: It can graph the lines associated with the system, shade in the appropriate sides of those lines (with the "Shade" command), and locate corner points. However, the "Shade" command is cumbersome to use; it is much easier to do the shading by hand and to use the calculator to graph the lines associated with the inequalities and to find the corner points. If you wish to explore the use of the "Shade" command, consult your calculator's operating manual.

Graphing a Linear Inequality

EXAMPLE 7 Graph $3x - 2y < 12$.

Solution

Step 1 *Solve the linear inequality for y.*

$$3x - 2y < 12 \rightarrow -2y < -3x + 12 \rightarrow y > \tfrac{3}{2}x - 6$$

Multiplying or dividing by a negative reverses the direction of the inequality.

Step 2 *Graph the line*, using the procedure discussed in Appendix C. The line associated with the inequality is $y = \frac{3}{2}x - 6$. Enter 3/2*x − 6 for Y_1. (Be sure to include the "*." Some calculators would interpret $3/2x - 6$ as $\frac{3}{2x} - 6$.) Be sure that no other equations are selected. Use the standard viewing window. The line's graph is shown in Figure 8.11.

Step 3 *Copy the line's graph onto paper.* The inequality is a ">" inequality, so use a dashed line.

Step 4 *By hand, shade in one side of the line.* To graph $y > \frac{3}{2}x - 6$, shade in the region above the line $y = \frac{3}{2}x - 6$. The region's graph is shown in Figure 8.12.

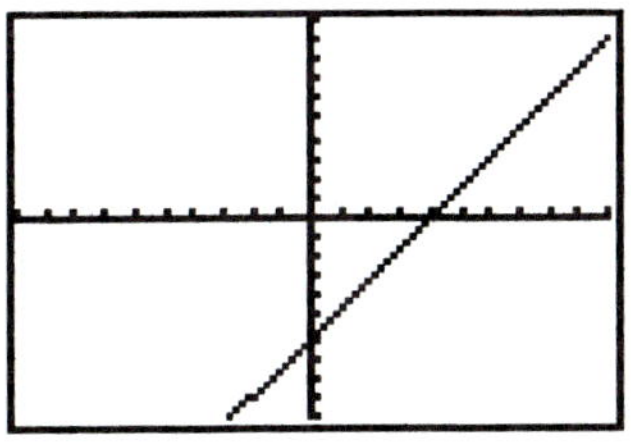

The graph of $y = \frac{3}{2}x - 6$

Figure 8.11

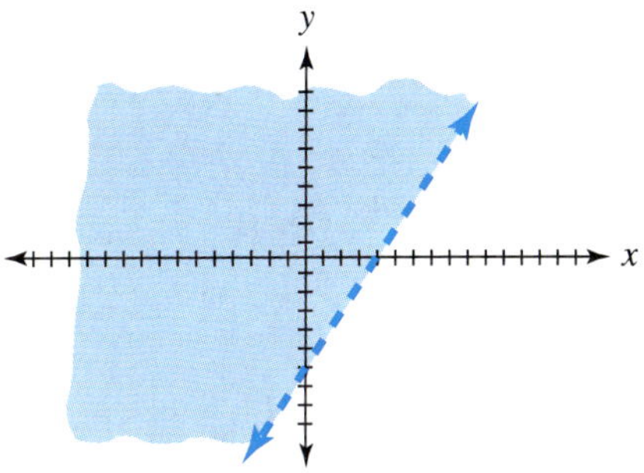

The graph of $y > \frac{3}{2}x - 6$

Figure 8.12 ■

Graphing a System of Linear Inequalities

EXAMPLE 8 Graph the system of inequalities:

$$x + y \geq 3$$
$$-x + 2y \leq 0$$

Solution

Step 1 *Solve the linear inequalities for y.*

$$x + y \geq 3 \rightarrow y \geq -x + 3$$
$$-x + 2y \leq 0 \rightarrow y \leq \tfrac{1}{2}x$$

Step 2 *Graph the lines*, using the procedure discussed in Appendix C. Enter $-x + 3$ for Y_1 and $1/2*x$ for Y_2. Be sure that no other equations are selected. Use the standard viewing window. See Figure 8.13.

Step 3 *Copy the lines' graphs onto paper*. The inequalities are both "$\leq$ or $\geq$" inequalities, so use solid lines.

Step 4 *By hand, shade in the region of solutions*. The region of solutions is above the line $Y_1 = -x + 3$, because the associated inequality is $y \geq -x + 3$, and y increases as you move up. The region of solutions is also below the line $Y_2 = \frac{1}{2}x$, because the associated inequality is $y \leq \frac{1}{2}x$, and y decreases as you move down. The region of solutions is the region that is both above the line $Y_1 = -x + 3$ and below the line $Y_2 = \frac{1}{2}x$. It is shown in Figure 8.14.

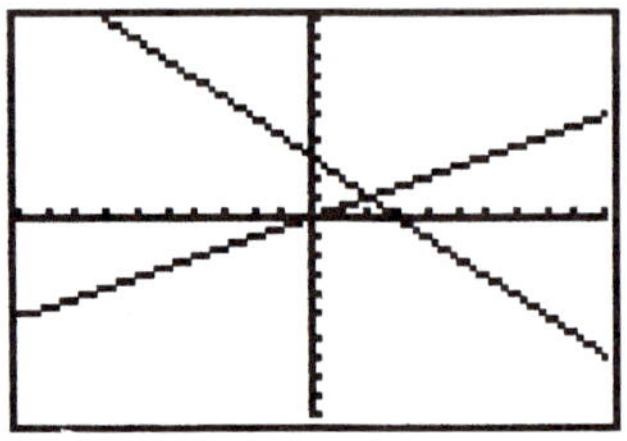

The graph of $x + y = 3$ and $-x + 2y = 0$

Figure 8.13

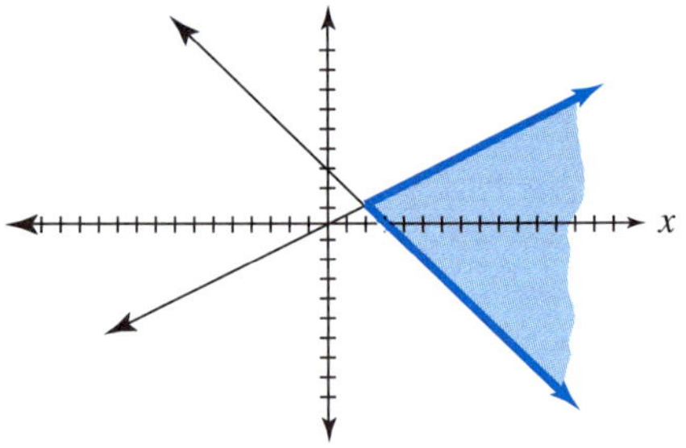

The graph of $x + y \geq 3$ and $-x + 2y \leq 0$

Figure 8.14 ■

Nonstandard Viewing Windows The previous examples were chosen so that the standard viewing window would be appropriate. In the following example, we must determine an appropriate viewing window.

EXAMPLE 9 Graph the system of inequalities:

$$2x + y \leq 8$$
$$x + 2y \leq 10$$
$$x \geq 0$$
$$y \geq 0$$

Solution

$$2x + y \leq 8 \rightarrow y \leq -2x + 8$$
$$x + 2y \leq 10 \rightarrow 2y \leq -x + 10 \rightarrow y \leq -\tfrac{1}{2}x + 5$$

$x \geq 0$ cannot be solved for y, and $y \geq 0$ is already solved for y.

Enter $-2x + 8$ for Y_1 and $-1/2*x + 5$ for Y_2. Be sure that no other equations are selected. The last pair of inequalities tells us that the region of solutions is in the first quadrant; if we set *X*min and *Y*min equal to -1, we'll leave ourselves a little extra room. By inspecting the two lines' equations, we can tell that the largest y-intercept is 8, so we'll set *Y*max equal to 8. Substituting 0 for y in each equation gives x-intercepts of (4, 0) and (10, 0), so we'll set *X*max equal to 10. The resulting graph is shown in Figure 8.15.

The region of solutions is below the line $Y_1 = -2x + 8$, because the associated inequality is $y \leq -2x + 8$, and y decreases as you move down. The region of solutions is also below the line $Y_2 = -\frac{1}{2}x + 5$, because the associated inequality is $y \leq -\frac{1}{2}x + 5$. The region of solutions is the part of the first quadrant that is both on or below the line $Y_1 = -2x + 8$ and on or below the line $Y_2 = -\frac{1}{2}x + 5$. It is shown in Figure 8.16.

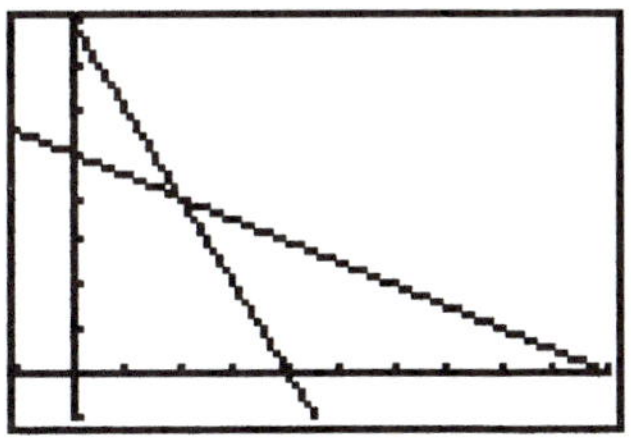

The graph of $2x + y = 8$ and $x + 2y = 10$

Figure 8.15

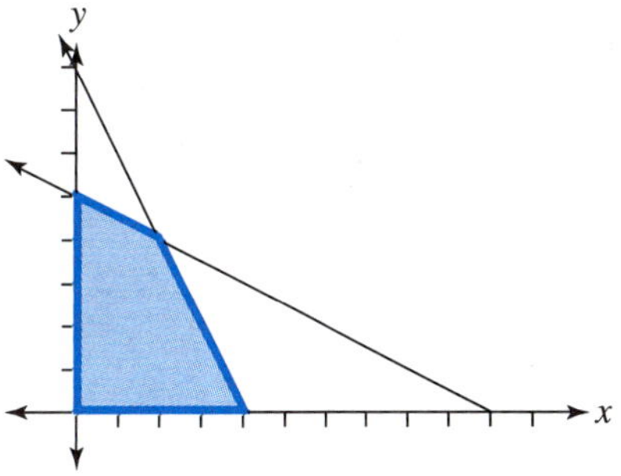

The system's region of solutions

Figure 8.16 ■

Finding Corner Points After using a graphing calculator to graph a system's region of solutions, you can use the calculator to find the corner points as well.

EXAMPLE 10 Find the corner points for the region graphed in Example 9.

Solution This region has four corner points. One is clearly located at (0, 0). A second is the y-intercept of the line $y = -\frac{1}{2}x + 5$; by inspecting the equation, we can tell that it is (0, 5).

A third corner point is the intersection of the two lines; it can be found using the procedure discussed in Appendix D (summarized below).

TI-80/82/83	Select "intersect" from the "CALC" menu.
TI-81	Use the ZOOM and TRACE buttons.
TI-85/86	Select "MATH" from the "GRAPH" menu, and then "ISECT" from the "GRAPH MATH" menu.
Casio	PRESS G-SOLV and then ISCT.

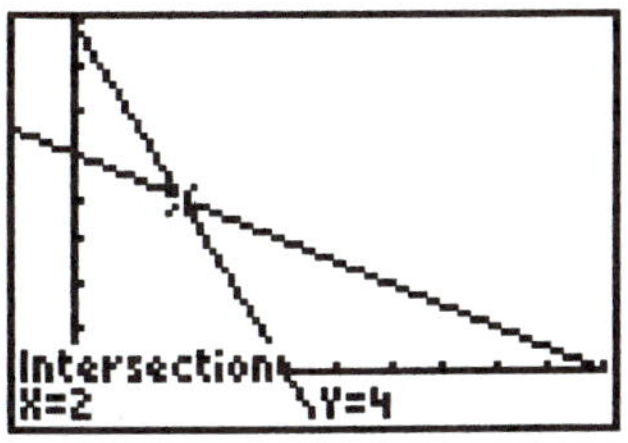

Figure 8.17

This third corner point is (2, 4), as shown in Figure 8.17.

The fourth corner point is the x-intercept or root of the line $y = -2x + 8$; it can be found using the procedure discussed in Appendix C, Exercises 8 and 9 (summarized below).

TI-80/82/83	Select "root" or "zero" from the "CALC" menu.
TI-81	Use the ZOOM and TRACE buttons.
TI-85/86	Select "MATH" from the "GRAPH" menu, and then "ROOT" from the "GRAPH MATH" menu.
Casio	PRESS G-SOLV and then ROOT.

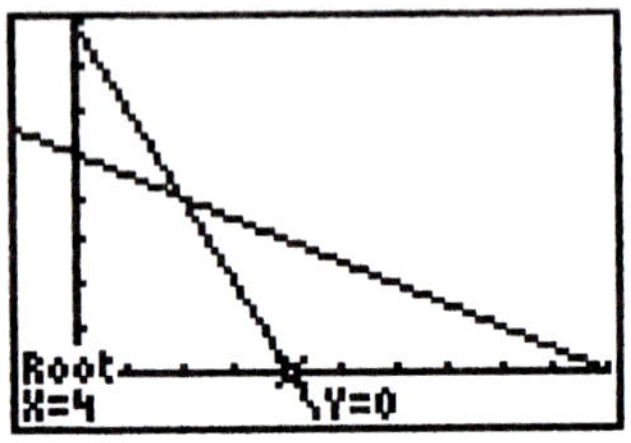

Figure 8.18

This fourth corner point is (4, 0), as shown in Figure 8.18. ■

Exercises

In Exercises 29–36, use a graphing calculator to graph the region of solutions of the inequality given earlier in this section.

29. Exercise 1
30. Exercise 2
31. Exercise 3
32. Exercise 4
33. Exercise 5
34. Exercise 6
35. Exercise 7
36. Exercise 8

In Exercises 37–49, use a graphing calculator to graph the region of solutions of the system of inequalities given earlier in this section, and to find the corner points.

37. Exercise 9
38. Exercise 10
39. Exercise 11
40. Exercise 12
41. Exercise 13
42. Exercise 14
43. Exercise 15
44. Exercise 16
45. Exercise 17
46. Exercise 18
47. Exercise 19
48. Exercise 20
49. Exercise 21

8.1 The Geometry of Linear Programming

As we stated in the introduction to this chapter, linear programming is a method for solving problems in which a quantity is to be maximized or minimized, when that quantity is subject to various restrictions. The following is a typical linear programming problem.

A craftsman produces two products, coffee tables and end tables. Production of one coffee table requires six hours of his labor, and the materials cost him \$200. Production of one end table requires five hours of labor, and the materials cost him \$100. The craftsman wants to work no more than 40 hours each week, and his financial resources allow him to pay no more than \$1,000 for materials each week. If he can sell as many tables as he can make and if his profit is \$240 per coffee table and \$160 per end table, how

Rick McClain/Gamma Liaison

How should a craftsman allocate his time and money to maximize profit?

many coffee tables and how many end tables should he make each week to maximize weekly profit?

Any linear programming problem has three features: *variables*, an *objective*, and *constraints*. In the problem above, the **variables** (or quantities that can vary) are the following:

- the number of coffee tables made each week
- the number of end tables made each week
- the number of hours the craftsman works each week
- the amount of money he spends on materials each week
- the weekly profit

The last three variables depend on the first two, so they are called the **dependent variables,** whereas the first two are called the **independent variables.**

The craftsman's objective is to maximize profit. The **objective function** is a function that mathematically describes the profit.

The **constraints** (or restrictions) are as follows:

- the craftsman's weekly hours ≤ 40
- the craftsman's weekly expenses $\leq \$1{,}000$

The constraints form a system of inequalities. To analyze the effect of these constraints on the craftsman's profit, we must graph the system of inequalities. The resulting graph is called the **region of possible solutions,** because it contains all the points that could *possibly* solve the craftsman's problem.

Creating a Model

A **model** is a mathematical description of a real-world situation. In this section, we discuss how to model a linear programming problem (that is, how to translate it into mathematical terms), how to find and graph the region of possible solutions, and how to analyze the effect of the constraints on the objective and solve the problem.

EXAMPLE 1 Model the linear programming problem from the beginning of this section and graph the region of possible solutions. The problem is summarized as follows:

A craftsman produces two products: coffee tables and end tables. Production data are given in Figure 8.19. If the craftsman wants to work no more than 40 hours each week and if his financial resources allow him to pay no more than $1,000 for materials each week, how many coffee tables and how many end tables should he make each week to maximize weekly profit?

	Labor (per table)	Cost of Materials (per table)	Profit (per table)
coffee tables	6 hours	$200	$240
end tables	5 hours	$100	$160

Figure 8.19

Solution

Step 1 *List the independent variables.* We have already done this. If we call them x and y, the independent variables are

x = number of coffee tables made each week

y = number of end tables made each week

Step 2 *List the constraints and translate them into linear inequalities.* We have already determined that the constraints (or restrictions) are as follows:

the craftsman's weekly hours ≤ 40

the craftsman's weekly expenses $\leq \$1{,}000$

We need to translate these constraints into linear inequalities. First, let's translate the time constraint.

$$\text{hours} \leq 40$$

$$(\text{coffee table hours}) + (\text{end table hours}) \leq 40$$

$$\begin{pmatrix}\text{6 hours per}\\ \text{coffee table}\end{pmatrix} \cdot \begin{pmatrix}\text{number of}\\ \text{coffee tables}\end{pmatrix} + \begin{pmatrix}\text{5 hours per}\\ \text{end table}\end{pmatrix} \cdot \begin{pmatrix}\text{number of}\\ \text{end tables}\end{pmatrix} \leq 40$$

$$6 \quad x \quad + \quad 5 \quad y \quad \leq 40$$

Next, we'll translate the money constraint.

$$\text{money spent} \leq 1{,}000$$

$$(\text{coffee table money}) + (\text{end table money}) \leq 1{,}000$$

$$\begin{pmatrix}\text{\$200 per}\\ \text{coffee table}\end{pmatrix} \cdot \begin{pmatrix}\text{number of}\\ \text{coffee tables}\end{pmatrix} + \begin{pmatrix}\text{\$100 per}\\ \text{end table}\end{pmatrix} \cdot \begin{pmatrix}\text{number of}\\ \text{end tables}\end{pmatrix} \leq 1{,}000$$

$$200 \quad x \quad + \quad 100 \quad y \quad \leq 1{,}000$$

There are two more constraints. Both x and y count things (the number of tables), so neither can be negative; therefore,

$$x \geq 0 \qquad \text{and} \qquad y \geq 0$$

Our constraints are

$$6x + 5y \leq 40$$

$$200x + 100y \leq 1{,}000$$

$$x \geq 0 \qquad \text{and} \qquad y \geq 0$$

Step 3 *Find the objective and translate it into a linear equation.* The objective is to maximize profit. If we let z = profit, we get

$$\begin{aligned} z &= (\text{coffee table profit}) + (\text{end table profit}) \\ &= (\$240 \text{ per coffee table})(\text{number of coffee tables}) \\ &\quad + (\$160 \text{ per end table})(\text{number of end tables}) \\ &= 240x + 160y \end{aligned}$$

This equation is our objective function.

Steps 1 through 3 yield the model, or mathematical description, of the problem.

Independent Variables

x = number of coffee tables
y = number of end tables

Constraints

$6x + 5y \leq 40$ — the time constraint
$200x + 100y \leq 1{,}000$ — the money constraint
$x \geq 0$ and $y \geq 0$

Objective Function

$z = 240x + 160y$ — z measures profit

Step 4 *Graph the region of possible solutions*. With the aid of the accompanying chart, we can graph the two inequalities on the same axes, as shown in Figure 8.20.

Original Inequality	Slope-Intercept Form	Associated Equation	Graph of the Inequality
$6x + 5y \leq 40$	$5y \leq -6x + 40$ $\rightarrow y \leq -\frac{6}{5}x + 8$	$y = -\frac{6}{5}x + 8$	all points on or below the line with slope $-6/5$ and y-intercept 8
$200x + 100y \leq 1{,}000$	$100y \leq -200x + 1{,}000$ $\rightarrow y \leq -2x + 10$	$y = -2x + 10$	all points on or below the line with slope -2 and y-intercept 10
$x \geq 0$	(not applicable)	$x = 0$	all points on or to the right of the y-axis
$y \geq 0$	$y \geq 0$	$y = 0$	all points on or above the x-axis

The last two constraints tell us that the region of possible solutions is in the first quadrant. The region of possible solutions is shown in Figure 8.20.

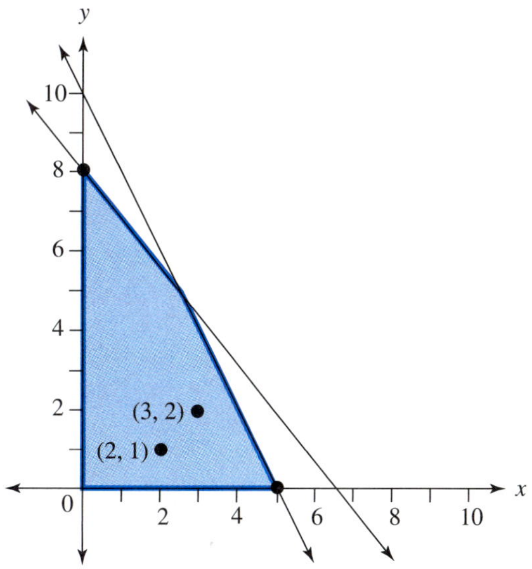

The graph of $y = -\frac{6}{5}x + 8$ and $y = -2x + 10$

Figure 8.20

Analyzing the Model

The region of possible solutions graphed in Example 1 consists of all the points that satisfy the constraints, that is, all the points at which the craftsman's weekly hours are no more than 40 and his weekly expenses are no more than \$1,000. Let's arbitrarily select two points in the region, verify that the constraints are satisfied, and find the craftsman's profit at those points. By inspecting the graph in Figure 8.20 on page 587, we can see that (2, 1) and (3, 2) are clearly in the region.

Recall that x measures the number of coffee tables to be made each week, y measures the number of end tables, and z measures weekly profit. The first constraint is that the time used, $6x + 5y$, be no more than 40 hours per week. The second constraint is that the money used, $200x + 100y$, be no more than \$1,000 each week. The weekly profit is $z = 240x + 160y$. Figure 8.21 gives the time used, the money spent, and the profit at our two points.

Point	(2, 1)	(3, 2)
Time Used	$6 \cdot 2 + 5 \cdot 1 = 17$	$6 \cdot 3 + 5 \cdot 2 = 28$
Money Spent	$200 \cdot 2 + 100 \cdot 1 = 500$	$200 \cdot 3 + 100 \cdot 2 = 800$
Profit	$240 \cdot 2 + 160 \cdot 1 = 640$	$240 \cdot 3 + 160 \cdot 2 = 1{,}040$

Figure 8.21

The table in Figure 8.21 shows that if the craftsman makes two coffee tables and one end table each week, he will use 17 hours (of 40 available hours), spend \$500 (of the \$1,000 available), and profit \$640. If he makes three coffee tables and two end tables each week, he will use 28 hours, spend \$800, and profit \$1,040. Each of these points represents a *possible* solution to the craftsman's problem, because each satisfies the time constraint and the money constraint. Neither represents the *actual* solution, because neither maximizes his profit; he has both money and time left over, so he should be able to increase his profit by building more tables. This is why the region in Figure 8.20 is called the region of *possible* solutions.

Common sense tells us that in order to maximize profit our craftsman must use all his time and/or money and make more tables. To make more tables means to increase the value of x and/or y, which implies that we should choose points on the boundary of the region of feasible solutions. There are quite a few points on the boundary—too many to find and substitute into the equation for profit. Fortunately, the **Corner Principle** comes to our rescue. (We discuss why the Corner Principle is true later in this section.)

Corner Principle

The maximum and minimum values of an objective function occur at corner points of the region of possible solutions if that region is bounded.

Our region is a bounded region, so the Corner Principle applies. The region has four corner points (see Figure 8.22):

- P_1 at the origin
- P_2 where $y = -\frac{6}{5}x + 8$ intersects the y-axis
- P_3 where $y = -\frac{6}{5}x + 8$ intersects $y = -2x + 10$
- P_4 where $y = -2x + 10$ intersects the x-axis

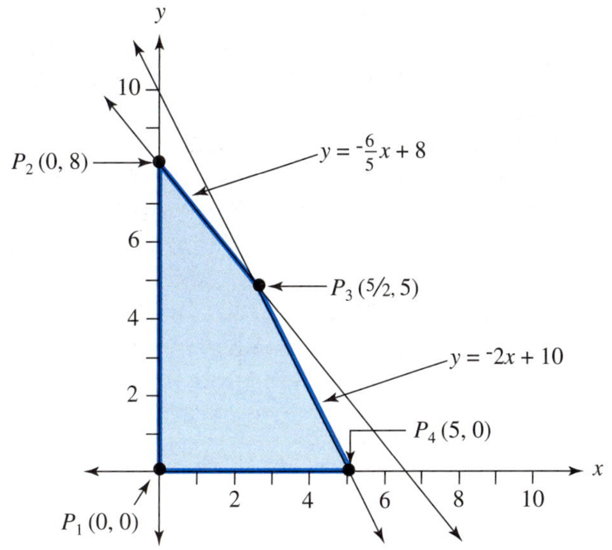

Figure 8.22

We can find each of these points by solving a system of equations. For example, P_3 can be found by solving the following system:

$$y = -\tfrac{6}{5}x + 8$$
$$y = -2x + 10$$

P_4 can be found by solving the following system:

$$y = -2x + 10$$
$$y = 0 \qquad \text{the } x\text{-axis}$$

The corner points, shown in Figure 8.22 are as follows:

$P_1\ (0, 0)$
$P_2\ (0, 8)$
$P_3\ (\tfrac{5}{2}, 5\ )$
$P_4\ (5, 0)$

Let's verify that the constraints are satisfied at each of these points and, more important, find the profit at each point.

Point	P_1 **(0, 0)**	P_2 **(0, 8)**	$P_3\left(\frac{5}{2}, 5\right)$	P_4 **(5, 0)**
Time Used	$6 \cdot 0 + 5 \cdot 0 = 0$	$6 \cdot 0 + 5 \cdot 8 = 40$	$6 \cdot \frac{5}{2} + 5 \cdot 5 = 40$	$6 \cdot 5 + 5 \cdot 0 = 30$
Money Spent	$200 \cdot 0 + 100 \cdot 0 = 0$	$200 \cdot 0 + 100 \cdot 8 = 800$	$200 \cdot \frac{5}{2} + 100 \cdot 5 = 1{,}000$	$200 \cdot 5 + 100 \cdot 0 = 1{,}000$
z **= Profit**	$240 \cdot 0 + 160 \cdot 0 = 0$	$240 \cdot 0 + 160 \cdot 8 = 1{,}280$	$240 \cdot \frac{5}{2} + 160 \cdot 5 = 1{,}400$	$240 \cdot 5 + 160 \cdot 0 = 1{,}200$

At each corner point, each constraint is satisfied: the time used is at most 40 hours, and the money spent is at most \$1,000. The highest profit, \$1,400, occurs at P_3 $(\frac{5}{2}, 5)$. The Corner Principle tells us that this is the point in the region of possible solutions at which the highest profit occurs. The craftsman will maximize his profit if he makes $2\frac{1}{2}$ coffee tables and 5 end tables each week (finishing that third coffee table during the next week), working his entire 40 hours per week and spending his entire \$1,000 per week on materials.

Some Graphing Tips

In solving a linear programming problem, an accurate graph of the region of possible solutions is important. A less-than-accurate graph can be the source of errors and frustration. To avoid difficulty if you are graphing without the aid of technology, follow these recommendations:

- Find the x-intercept of each line and use it (as well as the slope and y-intercept) to graph the line.
- Use graph paper and a straightedge.
- Use your graph to check your corner point computations—the computed location of a corner point should fit with the graph.

Why the Corner Principle Works

Each point in the region of possible solutions has a value of z associated with it. For example, we found that P_3 $(\frac{5}{2}, 5)$ has a z-value of 1,400. Think of each point in the region as being a light bulb, with the brightness of the light given by the z-value. The point at which the craftsman's profit is maximized is the point with the brightest light.

Some points are just as bright as others. For example, all points that satisfy the equation $320 = 240x + 160y$ have a brightness of $z = 320$. If we solve this equation for y, we get

$$320 = 240x + 160y$$
$$-160y = 240x - 320$$
$$y = \frac{-3}{2}x + 2$$

This is a line with slope $\frac{-3}{2}$ and y-intercept 2. Points that satisfy the equation $640 = 240x + 160y$ have a brightness of $z = 640$. Solving this equation for y gives $y = (\frac{-3}{2})x + 4$. Points that satisfy the equation $960 = 240x + 160y$ have a brightness of 960; solving for y gives $y = (\frac{-3}{2})x + 6$. These three lines have the same slope, so they are parallel, as shown in Figure 8.23.

The light bulbs that fill the region of possible solutions form parallel rows, and all bulbs in any one row are equally bright. Rows closer to the upper right corner of the region are brighter, and rows closer to the lower left corner are dimmer. The brightest bulb is at corner point P_3 $(\frac{5}{2}, 5)$, and the dimmest is at corner point P_1 $(0, 0)$.

Any linear programming problem (including our problem about the craftsman) must have constraints that are expressible as linear inequalities and an objective function that is expressible as a linear equation. This means that when a problem has two

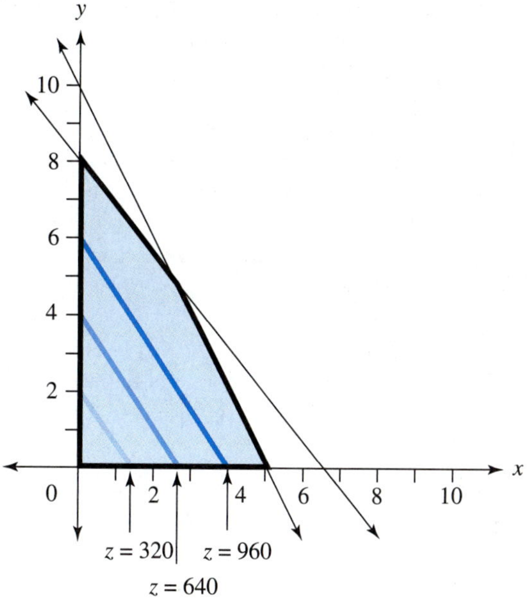

Figure 8.23

independent variables, the region of possible solutions is bounded by lines, and each value of z will correspond to a line. Thus, the graph must always look something like one of the three possibilities shown in Figure 8.24.

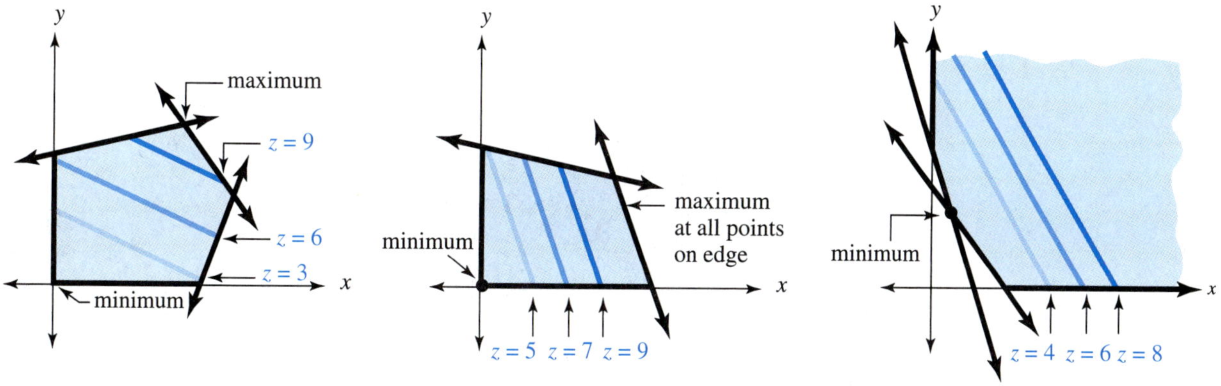

Figure 8.24

If the region is bounded, the maximum and minimum must be at corner points or at all points on an edge. Therefore, we can find all corner points, substitute them into the objective function, and choose the biggest or smallest. If two corners yield the same maximum z-value, we know that the maximum occurs at all points on the boundary line between those corners.

An unbounded region does not necessarily have a maximum or a minimum. Unbounded regions are explored further in the exercises.

Linear Programming Steps

1. *List the independent variables.*
2. *List the constraints and translate them into linear inequalities.*
3. *Find the objective and translate it into a linear equation.* This equation is called the objective function.
4. *Graph the region of possible solutions.* This is the region described by the constraints. Graph each line carefully, using the x-intercept as well as the slope and y-intercept, if you are graphing without the aid of technology.
5. *Find all corner points and the z-values associated with these points.* ✓ Check your corner point computations by verifying that a point's computed location fits with the graph, if you are graphing without the aid of technology.
6. *Find the maximum/minimum.* For a bounded region, the maximum occurs at the corner with the largest z-value, and the minimum occurs at the corner with the smallest z-value. If two corners give the same maximum (or minimum) value, then the maximum (or minimum) occurs at all points on the boundary line between those corners.

In Example 1, the craftsman maximized his profit by exhausting all of his resources. In Example 2, we will find that maximizing profit does not necessarily entail exhausting all resources.

EXAMPLE 2 Pete's Coffees sells two blends of coffee beans, Rich Blend and Hawaiian Blend. Rich Blend is one-half Colombian beans and one-half Kona beans, and Hawaiian Blend is one-quarter Colombian beans and three-quarters Kona beans. Profit on the Rich Blend is \$2 per pound, while profit on the Hawaiian Blend is \$3 per pound. Each day, the shop can obtain 200 pounds of Colombian beans and 60 pounds of Kona

Cindy Charles/PhotoEdit

How should Pete's Coffees blend its beans to maximize its profit?

beans, and it uses that coffee only in the two blends. If the shop can sell all that it makes, how many pounds of Rich Blend and of Hawaiian Blend should Pete's Coffees prepare each day to maximize profit?

Solution

Step 1 *List the independent variables.* The variables are as follows:

- the amount of Rich Blend to be prepared each day
- the amount of Hawaiian Blend to be prepared each day
- the daily profit

Profit depends on the amount of the two blends prepared, so profit is the dependent variable, and the amounts of Rich Blend and Hawaiian Blend are the independent variables. If we call them x and y, then

x = pounds of Rich Blend to be prepared each day

y = pounds of Hawaiian Blend to be prepared each day

Step 2 *List the constraints and translate them into linear inequalities.* The constraints, or restrictions, are that there are 200 pounds of Colombian beans available each day and only 60 pounds of Kona beans. In the blends, no more than the amount available can be used. First, let's translate the Colombian bean constraint:

Colombian beans used $\leq$ 200

(Colombian in Rich Blend) + (Colombian in Hawaiian Blend) $\leq$ 200

(one-half of Rich Blend) + (one-fourth of Hawaiian Blend) $\leq$ 200

$$\frac{1}{2}x \quad + \quad \frac{1}{4}y \quad \leq 200$$

Next, we'll translate the Kona bean constraint:

Kona beans used $\leq$ 60

(Kona in Rich Blend) + (Kona in Hawaiian Blend) $\leq$ 60

(one-half of Rich Blend) + (three-quarters of Hawaiian Blend) $\leq$ 60

$$\frac{1}{2}x \quad + \quad \frac{3}{4}y \quad \leq 60$$

Also, x and y count things (pounds of coffee), so neither can be negative.

$$x \geq 0 \quad \text{and} \quad y \geq 0$$

There is another implied constraint in this problem. Pete's sells its coffee in 1-pound bags, so x and y must be whole numbers. Special methods are available for handling such constraints, but these methods are beyond the scope of this class. We will ignore such constraints and accept fractional answers should they occur.

Step 3 *Find the objective and translate it into a linear equation.* The objective is to maximize profit. If we let z = profit, we get

z = (Rich Blend profit) + (Hawaiian Blend profit)

= ($2 per pound) (pounds of Rich Blend) + ($3 per pound) (pounds of Hawaiian Blend)

$= 2x + 3y$

Steps 1 through 3 yield the mathematical model:

Independent Variables

x = pounds of Rich Blend to be prepared each day

y = pounds of Hawaiian Blend to be prepared each day

Constraints

$$\frac{1}{2}x + \frac{1}{4}y \leq 200$$ the Colombian bean constraint

$$\frac{1}{2}x + \frac{3}{4}y \leq 60$$ the Kona bean constraint

$x \geq 0$ and $y \geq 0$

Objective Function

$$z = 2x + 3y$$

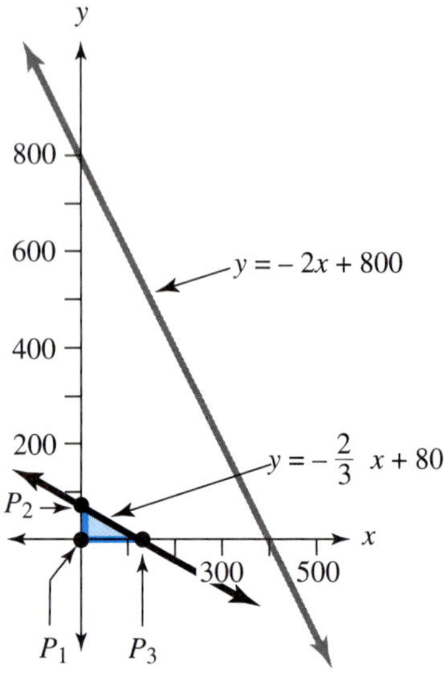

Figure 8.25

Step 4 *Graph the region of possible solutions.* With the aid of the accompanying chart, we can graph the two inequalities on the same axes, as shown in Figure 8.25.

Original Inequality	Slope-Intercept Form	Associated Equation	Graph of the Inequality
$\frac{1}{2}x + \frac{1}{4}y \leq 200$	$\frac{1}{4}y \leq -\frac{1}{2}x + 200 \rightarrow$ $y \leq -2x + 800$	$y = -2x + 800$	all points on or below the line with slope -2 and y-intercept 800
$\frac{1}{2}x + \frac{3}{4}y \leq 60$	$\frac{3}{4}y \leq -\frac{1}{2}x + 60 \rightarrow$ $y \leq -\frac{2}{3}x + 80$	$y = -\frac{2}{3}x + 80$	all points on or below the line with slope $-2/3$ and y-intercept 80
$x \geq 0$	(not applicable)	$x = 0$	all points on or to the right of the y-axis
$y \geq 0$	$y \geq 0$	$y = 0$	all points on or above the x-axis

If you are graphing the region without the aid of technology, it's easier to get an accurate graph if you find the x-intercept of each line and use it (as well as the slope and y-intercept) to graph the line. To find the x-intercept of $y = -\frac{2}{3}x + 80$, solve the system

$$y = -\frac{2}{3}x + 80$$

$$y = 0$$ the equation of the x-axis

The x-intercept is the point (120, 0). The region of possible solutions is shown in Figure 8.25. It is a bounded region. Notice that $y = -2x + 800$ is not a boundary of the region of possible solutions; if we were not given the first constraint, we would have the same region of possible solutions. That constraint, which describes the limited amount of Colombian beans available to Pete's Coffees, is not really a limitation.

Step 5 *Find all corner points and the z-values associated with these points.* Clearly, $P_1 = (0, 0)$. P_2 and P_3 have already been found; they are the y- and x-intercepts of $y = -\frac{2}{3}x + 80$. P_2 is (0, 80), and P_3 is (120, 0).

Step 6 *Find the maximum.* The result of substituting the corner points into the objective function $z = 2x + 3y$ is given in Figure 8.26.

Point	Value of $z = 2x + 3y$
P_1 (0, 0)	$z = 2 \cdot 0 + 3 \cdot 0 = 0$
P_2 (0, 80)	$z = 2 \cdot 0 + 3 \cdot 80 = 240$
P_3 (120, 0)	$z = 2 \cdot 120 + 3 \cdot 0 = 240$

Figure 8.26

The two corner points P_2 and P_3 give the same maximum value, $z = 240$. Thus, the maximum occurs at P_2 and P_3 and at all points between them on the boundary line $y = -\frac{2}{3}x + 80$, as shown in Figure 8.27. This line includes points such as (30, 60), (60, 40), and (90, 20). (Find points like these by substituting appropriate values of x in the equation.) The meaning of these points is given in Figure 8.28. Pete's Coffees can choose to produce its Rich Blend and Hawaiian Blend in any of the amounts given in Figure 8.28 (or any other amount given by a point on the line). Because each of these choices will maximize Pete's profit at $240 per day, the choice must be made using criteria other than profit. Perhaps the Hawaiian Blend tends to sell out earlier in the day than the Rich Blend. In this case, Pete might choose to produce 30 pounds of Rich Blend and 60 pounds of Hawaiian Blend.

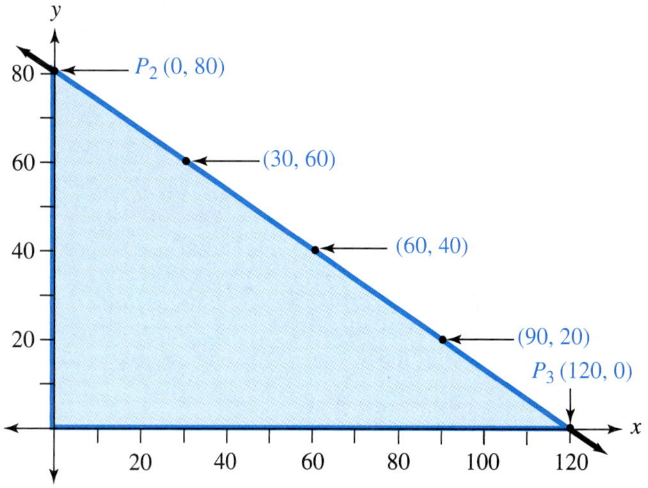

Figure 8.27

Point	Interpretation
P_2 (0, 80)	Each day, prepare no Rich Blend and 80 pounds of Hawaiian Blend for a profit of $240.
P_3 (120, 0)	Each day, prepare 120 pounds of Rich Blend and no Hawaiian Blend for a profit of $240.
(30, 60)	Each day, prepare 30 pounds of Rich Blend and 60 pounds of Hawaiian Blend for a profit of $240.
(60, 40)	Each day, prepare 60 pounds of Rich Blend and 40 pounds of Hawaiian Blend for a profit of $240.
(90, 20)	Each day, prepare 90 pounds of Rich Blend and 20 pounds of Hawaiian Blend for a profit of $240.

Figure 8.28 ■

HISTORICAL NOTE

George Dantzig

1914–

Edward W. Souza/News Service, Stanford University

George Bernard Dantzig's father was both a writer and a mathematician. He had hoped that his first son would be a writer and his second a mathematician, so he named his first son after the playwright and critic George Bernard Shaw and his second after Henri Poincaré, a famous French mathematician under whom the father had studied. As it happened, both sons became mathematicians.

In 1939, George Dantzig was a graduate student at the University of California at Berkeley. One day he arrived late at a statistics class. He copied the two problems that were written on the blackboard, assuming that they were homework problems. About six weeks after turning the homework in, he was awakened early one Sunday morning by his excited professor, who wanted to send off Dantzig's work right away for publication. The two problems were not homework but famous unsolved problems in statistics. Their solutions became Dantzig's Ph.D. thesis in mathematics.

During World War II, Dantzig was hired by the Air Force to find practical ways to distribute personnel, weapons, and supplies to the various fronts. Shortly after the end of the war, he became mathematics advisor to the U.S. Air Force Comptroller at the Pentagon, where he was responsible for finding a way to mechanize this planning process. The result was linear programming. The procedure was quickly applied to a wide variety of business, economic, and environmental topics. Tjalling Koopmans of the United States and Leonid Kantorovich of the Soviet Union received the 1975 Nobel Prize in economics for their use of linear programming in developing the theory of allocation of resources. Surprisingly, Dantzig himself was not honored. George Dantzig is now Professor Emeritus at Stanford University.

Source: Donald J. Albers and Constance Reid, "An Interview with George B. Dantzig. The Father of Linear Programming," *College Mathematics Journal,* vol. 17, no. 4, September 1986.

8.1 Exercises

In Exercises 1–6, convert the information to a linear inequality. Give the meaning of each variable used.

1. A landscape architect wants his project to use no more than 100 gallons of water per day. Each shrub requires 1 gallon of water per day, and each tree requires 3 gallons of water per day.

2. A shopper wishes to spend no more than $150. Each pair of pants costs $25 and each shirt costs $21.

3. A bookstore owner wishes to generate at least $5,000 in profit this month. Each hardback book generates $4.50 in profit, and each paperback generates $1.25 in profit.

4. Dick Rudd wants to take at least 1,000 mg of vitamin C. Each tablet of Megavite has 30 mg of vitamin C, and each tablet of Healthoboy has 45 mg of vitamin C.

5. A warehouse has 1,650 cubic feet of unused storage space. Refrigerators take up 63 cubic feet each, and dishwashers take up 41 cubic feet each.

6. A coffee shop owner has 1,000 pounds of Java beans. In addition to selling pure Java beans, the shop also sells Blend Number 202, which is 32% Java beans.

In Exercises 7–16, use the method of linear programming to solve the problem.

7. A craftswoman produces two products, floor lamps and table lamps. Production of one floor lamp requires 75 minutes of her labor and materials that cost $25. Production of one table lamp requires 50 minutes of labor, and the materials cost $20. The craftswoman wishes to work no more than 40 hours each week, and her financial resources allow her to pay no more than $900 for materials each week. If she can sell as many lamps as she can make and if her profit is $39 per floor lamp and $33 per table lamp, how many floor lamps and how many table lamps should she make each week to maximize her weekly profit? What is that maximum profit?

8. Five friends, all of whom are experienced bakers, form a company that will make bread and cakes and sell them to local restaurants and specialty stores. Each loaf of bread requires 50 minutes of labor and ingredients costing $0.90 and can be sold for $1.20 profit. Each cake requires 30 minutes of labor and ingredients costing $1.50 and can be sold for $4.00 profit. The partners agree that no one will work more than 8 hours a day. Their financial resources do not allow them to spend more than $190 per day on ingredients. How many loaves of bread and how many cakes should they make each day to maximize their profit? What is that maximum profit?

Ed Lallo/Gamma Liaison

9. Pete's Coffees sells two blends of coffee beans: Morning Blend and South American Blend. Morning Blend is one-third Mexican beans and two-thirds Colombian beans, and South American Blend is two-thirds Mexican beans and one-third Colombian beans. Profit on the Morning Blend is $3 per pound, while profit on the South American Blend is $2.50 per pound. Each day, the shop can obtain 100 pounds of Mexican beans and 80 pounds of Colombian beans, and it uses that coffee only in the two blends. If the shop can sell all that it makes, how many pounds of Morning Blend and of South American Blend should Pete's Coffees prepare each day to maximize profit? What is that maximum profit?

10. Pete's Coffees sells two blends of coffee beans: Yusip Blend and Exotic Blend. Yusip Blend is one-half Costa Rican beans and one-half Ethiopian beans, and Exotic Blend is one-quarter Costa Rican beans and three-quarters Ethiopian beans. Profit on the Yusip Blend is $3.50 per pound, while profit on the Exotic Blend is $4 per pound. Each day, the shop can obtain 200 pounds of Costa Rican beans and 330 pounds of Ethiopian beans, and it uses that coffee only in the two blends. If the shop can sell all that it makes, how many pounds of Yusip Blend and of Exotic Blend should Pete's Coffees prepare each day to maximize profit? What is that maximum profit?

11. Bake-Em-Fresh sells its bread to supermarkets. Shopgood Stores needs at least 15,000 loaves each week, and Rollie's Markets needs at least 20,000 loaves each week. Bake-Em-Fresh can ship at most 45,000 loaves to these two stores each week if it wishes to satisfy its other customers' needs. If shipping costs an average of 8¢ per loaf to Shopgood Stores and 9¢ per loaf to Rollie's Markets, how many loaves should Bake-Em-Fresh allot to Shopgood and to Rollie's each week? What shipping costs would this entail?

12. Notel Chips manufactures computer chips. Its two main customers, HAL Computers and Peach Computers, just submitted orders that must be filled immediately. HAL needs at least 130 cases of chips, and Peach needs at least 150 cases. Owing to a limited supply of silicon, Notel cannot send more than a total of 300 cases. If shipping costs $100 per case for shipments to HAL and $90 per case for shipments to Peach, how many cases should Notel send to each customer to minimize shipping costs? What shipping costs would this entail?

13. Global Air Lines has contracted with a tour group to transport a minimum of 1,600 first-class passengers and 4,800 economy-class passengers from New York to London during a 6-month time period. Global Air has two types of airplanes, the Orville 606 and the Wilbur W-1112. The Orville 606 carries 20 first-class passengers and 80 economy-class passengers and costs $12,000 to operate. The Wilbur W-1112 carries 80 first-class passengers and 120 economy-class passengers and costs $18,000 to operate. During the time period involved, Global Air can schedule no more than 52 flights on Orville 606s and no more than 30 flights on Wilbur W-1112s. How should Global Air Lines schedule its flights? What operating costs would this schedule entail?

14. Compucraft sells personal computers and printers made by Peach Computers. The computers come in 12-cubic-foot boxes, and the printers come in 8-cubic-foot boxes. Compucraft's owner estimates that at least 30 computers can be sold each month and that the number of computers sold will be at least 50% more than the number of printers. The computers cost Compucraft \$1,000 each and can be sold at a \$1,000 profit, while the printers cost \$300 each and can be sold for a \$350 profit. Compucraft has 1,000 cubic feet of storage available for the Peach personal computers and printers and sufficient financing to spend \$70,000 each month on computers and printers. How many computers and printers should Compucraft order from Peach each month to maximize profit? What is that maximum profit?

15. The Appliance Barn has 2,400 cubic feet of storage space for refrigerators. The larger refrigerators come in 60-cubic-foot packing crates, and the smaller ones come in 40-cubic-foot crates. The larger refrigerators can be sold for a \$250 profit, while the smaller ones can be sold for a \$150 profit.
    a. If the manager is required to sell at least 50 refrigerators each month, how many large refrigerators and how many small refrigerators should he order each month to maximize profit?
    b. If the manager is required to sell at least 40 refrigerators each month, how many large refrigerators and how many small refrigerators should he order each month to maximize profit?
    c. Should the Appliance Barn owner require his manager to sell 40 or 50 refrigerators per month?

16. City Electronics Distributors handles two lines of televisions, the Packard and the Bell. It purchases up to \$57,000 worth of television sets from the manufacturers each month and stores them in a 9,000-cubic-foot warehouse. The Packards come in 36-cubic-foot packing crates, and the Bells come in 30-cubic-foot crates. The Packards cost City Electronics \$200 each and can be sold to a retailer for a \$200 profit, while the Bells cost \$250 each and can be sold for a \$260 profit. City Electronics must stock enough sets to meet its regular customers' standing orders.
    a. If City Electronics has standing orders for 250 sets in addition to orders from other retailers, how many sets should City Electronics order each month to maximize profit?
    b. If City Electronics' standing orders increase to 260 sets, how many sets should City Electronics order each month to maximize profit?

17. How much of their available Mexican beans would be unused if Pete's Coffees of Exercise 9 maximizes its profit? How much of the available Colombian beans would be unused?

18. How much of their available time would be unused if the five friends of Exercise 8 maximize their profit? How much of their available money would be unused?

19. How much of her available time would be unused if the craftswoman of Exercise 7 maximizes her profit? How much of her available money would be unused?

20. How much of the available Costa Rican beans would be unused if Pete's Coffees of Exercise 10 maximizes its profit? How much of the available Ethiopian beans would be unused?

In Exercises 21–24, the region of possible solutions is not bounded: thus, there may not be both a maximum and a minimum. After graphing the region of possible solutions and finding each corner point, you can determine if both a maximum and a minimum exist by choosing two arbitrary values of z and graphing the corresponding lines, as discussed in the section "Why the Corner Principle Works."

21. The objective function $z = 2x + 3y$ is subject to the constraints

$$3x + y \geq 12$$
$$x + y \geq 6$$
$$x \geq 0 \quad \text{and} \quad y \geq 0$$

Find the following.
    a. the point at which the maximum occurs (if there is such a point)
    b. the maximum value
    c. the point at which the minimum occurs (if there is such a point)
    d. the minimum value

22. The objective function $z = 3x + 4y$ is subject to the constraints

$$2x + y \geq 10$$
$$3x + y \geq 12$$
$$x \geq 0 \quad \text{and} \quad y \geq 0$$

Find the following.
    a. the point at which the maximum occurs (if there is such a point)
    b. the maximum value

c. the point at which the minimum occurs (if there is such a point)

d. the minimum value

23. U.S. Motors manufactures quarter-ton, half-ton, and three-quarter-ton panel trucks. United Delivery Service has placed an order for at least 300 quarter-ton, 450 half-ton, and 450 three-quarter-ton panel trucks. U.S. Motors builds the trucks at two plants, one in Detroit and one in Los Angeles. The Detroit plant produces 30 quarter-ton trucks, 60 half-ton trucks, and 90 three-quarter-ton trucks each week, at a total cost of $540,000. The Los Angeles plant produces 60 quarter-ton trucks, 45 half-ton trucks, and 30 three-quarter-ton trucks each week, at a total cost of $360,000. How should U.S. Motors schedule its two plants so that it can fill this order at minimum cost? What is that minimum cost?

24. Eaton's Chocolates produces semisweet chocolate chips and milk chocolate chips at its plants in Bay City and Estancia. The Bay City plant produces 3,000 pounds of semisweet chips and 2,000 pounds of milk chocolate chips each day at a cost of $1,000, while the Estancia plant produces 1,000 pounds of semisweet chips and 6,000 pounds of milk chocolate chips each day at a cost of $1,500. Eaton's has an order from SafeNShop Supermarkets for at least 30,000 pounds of semisweet chips and 60,000 pounds of milk chocolate chips. How should it schedule its production so it can fill the order at minimum cost? What is that minimum cost?

Answer the following questions using complete sentences and your own words.

Concept Questions

25. A bicycle manufacturing firm hired a consultant who used linear programming to determine that the firm should discontinue manufacturing some of its product line. What negative ramifications might occur if this advice is followed? How could linear programming be used if the firm decided not to follow this advice?

26. Is it possible for an unbounded region to have both a maximum and a minimum? Draw a number of graphs (similar to those in Figure 8.24) to see. Justify your answer with graphs.

27. Discuss why it is that, when two corners give the same maximum (or minimum) value, the maximum (or minimum) occurs at all points on the boundary line between those corners.

28. In Example 2, Pete's Coffees mixes Colombian beans and Kona beans to make Rich Blend and Hawaiian Blend. It may be that Pete's could make slight variations in the proportions of Colombian and Kona beans in these two blends without affecting the taste of the coffee. Why would Pete's consider doing this? What effect would it have on the objective function, the constraints, the region of possible solutions, and the solution to the example?

29. In an open market, the selling price of a product is determined by the supply of the product and the demand for the product. In particular, the price falls if a larger supply of the product becomes available, and the price increases if there is suddenly a larger demand for the product. Any change in the selling price could affect the profit. However, when modeling a linear programming problem, we assume a constant profit. How could linear programming be adjusted to allow for the fact that prices change with supply and demand?

30. Make up a linear programming problem (a math problem, not a word problem) that has no maximum.

31. Make up a linear programming problem (a math problem, not a word problem) that has no minimum.

32. Make up a word problem that results in the following linear programming problem:

$$\text{maximize } z = 5x + 4y$$

subject to the constraints

$$10x + 20y \le 1{,}000$$
$$30x + 10y \le 2{,}000$$
$$x \ge 0 \quad \text{and} \quad y \ge 0$$

History Questions

33. For whom was George Dantzig named?

34. How did Dantzig choose the topic of his Ph.D. thesis?

35. What did Dantzig do during World War II?

36. What was surprising about the awarding of a Nobel Prize to two economists for their use of linear programming?

Project

37. Do some research on the Nobel Prize and find out why Dantzig was not eligible to receive a Nobel Prize for his invention of linear programming.

8.2 Introduction to the Simplex Method

In section 8.1, we discussed how to solve a linear programming problem by graphing the region of possible solutions, finding the corner points, and selecting the optimal corner point. This method, called the **geometric method,** is not the only method of solving a linear programming problem. Although it is an excellent method for solving problems, like those in section 8.1, that have two independent variables and only a handful of constraints, it is quite cumbersome for solving more complex problems.

Usually, a real-life problem has many variables. For example, the craftsman problem would be more realistic if the craftsman had two types of coffee tables, two types of end tables, and three different types of chairs, for a total of seven independent variables. Modern linear programming problems can involve thousands of variables and constraints and thousands of corner points. Finding all the corner points and the value of the objective function at each point can be an overwhelming task, even for the fastest of computers.

After inventing the geometric method, George Dantzig invented the **simplex method,** a method of linear programming for two *or more* variables that uses matrices and a procedure very similar to the Gauss-Jordan method of solving systems of equations. This method works well with more complex problems that have a large number of variables and constraints. It is also more amenable to graphing calculator or computer use than is the geometric method.

The simplex method involves a number of steps. First, the problem is modeled, in exactly the same way that problems are modeled in the geometric method. Next, the constraint inequalities are converted to equations, and the model is converted to a matrix problem, as we will show in this section. Finally, the matrix problem is solved using the row operations and a Gauss-Jordan–like procedure, as we will show in Section 8.3. We will discuss only problems in which each constraint is a $\leq$ inequality and in which the objective is to maximize some quantity.

We'll use the craftsman problem from the previous two sections, summarized below, to introduce the procedure and explain why it works.

A craftsman produces two products, coffee tables and end tables. Production data are given in Figure 8.29. If the craftsman wants to work no more than 40 hours each week and if his financial resources allow him to pay no more than $1,000 for materials each week, how many coffee tables and how many end tables should he make each week to maximize his weekly profit?

	Labor (per table)	**Cost of Materials (per table)**	**Profit (per table)**
Coffee Tables	6 hours	$200	$240
End Tables	5 hours	$100	$160

Figure 8.29

Step 1 *Set up the model.* This work was done previously. Because there are frequently more than two variables, the simplex method uses x_1, x_2, and so on rather than x and y to represent the variables.

Independent Variables

x_1 = number of coffee tables

x_2 = number of end tables

Constraints

C_1: $6x_1 + 5x_2 \le 40$ the time constraint

C_2: $200x_1 + 100x_2 \le 1{,}000$ the money constraint

C_3: $x_1 \ge 0$

C_4: $x_2 \ge 0$

Notice that we've labeled the constraints C_1 through C_4, for easy reference.

Objective Function

$z = 240x_1 + 160x_2$ z measures profit

Step 2 *Convert the constraint inequalities to equations.* Recall that the constraint

$$C_1\text{: } 6x_1 + 5x_2 \le 40$$

was derived from the data

$$(\text{coffee table hours}) + (\text{end table hours}) \le 40$$

If the total number of hours is less than 40, then some "slack" or unused hours are left, and we could say

$$(\text{coffee table hours}) + (\text{end table hours}) + (\text{unused hours}) = 40$$

C_1 can be rewritten as

$$C_1\text{: } 6x_1 + 5x_2 + s_1 = 40$$

where s_1 measures unused hours and $s_1 \ge 0$.

The variable s_1 is called a **slack variable** because it "takes up the slack" between the hours used (which can be less than 40) and the hours available (which are exactly 40). The introduction of slack variables converts the constraint inequalities to equations, allowing us to use matrices and a Gauss-Jordan–like procedure to solve the problem.

The constraint

$$C_2\text{: } 200x_1 + 100x_2 \le 1{,}000$$

was derived from the idea

$$(\text{coffee table money}) + (\text{end table money}) \le 1{,}000$$

which could be rephrased as

$$(\text{coffee table money}) + (\text{end table money}) + (\text{unused money}) = 1{,}000$$

Thus, C_2 can be rewritten as

$$C_2\text{: } 200x_1 + 100x_2 + s_2 = 1{,}000$$

where s_2 is a slack variable that measures unused money and $s_2 \ge 0$. This new variable "takes up the slack" between the money used (which can be less than \$1,000) and the money available (which is exactly \$1,000).

Constraints C_3: $x_1 \ge 0$ and C_4: $x_2 \ge 0$ remind us that variables x_1 and x_2 count things that cannot be negative (coffee tables and end tables, respectively); there is no need to convert these constraints to equations.

Step 3 *Rewrite the objective function with all variables on the left side.* The objective function

$$z = 240x_1 + 160x_2$$

becomes

$$-240x_1 - 160x_2 + z = 0$$

or, using *all* the variables,

$$-240x_1 - 160x_2 + 0s_1 + 0s_2 + 1z = 0$$

We have rewritten our model:

Independent Variables	**Slack Variables**
x_1 = number of coffee tables	s_1 = unused hours
x_2 = number of end tables	s_2 = unused money

$(x_1, x_2, s_1, s_2 \geq 0)$

Constraints

C_1: $6x_1 + 5x_2 + s_1 = 40$ the time constraint

C_2: $200x_1 + 100x_2 + s_2 = 1{,}000$ the money constraint

Objective Function

$-240x_1 - 160x_2 + 0s_1 + 0s_2 + 1z = 0$ z measures profit

Step 4 *Make a matrix out of the rewritten constraints and the rewritten objective function.* Basically, we copy the constraints and the objective function without the variables, as with the Gauss-Jordan method. The constraints go in the first rows, and the objective function goes in the last row.

The first constraint is

$$C_1: 6x_1 + 5x_2 + s_1 = 40$$

or, using *all* the variables,

$$C_1: 6x_1 + 5x_2 + 1s_1 + 0s_2 + 0z = 40$$

This equation becomes the first row of our matrix:

6 5 1 0 0 40

Similar alterations of the second constraint and the objective function yield the following matrix:

$$\begin{array}{c} \\ \\ \\ \end{array}\begin{array}{cccccc} x_1 & x_2 & s_1 & s_2 & z & \\ \end{array}$$

$$\left[\begin{array}{rrrrrr} 6 & 5 & 1 & 0 & 0 & 40 \\ 200 & 100 & 0 & 1 & 0 & 1{,}000 \\ -240 & -160 & 0 & 0 & 1 & 0 \end{array}\right] \begin{array}{l} \leftarrow C_1 \\ \leftarrow C_2 \\ \leftarrow C \text{ objective function} \end{array}$$

This matrix is called the **first simplex matrix.** The simplex method requires us to explore a series of matrices, just as the geometric method requires us to explore a series of corner points. Each simplex matrix will provide us with a corner point of the region of possible solutions, *without our actually graphing that region.* The last simplex matrix will provide us with the optimal corner point, that is, the point that solves the problem.

Step 5 *Determine the possible solution that corresponds to the matrix.* The possible solution is found by using a method very similar to the Gauss-Jordan method. Recall that with the Gauss-Jordan method we pivot until we obtain an identity matrix augmented with one extra column. We can find the value of the variable heading each column by reading down the column, turning at 1, and stopping at the end of the row.

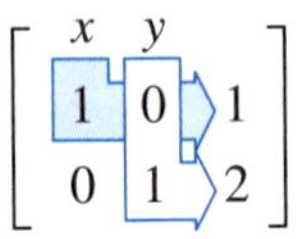

In the situation above, we do not have an identity matrix augmented with one extra column. Some of the columns (the "s_1," "s_2," and "z" columns) are identity matrix columns; we can find the values of the variables heading these columns by reading down the column, turning at 1, and stopping at the end of the row. Some of the columns (the "x_1" and "x_2" columns) are *not* identity matrix columns; the value of the variables heading these columns is zero.

$$\begin{array}{c} \begin{matrix} x_1 & x_2 & s_1 & s_2 & z & \end{matrix} \\ \left[\begin{array}{ccccc|c} 6 & 5 & 1 & 0 & 0 & 40 \\ 200 & 100 & 0 & 1 & 0 & 1{,}000 \\ -240 & -160 & 0 & 0 & 1 & 0 \end{array}\right] \begin{array}{l} \rightarrow s_1 = 40 \\ \rightarrow s_2 = 1{,}000 \\ \rightarrow z = 0 \end{array} \end{array}$$

$\uparrow$ $\uparrow$

not identity matrix columns, so $x_1 = 0, x_2 = 0$

The possible solution that corresponds to the matrix is

$(x_1, x_2, s_1, s_2) = (0, 0, 40, 1000)$ with $z = 0$

Why do the numbers read from the matrix in this manner yield a possible solution? Clearly, $(x_1, x_2) = (0, 0)$ satisfies all four constraints, so it is a possible solution. If we substitute this solution into C_1 and C_2, and the objective function, we get the following:

$$C_1\text{: } 6x_1 + 5x_2 + s_1 = 40 \qquad C_2\text{: } 200x_1 + 100x_2 + s_2 = 1{,}000$$
$$6 \cdot 0 + 5 \cdot 0 + s_1 = 40 \qquad 200 \cdot 0 + 100 \cdot 0 + s_2 = 1{,}000$$
$$s_1 = 40 \qquad s_2 = 1{,}000$$
$$z = 240x_1 + 160x_2 = 240 \cdot 0 + 160 \cdot 0 = 0$$

Thus, $(x_1, x_2, s_1, s_2) = (0, 0, 40, 1{,}000)$ with $z = 0$ satisfies all the constraints and the objective function and is a possible solution.

The possible solution means "make no coffee tables, make no end tables, have 40 unused hours and 1,000 unused dollars, and make no profit." When we solved this same problem with the geometric method in Section 8.1, one of the corner points was (0, 0) (see Figure 8.30). We have just found that corner point with the

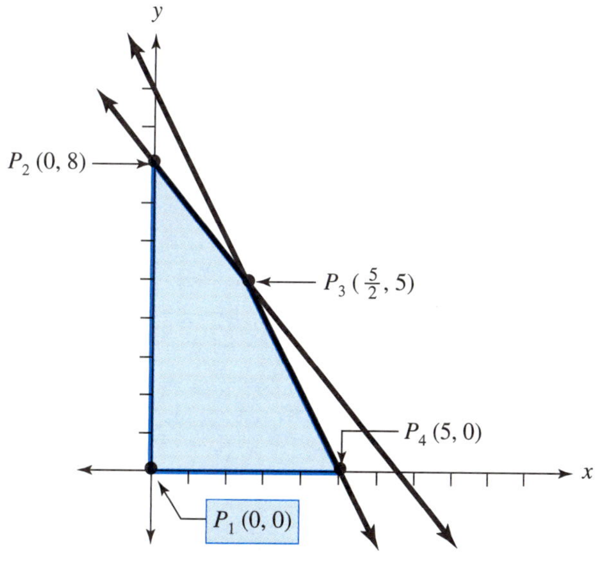

Figure 8.30

simplex method rather than the geometric method (the geometric method does not find values of the slack variables). The simplex method continues to locate corner points until the optimal corner point is found. The possible solution is only one of many corner points; it is not the optimal corner point, and it does not maximize profit. In the next section, we will see how to build on this work to find the maximal solution.

Simplex Method Steps

1. *Model the problem.*
   a. List the independent variables.
   b. List the constraints and translate them into linear inequalities.
   c. Find the objective and translate it into a linear equation.
2. *Convert each constraint from an inequality to an equation.*
   a. Use one slack variable for each constraint.
   b. Determine what each slack variable measures.
3. *Rewrite the objective function with all variables on the left side.*
4. *Make a matrix out of the rewritten constraints and the rewritten objective function.* This is the first simplex matrix.
5. *Determine the possible solution that corresponds to the matrix.*
   a. Identity matrix columns—turn at 1.
   b. Other columns—the value of the variable is zero.

The remaining steps, which build on this work to find the maximal solution, will be discussed in Section 8.3.

EXAMPLE 1 Find the first simplex matrix and the corresponding solution for the objective function

$$z = \frac{1}{2}x_1 + x_2$$

subject to the following constraints:

$$C_1: 2x_1 + x_2 \leq 15$$
$$C_2: x_1 + x_2 \leq 10$$
$$C_3: x_1 \geq 0$$
$$C_4: x_2 \geq 0$$

Solution

Step 1 *Model the problem.* No modeling needs to be done, because the problem is stated in the language of mathematics rather than as a real-world problem.

Step 2 *Convert each constraint from an inequality to an equation.*

$$C_1: 2x_1 + x_2 \leq 15 \rightarrow 2x_1 + x_2 + s_1 = 15$$
$$C_2: \ x_1 + x_2 \leq 10 \rightarrow \ x_1 + x_2 + s_2 = 10$$

Step 3 *Rewrite the objective function with all variables on the left side.*

$$z = \frac{1}{2}x_1 + x_2 \rightarrow -\frac{1}{2}x_1 - 1x_2 + 0s_1 + 0s_2 + 1z = 0$$

The model is rewritten as follows:

Constraints

C_1: $2x_1 + x_2 + s_1 = 15$
C_2: $x_1 + x_2 + s_2 = 10$
$(x_1, x_2, s_1, s_2 \geq 0)$

Objective Function

$$-\frac{1}{2}x_1 - 1x_2 + 0s_1 + 0s_2 + 1z = 0$$

Step 4 *Make a matrix out of the rewritten constraints and the rewritten objective function.* The first simplex matrix is as follows:

$$\begin{array}{c} \\ \\ \\ \end{array}\begin{matrix} x_1 & x_2 & s_1 & s_2 & z & \end{matrix}$$
$$\begin{bmatrix} 2 & 1 & 1 & 0 & 0 & 15 \\ 1 & 1 & 0 & 1 & 0 & 10 \\ -1/2 & -1 & 0 & 0 & 1 & 0 \end{bmatrix} \begin{matrix} \leftarrow C_1 \\ \leftarrow C_2 \\ \leftarrow \text{objective function} \end{matrix}$$

Step 5 *Determine the possible solution that corresponds to the matrix.* The x_1 and x_2 columns are not identity matrix columns, so the value of those variables is zero. The s_1, s_2, and z columns are identity matrix columns, so we find the values of those variables by reading down the column and turning at 1. The possible solution is $(x_1, x_2, s_1, s_2) = (0, 0, 15, 10)$ and $z = 0$.

Remember that this is only a *possible* solution; it is not the *maximal* solution. In the next section, we will see how to build on this work to find the maximal solution. ■

Comparison of the Gauss-Jordan and Simplex Methods

The Gauss-Jordan method solves systems of linear equations. The simplex method solves linear programming problems by optimizing linear objective functions subject to linear constraints.

The Gauss-Jordan method begins with a system of equations. The simplex method begins with a system of inequalities that is converted to a system of equations by the introduction of slack variables.

The Gauss-Jordan method does not produce a solution until you reach the final Gauss-Jordan matrix; none of the matrices that precede the final matrix have their own solution. The simplex method produces a series of possible solutions, one with each simplex matrix. Each possible solution would be a corner point of the region of possible solutions if we were using the geometric method.

We find the solution of a system of equations in the final Gauss-Jordan matrix by reading down each variable's column and turning at 1. Possible solutions of a linear programming problem are found in the same way *when a variable heads an identity matrix column*. The value of a variable that does not head an identity matrix column is zero.

The final Gauss-Jordan matrix is always an identity matrix augmented with one extra column; it always has the form

$$\begin{array}{cc} x & y \\ \end{array}$$
$$\left[\begin{array}{cc|c} 1 & 0 & 33 \\ 0 & 1 & 6 \end{array}\right]$$

The solution corresponding to this matrix is $(x, y) = (33, 6)$. A simplex matrix has some columns that would be found in an identity matrix and some that would not; it could have the form

$$\begin{array}{ccccc} x_1 & x_2 & s_1 & s_2 & z \end{array}$$
$$\left[\begin{array}{ccccc|c} 5 & 0 & 3 & 1 & 0 & 11 \\ -8 & 1 & 1 & 0 & 0 & 9 \\ 2 & 0 & 7 & 0 & 1 & 15 \end{array}\right]$$

The solution corresponding to this matrix is $(x_1, x_2, s_1, s_2) = (0, 9, 0, 11)$ with $z = 15$.

Karmarkar's New Method

Narendra Karmarkar was born in 1956 in Gwalior, India, and grew up in Poona, near Bombay. He is a current example of the Hindu tradition of excellence in mathematics. Both his father and his uncle were mathematicians. He attended the California Institute of Technology and received his doctorate from the University of California at Berkeley.

In 1984, Narendra Karmarkar, a mathematician at AT&T Bell Laboratories, invented a new method of linear programming that is an alternative to both the geometric method and the simplex method. The simplex method, as we saw in this section, starts at the corner point (0, 0) and methodically locates other corner points until the optimal corner point is found. A modern linear programming problem can involve thousands of corner points, and the simplex method involves locating a large proportion of these points. **Karmarkar's method,** on the other hand, makes use of quicker search routes through the interior of the region of possible solutions. This method has been found to be much quicker than the simplex method with some kinds of problems (see Figure 8.31).

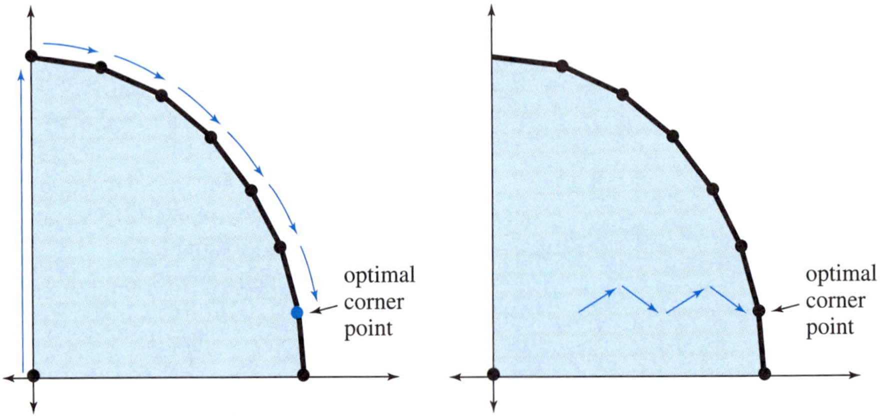

(a) a possible simplex method path through a region of possible solutions with many corner points

(b) a possible Karmarkar method path through the same region

Figure 8.31

The Granger Collection, NY

Srinivasa Ramanujan

HISTORICAL NOTE

Hindu Mathematics

The origins of Hindu mathematics are similar to those of Egyptian mathematics discussed in Chapter 6. The Hindus used geometry to construct buildings, and they stretched knotted ropes to survey land, measure buildings, and create right angles—as did the Egyptians. They knew the "Pythagorean" theorem before the time of Pythagoras.

In the fourth century A.D., Hindu scientists wrote the *Siddhantas,* a series of books on astronomy and trigonometry. It is not known whether this work was original or derived from Greek sources. The Hindus' trigonometry, however, was markedly superior to that of the Greeks; it is quite similar to modern trigonometry. In fact, the trigonometric word *sine* is a mistranslation of the Hindu word *jiva.* The *Siddhantas* included the earliest known tables of the sine of an angle.

The most important achievement of Hindu mathematics was the development of the numeration system that we use today, a place system based on powers of 10 that includes a symbol for zero. It is thought that the Hindu system was based on the abacus, which has columns of beads, just as our numeration system has columns of numbers. A place system is a system in which the value of a digit is determined by its place; for example, 30 is a different number than 3 because the zero puts the three in a different place. Roman numerals do not make use of a place system; V means "five" regardless of where the V is placed. It is quite likely that the Hindus invented the zero symbol, and this invention made the place system work.

The Hindu numeration system made arithmetic calculations much easier—imagine what multiplication or long division would be like with a nonplace system such as Roman numerals! Our modern methods of multiplication and long division differ only in appearance from the methods invented by the Hindus; the principles are the same. Hindu numbers, as well as their methods of multiplication and division, were used quite profitably by the Arabs, as discussed in Chapter 10. Europe acquired these inventions from Arabic books translated into Latin in the twelfth century.

There have been a number of prominent Indian mathematicians. Perhaps the most prominent was Srinivasa Ramanujan (1889–1920), a self-taught genius in pure mathematics. He became known to the Western world when he sent a letter to G.H. Hardy, a leading mathematician at Cambridge University in England. That letter contained 120 theorems. Some of the theorems were known to Hardy, but most were novel. Hardy said that he "had never seen anything in the least like them before. A single look at them is enough to show that they could only be written down by a mathematician of the highest class. They must be true, because, if they were not true, no one would have had the imagination to invent them." Hardy arranged to bring Ramanujan to Cambridge, where the two became good friends and collaborators.

A page from the *Siddhantas*

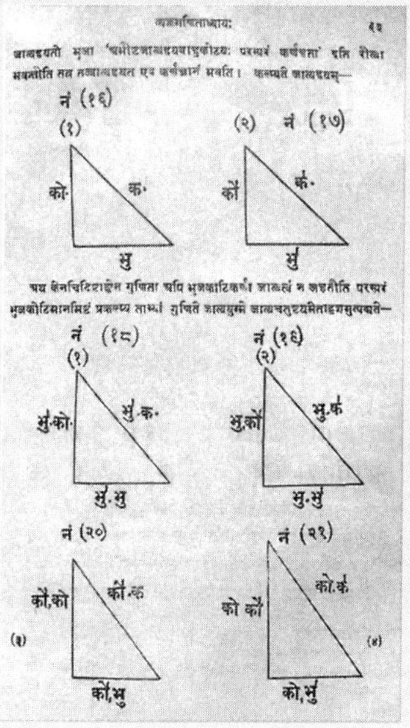

New York City Public Library, Photo by David Grossman

Scientists at Bell Labs used Karmarkar's method to find the most economical way to build a telephone network that links United States cities and to find the most economical way to route calls through the network once it was built. This involved a linear programming problem with 800,000 variables. The problem was not attempted with the simplex method because it would have taken weeks of computer time. It was solved with Karmarkar's method in less than one hour. Its application was quite profitable: AT&T estimates that its $15 billion system has an extra 9–10% of capacity because of the use of Karmarkar's method.

Airlines are currently attempting to use Karmarkar's method to quickly reroute airplanes and reschedule pilots and crews in the event of a major weather disruption. The simplex method is inappropriate here; it might necessitate a number of hours of computer time to find the answer. The Karmarkar method may yield the answer in a few minutes. Sometime in the not-so-distant future, you may be pleasantly surprised when your airline responds quickly and effectively to the closure of the airport at which you were to catch a connecting flight. When that happens, thank Narendra Karmarkar!

8.2 Exercises

In Exercises 1–4, convert the linear inequality to a linear equation by introducing a slack variable.

1. $3x_1 + 2x_2 \le 5$
2. $512x_1 + 339x_2 \le 254$
3. $3.41x_1 + 9.20x_2 + 6.16x_3 \le 45.22$
4. $99.52x_1 + 21.33x_2 + 102.15x_3 \le 50.14$

In Exercises 5–10, (a) translate the information into a linear inequality, (b) convert the linear inequality into a linear equation by introducing a slack variable, and (c) determine what each variable (including the slack variable) measures.

5. A shopper has $42.15 in cash. She wants to purchase meat at $6.99 a pound, cheese at $3.15 a pound, and bread at $1.98 a loaf. The shopper must pay for these items in cash.
6. A sharp dresser is at his favorite store, where everything is on sale. His budget will allow him to spend up to $425. Shirts are on sale for $19.99 each, slacks for $74.98 a pair, and sweaters for $32.98 each.
7. A plumber is starting an eight-hour day. Laying pipe takes her five minutes per foot, and installing elbows takes her four minutes per elbow.
8. An electrician is starting a $7\frac{1}{2}$-hour day. Installing conduit takes him 2.5 minutes per foot, connecting lines takes him 1 minute per connection, and installing circuit breakers takes him 4.5 minutes per circuit breaker.
9. A mattress warehouse has 50,000 cubic feet of space. Twin beds are 24 cubic feet, double beds are 36 cubic feet, queen-size beds are 56 cubic feet, and king-size beds are 72 cubic feet.
10. A nursery has 1,000 square feet of outdoor display space. Trees need 16 square feet each, shrubs need 10 square feet each, and flowering plants need 1 square foot each.

In Exercises 11–16, find the possible solution that corresponds to the given matrix.

11. $$\begin{array}{c} \begin{matrix} x_1 & x_2 & s_1 & s_2 & z & \end{matrix} \\ \left[\begin{array}{rrrrr|r} 5 & 0 & -3 & 1 & 0 & 12 \\ 3 & 0 & 19 & 0 & 1 & 22 \\ -21 & 1 & 48 & 0 & 0 & 19 \end{array}\right] \end{array}$$

12. $$\begin{array}{c} \begin{matrix} x_1 & x_2 & s_1 & s_2 & z & \end{matrix} \\ \left[\begin{array}{rrrrr|r} 1 & 31 & 0 & 9 & 0 & 7.4 \\ 0 & 0 & 0 & 15 & 1 & 4.9 \\ 0 & 1 & 1 & 0 & 0 & 20 \end{array}\right] \end{array}$$

13. $$\left[\begin{array}{rrrrrr|r} x_1 & x_2 & s_1 & s_2 & s_3 & z & \\ 1.9 & 0.3 & 0 & 0 & 1 & 0 & 0.5 \\ 3.2 & 0.7 & 0 & 1 & 0 & 0 & 9.3 \\ -5.5 & 0.8 & 1 & 0 & 0 & 0 & 7.8 \\ -2.1 & 3.2 & 0 & 0 & 0 & 1 & 9.6 \end{array}\right]$$

14. $$\left[\begin{array}{rrrrrr|r} x_1 & x_2 & s_1 & s_2 & s_3 & z & \\ 0 & 1 & 7 & 0 & 62 & 0 & 0.5 \\ 0 & 0 & 5 & 0 & 5 & 1 & 9.3 \\ 1 & 0 & 0 & 0 & 0 & 0 & 7.8 \\ 0 & 0 & 9 & 1 & 1 & 0 & 9.6 \end{array}\right]$$

15.

x_1	x_2	s_1	s_2	s_3	s_4	z	
1	0	6	0	0	76	0	25
0	0	8	0	1	13	0	46
0	0	3	1	0	50	0	32
0	1	4	0	0	−9	0	73
0	0	4	0	0	12	1	63

16.

x_1	x_2	s_1	s_2	s_3	s_4	z	
32	0	1	0	0	7	0	25
7	0	0	0	1	3	0	46
10	0	0	1	0	5	0	32
65	1	0	0	0	−9	0	73
24	0	0	0	0	2	1	44

In Exercises 17–22, find the first simplex matrix and the possible solution that corresponds to it.

17. *Objective Function:*
$z = 2x_1 + 4x_2$
Constraints:
$3x_1 + 4x_2 \leq 40$
$4x_1 + 7x_2 \leq 50$
$x_1 \geq 0$
$x_2 \geq 0$

18. *Objective Function:*
$z = 2.4x_1 + 1.3x_2$
Constraints:
$6.4x_1 + x_2 \leq 360$
$x_1 + 9.5x_2 \leq 350$
$x_1 \geq 0$
$x_2 \geq 0$

19. *Objective Function:*
$z = 12.10x_1 + 43.86x_2$
Constraints:
$112x_1 - 3x_2 \leq 370$
$x_1 + x_2 \leq 70$
$47x_1 + 19x_2 \leq 512$
$x_1 \geq 0$
$x_2 \geq 0$

20. *Objective Function:*
$z = -3.52x_1 + 4.72x_2$
Constraints:
$2x_1 + 4x_2 \leq 7{,}170$
$32x_1 - 19x_2 \leq 1{,}960$
$x_1 \leq 5$
$x_1 \geq 0$
$x_2 \geq 0$

21. *Objective Function:*
$z = 4x_1 + 7x_2 + 9x_3$
Constraints:
$5x_1 + 3x_2 + 9x_3 \leq 10$
$12x_1 + 34x_2 + 100x_3 \leq 10$
$52x_1 + 7x_2 + 12x_3 \leq 10$
$x_1 \geq 0$
$x_2 \geq 0$
$x_3 \geq 0$

22. *Objective Function:*
$z = 9.1x_1 + 3.5x_2 + 8.22x_3$
Constraints:
$x_1 + 16x_2 + 9.5x_3 \leq 1{,}210$
$72x_1 + 3.01x_2 + 50x_3 \leq 1{,}120$
$57x_1 + 87x_2 + 742x_3 \leq 309$
$x_1 \geq 0, x_2 \geq 0, x_3 \geq 0$

In Exercises 23–26, find the first simplex matrix and the possible solution that corresponds to it. Do not try to solve the problem.

23. Five friends, each of whom is an experienced baker, form a company that will make bread and cakes and sell them to local restaurants and specialty stores. Each loaf of bread requires 50 minutes of labor and ingredients costing \$0.90 and can be sold for a \$1.20 profit. Each cake requires 30 minutes of labor and ingredients costing \$1.50 and can be sold for a \$4.00 profit. The partners agree that no one will work more than 8 hours a day. Their financial resources do not allow them to spend more than \$190 per day on ingredients. How many loaves of bread and how many cakes should they make each day to maximize their profit?

24. A craftswoman produces two products, floor lamps and table lamps. Production of one floor lamp requires 75 minutes of her labor and materials that cost \$25. Production of one table lamp requires 50 minutes of labor and materials that cost \$20. The craftswoman wants to work no more than 40 hours each week, and her financial resources allow her to pay no more than \$900 for materials each week. If she can sell as many lamps as she can make and if her profit is \$40 per floor lamp and \$32 per table lamp, how many floor lamps and how many table lamps should she make each week to maximize her weekly profit?

25. Pete's Coffees sells two blends of coffee beans: Yusip Blend and Exotic Blend. Yusip Blend is one-half Costa Rican beans and one-half Ethiopian beans, and Exotic Blend is one-quarter Costa Rican beans and three-quarters Ethiopian beans. Profit on the Yusip Blend is \$3.50 per pound, while profit on the Exotic Blend is \$4 per pound. Each day, the shop can obtain 200 pounds of Costa Rican beans and 330 pounds of Ethiopian beans, and it uses those beans only in the two blends. If it can sell all that it makes, how many pounds of Yusip Blend and of Exotic Blend should Pete's Coffees prepare each day to maximize profit?

26. Pete's Coffees sells two blends of coffee beans: Morning Blend and South American Blend. Morning Blend is one-third Mexican beans and two-thirds Colombian beans, and South American Blend is two-thirds Mexican beans and one-third Colombian beans. Profit on the Morning Blend is \$3 per pound, while profit on the South American Blend is \$2.50 per pound. Each day, the shop can obtain 100 pounds of Mexican beans and 80 pounds of Colombian beans, and it uses those beans only in the two blends. If it can sell all that it makes, how many pounds of Morning Blend and of South American Blend should Pete's Coffees prepare each day to maximize profit?

Answer the following questions using complete sentences and your own words.

History Questions

27. What was one of the earliest uses of Karmarkar's method of linear programming?
28. What motivated Karmarkar's invention?
29. Why is Karmarkar's method an improvement?

8.3 The Simplex Method: Complete Problems

In Section 8.2, we saw how to convert a linear programming problem to a matrix problem. In this section, we will see how to use the simplex method to solve a linear programming problem. We'll start by continuing where we left off with the craftsman problem. We have modeled the problem, introduced slack variables, and rewritten the objective function so all variables are on the left side.

Independent Variables

x_1 = number of coffee tables

x_2 = number of end tables

$(x_1, x_2, s_1, s_2 \geq 0)$

Slack Variables

s_1 = unused hours

s_2 = unused money

Constraints

C_1: $6x_1 + 5x_2 + s_1 = 40$ — the time constraint

C_2: $200x_1 + 100x_2 + s_2 = 1{,}000$ — the money constraint

Objective Function

$-240x_1 - 160x_2 + 0s_1 + 0s_2 + 1z = 0$ — z measures profit

We have also made a matrix out of the rewritten constraints and the rewritten objective function (the "first simplex matrix") and determined the possible solution that corresponds to that matrix. The first simplex matrix was as follows:

$$\begin{array}{c} \\ \\ \\ \\ \end{array}\begin{matrix} x_1 & x_2 & s_1 & s_2 & z & \end{matrix}$$

$$\begin{bmatrix} 6 & 5 & 1 & 0 & 0 & 40 \\ 200 & 100 & 0 & 1 & 0 & 1{,}000 \\ -240 & -160 & 0 & 0 & 1 & 0 \end{bmatrix} \begin{matrix} \leftarrow C_1 \\ \leftarrow C_2 \\ \leftarrow \text{objective function} \end{matrix}$$

The corresponding possible solution was

$$(x_1, x_2, s_1, s_2) = (0, 0, 40, 1{,}000) \quad \text{with } z = 0$$

This solution is a corner point of the region of possible solutions (see Figure 8.32); it is point P_1 (0, 0) (the geometric method does not find values of the slack variables). However, it is not the optimal corner point; it is a possible solution but not the maximal solution. Our goal in the remaining steps of the simplex method is to find the maximal solution. *We do this with a procedure identical to that used in the Gauss-Jordan method except for how we tell where to pivot.* Recall that with the Gauss-Jordan method, pivoting is used to obtain a matrix of the form

$$\begin{array}{cc} x & y \end{array}$$
$$\begin{bmatrix} 1 & 0 & 33 \\ 0 & 1 & 6 \end{bmatrix}$$

This is not the case with the simplex method.

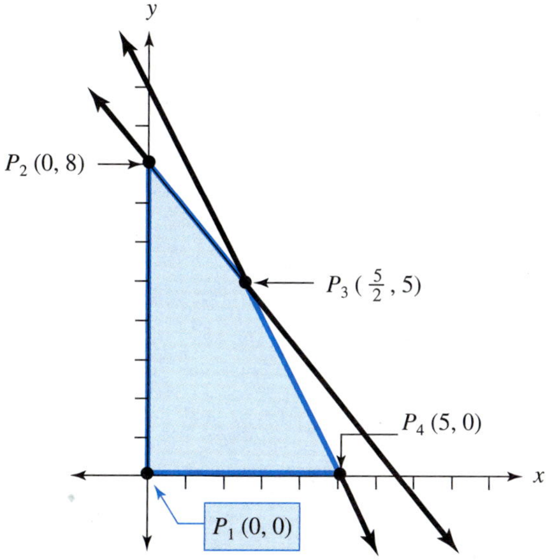

Figure 8.32
Each simplex matrix has a corresponding possible solution. Each of these possible solutions is a corner point of the region of possible solutions. The possible solution $(x_1, x_2, s_1, s_2) = (0, 0, 40, 1{,}000)$ is the corner point P_1 (0, 0).

Pivoting with the Simplex Method

Step 1 *Look at the last row, the objective function row. Select the most negative entry in that row. The column containing that entry is the* ***pivot column.*** If the last row contains no negative entries, then no pivoting is necessary; the possible solution that corresponds to the matrix is the maximal solution.

$$\begin{array}{cccccc} x_1 & x_2 & s_1 & s_2 & z & \\ \end{array}$$
$$\begin{bmatrix} 6 & 5 & 1 & 0 & 0 & 40 \\ 200 & 100 & 0 & 1 & 0 & 1{,}000 \\ -240 & -160 & 0 & 0 & 1 & 0 \end{bmatrix} \leftarrow \text{objective function}$$

The only negative entries in the last row are −240 and −160; the most negative is −240. This entry is in the first column, so our pivot column is the first column. We will pivot on one of the entries of the pivot column.

$$\begin{array}{c} \\ \\ \\ \\ \end{array}\begin{array}{cccccc} x_1 & x_2 & s_1 & s_2 & z & \\ \end{array}$$
$$\left[\begin{array}{cccccc} 6 & 5 & 1 & 0 & 0 & 40 \\ 200 & 100 & 0 & 1 & 0 & 1{,}000 \\ -240 & -160 & 0 & 0 & 1 & 0 \end{array}\right]$$
$$\begin{array}{l} \uparrow \\ \text{pivot column} \end{array}$$

Step 2 *Divide the last entry in each constraint row by the corresponding entry in the pivot column. The row that yields the smallest nonnegative such quotient is the* ***pivot row.***

$$\left[\begin{array}{cccccc} 6 & 5 & 1 & 0 & 0 & 40 \\ 200 & 100 & 0 & 1 & 0 & 1{,}000 \\ -240 & -160 & 0 & 0 & 1 & 0 \end{array}\right]\begin{array}{l} \leftarrow 40/6 \approx 6.67 \\ \leftarrow 1{,}000/200 = 5 \\ \leftarrow \text{not a constraint row} \end{array}$$
$$\begin{array}{l} \uparrow \\ \text{pivot column} \end{array}$$

Our quotients are 6.67 and 5; the smallest nonnegative quotient is 5, which is in row two. Our pivot row is the second row.

Step 3 *Pivot on the entry in the pivot row and pivot column.*

$$\left[\begin{array}{cccccc} \boxed{6} & 5 & 1 & 0 & 0 & 40 \\ \boxed{200} & \boxed{100} & \boxed{0} & \boxed{1} & \boxed{0} & \boxed{1{,}000} \\ \boxed{-240} & -160 & 0 & 0 & 1 & 0 \end{array}\right]\begin{array}{l} \\ \boxed{\leftarrow \textit{pivot row}} \\ \\ \end{array}$$
$$\begin{array}{l} \uparrow \\ \boxed{\textit{pivot column}} \end{array}$$

We will pivot on the entry "200." Now that we've determined *where* to pivot, we use the row operations as we would with the Gauss-Jordan method.

R2 ÷ 200

$$\left[\begin{array}{cccccc} 6 & 5 & 1 & 0 & 0 & 40 \\ 1 & 0.5 & 0 & 0.005 & 0 & 5 \\ -240 & -160 & 0 & 0 & 1 & 0 \end{array}\right]$$

R1 − 6R2:R1
240R2 + R3:R3

$$\left[\begin{array}{cccccc} 0 & 2 & 1 & -0.03 & 0 & 10 \\ 1 & 0.5 & 0 & 0.005 & 0 & 5 \\ 0 & -40 & 0 & 1.2 & 1 & 1{,}200 \end{array}\right]$$

Of course, these calculations could be performed on a graphing calculator or a computer equipped with the Amortrix software, as discussed in Chapter 7. We have finished our first pivot. The above matrix, the result of the first pivot, is called the **second simplex matrix.**

When to Stop Pivoting

The simplex method requires us to explore a series of matrices, just as the geometric method requires us to explore a series of corner points. Each simplex matrix provides us with a corner point of the region of possible solutions; the final simplex matrix will provide us with the optimal corner point, the point that solves the problem.

The second simplex matrix

$$\begin{array}{cccccc} x_1 & x_2 & s_1 & s_2 & z & \\ \end{array}$$
$$\begin{bmatrix} 0 & 2 & 1 & -0.03 & 0 & 10 \\ 1 & 0.5 & 0 & 0.005 & 0 & 5 \\ 0 & -40 & 0 & 1.2 & 1 & 1{,}200 \end{bmatrix}$$

provides us with a corner point. Because the "x_1," "s_1," and "z" columns are identity matrix columns, their values are found as in the Gauss-Jordan method. The "x_2" and "s_2" columns are not identity matrix columns, so their values are zero. Thus, the corner point that corresponds to the second simplex matrix is

$$(x_1, x_2, s_1, s_2) = (5, 0, 10, 0) \text{ with } z = 1{,}200$$

This is corner point P_4 (see Figure 8.33).

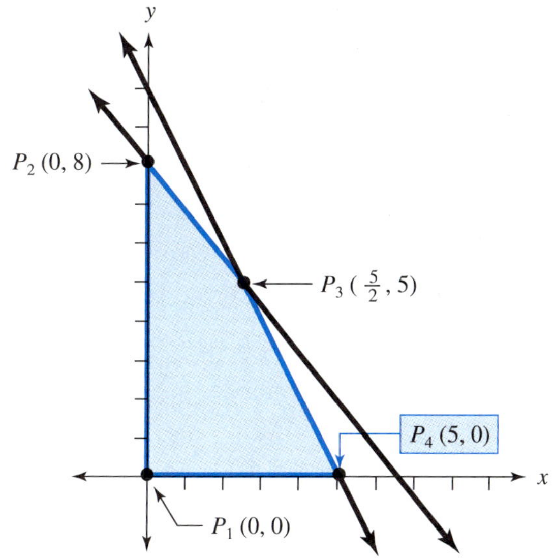

Figure 8.33

Because x_1 is the number of coffee tables, x_2 the number of end tables, s_1 the number of unused hours, and s_2 the amount of unused money, this possible solution means "make five coffee tables and no end tables, have ten unused hours and no unused money, and make a profit of $1,200."

Is this the optimal corner point, the point that solves the craftsman's problem? We answer this question by determining whether it is possible to pivot again. We must apply the steps in pivoting with the simplex method to the second simplex matrix.

Step 1 *Select the most negative entry in the last row, the objective function row. The column containing that entry is the pivot column.* If the last row contains no negative entries, then no pivoting is necessary; the possible solution that corresponds to the matrix is the maximal solution.

$$\begin{array}{cccccc} x_1 & x_2 & s_1 & s_2 & z & \\ \end{array}$$
$$\begin{bmatrix} 0 & 2 & 1 & -0.03 & 0 & 10 \\ 1 & 0.5 & 0 & 0.005 & 0 & 5 \\ 0 & -40 & 0 & 1.2 & 1 & 1{,}200 \end{bmatrix} \begin{array}{l} \\ \leftarrow \text{objective function} \\ \\ \end{array}$$

Because there is a negative entry in the objective function row, we must pivot again. This negative entry is in the "x_2" column, so our pivot column is the "x_2" column.

$$\begin{array}{cccccc} x_1 & x_2 & s_1 & s_2 & z & \\ \end{array}$$
$$\begin{bmatrix} 0 & 2 & 1 & -0.03 & 0 & 10 \\ 1 & 0.5 & 0 & 0.005 & 0 & 5 \\ 0 & -40 & 0 & 1.2 & 1 & 1{,}200 \end{bmatrix}$$

↑ pivot column

Step 2 *Divide the last entry in each constraint row by the corresponding entry in the pivot column. The row that yields the smallest nonnegative such quotient is the pivot row.*

$$\begin{bmatrix} 0 & 2 & 1 & 0.03 & 0 & 10 \\ 1 & 0.5 & 0 & 0.005 & 0 & 5 \\ 0 & -40 & 0 & 1.2 & 1 & 1{,}200 \end{bmatrix} \begin{array}{l} \leftarrow 10/2 = 5 \\ \leftarrow 5/0.5 = 10 \\ \leftarrow \text{not a constraint row} \end{array}$$

↑ pivot column

Our quotients are 5 and 10; the smallest nonnegative quotient is 5, which is in row one. Our pivot row is the first row.

Step 3 *Pivot on the entry in the pivot row and pivot column.*

$$\begin{bmatrix} \boxed{0} & \boxed{2} & \boxed{1} & \boxed{-0.03} & \boxed{0} & \boxed{10} \\ 1 & \boxed{0.5} & 0 & 0.005 & 0 & 5 \\ 0 & \boxed{-40} & 0 & 1.2 & 1 & 1{,}200 \end{bmatrix} \begin{array}{l} \leftarrow \text{pivot row} \\ \\ \\ \end{array}$$

↑ pivot column

R1 ÷ 2

$$\begin{bmatrix} 0 & 1 & 0.5 & -0.015 & 0 & 5 \\ 1 & 0.5 & 0 & 0.005 & 0 & 5 \\ 0 & -40 & 0 & 1.2 & 1 & 1{,}200 \end{bmatrix}$$

−0.5R1 + R2:R2

40R1 + R3:R3

$$\begin{array}{cccccc} x_1 & x_2 & s_1 & s_2 & z & \\ \end{array}$$
$$\begin{bmatrix} 0 & 1 & 0.5 & -0.015 & 0 & 5 \\ 1 & 0 & -0.25 & 0.0125 & 0 & 2.5 \\ 0 & 0 & 20 & 0.6 & 1 & 1{,}400 \end{bmatrix}$$

This matrix is our third *and last* simplex matrix, because it is not possible to pivot further—the bottom row contains no negative entries. The solution that corresponds to this matrix is the optimal corner point:

$$(x_1, x_2, s_1, s_2) = (2.5, 5, 0, 0) \text{ with } z = 1{,}400$$

meaning, "each week make 2.5 coffee tables and 5 end tables, have no unused hours and no unused money, and make a profit of $1,400" (see Figure 8.34).

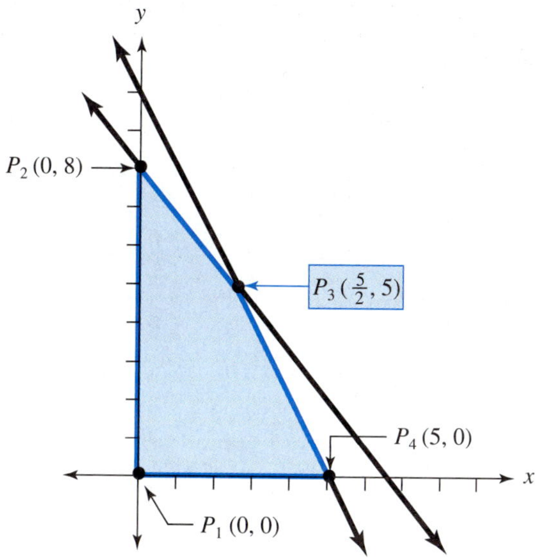

Figure 8.34

Why the Simplex Method Works

In the craftsman problem, our first simplex matrix was

$$\begin{array}{c} \begin{array}{cccccc} x_1 & x_2 & s_1 & s_2 & z & \end{array} \\ \begin{bmatrix} 6 & 5 & 1 & 0 & 0 & 40 \\ 200 & 100 & 0 & 1 & 0 & 1{,}000 \\ -240 & -160 & 0 & 0 & 1 & 0 \end{bmatrix} \end{array}$$

and its corresponding possible solution was

$$(x_1, x_2, s_1, s_2) = (0, 0, 40, 1{,}000) \quad \text{with } z = 0$$

This means "make no tables and generate no profit." We selected the first column as our pivot column because it was the column with the most negative number in the bottom. Why do we select the column with the most negative number in the bottom?

The possible solution corresponding to the first simplex matrix involves no profit ($z = 0$). Certainly, the profit can be increased by making some tables. If the craftsman were to make only one type of table, which table would be best? Coffee tables generate more profit per table ($240 versus $160), so the craftsman would be better off making coffee tables. This choice of coffee tables over end tables, or of $240 over $160, was made in the pivoting process when we chose the first column as the pivot column. That choice was made because 240 is larger than 160.

After selecting the pivot column, we selected the second row as our pivot row because 5 is the smallest nonnegative quotient. Why do we select the row with the smallest nonnegative quotient?

$$\begin{bmatrix} 6 & 5 & 1 & 0 & 0 & 40 \\ 200 & -100 & 0 & 1 & 0 & 1{,}000 \\ -240 & -160 & 0 & 0 & 1 & 0 \end{bmatrix} \begin{array}{l} \leftarrow 40/6 = 6\frac{2}{3} \\ \leftarrow 1{,}000/200 = 5 \\ \leftarrow \text{not a constraint row} \end{array}$$

If the craftsman were to make only coffee tables, he should make the largest amount allowed by the constraints when 0 is substituted for x_2 (that is, when no end tables are made).

Time Constraint	Money Constraint
C_1: $6x_1 + 5x_2 \leq 40$	C_2: $200x_1 + 100x_2 \leq 1{,}000$
$6x_1 + 5 \cdot 0 \leq 40$	$200x_1 + 100 \cdot 0 \leq 1{,}000$
$6x_1 \leq 40$	$200x_1 \leq 1{,}000$
$x \leq \frac{40}{6} = \frac{20}{3} = 6\frac{2}{3}$	$x_1 \leq \frac{1{,}000}{200} = 5$

The time constraint will not be violated as long as $x_1 \leq 6\frac{2}{3}$; the money constraint will not be violated as long as $x_1 \leq 5$. The craftsman can make at most $x_1 = 5$ coffee tables and not violate either constraint. This choice of 5 over $6\frac{2}{3}$ was made in the pivoting process when we chose the second row as the pivot row. That choice was made because 5 is less than $6\frac{2}{3}$.

We have just used logic to determine that if the craftsman makes only one type of table, he should make coffee tables, and that he could make at most five coffee tables without violating his constraints. The simplex method leads to the same conclusion. The second simplex matrix was

$$\begin{array}{cccccc} x_1 & x_2 & s_1 & s_2 & z & \\ \end{array}$$
$$\begin{bmatrix} 0 & 2 & 1 & -0.03 & 0 & 10 \\ 1 & 0.5 & 0 & 0.005 & 0 & 5 \\ 0 & -40 & 0 & 1.2 & 1 & 1{,}200 \end{bmatrix}$$

and its corresponding possible solution was

$$(x_1, x_2, s_1, s_2) = (5, 0, 10, 0) \quad \text{with } z = 1{,}200$$

This does suggest that the craftsman make five coffee tables and no end tables. This matrix was not the final simplex matrix, so this possible solution was not the optimal solution. That is, the craftsman will increase his profit if he makes more than one type of table.

The last row of this matrix, the objective function row, is

$$0 \qquad -40 \qquad 0 \qquad 1.2 \qquad 1 \qquad 1{,}200$$

We pivoted again because of the presence of a negative number in this row. This pivoting resulted in an increased profit. Why does the presence of a negative number imply that if we pivot again, the profit will be increased? The last row represents the equation

$$0x_1 + -40x_2 + 0s_1 + 1.2s_2 + 1z = 1{,}200$$

which can be rewritten as

$$z = 1{,}200 + 40x_2 - 1.2s_2$$

by solving for z. According to this last equation, the profit is $z = 1{,}200$ if x_2 and s_2 are both zero. If x_2 is positive, then the profit could be larger than 1,200. Thus, it is possible to achieve a larger profit by changing x_2 from zero to a positive number. Because x_2 measures the number of end tables, this means that the craftsman should make some end tables. Our last pivot did in fact result in his making some end tables.

The following list of steps includes those steps developed in Section 8.2, as well as those developed in this section. Notice that you can check your work at step 7; if you have made an arithmetic error during your pivot, catch it here before you go any further.

Simplex Method Steps

1. *Model the problem.*
   a. List the independent variables.
   b. List the constraints and translate them into linear inequalities.
   c. Find the objective and translate it into a linear equation.
2. *Convert each constraint from an inequality to an equation.*
   a. Use one slack variable for each constraint.
   b. Determine what each slack variable measures.
3. *Rewrite the objective function with all variables on the left side.*
4. *Make a matrix out of the rewritten constraints and the rewritten objective function.* This is the first simplex matrix.
5. *Determine the possible solution that corresponds to the matrix.*
   a. Identity matrix columns—turn at 1.
   b. Other columns—the value of the variable is zero.
6. *Pivot to find a better possible solution.*
   a. The pivot column is the column with the most negative entry in the last row. In case of a tie, choose either.
   b. Divide the last entry of each constraint row by the entry in the pivot column. The pivot row is the row with the smallest nonnegative such quotient. In case of a tie, choose either.
   c. Pivot on the selected row and column; that is, use the row operations to transform the selected column into an identity matrix column with 1 in the selected row. This gives a new simplex matrix.
7. *Determine the possible solution that corresponds to the matrix* (as in step 5).
   ✔ Check your work by seeing if the solution substitutes into the objective function and constraints.
8. *Determine whether the current possible solution maximizes the objective function.*
   a. If the last row of the new simplex matrix has no negative entries, then the problem is solved, and the current solution is the maximal solution.
   b. If the last row of the new simplex matrix has one or more negative entries, then pivot again. Return to step 6.
9. *Interpret the final solution.* Express the solution in words, as a solution to a real-world problem rather than a mathematical problem.

Example 1 The Leather Factory has one sewing machine with which it makes coats and vests. Each coat requires 50 minutes on the sewing machine and uses 12 square feet of leather. Each vest requires 30 minutes on the sewing machine and uses 8 square feet of leather. The sewing machine is available 8 hours a day, and the Leather Factory can obtain 118 square feet of leather a day. The coats sell for \$175 each, and the vests sell for \$100 each. Find the daily production level that would yield the maximum revenue.

Solution

Step 1 *Model the problem.*

Independent Variables

x_1 = number of coats made each day

x_2 = number of vests made each day

Constraints

C_1: sewing machine hours $\leq$ 8 hours $= 480$ minutes

(coat sewing time) + (vest sewing time) ≤ 480 minutes

$$\begin{pmatrix}\text{time}\\\text{per}\\\text{coat}\end{pmatrix}\cdot\begin{pmatrix}\text{number}\\\text{of}\\\text{coats}\end{pmatrix}+\begin{pmatrix}\text{time}\\\text{per}\\\text{vest}\end{pmatrix}\cdot\begin{pmatrix}\text{number}\\\text{of}\\\text{vests}\end{pmatrix}\leq 480 \text{ minutes}$$

$$50 \quad x_1 \quad + \quad 30 \quad x \quad \leq 480 \text{ minutes}$$

C_2: leather used $\leq$ 118 square feet

(coat leather) + (vest leather) $\leq$ 118 square feet

$$\begin{pmatrix}\text{leather}\\\text{per}\\\text{coat}\end{pmatrix}\cdot\begin{pmatrix}\text{number}\\\text{of}\\\text{coats}\end{pmatrix}+\begin{pmatrix}\text{leather}\\\text{per}\\\text{vest}\end{pmatrix}\cdot\begin{pmatrix}\text{number}\\\text{of}\\\text{vests}\end{pmatrix}\leq 118 \text{ square feet}$$

$$12 \quad x_1 \quad + \quad 8 \quad x_2 \quad \leq 118$$

Objective Function

maximize z = revenue = coat revenue + vest revenue

$$z=\begin{pmatrix}\text{price}\\\text{per}\\\text{coat}\end{pmatrix}\cdot\begin{pmatrix}\text{number}\\\text{of}\\\text{coats}\end{pmatrix}+\begin{pmatrix}\text{price}\\\text{per}\\\text{vest}\end{pmatrix}\cdot\begin{pmatrix}\text{number}\\\text{of}\\\text{vests}\end{pmatrix}$$

$$= 175 \quad x_1 \quad + \quad 100 \quad x_2$$

Step 2 *Convert each constraint from an inequality to an equation.*

$$C_1\colon 50x_1 + 30x_2 \leq 480$$
$$50x_1 + 30x_2 + s_1 = 480$$

where s_1 = unused sewing machine time.

$$C_2\colon 12x_1 + 8x_2 \leq 118$$
$$12x_1 + 8x_2 + s_2 = 118$$

where s_2 = unused leather.

Step 3 *Rewrite the objective function with all variables on the left side.*

$$z = 175x_1 + 100x_2$$
$$-175x_1 - 100x_2 + 1z = 0$$

or, using *all* the variables,

$$-175x_1 - 100x_2 + 0s_1 + 0s_2 + 1z = 0$$

Step 4 *Make a matrix out of the rewritten constraints and the rewritten objective function.* The first simplex matrix is as follows:

$$\begin{array}{cccccc} x_1 & x_2 & s_1 & s_2 & z & \\ \end{array}$$
$$\left[\begin{array}{rrrrrr} 50 & 30 & 1 & 0 & 0 & 480 \\ 12 & 8 & 0 & 1 & 0 & 118 \\ -175 & -100 & 0 & 0 & 1 & 0 \end{array}\right]$$

Step 5 *Determine the possible solution that corresponds to the first simplex matrix.* The solution is $(x_1, x_2, s_1, s_2) = (0, 0, 480, 118)$ with $z = 0$. This means "make no product and make no profit."

Step 6 *Pivot to find a better possible solution.*

$$\begin{array}{cccccc} x_1 & x_2 & s_1 & s_2 & z & \\ \end{array}$$
$$\begin{bmatrix} 50 & 30 & 1 & 0 & 0 & 480 \\ 12 & 8 & 0 & 1 & 0 & 118 \\ -175 & -100 & 0 & 0 & 1 & 0 \end{bmatrix}$$

↑ pivot column

Our pivot column is the first column, because -175 is the most negative entry in the bottom row. To select the pivot row, divide the last entry in each constraint row by the entry in the pivot column and select the row with the smallest nonnegative such quotient.

$$\begin{bmatrix} 50 & 30 & 1 & 0 & 0 & 480 \\ 12 & 8 & 0 & 1 & 0 & 118 \\ -175 & -100 & 0 & 0 & 1 & 0 \end{bmatrix} \begin{array}{l} \leftarrow 480/50 = 9.6 \\ \leftarrow 118/12 \approx 9.8333 \\ \leftarrow \text{not a constraint row} \end{array}$$

Because 9.6 is the smallest nonnegative quotient, our pivot row is the first row.

$$\begin{bmatrix} \boxed{50} & \boxed{30} & \boxed{1} & \boxed{0} & \boxed{0} & \boxed{480} \\ \boxed{12} & 8 & 0 & 1 & 0 & 118 \\ \boxed{-175} & -100 & 0 & 0 & 1 & 0 \end{bmatrix} \leftarrow \text{pivot row}$$

↑ pivot column

We will pivot on the 50.

R1 ÷ 50
$$\begin{bmatrix} 1 & 0.6 & 0.02 & 0 & 0 & 9.6 \\ 12 & 8 & 0 & 1 & 0 & 118 \\ -175 & -100 & 0 & 0 & 1 & 0 \end{bmatrix}$$

−12R1 + R2:R2
175R1 + R3:R3
$$\begin{array}{cccccc} x_1 & x_2 & s_1 & s_2 & z & \\ \end{array}$$
$$\begin{bmatrix} 1 & 0.6 & 0.02 & 0 & 0 & 9.6 \\ 0 & 0.8 & -0.24 & 1 & 0 & 2.8 \\ 0 & 5 & 3.5 & 0 & 1 & 1{,}680 \end{bmatrix}$$

Step 7 *Determine the possible solution that corresponds to the matrix.* The possible solution is $(x_1, x_2, s_1, s_2) = (9.6, 0, 0, 2.8)$ with $z = 1{,}680$.

> ✔ Check your work by seeing if the solution substitutes into the objective function and constraints:
>
> C_1: $50x_1 + 30x_2 + s_1 = 50 \cdot 9.6 + 30 \cdot 0 + 0 = 480$
>
> C_2: $12x_1 + 8x_2 + s_2 = 12 \cdot 9.6 + 8 \cdot 0 + 2.8 = 118$
>
> objective function: $z = 175x_1 + 100x_2 = 175 \cdot 9.6 + 100 \cdot 0$
> $= 1{,}680$

If any of these substitutions failed, we would know to stop and find our error.

Step 8 *Determine whether the current possible solution maximizes the objective function.* The last row contains no negative entries, so we don't need to pivot again, and the problem is finished. Some linear programming problems involve only one pivot, some involve two, and some involve more. This problem involved only one pivot.

Step 9 *Interpret the final solution.* Recall that x_1 measures the number of coats made per day, x_2 measures the number of vests, s_1 measures the number of slack machine hours, and s_2 measures the amount of unused leather. Thus, the Leather Factory should make 9.6 coats per day and no vests. (If workers spend all day making coats, they will make 9.6 coats. They'll finish the tenth coat the next day.) This will result in no unused machine hours and 2.8 square feet of unused leather and will generate a maximum revenue of $1,680 per day. These profit considerations indicate that the Leather Factory should increase the cost of its leather vests if it wants to sell vests. ■

8.3 Exercises

In Exercises 1–4, determine where to pivot.

1.

x_1	x_2	s_1	s_2	z	
5	0	3	1	0	12
3	0	19	0	10	22
−21	1	−48	0	0	19

2.

x_1	x_2	s_1	s_2	z	
1	31	0	9	0	7.4
0	20	0	15	1	4.9
0	−1	1	0	0	20

3.

x_1	x_2	s_1	s_2	s_3	z	
19	3	0	0	1	0	5
32	7	0	1	0	0	93
−55	8	1	0	0	0	8
−21	32	0	0	0	1	96

4.

x_1	x_2	s_1	s_2	s_3	z	
0	1	7	0	62	0	5
0	0	5	1	5	0	93
1	0	0	0	0	0	78
0	0	−9	1	−1	1	96

In Exercises 5–8, (a) determine where to pivot, (b) pivot, and (c) determine the solution that corresponds to the resulting matrix.

5.

x_1	x_2	s_1	s_2	z	
1	2	1	0	0	3
4	1	0	1	0	2
−6	−4	0	0	1	0

6.

x_1	x_2	s_1	s_2	z	
8	2	1	0	0	4
5	1	0	1	0	3
−2	4	0	0	1	0

7.

x_1	x_2	s_1	s_2	z	
1	1	5	0	0	3
3	0	1	1	0	12
−6	0	−2	0	1	6

8.

x_1	x_2	s_1	s_2	z	
0	2	1	5	0	8
1	3	0	9	0	6
0	−4	0	2	2	12

Use the simplex method to solve Exercises 9–16. (Some of these exercises were started in Section 8.2.)

9. Five friends, all of whom are experienced bakers, form a company that will make bread and cakes and sell them to local restaurants and specialty stores. Each loaf of bread requires 50 minutes of labor and ingredients costing $0.90 and can be sold for $1.20 profit. Each cake requires 30 minutes of labor and ingredients costing $1.50 and can be sold for $4.00 profit. The partners agree that no one will work more than eight hours a day. Their financial resources do not allow them to spend more than $190 per day on ingredients. How many loaves of bread and how many cakes should they make each day to maximize their profit? What is the maximum profit? Will this leave any extra time or money? If so, how much?

10. A craftswoman produces two products, floor lamps and table lamps. Production of one floor lamp requires 75 minutes of her labor and materials that cost $25. Production of one table lamp requires 50 minutes of labor and materials that cost $20. The craftswoman wants to work no more than 40 hours each week, and her financial resources allow her to pay no more than $900 for materials each week. If she can sell as many lamps as she can make and if her profit is $40 per floor lamp and $32 per table lamp, how many floor lamps and how many table lamps should she make each week to maximize her weekly profit? What is that maximum profit? Will this leave any unused time or money? If so, how much?

11. A furniture manufacturing firm makes sofas and chairs, each of which is available in several styles. Each sofa, regardless of style, requires eight hours in the upholstery shop and four hours in the carpentry shop and can be sold for a profit of $450. Each chair requires six hours in the upholstery shop and 3.5 hours in the carpentry shop and can be sold for a profit of $375. There are nine people working in the upholstery shop and five in the carpentry shop, each of whom can work no more than 40 hours per week. How many sofas and how many chairs should the firm make each week to maximize its profit? What is that maximum profit? Would this leave any extra time in the upholstery shop or the carpentry shop? If so, how much?

12. City Electronics Distributors handles two lines of televisions, the Packard and the Bell. The company purchases up to $57,000 worth of television sets from the manufacturers each month, which it stores in its 9,000-cubic-foot warehouse. The Packards come in 36-cubic-foot packing crates, and the Bells come in 30-cubic-foot crates. The Packards cost City Electronics $200 each and can be sold to a retailer for a $200 profit, while the Bells cost $250 each and can be sold for a $260 profit. How many sets should City Electronics order each month to maximize profit? What is that maximum profit? Would this leave any unused storage space or money? If so, how much?

13. J & M Winery makes two jug wines, House White and Premium White, which it sells to restaurants. House White is a blend of 75% French colombard grapes and 25% sauvignon blanc grapes, and Premium White is 75% sauvignon blanc grapes and 25% French colombard grapes. J & M also makes a Sauvignon Blanc, which is 100% sauvignon blanc grapes. Profit on the House White is $1.00 per liter, profit on the Premium White is $1.50 per liter, and profit on the Sauvignon Blanc is $2.00 per liter. This season, J & M can obtain 30,000 pounds of French colombard grapes and 20,000 pounds of sauvignon blanc grapes. It takes 2 pounds of grapes to make 1 liter of wine. If J & M can sell all that it makes, how many liters of House White, of Premium White, and of Sauvignon Blanc should J & M prepare to maximize profit? What is that maximum profit? Would this leave any extra grapes? If so, what amount?

14. J & M Winery makes two jug wines, House Red and Premium Red, which it sells to restaurants. House Red is a blend of 20% cabernet sauvignon grapes and 80% gamay grapes. Premium Red is 60% cabernet sauvignon grapes and 40% gamay grapes. J & M also makes a Cabernet Sauvignon, which is 100% cabernet sauvignon grapes. Profit on the House Red is $0.90 per liter, profit on the Premium Red is $1.60 per liter, and profit on the Cabernet Sauvignon is $2.50 per liter. This season, J & M can obtain 30,000 pounds of gamay grapes and 22,000 pounds of cabernet sauvignon grapes. It takes 2 pounds of grapes to make 1 liter of wine. If J & M can sell all that it makes, how many liters of House Red, of Premium Red, and of Cabernet Sauvignon should J & M prepare to maximize profit? What is that maximum profit? Would this leave any extra grapes? If so, what amount?

15. Pete's Coffees sells two blends of coffee beans: Smooth Sipper and Kona Blend. Smooth Sipper is composed of equal amounts of Kona, Colombian, and Arabian beans, and Kona Blend is one-half Kona beans and one-half Colombian beans. Profit on the Smooth Sipper is $3 per pound, while profit on the Kona Blend is $4 per pound. Each day, the shop can obtain 100 pounds of Kona beans, 200 pounds of Colombian beans, and 200 pounds of Arabian beans. It uses those beans only in the two blends. If it can sell all that it makes, how many pounds of Smooth Sipper and of Kona Blend should Pete's Coffees prepare each day to maximize profit? What is that maximum profit? Would this leave any extra beans? If so, what amount?

16. Pete's Coffees sells two blends of coffee beans: African Blend and Major Thompson's Blend. African Blend is one-half Tanzanian beans and one-half Ethiopian beans, and Major Thompson's Blend is one-quarter Tanzanian beans and three-quarters Colombian beans. Profit on the African Blend is $4.25 per pound, while profit on Major Thompson's Blend is $3.50 per pound. Each day, the shop can obtain 300 pounds of Tanzanian beans, 200 pounds of Ethiopian beans, and 450 pounds of Colombian beans. It uses those beans only in the two blends. If it can sell all that it makes, how many pounds of African Blend and of Major Thompson's Blend should Pete's Coffees prepare each day to maximize profit? What is that maximum profit? Would this leave any extra beans? If so, what amount?

Answer the following questions using complete sentences and your own words.

Concept Questions

17. Compare and contrast the geometric method of linear programming with the simplex method of linear programming. Give advantages and disadvantages of each.

18. Compare and contrast the simplex method of linear programming with Karmarkar's method of linear programming. Give advantages and disadvantages of each.

19. Explain how the maximal value of the objective function can be found without looking at the number in the lower right corner of the final simplex matrix.

20. Explain why the entries in the last row have had their signs changed, but the entries in all of the other rows have not.

21. Why not solve all linear programming problems with the geometric method, and never use the simplex method?

22. Find at least one exercise in Section 8.1 that could not be solved with the methods of Sections 8.2 and 8.3, and explain why it could not be solved with those methods.

23. Select an assigned exercise in this section and do that problem with the geometric method.

WARNING: Not all problems can be done with the geometric method.

History Questions

24. Who invented the simplex method? What motivated its invention?

Technology and Simplex Method

The simplex method has two major advantages over the geometric method: It will work with problems that have more than two independent variables, and it is amenable to the use of a computer or a graphing calculator. Frequently, real-world linear programming problems have so many constraints and variables that the calculations involved in pivoting can be overwhelming if done by hand. This difficulty is eliminated by use of either the Amortrix computer software introduced in Chapter 7 or a graphing calculator.

EXAMPLE 2 Our friend the craftsman now owns his own shop. He still makes coffee tables and end tables, but he makes each in three different styles: antique, art deco, and modern. The amount of labor and the cost of the materials required by each product, along with the profit they generate, are shown in Figure 8.35. The craftsman now has two employees, each of whom can work up to 40 hours a week. The craftsman himself frequently has to work more than 40 hours a week, but he will not allow himself to exceed 50 hours a week. His new bank loan allows him to spend up to $4,000 a week on materials. How many coffee tables and end tables of each style should he make each week to maximize his profit?

	Style	Hours of Labor	Cost of Materials	Profit
Coffee Table	antique	6.00	$230	$495
Coffee Table	art deco	6.25	$220	$500
Coffee Table	modern	5.00	$190	$430
End Table	antique	5.25	$125	$245
End Table	art deco	5.75	$120	$250
End Table	modern	4.5	$105	$250

Figure 8.35

Solution **Independent Variables**

x_1 = number of antique coffee tables

x_2 = number of art deco coffee tables

x_3 = number of modern coffee tables

x_4 = number of antique end tables

x_5 = number of art deco end tables

x_6 = number of modern end tables

Constraints

C_1: hours worked $\leq 2 \cdot 40 + 50 = 130$

$$\begin{pmatrix}\text{antique}\\ \text{coffee}\\ \text{table}\\ \text{hours}\end{pmatrix} + \cdots + \begin{pmatrix}\text{modern}\\ \text{end}\\ \text{table}\\ \text{hours}\end{pmatrix} \leq 130$$

$$6x_1 + 6.25x_2 + 5x_3 + 5.25x_4 + 5.75x_5 + 4.5x_6 \leq 130$$

$$6x_1 + 6.25x_2 + 5x_3 + 5.25x_4 + 5.75x_5 + 4.5x_6 + s_1 = 130$$

C_2: money spent $\leq 4{,}000$

$$\begin{pmatrix}\text{antique}\\ \text{coffee}\\ \text{table}\\ \text{money}\end{pmatrix} + \cdots + \begin{pmatrix}\text{modern}\\ \text{end}\\ \text{table}\\ \text{money}\end{pmatrix} \leq 4{,}000$$

$$230x_1 + 220x_2 + 190x_3 + 125x_4 + 120x_5 + 105x_6 \leq 4{,}000$$

$$230x_1 + 220x_2 + 190x_3 + 125x_4 + 120x_5 + 105x_6 + s_2 = 4{,}000$$

where s_1 = unused hours and s_2 = unused money.

Objective Function

maximize z = profit

$$= \begin{pmatrix}\text{antique coffee}\\ \text{table profit}\end{pmatrix} + \cdots + \begin{pmatrix}\text{modern end}\\ \text{table profit}\end{pmatrix}$$

$$= 495x_1 + 500x_2 + 430x_3 + 245x_4 + 250x_5 + 250x_6$$

$$-495x_1 - 500x_2 - 430x_3 - 245x_4 - 250x_5 - 250x_6 + 0s_1 + 0s_2 + 1z = 0$$

The first simplex matrix:

$$\begin{array}{c} \begin{matrix} x_1 & x_2 & x_3 & x_4 & x_5 & x_6 & s_1 & s_2 & z & \end{matrix} \\ \begin{bmatrix} 6.00 & 6.25 & 5.00 & 5.25 & 5.75 & 4.50 & 1 & 0 & 0 & 130 \\ 230 & 220 & 190 & 125 & 120 & 105 & 0 & 1 & 0 & 4{,}000 \\ -495 & -500 & -430 & -245 & -250 & -250 & 0 & 0 & 1 & 0 \end{bmatrix} \end{array}$$

Its corresponding possible solution:

$(x_1, x_2, x_3, x_4, x_5, x_6, s_1, s_2) = (0, 0, 0, 0, 0, 0, 130, 4000)$ with $z = 0$

At this point, you would enter the first simplex matrix into the computer or graphing calculator. You would divide to determine where to pivot.

$$\begin{array}{c} \begin{matrix} x_1 & x_2 & x_3 & x_4 & x_5 & x_6 & s_1 & s_2 & z & \end{matrix} \\ \begin{bmatrix} 6.00 & \boxed{6.25} & 5.00 & 5.25 & 5.75 & 4.50 & 1 & 0 & 0 & 130 \\ \boxed{230} & \boxed{220} & \boxed{190} & \boxed{125} & \boxed{120} & \boxed{105} & \boxed{0} & \boxed{1} & \boxed{0} & \boxed{4000} \\ -495 & \boxed{-500} & -430 & -245 & -250 & -250 & 0 & 0 & 1 & 0 \end{bmatrix} \end{array} \begin{array}{l} \leftarrow 130/6.25 = 20.8 \\ \leftarrow 4000/220 \approx 18.2 \\ \\ \end{array}$$

After determining that the appropriate pivot is row two, column two, you would instruct the computer or graphing calculator to divide row two by 220. After completing this pivot and performing all further pivots, check that your answer is reasonable by substituting it into the constraints and the objective function. ■

Decimal Answers In some linear programming problems, decimal answers make sense. In the example on the previous page, x_1 measures the number of antique coffee tables made each week. A decimal answer such as 4.5 would make sense here, because the fifth table would be halfway done at the end of the week and finished at the beginning of the next week.

In some linear programming problems, decimal answers do not make sense. If x_1 measured the number of loaves of bread a bakery made per day, then x_1 would have to be a whole number. A baker cannot make a portion of a loaf of bread. There are specific linear programming methods for problems in which the solution must be a whole number, but such methods are beyond the scope of this book. For our purposes, we will round off decimal answers to the nearest whole number when a decimal answer would not be reasonable.

Exercises

Use Amortrix or a graphing calculator to solve the linear programming problems in Exercises 25–28.

25. *Objective Function:*
maximize $z = 25x_1 + 53x_2 + 18x_3 + 7x_4$

Constraints:
$3x_1 + 2x_2 + 5x_3 + 12x_4 \leq 28$
$4x_1 + 5x_2 + x_3 + 7x_4 \leq 32$
$x_1 + 7x_2 + 9x_3 + 10x_4 \leq 25$

26. *Objective Function:*
maximize $z = 275x_1 + 856x_2 + 268x_3 + 85x_4$

Constraints:
$5.2x_1 + 9.8x_2 + 7.2x_3 + 3.7x_4 \leq 33.6$
$3.9x_1 + 5.3x_2 + 1.4x_3 + 2.5x_4 \leq 88.3$
$5.2x_1 + 7.7x_2 + 4.6x_3 + 4.6x_4 \leq 24.7$

27. *Objective Function:*
maximize $z = 37x_1 + 19x_2 + 53x_3 + 49x_4$

Constraints:
$6.32x_1 + 7.44x_2 + 8.32x_3 + 1.46x_4 + 9.35x_5 \leq 63$
$8.36x_1 + 5.03x_2 + x_3 + 5.25x_5 \leq 32$
$1.14x_1 + 9.42x_2 + 9.39x_3 + 10.42x_4 + 9.32x_5 \leq 14.7$

28. *Objective Function:*
maximize $z = 17x_1 + 26x_2 + 85x_3 + 63x_4 + 43x_5$

Constraints:
$72x_1 + 46x_2 + 73x_3 + 26x_4 + 54x_5 \leq 185$
$37x_1 + 84x_2 + 45x_3 + 83x_4 + 85x_5 \leq 237$

Use Amortrix or a graphing calculator to solve Exercises 29–36. Interpret each solution; that is, explain the values of each slack variable and z in addition to answering the question.

29. Finish Example 2 from this section.

30. The Leather Factory has one sewing machine with which it makes coats and vests. Each coat requires 50 minutes on the sewing machine and uses 12 square feet of leather. Each vest requires 30 minutes on the sewing machine and uses 8 square feet of leather. The sewing machine is available 8 hours a day, and the Leather Factory can obtain 118 square feet of leather a day. The coats generate a profit of \$175 each, and the vests generate a profit of \$100

Kathleen Olson

	Labor	Cost of Ingredients	Profit
Nine-Grain Bread	50 min. per loaf	$1.05 per loaf	$0.60 per loaf
Sourdough Bread	50 min. per loaf	$0.95 per loaf	$0.70 per loaf
Chocolate Cake	35 min. per cake	$2.00 per cake	$2.50 per cake
Poppyseed Cake	30 min. per cake	$1.55 per cake	$2.00 per cake
Blueberry Muffins	15 min. per dozen	$1.60 per dozen	$16.10 per dozen
Apple-Cinnamon Muffins	15 min. per dozen	$1.30 per dozen	$14.40 per dozen

Figure 8.36

each. Find the daily production level that would yield maximum profit.

31. The five friends have been so successful with their baking business that they have opened their own shop—the Five Friends Bakery. The bakery sells two types of breads: Nine-Grain and Sourdough; two types of cakes: Chocolate and Poppyseed; and two types of muffins: Blueberry and Apple-Cinnamon. Production data are given in Figure 8.36. The five partners have hired four workers. No one (neither partner nor worker) works more than 40 hours a week. Their business success (together with their new bank loan) allows them to spend up to $2,000 per week on ingredients. What weekly production schedule should they follow to maximize their profit?

32. Fiat Lux, Inc., manufactures floor lamps, table lamps, and desk lamps. Production data are given in Figure 8.37.

	Wood Shop	Metal Shop	Electrical Shop	Testing	Profit
Floor Lamps	20 min.	30 min.	15 min.	5 min.	$55
Table Lamps	25 min.	15 min.	12 min.	5 min.	$45
Desk Lamps	30 min.	10 min.	11 min.	4 min.	$40

Figure 8.37

There are seven employees in the wood shop, five in the metal shop, three in the electrical shop, and one in testing; each works no more than 40 hours per week. If Fiat Lux, Inc., can sell all the lamps it produces, how many should it produce each week?

33. A furniture manufacturing firm makes sofas, love seats, easy chairs, and recliners. Each piece of furniture is constructed in the carpentry shop, then upholstered, and finally coated with a protective coating. Production data are given in Figure 8.38.

	Carpentry	Upholstery	Coating	Profit
Sofas	4 hours	8 hours	20 min.	$450
Love Seats	4 hours	7.5 hours	20 min.	$350
Easy Chairs	3 hours	6 hours	15 min.	$300
Recliners	6 hours	6.5 hours	15 min.	$475

Figure 8.38

Fifteen people work in the upholstery shop, nine work in the carpentry shop, and one person applies the protective coating, and each person can work no more than 40 hours per week. How many pieces of furniture should the firm manufacture each week to maximize its profit?

34. City Electronics Distributors handles two lines of televisions, the Packard and the Bell. The Packard comes in three sizes: 18-inch, 24-inch, and rear projection; the Bell comes in two sizes: 18-inch and 30-inch. City Electronics purchases up to $57,000 worth of television sets from manufacturers each month and stores them in its 9,000-cubic-foot warehouse. Storage and financial data are given in Figure 8.39. How many sets should City Electronics order each month to maximize profit?

	Size	Cost	Profit
Packard 18"	30 cubic ft	$150	$175
Packard 24"	36 cubic ft	$200	$200
Packard Rear Projection	65 cubic ft	$450	$720
Bell 18"	28 cubic ft	$150	$170
Bell 30"	38 cubic ft	$225	$230

Figure 8.39

Projects

35. J & M Winery makes two jug wines, House White and Premium White, and two higher-quality wines, Sauvignon Blanc and Chardonnay, which it sells to restaurants, supermarkets, and liquor stores. House White is a blend of 40% French colombard grapes, 40% chenin blanc grapes, and 20% sauvignon blanc grapes, while Premium White is 75% sauvignon blanc grapes and 25% French colombard grapes. J & M's Sauvignon Blanc is 100% sauvignon blanc grapes, and its Chardonnay is 90% chardonnay grapes and 10% chenin blanc grapes. Profit on the House White is $1.00 per liter, profit on the Premium White is $1.50 per liter, profit on the Sauvignon Blanc is $2.25 per liter, and profit on the Chardonnay is $3.00 per liter. This season, J & M can obtain 30,000 pounds of French colombard grapes, 25,000 pounds of chenin blanc grapes, and 20,000 pounds each of sauvignon blanc grapes and chardonnay grapes. It takes 2 pounds of grapes to make 1 liter of wine. If the company can sell all it makes, how many liters of its various products should J & M prepare to maximize profit?

36. J & M Winery makes two jug wines, House Red and Premium Red, and two higher-quality wines, Cabernet Sauvignon and Zinfandel, which it sells to restaurants, supermarkets, and liquor stores. House Red is a blend of 20% pinot noir grapes, 30% zinfandel grapes, and 50% gamay grapes. Premium Red is 60% cabernet sauvignon grapes and 20% each pinot noir and gamay grapes. J & M's Cabernet Sauvignon is 100% cabernet sauvignon grapes, and its Zinfandel is 85% zinfandel grapes and 15% gamay grapes. Profit on the House Red is $0.90 per liter, profit on the Premium Red is $1.60 per liter, profit on the Zinfandel is $2.25 per liter, and profit on the Cabernet Sauvignon is $3.00 per liter. This season, J & M can obtain 30,000 pounds each of pinot noir, zinfandel, and gamay grapes and 22,000 pounds of cabernet sauvignon grapes. It takes 2 pounds of grapes to make 1 liter of wine. If the company can sell all that it makes, how many liters of its various products should J & M prepare to maximize profit?

Chapter 8 Review

TERMS

bounded versus unbounded regions
constraint
corner point
corner principle
dependent variable
geometric method
independent variable
linear equation
linear inequality
linear programming model
objective function
optimization
pivoting
region of possible solutions
region of solutions
simplex method
slack variable
system of linear inequalities

REVIEW EXERCISES

Graph the region of solutions of the inequalities in Exercises 1–4.

1. $4x - 5y > 7$
2. $3x + 4y < 10$
3. $3x + 6y \leq 9$
4. $6x - 8y \geq 12$

Graph the region of solutions of the systems of inequalities in Exercises 5–10, and find all corner points.

5. $8x - 4y < 10$
$3x + 5y \geq 7$
6. $x - 5y < 7$
$3x + 2y > 6$

7. $5x - y > 8$
$x \geq -3$
8. $6x + 4y \leq 7$
$y < 4$
9. $x - y \geq 7$
$5x + 3y \leq 9$
$x \geq 0$
$y \geq 0$
10. $3x - 2y \leq 12$
$5x + 3y \leq 15$
$x \geq 0$
$y \geq 0$

Use the geometric method to solve the linear programming problems in Exercises 11 and 12.

11. Mowson Audio Co. makes stereo speaker assemblies. It purchases speakers from a speaker manufacturing firm and installs them in its own cabinets. Mowson's model 110 speaker assembly, which sells for \$200, has a tweeter and a midrange speaker. The model 330 assembly, which sells for \$350, has two tweeters, a midrange speaker, and a woofer. Mowson currently has in stock 90 tweeters, 60 midrange speakers, and 44 woofers. How many speaker assemblies should Mowson make to maximize its income? What is that maximum income?
12. The Stereo Guys store sells two lines of personal stereos, the Sunny and the Iwa. The Sunny comes in a 12-cubic-foot box and can be sold for a \$220 profit, and the Iwa comes in an 8-cubic-foot box and can be sold for a \$200 profit. The Stereo Guys marketing department estimates that at least 600 personal stereos can be sold each month and that, due to Sunny's reputation for quality, the demand for the Sunny unit is at least twice that for the Iwa. If the Stereo Guys warehouse has 12,000 cubic feet of space available for personal stereos, how many Sunnys and Iwas should the company stock each month to maximize profit? What is that maximum profit?
13. Use the simplex method to solve Exercise 11. Interpret the solution; that is, explain the values of each slack variable and z in addition to answering the question.
14. Mowson Audio Co. (see Exercise 11) has introduced a new speaker assembly. The new model 220 has a tweeter, a midrange speaker, and a woofer and sells for \$280. If Mowson Audio has 140 tweeters, 90 midrange speakers, and 66 woofers in stock, how many speaker assemblies should it make to maximize its income? Interpret the solution in addition to answering the question.

Answer the following questions using complete sentences.

Concept and History Questions

15. Who invented the geometric method of linear programming?
16. Who invented the simplex method of linear programming? Under what circumstances was it invented?
17. How does the simplex method differ from the geometric method of linear programming?
18. How does Karmarkar's method differ from the simplex method of linear programming?

Exponential and Logarithmic Functions

9

In this chapter, we will study a wide assortment of seemingly unrelated topics:

- the prediction of the size of human and animal populations
- the effect of inflation on home prices
- the decay of radioactive materials
- radiocarbon dating of archeological artifacts
- the Richter scale (which rates earthquakes)
- the decibel scale (which rates volumes of sounds)

We will answer questions such as the following:

- When will the world's population exceed the earth's ability to support it?
- How long will it take for my home to double its value?
- How long must the radioactive waste from a nuclear plant be stored?
- What does radiocarbon dating say about the authenticity of the Dead Sea Scrolls and the Shroud of Turin?
- How does a 7.0 earthquake compare with an 8.0 earthquake?

These diverse topics and questions are all related mathematically. Each involves the use of exponential and logarithmic functions.

In Sections 9.0A and 9.0B, we will discuss the algebra of exponential and logarithmic functions and the related use of calculators. In Sections 9.1 through 9.3, we will discuss the topics and questions listed above.

9.0A Review of Exponentials and Logarithms

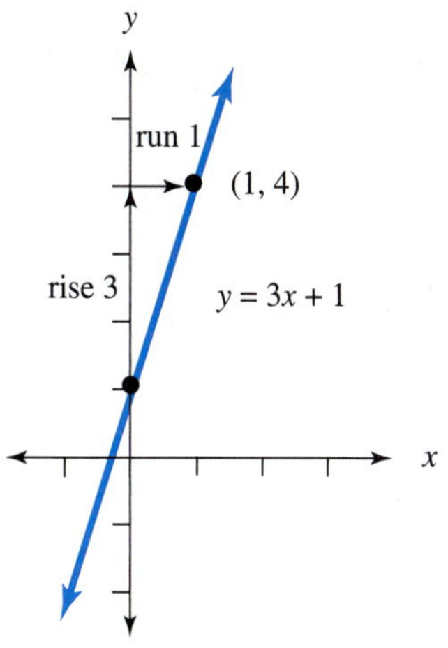

Figure 9.1

Functions

An equation is said to be a **function** if to each value of x there corresponds one and only one value of y. For example, consider the equation $y = 3x + 1$; $x = 1$ corresponds to one and only one value of y (in particular, to $y = 4$). Other values of x also correspond to one and only one value of y, so the equation $y = 3x + 1$ is a function. When the value of y depends in this manner on the value of x, we say that y is a function of x; x is called the **independent variable,** and y is called the **dependent variable.**

Our equation $y = 3x + 1$ is called a **linear function,** because it is a function whose graph is a line. The graph of the equation $y = 3x + 1$ is shown in Figure 9.1. The slope of this line is 3; for every one-unit increase in x, the value of y increases by three units [slope = rise/run = (change in y)/(change in x)].

Exponential Functions

A function in which x appears only in the exponent is called an **exponential function.** For example, $y = 2^x$ is an exponential function. Again, y is the dependent variable and x is the independent variable.

> **Exponential Function**
>
> An equation of the form $y = b^x$, where b is a positive constant, $(b \neq 1)$ is called an **exponential function.** The positive constant b is called the **base.**

As Example 1 shows, the graph of an exponential function is quite different from the graph of a linear function.

EXAMPLE 1 Sketch the graph of $y = 2^x$.

Solution We can graph this exponential function by finding several ordered pairs, as in Figure 9.2.

The graph of all ordered pairs (x, y) satisfying the equation $y = 2^x$ is the curve shown in Figure 9.3. ■

x	$y = 2^x$	Ordered Pair (x, y)
3	$2^3 = 8$	(3, 8)
2	$2^2 = 4$	(2, 4)
1	$2^1 = 2$	(1, 2)
0	$2^0 = 1$	(0, 1)
−1	$2^{-1} = \frac{1}{2}$	$\left(-1, \frac{1}{2}\right)$
−2	$2^{-2} = \frac{1}{4}$	$\left(-2, \frac{1}{4}\right)$
−3	$2^{-3} = \frac{1}{8}$	$\left(-3, \frac{1}{8}\right)$

Figure 9.2

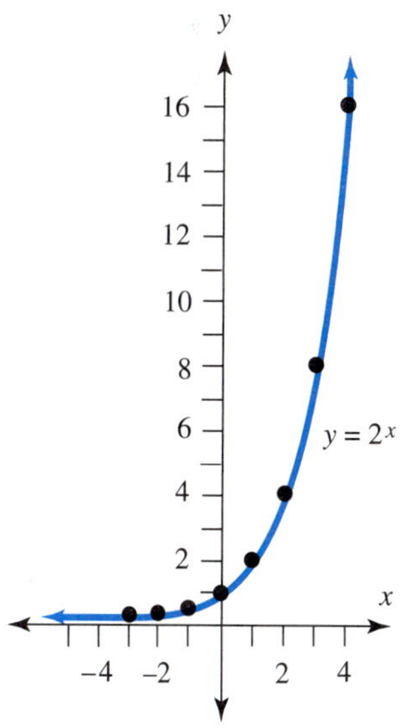

Figure 9.3

The graph of $y = 2^x$ is not a straight line. Typically, the graph of an exponential function is nearly horizontal over a large portion of the x-axis and then turns upward rather abruptly, increasing without bound. Because of this curvature, the slope is not constant as it is with a linear function; the vertical change is minimal where the graph is nearly horizontal, whereas the vertical change is quite extreme where the graph is steep.

In this chapter, we will study various applications of exponential functions. These applications include the prediction of the size of human and animal populations, inflation, the decay of radioactive materials, the dating of archeological artifacts, the Richter scale (which rates earthquakes), and the decibel scale (which rates volumes of sounds).

Rational and Irrational Numbers

Real numbers are either rational or irrational. A real number that is a terminating decimal, such as $\frac{3}{4} = 0.75$, or a repeating decimal, such as $\frac{2}{3} = 0.6666\ldots$, is called a ***ratio*nal number** because it can be written as the *ratio* of two integers—that is, as a fraction. A real number that is neither a terminating decimal nor a repeating decimal is called an **irrational number.** The irrational numbers with which you are probably most familiar are square roots. For example, $\sqrt{2} = 1.414213562\ldots$ and $\sqrt{23} = 4.795831523\ldots$ are irrational.

One important irrational number is π. Whenever the circumference of any circle is divided by its diameter, the resulting number is *always* $\pi = 3.141592654\ldots$. Another "famous" irrational number is $e = 2.71828182\ldots$. The number e is used in most of the applications of mathematics in this chapter, including population prediction, inflation, radioactive decay, and the dating of archeological artifacts.

The Natural Exponential Function

The **natural exponential function** is $y = e^x$. Calculations involving e^x will be performed on a calculator. Some scientific calculators have a button that is labeled "e^x" on the button itself and "ln" or "ln x" above the button; to calculate e^1 with such a calculator, press

1 $\boxed{e^x}$

Other scientific calculators have a button that is labeled "ln" or "ln x" on the button itself and "e^x" above the button; to calculate e^1 with such a calculator, press either

1 $\boxed{\text{2nd}}$ $\boxed{e^x}$ or 1 $\boxed{\text{shift}}$ $\boxed{e^x}$ or 1 $\boxed{\text{INV}}$ $\boxed{e^x}$

In the future, we will simply write $\boxed{e^x}$ to refer to each of these sequences of keystrokes.

Most graphing calculators have a button that is labeled "LN" on the button itself and "e^x" above the button; to calculate e^1 with such a calculator, press

$\boxed{\text{2nd}}$ $\boxed{e^x}$ 1 $\boxed{\text{ENTER}}$

EXAMPLE 2 Use a calculator to find the following values:

a. e **b.** $1/e$ **c.** e^2 **d.** $1/e^2$ **e.** e^3

f. Using the above values, sketch the graph of $y = e^x$.

Solution *Finding the Values*

a. $e = e^1 = 2.718281828\ldots$

1 [e^x]
[2nd] [e^x] 1 [ENTER]

b. $1/e = 0.367879441\ldots$

1 [÷] 1 [e^x] [=]
1 [÷] [2nd] [e^x] 1 [ENTER]

An alternative way to find $1/e$ is to realize that $1/e = e^{-1}$.

1 [+/−] [e^x]
[2nd] [e^x] [(−)] 1 [ENTER]

c. $e^2 = 7.389056099\ldots$

2 [e^x]
[2nd] [e^x] 2 [ENTER]

d. $1/e^2 = 0.135335283\ldots$

(scientific calculator)	1 [÷] 2 [e^x] [=] or 2 [+/−] [e^x]
(graphing calculator)	1 [÷] [2nd] [e^x] 2 [ENTER] or [2nd] [e^x] [(−)] 2 [ENTER]

e. $e^3 = 20.08553692\ldots$

(scientific calculator)	3 [e^x]
(graphing calculator)	[2nd] [e^x] 3 [ENTER]

f. *Sketching the graph of* $y = e^x$ First, express the above values as ordered pairs, rounding off y-coordinates to the nearest tenth (which is sufficient for plot plotting), as shown in Figure 9.4. Then plot the ordered pairs and connect them with a smooth curve, as shown in Figure 9.5. ■

x	$y = e^x$	Ordered Pair (x, y)
3	$e^3 \approx 20.1$	(3, 20.1)
2	$e^2 \approx 7.4$	(2, 7.4)
1	$e^1 \approx 2.7$	(1, 2.7)
0	$e^0 = 1$	(0, 1)
−1	$e^{-1} \approx 0.4$	(−1, 0.4)
−2	$e^{-2} \approx 0.1$	(−2, 0.1)

Figure 9.4

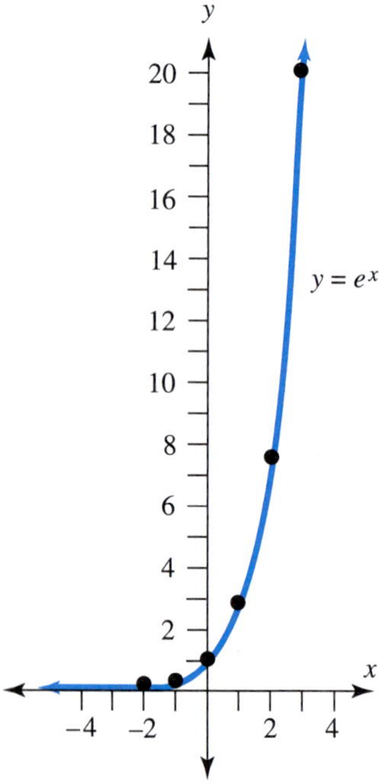

Figure 9.5

The graph of $y = e^x$ has the same shape as the graph of $y = 2^x$. Both are nearly horizontal over a portion of the x-axis and then turn upward, increasing without bound; both graphs are entirely above the x-axis. Exponential functions (functions of the form $y = b^x$) always have these qualities. The independent variable x can have any value, but the dependent variable y can have only a positive value.

Because $y = e^x$ is an exponential function, its graph is not a straight line, and the slope is not constant; going from $x = -2$ to $x = -1$, the slope is

$$m \approx \frac{0.4 - 0.1}{-1 - (-2)} = 0.3 \quad \text{using the ordered pairs generated in Figure 9.4}$$

However, going from $x = 2$ to $x = 3$, the slope is

$$m \approx \frac{20.1 - 7.4}{3 - 2} = 12.7$$

This means that e^x grows much more quickly when x is a larger number. When we are dealing with population growth, this trait can become quite disturbing; the larger the population, the faster it grows. The use of exponential functions to analyze population growth is explored in Section 9.1.

Logarithms

Ten raised to what power gives 100? 1,000? 346? The answers to the first and second questions are easily found; the solution to $10^x = 100$ is $x = 2$, and the solution to $10^x = 1{,}000$ is $x = 3$. However, the solution to $10^x = 346$ is not so obvious! We could safely say that because $10^2 = 100$ and $10^3 = 1{,}000$, and because 346 is between 100 and 1,000, the solution to $10^x = 346$ must be between 2 and 3. To be more accurate would be difficult.

The idea of finding the exponent to which a number must be raised in order to get some particular number is the central concept of a **logarithm.** The question "3 raised to what power gives 9?" is the same question as "What is the logarithm (base 3) of 9?" (The answer to either question is 2.) Algebraically, the question "3 raised to what power gives 9?" is written as "$3^x = 9$," and the question "What is the logarithm (base 3) of 9?" is written "$\log_3 9 = x$." When we say $x = \log_3 9$, we are saying that x is the exponent to which we must raise 3 in order to get 9.

Logarithm Definition

$$\log_b u = v \text{ (or the } \textbf{logarithm} \text{ of } u)$$

means the same as

$$b^v = u$$

In either equation, b is called the **base** and must be a positive number.

This definition of a logarithm allows us to rewrite a logarithmic equation as an exponential equation. The next example uses this fact.

EXAMPLE 3 **a.** Find v if $v = \log_2 8$. **b.** Find u if $\log_5 u = -2$. **c.** Find b if $\log_b 9 = 2$.

Solution

a. $v = \log_2 8$ can be rewritten as $2^v = 8$. By inspection, $v = 3$.

b. $\log_5 u = -2$ can be rewritten as $5^{-2} = u$. Therefore, $u = 1/5^2 = 1/25$.

c. $\log_b 9 = 2$ can be rewritten as $b^2 = 9$. By inspection, $b = \pm 3$. However, the base b of a logarithm is required to be positive. Thus, $b = 3$. ■

HISTORICAL NOTE

Leonhard Euler
1707–1783

Leonhard Euler (pronounced "Oiler") was probably the most versatile and prolific writer in the history of mathematics. Euler produced over 700 books and papers, many of which were created during the last seventeen years of his life while he was totally blind. He wrote an average of 800 pages of mathematics each year. Euler's writings spanned pure mathematics, astronomy, annuities, life expectancy, lotteries, and music.

Euler was the son of a Swiss minister and mathematician who studied under Jacob Bernoulli. Euler was schooled in theology and mathematics and was to become a minister. However, his mathematical curiosity and ability prevailed, and his father eventually allowed him to concentrate on mathematics. He was awarded his master's degree at the age of sixteen! At the age of twenty-six, at the invitation of Catherine I, he was appointed to the Academy of St. Petersburg, a major center of scientific research. Later he was invited to Berlin by Frederick the Great.

Euler's phenomenal memory has been the subject of many legends. Having memorized Virgil's *Aeneid,* Euler could recite the first and last lines on any page of his copy. When confronted by two students who asked him to settle a disputed mathematical calculation in the fiftieth place, Euler successfully calculated the true result in his head!

One of Euler's skills was the creation of notations that were useful and compact. His writings were very popular, and the notations he introduced have endured. He introduced the notation $f(x)$ for a function, and he was the first to use π for 3.14159 . . . , i for $\sqrt{-1}$, and e for 2.718281828. . . .

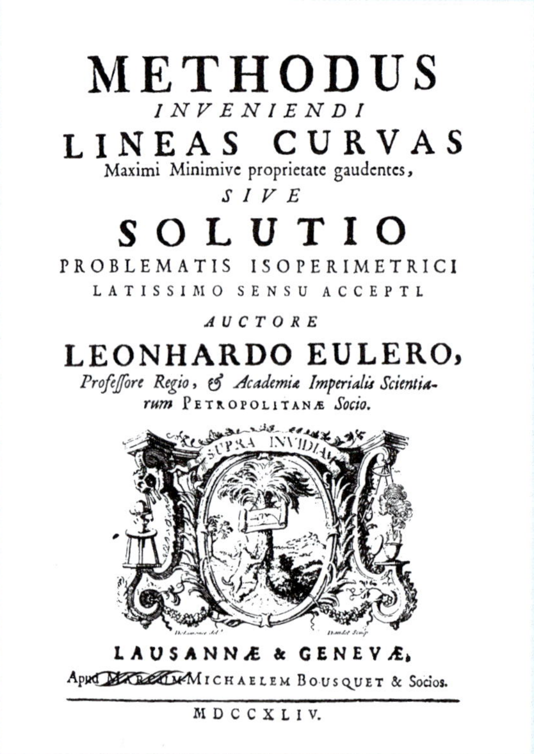

METHODUS
INVENIENDI
LINEAS CURVAS
Maximi Minimive proprietate gaudentes,
SIVE
SOLUTIO
PROBLEMATIS ISOPERIMETRICI
LATISSIMO SENSU ACCEPTI.
AUCTORE
LEONHARDO EULERO,
Professore Regio, & Academiæ Imperialis Scientiarum PETROPOLITANÆ *Socio.*

LAUSANNÆ & GENEVÆ,
Apud MARCUM-MICHAELEM BOUSQUET & Socios.
MDCCXLIV.

Common Logarithms

$\log_{10} x$, in which 10 is the base, is called the **common logarithm.** Base 10 logarithms are used so commonly that the base is understood to be 10 when no base is written. Thus, log 346 is an abbreviation for $\log_{10} 346$; log 346 is the exponent to which 10 must be raised in order to get 346.

Common Logarithm Definition

$$y = \log x \qquad (\text{or} \quad y = \log_{10} x)$$

means the same as

$$10^y = x$$

EXAMPLE 4 Find the following values by using a calculator:

a. log 346 **b.** log(0.82) **c.** log(−10) **d.** $\log(10^5)$

Solution **a.** The [log] button on your calculator is used to find the common logarithm of a number. Simply enter the number you want to find the logarithm of and press [log]. Some calculators require that you press [log] first and then enter the desired number.

$$\log 346 = 2.539076099\ldots$$

346 [log]

[LOG] 346 [ENTER]

Therefore, $10^{2.539076099} \approx 346$. (This is the answer to the earlier question "Ten raised to what power gives 346?" We had estimated that the power must be between 2 and 3; now we know that the power 2.539076099 gives 346.)

b. $\log(0.82) = -0.086186147\ldots$

0.82 [log]

[LOG] 0.82 [ENTER]

Therefore, $10^{-0.086186147} \approx 0.82$

c. $\log(-10) =$ ERROR

10 [+/−] [log]

[LOG] [(] [(−)] 10 [)] [ENTER]

Your calculator cannot find the common logarithm of −10 because negative numbers do not have logarithms! To see why, let $x = \log(-10)$. The corresponding exponential equation would be $10^x = -10$. This equation has no solution, because 10 raised to *any* power *always* gives a positive result; it is impossible for 10^x to equal −10. Consequently, log(−10) is undefined.

d. $\log(10^5) = \log 100{,}000 = 5$

(scientific calculator)	10 [y^x] 5 [=] [log]
(graphing calculator)	[LOG] [(] 10 [^] 5 [)] [ENTER]

■

The values of the independent variable x in the common logarithm function $y = \log x$ must always be positive, as discussed in part (c) above, and the values of the dependent variable y can be either positive or negative, as shown in parts (a) and (b). This situation is exactly the reverse of that for an exponential function, where the independent variable can have any value and the dependent variable must be positive.

It should be noted that the explanation given in part (c) of Example 4 is an oversimplification; technically, it should be stated that negative numbers do not have logarithms that are *real numbers*. Negative numbers *do* have logarithms; however, the logarithm of a negative number is a complex (or imaginary) number. Some graphing calculators will compute the (complex) logarithm of a negative number; however, we will not study complex numbers in this textbook.

EXAMPLE 5 Sketch the graph of $y = \log x$.

Solution Rather than computing the value of y for specific values of x, recall that $y = \log x$ means the same as $x = 10^y$. This exponential form allows for easier computations than does the original logarithmic form. Therefore, we compute the value of x for specific integer values of y. See Figure 9.6. Now plot the ordered pairs and connect them with a smooth curve, as shown in Figure 9.7.

y	$x = 10^y$	(x, y)
1	$10^1 = 10$	(10,1)
0	$10^0 = 1$	(1,0)
−1	$10^{-1} = \frac{1}{10}$	$(\frac{1}{10}, -1)$
−2	$10^{-2} = \frac{1}{100}$	$(\frac{1}{100}, -2)$

Figure 9.6

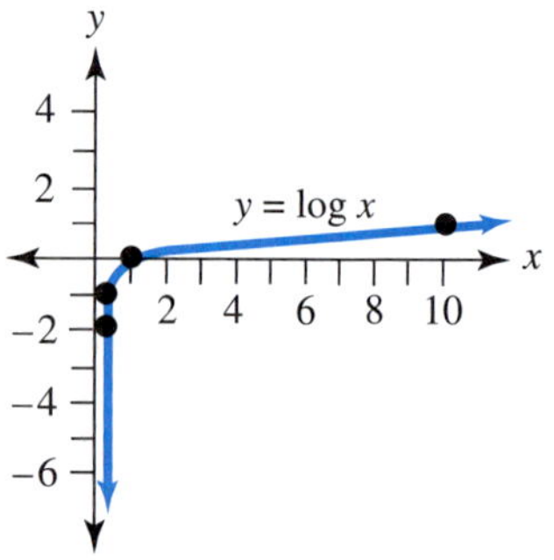

Figure 9.7 ■

Notice that when x is between 0 and 1, the graph rises very steeply and the slope is large. In this interval, small changes in x produce large changes in y; for example, as x changes from $x = \frac{1}{100}$ to $x = \frac{1}{10}$, y changes from $y = -2$ to $y = -1$. On the other hand, when x is greater than 1, the graph rises very slowly and the slope is small. Here, large changes in x produce very small changes in y; for example, as x changes from $x = 1$ to $x = 10$, y changes only from $y = 0$ to $y = 1$. Consequently, logarithms can be used to expand small variations and compress large ones (see Figure 9.8). This characteristic will be important in our study of the Richter scale and the decibel scale in Section 9.3.

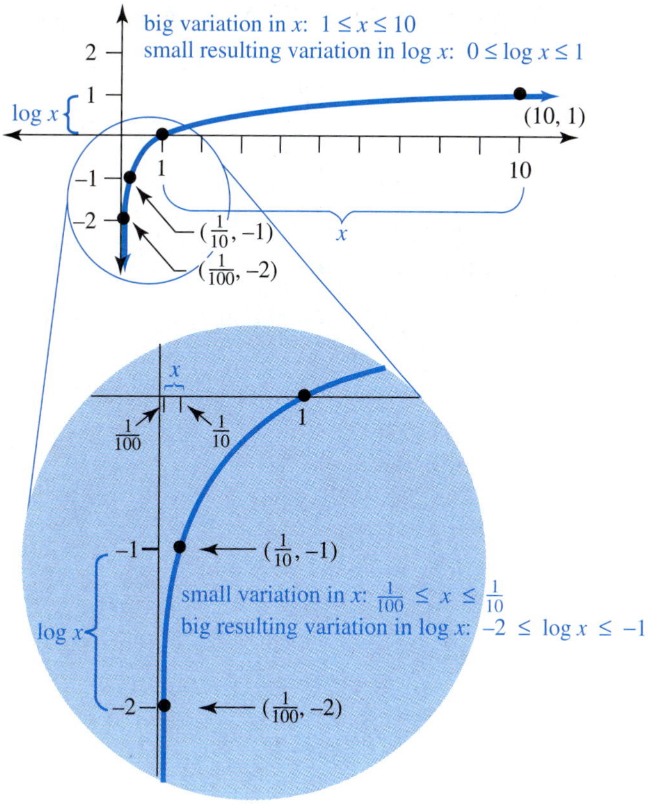

Figure 9.8

The Natural Logarithm Function

The number e is often used as the base of a logarithm. $\text{Log}_e\ x$ is called the **natural logarithm function** and is abbreviated $\ln x$. Thus, $\ln 2$ is an abbreviation for $\log_e 2$; $\ln 2$ is the exponent to which e must be raised in order to get 2.

> **Natural Logarithm Definition**
>
> $$y = \ln x \qquad (\text{or} \qquad y = \log_e x)$$
>
> means the same as
>
> $$e^y = x$$

The common logarithm (base 10) and the natural logarithm (base e) are the most frequently used types of logarithms.

EXAMPLE 6 Find the following values by using a calculator:

a. $\ln 5.2$ **b.** $\ln 0.4$ **c.** $\ln(-1.2)$ **d.** $\ln(e^2)$

Solution **a.** The $\boxed{\ln x}$ button on your calculator is used to find the natural logarithm of a number. Simply enter the number you want to find the logarithm of and press [ln x]. On a graphing calculator, you must press [LN] first and then enter the desired number.

$$\ln 5.2 = 1.648658626\ldots$$

5.2 [ln x]

[LN] 5.2 [ENTER]

Therefore, $e^{1.648658626} \approx 5.2$.

b. $\ln 0.4 = -0.916290731\ldots$

0.4 [ln x]

[LN] 0.4 [ENTER]

Therefore, $e^{-0.916290731} \approx 0.4$.

c. $\ln(-1.2) =$ ERROR

1.2 [+/−] [ln x]

[LN] [(] [(−)] 1.2 [)] [ENTER]

Your scientific calculator cannot find the natural logarithm of -1.2 because negative numbers do not have logarithms (in the real number system). If we let $y = \ln(-1.2)$, then the corresponding exponential equation would be $e^y = -1.2$. However, this equation has no solution (in the real number system) because e raised to any power always gives a positive result. Because we are working with real numbers only, $\ln(-1.2)$ is undefined.

d. To find $\ln(e^2)$, you must first calculate e^2: $e^2 = 7.389056099\ldots$. Now find the natural logarithm of this number.

$$\ln(e^2) = \ln(7.389056099\ldots) = 2$$

2 [e^x] [ln x]

This part gives e^2.

[LN] [( ] [2nd] [e^x] 2 [ )] [ENTER]

TI-83 users should not type the left parenthesis; it's automatically inserted.

9.0A Exercises

In Exercises 1–12, find the value of u, v, or b.

1. $v = \log_2 8$
2. $v = \log_2 16$
3. $v = \log_2(\frac{1}{4})$
4. $v = \log_2(\frac{1}{8})$
5. $\log_5 u = -2$
6. $\log_5 u = -3$
7. $\log_5 u = 0$
8. $\log_5 u = 1$
9. $\log_b 9 = 2$
10. $\log_b 4 = 2$
11. $\log_b 27 = -3$
12. $\log_b 25 = -2$

In Exercises 13–16, rewrite the logarithm as an exponential.

13. $K = \log_b H$
14. $W = \log_b(T + 1)$
15. $(G + 5) = \log_b(F - E)$
16. $\frac{D}{2} = \log_b C$

In Exercises 17–20, rewrite the exponential as a logarithm.

17. $b^T = S$
18. $b^{G+1} = H$
19. $b^{MN} = P - Q$
20. $b^{4-K} = L + 8$

In Exercises 21–46, use a calculator to find each value. Give the entire display of your calculator.

21. **a.** $e^{1.2}$ **b.** $10^{1.2}$
22. **a.** $e^{3.4}$ **b.** $10^{3.4}$
23. **a.** $2\,e^{0.08}$ **b.** $2(10)^{0.08}$
24. **a.** $3\,e^{0.05}$ **b.** $3(10)^{0.05}$
25. **a.** $\dfrac{1}{e^{1.5}}$ **b.** $\dfrac{1}{10^{1.5}}$
26. **a.** $\dfrac{1}{e^{2.5}}$ **b.** $\dfrac{1}{10^{2.5}}$
27. **a.** $\dfrac{4}{e^{0.24}}$ **b.** $\dfrac{4}{10^{0.24}}$
28. **a.** $\dfrac{12}{e^{0.98}}$ **b.** $\dfrac{12}{10^{0.98}}$
29. **a.** $\dfrac{e^{4.7}}{2\,e^{5.1}}$ **b.** $\dfrac{10^{4.7}}{2(10)^{5.1}}$
30. **a.** $\dfrac{5\,e^{0.48}}{e^{0.84}}$ **b.** $\dfrac{5(10)^{0.48}}{10^{0.84}}$
31. **a.** $\ln 2.15$ **b.** $\log 2.15$
32. **a.** $\ln 1.62$ **b.** $\log 1.62$
33. **a.** In 0.58 **b.** $\log 0.58$
34. **a.** $\ln 0.23$ **b.** $\log 0.23$
35. **a.** $\ln(e^{3.4})$ **b.** $\log(10^{3.4})$
36. **a.** $\ln(e^{1.28})$ **b.** $\log(10^{1.28})$
37. **a.** $\ln(4\,e^{0.02})$ **b.** $\log[4(10)^{0.02}]$
38. **a.** $\ln(3\,e^{0.56})$ **b.** $\log[3(10)^{0.56}]$
39. **a.** $e^{\ln 3.6}$ **b.** $10^{\log 3.6}$
40. **a.** $e^{\ln 5.8}$ **b.** $10^{\log 5.8}$
41. **a.** $e^{2\ln 5}$ **b.** $10^{2\log 5}$
42. **a.** $e^{4\ln 3}$ **b.** $10^{4\log 3}$
43. **a.** $\ln(e^2)$ **b.** $\log(e^2)$
44. **a.** $\ln(e^3)$ **b.** $\log(e^3)$
45. **a.** $\ln(10^{2.47})$ **b.** $\log(10^{2.47})$
46. **a.** $\ln(10^{1.81})$ **b.** $\log(10^{1.81})$
47. Put the following in numerical order, from smallest to largest: $e^{2.8}$, $\ln 2.8$, $10^{2.8}$, $\log 2.8$.
48. Put the following in numerical order, from smallest to largest: $e^{1.6}$, $\ln 1.6$, $10^{1.6}$, $\log 1.6$.

Answer the following questions using complete sentences and your own words.

Concept Questions

49. What is an exponential function?

50. What is a logarithmic function?

51. If $x > 0$, which value is larger, 10^x or $\log x$? Why?

52. If $x > 0$, which value is larger, e^x or $\ln x$? Why?

53. If $x > 1$, which value is larger, $\log x$ or $\ln x$? Why?

54. If $0 < x < 1$, which value is larger, $\log x$ or $\ln x$? Why?

55. If $x > 0$, which value is larger, e^x or 10^x? Why?

56. If $x < 0$, which value is larger, e^x or 10^x? Why?

History Question

57. Who invented the symbol e? What other symbols did this person introduce?

9.0B Review of Properties of Logarithms

Logarithms have several properties that can be used in simplifying complicated expressions and solving equations. We will use these properties extensively in the applications of logarithms and exponentials in this chapter. Perhaps the most important of these properties are the *Inverse Properties*.

The Inverse Properties

In Section 9.0A, we used the calculator to compute $\log(10^5) = 5$ and $\ln(e^2) = 2$ (Examples 4d and 6d). Note that in either case, the base of the logarithm is the same as the base of the exponential and that the logarithm appears to have "canceled" the exponential. Does this "canceling" always work? Does $\log_b(b^v) = v$ for all bases b and all numbers v?

Recall that by the definition of a logarithm, $\log_b u = v$ means the same as $b^v = u$. Since b^v is equal to u, we can substitute b^v for u in the equation $\log_b u = v$. This substitution results in $\log_b(b^v) = v$, which indicates that a base b logarithm will always "cancel" a base b exponential. *Cancel* is not really the right word here; "canceling" refers to what happens when you reduce a fraction. Instead, we say that a base b logarithm is the **inverse** of a base b exponential; whatever an exponential function does to a number, a logarithm undoes it and gives back the number.

Because $\log_b(b^x) = x$ for *any* base b, it is true for bases $b = 10$ and $b = e$. In other words, $\log_{10}(10^x) = x$ and $\log_e(e^x) = x$. Using the alternate notations, $\log(10^x) = x$ and $\ln(e^x) = x$.

Inverse Properties

$$\log(10^x) = x \qquad (\text{or } \log_{10}(10^x) = x)$$
$$\ln(e^x) = x \qquad (\text{or } \log_e(e^x) = x)$$

EXAMPLE 1 Simplify the following by using the Inverse Properties:

a. $\log(10^{3x})$ **b.** $\ln(e^{-0.012x})$ **c.** $\ln(10^{5x})$

Solution **a.** We are applying a base 10 logarithm to a base 10 exponential, so we can apply the Inverse Property $\log(10^x) = x$ and obtain

$$\begin{aligned}\log(10^{3x}) &= \log_{10}(10^{3x}) \\ &= 3x\end{aligned}$$

b. $$\begin{aligned}\ln(e^{-0.012x}) &= \log_e(e^{-0.012x}) \\ &= -0.012x \quad \text{using the Inverse Property } \ln(e^x) = x\end{aligned}$$

c. The Inverse Properties do *not* apply to $\ln(10^{5x}) = \log_e(10^{5x})$, because the logarithm and the exponential are different bases (base e and base 10, respectively). Thus, $\ln(10^{5x})$ cannot be simplified by using the Inverse Properties. ■

We have just seen that a logarithm "undoes" an exponential when the logarithm is applied to the exponential and the bases are the same. Will the same thing happen when an exponential is applied to a logarithm? Does $b^{(\log_b x)}$ simplify to just x?

Recall that, by the definition of a logarithm, $b^v = u$ means the same as $\log_b u = v$. Because $\log_b u$ is equal to v, we can substitute it for v in the equation $b^v = u$. This substitution results in $b^{(\log_b u)} = u$. This means that a base b exponential will always "undo" a base b logarithm; an exponential (of the same base) is the **inverse** of a logarithm.

Because $b^{(\log_b x)} = x$ for *any* base b, it is true for bases $b = 10$ and $b = e$. In other words, $10^{(\log_{10} x)} = x$ and $e^{(\log_e x)} = x$. Using the alternate notations, $10^{(\log x)} = x$ and $e^{(\ln x)} = x$.

Inverse Properties

$$10^{\log x} = x \qquad (\text{or } 10^{\log_{10} x} = x)$$
$$e^{\ln x} = x \qquad (\text{or } e^{\log_e x} = x)$$

EXAMPLE 2 Simplify the following.

a. $10^{\log(2x-3)}$ **b.** $e^{\ln(0.023x)}$

Solution **a.** We are raising 10 to a common log (base 10) power, so we can apply an Inverse Property:

$$10^{\log(2x-3)} = 2x - 3 \quad \text{using the Inverse Property } 10^{\log x} = x$$

b. We are raising e to a natural log (base e) power, so we can apply an Inverse Property:

$$e^{\ln(0.023x)} = 0.023x \qquad \text{using the Inverse Property } e^{\ln x} = x$$

■

Solving Exponential Equations

Recall that an exponential equation is one in which x appears only in the exponent. If the variable can be removed from the exponent, then the resulting equation can be solved in a manner that is familiar to you from algebra. To solve an exponential equation, apply a logarithm and use the properties of logarithms to remove the variable from the exponent.

EXAMPLE 3 Solve $10^x = 0.47$.

Solution We have a base 10 exponential, so we apply the common logarithm to both sides and simplify:

$$\begin{aligned} 10^x &= 0.47 \\ \log(10^x) &= \log(0.47) \\ x &= \log 0.47 \qquad \text{using the Inverse Property } \log(10^x) = x \\ x &= -0.327902142\ldots \end{aligned}$$

0.47 [log]

[LOG] 0.47 [ENTER]

✔ As a rough check, we know that $x = -0.33$ is between 0 and -1, so we should expect $10^x = 10^{-0.33}$ to be between $10^0 = 1$ and $10^{-1} = \frac{1}{10} = 0.1$. As a more accurate check, use your calculator to verify that $10^{-0.327902142} \approx 0.47$.

10 [y^x] 0.327902142 [+/−] [=]

10 [^] [(] [(−)] 0.327902142 [)] [ENTER]

■

EXAMPLE 4 Solve $5e^{0.01x} = 8$.

Solution Before we apply a logarithm, we must first "isolate" the exponential by dividing by 5:

$$5e^{0.01x} = 8$$

$$e^{0.01x} = \frac{8}{5}$$

$$\ln(e^{0.01x}) = \ln\frac{8}{5} \quad \text{taking ln of each side}$$

$$0.01x = \ln\frac{8}{5} \quad \text{using the Inverse Property } \ln(e^x) = x$$

$$x = \frac{\ln(8/5)}{0.01}$$

$$x = 47.00036292\ldots$$

8 [÷] 5 [=] [ln x] [÷] 0.01 [=]
[LN] [(] 8 [÷] 5 [)] [÷] 0.01 [ENTER]
TI-83 users should not type the left parenthesis; it's automatically inserted.

Note: A common mistake is to improperly apply the natural logarithm at the first step; that is, if $5e^{0.01x} = 8$, then $5\ \ln(e^{0.01x}) \neq \ln 8$. You should always isolate the exponential as your first step!

✔ Use your calculator to verify that $5e^{(0.01)(47.00036292)}$ is 8.

0.01 [×] 47.00036292 [=] [e^x] [×] 5 [=]
5 [×] [2nd] [e^x] [(] 0.01 [×] 47.00036292 [)] [ENTER]
TI-83 users should not type the left parenthesis; it's automatically inserted.

■

HISTORICAL NOTE

John Napier

1550–1617

The Granger Collection, NY

John Napier was a Scottish landowner and member of the upper class. As such, he had a great deal of leisure time, much of which he devoted to mathematics, politics, and religion.

As is the case today, scientists of Napier's time frequently had to multiply and divide large numbers. Of course, at that time no calculators (or even slide rules) existed; a scientist had to make all calculations by hand. Such work was tedious and prone to errors. Napier invented logarithms as a system that would allow the relatively easy calculation of products and quotients, as well as powers and roots.

With this system, the product of two numbers was calculated by finding the logarithms of those numbers in a book of tables and adding the results. Napier spent *twenty years* creating these tables. This procedure represented a real shortcut: Addition replaced multiplication, and addition is much simpler than multiplication when done by hand. This method of multiplying by adding was an especially useful application of what we know as the Multiplication-Becomes-Addition Property, $\log(A \cdot B) = \log A + \log B$. Napier did not invent the notation $\log_b x$; this notation was invented much later by Leonhard Euler. Napier did not use any notation in his writings on logarithms; he wrote everything out verbally.

Although Napier did not invent the decimal point, he is responsible for its widespread use. Napier's calculating system was very popular, and the logarithms in his tables were decimal numbers written with a decimal point.

A similar system was developed independently by Joost Bürgi, a Swiss mathematician and watchmaker. Napier is generally credited with the invention of logarithms, because he published his work before Bürgi.

As a member of the Scottish aristocracy, Napier was an active participant in local and national affairs. He was also a very religious man and belonged to the Church of Scotland. These two interests led to an interesting story about Napier.

In sixteenth-century Scotland, politics and religion were inexorably entwined. Mary, Queen of Scots, was a Roman Catholic; Elizabeth, Queen of England, was a Protestant; and both the Church of Scotland and the Church of England were Protestant churches.

Steps for Solving Exponential Equations

1. Isolate the exponential; that is, rewrite the problem in the form $e^A = B$.
2. Take the natural logarithm of each side.
3. Use the Inverse Property $\ln(e^x) = x$ to simplify.
4. Solve.
5. Use your calculator to check the answer.

(These steps also apply to solving a base 10 exponential equation. However, instead of applying the natural logarithm to each side, apply the common log.)

The Exponent-Becomes-Multiplier Property

A property of logarithms that is useful in solving exponential equations is the *Exponent-Becomes-Multiplier* Property.

Furthermore, Mary had a strong claim to the throne of England. Catholic factions wanted Scotland to form an alliance with France; Protestant factions wanted an alliance with England. After a series of power struggles, Mary was forced to abdicate her throne to her son, who then became James VI, King of Scotland. Mary was eventually beheaded for plotting against the English throne.

It was well known that James VI wished to succeed Elizabeth to the English throne. It was suspected that he had enlisted the help of Phillip II, King of Spain and a Catholic, to attain this goal. It was also suspected that James VI was arranging an invasion of Scotland by Spain. John Napier was a member of the committee appointed by the Scottish church to express its concern to James.

Napier was not content with expressing his concerns through the church. He wrote one of the earliest Scottish interpretations of the scriptures, *A Plaine Discovery of the Whole Revelation of Saint John,* which was clearly calculated to influence contemporary events. In this work, Napier urged James VI to see that "justice be done against the enemies of God's Church." It also declared, "Let it be your Majesty's continuall study to reform . . . your country, and first to begin at your Majesty's owne house, familie and court, and purge the same of all suspicion of Papists and Atheists and Newtrals."

This tract was widely read in Europe as well as in Scotland, and Napier earned a considerable reputation as a scholar and theologian. It has been suggested that the tract saved him from persecution as a warlock; Napier had previously been suspected of being in league with the devil.

Napier also invented several secret weapons for the defense of his faith and country against a feared Spanish invasion. These inventions included two kinds of burning mirrors designed to set fire to enemy ships at a distance and an armored chariot that would allow its occupants to fire in all directions. Whether any of these devices were ever constructed is not known.

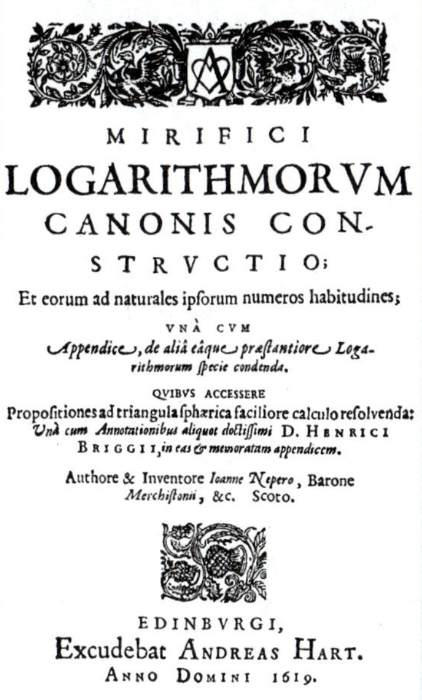

MIRIFICI
LOGARITHMORVM
CANONIS CON-
STRVCTIO;
Et eorum ad naturales ipsorum numeros habitudines;
VNÀ CVM
Appendice, de aliâ eâque præstantiore Logarithmorum specie condenda.
QVIBVS ACCESSERE
Propositiones ad triangula sphærica faciliore calculo resolvenda:
Unà cum Annotationibus aliquot doctissimi D. HENRICI BRIGGII, *in eas & memoratam appendicem.*
Authore & Inventore *Ioanne Nepero*, Barone *Merchistonii*, &c. Scoto.

EDINBVRGI,
Excudebat ANDREAS HART.
ANNO DOMINI 1619.

The Granger Collection, NY

Exponent-Becomes-Multiplier Property

$$\log(A^n) = n \cdot \log A$$
$$\ln(A^n) = n \cdot \ln A$$

When we take the logarithm of a base raised to a power, the exponent can be brought down in front of the logarithm, thus becoming a multiplier of the logarithm. Of course, these equations can be "reversed" to obtain $n \cdot \log A = \log(A^n)$ and $n \ln A = \ln(A^n)$; a multiplier of a logarithm can become an exponent.

This property seems rather strange—after all, exponents don't normally turn into multipliers. Or do they? Consider the Exponent Property:

$$(x^m)^n = x^{m \cdot n}$$

In the left-hand side of the equation, n is an exponent; in the right-hand side, n is a multiplier. This property is actually the basis of the Exponent-Becomes-Multiplier Property.

To see why $\log(A^n) = n \cdot \log A$, let $\log A = a$ and rewrite this logarithm as an exponential:

$$\log A = a$$
$$A = 10^a \quad \text{using the Common Logarithm Definition}$$

Raising each side to the nth power, we get

$$\begin{aligned} A^n &= (10^a)^n \\ &= 10^{an} \qquad \text{using the exponent law } (x^m)^n = x^{mn} \end{aligned}$$

Taking the log of each side and simplifying, we have

$$\begin{aligned} \log(A^n) &= \log(10^{an}) \\ &= an \qquad \text{using the Inverse Property } \log(10^x) = x. \end{aligned}$$

Finally, we substitute the original expression $\log A = a$ to get the desired result:

$$\log(A^n) = n \cdot \log A$$

A similar method can be used to obtain the natural logarithm version of the Exponent-Becomes-Multiplier Property. (See Exercise 50.)

EXAMPLE 5 **a.** Rewrite $\log(1.0125^x)$ so that the exponent is eliminated.
b. Rewrite $\ln(1 + 0.09/12)^x$ so that the exponent is eliminated.

Solution **a.** We are taking the log of an exponential, so we can apply the Exponent-Becomes-Multiplier Property:

$$\begin{aligned} \log(A^n) &= n \cdot \log A \qquad \text{Exponent-Becomes-Multiplier Property} \\ \log(1.0125^x) &= x \cdot \log 1.0125 \qquad \text{substituting 1.0125 for } A \text{ and } x \text{ for } n \end{aligned}$$

b. The Exponent-Becomes-Multiplier Property applies to natural logs too:

$$\begin{aligned} \ln(A^n) &= n \cdot \ln A \qquad \text{Exponent-Becomes-Multiplier Property} \\ \ln\left(1 + \frac{0.09}{12}\right)^x &= x \cdot \ln\left(1 + \frac{0.09}{12}\right) \qquad \text{substituting } \frac{0.09}{12} \text{ for } A \text{ and } x \text{ for } n \\ &= x \cdot \ln(1 + 0.0075) \\ &= x \cdot \ln(1.0075) \end{aligned}$$

■

EXAMPLE 6 Solve $1.03^x = 2$.

Solution This equation is different from those in the previous examples in that the base is neither 10 nor e, so the Inverse Properties cannot be used to simplify the equation. However, by taking the log (or ln) of each side, we can use the Exponent-Becomes-Multiplier Property of logarithms to change the exponent x into a multiplier, which gives us an easier equation to work with.

$$\begin{aligned} 1.03^x &= 2 \\ \log(1.03^x) &= \log 2 \qquad \text{taking the log of each side} \\ x \cdot \log 1.03 &= \log 2 \qquad \text{using the Exponent-Becomes-Multiplier Property } \log(A^n) = n \cdot \log A \\ x &= \frac{\log 2}{\log 1.03} \qquad \text{dividing by log 1.03} \\ x &= 23.44977225\ldots \end{aligned}$$

2 [log] [÷] 1.03 [log] [=]

[LOG] 2 [÷] [LOG] 1.03 [ENTER]

TI-83: [LOG] 2 [)] [÷] [LOG] 1.03 [)] [ENTER]
(A left parenthesis is automatically inserted when the [LOG] button is pushed.)

✔ Use your calculator to verify that $1.03^{23.44977225}$ is 2.

1.03 [y^x] 23.44977225 [=]

1.03 [^] 23.44977225 [ENTER]

■

The Division-Becomes-Subtraction Property

There are times when rewriting one logarithm as two logarithms, or vice versa, is advantageous. To this end, we have the following property.

Division-Becomes-Subtraction Property

$$\log \frac{A}{B} = \log A - \log B$$

$$\ln \frac{A}{B} = \ln A - \ln B$$

The logarithm of a quotient can be rewritten as the logarithm of the numerator *minus* the logarithm of the denominator, hence the title *Division-Becomes-Subtraction*. Of course, these equations can be "reversed" to obtain

$$\log A - \log B = \log \frac{A}{B} \qquad \text{and} \qquad \ln A - \ln B = \ln \frac{A}{B}$$

These properties seem rather strange; after all, division doesn't normally turn into subtraction. Or does it? Consider the exponent property

$$\frac{x^m}{x^n} = x^{m-n}$$

When we use exponents, division *does* become subtraction. Furthermore, logarithms are closely related to exponents, so it shouldn't be surprising that both logarithms and exponents have a Division-Becomes-Subtraction Property.

To see why $\log A - \log B = \log \frac{A}{B}$, let $\log A = a$ and $\log B = b$. Now rewrite each expression as an exponential:

$$A = 10^a \quad \text{and} \quad B = 10^b \quad \text{using the Common Logarithm Definition}$$

Dividing the first equation by the second yields the following:

$$\frac{A}{B} = \frac{10^a}{10^b}$$

$$\frac{A}{B} = 10^{a-b} \quad \text{using the exponent law } \frac{x^m}{x^n} = x^{m-n}$$

$$\log \frac{A}{B} = \log(10^{a-b}) \quad \text{taking the log of each side}$$

$$\log \frac{A}{B} = a - b \quad \text{using the Inverse Property } \log(10^x) = x$$

Finally, substitute the original expressions for a and b to get the desired result:

$$\log \frac{A}{B} = \log A - \log B$$

A similar method can be used to obtain the natural logarithm version of the Division-Becomes-Subtraction Property. (See Exercise 49.)

EXAMPLE 7 **a.** Rewrite $\log \frac{x}{3}$ so that the fraction is eliminated.
b. Rewrite $\log(2x) - \log 8$ as one logarithm.

Solution **a.** We have the log of a fraction, so we can apply the Division-Becomes-Subtraction Property.

$$\log \frac{A}{B} = \log A - \log B \quad \text{Division-Becomes-Subtraction Property}$$

$$\log \frac{x}{3} = \log x - \log 3 \quad \text{substituting } x \text{ for } A \text{ and 3 for } B$$

b. We are subtracting two logs, so we can apply the reverse of the Division-Becomes-Subtraction Property.

$$\log A - \log B = \log \frac{A}{B} \quad \text{Division-Becomes-Subtraction Property}$$

$$\log 2x - \log 8 = \log \frac{2x}{8} \quad \text{substituting } 2x \text{ for } A \text{ and 8 for } B$$

$$= \log \frac{x}{4} \quad \text{canceling}$$

■

In the last step of Example 6, we had

$$x = \frac{\log 2}{\log 1.03}$$

Frequently, students will attempt to rewrite this as

$$x = \log(2 - 1.03)$$

This is incorrect; it is the result of misremembering the Division-Becomes-Subtraction Property. That property says that $\log \frac{A}{B} = \log A - \log B$; it does *not* refer to $\frac{\log A}{\log B}$ or to $\log(A - B)$.

Common Errors

You might be tempted to "simplify" the expression $\dfrac{\log A}{\log B}$. However,

1. $\dfrac{\log A}{\log B} \neq \log A - \log B$
   (Subtracting two logs is not the same as dividing two logs.)
2. $\dfrac{\log A}{\log B} \neq \dfrac{A}{B}$
   (You cannot cancel the logs in the numerator and denominator of a fraction.)
3. $\dfrac{\log A}{\log B} \neq \log(A - B)$
   (The two logs on the left cannot be reduced into one log.)

The Multiplication-Becomes-Addition Property

We have seen that when logarithms are subtracted, they can be combined into one logarithm via the Division-Becomes-Subtraction Property. There is a similar property when logarithms are added.

Multiplication-Becomes-Addition Property

$$\log(A \cdot B) = \log A + \log B$$
$$\ln(A \cdot B) = \ln A + \ln B$$

The logarithm of a product can be rewritten as the sum of two logarithms, hence the title *Multiplication-Becomes-Addition.* Of course, these equations can be "reversed" to obtain $\log A + \log B = \log(A \cdot B)$ and $\ln A + \ln B = \ln(A \cdot B)$.

EXAMPLE 8

a. Rewrite $\log(3x)$ as two logarithms.
b. Rewrite $\ln(2x) + \ln 8$ as one simplified logarithm.

Solution

a. We have the log of a product, so we can apply the Multiplication-Becomes-Addition Property.

$$\log(A \cdot B) = \log A + \log B \quad \text{Multiplication-Becomes-Addition Property}$$
$$\log(3x) = \log 3 + \log x \quad \text{substituting 3 for } A \text{ and } x \text{ for } B$$

b. We are adding two logs, so we can apply the reverse of the Multiplication-Becomes-Addition Property.

$$\ln A + \ln B = \ln(A \cdot B) \quad \text{Multiplication-Becomes-Addition Property}$$

$$\ln(2x) + \ln 8 = \ln(2x \cdot 8) \quad \text{substituting } 2x \text{ for } A \text{ and } 8 \text{ for } B$$

$$= \ln(16x) \quad \text{multiplying}$$

■

Solving Logarithmic Equations

An equation in which x appears "inside" a logarithm is called a **logarithmic equation.** To solve an equation of this type, we apply an exponential function to each side and use the various properties that we have developed.

EXAMPLE 9 Solve $\log x = 5 + \log 3$.

Solution To solve for x, we have to eliminate the common logarithm. We can accomplish this by using the Inverse Property $10^{\log x} = x$. However, for this property to apply, the equation must contain only one logarithm. Therefore, our first step is to get all the log terms on one side and then combine them into one logarithm:

$$\log x = 5 + \log 3$$

$$\log x - \log 3 = 5$$

$$\log\left(\frac{x}{3}\right) = 5 \quad \text{using the Division-Becomes-Subtraction Property}$$

We can now apply the base 10 exponential function to each side—that is, exponentiate each side (base 10), and simplify:

$$10^{\log(x/3)} = 10^5$$

$$\frac{x}{3} = 100{,}000 \quad \text{using the Inverse Property } 10^{\log x} = x$$

$$x = 300{,}000$$

✔ Use your calculator to verify that $\log 300{,}000 = 5 + \log 3$ by computing the right and left sides separately and verifying that they are equal.

the left side: 300000 [log]
the right side: 5 [+] 3 [log] [=]

the left side: [LOG] 300000 [ENTER]
the right side: 5 [+] [LOG] 3 [ENTER]

■

pH: An Application of Logarithms

Chemists define pH by the formula $\text{pH} = -\log[H^+]$, where $[H^+]$ is the hydrogen ion concentration measured in moles per liter. Solutions with a pH of 7 are said to be neutral, whereas a pH less than 7 is classified as an acid and a pH greater than 7 is classified as a base.

EXAMPLE 10

a. An unknown substance has a hydrogen ion concentration of $[H^+] = 1.7 \times 10^{-5}$ moles per liter. Determine the pH and classify the substance as an acid or a base.

b. If a solution has a pH of 6.2, find the hydrogen ion concentration of the solution.

Solution

a. Because we know the hydrogen ion concentration, we substitute it into the pH formula and use a calculator.

$$\begin{aligned} pH &= -\log[H^+] && \text{pH formula} \\ &= -\log(1.7 \times 10^{-5}) && \text{substituting } 1.7 \times 10^{-5} \text{ for } [H^+] \\ &= 4.769551079 && \text{using a calculator} \\ &= 4.8 && \text{rounding to one decimal place} \end{aligned}$$

The pH of the substance is 4.8, so the substance is classified as an acid.

b. Because we know the pH, we substitute it into the pH formula and solve for $[H^+]$, the hydrogen ion concentration.

$$\begin{aligned} pH &= -\log[H^+] && \text{pH formula} \\ 6.2 &= -\log[H^+] && \text{substituting 6.2 for pH} \\ -6.2 &= \log[H^+] && \text{multiplying by } -1 \\ 10^{-6.2} &= 10^{\log[H^+]} && \text{exponentiating with base 10} \\ 10^{-6.2} &= [H^+] && \text{using the Inverse Property } 10^{\log x} = x \\ [H^+] &= 6.309573445 \times 10^{-7} && \text{using a calculator} \end{aligned}$$

The hydrogen ion concentration is 6.3×10^{-7} moles per liter. ■

Steps for Solving Logarithmic Equations

1. Get all the log terms on one side and all the nonlog terms on the other.
2. Combine the log terms into one term, using the **Division-Becomes-Subtraction Property**

$$\log\left(\frac{A}{B}\right) = \log A - \log B$$

and the **Multiplication-Becomes-Addition Property**

$$\log(A \cdot B) = \log A + \log B$$

3. Exponentiate each side (base 10).
4. Use the **Inverse Property** $10^{\log x} = x$ and simplify.
5. Solve.

(These steps also apply to solving an equation that contains natural logarithms. However, instead of exponentiating base 10, use the base e exponential and the inverse property $e^{\ln x} = x$ to simplify.)

For Those Who Have Read Section 6.8: Fractal Geometry

EXAMPLE 11 In Example 1 of Section 6.8 we found that the dimension of a Koch snowflake is the number d that satisfies the equation $3^d = 4$. We used a trial-and-error process to approximate the value of d because we were unable to solve this equation.

a. Use logarithms to solve the equation and find the exact value of d.
b. Use the result of part (a) to approximate d to the nearest hundredth.

Solution **a.** *Finding the exact value of d*

$$3^d = 4$$

$$\log(3^d) = \log 4 \quad \text{taking the log of each side}$$

$$d \cdot \log 3 = \log 4 \quad \text{using the Exponent-Becomes-Multiplier Property}$$

$$d = \frac{\log 4}{\log 3} \quad \text{dividing by log 3}$$

The exact value of d is

$$d = \frac{\log 4}{\log 3}$$

b. *Approximating d* Using a calculator, we get

$$d = 1.26185950\ldots \approx 1.26$$

Notice that this result agrees with that found in Example 1 of Section 6.8.

4 [LOG] [÷] 3 [LOG] [=]
[LOG] 4 [÷] [LOG] 3 [ENTER]
TI-83: [LOG] 4 [)] [÷] [LOG] 3 [)] [ENTER] (A left parenthesis is automatically inserted when the [LOG] button is pushed.)

Notice that in Example 11, you can use your calculator to check your answer by verifying that $3^{1.26185950\ldots}$ is 4.

3 [y^x] 1.26185950 [=]
3 [^] 1.26185950 [ENTER]

9.0B Exercises

In Exercises 1–8, simplify by using the Inverse Properties.

1. $\log(10^{7.5x})$
2. $\log(10^{-x})$
3. $\ln(e^{-0.058x})$
4. $\ln(e^{0.135x})$
5. $10^{\log(3x+1)}$
6. $10^{\log(3-5x)}$
7. $e^{\ln(8-x)}$
8. $e^{\ln(1.5x)}$

In Exercises 9–18, rewrite the given logarithm so that all products, quotients, and exponents are eliminated.

9. $\log(\frac{x}{5})$
10. $\ln(\frac{x}{8})$
11. $\ln(1.4x)$
12. $\log(10x)$
13. $\log(1.0625^x)$
14. $\ln(1.25^x)$
15. $\log(4x^3)$
16. $\ln(2x^5)$
17. $\ln(\frac{2x}{3})$
18. $\log(\frac{x^2}{7})$

In Exercises 19–28, rewrite as one simplified logarithm.

19. $\log(3x) - \log 6$
20. $\ln(2x) - \ln 8$
21. $\ln(5x) + \ln 2$
22. $\log(\frac{2x}{3}) + \log 6$
23. $\log x - \log 4 + \log 8$
24. $\ln x - \ln 5 + \ln 20$
25. $2\ln(3x) - \ln 9$
26. $3\log(2x) - \log 4$
27. $\log(9x) + \log(4x) - 2\log(6x)$
28. $\ln(4x) + \ln(16x) - 2\ln(8x)$

In Exercises 29–48, solve the given equation. (The solution of every other odd-numbered exercise is given in the back of the book; check the other exercises as shown in the text.)

29. a. $e^x = 0.25$
b. $10^x = 0.25$
30. a. $e^x = 0.65$
b. $10^x = 0.65$
31. a. $125\,e^{0.038x} = 250$
b. $125(10)^{0.038x} = 250$
32. a. $74\,e^{0.042x} = 148$
b. $74(10)^{0.042x} = 148$
33. a. $1{,}000\,e^{0.009x} = 1{,}500$
b. $1{,}000(10)^{0.009x} = 1{,}500$
34. a. $7{,}500\,e^{0.12x} = 10{,}000$
b. $7{,}500(10)^{0.12x} = 10{,}000$
35. a. $20\,e^{-0.015x} = 10$
b. $20(10)^{-0.015x} = 10$
36. a. $15\,e^{-0.005x} = 7.5$
b. $15(10)^{-0.005x} = 7.5$
37. a. $50\,e^{-0.0016x} = 40$
b. $50(10)^{-0.0016x} = 40$
38. a. $36\,e^{-0.0026x} = 27$
b. $36(10)^{-0.0026x} = 27$
39. a. $\log x = 0.12$
b. $\ln x = 0.12$
40. a. $\log x = 0.35$
b. $\ln x = 0.35$
41. a. $\log x = 1.85$
b. $\ln x = 1.85$
42. a. $\log x = 3.14$
b. $\ln x = 3.14$
43. a. $\log x + \log 4 = 1$
b. $\ln x - \ln 4 = 1$
44. a. $\log x + \log 1.4 = 2$
b. $\ln x - \ln 1.4 = 2$
45. a. $\log x = 1.8 + \log 3.6$
b. $\ln x = 1.8 - \ln 3.6$
46. a. $\log x = 2.1 + \log 1.7$
b. $\ln x = 2.1 - \ln 1.7$
47. a. $\log(2x) - \log 0.6 = 0.8$
b. $\ln(2x) + \ln 0.6 = 0.8$
48. a. $\log(3x) - \log 1.3 = 2.4$
b. $\ln(3x) + \ln 1.3 = 2.4$

49. Show that $\ln \frac{A}{B} = \ln A - \ln B$.

HINT: Use the same method as in the proof of $\log \frac{A}{B} = \log A - \log B$.

50. Show that $\ln(A^n) = n \cdot \ln A$.

HINT: Use the same method as in the proof of $\log(A^n) = n \cdot \log A$.

51. Show that $\log(A \cdot B) = \log A + \log B$.

HINT: Use a method similar to that used in Exercise 49.

52. Show that $\ln(A \cdot B) = \ln A + \ln B$.

Exercises 53–64 refer to Example 10.

53. For a certain fruit juice, $[H^+] = 3.0 \times 10^{-4}$. Determine the pH and classify the juice as acid or base.

54. For milk, $[H^+] = 1.6 \times 10^{-7}$. Determine the pH and classify milk as acid or base.

55. An unknown substance has a hydrogen ion concentration of 3.7×10^{-8} moles per liter. Determine the pH and classify the substance as acid or base.

56. An unknown substance has a hydrogen ion concentration of 2.4×10^{-5} moles per liter. Determine the pH and classify the substance as acid or base.

57. Fresh-brewed coffee has a hydrogen ion concentration of about 1.3×10^{-5} moles per liter. Determine the pH of fresh-brewed coffee.

58. Normally, human blood has a hydrogen ion concentration of about 3.98×10^{-8} moles per liter. Determine the normal pH of human blood.

59. When the pH of a person's blood drops below 7.4, a condition called *acidosis* sets in. Acidosis can result in death if pH reaches 7.0. What would the hydrogen ion concentration of a person's blood be at that point?

60. When the pH of a person's blood rises above 7.4, a condition called *alkalosis* sets in. Alkalosis can result in death if pH reaches 7.8. What would the hydrogen ion concentration of a person's blood be at that point?

61. You want to plant tomatoes in your backyard. Tomatoes prefer soil that has a pH range of 5.5 to 7.5. Using a testing kit, you determine the hydrogen ion concentration of your soil to be
   a. 3.5×10^{-7} moles per liter. Should you plant tomatoes? Why or why not?
   b. 3.5×10^{-4} moles per liter. Should you plant tomatoes? Why or why not?

62. You want to plant potatoes in your backyard. Potatoes prefer soil that has a pH range of 4.5 to 6.0. Using a testing kit, you determine the hydrogen ion concentration of your soil to be
   a. 2.1×10^{-7} moles per liter. Should you plant potatoes? Why or why not?
   b. 2.1×10^{-5} moles per liter. Should you plant potatoes? Why or why not?

63. Paprika prefers soil that has a pH range of 7.0 to 8.5. What range of hydrogen ion concentration does paprika prefer?

64. Spruce prefers soil that has a pH range of 4.0 to 5.0. What range of hydrogen ion concentration does spruce prefer?

For those who have read Section 6.8: Fractal Geometry. In Exercises 65–69, (a) Use logarithms to find the exact value of the dimension of the given fractal. (b) Use the result of part (a) to approximate d to the nearest hundredth. (The solution to part (b) is not given in the back of the book; instead, check as shown in the text.)

65. The Sierpinski gasket (See Exercise 9 in Section 6.8.)

66. The Sierpinski carpet (See Exercises 2 and 10 in Section 6.8.)

67. The Mitsubishi gasket (See Exercise 3 and 11 in Section 6.8.)

68. The Menger sponge (See Exercise 12 in Section 6.8.)

69. The square snowflake (See Exercises 4 and 13 in Section 6.8.)

Answer the following questions using complete sentences and your own words.

History Questions

70. Who invented logarithms?

71. What motivated the invention of logarithms?

72. Historically, how were logarithms used to calculate quotients?

    HINT: Alter the discussion of how logarithms were used to calculate products.

73. Was Napier in favor of or opposed to the attempt of James VI to succeed Elizabeth to the English throne? Why?

74. What possible catastrophe did Napier's political pursuits save him from?

9.1 Exponential Growth

In sections 9.0A and 9.0B, we reviewed the algebra used in connection with the exponential function $y = ae^{bx}$. In this and the following section, we will see how this function is used in topics as diverse as population growth, inflation, the decay of nuclear wastes, and radiocarbon dating. Each of these quantities grows (or decays) at a rate that reflects its current size: A larger population grows faster than a smaller one, and a larger amount of uranium decays faster than a smaller amount. This common relationship yields a powerful model that allows us to predict what the world population will be in the year 2100, how long it takes a home to double in value, and how long nuclear wastes must be stored. Or, looking back in time, we can use the model to determine the age of the Dead Sea Scrolls.

The first thing we need to do is determine why some things grow at a rate that reflects their current size.

EXAMPLE 1 A farmer has been studying the aphids in her alfalfa so that she can determine the best time to spray or release natural predators. In each of her two fields, she marked off one square foot of alfalfa and counted the number of aphids in that square foot. In her first field, she counted 100 aphids. One week later, the population had increased to 140; that is, its average growth rate is 40 aphids per week. In her second field, she counted 200 aphids. If the two fields provide the aphids with similar living conditions (favorable temperature, an abundance of food and space, and so on), what would be the most likely average growth rate for the second field?

Solution Forty aphids per week would not be realistic. Certainly, the larger population would have more births per week than the smaller one. The most likely average growth rate for the second field would be 80 aphids per week; this reflects the difference in size. Either population, then, would grow by the same percent; that is, the growth rate would be the same percentage of the population (see Figure 9.9).

	Average Growth Rate	Population	Growth as a Percent of (or Relative to) the Population
First Field	40 aphids/week	100 aphids	$\frac{40}{100} = 0.40 = 40\%$
Second Field	80 aphids/week	200 aphids	$\frac{80}{200} = 0.40 = 40\%$

Figure 9.9

■

Delta Notation

Delta (Δ) notation is frequently used to describe changes in quantities, with the symbol Δ meaning "change in." Thus, in Example 1, the change in population p could be written Δp, and the change in time t could be written Δt. Expressed in delta notation, the **average growth rate** of 40 aphids per week would be written as

$$\frac{\Delta p}{\Delta t} = \frac{40 \text{ aphids}}{1 \text{ week}}$$

while the **relative growth rate** (that is, the growth rate relative to, or as a percent of, the population) would be written as

$$\frac{\Delta p / \Delta t}{p} = \frac{40 \text{ aphids}/1 \text{ week}}{100 \text{ aphids}} = 40\% \text{ per week}$$

using dimensional analysis, as discussed in Appendix E, to cancel "aphids" with "aphids"

A **rate of change** always means one change divided by another change; the average growth rate

$$\frac{\Delta p}{\Delta t} = \frac{40 \text{ aphids}}{1 \text{ week}} = 40 \text{ aphids/week}$$

is an example of a rate of change.

EXAMPLE 2 Two years ago, Anytown, U.S.A., had a population of 30,000. Last year, there were 900 births and 300 deaths, for a net growth of 600. There was no immigration to or emigration from the town, so births and deaths were the only sources of population change. The average growth rate was therefore

$$\frac{\Delta p}{\Delta t} = \frac{600 \text{ people}}{1 \text{ year}}$$

What is the most likely average growth rate for Anytown this year, assuming no change in living conditions?

Solution Anytown should grow by the same percent each year, but by a percent of a changing amount. Because Anytown's population has increased slightly, the average growth rate would be a percent of a larger amount and would therefore be slightly higher. During the first year, the relative growth rate was

$$\frac{\Delta p/\Delta t}{p} = \frac{600 \text{ \sout{people}}/1 \text{ year}}{30{,}000 \text{ \sout{people}}} = 2\% \text{ per year}$$

using dimensional analysis to cancel "people" with "people"

In the following year, $\dfrac{\Delta p/\Delta t}{p}$ should also be 2%, but p is now 30,600:

$$\frac{\Delta p/\Delta t}{p} = 2\%$$

$$\frac{\Delta p/\Delta t}{30{,}600} = 2\%$$

$$\Delta p/\Delta t = 2\% \text{ of } 30{,}600 = 612$$

This means that the most likely average growth rate for Anytown this year would be 612 people/year (see Figure 9.10). It is important to realize that this is only a prediction; it might be a good prediction, but it is not a guarantee. For example, there might be fewer residents in their childbearing years this year than previously. This would have a significant effect on the average growth rate, as would changes in living conditions.

	Average Growth Rate $\Delta p/\Delta t$	**Population** p	**Relative Growth Rate** $\dfrac{\Delta p/\Delta t}{p}$
First Year	600 people/year	30,000 people	$\dfrac{600}{30{,}000} = 0.02 = 2\%$
Second Year	612 people/year	30,600 people	$\dfrac{612}{30{,}600} = 0.02 = 2\%$

Figure 9.10

In Examples 1 and 2, we saw that populations tend to grow at rates that reflect their size; it is most likely that a population will have an average growth rate that is a fixed percent of the current population size. In the next example, we will see that populations are not the only things that grow in this manner.

EXAMPLE 3 A house is purchased for \$140,000 in January 2003. A year later, the house next door is sold for \$149,800. The two houses are of the same style and size and are in similar condition, so they should have equal value. What would either house be worth in January 2005?

Solution Another increase of \$9,800, for a total value of \$159,600, would not be realistic. Experience has shown that more expensive homes tend to increase in value by a larger amount than do less expensive homes. The value would be more likely to increase by the same percent each year than by the same dollar amount. During 2003, the relative growth rate was

$$\frac{\Delta v/\Delta t}{v} = \frac{\$9{,}800/1 \text{ year}}{\$140{,}000} = 7\% \text{ per year}$$

In the following year, $\frac{\Delta v/\Delta t}{v}$ should also be 7% per year, but v is now \$149,800:

$$\frac{\Delta v/\Delta t}{v} = 7\% \text{ per year}$$

$$\frac{\Delta v/\Delta t}{149{,}800} = 7\% \text{ per year}$$

$$\Delta v/\Delta t = 7\% \text{ of } \$149{,}800 \text{ per year} = \$10{,}486 \text{ per year}$$

Therefore, the average growth rate in 2004 should be

$$\frac{\Delta v}{\Delta t} = \frac{\$10{,}486}{1 \text{ year}} = \$10{,}486 \text{ per year}$$

and the best guess of the value in January 2005 would be

$$\$149{,}800 + \$10{,}486 = \$160{,}286.$$

Remember that this is only a prediction, based on the assumption that the pattern set in 2003 will continue in 2004. This is a common assumption, unless economic forecasts indicate otherwise. ■

The Exponential Model

In the examples on populations, the average growth rate $\Delta p/\Delta t$ was a constant percent of the current population. In the example on inflation, the average growth rate $\Delta v/\Delta t$ was a constant percent of the current value. More generally, we are looking at situations in which *the average growth rate* $\Delta y/\Delta t$ *is a constant percent of the current value of* y, that is, in which

$$\frac{\Delta y}{\Delta t} = k \cdot y$$

(In the examples above, we used p for population and v for value in place of y, as a memory device, and k was a specific number, such as $7\% = 0.07$.) In calculus, it is determined that if a quantity y behaves in such a way that the average growth rate $\Delta y/\Delta t$ is a constant percent of the current value of y, then the size of y at some later time can be predicted using the equation

$$y = ae^{bt}$$

where t is time and a and b are constants. We used equations like this in Sections 9.0A and 9.0B, with specific constants such as 3 and 5 in place of a and b.

A **mathematical model** is an equation that is used to describe a real-world subject. The exponential equation

$$y = ae^{bt}$$

is called the **exponential model** because it is an exponential equation and is used to describe subjects such as populations and real estate appreciation. This model is extremely powerful—it allows us to predict the value of y, and its growth rate, with just a few data. Because many different quantities have a growth rate proportional to their size, the applications of this model are wide and varied. The model's use is illustrated in Examples 4–6, which are continuations of Examples 1–3.

EXAMPLE 4 One square foot of alfalfa has a population of 100 aphids. In one week, the population increases to 140; that is, its average growth rate is 40 aphids per week.

a. Develop the model that represents the population of aphids.
b. Predict the aphid population after three weeks.

Solution **a.** *Developing the model* As we discovered in Examples 1 and 2, the rate at which a population changes is a constant percent of the size of the population (as long as the population's growth is not limited by a lack of food, space, or other constraints). Thus, we can use the model $y = ae^{bt}$ or, with a more appropriate letter, $p = ae^{bt}$. To develop this model for our situation, we need to find the constants a and b.

Step 1 *Write the given information as two ordered pairs (t, p).* At the beginning of the experiment (at "time 0"), there were 100 aphids; that is, when $t = 0, p = 100$:

$$(t, p) = (0, 100)$$

After one week, there were 140 aphids, that is, when $t = 1, p = 140$:

$$(t, p) = (1, 140)$$

Step 2 *Substitute the first ordered pair into the model $p = ae^{bt}$ and simplify:*

$$\begin{aligned}(t, p) &= (0, 100)\\ 100 &= ae^{b \cdot 0}\\ 100 &= ae^{0}\\ 100 &= a \cdot 1\\ a &= 100\end{aligned}$$

Because a is a constant, we can rewrite our model $p = ae^{bt}$ as $p = 100\,e^{bt}$. To totally develop this model to fit our situation, we need to determine the value of the remaining constant b.

Step 3 *Substitute the second ordered pair into the model $p = 100\,e^{bt}$ and simplify:*

$$\begin{aligned}(t, p) &= (1, 140)\\ 140 &= 100\,e^{b \cdot 1}\\ 140 &= 100\,e^{b}\\ \frac{140}{100} &= e^{b}\\ 1.40 &= e^{b}\\ \ln 1.40 &= \ln e^{b}\\ \ln 1.40 &= b \quad \text{using the Inverse Property } \ln(e^x) = x\\ b &\approx 0.3364722366\end{aligned}$$

1.40 [ln x]
[LN] 1.40 [ENTER]

We have now developed the model $p = ae^{bt}$ to fit our situation. After substituting $a = 100$ and $b = 0.3364722366$, we get

$$p = 100\, e^{0.3364722366t}$$

where t is the number of weeks after the beginning of the experiment. Notice that a is the initial size of the population and that b is somewhat close to the relative growth rate of 40% = 0.40. The values of a and b always have these characteristics. The graph of the model is shown in Figure 9.11.

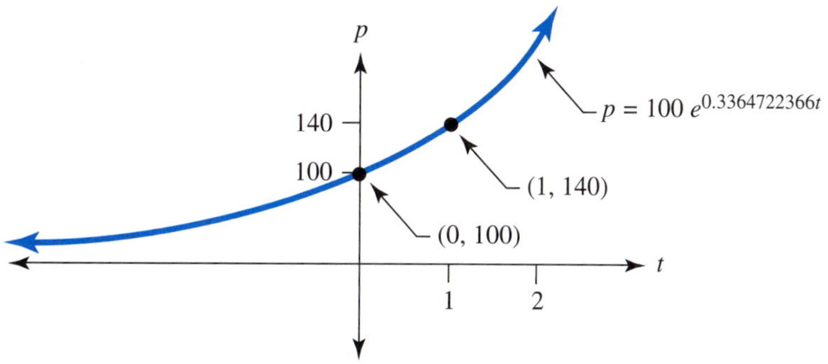

Figure 9.11
A graph showing the population of aphids

Our model will be most accurate if we do not round off $b = 0.3364722366$ at this point—that is, if we use all the decimal places that the calculator displays (your calculator might display more or fewer decimal places than shown here). One way to do this is to store the number in your calculator's memory. To do so, press

[STO] or [x→M] or [M in]
TI-81/85/86 [STO→] [B] [ENTER]
TI-80/82/83 [STO→] [ALPHA] [B] [ENTER]
Casio [→] [ALPHA] [B] [EXE]

when the number is on your calculator's screen.

b. *Predicting the population after three weeks* Now that the model has been developed, predictions can be made. At the end of three weeks, the aphid population would probably be as follows:

$$p = 100\, e^{(0.3364722366)(3)}$$
$$= 274.3999\ldots$$
$$\approx 274 \quad \text{rounding off to the nearest whole number of aphids}$$

Therefore, at the end of three weeks, we would expect 274 aphids.

> RCL [×] 3 [=] [e^x] [×] 100 [=]
>
> 100 [2nd] [e^x] [(] [ALPHA] [B] [×] 3 [)] [ENTER]
>
> TI-83 users should not type the left parenthesis—it's automatically inserted.

■

EXAMPLE 5 In the year 2003, Anytown, U.S.A., had a population of 30,000. In the following year, there were 900 births and 300 deaths, for a net growth of 600. There was no immigration to or emigration from the town, so births and deaths were the only sources of population change.

a. Develop the model that represents Anytown's population.
b. Predict Anytown's population in the year 2009.

Solution **a.** *Developing the model* As we discovered, the rate at which a population changes is a constant percent of the size of the population. Thus, we can use the model $p = ae^{bt}$. To develop this model for our situation, we need to find the constants a and b.

Step 1 *Write the given information as two ordered pairs* (t, p). For ease in calculating, let 2003 be $t = 0$; then 2004 is $t = 1$ and 2009 is $t = 6$. Also, we will count the population in thousands. The ordered pairs are (0, 30) and (1, 30.6).

Step 2 *Substitute the first ordered pair into the model* $p = ae^{bt}$ *and simplify*:

$$(t, p) = (0, 30)$$
$$30 = ae^{b \cdot 0}$$
$$30 = ae^{0}$$
$$30 = a \cdot 1$$
$$a = 30$$

As a shortcut, recall that a is always the initial population.

Step 3 *Substitute the second ordered pair into the model* $p = 30e^{bt}$ *and simplify*:

$$(t, p) = (1, 30.6)$$
$$30.6 = 30\, e^{b \cdot 1}$$
$$\frac{30.6}{30} = e^{b}$$
$$1.02 = e^{b}$$
$$\ln 1.02 = \ln e^{b}$$
$$\ln 1.02 = b \quad \text{using the Inverse Property } \ln(e^{x}) = x$$
$$b \approx 0.0198026273$$

Store this value of b in your calculator's memory.

✔ Notice that the value of b is close to the relative growth rate of $2\% = 0.02$, as it should be.

Thus, our model for Anytown's growth is

$$p = 30\, e^{0.0198026273t}$$

where t is years beyond 2003 and p is population in thousands.

b. *Predicting Anytown's population in the year 2009* Now that the model has been determined, predictions can be made. In the year 2009, $t = 6$ and

$$p = 30\, e^{(0.0198026273)(6)}$$
$$\approx 33.78487258$$

RCL ⊠ 6 = e^x ⊠ 30 =

30 2nd e^x (ALPHA B ⊠ 6) ENTER

TI-83 users should not type the left parenthesis—it's automatically inserted.

The population would probably be 33,785. ■

At this point, you might ask, "Why can't I just substitute the relative growth rate for b rather than use the second ordered pair to actually calculate b? After all, b is close to the relative growth rate." You can, but the resulting calculation will *not* necessarily be close to the true figure. In Example 5, if we use the relative growth rate for b, the resulting calculation is off by 40 people. Using the relative growth rate for b is equivalent to rounding off b; whenever you round off early, you lose accuracy.

Furthermore, while b is always somewhat close to the relative growth rate, it is not necessarily extremely close to that rate. In Example 4 on aphids, b was 0.3364722, and the relative growth rate was somewhat close at $40\% = 0.40$. In Example 5 on Anytown, b was 0.0198026273, and the relative growth rate was extremely close at $2\% = 0.02$. In general, b is always close to the relative growth rate, and the smaller these two numbers are, the closer they are.

In the previous two examples, we saw how to use the exponential model to predict the size of a population at some future date. The model can also be used to determine the time at which a quantity will have grown to any specified value.

EXAMPLE 6 A house is purchased for \$140,000 in January 2003. A year later, the house next door is sold for \$149,800. The two houses are of the same style and size and are in similar condition, so they should have equal value.

a. Develop the mathematical model that represents the home's value.
b. Find when the house would be worth \$200,000 (assuming that the rate of appreciation for houses continues unchanged).

Solution **a.** *Developing the model*

Step 1 *Write the given information as two ordered pairs* (t, v). If we let t measure years after January 2003 and v measure value in thousands of dollars, then the ordered pairs are $(t, v) = (0, 140)$ and $(t, v) = (1, 149.8)$.

Step 2 *Substitute the first ordered pair into the model* $y = ae^{bt}$ *and simplify to find* a. *(Or remember that* a *is always the initial value of* y.) Since a is the initial value of y (or here, v), $a = 140$. Alternatively, you could substitute (0, 140) into $v = ae^{bt}$:

$$140 = ae^{b \cdot 0}$$
$$140 = ae^{0}$$
$$140 = a \cdot 1$$
$$a = 140$$

The model is now

$$v = 140e^{bt}$$

Step 3 *Substitute the second ordered pair into the model and simplify to find* b:

$$(t, v) = (1, 149.8)$$
$$149.8 = 140\, e^{b \cdot 1}$$
$$149.8 = 140\, e^{b}$$
$$\frac{149.8}{140} = e^{b}$$
$$1.07 = e^{b}$$
$$\ln 1.07 = \ln e^{b}$$
$$\ln 1.07 = b$$
$$b \approx 0.0676586485$$

Store this value of b in your calculator's memory.

✔ Notice that b is close to the relative growth rate of $7\% = 0.07$.

The model is

$$v = 140\, e^{0.0676586485t}$$

where t is the number of years after January 2003 and v is the value in thousands of dollars.

b. *Finding when the house will be worth* \$200,000 The problem is to find t when $v = 200$, so we substitute 200 for v in our model:

$$200 = 140\, e^{0.0676586485t}$$
$$\frac{200}{140} = e^{0.0676586485t}$$
$$\ln \frac{200}{140} = \ln(e^{0.0676586485t})$$
$$\ln \frac{200}{140} = 0.0676586485t$$

$$t = \frac{\ln \frac{200}{140}}{0.0676586485}$$

$$t = 5.27168\ldots$$

200 [÷] 140 [=] [ln x] [÷] [RCL] [=]

[LN] [(] 200 [÷] 140 [)] [÷] [ALPHA] [B] [ENTER]

TI-83 users should not type the left parenthesis; it's automatically inserted.

Therefore, our prediction is that the house will be worth \$200,000 about $5\frac{1}{4}$ years after January 2003—that is, in April 2008. ■

Does the prediction above mean that the house definitely will be worth \$200,000 in April 2008? No; this calculation is based on the assumption that real estate appreciation continues at 7%. However, real estate values are affected by many things, and the rate changes frequently. The model $y = ae^{bt}$ allows us to make predictions, and it is important to remember that they are *only* predictions.

Steps in Developing an Exponential Model

1. *Write the information given as to the value of the quantity at two points in time as two ordered pairs* (t, y). (You might want to use another letter in place of y as a memory device.)
2. *Substitute the first ordered pair into the model* $y = ae^{bt}$ *and simplify to find* a. (Or remember that a is always the initial value of y.)
3. *Substitute the second ordered pair into the model and simplify to find* b. (As a check, recall that b is close to the relative growth rate.)

Exponential Growth and Compound Interest (For Those Who Have Read Chapter 5: Finance)

You might be wondering whether there is a relationship between the inflation of real estate, discussed in Examples 3 and 6, and compound interest, as discussed in Section 5.2. In Example 3, we found that a \$140,000 home increased in value by 7% from January 2003 to January 2004 (that is, its relative growth rate is 7%). Is this equivalent to depositing \$140,000 in an account that earns 7% interest compounded annually?

EXAMPLE 7 In January 2003, \$140,000 is deposited in an account that earns 7% interest compounded annually.

a. Find the future value in January 2004.
b. Find when the account would hold \$200,000.
c. Develop a compound interest model that can be used to answer questions involving the future value of the account.

Solution

a. $FV = P(1 + i)^n$ — the Compound Interest Formula

$= 140{,}000(1 + 0.07)^1$ — substituting

$= \$149{,}000$

b. $FV = P(1 + i)^n$ — the Compound Interest Formula

$200{,}000 = 140{,}000(1 + 0.07)^n$ — substituting

We will solve this exponential equation by following the steps developed in Section 9.0B. First, isolate the exponential by dividing by 140,000.

$$\frac{200000}{140000} = (1.07)^n$$

$$\frac{20}{14} = (1.07)^n \quad \text{reducing}$$

$$\ln(20/14) = \ln 1.07^n \quad \text{taking ln of each side}$$

$$\ln(20/14) = n \ln 1.07 \quad \text{using the Exponent-Becomes-Multiplier Property } \ln(A^n) = n \cdot \ln A \text{ dividing by } \ln 1.07$$

$$\frac{\ln(20/14)}{\ln 1.07} = n$$

$$n = 5.27168\ldots$$

20 [÷] 14 [=] [ln x] [÷] 1.07 [ln x] [=]

[LN] [(] 20 [÷] 14 [)] [÷] [LN] 1.07 [ENTER]

TI-83 users should not type the left parenthesis—it's automatically inserted.

Since n is the number of compounding periods and interest is compounded annually, $n = 5.27168\ldots$ means that it will take $5.27168\ldots$ years for the account to hold \$200,000.

This solution is mathematically correct. Practically speaking, however, the ".27168 . . ." part of the solution doesn't make sense. If interest is compounded annually, then at the end of each year, your account is credited with that year's interest. After five years, your account balance would be $140{,}000(1 + 0.07)^5 \approx \$196{,}357.24$. After $5.27168\ldots$ years, your account balance would not have changed, since interest won't be credited until the end of the year. After six years, your account balance would be $140{,}000(1 + 0.07)^6 \approx \$210{,}102.25$. Thus, the best answer to the question is that the account will never hold exactly \$200,000, but after six years (i.e., in January of 2009) the account will hold more than \$200,000.

c. To develop a compound interest model that can be used to answer questions involving the future value of the account, substitute 140,000 for P and $7\% = 0.07$ for i into the Compound Interest Formula.

$FV = P(1 + i)^n$ — the Compound Interest Formula

$FV = 140{,}000(1.07)^n$ — substituting

In this model, n must be a whole number of years. ■

Notice that the answer to part (a) of Example 7 matches Example 3, and that the mathematically correct (but practically incorrect) answer to part (b) of Example 7 matches that of part (b) of Example 6. This certainly implies that $v = 140\,e^{0.0676586485t}$, the exponential model developed in Example 6, and $FV = 140{,}000(1.07)^n$, the compound interest model developed in Example 7, are mathematically interchangeable. Regardless of which model is used, only a whole number of years makes sense in a compound interest problem (if the interest is compounded annually).

EXAMPLE 8 Show that the exponential model $v = 140\,e^{0.0676586485t}$ from Example 6 and the compound interest model $v = 140{,}000(1.07)^n$ from Example 7 are mathematically interchangeable.

Solution We will use the properties of logarithms to convert the exponential model to the compound interest model. In Example 6, we modeled the value of a house with

$$v = 140\,e^{0.0676586485t}$$

where v is the value in thousands of dollars. To convert this to dollars, as used in Example 7, multiply by 1,000:

$$\begin{aligned} v &= 1{,}000 \cdot 140\,e^{0.0675586485t} \\ &= 140{,}000\,e^{0.0676586485t} \end{aligned}$$

In Example 6, we found that $b = \ln 1.07 \approx 0.0676586485$, so we can replace 0.0676586485 with ln 1.07. This gives us

$$\begin{aligned} v &= 140{,}000\,e^{0.0676586485t} \\ &= 140{,}000\,e^{(\ln 1.07)(t)} \end{aligned}$$

For the moment, focus on e's exponent, $(\ln 1.07)(t)$:

$$\begin{aligned} (\ln 1.07)(t) &= t \cdot \ln 1.07 \\ &= \ln(1.07^t) \qquad \text{using the Exponent-Becomes-Multiplier Property } n \cdot \ln A = \ln(A^n) \end{aligned}$$

Thus, we can replace e's exponent, $(\ln 1.07)(t)$, with $\ln(1.07^t)$:

$$\begin{aligned} v &= 140{,}000\,e^{(\ln 1.07)(t)} \\ &= 140{,}000\,e^{\ln(1.07^t)} \\ &= 140{,}000(1.07^t) \qquad \text{using the Inverse Property } e^{\ln x} = x \\ &= 140{,}000(1.07^n) \qquad t \text{ and } n \text{ both measure number of years} \end{aligned}$$

Since we were able to convert the exponential model $v = 140\,e^{0.0676586485t}$ to the compound interest model $v = 140{,}000(1.07)^n$, we know that they are algebraically equivalent and therefore are mathematically interchangeable. Regardless of which model is used, only a whole number of years makes sense in a compound interest problem (if the interest is compounded annually). ■

9.1 Exercises

1. Use the model $p = 30\,e^{0.0198026273t}$ developed in Example 5 to predict the population of Anytown in the year 2007.
2. Use the model $p = 30\,e^{0.0198026273t}$ developed in Example 5 to predict the population of Anytown in the year 2026.
3. Use the model $v = 140\,e^{0.0676586485t}$ developed in Example 6 to predict when the house would be worth \$250,000.
4. Use the model $v = 140\,e^{0.0676586485t}$ developed in Example 6 to predict when the house would be worth \$300,000.

Exercises 5–12 deal with data from the U.S. Bureau of the Census on populations of metropolitan areas. These data allow us to find how fast the population is growing and when it will reach certain levels. Such calculations are very important, because they indicate the future needs of the population for goods and services and how well the area can support the population.*

5. The third largest metropolitan area in the United States is the Chicago/Gary/Kenosha metropolitan area. Its population in 1996 was 8,240 (in thousands); in 2000, it was 9,258.
   a. Convert this information to two ordered pairs (t, p), where t measures years since 1990 and p measures population in thousands.
   b. Find Δt, the change in time.
   c. Find Δp, the change in population.
   d. Find $\Delta p/\Delta t$, the average growth rate.
   e. Find $(\Delta p/\Delta t)/p$, the relative growth rate.

Chicago Board of Tourism

Chicago is the third largest metropolitan area in the United States.

6. The second largest metropolitan area in the United States is the Los Angeles/Anaheim/Riverside metropolitan area. Its population in 1990 was 14,532 (in thousands); in 2000, it was 16,374.
   a. Convert this information to two ordered pairs (t, p), where t measures years since 1990 and p measures population in thousands.
   b. Find Δt, the change in time.
   c. Find Δp, the change in population.
   d. Find $\Delta p/\Delta t$, the average growth rate.
   e. Find $(\Delta p/\Delta t)/p$, the relative growth rate.

7. The largest metropolitan area in the United States is the New York/northern New Jersey/Long Island metropolitan area. Its population in 1990 was 19,342 (in thousands); in 2000, it was 21,200.

©NYC & Company, Inc.

New York is the largest metropolitan area in the United States.

   a. Convert this information to two ordered pairs (t, p), where t measures years since 1990 and p measures population in thousands.
   b. Find Δt, the change in time.
   c. Find Δp, the change in population.
   d. Find $\Delta p/\Delta t$, the average growth rate.
   e. Find $(\Delta p/\Delta t)/p$, the relative growth rate.

8. The fifth largest metropolitan area in the United States is the San Francisco/Oakland/San Jose metropolitan area. Its population in 1990 was 6,253 (in thousands); in 2000, it was 7,039.
   a. Convert this information to two ordered pairs (t, p), where t measures years since 1990 and p measures population in thousands.
   b. Find Δt, the change in time.
   c. Find Δp, the change in population.
   d. Find $\Delta p/\Delta t$, the average growth rate.
   e. Find $(\Delta p/\Delta t)/p$, the relative growth rate.

9. **a.** Develop the model that represents the population of the Chicago/Gary/Kenosha metropolitan area (see Exercise 5).
   b. Predict the population in 2005.
   c. Predict the population in 2010.
   d. Predict when the population will be double its 1990 population.

10. **a.** Develop the model that represents the population of the Los Angeles/Anaheim/Riverside metropolitan area (see Exercise 6).
   b. Predict the population in 2006.

* *Statistical Abstract of the United States,* U.S. Department of Commerce, Bureau of the Census.

c. Predict the population in 2010.
d. Predict when the population will be 50% more than its 1990 population.

11. a. Develop the model that represents the population of the New York/northern New Jersey/Long Island metropolitan area (see Exercise 7).
b. Predict the population in 2003.
c. Predict the population in 2010.
d. Predict when the population will be 50% more than it was in 1990.

12. a. Develop the model that represents the population of the San Francisco/Oakland/San Jose metropolitan area (see Exercise 8).
b. Predict the population in 2004.
c. Predict the population in 2020.
d. Predict when the population will be double what it was in 1990.

13. A biologist is conducting an experiment that involves a colony of fruit flies. (Biologists frequently study fruit flies because their short life span allows the experimenters to easily study several generations.) One day, there were 2,510 flies in the colony. Three days later, there were 5,380.
a. Develop the mathematical model that represents the population of flies.
b. Use the model to predict the population after one week.
c. Use the model to predict when the population will be double its initial size.

14. A university keeps a number of mice for psychology experiments. One day, there were 89 mice. Three weeks later, there were 127.
a. Develop the mathematical model that represents the population of mice.
b. Use the model to predict the population after two months.
c. Use the model to predict when the population will be double its initial size.

15. In August 2000, Buck Meadows bought a house for \$230,000. In February 2001, a nearby house with the same floor plan was sold for \$310,000.
a. Develop the mathematical model that represents the value of the house.
b. Use the model to predict when the house will double its 2000 value.
c. Use the model to determine the value of the house one year after Mr. Meadows purchased it.
d. Use part (c) to determine $\Delta v/\Delta t$, that is, the rate at which the value of the house grew.

16. In July 1997, Alvarado Niles bought a house for \$189,000. In September 1998, a nearby house with the same floor plan was sold for \$207,000.
a. Develop the mathematical model that represents the value of the house.
b. Use the model to predict when the house will double its 1997 value.
c. Use the model to determine the value of the house one year after Mr. Niles purchased it.
d. Use part (c) to determine $\Delta v/\Delta t$, that is, the rate at which the value of the house grew.

Exercises 17–20 deal with data from the U.S. Bureau of the Census on energy consumption and the Gross Domestic Product (the total value of the nation's output of goods and services). These data allow us to predict future energy needs and to monitor economic growth.*

17. In 1995, 90.9 quadrillion British Thermal Units (Btu's) of energy were consumed in the United States. In 1999, 96.9 quadrillion Btu's were consumed.
a. Develop the exponential model that represents the amount of energy consumed.
b. Use the model to predict the amount of energy consumed in 2000. (The actual 2000 consumption was 98.5 quadrillion Btu's.)
c. Discuss why an exponential model might be appropriate to this situation.

18. In 1996, 3,246.7 billion kilowatt-hours (kwh) of electricity were consumed in the United States. In 1997, 3,288.7 billion kwh were consumed.
a. Develop the exponential model that represents the amount of electricity consumed.
b. Use the model to predict the amount of electricity consumed in 1998. (The actual 1998 consumption was 3,399.6 billion kwh.)
c. Discuss why an exponential model might be appropriate to this situation.

19. In 1993, the Gross Domestic Product (GDP) of the United States was \$6,343 billion. In 1995, it was \$7,401 billion.
a. Develop the exponential model that represents the nation's GDP.
b. Use the model to predict the GDP in 1999. (The actual 1999 GDP was \$9,269 billion.)
c. Discuss why an exponential model might be appropriate to this situation.

20. In 1998, there were 69,209 thousand cellular telephone subscriptions in the United States. In 1999, there were 86,047.
a. Develop the exponential model that represents the nation's cellular telephone subscriptions.

---

* *Source: 2001 Statistical Abstract of the United States,* U.S. Department of Commerce, Bureau of the Census.

b. Use the model to predict the number of subscriptions in 2000. (The actual number in 2000 was 109,478.)
c. Use the model to predict the number of subscriptions in 2010.

21. The first census of the United States was taken on August 2, 1790; the population then was 3,929,214. Since then, a census has been taken every ten years. The next census gave a U.S. population of 5,308,483.*
    a. Use these data to develop the mathematical model that represents the population of the United States.
    b. Which prediction would be more accurate, U.S. population in 1810 or in 1990? Why?
    c. Use the model to predict the U.S. population in 1810 and in 2000. (The actual populations were 7,239,881, and 281,421,906.)
22. In 1969, the National Academy of Sciences published a study titled "Resources and Man." That study places "the earth's ultimate carrying capacity at about 30 billion people, at a level of chronic near-starvation for the great majority (and with massive immigration to the now less-densely populated lands)!" The study goes on to state that 10 billion people is "close to (if not above) the maximum that an intensively managed world might hope to support with some degree of comfort and individual choice." The world population in 1980 was 4.478 billion; in 1991, it was 5.423 billion.
    a. Develop the mathematical model that represents world population.
    b. Use the model to predict when world population would reach the "somewhat comfortable" level of 10 billion.
    c. Use the model to predict when world population would reach "the earth's ultimate carrying capacity" of 30 billion.
23. Perhaps the most intuitive way to compare how fast different populations are growing is to calculate their "doubling times," that is, how long it takes for the populations to double. Find the doubling times for Africa, North America, and Europe, using the data in Figure 9.12.

	Africa	North America	Europe
1980 population	491 million	252 million	484 million
1990 population	643 million	277 million	509 million

Figure 9.12

24. China's population in 1980 was 983 million; in 1990, it was 1,154 million.
    a. Develop the mathematical model that represents China's population.
    b. Use the model developed in Exercise 23 that represents Africa's population and the model in part (a) to predict when Africa's population will exceed China's.
25. Use the data in Exercise 22 to compute the doubling time for the world's population. Compare it to the doubling time given in the article in Figure 9.13.
26. Use the data in Exercise 24 to compute the doubling time for China's population. Compare it to the doubling time given in the article in Figure 9.13.
27. A real estate investor bought a house for \$125,000. He estimated that houses in that area increase their value by 10% per year.
    a. If that estimate is accurate, why would it not take five years for the house to increase its value by 50%?
    b. How long would it take for the house to increase its value by 50%?
28. The average home in Metropolis in 2000 was \$235,600. In 2001 it was \$257,400. The average home price in Smallville in 2000 was \$112,100. In 2001, it was \$137,600. If these trends continue, when would Smallville's average home price exceed that of Metropolis?

Exercises 29–37 are only for those who have read Chapter 5.

29. Develop a compound interest model that represents the population of the New York/northern New Jersey/Long Island metropolitan area. Use the information in Exercise 7 and the method of Example 7.
30. Develop a compound interest model that represents the population of the San Francisco/Oakland/San Jose metropolitan area. Use the information in Exercise 8 and the method of Example 7.
31. Develop a compound interest model that represents the population of the Chicago/Gary/Kenosha metropolitan area. Use the information in Exercise 5 and the method of Example 7.
32. Develop a compound interest model that represents the population of the Los Angeles/Anaheim/Riverside metropolitan area. Use the information in Exercise 6 and the method of Example 7.

* *Historical Statistics of the United States*—Colonial Times to 1970, U.S. Department of Commerce, Bureau of the Census.

World population growth rate rises

ASSOCIATED PRESS.
Reprinted with permission.

WASHINGTON — Mankind's growth is accelerating again as the world adds the equivalent of another Mexico every year, the Population Reference Bureau said yesterday.

The private research group said that the world's population growth, after having slowed in the 1970s, is speeding up once more. As of mid-1989, the world will contain about 5.24 billion people, nearly a quarter-billion more than in 1987.

Demographers Carl Haub and Mary Kent reported that the expansion from 5 billion to 6 billion could be attained in a record time—within less than a decade—if growth continues at current rates.

The world is adding about 90 million people per year, slightly more than the current Mexican population, estimated at 87 million.

"Even to reach a stable world population size of 10 billion, double the current total, birth rates will have to begin a steady descent soon," Haub and Kent said.

At current rates, the world's population will rise to 6.3 billion by the year 2000 and to 8.3 billion by 2020 the group estimated, adding that the planet's population will double in 39 years.

China remains the most populous country on Earth with 1.1 billion people and an estimated doubling time of 49 years.

Second is India with 835 million, and it has a faster growth rate. Expected doubling time there is 32 years.

Far behind those two population giants, the Soviet Union is third with 289 million people and a doubling time of 70 years.

The United States is fourth with 248 million people and a doubling time of 98 years.

Other nations with more than 100 million residents, and their anticipated doubling times, are Indonesia, 185 million, 35 years; Brazil, 147 million, 34 years; Japan, 123 million, 141 years, Nigeria, 115 million, 24 years, Bangladesh, 115 million, 25 years; Pakistan, 110 million, 24 years.

Figure 9.13

33. Show that the exponential model developed in Exercise 9 and the compound interest model developed in Exercise 31 are mathematically interchangeable.

34. Show that the exponential model developed in Exercise 10 and the compound interest model developed in Exercise 32 are mathematically interchangeable.

35. In Section 5.2, we found the doubling time of an account that pays 5% interest compounded daily by using a graphing calculator to solve the equation.

$$2P = P\left(1 + \frac{0.05}{365}\right)^n$$

Use logarithms to solve the same equation.

36. Compare and contrast the method used in Exercise 35 with the method used in Section 5.2 to find doubling time. Which method do you prefer? Why?

37. Use logarithms to solve Exercise 47a in Section 5.3.

Answer the following questions using complete sentences and your own words.

Concept Questions

38. The exponential model is a powerful predictor of population growth, but it is based on the assumption that the two ordered pairs used to find a and b describe a steady tendency in the population's growth. In fact, that tendency may not be so steady, and the model's prediction may not be a good one.

a. What information were you given in Example 1 that implied a steady tendency in the aphid population's growth?

b. What information were you given in Example 2 that implied a steady tendency in Anytown's growth?

c. Discuss the factors of an animal population's growth that affect the accuracy of a prediction based on the exponential model. What other factors would be involved if the population were human?

39. In which of the following four situations would the exponential model give the most accurate prediction? The least accurate? Why?

- p is the population of a specific country; the two ordered pairs used to find a and b cover a time span of five years; you are asked to predict the population five years later.
- p is the population of a specific country; the two ordered pairs used to find a and b cover a time span of 100 years; you are asked to predict the population five years later.
- p is the population of the world; the two ordered pairs used to find a and b cover a time span of

five years; you are asked to predict the population five years later.

- p is the population of the world; the two ordered pairs used to find a and b cover a time span of 100 years; you are asked to predict the population five years later.

Projects

40. Do some research and find out the population of your community at two fairly recent points in time. Use that data to

a. Find $\Delta p/\Delta t$, the average growth rate.
b. Find $(\Delta p/\Delta t)/p$, the relative growth rate.
c. Develop the model that represents the population of your community.
d. Predict the population of your community in 10 years.
e. Determine your community's doubling time.
f. Also, discuss the accuracy of your model. Be certain to include in your discussion the prominence of emigration and immigration in your community.

9.2 Exponential Decay

What does the article in Figure 9.14 mean when it says that tritium gas "decays at a rate of 5.5% a year"? Why do scientists claim that the frozen corpse found in 1991 in the Austrian Alps "is 5,000 to 5,500 years old"? (See Figure 9.15.) The answers have to do with radioactivity. Radioactive materials have become more and more prevalent and useful since Marie Curie introduced them to the world in 1898.

A radioactive substance is not stable; over time, it transforms itself into another substance. This is called **radioactive decay,** and it is due to the interaction between nuclear particles (protons, neutrons, and electrons) in the radioactive substance. Because a larger quantity of a radioactive substance has more nuclear particles (and hence more interactions), we might guess that it decays faster, just as a larger population produces more offspring. This, in fact, is the case; a larger amount of radioactive material does experience more decay. That is, *the rate of decay is proportional to the amount of radioactive substance present,* just as the rate of growth of a population is proportional to the size of the population. Consequently, we can use the exponential model $y = ae^{bt}$, developed in Section 9.1. As a memory device, we will use the variable Q instead of y to represent "quantity"; therefore, our model will be $Q = ae^{bt}$.

Radioactive gas used in A-bomb may be missing

WASHINGTON — The Energy Department and the Nuclear Regulatory Commission are investigating the possible loss of enough tritium to help make a nuclear bomb, and authorities have suspended sales of the radioactive gas during the probe.

Tritium is used in nuclear weapons to increase their power. Since it decays at the rate of 5.5 percent a year, the gas must be replenished if weapons are to maintain their full explosive potential.

The federal government also sells 200 to 300 grams of tritium a year to American and foreign companies for use in biological and energy research and to make self-luminous lights, signs and dials.

The incident has led to fears in Congress that some of the gas is missing and has fallen into unfriendly hands. But federal officials say that there are no signs of a diversion and that the discrepancy may have resulted from errors in measuring the tritium.

Figure 9.14

Tests on iceman make him part of Stone Age

By John Noble Wilford

NEW YORK — Carbon dating tests show that the well-preserved body of a prehistoric human hunter found in an Alpine glacier last year is 5,000 to 5,500 years old, scientists reported yesterday.

The first scientifically established age for the frozen corpse is more than 1,000 years older than original estimates. It means the man lived and presumably froze to death well before the Bronze Age replaced the late Stone Age in Europe.

"Now we know he's not from the Bronze Age, but much older," said Dr. Werner Platzer, head of the anatomy department at Innsbruck University in Austria, who is directing research on the mummified corpse. "It means, I believe, that this is the only corpse we have from the Stone Age."

The tests on bones and skin tissue were conducted by scientists at Oxford University in England and a Swiss physics institute in Zurich. The body is being kept in cold storage at Innsbruck University.

The corpse was found in September in a glacier 10,500 feet up in the Austrian Alps, close to the Italian border. The body was mummified, and its leather and fur clothing was badly deteriorated. But scholars of prehistoric people were fascinated by a leather quiver with 14 arrows and an ax found by the iceman's side.

On first examination, the ax was thought to be made of bronze, which led to the assumption that this was an early Bronze Age man who lived about 4,000 years ago. More careful study showed the ax to be made of copper, thus a product of a simpler technology that seemed to place the iceman in the late Stone Age. This seemed to be confirmed by previous tests on samples of grasses taken from the hunter's clothing, which indicated that the body might be 5,000 years old.

Figure 9.15

EXAMPLE 1 Hospitals use the radioactive substance iodine-131 in research. It is effective in locating brain tumors and in measuring heart, liver, and thyroid activity. A hospital purchased 20 grams of the substance. Eight days later, when a doctor wanted to use some of the iodine-131, he observed that only 10 grams remained (the rest had decayed).

a. Develop a mathematical model that represents the amount of iodine-131 present.
b. Predict the amount remaining two weeks after purchase.

Solution **a.** *Developing the model* Because we're using the exponential model $Q = ae^{bt}$, we follow the steps from Section 9.1.

Step 1 *Write the given information as two ordered pairs* (t, Q). Originally there were 20 grams, so we have $(t, Q) = (0, 20)$. Eight days later, there were 10 grams, so we have $(t, Q) = (8, 10)$.

Step 2 Recall that *a is always the initial value of Q, so* a = 20. Alternatively, we could *substitute* (0, 20) *into the model* $Q = ae^{bt}$ *and simplify*:

$$Q = ae^{bt}$$
$$20 = ae^{b(0)}$$
$$20 = ae^{0}$$
$$20 = a \cdot 1$$
$$a = 20$$

The model is now $Q = 20e^{bt}$.

Step 3 *Substitute the second ordered pair* (8, 10) *into this new model and simplify*:

$$Q = 20\,e^{bt}$$
$$10 = 20\,e^{b(8)}$$
$$\frac{10}{20} = e^{8b}$$
$$0.5 = e^{8b}$$
$$\ln 0.5 = \ln e^{8b} \quad \text{taking ln of each side}$$
$$\ln 0.5 = 8b \quad \text{using the Inverse Property } \ln(e^x) = x$$
$$b = \frac{\ln 0.5}{8}$$
$$b \approx -0.086643397$$

Store this value of b in your calculator's memory.

0.5 [ln x] [÷] 8 [=]

[LN] 0.5 [÷] 8 [ENTER]

TI-83: [LN] 0.5 [)] [÷] 8 [ENTER]

(A left parenthesis is automatically inserted when the [LN] button is pushed.)

Thus, our model for the amount Q of iodine-131 remaining t days after purchase is $Q = 20\,e^{-0.086643397t}$.

b. *Predictions* Now that the model has been determined, predictions can be made. Two weeks after purchasing the iodine-131, $t = 14$ days, and

$$Q = 20\,e^{-0.086643397t}$$
$$= 20\,e^{-0.086643397(14)}$$
$$= 5.946035575\ldots$$
$$\approx 5.9$$

14 [×] [RCL] [=] [e^x] [×] 20 [=]

20 [×] [2nd] [e^x] [(] [ALPHA] [B] [×] 14 [)] [ENTER]

TI-83 users should not type the left parenthesis (it's automatically inserted).

After two weeks (14 days), we would expect approximately 5.9 grams to remain. ■

Notice that b is negative in Example 1. This indicates that the amount of iodine-131 is *decreasing*. In contrast, b was positive in Section 9.1, because the quantities under study were *increasing*.

> **Exponential Growth or Exponential Decay**
>
> In general, any quantity for which *the rate of change is proportional to the amount present* can be modeled by the formula $y = ae^{bt}$.
>
> 1. If $b > 0$, then y is growing exponentially.
> 2. If $b < 0$, then y is decaying exponentially.

Half-Life

A radioactive substance is not stable. One way to measure its instability is to determine the **half-life** of the substance, that is, the amount of time required for a quantity to reduce to one-half its initial size. In the previous example, the half-life of iodine-131 was 8 days; it took 8 days for 20 grams of the substance to decay into 10 grams. As we will see, it will take an additional 8 days for those 10 grams to decay into 5 grams.

EXAMPLE 2 Use the model $Q = 20\,e^{-0.086643397t}$ from Example 1 to calculate the amount of iodine-131 remaining at the following times.

a. 16 days after purchase **b.** 24 days after purchase

Solution **a.** Sixteen days after purchasing the 20 grams of iodine-131 (or eight days after observing that half of it had decayed), $t = 16$; therefore,

$$\begin{aligned} Q &= 20\,e^{-0.086643397t} \\ &= 20\,e^{-0.086643397(16)} \\ &= 5 \end{aligned}$$

> 16 [×] [RCL] [=] [e^x] [×] 20 [=]
>
> 20 [×] [2nd] [e^x] [(] [ALPHA] [B] [×] 16 [)] [ENTER]
>
> TI-83 users should not type the left parenthesis (it's automatically inserted).

Thus, 16 days after purchase, 5 grams of the initial 20 grams of iodine-131 will remain. Notice that 16 days is the same as two half-life periods. After one half-life period (8 days), 10 grams remained; after two half-life periods (16 days), 5 grams remained.

b. Twenty-four days (or three half-life periods) after purchasing the 20 grams of iodine-131, $t = 24$; therefore,

$$\begin{aligned} Q &= 20\,e^{-0.086643397t} \\ &= 20\,e^{-0.086643397(24)} \\ &= 2.5 \end{aligned}$$

Thus, 24 days after purchase, 2.5 grams of the initial 20 grams of iodine-131 will remain. After another 8 days (or a total of 32 days), only 1.25 grams will remain. Every 8 days, half of the material decays. ■

We can obtain a graph showing the exponential decay of the initial 20 grams of iodine-131 from the example above by plotting the ordered pairs (0, 20), (8, 10), (16, 5), and (24, 2.5) and continuing in this manner (after each 8-day period, only half the previous quantity remains). This graph is shown in Figure 9.16.

The half-life of a radioactive substance does not depend on the amount of substance present. For any given radioactive substance, the half-life is intrinsic to the substance itself; each radioactive substance has its own half-life. Figure 9.17 lists the half-life for various substances. Notice that half-lives have an incredibly wide range of values—from a few seconds to well over 20,000 years.

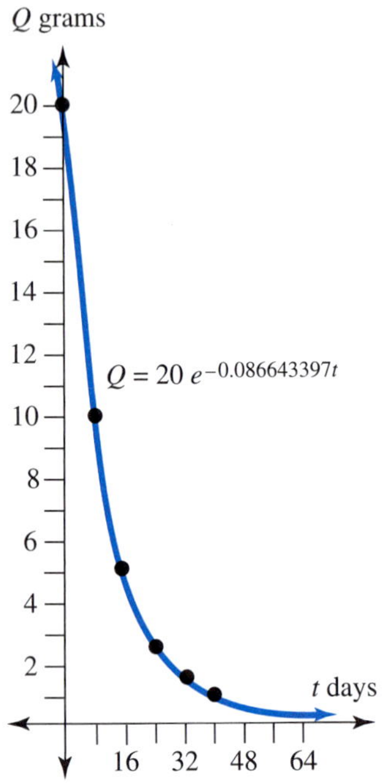

Figure 9.16
A graph showing the amount of iodine-131 remaining t days after purchasing 20 grams

EXAMPLE 3 Professor Frank Stein received 8.2 grams of plutonium-241 and stored it in his laboratory for future use in experiments. If he does not use any of the substance for experimental purposes, how much will remain one year later?

Solution Rather than focusing on the *specific* initial amount of 8.2 grams, we will determine the *general* model $Q = ae^{bt}$ for *any* initial amount and substitute our given values later. We do this here to emphasize the general approach of developing a model, rather than dwell on the specific numerical calculations.

Radioactive Substance	Half-life
krypton-91	10 seconds
silicon-31	2.6 hours
cobalt-55	18.2 hours
magnesium-28	21.0 hours
iodine-124	4.5 days
iodine-131	8.0 days
cobalt-60	5.3 years
plutonium-241	13 years
plutonium-238	86 years
carbon-14	5,730 years
plutonium-239	24,400 years

Figure 9.17

Step 1 *Express the information as ordered pairs* (t, Q). Because a is the initial amount of plutonium-241, $(t, Q) = (0, a)$. To obtain a second ordered pair, we find the half-life of plutonium-241 in Figure 9.17 (the half-life is $t = 13$ years). In 13 years, half of a, or $\frac{a}{2}$, will remain, and we obtain $(t, Q) = (13, \frac{a}{2})$.

Step 2 *Substitute the first ordered pair into the model, and simplify to find a.* Because we are not using a specific value for the initial amount a, we can omit this step. (Our model is still $Q = ae^{bt}$.)

Step 3 *Substitute* $(t, Q) = (13, \frac{a}{2})$ *into the model and simplify*:

$$Q = ae^{bt}$$

$$\frac{a}{2} = ae^{b(13)}$$

$$\frac{1}{2} = e^{13b} \quad \text{dividing by } a$$

$$0.5 = e^{13b}$$

$$\ln 0.5 = \ln e^{13b} \quad \text{taking the natural log}$$

$$\ln 0.5 = 13b \quad \text{using the Inverse Property } \ln(e^x) = x$$

$$b = \frac{\ln 0.5}{13}$$

$$b \approx -0.053319013$$

Store this value of b in your calculator's memory.

0.5 [ln x] [÷] 13 [=]

[LN] 0.5 [÷] 13 [ENTER]

TI-83: [LN] 0.5 [ )] [÷] 13 [ENTER]

(A left parenthesis is automatically inserted when the [LN] button is pushed.)

Thus, our model for the amount Q of plutonium-241 remaining t years after receiving an initial quantity of a grams is $Q = a\, e^{-0.053319013t}$. To predict the amount of plutonium-241 that will remain one year after Professor Stein receives his 8.2 grams, we substitute 8.2 for a and 1 for t:

$$\begin{aligned} Q &= a\, e^{-0.053319013t} \\ &= 8.2 \cdot e^{-0.053319013(1)} \\ &= 7.774235618\ldots \\ &\approx 7.8 \end{aligned}$$

RCL e^x × 8.2 =

8.2 × 2nd e^x ALPHA B ENTER

On the basis of the model $Q = 8.2\,e^{-0.053319013t}$, Professor Stein should expect approximately 7.8 grams of plutonium-241 to remain one year after receiving the 8.2 grams. ■

Relative Decay Rate

In Section 9.1, we discussed the average growth rate $\Delta p/\Delta t$, which indicates how fast a quantity is growing, and the relative growth rate $(\Delta p/\Delta t)/p$, which indicates how fast the quantity is growing *as a percent*. When calculating the constant b in the model $p = ae^{bt}$, we were able to check our work by seeing whether b was close to the relative growth rate.

This same concept can be applied to exponential decay. When using the exponential decay model $Q = a\,e^{bt}$, we can compute the **average decay rate** $\Delta Q/\Delta t$ by dividing the change in the amount of radioactive material by the corresponding change in time. We can also compute the **relative decay rate** by calculating $(\Delta Q/\Delta t)/Q$. Because Q is decreasing, ΔQ will be negative, as will the relative decay rate.

EXAMPLE 4 Find (a) the average decay rate and (b) the relative decay rate for the data given in Example 3.

Solution

a. *Finding the average decay rate* The initial quantity was 8.2 grams, and 7.774235625 grams remained after $t = 1$ year. (Using the rounded-off figure of 7.8 in place of 7.774235625 would result in a less accurate answer.)

$$\begin{aligned}\Delta Q &= \text{final quantity} - \text{initial quantity}\\ &= 7.774235625 - 8.2\\ &= -0.425764375 \text{ gram}\\ \Delta t &= 1 \text{ year}\end{aligned}$$

Therefore, the average decay rate is

$$\frac{\Delta Q}{\Delta t} = \frac{-0.425764375 \text{ gram}}{1 \text{ year}} \approx -0.4 \text{ gram per year}$$

The following year, there will not be as much material left, and what remains will decay at a slower rate. The average decay rate will not remain constant.

b. *Finding the relative decay rate*

$$\begin{aligned}\frac{\Delta Q/\Delta t}{Q} &= \frac{-0.425764375 \text{ gram per year}}{8.2 \text{ grams}}\\ &= -0.051922484 \text{ per year}\\ &= -5.1922484\% \text{ per year}\end{aligned}$$

This implies that the amount of radioactive material is decreasing by about 5.2% per year. While the average decay rate will change from year to year (because the material decays more slowly as the amount decreases), the relative decay rate will remain constant. Notice that the relative decay rate is close to b, which was -0.053319013. ■

EXAMPLE 5 Plutonium-239 is a waste product of nuclear reactors. How long will it take for this waste to lose 99.9% of its radioactivity and therefore be considered relatively harmless to the biosphere?*

Solution Regardless of the initial amount a of plutonium-239, we need to determine the time t required for 0.1% of the radioactivity to *remain*. (Remember, the model $Q = ae^{bt}$ determines the amount Q *remaining* after a time period t.) We first need to determine the model for plutonium-239.

Step 1 *Express the initial data.* Let a = the initial amount of plutonium-239, and note that the half-life is 24,400 years (Figure 9.16). *Express this information as ordered pairs* (t, Q). We have $(0, a)$ and $(24400, \frac{a}{2})$.

Step 2 *Substitute the first ordered pair into the model, and simplify to find a.* Because we are not using a specific value for the initial amount a, we can omit this step. (Our model is still $Q = ae^{bt}$.)

Step 3 *Substitute* $(t, Q) = (24400, \frac{a}{2})$ *into the model and simplify*:

$$
\begin{aligned}
Q &= ae^{bt} \\
\frac{a}{2} &= ae^{b(24,400)} \\
\frac{1}{2} &= e^{24,400b} \\
0.5 &= e^{24,400b} \\
\ln 0.5 &= \ln(e^{24,400b}) && \text{Taking ln of each side} \\
\ln 0.5 &= 24,400b && \text{using an Inverse Property} \\
b &= \frac{\ln 0.5}{24,400} \\
b &\approx -0.000028407
\end{aligned}
$$

Store this value of b in your calculator's memory.

For plutonium-239, therefore, our model is $Q = ae^{-0.000028407t}$. Because the initial amount of plutonium-239 is a, we need to determine the time t when 0.1% of a remains—that is, when the amount Q of plutonium-239 will equal $0.001a$. We substitute $0.001a$ for Q in the model $Q = ae^{-0.000028407t}$ and solve:

$$
\begin{aligned}
Q &= ae^{-0.000028407t} \\
0.001a &= ae^{-0.000028407t} \\
0.001 &= e^{-0.000028407t} && \text{dividing by } a \\
\ln 0.001 &= \ln(e^{-0.000028407t}) && \text{taking ln of each side} \\
\ln 0.001 &= -0.000028407t && \text{using an Inverse Property} \\
t &= \frac{\ln 0.001}{-0.000028407} \\
&= 243,165.1365\ldots \\
&\approx 240,000
\end{aligned}
$$

* H. A. Bethe, "The Necessity of Fission Power," *Scientific American*, January 1976.

0.001 [ln x] [÷] [RCL] [=]

[LN] 0.001 [÷] [ALPHA] [B] [ENTER]

TI-83: [LN] 0.001 [)] [÷] [ALPHA] [B] [ENTER]

(A left parenthesis is automatically inserted when the [LN] button is pushed.)

It will take approximately 240,000 years for any given quantity of plutonium-239 to lose 99.9% of its radioactivity and thus be considered relatively harmless to the biosphere. How long a time span is this, relative to all of human history? ■

The United States has assembled a vast stockpile of about 50 tons of plutonium. Because plutonium is so deadly, its storage and disposal are of major concern, as indicated by the 1993 news article shown in Figure 9.18. However, not every country in the world shares the United States' concern over the perils of plutonium. The Japanese Nuclear Agency has been promoting the use of

Plutonium stockpile could last forever

U.S. has 50 tons to think about disposing of

By Matthew L. Wald

AMARILLO, TEXAS — In 16 unremarkable concrete bunkers built by the Army for a war with Hirohito and Hitler, the United States has begun assembling about 50 tons of plutonium, a vast stockpile of one of the most expensive materials ever produced and perhaps the most important to safeguard.

The Energy Department says the bunkers, each about the size of a two-car garage, are going to be used for interim storage, meaning six or seven years.

But plutonium, which was invented by the Energy Department's predecessor, the Manhattan Project, may turn out to be the hardest thing on Earth to dispose of. And at the Energy Department, "interim" can have an elastic meaning. "Immediate" tends to mean several years. "Several years" can mean never.

President Clinton announced the formation of an inter-agency task force in September to consider how much plutonium the nation needs in the post-Soviet era and how to dispose of what is surplus. The Energy Department is also drawing up a plan for its weapons production complex for the next century.

But officials are choosing among a short list of unattractive options.

At the Pantex plant near Amarillo, where the plutonium is piling up, general manager Rich Loghry said when asked what would happen next, "I don't think people have a really good answer for what is going to happen to the plutonium."

Storage may be the leading option, but even this is tricky. Plutonium loses half its radioactivity every 24,000 years, so it will reach background levels of radioactivity in 10 "half-lives," or 240,000 years; the U.S. political system is focused largely on problems that can be solved in less than four.

Figure 9.18

Japanese make plutonium cute

Video claims it's safe to drink

By Seth Sutel

ASSOCIATED PRESS

Reprinted with permission.

TOKYO — Meet Mr. Pluto, the Japanese nuclear agency's round-faced, rosy-cheeked, animated answer to the public's concern about its plan to import 30 tons of plutonium as fuel for power plants.

In the country that best knows the dark side of atomic energy, not everyone is charmed by Mr. Pluto, who is featured in a promotional videotape prepared by the Power Reactor and Nuclear Fuel Development Corp.

Anti-nuclear groups said yesterday that they will campaign against distribution of the video, entitled "The Story of Plutonium: That Dependable Fellow, Mr. Pluto." They contend that it irresponsibly plays down the dangers plutonium poses.

Perky and pint-sized, Mr. Pluto is childlike, with cute red boots and a green helmet with antennae. On the front of the helmet is the chemical symbol for plutonium, Pu.

In one scene in the video, he shakes the hand of a cheerful youngster who is drinking a mug of plutonium-laced soda. The narration says that if plutonium were ingested, most of it would pass through the body without harm.

"The most fundamental lie in this video is the idea that plutonium is not dangerous," said Jinzaburo Takagi, a former nuclear chemist who heads the Citizens' Nuclear Information Center.

"Of course, it's very dangerous to drink plutonium," Takagi said. "To say otherwise, as they do in this video, is completely outrageous."

BY ASSOCIATED PRESS

The agency's video shows 'Mr. Pluto' greeting a boy drinking soda dosed with plutonium.

The highly radioactive, silvery metal is toxic to humans because it is absorbed by bone marrow. The inhalation of .0001 of a gram can induce lung cancer.

Figure 9.19

plutonium in its power plants, as indicated by the 1994 news article shown in Figure 9.19.

Radiocarbon Dating

Radioactive substances are used to determine the age of fossils and artifacts. The procedure is based on the fact that two types of carbon occur naturally: carbon-12, which is stable, and carbon-14, which is radioactive. The total amount of carbon-14 in the environment is rather small; there is only one atom of carbon-14 for every 1 trillion atoms of carbon-12! Living organisms maintain this ratio due to their intake of water, air, and nutrients. However, when an organism dies, the amount of carbon-14 decreases exponentially due to radioactive decay. By making extremely delicate measurements of the amounts of carbon-14 and carbon-12 in a fossil or artifact, scientists are able to estimate the age of the item under investigation.

EXAMPLE 6 Determine the model representing the amount Q of carbon-14 remaining t years after the death of an organism.

Solution

Step 1 Let a represent the original quantity of carbon-14 present in a living organism. The half-life of carbon-14 is 5,730 years (from Figure 9.15), so at time $t = 5{,}730$,

HISTORICAL NOTE

Marie Curie
1867–1934

The Granger Collection, NY

Born Marie Sklodowska in Warsaw, Poland, Marie Curie moved to Paris and enrolled as a student of physics at the Sorbonne in 1891. While researching the magnetic properties of various steel alloys, she met Pierre Curie, and they married in 1895. In the following year, Antoine Henri Becquerel discovered radioactivity in uranium. As a team, the Curies further investigated uranium and discovered the elements radium and polonium (named after Marie's native country). They also discovered that diseased, tumor-forming cells were destroyed faster than healthy cells when exposed to radium, laying the groundwork for modern radiation therapy. The word *radioactivity* was coined by Madame Curie.

In addition to publishing many important scientific papers and books, Madame Curie received many honors. She was the first person to win two Nobel Prizes, one in 1903 for the discovery of radioactivity (which she shared with her husband and Becquerel) and one in 1911 in chemistry. In 1908, at the University of Paris, she taught the first course on radioactivity ever offered. The Sorbonne created a special chair in physics for Pierre Curie; Marie was appointed his successor after he died in a street accident.

During World War I, Madame Curie devoted much of her time to providing radiological services to hospitals. On her death, Albert Einstein said, "Marie Curie is, of all celebrated beings, the only one whom fame has not corrupted."

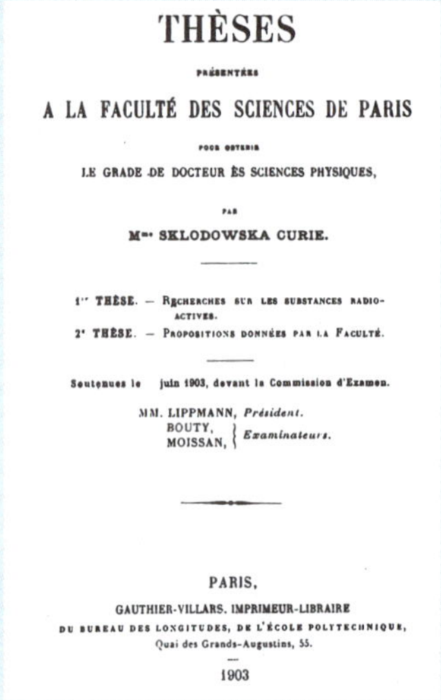
THÈSES

PRÉSENTÉES

A LA FACULTÉ DES SCIENCES DE PARIS

POUR OBTENIR

LE GRADE DE DOCTEUR ÈS SCIENCES PHYSIQUES,

PAR

Mme SKLODOWSKA CURIE.

1re THÈSE. — RECHERCHES SUR LES SUBSTANCES RADIO-ACTIVES.

2e THÈSE. — PROPOSITIONS DONNÉES PAR LA FACULTÉ.

Soutenues le juin 1903, devant la Commission d'Examen.

MM. LIPPMANN, *Président.*
BOUTY, MOISSAN, *Examinateurs.*

PARIS,

GAUTHIER-VILLARS, IMPRIMEUR-LIBRAIRE

DU BUREAU DES LONGITUDES, DE L'ÉCOLE POLYTECHNIQUE,

Quai des Grands-Augustins, 55.

1903

The Granger Collection, NY

the quantity of carbon-14 present would be $\frac{a}{2}$. Expressed as ordered pairs (t, Q), we have $(0, a)$ and $(5730, \frac{a}{2})$.

Step 2 Because we are not using a specific value for the initial amount a, we can omit this step. (Our model is still $Q = a\,e^{bt}$.)

Step 3 *We substitute* $(t, Q) = (5730, \frac{a}{2})$ *into the model and simplify.*

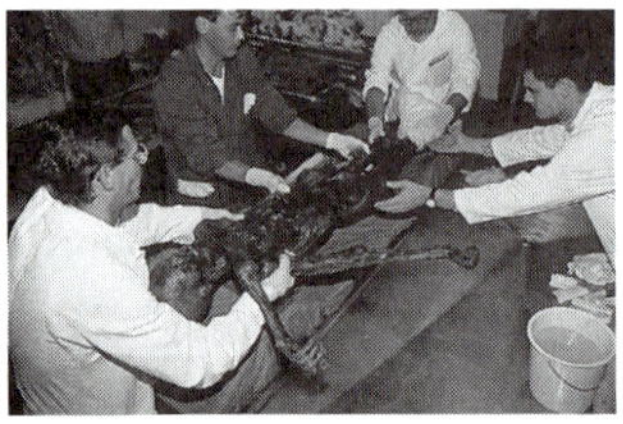
G. Hinterleitner/Liaison Agency

$$
\begin{aligned}
Q &= ae^{bt} \\
\frac{a}{2} &= ae^{b(5,730)} \\
\frac{1}{2} &= e^{5,730b} && \text{dividing by } a \\
0.5 &= e^{5,730b} \\
\ln 0.5 &= \ln(e^{5,730b}) && \text{Taking ln of each side} \\
\ln 0.5 &= 5,730b && \text{Using an Inverse Property} \\
b &= \frac{\ln 0.5}{5,730} \\
b &\approx -0.000120968
\end{aligned}
$$

HISTORICAL NOTE

Willard Frank Libby

1908–1980

Willard Frank Libby developed the radiocarbon dating technique in the mid-1940s. Carbon-14 was known to exist in nature, but little was known of its origins and properties. In 1939, Libby discovered that cosmic rays interacting with nitrogen at high altitudes produced a rapid formation of carbon-14. This high-altitude formation is the basis of the claim that the current ratio of carbon-14 to carbon-12 has been constant throughout history.

While working at the Enrico Fermi Institute of Nuclear Studies in Chicago, Libby was able to artificially produce carbon-14 and accurately determine its half-life. In addition, he devised a relatively simple device that measures the amount of carbon-14 in an organic sample. Before the creation of this device, measuring carbon-14 was a very expensive and difficult process. Libby's method made radiocarbon dating a practical possibility and revolutionized the fields of archeology and geology.

Libby received his doctorate degree in chemistry from the University of California at Berkeley in 1933 and taught there until 1945. During World War II, Libby also worked on the Manhattan Project, which developed the atomic bomb. During the years 1955–1959, he served on the U.S. Atomic Energy Commission, where he was instrumental in the formulation of many aspects of the commission.

After many years of dedicated research and numerous discoveries, Libby was awarded the Nobel Prize in chemistry in 1960 "for his method of using carbon-14 as a measurer of time in archeology, geology, geophysics, and other sciences."

Store this value of b in your calculator's memory.
We now have our model, $Q = ae^{-0.000120968t}$

■

The calculations in Example 6 produce the following general model.

Radiocarbon Dating Model

The quantity Q of carbon-14 remaining t years after the death of an organism (that had an initial amount a) is

$$Q = ae^{-0.000120968t} \quad \text{or} \quad Q = ae^{bt} \quad \text{where } b = \frac{\ln 0.5}{5{,}730}$$

EXAMPLE 7 The first of the Dead Sea Scrolls were discovered in 1947 in caves near the northwestern shore of the Dead Sea in the Middle East. The parchment scrolls, which were wrapped in linen and leather, contained all the books of the Old Testament (except Esther). An analysis showed that the scrolls contained approximately 79% of the expected amount of carbon-14 found in a living organism. Determine the age of the scrolls.

Solution Using the model $Q = ae^{-0.000120968t}$, we need to determine the amount of time t required so that 79% of a remains. To do this, we substitute $0.79a$ for Q in the model and solve.

Early archaeologists

People have always been curious as to the age of artifacts.

A fragment of the Habbakuk Commentary of the Old Testament

$$Q = ae^{-0.000120968t}$$

$$0.79a = ae^{-0.000120968t} \quad \text{substituting } Q = 0.79a$$

$$0.79 = e^{-0.000120968t} \quad \text{dividing by } a$$

$$\ln 0.79 = \ln(e^{-0.000120968t}) \quad \text{taking ln of both sides}$$

$$\ln 0.79 = -0.000120968t \quad \text{using an Inverse Property}$$

$$t = \frac{\ln 0.79}{-0.000120968} \quad \text{dividing by } -0.000120968$$

$$t = 1{,}948.22\ldots$$

$$t \approx 1{,}950$$

Thus, the Dead Sea Scrolls were approximately 1,950 years old when they were discovered and were therefore apparently created around the time when Jesus was alive. This result has been used to support the authenticity of the scrolls. ■

It should be noted that an underlying assumption in radiocarbon dating is that the current ratio of carbon-14 to carbon-12 in the biosphere remains constant over time and location. There have been disagreements within the scientific community over the validity of this assumption. Therefore, the dates obtained from radiocarbon dating are not guaranteed to be 100% reliable. Some of the results of radiocarbon dating have been controversial, as the article in Figure 9.20 indicates.

Shroud of Turin was created in 14th century, official says

ASSOCIATED PRESS

Reprinted with permission.

ROME — Laboratory tests show that the Shroud of Turin was made in the 14th century and could not be the burial cloth of Christ, the scientific adviser to the archbishop of Turin said he learned yesterday.

Professor Luigi Gonella said he has not yet seen the official report from the three laboratories that conducted the carbon-14 dating tests, but that all the leaks to the press dated it to the 14th century, and "somebody let me understand that the rumors were right."

He refused to identify who had told him about the results of the tests at Oxford University, the University of Arizona and the Swiss Federal Institute of Technology at the University of Zurich.

Figure 9.20

9.2 Exercises

1. Using the model $Q = 20\,e^{-0.086643397t}$ developed in Example 1, predict how much iodine-131 will remain after three weeks.
2. Using the model $Q = 20\,e^{-0.086643397t}$ developed in Example 1, predict how much iodine-131 will remain after thirty days.
3. Using the model $Q = 8.2\,e^{-0.053319013t}$ developed in Example 3, predict how much plutonium-241 will remain after two years.
4. Using the model $Q = 8.2\,e^{-0.053319013t}$ developed in Example 3, predict how much plutonium-241 will remain after ten years.

In Exercises 5–16, use Figure 9.16.

5. Silicon-31 is used to diagnose certain medical ailments. Suppose a patient is given 50 milligrams.
   a. Develop the mathematical model that represents the amount of silicon-31 present at time t.
   b. Predict the amount of silicon-31 remaining after one hour.
   c. Predict the amount of silicon-31 remaining after one day.
   d. Find $\Delta Q/\Delta t$, the average decay rate, for the first hour.
   e. Find $(\Delta Q/\Delta t)/Q$, the relative decay rate, for the first hour.
   f. Find $\Delta Q/\Delta t$, the average decay rate, for the first day.
   g. Find $(\Delta Q/\Delta t)/Q$, the relative decay rate, for the first day.
   h. Why is the absolute value of the answer to part (e) greater than the answer to part (g)?
6. Plutonium-238 is used as a compact source of electrical power in many applications, ranging from pacemakers to spacecraft. Suppose a power cell initially contains 1.6 grams of plutonium-238.

a. Develop the mathematical model that represents the amount of plutonium-238 present at time t.
b. Predict the amount of plutonium-238 remaining after one year.
c. Predict the amount of plutonium-238 remaining after 20 years.
d. Find $\Delta Q/\Delta t$, the average decay rate, for the first year.
e. Find $(\Delta Q/\Delta t)/Q$, the relative decay rate, for the first year.
f. Find $\Delta Q/\Delta t$, the average decay rate, for the first 20 years.
g. Find $(\Delta Q/\Delta t)/Q$, the relative decay rate, for the first 20 years.
h. Why is the absolute value of the answer to part (e) greater than the answer to part (g)?

7. How long will it take 64 grams of magnesium-28 to decay into the following amounts?
   a. 32 grams
   b. 16 grams
   c. 8 grams
8. How long will it take 56 milligrams of cobalt-55 to decay into the following amounts?
   a. 28 milligrams
   b. 14 milligrams
   c. 7 milligrams
9. How long will it take 500 grams of plutonium-241 to decay into 100 grams?
10. How long will it take 300 grams of cobalt-60 to decay into 10 grams?
11. How long will it take 30 milligrams of plutonium-238 to decay into 20 milligrams?
12. How long will it take 900 milligrams of silicon-31 to decay into 700 milligrams?
13. How long will it take a given quantity of plutonium-239 to lose 90% of its radioactivity?
14. How long will it take a given quantity of plutonium-239 to lose 95% of its radioactivity?
15. How long will it take a given quantity of krypton-91 to lose 99.9% of its radioactivity?
16. How long will it take a given quantity of carbon-14 to lose 99.9% of its radioactivity?
17. The article on the "Iceman" in Figure 9.15 stated that the frozen corpse was 5,000 to 5,500 years old. Assuming the corpse was 5,250 years old, how much carbon-14 should it contain?
18. The article on radioactive gas in Figure 9.14 at the beginning of this section stated that tritium decays at the rate of 5.5% a year; that is, 94.5% of any given quantity will remain at the end of 1 year.
   a. Find the value of b in the mathematical model for the exponential decay of tritium.
   b. Determine the half-life of tritium gas.
19. A lab technician had 58 grams of a radioactive substance. Ten days later, only 52 grams remained. Find the half-life of the radioactive substance.
20. A lab technician had 32 grams of a radioactive substance. Eight hours later, only 30 grams remained. Find the half-life of the radioactive substance.
21. In 1989, a Mayan codex (a remnant of ancient written records) was found in the remains of a thatched hut that had been buried under 15 feet of volcanic ash after a prehistoric eruption of Laguna Caldera Volcano, just north of San Salvador. An analysis of the roof material concluded that the material contained 84% of the expected amount of carbon-14 found in a living organism. Determine the age of the roof material and hence the age of the codex.
22. Two Ohlone Indian skeletons, along with burial goods such as quartz crystals, red ocher, mica ornaments, and olivella shell beads, were dug up at a construction site in San Francisco in 1989. An analysis of the skeletons revealed that they contained 88% of the expected amount of carbon-14 found in a living person. Determine the age of the skeletons.
23. In 1940, beautiful prehistoric cave paintings of animals, hunters, and abstract designs were found in the Lascaux cave near Montignac, France. Analyses revealed that charcoal found in the cave had lost 83% of the expected amount of carbon-14 found in living plant material. Determine the age of the charcoal and the paintings.

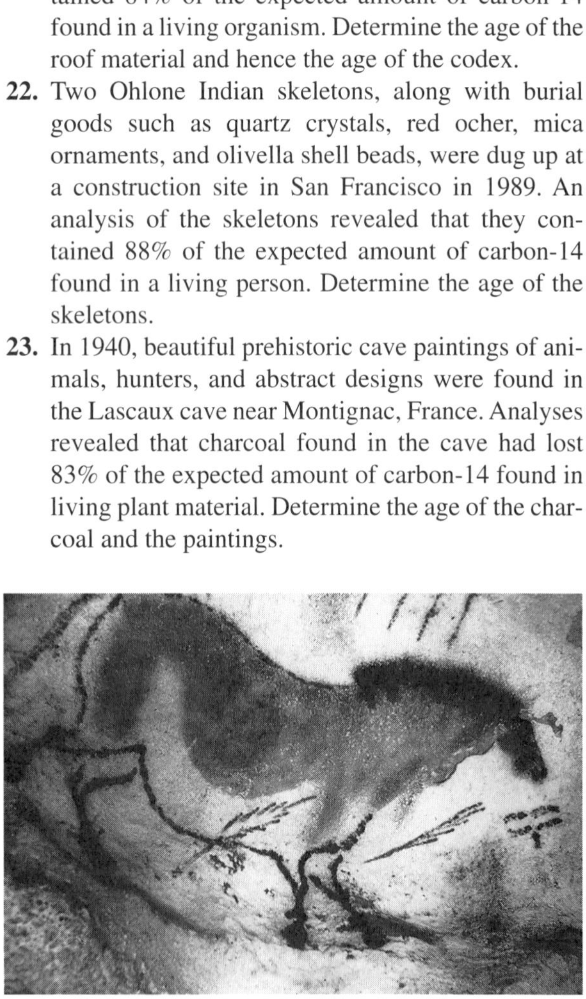
The Granger Collection, NY

24. To determine the onset of the last Ice Age, scientists analyzed fossilized wood found in Two Creeks Forest, Wisconsin. They theorized that the trees from which the wood was taken had been killed by the glacial advance. If the wood had lost 75% of its carbon-14, determine its age.

25. An ancient parchment contained 70% of the expected amount of carbon-14 found in living matter. Estimate the age of the parchment.

26. When analyzing wood found in an Egyptian tomb, scientists determined that the wood contained 55% of the expected amount of carbon-14 found in living matter. Estimate the age of the wood.

27. As noted in the article on the Shroud of Turin in Figure 9.20, radiocarbon dating analysis concluded that the shroud was made in the fourteenth century. Assuming the shroud was made in 1350 A.D., how much carbon-14 did it contain when it was tested in 1988? If the shroud was made at the time of Christ's death in 33 A.D., how much carbon-14 should it have contained when it was tested in 1988?

28. In 1993, archeologists digging in southern Turkey found an ancient piece of cloth. Radiocarbon dating determined that the cloth was 9,000 years old. How much carbon-14 did the cloth have when it was tested?

29. How much carbon-14 would you expect to find in a 5,730-year-old relic?

30. How much carbon-14 would you expect to find in an 11,460-year-old relic?

31. A museum claims that one of its mummies is 5,000 years old. An analysis reveals that the mummy contains 62% of the expected amount of carbon-14 found in living organisms. Is the museum's claim justified?

32. A museum claims that one of its skeletons is 9,000 years old. An analysis reveals that the skeleton contains 42% of the expected amount of carbon-14 found in living organisms. Is the museum's claim justified?

33. A flute carved from the wing bone of a crane was discovered in Jiahu, an excavation site of Stone Age artifacts in China's Yellow River Valley in the 1980s. Scientists estimate that the artifact has lost 63.5% of its carbon-14. Estimate the age of the flute.

Answer the following questions using complete sentences and your own words.

Concept Questions

34. What does the phrase *exponential decay* mean?

35. What does the phrase *half-life* mean?

36. In terms of mathematical models, what is the basic difference between exponential growth and exponential decay?

37. In terms of units, what is the difference between the average decay rate and the relative decay rate?

38. How does radiocarbon dating work?

39. What aspect of radiocarbon dating is controversial?

History Questions

40. Who invented radiocarbon dating? When? What other famous project did this person work on?

41. Who received a Nobel Prize for the discovery of radioactivity? When?

42. Who was the first person to receive two Nobel Prizes?

9.3 Logarithmic Scales

Until now, we have been studying exponential models that are based on the function $y = ae^{bt}$. Now let's look at another topic closely related to exponential models: logarithmic scales. A **logarithmic scale** is a scale in which logarithms serve to make data more manageable by expanding small variations and compressing large ones, as shown in Figure 9.21 on page 686. We will look at the Richter scale, which is used to rate earthquakes, and the decibel scale, which is used to rate the loudness of sounds.

Earthquakes

Most earthquakes are mild and cause little or no damage, but some are incredibly devastating. San Francisco was almost totally destroyed in 1906 by a large earthquake (and the fires it caused). Thousands of people were killed by a large quake in Armenia in 1988. However, engineers have made great progress in designing structures that can withstand major earthquakes. San Francisco survived its 1989 quake with very limited damage; although there was some major

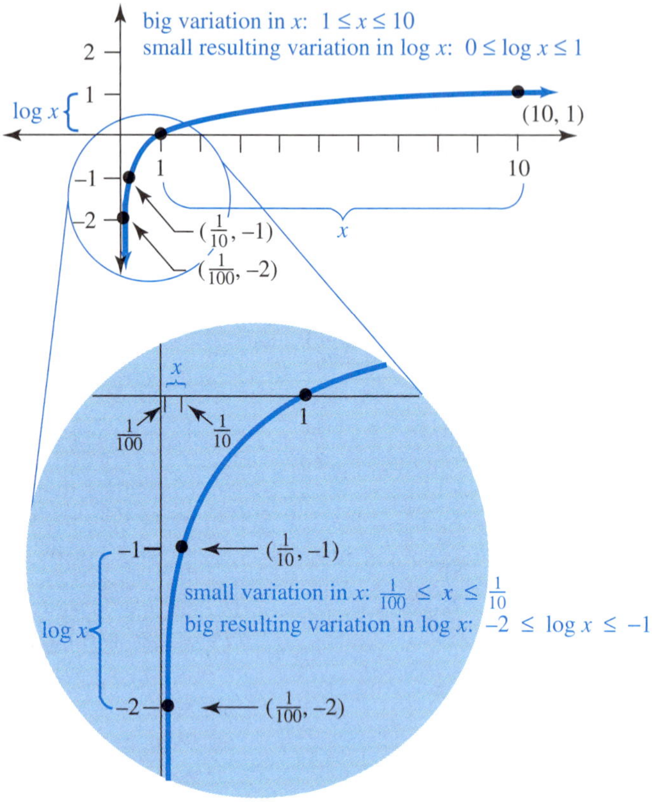

Figure 9.21

damage, most of the city survived without a scratch (contrary to many sensationalistic news reports).

Most earthquakes in recent times have been in the Middle East and along the Pacific Rim, but they are not limited to those regions. The most powerful earthquakes in North America's history were a series of three temblors in 1811–1812

Bettman/Corbis

David Weintraub/Stock Boston

San Francisco was almost totally destroyed by the great quake of 1906 (above left). Only limited damage was sustained in the quake of 1989 (above right). The few buildings that collapsed were old ones, and most were made of unreinforced masonry or stucco or were built over first-floor garages, as was the building shown above.

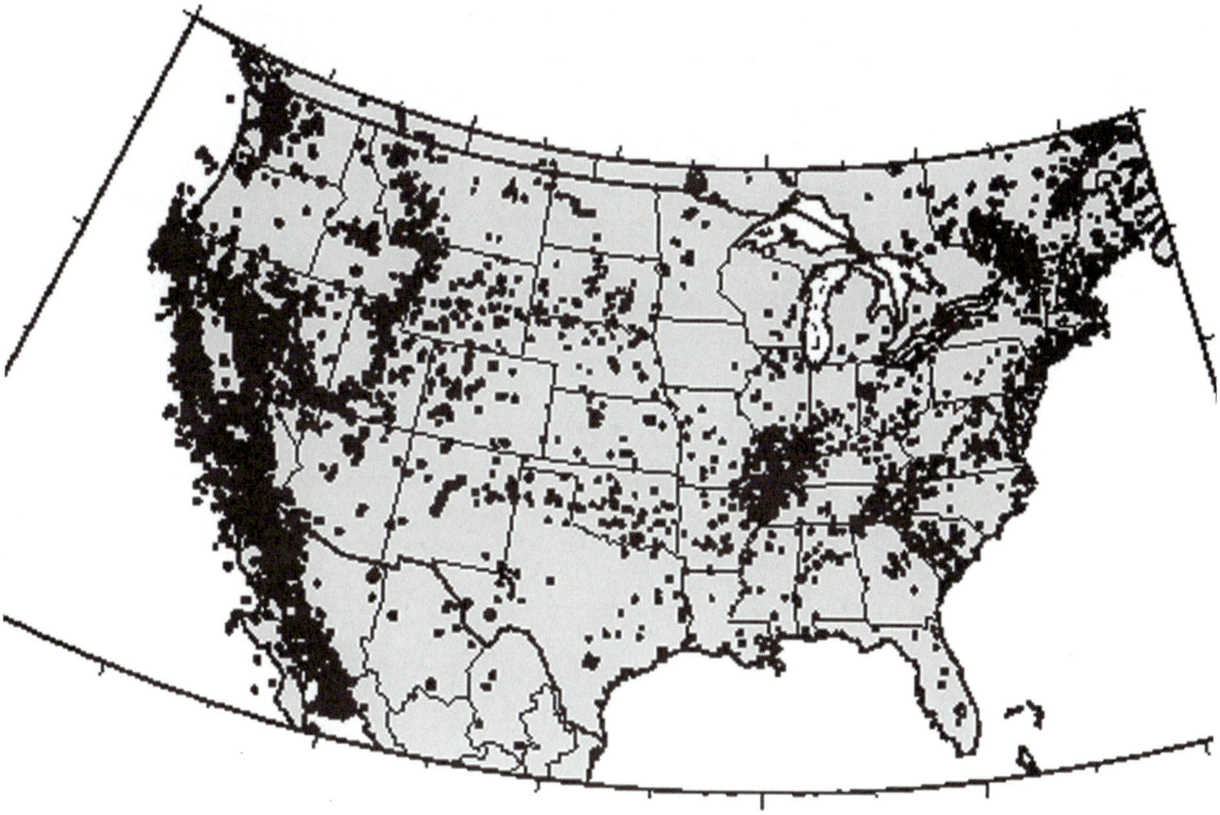

Courtesy of United States Geological Survey, 1990.

Figure 9.22
California is not the only state with tectonic stresses and thus prone to earthquakes. Each dot represents a recent earthquake.

on the New Madrid fault in Missouri. These quakes rerouted the Mississippi River, cracked plaster in Boston, and made church bells ring in Montreal. The U.S. Geological Survey now lists 39 states as having earthquake potential. (See Figure 9.22.) Geologists say that the eastern United States is certain to receive a violent jolt by 2010.* Perhaps the greatest current threat is posed by the New Madrid fault, where experts say there is a 50% chance of a serious quake by 2000. (Obviously, there was no serious quake in that region as of 2000; however, the possibility remains significant.) Because of the region's geology, such a quake would cause damage over a huge area, including parts of more than a dozen states. (See Figure 9.23.) Because of a lack of seismic building codes and other forms of preparedness, the effect would be much more devastating than necessary. David Stewart, director of the Center for Earthquake Studies at Southeast Missouri State University, stated, "Earthquake preparedness? On a scorecard of 100 I'd give San Francisco a 90, Armenia a 0 and Missouri about a 10."†

Most earthquakes are related to **tectonic stress** (that is, stress in the earth's crust). A group of more than 30 scientists from 19 different countries is compiling a global data base of stresses. A result of their efforts is the *World Stress Map*, shown in

* *Newsweek,* 30 October 1989, "East of the Rockies: A Lot of Shaking Going On."

† *San Francisco Examiner,* 26 November 1989, "Missouri Fault May See Next Big One."

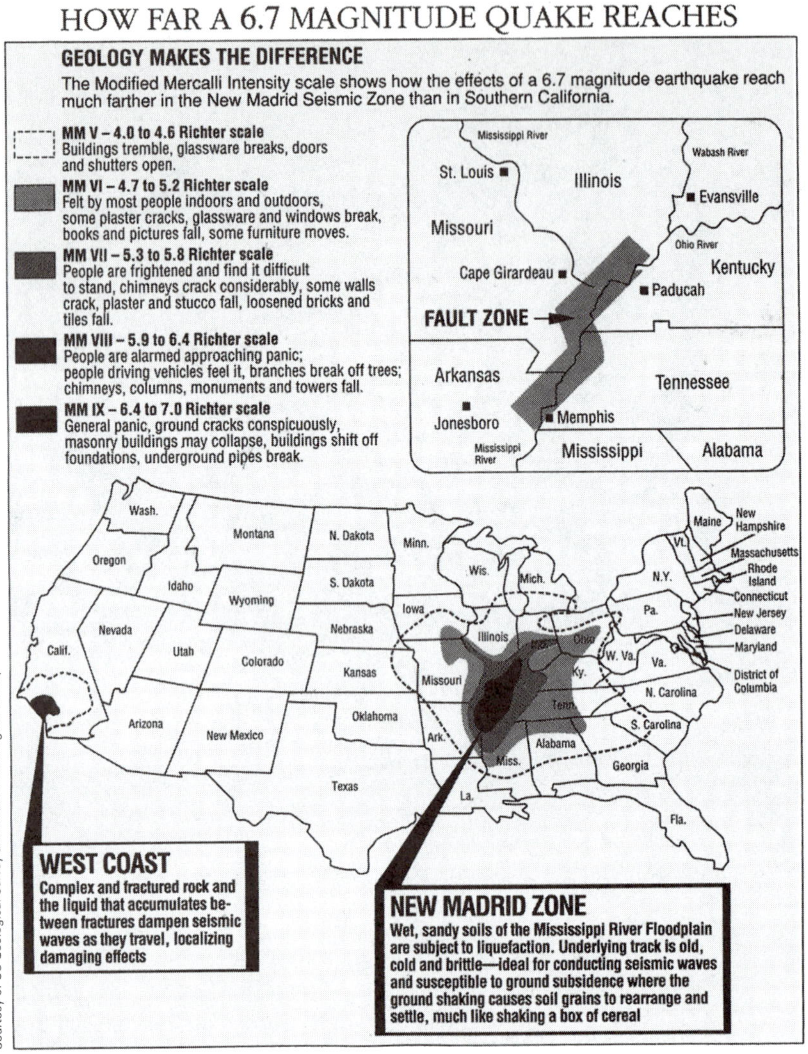

Figure 9.23

Figure 9.24, which indicates the presence of tectonic stresses throughout most of North America as well as much of Asia and Europe.

A **seismograph** (from *seismos*, a Greek word meaning "earthquake") is an instrument that records the amount of earth movement generated by an earthquake's seismic wave; the recording is called a **seismogram.** The **amplitude** of a seismogram is the vertical distance between the peak or valley of the recording of the seismic wave and a horizontal line formed if there is no earth movement; the amplitude is usually measured in micrometers (μm).

The Richter Scale

The most common method of comparing earthquakes was developed by Charles F. Richter, a seismologist at the California Institute of Technology, in 1935. Prior to Richter, it was known that the amplitude of a recording of a seismic wave is affected by the strength of the earthquake and by the distance between the earthquake and the seismograph. Richter wanted to develop a scale that would reflect

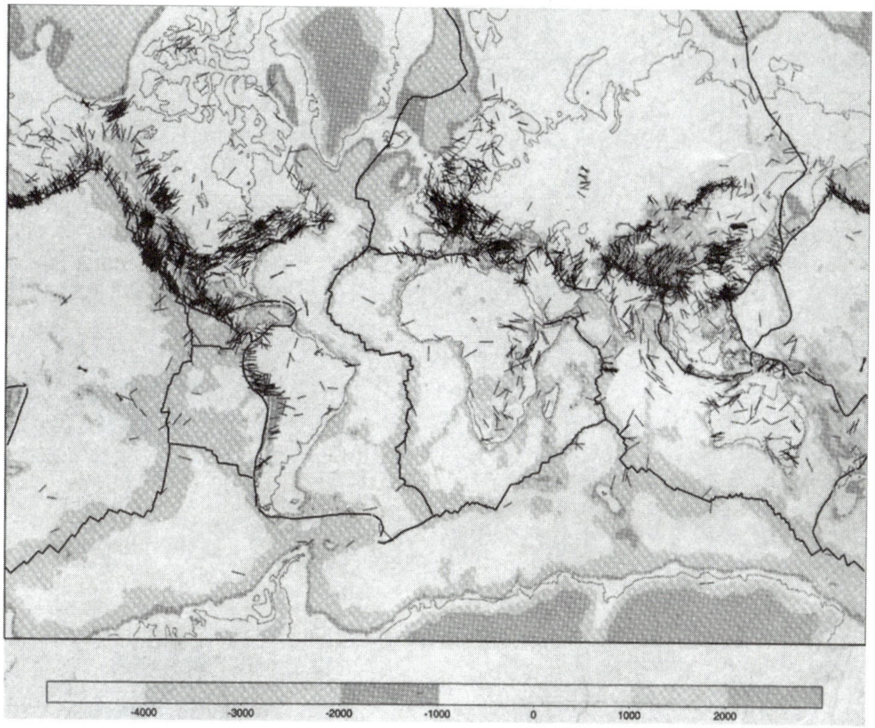

Nature, 28 September 1989

Figure 9.24
The *World Stress Map* shows the pressure of tectonic stresses in most of North America, Europe, and Asia.

only the actual strength of the earthquake and not the distance between the source (or **epicenter**) and the seismograph. Because larger earthquakes have amplitudes millions of times greater than those of smaller quakes, Richter used the common logarithm of the amplitude of the quake in developing his scale in order to compress the enormous variation inherent in earth movement down to a more manageable range of numbers.

John T. Barr/Liaison Agency

Because the epicenter of the 1994 Northridge quake was within Los Angeles itself, the city suffered more extensive damage than did San Francisco in 1989.

Examining data from many earthquakes, Richter discovered an interesting pattern: If A_{10} and A_{20} are the amplitudes of one earthquake measured 10 and 20 kilometers, respectively, from the epicenter and if B_{10} and B_{20} are similar 10- and 20-kilometer measurements for a second earthquake, as shown in Figure 9.25, then

$$\log A_{10} - \log B_{10} = \log A_{20} - \log B_{20}$$

or, equivalently, by the Division-Becomes-Subtraction Property,

$$\log \frac{A_{10}}{B_{10}} = \log \frac{A_{20}}{B_{20}}$$

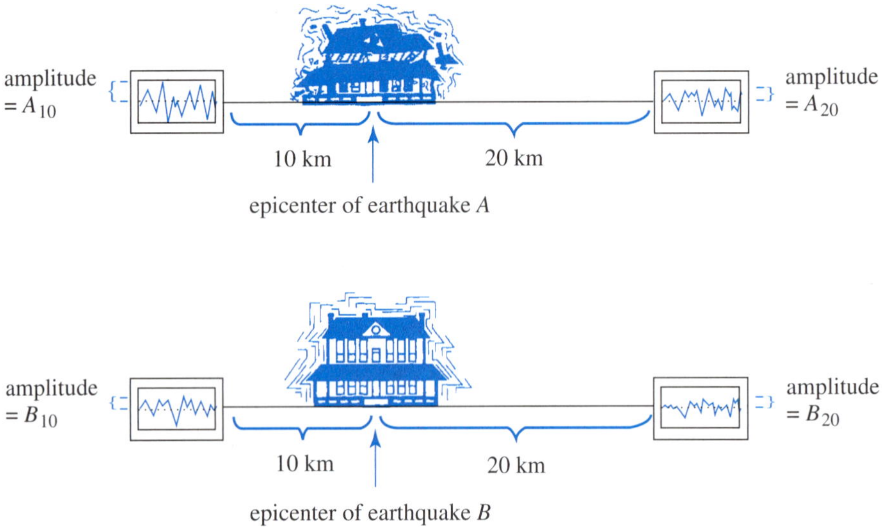

Figure 9.25

This pattern continues *regardless of the distance*. That is, if A_{95} and B_{95} are the amplitudes of two quakes measured 95 kilometers from their epicenters, then

$$\log A_{10} - \log B_{10} = \log A_{20} - \log B_{20} = \log A_{95} - \log B_{95}$$

Richter's goal was to develop a scale that would measure the actual strength of an earthquake *without being biased by the distance between the quake's epicenter and the seismograph*. This pattern went a long way toward fulfilling Richter's goal; the only difficulty was that the pattern is a comparison of two quakes, not a measurement of a single quake. Richter remedied that problem by creating a "standard earthquake" of a certain fixed strength; this standard earthquake would be a yardstick against which all future earthquakes would be measured. Richter's standard earthquake is an average of a large number of extremely small southern California earthquakes, and the magnitude of an earthquake is a measure of how much stronger than the standard a given earthquake is. Figure 9.26 gives logs of amplitudes at different distances for the standard quake of "magnitude 0."

distance (km)	20	60	100	140	180	220	260	300	340	380	420	460	500	540	580
log of amplitude (πm)	−1.7	−2.8	−3.0	−3.2	−3.4	−3.6	−3.8	−4.0	−4.2	−4.4	−4.5	−4.6	−4.7	−4.8	−4.9

Figure 9.26
Richter's standard earthquake

Richter's Definition of Earthquake Magnitude

The magnitude M of an earthquake of amplitude A is

$$M = \log A - \log A_0$$

where A_0 is the amplitude of the "standard earthquake" measured at the same distance.

EXAMPLE 1 A seismograph 20 kilometers from an earthquake's epicenter recorded a maximum amplitude of 5.0×10^6 micrometers. Find the earthquake's magnitude.

Solution

$$\begin{aligned} M &= \log A - \log A_0 \\ &= \log(5.0 \times 10^6) - \log A_0 \\ &= \log(5.0 \times 10^6) - (-1.7) \quad \text{log } A_0 \text{ is found in Figure 9.26} \\ &\approx 6.699 + 1.7 \\ &= 8.399 \\ &\approx 8.4 \end{aligned}$$

The earthquake was 8.4 on the Richter scale. (By tradition, earthquake magnitudes are rounded to the nearest tenth.) ■

An earthquake is usually measured by seismographs in at least three different locations in order to accurately determine its Richter scale rating and the location of its epicenter. The Richter scale ratings of some major earthquakes are given in Figure 9.27.

1811–1812	New Madrid, MO	8.7	1990	Philippines	7.7
1906	San Francisco	8.3	1991	India and Nepal	7.7
1933	Japan	8.9	1992	Yucca Valley, CA	7.4
1950	India	8.7	1994	Los Angeles (Northridge)	6.8
1960	Chile (the greatest earthquake ever recorded)	9.5	1995	Kobe, Japan	7.2
1964	Alaska	8.5	1994	Kuril Islands (between Russia and Japan)	8.2
1968	22 midwestern states	5.5	1997	Western Pakistan	7.3
1976	China	8.0	1999	Turkey	7.8
1983	Coalinga, CA	6.5	1999	Taiwan	7.6
1983	Hilo, Hawaii	6.3	1999	Oaxaca, Mexico	7.6
1985	Mexico City	8.1	2001	El Salvador	7.7
1988	Armenia	6.9	2001	India	7.7
1989	San Francisco/Loma Prieta (the "world series quake")	7.1	2001	Peru	8.4
1990	Iran	7.7			

Figure 9.27
Magnitudes of some major earthquakes

In addition to comparing the amplitude of an earthquake with that of Richter's artificial "standard earthquake," we can also use Richter's definition to compare the amplitudes of two actual earthquakes.

Magnitude Comparison Formula

If M_1 and M_2 are the magnitudes of two earthquakes, and if A_1 and A_2 are their amplitudes measured at equal distances, then

$$M_1 - M_2 = \log\left(\frac{A_1}{A_2}\right)$$

The Magnitude Comparison Formula is true because

$$\begin{aligned} M_1 - M_2 &= (\log A_1 - \log A_0) - (\log A_2 - \log A_0) && \text{using Richter's definition} \\ &= \log A_1 - \log A_2 \\ &= \log\left(\frac{A_1}{A_2}\right) && \text{using the Division-Becomes-Subtraction Property} \end{aligned}$$

Notice that if the second earthquake in the formula above is Richter's standard earthquake, $M_2 = 0$ (Richter's standard quake is one of magnitude zero), and A_2 is A_0, the magnitude comparison formula turns into Richter's definition of earthquake magnitude.

EXAMPLE 2 If one earthquake has magnitude 6 and another has magnitude 3, it does not mean that the first caused twice as much earth movement, as many people believe. Find how the stronger earthquake actually compares to the weaker one.

Solution We use the Magnitude Comparison Formula:

$$\begin{aligned} M_1 - M_2 &= \log\left(\frac{A_1}{A_2}\right) \\ 6 - 3 &= \log\left(\frac{A_1}{A_2}\right) \\ 3 &= \log\left(\frac{A_1}{A_2}\right) \\ 10^3 &= 10^{\log(A_1/A_2)} \\ 10^3 &= \frac{A_1}{A_2} && \text{using the Inverse Property } 10^{\log x} = x \\ A_1 &= 10^3 A_2 \end{aligned}$$

The stronger earthquake's amplitude is actually $10^3 = 1{,}000$ times that of the weaker earthquake; thus, the stronger earthquake would cause about 1,000 times as much earth movement (not twice as much)! ■

Earthquake Magnitude and Energy

The magnitude of an earthquake is determined by the amplitude of the earthquake's seismic wave and thus by the amount of ground movement caused by the quake. The amount of ground movement does not depend exclusively on the amount of energy radiated by the earthquake, because different geological compositions will transmit the energy in different ways. However, ground movement and energy are closely related.

> **Energy Formula**
>
> The energy E (in ergs) released by an earthquake of magnitude M is approximated by
>
> $$\log E \approx 11.8 + 1.45M$$

EXAMPLE 3 Find approximately how much more energy is released by an earthquake of magnitude 6 than by an earthquake of magnitude 3.

Solution

$$\begin{aligned} M_1 &= 6 & M_2 &= 3 \\ \log E_1 &\approx 11.8 + 1.45M_1 & \log E_2 &\approx 11.8 + 1.45M_2 \\ &= 11.8 + 1.45 \cdot 6 & &= 11.8 + 1.45 \cdot 3 \\ &= 20.5 & &= 16.15 \\ 10^{\log E_1} &\approx 10^{20.5} & 10^{\log E_2} &\approx 10^{16.15} \\ E_1 &\approx 10^{20.5} & E_2 &\approx 10^{16.15} \end{aligned}$$

$$\begin{aligned} \frac{E_1}{E_2} &\approx \frac{10^{20.5}}{10^{16.15}} \\ &= 10^{20.5-16.15} \\ &= 10^{4.35} \\ &= 22{,}387.21139\ldots \\ &\approx 22{,}000 \\ E_1 &\approx 22{,}000 \cdot E_2 \end{aligned}$$

The earthquake of magnitude 6 releases approximately 22,000 times as much energy as the earthquake of magnitude 3. (Recall from Example 2 that the magnitude 6 quake causes 1,000 times as much earth movement as the magnitude 3 quake.) ■

The Decibel Scale

A sound is a vibration received by the ear and processed by the brain. The **intensity of a sound** is a measure of the "strength" of the vibration; it is determined by placing a surface in the path of the sound and measuring the amount of energy in that surface per unit of area per second. This surface acts like an eardrum.

Intensity is not a good measure of loudness as perceived by the human brain, for two reasons. First, experiments have shown that humans perceive loudness on the basis of the ratio of intensities of two different sounds. For example, in order for the human ear and brain to distinguish an increase in loudness, one sound must typically have 25% more energy than another. If two sounds have intensities I_1 and I_2 and if I_1 is 25% greater than I_2, then

$$\begin{aligned} I_1 &= 125\% \text{ of } I_2 \\ &= 1.25 I_2 \end{aligned}$$

and the ratio of I_1 to I_2 is

$$\frac{I_1}{I_2} = 1.25$$

and the first sound would be perceived as slightly louder than the second. To determine loudness (as humans perceive it), we must look at the ratio of sound intensities.

The second reason why intensity is not a good measure of loudness has to do with how the brain processes sound. The human ear registers an amazing range of sound intensities. The intensity of a painfully loud sound is 100 trillion (10^{14} = 100,000,000,000,000) times that of a barely audible sound. More precisely, if I_1 is the intensity of a painfully loud sound and I_2 is the intensity of a barely audible sound, then $I_1 = 10^{14} \cdot I_2$, or $I_1/I_2 = 10^{14}$. The ear registers this range of sound intensities, but the brain condenses it to a smaller, more manageable range. The brain processes sound in a roughly logarithmic fashion.

The **decibel scale** approximates loudness as perceived by the human brain. It is based on the ratio of intensities of sounds, and its range is roughly on a par with the range in loudness that a human perceives rather than with the range in energy intensities that the ear or a mechanical device would actually measure.

Decibel Rating Definition

If a sound has intensity I (in watts per square centimeter, measured at a standard distance), then its decibel rating is

$$D = 10 \log\left(\frac{I}{I_0}\right)$$

where I_0 is a "standard intensity" ($I_0 \approx 10^{-16}$ watts/cm^2, the intensity of a barely audible sound).

The word *decibel* is abbreviated *dB*. The prefix *deci* is the metric system's prefix meaning "tenth." One decibel, therefore, is one-tenth of a bel. The *bel* is named for Alexander Graham Bell, the inventor of the telephone.

EXAMPLE 4 The background noise of a quiet library has a measured intensity of 10^{-12} watts/cm^2. Find the decibel rating of the sound.

Solution

$$I_1 = 10^{-12}$$

$$D_1 = 10 \log\left(\frac{I_1}{I_0}\right)$$

$$= 10 \log\left(\frac{10^{-12}}{10^{-16}}\right)$$
$$= 10 \log (10^{-12-(-16)})$$
$$= 10 \log 10^4$$
$$= 10 \cdot 4 \quad \text{using the Inverse Property } \log 10^x = x$$
$$= 40$$

The decibel rating of the background noise is 40 dB. ■

EXAMPLE 5 When two students started a conversation in the library (see Example 4), the intensity shot up to 10^{-10} watts/cm^2. Find the increase in decibels.

Solution

$$I_2 = 10^{-10}$$
$$D_2 = 10 \log\left(\frac{I_2}{I_0}\right)$$
$$= 10 \log\left(\frac{10^{-10}}{10^{-16}}\right)$$
$$= 10 \log(10^{-10-(-16)})$$
$$= 10 \log 10^6$$
$$= 10 \cdot 6 \quad \text{using the Inverse Property } \log 10^x = x$$
$$= 60$$
$$D_2 - D_1 = 60 - 40 \quad D_1 = 40, \text{ from Example 4}$$
$$= 20$$

The increase in decibels, or **dB gain,** is 20 dB. ■

Mark Burnett/Stock Boston

Recording engineers monitor the music's decibel level so as to get a high-quality recording.

An alternative approach to finding the dB gain is to use a slightly altered version of the Decibel Formula.

dB Gain Formula

If I_1 and I_2 are the intensities of two sounds, then the dB gain is

$$D_1 - D_2 = 10 \log\left(\frac{I_1}{I_2}\right)$$

The dB Gain Formula is true because

$$\begin{aligned}
D_1 - D_2 &= 10 \log\left(\frac{I_1}{I_0}\right) - 10 \log\left(\frac{I_2}{I_0}\right) && \text{Decibel Rating Definition}\\
&= 10\left[\log\left(\frac{I_1}{I_0}\right) - \log\left(\frac{I_2}{I_0}\right)\right] && \text{factoring}\\
&= 10[(\log I_1 - \log I_0) - (\log I_2 - \log I_0)] && \text{Division-to-Subtraction Property}\\
&= 10[\log I_1 - \log I_2]\\
&= 10 \log\left(\frac{I_1}{I_2}\right) && \text{Division-to-Subtraction Property}
\end{aligned}$$

EXAMPLE 6 Use the dB Gain Formula to find the increase in decibels when the quiet library in Example 4 became somewhat less quiet due to the students' conversation in Example 5.

Solution

$$\begin{aligned}
I_1 &= 10^{-10} \text{ watts/cm}^2 && \text{(conversation)}\\
I_2 &= 10^{-12} \text{ watts/cm}^2 && \text{(library background noise)}
\end{aligned}$$

$$\begin{aligned}
D_1 - D_2 &= 10 \log\left(\frac{I_1}{I_2}\right)\\
&= 10 \log\left(\frac{10^{-10}}{10^{-12}}\right)
\end{aligned}$$

Approximate Levels of Sounds

0 dB	threshold of hearing	80 dB	auto interior
10 dB	soundproof room	90 dB	bus or truck interior
20 dB	radio, television, or recording studio	100 dB	electric saw
30 dB	quiet lecture hall; bedroom	110 dB	loud orchestral music (in audience)
40 dB	quiet room in home	120 dB	amplified rock music; nearby jet engine
50 dB	restaurant; private office	130 dB	artillery fire at close proximity (threshold of pain)
60 dB	normal conversation; business office		
70 dB	average street noise; loud telephone bell		

The Granger Collection, NY

HISTORICAL NOTE

Alexander Graham Bell

1847–1922

Alexander Melville Bell lectured at the University of Edinburgh (Scotland), at the University of London, and in Boston. He developed a system of visible speech for the deaf, with symbols for every sound of the human voice.

Alexander Graham Bell, the son of Alexander Melville Bell and the husband of a deaf woman, had his own school of vocal physiology in Boston and was very active in issues related to education of the deaf. In 1875, he conceived the idea of the telephone. On March 10, 1876, he used his experimental apparatus to transmit the now famous "Watson, come here; I want you" to his assistant. Later that same year, the telephone was introduced to the world at the Philadelphia Centennial Exposition. The Bell Telephone Company was organized a year later.

Bell also established the Volta Laboratory in Washington, D.C., where the first successful phonograph record was produced. He invented both the flat and cylindrical wax recorders for phonographs, as well as the photophone, which transmits speech by light rays, and the audiometer, which measures a person's hearing ability. He investigated the nature and causes of deafness and studied its heredity. He helped found the magazine *Science*, was president of the National Geographic Society, and was a regent of the Smithsonian Institution.

The Granger Collection, NY

Bell's telephone, as described in an 1877 newspaper. The first two figures are the transmitter and receiver that were shown at the Philadelphia Exposition.

$$= 10 \log(10^{-10-(-12)})$$
$$= 10 \log 10^2$$
$$= 10 \cdot 2 = 20 \text{ dB gain}$$

Notice that $I_1/I_2 = 10^{-10}/10^{-12} = 10^2$, so the conversation had a sound intensity $10^2 = 100$ times that of the background noise. ■

EXAMPLE 7 The noise on a busy freeway varies from 81 dB to 92 dB. Find the corresponding variation in intensities.

Solution We are given $D_1 = 92$ dB and $D_2 = 81$ dB, and we are asked to compare I_1 and I_2.

$$D_1 - D_2 = 10 \log\left(\frac{I_1}{I_2}\right)$$
$$92 - 81 = 10 \log\left(\frac{I_1}{I_2}\right)$$
$$11 = 10 \log\left(\frac{I_1}{I_2}\right)$$
$$1.1 = \log\left(\frac{I_1}{I_2}\right)$$
$$10^{1.1} = 10^{\log (I_1/I_2)}$$
$$10^{1.1} = \frac{I_1}{I_2} \qquad \text{using the Inverse Property } 10^{\log x} = x$$
$$I_1 = 10^{1.1} I_2 \approx 12.5893 I_2$$

The louder freeway noise has almost 13 times the sound intensity of the quieter freeway noise. ■

9.3 Exercises

In Exercises 1–4, find the magnitude of the given earthquake.

1. An earthquake was measured by a seismograph 100 kilometers from the epicenter that recorded a maximum amplitude of 3.9×10^4 μm.

2. An earthquake was measured by a seismograph 60 kilometers from the epicenter that recorded a maximum amplitude of 6.3×10^5 μm.

3. An earthquake was measured by three seismographs. The closest, 60 kilometers from the epicenter, recorded a maximum amplitude of 250 μm. A second seismograph 220 kilometers from the epicenter recorded a maximum amplitude of 40 μm. The third seismograph, 460 kilometers from the epicenter, recorded a maximum amplitude of 4 μm.

4. An earthquake was measured by three seismographs. The closest, 20 kilometers from the epicenter, recorded a maximum amplitude of 25 μm. A second seismograph 60 kilometers from the epicenter recorded a maximum amplitude of 2 μm. The third seismograph, 180 kilometers from the epicenter, recorded a maximum amplitude of 0.50 μm.

In Exercises 5–12, use the information in Figure 9.27 (page 691) to compare the two earthquakes given by (a) finding how much more earth movement was caused by the larger and (b) finding how much more energy was released by the larger.

5. San Francisco (1989) and San Francisco (1906)

6. New Madrid and Los Angeles

7. San Francisco (1906) and Los Angeles

8. San Francisco (1989) and Los Angeles

9. Iran and Turkey

10. Turkey and Taiwan

11. New Madrid and Coalinga

12. Yucca Valley and Los Angeles

13. Shortly after the 1989 San Francisco quake, it was announced that the quake was of magnitude 7.0. Later, after the data from more seismographs had been analyzed, the rating was increased to 7.1.

a. How large an increase of earth movement corresponds to this increase in magnitude?

b. How large an increase of energy released corresponds to this increase in magnitude?

14. Find a rule of thumb for the increase in earth movement that corresponds to an increase in Richter magnitude of one unit.

HINT: Follow Example 2, using magnitudes $x + 1$ and x.

15. Find a rule of thumb for the increase in energy released that corresponds to an increase in Richter magnitude of one unit.

HINT: Follow Example 3, using magnitudes $x + 1$ and x.

In Exercises 16–20, find the decibel rating of the given sound.

16. a very faint whisper at 10^{-14} watts/cm^2
17. a television at 10^{-9} watts/cm^2
18. a quiet residence at 3.2×10^{-12} watts/cm^2
19. a vacuum cleaner at 2.5×10^{-9} watts/cm^2
20. an outboard motor at 1.6×10^{-6} watts/cm^2

In Exercises 21–24, find the dB gain for the given sound.

21. a whisper increasing from 10^{-14} watts/cm^2 to 10^{-13} watts/cm^2
22. noise in a dormitory increasing from 3.2×10^{-12} watts/cm^2 to 2.1×10^{-11} watts/cm^2
23. a stereo increasing from 3.9×10^{-9} watts/cm^2 to 3.2×10^{-8} watts/cm^2
24. a motorcycle increasing from 6.3×10^{-8} watts/cm^2 to 3.1×10^{-6} watts/cm^2
25. If a single singer is singing at 74 dB, how many singers have joined him if the level increases to 81 dB and each singer is equally loud?
26. If a single singer is singing at 74 dB, how many singers have joined him if the level increases to 83 dB and each singer is equally loud?
27. If a single singer is singing at 74 dB, how many singers have joined her if the level increases to 77 dB and each singer is equally loud?
28. If a single singer is singing at 74 dB, how many singers have joined her if the level increases to 82.5 dB and each singer is equally loud?
29. If a single trumpet is playing at 78 dB, how many trumpets have joined in if the level increases to 85.8 dB and each trumpet is equally loud?
30. If a single trumpet is playing at 78 dB, how many trumpets have joined in if the level increases to 88 dB and each trumpet is equally loud?
31. Find a rule of thumb for the dB gain if the number of sound sources doubles (where each source produces sounds at the same level).
32. Find a rule of thumb for the dB gain if the number of sound sources increases tenfold (where each source produces sounds at the same level).
33. According to the articles in Figures 9.28 and 9.29, the rock group Pink Floyd reduced its volume to protect Venice's buildings. How much effect would the stated decrease in volume have?
34. According to the article in Figure 9.30, the Navy will restrict its sonar operators to a lower sound level to protect marine mammals. How much effect would the stated decrease in volume have?

Answer the following questions using complete sentences and your own words.

Concept Question

35. Compare and contrast the Richter scale and the decibel scale. Why are the two formulas so similar? Why does each formula involve logarithms?

History Questions

36. Name three of Bell's other inventions besides the telephone.
37. The inventor of the Richter scale performed his research at what school?
38. In addition to electronics, what field of study interested Alexander Graham Bell?

Venice balks at concert by floating Pink Floyd

UNITED PRESS INTERNATIONAL
Reprinted with permission.

VENICE — The British rock band Pink Floyd arrived yesterday to prepare for a weekend world television concert as promoters and city leaders argued over whether the group's decibels would damage ancient monuments.

Pink Floyd originally was scheduled to perform tomorrow on a giant floating platform moored alongside the lagoon city's world famous St. Mark's Square.

But the city's superintendent for architectural and environmental property vetoed the idea on grounds that vibrations from the group's amplified music could damage the structures of such historic buildings as St. Mark's Basilica and the Ducal Palace.

Yesterday morning, the organizers and city officials reached agreement that the concert could go ahead provided the floating platform is moored far enough out in the Venice lagoon to restrict the sound hitting the famed buildings to 60 decibels.

Scientists calculated that could be achieved if the huge platform—which measures 295 feet by 88 feet—is anchored 150 yards away from the St. Mark's Square shore.

Figure 9.28

Pink Floyd's Venice concert draws 200,000

ASSOCIATED PRESS

Reprinted with permission.

VENICE — Fans of the rock group Pink Floyd, including Woody Allen and Tom Cruise, more than doubled the population of this canal city to see the band perform on a floating stage near St. Mark's Square.

100 Million Listeners

It was broadcast live in Italy and 23 other countries, reaching a total audience of 100 million people, according to the state-run RAI network. It was not broadcast in the United States.

The concert set an attendance record as Italy's largest rock event and attracted more visitors to the historic city than did the annual Carnival festival.

City preservation officials had threatened to deny permission for the concert. But on Thursday they dropped their objections when concert organizers agreed to measures to reduce potential damage to the surrounding buildings.

A Quieter Sound

The stage was pulled farther from the shore than previously planned and the rock group reduced its volume to a maximum of 60 decibels, instead of 105 as planned.

Italy's vice premier, Gianni De Michelis, spoke out in favor of the concert.

"Venice must give space to all cultural languages, including that of rock music," he said.

Figure 9.29

Suit filed to halt Navy sonar system

SAN FRANCISCO CHRONICLE,
8 AUGUST 2002

Environmental groups sued . . . to stop the U.S. Navy from deploying a new anti-submarine sonar system that they say could harm sea creatures wherever it's used in the world's oceans. The federal lawsuit seeks an injunction against using the low-frequency active sonar system, which critics fear will threaten entire populations of whales, dolphins, seals and other marine mammals. . . . The Navy says active sonar arrays are capable of generating 215 decibels of sound—dangerous for marine mammals—but that operators will be restricted to 180 decibels, a sound level emitted by some whales. . . . Naomi Rose, marine mammal scientist for the New York-based Humane Society of the United States, . . . called a safety exposure level of 180 decibels "completely unacceptable and not substantiated by data." . . . "It's not precautionary, as the Marine Mammal Protection Act demands," Rose said, "and is out of step with the standards that other countries are setting."

Figure 9.30

Chapter 9 Review

TERMS

amplitude of a seismogram
average decay rate
average growth rate
base of a logarithmic function
base of an exponential function
common logarithm
dB gain
decibel
delta notation
dependent and independent variables
epicenter
exponential function
exponential model
function
half-life
intensity of a sound (versus its perceived loudness)
logarithm
logarithmic scale
mathematical model
natural logarithm
radioactive decay
rate of change
relative decay rate
relative growth rate
seismogram
seismograph
tectonic stress

REVIEW EXERCISES

In Exercises 1–3, find the value of x.

1. $x = \log_3 81$
2. $2 = \log_x 16$
3. $5 = \log_5 x$
4. State the Inverse Properties of the common logarithm.
5. State the Inverse Properties of the natural logarithm.
6. State the Division-Becomes-Subtraction Property of the common logarithm.
7. State the Division-Becomes-Subtraction Property of the natural logarithm.
8. State the Exponent-Becomes-Multiplier Property of the common logarithm.
9. State the Exponent-Becomes-Multiplier Property of the natural logarithm.
10. State the Multiplication-Becomes-Addition Property of the common logarithm.
11. State the Multiplication-Becomes-Addition Property of the natural logarithm.

In Exercises 12–14, rewrite each of the expressions (if possible) using properties of logarithms.

12. $\log 3x - \log 4 + \log 7x^2$
13. $\log(x + 2)$
14. $\log(\frac{x}{2})$
15. Solve $520\, e^{0.03x} = 730$.
16. Solve $\ln x - \ln 4 = \ln 2$.
17. Solve $\log 5x + \log x^2 - \log x = 12$.
18. A bacteria culture had a population of about 10,000 at 12 noon. At 2 P.M., the population had grown to 25,000.
    a. Develop the mathematical model that represents the population of the bacteria culture.
    b. Use the model to predict the population at 6 A.M. the next day.
    c. Use the model to predict when the population will double its original size.
    d. On what assumptions is the model based?
19. Cobalt-60 is used by hospitals in radiation therapy. Suppose a hospital purchases 300 grams.
    a. Develop the mathematical model that represents the amount of cobalt-60 at time *t*.
    b. Predict the amount *lost* if delivery takes two weeks.
    c. Predict the amount *remaining* after one year (if none is used).
    d. Determine how long it will take for 90% of the cobalt-60 to decay.
20. Archeologists found a wooden bowl in an Indian burial mound. The bowl contained 73% of the amount of carbon-14 found in living matter. Estimate the age of the burial mound.
21. An earthquake was measured by two seismographs. The closest, 20 kilometers from the epicenter, recorded a maximum amplitude of 25 μm. The second seismograph, 60 km from the epicenter, recorded a maximum amplitude of 2 μm. Find the magnitude of the earthquake.
22. A 1989 earthquake in Newcastle, Australia measured 5.5 on the Richter scale. The 1989 San Francisco earthquake measured 7.1.
    a. How much more earth movement was caused by the San Francisco quake?
    b. How much more energy was released by the San Francisco quake?
23. Find the decibel rating of a scooter at 1.6×10^{-6} watts/cm^2.
24. The scooter in Exercise 23 increases to 1.3×10^{-5} watts/cm^2. Find the dB gain.
25. If a single trumpet is playing at 78 dB, how many trumpets have joined in if the level increases to 84 dB?

Answer the following questions using complete sentences and your own words.

Concept Questions

26. What is an exponential function?
27. What is a logarithmic function?
28. In terms of mathematical models, what is the basic difference between exponential growth and exponential decay?
29. How does radiocarbon dating work? What aspect of radiocarbon dating is controversial?

History Questions

30. What underlying theme unites all of Alexander Graham Bell's professional interests?
31. Name three of Bell's other inventions besides the telephone.
32. Who introduced the symbol *e*? What other symbols did this person invent?
33. Who invented logarithms? What motivated their invention?
34. Who invented radiocarbon dating?
35. Who was the first person to receive two Nobel Prizes? For what was the first prize awarded?

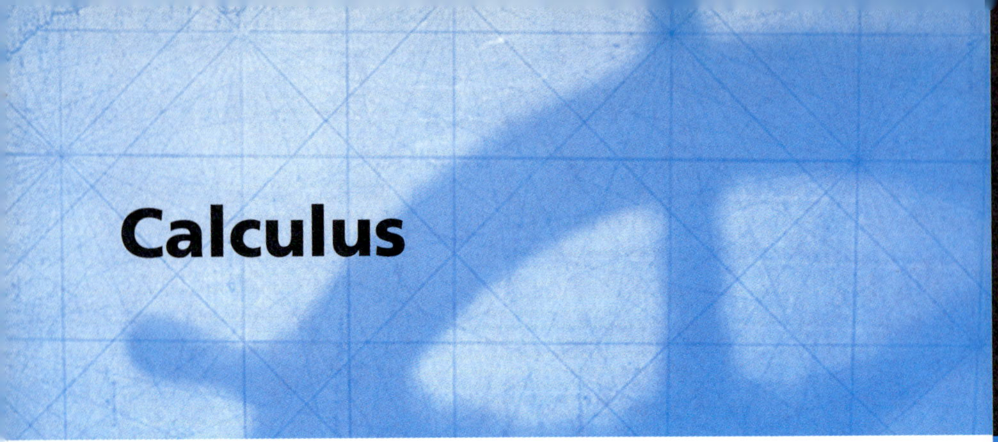

10 Calculus

The word *calculus* comes from a Latin word that means "pebble." The manipulation of piles of pebbles was one of the earliest methods of calculation. The Roman abacus (or "calculator") used pebbles instead of beads. Physicians use the word *calculus* to refer to unwelcome stonelike substances in the body, such as kidney stones and gallstones. Dentists clean calculus off their patients' teeth.

In modern mathematics, calculus is the name for the mathematical tool that is used to study how things change and the rate at which they change. Algebra, geometry, and all other branches of mathematics that preceded calculus either are useful only in the study of unchanging quantities or allow for change but do not study the change itself. Geometry studies unchanging shapes, such as a circle with a radius of 4 feet or a triangle with sides of length 3 inches and 4 inches and a hypotenuse of length 5 inches. Algebra allows for change; for example, in the algebraic equation $y = 3x + 1$, the value of the variable x could change from 5 to 6. However, algebra does not focus on the change itself. Calculus provides the tools we need to focus on change.

We live in a world of change, so it should not be surprising that calculus is an important subject. Practically every development in science and technology from the late 1600s to modern times is connected with calculus. Calculus is used to study gravity, artillery, flight, the movements of the planets, heat, light, sound, electricity, magnetism, the design of structures and machines, water flow, population growth, efficient resource allocation, and production cost minimization. College students majoring in physics, engineering, chemistry, biology, economics, architecture, and business are routinely required to study calculus. Calculus has had a greater impact than any other branch of mathematics.

It is said that Isaac Newton and Gottfried Wilhelm Leibniz invented calculus, each working independently of the other. It is generally agreed that the two mathematicians share equal "billing"; Newton had the calculus first (1665–1666, compared to Leibniz, 1673–1676), but Leibniz published it first (1684–1686, compared to Newton, 1711–1736) and formulated it in a more useful manner. However, to say that these two men invented calculus is overly simplistic. While their achievements were tremendous, they did not work in a vacuum. Many of the ideas of calculus came from others. Newton and Leibniz absorbed these ideas, saw new features hidden in them and new relationships between them, perfected them, and added their own ideas.

In this chapter, we explore the history of calculus and its place in our culture, and we investigate its main concepts.

10.0 Review of Ratios, Parabolas, and Functions

This chapter draws on a diverse set of algebraic and geometric topics. Because calculus is a study of how things change and the rate at which they change, we will review rates and two closely related topics: ratios and similar triangles. One of the questions that originally motivated the invention of calculus involves parabolas, so we will also review parabolas. Finally, because the central concept of calculus involves secant lines and their slopes and uses functional notation, we will review these topics.

Ratios, Proportions, and Rates

A **ratio** is a comparison of two quantities, usually expressed as a fraction. In fact, a fraction is frequently called a "*ratio*nal number," because one meaning of the word *rational* is "having to do with ratios." A college might have a ratio of two men to three women, that is, of (2 men)/(3 women). If 3,000 women are enrolled at the school, then there should be 2,000 men, because (2,000 men)/(3,000 women) = (2 men)/(3 women). This equation is an example of a **proportion,** which is an equality of two ratios.

A **rate** is a ratio that is used as a form of measurement. If someone drives 45 miles in one hour, then the ratio of distance to time is distance/time = (45 miles)/(1 hour) = 45 miles per hour. This is the driver's average speed, or average *rate*. If a $2,000 deposit earns $100 interest in 1 year, then the ratio of interest to principal is interest/principal = $100/$2000 = 1/20 = 0.05 = 5%. This is called the interest *rate*.

EXAMPLE 1 A car's odometer read 57,362.5 at 2 P.M. and 57,483.8 at 4 P.M.

a. Find the car's average rate.
b. Use proportions to find the distance that would be traveled in 3 more hours at the same average rate.

Solution **a.** The average rate is the ratio of the change in distance to the change in time:

$$\text{change in distance} = 57{,}483.8 - 57{,}362.5 = 121.3 \text{ miles}$$
$$\text{change in time} = 4 - 2 = 2 \text{ hours}$$
$$\text{average rate} = \frac{\text{change in distance}}{\text{change in time}} = \frac{121.3 \text{ mi}}{2 \text{ hr}} = \frac{60.65 \text{ mi}}{1 \text{ hr}}$$
$$= 60.65 \text{ miles per hour}$$

b. To find the distance traveled in three more hours, we solve the proportion

$$\frac{60.65 \text{ mi}}{1 \text{ hr}} = \frac{x}{3 \text{ hr}}$$
$$x = 3 \cancel{\text{hr}} \cdot 60.65 \text{ mi}/\cancel{\text{hr}} = 181.95 \text{ mi}$$ using dimensional analysis to cancel "hours," as discussed in Appendix E ■

The average speed, or average rate, of a car is the ratio of distance to time; speed = distance/time. If we multiply each side of this equation by time, we get distance = speed · time (or distance = rate · time). In part (b) of Example 1, we used a proportion to find a distance. Instead, we could use this formula.

$$\text{distance} = \text{rate} \cdot \text{time} = 60.65 \text{ mi/hr} \cdot 3 \text{ hr} = 181.95 \text{ mi}$$

Delta Notation

Delta notation is frequently used to describe changes in quantities. Delta notation consists of the symbol Δ (the Greek letter delta), which means "change in," followed by a letter that refers to the quantity that changes. Thus, in Example 1, the change in distance could be written Δd, and the change in time could be written Δt (read these as "delta d" and "delta t"). Expressed in delta notation, the car's average rate would be written as

$$\text{average rate} = \frac{\Delta d}{\Delta t} = \frac{121.3 \text{ mi}}{2 \text{ hr}} = 60.65 \text{ miles per hour}$$

This average rate is the ratio of the change in distance ($\Delta d = 121.3$ mi) to the change in time ($\Delta t = 2$ hr). Thus, it is a ratio of two changes, or a **rate of change.**

EXAMPLE 2 A karat is a unit for measuring the fineness of gold. Pure gold is 24 karats fine; gold that is less than 24 karats is an alloy rather than pure gold. The number of karats is the numerator of a ratio between the amount of pure gold and the total amount of material; the denominator of that ratio is 24.

a. If a ring is 18-karat gold, find what portion of the ring is pure gold.
b. If the ring weighs 7 grams, find the amount of pure gold in the ring.

Solution **a.**

$$\frac{\text{amount of pure gold}}{\text{total amount of material}} = \frac{18 \text{ units of pure gold}}{24 \text{ units of material}} = \frac{3 \text{ units of pure gold}}{4 \text{ units of material}}$$

Therefore, three-fourths of the ring is pure gold.

b. To find the amount of pure gold, we solve the following proportion:

$$\frac{3 \text{ units of pure gold}}{4 \text{ units of material}} = \frac{x \text{ grams pure gold}}{7 \text{ grams of material}}$$

$$x = 7 \cdot \frac{3}{4} = \frac{21}{4} = 5.25 \text{ grams}$$

The ring has 5.25 grams of pure gold in it. ■

Similar Triangles

If a triangle is magnified, the sides will enlarge, but *the angles do not enlarge*. For example, if we have a triangle with sides of length 1, 1, and $\sqrt{2}$, then its angles would be 45°, 45°, and 90°. If we magnified that triangle by a factor of 2, the lengths of the sides of the new triangle would be 2, 2, and $2\sqrt{2}$, but the angles would still be

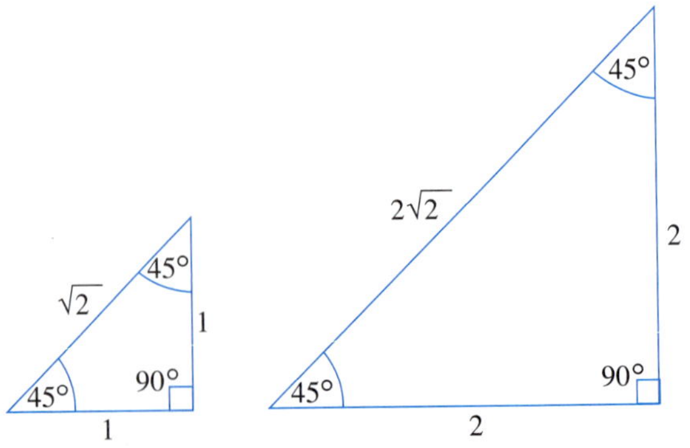

Figure 10.1

45°, 45°, and 90°. These two triangles are called **similar triangles** because they have the same angles. Similar triangles have the same shape but different sizes, as shown in Figure 10.1.

Notice that the ratios of corresponding sides are always equal to 2. In particular,

- The ratio of the bottom side of the larger triangle to the bottom side of the smaller triangle is $2/1 = 2$.
- The ratio of the right side of the larger triangle to the right side of the smaller triangle is $2/1 = 2$.
- The ratio of the top side of the larger triangle to the top side of the smaller triangle is $2\sqrt{2}/\sqrt{2} = 2$.

Similar Triangles and the Great Pyramid

The fact that the ratios of corresponding sides of similar triangles are always equal was first demonstrated by the Greek mathematician Thales around 600 B.C. He used this fact to measure the height of the Great Pyramid of Cheops. He stuck a pole into the ground right at the tip of the shadow of the pyramid. This created two similar triangles, one formed by the pyramid and its shadow, the other formed by the pole and its shadow. Suppose we call the height of the pyramid h, its base b, the length of its shadow S, the height of the pole p, and the length of its shadow s, as shown in Figure 10.2. Then

$$\frac{h}{p} = \frac{\frac{1}{2}b + S}{s}$$

$$h = \frac{\frac{1}{2}b + S}{s} \cdot p$$

The lengths of the pole, the two shadows, and the base of the pyramid were easily measured, so the height of the pyramid could be calculated.

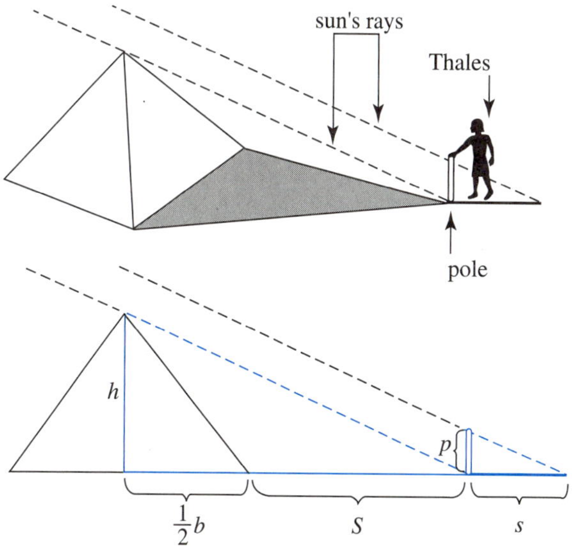

Figure 10.2

Similar Triangles

If two triangles are similar, the ratios of their corresponding sides are always equal.

$$\frac{a}{c} = \frac{b}{d}$$

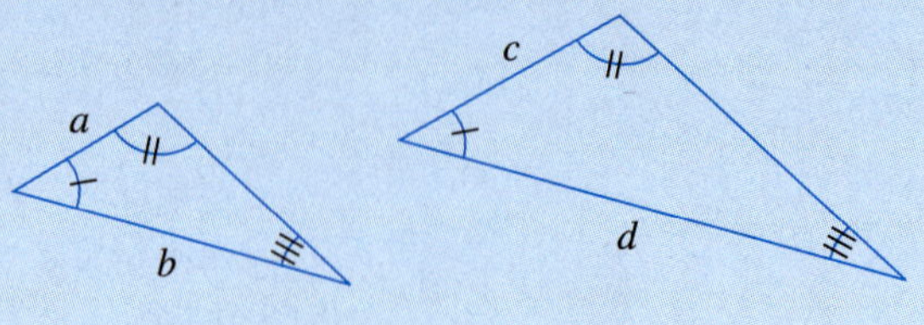

In the above illustrations, the two triangles are similar because they have the same angles. In particular, the single slash mark in the left angles of the two triangles means that those two angles are equal. Similarly, the double slash mark in the top angles means that those two angles are equal, and the triple slash mark in the third angles means that those two angles are equal.

EXAMPLE 3 The two triangles in Figure 10.3 are similar because they have the same angles, as indicated by the marks inside their angles. Find the lengths of the unknown sides.

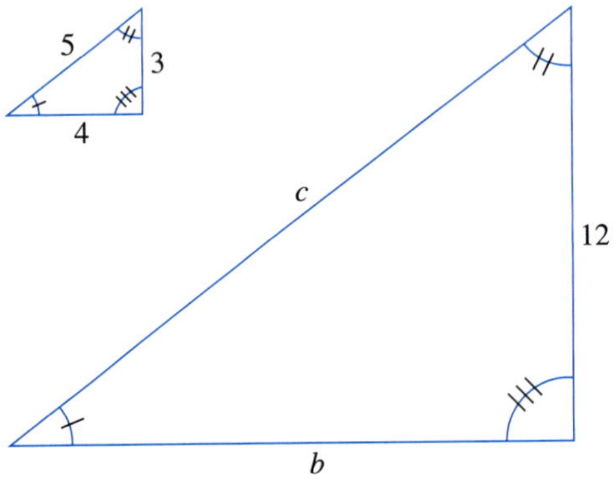

Figure 10.3

Solution The two triangles are similar, so the ratios of their corresponding sides are equal:

$$\frac{b}{4} = \frac{12}{3} \qquad b \text{ corresponds to 4 and 12 corresponds to 3}$$

$$4 \cdot \frac{b}{4} = 4 \cdot \frac{12}{3} \qquad \text{solving for } b$$

$$b = 4 \cdot 4 = 16$$

We can find c in a similar manner:

$$\frac{c}{5} = \frac{12}{3} \qquad c \text{ corresponds to 5 and 12 corresponds to 3}$$

$$5 \cdot \frac{c}{5} = 5 \cdot \frac{12}{3} \qquad \text{solving for } c$$

$$c = 5 \cdot 4 = 20$$

■

Parabolas

A parabola always has this shape:

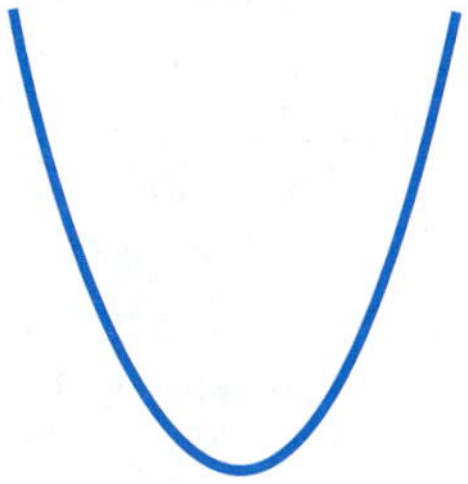

A parabola can be fatter or thinner than shown above, or it can be upside down or sideways, as shown in Figure 10.4, but it will always have the same basic shape.

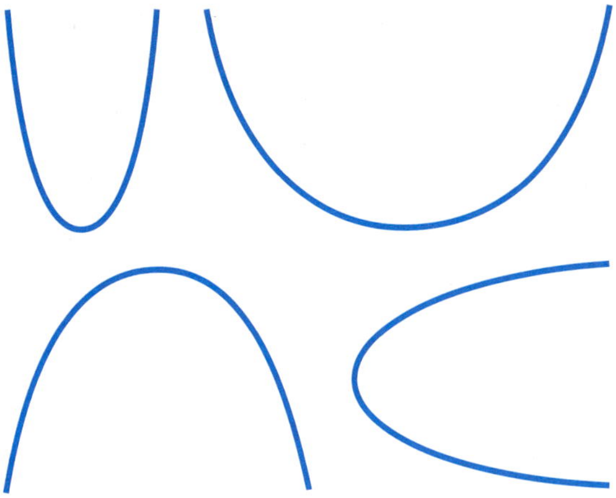

Figure 10.4

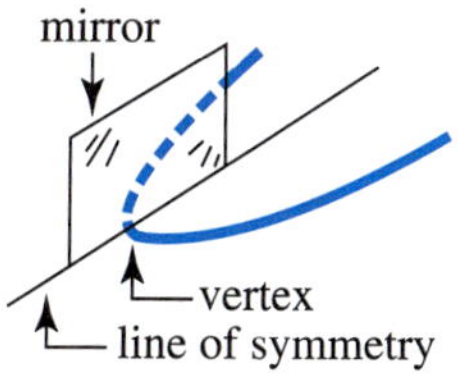

Figure 10.5

One important aspect of a parabola's shape is that it is **symmetrical;** that is, there is a line along which a mirror could be placed so that one side of the line is the mirror image of the other. That line is called the **line of symmetry.** The point at which the parabola intersects the line of symmetry is called the **vertex** (see Figure 10.5).

The equation of a parabola always involves a quadratic expression. Typically, it is either of the form $y = ax^2 + bx + c$ or of the form $x = ay^2 + by + c$. We can graph a parabola by plotting points, as long as we find enough points that the vertex is included.

EXAMPLE 4 Graph the parabola $y = x^2 - 6x + 5$ by plotting points until the vertex is found. Find the vertex and the equation of the line of symmetry.

Solution Substituting for x yields the following points:

x	0	1	2	3	4	5	6
y	5	0	−3	−4	−3	0	5

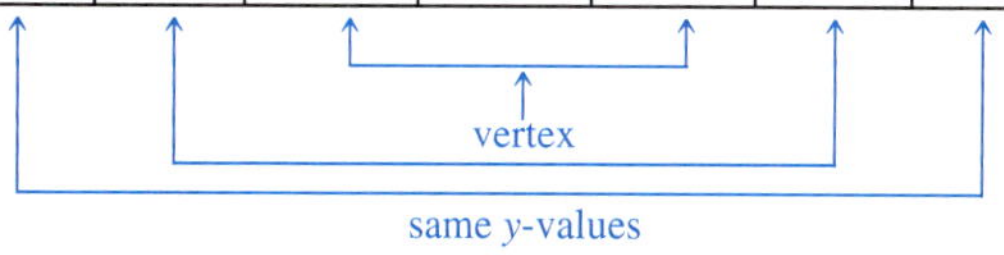

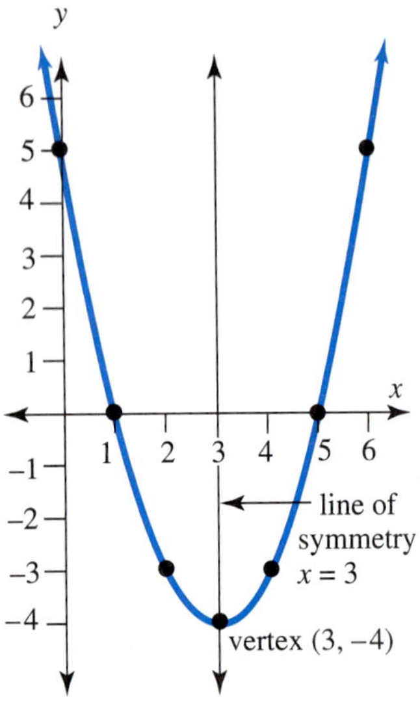

Figure 10.6

As shown on page 707, the points $(2, -3)$ and $(4, -3)$ have the same y-values, as do the points $(1, 0)$ and $(5, 0)$, and the points $(0, 5)$ and $(6, 5)$. If we place a mirror halfway between each of these pairs of points, then the points would reflect onto each other. The line of symmetry is the vertical line where we would place the mirror, as shown in Figure 10.6. This line goes through $(3, -4)$ and other points with an x-coordinate of 3, such as $(3, 0)$ and $(3, 1)$, so the equation of the line of symmetry is $x = 3$. The vertex is where the line of symmetry intersects the parabola, so it is the point $(3, -4)$. The parabola resulting from these points is shown in Figure 10.6.

Notice that the following shorter list of points would have been sufficient for graphing our parabola.

x	0	1	2	3	4
y	5	0	-3	-4	-3

same y-value

The locations of the line of symmetry and the vertex are apparent as soon as we see that if x is either 2 or 4, the y-value is the same. We don't have to list any more points, because symmetry tells us that the y-value at $x = 5$ (2 units to the right of the vertex) will be the same as the y-value at $x = 1$ (2 units to the left of the vertex). When you graph a parabola, list points until you can see the symmetry. ■

Secant Lines and Their Slopes

If a line goes through the points $(1, 1)$ and $(2, 3)$, then as you move from the first point to the second point, the x-value changes from 1 to 2—for a change of $2 - 1 = 1$. The mathematical notation for "change in x" is Δx, so we could write $\Delta x = 1$. This is a horizontal change, frequently called the *run*. Similarly, as you move from $(1, 1)$ to $(2, 3)$, the y-value changes from 1 to 3, for a change of $\Delta y = 3 - 1 = 2$. This is a vertical change, frequently called the *rise*. See Figure 10.7.

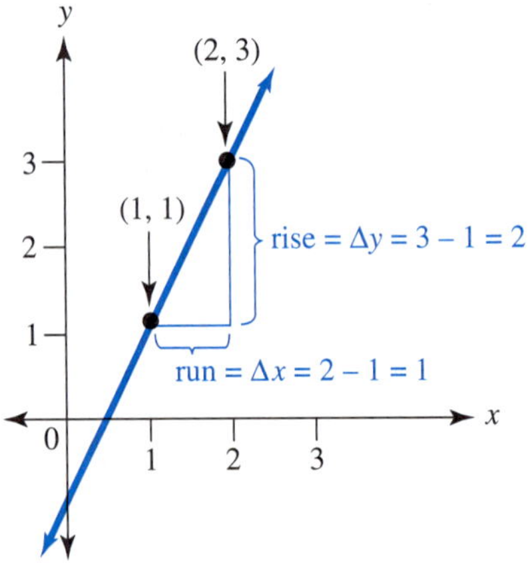

Figure 10.7

The **slope** of a line is a measure of its steepness; it is the ratio of the rise to the run. The slope of the line that goes through the points $(1, 1)$ and $(2, 3)$ is

$$\frac{\text{rise}}{\text{run}} = \frac{2}{1} = 2$$

The slope of a line through the points (x_1, y_1) and (x_2, y_2) is given by

$$m = (y_2 - y_1)/(x_2 - x_1).$$

The numerator is a change in y, and the denominator is a change in x, so we could say that the slope is the ratio $\Delta y/\Delta x$.

Slope of a Line

The slope of a line through (x_1, y_1) and (x_2, y_2) is

$$m = \frac{\Delta y}{\Delta x} = \frac{y_2 - y_1}{x_2 - x_1} = \frac{\text{rise}}{\text{run}}$$

A line is **secant** to a curve if the line intersects the curve in at least two points. Later in this chapter, we will discuss secants as lines that help describe the slope of a curve.

EXAMPLE 5

a. Sketch the curve $y = x^2$.
b. Sketch the line secant to the curve through $x = 2$ and $x = 3$.
c. Find the slope of the secant line.

Solution

a. The equation is of the form $y = ax^2 + bx + c$ (with $a = 1$, $b = 0$, and $c = 0$), so the curve is a parabola. Substituting for x yields the following points:

x	0	1	2	3
y	0	1	4	9

The symmetry is not yet apparent, since no two points have the same y-value. We can continue to substitute larger values for x, or we can try some negative values.

x	−1	0	1	2	3
y	1	0	1	4	9

same y-value

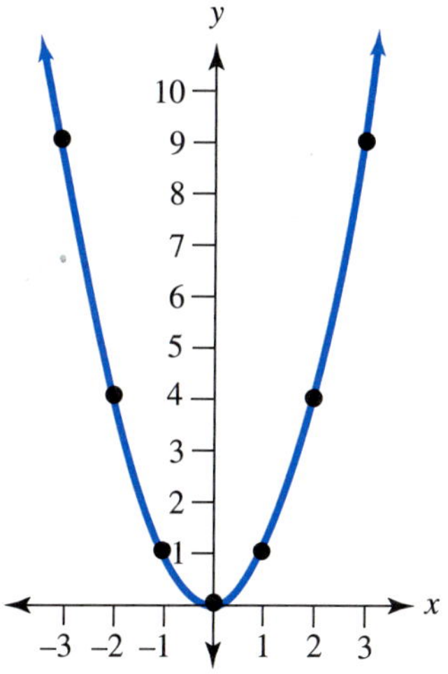

Figure 10.8

The graph is sketched as shown in Figure 10.8.

b. The secant line intersects the curve at the points (2, 4) and (3, 9). (See Figure 10.9.)

c. The slope of the secant line is

$$m = \frac{\Delta y}{\Delta x} = \frac{9 - 4}{3 - 2} = \frac{5}{1} = 5$$

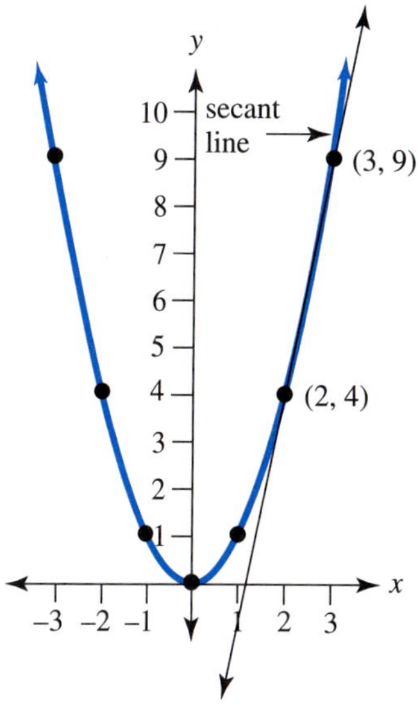

Figure 10.9

■

Locating the Vertex

The tricky part of graphing a parabola is knowing which values to substitute for x so that the vertex can be located. The key is to look at the secant lines and their slopes. Consider the sketching of $y = x^2$ in Example 5. Initially, we substituted 0, 1, 2, and 3 for x, without locating the vertex. If we had calculated the slopes of some secant lines, the direction in which the vertex lies would have been clear. (See Figure 10.10.)

Points	Slope of Secant Line	Slope Observations
(0, 0) and (1, 1)	$m = \frac{\Delta y}{\Delta x} = \frac{1-0}{1-0} = 1$	slope is closer to 0 and line is less steep
(1, 1) and (2, 4)	$m = \frac{\Delta y}{\Delta x} = \frac{4-1}{2-1} = 3$	
(2, 4) and (3, 9)	$m = \frac{\Delta y}{\Delta x} = \frac{9-4}{3-2} = 5$	slope is larger and line is steeper

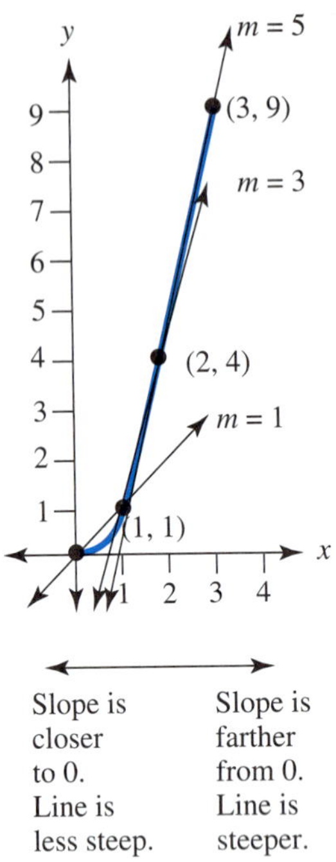

Figure 10.10

As we move to the right along the curve, the slope becomes larger and the secant line becomes steeper; as we move to the left along the curve, the slope becomes closer to 0 and the line becomes less steep. For this reason, the vertex must be farther to the left.

> **Sketching Parabolas**
>
> 1. Substitute 0, 1, and 2 for x, compute the corresponding y-values, and plot the corresponding points.
> 2. Look at the slope of the secant line between $x = 0$ and $x = 1$, and the slope of the secant line between $x = 1$ and $x = 2$. The vertex lies in the direction in which the slope is closer to 0 and the line is less steep.
> 3. Substitute more values for x, using the information from step 2. Continue to plot points until the symmetry is apparent.

EXAMPLE 6 Sketch the curve $y = -x^2 - 2x + 3$. Find the vertex and the equation of the line of symmetry.

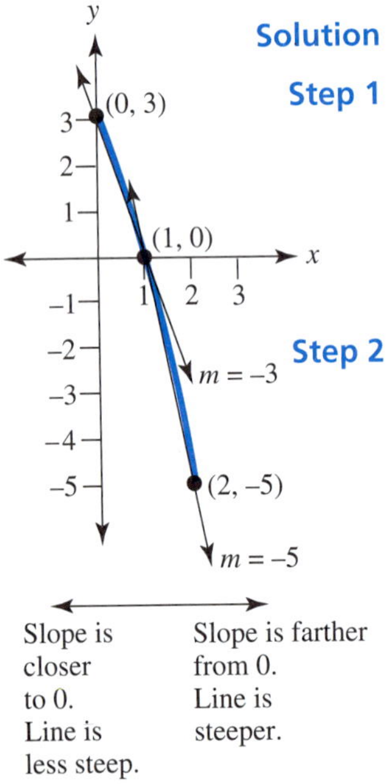

Figure 10.11

Solution

Step 1 *Substitute* 0, 1, *and* 2 *for x, compute the corresponding y-values, and plot the corresponding points, as shown in Figure 10.11.*

x	0	1	2
y	$0 - 0 + 3 = 3$	$-1 - 2 + 3 = 0$	$-4 - 4 + 3 = -5$

Step 2 *Look at the slope of the secant line between $x = 0$ and $x = 1$, and the slope of the secant line between* $x = 1$ and $x = 2$. *The vertex lies in the direction in which the slope is closer to* 0 *and the line is less steep.*

Points	Slope of Secant Line
(0, 3) and (1, 0)	$m = \frac{\Delta y}{\Delta x} = \frac{0-3}{1-0} = -3$
(1, 0) and (2, −5)	$m = \frac{\Delta y}{\Delta x} = \frac{-5-0}{2-1} = -5$

The vertex lies to the left, because the slope is closer to 0 and the line is less steep in that direction. See Figure 10.11.

Step 3 *Substitute more values for x, using the information from step 2. Continue to plot points until the symmetry is apparent.*

x	-2	-1	0	1	2
y	$-4 + 4 + 3 = 3$	$-1 + 2 + 3 = 4$	3	0	-5

same y-value

The graph is sketched in Figure 10.12. The vertex is at $(-1, 4)$. The line of symmetry is the vertical line through $(-1, 4)$ and other points with an x-coordinate of -1, such as $(-1, 0)$ and $(-1, 1)$, so its equation is $x = -1$.

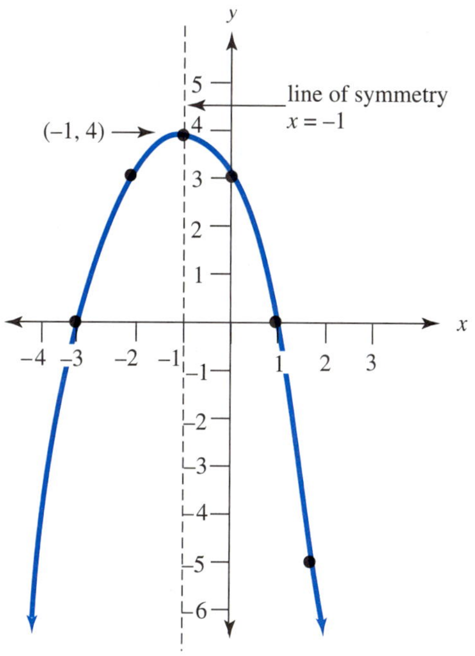

Figure 10.12

Functions

A **function** from x to y is a rule that associates a single value of y to each value of x. The equation

$$y = 3x - 5$$

is a function, because it associates a single value of y to each value of x. For example, to the x-value 0 it associates the *single* y-value $3 \cdot 0 - 5 = -5$, and to the x-value 12 it associates the *single* y-value $3 \cdot 12 - 5 = 31$.

When an equation is a function, it can be rewritten with functional notation. **Functional notation** involves giving the function a name. Usually, the name is the letter f (which stands for "function") or some other letter near f in the alphabet. If we named the above function f, then we would write

$$f(x) = 3x - 5$$

instead of

$$y = 3x - 5$$

The notation $f(x)$ looks like multiplication, but it is not. It is read "f of x" rather than "f times x."

The instructions "find $f(0)$" mean "substitute 0 for x in the function named f." Thus, for the function $f(x) = 3x - 5$,

$$f(0) = 3 \cdot 0 - 5 = -5$$

EXAMPLE 7 Given $g(x) = 5x + 2$ and $h(x) = x^2 - 1$, find the following:

a. $g(7)$
b. $h(4)$
c. $g(3x - 1)$
d. $h(x + \Delta x)$

Solution **a.** "Find $g(7)$" means "substitute 7 for x in the function named g."

$$g(x) = 5x + 2$$
$$g(7) = 5 \cdot 7 + 2 = 37$$

b. "Find $h(4)$" means "substitute 4 for x in the function named h."

$$h(x) = x^2 - 1$$
$$h(4) = 4^2 - 1 = 15$$

c. "Find $g(3x - 1)$" means "substitute $3x - 1$ for x in the function named g."

$$g(x) = 5x + 2$$
$$g(3x - 1) = 5 \cdot (3x - 1) + 2 = 15x - 5 + 2 = 15x - 3$$

d. "Find $h(x + \Delta x)$" means "substitute $x + \Delta x$ for x in the function named h."

$$h(x) = x^2 - 1$$
$$h(x + \Delta x) = (x + \Delta x)^2 - 1 = x^2 + 2x\Delta x + \Delta x^2 - 1$$

■

10.0 Exercises

In Exercises 1–8, use ratios to find the unknown lengths.

1.

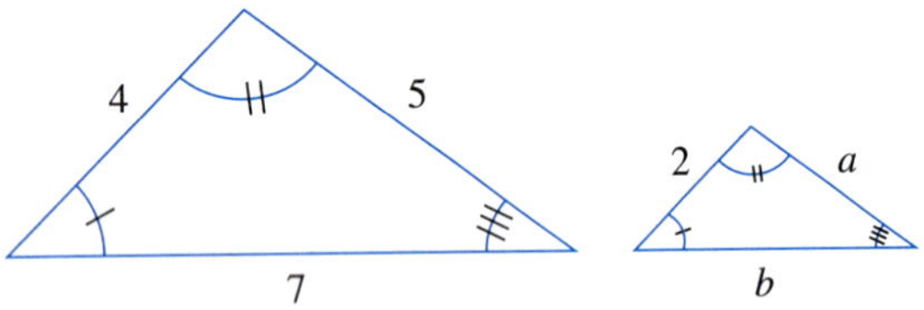

2.

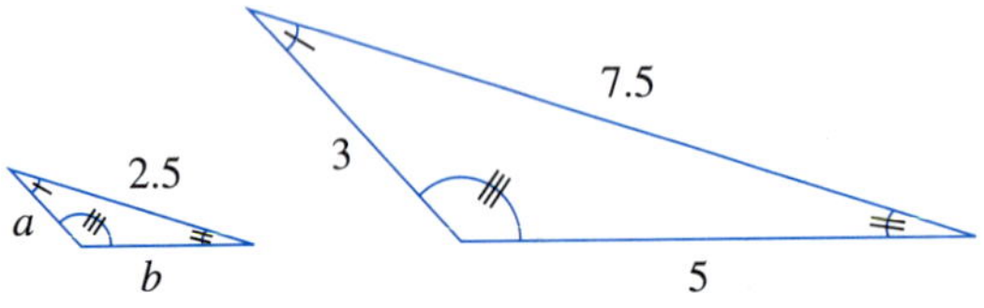

3.

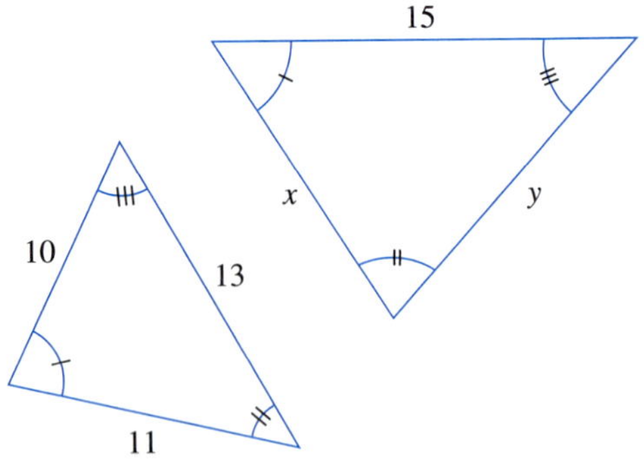

4.

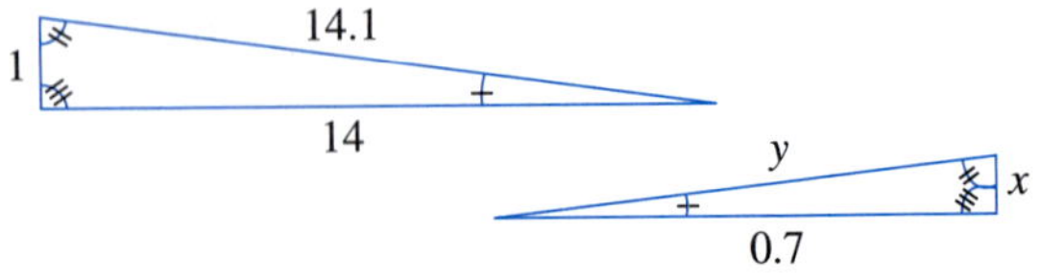

5.

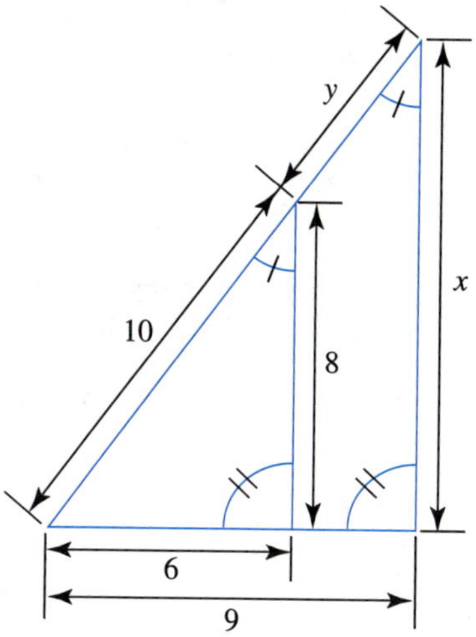

6.

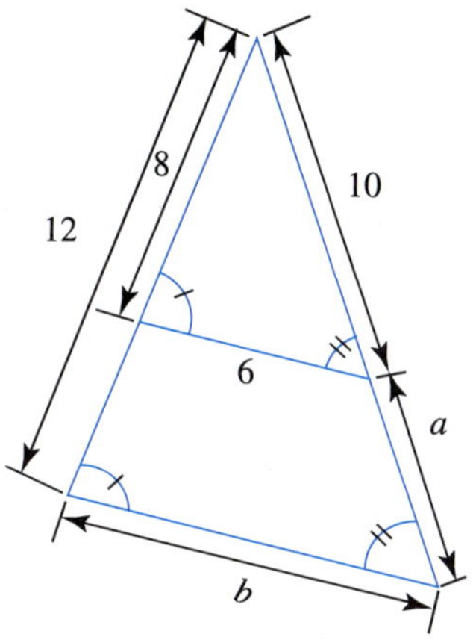

7.

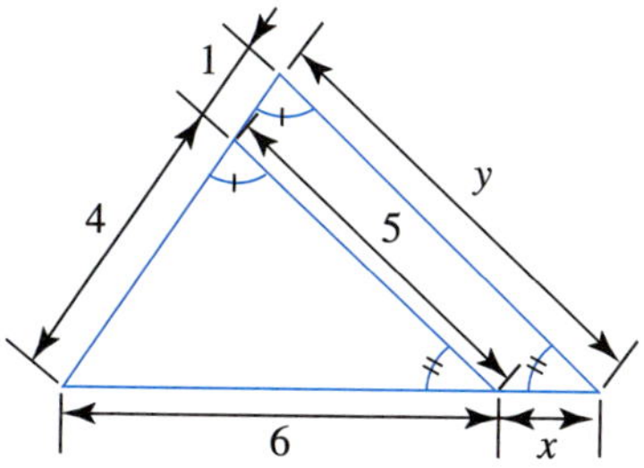

8.

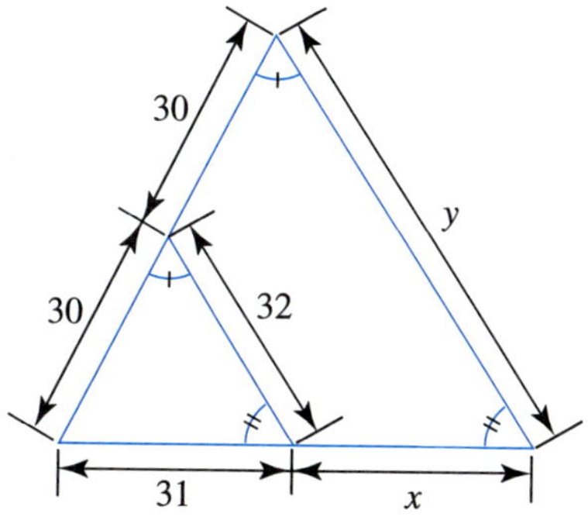

9. The University of Utopia has a student/teacher ratio of 15 to 1.

a. Write this ratio as a fraction.

b. How large is the University of Utopia faculty if the university has 5,430 students?

10. Sylvia Silver received \$2,330 in interest on an investment of \$29,125.

a. Find the ratio of interest to principal.

b. Express this ratio as an interest rate.

11. In one year, Gordon Gotrox received \$1,575 in interest on an investment of \$17,500.

a. Find the ratio of interest to principal.

b. Express this ratio as an interest rate.

12. A car's odometer read 33,482.4 at 9 A.M. and 33,812.9 at 4 P.M.

a. Find the car's average rate.

b. Use proportions to find the distance that would be traveled in four more hours at the same average rate.

c. Use a formula to find the distance that would be traveled in four more hours at the same average rate.

13. A car's odometer read 101,569.3 at 10 A.M. and 101,633.5 at 12 noon.

a. Find the car's average rate.

b. Use proportions to find the distance that would be traveled in four more hours at the same average rate.

c. Use a formula to find the distance that would be traveled in four more hours at the same average rate.

14. A car's odometer read 101,569.3 at 10:00 A.M. and 101,633.5 at 11:30 A.M.

a. Find the car's average rate.

b. Use proportions to find the distance that would be traveled in four more hours at the same average rate.

c. Use a formula to find the distance that would be traveled in four more hours at the same average rate.

15. A car's odometer read 72,938.0 at 9:00 A.M. and 73,130.3 at 11:30 A.M.

a. Find the car's average rate.

b. Use proportions to find the distance that would be traveled in four more hours at the same average rate.

c. Use a formula to find the distance that would be traveled in four more hours at the same average rate.

16. Phyllis Peterson has a new Toyonda Suppord. When she filled the tank with gas, the odometer read 37.6 miles. Later, she added 7.3 gallons, and the odometer read 274.1 miles.

a. Find the ratio of miles traveled to gallons.
b. Use a proportion to predict the number of miles she could travel on a full tank if her tank holds 14 gallons.
c. Use a formula to predict the number of miles she could travel on a full tank if her tank holds 14 gallons.

17. Rick Mixter has a new Nissota Suppette. When he filled the tank with gas, the odometer read 5.4 miles. Later, he added 13.3 gallons, and the odometer read 332.5 miles.
a. Find the ratio of miles traveled to gallons consumed.
b. Use a proportion to predict the number of miles he could travel on a full tank if his tank holds 15 gallons.
c. Use a formula to predict the number of miles he could travel on a full tank if his tank holds 15 gallons.

18. If a ring is 18-karat gold and weighs 4 grams, find the amount of pure gold in the ring.
19. If a ring is 12-karat gold and weighs 8 grams, find the amount of pure gold in the ring.
20. If a ring is 18-karat gold and weighs 8 grams, find the amount of pure gold in the ring.
21. When Thales determined the height of the Great Pyramid of Cheops, he found that the base of the pyramid was 756 feet long, the pyramid's shadow was 342 feet long, the pole was 6 feet tall, and the pole's shadow was 9 feet long. (Naturally, the unit of measurement was not feet.) Find the height of the Great Pyramid.
22. Draw the two similar triangles that are a part of Figure 10.2. List the two pairs of corresponding sides.

In Exercises 23–32, (a) graph the parabola, (b) give the equation of its line of symmetry, and (c) find its vertex.

23. $y = 3x^2 + 4$
24. $y = 2 - 4x^2$
25. $y = -x^2 - 4x + 3$
26. $y = x^2 + 2x + 1$
27. $y = 2x + x^2 + 3$
28. $y = x^2 - 8x + 10$
29. $y = 8x - x^2 - 14$
30. $y = 4x - x^2$
31. $y = 12x - 2x^2$
32. $y = 2x^2 - 8x + 4$

In Exercises 33–36, (a) find the slope of the secant line that intersects the given parabola at the given points and (b) graph the parabola and the secant line.

33. $y = 2 - 3x^2, x = 1, x = 3$
34. $y = x^2 - 2x + 2, x = -2, x = 0$
35. $y = x^2 - 6x + 11, x = 1, x = 5$
36. $y = 3 - 4x^2, x = 1, x = 2$

In Exercises 37–48, use the following functions:

$$f(x) = 8x - 11$$
$$g(x) = 3x^2$$
$$h(x) = x^2 - 3$$
$$k(x) = x^2 + 2x - 4$$

37. Find $f(4)$.
38. Find $g(-2)$.
39. Find $h(-3)$.
40. Find $k(-5)$.
41. Find $f(x + 3)$.
42. Find $g(x - 2)$.
43. Find $h(x - 7)$.
44. Find $k(x + 1)$.
45. Find $f(x + \Delta x)$.
46. Find $g(x + \Delta x)$.
47. Find $h(x + \Delta x)$.
48. Find $k(x + \Delta x)$.

10.1 The Antecedents of Calculus

To gain some perspective on why calculus was invented, we'll start with a brief history of the mathematics antecedent to calculus—algebra, geometry, and, most important, analytic geometry.

Analytic geometry is a blending together of algebra and geometry. The graphing of $y = 2x + 1$ as a line with slope 2 and y intercept 1, shown in Figure 10.13 on page 715, is a modern example of analytic geometry; it combines algebra (the equation $y = 2x + 1$) with geometry (the line).

What follows is not a history of mathematics in general but a history of the people and discoveries that had a major impact on the invention of calculus.

Ancient Mathematics

One of the earliest applications of mathematics was to the study of the moon, planets, and stars. These celestial bodies were worshiped as gods, and the study of their perfect motions was a primary duty of priests. This study also resulted in the ability to predict the seasons; priests were thus able to tell the people when to plant and when to harvest. It also led to the ability to navigate by the stars. Such knowledge was one source of the priests' power.

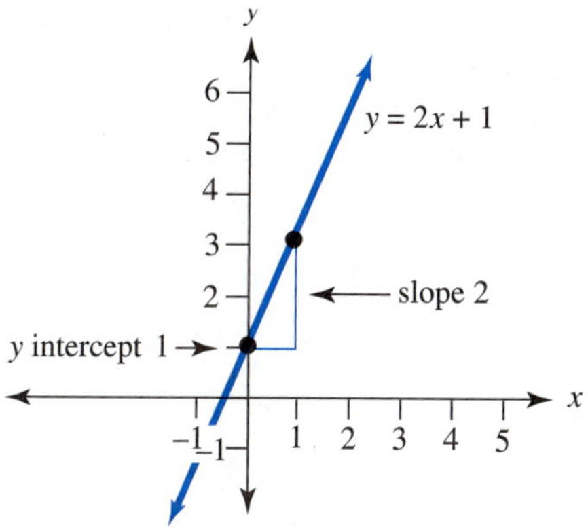

Figure 10.13

The Egyptians probably had a calendar by 4241 B.C. and had developed an arithmetic for use in agriculture and commerce by 3000 B.C. At the same time, the Babylonians had developed a more advanced arithmetic as well as a rudimentary form of algebra that could solve some quadratic equations and systems of first-degree equations. The Ahmes Papyrus (or Rhind Papyrus), written in Egypt in 1650 B.C., contains methods of solving linear equations, a detailed treatment of fractions, and methods for computing some areas and volumes. (See Chapter 6 for more information on Egyptian mathematics.)

Greek Mathematics

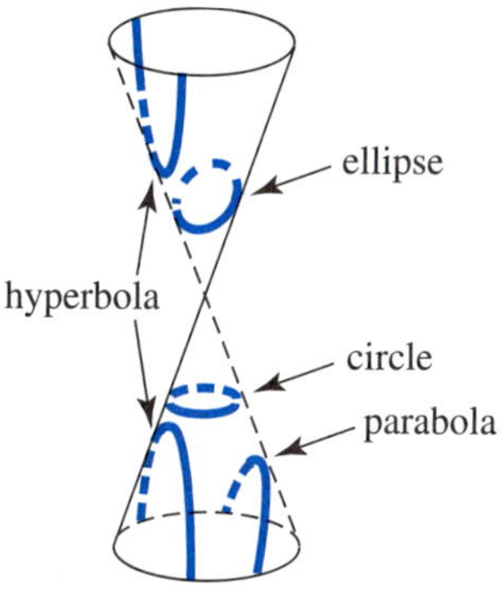

Figure 10.14
The conic sections—cross sections of a double cone

Greek geometry began in 585 B.C. with Thales of Miletus, a Greek merchant whose travels brought him in contact with the mathematics of Babylonia and Egypt. Until this time, geometry had consisted strictly of measuring techniques (in fact, the word *geometry* comes from the Greek and means "earth measurement"). However, Thales made abstract, general statements—such as "when two lines intersect, they create pairs of equal angles"—and attempted to justify those statements logically. Three hundred years later, the Greek mathematician Euclid organized geometry into a systematic, logical structure. (See Chapter 6 for more information on the geometry of Thales and Euclid.)

Around 220 B.C., the Greek mathematician Apollonius of Perga wrote an eight-volume work in which he investigated the curves obtained by taking cross sections of a double cone. These curves, the circle, ellipse, parabola, and hyperbola, are called conic sections. (See Figure 10.14.) Apollonius's work was the standard reference on conic sections for almost 2,000 years, even though several volumes were lost.

Apollonius studied parabolas by superimposing line segments and squares over the curves. Apollonius arbitrarily selected a point on a parabola. As shown in Figure 10.15, he then drew a square with one side on the parabola's line of symmetry and a second side connecting the line of symmetry and the selected point. He also drew a line segment along the line of symmetry, connecting the vertex and the square.

Apollonius of Perga

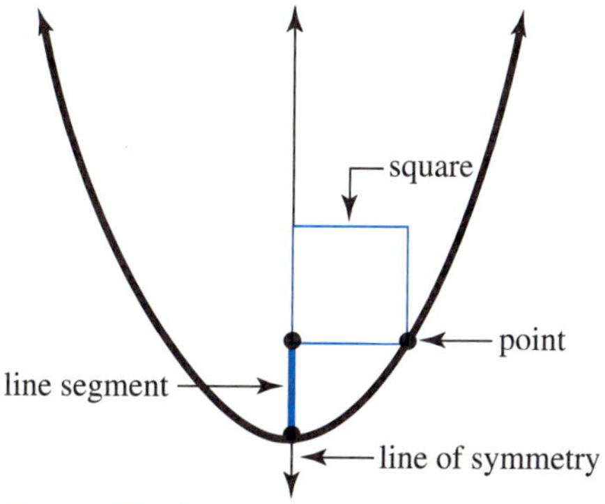

Figure 10.15

Apollonius then selected a second point on the parabola and drew a second square and a second line segment, just as he had done with the first point (see Figure 10.16). He found that the ratio of the lengths of the line segments is always equal to the ratio of the areas of the squares. In other words,

$$\frac{\text{length of line segment 1}}{\text{length of line segment 2}} = \frac{\text{area of square 1}}{\text{area of square 2}}$$

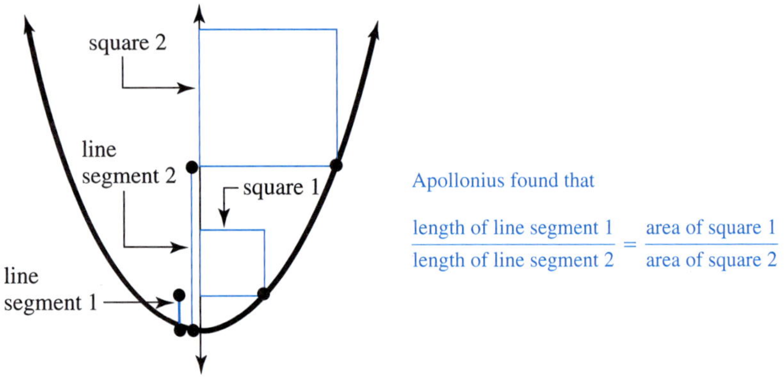

Figure 10.16

If Apollonius had had the more modern idea of x- and y-coordinates, had made the y-axis the line of symmetry and the origin the vertex, and had labeled the two points on the parabola (x_1, y_1) and (x_2, y_2), he might have realized that the lengths of the line segments are y_1 and y_2 and the bases of the squares are x_1 and x_2 (see Figure 10.17).

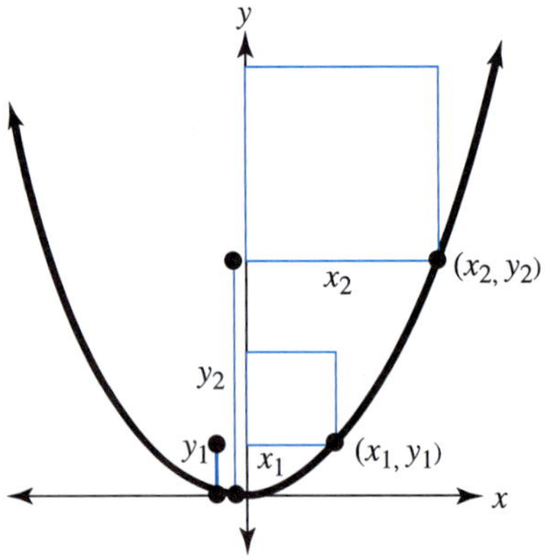

Figure 10.17

Thus, the areas of the squares are x_1^2 and x_2^2. The modern version of Apollonius's conjecture that

$$\frac{\text{length of line segment 1}}{\text{length of line segment 2}} = \frac{\text{area of square 1}}{\text{area of square 2}}$$

would be

$$\frac{y_1}{y_2} = \frac{x_1^2}{x_2^2}$$

EXAMPLE 1 The parabola $y = 2x^2$ goes through the points (1, 2) and (2, 8).

a. At each of these two points, compute the length of the line segment and the area of the square described by Apollonius.

b. Compute (length of line segment 1)/(length of line segment 2) and (area of square 1) /(area of square 2).

Solution **a.** The parabola is graphed in Figure 10.18.

At the point (1, 2), the line segment has length $y_1 = 2$. The base of the square has length $x_1 = 1$, so the area of the square is $1^2 = 1$.

At the point (2, 8), the line segment has length $y_2 = 8$. The base of the square has length $x_2 = 2$, so the area of the square is $2^2 = 4$.

b. $$\frac{\text{length of line segment 1}}{\text{length of line segment 2}} = \frac{2}{8} = \frac{1}{4}$$

$$\frac{\text{area of a square 1}}{\text{area of a square 2}} = \frac{1^2}{2^2} = \frac{1}{4}$$

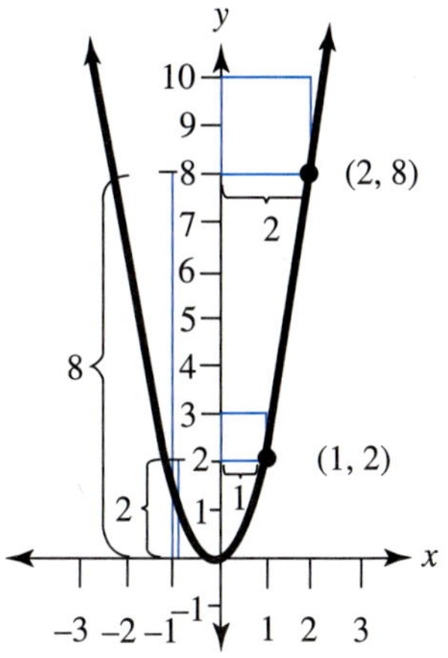

Figure 10.18

As Apollonius observed, the ratio of the length of the line segments is equal to the ratio of the areas of the squares—both equal 1/4. ■

It's important to remember that Apollonius studied parabolas, ellipses, and hyperbolas *without* any of the tools used in Example 1. He did not have x- and y-axes, ordered pairs, or equations of parabolas.

Analytic geometry has its origins in Apollonius's work on conic sections. His work involved both geometry and algebra: The parabola, line segments, and squares are geometrical, while the comparisons of lengths and areas are essentially algebraic. This type of algebra is typical of the algebra of the Greeks. The algebra is applied to geometrical objects and is based on ratios and proportions rather than on equations. Furthermore, it is verbal rather than symbolic. That is, the Greeks actually wrote out a verbal description, such as "the area of a square with base perpendicular to the line of symmetry," rather than the equivalent symbolic description x^2. Symbolic, equation-based algebra is much more powerful than verbal, proportion-based algebra, and calculus could not be invented until algebra had evolved.

Hindu Mathematics

Hindu mathematicians in India expanded Babylonian and Greek ideas in arithmetic, geometry, and algebra. The Hindus also developed the numeration system that we use today, a place system based on powers of 10 that includes a symbol for zero. It is thought that the Hindu system was based on the abacus, which has columns of beads, just as our numeration system has columns of numbers (the ones column, the tens column, and so on). A place system is a system in which the value of a digit is determined by its place; for example, 30 is a different number than 3 because the zero puts the three in a different place. Roman numerals do not comprise a place system; V means "five" regardless of where the V is placed. The place system made arithmetic calculations much easier—imagine what multiplication or long division would be like with a nonplace system such as Roman numerals! (See Chapter 8 for more information on Hindu mathematics.)

A comparison of two different methods of computation, featuring Boethius using Hindu numerals and Pythagoras using an abacus.

Bettmann/Corbis

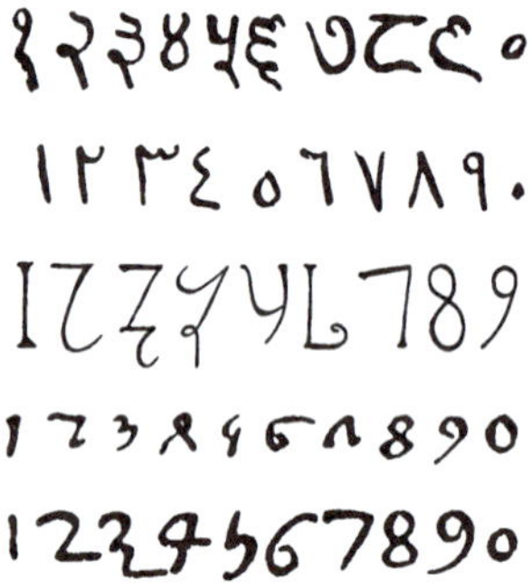

Hindu, Arabic, and European numerals

Arabic Mathematics

Mathematics and the sciences entered an extremely long period of stagnation when the Greek civilization fell and was replaced by the Roman Empire. This inactivity was uninterrupted until after the Islamic religion and the resulting Islamic culture were founded by the prophet Muhammad in A.D. 622. Within a century, the Islamic empire stretched from Spain, Sicily, and Northern Africa to India.

Unlike the Roman Empire, Islamic culture encouraged the development of the sciences as well as the arts. Arab scholars translated many Greek and Hindu works in mathematics and the sciences, including Apollonius's work on conic sections. It's likely that much of the Greeks' work in science and mathematics would have been lost if not for these Arab scholars.

The Arab mathematician Mohammed ibn Musa al-Khowarizmi wrote two important books around A.D. 830, each of which was translated into Latin in the twelfth century. Much of the mathematical knowledge of medieval Europe was derived from the Latin translations of al-Khowarizmi's two works.

Al-Khowarizmi's first book, on arithmetic, was titled *Algorithmi de numero Indorum* (or *al-Khowarizmi on Indian Numbers*). The Latin translation of this book introduced to Europe the Hindu number system and the simpler calculation techniques (such as the procedures for multiplication and long division) that system allows. This system is now called the *Arabic* number system, because Europeans learned about the system from Arabs. The book's title is the origin of the word *algorithm*, which means a procedure for solving a certain type of problem, such as the procedure for long division.

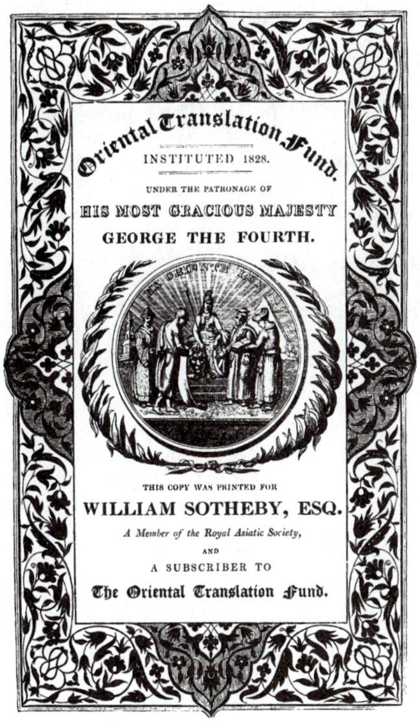
Oriental Translation Fund.
INSTITUTED 1828.
UNDER THE PATRONAGE OF
HIS MOST GRACIOUS MAJESTY
GEORGE THE FOURTH.
THIS COPY WAS PRINTED FOR
WILLIAM SOTHEBY, ESQ.
A Member of the Royal Asiatic Society,
AND
A SUBSCRIBER TO
The Oriental Translation Fund.

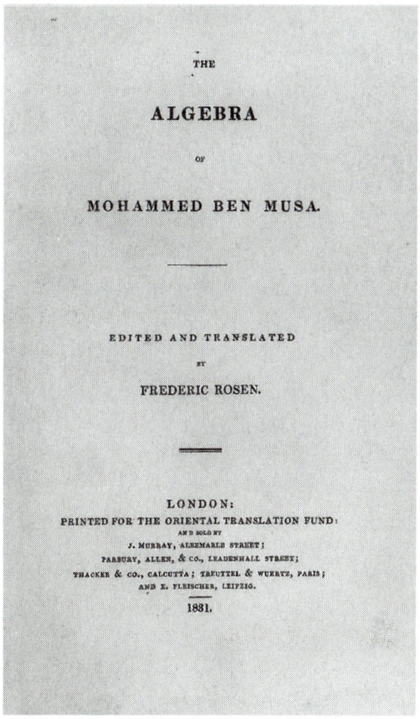
THE
ALGEBRA
OF
MOHAMMED BEN MUSA.
EDITED AND TRANSLATED
BY
FREDERIC ROSEN.
LONDON:
PRINTED FOR THE ORIENTAL TRANSLATION FUND:
AND SOLD BY
J. MURRAY, ALBEMARLE STREET;
PARBURY, ALLEN, & CO., LEADENHALL STREET;
THACKER & CO., CALCUTTA; TREUTTEL & WUERTZ, PARIS;
AND E. FLEISCHER, LEIPZIG.
1831.

The title pages of an 1831 edition of a translation of al-Khowarizmi's book, *Al-Jabr w'al Muqabalah*

Al-Khowarizmi's second book, *Al-Jabr w'al Muqabalah*, discussed linear and quadratic equations. In fact, the word *algebra* comes from the title of this second book. This title, which translates literally as *Restoration and Opposition*, refers to the solving of an equation by adding the same thing to each side of the equation (which "restores the balance" of the equation) and simplifying the result by canceling opposite terms (which is the title's "opposition"). For example (using modern symbolic algebra),

$$\begin{aligned} 6x &= 5x + 11 & \\ 6x + -5x &= 5x + 11 + -5x & \text{"al-jabr" or restoration of balance} \\ x &= 11 & \text{"al-muqabalah" or opposition} \end{aligned}$$

The quote on page 720 is from a translation of *Al-Jabr w'al Muqabalah*,* demonstrates several important features of al-Khowarizmi's algebra. First, it is entirely verbal, as was the algebra of Apollonius—there is no symbolic algebra at all. Second, this algebra differs from that of Apollonius in that it is not based on proportions. Third, the terminology betrays the algebra's connections with geometry. When al-Khowarizmi refers to "a square," he is actually referring to the area of a square; when he refers to "a root," he is actually referring to the length of one side of the square (hence the modern phrase "square root"). Modern symbolic algebra uses the notations x^2 and x in place of "a square" and "a root."

The quote from *Al-Jabr w'al Muqabalah* is on the left; a modern version of the same instructions is on the right. You might recognize this modern version from intermediate algebra, where it is called "completing the square."

* From L. C. Karpinski. *Robert of Chester's Latin translation of the Algebra of Khowarizmi* (New York: Macmillan, 1915).

The following is an example of squares and roots equal to numbers: a square and 10 roots are equal to 39 units. The question there-	$x^2 + 10x = 39$
fore in this type of equation is about as follows: what is the square which combined	Solve for x^2.
with ten of its roots will give a sum total of 39? The manner of solving this type of equation is to take one-half of the roots just	$\frac{1}{2} \cdot 10 = 5$
mentioned. Now the roots in the problem before us are 10. Therefore take 5, which	$5^2 = 25$
multiplied by itself gives 25, an amount	$x^2 + 10x + 25 = 39 + 25$
which you add to 39, giving 64. Having	$x^2 + 10x + 25 = 64$
taken then the square root of this which is 8,	$(x + 5)^2 = 8^2 x + 5 = 8$
subtract from it the half of the roots, 5, leaving 3. The number three therefore represents	$x + 5 - 5 = 8 - 5x = 3$
one root of this square, which itself, of course, is 9. Nine therefore gives that	$x^2 = 9$
square.	

Al-Khowarizmi, like Apollonius, understood numbers to be lengths of line segments, areas, and volumes. He did not recognize negative numbers, because neither a line nor an area nor a volume can be represented by a negative number. Modern algebra recognizes negative numbers and gives two solutions to the above equation:

$$(x + 5)^2 = 64$$

$$x + 5 = 8 \quad \text{or} \quad x + 5 = -8$$

$$x = 3 \quad \text{or} \quad x = -13$$

Al-Khowarizmi justifies his verbal algebraic solution with a geometrical demonstration. Apparently, he felt that only a geometrical demonstration in the style of the ancient Greeks would be sufficiently convincing. Al-Khowarizmi continues: "Now, however, it is necessary that we should demonstrate geometrically the truth of the same problems which we have explained in numbers. Therefore our first proposition is this, that a square and 10 roots equal 39 units. The proof is that we construct a square of unknown sides and let this square figure represent the square which together with its root you wish to find." (He is describing a square with sides of length x.)

He next instructs the reader to construct a rectangle with one side of length 10 and the other of length equal to that of the square. If the square's side is of length x, then the area of the square is x^2 and the area of the rectangle is $10x$, so the total area is $x^2 + 10x$. We are given that "a square and 10 roots equals 39 units," so we know that $x^2 + 10x = 39$, and thus the combined areas of the square and the rectangle must be 39, as shown in Figure 10.19.

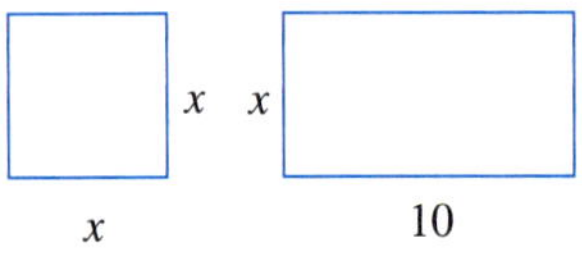

Combined area is $x^2 + 10x = 39$.

Figure 10.19

HISTORICAL NOTE

Arabic Mathematics

The Granger Collection, NY

Arab astronomers

The prophet Muhammad established an Arab state that provided protection and freedom of worship to Moslems, Jews, and Christians. As a result, scholars from throughout the Middle East were attracted to Baghdad. Thus, the Arab academic community was multiethnic and cosmopolitan and owed its inception to Arab initiative and patronage. The city became an important center of learning under three great patrons—al-Mansur, al-Raschid and al-Mamun. Al-Mamun had a dream in which he was visited and advised by the Greek scientist Aristotle. As a result of this dream, al-Mamun had Arabic translations made of all the ancient Greek works available, and he established a library like the old one at Alexandria.

The Arabs preserved, translated, and assimilated Greek, Hindu, and Babylonian science. Having digested what they learned from their predecessors, they were able to enrich it with new observations, new results, and new techniques.

Among the faculty at al-Mamun's library was the mathematician and astronomer Mohammed ibn Musa al-Khowarizmi. His work typified Arab science in that his book on arithmetic was a handbook on ideas obtained from another culture, while his book on algebra appears to be a unique work with no significant predecessor. Its prominence in Europe came to rival that of Euclid's *Elements.* Al-Khowarizmi's algebra book was a - systematic and exhaustive discussion of the solving of linear and quadratic equations. Later, a book by the Persian poet, mathematician, and astronomer Omar Khayyam went beyond that of al-Khowarizmi to include third-degree equations. This Arab interest in algebra seems to have been originally motivated by their complicated system of inheritance laws; the division of an estate involved the solving of some rather sophisticated equations involving areas.

There were four parts to Arabic mathematics. Their numeration system and arithmetic were derived from Hindu sources, but the Arabs added to it with their invention of decimal fractions. Their algebra was a unique work with no significant predecessor. Their trigonometry had Greek and Hindu sources; the Arabs combined these and added new functions and formulas. Finally, their geometry was quite Greek in form; it combined the contents and methods of Euclid, Apollonius, and Archimedes.

Many Arab mathematicians attempted to prove Euclid's parallel postulate, discussed in Section 6.7 (Non-Euclidean Geometry). Girolamo Saccheri's familiarity with one Arab "proof" motivated his own "proof."

Arab science was not limited to mathematics. It also stressed astronomy, optics, anatomy, medicine, pharmacology, botany, and agriculture.

From an Arab translation of Apollonius' *Conics.*

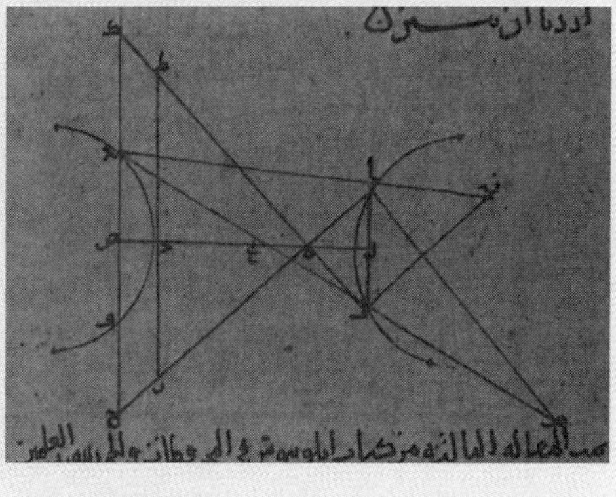

The Granger Collection, NY

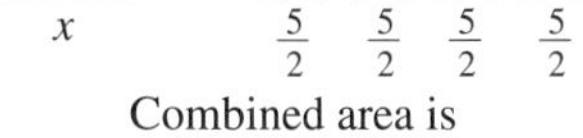

Combined area is

$x^2 + 4 \cdot \frac{5}{2}x = x^2 + 10x = 39$

Figure 10.20

The rectangle can be cut into four strips, each with one side of length x and the other of length $\frac{10}{4} = \frac{5}{2}$, as shown in Figure 10.20. Each strip can be glued to the edge of the square, resulting in a figure whose area is still $x^2 + 4 \cdot \frac{5}{2}x = x^2 + 10x = 39$, as shown in Figure 10.21 on page 722.

If we filled in the corners of this figure, the result would be a new square that is larger than the original square. (This step explains the name of the modern version of al-Khowarizmi's method—"completing the square.") The corners are each squares of area $\left(\frac{5}{2}\right)^2 = \frac{25}{4}$, so the area of the four corners is $4 \cdot \frac{25}{4} = 25$. The area of

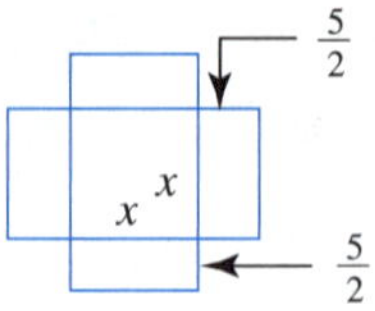

Combined area is still
$x^2 + 4 \cdot \frac{5}{2}x = x^2 + 10x = 39$

Figure 10.21

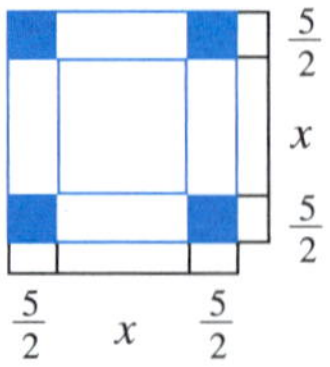

Total area is $39 + 25 = 64$.

Figure 10.22

the figure with its corners filled in, shown in Figure 10.22, is $39 + 25 = 64$. Because the area of the large square is 64, it must have sides of length 8. To find the length of a side of the original square and to solve the problem, we would have to subtract $2 \cdot \frac{5}{2} = 5$ from 8 and get $8 - 5 = 3$. Thus, the area of the smaller, original square is $3 \cdot 3 = 9$.

The Moors (Moslems from North Africa) entered Spain in A.D. 711 and built universities in Toledo, Cordoba, and Seville. This culture was the only major exception to the mathematical and scientific stagnation in Europe that started with the Roman Empire and continued through the Middle Ages. In 1085, the Christians reconquered Toledo; many Moors were killed or expelled, to Spain's great loss.

During the twelfth century, many Arabic mathematical and scientific works (including al-Khowarizmi's two books), as well as Greek and Hindu works that had been translated into Arabic, were translated into Latin, often by Jewish scholars in Spain. Greek works of literature and philosophy were also translated.

The Arabic number system and the recently invented Italian double-entry bookkeeping system revolutionized trade. As trade expanded, Arab and Greek knowledge was transmitted throughout Europe. European merchants began to visit the East in order to acquire more scientific information. Many of the famous European universities, including Oxford, Cambridge, Paris, and Bologna, were built.

Mathematics during the Middle Ages

During the Middle Ages, the Greek philosopher Aristotle's concepts of change and motion were favorite topics at the universities, especially at Oxford and Paris. Around 1360, the Parisian scholar Nicole Oresme drew a picture to describe the speed of a moving object, which he described as the "intensity of motion." He drew a horizontal line, along which he marked points representing different points in time. At each of these points, he drew a vertical line, the length of which represented the speed of the object at that point in time, as shown in Figure 10.23. Oresme called his horizontal and vertical lines *longitudo* and *latitudo*, which indicate the concept's origins in mapmaking. If the speed of the object increased in a uniform way, the endpoints of these vertical lines would themselves lie in a straight line.

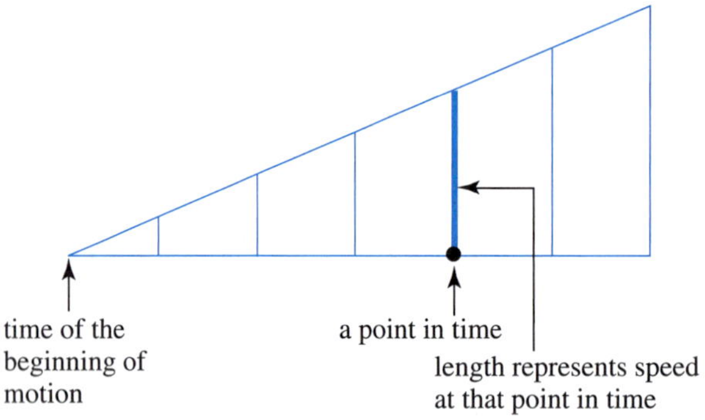

Figure 10.23

Using modern terminology, Oresme's horizontal line was the x-axis, the points along the line were x-coordinates, the length of the vertical lines were y-coordinates, and the diagonal line was the graph of the speed. Oresme's work is considered one of the earliest appearances of the idea of x- and y-coordinates and of the graphing of a variable quantity; for this reason, his work is a cornerstone of analytic geometry.

Mathematics during the Renaissance

In the 1400s, Europe was undergoing a massive transformation. The Black Death, a plague that killed as much as three-quarters of the population of Europe, was over, and Europe was becoming revitalized. In 1453, the Turks conquered Constantinople, the last remaining center of Greek culture. Many Eastern scholars moved from Constantinople to Europe, bringing Greek knowledge and manuscripts with them. Around the same time, Gutenberg invented the movable-type printing press, which greatly increased the availability of scientific information in the form of both new works and translations of ancient works. Translations of Euclid's work and some of Apollonius's work on geometry were printed, as was the Franciscan monk Luca Pacioli's *Summa de arithmetica, geometrica, proportioni et proportionalita*, which was a summary of the arithmetic, geometry, algebra, and double-entry bookkeeping known at that time. The book's title makes it evident that algebra was still based on proportions rather than equations.

The ready availability of Greek and Arab knowledge, together with information about the Greek way of life, created a pointed contrast in the medieval European mind. The Greeks attempted to understand the physical world, whereas the Church suppressed any theory that was opposed to established doctrines. (Galileo was imprisoned for suggesting that the earth rotates around the sun, rather than the sun rotating around the earth.)

A general spirit of skepticism replaced the acceptance of old ideas, and knowledge expanded in all directions. Many libraries and universities were established, and scholars, poets, artists, and craftsmen received encouragement and material support from wealthy benefactors. The religious reformation championed by Martin Luther in Germany and by Henry VIII and his daughter Elizabeth I in England shattered the existing order in Europe. This revolt against papal authority inspired an intellectual revolt against authority and tradition.

Technology grew quickly. The introduction of gunpowder in the thirteenth century had revolutionized warfare. Europe was engulfed in an arms race; a nation's fate hinged on its advances in the design and use of cannons, muskets, and fortifications. Some of Europe's best minds—including Galileo, Leonardo da Vinci, Michelangelo, and Albrecht Dürer—were devoted to various aspects of the problem. It was necessary to learn about the trajectories of projectiles, their range, the heights they could reach, and the effect of muzzle velocity. In fact, it became necessary to investigate the principles of motion itself.

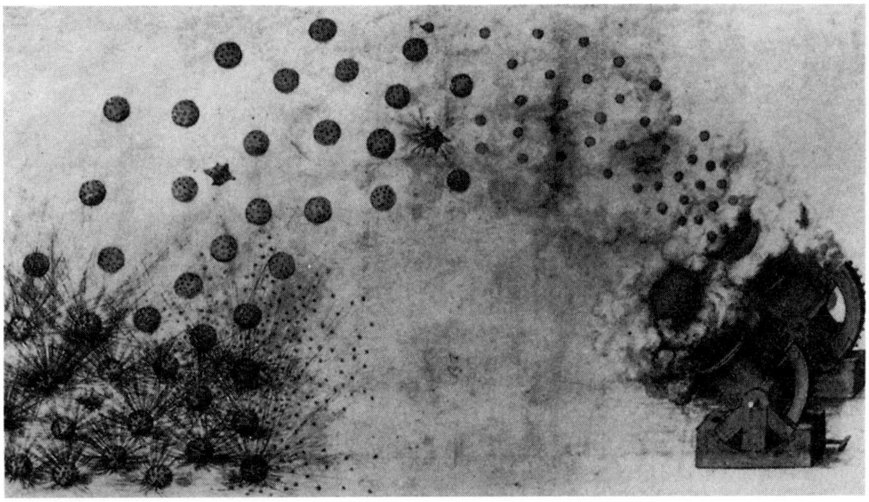

A Leonardo da Vinci illustration showing the trajectories of cannonballs

The algebra of the sixteenth century had as much in common with that of the ancient Greeks as it did with our modern algebra. To a large extent, equations had replaced proportions, but algebra was still very verbal, with only a limited use of symbols. For example, the Latin phrase "*A* cubus p. 6 in *A* planum aequalis 20" (where "p." means "plus") would be written instead of the modern $A^3 + 6A^2 = 20$.

The terminology indicates that algebra was still connected with geometry, like the algebras of Apollonius and al-Khowarizmi. "*A* cubus" refers to a cube whose edges are of length *A*, and "6 in *A* planum" refers to a rectangular box with two edges of length *A* and a third edge of length 6. "*A* cubus p. 6 in *A* planum aequalis 20" means that the volumes of the cube and the box add to 20 (see Figure 10.24).

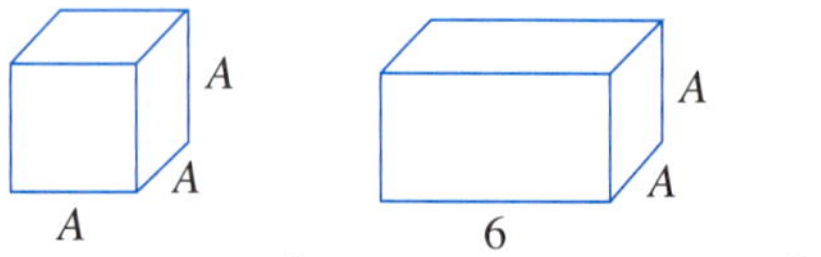

Figure 10.24

A variable referred to the length of a line segment, the square of a variable referred to the area of a square, and the cube of a variable referred to the volume of a cube. From this perspective, both negative numbers and powers higher than cubes were impossible. Furthermore, a square could not be added to a cube (that is, one could not add $x^2 + x^3$), because areas and volumes are not like quantities and cannot be combined.

Algebra was still a set of specific techniques that could be used to solve specific equations. There was little generalization, and there was no way to write an equation to represent an entire class of equations, as we would now write $ax^2 + bx + c = 0$ to represent all quadratic equations. There were only ways to write specific equations such as $3x^2 + 5x + 7 = 0$. Thus, it was impossible to write a formula like the quadratic formula [if $ax^2 + bx + c = 0$, then $x = (-b \pm \sqrt{b^2 - 4ac})/2a$]. It was only possible to give an example, such as al-Khowarizmi's example of completing the square.

In the late sixteenth century, algebra matured into a much more powerful tool. It became more symbolic. Exponents were introduced; what had been written as "cubus," "*A* cubus" or "*AAA*" could now be written as "A^3." The symbols $+$, $-$, and $=$ were also introduced.

François Viète, a French lawyer who studied mathematics as a hobby, began using vowels to represent variables and consonants to represent constants. (The more modern convention of using letters from the end of the alphabet, such as *x* and *y*, to represent variables and letters from the beginning of the alphabet, such as *a* and *b*, to represent constants was introduced later by Descartes.) This allowed mathematicians to represent the entire class of quadratic equations by writing "$A^2 + BA = C$" (where the vowel *A* is the variable and the consonants *B* and *C* are the constants) and made it possible to discuss general techniques that could be used to solve classes of equations. All these notational changes were slow to gain acceptance. No one mathematician adopted all the new notations. Viète's algebra was quite verbal; he did not even adopt the symbol $+$ until late in his life.

François Viète

In 1637, the famous French philosopher and mathematician René Descartes published *Discours de la méthode pour bien conduire sa raison et chercher la vérité dans les sciences* (*Discourse on the Method of Reasoning Well and Seeking Truth in the Sciences*). This work, originally published anonymously due to Descartes's fear of negative reaction from the Church, was a revolutionary departure from the old accepted ideas; in fact, his method of reasoning was based on systematic doubt. His doubt was so systematic that he doubted his own existence.

He decided that the only thing he couldn't doubt was doubt itself. This led to the conclusion "I think; therefore I exist." He hoped that through his doubting process he would be able to reach clear and distinct ideas. His approach led him to a scientific view that was predominant for about 100 years, after which it was replaced by Isaac Newton's theories.

Descartes's *Discours* contained three appendices in which he illustrated his general philosophical method. One of these appendices, *La géométrie*, explored the relationship between algebra and geometry in a way unforeseen by Apollonius, al-Khowarizmi, and Oresme. Descartes shared the Greek view that a variable corresponds to the length of a line segment, but he showed that a variable raised to *any* whole-number power (including powers larger than the third power) corresponds to the length of another line segment that could be constructed with a compass and straightedge. This new view freed algebra from some of its geometrical constraints by allowing for expressions such as x^4 and $x^2 + x^3$. Descartes was able to solve a variety of quadratic equations by using a compass and straightedge to construct lines of the correct length.

After showing how to interpret algebraic operations and solve quadratic equations geometrically, Descartes went on to show that algebra could be applied to geometric problems. His typical procedure was to begin with a geometrical problem, convert it to an algebraic equation, simplify it algebraically, and solve the simplified equation geometrically. We will see an example of this procedure in Section 10.2.

The analytic geometry of René Descartes was an amazing approach that combined algebra and geometry in new and unique ways. However, it did not especially resemble our modern analytic geometry, which consists of ordered pairs, x- and y-axes, and a correspondence between algebraic equations and their graphs. Descartes used an x-axis, but he did not have a y-axis. (So Descartes did not use what we now call Cartesian coordinates in his honor!) Although he knew that an equation in two unknowns determines a curve, he had very little interest in sketching curves; he never plotted a new curve directly from its equation.

Descartes's algebra, on the other hand, was more modern than that of any of his predecessors or contemporaries. Algebra had been advancing steadily since the Renaissance, and it found its culmination in Descartes's *La géométrie*. Here we see a symbolic, equation-based algebra rather than a verbal, proportion-based algebra. With Descartes, algebra had finally evolved enough so that calculus could be invented.

In 1629, eight years before Descartes's *La géométrie*, the French lawyer and amateur mathematician Pierre de Fermat attempted to recreate one of the lost works of Apollonius on conic sections using references to that work made by other Greek mathematicians. The restoration of lost works of antiquity was a popular pastime of the upper class at that time. Fermat applied Viète's algebra to Apollonius's analytic geometry and created an analytic geometry much more similar to the modern one than was Descartes's. Fermat emphasized the sketching of graphs of equations. He showed a parallelism between certain types of equations and certain types of graphs; for example, he showed that the graph of "*d* planum p. *a* planum aequetur *b* in *e*" ($d^2 + a^2 = be$) is always a parabola. Notice that he used vowels for variables and consonants for constants.

Fermat's analytic geometry was more modern than that of Descartes, but his algebra was less modern. It was verbal and did not allow the addition of "unlike quantities." The expression "*d* planum p. *a* planum aequetur *b* in *e*" combines areas with areas; "*d* planum" refers to the area of a square with side d, "*a* planum" refers to the area of a square with side a, and "*b* in *e*" refers to the area of a rectangle with sides b and e. Descartes would have written "$c + x^2 = by$."

HISTORICAL NOTE

René Descartes

1596–1650

North Wind Picture Archives

René Descartes's family was rather well off, and he inherited enough money to be able to afford a life of study and travel. At the age of eight, he was sent away to a Jesuit school where, at first due to health problems, he was allowed to stay in bed all morning. He maintained this habit all his life and felt that his morning meditative hours were his most productive.

At the age of sixteen, he left school and went to Paris, where he studied mathematics. Four years later, he became a professional soldier, enlisting in the army of Prince Maurice of Nassau and later in the Bavarian army. He found that a soldier's life, though busy and dangerous at times, provided him with sufficient leisure time to continue his studies.

After quitting the army, he spent several years traveling in Denmark, Switzerland, Italy, and Holland. He eventually settled in Holland for 20 years, where he studied science, philosophy, and mathematics and spent four years writing a book on the workings of the physical world. Holland allowed much more freedom of thought than did most of Europe at that time. Even so, when he heard of Galileo's condemnation by the Church and his imprisonment for writing that the earth revolves around the sun, Descartes prudently abandoned his work and instead wrote on his philosophy of science. The resulting work, *Discourse on the Method of Rightly Conducting Reason and Seeking Truth in the Sciences,* was followed by two more books—*Meditationes,* in which he further explains his philosophical views, and *Principia philosophiae* (*Philosophical Principles*), in which he details a theory on the workings of the universe that became the accepted scientific view for about 100 years.

Holland's liberal attitude toward new views was not without exception. At one time, it was forbidden to print or sell any of Descartes's works. Another time, he was brought before a judge on charges of atheism.

Although he made numerous advances in optics and wrote on physics, physiology, and psychology, Descartes is most famous for his philosophical and mathematical works. He has been called the father of modern philosophy, because he attempted to build a completely new system of thought.

When Queen Christina of Sweden invited him to become her tutor, he accepted. Stockholm was cold and miserable, and the queen forced Descartes to break his habit of staying in bed by requiring him to instruct her daily at 5:00 A.M. After four months in Sweden, Descartes came down with pneumonia and died at the age of fifty-four.

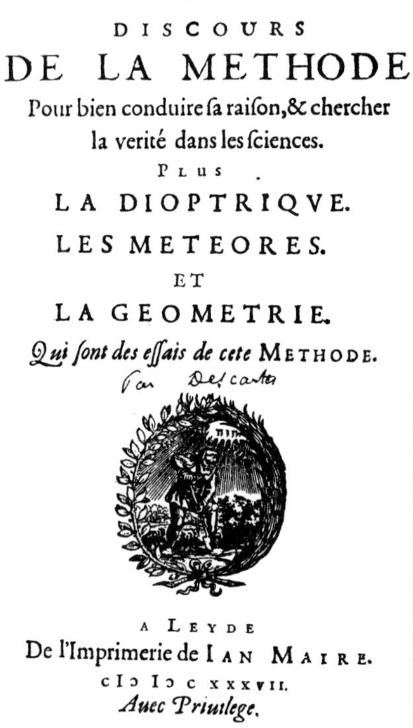
DISCOURS
DE LA METHODE
Pour bien conduire ſa raiſon,& chercher la verité dans les ſciences.
PLUS
LA DIOPTRIQVE.
LES METEORES.
ET
LA GEOMETRIE.
Qui ſont des eſſais de cete METHODE.
A LEYDE
De l'Imprimerie de IAN MAIRE.
cIↃ IↃ c xxxvii.
Auec Priuilege.

Brown University Library

Analytic geometry is usually considered to be an invention of Descartes, because Descartes published before Fermat. However, Fermat had written his work well before Descartes wrote his. Unfortunately, Fermat never published his work; it was released only after his death, almost 50 years after it was written. It would be most correct to say that analytic geometry was invented by Apollonius, al-Khowarizmi, Oresme, Descartes, and Fermat.

10.1 Exercises

In Exercises 1–4, do the following.

a. Sketch the parabola given by the equation.

b. At $x = 1$ and $x = 2$, compute the length of the line segment and the area of the square described by Apollonius. Sketch these two line segments and two squares in their proper places, superimposed over the parabola.

c. Compute the ratio of the lengths of the two line segments and the ratio of the areas of the two squares in part (b).

d. At $x = 3$ and $x = 4$, compute the length of the line segment and the area of the square described by Apollonius. Sketch these two line segments and two squares in their proper places, superimposed over the parabola.

e. Compute the ratio of the lengths of the two line segments and the ratio of the areas of the two squares in part (d).

1. $y = \frac{1}{4}x^2$ **2.** $y = \frac{1}{2}x^2$

3. $y = 3x^2$ **4.** $y = 10x^2$

In Exercises 5–8, do the following.

a. Sketch the parabola given by the equation.

b. At $= 1$ and $x = 2$, compute the length of the line segment and the area of the square described by Apollonius. Sketch these two line segments and two squares in their proper places, superimposed over the parabola.

c. Compute the ratio of the lengths of the two line segments and the ratio of the areas of the two squares in part (b).

5. $y = x^2 + 1$ **6.** $y = x^2 + 3$

7. $y = x^2 + 2x + 1$ **8.** $y = x^2 + 6x + 9$

In Exercises 9–14, do the following.

a. Solve the problem using the method of al-Khowarizmi. Use language similar to his, but don't just copy the example given in this section. Do not use modern terminology and notation.

b. Illustrate your solution to the problem in the manner of al-Khowarizmi.

c. Give a modern version of your solution from part (a).

9. A square and 12 roots are equal to 45 units.

10. A square and 8 roots are equal to 48 units.

11. A square and 6 roots are equal to 7 units.

12. A square and 6 roots are equal to 55 units.

13. A square and 2 roots are equal to 80 units.

14. A square and 4 roots are equal to 96 units.

15. Al-Khowarizmi's verbal description of the solution of $x^2 + 10x = 39$ differs from his geometrical description in one area. In his verbal description, he says to take one-*half* of the roots in the problem ($\frac{1}{2} \cdot 10 = 5$), to multiply the result by itself ($5 \cdot 5 = 25$), and to add this to 39 ($25 + 39 = 64$). In his geometrical description, he takes one-*fourth* of the ten roots ($\frac{1}{4} \cdot 10 = \frac{5}{2}$) and fills in the four corners of a figure with small squares where each square has a side of length $\frac{5}{2}$. This gives a total area of $39 + 4 \cdot (\frac{5}{2})^2 = 39 + 25 = 64$. By answering the following questions, show that these seemingly different sets of instructions always have the same result.

a. Consider the more general problem $x^2 + bx = c$ (rather than al-Khowarizmi's specific example $x^2 + 10x = 39$), and find the result of taking half of the b roots, squaring the result, and adding this to the given area c.

b. In the more general problem $x^2 + bx = c$, find the total area that results from filling in the four corners of a figure with small squares, each of length of one-fourth of the b roots.

16. Apply al-Khowarizmi's method to the problem $x^2 + bx = c$ and obtain a formula for the solution of all equations of this type.

HINT: Follow the modern version of the example in the text but with b and c in place of 10 and 39. You will need to use the $\sqrt{}$ symbol in the next-to-last step.

17. Apply al-Khowarizmi's method to the problem $ax^2 + bx = c$ and obtain a formula for the solution of all equations of this type.

HINT: Divide each side by a and get $x^2 + (b/a)x = c/a$. Then pattern your work after what you did in Exercise 16.

Answer the following questions using complete sentences and your own words.

Concept and History Questions

18. Al-Khowarizmi described his method of solving quadratic equations with an example; he did not generalize his method into a formula, as shown in Exercise 16. What characteristic of the mathematics of his time limited him to this form of a description? What change in mathematics lifted this limitation? To whom is that change due? Approximately how many years after al-Khowarizmi did this change occur?

19. What were some of the earliest uses of mathematics?

20. What type of mathematics did the Egyptians and the Babylonians have?

21. What type of mathematics are the Greeks known for?

22. What are the conic sections?

23. Which Greek mathematician is known for his work on conic sections?

24. Why is our modern number system called the Arabic number system? Which culture created this system? What is important about this system?

25. The mathematics and science of the Greeks could well have been lost if it were not for a certain culture. Which culture saved this Greek knowledge, expanded it, and reintroduced it to Europe?

26. What were the subjects of al-Khowarizmi's two books?

27. What is analytic geometry?

28. What did Apollonius contribute to analytic geometry?

29. What did Oresme contribute to analytic geometry?

30. What did Descartes contribute to analytic geometry?

31. What did Fermat contribute to analytic geometry?

32. What did Viète contribute to algebra?

33. What did Descartes contribute to algebra?

34. What subjects did Descartes study besides mathematics?

35. Why has Descartes been called the father of modern philosophy?

36. Describe Descartes's philosophy.

37. What habit did Descartes maintain all his life?

38. Why did Descartes abandon his book on the workings of the physical world? What did he write about instead?

39. Why was the idea of a number raised to the fourth power impossible in earlier algebra? Why was it not possible to add squares and cubes? Who removed these restrictions from algebra? How did he remove them?

Projects

40. In the next-to-last paragraph in this section, it is stated that Descartes would have written "$c + x^2 = by$" instead of "*d* planum *p*. *a* planum aequetur *b* in *e*." Write several paragraphs in which you explain in detail the transition in language seen here. In your analysis, include the change in conventions regarding the representation of variables and constants and the change in conventions requiring that only "like quantities" can be added. Also discuss the use of verbal algebra in one statement and symbolic algebra in the other.

41. Write several paragraphs tracing the evolution of algebra as well as the evolution of analytic geometry, as described in this section. Use this text as well as other sources.

42. Write an essay summarizing the contributions of Apollonius, Oresme, Descartes, and Fermat to analytic geometry. Use this text as well as other sources.

43. Algebra underwent several qualitative changes before calculus was invented. Write an essay in which you compare and contrast the algebras of Apollonius, al-Khowarizmi, and Descartes. Use this text as well as other sources.

44. Pierre de Fermat once wrote in the margin of a book a brief mathematical statement involving the Pythagorean theorem, along with a claim that he had found a proof of that statement. He never actually wrote the proof. For 350 years, mathematicians were unsuccessful at proving what became known as Fermat's Last Theorem, until a Princeton mathematician is thought to have succeeded in 1993. Write a research paper on Fermat's Last Theorem and the reaction to its recent proof.

45. Write a research paper on a historical topic referred to in this section, or a related topic. Following is a partial list of topics from this section. (Topics in italics were not specifically mentioned in the section.)

- the Ahmes Papyrus (or Rhind Papyrus)
- al-Khowarizmi and his arithmetic and algebra
- Apollonius of Perga and his geometry
- Babylonian algebra
- Constantinople's fall and its effect on Europe
- *Descartes's theory of vortices*
- Egyptian algebra and its use in agriculture and commerce
- Euclid and his geometry
- Pierre de Fermat and his analytic geometry
- Gutenberg and/or his press
- Hindu and Arabic numerals
- Omar Khayyam, Persian poet and mathematician, author of *The Rubaiyat*
- the Moors
- the Islamic Empire
- Islamic mathematicians
- Nicole Oresme and his analytic geometry
- *Pythagoras, Greek mystic and mathematician*
- Thales of Miletus, Greek businessman and mathematician
- François Viète and his algebra

10.2 Four Problems

Four problems that were instrumental in the invention of calculus came together in the 1600s. The earliest of these problems is to find the area of any shape. This is one of the most ancient of all mathematical problems. The Ahmes Papyrus, written in Egypt in 1650 B.C., contains methods for computing some areas. The Greeks had a great deal of success with finding areas; they developed methods for finding areas of circles, rectangles, triangles, and other shapes. However, they were not able to find the area of *any shape*.

The second problem is to find the line tangent to a given curve at any point on that curve. This problem is almost as old as the area problem; it, too, was a subject of Greek geometry. The Greeks attempted to find methods of constructing a line tangent to a curve at a point using a compass and a straightedge. If you've had a class in geometry, you've constructed a line tangent to a circle. The construction involves drawing a radius and then using a compass to construct a line perpendicular to the radius at the point where the radius touches the circle. This line is tangent to the circle (see Figure 10.25).

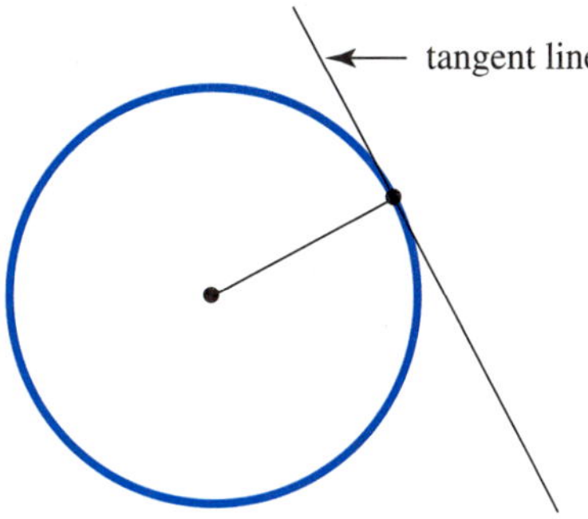

Figure 10.25

What does it mean to say that a line is tangent to a curve? The line tangent to a circle is frequently described as a line that intersects the circle once. However, if the curve is not a circle, the tangent line might hit the curve more than once. The **tangent line** is the straight line to which the curve is most similar in orientation at the point of tangency. If you take that part of the curve that is near the point of tangency and magnify it, as shown in Figure 10.26, the curve will look as if it's almost a straight line. That line is the tangent line.

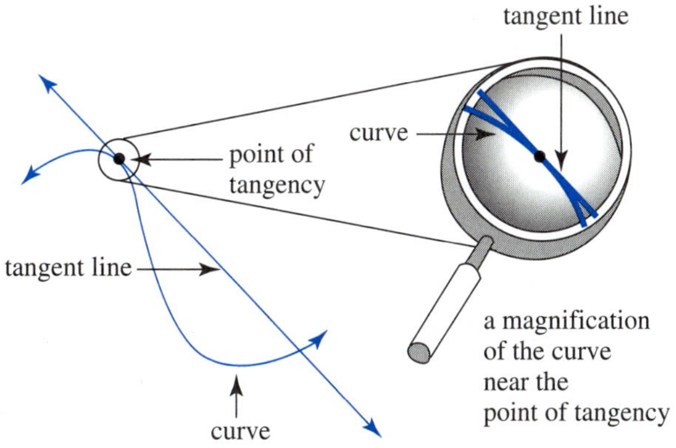

Figure 10.26

The third problem has to do with gravity. In the early 1600s, the Italian scientist Galileo Galilei was attempting to describe the motion of a falling object. In particular, he was attempting to find the distance traveled by a falling object, as well as its speed. He did pursue this matter in the Italian town of Pisa, but he did not actually drop balls from the Leaning Tower of Pisa, as the story claims.

The fourth problem is related to arms and war. In the 1500s and 1600s, Europe was the scene of almost constant warfare. Ever since the introduction of gunpowder in the thirteenth century, as stated earlier, Europe was engulfed in an arms race; a nation's fate hinged on its advances in the design and use of cannons, muskets, and fortifications. Aiming a cannon so that the ball would hit the target was difficult. If one simply aimed the cannon directly at the target, the ball would fall short. If one pointed the cannon too high, the ball would either never reach the target or pass completely over it. Aiming a cannon so that the ball would go over a fortification wall and into the enemy's stronghold was even more difficult. It became necessary to *find the path* (*or trajectory*) *of a cannonball* and determine how to control that **trajectory.** In fact, it became necessary to investigate the principles of motion itself. The first country that found how to do this would have a significant advantage over other countries. Some of Europe's best minds—including Galileo, Leonardo da Vinci, Michelangelo, and Albrecht Dürer—were devoted to various aspects of the problem. Calculus was, to some extent, the result of this arms race.

These and other problems motivated the invention of calculus. (A fifth problem, which we will not specifically investigate, involved finding the orbits of the planets.) In this section, we discuss the partial solutions to these problems that preceded Newton and Leibniz. In an attempt to keep things in their proper historical perspective, we discuss these partial solutions in the order of their solution, rather than in the order of their inception, given at the beginning of this section.

This attacking army is unsuccessful in its siege, due to a lack of knowledge of trajectories. The cannonballs are glancing off the town's walls without causing any damage.

Problem 1: Find the Distance Traveled by a Falling Object

Galileo's work on the motion of a falling object, *Discorsi e dimostriazioni matematiche intorno a due nuove scienze* (*Dialogues Concerning Two New Sciences*) was completed in 1636. However, the story of Galileo's investigations of gravity begins almost 300 years earlier with the Parisian scholar Nicole Oresme.

Recall that in about 1360, Oresme drew the graph of the speed of a moving object (Figure 10.27), which he thought of as the "intensity" of the object's motion. In this graph, point A represents the time at which the motion started, and points B and D represent later times. The length of a horizontal line segment, such as AB, represents the amount of time from the beginning of motion until time B, and the length of a vertical line segment, such as BC, represents the speed of the object at time B. These lengths are increasing in a uniform way, so Oresme's graph was of an object whose speed is increasing in a uniform way. Oresme somehow realized that the area of triangle ABC gives the distance traveled by the object from time A until time B, although he never explained why this would be the case.

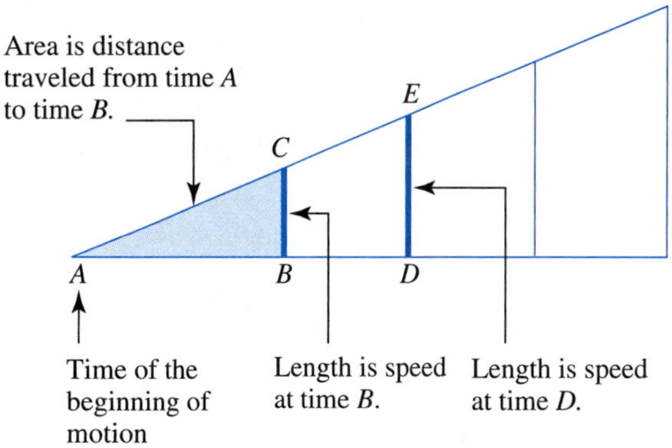

Figure 10.27

Oresme used this triangle to show that if an object's speed is increasing uniformly, then that speed is proportional to the time in motion. His reasoning, illustrated in Figure 10.28, went like this:

ΔABC is similar to ΔADE — the triangles have equal angles

$$\frac{BC}{DE} = \frac{AB}{AD}$$ ratios of corresponding sides are equal

$$\frac{\text{speed at time } B}{\text{speed at time } D} = \frac{\text{time in motion until time } B}{\text{time in motion until time } D}$$

The speed of the object is proportional to its time in motion.

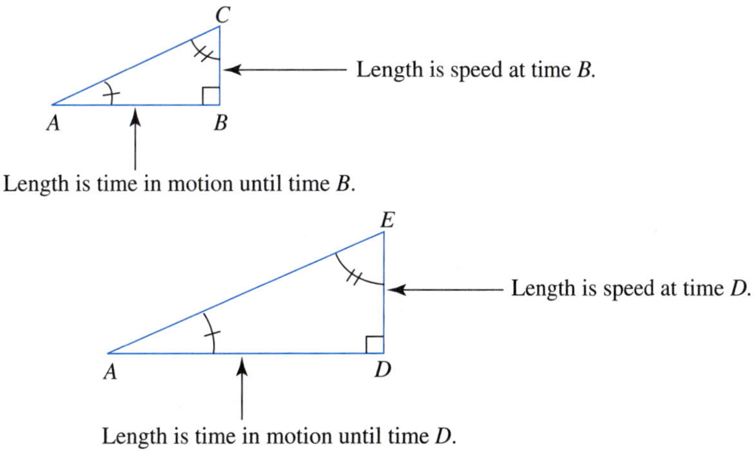

Figure 10.28

If the algebra of Oresme's time had been based on equations rather than proportions, he might have reasoned further and obtained a formula for speed. However, algebra didn't reach that point in its development until the time of Descartes and Galileo, around 300 years later.

Oresme never applied his triangle to the motion of a falling object. However, Galileo was familiar with Oresme's work and applied the triangle to this type of motion, reasoning that a falling object moves in such a way that its speed increases uniformly. This allowed him to conclude, like Oresme, that speed is proportional to time.

Galileo lived in a time when mathematics was rapidly advancing. Descartes's *La géométrie*, Fermat's (unpublished) work on conic sections, and Galileo's work on the motion of a falling object were finished within a few years of each other. Unfortunately, Galileo did not use the algebra or analytic geometry of Descartes and Fermat, presumably because it was too new and he was not familiar with it; instead, he used the 2,000-year-old analytic geometry of Apollonius and an algebra that was verbal and based on proportions rather than equations. Thus, like Oresme, he was unable to express his conclusion that speed is proportional to time as a formula. Instead, he used language such as "the ratio of the velocities of an object falling from rest is the same as the ratio of the time intervals employed in traversing their distances." This language is cumbersome and makes it difficult to apply the relationship between speed and time in motion discovered by Galileo. A formula of the form

$$\text{speed} = ???$$

would have made it easier to apply that relationship and might have led to new results and discoveries.

In addition to concluding that the speed of a falling object is proportional to its time in motion, Galileo was able to use Oresme's triangle to make an observation about the distance traveled by a falling object. Oresme had realized that the area of his triangle gives the distance traveled by the object. Galileo observed that if a falling object covered a certain distance in the first interval of time measuring its fall, then it would cover three times that distance in the second interval of time and five times that distance in the third interval of time (and similarly with following intervals of time). Figures 10.29 and 10.30 on page 733 illustrate his observations.

Galileo had established the following pattern:

Interval of Time	Distance Traveled during That Interval	Total Distance Traveled
first	1	1
second	3	$1 + 3 = 4$
third	5	$1 + 3 + 5 = 9$

Notice that in each case, the total distance traveled is the square of the amount of time in motion. Galileo concluded that the distance traveled by a falling object was proportional to the square of its time in motion, or, as he phrased it, "the spaces described by a body falling from rest with a uniformly accelerated motion are as to each other as the squares of the time-intervals employed in traversing these distances."

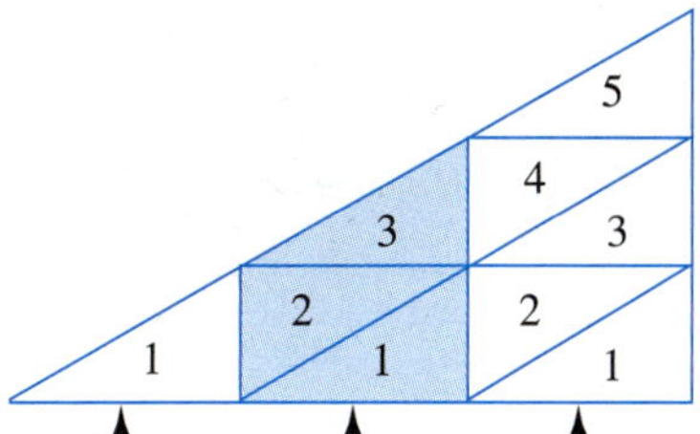

Figure 10.29

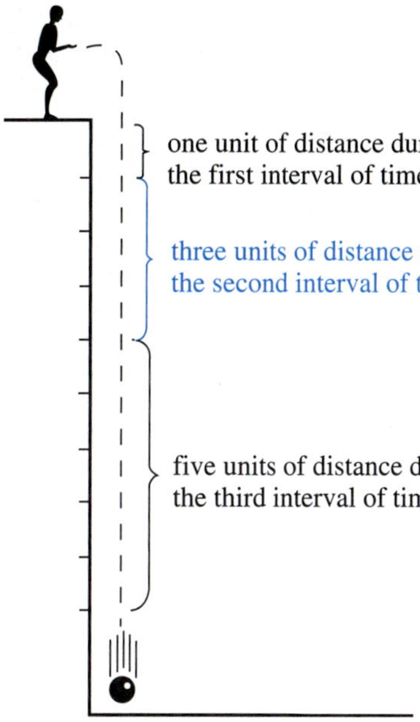

Figure 10.30

In a roundabout way, Galileo verified his conclusions regarding falling bodies by timing a ball rolling down a ramp. He was unable to experiment effectively with falling objects, because things fell too fast to be timed with the water clock with which he was obliged to measure intervals. His water clock consisted of a vessel of water that was allowed to drain through a pipe while the ball was rolling. The drained water was weighed at the end of the trip. If a ball was timed during two different trips, he found that

$$\frac{\text{distance from first trip}}{\text{distance from second trip}} = \frac{(\text{weight of water from first trip})^2}{(\text{weight of water from second trip})^2}$$

Galileo was never able to actually determine how far a ball would fall (or roll down a ramp) for a given period of time because of his use of proportional, verbal algebra and the quality of his clock.

Problem 2: Find the Trajectory of a Cannonball

Galileo thought of the motion of a cannonball as having two separate components—one due to the firing of the cannon, in the direction in which the cannon is pointed, and the other due to gravity, in a downward direction. He used his result regarding the distance traveled by a falling object and Apollonius's description of a parabola to show that the path of the cannonball is actually a parabola.

Galileo reasoned that if a cannon was at the top of a cliff and pointed horizontally, then the cannonball's motion could be resolved into a horizontal motion due to the firing of the cannon and a vertical motion due to gravity. Consider the cannonball's position at times t_1 and t_2 and let the horizontal distances traveled be x_1 and x_2 and the vertical distances be y_1 and y_2, as shown in Figure 10.31 on page 734.

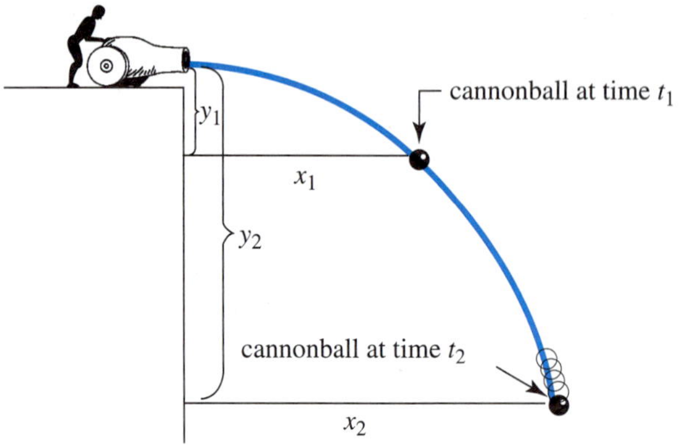

x_1 and x_2 are the horizontal distances traveled by the cannonball

y_1 and y_2 are the vertical distances traveled by the cannonball

Figure 10.31

The horizontal motion has constant speed, because there is nothing to slow it down (except a very small amount of air resistance, which Galileo chose to ignore). In twice the time, the cannonball would cover twice the horizontal distance. Thus, the horizontal distance traveled is proportional to the time in motion. In other words,

$$\frac{x_1}{x_2} = \frac{t_1}{t_2}$$

The vertical motion is due to gravity, so Galileo's conclusion that "the spaces described by a body falling from rest with a uniformly accelerated motion are as to each other as the squares of the time-intervals employed in traversing these distances" applies—or, using modern algebra,

$$\frac{y_1}{y_2} = \frac{t_1^2}{t_2^2}$$

Combining these two results yields

$$\begin{aligned}\frac{y_1}{y_2} &= \frac{t_1^2}{t_2^2}\\ &= \left(\frac{t_1}{t_2}\right)^2\\ &= \left(\frac{x_1}{x_2}\right)^2 \quad \text{substituting}\\ &= \frac{x_1^2}{x_2^2}\end{aligned}$$

Note: Symbolic algebra is used for the convenience of the reader. Galileo used verbal algebra.

This told Galileo that the path of a cannonball must be parabolic, because Galileo knew that Apollonius had shown the following to be true for any parabola:

$$\frac{\text{length of line segment 1}}{\text{length of line segment 2}} = \frac{\text{area of square 1}}{\text{area of square 2}}$$

That is,

$$\frac{y_1}{y_2} = \frac{x_1^2}{x_2^2}$$

as shown in Figure 10.32.

Thus, Galileo used his own work on a falling body and Apollonius's description of a parabola to show that a horizontally fired cannonball follows a parabolic path. He was also able to show that if a cannonball is fired at an angle, it will still follow a parabolic path. Naturally, this conclusion applies to all other projectiles, including balls that are thrown rather than fired.

Unfortunately, Galileo was hampered by his use of ancient mathematics, so he was unable to translate this discovery into an equation for the path of a cannonball, and he was unable to determine how to aim a cannon so the ball would hit the target.

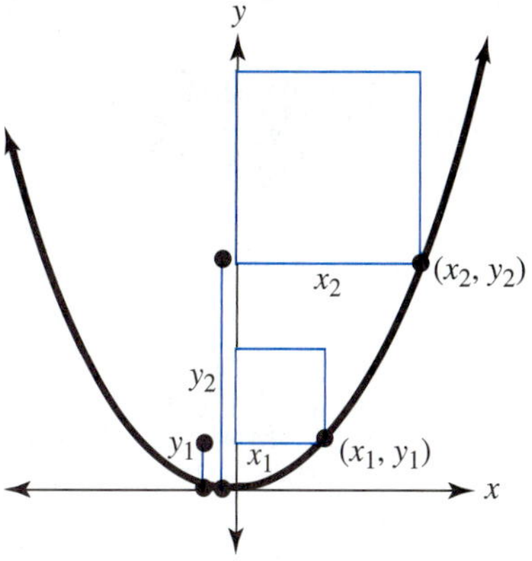

Figure 10.32

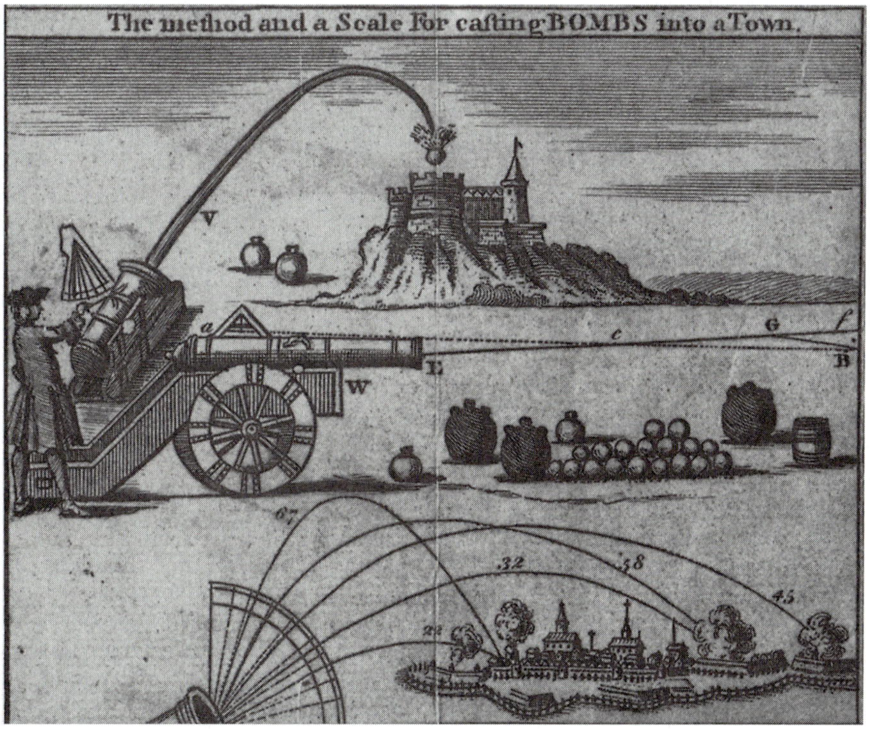

Scientists were able to use Galileo's discovery that a cannonball follows a parabolic path to determine how to aim a cannon, as shown in this engraving. Notice the use of angles in determining the ball's path; this will be discussed in Section 10.5.

HISTORICAL NOTE

Galileo Galilei

1564–1642

Galileo Galilei was born in Pisa, Italy, in 1564. His father was a noble of Florence, and Galileo was well-educated. He studied medicine at the University of Pisa until he became distracted by two different events. One day while daydreaming in church, he watched a lamp swing back and forth as a result of its being pulled aside for lighting. He timed the swings with his pulse and noticed that each lasted the same amount of time, even though the swings were decreasing in size. Later, he attended a lecture on geometry at the university. As a result, Galileo's interest in physics and mathematics was aroused, and he changed his major.

At age twenty-five, Galileo was appointed professor of mathematics at the University of Pisa, where he began experimenting with motion, especially motion due to gravity. His findings contradicted the accepted teachings of Aristotle. Other professors were shocked that Galileo would consider contradicting Aristotle and refused to even consider Galileo's experimental evidence. Science at that time was based solely on "logical" thinking, with no attempt to support conclusions by experimentation. Galileo was forced to resign his position, and he became a professor at the University of Padua, where his pursuits found greater acceptance.

Meanwhile, in Holland, an apprentice lens grinder had discovered that if he looked through two lenses held an appropriate distance apart, objects appeared to be closer. His master used this principle to make a toy that he then displayed in his window. The toy came to the attention of Prince Maurice of Nassau, commander of the armed forces of the Netherlands (and, coincidentally, Descartes's commander), who envisioned military uses for it.

Galileo heard of this device two years later and started building telescopes. His fifth telescope was large enough and accurate enough to enable him to observe sunspots and the mountains on the moon and to see that Jupiter had four moons that seemed to circle around it. These observations again contradicted the accepted science of Aristotle, which held that the sun is without blemish and that all celestial bodies revolve around the earth. Galileo chose to publicly support the Copernican system, which stated that the planets revolve around the sun. His foes would not accept his evidence; he was even accused of falsifying his evidence by placing the four moons of Jupiter inside his telescope. The Church denounced the Copernican system as being dangerous to the faith and summoned Galileo to Rome, warning him not to teach or uphold that system.

Galileo responded by writing a masterpiece, *Dialogue on the Two Chief Systems of the World,* in which he compared the strengths and weaknesses of the two theories of celestial motion. This work was written so that the nonspecialist

Problem 3: Find the Line Tangent to a Given Curve

The development of analytic geometry by Fermat and Descartes made the solution of this tangent problem much more likely, because mathematicians could apply algebra to what had been a strictly geometrical problem.

Descartes typically began with a geometrical problem, converted it to an algebraic equation, simplified it algebraically, and solved the resulting equation geometrically. He applied this procedure to the tangent line problem. He would start with the equation of the curve and the location on the curve of the point P at which he wished to find the tangent line. Using algebra, he would find the equation of the circle through point P, with center on the x-axis, that had the same tangent line. Then he would construct the line tangent to the circle at the point P, using a compass and straightedge, as shown in Figure 10.33. Unfortunately, the algebra involved could be quite overwhelming.

could read it, and it became quite popular. Galileo was again summoned to Rome, where he was tried by the Inquisition and forced to recant his findings under threat of torture. It is said that after his forced denial of the earth's motion, he muttered "e pur si muove" ("nevertheless it does move"). He was imprisoned in his home and died nine years later. As a result of Galileo's experience, most scientists left Italy and went to Holland, where new scientific views were viewed with a more tolerant attitude.

A note in Galileo's handwriting in the margin of his personal copy of the *Dialogue* states, "In the matter of introducing novelties. And who can doubt that it will lead to the worst disorders when minds created free by God are compelled to submit slavishly to an outside will? When we are told to deny our senses and subject them to the whim of others? When people devoid of whatsoever competence are made judges over experts and are granted authority to treat them as they please? These are the novelties which are apt to bring about the ruin of commonwealths and the subversion of the state."

It is interesting to note that in 1992, the Roman Catholic Church admitted that it was wrong in having condemned Galileo and opposed the Copernican system.

Galileo is called the father of modern science because of his emphasis on experimentation and his interest in determining how things work rather than what causes them to work as they do. He insisted that a theory was unsound, no matter how logical it seemed, if observation did not support it. He invented the pendulum clock and the thermometer; he constructed one of the first compound microscopes; and he greatly improved the design of telescopes. His last work, *Dialogues Concerning Two New Sciences,* contains many of his contributions to science, including those concerning motion and the strength of materials. In that work, he showed that a projectile follows a parabolic path and came to conclusions that foreshadowed Newton's laws of motion. He also held that motion does not require a force to maintain it (as Aristotle claimed), but rather the "creation or destruction of motion" (that is, acceleration or deceleration) requires the application of force. Thus, he was the first to appreciate the importance of the concept of acceleration.

The frontispiece of Galileo's *Dialogue on the Two Chief Systems of the World,* showing Galileo (left) conversing with Ptolemy (center) and Copernicus. Ptolemy was the main historical proponent of the "sun revolves around the earth" theory.

North Wind Picture Archives

About 30 years after Descartes's work was published, an English mathematician named Isaac Barrow published a method of finding the slope of the tangent line in his book *Lectiones opticae et geometricae* (*Optical and Geometrical Lectures*). His method was very similar to an earlier one due to Fermat and might have been based on that method. Barrow was the first holder of the Lucas chair in mathematics at Cambridge, a prestigious professorship funded by member of parliament Henry Lucas. Isaac Newton was the second holder of the Lucas chair; Stephen Hawking, author of *A Brief History of Time*, is its current holder.

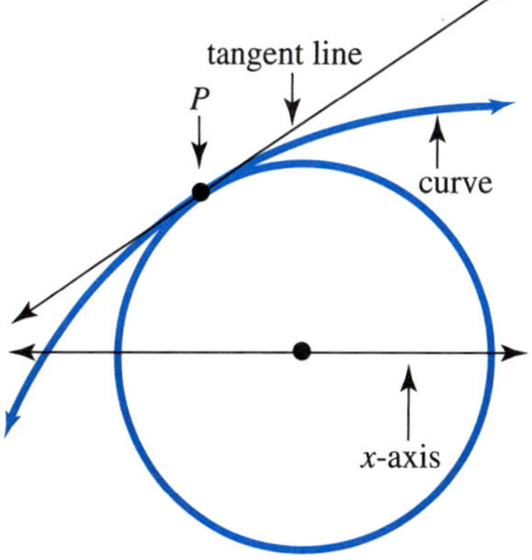

Figure 10.33

The Granger Collection, NY

Isaac Barrow, first holder of the Lucas chair

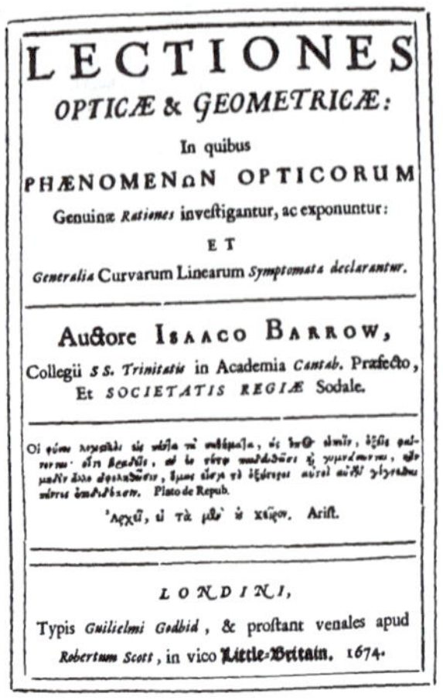
LECTIONES
OPTICÆ & GEOMETRICÆ:
In quibus
PHÆNOMENΩN OPTICORUM
Genuinæ Rationes investigantur, ac exponuntur:
ET
Generalia Curvarum Linearum Symptomata declarantur.

Auctore ISAACO BARROW,
Collegii S S. Trinitatis in Academia Cantab. Præfecto,
Et SOCIETATIS REGIÆ Sodale.

Plato de Repub.
Arist.

LONDINI,
Typis Gulielmi Godbid, & prostant venales apud
Robertum Scott, in vico Little-Britain. 1674.

Brown University Library

James D. Wilson/Liaison Agency

Stephen Hawking, current holder of the Lucas chair

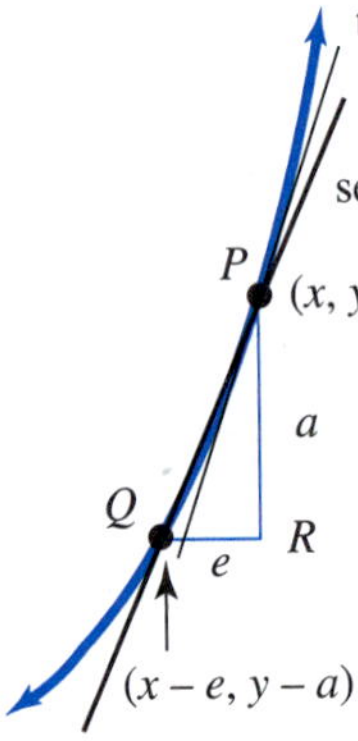

Figure 10.34

Barrow's method involved two points on the curve, P and Q, with P the point at which he wanted to find the tangent line and Q a point very close to P, as shown in Figure 10.34. The secant line PQ is very close to the desired tangent line. If a and e are the lengths of the vertical and horizontal sides of triangle PQR, respectively, then the slope of the secant line PQ is

$$\frac{\text{rise}}{\text{run}} = \frac{a}{e}$$

The tangent line has almost the same slope as the secant line PQ. As we can see in Figure 10.34, both have a rise of a, but the tangent's run is not quite the same as that of the secant.

If we let point P be the ordered pair (x, y), then point Q is $(x - e, y - a)$. (*Note*: Barrow did not use ordered pairs; they are used here for the convenience of the reader.) Because point Q is a point on the curve, $(x - e, y - a)$ would successfully substitute into the equation of the curve.

After making this substitution and simplifying the result, Barrow would "omit all terms containing a power of a or e, or products of these (for these terms have no value)." If the result were solved for a/e, the slope of the tangent line would be obtained. Why these terms should be omitted, or why their omission changes a/e from the slope of the secant to the slope of the tangent, is not clear, and Barrow did not attempt an explanation. However, the method does work.

EXAMPLE 1 Use Barrow's method to find the slope of the line tangent to the parabolic curve $y = x^2$ at the point (2, 4), as shown in Figure 10.35.

Solution P is at (2, 4), so Q is at $(2 - e, 4 - a)$.

$4 - a = (2 - e)^2$	substituting $2 - e$ for x and $4 - a$ for y in $y = x^2$
$4 - a = 4 - 4e + e^2$	multiplying
$e^2 - 4e + a = 0$	collecting like terms
$-4e + a = 0$	omitting all terms containing a power of a or e
$a = 4e$	
$\frac{a}{e} = 4$	solving for a/e

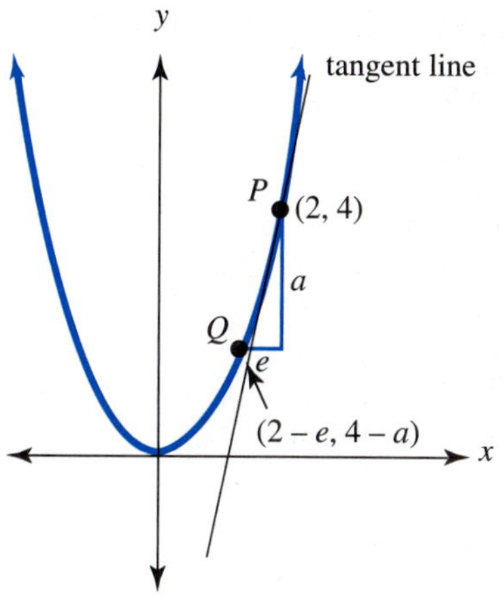

Figure 10.35

The slope of the tangent line at (2, 4) is 4. This result fits with the sketch of the tangent line; its slope could reasonably be 4. ■

> **Barrow's Method of Finding the Slope of the Tangent Line**
>
> 1. *Find Q.* Get Q's x-coordinate by subtracting e from P's x-coordinate. Get Q's y-coordinate by subtracting a from P's y-coordinate.
> 2. *Substitute the ordered pair describing Q into the equation of the curve.* Simplify the result.
> 3. *Omit all terms containing a power of a or e, or products of these.*
> 4. *Solve the result for a/e.* This is the slope of the tangent line.

Barrow's method is not nearly as geometrical as that of Descartes; there is no construction of a line with compass and straightedge. The advantage of Barrow's method is that the algebra tends to be much easier, but the disadvantage is that why Barrow's method works is not clear.

Once the slope of the tangent line has been found, finding the equation of the tangent line is easy. To do so, use the point-slope formula from intermediate algebra.

> **Point-Slope Formula**
>
> A line through (x_1, y_1) with slope m has equation
>
> $$y - y_1 = m(x - x_1)$$

EXAMPLE 2 Find the equation of the line tangent to the curve $y = x^2$ at the point (2, 4).

Solution We are given $(x_1, y_1) = (2, 4)$. In Example 1, we found that the slope of this line is $m = 4$.

$$\begin{aligned} y - y_1 &= m(x - x_1) && \text{Point-Slope Formula} \\ y - 4 &= 4(x - 2) && \text{substituting} \\ y - 4 &= 4x - 8 && \text{distributing} \\ y &= 4x - 4 \end{aligned}$$

■

Problem 4: Find the Area of Any Shape

Some of the ancient Greek geometers developed clever ways of finding areas and volumes. Eudoxus and Archimedes found areas of curved shapes by filling in the region with a sequence of successively smaller triangles and finding the sum of the areas of those triangles (see Figure 10.36).

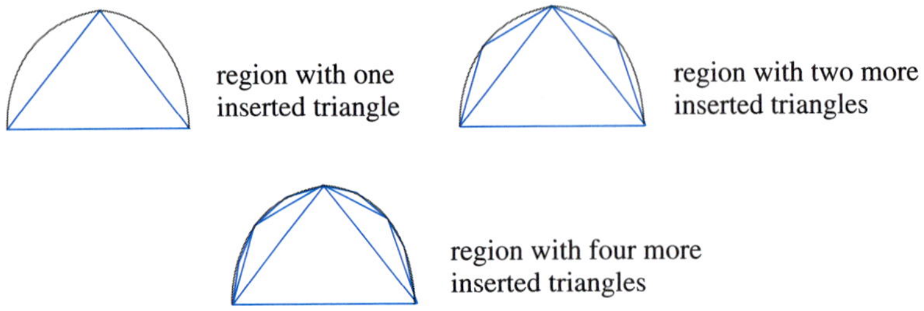

Figure 10.36

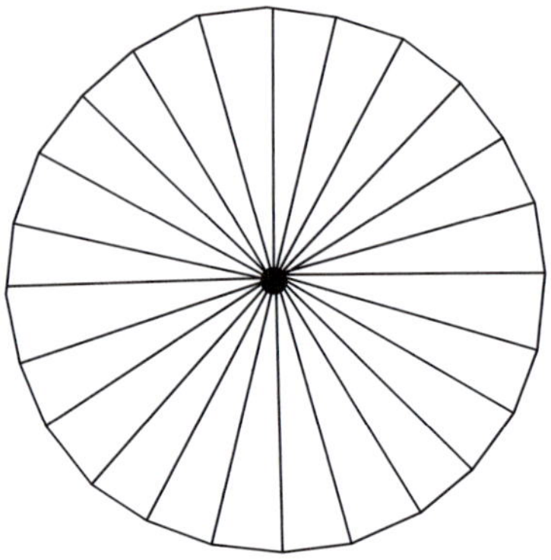

Figure 10.37

Around 1450, European mathematicians learned of these methods when a manuscript was found at Constantinople. In the early 1600s, the astronomer and mathematician Johann Kepler adapted Archimedes' methods to find areas and volumes. While his main use for these techniques was in his studies of the planets' orbits, he also used them in a paper on the volumes of wine barrels. We will illustrate Kepler's method by showing that the area of a circle is πr^2.

Kepler envisioned the circumference of a circle as being composed of an infinite number of very short straight lines, as illustrated in Figure 10.37.

If each of these short lines is taken as the base of a triangle with vertex at the center of the circle, and if the radius of the circle is r, then the area of the circle would be as follows:

$$\text{(area of first triangle)} + \text{(area of second triangle)} + \cdots$$

$$= \left(\frac{1}{2} \cdot \text{base} \cdot \text{height}\right) + \left(\frac{1}{2} \cdot \text{base} \cdot \text{height}\right) + \cdots$$

$$= \frac{1}{2}\,\text{height (base} + \text{base} + \cdots)$$ because the heights are all equal

$$= \frac{1}{2}\,r(\text{base} + \text{base} + \cdots)$$ because the height goes from the circumference to the center

$$= \frac{1}{2}\,r(\text{circumference of the circle})$$

$$= \frac{1}{2}\,r(2\pi r)$$

$$= \pi r^2$$

Twenty years after Kepler published his method of computing areas, a student of Galileo named Bonaventura Cavalieri published a popular book that showed how to compute areas by thinking of a region as being composed of an infinite number of rectangles rather than an infinite number of triangles (see Figure 10.38). Galileo himself had used a similar approach in analyzing Oresme's triangle.

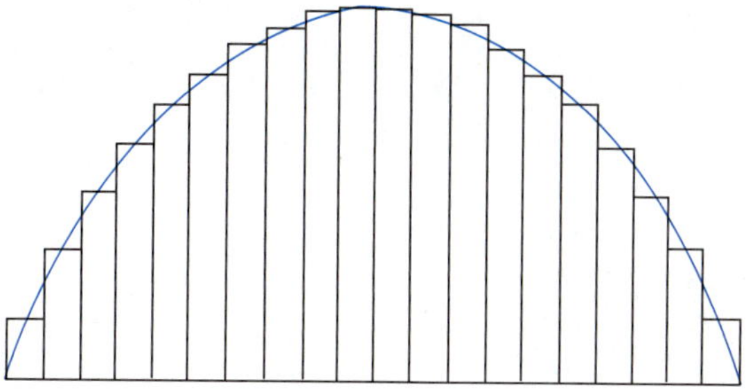

Figure 10.38

10.2 Exercises

Reread the discussion of Oresme's triangle at the beginning of this section. Use that triangle to answer the questions in Exercises 1–4 about an object whose speed increases uniformly from the moment motion begins.

1. After 10 seconds, an object is moving at a rate of 4 feet per second.
   - **a.** Find the distance it has traveled in those 10 seconds.

     HINT: According to Oresme, distance = area of triangle.
   - **b.** Find its speed after 15 seconds.

     HINT: Use similar triangles.
   - **c.** Find the distance it has traveled in 15 seconds.
   - **d.** Find its average speed during its first 15 seconds of travel.

     HINT: Average speed = distance/time
2. After 15 minutes, an object is moving at a rate of 6 feet per minute.
   - **a.** Find the distance it has traveled in those 15 minutes.
   - **b.** Find its speed after 10 minutes.
   - **c.** Find the distance it has traveled in 10 minutes.
   - **d.** Find its average speed during its first 10 minutes of travel.
3. After 5 minutes, an object has traveled 100 feet.
   - **a.** Find its speed at the moment it has traveled 100 feet.
   - **b.** Find its speed after 7 minutes of motion.
   - **c.** Find the distance it has traveled in 7 minutes.
   - **d.** Find its average speed for those 7 minutes.
4. After 3 seconds, an object has traveled 2 feet.
   - **a.** Find its speed at the moment it has traveled 2 feet.
   - **b.** Find its speed after 10 seconds of motion.
   - **c.** Find the distance it has traveled in 10 seconds.
   - **d.** Find its average speed for those 10 seconds.
5. Galileo used Oresme's triangle (refer back to Figure 10.29) to observe the following pattern in the distance traveled by a falling object.

Interval of Time	Distance Traveled during That Interval	Total Distance Traveled
first	1	1
second	3	$1 + 3 = 4$
third	5	$1 + 3 + 5 = 9$

Expand the illustration in Figure 10.29 and the accompanying chart to include the fourth through sixth intervals of time.

In Exercises 6–9, use the following information: Using a water clock to time a ball rolling down a ramp, Galileo found that if a ball was timed during two different trips, the following relationship held:

$$\frac{\text{distance from first trip}}{\text{distance from second trip}} = \frac{(\text{weight of water from first trip})^2}{(\text{weight of water from second trip})^2}$$

6. If a ball would roll 25 feet while 8 ounces of water flowed out of a water clock, how far would it roll while 4 ounces of water flowed from the clock?
7. If a ball would roll 32 feet while 6 ounces of water flowed out of a water clock, how far would it roll while 12 ounces of water flowed from the clock?
8. If a ball would roll 25 feet while 8 ounces of water flowed out of a water clock, how much water would flow out of the clock while the ball rolled 50 feet?
9. If a ball would roll 40 feet while 7 ounces of water flowed out of a water clock, how much water would flow out of the clock while the ball rolled 10 feet?

In Exercises 10–13, use the following information: Galileo found that "the spaces described by a body falling from rest with a uniformly accelerated motion are as to each other as the squares of the time-intervals employed in traversing these distances." In other words,

$$\frac{\text{distance from first trip}}{\text{distance from second trip}} = \frac{(\text{time of first trip})^2}{(\text{time of second trip})^2}$$

Galileo was unable to time a falling object, due to the inaccuracy of his water clock. Suppose Galileo had a stopwatch and used it to find that an object falls 1,600 feet in 10 seconds. How far would that object fall in the times given?

10. 3 seconds
11. 12 seconds
12. 20 seconds
13. 1 second
14. A cannon is 140 feet away from a stone wall that is 25 feet tall. If a cannonball is fired so that it barely passes over the wall and begins to descend immediately after passing over the wall, at approximately what distance beyond the wall will the cannonball hit the ground?

    HINT: Draw a picture and use Galileo's conclusion regarding the path of a projectile.

15. A woman throws a ball over an 8-foot fence that is 30 feet away. The ball barely passes over the fence and begins to descend immediately after passing over the fence. Approximately how far from the fence should her friend stand in order to catch the ball?

    HINT: Draw a picture and use Galileo's conclusion regarding the path of a projectile.

In Exercises 16–25, (a) use Barrow's method to calculate the slope of the line tangent to the given curve at the given point. Also (b) find the equation of the tangent line, and (c) include an accurate sketch of the curve and the tangent line.

16. $y = 2x^2$ at $(1, 2)$
17. $y = 2x^2$ at $(3, 18)$
18. $y = x^3$ at $(2, 8)$
19. $y = x^3$ at $(3, 27)$
20. $y = x^2 - 2x + 1$ at $(1, 0)$
21. $y = x^2 - 2x + 1$ at $(3, 4)$
22. $y = x^2 + 4x + 7$ at $x = -1$
23. $y = x^2 + 6x + 11$ at $x = -1$
24. $y = x^2 + 6x + 11$ at $x = 0$
25. $y = x^2 + 4x + 6$ at $x = 0$
26. Oresme thought that the area of his triangle equaled the distance the object travels, but he never explained why this was so. This exercise outlines a possible explanation of the relationship.
    a. Why would Oresme use a rectangle instead of a triangle to describe the motion of an object whose speed is constant?
    b. If an object moves with constant speed, then

    $$\text{distance traveled} = \text{speed} \cdot \text{time}$$

    Why is the area of Oresme's rectangle equal to the distance the object travels?
    c. Oresme described the motion of an object whose speed is increasing uniformly with a triangle. Such a triangle has the same area as a rectangle whose height is half that of the triangle, as shown in Figure 10.39. What does this imply about the average speed of an object whose speed is increasing uniformly?

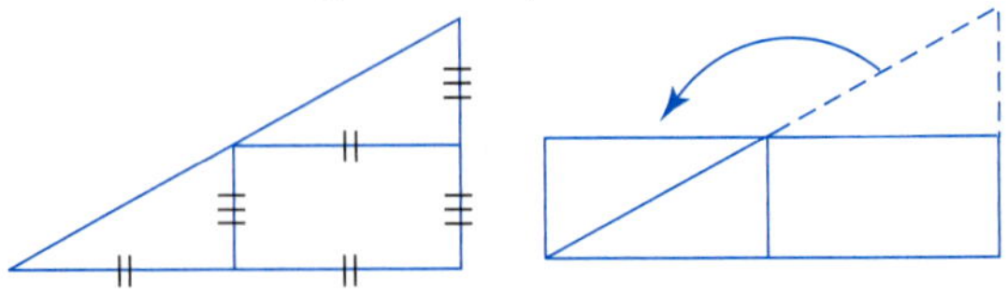

Figure 10.39

    d. Why is the area of Oresme's triangle equal to the distance the object travels?
27. Kepler showed that the area of a circle is πr^2. Use similar reasoning to show that the volume of a

sphere is $\frac{4}{3}\pi r^3$. You will need to envision the sphere as being composed of an infinite number of very thin cones (see Figure 10.40). You will need the following two formulas: the volume of a cone is $\frac{1}{3}\pi r^2 h$, and the surface area of a sphere is $4\pi r^2$.

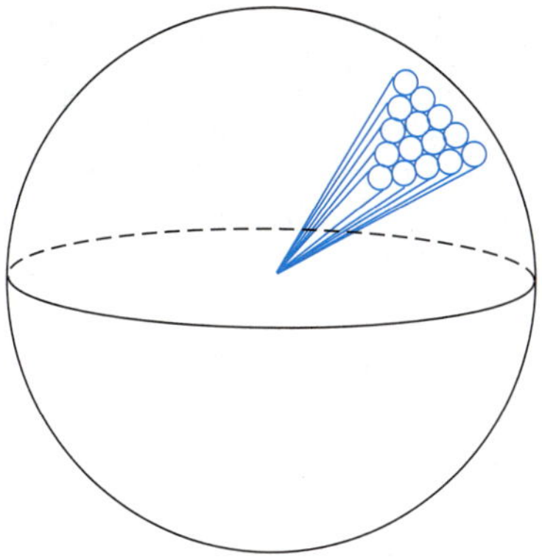

Figure 10.40

Answer the following questions using complete sentences and your own words.

Concept Questions

28. Why did Galileo conclude that the distance traveled by a falling object is *proportional to* the square of its time in motion, even though the chart on page 732 shows that the total distance traveled is *equal to* the square of the amount of time in motion?

29. Explain why "in twice the time, the cannonball would cover twice the horizontal distance" implies that

$$\frac{x_1}{x_2} = \frac{t_1}{t_2}$$

as discussed on page 734.

30. In using Kepler's method to find the area of a circle, "circumference of the circle" was substituted for "base + base + $\cdots$." Explain why these two quantities are equal.

31. How did Galileo conclude that the distance traveled by a falling object is proportional to the square of its time in motion?

32. Compare and contrast Descartes' and Barrow's methods of finding a line tangent to a curve at a given point. Be certain to discuss both their similarities and their differences.

History Questions

33. What role did Oresme play in the "distance traveled by a falling object" problem? What role did Galileo play?

34. Why was Galileo unable to verify experimentally his conclusion that the distance traveled by a falling object is proportional to the square of its time in motion? How did he partially verify it?

35. How did Galileo conclude that a cannonball fired horizontally from the top of a cliff would have a parabolic trajectory?

36. Why was Galileo unable to actually find the equation of the trajectory of a cannonball?

37. What was Galileo studying at the university when he changed his major to science and mathematics? What two events prompted that change?

38. What reaction did his colleagues have to Galileo's experiments concerning motion? How did he respond?

39. What reaction did his colleagues have to Galileo's studies of the heavens with his telescope? How did he respond?

40. What reaction did the Church have to Galileo's studies of the heavens with his telescope? How did he respond?

41. Why is Galileo called the father of modern science?

42. Why was Kepler interested in computing areas and volumes?

Projects

43. The four problems discussed in this section were not the *only* problems that motivated the invention of calculus. A fifth problem involved finding the orbits of the planets. Write a research paper on the history of this problem. Start your research by reading about Copernicus and Kepler.

44. Write a research paper on a historical topic referred to in this section, or a related topic. Following is a partial list of topics from this section. (The topics in italics were not mentioned in the section.)

- Archimedes, Greek inventor and mathematician
- Isaac Barrow, English scientist and mathematician
- Bonaventura Cavalieri, Italian mathematician
- Nicholas Copernicus and the Copernican system
- Eudoxus, Greek astronomer, mathematician, and physicist
- the impact of gunpowder on European warfare
- the Inquisition and the punishment of heretics
- Johann Kepler, German astronomer and mathematician
- the Roman Catholic church, Galileo, and the Copernican system
- *Evangelista Torricelli, Italian physicist and mathematician*
- *John Wallis, English mathematician*
- the water clock and other ancient timekeeping devices

10.3 Newton and Tangent Lines

Mathematics had made tremendous advances by the late 1600s. Algebra had become much more powerful due to Viète, Descartes, and many others. In particular, algebra was based more on equations and less on ratios and proportions, and it had shed its verbal style for a simpler, easier-to-use symbolic notation. Analytic geometry had been developed by Fermat and Descartes. Galileo had accomplished much in his study of the speed and motion of both a falling object and a cannonball. Several mathematicians—including Descartes, Fermat, and Barrow—had developed methods for finding the slope of a tangent line. Isaac Newton, a young student of Barrow at Cambridge University, built on this foundation and created calculus.

Newton improved Barrow's method of finding the slope of a tangent line and in doing so found the key that unlocked all four of the problems. In creating a more logical method of finding the slope of a tangent line, Newton also created a method of finding the speed of any moving object (including a falling object and a cannonball), as well as a method of finding the area of any shape. We will explore Newton's ideas about tangent lines in this section and his ideas on speed and area in the following sections.

Newton was not only a student of Barrow but also a close friend. He helped Barrow prepare *Optical and Geometrical Lectures* for publication. In the preface of the book, Barrow acknowledges indebtedness to Newton for some of the material. It should not be surprising, then, that Newton's method of finding the slope of the tangent line is very similar to Barrow's. The method given here is a later version found in Newton's *De quadratura curvarum* (*Quadrature of Curves*).

Newton's method involves Barrow's two points P and Q, except that Q is thought of as being to the right of P by a small amount o, as shown in Figure 10.41. (Barrow's point Q was to the left of P by an amount e). If point P has an x-coordinate of x, then point Q would have an x-coordinate of $x + o$ (o is the letter "oh," not the number "zero"). Newton called o an **evanescent increment**—that is, an imperceptibly small increase in x. Newton's method consisted of computing the slope of the secant line PQ and then replacing the evanescent increment o with the number zero. He called the changes in x and y **fluxions** and the ratio of their changes (that is, the slope) the **ultimate ratio of fluxions.**

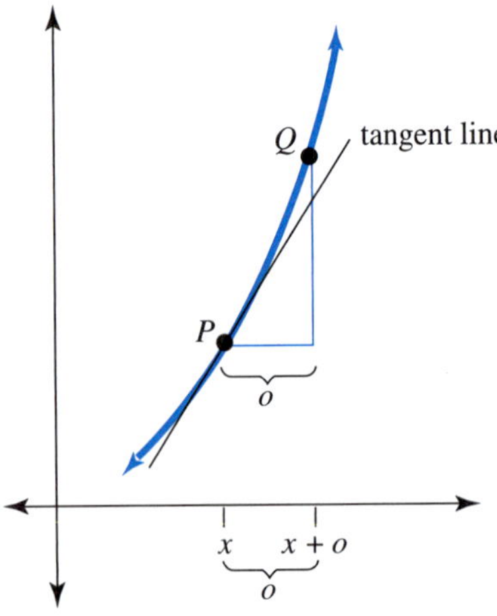

Figure 10.41

EXAMPLE 1 Use Newton's method to find the slope of the line tangent to the curve $y = x^2$ at the point (2, 4), as shown in Figure 10.42.

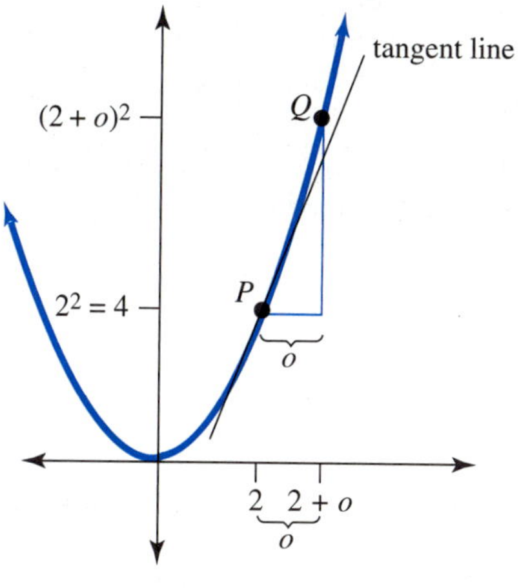

Figure 10.42

Solution P is at $x = 2$, with a y-coordinate of $x^2 = 2^2 = 4$.
Q is at $x = 2 + o$ with a y-coordinate of $(2 + o)^2$.

$$\begin{aligned}
\text{slope of } PQ &= \frac{\Delta y}{\Delta x} = \frac{y_2 - y_1}{x_2 - x_1} \\
&= \frac{(2 + o)^2 - 4}{(2 + o) - 2} \\
&= \frac{4 + 4 \cdot o + o^2 - 4}{o} \\
&= \frac{4 \cdot o + o^2}{o} \\
&= \frac{o(4 + o)}{o} && \text{factoring} \\
&= 4 + o && \text{canceling} \\
&= 4 + 0 = 4 && \text{since } o \text{ is imperceptibly small}
\end{aligned}$$

The slope of the tangent line at (2, 4) is 4. ■

> **Newton's Method of Finding the Slope of the Tangent Line**
>
> 1. *Find Q.* Get Q's x-coordinate by adding o to P's x-coordinate. Use the equation of the curve to get Q's y-coordinate.
> 2. *Find the slope of PQ.* Simplify the result by factoring and canceling.
> 3. *To find the slope of the tangent line, replace o with* 0.

Compare this method with Barrow's method as illustrated in Example 1 of Section 10.2. Newton's method is very similar to Barrow's, except that Newton filled some of the logical holes left by Barrow. In particular, Newton's method included nothing like Barrow's unexplained instructions to "omit all terms containing a power of a or e, or products of these (for these terms have no value)." Instead, Newton's procedure involved a normal factoring and canceling followed by equating the evanescent increment o with zero.

Unfortunately, this attempted explanation creates as many problems as it solves. If the evanescent increment o is so small that it can be replaced with the number zero in the last step of Example 1, then why not replace o with zero a few steps earlier and obtain

$$\frac{4 \cdot o + o^2}{o} = \frac{4 \cdot 0 + 0^2}{0} = \frac{0}{0}$$

which is undefined? Furthermore, if P is at $x = 2$ and Q is at $x = 2 + o$, and if o can be replaced with zero, then aren't P and Q the same point? How can one find the slope of a line between two points when the points are the same?

Bishop George Berkeley

Gottfried Wilhelm Leibniz, working independently in Europe, had also developed a calculus based on evanescent increments (which he called **infinitesimals**), and his theory had the same difficulties. (It is interesting to note that Leibniz had previously obtained a copy of Barrow's *Optical and Geometrical Lectures*.) Both Leibniz and Newton repeatedly changed their explanations of the concept of the infinitely small and repeatedly failed to offer an explanation that had no logical holes. It is probably most accurate to say that each man understood the concept but not well enough to explain it in a logically valid manner. It is said that Leibniz once attempted to explain the concept of the imperceptibly small to the queen of Prussia, who responded that she needed no instruction on the subject, because the behavior of the members of her court had made her thoroughly familiar with it.

Newton accepted his theory because it fit with his intuition and because it worked. His contemporaries, however, were far from unanimous in accepting it. Some of the strongest criticisms came from religious leaders, perhaps because mathematics was the principal avenue by which science was invading the physical universe, which had been the exclusive domain of the Church.

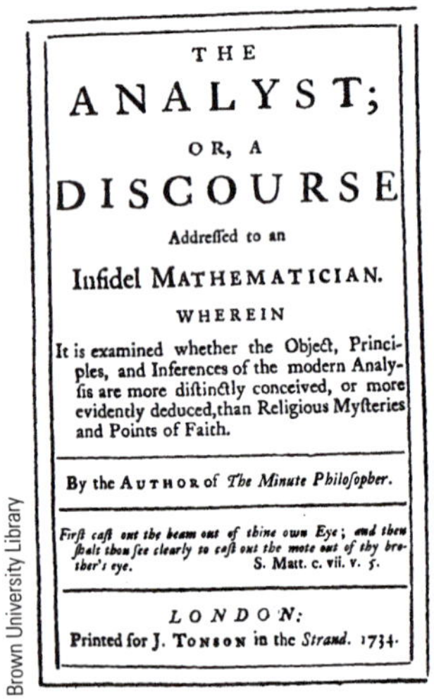

THE

ANALYST;

OR, A

DISCOURSE

Addreffed to an

Infidel MATHEMATICIAN.

WHEREIN

It is examined whether the Object, Principles, and Inferences of the modern Analyfis are more diftinctly conceived, or more evidently deduced, than Religious Myfteries and Points of Faith.

By the AUTHOR of *The Minute Philofopher*.

Firft caft out the beam out of thine own Eye; and then fhalt thou fee clearly to caft out the mote out of thy brother's eye. S. Matt. c. vii. v. 5.

LONDON:

Printed for J. TONSON in the *Strand*. 1734.

The essay in which Berkeley criticizes Newton's evanescent increments

The most famous criticism came from Bishop George Berkeley. In his essay "The Analyst: A Discourse Addressed to an Infidel Mathematician," Berkeley states, "And what are these same evanescent increments? They are neither finite quantities, or quantities infinitely small, nor yet nothing. May we not call them the ghosts of departed quantities? Certainly . . . he who can digest a second or third fluxion . . . need not, methinks, be squeamish about any point in Divinity."

Cauchy's Reformulation of Newton's Method

A reformulation of Newton's method that avoids the difficulties caused by evanescent increments or infinitesimals was developed about 150 years later by the French mathematician Augustin Louis Cauchy. This formulation uses the symbol Δx in place of o, where Δx means "change in x" and is *not* meant to be infinitely small. With this notation, point Q would have an x-coordinate of $x + \Delta x$ rather than $x + o$. The smaller the value of Δx, the closer the line PQ is to the desired tangent line, and the closer the slope of the line PQ is to the slope of the tangent line (Figure 10.43). However, we cannot let Δx equal zero, or we will encounter all the logical holes of Newton's method. Rather, the slope of the tangent line is obtained by calculating the slope of the secant line and observing the effect of allowing Δx to *approach zero without reaching zero*.

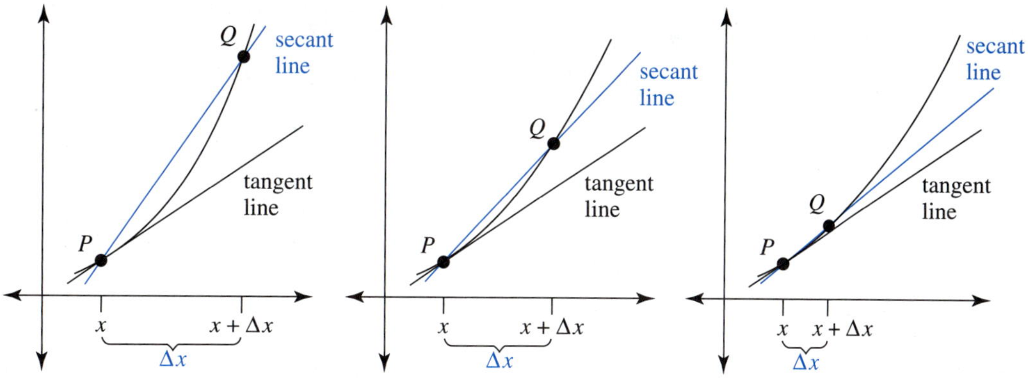

Figure 10.43
As we move from the left graph to the middle graph to the right graph, Δx gets smaller, line PQ gets closer to the tangent line, and the slope of line PQ gets closer to the slope of the tangent line.

EXAMPLE 2 Use Cauchy's method to find the slope of the line tangent to the curve $y = x^2$ at the point $x = 2$, as shown in Figure 10.44.

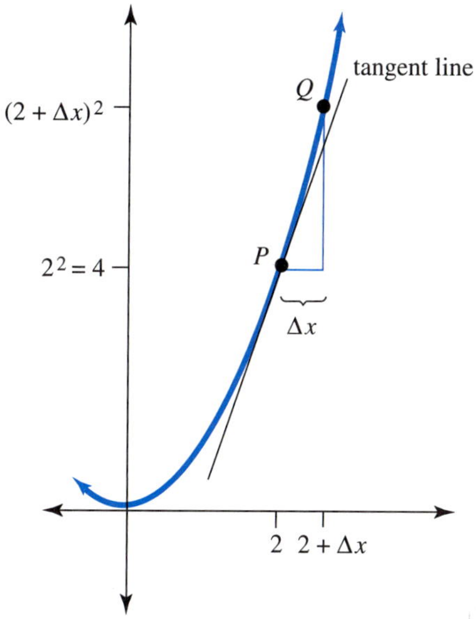

Figure 10.44

Solution P is at $x = 2$, with a y-coordinate of $2^2 = 4$. Q is at $x = 2 + \Delta x$, with a y-coordinate of $(2 + \Delta x)^2$:

$$\begin{aligned}
\text{slope of secant line } PQ &= \frac{\Delta y}{\Delta x} = \frac{y_2 - y_1}{x_2 - x_1} \\
&= \frac{(2 + \Delta x)^2 - 4}{(2 + \Delta x) - 2} \\
&= \frac{4 + 4 \cdot \Delta x + \Delta x^2 - 4}{\Delta x} \\
&= \frac{4 \cdot \Delta x + \Delta x^2}{\Delta x} \\
&= \frac{\Delta x(4 + \Delta x)}{\Delta x} && \text{factoring} \\
&= 4 + \Delta x && \Delta x \neq 0 \text{, so it can be canceled}
\end{aligned}$$

Next, observe the effect of allowing Δx to approach zero without reaching zero:

If P is at $x =$	and Q is at $x =$	then $\Delta x =$	and the slope of PQ is $4 + \Delta x =$
2	3	$3 - 2 = 1$	$4 + \Delta x = 4 + 1 = 5$
2	2.5	$2.5 - 2 = 0.5$	$4 + \Delta x = 4 + 0.5 = 4.5$
2	2.1	$2.1 - 2 = 0.1$	$4 + \Delta x = 4 + 0.1 = 4.1$

Reading from top to bottom in the above table, we can see that

- Q's x-coordinate gets closer to 2, without reaching 2.
- Δx gets closer to 0, without reaching 0.
- The slope of PQ gets closer to 4, without reaching 4.

This is illustrated in Figure 10.45.

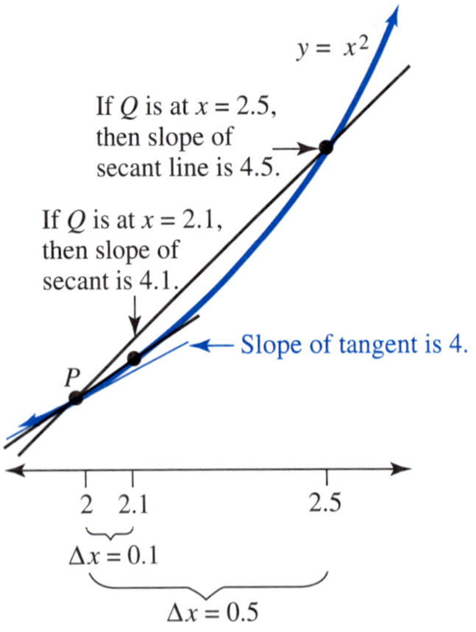

Figure 10.45

HISTORICAL NOTE

Isaac Newton

1642–1727

Where the statue stood
Of Newton, with his prism and
silent face,
The marble index of a mind forever
Voyaging through strange seas of
thought alone.
William Wordsworth

Isaac Newton, considered by many to be a creative genius and the greatest scientist who ever lived, was born in 1642, the year of Galileo's death, to a family of farmers in Woolsthorpe, England. His father had died before his birth, and his mother soon remarried and moved to a neighboring town, leaving young Isaac in the care of his grandmother. Although he was not an especially good student in his youth, he did show a real mechanical aptitude. He constructed perfectly functioning mechanical toys, including a water-powered wooden clock and a miniature wheat mill with a fat mouse acting as both power source and product consumer. Perhaps owing to this mechanical skill and the fact that he was a better student than farmer, it was decided to send him to Cambridge University when he was nineteen. Before leaving Woolsthorpe, he became engaged to a local girl. However, even though he remembered her affectionately all his life, he withdrew from her and never married.

At Cambridge Newton became interested in mathematics. He read Euclid's *Elements,* Descartes's *La géométrie,* and Viète's work on algebra. He received his bachelor's degree in 1664, after an undistinguished four years of study. Later that same year, the university was closed for two years because of the Great Plague (a bubonic plague), and Newton returned to the family farm in Woolsthorpe to avoid exposure to the plague. In those two years, he invented calculus (which he called "the method of fluxions"), proved experimentally that white light is composed of all colors, and discovered the law of universal gravitation, in which he provided a single explanation of both falling bodies on earth and the motion of planets and comets. In his later years, Newton talked about discovering the law of universal gravitation while sitting under an apple tree at the farm. He said that he was wondering what force could hold the moon in its path when the fall of an apple made him think that it might be the same gravitational force, diminished by distance, that acted on the apple. Unfortunately, he chose not to publish any of his work from this period for many years.

He returned to Cambridge when the danger from the plague was over and studied optics and mathematics under Isaac Barrow, holder of Cambridge's Lucas chair in mathematics. Newton communicated some of his discoveries to Barrow, including part of his method of fluxions, and helped him prepare his book *Optical Lectures* for publication. After several years, Barrow retired and recommended Newton as his successor to the Lucas chair. Newton sent a paper on optics to the Royal Society (a professional scientific organization), where some found his ideas interesting and others attacked them vehemently. Newton disliked the ensuing argument so much that he vowed never to publish again.

North Wind Picutre Archives
Newton analyzing the ray of light

Newton became quite neurotic, paralyzed by fear of exposing his discoveries and beliefs to the world. A colleague said that he was "of the most fearful, cautious, and suspicious temper that I ever knew." Early in life, Newton abandoned the orthodox Christian belief in the Trinity. He considered this to be a

The Granger COllection, NY

Newton sitting under an apple tree, discovering the law of universal gravitation

dreadful secret and tried to conceal it his whole life. He published his theories only under extreme pressure from friends. Once Edmond Halley, a famous astronomer and mathematician (for whom Halley's comet is named), visited Newton at Cambridge and asked him what law of force would cause the planets to move in elliptical orbits. Newton immediately told him, to which Halley responded, "Yes, but how do you know that? Have you proved it?" Newton answered, "Why, I've known it for years. If you'll give me a few days, I'll certainly find you a proof of it." This exchange rearoused his interest in physics and astronomy, and he started writing up his discoveries, including his law of universal gravitation. The result was his first mas-

North Wind Picutre Archives

Edmond Halley, Newton's benefactor

terpiece, the three-volume *Philosophiae Naturalis Principia Mathematica* (*Mathematical Principals of Natural Philosophy*), published in 1687. This revolutionary work, published at Halley's expense, made an enormous impression throughout Europe. It totally altered the nature of scientific thought. Newton had used calculus extensively in the development of this work, but he wrote it without using any calculus because he wished to keep his calculus a secret.

Newton wrote three major papers on his calculus between

The closer Δx is to zero, the closer Q is to P and the closer the slope of the secant line PQ is to the slope of the tangent line. If Δx is allowed to approach zero without reaching zero, then $4 + \Delta x$ will approach 4 without reaching 4. The slope of the tangent line at (2, 4) is 4. ■

Cauchy's Method of Finding the Slope of the Tangent Line

1. *Find Q.* Get Q's x-coordinate by adding Δx to P's x-coordinate. Use the equation of the curve to get Q's y-coordinate.
2. *Find the slope of PQ.*
3. *To find the slope of the tangent line, allow Δx to approach zero.*

You should review Newton's original method in Example 1 and compare it to Cauchy's version in Example 2. Notice that in finding the slope of line PQ in Example 2, the Δx's can be canceled, because Δx *does not equal zero*. In finding

1669 and 1676. He did not publish these works but merely circulated them among his friends. In 1693, Newton learned that calculus was becoming well known on the continent and that it was being attributed to Leibniz. At the insistence of his friends, Newton slowly began to publish his three papers, which appeared between 1711 and 1736.

By 1712, the question of who really invented calculus and whether either mathematician had plagiarized the other had become matters of consuming public interest, and Newton, Leibniz, and their backers began to attack each other. Leibniz and his followers went on to perfect and expand Leibniz's calculus. England disregarded this and all other work from the continent out of loyalty to Newton, and as a result, failed to progress mathematically for 100 years.

In his later years, Newton retired from mathematics and physics and turned his attention to alchemy, chemistry, theology, and history. He became Cambridge's representative in Parliament and then Master of the Mint. Occasionally, though, Newton returned to mathematics. Johann Bernoulli (a follower of Leibniz) once posed a challenging problem to all the mathematicians of Europe. Newton heard of the problem about six months after it had been posed, during which time no one had solved it. Newton solved the problem after dinner that same day and sent the solution to Bernoulli anonymously. Despite the anonymity, Bernoulli knew its source, saying, "I recognize the lion by his claw."

Newton was well known for his ability to focus his concentration. Such ability was undoubtedly necessary, considering the fact that he totally reshaped the face of mathematics and physics. Stories of the effects of this concentration include the story of a chicken dinner to which Newton was invited by a friend. Newton forgot about the dinner and did not show up. Eventually, his host ate the chicken, leaving the bones on a covered platter. Newton recalled the engagement later in the evening, arrived at the friend's house, lifted the cover to discover a consumed carcass, and exclaimed, "Dear me, I had forgotten that we had already dined."

Newton was knighted for his work at the mint and for his scientific discoveries. Near the end of his life, he appraised his efforts, saying, "I do not know what I may appear to the world; but to myself I seem to have been only like a boy playing on the seashore, and diverting myself in now and then finding a smoother pebble or a prettier shell than ordinary, whilst the great ocean of truth lay all undiscovered before me." He was buried with great ceremony at Westminster Abbey.

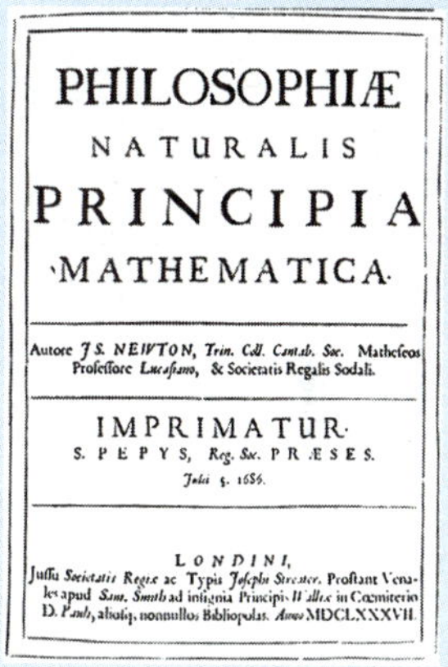
PHILOSOPHIÆ

NATURALIS

PRINCIPIA

MATHEMATICA.

Autore JS. NEWTON, Trin. Coll. Cantab. Soc. Matheseos Professore Lucasiano, & Societatis Regalis Sodali.

IMPRIMATUR.

S. PEPYS, Reg. Soc. PRÆSES.

Julii 5. 1686.

LONDINI,

Jussu Societatis Regiæ ac Typis Josephi Streater. Prostant Venales apud Sam. Smith ad insignia Principis Walliæ in Cœmiterio D. Pauli, aliosq; nonnullos Bibliopolas. Anno MDCLXXXVII.

The Granger Collection, NY

the slope of the tangent line in Example 2, we are still not allowing Δx to equal zero; we are observing the result of allowing Δx to *approach zero*. Realize that this is a very sophisticated concept that is difficult to grasp. Newton and Leibniz each had a great deal of difficulty explaining it, and it was 150 years before mathematicians were able to make it logically correct. So if you do not achieve understanding overnight, you're not alone!

10.3 Exercises

In Exercises 1–10, do the following.
a. Use Newton's method to find the slope of the line tangent to the given curve at the given point.
b. Use Cauchy's method to find the slope of the line tangent to the given curve at the given point.
c. Find the equation of the tangent line.

1. $y = 3x^2$ at $(4, 48)$
2. $y = 4x^2$ at $(1, 4)$
3. $y = x^2 + 2$ at $x = 3$
4. $y = x^2 + 5$ at $x = 4$
5. $y = 2x^2 - 5x + 1$ at $x = 7$
6. $y = 3x^2 + 4x - 3$ at $x = 5$
7. $y = x^3$ at $x = 2$
8. $y = x^3$ at $x = 5$
9. $y = x^3 - x^2$ at $x = 1$
10. $y = x^3 - x - 1$ at $x = 3$

Answer the following questions using complete sentences and your own words.

Concept Questions

11. How does Cauchy's method fill the logical holes that exist in Newton's method of finding the slope of a tangent line?

12. How does Cauchy's method of allowing Δx to approach zero resemble Eudoxus's, Archimedes', and Kepler's methods of computing areas?

13. Compare and contrast Barrow's, Newton's, and Cauchy's methods of finding the slope of a tangent line. Be certain to discuss both their similarities and their differences.

History Questions

14. Of what aspect of Newton's calculus was Bishop Berkeley especially critical?

15. Why did Newton leave Cambridge for two years? What did he accomplish during that time?

16. What prompted Newton to write his *Philosophiae Naturalis Principia Mathematica* (*Mathematical Principals of Natural Philosophy*)?

17. Who paid for the publication of Newton's *Philosophiae Naturalis Principia Mathematica*?

18. What kind of impression did *Philosophiae Naturalis Principia Mathematica* make on English and European scientists?

19. What subject was deliberately omitted from *Philosophiae Naturalis Principia Mathematica*? Why?

20. Why did Newton finally start to publish his works on calculus?

Projects

21. Write a research paper on a historical topic referred to in this section or on a related topic. Following is a partial list of topics from this section:

- Bishop George Berkeley, Newton's critic
- any or all of the Bernoullis, a family of mathematicians likened to the Bachs in music
- Augustin Louis Cauchy, French mathematician and baron
- the controversy over who created calculus, Newton or Leibniz?
- Edmond Halley, Newton's friend and benefactor, who used Newton's work to calculate the orbit of Halley's comet
- Gottfried Wilhelm Leibniz, German philosopher, diplomat, logician, and mathematician
- the Lucas chair in mathematics at Cambridge
- the Royal Society, oldest scientific organization in Great Britain

10.4 Newton on Falling Objects and the Derivative

Average Speed and Instantaneous Speed

Before further investigating Newton's solutions to the four problems (area of any shape, tangent of a curve, speed of a falling object, and path of a cannonball), we need to discuss instantaneous speed. Suppose you're driving in your car and after one hour have gone 45 miles; then, after three hours, you have gone 155 miles. Your **average speed** in the last two hours of driving would be as follows:

$$\begin{aligned}\text{average speed} &= \frac{\text{change in distance}}{\text{change in time}} = \frac{\Delta d}{\Delta t}\\ &= \frac{155\text{ mi} - 45\text{ mi}}{3\text{ hr} - 1\text{ hr}}\\ &= \frac{110\text{ mi}}{2\text{ hr}}\\ &= 55\text{ mi/hr}\end{aligned}$$

This does not mean that your speedometer would show a speed of 55 miles/hour at every moment during those two hours of driving; the speedometer shows not average speed but instantaneous speed. **Instantaneous speed** means the speed during an instant, rather than during two hours. Clearly, instantaneous speed is more useful to the driver than average speed; few cars have average speedometers, but all have instantaneous speedometers.

Newton on Gravity

Newton's method of fluxions and evanescent increments was motivated as much by the question of the speed of a falling object as by the tangent line problem. In fact, his word *fluxion* has the same Latin root as the word *flux*, which means continual change.

Recall that Galileo had determined that the distance traveled by a falling object is proportional to the square of the time in motion. After Galileo's death, others were able to use the new algebra to phrase these facts as equations rather than proportions.

$$\frac{d_1}{d_2} = \frac{t_1^2}{t_2^2}$$

$$d_1 t_2^2 = d_2 t_1^2 \quad \text{cross multiplying}$$

$$\frac{d_1}{t_1^2} = \frac{d_2}{t_2^2} \quad \text{dividing by } t_1^2 t_2^2$$

This means that dividing d by t^2 will always result in the same number, regardless of how long or how far an object has been falling.

The pendulum clock, which was much more accurate than Galileo's water clock, was invented by Galileo and perfected by the Dutch scientist Christiaan Huygens. It allowed scientists to time falling objects accurately enough to determine that dividing d by t^2 will always result in the number 16 (if d is measured in feet, t is measured in seconds, and the object is shaped so that it meets little wind resistance). Thus,

$$\frac{d}{t^2} = 16$$

$$d = 16t^2 \quad \text{multiplying by } t^2$$

EXAMPLE 1 **a.** Find the distance a falling object has traveled after 2 seconds.
b. Find the distance a falling object has traveled after 3 seconds.
c. Find the falling object's average speed during the time interval from 2 to 3 seconds after the motion starts.

Solution **a.**
$$\begin{aligned} d &= 16t^2 \\ &= 16 \cdot 2^2 \\ &= 64 \text{ ft} \end{aligned}$$

b.
$$\begin{aligned} d &= 16t^2 \\ &= 16 \cdot 3^2 \\ &= 144 \text{ ft} \end{aligned}$$

c.
$$\begin{aligned} \text{average speed} &= \frac{\text{change in distance}}{\text{change in time}} = \frac{\Delta d}{\Delta t} \\ &= \frac{144 \text{ ft} - 64 \text{ ft}}{3 \text{ sec} - 2 \text{ sec}} \\ &= 80 \text{ ft/sec} \end{aligned}$$

■

EXAMPLE 2 **a.** Use the formula $d = 16t^2$ to find a formula for the average speed of a falling object from an earlier time t to a later time $t + \Delta t$.
b. Use the result of part (a) to find the average speed of a falling object from an earlier time $t = 2$ to a later time $t + \Delta t = 3$.
c. Use the result of part (a) to find the instantaneous speed of a falling object at time t.

Solution **a.** The distance at the earlier time t is $16t^2$, and the distance at the later time $t + \Delta t$ is $16(t + \Delta t)^2$.

$$\begin{aligned}
\text{average speed} &= \frac{\text{change in distance}}{\text{change in time}} = \frac{\Delta d}{\Delta t} \\
&= \frac{\text{distance at time } (t + \Delta t) - \text{distance at time } t}{(t + \Delta t) - t} \\
&= \frac{16(t + \Delta t)^2 - 16t^2}{\Delta t} \\
&= \frac{16(t^2 + 2t\Delta t + \Delta t^2) - 16t^2}{\Delta t} && \text{squaring} \\
&= \frac{(16t^2 + 32t\Delta t + 16\Delta t^2) - 16t^2}{\Delta t} \\
&= \frac{32t\Delta t + 16\Delta t^2}{\Delta t} \\
&= \frac{\Delta t(32t + 16\Delta t)}{\Delta t} && \text{factoring} \\
&= 32t + 16\Delta t
\end{aligned}$$

b. The earlier time is $t = 2$, the later time is $t + \Delta t = 3$, and $\Delta t = 3 - 2 = 1$. Using the new formula, we have

$$\begin{aligned}
\text{average speed} &= 32t + 16\Delta t \\
&= 32 \cdot 2 + 16 \cdot 1 \\
&= 80 \text{ ft/sec}
\end{aligned}$$

This answer agrees with that obtained in Example 1 but involves less work.

c. The speed that Newton and Galileo were most interested in was not average speed but rather instantaneous speed. Instantaneous speed can be thought of as the average speed during an instant, or, as Newton would have said, during an "evanescent increment" of time. Thus,

$$\begin{aligned}
\text{instantaneous speed} &= \text{average speed during an instant} \\
&= 32t + 16\Delta t \\
&= 32t + 16 \cdot o \\
&= 32t
\end{aligned}$$

To use Cauchy's language, if Δt is allowed to approach zero without reaching zero, then average speed approaches instantaneous speed, and $32t + 16\Delta t$ approaches $32t$.

Thus, Newton found that the instantaneous speed of a falling object is given by $s = 32t$. Newton called this the "fluxion of gravity." ■

Falling Object Formulas

If an object falls for t seconds, then its distance d (in feet) and its speed s (in feet per second) are given by the equations

$$d = 16t^2$$

$$s = 32t$$

Technically, these falling object formulas have limited accuracy. At a certain speed (that depends on the shape of the object), air resistance will become so great that a falling object will stop accelerating, and the speed will no longer be given by $s = 32t$. We will ignore this, as did Galileo and Newton.

EXAMPLE 3 **a.** Find the instantaneous speed of a falling object after 3 seconds.
b. Find the average speed of a falling object during its first 3 seconds of descent.

Solution **a.** The instantaneous speed is

$$\begin{aligned} s &= 32t \\ &= 32 \cdot 3 \\ &= 96 \text{ ft/sec} \end{aligned}$$

b. The average speed is

$$\begin{aligned} \text{average speed} &= \frac{\text{change in distance}}{\text{change in time}} = \frac{\Delta d}{\Delta t} \\ &= \frac{\text{distance at time 3} - \text{distance at time 0}}{3 - 0} \\ &= \frac{16 \cdot 3^2 - 16 \cdot 0^2}{3} \\ &= 48 \text{ ft/sec} \end{aligned}$$

Alternatively, the earlier time is $t = 0$, the later time is $t + \Delta t = 3$, and $\Delta t = 3 - 0 = 3$; thus,

$$\begin{aligned} \text{average speed} &= 32t + 16\Delta t \\ &= 32 \cdot 0 + 16 \cdot 3 \\ &= 48 \text{ ft/sec} \end{aligned}$$

This example is illustrated in Figure 10.46.

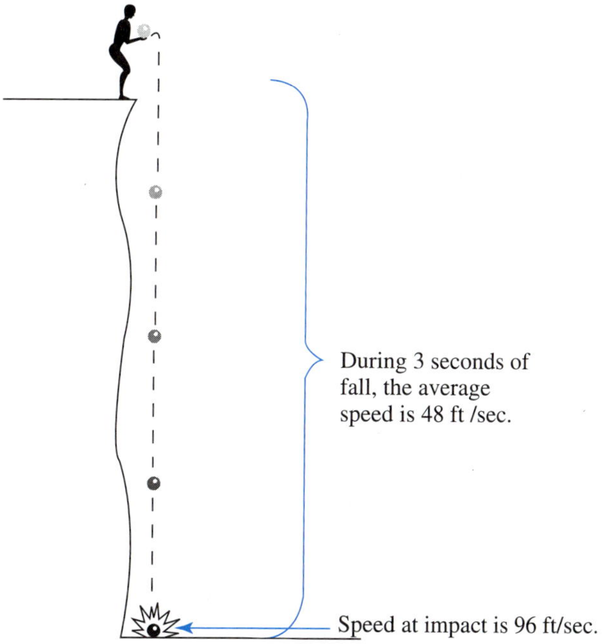

Figure 10.46

The Derivative

The procedure used in Examples 2 and 3 to find the instantaneous speed of a falling object is extremely similar to that used in Section 10.3 to find the slope of a tangent line. To see this important similarity, compare the two procedures' steps.

To Find the Instantaneous Speed of a Falling Object (as discussed in Example 2)	To Find the Slope of the Tangent Line (as discussed in Section 10.3)
1. Find the later time and the distance at that later time. The later time is $t + \Delta t$. Use the equation $d = 16t^2$ to get the distance at that later time.	1. Find Q's x-coordinate and y-coordinate. The x-coordinate is $x + \Delta x$. Use the equation of the curve to get Q's y-coordinate.
2. Find the average speed $\Delta d/\Delta t$.	2. Find the slope $\Delta y/\Delta x$.
3. To find the instantaneous speed, allow Δt to approach zero.	3. To find the slope of the tangent line, allow Δx to approach zero.

Newton's concept of fluxions and evanescent increments is the main idea of calculus. It can be used to find the slope of a tangent line or the speed of a moving object. What Newton called the ultimate ratio of fluxions is now called the **derivative**. To find the derivative of a function $f(x)$, we use the procedure that was used both in Section 10.3 to find the slope of a tangent line and in this section to find the instantaneous speed of a falling object.

If point P has an x-coordinate of x and a y-coordinate of $f(x)$, then point Q has an x-coordinate of $x + \Delta x$ and a y-coordinate of $f(x + \Delta x)$, as shown in Figure 10.47. The slope of line PQ is

$$\frac{\Delta y}{\Delta x} = \frac{y_2 - y_1}{x_2 - x_1} = \frac{f(x + \Delta x) - f(x)}{(x + \Delta x) - (x)} = \frac{f(x + \Delta x) - f(x)}{\Delta x}$$

The slope of the tangent line is the limiting result of allowing Δx to approach zero without reaching zero. This is written as

$$\text{slope of tangent} = \lim_{\Delta x \to 0} \frac{f(x + \Delta x) - f(x)}{\Delta x}$$

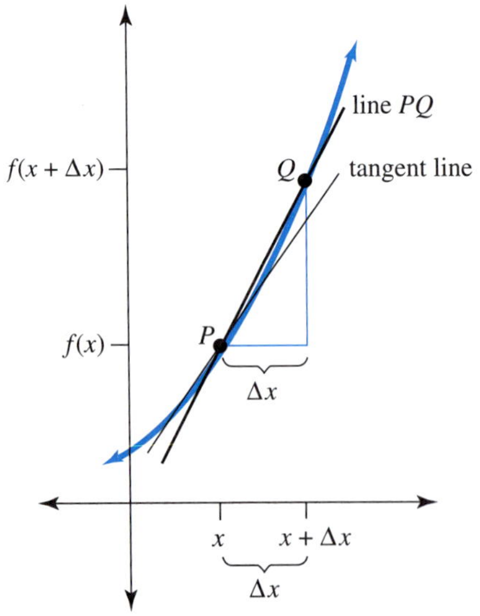

Figure 10.47

where

$$\lim_{\Delta x \to 0}$$

is read "the limit as Δx approaches zero."

The derivative of a function can be used to find the speed of a moving object as well as to find the slope of a tangent line, so it would be inappropriate to always refer to it as "slope of tangent," as we did above. Instead, it is referred to as either $f'(x)$ or df/dx. The former notation is Newton's, and the latter notation is Leibniz's. Leibniz's notation df/dx is derived from

$$\frac{\Delta f}{\Delta x} = \frac{\text{change in } f}{\text{change in } x}$$

(The Δs are merely changed to ds.)

The Derivative

The derivative of a function $f(x)$ is the function

$$f'(x) \quad \text{or} \quad \frac{df}{dx} = \lim_{\Delta x \to 0} \frac{f(x + \Delta x) - f(x)}{\Delta x}$$

EXAMPLE 4 Find the derivative of $f(x) = mx + b$, the line with slope m and y-intercept b.

Solution

$$\begin{aligned}
\frac{df}{dx} &= \lim_{\Delta x \to 0} \frac{f(x + \Delta x) - f(x)}{\Delta x} \\
&= \lim_{\Delta x \to 0} \frac{[m(x + \Delta x) + b] - [mx + b]}{\Delta x} && \text{since } f(x + \Delta x) = m(x + \Delta x) + b \\
&= \lim_{\Delta x \to 0} \frac{mx + m\Delta x + b - mx - b}{\Delta x} \\
&= \lim_{\Delta x \to 0} \frac{m\Delta x}{\Delta x} && \text{combining like terms} \\
&= \lim_{\Delta x \to 0} m && \text{canceling} \\
&= m
\end{aligned}$$

Thus, the slope of the line tangent to $f(x) = mx + b$ is m. This result should make sense to you, because $f(x) = mx + b$ is itself a line with slope m. ■

EXAMPLE 5

a. Find the derivative of $f(x) = x^2$.

b. Use the derivative of $f(x)$ to find the slope of the tangent line at $x = 1$ and at $x = 2$.

c. Graph $f(x)$ and the tangent lines at $x = 1$ and $x = 2$. Show the slopes on the tangent lines' graphs.

Solution

a. The derivative of $f(x)$ is

$$\begin{aligned}
\frac{df}{dx} &= \lim_{\Delta x \to 0} \frac{f(x + \Delta x) - f(x)}{\Delta x} \\
&= \lim_{\Delta x \to 0} \frac{(x + \Delta x)^2 - x^2}{\Delta x} && \text{since } f(x + \Delta x) = (x + \Delta x)^2
\end{aligned}$$

$$= \lim_{\Delta x \to 0} \frac{x^2 + 2x\Delta x + \Delta x^2 - x^2}{\Delta x} \quad \text{multiplying}$$

$$= \lim_{\Delta x \to 0} \frac{2x\Delta x + \Delta x^2}{\Delta x} \quad \text{combining like terms}$$

$$= \lim_{\Delta x \to 0} \frac{\Delta x(2x + \Delta x)}{\Delta x} \quad \text{factoring}$$

$$= \lim_{\Delta x \to 0} (2x + \Delta x) \quad \text{canceling}$$

$$= 2x + 0 = 2x \quad \text{since } \Delta x \text{ approaches 0}$$

The derivative of $f(x) = x^2$ is $df/dx = 2x$.

b. To find the slope of the tangent line at $x = 1$, substitute 1 for x in df/dx:

$$\frac{df}{dx} = 2x = 2 \cdot 1 = 2$$

To find the slope of the tangent line at $x = 2$, substitute 2 for x in df/dx:

$$\frac{df}{dx} = 2x = 2 \cdot 2 = 4$$

c. The graph of $f(x)$ and the tangent lines is shown in Figure 10.48. Notice that the calculated slopes perfectly match the tangent lines' graphs; that is, the tangent lines do in fact seem to be of slopes 2 and 4.

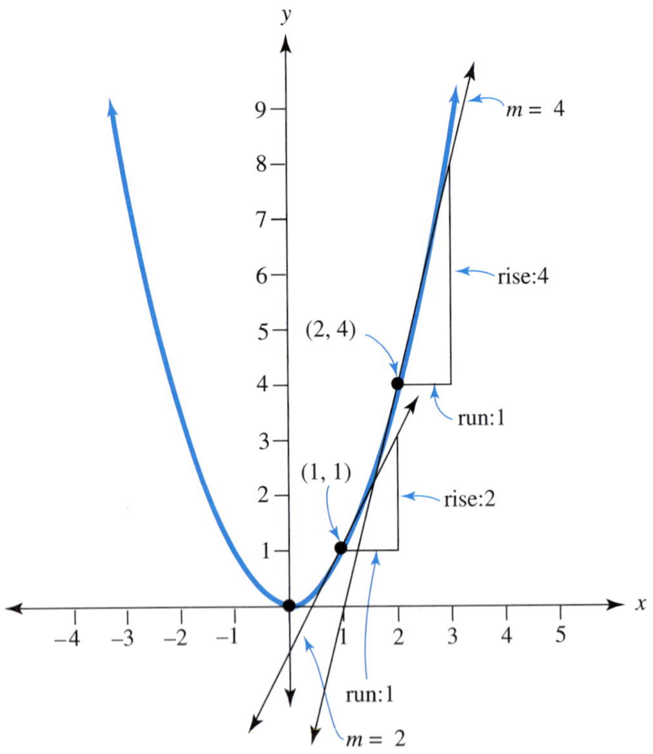

Figure 10.48

EXAMPLE 6 Find the derivative of $f(x) = x^2 - 6x + 7$.

Solution

$$\frac{df}{dx} = \lim_{\Delta x \to 0} \frac{f(x + \Delta x) - f(x)}{\Delta x}$$

$$= \lim_{\Delta x \to 0} \frac{[(x + \Delta x)^2 - 6(x + \Delta x) + 7] - [x^2 - 6x + 7]}{\Delta x}$$ since $f(x + \Delta x) = (x + \Delta x)^2 - 6(x + \Delta x) + 7$

$$= \lim_{\Delta x \to 0} \frac{[x^2 + 2x\Delta x + \Delta x^2 - 6x - 6\Delta x + 7] - [x^2 - 6x + 7]}{\Delta x}$$ multiplying

$$= \lim_{\Delta x \to 0} \frac{x^2 + 2x\Delta x + \Delta x^2 - 6x - 6\Delta x + 7 - x^2 + 6x - 7}{\Delta x}$$ distributing

$$= \lim_{\Delta x \to 0} \frac{2x\Delta x + \Delta x^2 - 6\Delta x}{\Delta x}$$ combining like terms

$$= \lim_{\Delta x \to 0} \frac{\Delta x(2x + \Delta x - 6)}{\Delta x}$$ factoring

$$= \lim_{\Delta x \to 0} (2x + \Delta x - 6)$$ canceling

$$= 2x + 0 - 6 = 2x - 6$$ since Δx approaches 0

The derivative of $f(x) = x^2 - 6x + 7$ is $df/dx = 2x - 6$. ■

Interpreting a Derivative

In any situation, Δx means the change in x, and $\Delta f = f(x + \Delta x) - f(x)$ means the change in $f(x)$. A **rate of change** refers to one change divided by another change, so

$$\frac{\Delta f}{\Delta x} = \frac{f(x + \Delta x) - f(x)}{\Delta x}$$

is a rate of change; it is called the **average rate of change of f**. If we allow Δx to approach zero, then the result,

$$\frac{df}{dx} = \lim_{\Delta x \to 0} \frac{f(x + \Delta x) - f(x)}{\Delta x}$$

is called the **instantaneous rate of change of f**. The word *instantaneous* is used here because x often measures time, so Δx measures the change in time; if Δx approaches zero, then the amount of time becomes an instant.

We've already seen that if x measures time and $f(x)$ measures distance traveled, then

$$\frac{\Delta f}{\Delta x} = \frac{f(x + \Delta x) - f(x)}{\Delta x}$$

measures the average rate of change of distance, or average speed, and

$$\frac{df}{dx} = \lim_{\Delta x \to 0} \frac{f(x + \Delta x) - f(x)}{\Delta x}$$

measures the instantaneous rate of change of distance, or instantaneous speed. Furthermore, if time is measured in seconds and distance in feet, then Δx would also be in seconds and Δf would also be in feet; therefore, both $\Delta f/\Delta x$ and df/dx would be in feet/second, or feet per second.

What do

$$\frac{\Delta f}{\Delta x} = \frac{f(x + \Delta x) - f(x)}{\Delta x} \quad \text{and} \quad \frac{df}{dx} = \lim_{\Delta x \to 0} \frac{f(x + \Delta x) - f(x)}{\Delta x}$$

measure under different circumstances? We can answer this question by replacing the symbols $\Delta f/\Delta x$ with the words

$$\frac{\text{change in (whatever } f \text{ measures)}}{\text{change in (whatever } x \text{ measures)}}$$

For example, if x measures time and $f(x)$ measures speed, then

$$\frac{\Delta f}{\Delta x} \quad \text{means} \quad \frac{\text{change in speed}}{\text{change in time}}$$

It measures the average rate of change of speed, which is called average acceleration (*acceleration* means a change in speed). Also,

$$\frac{df}{dx} = \lim_{\Delta x \to 0} \frac{f(x + \Delta x) - f(x)}{\Delta x}$$

measures the *instantaneous* rate of change of speed, or *instantaneous* acceleration, because Δx means "change in time," and if Δx approaches zero, then the amount of time becomes an instant.

If x measures time and $f(x)$ measures the population of a state, then $\Delta f/\Delta x$ means (change in population)/(change in time); it measures the average rate of change of population, which is called **average growth rate**. Also, df/dx measures the instantaneous rate of change of population, or **instantaneous growth rate**. Growth rates are commonly used in biology and demographics.

If x measures the number of items manufactured by a factory each day and $f(x)$ measures the profit obtained from selling those items, then $\Delta f/\Delta x$ means (change in profit)/(change in number of items); it measures the average rate of change of profit, which is called **average marginal profit**. Also, df/dx measures the instantaneous rate of change of profit, or **instantaneous marginal profit**, and is used by manufacturers to determine whether production should be increased. The meanings of $\Delta f/\Delta x$ and df/dx under these different circumstances are summarized in Figure 10.49.

Given a function that measures some physical quantity, the derivative of that function measures the rate at which that quantity changes. Calculus is the mathematical tool used to study how things change. Calculus has given society, for better or worse, the power to control and predict the physical universe. For this reason, the invention of calculus may well be the single most important intellectual achievement of the Renaissance.

x	$f(x)$	$\frac{\Delta f}{\Delta x} = \frac{f(x + \Delta x) - f(x)}{\Delta x}$	$\frac{df}{dx} = \lim_{\Delta x \to 0} \frac{f(x + \Delta x) - f(x)}{\Delta x}$
the x-value of a point	the y-value of a point	the slope of the secant line through x and $x + \Delta x$	the slope of the tangent line at x
time in seconds	distance in feet	average speed in feet per second	instantaneous speed in feet per second
time	instantaneous speed	average acceleration	instantaneous acceleration
time in years	population of a town in thousands	average growth rate in thousands of people per year	instantaneous growth rate in thousands of people per year
number of items manufactured by a factory per day	profit from selling those items	average marginal profit	instantaneous marginal profit

Figure 10.49

10.4 Exercises

In Exercises 1–4, find (a) the distance an object will fall in the given amount of time, (b) the instantaneous speed of an object after it has fallen for the given amount of time, and (c) the average speed of an object that falls for the given amount of time.

1. 1 second
2. 10 seconds
3. 5 seconds
4. 15 seconds
5. A parachutist falls for 8 seconds before she opens her parachute.
   a. How far does she fall in those 8 seconds?
   b. At what speed is she falling when she opens the parachute?
   c. What is her average speed while she is in free-fall?
6. A parachutist falls for 20 seconds before he opens his parachute.
   a. How far does he fall in those 20 seconds?
   b. At what speed is he falling when he opens the parachute?
   c. What is his average speed while he is in free-fall?
7. A rock is dropped from a six-story building. It falls 64 feet.
   a. How long will it take the rock to hit the ground?
   HINT: distance = 64, time = ?
   b. At what speed will the rock be falling when it hits the ground?
   c. What is the average speed of the rock during its journey?
8. A tourist drops a penny from the observation platform on the top of the Empire State Building in New York City. It falls 1,248 feet.
   a. How long will it take the penny to hit the ground?
   b. At what speed will the penny be falling when it hits the ground?
   c. What is the average speed of the penny during its journey?
9. Newton took the derivative of $d = 16t^2$ to find $s = 32t$, which he called the "fluxion of gravity." Find Newton's "second fluxion of gravity" by taking the derivative of $s = 32t$. Determine what this second fluxion measures.

In Exercises 10–13, find the derivative of the given function by calculating

$$\frac{df}{dx} = \lim_{\Delta x \to 0} \frac{f(x + \Delta x) - f(x)}{\Delta x}$$

10. $f(x) = 3x - 1$
11. $f(x) = 2x + 12$
12. $f(x) = 19x + 112$
13. $f(x) = -7x + 42$

In Exercises 14–17, use the result of Example 4 to find the derivative of the given function. (These exercises have the same answers as Exercises 10–13.)

14. $f(x) = 3x - 1$
15. $f(x) = 2x + 12$
16. $f(x) = 19x + 112$
17. $f(x) = -7x + 42$

In Exercises 18–23, find the derivative of the given function by calculating

$$\frac{df}{dx} = \lim_{\Delta x \to 0} \frac{f(x + \Delta x) - f(x)}{\Delta x}$$

18. $f(x) = 5x^2$
19. $f(x) = 11x^2$
20. $f(x) = 4x^2 + 7$
21. $f(x) = 3x^2 - 11$
22. $f(x) = 2x^2 + 5x - 3$
23. $f(x) = 3x^2 - 2x + 7$
24. Show that the derivative of $f(x) = ax^2 + bx + c$ is

$$\frac{df}{dx} = 2ax + b$$

HINT: Use the procedure used in Example 4.

In Exercises 25–29, use the result of Exercise 24 to find the derivative of the given function. (These exercises have the same answers as Exercises 19–23.)

25. $f(x) = 11x^2$
26. $f(x) = 4x^2 + 7$
27. $f(x) = 3x^2 - 11$
28. $f(x) = 2x^2 + 5x - 3$
29. $f(x) = 3x^2 - 2x + 7$

In Exercises 30–33, do the following.

a. *Find the derivative of the given function.*
b. *Use part (a) to find the slopes of the tangent lines at $x = 1$ and at $x = 2$.*
c. *Accurately graph f(x) and the tangent lines at $x = 1$ and $x = 2$. Show the slopes of the tangent lines, as in Example 5.*

30. $f(x) = x^2 - 2x + 2$
31. $f(x) = x^2 + 4x + 5$
32. $f(x) = x^2 - 4x + 2$
33. $f(x) = x^2 - 2x + 3$

In Exercises 34–37, find the derivative of the given function by calculating

$$\frac{df}{dx} = \lim_{\Delta x \to 0} \frac{f(x + \Delta x) - f(x)}{\Delta x}$$

34. $f(x) = x^3$
35. $f(x) = x^3 + 3$
36. $f(x) = 2x^3 + x^2$
37. $f(x) = 3x^3 + 4x - 1$
38. **a.** Show that the derivative of $f(x) = ax^3 + bx^2 + cx + d$ is $df/dx = 3ax^2 + 2bx + c$.

HINT: Use the procedure used in Example 4.

b. Use the result of part (a) to guess the formulas for the derivative of $f(x) = ax^n$ and of $f(x) = ax^n + c$.

In Exercises 39–46, use the result of Exercise 38 to find the derivative of the given function. (Exercises 39–42 have the same answers as Exercises 34–37.)

39. $f(x) = x^3$
40. $f(x) = x^3 + 3$
41. $f(x) = 2x^3 + x^2$
42. $f(x) = 3x^3 + 4x - 1$
43. $f(x) = 8x^5 + 4$
44. $f(x) = 9x^6 - 3$
45. $f(x) = 5x^4 - 3x^2 + 2x - 7$
46. $f(x) = 3x^5 - 4x^3 + x + 9$

Concept Questions

In Exercises 47–53, determine what is measured by each of the following. (Include units with each answer.)

a. $\frac{\Delta f}{\Delta x} = \frac{f(x + \Delta x) - f(x)}{\Delta x}$

b. $\frac{df}{dx} = \lim_{\Delta x \to 0} \frac{f(x + \Delta x) - f(x)}{\Delta x}$

47. x measures time in seconds, and $f(x)$ measures the distance from an Atlas rocket to its launching pad in miles.

48. x measures time in minutes, and $f(x)$ measures the volume of water flowing out of a reservoir in gallons.

49. x measures time in hours, and $f(x)$ measures the volume of water flowing through a river in gallons.

50. x measures time in years, and $f(x)$ measures the height of a certain tree in inches.

51. x measures time in years, and $f(x)$ measures the length of a certain shark in centimeters.

52. x measures time in days, and $f(x)$ measures the population of a certain beehive.

53. x measures time in years, and $f(x)$ measures the population of Metropolis in thousands.

54. At the end of Example 4, we stated that $\lim_{\Delta x \to 0} m = m$. Explain why this is true.

55. *For those who have read Section 6.9: The Perimeter and Area of a Fractal.* What is the relationship between P_n, the perimeter at step n, and P, the perimeter of the fractal? Your answer should involve a key concept from calculus.

10.5 The Trajectory of a Cannonball

Galileo showed that the motion of a cannonball could be resolved into two simultaneous motions, one due to the explosion and the other due to gravity. The explosion causes a motion in the direction in which the cannon is pointed; this motion has constant speed, because there is nothing to slow it down (ignoring air resistance). Gravity causes a vertical motion, with speed changing in a uniform way like that of a falling body. Galileo used this idea, along with Apollonius's description of a parabola, to show that a cannonball follows a parabolic path. He was unable to get an equation for the path (and thus unable to determine how to aim a cannon so the ball would hit the target) because he used ancient, proportion-based algebra rather than modern, equation-based algebra.

After Galileo died, other mathematicians used the modern analytic geometry of Fermat and Descartes, the modern algebra of Viète and Descartes, and Newton's work on gravity to find an equation for the path of a cannonball. We will find such an equation for the situation in which the cannon is pointed at a 30° angle, a 45° angle, or a 60° angle.

Our goal is to obtain an equation for the cannonball's motion. Since an equation is algebraic and the motion it describes is geometric, we must use analytic geometry to obtain that equation. To use analytic geometry, we will place an x-axis and a y-axis such that the x-axis runs along the ground and the y-axis runs through the mouth of the cannon (as illustrated in Figure 10.50). This means that the position of the cannonball at a time t is (x, y), where x measures the horizontal distance from the cannon to the

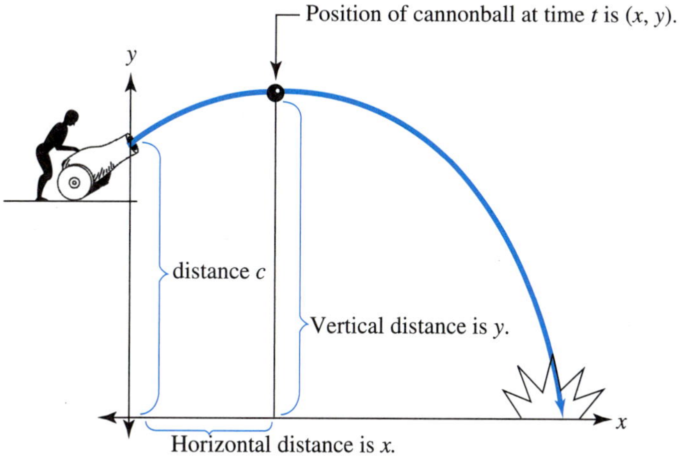

Figure 10.50

cannonball and y measures the vertical distance from the cannonball to the ground. We will be able to obtain our equation if we can find these horizontal and vertical distances.

Galileo resolved a cannonball's motion into a diagonal motion due to the explosion and a downward motion due to gravity. We will follow Galileo's lead and separately consider the motion due to the explosion and the motion due to gravity. In considering the motion due to the explosion, we will not need to take gravity into account, and in considering the motion due to gravity, we will not need to take the explosion into account. Later, we will combine the results of our two analyses and get one equation that describes the cannonball's motion.

The motion due to gravity fits naturally into an analytic geometry perspective, since it is in the y-direction. The motion due to the explosion fits less naturally, since it is both forward and upward—that is, in both the x- and y-directions. We can break this diagonal motion down into an x-component and a y-component by using similar triangles.

The Motion Due to the Explosion

Think of a cannonball that's fired from a cannon pointed at a 45° angle. That cannonball does not move in a straight line at a 45° angle, because gravity pulls it downward. For the time being, however, we are considering only the motion due to the explosion, without the effects of gravity, so we must view the cannonball as moving along a straight line at a 45° angle, as shown in Figure 10.51. If the cannonball's speed is s feet per second, then in 1 second, the ball would move s feet along that line. The horizontal component h of that motion is represented by the horizontal side of the triangle in Figure 10.52; the vertical component v of that motion is represented by the vertical side of the triangle. The upper angle of that triangle must be 45°, since the three angles of a triangle always add to 180°.

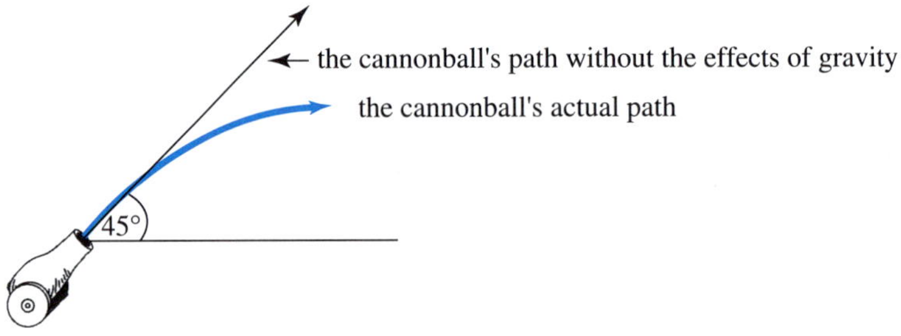

Figure 10.51

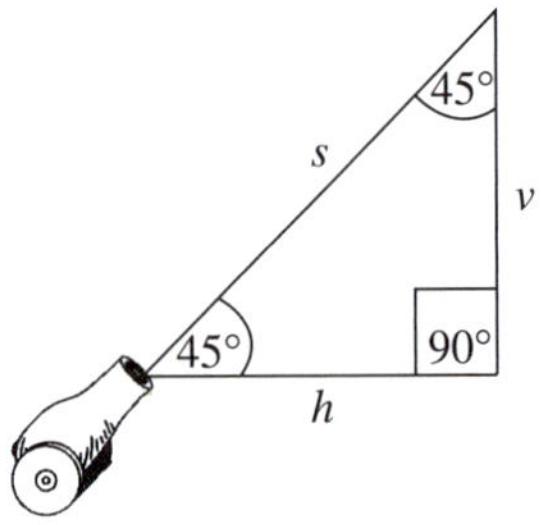

Figure 10.52

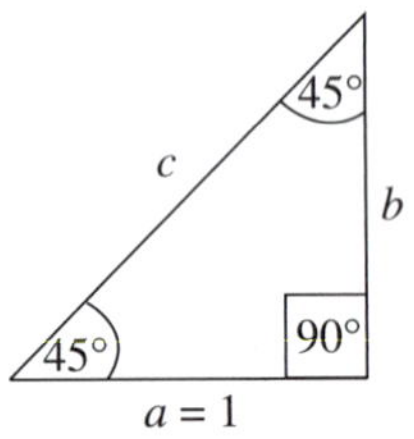

Figure 10.53

The triangle in Figure 10.52 is similar to any other triangle with the same angles. In particular, it is similar to a triangle with a horizontal side of length $a = 1$, as shown in Figure 10.53. That triangle's vertical side must be of the same length, since the horizontal and vertical sides are across from angles of the same size. The triangle is a right triangle, so we can apply the Pythagorean Theorem to find the length of the diagonal side.

$$a^2 + b^2 = c^2 \quad \text{Pythagorean Theorem}$$
$$1^2 + 1^2 = c^2 \quad a = 1 \text{ and } b = 1$$
$$1 + 1 = c^2$$
$$c^2 = 2$$
$$c = \sqrt{2}$$

Since the two triangles are similar, the ratios of their corresponding sides are equal, as shown in Figure 10.54. In particular:

$$\frac{h}{1} = \frac{s}{\sqrt{2}} \quad \text{the horizontal sides' ratio equals the diagonal sides' ratio}$$

$$\frac{v}{1} = \frac{s}{\sqrt{2}} \quad \text{the vertical sides' ratio equals the diagonal sides' ratio}$$

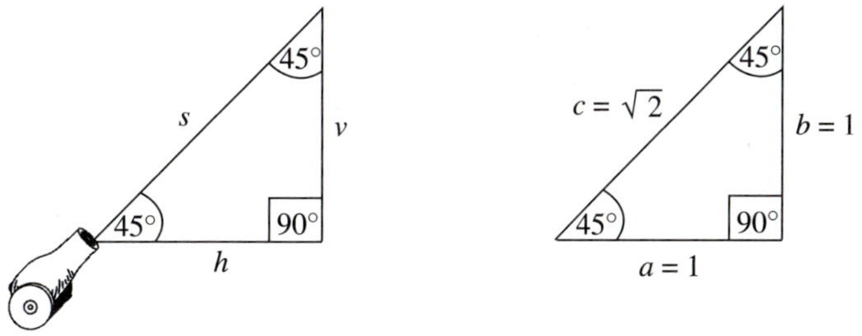

Figure 10.54

Recall that s is the cannonball's speed, h is its horizontal component, and v is its vertical component. Thus, the horizontal speed (or rate) due to the explosion is $h = s/\sqrt{2}$, and the vertical speed (or rate) due to the explosion is $v = s/\sqrt{2}$. And we can use the formula "distance = rate · time" to find the horizontal and vertical distances traveled due to the explosion. See Figure 10.55.

Motion Due to the Explosion			
Direction	**Rate**	**Time**	**Distance = Rate · Time**
horizontal	$\frac{s}{\sqrt{2}}$	t	$\frac{s}{\sqrt{2}} \cdot t$
vertical	$\frac{s}{\sqrt{2}}$	t	$\frac{s}{\sqrt{2}} \cdot t$

Figure 10.55

The Motion Due to Gravity

As Galileo discovered, a cannonball moves along a parabola, unlike a falling object. However, we are now considering only the motion due to gravity, without the effects of the explosion. If there were no explosion, the cannonball would fall straight down, just like any other falling object; in t seconds, it would fall $16t^2$ feet. And since this motion is due to gravity, it is all vertical motion.

The Equation for the Cannonball's Motion

Figure 10.56 summarizes our work so far.

Motion Due to	Direction	Distance
the explosion	forward (x)	$\frac{s}{\sqrt{2}} \cdot t$
the explosion	upward (y)	$\frac{s}{\sqrt{2}} \cdot t$
gravity	downward (y)	$16t^2$

Figure 10.56

Recall that we placed the x-axis and y-axis such that the position of the cannonball at a time t is (x, y), where x measures the horizontal distance from the cannonball to the cannon and y measures the vertical distance from the cannonball to the ground, as shown in Figure 10.50. The horizontal motion is due solely to the explosion, and we found that distance to be $(s/\sqrt{2})t$. Thus, $x = (s/\sqrt{2})t$.

The vertical motion is due both to the explosion and to gravity. Initially, the y-coordinate is just the distance from the cannon to the origin—that is, the height of the cannon's muzzle. Call this distance c. The cannonball's height is increased by the explosion and decreased by gravity, so to the initial height, we add $(s/\sqrt{2})t$ and subtract $16t^2$. Thus,

$$y = c + \frac{s}{\sqrt{2}} \cdot t - 16t^2$$

To get the equation of the cannonball's path, we combine the formula for the cannonball's x-coordinate $[x = (s/\sqrt{2})t]$ and the formula for the cannonball's y-coordinate $[y = c + (s/\sqrt{2})t - 16t^2]$.

$$x = \frac{s}{\sqrt{2}}t \qquad \text{the } x\text{-equation}$$

$$\frac{x \cdot \sqrt{2}}{s} = t \qquad \text{solving for } t$$

$$y = c + \frac{s}{\sqrt{2}} \cdot t - 16t^2 \qquad \text{the } y\text{-equation}$$

$$= c + \frac{s}{\sqrt{2}} \cdot \frac{x \cdot \sqrt{2}}{s} - 16\left(\frac{x \cdot \sqrt{2}}{s}\right)^2 \qquad \text{substituting } \frac{x \cdot \sqrt{2}}{s} \text{ for } t$$

$$= c + x - 16\left(\frac{x^2 \cdot 2}{s^2}\right) \qquad \text{canceling and squaring}$$

$$= -\frac{32x^2}{s^2} + x + c$$

This is the equation of the trajectory of a cannonball that is fired at a 45° angle.

Similar work would allow us to find the equation of the trajectory of a cannonball that is fired at a 30° angle and one that is fired at a 60° angle. Those equations are given below.

Although the following equations were discovered in an intensive study of cannons, they also apply to other projectiles, such as a ball, a rock, an arrow from a bow or crossbow, and a shell from a rifle or a musket. Thus, while the following example involves a cannonball shot with a certain initial speed at a certain angle, it could just as well involve a football kicked at the same speed and angle. The problem would be solved in the same manner, and the solutions would be the same.

Trajectory of a Projectile

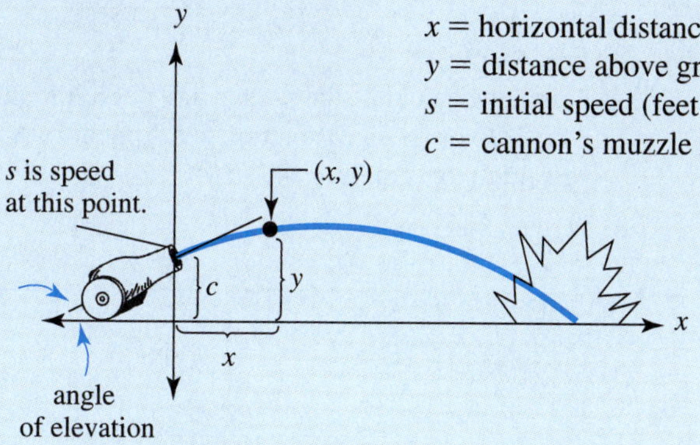

x = horizontal distance from cannon (feet)
y = distance above ground (feet)
s = initial speed (feet per second)
c = cannon's muzzle height

$y = -\dfrac{32x^2}{s^2} + x + c$ if the cannon's angle of elevation is 45°

$y = -\dfrac{64x^2}{3s^2} + \dfrac{x}{\sqrt{3}} + c$ if the cannon's angle of elevation is 30°

$y = -\dfrac{64x^2}{s^2} + \sqrt{3}x + c$ if the cannon's angle of elevation is 60°

EXAMPLE 1 A cannonball is shot with an initial speed of 200 feet per second from a cannon whose muzzle is 3 feet above the ground and at a 30° angle.

a. Find the equation of the cannonball's trajectory.
b. How far from the cannon will the ball strike the ground?
c. Will the ball clear an 85-foot-high wall that is 1,000 feet from the cannon?
d. Sketch the cannonball's path.

Solution **a.** $y = -\dfrac{64x^2}{3s^2} + \dfrac{x}{\sqrt{3}} + c$ using the 30° angle formula

$= -\dfrac{64x^2}{3(200)^2} + \dfrac{x}{\sqrt{3}} + 3$ substituting 200 for s and 3 for c

$= -\dfrac{64x^2}{120{,}000} + \dfrac{x}{\sqrt{3}} + 3$

$= -0.0005x^2 + 0.5774x + 3$ rounding to four decimal places

b. Finding how far the ball will travel means finding the value of x that will make $y = 0$, because y measures the distance above the ground.

$$y = -0.0005x^2 + 0.5774x + 3$$

$$0 = -0.0005x^2 + 0.5774x + 3 \quad \text{substituting 0 for } y$$

To solve this equation, we need to use the Quadratic Formula:

if $ax^2 + bx + c = 0$, then

$$x = \frac{-b \pm \sqrt{b^2 - 4ac}}{2a}$$

$$x = \frac{-0.5774 \pm \sqrt{0.5774^2 - 4(-0.0005)(3)}}{2(-0.0005)}$$

First, we calculate the radical and put it into the calculator's memory (usually with a [STO] or [$x \rightarrow$ M] button; see Appendix A or B for more information on using a calculator's memory).

.5774 [x^2] [+] 4 [×] .0005 [×] 3 [=] [$\sqrt{x}$] [STO]

TI-80/82/83 [2nd] [$\sqrt{\ }$] [(] .5774 [x^2] [+] 4 [×] .0005 [×] 3 [)] [STO→] [ALPHA] [A] [ENTER]

TI-83 users should not type the left parenthesis—it's automatically inserted.

TI-81/85/86 [2nd] [$\sqrt{\ }$] [(] .5774 [x^2] [+] 4 [×] .0005 [×] 3 [)] [STO→] [A] [ENTER]

Casio [SHIFT] [$\sqrt{\ }$] [(] .5774 [x^2] [+] 4 [×] .0005 [×] 3 [)] [→] [ALPHA] [A] [EXE]

Our last equation for x becomes

$$x = \frac{-0.5774 \pm \text{memory}}{-0.001}$$

$$= \frac{-0.5774 + \text{memory}}{-0.001} \quad \text{or} \quad \frac{-0.5774 - \text{memory}}{-0.001}$$

We want to combine -0.5774 with the number that is in the calculator's memory. On most scientific calculators, this is done with a [RCL] or [RM] button.

.5774 [+/−] [+] [RCL] [=] [÷] .001 [+/−] [=] and

.5774 [+/−] [−] [RCL] [=] [÷] .001 [+/−] [=]

[(] [(−)] .5774 [+] [ALPHA] [A] [)] [÷] [(−)] .001 [ENTER] and

[(] [(−)] .5774 [−] [ALPHA] [A] [)] [÷] [(−)] .001 [ENTER]

This gives $x = -5.172536\ldots$ or $1{,}159.972536\ldots \approx -5$ or 1,160. The parabola intersects the x-axis at approximately $x = -5$ and at $x = 1{,}160$. The negative value for x does not fit our situation. The cannonball will travel approximately 1,160 feet.

c. The ball will clear an 85-foot-high wall that is 1,000 feet away if $y > 85$ when $x = 1{,}000$.

$$\begin{aligned} y &= -0.0005x^2 + 0.5774x + 3 \\ &= -0.0005(1{,}000)^2 + 0.5774(1{,}000) + 3 \\ &= 80.4 \end{aligned}$$

Thus, the ball will be about 80 feet high when it is 1,000 feet from the cannon; it will *not* clear an 85-foot-high wall. The cannon's angle of elevation would have to be increased a little.

d. The path is sketched in Figure 10.57.

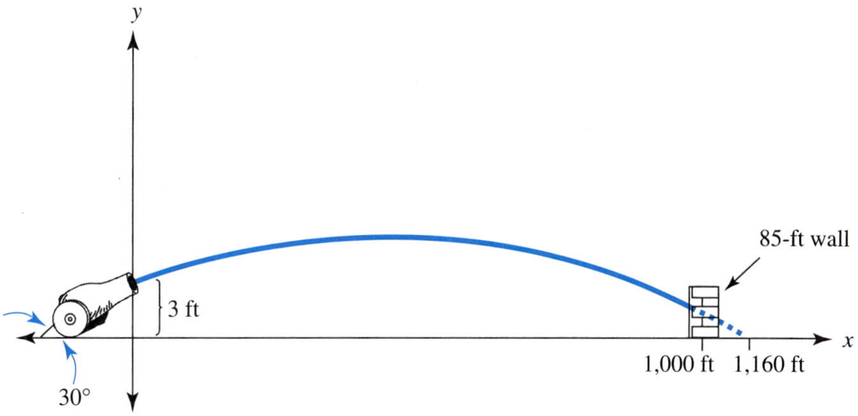

Figure 10.57

EXAMPLE 2 A cannonball is shot with an initial speed of 200 feet per second from a cannon whose muzzle is 4 feet above the ground and at a 45° angle.

a. Find the equation of the cannonball's trajectory.

b. What is the maximum height the ball will attain?

Solution a. $y = -\dfrac{32x^2}{s^2} + x + c$ using the 45° angle formula

$= -\dfrac{32x^2}{200^2} + x + 4$ substituting

$= -0.0008x^2 + x + 4$

b. Finding the ball's maximum height means finding the vertical distance from the ground at the highest point; that is, it asks for the value of y at the vertex. In Section 10.0, we were able to find vertices of parabolas by plotting a handful of points in the region where the slope of the secant line is close to 0. In our current situation, such an approach would involve finding hundreds of points. Instead, observe that the cannonball is at its highest point when the slope of the tangent line is 0, as shown in Figure 10.58. Thus, we will first find the slope of the tangent line at an arbitrary point P, by taking the derivative of the equation of the path. Next, we will set that slope equal to 0 and solve for x. This will give us the location of the point P at which the slope of the tangent line is 0.

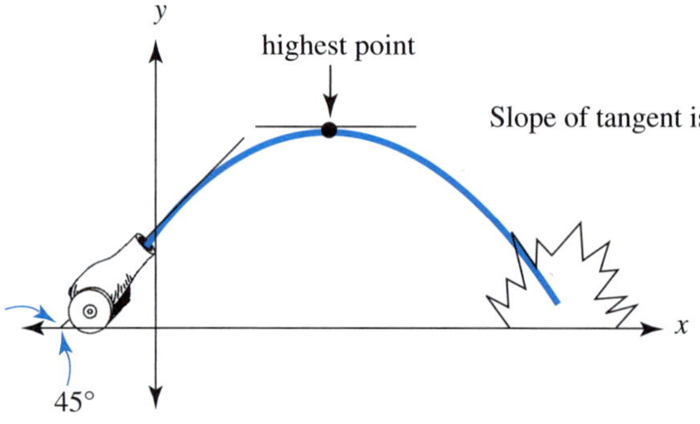

Figure 10.58

Step 1 *Find the slope of the tangent line at an arbitrary point P by taking the derivative.*

Rather than taking the derivative of $f(x)$ with Newton's or Cauchy's method, we can use a shortcut. Recall from Exercise 24 of Section 10.4 that if $f(x) = ax^2 + bx + c$, then $df/dx = 2ax + b$.

If $f(x) = \boxed{a}\, x^2 + \boxed{b}\, x + c$ then $\dfrac{df}{dx} = 2\boxed{a}\, x + \boxed{b}$

If $f(x) = \boxed{-0.0008}\, x^2 + \boxed{1}\, x + 4$ then $\dfrac{df}{dx} = 2(\boxed{-0.0008})\, x + \boxed{1}$

By comparing the formula with our problem, we can see that the "a" in our formula matches "– 0.0008" in our problem, and "b" matches "1." Thus, the derivative of $f(x) = -0.0008x^2 + 1x + 4$ is $2(-0.0008)x + 1 = -0.0016x + 1$.

Step 2 *The cannonball is at its highest point when the slope of the tangent line is 0.*

$$-0.0016x + 1 = 0$$

$$-0.0016x = -1$$

$$x = \frac{-1}{-0.0016} = 625$$

This is the x-coordinate of the point P, where the ball is at its highest point. The question asked for the vertical distance from the ground at the highest point; that is, it asks for the y-coordinate of P.

$$
\begin{aligned}
y &= -0.0008x^2 + x + 4 \\
&= -0.0008(625)^2 + 625 + 4 \quad \text{substituting 625 for } x \\
&= 316.5 \approx 317 \quad \text{rounding to the nearest foot}
\end{aligned}
$$

The cannonball will travel to a maximum height of 317 feet (rounded off), as shown in Figure 10.59.

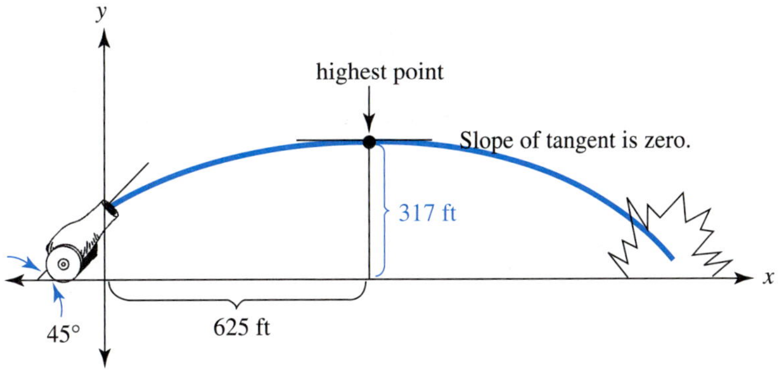

Figure 10.59

■

The Generalized Equation for the Cannonball's Motion

(for those who have read Section 6.5: Trigonometry)

Earlier in this section, we derived the equation of the trajectory of a cannonball shot from a cannon whose angle of elevation is 45°. In the exercises you will derive equations for two other angles of elevation—30° and 60°. There are quite a few angles other than these three; clearly, it's impractical to have separate equations for 1°, 2°, and so on. Trigonometry allows us to generate one equation that will work for any angle.

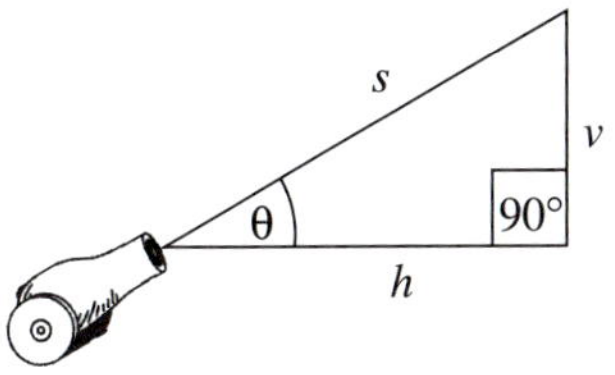

Figure 10.60

If a cannon's angle of elevation is θ (the Greek letter "theta"), its speed is s feet per second, and that speed's horizontal and vertical components are h and v, respectively, then we can draw the triangle shown in Figure 10.60, a generalized version of the specific triangle drawn for a 45° angle of elevation in Figure 10.54.

Since the opposite side is v, the adjacent side is h, and the hypotenuse is s, we have

$$\sin\theta = \frac{\text{opp}}{\text{hyp}} = \frac{v}{s} \quad \text{and} \quad \cos\theta = \frac{\text{adj}}{\text{hyp}} = \frac{h}{s}$$

We can solve the first equation for v and the second equation for h by multiplying by s:

$$s \cdot \sin\theta = \cancel{s} \cdot \frac{v}{\cancel{s}} \quad \text{and} \quad s \cdot \cos\theta = \cancel{s} \cdot \frac{h}{\cancel{s}}$$

$$v = s\sin\theta \quad \text{and} \quad h = s\cos\theta$$

Since h is the horizontal speed (or rate), and distance = rate · time, the horizontal distance is rate · time = $s \cos\theta \cdot t = st \cos\theta$. Similar work gives us the vertical distance, as shown in Figure 10.61.

Motion Due to the Explosion			
Direction	**Rate**	**Time**	**Distance = Rate · Time**
horizontal	$h = s \cos\theta$	t	$(s \cos\theta) \cdot t = st \cos\theta$
vertical	$h = s \sin\theta$	t	$(s \sin\theta) \cdot t = st \sin\theta$

Figure 10.61

The motion due to gravity is not affected by the angle of elevation; the cannonball still falls $16t^2$ feet in t seconds. Naturally, this motion is all vertical.

Figure 10.62 summarizes our work so far.

Motion Due to	Direction	Distance
the explosion	forward (x)	$st \cos\theta$
the explosion	upward (y)	$st \sin\theta$
gravity	downward (y)	$16t^2$

Figure 10.62

The horizontal motion is due solely to the explosion, and we found that distance to be $st \cos\theta$; thus, $x = st \cos\theta$. The vertical motion is due both to the explosion and to gravity. Initially, the y-coordinate is just the distance c from the cannon to the origin—that is, the height of the cannon's muzzle. The cannonball's height is increased by the explosion and decreased by gravity, so to the initial height, we must add $st \sin\theta$ and subtract $16t^2$. Thus,

$$y = c + st \sin\theta - 16t^2$$

To get the equation of the cannonball's path, we solve the x-equation for t and substitute it into the y-equation, as we did for the 45° equation earlier in this section:

$$x = st \cos\theta \qquad \text{the } x\text{-equation}$$

$$\frac{x}{s \cos\theta} = \frac{\cancel{s}t\,\cancel{\cos\theta}}{\cancel{s}\,\cancel{\cos\theta}} \qquad \text{solving for } t$$

$$\frac{x}{s \cos\theta} = t$$

$$y = c + s \cdot t \sin\theta - 16 t^2 \qquad \text{the } y\text{-equation}$$

$$y = c + s \cdot \frac{x}{s \cos\theta} \sin\theta - 16\left(\frac{x}{s \cos\theta}\right)^2 \qquad \text{substituting } \frac{x}{s \cos\theta} \text{ for } t$$

$$y = c + \frac{x \sin\theta}{\cos\theta} - 16\frac{x^2}{(s \cos\theta)^2} \qquad \text{canceling and simplifying}$$

This is the equation of the trajectory of a cannonball shot from a cannon whose angle of elevation is θ, but it can be further simplified. Recall that $\sin\theta = \text{opp/hyp}$, $\cos\theta = \text{adj/hyp}$, and $\tan\theta = \text{opp/adj}$. Thus,

$$\frac{\sin\theta}{\cos\theta} = \frac{\dfrac{\text{opp}}{\text{hyp}}}{\dfrac{\text{adj}}{\text{hyp}}}$$

$$= \frac{\dfrac{\text{opp}}{\cancel{\text{hyp}}}}{\dfrac{\text{adj}}{\cancel{\text{hyp}}}} \cdot \frac{\cancel{\text{hyp}}}{\cancel{\text{hyp}}} \qquad \text{since hyp/hyp} = 1$$

$$= \frac{\text{opp}}{\text{adj}} \qquad \text{canceling}$$

$$= \tan\theta$$

This means that we can replace the $\sin\theta/\cos\theta$ in the above equation with $\tan\theta$. Thus, the equation is

$$y = c + x\tan\theta - 16\frac{x^2}{(s\cos\theta)^2}$$

$$y = -16\frac{x^2}{(s\cos\theta)^2} + x\tan\theta + c$$

Trajectory of a Projectile

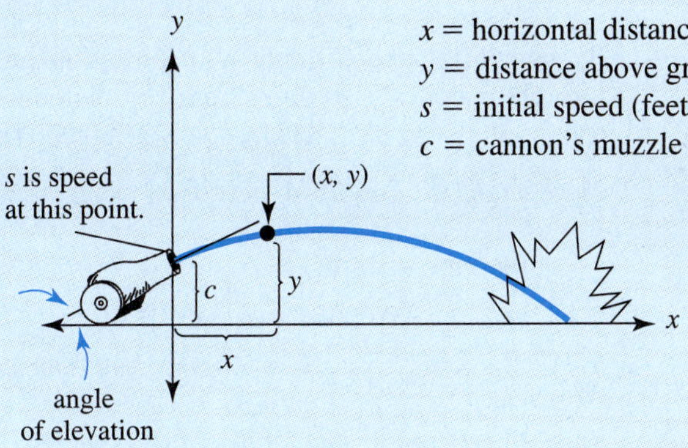

x = horizontal distance from cannon (feet)
y = distance above ground (feet)
s = initial speed (feet per second)
c = cannon's muzzle height

$$y = -16\frac{x^2}{(s\cos\theta)^2} + x\tan\theta + c \qquad \text{if the cannon's angle of elevation is } \theta$$

EXAMPLE 3 A cannonball is shot with an initial speed of 250 feet per second from a cannon whose muzzle is 4 feet above the ground and at a 40° angle.

a. Find the equation of the cannonball's trajectory.
b. What is the maximum height the ball will attain?

Solution **a.** $y = -16\dfrac{x^2}{(s\cos\theta)^2} + x\tan\theta + c$

$= -16\dfrac{x^2}{(250\cos 40°)^2} + x\tan 40° + 4$ substituting 250 for s and 40 for θ

$= -0.0004362465769x^2 + 0.8390996312x + 4$

To compute the coefficient of x^2, type

16 [+/−] [÷] [(] 250 [×] 40 [cos] [)] [x^2] [=]

[(−)] 16 [÷] [(] 250 [×] [cos] 40 [)] [x^2] [ENTER]

$\approx -0.0004x^2 + 0.8391x + 4$ rounding to 4 decimal places

b. We will follow the steps used in Example 2 to find the ball's maximum height.

Step 1 *Find the slope of the tangent line at an arbitrary point P by taking the derivative.*

$$\text{If } f(x) = a\,x^2 + b\,x + c \text{ then } \frac{df}{dx} = 2\,a\,x + b$$

$$\text{If } f(x) = -0.0004x^2 + 0.8391x + 4 \text{ then } \frac{df}{dx} = 2(-0.0004)x + 0.8391$$

By comparing the formula with our problem, we can see that the "a" in our formula matches "−0.0004" and "b" matches "0.8391." Thus, the derivative of $f(x) = -0.0004x^2 + 0.8391x + 4$ is $2(-0.0004)x + 0.8391 = -0.0008x + 0.8391$.

Step 2 *The cannonball is at its highest point when the slope of the tangent line is* 0.

$$-0.0008x + 0.8391 = 0$$
$$-0.0008x = -0.8391$$
$$x = -0.8391/-0.0008 = 1048.875 \approx 1049$$

This is the x-coordinate of the point P, where the ball is at its highest point. The question asked for the vertical distance from the ground at the highest point; that is, it asks for the y-coordinate of P.

$$\begin{aligned} y &= -0.0004x^2 + 0.8391x + 4 \\ &= -0.0004(1048.875)^2 + 0.8391(1048.875) + 4 \\ &= 1{,}492.930506 \approx 1{,}493 \text{ feet} \end{aligned}$$

The cannonball will travel to a maximum height of 1,493 feet, as shown in Figure 10.63.

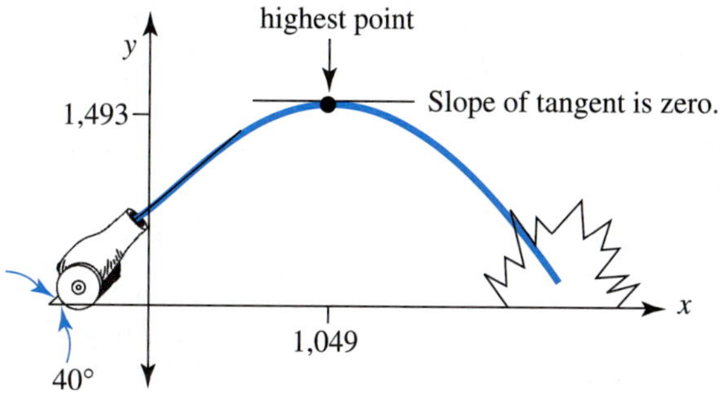

Figure 10.63

10.5 Exercises

In Exercises 1–6, answer the following questions.

a. Find the equation of the trajectory of a cannonball shot with the given initial speed from a cannon whose muzzle is at the given orientation.
b. How far will the ball travel?
c. What is the maximum height the ball will attain?
d. Will the ball clear an 80-foot wall that is 1,900 feet from the cannon?
e. Sketch the cannonball's path. In your sketch, include everything that is known about the path.

(Round all decimals to four decimal places, and round distance answers to the nearest foot, as in the examples.)

1. initial speed: 250 feet per second
muzzle height: 4 feet
angle of elevation: 30°

2. initial speed: 300 feet per second
muzzle height: 4 feet
angle of elevation: 30°

3. initial speed: 300 feet per second
muzzle height: 4 feet
angle of elevation: 45°

4. initial speed: 250 feet per second
muzzle height: 4 feet
angle of elevation: 45°

5. initial speed: 325 feet per second
muzzle height: 5 feet
angle of elevation: 60°

6. initial speed: 275 feet per second
muzzle height: 4 feet
angle of elevation: 60°

In Exercises 7–12, answer the following questions.

a. Find the equation of the trajectory of a cannonball shot with the given initial speed from a cannon whose muzzle has the given orientation.
b. How far will the ball travel?
c. What is the maximum height the ball will attain?
d. Will the ball clear a 70-foot wall that is 1,100 feet from the cannon?
e. Sketch the cannonball's path. In your sketch, include everything that is known about the path.

(Round all decimals to four decimal places, and round distance answers to the nearest foot, as in the examples.)

7. initial speed: 200 feet per second
muzzle height: 5 feet
angle of elevation: 30°

8. initial speed: 225 feet per second
muzzle height: 4 feet
angle of elevation: 30°

9. initial speed: 250 feet per second
muzzle height: 4 feet
angle of elevation: 45°

10. initial speed: 300 feet per second
muzzle height: 5 feet
angle of elevation: 45°

11. initial speed: 310 feet per second
muzzle height: 3 feet
angle of elevation: 60°

12. initial speed: 360 feet per second
muzzle height: 5 feet
angle of elevation: 60°

In Exercises 13–16, answer the following questions.

a. Find the equation of the trajectory of a cannonball shot with the given initial speed from a cannon whose muzzle is at the given orientation.
b. How far from the target must this cannon be placed, if the target is placed as indicated? (Choose the answer that allows the cannon crew to be as far as possible from the enemy.)
c. Sketch the cannonball's path. In your sketch, include everything that is known about the path.

13. initial speed: 240 feet per second
muzzle height: 4 feet
angle of elevation: 45°
target: 20 feet above the ground
14. initial speed: 200 feet per second
muzzle height: 3 feet
angle of elevation: 30°
target: 50 feet above the ground
15. initial speed: 300 feet per second
muzzle height: 4 feet
angle of elevation: 30°
target: 20 feet above the ground
16. initial speed: 200 feet per second
muzzle height: 5 feet
angle of elevation: 45°
target: 40 feet above the ground

17. A baseball player hits a home run that just clears the 10-foot-high fence 400 feet from home plate. He hit the ball at a 30° angle of elevation, 3 feet above the ground.
   a. Find the ball's initial speed. (Round to the nearest foot per second.)

      HINT: Substitute 400 for x and 10 for y in the 30° equation.

   b. Find the equation of the ball's trajectory. (Round all decimals to four decimal places.)
   c. What is the maximum height the ball attained? (Round to the nearest foot.)
   d. Find the ball's initial speed if the angle of elevation were 45°. (Round to the nearest foot per second.)

18. A football player kicks a field goal at a 45° angle of elevation from a distance of 50 yards from the goal posts.
   a. Find the ball's initial speed if the ball hits the goal post's crossbar, 3 yards above ground. (Round to the nearest foot per second.)

      HINT: See Exercise 17.

   b. Find the equation of the ball's trajectory. (Round all decimals to four decimal places.)
   c. What is the maximum height the ball attained? (Round to the nearest foot.)
   d. What would the football player have to do differently to get the ball over the crossbar?

19. A 6-foot-tall hiker is walking on a 100-foot-high cliff overlooking the ocean. He throws a rock at a 45° angle of elevation with an initial speed of 90 feet per second.
   a. Find the equation of the trajectory of the rock.
   b. How far does the rock travel?
   c. How long does it take for the rock to land in the water below?

      HINT: Use the answer to part (b) and the "x-equation" that was obtained when we derived the equation of a projectile with a 45° angle of elevation.

20. A firecracker is launched at a 45° angle of elevation from ground level with an initial speed of 120 feet per second.
   a. Find the equation of the trajectory of the firecracker.
   b. Find the firecracker's maximum height and the horizontal distance between the launch point and the point of maximum height.
   c. Find the amount of time that the fuse should burn if the firecracker is to detonate at the peak of its trajectory.

      HINT: Use the answer to part (b) and the "x-equation" that was obtained when we derived the equation of a projectile with a 45° angle of elevation.

21. In finding the equation of a cannonball's trajectory, we used a right triangle with a 45° angle, because our equation was for a cannon pointed at a 45° angle. That triangle had sides of length 1, 1, and $\sqrt{2}$. If the cannon is pointed at a 30° angle, we use a right triangle with a 30° angle. In this exercise, we will find the lengths of such a triangle's sides. Explain the reasoning that you use to answer each of the following questions.
   a. What is the size of the triangle's third angle?
   b. Make a larger triangle by attaching a duplicate of our 30° triangle underneath it, as shown in Figure 10.64. What are the sizes of that larger triangle's three angles?
   c. If the original 30° triangle's vertical side is of length 1, what is the length of the larger triangle's vertical side?
   d. What are the lengths of the larger triangle's other two sides?

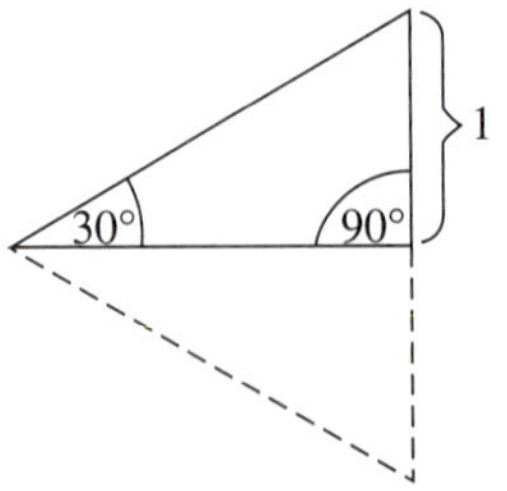

Figure 10.64

e. You now know the length of two of the 30° triangle's sides. Use the Pythagorean Theorem to find the length of its third side.

22. Use Exercise 21 and a procedure similar to that used in this section (in finding the trajectory equation for a cannon whose angle of elevation is 45°) to find the trajectory equation for a cannon whose angle of elevation is 30°.

23. Use Exercise 21 and a procedure similar to that used in this section (in finding the trajectory equation for a cannon whose angle of elevation is 45°) to find the trajectory equation for a cannon whose angle of elevation is 60°.

24. Use the "x-equation" that was obtained in Exercise 22 to find how long it takes for the cannonball in Exercise 2 to hit the ground.

25. Use the "x-equation" that was obtained in Exercise 22 to find how long it takes for the baseball in Exercise 17 to hit the ground.

26. Find how long it takes for the football in Exercise 18 to hit the crossbar.

27. Use the "x-equation" that was obtained in Exercise 23 to find how long it takes for the cannonball in Exercise 5 to hit the ground.

Exercises 28–31 are for those who have read Section 6.5: Trigonometry. In Exercises 28–31, answer the following questions.

a. Find the equation of the trajectory of a cannonball shot with the given initial speed from a cannon whose muzzle has the given orientation.

b. How far will the ball travel?

c. What is the maximum height that the ball will attain?

d. Sketch the cannonball's path. In your sketch, include everything that is known about the path.

(Round all decimals to four decimal places, and round distance answers to the nearest foot, as in the examples.)

28. initial speed: 200 feet per second
muzzle height: 5 feet
angle of elevation: 10°

29. initial speed: 200 feet per second
muzzle height: 5 feet
angle of elevation: 20°

30. initial speed: 200 feet per second
muzzle height: 5 feet
angle of elevation: 80°

31. initial speed: 200 feet per second
muzzle height: 5 feet
angle of elevation: 70°

Answer the following questions using complete sentences and your own words.

Concept Questions

32. Why do the equations of the trajectory of a projectile apply equally well to cannonballs, baseballs hit by bats, and thrown objects?

33. Would the equations of the trajectory of a projectile apply to missiles? Why or why not?

10.6 Newton and Areas

Find the Area of Any Shape

Recall that the Greek geometers had found areas of specific shapes such as circles, rectangles, and triangles and that Archimedes and Kepler had found areas of regions by filling those regions with an infinite number of triangles or rectangles. Newton used his calculus to find areas in a different manner. We will use modern notation in explaining Newton's concept.

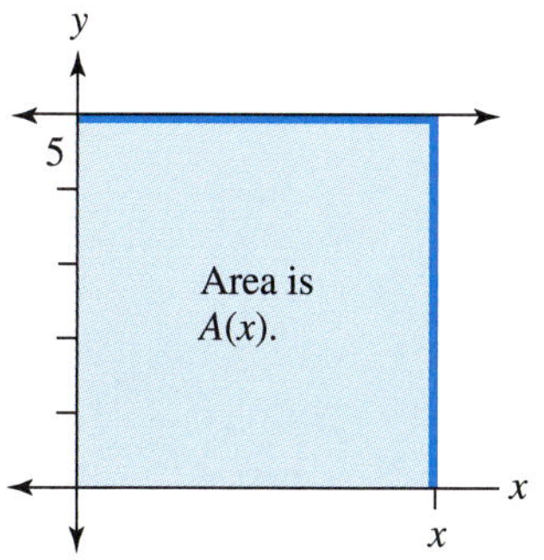

Figure 10.65

The Area Function $A(x)$

Consider the region bounded below by the x-axis, above by the horizontal line $y = 5$, on the left by the y-axis, and on the right by a vertical line, as illustrated in Figure 10.65. Different locations of a right-hand boundary produce different regions with different areas.

Let $A(x)$ be a function whose input x is the specific location of the right-hand boundary and whose output is the area of the resulting region. Thus, as shown in Figure 10.66, $A(2)$ would be the area of a rectangle with height 5 and base 2; $A(2) = 5 \cdot 2 = 10$. Similarly, $A(8)$ would be the area of a rectangle with height 5 and base 8; $A(8) = 5 \cdot 8 = 40$. More generally, $A(x)$ would be the area of a rectangle with height 5 and base x; $A(x) = 5x$, as shown in Figure 10.67. This is the area function for a rectangle of height 5; that is, it is the area function associated with $f(x) = 5$.

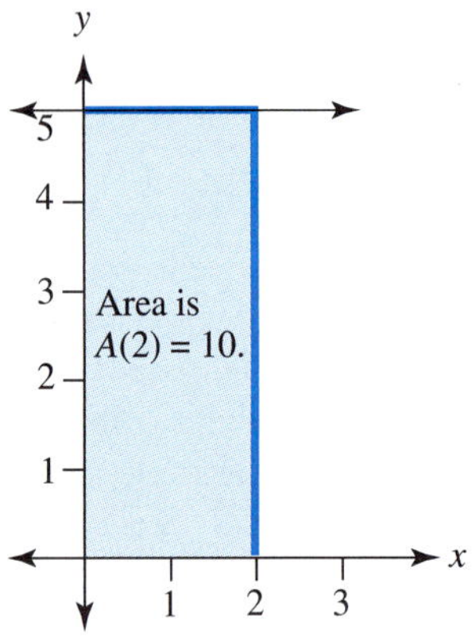

Figure 10.66

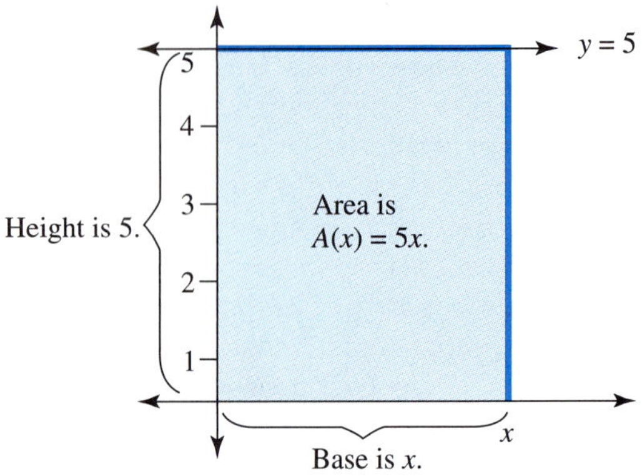

Figure 10.67

Every positive function $f(x)$ has an area function $A(x)$ associated with it. The area is that of the region bounded below by the x-axis, above by the curve or line given by $f(x)$, on the left by the y-axis, and on the right by a vertical line at x, as shown in Figure 10.68. We just found that the function $f(x) = 5$ (or $y = 5$) has an area function $A(x) = 5x$ associated with it.

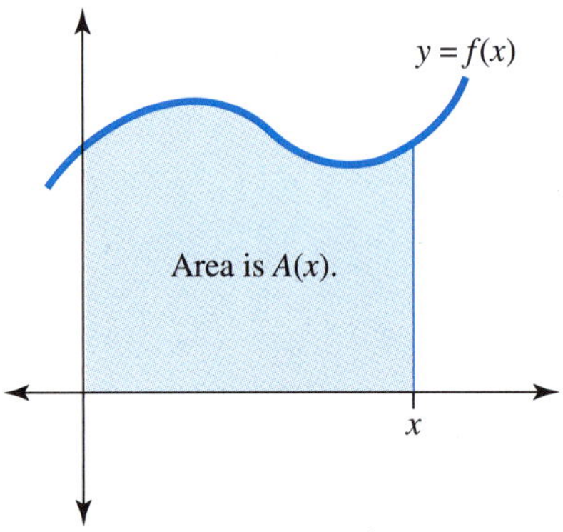

Figure 10.68

EXAMPLE 1

a. Find the area function associated with $f(x) = 3x$.
b. Sketch the region whose area is found by that area function.
c. Find $A(6)$ and describe the region whose area is $A(6)$.
d. Find $A(4)$ and describe the region whose area is $A(4)$.
e. Find $A(6) - A(4)$ and describe the region whose area is $A(6) - A(4)$.

Solution

a. $f(x) = 3x$ (or $y = 3x + 0$) is a line whose slope is 3 and whose intercept is 0. Its area function $A(x)$ gives the area of the region bounded below by the x-axis, above

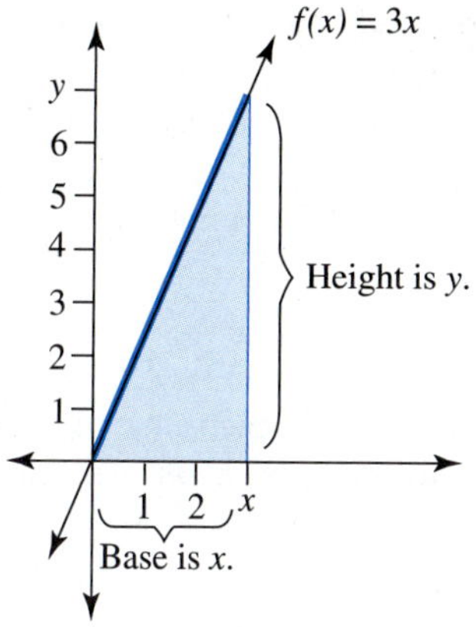

Figure 10.69

by the line $f(x) = 3x$, on the left by the y-axis, and on the right by a vertical line at x, as shown in Figure 10.69. That region is a triangle, so its area is given by the formula "area $= \frac{1}{2} \cdot$ base $\cdot$ height." Since the base is x and the height is y, the area is

$$\begin{aligned} A(x) &= \frac{1}{2} \cdot \text{base} \cdot \text{height} \\ &= \frac{1}{2} \cdot \text{x} \cdot \text{y} \\ &= \frac{1}{2} \cdot x \cdot 3x \qquad \text{since } y = 3x \\ &= \frac{3x^2}{2} \end{aligned}$$

Thus, the area function associated with $f(x) = 3x$ is $A(x) = \dfrac{3x^2}{2}$.

b. The region is shown in Figure 10.69.

c. $A(x) = \dfrac{3x^2}{2}$ so $A(6) = \dfrac{3 \cdot 6^2}{2} = 54$

This is the area of the region bounded below by the x-axis, above by the line $f(x) = 3x$, on the left by the y-axis, and on the right by a vertical line at $x = 6$.

d. $A(x) = \dfrac{3x^2}{2}$ so $A(4) = \dfrac{3 \cdot 4^2}{2} = 24$

This is the area of the region bounded below by the x-axis, above by the line $f(x) = 3x$, on the left by the y-axis, and on the right by a vertical line at $x = 4$.

e. $A(6) - A(4) = 54 - 24 = 30$

This is the area of the region bounded below by the x-axis, above by the line $f(x) = 3x$, on the left by a vertical line at $x = 4$, and on the right by a vertical line at $x = 6$, as shown in Figure 10.70.

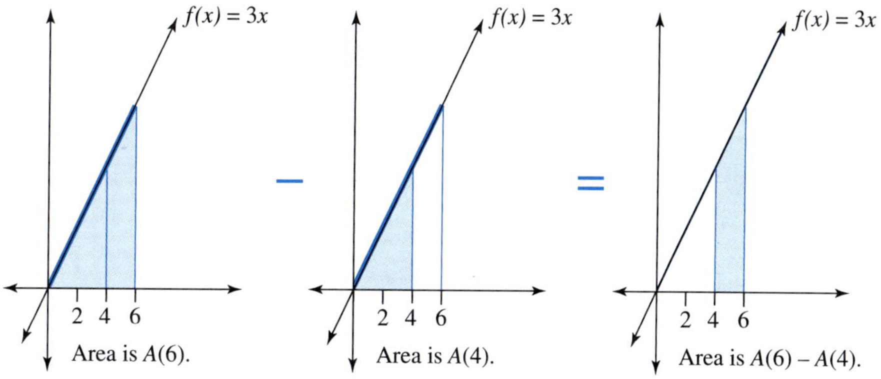

Figure 10.70

Newton and the Area Function

We can find areas of rectangles, triangles, and other basic shapes quite easily, because the ancient Greeks found formulas for those areas centuries ago (as was discussed in Chapter 6). These formulas are of little value in finding the area bounded by curves (unless, like Kepler and Archimedes, you fill the region with an infinite number of triangles).

Newton's goal was to find the area of a region bounded below by the x-axis, above by a *curve* $y = f(x)$, on the left by the y-axis, and on the right by a vertical line at x, as shown in Figure 10.68. He wanted to find the area function associated with the *curve* $y = f(x)$.

The key to the area problem is to find the derivative of the area function $A(x)$:

$$\frac{dA}{dx} = \lim_{\Delta x \to 0} \frac{A(x + \Delta x) - A(x)}{\Delta x}$$

We can interpret this derivative geometrically by analyzing it piece by piece. $A(x + \Delta x)$ refers to the area of the region bounded on the right by a vertical line at $x + \Delta x$, slightly to the right of x, as shown in Figure 10.71. $A(x + \Delta x) - A(x)$ then refers to the difference between this area and the area given by $A(x)$, as shown in Figure 10.72. This is the area of the narrow strip bounded on the left by a vertical line at x and on the right by a vertical line at $x + \Delta x$. Because Δx is allowed to approach zero, the strip is quite narrow; it is so narrow that we can think of it as a rectangle whose height is y, or $f(x)$, and whose width is Δx. Thus, the numerator can be rewritten as

$$\begin{aligned} A(x + \Delta x) - A(x) &= \text{area of thin rectangle} \\ &= \text{base} \cdot \text{height} \\ &= \Delta x \cdot f(x) \end{aligned}$$

and the entire fraction can be rewritten as

$$\frac{A(x + \Delta x) - A(x)}{\Delta x} = \frac{\Delta x \cdot f(x)}{\Delta x} = f(x)$$

This means that the derivative of the area function is $dA/dx = f(x)$ or, to put it another way, that the area function $A(x)$ is that function whose derivative is $f(x)$. For this reason, $A(x)$ is called the **antiderivative** of $f(x)$; it is denoted by $A(x) = \int f(x)\, dx$.

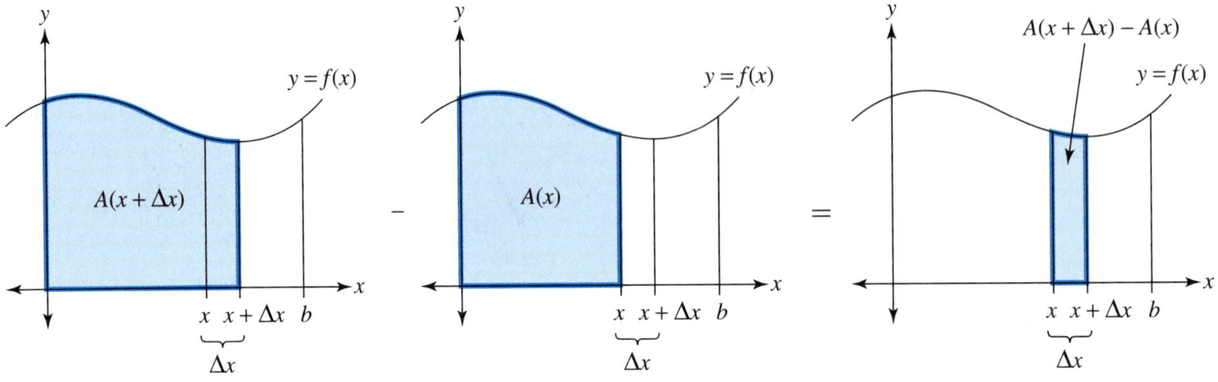

Figure 10.71

Figure 10.72

The antiderivative symbol $\int$ has its origins in Archimedes' and Kepler's methods of finding areas. Each of these methods involves breaking down a region into very small pieces and finding the sum of the areas of those pieces. The symbol $\int$ is a stylized S and stands for *sum*.

Antiderivatives

Before we continue with the area problem, we must pause and discuss antiderivatives.

Antiderivative Definition

If the derivative of $f(x)$ is $g(x)$, then the **antiderivative** of $g(x)$ is $f(x)$.

$$\text{If } \frac{df}{dx} = g(x), \text{ then } \int g(x)\,dx = f(x).$$

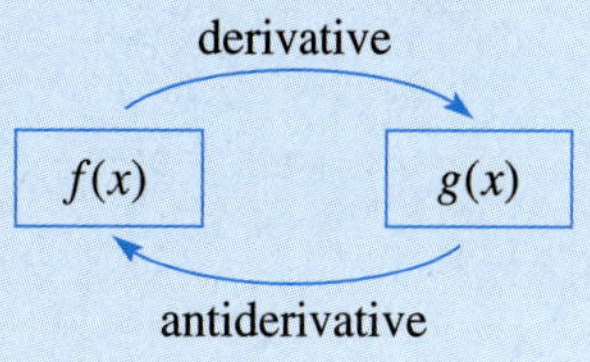

EXAMPLE 2 **a.** Find the derivative of $f(x) = 3x + c$ (where c is any constant).
b. Rewrite the derivative found in part (a) as an antiderivative.

Solution **a.** Although we could find the derivative of $f(x) = 3x + c$ by computing

$$\frac{df}{dx} = \lim_{\Delta x \to 0} \frac{f(x + \Delta x) - f(x)}{\Delta x}$$

it is much easier to use a shortcut. In Example 4 of Section 10.4, we found that the derivative of $f(x) = mx + b$ is $df/dx = m$.

$$\text{If } f(x) = \boxed{m}\,x + \boxed{b} \text{ then } \frac{df}{dx} = m \qquad \text{the formula (from Example 4)}$$

$$\text{If } f(x) = \boxed{3}\,x + \boxed{c} \text{ then } \frac{df}{dx} = ? \qquad \text{our problem}$$

By comparing the formula with our problem, we can see that "m" in the formula matches "3" in our problem, and "b" in the formula matches "c" in our problem. Thus, the derivative of $f(x) = 3x + c$ is $df/dx = 3$.

Because c is *any* constant, the derivative of $f(x) = 3x + 7$ is $df/dx = 3$, and the derivative of $f(x) = 3x + 107$ is $df/dx = 3$. This should make sense to you, because the derivative gives the slope, and $f(x) = 3x + 7$ and $f(x) = 3x + 107$ are both lines with slope 3.

b. In part (a), we found that the derivative of $f(x) = 3x + c$ is $df/dx = 3$. This means that the antiderivative of 3 is $3x + c$. Using the antiderivative symbol, we have

$$\int 3\,dx = 3x + c$$

The relationship between the above derivative and antiderivative statements is illustrated in Figure 10.73. ■

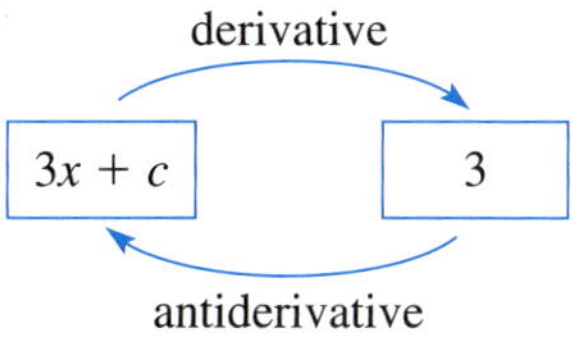

Figure 10.73

EXAMPLE 3 Find $\int 8x^3\,dx$. That is, find the antiderivative of $8x^3$.

Solution To find the antiderivative of $8x^3$ is to find the function whose derivative is $8x^3$. What could we take the derivative of and get $8x^3$? Finding it involves a series of successive guesses.

First Guess It would have to involve an x^4, because the derivative of x^4 is $4x^3$ (using the result from part (b) of Exercise 38 in Section 10.4). However, we don't want the derivative to be $4x^3$; we want it to be $8x^3$. We've got the correct power but an incorrect coefficient.

Second Guess It must be $2x^4$, because the derivative of $2x^4$ is $2 \cdot 4x^3 = 8x^3$ (using the result from part (b) of Exercise 38 in Section 10.4). This seems to be the right answer. However, the derivative of $2x^4 + 4$ is $8x^3$, and the derivative of $2x^4 + 19$ is $8x^3$, because the derivative of any constant is 0.

Answer It must be $2x^4 + c$, because the derivative of $2x^4 + c$ is $8x^3 + 0 = 8x^3$. Thus, $\int 8x^3\,dx = 2x^4 + c$. ■

EXAMPLE 4 Find $\int 5x^7 + 3x + 7\,dx$. That is, find the antiderivative of $5x^7 + 3x + 7$.

Solution To find the antiderivative of $5x^7 + 3x + 7$ is to find the function whose derivative is $5x^7 + 3x + 7$. What could we take the derivative of and get $5x^7 + 3x + 7$?

First Guess The goal of the first guess is to get the powers correct. Since taking a derivative decreases the powers by 1, our antiderivative would have to involve $x^8 + x^2 + x^1$, because the derivative of $x^8 + x^2 + x^1$ is

$$8x^7 + 2x^1 + 1x^0 = 8x^7 + 2x + 1$$

However, we don't want the derivative to be

$$8x^7 + 2x + 1$$

we want it to be

$$5x^7 + 3x + 7$$

We've got the correct powers but incorrect coefficients.

Second Guess The goal of the second guess is to make the coefficients correct. If we put a $\frac{5}{8}$ in front of the x^8, then the eights will cancel and the 5 will remain: The derivative of $\frac{5}{8}x^8$ is $\frac{5}{8} \cdot 8x^7 = 5x^7$. Similarly, we must put a $\frac{3}{2}$ in front of the x^2 and a 7 in front of the x^1. Thus, the antiderivative of $5x^7 + 3x + 7$ must be $\frac{5}{8}x^8 + \frac{3}{2}x^2 + 7x$, since the derivative of $\frac{5}{8}x^8 + \frac{3}{2}x^2 + 7x$ is

$$\frac{5}{8} \cdot 8x^7 + \frac{3}{2} \cdot 2x + 7 = 5x^7 + 3x + 7$$

Answer The antiderivative must be $\frac{5}{8}x^8 + \frac{3}{2}x^2 + 7x + c$, because the derivative of $\frac{5}{8}x^8 + \frac{3}{2}x^2 + 7x + c$ is $5x^7 + 3x + 7 + 0 = 5x^7 + 3x + 7$. Thus,

$$\int 5x^7 + 3x + 7\,dx = \frac{5}{8}x^8 + \frac{3}{2}x^2 + 7x + c$$ ■

Antiderivatives and Areas

Newton found that the derivative of the area function $A(x)$ is $f(x)$. In other words, he found that $A(x) = \int f(x)\,dx$ [that the area function is the antiderivative of $f(x)$].

EXAMPLE 5

a. Use antiderivatives to find the area function associated with $f(x) = x + 2$.

b. Find the area of a region bounded below by the x-axis, above by $f(x) = x + 2$, on the left by a vertical line at $x = 1$, and on the right by a vertical line at $x = 2$.

Solution **a.** $A(x) = \int x + 2\,dx = ?$ What can we take the derivative of and get $x + 2$?

First Guess Since taking a derivative decreases the power by 1, and since $x + 2 = x^1 + 2x^0$, we would have to take the derivative of $x^2 + x^1$. However, the derivative of $x^2 + x^1$ is $2x^1 + 1x^0 = 2x + 1$.

Second Guess To make the coefficients right, we need a $\frac{1}{2}$ in front of the x^2 and a 2 in front of the x^1. The derivative of $\frac{1}{2}x^2 + 2x^1$ is

$$\frac{1}{2} \cdot 2x + 2 \cdot 1x^0 = x + 2$$

Answer The antiderivative of $f(x) = x + 2$ is $\frac{1}{2}x^2 + 2x + c$:

$$A(x) = \int x + 2\,dx = \frac{1}{2}x^2 + 2x + c$$

b. The area of a region bounded below by the x-axis, above by $f(x) = x + 2$, on the left by a vertical line at $x = 1$, and on the right by a vertical line at $x = 2$, is $A(2) - A(1)$, as shown in Figure 10.74.

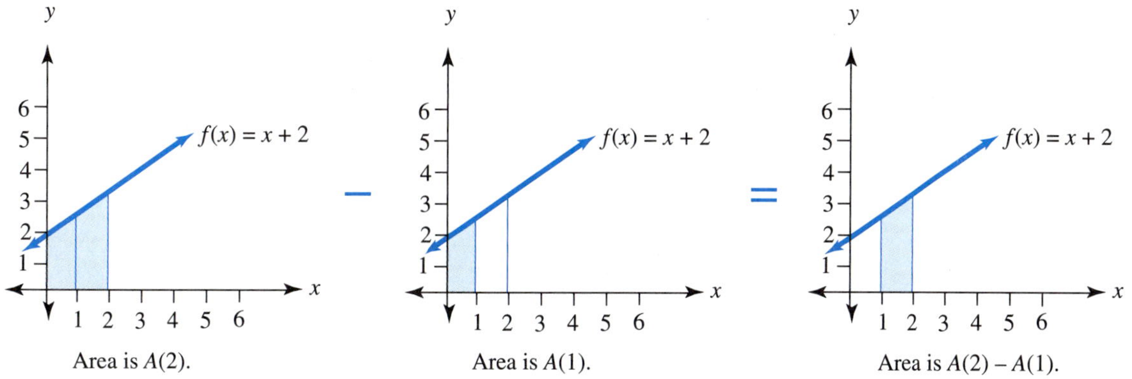

Figure 10.74

$$\begin{aligned} A(2) - A(1) &= \left(\frac{1}{2} \cdot 2^2 + 2 \cdot 2 + c\right) - \left(\frac{1}{2} \cdot 1^2 + 2 \cdot 1 + c\right) \\ &= 2 + 4 + c - \frac{1}{2} - 2 - c \\ &= 3\frac{1}{2} \end{aligned}$$

The area of the region is $3\frac{1}{2}$. ■

We could find the area of the region in Example 5 without calculus, since the region is a rectangle topped with a triangle, as shown in Figure 10.75 on page 784. The rectangle's base is $2 - 1 = 1$. Its height is the length of the vertical line at $x = 1$; that is, its height is $f(1)$. And, since $f(x) = x + 2$, $f(1) = 1 + 2 = 3$. Thus, the rectangle's area is

$$\text{rectangle's area} = b \cdot h = 1 \cdot 3 = 3$$

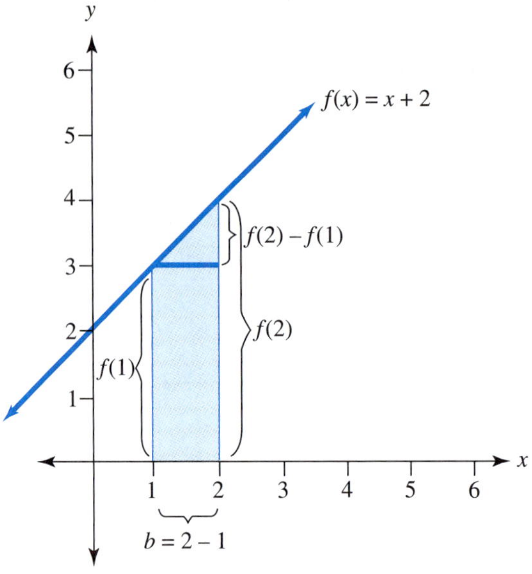

Figure 10.75

The triangle's base is $2 - 1 = 1$. Its height is the length of the top part of the vertical line at $x = 2$; that is, its height is $f(2) - f(1) = 4 - 3 = 1$. Thus, the triangle's area is

$$\text{triangle's area} = \frac{1}{2} \cdot b \cdot h = \frac{1}{2} \cdot 1 \cdot 1 = \frac{1}{2}$$

The area of our region is then

$$\text{rectangle's area} + \text{triangle's area} = 3 + \frac{1}{2} = 3\frac{1}{2}$$

In the following example, $f(x)$ is a curve rather than a line, and we would be unable to find the area without calculus.

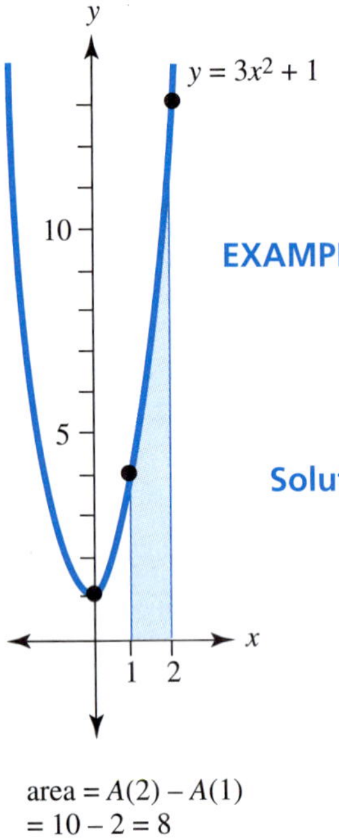

Figure 10.76

EXAMPLE 6 Find the area of the region bounded above by the curve $f(x) = 3x^2 + 1$, below by the x-axis, and on the sides by

a. a vertical line at $x = 1$ and a vertical line at $x = 2$
b. the y-axis and a vertical line at $x = 2$

Solution **a.** The curve is a parabola; the region is illustrated in Figure 10.76. To find the area, we must first find the area function $A(x)$ associated with $f(x) = 3x^2 + 1$. This antiderivative is $x^3 + x + c$:

$$A(x) = \int 3x^2 + 1\, dx = x^3 + x + c$$

Our region has a right-hand boundary at $x = 2$ and a left-hand boundary at $x = 1$, so we want $A(2) - A(1)$, as shown in Figure 10.77.

$$A(x) = x^3 + x + c$$
$$A(2) - A(1) = (2^3 + 2 + c) - (1^3 + 1 + c) = (10 + c) - (2 + c) = 8$$

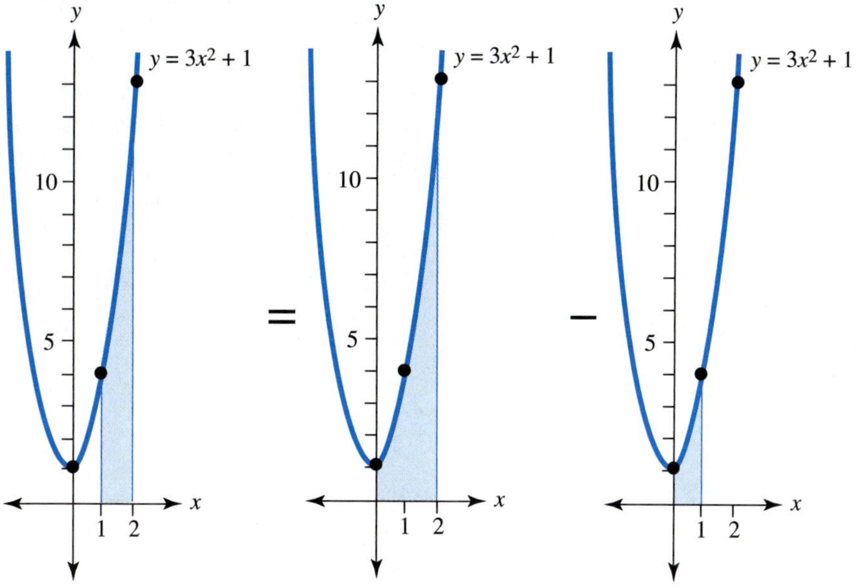

Figure 10.77

area = $A(2) - A(1)$ area = $A(2) = 10 + c$ area = $A(1) = 2 + c$

b. Our region has a right-hand boundary at $x = 2$ and a left-hand boundary at the y-axis, as shown in Figure 10.78. Its area is $A(2)$:

$$A(x) = x^3 + x + c$$
$$A(2) = 2^3 + 2 + c = 10 + c$$

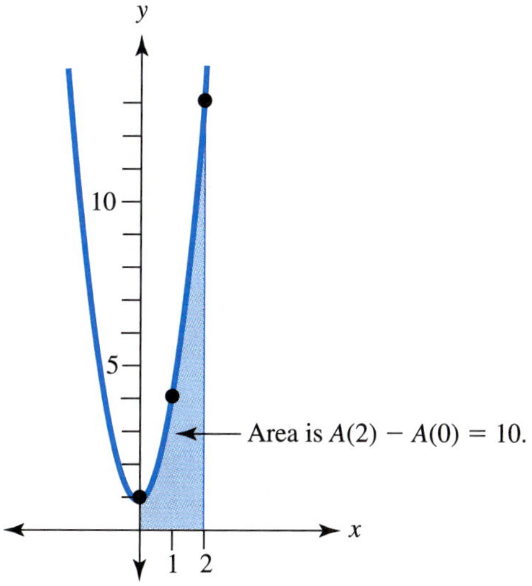

Figure 10.78

This doesn't do us any good because *we don't know what c is*! Rather than trying to find the value of c, we realize that we can view our left-hand boundary as the *y axis* or as a line at $x = 0$, so we want $A(2) - A(0)$:

$$\begin{aligned} A(x) &= x^3 + x + c \\ A(2) - A(0) &= (2^3 + 2 + c) - (0^3 + 0 + c) \\ &= 8 + 2 + c - c \\ &= 10 \end{aligned}$$

■

Area Steps

To find the area of a region bounded below by the x-axis, above by the curve $y = f(x)$, and on the left and right by vertical lines at a and at b:

1. Find $A(x) = \int f(x)\,dx$, the antiderivative of $f(x)$.
2. Find $A(b) - A(a) = A(\text{right boundary}) - A(\text{left boundary})$.

10.6 Exercises

In Exercises 1–8, do the following.

a. Find the area function associated with the given function.

b. Sketch the region whose area is found by that area function.

c. Find A(3), and describe the region whose area is A(3).

d. Find A(4), and describe the region whose area is A(4).

e. Find A(4) − A(3), and describe the region whose area is A(4) − A(3).

1. $f(x) = 2$

2. $f(x) = 6$

3. $f(x) = 2x$

4. $f(x) = 4x$

5. $f(x) = x + 2$

6. $f(x) = x + 1$

7. $f(x) = 2x + 3$

8. $f(x) = 3x + 1$

In Exercises 9–12, find the antiderivative of the given function.

9. $f(x) = 5$

10. $f(x) = 9$

11. $f(x) = 6x + 4$

12. $f(x) = 10x - 13$

In Exercises 13–16,

a. Use the results of Exercises 9–12 to find the area of a region bounded below by the x-axis, above by the line y 5 f(x), and on each side by the given vertical lines.

b. Use triangles and rectangles to find the area of the same region.

13. $f(x) = 5$, vertical lines at 2 and 7.

14. $f(x) = 9$, vertical lines at 3 and 11.

15. $f(x) = 6x + 4$, vertical lines at 0 and 5.

16. $f(x) = 10x - 13$, vertical lines at 4 and 20.

In Exercises 17–20, find the antiderivative of the given function.

17. $g(x) = 8x + 7$

18. $g(x) = 9x - 3$

19. $g(x) = 3x^2 - 2x + 5$

20. $g(x) = 4x^2 + 2x + 1$

In Exercises 21–24, use the results of Exercises 17–20 to find the area of a region bounded below by the x-axis, above by the curve y = g(x), and on each side by the given vertical lines.

21. $g(x) = 8x + 7$, vertical lines at 8 and 11.

22. $g(x) = 9x - 3$, vertical lines at 2 and 5.

23. $g(x) = 3x^2 - 2x + 5$, vertical lines at 0 and 7.

24. $g(x) = 4x^2 + 2x + 1$, vertical lines at 1 and 12.

In Exercises 25–32, find (a) the derivative of the given function, and (b) the antiderivative of the given function.

25. $f(x) = x$

26. $g(x) = 9$

27. $h(x) = 5$

28. $k(x) = 2x$

29. $m(x) = 8x$

30. $n(x) = 9x^8 - 7$

31. $p(x) = 3x^5 + 4$

32. $q(x) = 4x^3 - 3x$

In Exercises 33–39, find the area of the region bounded below by the x-axis, above by the given function, and on each side by the given vertical lines.

33. $f(x) = 3x^2$, vertical lines at

a. 1 and 4

b. 3 and 100

34. $f(x) = 3x^2 + 5$, vertical lines at
a. 0 and 6
b. 2 and 12

35. $f(x) = 6x^2 + 4$, vertical lines at
a. 2 and 6
b. 5 and 17

36. $f(x) = 9x^2 + 7$, vertical lines at
a. 1 and 2
b. 3 and 20

37. $f(x) = 5x^6 + 3$, vertical lines at
a. 4 and 7
b. 3 and 10

38. $f(x) = 4x^7 + 2$, vertical lines at
a. 3 and 11
b. 4 and 10

39. $f(x) = 3x^4 - \frac{2}{3}x^3 - 5x + 11$,
a. vertical lines at 2 and 19
b. the y-axis and a vertical line at 5

40. **a.** Use the procedure from Example 3 to determine a formula for $\int x^n \, dx$.

HINT: Follow the steps used in Example 3, except use an n in place of 3.

b. Check your answer to part a by taking its derivative.
c. Use the procedure from Example 3 to determine a formula for $\int ax^n \, dx$.
d. Check your answer to part c by taking its derivative.

41. Use the formula developed in Exercise 40 to find the antiderivative of the function in Exercise 17.

42. Use the formula developed in Exercise 40 to find the antiderivative of the function in Exercise 18.

43. Use the formula developed in Exercise 40 to find the antiderivative of the function in Exercise 19.

44. Use the formula developed in Exercise 40 to find the antiderivative of the function in Exercise 20.

45. Use the formula developed in Exercise 40 to find the antiderivative of $f(x) = 5x^6 - 3x^2 + 13$.

46. Use the formula developed in Exercise 40 to find the antiderivative of $g(x) = 9x^7 + 13x^4 + 7$.

47. Use the formula developed in Exercise 40 to find the antiderivative of $h(x) = 18/x^2$.

HINT: Use exponent rules to rewrite the function without a quotient.

48. Use the formula developed in Exercise 40 to find the antiderivative of $k(x) = \frac{5}{x} + 3$.

Answer the following question using complete sentences and your own words.

Concept Question

49. Of Exercises 21–24, which could be done with triangles and rectangles rather than antiderivatives? Which could not be done with triangles and rectangles? Why?

10.7 Conclusion

We have briefly investigated several of the more common applications of calculus. Each of the applications we explored represents only the tip of an iceberg.

Rates of Change

Newton's method of fluxions and evanescent increments was motivated in part by the problem of finding the speed of a falling object. One use of calculus is to determine the rate at which things change, and "speed" is the rate at which distance changes.

Calculus can be used to determine many other rates of change, including the rate at which a population grows ("growth rate"), the rate at which a manufacturing firm's profit changes as its production increases ("marginal profit"), and the rate at which water flows through a river ("flow rate").

Sketching Curves

Newton's method of fluxions was also motivated by the problem of finding the slope of a tangent line. The ability to find this slope leads to a powerful method of sketching curves that involves plotting certain key points.

In Section 10.5, we sketched parabolas to answer questions concerning the trajectory of a cannonball. We did this by plotting key points that were important in analyzing the trajectory. In particular, we plotted the points at which the curve intersects the x- and y-axes and the vertex, which we found by determining where the tangent line has a slope of zero (see Figure 10.79).

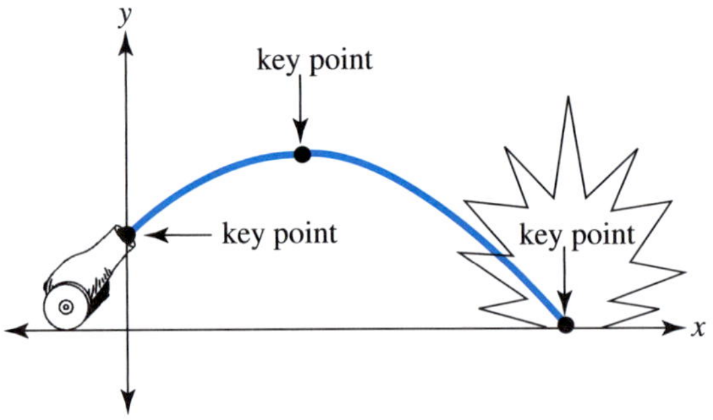

Figure 10.79

By expanding on this procedure, we can sketch the graph of almost any curve—for example, the one in Figure 10.80. This is an important skill in many fields, including business and economics, in which graphs are sketched to analyze a firm's profit and to analyze the relationship between the sales price of a product and its success in the marketplace.

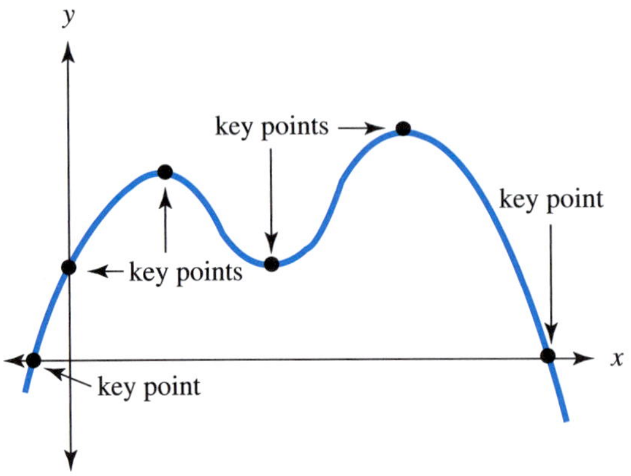

Figure 10.80

EXAMPLE 1 **a.** Find where the graph of $f(x) = 2x^3 + 3x^2 - 12x$ intersects the y-axis by substituting 0 for x.

b. Find where the graph of $f(x) = 2x^3 + 3x^2 - 12x$ intersects the x-axis by substituting 0 for $f(x)$ and solving for x.

c. Find where the graph of $f(x) = 2x^3 + 3x^2 - 12x$ has a vertex by determining where the tangent line has a slope of zero.

d. Use the results of parts (a), (b), and (c) to sketch the graph of $f(x) = 2x^3 + 3x^2 - 12x$.

Solution

a. $f(x) = 2x^3 + 3x^2 - 12x$

$f(0) = 2 \cdot 0^3 + 3 \cdot 0^2 - 12 \cdot 0 = 0$ substituting 0 for x

The graph of $f(x) = 2x^3 + 3x^2 - 12x$ intersects the y-axis at the point (0, 0), as shown in Figure 10.81 on page 790.

b.

$$f(x) = 2x^3 + 3x^2 - 12x$$

$$0 = 2x^3 + 3x^2 - 12x \qquad \text{substituting 0 for } y$$

$$0 = x(2x^2 + 3x - 12) \qquad \text{factoring}$$

$$\text{either } x = 0 \text{ or } 2x^2 + 3x - 12 = 0$$

$$\text{either } x = 0 \text{ or } x = \frac{-3 \pm \sqrt{3^2 - 4 \cdot 2 \cdot -12}}{2 \cdot 2} \qquad \text{using the quadratic formula as discussed in Section 10.5}$$

$$\text{either } x = 0 \text{ or } x = 1.81173769149\ldots \text{ or } x = -3.31173769149\ldots$$

$$\text{either } x = 0 \text{ or } x \approx 1.8 \text{ or } x \approx -3.3$$

The graph of $f(x) = 2x^3 + 3x^2 - 12x$ intersects the x-axis at the points (0, 0), (1.8, 0), and (−3.3, 0) (approximately), as shown in Figure 10.81 on page 790.

c. Rather than taking the derivative of $f(x)$ with Newton's or Cauchy's method, we can use a shortcut. Recall from Exercise 38 of Section 10.4 that if $f(x) = ax^3 + bx^2 + cx + d$, then $df/dx = 3ax^2 + 2bx + c$.

If $f(x) = ax^3 + bx^2 + cx + d$ then $df/dx = 3ax^2 + 2bx + c$

If $f(x) = 2x^3 + 3x^2 - 12x + 0$ then $df/dx = 3(2)x^2 + 2(3)x - 12$

By comparing the formula with our problem, we can see that the "a" in our formula matches "2" in our problem, the "b" matches "3", the "c" matches "−12," and the "d" matches "0." Thus, the derivative of $f(x) = 2x^3 + 3x^2 - 12x$ is $df/dx = 6x^2 + 6x - 12$.

The graph of $f(x) = 2x^3 + 3x^2 - 12x$ has a vertex where the tangent line has a slope of zero.

$$\text{slope of tangent} = 0$$

$$df/dx = 6x^2 + 6x - 12 = 0$$

$$x = \frac{-6 \pm \sqrt{6^2 - 4 \cdot 6 \cdot -12}}{2 \cdot 6} \qquad \text{using the quadratic formula as discussed in Section 10.5}$$

$$= 1 \text{ or } -2$$

The graph of $f(x) = 2x^3 + 3x^2 - 12x$ has two vertices: one at $x = 1$ and a second at $x = -2$. And since $f(1) = 2(1)^3 + 3(1)^2 - 12(1) = -7$ and $f(-2) = 2(-2)^3 + 3(-2)^2 - 12(-2) = 20$, the vertices are at the points (1, −7) and (−2, 20), as shown in Figure 10.81 on page 790.

d. The graph of $f(x) = 2x^3 + 3x^2 - 12x$ is shown in Figure 10.81 on page 790.

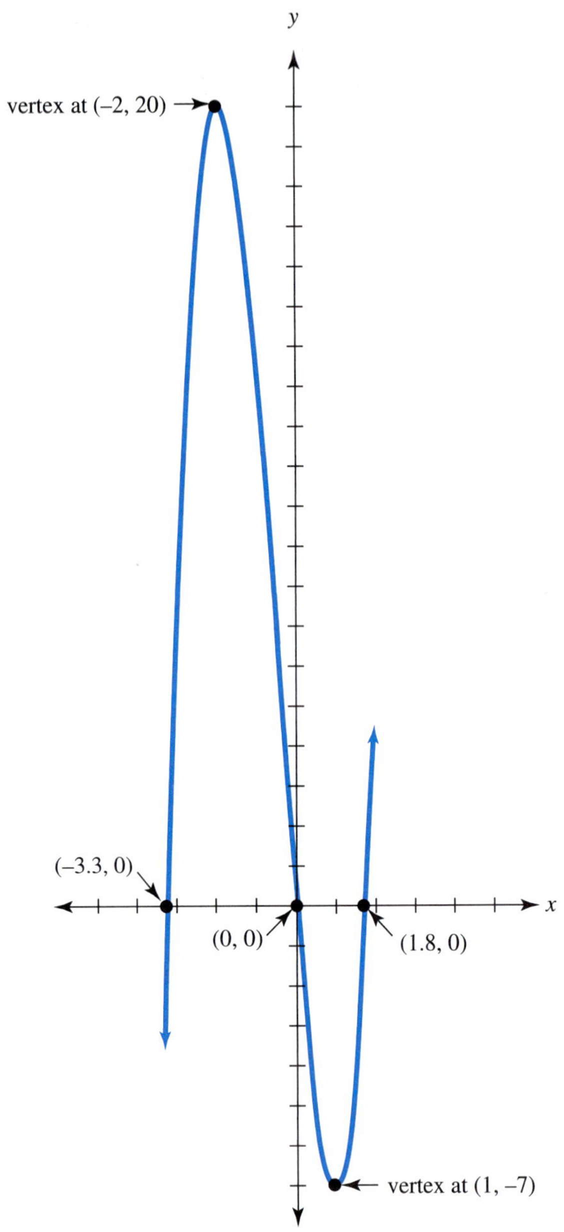

Figure 10.81
The graph of $f(x) = 2x^2 + 3x^2 - 12x$

Finding Maximum and Minimum Values

In Section 10.5, we found the maximum height attained by a cannonball. In many fields, finding where a function has its maximum or minimum value is important. In business, it's important to determine the circumstances under which a company's profit will be maximized and its costs minimized. In engineering, it is important to determine the maximum weight a structure can support.

The basic principle involved in finding the maximum or minimum value of a function is to find where the tangent line has a slope of zero, as we did when we found the maximum height attained by a cannonball. When we found that maximum height, we knew that the graph was that of a parabola; in most situations, the shape of the curve is not known, and it will be necessary to do more work than simply finding where the tangent line has a slope of zero.

Motion

Newton's method of fluxions was motivated in part by the problem of finding the path followed by a cannonball as well as by the problem of finding the paths followed by the planets. Calculus can be used to analyze any type of motion, including that of the space shuttle as it leaves the earth and goes into orbit and that of an automobile negotiating a tight curve. Such analyses are important in designing the space shuttle or the automobile in order to develop a vehicle that can withstand the pressures that will be exerted on it and perform as required.

Areas and Volumes

In Section 10.6, we found the areas of shapes bounded by one curve and some straight lines. We did not explore the techniques involved in finding areas of more complicated shapes. If the region is bounded by more than one function, you compute the antiderivative of each function.

In Figure 10.82, the area of the region is the sum of the area of the left side and the area of the right side. The left side's area is $A(3) - A(1)$, where $A(x) = \int f(x)\,dx$. The right side's area is $A(5) - A(3)$, where $A(x) = \int g(x)\,dx$.

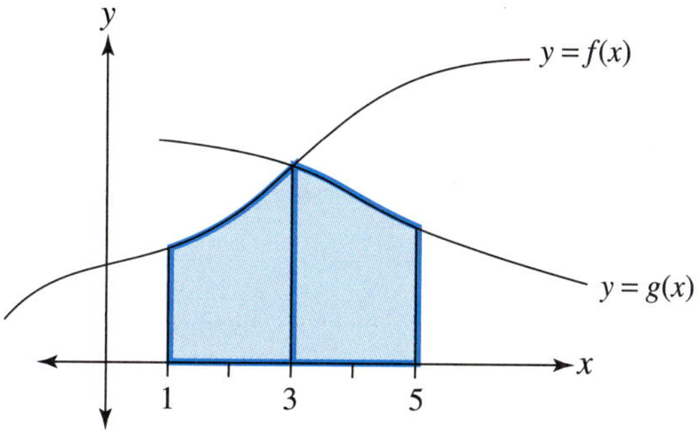

Figure 10.82

In Figure 10.83, the area of the region is the area of the top part minus the area of the bottom part. The top part's area is $A(8) - A(2)$, where $A(x) = \int f(x)\,dx$. The bottom part's area is $A(8) - A(2)$, where $A(x) = \int g(x)\,dx$.

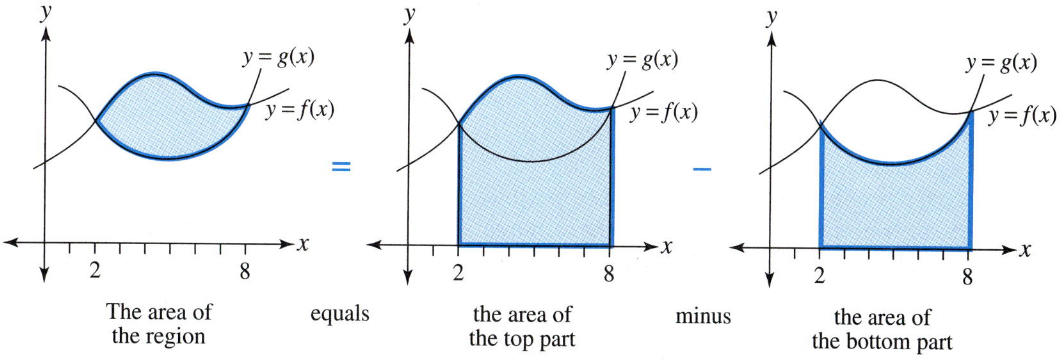

Figure 10.83

Summing Up

The point of this chapter is to explore concepts, not to develop a proficiency in calculus. We have investigated the main ideas of this subject, but of course there is much more to calculus than what we have seen here.

10.7 Exercises

Projects

1. The distance traveled by a falling object is $d = 16t^2$, if the distance d is in feet. Only the United States uses feet to measure distance; the rest of the world uses meters and the metric system. If the distance is measured in meters rather than feet, the formula is almost the same; the only difference is the multiplier is not 16. That is, the distance traveled by a falling object is $d = k \cdot t^2$, where k is some number other than 16, if the distance d is in meters. In this exercise we will find k.

a. Complete the following chart, using dimensional analysis (see Appendix E) and the fact that 1 meter ≈ 3.28 feet.

Time in Seconds	Distance Traveled in Feet	Distance Traveled in Meters
1 sec	$16 \cdot 1^2 = 16$ ft	(convert 16 feet to meters)
2 sec		
3 sec		

b. In the equation $d = k \cdot t^2$, substitute 1 for t and the appropriate distance in meters (from part a) for d, and solve for k.

c. In the equation $d = k \cdot t^2$, substitute 2 for t and the appropriate distance in meters (from part a) for d, and solve for k.

d. In the equation $d = k \cdot t^2$, substitute 3 for t and the appropriate distance in meters (from part a) for d, and solve for k.

e. Use the results of parts (b), (c), and (d) to find the metric system version of the formula $d = 16t^2$.

f. Use derivatives and the result of part (e) to find the metric system equation for the speed s of a falling object. What units would s be in?

2. a. Find where the graph of $f(x) = -x^4 + 8x^2$ intersects the y-axis by substituting 0 for x.

b. Find where the graph of $f(x) = -x^4 + 8x^2$ intersects the x-axis by substituting 0 for $f(x)$ and solving for x.

c. Find where the graph of $f(x) = -x^4 + 8x^2$ has a vertex by determining where the tangent line has a slope of zero.

d. Use the results of parts (a), (b), and (c) to sketch the graph of $f(x) = -x^4 + 8x^2$.

e. By inspecting the graph from part (d), determine the maximum value of $f(x)$.

In Exercises 3–6, use antiderivatives to find the indicated area.

3.

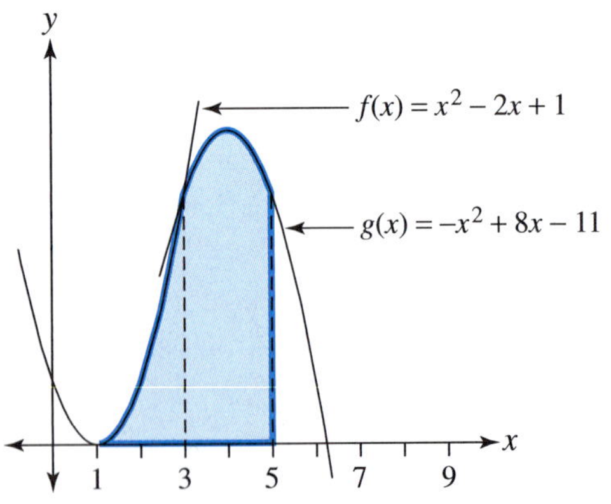

a. Find the area of the left side.

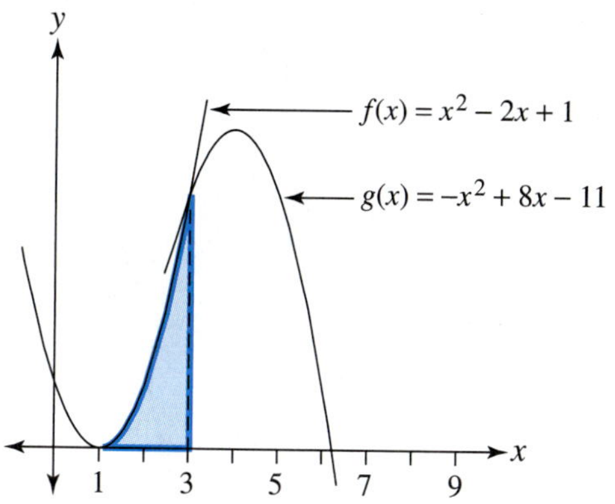

b. Find the area of the right side.

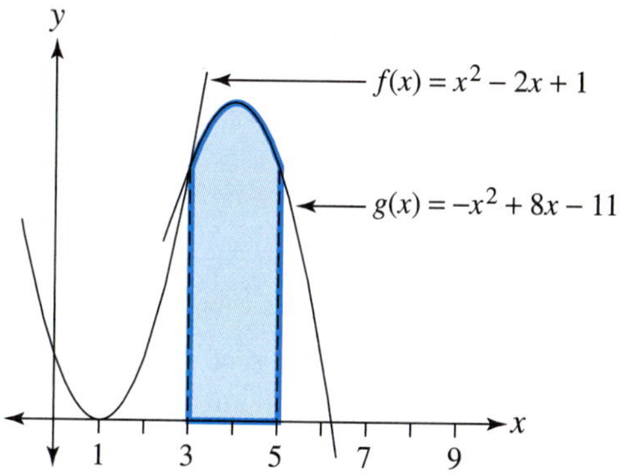

c. Find the area of the entire region by combining the answers to parts (a) and (b).

4.

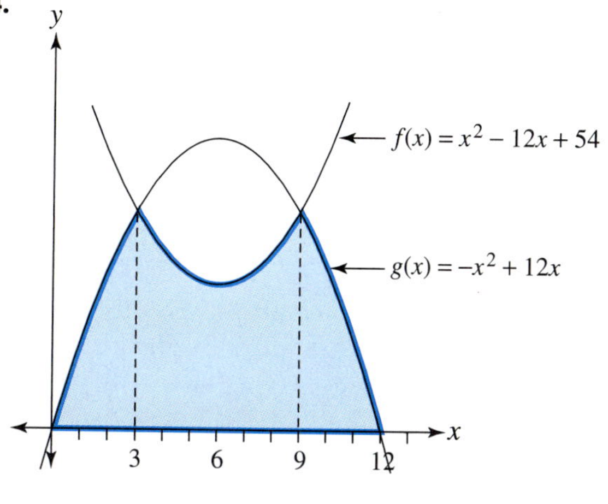

a. Find the area of the left part.
b. Find the area of the middle part.
c. Find the area of the right part.
d. Find the area of the entire region by combining the answers to parts (a), (b), and (c).

5.

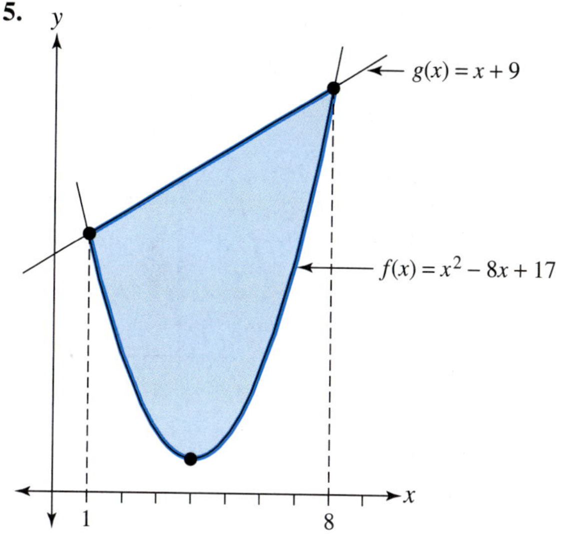

a. Find the area of the top part.

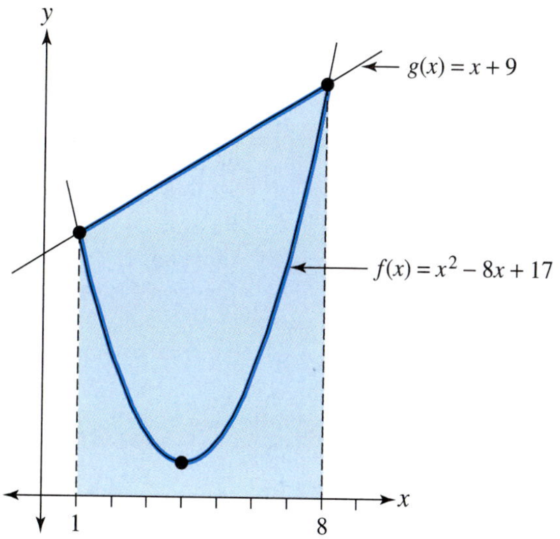

b. Find the area of the bottom part.

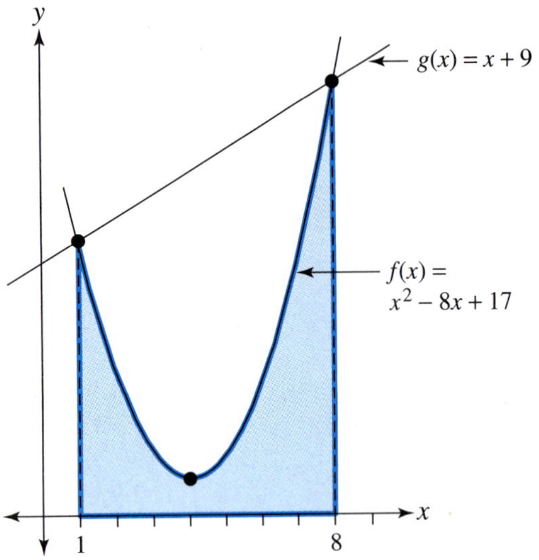

c. Find the area of the entire region by combining the answers to parts (a) and (b).

6.

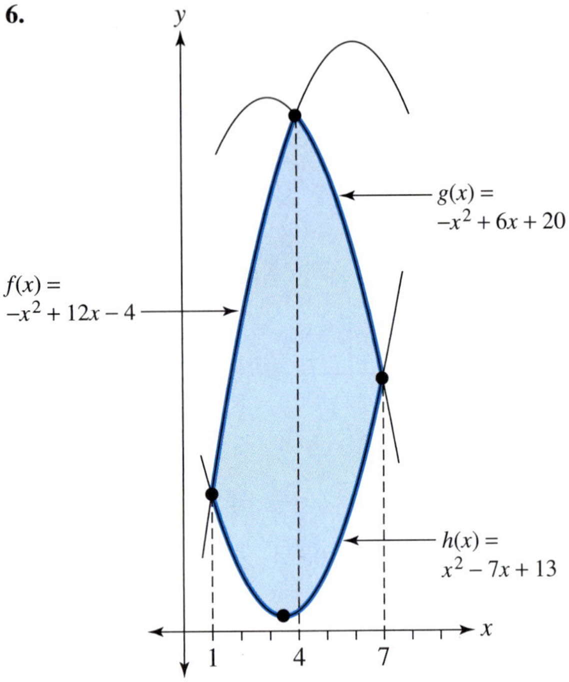

Chapter 10 Review

TERMS

analytic geometry
antiderivative
average versus instantaneous rate of change
average versus instantaneous speed
conic sections
delta notation
derivative
evanescent increment
fluxion
function
functional notation
infinitesimal
line of symmetry of a parabola
proportion
rate
rate of change
ratio
secant line
similar triangles
tangent line
trajectory
vertex of a parabola

REVIEW EXERCISES

1. What role did the following people play in the development of calculus?
- Apollonius of Perga
- Isaac Barrow
- George Berkeley
- Augustin Louis Cauchy
- René Descartes
- Pierre de Fermat
- Galileo Galilei
- Mohammed ibn Musa al-Khowarizmi
- Gottfried Leibniz
- Isaac Newton
- Nicole Oresme
- François Viète

2. Find the unknown lengths in Figure 10.84.

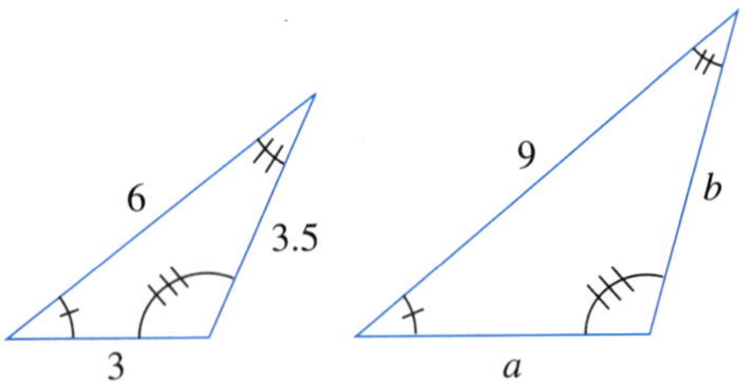

Figure 10.84

3. When Jorge Gonzalez filled his car's tank with gas, the odometer read 4,215.6 miles. Later, he added 8.2 gallons, and the odometer read 4,465.1 miles.

a. Find the ratio of miles traveled to gallons consumed.

b. Predict the number of miles that he could travel on a full tank if his tank holds 15 gallons.

4. $f(x) = 8x^2 - 7$

a. Find $f(-3)$

b. Find $f(x + \Delta x)$.

5. An object is moving in such a way that its speed increases uniformly from the moment that motion begins. After 30 seconds, the object is moving at a rate of 8 feet per second. Use Oresme's triangle to find the following:

a. the distance it has traveled in those 30 seconds

b. its speed after 25 seconds

c. the distance it has traveled in 25 seconds

d. its average speed during its first 25 seconds of travel

6. Using a water clock to time a ball rolling down a ramp, Galileo found that if a ball was timed during two different trips, the following relationship held:

$$\frac{\text{distance from first trip}}{\text{distance from second trip}} = \frac{(\text{weight of water from first trip})^2}{(\text{weight of water from second trip})^2}$$

If a ball would roll 100 feet while 5 ounces of water flowed out of a water clock, how far would it roll while 10 ounces of water flowed from the clock?

7. Given $y = 2x^2$, do the following.

a. Sketch the parabola given by the equation.

b. At $x = 1$ and $x = 2$, compute the length of the line segment and the area of the square described by Apollonius. Sketch these two line segments and two squares in their proper places, superimposed over the parabola.

c. Compute the ratio of the lengths of the two line segments and the ratio of the areas of the two squares in part (b).

d. At $x = 3$ and $x = 4$, compute the length of the line segment and the area of the square described by Apollonius. Sketch these two line segments and two squares in their proper places, superimposed over the parabola.

e. Compute the ratio of the lengths of the two line segments and the ratio of the areas of the two squares in part (d).

8. A square and 10 roots are equal to 11 units.

a. Solve the problem using the method of al-Khowarizmi. Do not use modern terminology and notation.

b. Illustrate your solution to the problem in the manner of al-Khowarizmi.

c. Give a modern version of your solution from part (a).

9. Given $y = x^2 + 4x + 11$, do the following.

a. Graph the parabola.

b. Give the equation of its line of symmetry.

c. Find its vertex.

d. Find the slope of the secant line that intersects the parabola at $x = 1$ and $x = 2$.

e. Find the slope of the tangent line at $x = 1$.

f. Find the equation of the tangent line at $x = 1$.

g. Graph the secant line from part (d) and the tangent line from parts (e) and (f).

10. Calculate the slope of the line tangent to $y = 3x^2 - 1$ at the point where $x = 2$, using the following methods. Find the equation of the tangent line and sketch the graph of the curve and the tangent line.

a. Barrow's method

b. Newton's method

c. Cauchy's method

11. You drop a rock off a bridge. After 2.5 seconds, you hear it hit the water below.

a. How far does the rock fall?

b. At what speed is the rock moving when it hits the water?

c. What is the average speed of the rock during its fall?

In Exercises 12–14, find the derivative of the given function, by calculating

$$\frac{df}{dx} = \lim_{\Delta x \to 0} \frac{f(x + \Delta x) - f(x)}{\Delta x}$$

12. $f(x) = 3$

13. $f(x) = 2x - 15$

14. $f(x) = x^2 - 3x + 5$

15. Craig James is dieting. If $f(x)$ measures Craig's weight x weeks after he started to diet, find what is measured by each of the following. (Include units with each answer.)

a. $\dfrac{\Delta f}{\Delta x} = \dfrac{f(x + \Delta x) - f(x)}{\Delta x}$

b. $\dfrac{df}{dx} = \lim_{\Delta x \to 0} \dfrac{f(x + \Delta x) - f(x)}{\Delta x}$

16. In October 1965, the state Department of Fish and Game started to keep records concerning the number of bass in Ocean Bay. A biologist has used these records to create a function $f(x)$ that measures the number of bass in the bay x weeks after October 1, 1965. Find what is measured by each of the following. (Include units with each answer.)

a. $\dfrac{\Delta f}{\Delta x} = \dfrac{f(x + \Delta x) - f(x)}{\Delta x}$

b. $\dfrac{df}{dx} = \lim_{\Delta x \to 0} \dfrac{f(x + \Delta x) - f(x)}{\Delta x}$

17. A cannonball is shot with an initial speed of 200 feet per second from a cannon whose muzzle is 4.5 feet above the ground at a 30° angle.
    a. Find the equation of the path of the cannonball.
    b. How far will the ball travel?
    c. What is the maximum height the ball will attain?
    d. Will the ball clear a 45-foot-high wall that is 100 feet from the cannon?
    e. Sketch the cannonball's path. In your sketch, include everything that is known about the path.
18. Find the antiderivative of $f(x) = 3x - 2$.
19. Find the area of a region bounded below by the x-axis, above by the curve $y = 3x^2 + 7$, and on the sides by vertical lines at 1 and 5.

Answer the following questions using complete sentences and your own words.

20. What were the four problems that led to the invention of calculus? Who worked on these problems prior to Newton? What did they discover?
21. How did Galileo use Oresme's triangle to investigate the distance traveled by a falling object? How did he use it to investigate the speed of a falling object?
22. Why was Galileo unable to find the equation of the trajectory of a cannonball?
23. What did Apollonius contribute to analytic geometry?
24. What did Oresme contribute to analytic geometry?
25. What did Descartes contribute to analytic geometry?
26. What did Fermat contribute to analytic geometry?
27. What did Viète contribute to algebra?
28. What did Descartes contribute to algebra?
29. What changes did mathematics need to undergo before calculus could be invented? Why were these changes necessary?
30. Why didn't Newton initially publish his works on calculus? What finally prompted him to publish them?

A Using a Scientific Calculator

What Kind of Calculator Do You Have?

There are three kinds of calculators: graphing calculators, scientific calculators, and basic calculators. Graphing calculators have bigger screens, which they use to display graphs. If you have one of these, read Appendix B.

Scientific calculators and basic calculators have smaller screens; they don't graph, so they don't need the extra space on the screen. If you don't have a graphing calculator, you need to determine whether you have a scientific calculator or a basic calculator because a basic calculator is insufficient for this text. Here's how you can tell. Type

2 [+] 3 [×] 4 [=]

If you get 14, you probably have a scientific calculator. Read this appendix to learn how to use it. If you get 20, you have a basic calculator, which is insufficient for this text.

The following discussion is meant to apply to any scientific calculator. Read it with your calculator (and its instruction booklet, if you have it) by your side. When a calculation is discussed, do that calculation on your calculator.

Some scientific calculators work like graphing calculators, except they don't graph (and thus they don't have bigger screens). If the instructions in this appendix don't always work with your calculator, try the instructions in Appendix B.

Unfortunately, scientific calculators don't all work exactly the same way. Even if you do everything correctly, your answer may have a few more or less decimal places than the one given, or may differ in the last decimal place. Occasionally, the way a calculation is performed on your calculator will differ slightly from that discussed below. If so, consult your instruction booklet, experiment a little, or consult your instructor. To paraphrase Abraham Lincoln, the following discussion applies to most scientific calculators all of the time and to all scientific calculators most of the time, but it doesn't apply to all scientific calculators all of the time.

The Equals Button

Some buttons perform operations on a pair of numbers. For example, the [+] button is used to add two numbers; it only makes sense to use this button in conjunction with a pair of numbers. *When using such a button, you must finish the typing with the* [=] *button.* That is, to add 3 and 2, type

3 [+] 2 [=]

Some buttons perform operations on a single number. For example, the [x^2] button is used to square a number; it only makes sense to use this button in conjunction with a single number. *When using such a button, you do **not** type the* [=] *button*. That is, to square 5, type

5 [x^2]

The reason for this distinction is that the calculator must be told when you are done entering information and you are ready for it to compute. If you typed

3 [+] 2

the calculator would have no way of knowing if you were ready for it to compute "3 + 2" or if you were in the middle of instructing it to compute "3 + 22." You must follow "3 + 2" with the [=] button to tell the calculator that you are done entering information and are ready for it to compute. On the other hand, when you type

5 [x^2]

the calculator knows that you are done entering information; if you meant to square 53, you would not have pressed the [x^2] button after the "5." There is no need to follow "5 [x^2]" with the [=] button.

The Clear Button

Most calculators have two "clear" buttons (buttons that erase): one that clears just the last number you typed and one that clears everything. Frequently, the button that clears just the last number you typed is labeled [CE] (for "clear entry") and the one that clears everything is labeled [CA] (for "clear all"). These buttons are used when you err in typing. For example, if you typed "5 × 2" and then realized you wanted "5 × 3," you could clear the incorrect "2" with the [CE] button by typing

5 [×] 2 [CE] 3 [=]

This would make the display read "15," since the [CE] button clears just the last entry (the "2"), and 5 × 3 = 15. However, if you typed

5 [×] 2 [CA] 3 [=]

the display would read "3," since the [CA] button clears everything that came before it (the "5 × 2").

Some calculators label their "clear entry" button [CE] and their "clear all" button [C]. Some calculators label their "clear entry" button [C] and their "clear all" button [CA]. Some calculators have a button labeled [C] that functions as a "clear entry" button if you press it once and as a "clear all" button if you press it twice. Some experimentation with your clear button(s) and the above two examples will show you how yours works.

The Subtraction Symbol and the Negative Symbol

If "3 − −2" were read aloud, you could say "three subtract negative two" or "three minus minus two," and you would be understood. The expression is understandable even if the distinction between the negative symbol and the subtraction symbol is not made clear. With a calculator, however, this distinction is crucial. The subtraction button is labeled [−]. There is no negative button; instead, there is a [+/−] button that changes the sign of whatever number is on the display. Typing

5 [+/−]

makes the display read "−5." Typing

5 [+/−] [+/−]

makes the display read "5," because two sign changes undo each other. Typing

[+/−] 5

makes the display read "5," not "−5." *You must press the sign-change button after the number itself.*

EXAMPLE 1 Calculate $3 - -2$, both (a) by hand and (b) with a calculator.

Solution

a. $3 - -2 = 3 + 2 = 5$

b. To do this, we must use the subtraction button, which is labeled [−], and the sign-change button, which is labeled [+/−]. Typing

3 [−] 2 [+/−] [=]

makes the display read "5." Remember that you must press the sign-change button *after* the number itself; you press "2" followed by [+/−], not [+/−] followed by "2." Also, you must finish the typing with the [=] button, because the [=] button performs an operation on a pair of numbers. ■

Order of Operations and Use of Parentheses

Scientific calculators are programmed so that they follow the standard order of operations. That is, they perform calculations in the following order:

1. **P**arentheses-enclosed work
2. **E**xponents
3. **M**ultiplication and
   Division, from left to right
4. **A**ddition and
   Subtraction, from left to right

This order can be remembered by remembering the word *PEMDAS*, which stands for

Parentheses/**E**xponents/**M**ultiplication/**D**ivision/**A**ddition/**S**ubtraction

Frequently, the fact that calculators are programmed to follow the order of operations means that you perform a calculation on your calculator in exactly the same way that it is written.

EXAMPLE 2 Calculate $2(3 + 4)$, both (a) by hand and (b) with a calculator.

Solution

a. $2(3 + 4) = 2(7)$ parentheses-enclosed work comes first

$= 14$

b. Type

2 [×] [(] 3 [+] 4 [)] [=]

and the display reads "14." Notice that this is typed on a calculator exactly as it is written, with two important exceptions:

- We must press the [×] button to mean multiplication; we cannot use parentheses to mean multiplication, as is done in part (a).
- We must finish the typing with the [=] button. ■

EXAMPLE 3 Calculate $2 \cdot 3^2$, both (a) by hand and (b) with a calculator.

Solution **a.** $2 \cdot 3^2 = 2 \cdot 9$ exponents come before multiplication
$= 18$

b. Type

2 $\boxed{\times}$ 3 $\boxed{x^2}$ $\boxed{=}$

and the display reads "18." Notice that this is typed on a calculator exactly as it is written, except that we must finish the typing with the $\boxed{=}$ button and we must press the $\boxed{\times}$ button to mean multiplication. The $\boxed{\bullet}$ button is a decimal-point button, not a multiplication-dot button. ■

EXAMPLE 4 Calculate $(2 \cdot 3)^2$, both (a) by hand and (b) with a calculator.

Solution **a.** $(2 \cdot 3)^2 = 6^2$ parentheses-enclosed work comes first
$= 36$

b. Type

$\boxed{(}$ 2 $\boxed{\times}$ 3 $\boxed{)}$ $\boxed{x^2}$

and the display reads "36." Notice that this is typed on a calculator exactly as it is written, except that we must use the $\boxed{\times}$ button to mean multiplication. Also, we do not use the $\boxed{=}$ button, because the $\boxed{x^2}$ button performs an operation on a single number. ■

EXAMPLE 5 Calculate $4 \cdot 2^3$, both (a) by hand and (b) with a calculator.

Solution **a.** $4 \cdot 2^3 = 4 \cdot 8$ exponents come before multiplication
$= 32$

b. To do this, we must use the exponent button, which is labeled either $\boxed{y^x}$ or $\boxed{x^y}$. Type

4 $\boxed{\times}$ 2 $\boxed{y^x}$ 3 $\boxed{=}$ or 4 $\boxed{\times}$ 2 $\boxed{x^y}$ 3 $\boxed{=}$

and the display reads "32." This is typed on a calculator exactly as it is written, except that we must press the $\boxed{\times}$ button to mean multiplication and the $\boxed{=}$ button to finish the calculation. ■

Sometimes, you don't perform a calculation on your calculator in the same way that it is written, even though calculators are programmed to follow the order of operations.

EXAMPLE 6 Calculate $\frac{2}{3 \cdot 4}$ with a calculator.

Solution **Wrong** It is incorrect to type

2 $\boxed{\div}$ 3 $\boxed{\times}$ 4 $\boxed{=}$

According to the order of operations, multiplication and division are done *from left to right*, so the above typing is algebraically equivalent to

$= \frac{2}{3} \cdot 4$ first dividing then multiplying, since division is on the left and multiplication is on the right

$= \frac{2}{3} \cdot \frac{4}{1} = \frac{2 \cdot 4}{3}$

which is not what we want. The difficulty is that the large fraction bar in the expression $\frac{2}{3 \cdot 4}$ groups the "3 · 4" together in the denominator; in the above typing, nothing groups the "3 · 4" together, and only the 3 ends up in the denominator.

Right The calculator needs parentheses inserted in the following manner:

$$\frac{2}{(3 \cdot 4)}$$

Thus, it is correct to type

2 [÷] [(] 3 [×] 4 [)] [=]

This makes the display read 0.166666667, the correct answer.

Also Right It is correct to type

2 [÷] 3 [÷] 4 [=]

According to the order of operations, multiplication and division are done from left to right, so the above typing is algebraically equivalent to

$$\frac{2}{3} \div 4 \quad \text{doing the left-hand division first}$$

$$= \frac{2}{3} \cdot \frac{1}{4} \quad \text{inverting and multiplying}$$

$$= \frac{2}{3 \cdot 4}$$

which is what we want. *When you're calculating something that involves only multiplication and division and you don't use parentheses, the* [×] *button places a factor in the numerator, and the* [÷] *button places a factor in the denominator.* ■

EXAMPLE 7 Calculate $\frac{2}{3/4}$ with a calculator.

Solution **Wrong** It is incorrect to type

2 [÷] 3 [÷] 4 [=]

even though that matches the way the problem is written algebraically. As discussed in Example 6, this typing is algebraically equivalent to

$$\frac{2}{3 \cdot 4}$$

which is not what we want.

Right The calculator needs parentheses inserted in the following manner:

$$\frac{2}{(3 \div 4)}$$

Thus, it is correct to type

2 [÷] [(] 3 [÷] 4 [)] [=]

since, according to the order of operations, parentheses-enclosed work is done first. This makes the display read 2.66666667, the correct answer. ■

EXAMPLE 8 Calculate $\dfrac{2+3}{4}$ with a calculator.

Solution **Wrong** It is incorrect to type

$$2 \ \boxed{+} \ 3 \ \boxed{\div} \ 4 \ \boxed{=}$$

According to the order of operations, division is done before addition, so this typing is algebraically equivalent to

$$2 + \frac{3}{4}$$

which is not what we want. The large fraction bar in the expression $\frac{2+3}{4}$ groups the "2 + 3" together in the numerator; in the above typing, nothing groups the "2 + 3" together, and only the 3 ends up in the numerator.

Right The calculator needs parentheses inserted in the following manner:

$$\frac{(2+3)}{4}$$

Thus, it is correct to type

$$\boxed{(} \ 2 \ \boxed{+} \ 3 \ \boxed{)} \ \boxed{\div} \ 4 \ \boxed{=}$$

This makes the display read 1.25, the correct answer.

Also Right It is correct to type

$$2 \ \boxed{+} \ 3 \ \boxed{=} \ \boxed{\div} \ 4 \ \boxed{=}$$

The first $\boxed{-}$ makes the calculator perform all prior calculations before continuing. This too makes the display read "1.25," the correct answer. ■

The Shift Button

A scientific calculator's function buttons each have two labels and two uses. For example, your calculator might have a button that is labeled $\boxed{x^2}$ on the button itself and "$\sqrt{x}$" above the button. If so, typing

$$4 \ \boxed{x^2}$$

makes the display read 16, since $4^2 = 16$. To take the square root of 4, you have to use the shift button. The shift button is so named because of its similarity to the shift key on a typewriter, which determines whether the letter typed is uppercase or lowercase. Your shift button may be labeled $\boxed{\text{SHIFT}}$, $\boxed{\text{2nd}}$, or $\boxed{\text{INV}}$. Typing

$$4 \ \boxed{\text{SHIFT}} \ \boxed{x^2} \qquad \text{or} \qquad 4 \ \boxed{\text{2nd}} \ \boxed{x^2} \qquad \text{or} \qquad 4 \ \boxed{\text{INV}} \ \boxed{x^2}$$

makes the display read 2, since $\sqrt{4} = 2$.

Some calculators have a button that is labeled $\boxed{\sqrt{x}}$ on the button itself and "x^2" above the button. If yours is like this, then typing

$$4 \ \boxed{\sqrt{x}}$$

makes the display read 2, and typing

$$4 \ \boxed{\text{INV}} \ \boxed{\sqrt{x}}$$

makes the display read 16. In either case, *the label on the button says what the button does without using the shift button, and the label above the button says what the button does using the shift button.*

Frequently, the two functions that share a function button are functions that "undo" each other. For example, typing

3 $\boxed{x^2}$

makes the display read 9, since $3^2 = 9$, and typing

9 $\boxed{\text{INV}}$ $\boxed{x^2}$

makes the display read 3, since $\sqrt{9} = 3$. This is done as a memory device; it is easier to find the various operations on the keyboard if the two operations that share a button also share a relationship.

Notice that this doesn't always work. For example, typing

3 $\boxed{+/-}$ $\boxed{x^2}$

makes the display read 9, since $(-3)^2 = 9$, but typing

9 $\boxed{\text{INV}}$ $\boxed{x^2}$

won't make the display read -3. Two functions that *always* undo each other are called **inverses.** The x^2 and $\sqrt{x}$ functions are not inverses, because of the above counterexample. However, there is an inverse-type relationship between the x^2 and $\sqrt{x}$ functions (they undo each other sometimes), and that is why the "shift" button is sometimes labeled $\boxed{\text{INV}}$. ■

Memory

The memory is a place to store a number for later use, without having to write it down. If a number is on your display, you can place it into the memory (or **store** it) by pressing the button labeled $\boxed{\text{STO}}$ or $\boxed{x \rightarrow \text{M}}$ or $\boxed{\text{M in}}$, and you can take it out of the memory (or **recall** it) by pressing the button labeled $\boxed{\text{RCL}}$ or $\boxed{\text{RM}}$ or $\boxed{\text{MR}}$. (WARNING: A button labeled $\boxed{\text{M+}}$ does *not* store; it adds the number on the display to whatever is currently stored.) In the above button labels, "M" stands for *memory* and "R" stands for *recall.*

Typing

5 $\boxed{\text{STO}}$ or 5 $\boxed{x \rightarrow \text{M}}$ or 5 $\boxed{\text{M in}}$

makes the calculator store a 5 in its memory. If you do other calculations or just clear your display and later press

$\boxed{\text{RCL}}$ or $\boxed{\text{RM}}$ or $\boxed{\text{MR}}$

then your display will read "5."

Some calculators have more than one memory. If yours does, then pressing the button labeled $\boxed{\text{STO}}$ won't do anything; pressing $\boxed{\text{STO}}$ and then "1" will store it in memory number 1; pressing $\boxed{\text{STO}}$ and then "2" will store it in memory number 2, and so on. Pressing $\boxed{\text{RCL}}$ and then "1" will recall what has been stored in memory number 1.

EXAMPLE 9 Use the quadratic formula and your calculator's memory to solve

$$2.3x^2 + 4.9x + 1.5 = 0$$

Solution The quadratic formula says that if $ax^2 + bx + c = 0$, then

$$x = \frac{-b \pm \sqrt{b^2 - 4ac}}{2a}$$

We have $2.3x^2 + 4.9x + 1.5 = 0$, so $a = 2.3$, $b = 4.9$, and $c = 1.5$. This gives

$$x = \frac{-4.9 \pm \sqrt{4.9^2 - 4 \cdot 2.3 \cdot 1.5}}{2 \cdot 2.3}$$

The quickest way to do this calculation is to calculate the radical, store it, and then calculate the two fractions.

Step 1 *Calculate the radical.* To do this, type

4.9 [x^2] [−] 4 [×] 2.3 [×] 1.5 [=] [$\sqrt{x}$] [STO]

This makes the display read "3.195309" and stores the number in the memory. Notice the use of the [=] button; this makes the calculator finish the prior calculation before taking a square root. If the [=] button were not used, the order of operations would require the calculator to take the square root of 1.5.

Step 2 *Calculate the first fraction.* To do this, type

4.9 [+/−] [+] [RCL] [=] [÷] 2 [÷] 2.3 [=]

This makes the display read "−0.3705849." Notice the use of the [=] button.

Step 3 *Calculate the second fraction.* To do this, type

4.9 [+/−] [−] [RCL] [=] [÷] 2 [÷] 2.3 [=]

This makes the display read "−1.7598498."

The solutions to $2.3x^2 + 4.9x + 1.5 = 0$ are $x = -0.3705849$ and $x = -1.7598498$. These are approximate solutions in that they show only the first seven decimal places.

Step 4 *Check your solutions.* These solutions can be checked by seeing if they satisfy the equation $2.3x^2 + 4.9x + 1.5 = 0$. To check the first solution, type

2.3 [×] .3705849 [+/−] [x^2] [+] 4.9 [×] .3705849 [+/−] [+] 1.5 [=]

and the display will read either "0" or a number very close to 0. ■

Scientific Notation

Typing

4000000 [×] 8000000 [=]

makes the display read "3.2   13" rather than "32000000000000." This is because the calculator does not have enough room on its display for "32000000000000." When the display shows "3.2   13," read it as "3.2×10^{13}," which is written in scientific notation. Literally, "3.2×10^{13}" means "multiply 3.2 by 10, thirteen times," but as a shortcut, you can interpret it as "move the decimal point in the '3.2' thirteen places to the right."

Typing

.0000005 [×] .0000007 [=]

makes the display read "3.5   −13" rather than "0.00000000000035," because the calculator does not have enough room on its display for "0.00000000000035." Read "3.5   −13" as "3.5×10^{-13}." Literally, this means "divide 3.5 by 10, thirteen times," but as a shortcut, you can interpret it as "move the decimal point in the '3.5' thirteen places to the left."

You can type a number in scientific notation by using the button labeled EXP (which stands for *exponent*) or EE (which stands for *enter exponent*). For example, typing

5.2 EXP 8 or 5.2 EE 8

makes the display read "5.2 8," which means "5.2×10^8," and typing

3 EXP 17 +/− or 3 EE 17 +/−

makes the display read "3 −17," which means "3×10^{-17}." Notice that the sign-change button is used to make the exponent negative.

Be careful that you don't confuse the y^x button with the EXP button. The EXP button does *not* allow you to type in an exponent; it allows you to type in scientific notation. For example, typing

3 EXP 4

makes the display read "3 4," which means "3×10^4," and typing

3 y^x 4

makes the display read "81," since $3^4 = 81$.

Exercises

Perform the following calculations. The correct answer is given in brackets [ ]. In your homework, write down what you type on your calculator to get that answer. Answers are not given in the back of the book.

1. $-3 - -5$ [2]
2. $-6 - 3$ [−9]
3. $4 - -9$ [13]
4. $-6 - -8$ [2]
5. $-3 - (-5 - -8)$ [−6]
6. $-(-4 - 3) - (-6 - -2)$ [11]
7. $-8 \cdot -3 \cdot -2$ [−48]
8. $-9 \cdot -3 -2$ [25]
9. $(-3)(-8) - (-9)(-2)$ [6]
10. $2(3 - 5)$ [−4]
11. $2 \cdot 3 - 5$ [1]
12. $4 \cdot 11^2$ [484]
13. $(4 \cdot 11)^2$ [1,936]
14. $4 \cdot (-11)^2$ [484]
15. $4 \cdot (-3)^3$ [−108]

WARNING: Some calculators will not raise a negative number to a power. If yours has this characteristic, how can you use your calculator on this exercise?

16. $(4 \cdot -3)^3$ [−1,728]
17. $\frac{3 + 2}{7}$ [0.7142857]
18. $\frac{3 \cdot 2}{7}$ [0.8571429]
19. $\frac{3}{2 \cdot 7}$ [0.2142857]
20. $\frac{3 \cdot 2}{7 \cdot 5}$ [0.1714286]
21. $\frac{3 + 2}{7 \cdot 5}$ [0.1428571]
22. $\frac{3 \cdot -2}{7 + 5}$ [−0.5]
23. $\frac{3}{7/2}$ [0.8571429]
24. $\frac{3/7}{2}$ [0.2142857]
25. 1.8^2 [3.24]
26. $\sqrt{1.8}$ [1.3416408]
27. $47{,}000{,}000^2$ [2.209×10^{15}]
28. $\sqrt{0.0000000000027}$ [1.643168×10^{-6}]
29. $(-3.92)^7$ [−14,223.368737]
30. $(5.72 \times 10^{19})^4$ [1.070494×10^{79}]
31. $(3.76 \times 10^{-12})^{-5}$ [1.330641×10^{57}]
32. $(3.76 \times 10^{-12}) -5$ [−5]
33. Solve $4.2x^2 + 8.3x + 1.1 = 0$ for x. Check your two answers by substituting them back into the equation.
34. Solve $5.7x^2 + 12.3x - 8.1 = 0$ for x. Check your two answers by substituting them back into the equation.
35. Which of the following buttons must be used in conjunction with the = button, and why?

+ − × ÷ x^2 $\sqrt{x}$ y^x $1/x$ +/−

B Using a Graphing Calculator

The following discussion was written specifically for Texas Instruments (or "TI") and Casio graphing calculators, but it frequently applies to other brands as well. Texas Instruments TI-81, TI-82, TI-83, TI-85, and TI-86 models and Casio CFX-9850 and CFX-9950 models are specifically addressed. Read this discussion with your calculator close at hand. When a calculation is discussed, do that calculation on your calculator.

To do any of the calculations discussed in this section on a Casio, start by pressing MENU. Then use the arrow buttons to select "RUN" on the main menu, and press EXE (which stands for "execute").

The Enter Button

A TI graphing calculator will never perform a calculation until the ENTER button is pressed; a Casio will never perform a calculation until the EXE button is pressed. A Casio's EXE button functions like a TI's ENTER button; often, we will refer to either of these buttons as the ENTER button.

To add 3 and 2, type

3 + 2 ENTER

and the display will read 5. To square 4, type

4 x^2 ENTER

and the display will read 16. If the ENTER button isn't pressed, the calculation will not be performed.

The 2nd and Alpha Buttons

Most calculator buttons have more than one label and more than one use; you select from these uses with the 2nd and ALPHA buttons. A Casio's SHIFT button functions like a TI's 2nd button; often, we will refer to either of these buttons as the 2nd button. For example, one button is labeled "x^2" on the button itself, "$\sqrt{\ }$" above the button, and either "I" or "K" above and to the right of the button. If it is used without the 2nd or ALPHA buttons, it will square a number. Typing

4 x^2 ENTER

makes the display read 16, since $4^2 = 16$. If it is used with the [2nd] button, it will take the square root of a number. Typing

[2nd] [$\sqrt{\ }$] 4 [ENTER]

makes the display read 2, since $\sqrt{4} = 2$. If it is used with the [ALPHA] button, it will display the letter I or K.

Notice that to square 4, you press the [x^2] button *after* the 4, but to take the square root of 4, you press the [$\sqrt{\ }$] button *before* the 4. This is because graphing calculators are designed so that the way you type something is as similar as possible to the way it is written algebraically. When you write 4^2, you write the 4 first and then the squared symbol; thus, on your graphing calculator, you press the 4 first and then the [x^2] button. When you write $\sqrt{4}$, you write the square root symbol first and then the 4; thus, on your graphing calculator, you press the [$\sqrt{\ }$] button first and then the 4.

Frequently, the two operations that share a button are operations that "undo" each other. For example, typing

3 [x^2] [ENTER]

makes the display read 9, since $3^2 = 9$, and typing

[2nd] [$\sqrt{\ }$] 9 [ENTER]

makes the display read 3, since $\sqrt{9} = 3$. This is done as a memory device; it is easier to find the various operations on the keyboard if the two operations that share a button also share a relationship.

Two operations that *always* undo each other are called **inverses.** The x^2 and $\sqrt{x}$ operations are not inverses because $(-3)^2 = 9$, but $\sqrt{9} \neq -3$. However, there is an inverse-type relationship between the x^2 and $\sqrt{x}$ operations—they undo each other sometimes. When two operations share a button, they are inverses or they share an inverse-type relation.

Correcting Typing Errors

If you've made a typing error *and you haven't yet pressed* [ENTER], you can correct that error with the [←] button. For example, if you typed "5 × 2 + 7" and then realized you wanted "5 × 3 + 7," you can replace the incorrect 2 with a 3 by pressing the [←] button until the 2 is flashing, and then press "3."

If you realize that you've made a typing error *after* you pressed [ENTER], just press [2nd] [ENTRY] to reproduce the previously entered line. (With a Casio, press [←].) Then correct the error with the [←] button, as described above.

The [INS] button allows you to insert a character. For example, if you typed "5 × 27" and you meant to type "5 × 217," press the [←] button until the 7 is flashing, and then insert 1 by typing

[2nd] [INS] 1

(With a TI-81, do not type [2nd].)

The [DEL] button allows you to delete a character. For example, if you typed "5 × 217" and you meant to type "5 × 27," press the [←] button until the 1 is flashing, and then press [DEL].

If you haven't yet pressed [ENTER], the [CLEAR] button erases an entire line. If you have pressed [ENTER], the [CLEAR] button clears everything off of the screen. With a Casio, the [AC] button functions in the same way. "AC" stands "all clear."

The Subtraction Symbol and the Negative Symbol

If you read "3 − -2" aloud, you could say "three subtract negative two" or "three minus minus two," and you would be understood. The expression is understandable even if the distinction between the negative symbol and the subtraction symbol is not made clear. With a calculator, however, this distinction is crucial. The subtraction button is labeled "−," and the negative button is labeled "(−)."

EXAMPLE 1 Calculate 3 − -2, both (a) by hand and (b) with a calculator.

Solution **a.** 3 − -2 = 3 + 2 = 5

b. Type

3 [−] [(−)] 2 [ENTER]

and the display will read 5. ■

If you had typed

3 [−] [−] 2 [ENTER] or 3 [(−)] [−] 2 [ENTER]

the calculator would have responded with an error message.

The Multiplication Symbol

In algebra, we do not use "x" for multiplication. Instead, we use "x" as a variable, and we use "·" for multiplication. However, Texas Instruments graphing calculators use "×" as the label on the multiplication button, and "*" for multiplication on the display screen. (The "variable x" button is labeled either "X,T," "X|T," "X,T,θ," "X,T,θ,n", or "x-VAR.") This is one of the few instances in which you don't type things in the same way that you write them algebraically.

Order of Operations and Use of Parentheses

Texas Instruments and Casio graphing calculators are programmed to follow the order of operations. That is, they perform calculations in the following order:

1. **P**arentheses-enclosed work
2. **E**xponents
3. **M**ultiplication and
   Division, from left to right
4. **A**ddition and
   Subtraction, from left to right

You can remember this order by remembering the word "**PEMDAS**," which stands for

Parentheses/**E**xponents/**M**ultiplication/**D**ivision/**A**ddition/**S**ubtraction

EXAMPLE 2 Calculate 2(3 + 4), both (a) by hand and (b) with a calculator.

Solution **a.** 2(3 + 4) = 2(7) parentheses-enclosed work comes first
= 14

b. Type

2 [×] [(] 3 [+] 4 [)] [ENTER]

and the display will read 14. ■

In the instructions to Example 2, notice that we wrote "2(3 + 4)" rather than "2 · (3 + 4)"; in this case, it's not necessary to write the multiplication symbol. Similarly, it's not necessary to type the multiplication symbol. Example 2b could be computed by typing

2 [( ] 3 [+] 4 [ )] [ENTER]

EXAMPLE 3 Calculate $2 \cdot 3^3$, both (a) by hand and (b) with a calculator.

Solution **a.** $2 \cdot 3^3 = 2 \cdot 27$ exponents come before multiplication

$= 54$

b. To do this, we must use the exponent button, which is labeled "∧." Type

2 [×] 3 [∧] 3 [ENTER]

and the display will read 54. ■

EXAMPLE 4 Calculate $(2 \cdot 3)^3$, both (a) by hand and (b) with a calculator.

Solution **a.** $(2 \cdot 3)^3 = 6^3$ parentheses-enclosed work comes first

$= 216$

b. Type

[( ] 2 [×] 3 [ )] [∧] 3 [ENTER]

and the display will read 216. ■

In Example 3, the exponent applies only to the 3, because the order of operations dictates that exponents come before multiplication. In Example 4, the exponent applies to the (2 · 3), because the order of operations dictates that parentheses-enclosed work comes before exponents. In each example, the way you type the problem matches the way it is written algebraically, because the calculator is programmed to follow the order of operations. Sometimes, however, the way you type a problem doesn't match the way it is written algebraically.

EXAMPLE 5 Calculate $\dfrac{2}{3 \cdot 4}$ with a calculator.

Solution **Wrong** It is incorrect to type

2 [÷] 3 [×] 4 [ENTER]

even though that matches the way the problem is written algebraically. According to the order of operations, multiplication and division are done *from left to right*, so the above typing is algebraically equivalent to

$= \dfrac{2}{3} \cdot 4$ first dividing and then multiplying, since division is on the left and multiplication is on the right

$= \dfrac{2}{3} \cdot \dfrac{4}{1} = \dfrac{2 \cdot 4}{3}$

which is not what we want. The difficulty is that the large fraction bar in the expression $\frac{2}{3 \cdot 4}$ groups the "3 · 4" together in the denominator; in the above typing, nothing groups the "3 · 4" together, and only the 3 ends up in the denominator.

Right The calculator needs parentheses inserted in the following manner:

$$\frac{2}{(3 \cdot 4)}$$

Thus, it is correct to type

2 [÷] [(] 3 [×] 4 [)] [ENTER]

This makes the display read 0.166666667, the correct answer.

Also Right It is correct to type

2 [÷] 3 [÷] 4 [ENTER]

According to the order of operations, multiplication and division are done from left to right, so the above typing is algebraically equivalent to

$$\frac{2}{3} \div 4 \qquad \text{doing the left-hand division first}$$

$$= \frac{2}{3} \cdot \frac{1}{4} \qquad \text{inverting and multiplying}$$

$$= \frac{2}{3 \cdot 4}$$

which is what we want. *When you're calculating something that involves only multiplication and division and you don't use parentheses, the* [×] *button places a factor in the numerator, and the* [÷] *button places a factor in the denominator.* ■

EXAMPLE 6 Calculate $\frac{2}{3/4}$ with a calculator.

Solution **Wrong** It is incorrect to type

2 [÷] 3 [÷] 4 [ENTER]

even though that matches the way the problem is written algebraically. As was discussed in Example 5, this typing is algebraically equivalent to

$$\frac{2}{3 \cdot 4}$$

which is not what we want.

Right The calculator needs parentheses inserted in the following manner:

$$\frac{2}{(3/4)}$$

Thus, it is correct to type

2 [÷] [(] 3 [÷] 4 [)] [ENTER]

since, according to the order of operations, parentheses-enclosed work is done first. This makes the display read 2.6666667, the correct answer. ■

EXAMPLE 7 Calculate $\dfrac{2 + 3}{4}$ with a calculator.

Solution **Wrong** It is incorrect to type

2 [+] 3 [÷] 4 [ENTER]

even though that matches the way the problem is written algebraically. According to the order of operations, division is done before addition, so this typing is algebraically equivalent to

$$2 + \frac{3}{4}$$

which is not what we want. The large fraction bar in the expression $\frac{2+3}{4}$ groups the "2 + 3" together in the numerator; in the above typing, nothing groups the "2 + 3" together, and only the 3 ends up in the numerator.

Right The calculator needs parentheses inserted in the following manner:

$$\frac{(2 + 3)}{4}$$

Thus, it is correct to type

[(] 2 [+] 3 [)] [÷] 4 [ENTER]

This makes the display read 1.25, the correct answer.

Also Right It is correct to type

2 [+] 3 [ENTER] [÷] 4 [ENTER]

The first [ENTER] makes the calculator perform all prior calculations before continuing. This too makes the display read 1.25, the correct answer. ■

Memory

The **memory** is a place to store a number for later use, without having to write it down. Graphing calculators have a memory for each letter of the alphabet; that is, you can store one number in memory A, a second number in memory B, and so on. Pressing

TI-81/85/86	5 [STO→] [A] [ENTER]
TI-80/82/83	5 [STO→] [ALPHA] [A] [ENTER]
Casio	5 [→] [ALPHA] [A] [EXE]

will store 5 in memory A. Similar keystrokes will store in memory B. Pressing [ALPHA] [A] will recall what has been stored in memory A.

EXAMPLE 8 Calculate $\dfrac{3 + \dfrac{5 + 7}{2}}{4}$ (a) by hand and (b) by first calculating $\frac{5+7}{2}$ and storing the result.

Solution **a.** $\dfrac{3 + \dfrac{5+7}{2}}{4} = \dfrac{3 + \dfrac{12}{2}}{4} = \dfrac{3+6}{4} = \dfrac{9}{4} = 2.25$

b. First, calculate $\frac{5+7}{2}$ and store the result in memory A. The large fraction bar in this expression groups the "5 + 7" together in the numerator; in our typing, we must group the "5 + 7" together with parentheses. The calculator needs parentheses inserted in the following manner:

$$\frac{(5+7)}{2}$$

Type

TI-81/85/86	[(] 5 [+] 7 [)] [÷] 2 [STO→] [A] [ENTER]
TI-80/82/83	[(] 5 [+] 7 [)] [÷] 2 [STO→] [ALPHA] [A] [ENTER]
Casio	[(] 5 [+] 7 [)] [÷] 2 [→] [ALPHA] [A] [EXE]

What remains is to compute $\frac{3+A}{4}$. Again, the large fraction bar groups the "3 + A" together, so the calculator needs parentheses inserted in the following manner:

$$\frac{(3+A)}{4}$$

Type

[(] 3 [+] [ALPHA] [A] [)] [÷] 4 [ENTER]

and the display will read 2.25. ■

EXAMPLE 9 Calculate $\dfrac{3 + \dfrac{5+7}{2}}{4}$ using one line of instructions and without using the memory.

Solution There are two large fractions bars, one grouping the "5 + 7" together and one grouping the "3 + $\frac{5+7}{2}$" together. In our typing, we must group each of these together with parentheses. The calculator needs parentheses inserted in the following manner:

$$\frac{\left(3 + \dfrac{(5+7)}{2}\right)}{4}$$

Type

[(] 3 [+] [(] 5 [+] 7 [)] [÷] 2 [)] [÷] 4 [ENTER]

and the display will read 2.25. ■

EXAMPLE 10 **a.** Use the quadratic formula and your calculator's memory to solve $2.3x^2 + 4.9x + 1.5 = 0$.

b. Check your answers.

Solution **a.** According to the Quadratic Formula, if $ax^2 + bx + c = 0$, then

$$x = \frac{-b \pm \sqrt{b^2 - 4ac}}{2a}$$

For our problem, $a = 2.3$, $b = 4.9$ and $c = 1.5$. This gives

$$x = \frac{-4.9 \pm \sqrt{4.9^2 - 4 \cdot 2.3 \cdot 1.5}}{2 \cdot 2.3}$$

One way to do this calculation is to calculate the radical, store it, and then calculate the two fractions.

Step 1 *Calculate the radical.* To do this, type

TI-80/82	[2nd] [√] [(] 4.9 [x^2] [−] 4 [×] 2.3 [×] 1.5 [)] [STO→] [ALPHA] [A] [ENTER]
TI-81/85/86	[2nd] [√] [(] 4.9 [x^2] [−] 4 [×] 2.3 [×] 1.5 [)] [STO→] [ALPHA] [A] [EXE]
TI-83	[2nd] [√] 4.9 [x^2] [−] 4 [×] 2.3 [×] 1.5 [)] [STO→] [ALPHA] [A] [ENTER]
Casio	[SHIFT] [√] [(] 4.9 [x^2] [−] 4 [×] 2.3 [×] 1.5 [)] [→] [ALPHA] [A] [EXE]

This makes the display read "3.195309062," and stores the number in the memory A. Notice the use of parentheses.

Step 2 *Calculate the first fraction.* The first fraction is

$$\frac{-4.9 \pm \sqrt{4.9^2 - 4 \cdot 2.3 \cdot 1.5}}{2 \cdot 2.3}$$

However, the radical has already been calculated and stored in memory A, so this is equivalent to

$$\frac{-4.9 + \text{A}}{2 \cdot 2.3}$$

Type

(calculator icon)	TI-80/82/83	((−) 4.9 + ALPHA A) ÷ (2 × 2.3) ALPHA B ENTER
	TI-81/85/86	((−) 4.9 + ALPHA A) ÷ (2 × 2.3) STO→ B ENTER
	Casio	((−) 4.9 + ALPHA A) ÷ (2 × 2.3) → ALPHA B EXE

This makes the display read "−0.3705849866" and stores the number in memory B.

Step 3 *Calculate the second fraction.* The second fraction is

$$\frac{-4.9 - \sqrt{4.9^2 - 4 \cdot 2.3 \cdot 1.5}}{2 \cdot 2.3} = \frac{-4.9 - A}{2 \cdot 2.3}$$

Memories A and B are already in use, so we will store this in memory C. Instead of retyping the line in step 2 with the "+" changed to a "−" and the "B" to a "C," press 2nd ENTRY (← with a Casio) to reproduce that line, and use the ← button to make these changes. Press ENTER and the display will read −1.759849796, and that number will be stored in memory C. The solutions to $2.3x^2 + 4.9x + 1.5 = 0$ are $x = -0.370584966$ and $x = -1.759849796$. These are approximate solutions in that they show only the first nine decimal places.

Step 4 *Check your solutions.* These two solutions are stored in memories B and C; they can be checked by seeing if they satisfy the equation $2.3x^2 + 4.9x + 1.5 = 0$. To check the first solution, type

2.3 × ALPHA B x^2 + 4.9 × ALPHA B + 1.5 ENTER

and the display should read either "0" or a number very close to 0. ■

Scientific Notation

Typing

4000000 × 8000000 ENTER

makes the display read "3.2E13" rather than "32000000000000." This is because the calculator does not have enough room on its display for "32000000000000." When the display shows "3.2E13," read it as "3.2×10^{13}," which is written in scientific notation. Literally, "3.2×10^{13}" means "multiply 3.2 by 10, thirteen times," but as a shortcut, you can interpret it as "move the decimal point in the '3.2' thirteen places to the right."

Typing

.0000005 × .0000007 ENTER

makes the display read "3.5E-13" rather than "0.00000000000035," because the calculator does not have enough room on its display for "0.00000000000035." Read "3.5E-13" as "3.5×10^{-13}." Literally, this means "divide 3.5 by 10, thirteen times," but as a shortcut, you can interpret it as "move the decimal point in the '3.5' thirteen places to the left."

You can type a number in scientific notation by using the TI button labeled "EE" (which stands for "Enter Exponent") or the Casio button labeled "EXP" (which stands for "Exponent"). For example, typing

5.2 [EE] 8 [ENTER]

makes the display read "52000000." (If the "EE" label is above the button, you will need to use the [2nd] button.) On a Casio, type

5.2 [EXP] 8 [EXE]

Be careful that you don't confuse the [EE] or [EXP] button with the [∧] button. The [EE] or [EXP] button does *not* allow you to type in an exponent; it allows you to type in scientific notation. For example, typing

3 [EE] 4 [ENTER]

makes the display read "30000," since $3 \times 10^4 = 30{,}000$. Typing

3 [∧] 4 [ENTER]

makes the display read "81," since $3^4 = 81$.

Exercises

In Exercises 1–32, use your calculator to perform the given calculation. The correct answer is given in brackets [ ]. In your homework, write down what you type to get that answer. Answers are not given in the back of the book.

1. $-3 - -5$ [2]
2. $-6 - 3$ [−9]
3. $4 - -9$ [13]
4. $-6 - -8$ [2]
5. $-3 - (-5 - -8)$ [−6]
6. $-(-4 - 3) - (-6 - -2)$ [11]
7. $-8 \cdot -3 \cdot -2$ [−48]
8. $-8 \cdot -3 - 2$ [22]
9. $(-3)(-8) - (-9)(-2)$ [6]
10. $2(3 - 5)$ [−4]
11. $2 \cdot 3 - 5$ [1]
12. $4 \cdot 11^2$ [484]
13. $(4 \cdot 11)^2$ [1,936]
14. $4 \cdot (-11)^2$ [484]
15. $4 \cdot (-3)^3$ [−108]
16. $(4 \cdot -3)^3$ [−1,728]
17. $\dfrac{3 + 2}{7}$ [0.7142857]
18. $\dfrac{3 \cdot 2}{7}$ [0.8571429]
19. $\dfrac{3}{2 \cdot 7}$ [0.2142857]
20. $\dfrac{3 \cdot 2}{7 \cdot 5}$ [0.1714286]
21. $\dfrac{3 + 5}{7 \cdot 2}$ [0.1428571]
22. $\dfrac{3 \cdot -2}{7 + 5}$ [−0.5]
23. $\dfrac{3}{7/2}$ [0.8571429]
24. $\dfrac{3/7}{2}$ [0.2142857]
25. 1.8^2 [3.24]
26. $\sqrt{1.8}$ [1.3416408]
27. $47{,}000{,}000^2$ [2.209×10^{15}]
28. $\sqrt{0.0000000000027}$ [1.643168×10^{-6}]
29. $(-3.92)^7$ [−14,223.368737]
30. $(5.72 \times 10^{19})^4$ [1.070494×10^{79}]
31. $(3.76 \times 10^{-12})^{-5}$ [1.330641×10^{57}]
32. $(3.76 \times 10^{-12}) - 5$ [−5]

In Exercises 33–36, perform the given calculation (a) by hand; (b) with a calculator, using the memory; and (c) with a calculator, using one line of instruction and without using memory, as shown in Examples 8 and 9. In your homework, for parts (b) and (c), write down what you type. Answers are not given in the back of the book.

33. $\dfrac{\dfrac{9 - 12}{5} + 7}{2}$

34. $\dfrac{\dfrac{4-11}{6}+8}{7}$

35. $\dfrac{\dfrac{7+9}{5}+\dfrac{8-14}{3}}{3}$

36. $\dfrac{\dfrac{4-16}{5}+\dfrac{7-22}{2}}{5}$

In Exercises 37–38, use your calculator to solve the given equation for x. Check your two answers, as shown in Example 10. In your homework, write down what you type to get the answers, and what you type to check the answers. Answers are not given in the back of the book.

37. $4.2x^2 + 8.3x + 1.1 = 0$
38. $5.7x^2 + 12.3x - 8.1 = 0$
39. Discuss the use of parentheses in Example 10a, step 1. Why are they necessary? What would happen if they were omitted?
40. Discuss the use of parentheses in Example 10a, step 2. Why are they necessary? What would happen if they were omitted?
41. a. Calculate $\dfrac{\dfrac{5+7.1}{3}+\dfrac{2-7.1}{5}}{7}$ using one line of instruction and without using the memory. In your homework, write down what you type as well as the solution.
   b. Use the 2nd ENTRY feature to calculate $\dfrac{\dfrac{5+7.2}{3}+\dfrac{2-7.2}{5}}{7}$
   c. Use the 2nd ENTRY feature to calculate $\dfrac{\dfrac{5+9.3}{3}+\dfrac{2-4.9}{5}}{7}$
42. a. Calculate $\dfrac{3+\dfrac{5+\dfrac{6-8.3}{2}}{3}}{9}$ using one line of instruction and without using the memory. In your homework, write down what you type as well as the solution.
   b. Use the 2nd ENTRY feature to calculate $\dfrac{3+\dfrac{5-\dfrac{6-8.3}{2}}{3}}{9}$
   c. Use the 2nd ENTRY feature to calculate $\dfrac{3-\dfrac{5+\dfrac{6-8.3}{2}}{3}}{9}$
43. a. What is the result of typing "8.1 EE 4"?
   b. What is the result of typing "8.1 EE 12"?
   c. Why do the instructions in part (b) yield an answer in scientific notation, while the instructions in part (a) yield an answer that's not in scientific notation?
   d. By using the MODE button, your calculator can be reset so that all answers will appear in scientific notation. Describe how this can be done.

C Graphing with a Graphing Calculator

Graphing with a Texas Instruments Graphing Calculator

The Graphing Buttons

The graphing buttons on a TI graphing calculator are all at the top of the keypad, directly under the screen. The labels on these buttons vary a little from model to model, but their uses are the same. The button labels and their uses are listed in Figure A.1.

TI-80/81/82/83	Y=	RANGE or WINDOW	ZOOM	TRACE	GRAPH
TI-85/86 labels on buttons	M1 F1	M2 F2	M3 F3	M4 F4	M5 F5
TI-85/86 labels on screen in graphing mode*	y(x) =	RANGE	ZOOM	TRACE	GRAPH
use this button to tell the calculator:	what to graph	what part of the graph to draw	to zoom in or out	to give the coordinates of a highlighted point	to draw the graph

Figure A.1

Graphing a Line

To graph $y = 2x - 1$ on a TI graphing calculator, follow these steps.

Step 1 **[For TI-85/86 calculators only]** *Put the calculator into graphing mode* by pressing the GRAPH button. This activates the "F" buttons and puts labels at the bottom of the screen, as shown in Figure A.2 on page A-22.

* **TI-85/86 users** Your calculator is different from the other TI models in that its graphing buttons are labeled "F1" through "F5" ("M1" through "M5" when preceded by the 2nd button). The use of these buttons varies, depending on what you're doing with the calculator. When these buttons are active, their uses are displayed at the bottom of the screen.

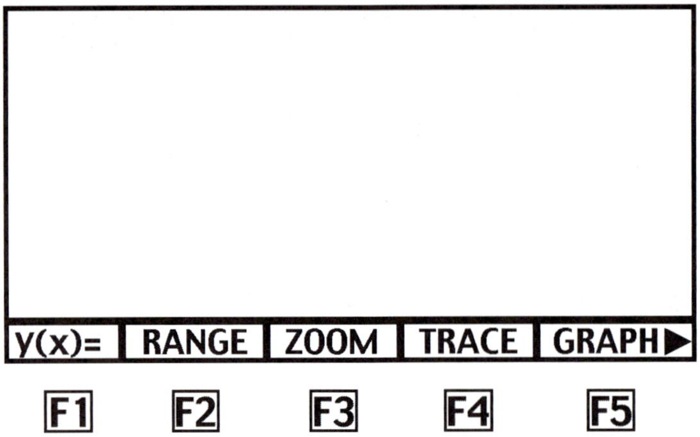

Figure A.2
The screen labels for a TI-85/86's "F" buttons

Step 2 *Set the calculator up for instructions on what to graph* by pressing [Y=] ([y(x)=] or [F1] on a TI-85/86). This produces the screen similar to that shown in Figure A.3. If your screen has things written after the equals symbols, use the [↑] and [↓] buttons along with the [CLEAR] button to erase them.

Step 3 *Tell the calculator what to graph* by typing "2*x* − 1" where the screen reads "Y_1=." To type the *x* symbol:

Figure A.3
A TI-82's "Y=" screen (other models' screens are similar)

TI-80/81	press [X.T] or [X|T]
TI-82	press [X.T.θ]
TI-83	press [X.T.θ.n]
TI-85/86	press either [x-VAR] or [F1] (which is now labeled "x" at the bottom of the screen)

After typing "2x − 1," press the [ENTER] button.

Step 4 *Set the calculator up for instructions on what part of the graph to draw.*

TI-80/81/82/83	press [RANGE] or [WINDOW]
TI-85/86	press [RANGE] or [WIND] (i.e., [2nd] [M2])*

* **TI-85/86 users** Your calculator now has a double row of labels at the bottom of the screen. The bottom row of labels refers to the current use of the [F1] through [F5] buttons, and the top row of labels refers to the current use of the [M1] through [M5] buttons (which require the use of the [2nd] button, as indicated by the orange lettering). Pressing the [EXIT] button removes one row of labels.

Step 5 *Tell the calculator what part of the graph to draw* by entering the values shown in Figure A.4 (If necessary, use the [↑] and [↓] buttons to move from line to line).

- "*x*min" and "*x*max" refer to the left and right boundaries of the graph, respectively
- "*y*min" and "*y*max" refer to the lower and upper boundaries of the graph, respectively
- "*x*scl" and "*y*scl" refer to the scales on the *x*- and *y*-axes (i.e., to the location of the tick marks on the axes)

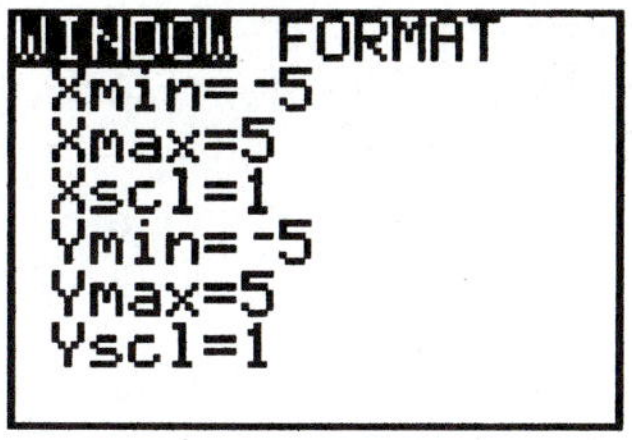

Figure A.4
A TI-82's "Window" screen (other models' screens are similar)

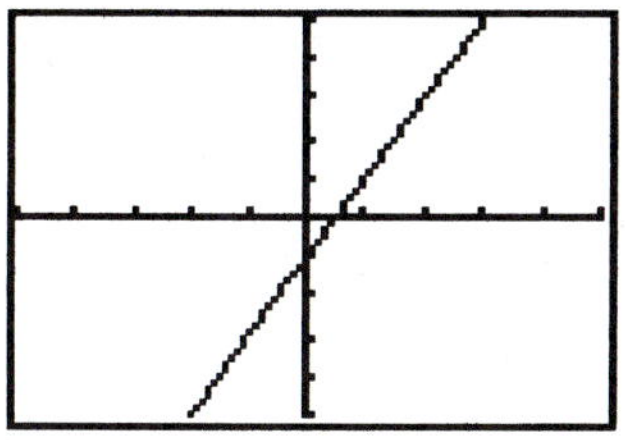

Figure A.5

Step 6 *Tell the calculator to draw a graph* by pressing the [GRAPH] button.* This produces the screen shown in Figure A.5.

* **TI-85/86 users** Your calculator has two different [GRAPH] buttons:

- One is labeled "GRAPH" on the button itself. This button puts the calculator in graphing mode, as discussed in step 1.
- One is labeled "F5" on the button itself, and "GRAPH" on the screen, when the calculator is in graphing mode. This tells the calculator to draw a graph, as discussed in step 6.

Step 7 *Discontinue graphing* by pressing [2nd] [QUIT].

Graphing with a Casio Graphing Calculator

To graph $y = 2x - 1$ on a Casio graphing calculator, follow these steps.

Step 1 *Put the calculator into graphing mode* by pressing Menu, using the arrow buttons to highlight "GRAPH" on the main menu, and then pressing [EXE].

Step 2 *Tell the calculator what to graph* by typing "$2x - 1$" where the screen reads "Y1=" (see Figure A.6). (If something is already there, press [F2], which now has the label "DEL" directly above it on the screen, and then press [F1], which is labeled "YES.") To type the *x* symbol, press [x,θ,T]. After typing "$2x - 1$," press the [EXE] button.

Step 3 *Tell the calculator to draw a graph* by pressing the [F6] button, which now has the label "DRAW" directly above it on the screen.

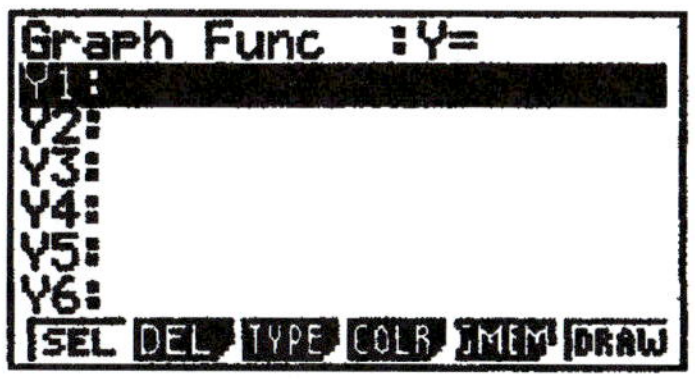

Figure A.6
A Casio's "Y=" screen

Step 4 *Set the calculator up for instructions on what part of the graph to draw* by pressing [SHIFT] and then the [F3] button, which now has the label "V-WIN" directly above it on the screen. ("V-WIN" is short for "viewing window.")

Step 5 *Tell the calculator what part of the graph to draw* by entering the values shown in Figure A.7. Press [EXE] after each entry. Then press [EXIT] to return to the "Y=" screen.

- "Xmin" and "Xmax" refer to the left and right boundaries of the graph, respectively.
- "Ymin" and "Ymax" refer to the lower and upper boundaries of the graph, respectively.
- "Xscale" and "Yscale" refer to scales on the *x*- and *y*-axes, respectively (i.e., to the location of the tick marks on the axes).

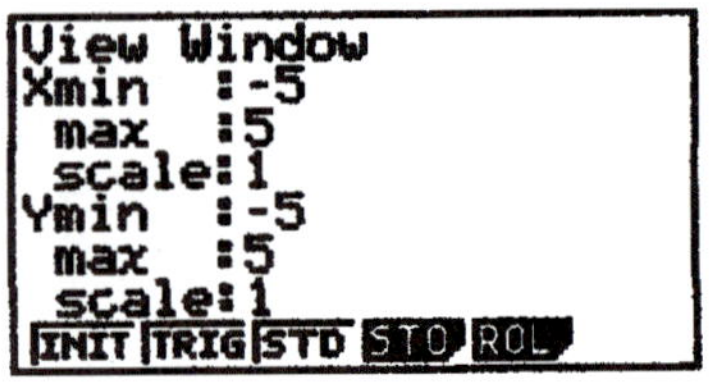

Figure A.7
A Casio's View Window

Step 6 *Tell the calculator to draw a graph* by pressing the [F6] button, which now has the label "DRAW" directly above it on the screen (see Figure A.8).

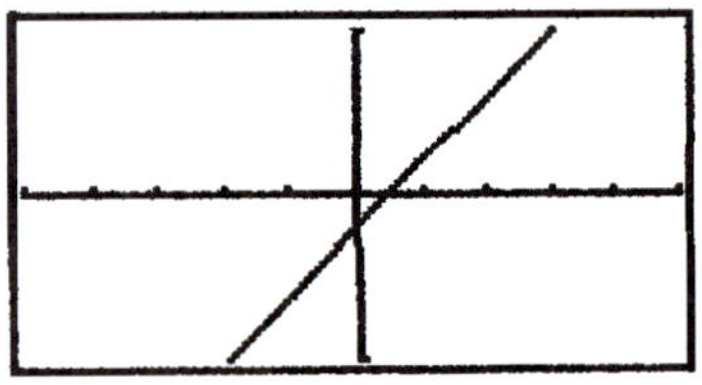

Figure A.8

Step 7 *Discontinue graphing* by pressing [MENU].

Exercises

In the following exercises, you will explore some of your calculator's graphing capabilities. Answers are not given in the back of the book.

1. *Exploring the "Zoom Standard" command.* Use your calculator to graph $y = 2x - 1$, as discussed in this section. When that graph is on the screen, select the TI "Zoom Standard" command from the "Zoom menu" or the Casio "STD" command from the "V-WIN" menu by doing the following:

TI-80/81/82/83	• press [ZOOM] • select option 6 "Standard" or "ZStandard" by either: • using the down arrow to scroll down to that option and pressing [ENTER], or • typing the number "6"
TI-85/86	• press [ZOOM] (i.e., [F3]) • press [ZSTD] (i.e., [F4])
Casio	• press [SHIFT] • press [V-WIN] (i.e., [F3]) • press [STD] (i.e., [F3])

   a. What is the result?
   b. How else could you accomplish the same thing without using any zoom commands?

2. *Exploring the "RANGE," "WINDOW," or "V-WIN" screen.* Use your calculator to graph $y = 2x - 1$ as discussed in this section. When that graph is on the screen, use the "RANGE," "WINDOW," or "V-WIN" screen described in steps 4 and 5 of this section to reset the following:

   - *x*min to 1
   - *x*max to 20
   - *y*min to 1
   - *y*max to 20

   a. Why are there no axes shown?
   b. Why does the graph start exactly in the lower left corner of the screen?

3. *Exploring the "RANGE," "WINDOW," or "V-WIN" screen.* Use your calculator to graph $y = 2x - 1$, as discussed in this section. When that graph is on the screen, use the "RANGE," "WINDOW," or "V-WIN" screen described in steps 4 and 5 of this section to reset the following:

   - *x*min to -1
   - *x*max to -20
   - *y*min to -1
   - *y*max to -20

Why did the TI calculator respond with an error message? What odd thing did the Casio calculator do? (Casio hint: use the TRACE button to investigate.)

4. *Exploring the "RANGE," "WINDOW," or "V-WIN" screen.* Use your calculator to graph $y = 2x - 1$, as discussed in this section. When that graph is on the screen, use the "RANGE" or "WINDOW" screen described in steps 4 and 5 of this section to reset the following:
   - *x*min to -5
   - *x*max to 5
   - *y*min to 10
   - *y*max to 25

   a. Why was only one axis shown?
   b. Why was no graph shown?

5. *Exploring the "TRACE" command.* Use your calculator to graph $y = 2x - 1$, as discussed in this section. When that graph is on the screen, press TRACE (F4 on a TI-85/86, F1 on a Casio). This causes two things to happen:
   - A mark appears at a point on the line. This mark can be moved with the left and right arrow buttons.
   - The corresponding ordered pair is printed out at the bottom of the screen.

   a. Use the "TRACE" feature to locate the line's x-intercept, the point at which the line hits the x-axis. (You may need to approximate it.)
   b. Use algebra, rather than the graphing calculator, to find the x-intercept.
   c. Use the "TRACE" feature to locate another ordered pair on the line.
   d. Use substitution to check that the ordered pair found in part (c) is in fact a point on the line.

6. *Exploring the "ZOOM BOX" command.* Use your calculator to graph $y = 2x - 1$ as discussed in this section. When that graph is on your screen, press ZOOM (F3 on a TI-85/86, F2 on a Casio) and select option 1, "BOX" or "ZBOX," in the manner described in Exercise 1. This seems to have the same result as TRACE, except the mark does not have to be a point on the line. Use the four arrow buttons to move to a point of your choice (that may be on or off the line). Press ENTER. Use the arrow buttons to move to a different point, so that the resulting box encloses a part of the line. Press ENTER again. What is the result of using the "Zoom Box" command?

7. *Exploring the "ZOOM IN" command.* Use your calculator to graph $y = 2x - 1$, as discussed in this section. When that graph is on your screen, press ZOOM (F3 on a TI-85/86, F2 on a Casio) and select "Zoom In" or "ZIN," in the manner described in Exercise 1. (Press IN or F3 on a Casio.) This causes a mark to appear on the screen. Use the four arrow buttons to move the mark to a point of your choice, either on or near the line. Press ENTER.
   a. What is the result of the "Zoom In" command?
   b. How could you accomplish the same thing without using any zoom commands?
   c. The "Zoom Out" command is listed right next to the "Zoom In" command? What does it do?

8. *Zooming in on x-intercepts.* Use the "Zoom In" command described in Exercise 7, and the "Trace" command described in Exercise 5, to approximate the location of the x-intercept of $y = 2x - 1$ as accurately as possible. You may need to use these commands more than once.
   a. Describe the procedure you used to generate this answer.
   b. According to the calculator, what is the x-intercept?
   c. Is this answer the same as that of Exercise 5(b)? Why or why not?

9. *Calculating x-intercepts.* The TI-82, TI-83, TI-85, TI-86, and Casio will calculate the x-intercept (also called a "root" or "zero") without using the "Zoom In" and "Trace" commands. First, use your calculator to graph $y = 2x - 1$ as discussed in this section. When that graph is on the screen, do the following.

On a TI-82/83:
- Press 2nd CALC and select option 2, "root" or "zero."
- The calculator responds by asking, "Lower Bound?" or "Left Bound?" Use the left and right arrow buttons to move the mark to a point slightly to the left of the x-intercept, and press ENTER.
- When the calculator asks, "Upper Bound?" or "Right Bound?", move the mark to a point slightly to the right of the x-intercept, and press ENTER.
- When the calculator asks, "Guess?", move the mark to a point close to the x-intercept, and press ENTER. The calculator will then display the location of the x-intercept.

On a TI-85/86:
- Press MORE until the "MATH" label appears above the F1 button, and then press that button.
- Press ROOT. This causes an ordered pair to appear on the screen and a mark to appear at the corresponding point on the line. The calculator is

asking you if it is on the correct line. However, there is only one line in the exercise, so press ENTER. (If there were graphs of several different lines on the screen, you could use the up and down arrow buttons to select the correct line). The calculator will then display the location of the x-intercept.

On a Casio:

- Press G-Solv (i.e., F5).
- Press ROOT. Wait and watch.

a. According to the calculator, what is the x-intercept of the line $y = 2x - 1$?

b. Is this answer the same as that of Exercise 5(b)? Why or why not?

D Finding Points of Intersection with a Graphing Calculator

The graphs of $y = x + 3$ and $y = -x + 9$ intersect on the standard viewing screen. To find the point of intersection with a TI or Casio graphing calculator, first enter the two equations on the "Y=" screen (one as Y_1 and one as Y_2), erase any other equations, and erase the "Y= " screen by pressing [2nd] [QUIT]. Then follow the following instructions.

On a TI-80/81

- Graph the two equations on the standard viewing screen.
- Use the "Zoom In" and "Trace" commands discussed in Appendix C Exercises 5–8 to approximate the location of the point of intersection as accurately as possible. You may need to use these commands more than once.
- Check your answer by substituting the ordered pair into each of the two equations.

On a TI-82/83

- Graph the two equations on the standard viewing screen.
- Press [2nd] [CALC] and select option 5, "intersect."
- When the calculator responds with "First curve?" and a mark on the first equation's graph, press [ENTER].
- When the calculator responds with "Second curve?" and a mark on the second equation's graph, press [ENTER].
- When the calculator responds with "Guess?", use the left and right arrows to place the mark near the point of intersection, and press [ENTER].
- Check your answer by substituting the ordered pair into each of the two equations.

On a TI-85/86

- Graph the two equations on the standard viewing screen.
- Press [MORE] until the "MATH" option appears, and select that option.
- Press [MORE] until the "ISECT" option appears, and select that option.
- When the calculator responds with a mark on the first equation's graph, press [ENTER].
- When the calculator responds with a mark on the second equation's graph, press [ENTER].
- Check your answer by substituting the ordered pair into each of the two equations.

On a Casio

- Graph the two equations on the standard viewing screen.
- Press [G-Solv] (i.e., [F5]).

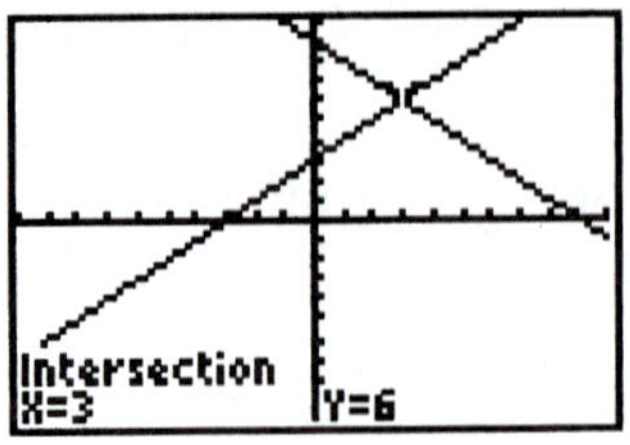

Figure A.9
Finding the point of intersection on a TI-82

- Press ISCT (i.e., F5). Watch and wait.
- After a pause, the calculator will display the location of the x-intercept.

Figure A.9 shows the results of computing the point of intersection of $y = x + 3$ and $y = -x + 9$.

The information on the screen indicates that the point of intersection is (3, 6). To check this, substitute 3 for x into each of the two equations; you should get 6.

$y = x + 3 = 3 + 3 = 6.$ ✔

$y = -x + 9 = -3 + 9 = 6.$ ✔

Exercises

In Exercises 1–6, do the following.

a. Use the graphing calculator to find the point of intersection of the given equations.

b. Check your solutions by substituting the ordered pair into each of the two equations. Answers are not given in the back of the book.

1. $y = 3x + 2$ and $y = 5x + 5$

2. $y = 2x - 6$ and $y = 3x + 4$

3. $y = 8x - 14$ and $g(x) = 11x + 23$

HINT: You will have to change xmin, xmax, ymin, and ymax to find the point.

4. $y = -7x + 12$ and $y = -12x - 71$

HINT: You will have to change xmin, xmax, ymin, and ymax to find the point.

5. $y = x^2 - 2x + 3$ and $y = -x^2 - 3x + 12$ (Find two answers.)

6. $f(x) = 8x^2 - 3x - 7$ and $y = 2x + 4$ (Find two answers.)

E Dimensional Analysis

Most people know how to convert 6 feet into yards (it's 6 ÷ 3 = 2 yards). Few people know how to convert 50 miles per hour into feet per minute. The former problem is so commonplace that people just remember how to do it; the latter problem is not so common, and people don't know how to do it. Dimensional analysis is an easy way of converting a quantity from one set of units to another; it can be applied to either of these two problems.

Dimensional analysis involves using a standard conversion (such as 1 yard = 3 feet) to create a fraction, including units (such as "feet" and "yards") in that fraction, and canceling units in the same way that variables are canceled. That is, in the fraction 2 feet/3 feet, we can cancel feet with feet and obtain 2/3, just as we can cancel x with x in the fraction $2x/3x$ and obtain 2/3.

EXAMPLE 1 Use dimensional analysis to convert 6 feet into yards.

Solution Start with the standard conversion:

$$1 \text{ yard} = 3 \text{ feet}$$

Create fractions by dividing each side by 3 feet:

$$\frac{1 \text{ yard}}{3 \text{ feet}} = \frac{\cancel{3 \text{ feet}}}{\cancel{3 \text{ feet}}}$$

$$\frac{1 \text{ yard}}{3 \text{ feet}} = 1$$

To convert 6 feet into yards, multiply 6 feet by the fraction 1 yard/3 feet. This is valid because that fraction is equal to 1, and multiplying something by 1 doesn't change its value.

$$6 \text{ feet} = 6 \text{ feet} \cdot 1$$

$$= 2\ \cancel{\text{feet}} \cdot \frac{1 \text{ yard}}{\cancel{3 \text{ feet}}}$$

$$= 2 \text{ yards}$$

It is crucial to include units in this work. If at the beginning of this example, we had divided by 1 yard instead of 3 feet, we would have obtained:

$$\begin{aligned} 1 \text{ yard} &= 3 \text{ feet} \\ \frac{\cancel{1 \text{ yard}}}{\cancel{1 \text{ yard}}} &= \frac{3 \text{ feet}}{1 \text{ yard}} \\ 1 &= \frac{3 \text{ feet}}{1 \text{ yard}} \end{aligned}$$

Multiplying 6 feet by 3 feet/1 yard would not allow us to cancel feet with feet and would not leave an answer in yards.

$$\begin{aligned} 6 \text{ feet} &= 6 \text{ feet} \cdot 1 \\ &= 6 \text{ feet} \cdot \frac{3 \text{ feet}}{1 \text{ yard}} \end{aligned}$$

It's important to include units in dimensional analysis, because it's only by looking at the units that we can tell that multiplying by 1 yard/3 feet is productive and multiplying by 3 feet/1 yard isn't. ■

EXAMPLE 2 Use dimensional analysis to convert 50 miles/hour to feet/minute.

Solution The appropriate standard conversions are

$$\begin{aligned} 1 \text{ mile} &= 5{,}280 \text{ feet} \\ 1 \text{ hour} &= 60 \text{ minutes} \end{aligned}$$

This problem has two parts, one for each of these two standard conversions.

Part 1 *Use the standard conversion "1 mile = 5,280 feet" to convert miles/hour to feet/hour*. The fraction 50 miles/hour has miles in the numerator, and we are to convert it to a fraction that has feet in the numerator. To replace miles with feet, first rewrite the standard conversion as a fraction that has miles in the denominator, and then multiply by that fraction. This will allow miles to cancel.

$$\begin{aligned} 1 \text{ mile} &= 5{,}280 \text{ feet} && \text{a standard conversion} \\ \frac{\cancel{1 \text{ mile}}}{\cancel{1 \text{ mile}}} &= \frac{5280 \text{ feet}}{1 \text{ mile}} && \text{placing miles in the denominator} \\ 1 &= \frac{5280 \text{ feet}}{1 \text{ mile}} \end{aligned}$$

Now multiply 50 miles/hour by this fraction and cancel:

$$\begin{aligned} \frac{50 \text{ miles}}{\text{hour}} &= \frac{50 \text{ miles}}{\text{hour}} \cdot 1 \\ &= \frac{50 \cancel{\text{ miles}}}{\text{hour}} \cdot \frac{5280 \text{ feet}}{1 \cancel{\text{ mile}}} \\ &= \frac{50 \cdot 5280 \text{ feet}}{\text{hour}} \end{aligned}$$

Part 2 *Use the standard conversion "1 hour = 60 minutes" to convert feet/hour to feet/minute*. To replace hours with minutes, first rewrite the standard conversion as a fraction that has hours in the numerator, and then multiply by that fraction. This will allow hours to cancel.

$$1 \text{ hour} = 60 \text{ minutes} \quad \text{a standard conversion}$$
$$\frac{1 \text{ hour}}{60 \text{ minutes}} = \frac{60 \cancel{\text{ minutes}}}{60 \cancel{\text{ minutes}}} \quad \text{placing hours in the numerator}$$
$$\frac{1 \text{ hour}}{60 \text{ minutes}} = 1$$

Continuing where we left off, multiply by this fraction and cancel:

$$\begin{aligned}
\frac{50 \text{ miles}}{\text{hour}} &= \frac{50 \cdot 5280 \text{ feet}}{\text{hour}} && \text{from part 1} \\
&= \frac{50 \cdot 5280 \text{ feet}}{\text{hour}} \cdot 1 \\
&= \frac{50 \cdot 5280 \text{ feet}}{\cancel{\text{hour}}} \cdot \frac{1 \cancel{\text{ hour}}}{60 \text{ minutes}} \\
&= \frac{4400 \text{ feet}}{1 \text{ minute}} && \text{since } 50 \cdot 5280/60 = 4400
\end{aligned}$$

Thus, 50 miles/hour is equivalent to 4,400 feet/minute. ■

EXAMPLE 3 How many feet will a car travel in half a minute if that car's rate is 50 miles/hour?

Solution This seems to be a standard algebra problem that uses the formula "distance = rate · time"; we're given the rate (50 miles/hour) and the time (1/2 minute), and we are to find the distance. However, the units are not consistent.

$$\begin{aligned}
\text{distance} &= \text{rate} \cdot \text{time} \\
&= \frac{50 \text{ miles}}{\text{hour}} \cdot \frac{1}{2} \text{ minute}
\end{aligned}$$

None of these units cancels, and we are not left with an answer in feet. If, however, the car's rate was in feet/minute rather than miles/hour, the units would cancel, and we would be left with an answer in feet:

$$\begin{aligned}
\text{distance} &= \text{rate} \cdot \text{time} \\
&= \frac{50 \text{ miles}}{\text{hour}} \cdot \frac{1}{2} \text{ minute} \\
&= \frac{4400 \text{ feet}}{1 \cancel{\text{ minute}}} \cdot \frac{1}{2} \cancel{\text{ minute}} && \text{from Example 2} \\
&= 2{,}200 \text{ feet}
\end{aligned}$$

The car would travel 2,200 feet in half a minute. ■

Exercises

In Exercises 1–6, use dimensional analysis to convert the given quantity.

1. **a.** 12 feet into yards
   b. 12 yards into feet
2. **a.** 24 feet into inches
   b. 24 inches into feet
3. **a.** 10 miles into feet
   b. 10 feet into miles (Round off to the nearest ten thousandth of a mile.)
4. **a.** 2 hours into minutes
   b. 2 minutes into hours (Round off to the nearest thousandth of an hour.)
5. 2 miles into inches

   HINT: Convert first to feet, then to inches.
6. 3 hours into seconds

   HINT: Convert first to minutes, then to seconds.

In Exercises 7–12, use the following information. The metric system is based on three different units:

- the gram (1 gram = 0.0022046 pound)
- the meter (1 meter = 39.37 inches)
- the liter (1 liter = 61.025 cubic inches)

Other units are formed by adding the following prefixes to these basic units:

- kilo-, which means one thousand (for example, 1 kilometer = 1,000 meters)
- centi-, which means one hundredth (for example, 1 centimeter = 1/100 meter)
- milli-, which means one thousandth (for example, 1 millimeter = 1/1,000 meter)

7. Use dimensional analysis to convert
   a. 50 kilometers to meters
   b. 50 meters to kilometers
8. Use dimensional analysis to convert
   a. 30 milligrams to grams
   b. 30 grams to milligrams
9. Use dimensional analysis to convert
   a. 2 centiliters to liters
   b. 2 liters to centiliters
10. Use dimensional analysis to convert
    a. 1 yard to meters (Round off to the nearest hundredth.)
    b. 20 yards to centimeters (Round off to the nearest centimeter.)
11. Use dimensional analysis to convert
    a. 1 pound to grams (Round off to the nearest gram.)
    b. 8 pounds to kilograms (Round off to the nearest tenth of a kilogram.)
12. Use dimensional analysis to convert
    a. 1 cubic inch to liters (Round off to the nearest thousandth.)
    b. 386 cubic inches to liters (Round off to the nearest tenth.)
13. **a.** Use dimensional analysis to convert 60 miles per hour into feet per second.
    b. Leadfoot Larry speeds through an intersection at 60 miles per hour, and he is ticketed for speeding and reckless driving. He pleads guilty to speeding but not guilty to reckless driving, telling the judge that at that speed, he would have plenty of time to react to any cross traffic. The intersection is 80 feet wide. How long would it take Larry to cross the intersection? (Round off to the nearest tenth of a second.)
14. In the United States, a typical freeway speed limit is 65 miles per hour.
    a. Use dimensional analysis to convert this to kilometers per hour. (Round off to the nearest whole number.)
    b. At this speed, how many miles can be traveled in 10 minutes? (Round off to the nearest whole number.)
15. In Germany, a typical autobahn speed limit is 130 kilometers per hour.
    a. Use dimensional analysis to convert this into miles per hour. (Round off to the nearest whole number.)
    b. How many more miles will a car traveling at 130 kilometers per hour go in one hour than a car traveling at 65 miles per hour? (Round off to the nearest whole number.)
16. Light travels 6×10^{12} miles per year.
    a. Convert this to miles per hour. (Round off to the nearest whole number.)
    b. How far does light travel in 1 second? (Round off to the nearest whole number.)
17. In December 1999, Massachusetts' Capital Crossing Bank offered a money market account with a 5.75% interest rate. This means that Capital Crossing Bank will pay interest at a rate of 5.75% per year.
    a. Use dimensional analysis to convert this to a percent per day.
    b. If you deposited $10,000 on September 1, how much interest would your account earn by

October 1? (There are 30 days in September.) (Round off to the nearest cent.)

c. If you deposited \$10,000 on October 1, how much interest would your account earn by November 1? (There are 31 days in October.) (Round off to the nearest cent.)

18. In December 1999, Kentucky's Republic Bank and Trust offered a money market account with a 5.36% interest rate. This means that Republic Bank will pay interest at a rate of 5.36% per year.

a. Use dimensional analysis to convert this to a percent per day.

b. If you deposited \$10,000 on September 1, how much interest would your account earn by October 1? (There are 30 days in September.) (Round off to the nearest cent.)

c. If you deposited \$10,000 on October 1, how much interest would your account earn by November 1? (There are 31 days in October.) (Round off to the nearest cent.)

19. You cannot determine whether a person is overweight by merely determining his or her weight; if a short person and a tall person weigh the same, the short person could be overweight and the tall person could be underweight. Body mass index (BMI) is becoming a standard way of determining if a person is overweight, since it takes both weight and height into consideration. BMI is defined as (weight in kilograms)/(height in meters)^2. According to the World Health Organization, a person is overweight if his or her BMI is 25 or greater. In October 1996, Katherine Flegal, a statistician for the National Center for Health Statistics, said that according to this standard, one out of every two Americans is overweight. (*Source: San Francisco Chronicle*, 16 October 1996, page A6.)

a. Lenny is 6 feet tall. Convert his height to meters. (Round off to the nearest hundredth.)

b. Lenny weighs 169 pounds. Convert his weight to kilograms. (Round off to the nearest tenth.)

c. Determine Lenny's BMI. (Round off to the nearest whole number.) Is he overweight?

d. Fred weighs the same as Lenny, but he is 5′5″ tall. Determine Fred's BMI. (Round off to the nearest whole number.) Is Fred overweight?

e. Why does BMI use the metric system (kilograms and meters) rather than the English system (feet, inches, and pounds)?

20. According to the 1990 census, California, the most populous state, had 29,758,000 people residing in 158,706 square miles of area, while Rhode Island, the smallest state, had 1,003,000 people residing in 1,212 square miles of area. Determine which state is more crowded by computing the number of square feet per person in each state.

21. A wading pool is 4 feet wide, 6 feet long, and 11 inches deep. There are 7.48 gallons of water per cubic foot. How many gallons of water does it take to fill the pool?

22. John ate a 2,000 calorie lunch and immediately felt guilty. Jogging for one minute consumes 0.061 calorie per pound of body weight. John weighs 205 pounds. How long will he have to jog to burn off all of the calories from lunch?

F Body Table for the Standard Normal Distribution

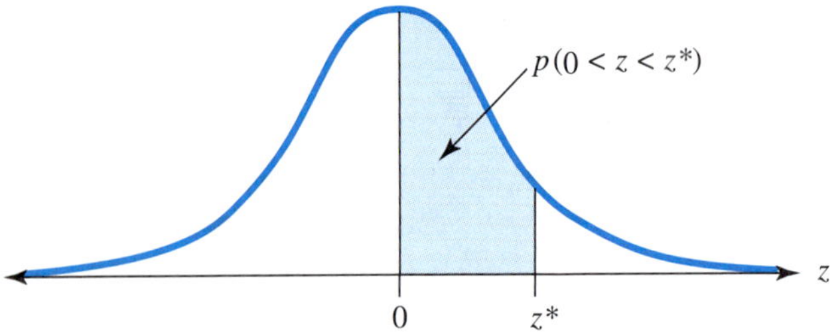

z*	0.00	0.01	0.02	0.03	0.04	0.05	0.06	0.07	0.08	0.09
0.0	0.0000	0.0040	0.0080	0.0120	0.0160	0.0199	0.0239	0.0279	0.0319	0.0359
0.1	0.0398	0.0438	0.0478	0.0517	0.0557	0.0596	0.0636	0.0675	0.0714	0.0753
0.2	0.0793	0.0832	0.0871	0.0910	0.0948	0.0987	0.1026	0.1064	0.1103	0.1141
0.3	0.1179	0.1217	0.1255	0.1293	0.1331	0.1368	0.1406	0.1443	0.1480	0.1517
0.4	0.1554	0.1591	0.1628	0.1664	0.1700	0.1736	0.1772	0.1808	0.1844	0.1879
0.5	0.1915	0.1950	0.1985	0.2019	0.2054	0.2088	0.2123	0.2157	0.2190	0.2224
0.6	0.2257	0.2291	0.2324	0.2357	0.2389	0.2422	0.2454	0.2486	0.2517	0.2549
0.7	0.2580	0.2611	0.2642	0.2673	0.2704	0.2734	0.2764	0.2794	0.2823	0.2852
0.8	0.2881	0.2910	0.2939	0.2967	0.2995	0.3023	0.3051	0.3078	0.3106	0.3133
0.9	0.3159	0.3186	0.3212	0.3238	0.3264	0.3289	0.3315	0.3340	0.3365	0.3389
1.0	0.3413	0.3438	0.3461	0.3485	0.3508	0.3531	0.3554	0.3577	0.3599	0.3621
1.1	0.3643	0.3665	0.3686	0.3708	0.3729	0.3749	0.3770	0.3790	0.3810	0.3830
1.2	0.3849	0.3869	0.3888	0.3907	0.3925	0.3944	0.3962	0.3980	0.3997	0.4015
1.3	0.4032	0.4049	0.4066	0.4082	0.4099	0.4115	0.4131	0.4147	0.4162	0.4177
1.4	0.4192	0.4207	0.4222	0.4236	0.4251	0.4265	0.4279	0.4292	0.4306	0.4319
1.5	0.4332	0.4345	0.4357	0.4370	0.4382	0.4394	0.4406	0.4418	0.4429	0.4441
1.6	0.4452	0.4463	0.4474	0.4484	0.4495	0.4505	0.4515	0.4525	0.4535	0.4545
1.7	0.4554	0.4564	0.4573	0.4582	0.4591	0.4599	0.4608	0.4616	0.4625	0.4633
1.8	0.4641	0.4649	0.4656	0.4664	0.4671	0.4678	0.4686	0.4692	0.4699	0.4706
1.9	0.4713	0.4719	0.4726	0.4732	0.4738	0.4744	0.4750	0.4756	0.4761	0.4767
2.0	0.4772	0.4778	0.4783	0.4788	0.4793	0.4798	0.4803	0.4808	0.4812	0.4817
2.1	0.4821	0.4826	0.4830	0.4834	0.4838	0.4842	0.4846	0.4850	0.4854	0.4857
2.2	0.4861	0.4864	0.4868	0.4871	0.4875	0.4878	0.4881	0.4884	0.4887	0.4890
2.3	0.4893	0.4896	0.4898	0.4901	0.4904	0.4906	0.4909	0.4911	0.4913	0.4916
2.4	0.4918	0.4920	0.4922	0.4925	0.4927	0.4929	0.4931	0.4932	0.4934	0.4936
2.5	0.4938	0.4940	0.4941	0.4943	0.4945	0.4946	0.4948	0.4949	0.4951	0.4952
2.6	0.4953	0.4955	0.4956	0.4957	0.4959	0.4960	0.4961	0.4962	0.4963	0.4964
2.7	0.4965	0.4966	0.4967	0.4968	0.4969	0.4970	0.4971	0.4972	0.4973	0.4974
2.8	0.4974	0.4975	0.4976	0.4977	0.4977	0.4978	0.4979	0.4979	0.4980	0.4981
2.9	0.4981	0.4982	0.4982	0.4983	0.4984	0.4984	0.4985	0.4985	0.4986	0.4986
3.0	0.4987	0.4987	0.4987	0.4988	0.4988	0.4989	0.4989	0.4989	0.4990	0.4990

G Answers to Selected Exercises

Chapter 1 Logic

Section 1.1 Deductive vs. Inductive Reasoning

1. a. valid

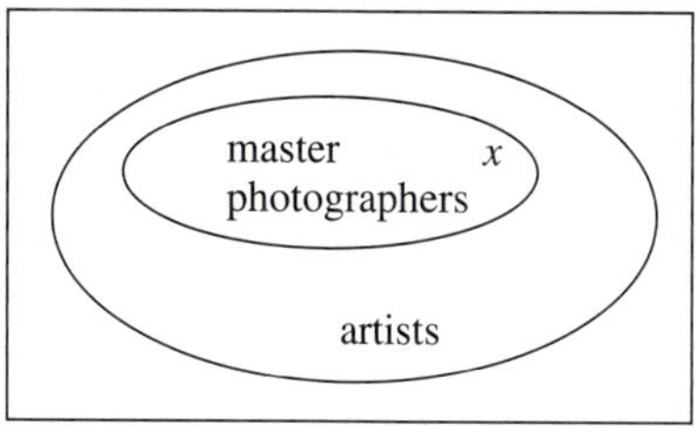

x = Ansel Adams

b. Not Valid

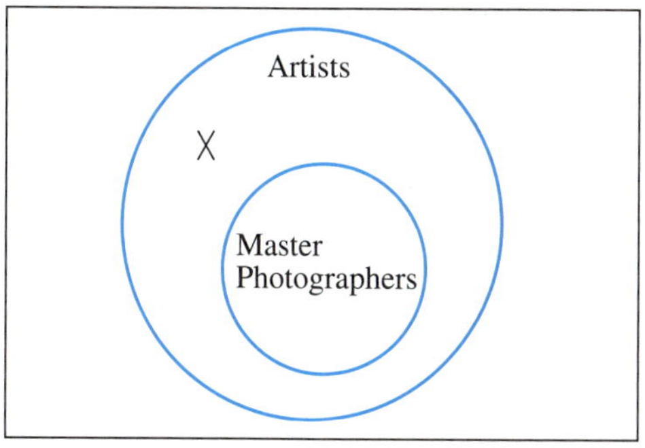

x = Ansel Adams

3. a. Not Valid

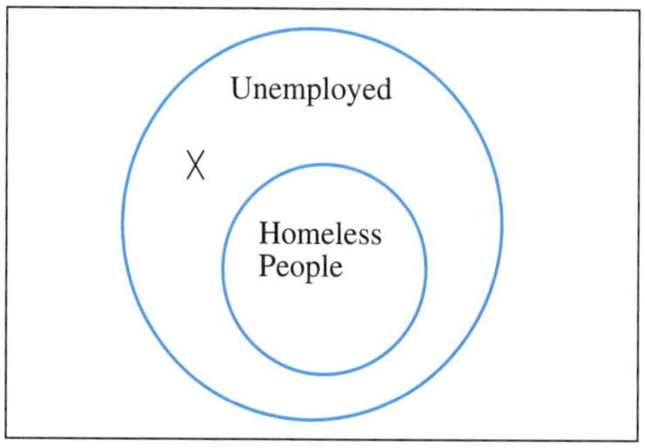

x = Bill Gates

b. Valid

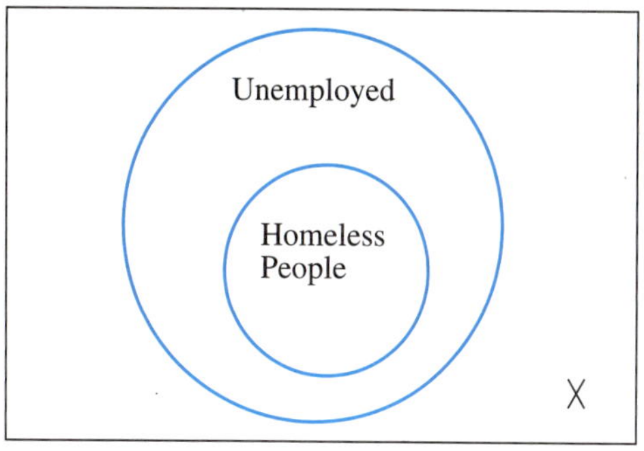

x = Bill Gates

5. invalid

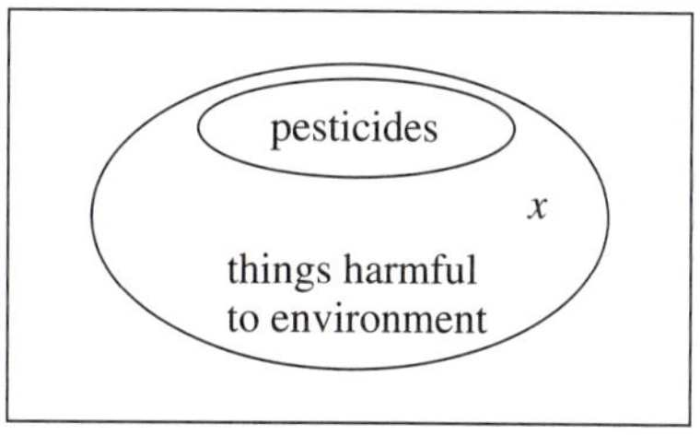

x = fertilizer

7. valid

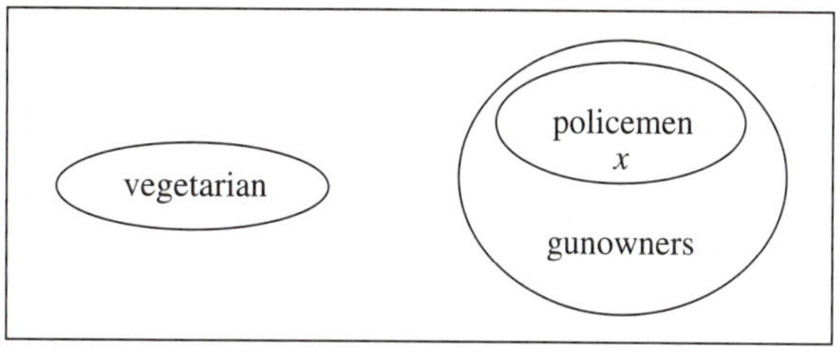

x = policeman

9. valid

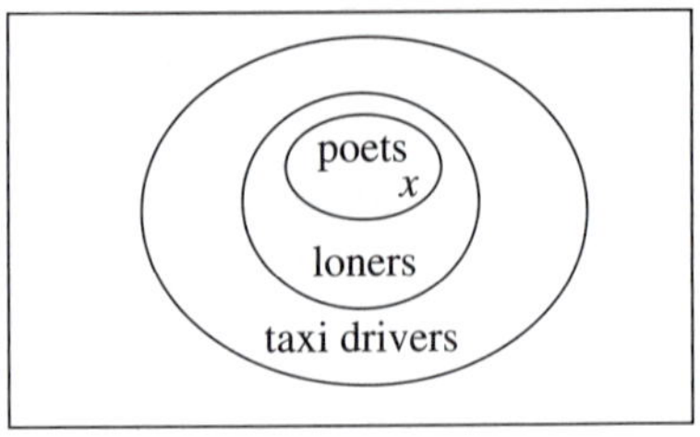

x = poet

11. valid

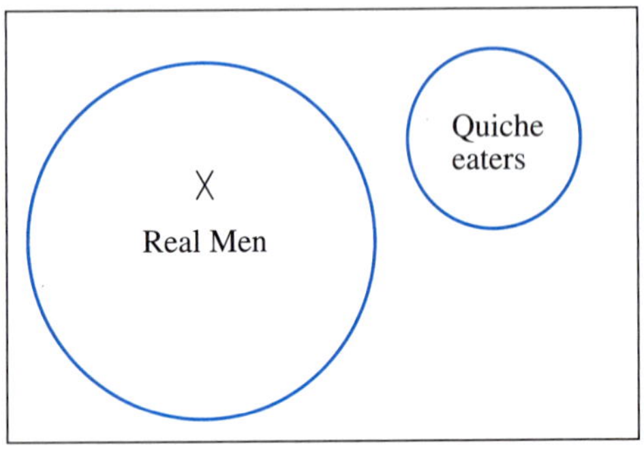

x = Clint Eastwood

13. valid

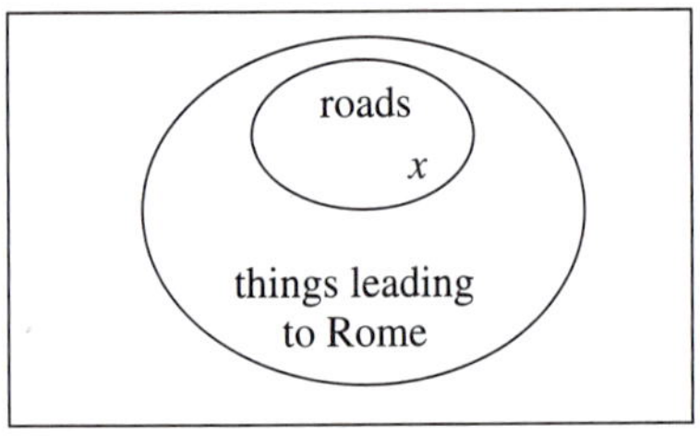

x = Route 66

15. invalid

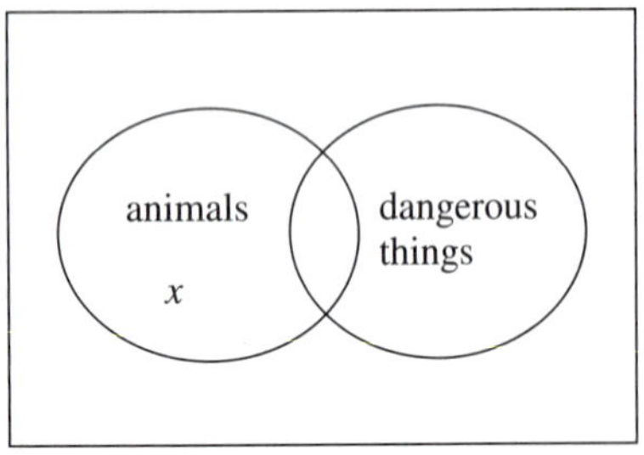

x = tiger

17. invalid

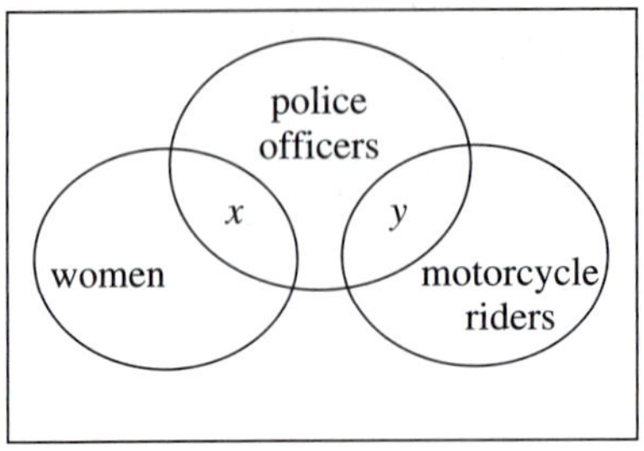

x = policewoman y = motorcycle officer

19. valid

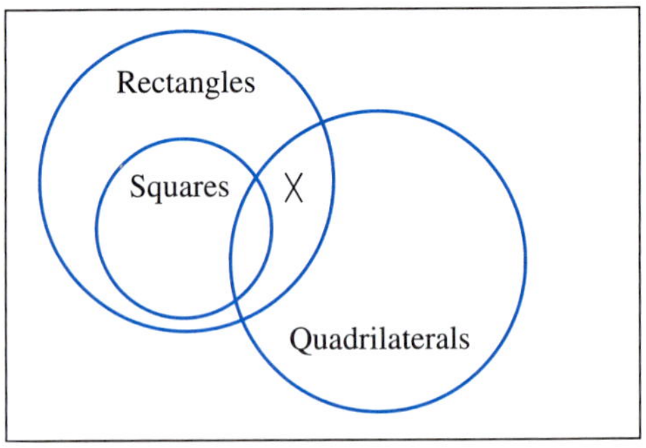

x = Quadrilateral

21. **a.** inductive **b.** deductive
23. 23. Add 5 each time
25. 20. Add 2 to the first term to get the second term, add 4 to the second term to get the third term, add 6 to the third term to get the fourth term, and so on.
27. 25. The first term is 1^2, the second term is 2^2, the third term is 3^2, and so on.
29. 13. Each term is a prime number.
31. 5. Think of each number in the sequence as a time (5:00, 8:00, etc.). Add three hours each time.
33. f, s; first letter of the natural numbers
35. T, W. Each member of the sequence is the first letter of the days of the week: Friday, Saturday, . . .
37. r. Each member of the sequence is the first letter of the corresponding word.
39. f. Each member of the sequence is the first letter after the first vowel in the corresponding word.
41. f. Look at the last letter in each word. Choose the next letter in the alphabet for the sequence.
43. 13. Add 3 each time.
1. Think of the numbers as times and add 3 hours each time.
47. $x = 3 \pm \sqrt{2}$

Section 1.2 Symbolic Logic

1. (a) and (b) are statements
3. (a) and (d) are negations; (b) and (c) are negations
5. **a.** Her dress is red.

b. No computer is priced under \$100. (All computers are priced \$100 or more.)
c. Some dogs are not four-legged animals.
d. Some sleeping bags are waterproof.

7. a. $p \wedge q$ b. $\sim p \rightarrow \sim q$
c. $\sim(p \vee q)$ d. $p \wedge \sim q$

9. a. $(p \vee q) \rightarrow r$ b. $q \wedge p \wedge r$
c. $r \wedge \sim(p \vee q)$ d. $q \rightarrow r$

11. p: a shape is a square
q: a shape is a rectangle
$p \rightarrow q$

13. p: a number is a whole number
q: a number is even
r: a number is odd
$p \rightarrow (q \vee r)$

15. p: a person is born in the U.S.
q: a person is an American citizen
$p \rightarrow q$

17. p: a car is a muscle car from the 60's
q: a car pollutes
$p \rightarrow q$

19. p: I sleep soundly
q: I drink coffee
r: I eat chocolate
$(q \vee r) \rightarrow (\sim p)$

21. p: your check is accepted
q: you have a driver's license
r: you have a credit card
$\sim(q \vee r) \rightarrow (\sim p)$

23. p: You do drink.
q: You do drive.
r: You are fined.
s: You go to jail.
$(p \wedge q) \rightarrow (r \vee s)$

25. p: you get a refund
q: you get store credit
r: the product is defective
$r \rightarrow (p \vee q)$

27. a. I am an environmentalist and I recycle my aluminum cans.
b. If I am an environmentalist, then I recycle my aluminum cans.
c. If I do not recycle my aluminum cans, then I am not an environmentalist.
d. I recycle my aluminum cans or I am not an environmentalist.

29. a. If I recycle my aluminum cans or newspapers, then I am an environmentalist.
b. If I am not an environmentalist, then I do not recycle my aluminum cans or newspapers.
c. I recycle my aluminum cans and newspapers or I am not an environmentalist.
d. If I recycle my newspapers and do not recycle my aluminum cans, then I am not an environmentalist.

Section 1.3 Truth Tables

1.

	p	q	$\sim q$	$p \vee \sim q$
1.	T	T	F	T
2.	T	F	T	T
3.	F	T	F	F
4.	F	F	T	T

3.

	p	$\sim p$	$p \vee \sim p$
1.	T	F	T
2.	F	T	T

5.

	p	q	$\sim q$	$p \rightarrow \sim q$
1.	T	T	F	F
2.	T	F	T	T
3.	F	T	F	T
4.	F	F	T	T

7.

	p	q	$\sim p$	$\sim q$	$\sim q \rightarrow \sim p$
1.	T	T	F	F	T
2.	T	F	F	T	F
3.	F	T	T	F	T
4.	F	F	T	T	T

9.

	p	q	$\sim p$	$p \vee q$	$(p \vee q) \rightarrow \sim p$
1.	T	T	F	T	F
2.	T	F	F	T	F
3.	F	T	T	T	T
4.	F	F	T	F	T

11.

	p	q	$p \vee q$	$p \wedge q$	$(p \vee q) \rightarrow (p \wedge q)$
1.	T	T	T	T	T
2.	T	F	T	F	F
3.	F	T	T	F	F
4.	F	F	F	F	T

13.

	p	q	r	$q \vee r$	$\sim(q \vee r)$	$p \wedge \sim(q \vee r)$
1.	T	T	T	T	F	F
2.	T	T	F	T	F	F
3.	T	F	T	T	F	F
4.	T	F	F	F	T	T
5.	F	T	T	T	F	F
6.	F	T	F	T	F	F
7.	F	F	T	T	F	F
8.	F	F	F	F	T	F

15.

	p	q	r	$\sim q$	$\sim q \wedge r$	$p \vee (\sim q \wedge r)$
1.	T	T	T	F	F	T
2.	T	T	F	F	F	T
3.	T	F	T	T	T	T
4.	T	F	F	T	F	T
5.	F	T	T	F	F	F
6.	F	T	F	F	F	F
7.	F	F	T	T	T	T
8.	F	F	F	T	F	F

17.

	p	q	r	$\sim r$	$\sim r \vee p$	$q \wedge p$	$(\sim r \vee p) \rightarrow (q \wedge p)$
1.	T	T	T	F	T	T	T
2.	T	T	F	T	T	T	T
3.	T	F	T	F	T	F	F
4.	T	F	F	T	T	F	F
5.	F	T	T	F	F	F	T
6.	F	T	F	T	T	F	F
7.	F	F	T	F	F	F	T
8.	F	F	F	T	T	F	F

19.

	p	q	r	$\sim r$	$p \vee r$	$q \wedge \sim r$	$(p \vee r) \rightarrow (q \wedge \sim r)$
1.	T	T	T	F	T	F	F
2.	T	T	F	T	T	T	T
3.	T	F	T	F	T	F	F
4.	T	F	F	T	T	F	F
5.	F	T	T	F	T	F	F
6.	F	T	F	T	F	T	T
7.	F	F	T	F	T	F	F
8.	F	F	F	T	F	F	T

21. p: It is raining.
q: The streets are wet.
$p \rightarrow q$

	p	q	$p \rightarrow q$
1.	T	T	T
2.	T	F	F
3.	F	T	T
4.	F	F	T

23. p: It rains.
q: The water supply is rationed.
$\sim p \rightarrow q$

	p	q	$\sim p$	$\sim p \rightarrow q$
1.	T	T	F	T
2.	T	F	F	T
3.	F	T	T	T
4.	F	F	T	F

25. p: a shape is a square
q: a shape is a rectangle
$p \rightarrow q$

p	q	$p \rightarrow q$
T	T	T
T	F	F
F	T	T
F	F	T

27. p: it is a square
q: it is a triangle
$p \rightarrow (\sim q)$

p	q	$\sim q$	$p \rightarrow (\sim q)$
T	T	F	F
T	F	T	T
F	T	F	T
F	F	T	T

29. p: you have a drivers license
q: you have a credit card
r: your check is approved
$(p \vee q) \rightarrow r$

p	q	r	$p \vee q$	$(p \vee q) \rightarrow r$
T	T	T	T	T
T	T	F	T	F
T	F	T	T	T
T	F	F	T	F
F	T	T	T	T
F	T	F	T	F
F	F	T	F	T
F	F	F	F	T

31. p: Leaded gasoline is used.
q: The catalytic converter is damaged.
r: The air is polluted.
$p \rightarrow (q \wedge r)$

	p	q	r	$q \wedge r$	$p \rightarrow (q \wedge r)$
1.	T	T	T	T	T
2.	T	T	F	F	F
3.	T	F	T	F	F
4.	T	F	F	F	F
5.	F	T	T	T	T
6.	F	T	F	F	T
7.	F	F	T	F	T
8.	F	F	F	F	T

33. p: I have a college degree.
q: I have a job.
r: I own a house.
$p \wedge \sim(q \vee r)$

	p	q	r	$q \vee r$	$\sim(q \vee r)$	$p \wedge \sim(q \vee r)$
1.	T	T	T	T	F	F
2.	T	T	F	T	F	F
3.	T	F	T	T	F	F
4.	T	F	F	F	T	T
5.	F	T	T	T	F	F
6.	F	T	F	T	F	F
7.	F	F	T	T	F	F
8.	F	F	F	F	T	F

35. p: Proposition A passes.
q: Proposition B passes.
r: Jobs are lost.
s: New taxes are imposed.
$(p \wedge \sim q) \rightarrow (r \vee s)$

	p	q	r	s	$\sim q$	$p \wedge \sim q$	$r \vee s$	$(p \wedge \sim q) \rightarrow (r \vee s)$
1.	T	T	T	T	F	F	T	T
2.	T	T	T	F	F	F	T	T
3.	T	T	F	T	F	F	T	T
4.	T	T	F	F	F	F	F	T
5.	T	F	T	T	T	T	T	T
6.	T	F	T	F	T	T	T	T
7.	T	F	F	T	T	T	T	T
8.	T	F	F	F	T	T	F	F
9.	F	T	T	T	F	F	T	T
10.	F	T	T	F	F	F	T	T
11.	F	T	F	T	F	F	T	T
12.	F	T	F	F	F	F	F	T
13.	F	F	T	T	T	F	T	T
14.	F	F	T	F	T	F	T	T
15.	F	F	F	T	T	F	T	T
16.	F	F	F	F	T	F	F	T

p: Proposition A passes.
q: The legislature raises taxes.
r: The quality of education is lowered.
s: Unemployment rises.

37. equivalent

	p	q	$\sim q$	$p \vee \sim q$	$q \rightarrow p$
1.	T	T	F	T	T
2.	T	F	T	T	T
3.	F	T	F	F	F
4.	F	F	T	T	T

39. equivalent

	p	q	$\sim p$	$\sim q$	$p \vee \sim q$	$\sim p \rightarrow \sim q$
1.	T	T	F	F	T	T
2.	T	F	F	T	T	T
3.	F	T	T	F	F	F
4.	F	F	T	T	T	T

41. not equivalent

	p	q	$p \rightarrow q$	$q \rightarrow p$
1.	T	T	T	T
2.	T	F	F	T
3.	F	T	T	F
4.	F	F	T	T

43. equivalent

	p	q	$\sim p$	$\sim q$	$p \rightarrow q$	$\sim q \rightarrow \sim p$
1.	T	T	F	F	T	T
2.	T	F	F	T	F	F
3.	F	T	T	F	T	T
4.	F	F	T	T	T	T

45. not equivalent

	p	q	$\sim p$	$\sim q$	$p \vee \sim q$	$q \wedge \sim p$
1.	T	T	F	F	T	F
2.	T	F	F	T	T	F
3.	F	T	T	F	F	T
4.	F	F	T	T	T	F

49. I do not have a college degree or I am employed.
51. The television set is not broken and there is not a power outage.
53. The building contains asbestos and the original contractor is not responsible.
55. The lyrics are censored and the First Amendment has not been violated.

Section 1.4 More on Conditionals

1. a. If she is a police officer, then she carries a gun.
b. If she carries a gun, then she is a police officer.
c. If she is not a police officer, then she does not carry a gun.
d. If she does not carry a gun, then she is not a police officer.
e. a and d; b and c

3. a. If I watch television, then I do not do my homework.
b. If I do not do my homework, then I watch television.
c. If I do not watch television, then I do my homework.
d. If I do my homework, then I do not watch television.
e. a and d; b and c

5. a. If you do not pass this mathematics course, then you do not fulfill a graduation requirement.
b. If you fulfill a graduation requirement, then you pass this mathematics course.
c. If you do not fulfill a graduation requirement, then you do not pass this mathematics course.

7. a. If the electricity is turned on, then the television set does work.
b. If the television set does not work, then the electricity is turned off.
c. If the television set does work, then the electricity is turned on.

9. a. If you eat meat, then you are not a vegetarian.
b. If you are a vegetarian, then you do not eat meat.
c. If you are not a vegetarian, then you do eat meat.

11. a. premise: I take public transportation.
conclusion: Public transportation is convenient.
b. If I take public transportation, then it is convenient.
c. false statement when I take public transportation and it is not convenient

13. a. premise: I buy foreign products.
conclusion: Domestic products are not available.
b. If I buy foreign products, then domestic products are not available.
c. false statement when I buy foreign products and domestic products are available

15. a. premise: You may become a United States senator.
conclusion: You are at least thirty years old and have been a citizen for nine years.
b. If you become a United States senator, then you are at least thirty years old and have been a citizen for 9 years.
c. false statement when you become a United States senator and either you are not at least thirty years old or have not been a citizen for nine years, or both

17. If you obtain a refund, then you have a receipt, and if you have a receipt, then you will obtain a refund.
19. If $ax^2 + bx + c = 0$ has two distinct real solutions, then $b^2 - 4ac > 0$ and if $b^2 - 4ac > 0$, then $ax^2 + bx + c = 0$ has two distinct real solutions.
21. If a polygon is a triangle, then the polygon has three sides, and if a polygon has three sides, then the polygon is a triangle.
23. equivalent
25. equivalent
27. equivalent
29. If I do not walk to work, then it is raining.
31. If it is not cold, then it is not snowing.
33. If you are a vegetarian, then you do not eat meat.
35. (i) and (iv); (ii) and (iii)
37. (i) and (iii); (ii) and (iv)
39. (i) and (iii); (ii) and (iv)

Section 1.5 Analyzing Arguments

1. 1. $p \rightarrow q$
2. p

$\therefore q$

3. 1. $p \rightarrow q$
2. $\sim q$

$\therefore \sim p$

5. 1. $p \rightarrow q$
2. $\sim p$

$\therefore \sim q$

7. valid

9. valid

11. invalid; the argument is invalid when you do not exercise regularly and you are healthy

13. invalid; the argument is invalid when the Democrats have a majority and Smith is appointed and student loans are funded

15. invalid; the argument is invalid when
(1) you do not argue with a police officer and you do get a ticket and you do break the speed limit, or
(2) you do not argue with a police officer and you do not get a ticket and you do break the speed limit

17. Rewriting the argument:
If it is a pesticide, then it is harmful to the environment.
If it is a fertilizer, then it is not a pesticide.

If it is a fertilizer, then it is not harmful to the environment.
p: it is a pesticide.
q: it is harmful to the environment.
r: it is a fertilizer.

$p \rightarrow q$
$r \rightarrow \sim p$

$r \rightarrow \sim q$

$\{(p \rightarrow q) \wedge (r \rightarrow p)\} \rightarrow (r \rightarrow p)$

					1	2		3	
p	**q**	**r**	**$\sim p$**	**$\sim q$**	**$p \rightarrow q$**	**$r \rightarrow \sim p$**	**$1 \wedge 2$**	**$r \rightarrow \sim q$**	**$(1 \wedge 2) \rightarrow 3$**
T	T	T	F	F	T	F	F	F	T
T	T	F	F	F	T	T	T	T	T
T	F	T	F	T	F	F	F	T	T
T	F	F	F	T	F	T	F	T	T
F	T	T	T	F	T	T	T	F	F
F	T	F	T	F	T	T	T	T	T
F	F	T	T	T	T	T	T	T	T
F	F	F	T	T	T	T	T	T	T

The argument is not valid if it is not a pesticide and if it is harmful to the environment and it is a fertilizer.

19. Rewriting the argument:
If you are a poet, then you are a loner.
If you are a loner, then you are a taxi driver.

If you are a poet, then you are a taxi driver.
p: you are a poet.
q: you are a loner.
r: you are a taxi driver.

$p \rightarrow q$
$q \rightarrow r$

$p \rightarrow r$

$\{(p \rightarrow q) \wedge (q \rightarrow r)\} \rightarrow (p \rightarrow r)$

			1	2		3	
p	**q**	**r**	**$p \rightarrow q$**	**$q \rightarrow r$**	**$1 \wedge 2$**	**$p \rightarrow r$**	**$(1 \wedge 2) \rightarrow 3$**
T	T	T	T	T	T	T	T
T	T	F	T	F	F	F	T
T	F	T	F	T	F	T	T
T	F	F	F	T	F	F	T
F	T	T	T	T	T	T	T
F	T	F	T	F	F	T	T
F	F	T	T	T	T	T	T
F	F	F	T	T	T	T	T

The argument is valid.

21. Rewriting the argument:

If you are a professor, then you are not a millionaire.
If you are a millionaire, then you are literate.
―――――――――――――――――――
If you are a professor, then you are literate.

p: you are a professor.
q: you are a millionaire.
r: you are literate.

$p \rightarrow \sim q$
$q \rightarrow r$
―――――
$p \rightarrow r$

$\{(p \rightarrow \sim q) \wedge (q \rightarrow r)\} \rightarrow (p \rightarrow r)$

				1	2		3	
p	q	r	$\sim q$	$p \rightarrow \sim q$	$q \rightarrow r$	$1 \wedge 2$	$p \rightarrow r$	$(1 \wedge 2) \rightarrow 3$
T	T	T	F	F	T	F	T	T
T	T	F	F	F	F	F	F	T
T	F	T	T	T	T	T	T	T
T	F	F	T	T	T	T	F	F
F	T	T	F	T	T	T	T	T
F	T	F	F	T	F	F	T	T
F	F	T	T	T	T	T	T	T
F	F	F	T	T	T	T	T	T

The argument is not valid if you are a professor who is not a millionaire and is illiterate.

23. valid
25. valid
27. invalid; the argument is invalid when
(1) you are in a hurry and you eat at Lulu's Diner and you do not eat good food, or
(2) you are not in a hurry and you eat at Lulu's Diner and you do not eat good food
29. valid
31. valid
33. valid
35. valid
37. valid

Chapter 1 Review

1. a. inductive b. deductive
2. g. The next letter in the sequence is the first consonant after the first vowel in the word.
3. 9. Think of each number in the sequence as a time (1:00, 6:00, etc.). Add five hours to get the next number in the sequence.
4. valid

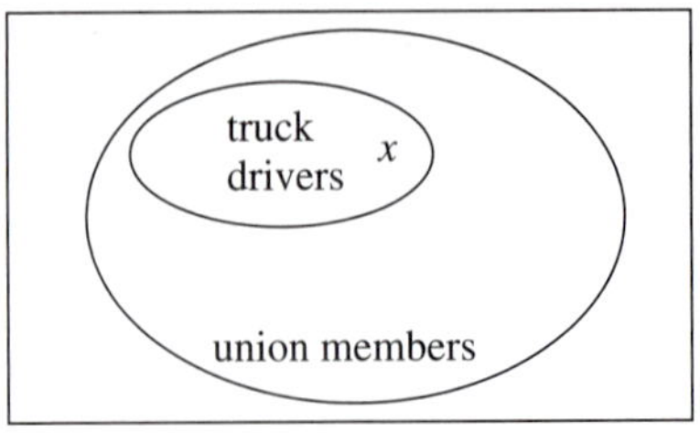

x = Rocky

5. invalid

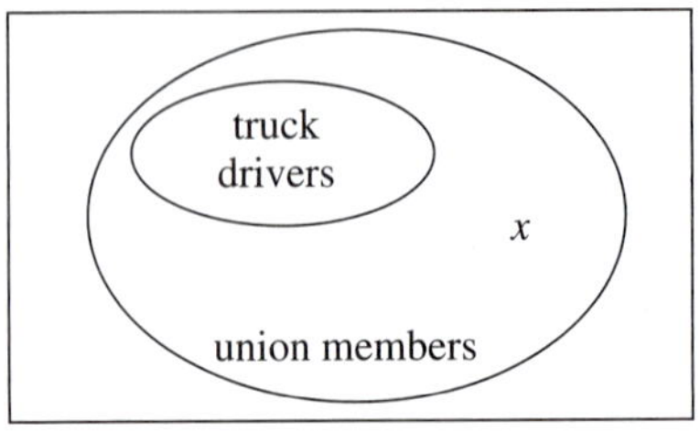

x = Rocky

6. invalid

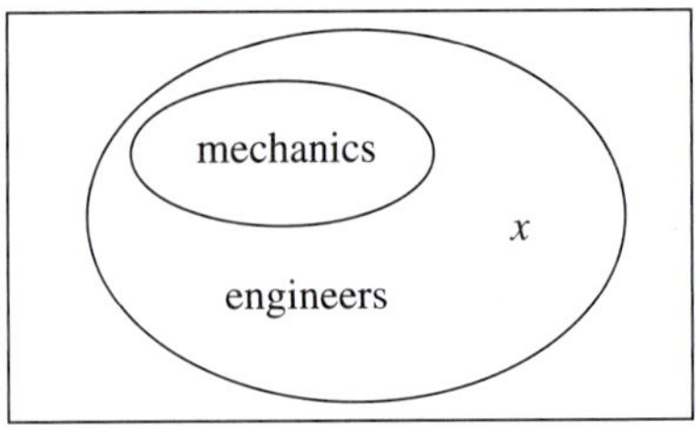

x = Casey Jones

7. valid

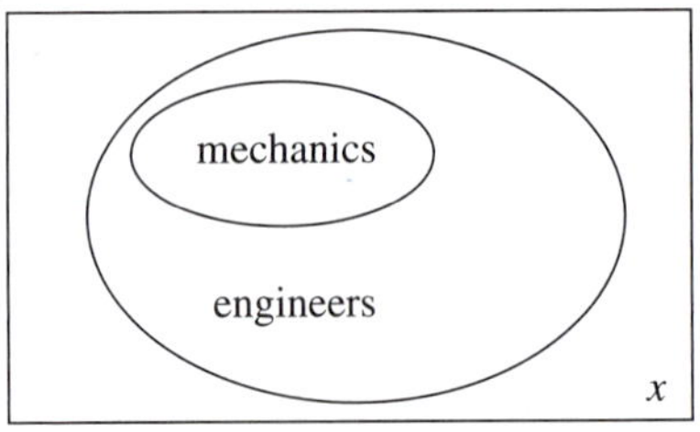

x = Casey Jones

8. invalid

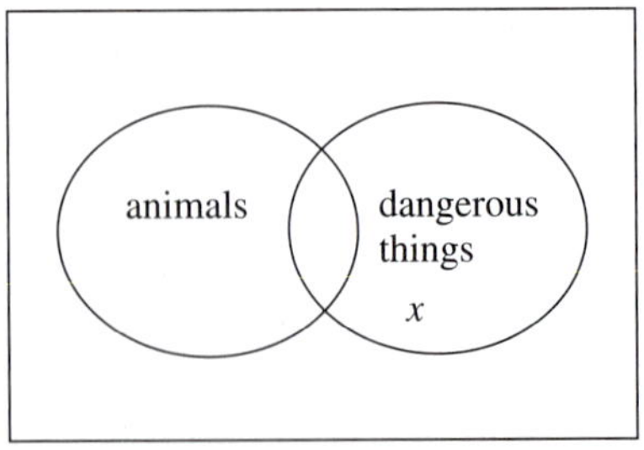

x = gun

9. valid

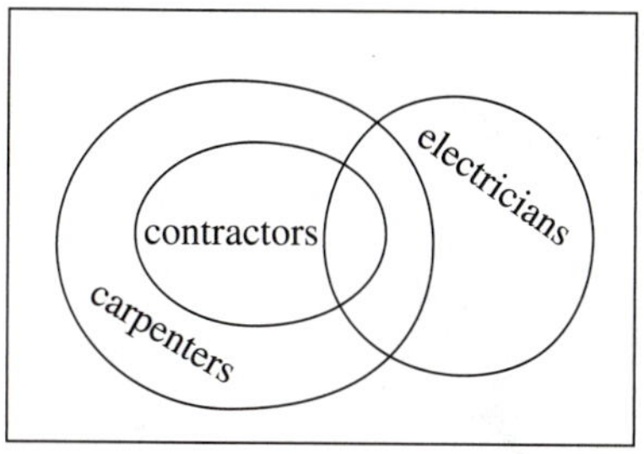

10. (a) and (b) are statements

11. (a) and (d) are negations; (b) and (c) are negations.

12. **a.** His car is new.
b. No building is earthquakeproof.
c. Some children do not eat candy.
d. Sometimes I cry in a movie theater.

13. **a.** $p \wedge q$ **b.** $\sim q \rightarrow \sim p$
c. $p \wedge \sim q$ **d.** $\sim(p \vee q)$

14. **a.** $q \rightarrow p$ **b.** $\sim r \wedge \sim q \wedge p$
c. $r \rightarrow \sim p$ **d.** $p \wedge \sim(q \vee r)$

15. **a.** If it is expensive, then it is desirable.
b. It is undesirable if and only if it is not expensive.
c. It is not expensive or undesirable.
d. It is expensive and it is desirable, or it is not expensive and it is undesirable.

16. **a.** If the movie is critically acclaimed or a box office hit, then the movie is available on videotape.
b. If the movie is critically acclaimed and it is not a box office hit, then the movie is not available on videotape.
c. The movie is not critically acclaimed or a box office hit and it is available on videotape.
d. If the movie is not available on videotape, then the movie is not critically acclaimed and it is not a box office hit.

17.

	p	q	$\sim q$	$p \vee \sim q$
1.	T	T	F	T
2.	T	F	T	T
3.	F	T	F	F
4.	F	F	T	T

18.

	p	q	$\sim q$	$p \wedge \sim q$
1.	T	T	F	F
2.	T	F	T	T
3.	F	T	F	F
4.	F	F	T	F

19.

	p	q	$\sim p$	$\sim p \rightarrow q$
1.	T	T	F	T
2.	T	F	F	T
3.	F	T	T	T
4.	F	F	T	F

20.

	p	q	$\sim q$	$p \wedge q$	$(p \wedge q) \rightarrow \sim q$
1.	T	T	F	T	F
2.	T	F	T	F	T
3.	F	T	F	F	T
4.	F	F	T	F	T

21.

	p	q	r	$p \vee r$	$\sim(p \vee r)$	$q \vee \sim(p \vee r)$
1.	T	T	T	T	F	T
2.	T	T	F	T	F	T
3.	T	F	T	T	F	F
4.	T	F	F	T	F	F
5.	F	T	T	T	F	T
6.	F	T	F	F	T	T
7.	F	F	T	T	F	F
8.	F	F	F	F	T	T

22.

	p	q	r	$\sim p$	$q \vee r$	$\sim p \rightarrow (q \vee r)$
1.	T	T	T	F	T	T
2.	T	T	F	F	T	T
3.	T	F	T	F	T	T
4.	T	F	F	F	F	T
5.	F	T	T	T	T	T
6.	F	T	F	T	T	T
7.	F	F	T	T	T	T
8.	F	F	F	T	F	F

23.

	p	q	r	$\sim r$	$q \wedge p$	$\sim r \vee p$	$(q \wedge p) \rightarrow (\sim r \vee p)$
1.	T	T	T	F	T	T	T
2.	T	T	F	T	T	T	T
3.	T	F	T	F	F	T	T
4.	T	F	F	T	F	T	T
5.	F	T	T	F	F	F	T
6.	F	T	F	T	F	T	T
7.	F	F	T	F	F	F	T
8.	F	F	F	T	F	T	T

24.

	p	q	r	$\sim r$	$p \vee r$	$q \wedge \sim r$	$(p \vee r) \rightarrow (q \wedge \sim r)$
1.	T	T	T	F	T	F	F
2.	T	T	F	T	T	T	T
3.	T	F	T	F	T	F	F
4.	T	F	F	T	T	F	F
5.	F	T	T	F	T	F	F
6.	F	T	F	T	F	T	T
7.	F	F	T	F	T	F	F
8.	F	F	F	T	F	F	T

25. equivalent
26. not equivalent
27. not equivalent
28. equivalent
29. Jesse did not have a party or somebody came.
30. You pay the fine and you go to jail.
31. I am not the winner and you are not blind.
32. He is employed or he did apply for financial assistance.
33. His application is ignored and the selection procedure has not been violated.
34. The jackpot is less than \$1 million.
35. **a.** If you are an avid jogger, then you are healthy.
b. If you are healthy, then you are an avid jogger.
c. If you are not an avid jogger, then you are not healthy.
d. If you are not healthy, then you are not an avid jogger.
e. You are an avid jogger if and only if you are healthy.
36. **a.** If he is not elected, then the country is not in big trouble.
b. If the country is in big trouble, then he is elected.
c. If the country is not in big trouble, then he is not elected.
37. **a.** premise: The economy improves.
conclusion: Unemployment goes down.
b. If the economy improves, then unemployment goes down.
38. **a.** premise: Unemployment goes down.
conclusion: The economy improves.
b. If unemployment goes down, then the economy improves.
39. If it is a computer, then it is repairable.
40. If it is a gemstone, then it is valuable.
41. not equivalent
42. equivalent
43. (i) and (iii); (ii) and (iv)
44. invalid
45. valid
46. invalid
47. valid
48. valid
49. invalid
50. invalid
51. invalid
52. p: a person is a professor
q: a person is educated
r: an animal is a monkey

$$\begin{array}{l} p \rightarrow q \\ \underline{r \rightarrow \sim q} \\ p \rightarrow \sim r \end{array}$$

$\{[p \rightarrow q] \wedge [r \rightarrow \sim q]\} \rightarrow (p \rightarrow \sim r)$
The argument is valid.

53. p: a person is a professor
q: a person is educated
r: an animal is a monkey

$$\begin{array}{l} p \rightarrow q \\ \underline{r \rightarrow \sim p} \\ r \rightarrow \sim q \end{array}$$

$\{[p \rightarrow q] \wedge [r \rightarrow \sim p]\} \rightarrow (r \rightarrow \sim q)$
The argument is invalid.

54. p: vehicles stop
q: the traffic light is red
r: there is an accident

$$\begin{array}{l} q \rightarrow p \\ p \rightarrow \sim r \\ \underline{r} \\ \sim q \end{array}$$

$\{(q \rightarrow p) \wedge (p \rightarrow \sim r) \wedge r\} \rightarrow \sim q$
The argument is valid.

Chapter 2 Sets and Counting

Section 2.1

1. **a.** well-defined
b. not well-defined
c. well-defined
d. not well-defined
3. proper: ∅, {Lennon}, {McCartney}
improper: {Lennon, McCartney}
5. proper: ∅, {yes}, {no}, {undecided}, {yes, no}, {yes, undecided}, {no, undecided}
improper: {yes, no, undecided}

7. a. {4, 5}
b. {1, 2, 3, 4, 5, 6, 7, 8}
c. {0, 6, 7, 8, 9}
d. {0, 1, 2, 3, 9}

9. a. { }
b. {0, 1, 2, 3, 4, 5, 6, 7, 8, 9}
c. {0, 2, 4, 6, 8}
d. {1, 3, 5, 7, 9}

11. {Friday}

13. {Monday, Tuesday, Wednesday, Thursday}

15. {Friday, Saturday, Sunday}

17.

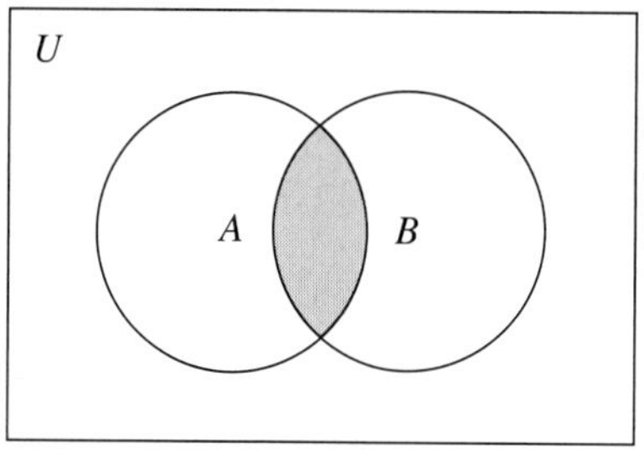

19.

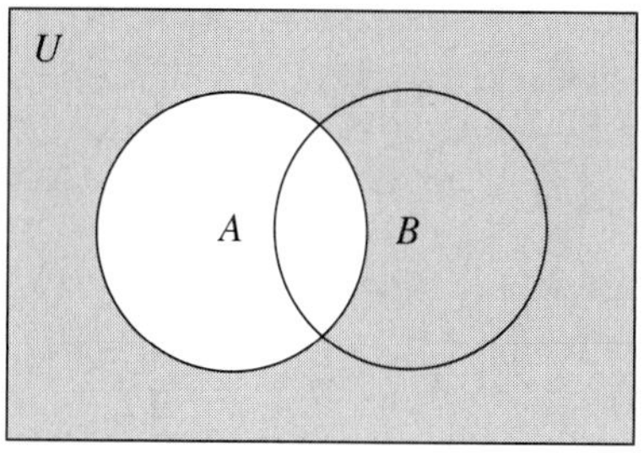

21.

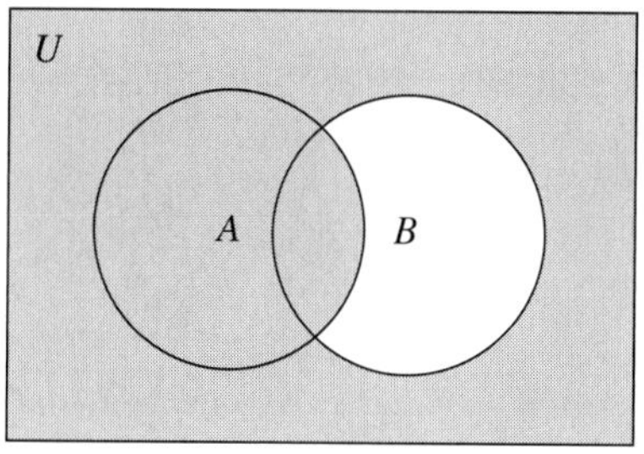

23.

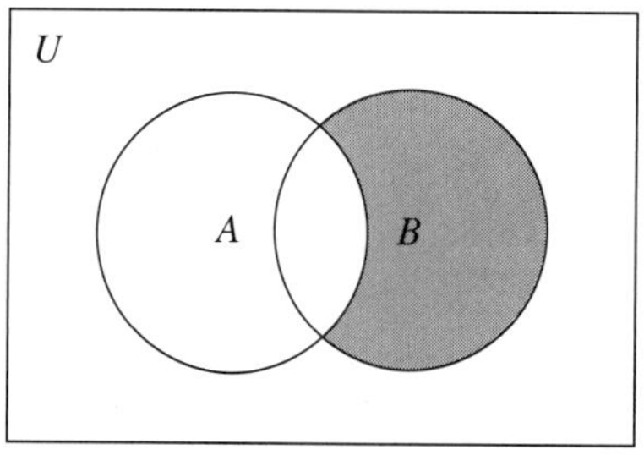

25.

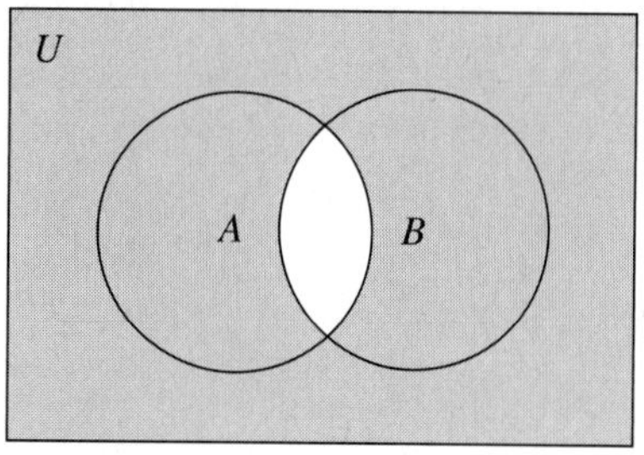

27. a. 21

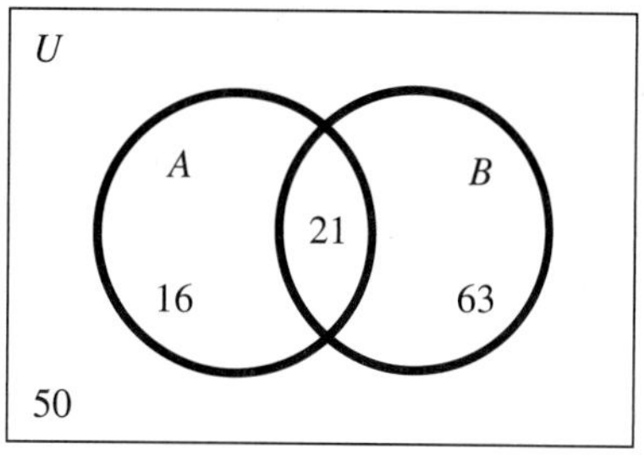

b. 0

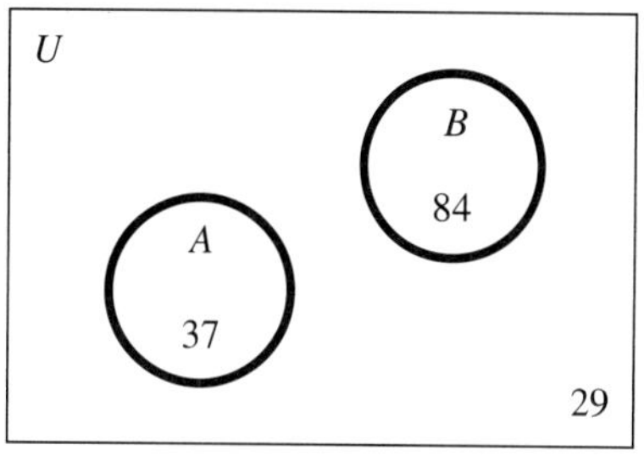

29. a.

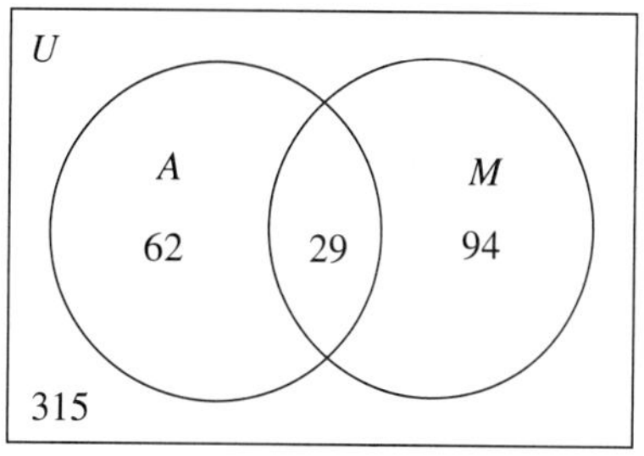

b. 37.0%

31. a.

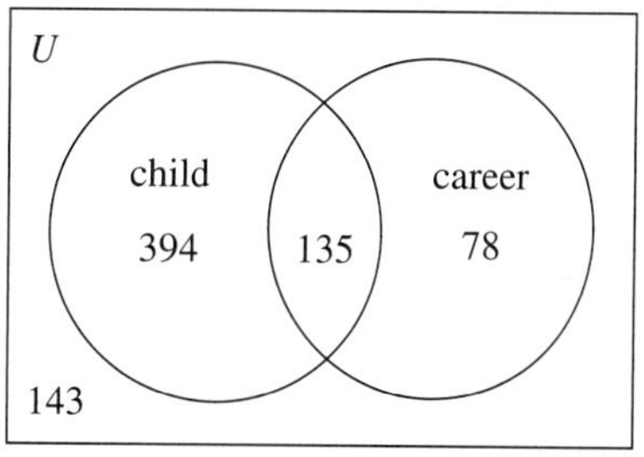

b. 18.0%

33. 42
35. 43
37. 8
39. 5
41. 16
43. 32
45. 6
47. 8
49. 0
51. **a.** {1, 2, 3} = A **b.** {1, 2, 3, 4, 5, 6} = B
c. E , F **d.** E , F
53. **a.** ∅, {a}; 2
b. ∅, {a}, {b}, {a, b}; 4
c. ∅, {a}, {b}, {c}, {a, b}, {a, c}, {b, c}, {a, b, c}; 8
d. ∅, {a}, {b}, {c}, {d}, {a, b}, {a, c}, {a, d}, {b, c}, {b, d}, {c, d}, {a, b, c}, {a, b, d}, {a, c, d}, {b, c, d}, {a, b, c, d}; 16
e. yes; $2^{n(A)}$
f. 64
65. (d)
67. (c)
69. (e)

Section 2.2 Applications of Venn Diagrams

1. **a.** 143 **b.** 16 **c.** 49 **d.** 57
3. **a.** 408 **b.** 1,343 **c.** 664 **d.** 149
5. **a.** 106 **b.** 448 **c.** 265 **d.** 159
7. **a.** $x + y - z$ **b.** $x - z$
c. $y - z$ **d.** $w - (x + y - z)$
9. **a.** 43.8% **b.** 10.8%
11. **a.** 77% **b.** 22%
13. **a.** 0% **b.** 54.1%
15. **a.** 44.0% **b.** 12.8%
17. **a.** 20.0% **b.** 58.8%
19. 16
21. {0, 4, 5}
23. {1, 2, 3, 6, 7, 8, 9}
25.

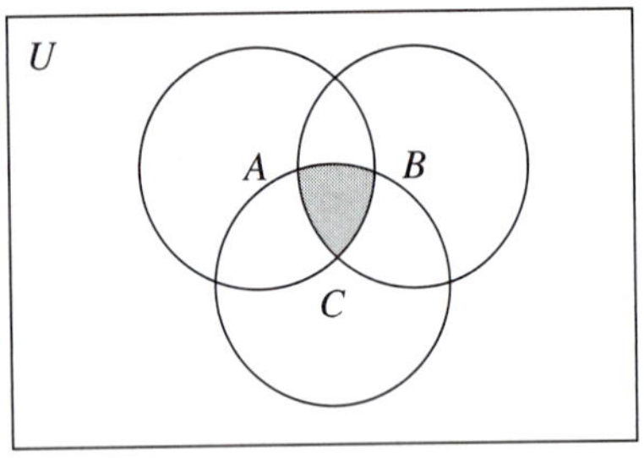

27.

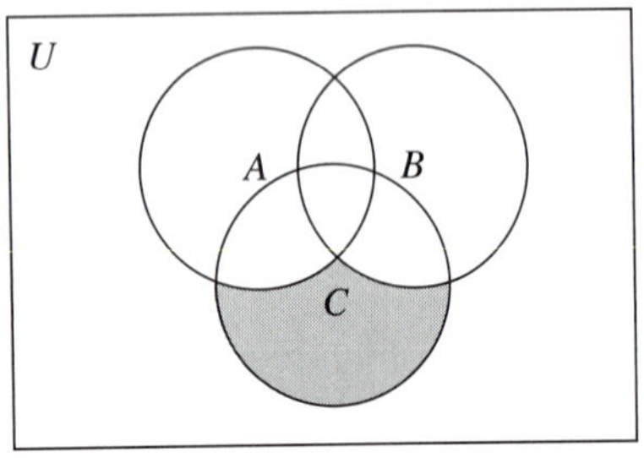

29.

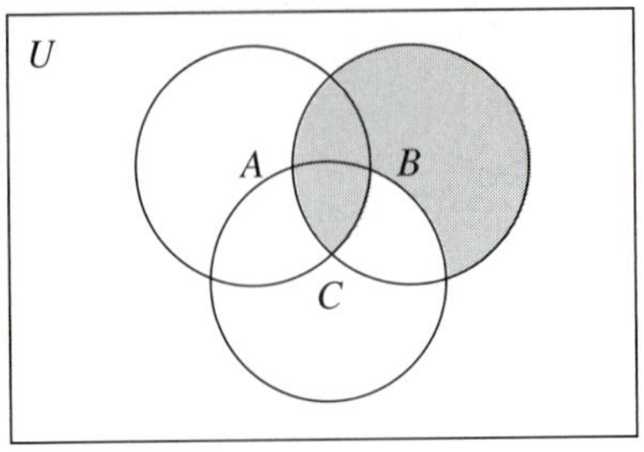

35. (d)
37. (d)
39. (b)
41. (e)

Section 2.3 Introduction to Combinatorics

1. **a.** 8

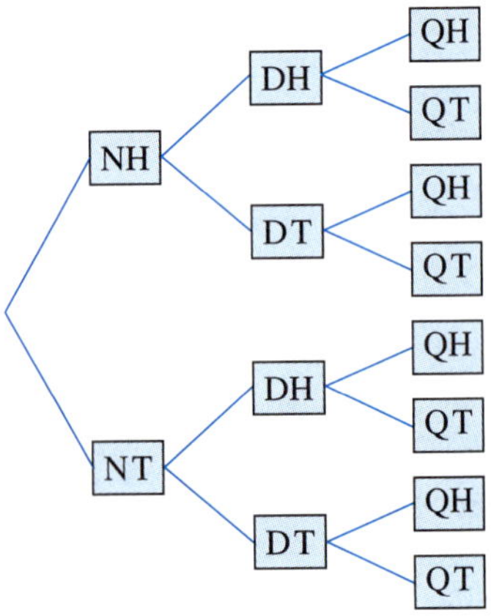

3. **a.** 12

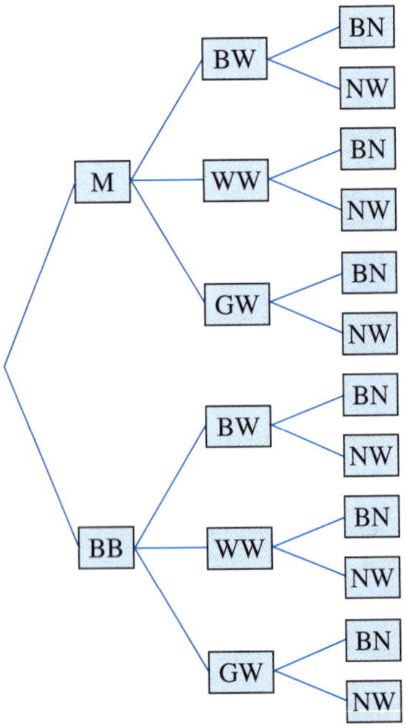

5. 24
7. 720
9. 2,646

11. 216
13. 1 billion
15. 10^{10} or 10 billion
17. a. 128 **b.** 800
19. a. 1.17936×10^9 **b.** 6.76×10^{10} **c.** 4.2×10^{10}
21. 540,000
23. 24
25. 3,628,800
27. $2.432902008 \times 10^{18}$
29. 17,280
31. a. 30 **b.** 360
33. 56
35. 70
37. 3,321
39. $1.046139494 \times 10^{13}$
41. 120
43. 35
45. 1
51. (b)
53. (d)

Section 2.4 Permutations and Combinations

1. a. 210 **b.** 35
3. a. 120 **b.** 1
5. a. 14 **b.** 14
7. a. 970,200 **b.** 161,700
9. a. $x!$ **b.** x
11. a. $x^2 - x$ **b.** $\frac{x^2 - x}{2}$
13. a. 6
b. $\{a, b\}, \{a, c\}, \{b, a\}, \{b, c\}, \{c, a\}, \{c, b\}$
15. a. 6
b. {a, b}, {a, c}, {a, d}, {b, c}, {b, d}, {c, d}
17. a. 479,001,600 **b.** 1
19. 24
21. 91
23. 2,184
25. a. 420 **b.** 1,001 **c.** 406
27. 2,598,960
29. a. 4,512 **b.** 58,656
31. 123,552
33. 22,957,480
35. 376,992
37. 5/36
39. The first five rows of triangle are:

1
1 1
1 2 1
1 3 3 1
1 4 6 4 1

a. 1 **b.** 2
c. 4 **d.** 8
e. Yes. The sum of each row is twice the previous sum.
f. We predict $2 \cdot 8 = 16$
g. The sum is 16. Our prediction was correct.
h. 2^{n-1}

41. a. 4th row **b.** nth row
c. no, the third **d.** yes
e. nth row, $(r + 1)$th number
45. (c)
47. (c)
49. (e)

Section 2.5 Voting Systems

1. a. 2,000 **b.** Cruz **c.** Yes
3. a. 10,351 **b.** Edelstein **c.** No
5. a. 30 **b.** Park **c.** 47%
d. Beach **e.** 53% **f.** Beach
g. 63 **h.** Beach **i.** 2
7. a. 65 **b.** Coastline **c.** 48%
d. Coastline **e.** 63% **f.** Coastline
g. 150 **h.** Coastline **i.** 2
9. a. 140 **b.** Shattuck **c.** 40%
d. Nirglotis **e.** 51% **f.** Shattuck
g. 284 **h.** Nirglotis **i.** 2
11. a. 1,342 **b.** Jones **c.** 44%
d. Jones **e.** 50.4% **f.** Jones
g. 3,960 **h.** Jones **i.** 3
13. a. 31,754 **b.** Darter **c.** 52%
d. Darter (He has a majority, so no run off is needed.)
e. 52% **f.** Darter **g.** 138,797
h. Darter **i.** 4
15. $6! = 720$
17. ${}_6C_2 = 15$
19. a. 75 **b.** 25
21. a. 21 **b.** 0

Section 2.6

1. $n(S) = 4$, $n(C) = 4$; equivalent; state $\leftrightarrow$ capital city
3. $n(R) = 3$, $n(G) = 4$, not equivalent
5. $n(C) = 22$, $n(D) = 22$; equivalent; $3n \leftrightarrow 4n$
7. $n(G) = 250$, $n(H) = 251$; not equivalent
9. $n(A) = 62$, $n(B) = 62$; equivalent, $2n - 1 \leftrightarrow 2n + 123$
11. a. $n \leftrightarrow 2n - 1$ **b.** 918 **c.** $\frac{x+1}{2}$
d. 1,563 **e.** $2n - 1$
13. a. $n \leftrightarrow 3n$ **b.** 312 **c.** $\frac{x}{3}$
d. 2,808 **e.** $3n$
15. a. -344 **b.** 248 **c.** 755 **d.** $\aleph_0$

Chapter 2 Review

1. a. Well defined.
b. Not well defined.
c. Not well defined.
d. Well defined.
2. a. {1, 3, 5, 7, 9}
b. {0, 2, 4, 6, 8}

c. {0, 1, 2, 3, 4, 5, 6, 7, 8, 9}
d. ∅

3. **a.** A < B 5 {Maria, Nobuko, Leroy, Mickey, Kelly, Rachel, Deanna}
b. A > B 5 {Leroy, Mickey}

4. proper: ∅, {Dallas}, {Chicago}, {Tampa}, {Dallas, Chicago}, {Dallas, Tampa}, {Chicago, Tampa}
improper: {Dallas, Chicago, Tampa}

5. **a.** 18
b.

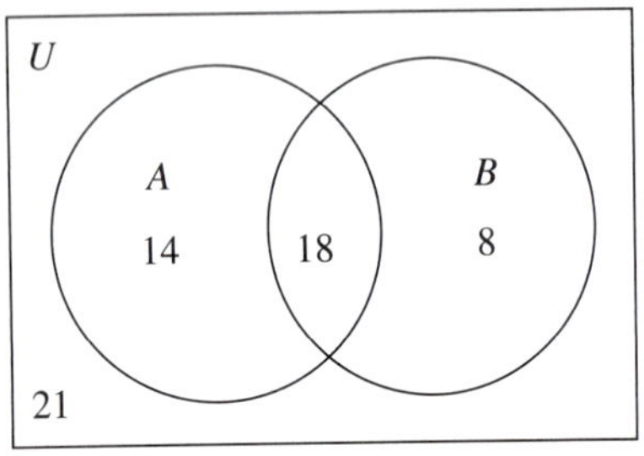

6. **a.** 1,670 **b.** 733 **c.** 346 **d.** 330
7. 29%
8. **a.** {b, f} **b.** {a, c, d, e, g, h, i}
9. **a.** 12 **b.**

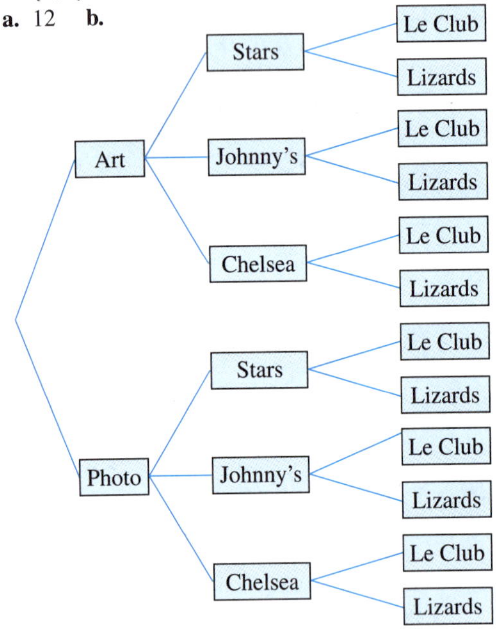

10. 180
11. 1,080,000
12. **a.** 3,628,800 **b.** 1 **c.** 531,360 **d.** 888,030
13. **a.** 165 **b.** 990
14. **a.** 32,760 **b.** 1,365 **c.** 54,486,432,000
15. **a.** 660 **b.** 1,540 **c.** 880
16. 45
17. 720
18. 133,784,560
19. 177,100,560
20. 5,245,786
22. **a.** row 7, number 4 **b.** row 7, number 5
c. same **d.** nth row, $(r + 1)$th number

23. **a.** 3 **b.** 3 **c.** 1 **d.** 1
e. 8 **f.** $2^{n(S)} = 2^3 = 8$
24. **a.** 74
b. Beethoven
c. 34%
d. Vivaldi
e. 59%
f. Beethoven
g. 197
h. Tie between Beethoven and Vivaldi
i. 2 points each
25. $n(A) = 5$, $n(B) = 5$; equivalent; Roman numeral ↔ number word
26. $n(C) = 450$, $n(D) = 450$; equivalent; $2n + 1 \leftrightarrow 2n$
27. $n(E) = 1$, $n(F) = 2$; not equivalent
28. **a.** $n \leftrightarrow n^2$ **b.** 29 **c.** $\sqrt{x}$
d. 20,736 **e.** n^2
29. **a.** 594 **b.** 264 **c.** 103 **d.** $\aleph_0$

Chapter 3 Probability

Section 3.1

5. **a.** win $85 **b.** win $85 **c.** lose $5
7. **a.** lose $20 **b.** win $160 **c.** lose $20
9. **a.** lose $10 **b.** win $50
11. **a.** win $50 **b.** lose $25
13. **a.** lose $50 **b.** win $50
15. **a.** lose $45 **b.** win $855 **c.** win $675
17. **a.** win $495 **b.** win $525 **c.** lose $45
d. lose $15
19. $3
21. $500
23. **a.** 13 **b.** $\frac{1}{4}$
25. **a.** 12 **b.** $\frac{3}{13}$
27. **a.** 4 **b.** $\frac{1}{13}$

Section 3.2 Basic Terms of Probability

1. selecting a jellybean
3. $\frac{12}{35}$
5. $\frac{18}{35}$
7. $\frac{5}{7}$
9. 0
11. 12:23
13. 18:17
15. selecting a card
17. **a.** $\frac{1}{2}$ **b.** 1:1
19. **a.** $\frac{1}{13}$ **b.** 1:12
21. **a.** $\frac{1}{52}$ **b.** 1:51
23. **a.** $\frac{3}{13}$ **b.** 3:10
25. **a.** $\frac{10}{13}$ **b.** 10:3
27. **a.** $\frac{3}{13}$ **b.** 3:10
29. **a.** $\frac{1}{38}$ **b.** 1:37

31. a. $\frac{3}{38}$ **b.** 3:35
33. a. $\frac{5}{38}$ **b.** 5:33
35. a. $\frac{6}{19}$ **b.** 6:13
37. a. $\frac{9}{19}$ **b.** 9:10
39. 1:4
41. $\frac{3}{5}$
43. a: $(b - a)$
45. a. $\frac{9}{19}$ **b.** 9:10
47. a. {(b, b), (b, g), (g, b), (g, g)} **b.** {(b, g), (g, b)}
c. {(b, g), (g, b), (g, g)} **d.** {(g, g)}
e. $\frac{1}{2}$ **f.** $\frac{3}{4}$ **g.** $\frac{1}{4}$
h. 1:1 **i.** 3:1 **j.** 1:3
49. a. {(b, b, b), (b, b, g), (b, g, b), (g, b, b), (b, g, g), (g, b, g), (g, g, b), (g, g, g)}
b. {(b, g, g), (g, b, g), (g, g, b)}
c. {(b, g, g), (g, b, g), (g, g, b), (g, g, g)}
d. {(g, g, g)} **e.** $\frac{3}{8}$ **f.** $\frac{1}{2}$
g. $\frac{1}{8}$ **h.** 3:5 **i.** 1:1
j. 1:7
51. a. $\frac{1}{4}$ **b.** $\frac{1}{2}$ **c.** $\frac{1}{4}$ **d.** equally likely
53. different
55. a. {(1, 1), (1, 2), (1, 3), (1, 4), (1, 5), (1, 6), (2, 1), (2, 2), (2, 3), (2, 4), (2, 5), (2, 6), (3, 1), (3, 2), (3, 3), (3, 4), (3, 5), (3, 6), (4, 1), (4, 2), (4, 3), (4, 4), (4, 5), (4, 6), (5, 1), (5, 2), (5, 3), (5, 4), (5, 5), (5, 6), (6, 1), (6, 2), (6, 3), (6, 4), (6, 5), (6, 6)}
b. {(1, 6), (2, 5), (3, 4), (4, 3), (5, 2), (6, 1)}
c. {(6, 5), (5, 6)}
d. {(1, 1), (2, 2), (3, 3), (4, 4), (5, 5), (6, 6)}
e. $\frac{1}{6}$ **f.** $\frac{1}{18}$ **g.** $\frac{1}{6}$ **h.** 1:5 **i.** 1:17 **j.** 1:5
57. a. $\frac{1}{4}$ **b.** $\frac{1}{4}$ **c.** $\frac{1}{2}$
59. a. $\frac{1}{4}$ **b.** $\frac{1}{2}$ **c.** $\frac{1}{4}$
61. a. 0 **b.** $\frac{1}{2}$ **c.** 1
63. a. $\frac{1}{2}$ **b.** 0 **c.** $\frac{1}{2}$

Section 3.3

1. not mutually exclusive
3. mutually exclusive
5. not mutually exclusive
7. mutually exclusive
9. mutually exclusive
11. a. $\frac{1}{26}$ **b.** $\frac{7}{13}$ **c.** $\frac{25}{26}$
13. a. $\frac{1}{52}$ **b.** $\frac{4}{13}$ **c.** $\frac{51}{52}$
15. a. $\frac{2}{13}$ **b.** $\frac{5}{13}$ **c.** 0 **d.** $\frac{7}{13}$
17. a. $\frac{9}{13}$ **b.** $\frac{8}{13}$ **c.** $\frac{4}{13}$ **d.** 1
19. $\frac{12}{13}$
21. $\frac{10}{13}$
23. $\frac{11}{13}$
25. $\frac{9}{13}$
27. 9:5
29. 2:5; 5:2
31. b:a
33. 12:1
35. 10:3
37. 10:3
39. a. $\frac{71}{175}$ **b.** $\frac{104}{175}$
41. a. $\frac{151}{700}$ **b.** $\frac{97}{140}$
43. a. $\frac{8}{25}$ **b.** $\frac{17}{25}$
45. a. $\frac{9}{25}$ **b.** $\frac{16}{25}$
47. a. $\frac{1}{6}$ **b.** $\frac{1}{9}$ **c.** $\frac{1}{18}$
49. a. $\frac{2}{9}$ **b.** $\frac{7}{18}$
51. a. $\frac{1}{6}$ **b.** $\frac{3}{4}$
53. a. $\frac{1}{6}$ **b.** $\frac{1}{2}$
55. a. $\frac{3}{4}$ **b.** $\frac{1}{2}$ **c.** 1
57. a. $\frac{1}{4}$ **b.** $\frac{1}{2}$ **c.** 0
59. a. 0.35 **b.** 0.25 **c.** 0.15 **d.** 0.6
61. a. 0.20 **b.** 0.95 **c.** 0.8

Section 3.4 Combinatorics and Probability

1. 0.7063
3. a. $\frac{1}{13,983,816}$ **b.** $\frac{1}{22,957,480}$
c. $\frac{1}{18,009,460}$ **d.** about 64% more likely
5. 0.00003
7. a. 0.000003080 **b.** 0.000462062
9. a. 0.000001737 **b.** 0.000295263
11. Easiest to win: 5/26, hardest to win: 6/53
13. a. 0.000004 **b.** 0.000160 **c.** 0.002367
d. 0.018303 **e.** 0.081504 **f.** 0.897662
15. a. 0.000495 **b.** 0.001981 **c.** 0.000015
d. 0.001966
17. 0.05
19. 0.48
21. 0.64
23. 0.36
25. a. 0.09 **b.** 0.42 **c.** 0.49
27. a. 0.0000000125 **b.** 0.0000000715
c. Winning the 6/49 is 5.7 times more likely.

Section 3.5 Expected Value

1. a. −\$0.053
3. a. −\$0.053
5. a. −\$0.053
7. a. −\$0.053
9. a. −\$0.053
11. \$549
13. 1.75
15. \$10.05
19. bank
21. 0.59 or more
23. a. 0 **b.** 1/16 **c.** 3/8
25. −\$0.28
27. −\$0.55
29. more than \$1,200/year
31. back yard

33. a. \$12.32 **b.** −\$1.34 **c.** −\$5.89

35. a. −\$.47
b. −\$2.74
c. −\$3.49

37. \$7.8 million

39. a. 13,983,816 **b.** \$13,983,816 **c.** 97 days

41. 6 successive losses; win \$1

Section 3.6 Conditional Probability

1. a. 0.23 **b.** 0.53 **c.** 0.14 **d.** 0.32 **e.** 0.08 **f.** 0.08

3. 0.20

5. 0.34

7. a. $\frac{1}{4}$ **b.** $\frac{4}{17}$ **c.** $\frac{1}{17}$

9. a. $\frac{1}{4}$ **b.** $\frac{13}{51}$ **c.** $\frac{13}{204}$

11. a. $p(B \mid A)$ **b.** $p(A')$ **c.** $p(C \mid A')$

13. a. $\frac{1}{16}$ **b.** $\frac{1}{3}$ **c.** 0 **d.** 1

15. a. $\frac{5}{36}$ **b.** $\frac{5}{18}$ **c.** 0 **d.** 1

17. a. $\frac{1}{12}$ **b.** $\frac{3}{10}$ **c.** 1

19. E_2 is most likely; E_3 is least likely

21. 0.46

23. 0.0005

25. 0.0020

27. 0.14

29. 0.20

31. 0.07

33. 0.00646

35. 0.01328

37. a. .2 **b.** .06

39. a. .105 **b.** .103

41. 92%

43. 39%

45. 0.05

47. a. 0.86 **b.** 0.34 **c.** 0.66 **d.** $N' \mid W$

49. $A' \mid B$

51. 0.14

Section 3.7 Independence; Trees in Genetics

1. a. dependent **b.** not mutually exclusive

3. a. dependent **b.** mutually exclusive

5. a. independent **b.** not mutually exclusive

7. a. dependent **b.** mutually exclusive

9. a. dependent **b.** mutually exclusive

11. no

13. dependent

15. dependent

17. dependent

19. dependent

21. a. 0.000001 **b.** 4

23. $0.0101 \approx 1\%$

25. a. 0.49 **b.** 0.999996 **c.** 0.51 **d.** 0.000004 **e.** a; c **f.** c; d

29. a. $\frac{1}{96} \approx 1\%$ **b.** $\frac{1}{144} \approx 0.7\%$

31. 0.7225, 0.2775

33. $\frac{1}{3}$

35. $\frac{1}{2}$

37. chestnut or shiny dark brown, each with $p = \frac{1}{2}$

39. light brown ($\frac{1}{8}$), reddish brown ($\frac{1}{4}$), dark red ($\frac{1}{8}$), medium brown ($\frac{1}{8}$), chestnut ($\frac{1}{4}$), or auburn ($\frac{1}{8}$)

41. light brown ($\frac{1}{16}$), reddish brown ($\frac{1}{8}$), dark red ($\frac{1}{16}$), medium brown ($\frac{1}{16}$), chestnut ($\frac{1}{8}$), auburn ($\frac{1}{16}$), dark brown ($\frac{1}{16}$), shiny dark brown ($\frac{1}{8}$), glossy dark brown ($\frac{1}{16}$), black ($\frac{1}{16}$), shiny black ($\frac{1}{8}$), glossy black ($\frac{1}{16}$)

43. a. $\frac{1}{6}$ **b.** $\frac{5}{6}$ **c.** $\frac{625}{1{,}296} \approx 0.48$ **d.** $\frac{671}{1{,}296} \approx 0.52$ **e.** \$0.04

Chapter 3 Review

2. a. $\frac{26}{52} = \frac{1}{2}$, 26:26 = 1:1
b. $\frac{4}{52} = \frac{1}{13}$, 4:48 = 1:12
c. $\frac{13}{52} = \frac{1}{4}$, 13:39 = 1:3
d. $\frac{1}{52}$, 1:51
e. $\frac{16}{52} = \frac{4}{13}$, 16:36 = 4:9
f. $\frac{48}{52} = \frac{12}{13}$, 48:4 = 12:1

3. a. $\frac{3}{38}$; 3:35 **b.** $\frac{6}{19}$; 6:13 **c.** $\frac{9}{19}$; 9:10

4. b. {(h, h, h), (h, h, t), (h, t, h), (t, h, h), (h, t, t), (t, h, t), (t, t, h), (t, t, t)}
c. {(h, t, t), (t, h, t), (t, t, h)}
d. {(h, t, t), (t, h, t), (t, t, h), (t, t, t)}
e. $\frac{3}{8}$; 3:5
f. $\frac{1}{2}$; 1:1

5. a. $\frac{1}{6}$ **b.** $\frac{1}{18}$ **c.** $\frac{7}{18}$ **d.** $\frac{1}{6}$ **e.** $\frac{11}{18}$ **f.** $\frac{7}{18}$

6. a. 0.013 **b.** 0.138 **c.** 0.151 **d.** 0.000008

7. a. 0.005 **b.** 0.069 **c.** 0.074

8. a. 0.14 **b.** no

9. a. $\frac{1}{2}$ **b.** $\frac{1}{2}$

10. a. $\frac{1}{2}$ **b.** $\frac{1}{2}$ **c.** $\frac{1}{4}$

11. a. $\frac{1}{4}$ **b.** $\frac{1}{2}$ **c.** $\frac{1}{4}$

12. a. $\frac{1}{2}$ **b.** $\frac{1}{2}$

13. a. 0 **b.** $\frac{1}{2}$ **c.** $\frac{1}{2}$

14. $\frac{1}{9}$

15. a. 0.56; 0.39 **b.** 0.40; 0.55 **c.** 0.67; 0.33 **d.** 0.50; 0.50 **e.** urban area; urban area **f.** O'Neill; Bell **g.** no **h.** O'Neill

16. a. dependent; not mutually exclusive
b. independent; not mutually exclusive
c. dependent; mutually exclusive
d. dependent; not mutually exclusive
e. independent; not mutually exclusive

17. a. 0.0000007 **b.** 0.00003

18. −\$0.053

19. −\$0.23

20. The low-number bet is the better bet.

21. $\frac{24}{2{,}598{,}960} \approx .0000092345$

Chapter 4 Statistics

Section 4.1 Population, Sample and Data

1. a.

Number of visits to the Library	Frequency
1	9
2	8
3	2
4	5
5	6
	Total = 30

b.

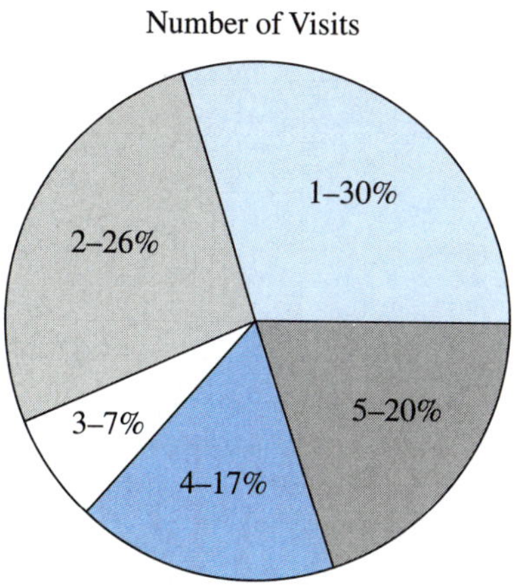

c.

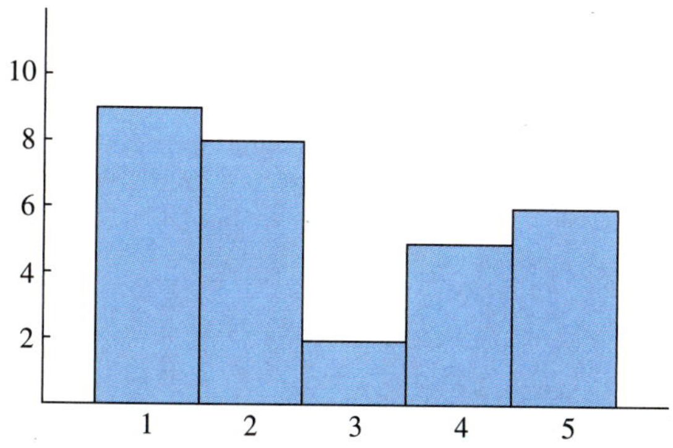

3. a.

Number of Children	Tally	Frequency	Relative Frequency	Central Angle
0	𝍸 III	8	0.2 = 20%	72°
1	𝍸 𝍸 III	13	0.325 = 32.5%	117°
2	𝍸 IIII	9	0.225 = 22.5%	81°
3	𝍸 I	6	0.15 = 15%	54°
4	III	3	0.075 = 7.5%	27°
5	I	1	0.025 = 2.5%	9°
		$n = 40$	1.000 = 100%	total = 360°

b.

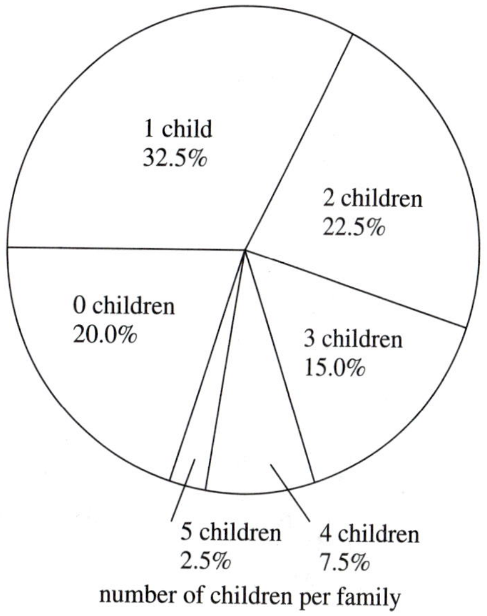

c.

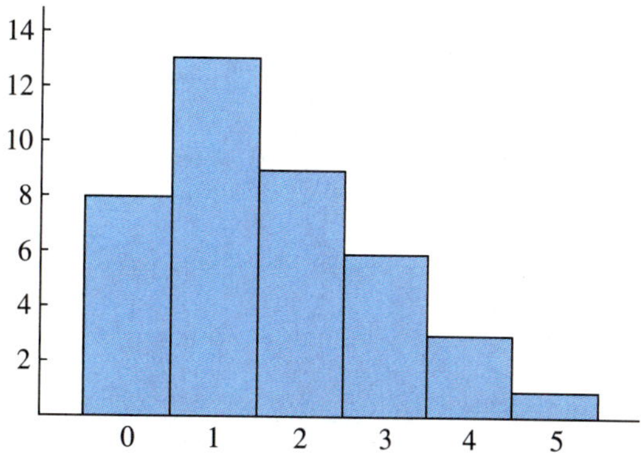

5. a.

Speed (in mph)	Tally	Frequency	Relative Frequency
$51 \le x < 56$	‖	2	$0.05 = 5\%$
$56 \le x < 61$	‖‖	4	$0.1 = 10\%$
$61 \le x < 66$	𝍸 ‖	7	$0.175 = 17.5\%$
$66 \le x < 71$	𝍸 𝍸	10	$0.25 = 25\%$
$71 \le x < 76$	𝍸 ‖‖	9	$0.225 = 22.5\%$
$76 \le x < 81$	𝍸 |||	8	$0.2 = 20\%$
		$n = 40$	$1.000 = 100\%$

b.

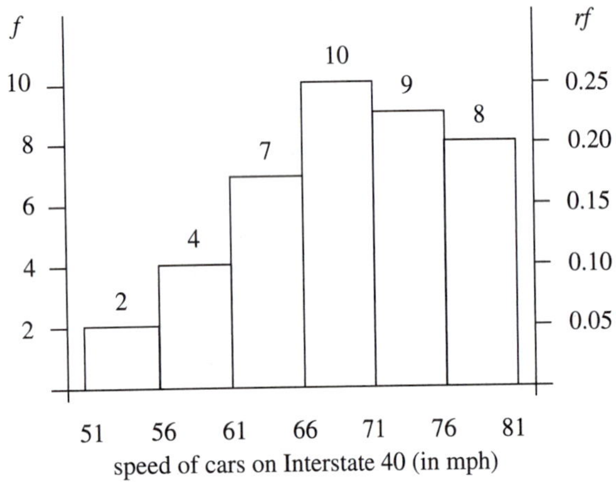

7.

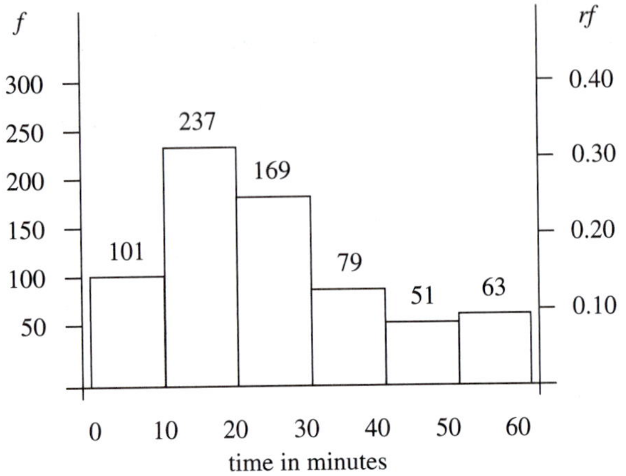

9.

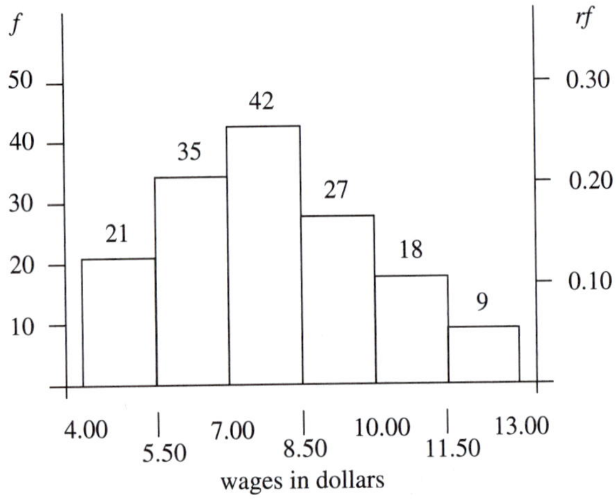

11.

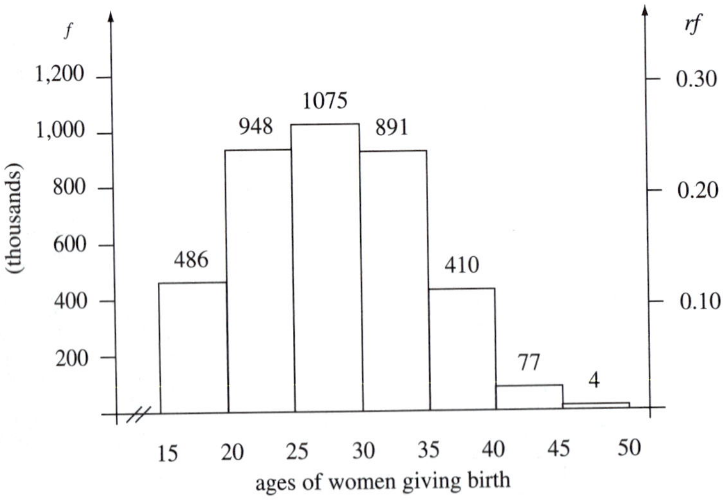

13. 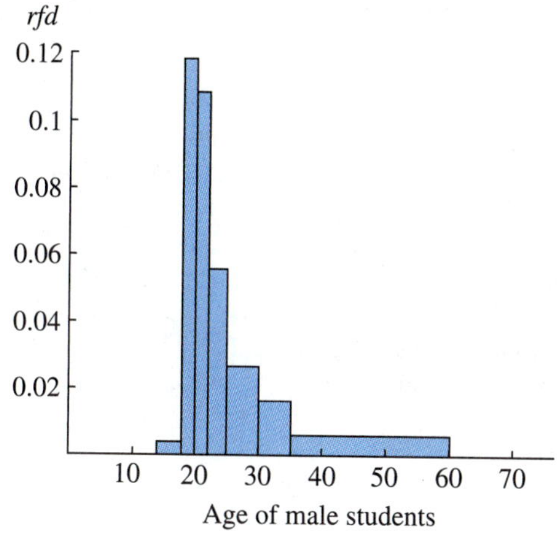

15. **a.** 28.5% **b.** 32% **c.** not possible
d. 97.5% **e.** 50% **f.** 52.5%

17.

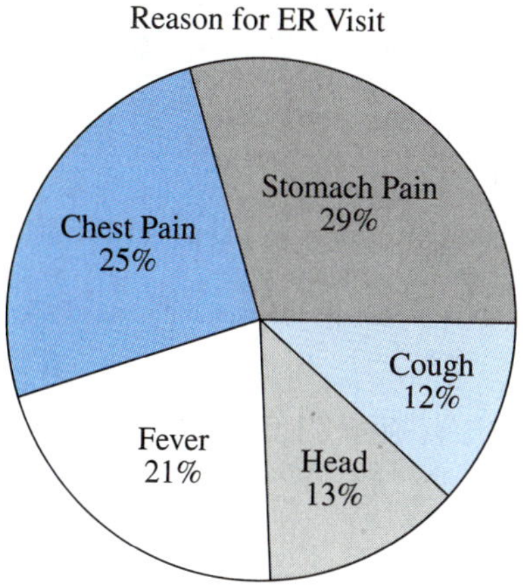

19. a.

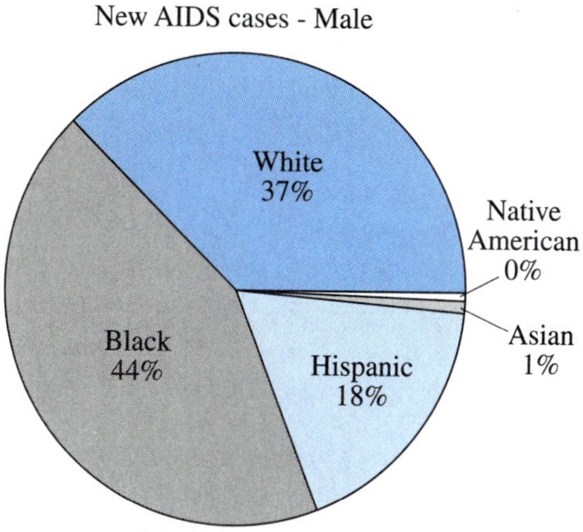

b.

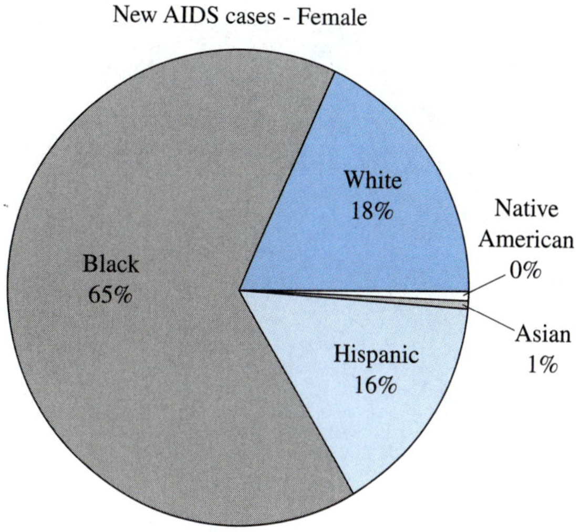

c.

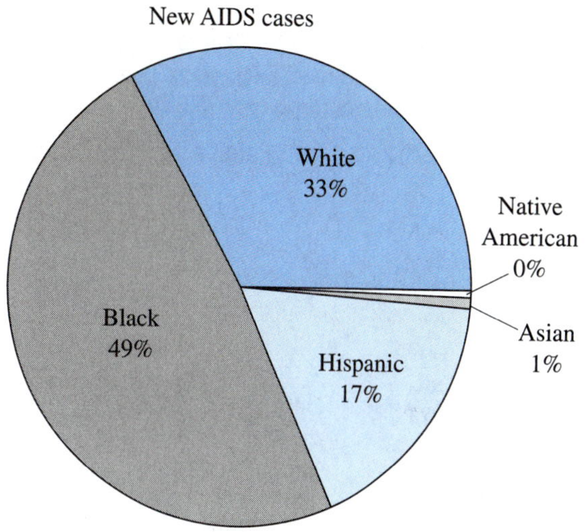

21. a.

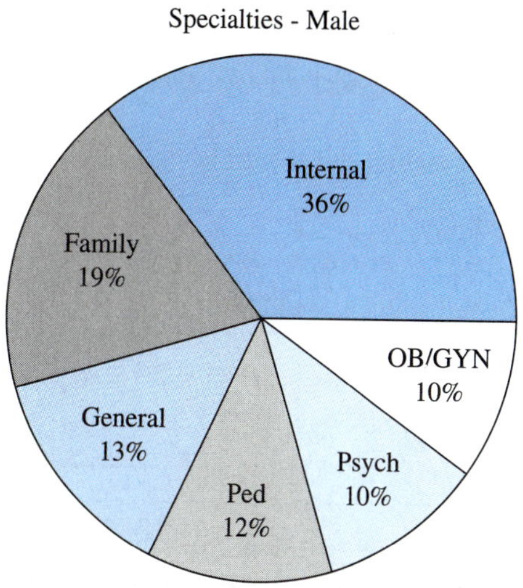

b.

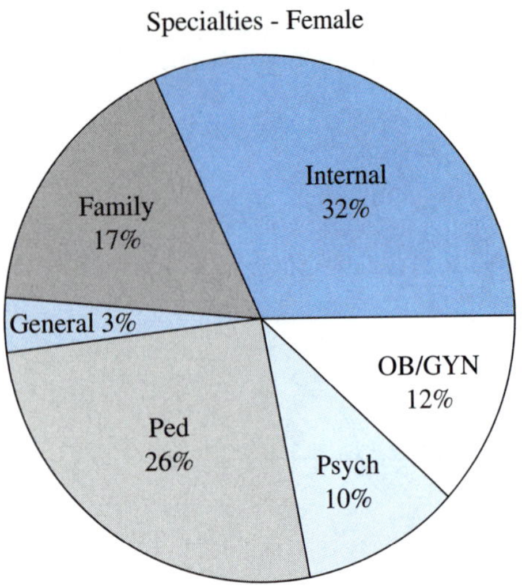

c.

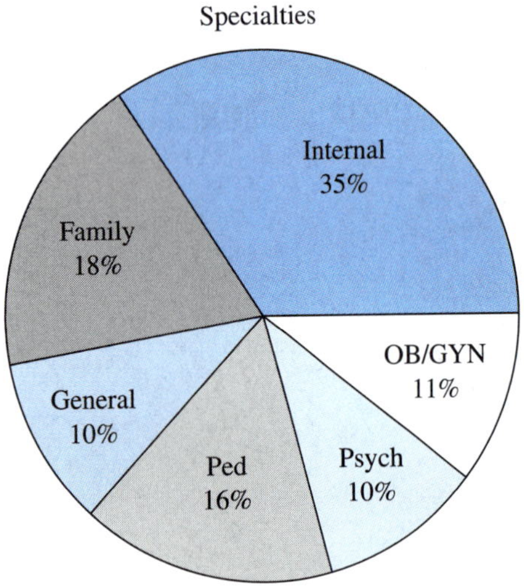

29.

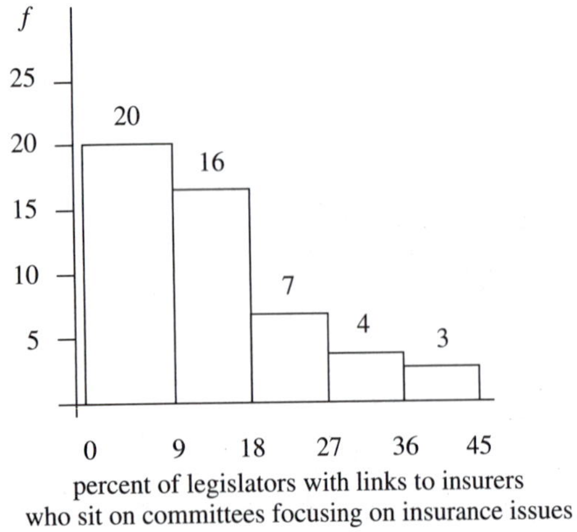

Section 4.2 Measures of Central Tendency

1. mean = 12.5; median = 11.5; mode = 9
3. mean = 1.45; median = 1.5; mode = 1.7
5. a. mean = 11; median = 10.5; mode = 9
b. mean = 25.5; median = 10.5; mode = 9
7. a. mean = 7; median = 7; mode = none
b. mean = 107; median = 107; mode = none
9. mean = 10.88, median = 10, mode = 9 and 15
11. mean = 36.1, median = 34, mode = none (many numbers have frequency 2)
13. mean = 7.6; median = 8; mode = 6 and 9
15. 16.028 oz
17. a. 42 **b.** 92
c. impossible (it would require a score of 142)
19. \$51,600
21. 51.4 mph
23. 39 years old
25. a. \$40,750 **b.** \$42,000 (it does not change)
27. a. \$3,583.09 **b.** \$4,274.41
29. 36.5 years old
31. a. 23 years old **b.** 28.5 years old

Section 4.3 Measures of Dispersion

1. a. 16.4; 4.0 **b.** 16.4; 4.0
3. a. 0 **b.** 0
5. a. 22; 7.5 **b.** 1100; 374.2
7. a. Joey, 168; Dee Dee, 167; Joey has the higher mean.
b. Joey, 30.9; Dee Dee, 18.2
c. Dee Dee is more consistent because his standard deviation is lower.
9. 13.4
11. a. 67.5; 7.1 **b.** 80% **c.** 90%
13. a. 3.175; 1.807 **b.** 58% **c.** 100%
15. a. 8; 1.4 **b.** 71% **c.** 94% **d.** 100%
17. 0.285
19. 6.30
25. 13.5; 11.3

Section 4.4 The Normal Distribution

1. a. 34.13% **b.** 34.13% **c.** 68.26%
3. a. 49.87% **b.** 49.87% **c.** 99.74%
5. a. one [22.4, 27.0]; two [20.1, 29.3]; three [17.8, 31.6]
b. 68.26%; 95.44%; 99.74%
c.

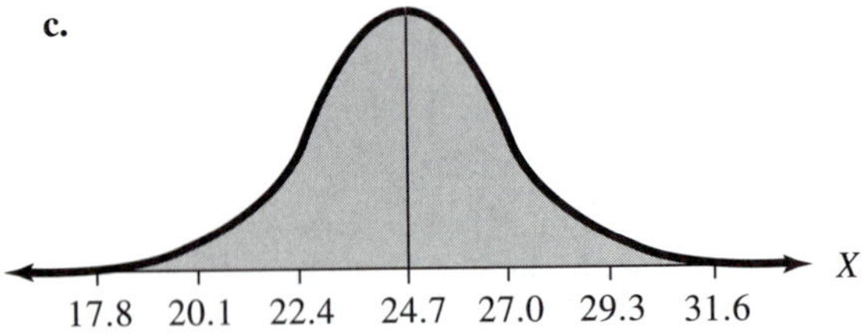

7. a. 0.4474 **b.** 0.0639 **c.** 0.5884
d. 0.0281 **e.** 0.0960 **f.** 0.8944
9. a. 0.34 **b.** −2.08 **c.** 0.62
d. −0.27 **e.** 1.64 **f.** −1.28

11. **a.** 0.42 **b.** −0.42 **c.** 2.08
d. −1.46 **e.** 2.96 **f.** −2.21
13. **a.** 0.3413 **b.** 0.6272 **c.** 0.8997
d. 0.9282 **e.** 0 **f.** 0.3300
15. **a.** 4.75% **b.** 49.72%
17. **a.** 0.1894 **b.** 0.7745
19. **a.** 95.25% **b.** 79.67%
21. **a.** 87 **b.** 62
23. 9.1 min

Section 4.5 Polls and Margin of Error

1. **a.** 0.68 **b.** 0.31
c. 1.39 **d.** 2.65
3. **a.** 1.44 **b.** 1.28
c. 1.15 **d.** 2.575
5. 1.75
7. 1.15
9. **a.** ±2.6% **b.** ±3.7%
11. **a.** 75.0% ±3.4% **b.** 75.0% ±4.0%
13. **a.** 79.0% **b.** 21.0%
c. ±1.6%
15. **a.** 86.0% **b.** 70.0%
c. ±2.8% for the men; ±2.5% for the women
17. **a.** 57.0% **b.** 43.0% **c.** ±2.2%
19. **a.** 36.0% **b.** ±1.6% **c.** ±2.3%
21. **a.** ±4.7% **b.** ±3.5% **c.** ±2.8%
23. 96.4%
25. 95.3%

Section 4.6 Linear Regression

1. **a.** $\hat{y} = 1.0x + 3.4$ **b.** 14.4
c. 15.6
d. 0.9529485 **e.** yes
3. **a.** $\hat{y} = 20.1x + 8.2$ **b.** 7.7
c. 12 **d.** 20.0951819
e. no
5. **a.** yes

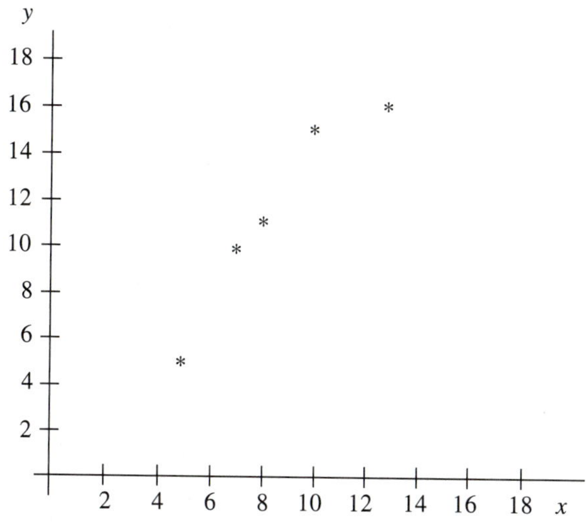

b. $\hat{y} = 1.4x - 0.3$ **c.** 12.3
d.

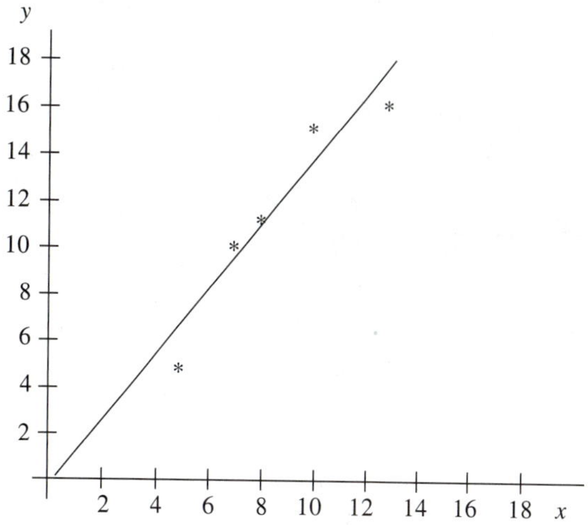

e. 0.9479459 **f.** yes
7. **a.** no

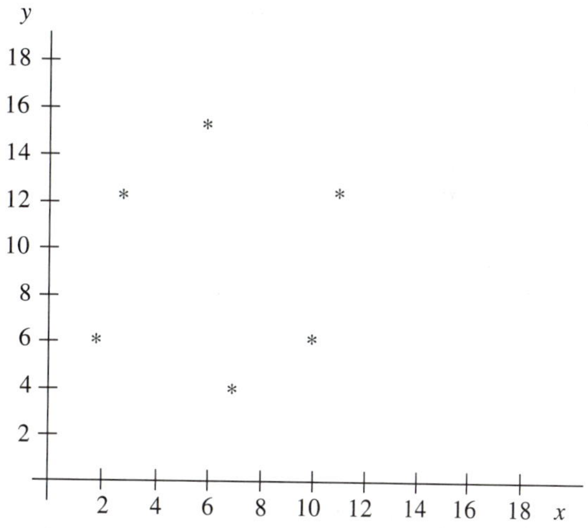

b. $\hat{y} = 0.0x + 9.1$ **c.** 9.1
d.

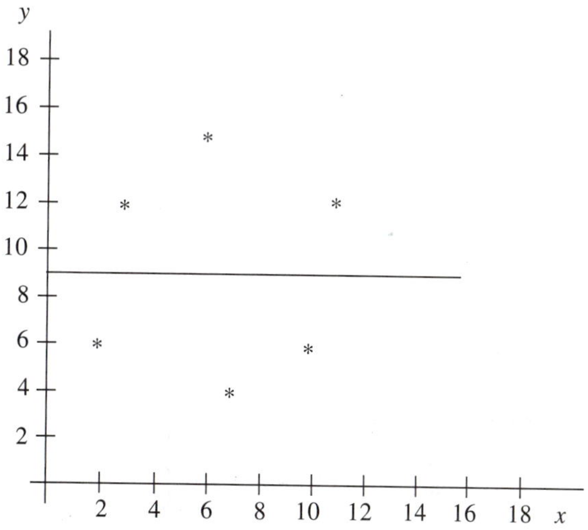

e. 0.0062782255
f. no

9. a. yes

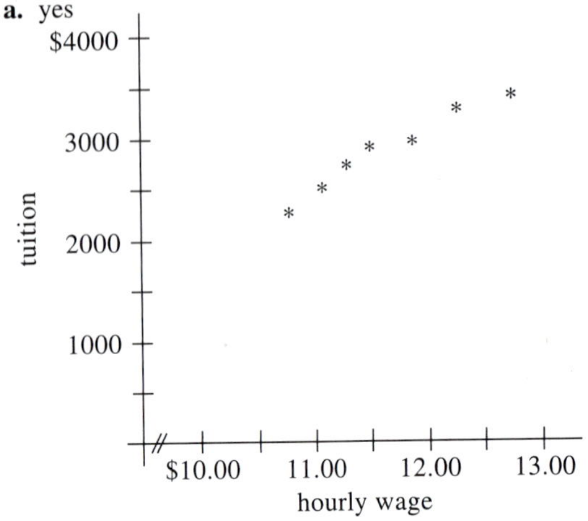

b. $\hat{y} = 21779.2 + 398.0x$
c. $3,195.80
d. $12.01
e. 0.9878803778
f. yes; r is close to 1

11. a. yes

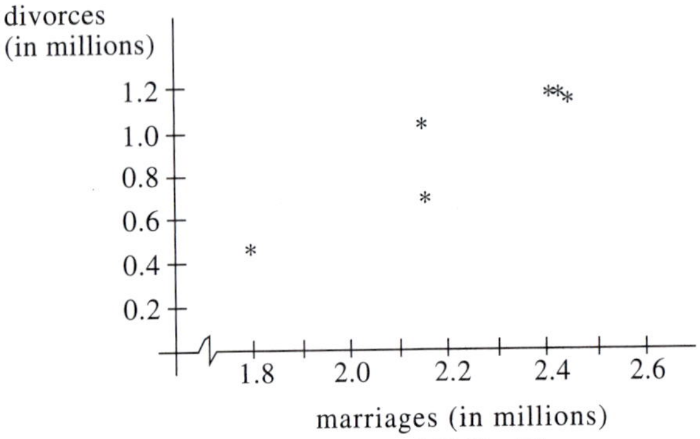

b. $\hat{y} = 1.1130x - 1.5239$ **c.** 1.5369 million
d. 2.7169 million **e.** 0.9338946
f. yes

Chapter 4 Review

1. a. 7.1 **b.** 7.5 **c.** 8 **d.** 2.1

2. a.

Number of Children	Tally	Frequency	Relative Frequency
0	𝍸 |||	8	0.2 = 20%
1	𝍸 𝍸	10	0.25 = 25%
2	𝍸 𝍸		11	0.275 = 27.5%
3	𝍸 ||	7	0.175 = 17.5%
4	|||	3	0.075 = 7.5%
5			1	0.025 = 2.5%
		$n = 40$	1.000 = 100%

b. 1.8 **c.** 2 **d.** 2 **e.** 1.3
f.

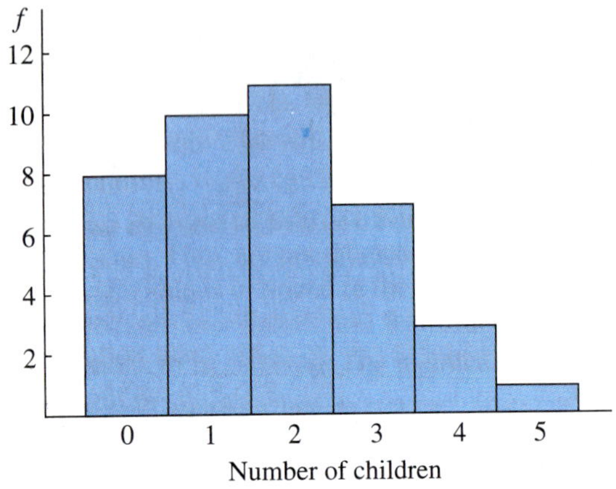

3. a. 40% **b.** 28% **c.** 86%
d. 14% **e.** 50% **f.** cannot determine

4. a. 10.17
b. 3.21
c.

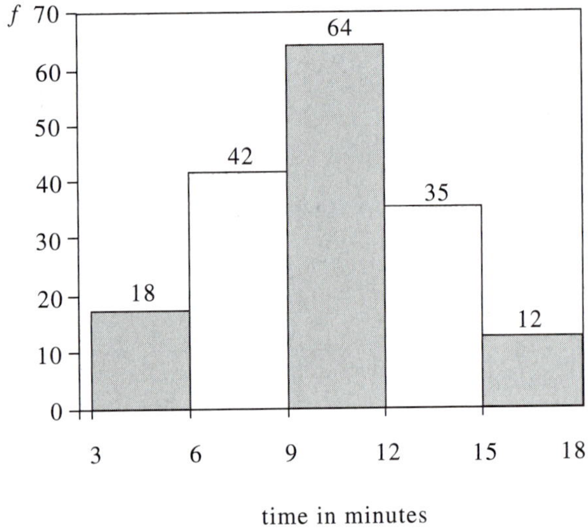

5. 100
6. $33,400
7. a. Timo, 98.8; Henke, 96.6; Henke has the lower mean.
b. Timo, 6.4; Henke, 10.4
c. Timo is more consistent because his standard deviation is lower.
8. a. $1.298; $0.131 **b.** 78%
c. 95% **d.** 100%
9. a. continuous **b.** neither
c. discrete **d.** neither
e. discrete **f.** continuous
10. a. 45.99% **b.** 45.99% **c.** 91.98%
11. a. one [71, 85]; two [64, 92]; three [57, 99]
b. 68.26%; 95.44%; 99.74%

c.

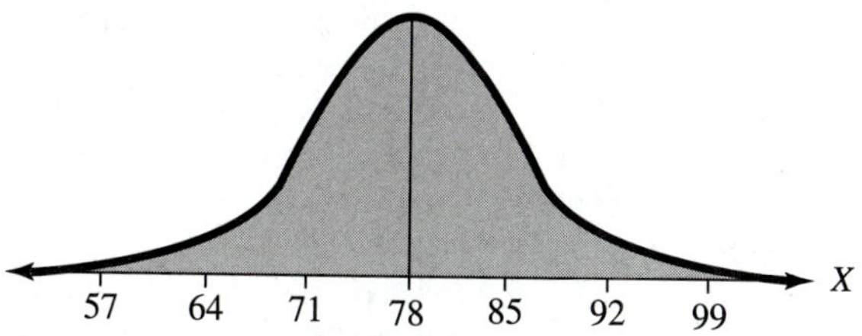

12. a. 0.0475 **b.** 0.4972

13. 401

14. a. 1.59 **b.** 1.645
c. 0.51 **d.** 2.81

15. a. 66.7% ± 2.4% **b.** 66.7% ± 2.8%

16. a. 75% ± 2.0% **b.** 75% ± 3.7%

17. a. ±4.1% **b.** ±3.1% **c.** ±2.5%

18. 81.6%

19. a. $\hat{y} = 3x + 15$ **b.** 63 **c.** 14
d. 1.0 **e.** yes

20. a. yes

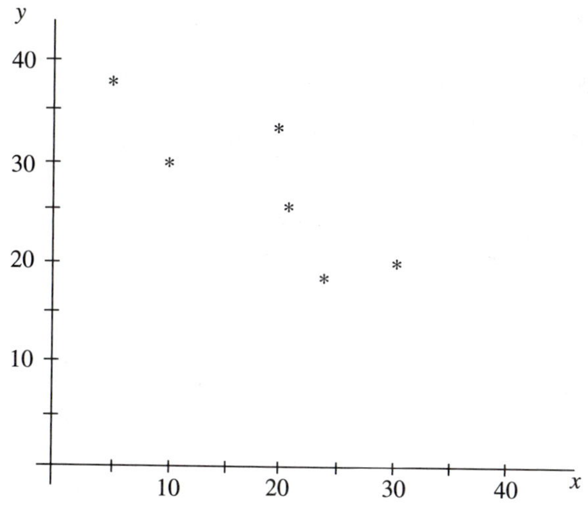

b. $\hat{y} = 20.70x + 40.25$
c. 29.75
d.

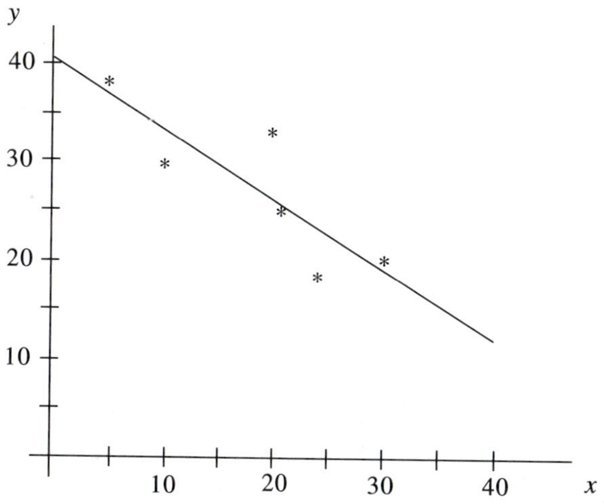

e. −0.8398392
f. yes

21. a. yes

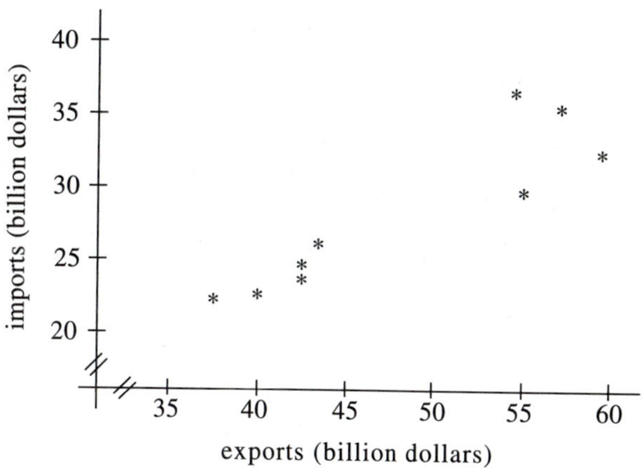

b. $\hat{y} = 20.719 + 0.607x$
c. \$29.631 billion
d. \$58.845 billion
e. 0.8979632779
f. yes, r is close to 1

Chapter 5 Finance

Section 5.1 Simple Interest

1. \$480
3. \$21.26
5. \$153.80
7. \$4,376.48
9. \$14,073.09
11. \$1,474.81
13. \$6,162.51
15. \$17,036.82
17. \$6,692.61
19. \$1,039.76
21. \$1,240.95
23. \$2,727.27
25. about $3\frac{1}{3}$ years
27. \$328.19
29. \$115.02
31. \$140.27; \$2.08
33. \$152.84; \$2.73
35. a. \$8,125 **b.** \$146,250
c. \$8,125 **d.** \$67.71
e. \$19,500.08 **f.** \$136,500
g. \$156,000.08
37. a. \$38,940 **b.** \$311,520
c. \$38,940 **d.** \$356.95
e. \$95,013.60 **f.** \$288,156
g. \$383,169.60

Section 5.2 Compound Interest

1. a. 0.03 b. 0.01
c. 0.000328767 d. 0.0046153846
e. 0.005
3. a. 0.00775 b. 0.002583333
c. 0.000084932 d. 0.001192307
e. 0.001291667
5. a. 0.02425 b. 0.008083333
c. 0.000265753 d. 0.003730769
e. 0.004041667
7. a. 34 quarters b. 102 months
c. 3,103 days
9. a. 120 quarters b. 360 months
c. 10,950 days
11. $7,189.67
13. $9,185.46
15. $3,951.74
17. 8.30%
19. 4.34%
21. a. 10.38% b. 10.47% c. 10.52%
23. $583.49
25. $470.54
27. First National: 9.55%; Citywide: 9.45%
29. verifies
31. does not verify—true yield is 9.74%
33. a. $9,664.97 b. $51.77
35. $19,741.51
37. a. $8,108.07 b. $690.65
39. a. 1.65% b. $1008.23 c. $8.23
d. 0.823% e. The CD is for one half year.
41. $r = (1 + i)^n - 1$
43. 7.45%
45. a. 5.70% b. 5.74% c. 5.77%
d. 5.79% e. 5.78% f. 5.78%
59. a. 14.2 years b. 167 months ≈ 13.92 years
c. 56 quarters = 14 years
d. 5,061 days ≈ 13.87 years
61. 1,822 days ≈ 4.99 years
10,346 days ≈ 28.35 years
63. 2,369 days ≈ 6.49 years
9,400 days ≈ 25.75 years

Section 5.3 Annuities

1. a. $1,478.56
3. a. $5,422.51
5. a. $770.00 b. $750.00 c. $20.00
7. a. $1,391.45 b. $1,350.00 c. $41.45
9. a. $283,037.86 b. $54,600.00 c. $228,437.86
11. a. $251,035.03 b. $69,600.00 c. $181,435.03
13. $1,396.14
15. $726.49
17. $18,680.03
19. $51.84
21. $33.93
23. $33.63
25. a. $537,986.93
b.

1	$537,986.93	$2,734.77	$650	$540,071.70
2	$540,071.70	$2,745.36	$650	$542,167.06
3	$542,167.06	$2,756.02	$650	$544,273.08
4	$544,273.08	$2,766.72	$650	$546,389.80
5	$546,389.80	$2,777.48	$650	$548,517.28

27. a. $175,384.62 b. $111.84
29. a. $1,484,909.47 b. $102,000
c. $1,382,909.47 d. $979,650.96
e. $643,631.10
31. $11,954.38
33. $45.82
35. a. $595.83 b. $5,367.01
c.

1	2	3	4	5
$5,367.01	$10,958.76	$16,784.67	$22,854.54	$29,178.58
6	7	8	9	10
$35,767.44	$42,632.21	$49,784.44	$57,236.17	$64,999.94

37. $P = \text{pymt}\dfrac{(1+i)^n - 1}{i(1+i)^n} = \text{pymt}\dfrac{1-(1+i)^{-n}}{i}$
39. $1,396.14
41. $726.49
47. a. 586 months = 48 years and 10 months
49. 1,171 payments = 48 years and 9.5 months

Section 5.4 Amortized Loans

1. a. $125.62 b. $1,029.76
3. a. $193.91 b. $1,634.60
5. a. $1,303.32 b. $314,195.20
7. a. $378.35 b. $3,658.36
c.

Payment Number	Principal Portion	Interest Portion	Total Payment	Balance Due
0	—	—	—	$14,502.44
1	$239.37	$138.98	$378.35	$14,263.07
2	$241.66	$136.69	$378.35	$14,021.41

9. a. $1,602.91 b. $407,047.60
c.

Payment Number	Principal Portion	Interest Portion	Total Payment	Balance Due
0	—	—	—	$170,000.00
1	$62.28	$1,540.63	$1,602.91	$169,937.72
2	$62.85	$1,540.06	$1,602.91	$169,874.87

d. $4,218.18 per month (assuming only home loan payment)
11. a. $18,950 b. $151,600
c. $18,950 d. $1,501.28
e. $189.50 f. $271.61
g. $1,962.39 h. $1,501.28
13. a. $404.72; $353.68
b. $4,597.24; $2,730.25
c. simple interest loan (why?)
15. a. $354.29; $1,754.44
b. $278.33; $2,359.84
c. $233.04; $2,982.40
17. a. $599.55; $115,838.00
b. $665.30; $139,508.00
c. $733.76; $164,153.60 d. $804.62; $189,663.20
e. $877.57; $215,925.20 f. $952.32; $242,835.20

19. Both verify, but not exactly. Their computations are actually more accurate than ours, since they are using more accurate round-off rules than we do.

21. **a.** $877.57; $215,925.20 **b.** $404.89; $215,814.20

23. **a.** $24,535.12

b.

Payment Number	Principal Portion	Interest Portion	Total Payment	Amount Due after Payment
0	—	—	—	$120,000.00
1	$23,647.62	$887.50	$24,535.12	$96,352.38
2	$23,822.51	$712.61	$24,535.12	$72,529.87
3	$23,998.70	$536.42	$24,535.12	$48,531.17
4	$24,176.19	$358.93	$24,535.12	$24,354.98
5	$24,354.98	$180.13	$24,535.11	$0.00

25. **a.** $12,232.14

b.

Payment Number	Principal Portion	Interest Portion	Total Payment	Amount Due after Payment
0	—	—	—	$48,000.00
1	$11,862.14	$370.00	$12,232.14	$36,137.86
2	$11,953.58	$278.56	$12,232.14	$24,184.28
3	$12,045.72	$186.42	$12,232.14	$12,138.56
4	$12,138.56	$93.57	$12,232.13	$0.00

27. **a.** $35.37 **b.** $1735.36

c.

Payment	Principal	Interest	Total Payment	Balance
0				$10288.00
1	$1699.99	$35.37	$1735.36	$8588.01
2	$1705.84	$29.52	$1735.36	$6882.17
3	$1711.70	$23.66	$1735.36	$5170.47
4	$1717.59	$17.77	$1735.36	$3452.88
5	$1723.67	$11.87	$1735.36	$1729.39
6	$1729.39	$5.94	$1735.33	$0

d. $124.13

29. add-on:

Payment Number	Principal Portion	Interest Portion	Total Payment	Amount Due after Payment
0	—	—	—	$14,829.32
1	$308.95	$95.77	$404.72	$14,520.37
2	$308.95	$95.77	$404.72	$14,211.42
3	$308.95	$95.77	$404.72	$13,902.47

simple interest:

Payment Number	Principal Portion	Interest Portion	Total Payment	Amount Due after Payment
0	—	—	—	$14,246.39
1	$248.32	$105.36	$353.68	$13,998.07
2	$250.15	$103.53	$353.68	$13,747.92
3	$252.00	$101.68	$353.68	$13,495.92

31. $3,591.73

33. $160,234.64

35. **a.** $1,735.74 **b.** $151,437.74 **c.** $1,204.91 **d.** $472,016.40 **e.** $343,404.24 **f.** yes (why?)

37. **a.** $1,718.80 **b.** $172,157.40 **c.** $240,354.60 **d.** $573,981.98 **e.** $1,200,543.86 **f.** The decision is between saving more for retirement or increasing their standard of living now. (Why?)

49. **b.** $1,511.44 **c.** $4,269.82 **e.** $270.64

51. **b.** $18,449.10 **c.** $18,156.11 **e.** $1,087.63

53. **b.** $50,702.84 **c.** $47,587.04 **e.** $3,818.14

55. **b.** $735.77 **c.** $2,353.64 **e.** $101.61

57.

Payment	Principal	Interest	Total Payment	Balance
0				$32,440.00
1	$2,638.90	$141.93	$2,780.82	$29,801.11
2	$2,650.44	$130.38	$2,780.82	$27,150.66
3	$2,662.04	$118.78	$2,780.82	$24,488.63
4	$2,673.68	$107.14	$2,780.82	$21,814.95
5	$2,685.38	$95.44	$2,780.82	$19,129.57
6	$2,697.13	$83.69	$2,780.82	$16,432.44
7	$2,708.93	$71.89	$2,780.82	$13,723.51
8	$2,720.78	$60.04	$2,780.82	$11,002.73
9	$2,732.68	$48.14	$2,780.82	$8,270.05
10	$2,744.64	$36.18	$2,780.82	$5,525.41
11	$2,756.65	$24.17	$2,780.82	$2,768.76
12	$2,768.76	$12.11	$2,780.87	$0.00

Section 5.5

1. 14.6%

3. 11.2%

5. **a.** $126.50 **b.** verifies

7. **a.** $228.77 **b.** does not verify

9. Either one could be less expensive, depending on the A.P.R.

11. Really Friendly S and L will have lower payments but higher fees and/or more points.

13. $8,109.54

15. **a.** $819.38 **b.** $801.40 **c.** $8,009.75 **d.** $9,496.12

Section 5.6

1. **a.** $143,465.15 **b.** $288,000.00

3. **a.** $179,055.64 **b.** $390,000.00

5. **a.** $96.26 **b.** $34,653.60; she receives $253,346.40 more than she paid

7. **a.** $138.30 **b.** $331.39

9. **a.** $185,464.46 **b.** $14,000.00 **c.** $14,560.00 **d.** $29,495.89

11. a. $36,488.32 b. 576,352.60 c. $454.10
d. $1,079.79
13. a. $12,565.57 b. 10.4713067% c. $138,977.90
d. $61.48
15. a. $17,502.53 b. 0.0927217 c. $245,973.09
d. $372.99
17. $490,907.37

Chapter 5 Review

1. $8,730.15
2. $15,547.72
3. $4,067.71
4. $4,104.26
5. about 1.9 years
6. $3,443.70; $57.03
7. a. $308.25 b. $70,692.00 c. $235,092.00
8. Bank of the South at 7.88% is better than Extremely Trustworthy Savings at 7.63%.
9. a. $4,580.78 b. $2,109.38
10. $37,546.62
11. $19,173.84
12. $37,738.26
13. a. $189,090.91 b. $88.22
14. a. $263.17 b. $3,651.62
c.

Period Number	Principal	Interest	Total	Amount Outstanding
0	—	—	—	$12,138.58
1	$153.16	$110.01	$263.17	$11,985.42
2	$154.55	$108.62	$263.17	$11,830.87

d. does not verify e. $6,885.51
15. a. $174.50 b. does not verify
16. a. $1,217.96 b. $7,299.98
17. a. $261,366.36 b. 255.25
c. payments = $91,890.00; receipts = $510,000.00
18. a. $58,409.10 b. $881,764.23
c. $556.55

Chapter 6 Geometry

Section 6.1 Perimeter and Area

1. 16.1 cm^2
3. 12.5 in.^{2}
5. 21.7 ft^2
7. 176 m^2
9. a. 33.2 in.^{2} b. 20.4 in.
11. a. 30 m^2 b. 30 m
13. a. 27.7 m^2 b. 24 m
15. a. 80 ft^2 b. 44 ft
17. a. 3,927.0 yd^2 b. 257.1 yd
19. a. 54 yd^2 b. 30 yd
21. a. 49.8 ft^2 b. 28.9 ft
23. a. 1,963.5 ft^2 b. 863.9 ft^2
c. 3,572.6 ft^2
25. a. 3,136 in.^{2} b. 21.8 ft^2
27. 8.0 ft^2
29. 3.9 mi
31. 8 ft
33. a. 5,256.6 yd^2 b. 325.7 yd
35. seven
37. large

Section 6.2 Volume and Surface Area

1. a. 38.22 m^3 b. 72.94 m^2
3. a. 785.40 in.^{3} b. 471.24 in.^{2}
5. a. 2.81 in.^{3} b. 9.62 in.^{2}
7. 16.76 ft^3
9. 21.33 ft^3
11. 36 ft^3
13. 47.12 ft^3
15. a. 136.5 in.^{3} b. 151 in.^{2}
17. 17,802.36 ft^3
19. a. hardball, 12.31 in.^{3}; softball, 29.18 in.^{3}; volume of softball is 137% more than that of hardball
b. hardball, 25.78 in.^{2}; softball, 45.83 in.^{2}; surface area of softball is 78% more than that of hardball
21. 49
23. 147
25. no; should have paid $118.75
27. 36.82 in.^{3}
29. $175.20
31. 91,445.760 ft^3
33. 301.59 ft^3

Section 6.3

1. a right triangle
3. not a right triangle
5. a. the regular pyramid (2,025 khar . 1,554 khar)
7. 36; 18; 9; 36; 18; 9; 63; 9; 3; 63; 189; 189
9. a square with sides of 21 units
11. $\frac{49}{16}$
13. a. 28.4444 palms^2 b. 28.2743 palms^2
c. 0.6%
15. a. 113.7778 palms^3
b. 113.0973 palms^3 c. 0.6%
17. 11 cubits
19. the khar (1 khar = 5887 in.^{3} > 1 ft^3 = 1,728 in.^{3})
21. 1.5 setats
23. a. 2 khet b. 200 cubits
25. a. 960 khar b. 954.2588 khar
c. 0.6%

Section 6.4 The Greeks

1. $x = 5$; $y = 60$
3. $x = 64.6$; $y = 2.5$
5. 36 ft
9. 5.3 ft (rounded down so the object will fit)
11. 7.5 in.
19. a. 3.061467459
b. 3.313708499
21. a. 3.121445152
b. 3.182597878

Section 6.5 Right Triangle Trigonometry

1. $\theta = 60°; x = 3; y = 3\sqrt{3}$
3. $\theta = 30°; x = 14; y = 7\sqrt{3}$
5. $\theta = 45°; x = 3; y = 3\sqrt{2}$
7. $\theta = 45°; x = \frac{7\sqrt{2}}{2}; y = \frac{7\sqrt{2}}{2}$
9. $B = 53°; b = 15.9; c = 19.9$
11. $B = 35.7°; a = 7.8; c = 9.6$
13. $A = 40.1°; a = 0.59; b = 0.70$
14. $a = .081; b = .028; A = 71.3°$
15. $A = 80.85°; b = 249; c = 1{,}556$
17. $B = 36.875°; a = 43.52; b = 32.64$
19. $A = 85.00°; a = 11.453; c = 11.497$
21. $A = 40.7°; B = 49.3°; b = 17.4$
23. $A = 40.6°; B = 49.4°; c = 9.2$
25. $A = 74.4°; B = 15.6°; a = 0.439$
27. **a.** 64.0 ft **b.** 92.3 ft
29. 150 ft
31. 18.4°
33. **a.** 176.1 ft **b.** 26.7 ft
35. 26.2 ft 37.167 ft
37. 167 ft
39. 630 ft
41. 1,454 ft
43. 24.6 ft

Section 6.6

1. center (0, 0); radius = 1

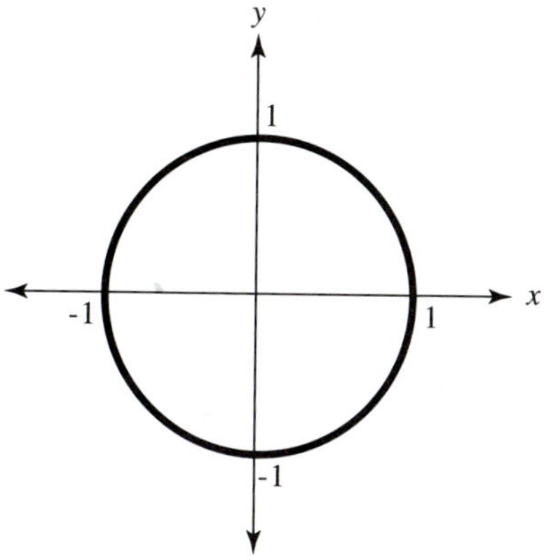

3. center (2, 0); radius = 3

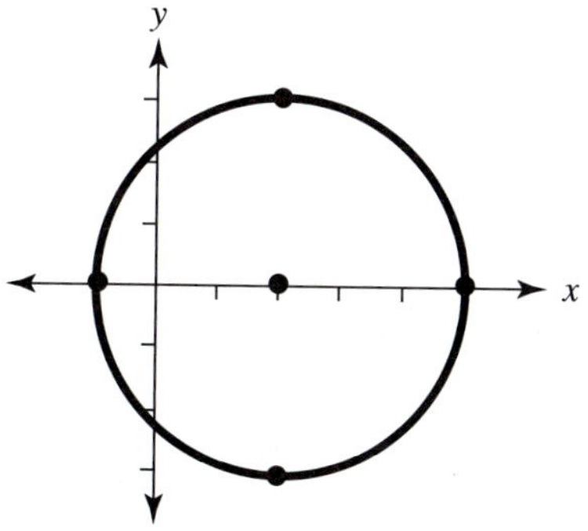

5. center (5, −2); radius = 4

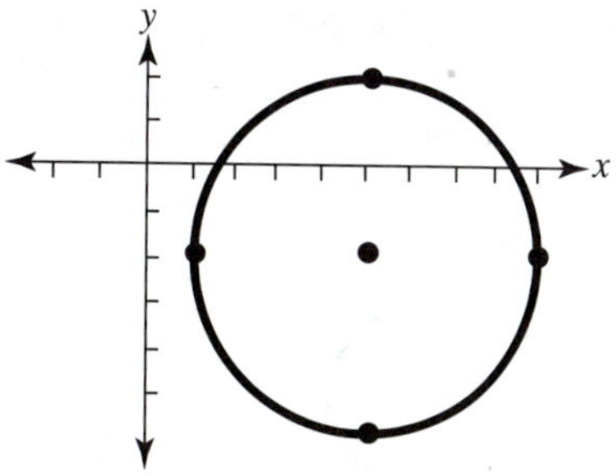

7. center (5, 5); radius = 5

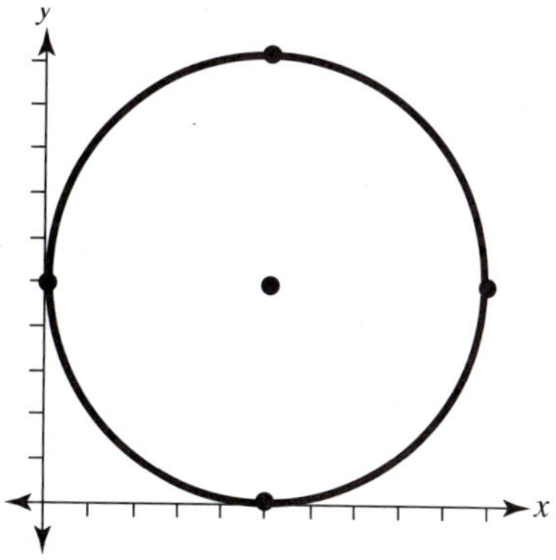

9. focus $(0, \frac{1}{4})$

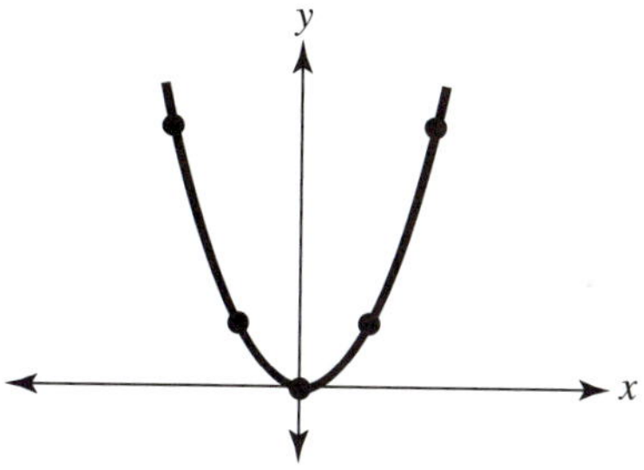

11. focus $(0, \frac{1}{2})$

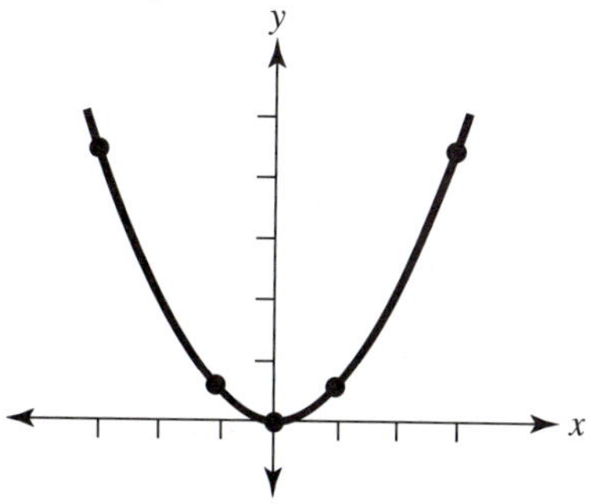

13. 2.9 ft above the bottom of the dish

15. $\frac{9}{16}$ inch above the bottom of the reflector

17. foci $(\sqrt{5}, 0)$ and $(-\sqrt{5}, 0)$

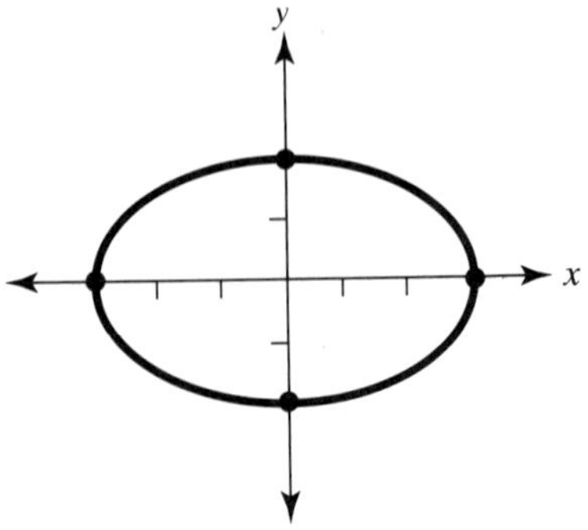

19. foci $(0, \sqrt{21})$ and $(0, -\sqrt{21})$

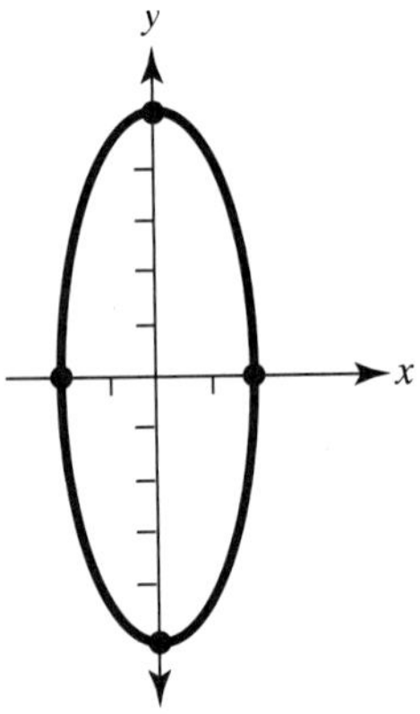

21. foci $(0, \sqrt{7})$ and $(0, -\sqrt{7})$

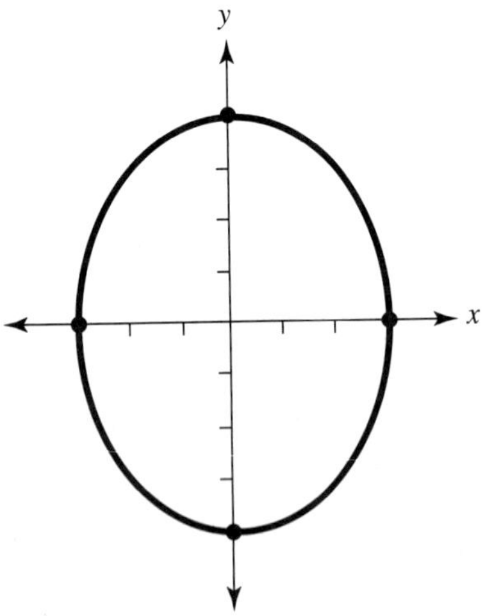

23. foci $(1, 2 + \sqrt{3})$ and $(1, 2 - \sqrt{3})$

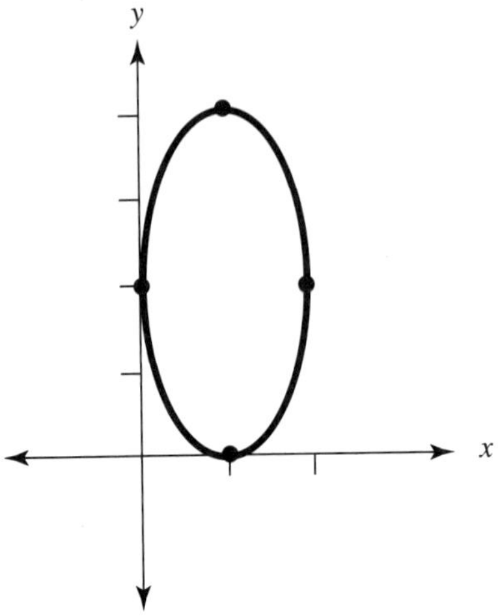

25. They should stand at the foci, which are located 9 ft from the center, in the long direction.

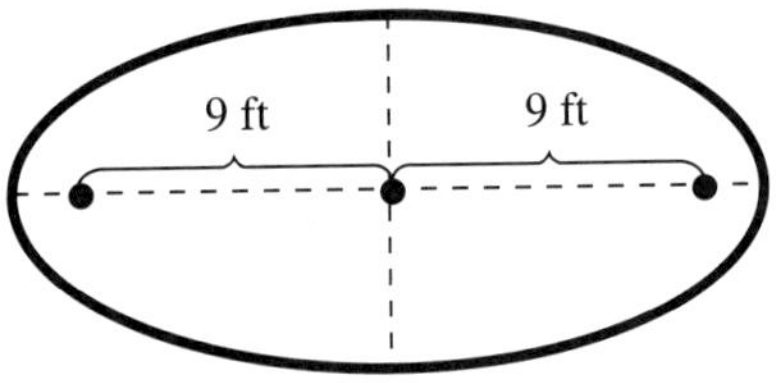

27. $\dfrac{x}{92.955^2} + \dfrac{y}{92.942^2} = 1$

29. a. foci $(\sqrt{2}, 0)$ and $(-\sqrt{2}, 0)$

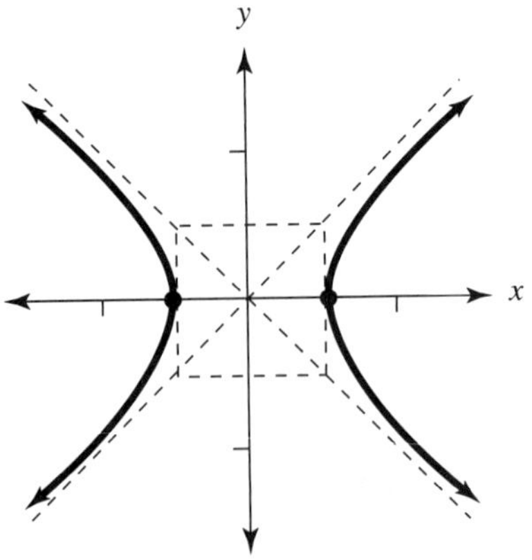

b. foci $(0, \sqrt{2})$ and $(0, -\sqrt{2})$

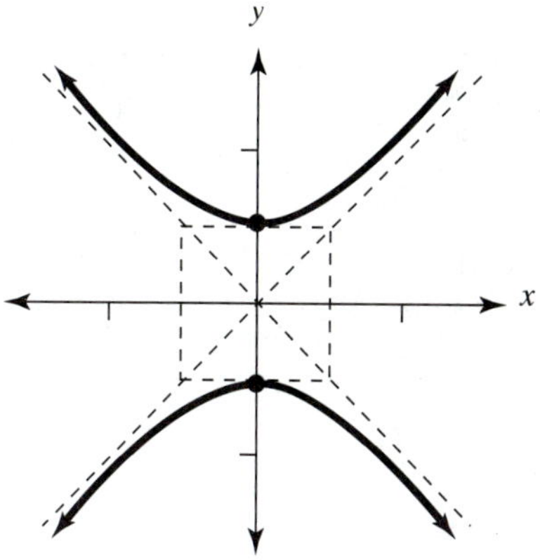

31. foci $(\sqrt{13}, 0)$ and $(-\sqrt{13}, 0)$

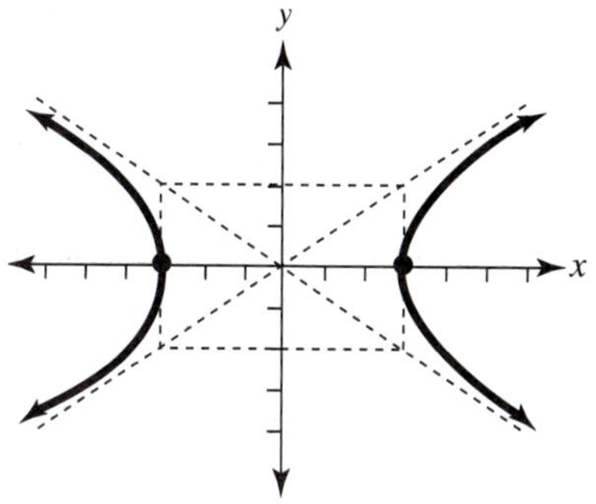

33. foci $(0, \sqrt{29})$ and $(0, -\sqrt{29})$

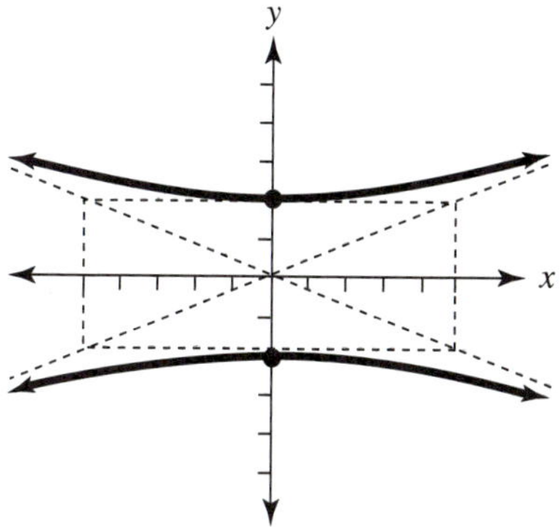

35. foci $(3 + \sqrt{5}, 0)$ and $(3 - \sqrt{5}, 0)$

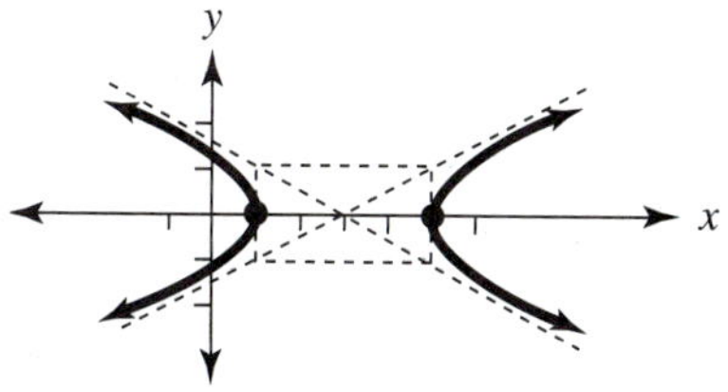

Section 6.7

1. one
3. zero or one
5. zero or one
7. none
9. zero, one, two, or three
11. two
13. infinitely many
15. zero or one
17. zero or one

Section 6.8 Fractal Geometry

1. b.

5. $d = 2$
7. $d = \frac{1}{3}$
9. $d \approx 1.6$ It's not 2-dimensional, as a regular triangle is.
11. $d \approx 1.6$ It's not 2-dimensional, as a regular triangle is.
13. $d = 1.5$ It's not 2-dimensional.
15. $d = 3$, the same dimension that we observed before.
17. $d = 2$

Section 6.9 The Perimeter and Area of a Fractal

1. a. In step 1, there is only one triangle. If each side is 1 ft. in length, then the perimeter is 3 ft. In step 2, the original triangle is modified by removing a center triangle. This results in three smaller triangles. Each side of the original triangle now has two triangles with sides of length $\frac{1}{2}$ ft. Thus, the perimeter of one of these smaller triangles is $\frac{1}{2}+\frac{1}{2}+\frac{1}{2} = 3 \cdot \frac{1}{2} = \frac{3}{2}$ ft. Since there are three such triangles, the total perimeter is $3 \cdot \frac{3}{2} = \frac{9}{2}$ ft.

1. b.

Step	Number of Triangles	Length of each side (in feet)	Perimeter of one triangle (in feet)	Total Perimeter of all triangles (in feet)
1	1	1	3	3
2	3	$1/2$	$3/2$	$9/2$
3	9	$1/4$	$3/4$	$27/4$
4	27	$1/8$	$3/8$	$81/8$
5	81	$1/16$	$3/16$	$243/16$
6	243	$1/32$	$3/32$	$729/32$

c. 3
d. $1/2$
e. $1/2$
f. $3/2$
g. $3(\frac{3}{2})^{n-1}$
h. The perimeter is infinite. The numbers get larger and larger without bound.

3. a. In step 1, there is only one square. If each side is 1 ft. in length, then the perimeter is 4 ft. In step 2, the original square is divided into nine smaller squares and modified by removing the center square. The removed square contributes to the perimeter. It has sides of length $1/3$ ft. Thus, the perimeter of this smaller square is $4 \cdot 1/3 = 4/3$ ft. The total perimeter is the sum of the previous perimeter and the new contribution: $4 + 4/3$ ft.

3. b.

Step	Number of new squares	Length of each side (in feet)	Perimeter of one new square (in feet)	Total perimeter of all new squares (in feet)	Total perimeter of all squares (in feet)
1	1	1	4	4	4
2	1	$1/3$	$4/3$	$4/3$	$4+4/3$
3	8	$1/9$	$4/9$	$32/9$	$4+4/3+32/9$
4	64	$1/27$	$4/27$	$256/27$	$4+4/3+32/9+256/27$
5	512	$1/81$	$4/81$	$2048/81$	$4+4/3+32/9+256/27+2048/81$
6	4096	$1/243$	$4/243$	$16384/243$	$4+4/3+32/9+256/27+2048/81+16384/243$

c. 8
d. $1/3$
e. $1/3$
f. $8/3$

g. $\frac{4}{3} \cdot \left(\frac{8}{3}\right)^{n-2}$, valid for n>1

h. The total perimeter is infinite. The numbers increase without bound.

5. a. $3 \cdot 2^{n-1}$ **b.** ∞ **c.** $\left(\frac{2}{3}\right)^{n-1}\frac{\sqrt{3}}{4}$ **d.** 0

Chapter 6 Review

1. Area = $x(x-2)$ square feet. Perimeter = $4x - 4$ feet.
2. 114 ft^2, 42 ft
3. 73.5 in.^{2}; 41 in.
4. 36 cm^2; 28 cm
5. 69 yd^2; 37.4 yd
6. 2,789.7 ft^3
7. 72.8 cm^3; 153.9 cm^2
8. 254.5 yd^3; 216.8 yd^2
9. 1,056 in.^{3}; 664 in.^{2}
10. 4.1 mi
11. nine
12. 672.1 yd^2
13. **a.** 224 in.^{3}
    b. 200 in.^{2}
14. **a.** hardball, 12.3 in.^{3}; tennis ball, 8.2 in.^{3}; volume of hardball is 50% more than that of tennis ball
    b. hardball, 25.8 in.^{2}; tennis ball, 19.6 in.^{2}; surface area of hardball is 32% more than that of tennis ball
15. **a.** 450.7 ft^3 **b.** 268.0 ft^2
16. **a.** 3.1605 cubits^2 **b.** 3.1416 cubits^2 **c.** 0.6%
17. **a.** 4.2140 cubits^3 **b.** 4.1888 cubits^3 **c.** 0.6%
18. the regular pyramid (40,000 khar . 28,140 khar)
19. a square with sides of 16 units

20. $\frac{256}{81}$

21. 6.5 ft (rounded down so the object will fit)

22. 4.7 ft

26. 3.313708499

27. 3.105828541

28. center (1, 1); radius = 1

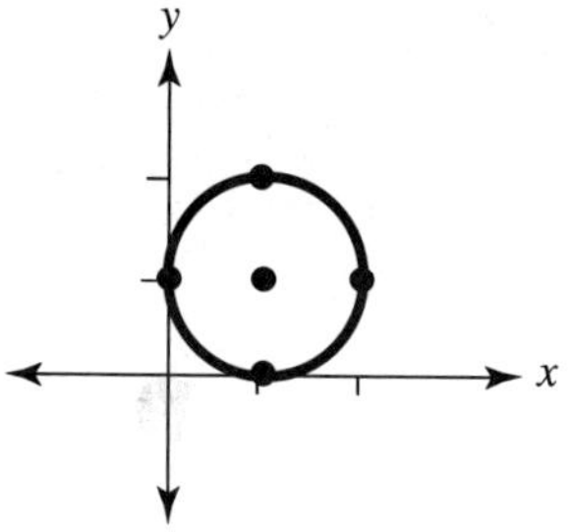

29. **a.** foci $(\sqrt{13}, 0)$ and $(-\sqrt{13}, 0)$

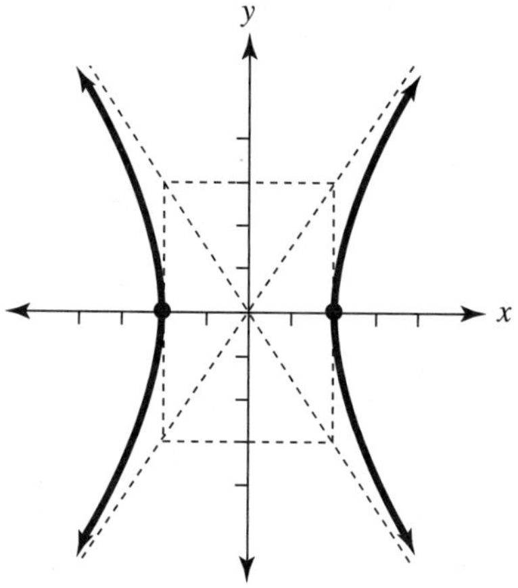

b. foci $(0, \sqrt{13})$ and $(0, -\sqrt{13})$

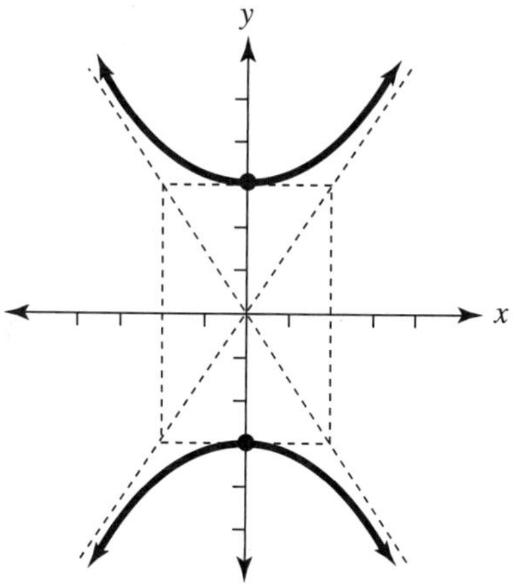

30. 2 ft above the bottom of the reflector

31. foci $(0, \sqrt{5})$ and $(0, -\sqrt{5})$

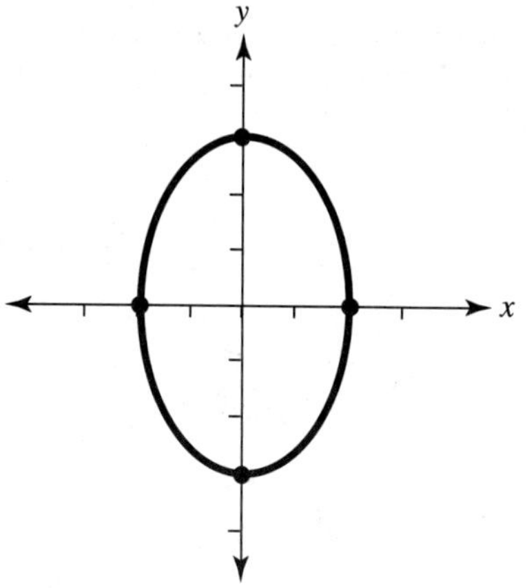

32. They should stand at the foci, which are located 17.3 ft from the center of the room, in the long direction.

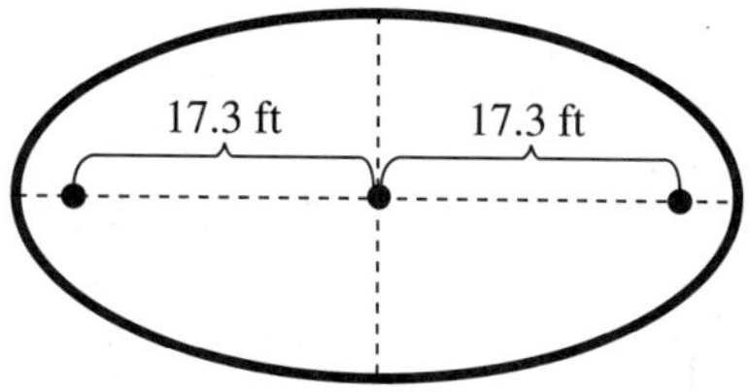

33. $B = 64.6°$; $a = 24.1$ ft; $b = 50.7$ ft

34. $A = 33.7°$; $B = 56.3°$; $c = 81.5$ cm

35. 213.2 ft

36. 983 ft

37. **a.** Given a line and a point not on that line, there is one and only one line through the point parallel to the original line.

b. 1. Given a line and a point not on that line, there are no lines through the point parallel to the original line (parallels don't exist).

2. Given a line and a point not on that line, there are at least two lines through the point parallel to the original line.

c. 1. a sphere (Riemannian geometry)

2. a Poincaré disk (Lobachevskian geometry)

d. see page 412, Figure 6.57

38. **a.** one **b.** more than one **c.** none

39. **a.** sum = 180° **b.** sum < 180° **c.** sum > 180°

40.

b. The total perimeter in steps 1–5: 4, $\frac{20}{3}$, $\frac{100}{9}$, $\frac{500}{27}$, $\frac{2500}{81}$

c. $4 \cdot \left(\frac{5}{3}\right)^{n-1}$

d. The perimeter is infinite.

e. $\left(\frac{5}{9}\right)^{n-1}$

f. 0

g. $d = 1.465$

h. If we focus in on any smaller square it will look like the entire box fractal.
i. For any square that is not removed, the recursive rule is to remove the same four squares in the middle of the edges.

Chapter 7 Matrices and Markov Chains

Section 7.0 Review of Matrices

1. a. 3×2 **b.** none
3. a. 2×1 **b.** column
5. a. 1×2 **b.** row
7. a. 3×3 **b.** square
9. a. 3×1 **b.** column
11. 22
13. 41
15. 3
17. −11
19. −3
21. a. $AC = \begin{bmatrix} 115 \\ 383 \\ 783 \end{bmatrix}$ **b.** *CA* does not exist.
23. a. *AD* does not exist. **b.** *DA* does not exist.
25. a. *CG* does not exist. **b.** *GC* does not exist.
27. a. *JB* does not exist. **b.** $BJ = \begin{bmatrix} 2{,}243 \\ 1{,}056 \\ 52 \end{bmatrix}$
29. a. $AF = \begin{bmatrix} -10 & 50 \\ -56 & 229 \\ 0 & 153 \end{bmatrix}$ **b.** *FA* does not exist.
31. \$130 on sale, \$170 regular price
33. [3.45 3.70] **35.** 3.07; 3; 3.50
37. $\begin{bmatrix} 106 \\ 68 \end{bmatrix}$ **39.** $\begin{bmatrix} 3 & -2 \\ 34 & 0 \end{bmatrix}$
41. does not exist
43. $\begin{bmatrix} 19 & 7 & 34 \\ 74 & 0 & -11 \\ 13 & -2 & 44 \end{bmatrix}$

Section 7.1 Markov Chains

1. a. *p*(current purchase is KickKola) = 0.14; *p*(current purchase is not KickKola) = 0.86
b. $P = [0.14\ 0.86]$
3. a. *p*(currently own) = 0.32; *p*(currently rent) = 0.68
b. $P = [0.32\ 0.68]$
5. a. *p*(Silver's) = 0.48; *p*(Fitness Lab) = 0.37; *p*(ThinNFit) = 0.15
b. $P = [0.48\ \ 0.37\ \ 0.15]$
7. a. *p*(next purchase is KickKola | current purchase is not Kick Kola) = 0.12; *p*(next purchase is KickKola | current purchase is KickKola)= 0.63; *p*(next purchase is not KickKola | current purchase is not KickKola) = 0.88; *p*(next purchase is not KickKola | current purchase is KickKola) = 0.37
b. $\begin{bmatrix} 0.63 & 0.37 \\ 0.12 & 0.88 \end{bmatrix}$
9. a. *p*(buy next residence | currently rent) = 0.12; *p*(rent next residence | currently own) = 0.03; *p*(rent next residence | currently rent) = 0.88; *p*(buy next residence | currently own) = 0.97
b. $\begin{bmatrix} 0.97 & 0.03 \\ 0.12 & 0.88 \end{bmatrix}$
11. a. *p*(Silver's next | currently Silver's) = 0.71; *p*(Fitness Lab next | currently Silver's) = 0.12; *p*(ThinNFit next | currently Silver's) = 0.17; *p*(Silver's next | currently Fitness Lab) = 0.32; *p*(Fitness Lab next | currently Fitness Lab) = 0.34; *p*(ThinNFit next | currently Fitness Lab) = 0.34; *p*(Silver's next | currently ThinNFit) = 0.02; *p*(Fitness Lab next | currently ThinNFit) = 0.02; *p*(ThinNFit next | currently ThinNFit) = 0.96
b. $\begin{bmatrix} 0.71 & 0.12 & 0.17 \\ 0.32 & 0.34 & 0.34 \\ 0.02 & 0.02 & 0.96 \end{bmatrix}$
13. a. 19% **b.** 22%
15. a. 46%; 19%; 35% **c.** 39%; 13%; 48%
d. 33%; 10%; 57%
17. 27%

Section 7.2 Systems of Linear Equations

1. yes
3. no
5. no
7. a. $-\frac{5}{2}, 2; \frac{6}{19}, -\frac{72}{19}$ **b.** one solution
9. a. $-\frac{4}{3}, 4; -\frac{4}{3}, 4$ **b.** infinite number of solutions
11. a. $-1, 7; -\frac{3}{2}, 4; -1, 7$ **b.** one solution
13. yes
15. no
17. no

Section 7.3 Long-Range Predictions with Markov Chains

1. $[0.1818\quad 0.8182]$
3. $L = [\frac{8}{15}\quad \frac{7}{15}]$
5. 24%
7. 80%; 20%; current trends continue, and residents' moving plans are realized
9. Sierra Cruiser will eventually control 26% of the market.

Section 7.4 Solving Larger Systems of Equations

1. $x = 4, y = -2, z = 1$
3. $x = 2, y = 5, z = 7$
5. $x = 5, y = 10, z = -2$
7. $\begin{bmatrix} 2 & 7 & 11 \\ 3 & -2 & 15 \end{bmatrix}$
9. $\begin{bmatrix} 2 & -1 & 5 \\ 3 & -1 & 41 \end{bmatrix}$
11. $\begin{bmatrix} 2 & 3 & -7 & 53 \\ 5 & -2 & 12 & 19 \\ 1 & 1 & 1 & 55 \end{bmatrix}$
13. $\begin{bmatrix} 5 & 0 & 1 & 2 \\ 3 & -1 & 0 & 15 \\ 2 & -2 & 2 & 53 \end{bmatrix}$
15. $\begin{bmatrix} 1 & 3 & 5 \\ 0 & 7 & 14 \end{bmatrix}$
17. $\begin{bmatrix} 1 & 0 & \frac{39}{5} \\ 0 & 1 & \frac{7}{5} \end{bmatrix}$
19. $\begin{bmatrix} 1 & 1 & 2 & 6 \\ 0 & -3 & 0 & -9 \\ 0 & 1 & -11 & -4 \end{bmatrix}$
21. (9, 2)
23. (0.5281, −0.6205, 0)
37. no solution
39. (4, 0, −7)

Section 7.5 More on Markov Chains

1. $[0.2115 \quad 0.5577 \quad 0.2308]$
3. 10.8%; 4.5%; 84.6%
5. 48.6%; 37.6%; 13.8%

Chapter 7 Review

1. neither; 3×2
2. square; 3×3
3. column; 4×1
4. row; 1×4
5. $\begin{bmatrix} 23 & -24 \\ -24 & 28 \end{bmatrix}$
6. does not exist
7. $\begin{bmatrix} -18 & 3 & 15 \\ 0 & -10 & -25 \end{bmatrix}$
8. $\begin{bmatrix} 35 & -19 & -72 & 24 \\ 12 & 34 & 51 & 24 \end{bmatrix}$
9. does not exist
10. \$43,000; \$41,300
11. no; yes
17. **a.** 21.5% **b.** 25.9% **c.** 29.6%
d. Current trends continue, and the survey accurately reflects future purchases.
18. **a.** If current trends continue, the percentages will be 22.1%, 19.6%, and 58.3%. Thus, more condominiums and townhouses will be needed.
b. If current trends continue, the percentages will be 17.6%, 23.9%, and 58.5%. Thus, significantly fewer apartments will be needed, and significantly more condominiums and townhouses will be needed.
c. emigration and immigration

Chapter 8 Linear Programming

Section 8.0 Review of Linear Inequalities

1.

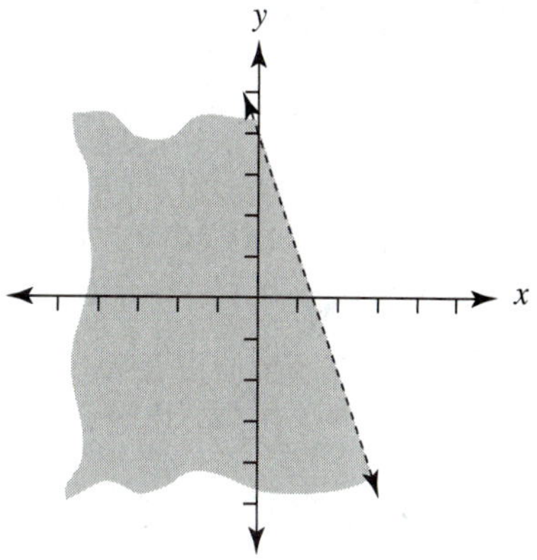

3.

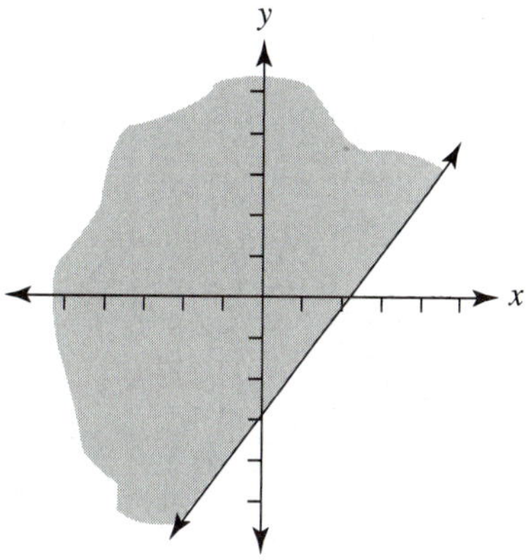

5.

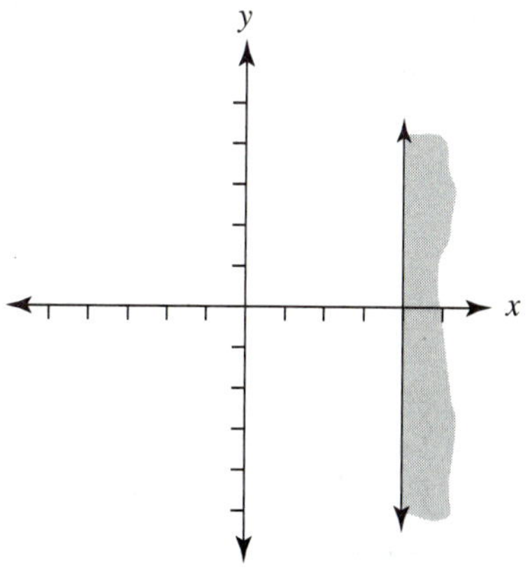

7.

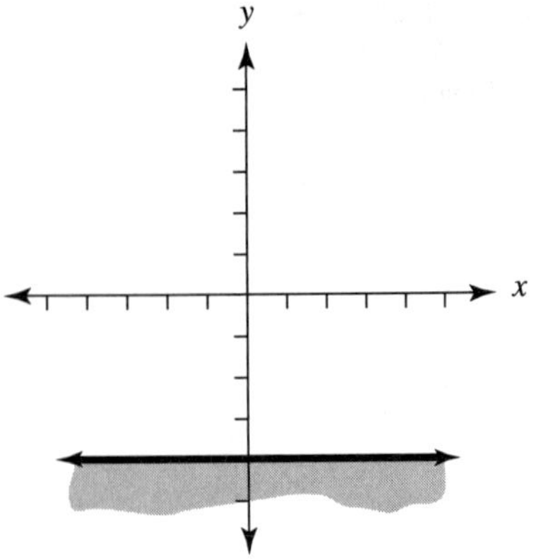

9. a.

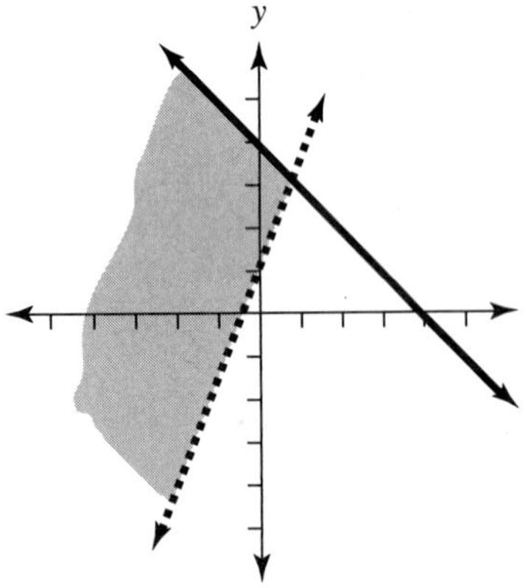

b. unbounded **c.** (1, 3)

11. a.

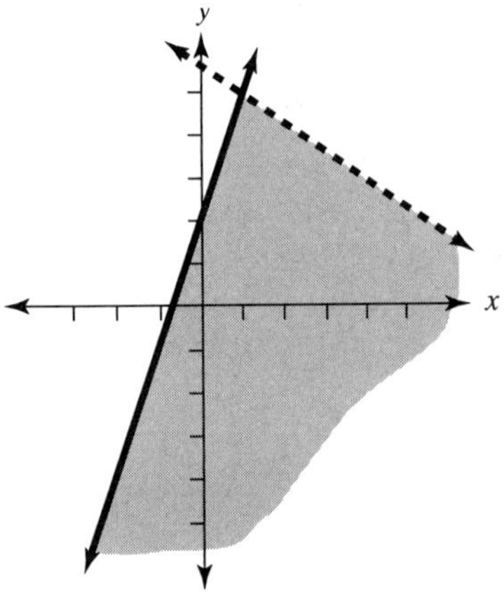

b. unbounded **c.** (1, 5)

13. a.

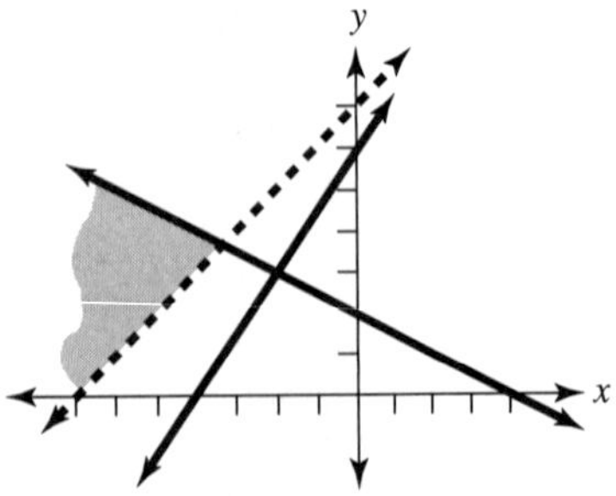

b. unbounded
c. $(-3\frac{1}{3}, 3\frac{2}{3})$

15. a.

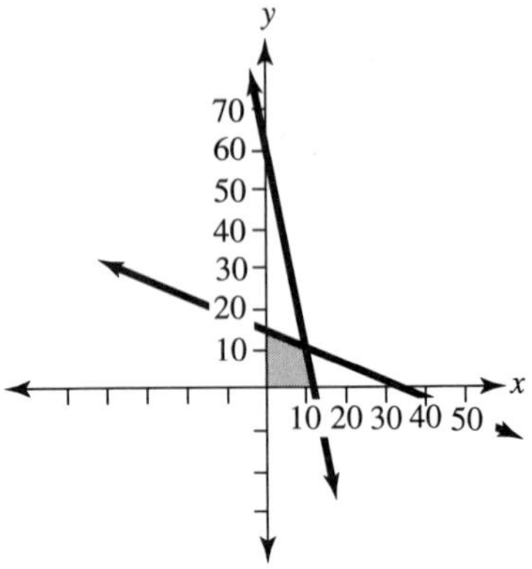

b. bounded
c. (0, 0), (0, 14), (10, 10), (12, 0)

17. a.

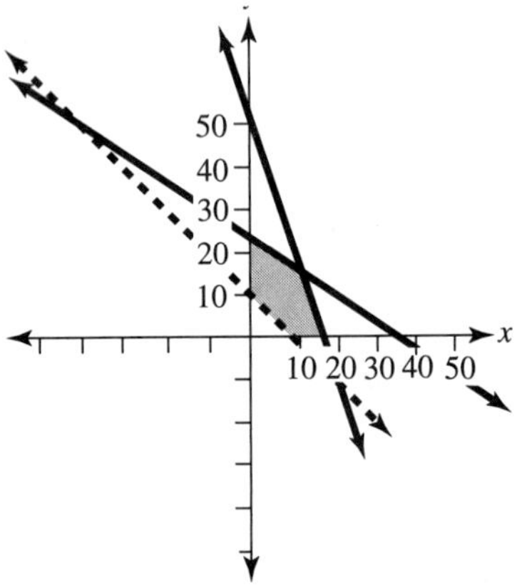

b. bounded
c. (0, 10), $(0, 23\frac{2}{11})$, (12, 15), $(17\frac{1}{7}, 0)$, (10, 0)

19. a.

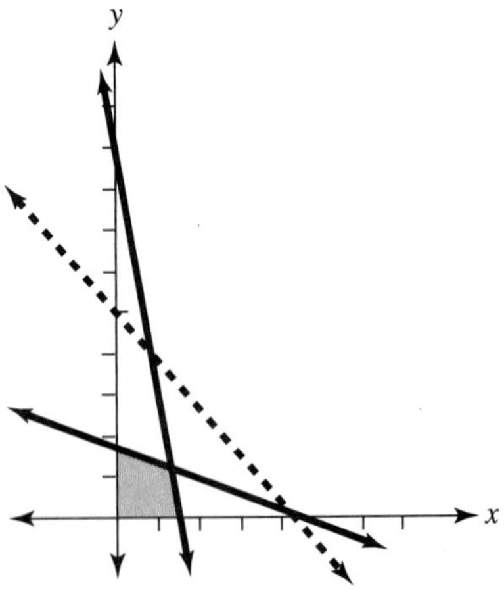

b. bounded
c. (0, 0), (0, 1.7), (1.3, 1.2), (1.5, 0)

21. a.

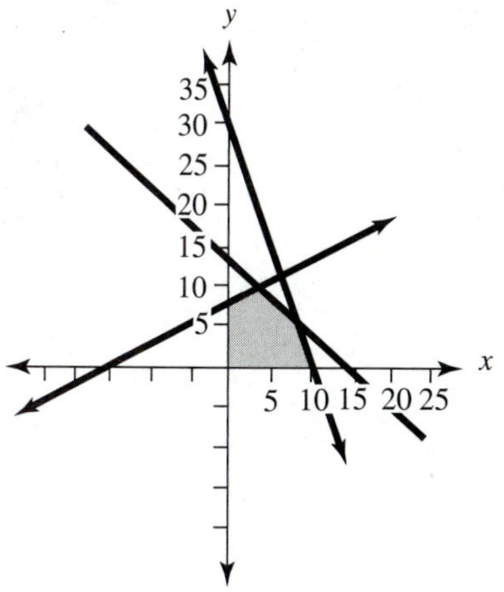

b. bounded **c.** (0, 0), (0, 8), (4, 10), (8, 6), (10, 0)

Section 8.1 The Geometry of Linear Programming

1. x = number of shrubs, y = number of trees; $1x + 3y \le 100$

3. x = number of hardbacks, y = number of paperbacks; $4.50x + 1.25y \ge 5{,}000$

5. x = number of refrigerators, y = number of dishwashers; $63x + 41y \le 1{,}650$

7. no floor lamps and 45 table lamps, for a profit of $1,485

9. 60 pounds of Morning Blend and 120 pounds of South American Blend each day for a profit of $480

11. 15,000 loaves to Shopgood and 20,000 loaves to Rollie's each week for shipping costs of $3,000

13. Global has many choices, including the following:
15 Orvilles and 30 Wilburs
18 Orvilles and 28 Wilburs
21 Orvilles and 26 Wilburs
Each of these generates a cost of $720,000.

15. a. 20 large and 30 small, for a profit of $9,500
b. 40 large and no small, for a profit of $10,000
c. 40

17. 0 Mexican; 0 Colombian

19. 150 minutes; $0

21. a. no maximum **b.** no maximum **c.** (6, 0) **d.** 12

23. 3 weeks at Detroit and 6 weeks at Los Angeles, for a minimum cost of $3,780,000

Section 8.2 Introduction to the Simplex Method

1. $3x_1 + 2x_2 + s_1 = 5$

3. $3.41x_1 + 9.20x_2 + 6.16x_3 + s_1 = 45.22$

5. a. $6.99x_1 + 3.15x_2 + 1.98x_3 \le 42.15$
b. $6.99x_1 + 3.15x_2 + 1.98x_3 + s_1 = 42.15$
c. x_1 = pounds of meat, x_2 = pounds of cheese, x_3 = loaves of bread, s_1 = unused money

7. a. $5x_1 + 4x_2 \le 480$
b. $5x_1 + 4x_2 + s_1 = 480$
c. x_1 = feet of pipe, x_2 = number of elbows, s_1 = unused time in minutes

9. a. $24x_1 + 36x_2 + 56x_3 + 72x_4 \le 50{,}000$
b. $24x_1 + 36x_2 + 56x_3 + 72x_4 + s_1 = 50{,}000$
c. x_1 = number of twin beds, x_2 = number of double beds, x_3 = number of queen-size beds, x_4 = number of king-size beds, s_1 = unused space in cubic feet

11. $(x_1, x_2, s_1, s_2) = (0, 19, 0, 12)$, $z = 22$

13. $(x_1, x_2, s_1, s_2, s_3) = (0, 0, 7.8, 9.3, 0.5)$, $z = 9.6$

15. $(x_1, x_2, s_1, s_2, s_3, s_4) = (25, 73, 0, 32, 46, 0)$, $z = 63$

17. $$\begin{bmatrix} 3 & 4 & 1 & 0 & 0 & 40 \\ 4 & 7 & 0 & 1 & 0 & 50 \\ -2 & -4 & 0 & 0 & 1 & 0 \end{bmatrix}$$
$(x_1, x_2, s_1, s_2) = (0, 0, 40, 50)$, $z = 0$

19. $$\begin{bmatrix} 112 & -3 & 1 & 0 & 0 & 0 & 370 \\ 1 & 1 & 0 & 1 & 0 & 0 & 70 \\ 47 & 19 & 0 & 0 & 1 & 0 & 512 \\ -12.10 & -43.86 & 0 & 0 & 0 & 1 & 0 \end{bmatrix}$$
$(x_1, x_2, s_1, s_2, s_3) = (0, 0, 370, 70, 512)$, $z = 0$

21. $$\begin{bmatrix} 5 & 3 & 9 & 1 & 0 & 0 & 0 & 10 \\ 12 & 34 & 100 & 0 & 1 & 0 & 0 & 10 \\ 52 & 7 & 12 & 0 & 0 & 1 & 0 & 10 \\ -4 & -7 & -9 & 0 & 0 & 0 & 1 & 0 \end{bmatrix}$$
$(x_1, x_2, x_3, s_1, s_2, s_3) = (0, 0, 0, 10, 10, 10)$, $z = 0$

23. $$\begin{bmatrix} 50 & 30 & 1 & 0 & 0 & 2{,}400 \\ 0.90 & 1.50 & 0 & 1 & 0 & 190 \\ -1.2 & -4 & 0 & 0 & 1 & 0 \end{bmatrix}$$
$(x_1, x_2, s_1, s_2) = (0, 0, 2400, 190)$, $z = 0$

25. $$\begin{bmatrix} \frac{1}{2} & \frac{1}{4} & 1 & 0 & 0 & 200 \\ \frac{1}{2} & \frac{3}{4} & 0 & 1 & 0 & 330 \\ -3.5 & -4 & 0 & 0 & 1 & 0 \end{bmatrix}$$
$(x_1, x_2, s_1, s_2) = (0, 0, 200, 330)$, $z = 0$

Section 8.3 The Simplex Method: Complete Problems

1. pivot on 19

3. pivot on 19

5. a. pivot on 4
b. $$\begin{bmatrix} 0 & \frac{7}{4} & 1 & -\frac{1}{4} & 0 & \frac{5}{2} \\ 1 & \frac{1}{4} & 0 & \frac{1}{4} & 0 & \frac{1}{2} \\ 0 & -\frac{5}{2} & 0 & \frac{3}{2} & 1 & 3 \end{bmatrix}$$
c. $(x_1, x_2, s_1, s_2) = (1/2, 0, 2\frac{1}{2}, 0)$, $z = 3$

7. a. pivot on 1 (row 1, column 1)
b. $$\begin{bmatrix} 1 & 1 & 5 & 0 & 0 & 3 \\ 0 & -3 & -14 & 1 & 0 & 3 \\ 0 & 6 & 28 & 0 & 1 & 24 \end{bmatrix}$$
c. $(x_1, x_2, s_1, s_2) = (3, 0, 0, 3)$, $z = 24$

9. Each day, make no bread and 80 cakes, for a profit of \$320. This uses all available time and leaves \$70 unspent.

11. Each week, make 15 sofas and 40 chairs, for a profit of \$21,750. This leaves no extra time in either shop.

13. Prepare 20,000 liters of House White, no Premium White, and 5,000 liters of Sauvignon Blanc, for a profit of \$30,000. This uses all the grapes.

15. Prepare 300 pounds of Smooth Sipper and no Kona Blend, for a profit of \$900. This leaves 100 pounds each of Colombian and Arabian beans.

25. (4.304, 2.956, 0, 0, 9.173, 0, 0), $z = 264.304$

27. (3.694, 0, 1.117, 0, 0, 30.359, 0, 0); $z = 195.884$

29. 13.2 modern coffee tables, 14.2 modern end tables; profit \$9,228.79; no slack

31. 427 dozen blueberry muffins, 1013 dozen apple-cin. muffins; profit \$21,461.33; no slack

33. 57.3 sofas, 21.8 recliners; \$36,140 profit; 15.5 slack coating hours

Chapter 8 Review

1.

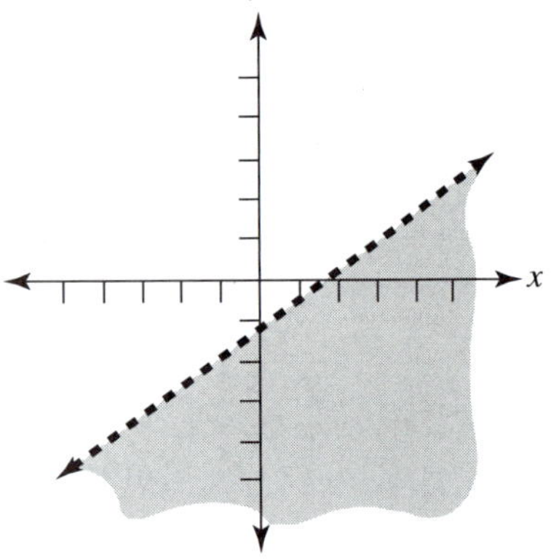

2.

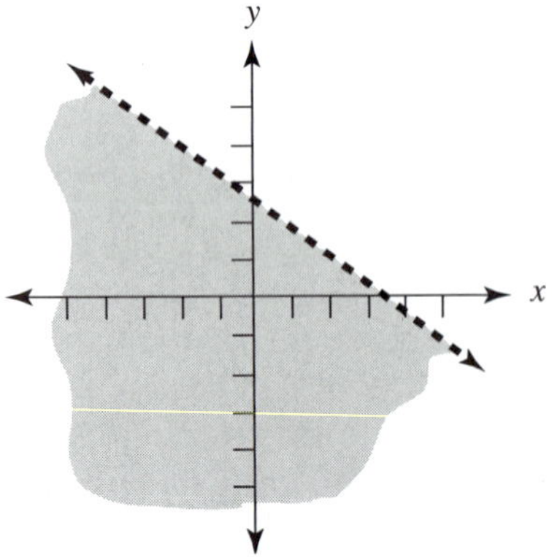

3.

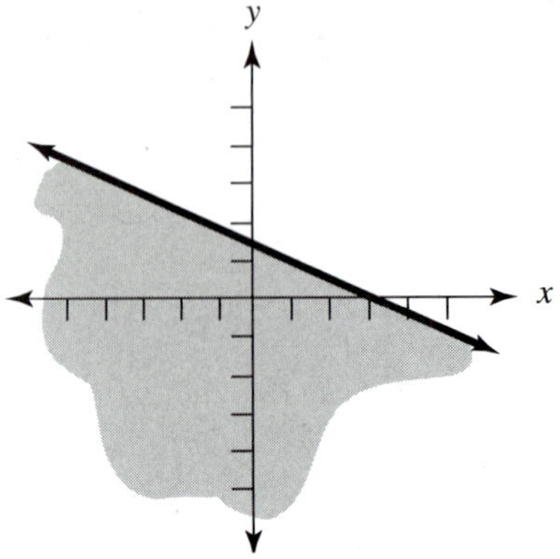

4.

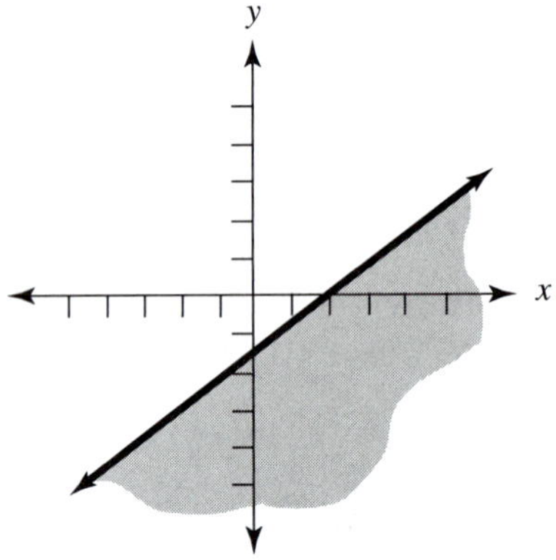

5. $(\frac{3}{2}, \frac{1}{2})$

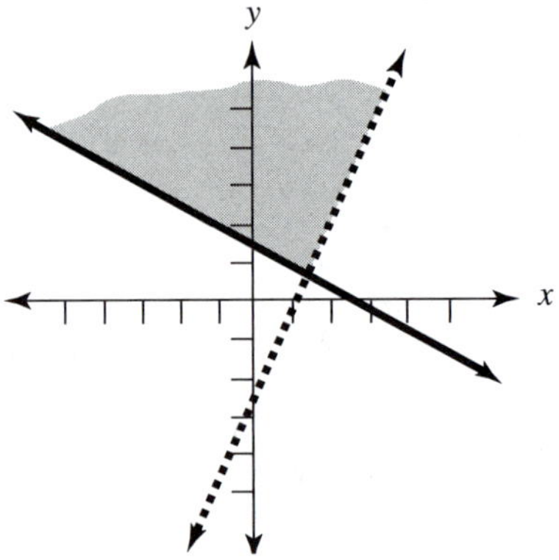

6. $(2\frac{10}{17}, -\frac{15}{17})$

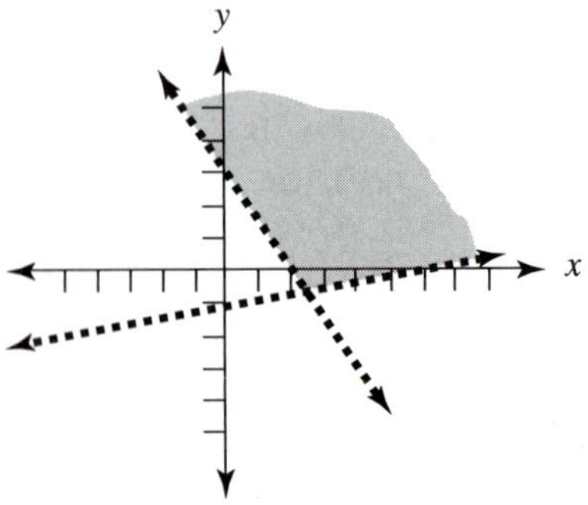

7. (−3, −23)

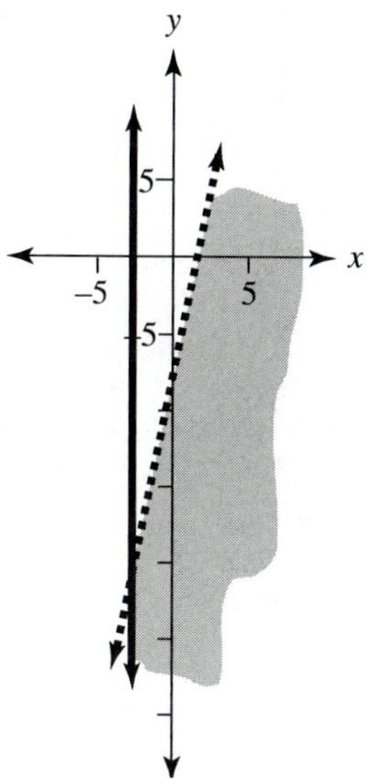

8. $(-1\frac{1}{2}, 4)$

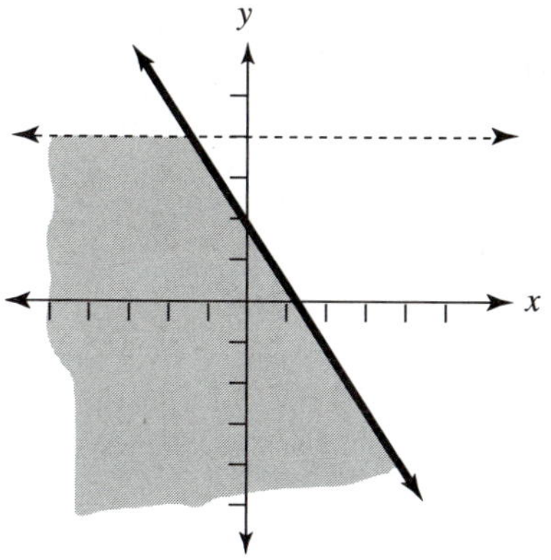

9. no solution

10. (0, 5), (3, 0), (4, 0)

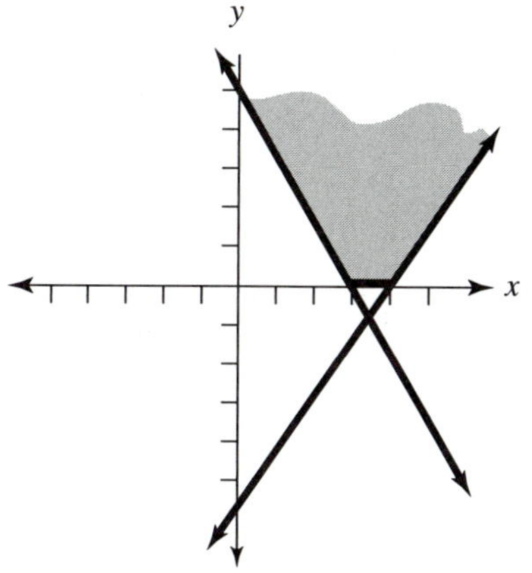

11. 30 of each assembly for a maximum income of $16,500

12. 750 Sunnys and 375 Iwas for a maximum profit of $240,000

13. Make 30 of each assembly, with no tweeters or midrange speakers left over and 14 woofers left over. This gives a maximum income of $16,500.

14. Make 24 model 110s, 16 model 220s, and 50 model 330s, with no tweeters, midrange speakers, or woofers left over. This gives a maximum income of $26,780.

Chapter 9 Exponential and Logarithmic Functions

Section 9.0 A Review of Exponentials and Logarithms

1. $v = 3$
3. $v = 22$
5. $u = \frac{1}{25}$
7. $u = 1$
9. $b = 3$
11. $b = \frac{1}{3}$
13. $b^k = H$
15. $b^{(G+5)} = F - E$
17. $T = \log_b S$
19. $MN = \log_b(P - Q)$
21. **a.** 3.320116923 **b.** 15.84893192
23. **a.** 2.166574135 **b.** 2.404528869
25. **a.** 0.2231301601 **b.** 0.0316227766
27. **a.** 3.146511444 **b.** 2.301759749
29. **a.** 0.335160023 **b.** 0.1990535853
31. **a.** 0.7654678421 **b.** 0.3324384599
33. **a.** −0.5447271754 **b.** −0.2365720064
35. **a.** 3.4 **b.** 3.4
37. **a.** 1.406294361 **b.** 0.622059913
39. **a.** 3.6 **b.** 3.6
41. **a.** 25 **b.** 25
43. **a.** 2 **b.** 0.8685889638
45. **a.** 5.68738518 **b.** 2.47
47. log 2.8, ln 2.8, $e^{2.8}$, $10^{2.8}$

Section 9.0 B Review of Properties of Logarithms

1. $7.5x$
3. $-0.058x$
5. $3x + 1$
7. $8 - x$
9. $\log x - \log 5$
11. $\ln 1.4 + \ln x$
13. $x \log 1.0625$
15. $\log 4 + 3 \log x$
17. $\ln 2 + \ln x - \ln 3$
19. $\log\frac{x}{2}$
21. $\ln(10x)$
23. $\log(2x)$
25. $\ln(x^2)$
27. 0
29. **a.** −1.386294361 **b.** −0.602059991
31. **a.** 18.24071528 **b.** 7.921841991

33. a. 45.05167868 **b.** 19.56569545
35. a. 46.20981204 **b.** 20.06866638
37. a. 139.4647196 **b.** 60.56875813
39. a. 1.318256739 **b.** 1.127496852
41. a. 70.79457844 **b.** 6.359819523
43. a. 2.5 **b.** 10.87312731
45. a. 227.144644 **b.** 1.680457629
47. a. 1.892872033 **b.** 1.85461744
53. 3.5; acid
55. 7.4; base
57. 4.9
59. 10^{-7} moles per liter
61. a. yes; pH = 6.5 is acceptable
b. no; pH = 3.5 is not acceptable
63. from 3.16×10^{-9} to 10^{-7} moles per liter
65. a. $\frac{\log 3}{\log 2}$ **b.** 1.58
67. a. $\frac{\log 6}{\log 3}$ **b.** 1.63
69. a. $\frac{\log 8}{\log 4}$ **b.** 1.50

Section 9.1 Exponential Growth

1. 32,473
3. July 2011
5. a. (0, 8240) and (10, 9258)
b. 10 years
c. 1,018 thousand people
d. 101.8 thousand people per year
e. 1.2% per year
7. a. (0, 19342) and (10, 21200)
b. 10 years
c. 1,858 thousand people
d. 185.8 thousand people per year
e. 1.0% per year
9. a. $p(t) = 8{,}240e^{.0116487699t}$ (in thousands)
b. 9,813.2 thousand
c. 10,401.8 thousand
d. Halfway through 2049
11. a. $p(t) = 19{,}342e^{.0091722284t}$ in thousands
b. 21,791.5 thousand
c. 23,236.5 thousand
d. early in 2034
13. a. $p = 2{,}510e^{0.2541352t}$ **b.** 14,870
c. 2.7 days
15. a. $p(t) = 230{,}000e^{.0497488314t}$
b. October 2001
c. $417,826
d. $15,652 per month
17. a. $p(t) = 90.9e^{.0159798794t}$
b. 98.5 quadrillion BTU's
19. a. $p(t) = 6{,}343e^{.0771316418t}$
b. $10,076 billion
21. a. $p(t) = 3{,}929{,}214e^{.0300866701t}$
b. 1810
c. In 1810: 7,171,916. In 2000: 2,179,040,956
23. Africa, 26 years; North America, 73 years; Europe, 138 years
25. 40 years
27. a. because it will not increase by the same amount each year
b. approximately 4.25 years
29. $p(t) = 19{,}342(1.009214422)^t$ thousand people
31. $p(t) = 8{,}240(1.011716881)^t$ thousand people
33. $8{,}240e^{.0116487699t} = 8{,}240(e^{.0116487699})^t$
$= 8{,}240(1.011716881)^t$
35. 5,060.3 days = 13.86 years
37. 586 months = 48 years and 10 months

Section 9.2 Exponential Decay

1. 3.2 g
3. 7.4 g
5. a. $Q = 50e^{-0.266595069t}$ (t in hours)
b. 38.3 mg
c. 0.08 mg
d. −11.7 mg per hour
e. −23.4% per hour
f. −2.1 mg per hour
g. −4.2% per hour
h. Radioactive substances decay faster when there is more substance present. The rate of decay is proportional to the amount present.
7. a. 21.0 hr
b. 42.0 hr
c. 63.0 hr
9. 30.2 years
11. 50.3 years
13. 81,055 years
15. 99.7 sec
17. 53% of the amount expected in a living organism
19. 63.5 days
21. 1,441 years
23. 14,648 years
25. 2,949 years
27. 93% of the amount expected in a living organism; 79% of the amount expected in a living organism
29. 50% of the amount expected in a living organism
31. The claim is not justified. (62% remaining carbon-14 would indicate approximately 3,950 years.)
33. approximately 8,300 years old

Section 9.3 Logarithmic Scales

1. 7.6
3. 5.2

5. **a.** 16 times as much earth movement
   b. 55 times as much energy released
7. **a.** 32 times as much earth movement
   b. 150 times as much energy released
9. **a.** 1.3 times as much earth movement
   b. 1.4 times as much energy released
11. **a.** 158 times as much earth movement
   b. 1,549 times as much energy released
13. **a.** 26%
   b. 40%
15. Energy released is magnified by a factor of 28.
17. 70 dB
19. 74 dB
21. 10 dB
23. 9.1 dB
25. four joined (five total)
27. one joined (two total)
29. five joined (six total)
31. a gain of 3 dB
33. The new sound would be 0.003% as intense as the old sound, so the energy hitting the buildings would be 0.003% as strong.

Chapter 9 Review

1. $x = 4$
2. $x = 4$
3. $x = 3{,}125$
4. $\log(10^x) = x$ and $10^{\log x} = x$
5. $\ln(e^x) = x$ and $e^{\ln x} = x$
6. $\log(\frac{A}{B}) = \log A - \log B$
7. $\ln(\frac{A}{B}) = \ln A - \ln B$
8. $\log(A^n) = n(\log A)$
9. $\ln(A^n) = n(\ln A)$
10. $\log(AB) = \log A + \log B$
11. $\ln(AB) = \ln A + \ln B$
12. $\log\left(\frac{21x^3}{4}\right)$
13. $\log(x + 2)$
14. $\log x - \log 2$
15. 11.30719075
16. 8
17. 447,213.6
18. **a.** $P = 10{,}000e^{0.458145365t}$ (t in hours)
   b. 38,147,000 **c.** 1:30 p.m.
   d. The growth conditions are unlimited, and the culture grows at a rate proportional to the population size.
19. **a.** $Q = 300e^{-0.130782486t}$ (t in years)
   b. 1.5 g **c.** 263.2 g **d.** 17.6 years
20. 2,600 years old
21. 3.1 on the Richter scale
22. **a.** 40 times more earth movement
   b. 209 times more energy
23. 102 dB
24. 9 dB gained
25. Three trumpets have joined in.

Chapter 10 Calculus

Section 10.0

1. $a = 5/2$, $b = 7/2$
3. $x = 16.5$, $y = 19.5$
5. $x = 12$, $y = 5$
7. $x = 1.5$, $y = 6.25$
9. **a.** 15 students/1 teacher
   b. 362 teachers
11. **a.** 9/100 **b.** 9%
13. **a.** 32.1 mph **b.** 128.4 mi
15. **a.** 76.9 mph **b.** 307.7 mi
17. **a.** ≈24.6 mpg **b.** ≈369 mi
   c. ≈369 mi
19. 4 g
21. 480 ft

23. **a.**

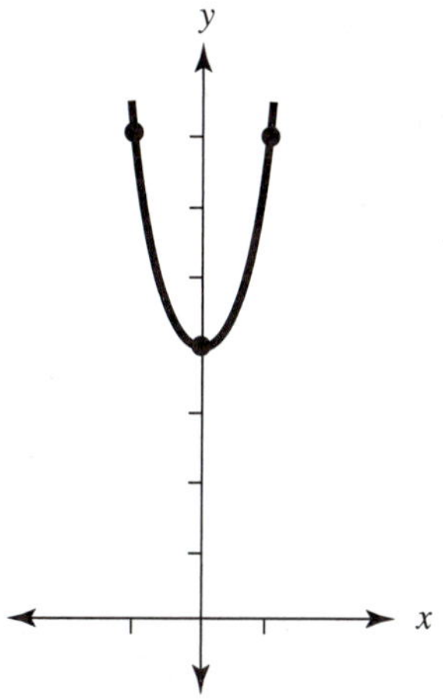

b. $x = 0$ **c.** (0, 4)

25. **a.**

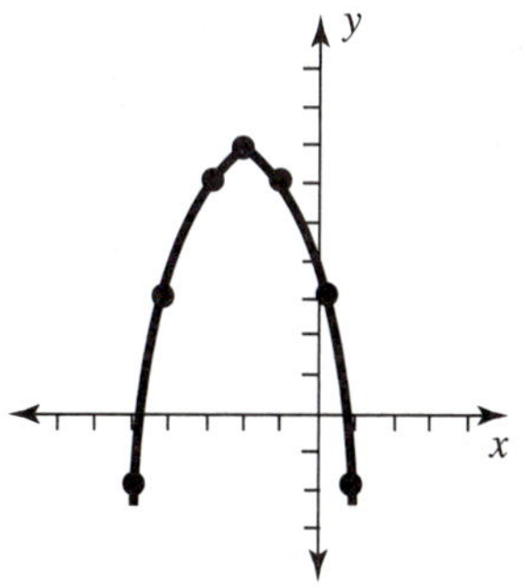

b. $x = -2$ **c.** (−2, 7)

27. a.

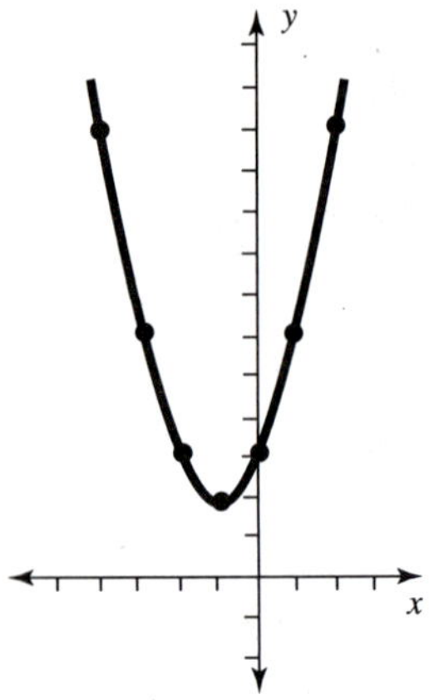

b. $x = -1$
c. $(-1, 2)$

29. a.

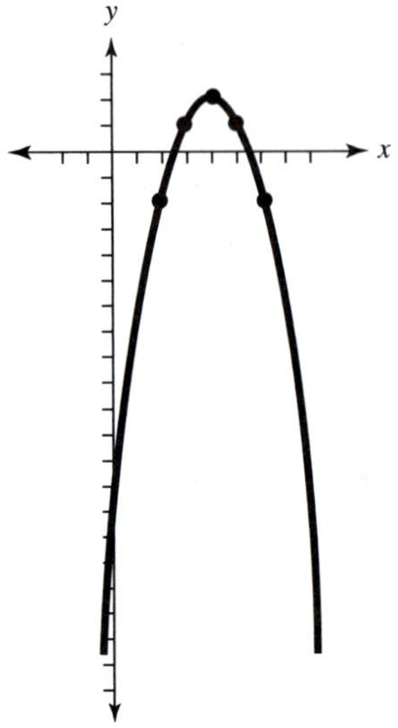

b. $x = 4$
c. $(4, 2)$

31. a.

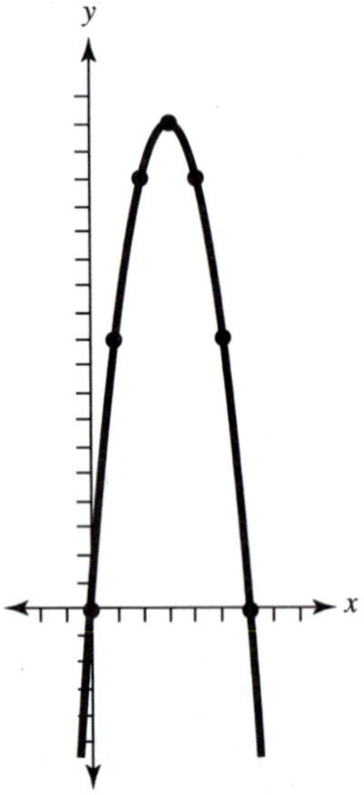

b. $x = 3$
c. $(3, 18)$

33. a. -12 **b.**

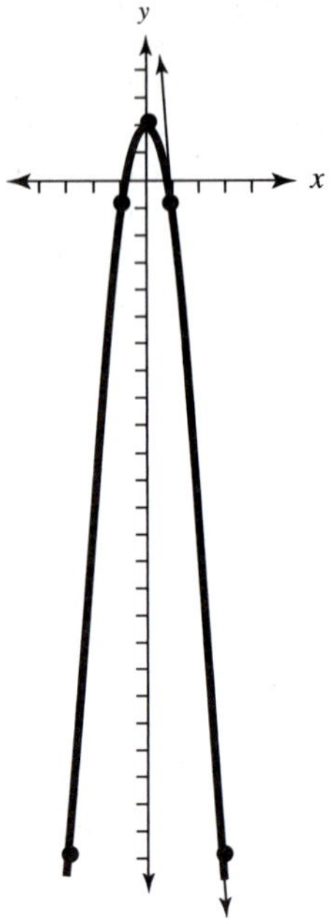

35. a. 0 **b.**

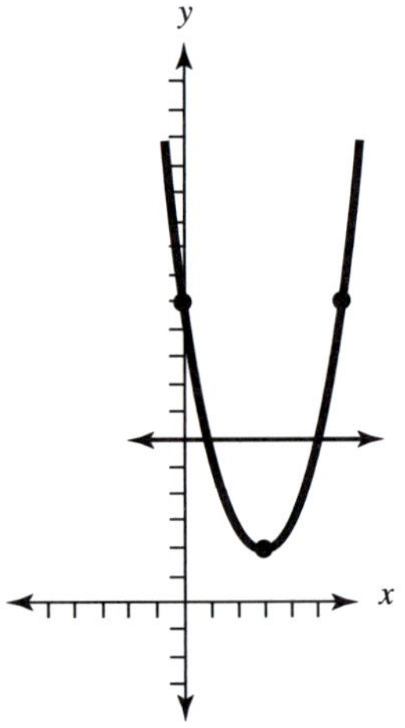

37. 21
39. 6
41. $8x + 13$
43. $x^2 - 14x + 46$
45. $8x + 8\Delta x - 11$
47. $x^2 + 2x\Delta x + (\Delta x)^2 - 3$

Section 10.1

1. a

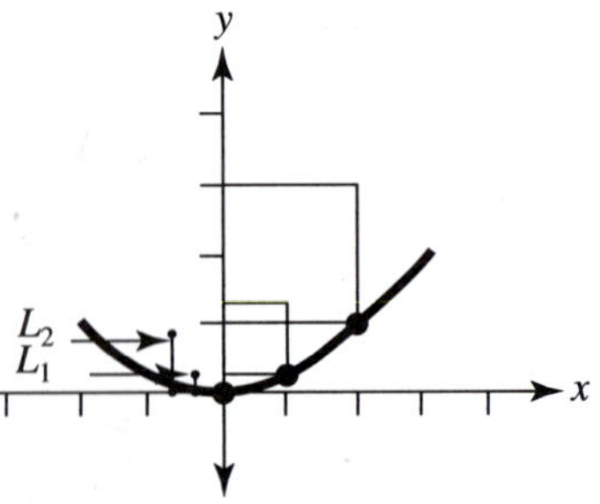

b. $L_1 = \frac{1}{4}, L_2 = 1, A_1 = 1, A_2 = 4$
c. $\frac{1}{4}, \frac{1}{4}$
d. $L_3 = \frac{9}{4}, L_4 = 4, A_3 = 9, A_4 = 16$
e. $\frac{9}{16}, \frac{9}{16}$

3.a.

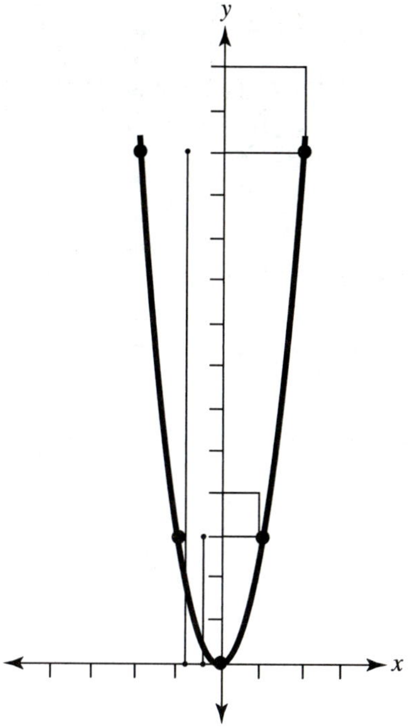

b. $L_1 = 3, L_2 = 12, A_1 = 1, A_2 = 4$
c. $\frac{1}{4}, \frac{1}{4}$
d. $L_3 = 27, L_4 = 48, A_3 = 9, A_4 = 16$
e. $\frac{9}{16}, \frac{9}{16}$

5. a.

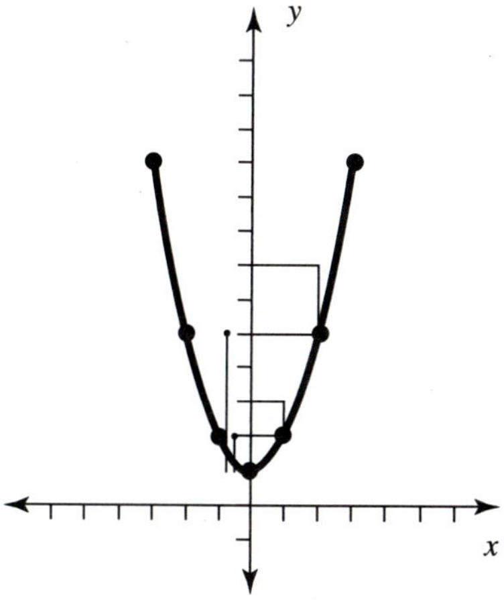

b. $L_1 = 1, L_2 = 4, A_1 = 1, A_2 = 4$
c. $\frac{1}{4}, \frac{1}{4}$

7. a.

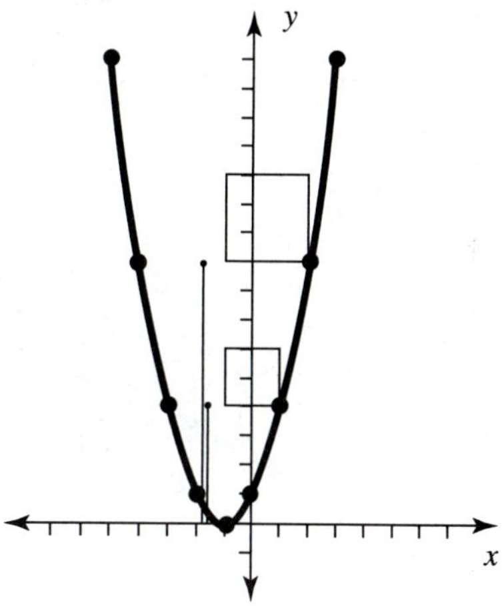

b. $L_1 = 4, L_2 = 9, A_1 = 4, A_2 = 9$
c. $\frac{4}{9}, \frac{4}{9}$

9. area of the square is 9
11. area of the square is 1
13. area of the square is 64
15. **a.** $c + \dfrac{b^2}{4}$ **b.** $c + \dfrac{b^2}{4}$
17. $x = \dfrac{-b \pm \sqrt{b^2 + 4ac}}{2a}$

10.2 Four Problems

1. **a.** 20 ft **b.** 6 ft/sec **c.** 45 ft **d.** 3 ft/sec
3. **a.** 40 ft/min **b.** 56 ft/min **c.** 196 ft. **d.** 28 ft/min
5. The last three entries under "total distance traveled" are 16, 25, and 36.
7. 128 ft
9. 3.5 oz
11. 2,304 ft
13. 16 ft
15. 30 ft
17. $m = 12; y = 12x - 18$
19. $m = 27; y = 27x - 54$
21. $m = 4; y = 4x - 8$
23. $m = 4; y = 4x + 10$
25. $m = 4; y = 4x + 6$

Section 10.3 Newton and Tangent Lines

1. $m = 24; y = 24x - 48$
3. $m = 6; y = 6x - 7$
5. $m = 23; y = 23x - 97$
7. $m = 12; y = 12x - 16$
9. $m = 1; y = x - 1$

Section 10.4 Newton on Falling Objects and the Derivative

1. **a.** 16 ft **b.** 32 ft/sec **c.** 16 ft/sec
3. **a.** 400 ft **b.** 160 ft/sec **c.** 80 ft/sec
5. **a.** 1,024 ft **b.** 256 ft/sec **c.** 128 ft/sec
7. **a.** 2 sec **b.** 64 ft/sec **c.** 32 ft/sec
9. The derivative is $\frac{ds}{dt} = 32$.
11. 2
13. -7
19. $22x$
21. $6x$

23. $6x - 2$

31. a. $f'(x) = 2x + 4$

b. Slope at $x = 1$ is 6. Slope at $x = 2$ is 8.

c.

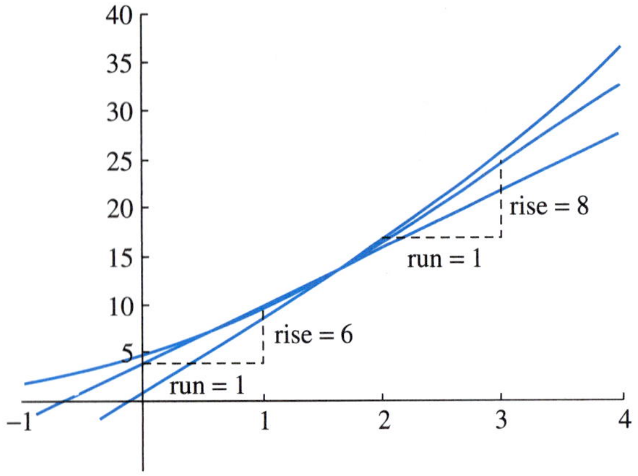

33. a. $f'(x) = 2x - 2$

b. Slope at $x = 1$ is 0. Slope at $x = 2$ is 2.

c.

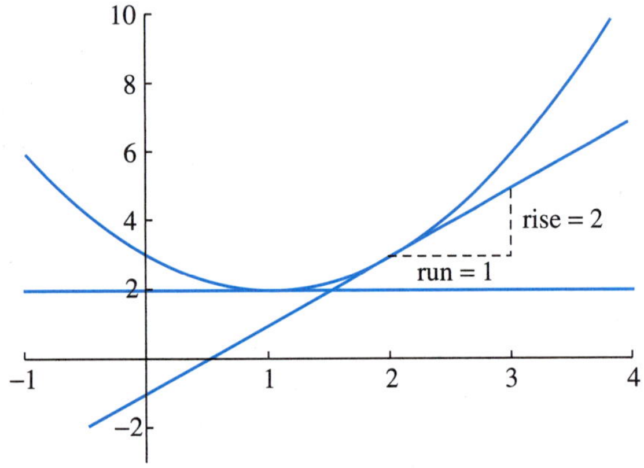

35. $3x^2$

37. $9x^2 + 4$

43. $40x^4$

45. $20x^3 - 6x + 2$

47. a. average speed in mi/sec

b. instantaneous speed in mi/sec

49. a. average flow rate in gal/hr

b. instantaneous flow rate in gal/hr

51. a. average growth rate in cm/yr

b. instantaneous growth rate in cm/yr

53. a. average growth rate in thousands of people per year

b. instantaneous growth rate in thousands of people per year

Section 10.5 The Trajectory of a Cannonball

1. a. $y = -0.0003x^2 + 0.5774x + 4$ **b.** 1,932 ft **c.** 282 ft **d.** no

3. a. $y = -0.0004x^2 + x + 4$ **b.** 2,504 ft **c.** 629 ft **d.** yes

5. a. $y = -0.0006x^2 + 1.7321x + 5$ **b.** 2,890 ft **c.** 1,225 ft **d.** yes

7. a. $y = -0.0005x^2 + 0.5774x + 5$ **b.** 1,163 ft **c.** 172 ft **d.** no

9. a. $y = -0.0005x^2 + x + 4$ **b.** 2,004 ft **c.** 504 ft **d.** yes

11. a. $y = -0.0007x^2 + 1.7321x + 3$ **b.** 2,476 ft **c.** 1,074 ft **d.** yes

13. a. $y = -0.0006x^2 + x + 4$ **b.** 1,651 ft

15. a. $y = -0.0002x^2 + 0.5774x + 4$ **b.** 2,859 ft

17. a. 123 ft/sec

b. $y = -0.0014x^2 + 0.5774x + 3$

c. 63 ft

d. 114 ft/sec

19. a. $y = -0.0040x^2 + x + 106$

b. 330 ft

c. 5.2 sec

21. a. 60° **b.** 60°, 60°, 60°

c. 2

d. 2 and 2

e. $\sqrt{3}$

23. $y = -\frac{64}{s^2}x^2 + \sqrt{3}x + C$

25. 3.9 secs

27. 17.8 secs

29. a. $y = -0.0005x^2 + 0.3640x + 5$ **b.** 741 ft

c. 71 ft

31. a. $y = -0.0034x^2 + 2.7475x + 5$ **b.** 810 ft

c. 560 ft

Section 10.6 Newton and Areas

1. a. $A(x) = 2x$

b.

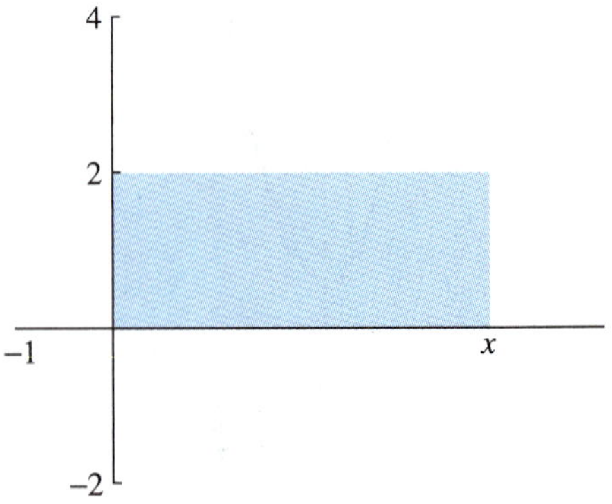

c. $A(3) = 6$

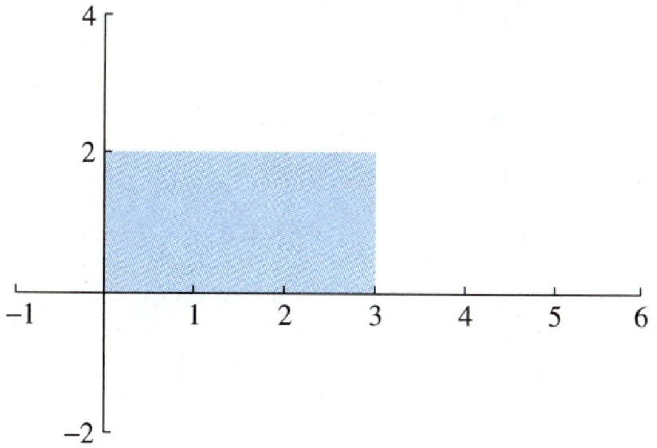

d. $A(4) = 8$

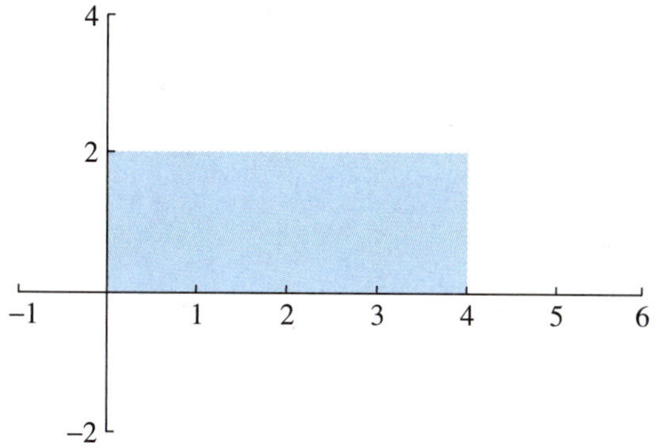

e. $A(4) - A(3) = 2$

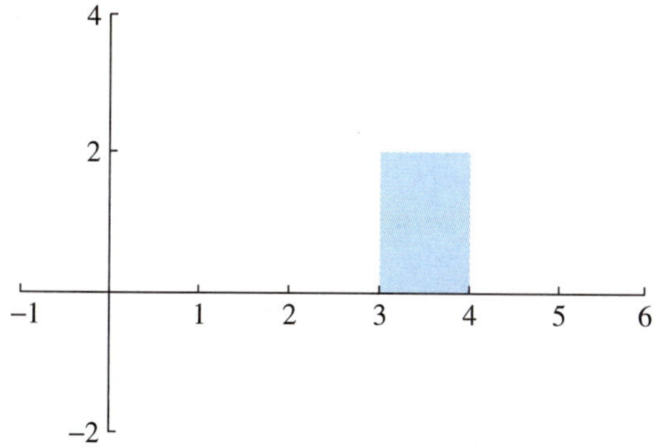

3. a. $A(x) = x^2$

b.

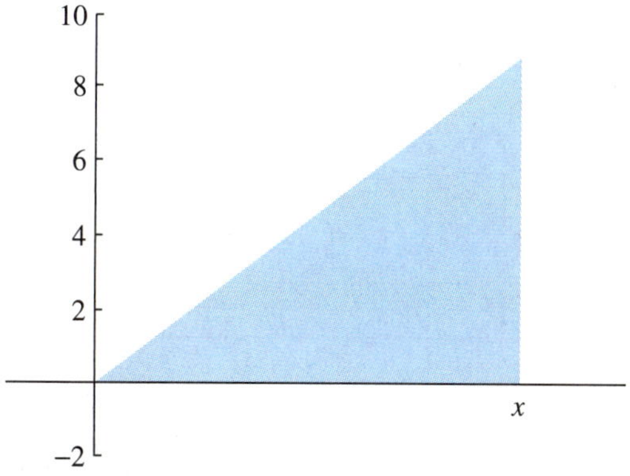

c. $A(3) = 9$

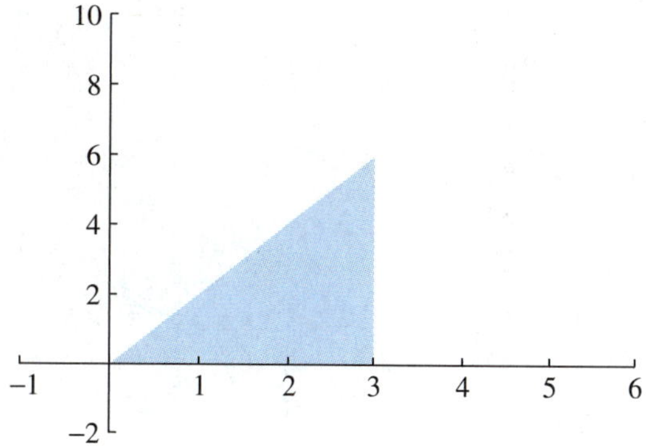

d. $A(4) = 16$

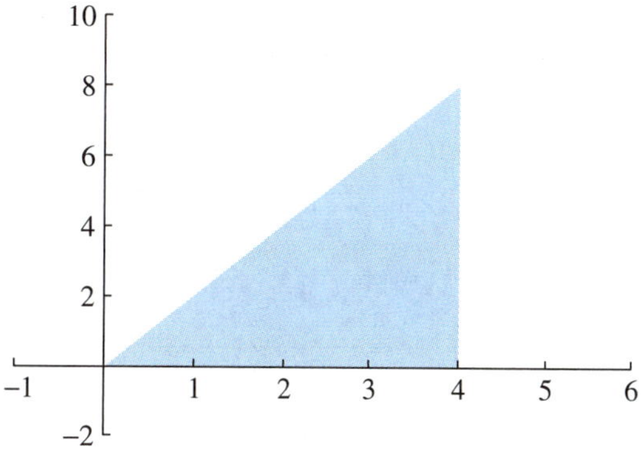

e. $A(4) - A(3) = 7$

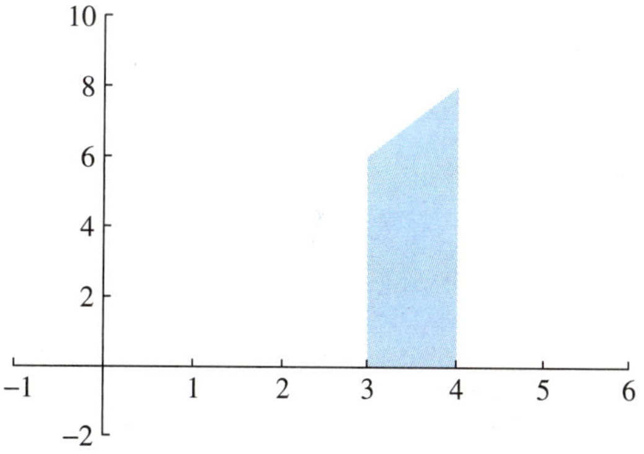

5. a. $A(x) = 2x + \frac{1}{2}x^2$

b.

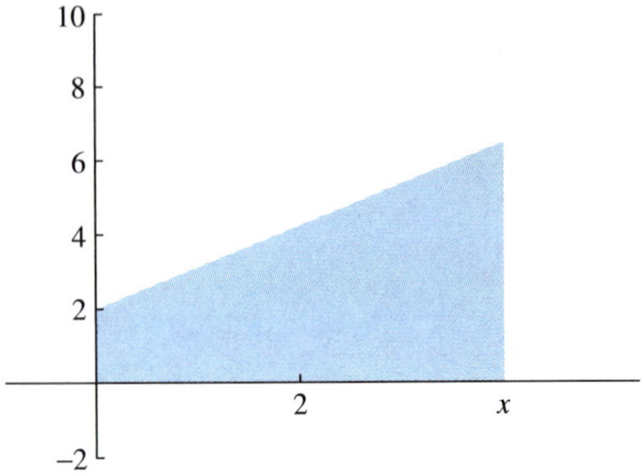

c. $A(3) = 10.5$

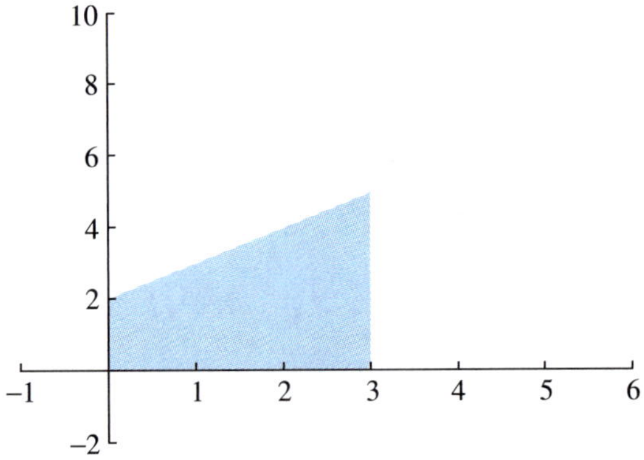

d. $A(4) = 16$

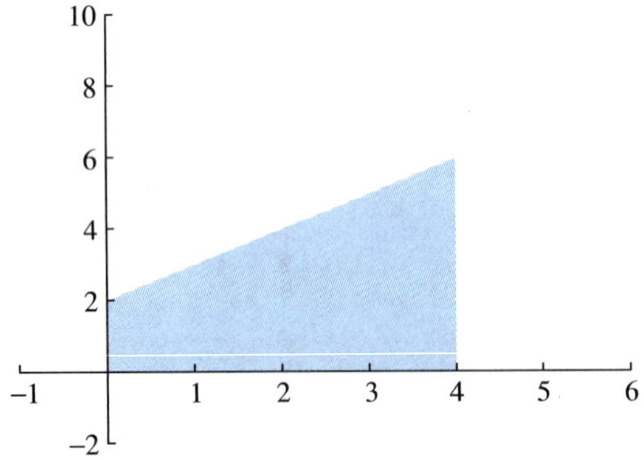

e. $A(4) - A(3) = 5.5$

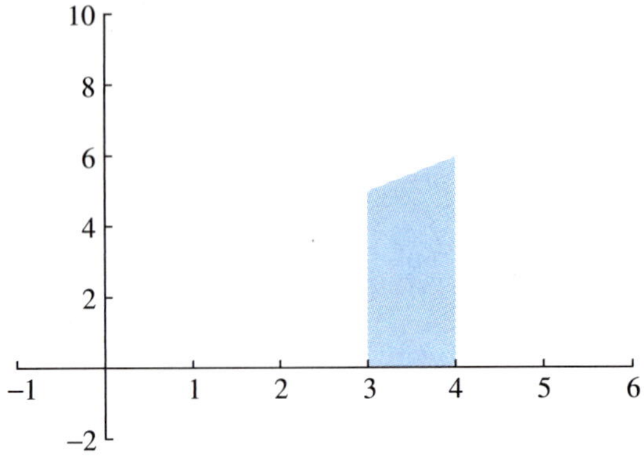

7. a. $A(x) = 3x + x^2$

b.

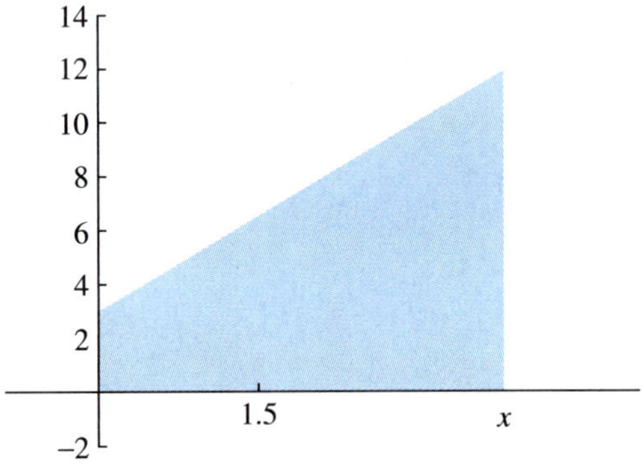

c. $A(3) = 18$

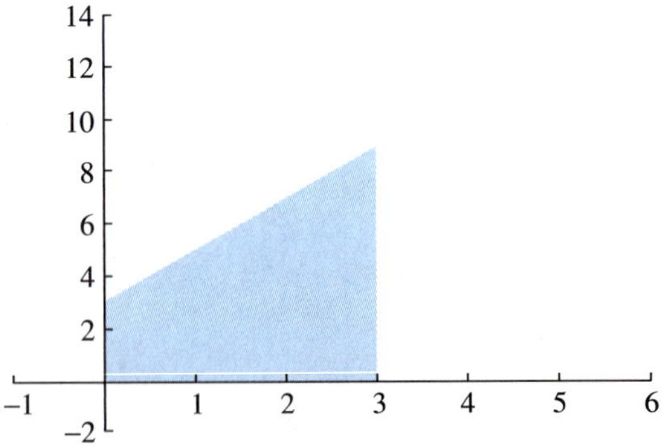

d. $A(4) = 28$

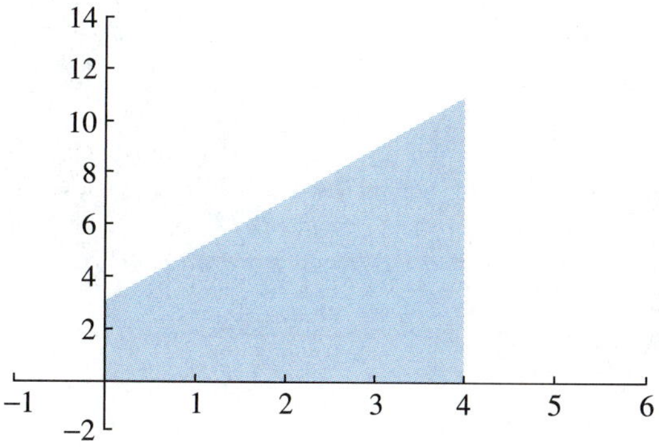

e. $A(4) - A(3) = 10$

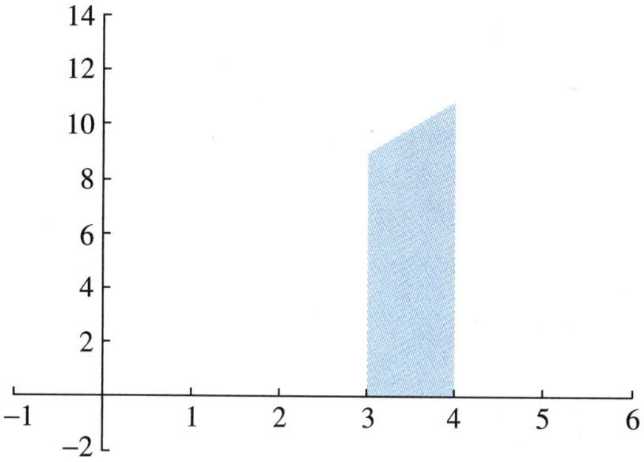

9. $5x + c$
11. $3x^2 + 4x + c$
13. area = 25
15. area = 95
17. $4x^2 + 7x + c$
19. $x^3 - x^2 + 5x + c$
21. 249
23. 329
25. **a.** 1 **b.** $\frac{x^2}{2} + c$
27. **a.** 0 **b.** $5x + c$
29. **a.** 8 **b.** $4x^2 + c$
31. **a.** $15x^4$ **b.** $\frac{x^2}{6} + 4x + c$
33. **a.** 63 **b.** 999,973
35. **a.** 432 **b.** 9,624
37. **a.** 576,551.14 **b.** 7,141,316
39. **a.** 1,463,217 **b.** 1,763
41. $4x^2 + 7x + c$
43. $x^3 - x^2 + 5x + c$
45. $\frac{5}{7}x^7 - x^3 + 13x + c$
47. $\frac{-18}{x} + c$

Chapter 10 Review

2. $a = 4.5, b = 5.25$
3. **a.** about 30.4 mpg
b. 456 mi
4. **a.** 65
b. $8x^2 + 16x(\Delta x) + 8(\Delta x)^2 - 7$
5. **a.** 120 ft
b. about 6.7 ft/sec
c. 83.3 ft
d. 3.3 ft/sec
6. 400 ft
7. **a.**

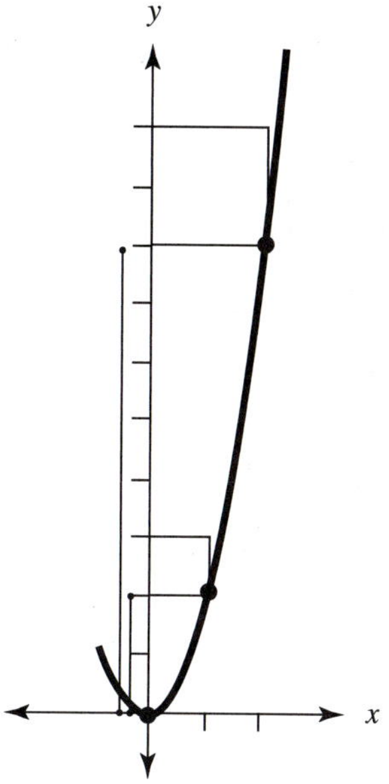

b. $L_1 = 2, L_2 = 8, A_1 = 1, A_2 = 4$
c. $\frac{1}{4}, \frac{1}{4}$
d. $L_3 = 18, L_4 = 32, A_3 = 9, A_4 = 16$
e. $\frac{9}{16}, \frac{9}{16}$
8. The root is 1, and the area of the square is 1.

9. a. and g.

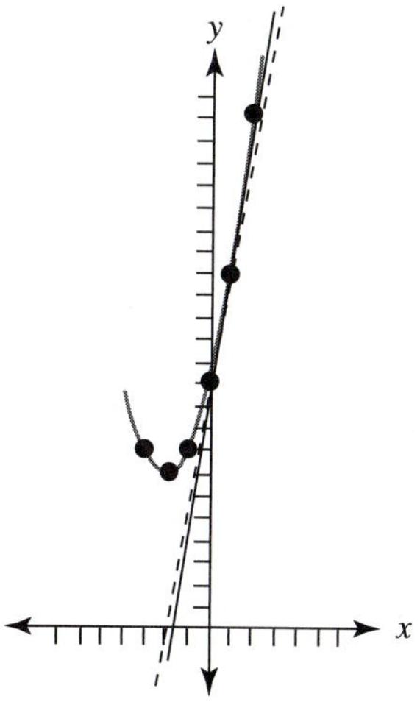

b. $x = -21$ **c.** $(-2, 7)$
d. 7 **e.** 6
f. $y = 6x + 10$

10. slope = 12;
$y = 12x - 13$

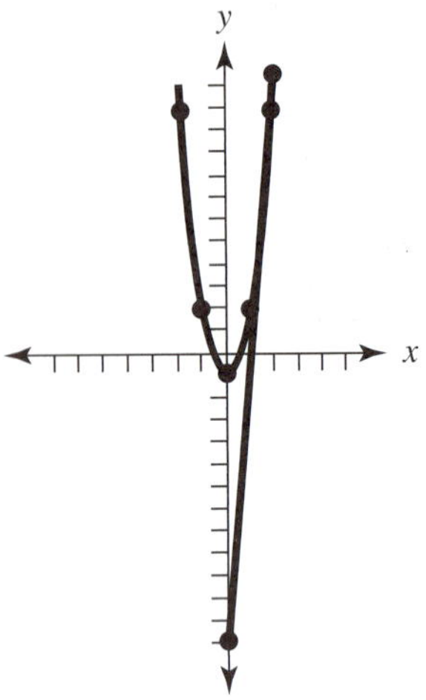

11. **a.** 100 feet **b.** 80 ft/sec **c.** 40 ft/sec
12. 0
13. 2
14. $2x - 3$
15. **a.** average rate of weight loss in pounds/week
b. instantaneous rate of weight loss in pounds/week
16. **a.** average growth rate in bass/week
b. instantaneous growth rate in bass/week
17. **a.** $y = -0.0005x^2 + 0.5774x + 4.5$ **b.** 1,163 ft
c. 171 ft **d.** yes
18. $\frac{3}{2}x^2 - 2x + c$
19. 152

Appendix E

1. **a.** 4 yd **b.** 36 ft
3. **a.** 52,800 ft **b.** 0.0019 mi
5. 126,720 in.
7. **a.** 50,000 m **b.** 0.05 km
9. **a.** 0.02 liters **b.** 200 cl
11. **a.** 454 g **b.** 3.6 kg
13. **a.** 88 ft/sec **b.** 0.9 sec
15. **a.** 81 mph **b.** 16 mi
17. **a.** 0.0157534247 **b.** \$47.26 **c.** \$48.84
19. **a.** 1.83 m **b.** 76.7 kg **c.** 23; no **d.** 28; yes
21. 164.56 gals

Index